Teacher's Edition

AGS

Geometry

by
Siegfried Haenisch

AGS®
American Guidance Service, Inc.
Circle Pines, Minnesota 55014-1796
1-800-328-2560

About the Author

Siegfried Haenisch, Ed.D., has taught mathematics at every level, from elementary to graduate school, most recently as Professor in the Department of Mathematics and Statistics at the College of New Jersey. The Mathematical Association of America granted him the 1995 Award for Distinguished Teaching of Mathematics. Dr. Haenisch was the site director for the training of teachers in the New Jersey Algebra Project. He was a member of the National Science Foundation Institutes in Mathematics at Rutgers University, Oberlin College, and Princeton University. At Yale University he was a member of the Seminar in the History of Mathematics, sponsored by the National Endowment in the Humanities. Dr. Haenisch currently serves as a mathematics curriculum consultant to school districts.

The publisher wishes to thank the following educators for their helpful comments during the review process for *Geometry.* Their assistance has been invaluable.

Candise Campbell
Special Education Teacher
Montclair High School
Montclair, California

Betty Davis
Academic Dean of Mathematics and Science
Jefferson High School
Shenandoah Junction, West Virginia

Reva Halloran
Learning Disabilities Teacher
Cabell Midland High School
Ona, West Virginia

Carolyn Hansen
Instructional Specialist for Mathematics and Technology
Williamsville Schools
Williamsville, New York

Deborah Horn
Special Education Teacher
Wayne City High School
Wayne City, Illinois

Sandee Hough
Program Manager
Adult Learning Center
University of Alaska, Anchorage
Anchorage, Alaska

Niki J. Pennington
Department Head
Exceptional Student Education
Durant High School
Plant City, Florida

Carolyn Scott
Secondary Eduational Coordinator
Terrebonne Parish School Board
Houma, Louisiana

Chet Singer
Supervisor of Mathematics, Business and Computer Education
Hicksville UFSD
Hicksville, New York

Elin Snyder Smith
Special Education Teacher
Muhlenberg South High School
Greenville, Kentucky

Marlinda Stull
Special Vocational Teacher and Special Education Instructor
Meade County High School
Brandenburg, Kentucky

Suzanne R. Swartz
Special Educator
Jefferson High School
Shenandoah Junction, West Virginia

Printed in the United States of America
ISBN 0-7854-2222-6 Product Number 91202
A 0 9 8 7 6 5 4 3

Table of Contents

Overview

Lesson Plans

Teacher's Resources

AGS Geometry

Teacher's Edition

Student Workbook

Student Text

Teacher's Resource Library on CD-ROM

AGS GEOMETRY teaches the basic concepts of geometry in a step-by-step approach and meets the standards set by most states and the National Council of Teachers of Mathematics (NCTM). This text is written for students and young adults who need extra help grasping new concepts. Because students learn in many different ways, emphasis is placed on including instruction using a variety of modalities. When appropriate, manipulatives are suggested to provide a tactile/ kinesthetic activity promoting concept mastery.

The text is organized into twelve discrete chapters. Chapters open with photographs depicting examples of how geometry is used and viewed in the world around us. These examples from real life will give students a better understanding of the relevance of geometry. Chapter openers also include Goals for Learning that introduce students to key objectives.

Short lessons with plenty of examples are included to illustrate and teach each new skill. Theorems and postulates are highlighted for quick reference. Math terms are defined in the side column and in the Glossary. Frequent sets of exercises allow students to practice their newly acquired skills. To further facilitate learning, sample solutions are given for the first problem in each new exercise set. In addition, the answers to all odd-numbered problems are in the Student Text. Combined with the many opportunities for practice within a lesson, supplementary problems in the back of the text offer extra reinforcement.

As recommended by NCTM, students are encouraged to write about mathematics. Students may record these writings in a notebook or portfolio. To encourage understanding and motivation, problem-solving exercises are integrated throughout the lessons. These exercises present problems that require geometric thinking and analysis. To extend this concept, application activities are included with each chapter. These activities help students relate geometric concepts to real-life situations, helping them to connect math instruction with the real world.

AGS Worktexts and Mathematics Texts

CHECK OUT AGS WORKTEXTS! AGS offers additional mathematics materials to help you tailor instruction to meet the diverse needs of your students. Each worktext contains 96 pages of up-to-date information and motivating skill lessons with lots of opportunities for practice and reinforcement. These may be used to accompany a basal program or as the core instructional tool.

ENHANCE YOUR MATH PROGRAM WITH AGS TEXTBOOKS!—an easy, effective way to teach students the practical skills they need. Each AGS textbook meets your math curriculum needs. These exciting, full-color books use student-friendly text and real-world examples to show students the relevance of math in their daily lives. Each presents a comprehensive coverage of skills and concepts. The short, concise lessons will motivate even your most reluctant students. With readabilities of all the texts below 4th grade reading level, your students can concentrate on learning the content. AGS is committed to making learning accessible to all students.

For more information on AGS worktexts and textbooks, call 1-800-328-2560, or visit our Web site: www.agsmath.com e-mail: agsmail@agsnet.com

Student Text Highlights

- Each lesson is clearly labeled to help students focus on the skill or concept to be learned.
- Vocabulary terms are boldfaced and then defined in the margin at the beginning of the lesson and in the Glossary.
- Reminder notes and tips help students recall and apply what they already know.

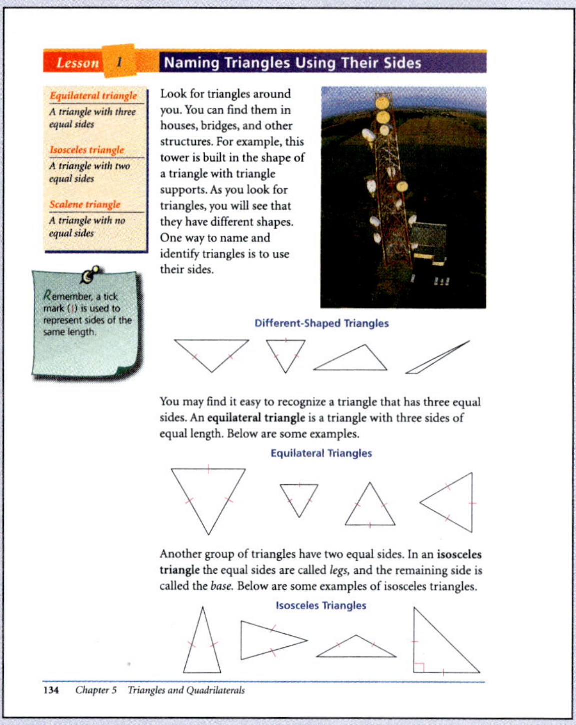

Lesson 1 — Naming Triangles Using Their Sides

Equilateral triangle — A triangle with three equal sides

Isosceles triangle — A triangle with two equal sides

Scalene triangle — A triangle with no equal sides

Remember, a tick mark (|) is used to represent sides of the same length.

Look for triangles around you. You can find them in houses, bridges, and other structures. For example, this tower is built in the shape of a triangle with triangle supports. As you look for triangles, you will see that they have different shapes. One way to name and identify triangles is to use their sides.

Different-Shaped Triangles

You may find it easy to recognize a triangle that has three equal sides. An **equilateral triangle** is a triangle with three sides of equal length. Below are some examples.

Equilateral Triangles

Another group of triangles have two equal sides. In an **isosceles triangle** the equal sides are called *legs*, and the remaining side is called the *base*. Below are some examples of isosceles triangles.

Isosceles Triangles

134 *Chapter 5 Triangles and Quadrilaterals*

- Goals for Learning at the beginning of each chapter identify learner outcomes.

Goals for Learning

- To construct triangles using sides
- To name triangles using their angles
- To name and distinguish quadrilaterals
- To find the measures of angles in quadrilaterals and triangles
- To define quadrilaterals and parallels and apply theorems to them
- To construct perpendiculars
- To construct triangles based on attributes
- To use the angle sum theorem to define triangles

Algebra Connection

- Algebra Connection shows students how skills they learned in algebra are used to solve problems in geometry. An Algebra Review section is also included in the back of the Student Text.

- Important rules, theorems, and postulates are highlighted for quick reference.

ANGLE SUM THEOREM: The sum of the angle measures in any triangle is 180°.

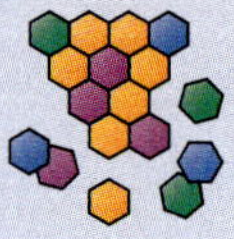

- Problem-solving exercises give students opportunities to apply lesson skills to solve practical problems.

- Examples, proofs, and constructions are clearly highlighted for emphasis and review.
- Exercise sets parallel the instruction and examples in each lesson.
- In each chapter, students are directed to the Selected Answers and Solutions found in the back of the Student Text.

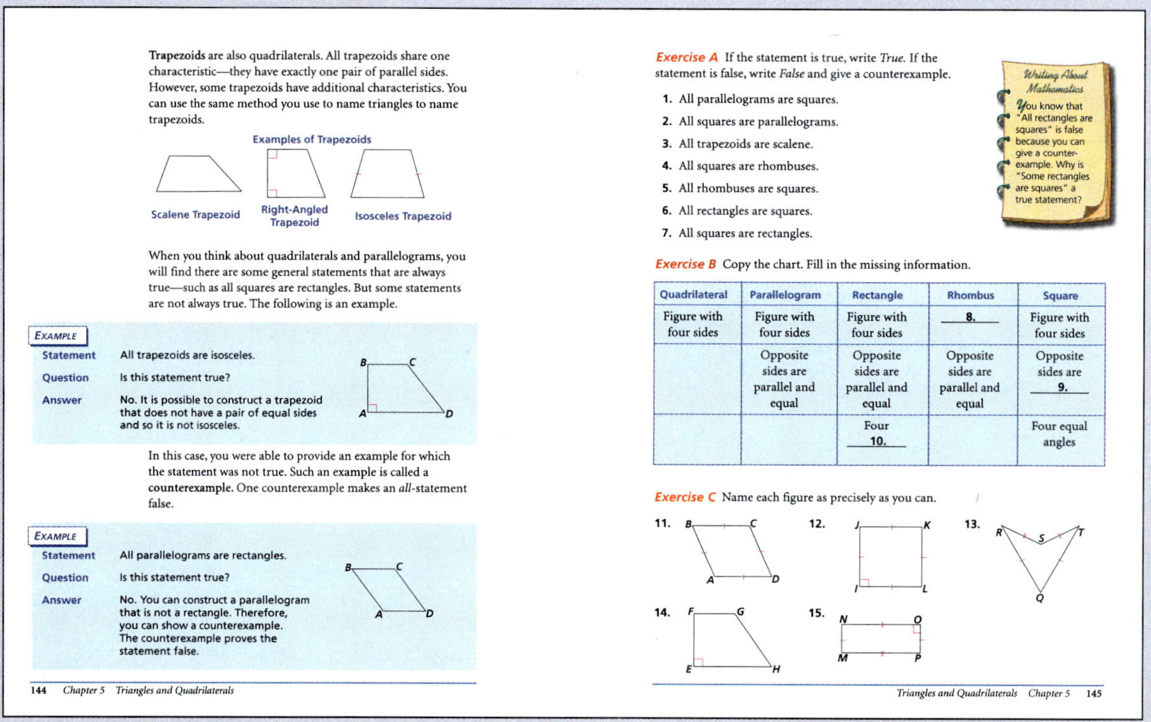

Trapezoids are also quadrilaterals. All trapezoids share one characteristic—they have exactly one pair of parallel sides. However, some trapezoids have additional characteristics. You can use the same method you use to name triangles to name trapezoids.

Examples of Trapezoids

When you think about quadrilaterals and parallelograms, you will find there are some general statements that are always true—such as all squares are rectangles. But some statements are not always true. The following is an example.

EXAMPLE

Statement All trapezoids are isosceles.

Question Is this statement true?

Answer No. It is possible to construct a trapezoid that does not have a pair of equal sides and so it is not isosceles.

In this case, you were able to provide an example for which the statement was not true. Such an example is called a **counterexample**. One counterexample makes an *all*-statement false.

EXAMPLE

Statement All parallelograms are rectangles.

Question Is this statement true?

Answer No. You can construct a parallelogram that is not a rectangle. Therefore, you can show a counterexample. The counterexample proves the statement false.

144 *Chapter 5 Triangles and Quadrilaterals*

Exercise A If the statement is true, write *True.* If the statement is false, write *False* and give a counterexample.

1. All parallelograms are squares.
2. All squares are parallelograms.
3. All trapezoids are scalene.
4. All squares are rhombuses.
5. All rhombuses are squares.
6. All rectangles are squares.
7. All squares are rectangles.

Writing About Mathematics

You know that "All rectangles are squares" is false because you can give a counterexample. Why is "Some rectangles are squares" a true statement?

Exercise B Copy the chart. Fill in the missing information.

Quadrilateral	Parallelogram	Rectangle	Rhombus	Square
Figure with four sides	Figure with four sides	Figure with four sides	8. ____	Figure with four sides
	Opposite sides are parallel and equal	Opposite sides are parallel and equal	Opposite sides are parallel and equal	Opposite sides are 9. ____
		Four 10. ____		Four equal angles

Exercise C Name each figure as precisely as you can.

11. B C A D
12. J K I L
13. R S T Q
14. F G E H
15. N O M P

Triangles and Quadrilaterals Chapter 5 145

■ Many features and optional activities reinforce and extend student learning beyond the lesson.

TECHNOLOGY CONNECTION

*A*ccurate measurements of distances and angles, such as flight paths and re-entry angles, are important to space exploration. The Mars Climate Orbiter, which traveled 416,000,000 miles to Mars, may have failed because it came about $12\frac{1}{2}$ miles too close to the planet's surface. That's an error of about 0.00003%!

Geometry Around Us

EXPERIENCING GEOMETRY

*C*ut a square from a sheet of paper and label the square *ABCD*. Draw diagonal $\overline{AC}$. Now fold the square in half along $\overline{AC}$, matching vertex *B* with vertex *D*. Label angles 1, 2, 3, and 4. Notice that ΔABC matches ΔADC exactly. What can you say about the measures of angles 1, 2, 3, and 4? Give reasons for your answer.

TRY THIS...

Try to construct a right triangle with three equal sides. Explain your observations.

Calculator Practice

Writing About Mathematics

*Y*ou know that "All rectangles are squares" is false because you can give a counter-example. Why is "Some rectangles are squares" a true statement?

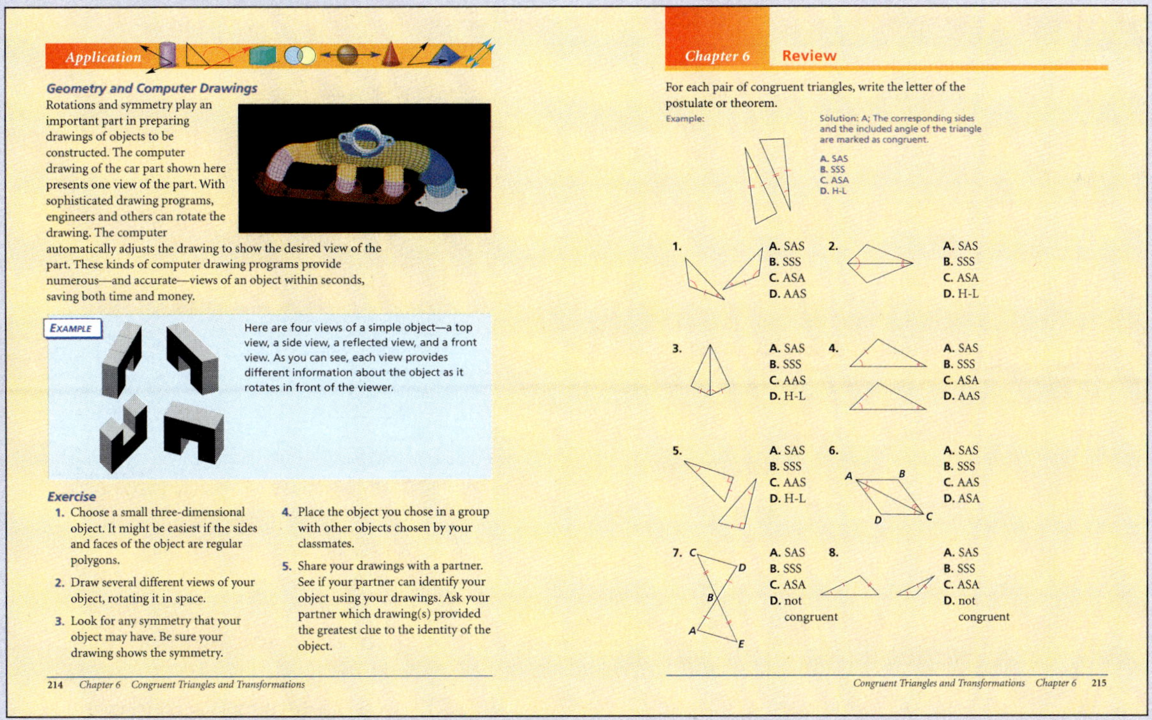

Application

Geometry and Computer Drawings

Rotations and symmetry play an important part in preparing drawings of objects to be constructed. The computer drawing of the car part shown here presents one view of the part. With sophisticated drawing programs, engineers and others can rotate the drawing. The computer automatically adjusts the drawing to show the desired view of the part. These kinds of computer drawing programs provide numerous—and accurate—views of an object within seconds, saving both time and money.

EXAMPLE

Here are four views of a simple object—a top view, a side view, a reflected view, and a front view. As you can see, each view provides different information about the object as it rotates in front of the viewer.

Exercise

1. Choose a small three-dimensional object. It might be easiest if the sides and faces of the object are regular polygons.
2. Draw several different views of your object, rotating it in space.
3. Look for any symmetry that your object may have. Be sure your drawing shows the symmetry.
4. Place the object you chose in a group with other objects chosen by your classmates.
5. Share your drawings with a partner. See if your partner can identify your object using your drawings. Ask your partner which drawing(s) provided the greatest clue to the identity of the object.

214 Chapter 6 Congruent Triangles and Transformations

Chapter 6 Review

For each pair of congruent triangles, write the letter of the postulate or theorem.

Example:

Solution: A; The corresponding sides and the included angle of the triangle are marked as congruent.

A. SAS
B. SSS
C. ASA
D. H-L

1. A. SAS B. SSS C. ASA D. AAS
2. A. SAS B. SSS C. ASA D. H-L
3. A. SAS B. SSS C. AAS D. H-L
4. A. SAS B. SSS C. ASA D. AAS
5. A. SAS B. SSS C. AAS D. H-L
6. A. SAS B. SSS C. AAS D. ASA
7. A. SAS B. SSS C. ASA D. not congruent
8. A. SAS B. SSS C. ASA D. not congruent

Congruent Triangles and Transformations Chapter 6 215

■ Application activities help students relate geometry to everyday life.

■ Chapter Reviews appear in the Student Text and allow students and teachers to check for understanding.

■ Sample problems and solutions are included to review key concepts. The review can be used for reteaching and to prepare for the chapter test.

■ In addition, Supplementary Problems for each chapter are located in the back of the Student Text.

■ Test-Taking Tips help reduce test anxiety and increase test scores.

TEST-TAKING TIP

When taking a multiple-choice test, read every choice before you answer a question. Put a line through choices you know are wrong. Then choose the best answer from the remaining choices.

Teacher's Edition Highlights

The comprehensive, wraparound Teacher's Edition provides instructional strategies at point of use. Everything from preparation guidelines to teaching tips and strategies is featured at point of use for your convenience.

Chapter 1

Planning Guide

Exploring Geometry: Points, Lines, and Angles in the Plane

Lesson	Student Text Lesson: Student Pages	Vocabulary	Proof	Practice Exercises	Problem Solving	Calculator Practice	Solutions Key
Lesson 1 Points and Lines in the Plane	2–6	✔		✔	6		✔
Lesson 2 Ruler Postulates	7–10	✔		✔			✔
Lesson 3 Copying and Bisecting Angles	11–14	✔		✔	14		✔
Lesson 4 Angle Measurement	15–19	✔		✔	19		✔
Lesson 5 Complementary and Supplementary Angles	20–23	✔		✔		22	✔
Lesson 6 Algebra and Angles	24–27			✔	27		✔
Application Drawing Networks	28			✔			✔

Lesson	Student Text Features: Algebra Connection	Constructions	Experiencing Geometry	Technology Connection	Try This	Writing About Mathematics
Lesson 1 Points and Lines in the Plane					6	
Lesson 2 Ruler Postulates				10		
Lesson 3 Copying and Bisecting Angles		12–14				
Lesson 4 Angle Measurement		18, 19				17
Lesson 5 Complementary and Supplementary Angles			23			23
Lesson 6 Algebra and Angles	24		27			24
Application Drawing Networks						

Lesson	Teaching Strategies: Warm-Up Activity	Common Error	Algebra Connection	Mental Math	Manipulatives	Calculator	Applications	Group Problem Solving
Lesson 1 Points and Lines in the Plane	2						4, 5	
Lesson 2 Ruler Postulates	7	8		9	8		10	10
Lesson 3 Copying and Bisecting Angles	11							
Lesson 4 Angle Measurement	15	16		19	18		18	17
Lesson 5 Complementary and Supplementary Angles	20			23		22		
Lesson 6 Algebra and Angles	24	25	24		25	26		26
Application Drawing Networks	28						28	

Lesson	Learning Styles: Auditory/Verbal	Visual/Spatial	Body/Kinesthetic	Logical/Mathematical	Interpersonal/Group Learning	LEP/ESL
Lesson 1 Points and Lines in the Plane		3, 5	3	6	4	
Lesson 2 Ruler Postulates	9					
Lesson 3 Copying and Bisecting Angles	13		13		14	12
Lesson 4 Angle Measurement	19	17				17
Lesson 5 Complementary and Supplementary Angles		22	23	23		
Lesson 6 Algebra and Angles						27
Application Drawing Networks						

Lesson	Teacher's Resource Library: Workbook Activities	Additional Practice	Self-Study Guide
Lesson 1 Points and Lines in the Plane	1	1	✔
Lesson 2 Ruler Postulates	2, 3	2	✔
Lesson 3 Copying and Bisecting Angles	4	3	✔
Lesson 4 Angle Measurement	5	4	✔
Lesson 5 Complementary and Supplementary Angles	6	5	✔
Lesson 6 Algebra and Angles	7	6	✔
Application Drawing Networks	8		✔

Chapter Activities

Teacher's Resource Library
Geometry on the Job 1: Surveyors
Community Connection 1: Your Town
Teacher's Edition
Chapter 1 Project

Assessment Options

Student Text
Chapter 1 Review
Teacher's Resource Library
Chapter 1 Mastery Tests A and B
Chapter 1 Cumulative Mastery Test
Teacher's Edition
Chapter 1 Alternative Assessment

Manipulatives

Factor Blocks
Angle Rule
Manipulatives Master 1 (Number Line)

xviA

xviB

Chapter Planning Guides

- The Planning Guide saves valuable preparation time by organizing all materials for each unit.
- A complete listing of lessons allows you to preview each chapter quickly.
- Assessment options are highlighted for easy reference. Options include:
  - Chapter Reviews
  - Chapter Mastery Tests, Forms A and B
  - Cumulative Mastery Tests
  - Midterm and Final Mastery Tests
  - Alternative Assessment items for each chapter
- Manipulatives used in teaching activities are listed to help you tailor instruction to meet individual student needs.
- Page numbers of Student Text and Teacher's Edition features help customize lesson plans to your students.
- A variety of teaching strategies and learning styles is listed to help include students with diverse needs.
- All activities from the Teacher's Resource Library CD-ROM are listed. These include:
  - Workbook Activities
  - Additional Practice
  - Geometry on the Job
  - Community Connections
  - Self-Study Guides
  - Mastery Tests
- The Solutions Key presents a step-by-step solution for each problem in the Student Text and shows students how to arrive at the correct answer. These reproducible pages are available on a separate CD-ROM.

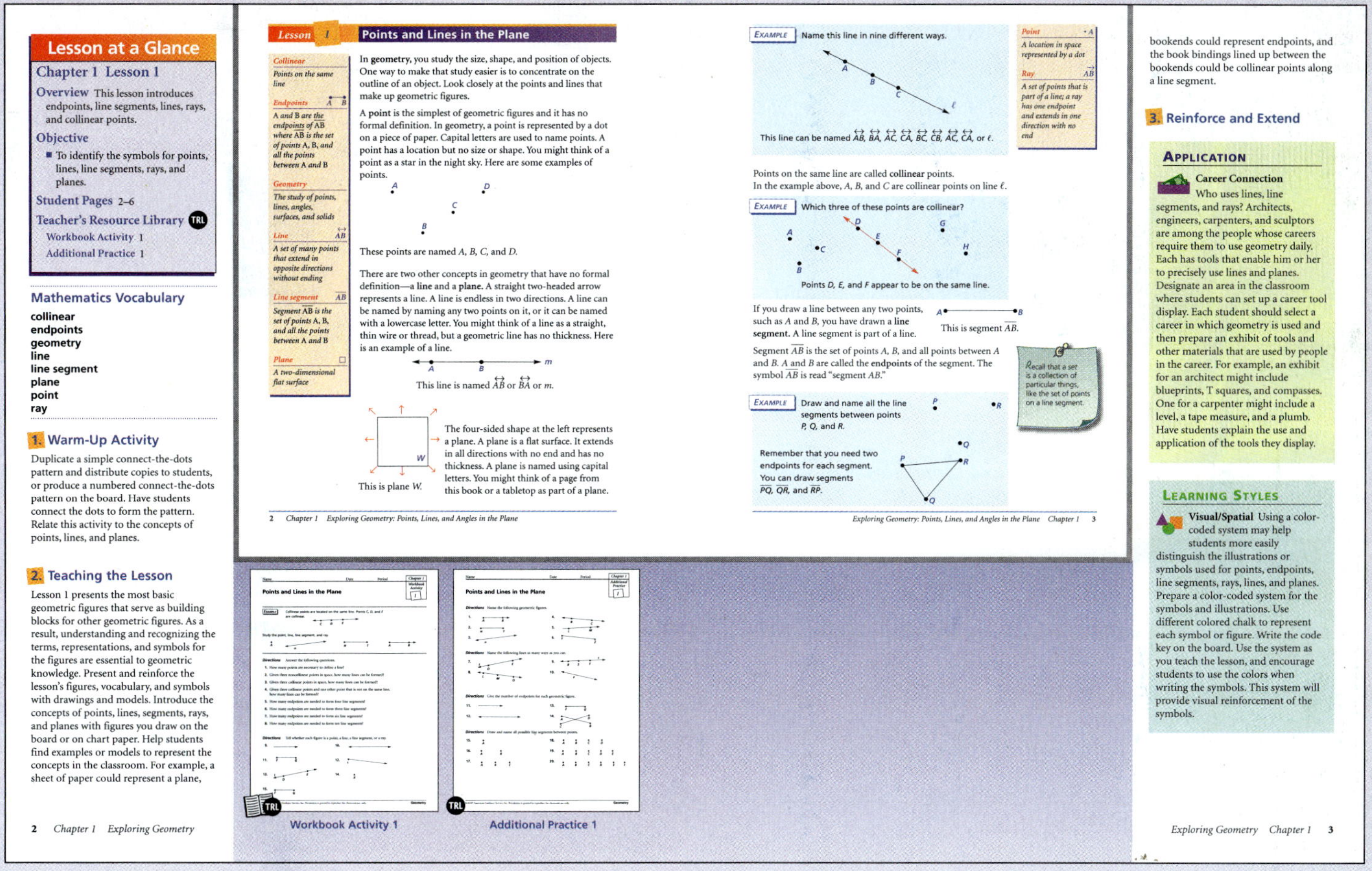

Lesson at a Glance

Chapter 1 Lesson 1

Overview This lesson introduces endpoints, line segments, lines, rays, and collinear points.

Objective

- To identify the symbols for points, lines, line segments, rays, and planes.

Student Pages 2–6

Teacher's Resource Library TRL

Workbook Activity 1

Additional Practice 1

Mathematics Vocabulary

collinear
endpoints
geometry
line
line segment
plane
point
ray

1. Warm-Up Activity

Duplicate a simple connect-the-dots pattern and distribute copies to students, or produce a numbered connect-the-dots pattern on the board. Have students connect the dots to form the pattern. Relate this activity to the concepts of points, lines, and planes.

2. Teaching the Lesson

Lesson 1 presents the most basic geometric figures that serve as building blocks for other geometric figures. As a result, understanding and recognizing the terms, representations, and symbols for the figures are essential to geometric knowledge. Present and reinforce the lesson's figures, vocabulary, and symbols with drawings and models. Introduce the concepts of points, lines, segments, rays, and planes with figures you draw on the board or on chart paper. Help students find examples or models to represent the concepts in the classroom. For example, a sheet of paper could represent a plane,

2 Chapter 1 Exploring Geometry

Lesson 1 Points and Lines in the Plane

Collinear — *Points on the same line*

Endpoints — *A and B are the endpoints of $\overline{AB}$ where $\overline{AB}$ is the set of points A, B, and all the points between A and B*

Geometry — *The study of points, lines, angles, surfaces, and solids*

Line $\overleftrightarrow{AB}$ — *A set of many points that extend in opposite directions without ending*

Line segment $\overline{AB}$ — *Segment $\overline{AB}$ is the set of points A, B, and all the points between A and B*

Plane — *A two-dimensional flat surface*

In **geometry,** you study the size, shape, and position of objects. One way to make that study easier is to concentrate on the outline of an object. Look closely at the points and lines that make up geometric figures.

A **point** is the simplest of geometric figures and it has no formal definition. In geometry, a point is represented by a dot on a piece of paper. Capital letters are used to name points. A point has a location but no size or shape. You might think of a point as a star in the night sky. Here are some examples of points.

These points are named *A, B, C,* and *D.*

There are two other concepts in geometry that have no formal definition—a **line** and a **plane.** A straight two-headed arrow represents a line. A line is endless in two directions. A line can be named by naming any two points on it, or it can be named with a lowercase letter. You might think of a line as a straight, thin wire or thread, but a geometric line has no thickness. Here is an example of a line.

This line is named $\overleftrightarrow{AB}$ or $\overleftrightarrow{BA}$ or *m.*

The four-sided shape at the left represents a plane. A plane is a flat surface. It extends in all directions with no end and has no thickness. A plane is named using capital letters. You might think of a page from this book or a tabletop as part of a plane.

This is plane *W.*

2 Chapter 1 Exploring Geometry: Points, Lines, and Angles in the Plane

EXAMPLE Name this line in nine different ways.

This line can be named $\overleftrightarrow{AB}$, $\overleftrightarrow{BA}$, $\overleftrightarrow{AC}$, $\overleftrightarrow{CA}$, $\overleftrightarrow{BC}$, $\overleftrightarrow{CB}$, $\overleftrightarrow{AC}$, $\overleftrightarrow{CA}$, or ℓ.

Point — *A location in space represented by a dot*

Ray $\overrightarrow{AB}$ — *A set of points that is part of a line; a ray has one endpoint and extends in one direction with no end*

Points on the same line are called **collinear** points.
In the example above, *A, B,* and *C* are collinear points on line ℓ.

EXAMPLE Which three of these points are collinear?

Points *D, E,* and *F* appear to be on the same line.

If you draw a line between any two points, such as *A* and *B,* you have drawn a **line segment.** A line segment is part of a line.

This is segment $\overline{AB}$.

Segment $\overline{AB}$ is the set of points *A, B,* and all points between *A* and *B. A* and *B* are called the **endpoints** of the segment. The symbol $\overline{AB}$ is read "segment *AB.*"

Recall that a set is a collection of particular things, like the set of points on a line segment.

EXAMPLE Draw and name all the line segments between points *P, Q,* and *R.*

Remember that you need two endpoints for each segment. You can draw segments $\overline{PQ}$, $\overline{QR}$, and $\overline{RP}$.

Exploring Geometry: Points, Lines, and Angles in the Plane Chapter 1 3

Points and Lines in the Plane

Workbook Activity 1

Points and Lines in the Plane

Additional Practice 1

bookends could represent endpoints, and the book bindings lined up between the bookends could be collinear points along a line segment.

3. Reinforce and Extend

APPLICATION

Career Connection Who uses lines, line segments, and rays? Architects, engineers, carpenters, and sculptors are among the people whose careers require them to use geometry daily. Each has tools that enable him or her to precisely use lines and planes. Designate an area in the classroom where students can set up a career tool display. Each student should select a career in which geometry is used and then prepare an exhibit of tools and other materials that are used by people in the career. For example, an exhibit for an architect might include blueprints, T squares, and compasses. One for a carpenter might include a level, a tape measure, and a plumb. Have students explain the use and application of the tools they display.

LEARNING STYLES

Visual/Spatial Using a color-coded system may help students more easily distinguish the illustrations or symbols used for points, endpoints, line segments, rays, lines, and planes. Prepare a color-coded system for the symbols and illustrations. Use different colored chalk to represent each symbol or figure. Write the code key on the board. Use the system as you teach the lesson, and encourage students to use the colors when writing the symbols. This system will provide visual reinforcement of the symbols.

Exploring Geometry Chapter 1 3

Lessons

- Quick overviews of chapters and lessons save planning time.
- Lesson objectives are listed for easy reference.
- Page references are provided for convenience.
- Instructional resources are clearly identified.
- Easy-to-follow lesson plan in three steps saves time: Warm-Up Activity, Teaching the Lesson, and Reinforce and Extend.
- Common Error alerts teachers to possible student errors before they are made.
- Algebra Connection reviews and applies algebraic skills and concepts to geometry.
- Mental Math activities encourage students to think independently.
- Manipulative activities are included to develop students' mathematical understanding at a concrete level.
- Calculator activities provide teaching suggestions for reinforcing and extending the lesson.
- Applications: Three areas of application—Career Connection, In the Community, and At Home—help students relate geometry to the world outside the classroom. Applications motivate students and make learning relevant.
- Group Problem Solving involves students in working together to solve a problem.
- Learning Styles provide teaching strategies to help meet the needs of students with diverse ways of learning. Modalities include Auditory/Verbal, Visual/Spatial, Body/Kinesthetic, Logical/Mathematical, and Interpersonal/Group Learning. Additional teaching activities are provided for LEP/ESL students.
- Answers are provided in the Teacher's Edition for all exercises in the Student Text. A separate Solutions Key for the student text is available on CD-ROM. Answers are also provided in the Teacher's Edition for the Workbook and other Teacher's Resource Library activities.
- Activity pages from the Teacher's Resource Library are shown at point of use.

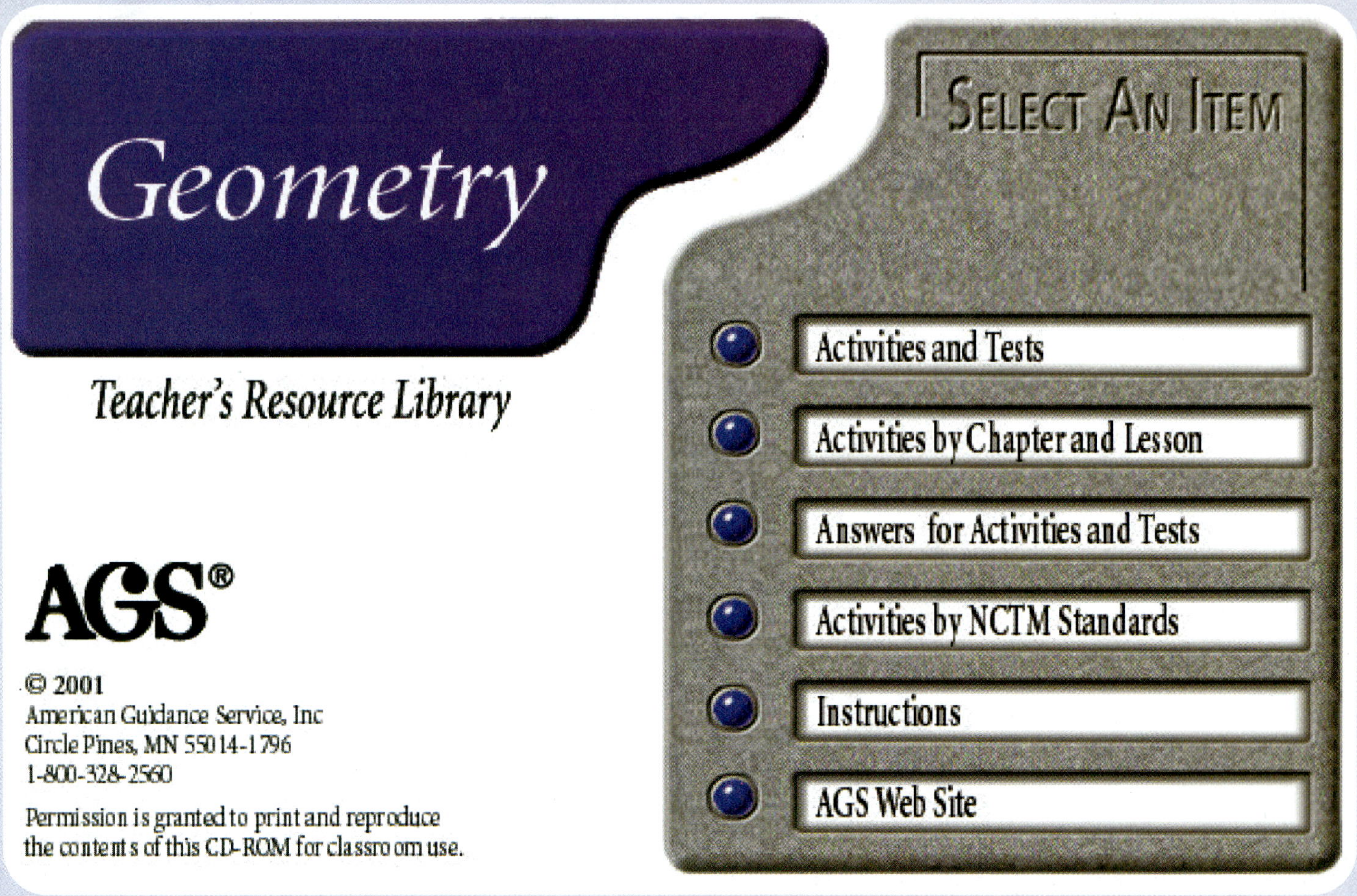

TRL All the activities you'll need to reinforce and extend the text are conveniently located on the AGS Teacher's Resource Library (TRL) CD-ROM. All of the reproducible activities featured in the Teacher's Edition are ready to view, select, and print. Additionally, you can preview other materials by directly linking to the AGS Web site.

Workbook Activities
Workbook activities may be reproduced as added reinforcement and practice for each lesson and application. These activities are also available in a bound workbook format.

Additional Practice
Additional reproducible activity pages offer more skill practice for students. The exercises parallel the practice exercises in the text.

Geometry on the Job
Career activities link geometry and mathematical concepts for each chapter to on-the-job applications in the workplace.

Community Connections
Relevant activities are designed to help students apply geometry skills and concepts outside the classroom and in the community.

Correlation to NCTM Standards
Selected activities are correlated to the standards set forth by the National Council of Teachers of Mathematics.

Self-Study Guide
An assignment guide provides the student with an outline for working through the text independently. The guide provides teachers with flexibility for individualized instruction or independent study.

Mastery Tests
Two forms of each Chapter Mastery Test allow use of one form as a pretest and the other form as a posttest. A Cumulative Mastery Test is available for each chapter, as well as a Midterm and Final Mastery Test.

Answer Key
An answer key for all of the reproducible activities and tests in the Teacher's Resource Library is included in the TRL and also in the Teacher's Edition.

■ **Workbook Activities:**
Workbook activities may be reproduced as added reinforcement and practice for lessons. These activities are also available in workbook format.

Name Date Period | Chapter 3 *Workbook Activity* 23

More Transversals

EXAMPLE A transversal creates eight angles. These angles are categorized as exterior, interior, corresponding, alternate interior, alternate exterior, or supplementary angles.

ℓ ∥ m

Directions Use the figure at the right for problems 1–5.

1. Name the interior angles.
2. Name all pairs of exterior angles.
3. Name eight pairs of supplementary angles.
4. Name all pairs of corresponding angles.
5. Name all pairs of alternate interior angles.

Directions Use the figure shown for problems 6–10. Name the angles as *exterior, interior, alternate exterior, alternate interior, corresponding,* or *supplementary.*

6. ∠*a* and ∠*e*
7. ∠*b* and ∠*h*
8. ∠*d* and ∠*f*
9. ∠*a* and ∠*d* and ∠*g* and ∠*f*
10. ∠*c* and ∠*b* and ∠*e* and ∠*h*

©AGS® American Guidance Service, Inc. Permission is granted to reproduce for classroom use only. Geometry

Name Date Period | Chapter 3 *Additional Practice* 12

Parallel Lines

Directions Use the figure below. Write *T* if you think the statement is true. Write *F* if you think the statement is false.

1. $\overline{AB}$ is parallel to $\overline{DC}$.
2. $\overline{AD}$ is parallel to $\overline{BC}$.
3. $\overline{DC}$ is parallel to $\overline{EF}$.
4. $\overline{CD}$ is parallel to $\overline{EF}$.
5. $\overline{EB}$ is parallel to $\overline{DF}$.
6. $\overline{AB}$ is parallel to $\overline{DF}$.
7. $\overline{ED}$ is parallel to $\overline{BF}$.
8. $\overline{AE}$ is not parallel to $\overline{FC}$.
9. $\overline{EB}$ is not parallel to $\overline{EF}$.
10. $\overline{DF}$ is parallel to $\overline{DC}$.

Directions Use symbols to describe each pair of lines.

11. 12. 13. 14. 15.

©AGS® American Guidance Service, Inc. Permission is granted to reproduce for classroom use only. Geometry

■ **Additional Practice:**
Reproducible pages offer additional skill practice and parallel the exercises from each lesson.

Name Date Period | Chapter 3 *Community Connection* 3

Construction Sites

Construction workers and carpenters w… The vertical beams holding up a wall m… beams in a building must be parallel wi… structure sits. The main tool that the w… are parallel is called a *level.* A level cons… water about a half of a centimeter wide… the water is an air bubble a few millime… with two parallel lines a few millimeters… its middle. Each line is exactly the same… the tube. This allows the air bubble to… lighter than the water, the air bubble wi… tube is more vertical.

Generally, two different tubes are placed… tool. One tube will be parallel to the str… perpendicular to the straightedge. When… surface, the parallel tube will show whe… tal. If the air bubble is exactly in betwee… surface is level. The perpendicular tube… such as a wall, is vertical.

Another tool used to check vertical surf… is a piece of thin string with a weight at… the weight straight down, so the string… can be held up to a vertical line and che… they line up, then the surface is vertical.

Directions

Step 1 Make arrangements to visit a c… workers or interview a contract… are used to make sure that line…

Step 2 Write a brief report of your fin… the class.

©AGS® American Guidance Service, Inc. Permission is gran…

Name Date Period | Chapter 3 *Geometry on the Job* 3

Drafting

Drafting is the art of making detailed drawings of an object. The drawings show the object from many different views. These drawings might be used by a machinist to manufacture an engine part or by a contractor to construct a building. A draftsperson uses angles, parallel lines, and perpendicular lines every time he or she sits down at a drafting board to work. Until very recently, draftspersons used a variety of tools to allow them to make accurate scale drawings that include angles, parallel lines, and perpendicular lines. Now, most drafting work is done using CAD (computer-aided design).

The tools that draftspersons used before CAD were fairly simple, yet very effective. A draftsperson worked at a large drawing board that could be tilted to a desired angle. The board included a sliding straightedge that could be moved up and down to produce lines parallel with the bottom of the board. Paper was taped to the board so that its bottom edge ran parallel with this straightedge. Vertical lines were drawn by using a tool called a T-square, which is essentially a straightedge with a perpendicular "T" attached so that it can run along the top of the straightedge. Another tool was the bevel. It is like a pair of straightedges with a protractor to measure the angle between them. It was used to create angled lines relative to the horizontal straightedge.

Try It Yourself
Find out more about drafting:

1. Visit the library to look for books about the subject.
2. Ask your teacher about contacting a local draftsperson or architect to speak to the class.
3. Arrange a visit to a drafting or CAD class at a local school or college. Observe how the tools are used to create accurate lines and angles.

Prepare a Report
Write a brief report of your observations and your research.

©AGS® American Guidance Service, Inc. Permission is granted to reproduce for classroom use only. Geometry

■ **Geometry on the Job and Community Connections:**
Reproducible pages that extend and apply geometric skills can be printed from the TRL.

Name Date Period | Chapter 1 *Mastery Test A* page 1

Chapter 1 Mastery Test A

Directions Name the following geometric figures as many ways as you can.

1. 2. 3. 4.

Directions Write the letter of the answer to each question.

A. none B. one C. two D. three

5. How many endpoints does a line have?
6. How many line segments can be created with three points?
7. How many endpoints does a line segment have?
8. How many endpoints does a ray have?

Directions Use the Ruler Placement Postulate to find the distance between the points.

9. 10. 11. 12.

Directions Use a ruler to measure these segments in centimeters or inches.

13. ______ centimeters
14. ______ inches
15. ______ centimeters
16. ______ inches

©AGS® American Guidance Service, Inc. Permission is granted to reproduce for classroom use only. Geometry

■ **Mastery Tests:**
Two forms of each Chapter Mastery Test, as well as Cumulative, Midterm, and Final Mastery Tests are conveniently referenced as assessment options.

Guidelines for Using Manipulatives

The manipulative activities in the AGS Geometry Teacher's Edition are intended to develop students' geometric understanding at a concrete level. Manipulatives are used to build visual representations, or models, of geometric figures, concepts, and relationships. Students introduced to geometric concepts through the use of manipulatives develop a grasp of the relationships among geometric figures and symbolic representations more quickly and in a more internalized manner. In other cases, the manipulatives can amplify or reinforce concepts for students with specific learning styles, e.g., Body/Kinesthetic or Visual/Spatial. For this reason, activities relying on manipulatives may appear in either *Teaching the Lesson* or *Reinforce and Extend* in the 3-step lesson plan.

The following general guidelines will help you to successfully integrate manipulatives in your geometry program.

- Spend as much time as necessary on guided practice when an activity or concept is first introduced. This will enable students to understand the use of the manipulatives before they begin to use them independently.

- Make manipulatives available at students' desks during group instruction so that they can execute the activity as you explain the geometric concept.

- Present activities in order. Students will develop their understanding of the use of manipulatives in small, incremental steps.

- Employ repetition as time permits. The suggested exercises represent the minimum amount of practice necessary for most students. For students who require additional practice, supplementary exercises are available in the Student Workbook.

- Point out to students that in some cases they can use manipulatives to check their work. Reviewing and evaluating one's own work is a metacognitive skill that improves problem-solving ability.

Manipulatives Used with AGS Geometry

All of the manipulatives used with AGS Geometry are available from ETA. Call 1-800-445-5985. For your information and planning, the chart below shows the manipulatives used within AGS Geometry. In addition, the Teacher's Resource Library (TRL) includes Manipulatives Masters 1–4 (shown below).

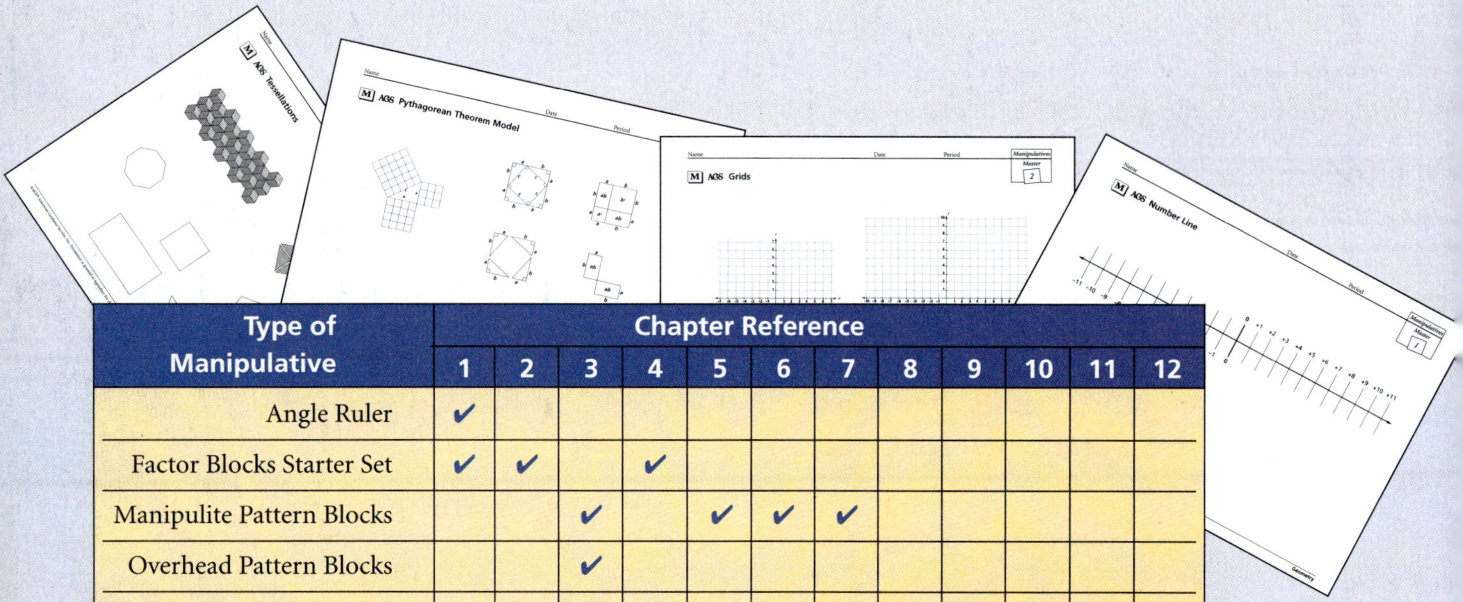

Type of Manipulative	Chapter Reference											
	1	2	3	4	5	6	7	8	9	10	11	12
Angle Ruler	✔											
Factor Blocks Starter Set	✔	✔		✔								
Manipulite Pattern Blocks			✔		✔	✔	✔					
Overhead Pattern Blocks			✔									
Polygons Power Pack					✔	✔	✔		✔	✔		
GeoReflector Mirror						✔						
Pattern Blocks Geoboard							✔					
Proofs of Pythagoras								✔				
Geometric Volume Set											✔	
GeoSolids											✔	

Manipulative Materials

The following describes the manipulative materials used in the instructional program in the Teacher's Edition.

Angle Ruler

The Angle Ruler will come in handy for measuring lengths to 30 cm (12 in.) as well as measuring angles. The device is particularly useful for measuring an initial angle and then confirming that the construction of an angle bisector resulted in two angles of equal measure. The Angle Ruler can also be pressed into service to show that the sum of angles in a triangle is 180°.

Factor Blocks

The dimensions of each Factor Block, given in prime numbers, can be used to compute each block's volume. Factor Blocks can provide useful visual models for exploration of factors, multiples, area, perimeter, and volume. In addition, students may find the blocks useful in determining Pythagorean Triples and in simplifying square root expressions.

Manipulite Pattern Blocks

Manipulite® Pattern Blocks provide students with a selection of polygons they can use to gain a hands-on familiarity with regular, similar, and congruent polygons including hexagons, squares, trapezoids, triangles, and rhombuses. The polygons can be used to emphasize the proportional characteristics of similar polygons. Further, the blocks can be helpful in demonstrating the formulas for areas of trapezoids and parallelograms. Polygon blocks can also be used to reinforce the difference between adjacent and nonadjacent vertices.

Overhead Pattern Blocks

The Overhead Pattern Blocks, which match the Manipulite Pattern blocks' colors and sizes, allow students to follow a demonstration on an overhead projector as they work with their own set of polygons at their desks. The blocks are useful for modeling congruent, similar, and regular polygons as well as developing a vocabulary for quadrilaterals.

Polygons Power Pack

The Polygons Power Pack can be employed to assist students in identifying and classifying a variety of polygons. Students can use polygons in the pack to strengthen their familiarity with triangles, quadrilaterals, and circles. Circles in the pack are particularly useful for comparing the radius, circumference, and area of circles.

GeoReflector Mirror

The clear plastic GeoReflector Mirror allows students to see the object on the near side of the mirror as well as its reflection on the far side of the mirror. This makes the device particularly helpful in supporting the logical proof that the distance from a mirror to an image is the same as the distance from the mirror to the object.

Pattern Blocks Geoboard

Students using the Pattern Blocks Geoboard can produce outline shapes of a variety of polygons. The resulting figures can be used to amplify the difference between area and perimeter as well as reinforce concepts such as congruence and similarity.

Proofs of Pythagorus

This set of squares and triangles allows students to model each of the proofs presented in AGS Geometry. Inner and outer square side lengths as well as length of sides of triangles can be measured and the algebraic statement $a^2 + b^2 = c^2$ can be confirmed. In addition, the Proofs of Pythagoras manipulatives allow students to produce an additional proof, not directly specified in the text.

Geometric Volume Set

Invite students to confirm the formulas in this lesson using the Geometric Volume Set and sand, dried beans, or rice to compare the volumes of pyramids, prisms, cones, and cylinders.

Relational GeoSolids

The clear plastic models of a variety of the solids are invaluable in helping students visualize the differences between solids such as rectangular prisms and triangular prisms or square pyramids and triangular pyramids. The GeoSolids are also useful for distinguishing among surface area, lateral faces, base, volume, and altitude as well as radius, diameter, and chords.

To order the manipulatives used with AGS Geometry, call ETA at 1-800-445-5985.

Learning Styles

The learning style activities in the AGS Geometry Teacher's Edition provide activities to help students with special needs understand the lesson. These activities focus on the following learning styles: Visual/Spatial, Auditory/Verbal, Body/Kinesthetic, Logical/ Mathematical, Interpersonal/Group Learning, LEP/ESL. These styles reflect Howard Gardner's theory of multiple intelligences. The writing activities suggested in the Student Text are appropriate for students who fit Gardner's description of Verbal/Linguistic Intelligence.

The activities are designed to help teachers capitalize on students' individual strengths and dominant learning styles. The activities reinforce the lesson by teaching or expanding upon the content in a different way.

Following are examples of activities featured in the AGS Geometry Teacher's Edition:

Visual/Spatial students benefit from seeing illustrations or demonstrations beyond what is in the text.

Learning Styles

Visual/Spatial Help students draw sketches and write statements on index cards to identify the theorems and other axioms, definitions, and properties they need to review. For example, they might draw parallel lines with alternate interior angles labeled x and y and write the statement $m\angle x = m\angle y$. Working with a partner, one student holds up a card and the other recites the theorem, property, etc., that justifies the statement on the card. Students can take turns holding the cards and reciting the theorems.

Body/Kinesthetic learners benefit from activities that include physical movement or tactile experiences.

Learning Styles

Body/Kinesthetic Ask pairs of students to model parallel, intersecting, and skew lines in various ways around the room. For example, the pair could stand against the same wall parallel to each other to model parallel lines, or one could stand leaning slightly to model intersecting lines. Ask for volunteers to display their line models to the class.

Interpersonal/Group learners benefit from working with at least one other person on activities that involve a process and an end product.

Learning Styles

Interpersonal/Group Learning Have individuals write several conditionals. Tell students to write conditionals for a variety of situations, not just mathematical statements. Then in groups of 3 or 4, students share their conditionals and the group decides whether they are true or false. As a group, students should then write the converse of each conditional and decide whether it is true or false. Encourage students to use counterexamples to illustrate why a converse is false. Ask the groups to turn in one paper showing their work, including their reasoning for determining whether a conditional or its converse is true or false.

Auditory/Verbal students benefit from having someone read the text aloud or listening to the text on audiocassette. Musical activities appropriate for the lesson may help auditory learners.

Learning Styles

Auditory/Verbal Have students compare the two methods for finding the midpoint of a segment. Ask them to describe different situations in which one of the methods seems easier to use than the other and to explain why.

Logical/Mathematical students learn by using logical/mathematical thinking in relation to the lesson content.

Learning Styles

Logical/Mathematical Ask students to explain how to find the slope of an escalator. They should identify the measurements needed to calculate the slope. Then ask them to estimate what the slope of an average escalator might be and to compare its slope to that of an elevator.

LEP/ESL students benefit from activities that promote English language acquisition and interaction with English-speaking peers.

Learning Styles

LEP/ESL Make examples of pairs of triangles that are congruent by each of the postulates on one side of a large index card and write the postulate abbreviation on the other side. Students work in pairs, with one partner describing the pair of triangles and naming the postulate that is illustrated while the other holds the card and verifies the answer. Then partners switch roles and repeat. The students should use both sides of the cards as cues.

Correlation of the NCTM Standards to AGS Geometry

Standard 1: Number and Operations	
Mathematics instructional programs should foster the development of number and operation sense to enable all students to— ■ understand numbers, ways of representing numbers, relationships among numbers, and number systems; ■ understand meanings of operations and how they relate to one another; ■ compute fluently and make reasonable estimates.	**AGS GEOMETRY:** Representation, relationships, and operations: **Chapter 1: Lesson 6; Chapter 7: Lessons 1, 3–5, Application.** Computational tools, strategies, and estimation: **See Problem-Solving Strategies on pages xii–xiii; Calculator Practice exercises on pages 22, 52, 72, 111, 126, 171, 199, 224, 253, 260, 274, 307, 330, 382; Chapter 8: Lesson 4; Chapter 10: Lessons 2, 3.**

Standard 2: Algebra	
Mathematics instructional programs should include attention to patterns, functions, symbols, and models to enable all students to— ■ understand patterns, relations, and functions; ■ represent and analyze mathematical situations and structure using algebraic symbols; ■ use mathematical models to represent and understand quantitative relationships; ■ analyze change in various contexts.	**AGS GEOMETRY:** Patterns and functional relationships: **Chapter 4: Lessons 4, 5; Chapter 5: Lesson 6; Chapter 6: Lesson 7; Chapter 7: Lesson 3; Chapter 8: Lessons 1–3, 5, 6; Chapter 9: Lessons 1, 2, 4, 5, Application; Chapter 10: Lesson 2; Chapter 12: Application.** Use of symbols: Universal symbols for geometric figures such as points, rays, angles, lines, line segments, perpendiculars, triangles, and so on are used throughout the text. Mathematical models: **Chapter 1: Application; Chapter 4: Application.**

Standard 3: Geometry	
Mathematics instructional programs should include attention to geometry and spatial sense to enable all students to— ■ analyze characteristics and properties of two- and three-dimensional geometric shapes and develop mathematical arguments about geometric relationships; ■ specify locations and describe spatial relationships using coordinate geometry and other represental systems; ■ apply transformations and use symmetry to analyze mathematical situations; ■ use visualization, spatial reasoning, and geometric modeling to solve problems.	**AGS GEOMETRY:** Analyze two- and three-dimensional objects: **Chapter 1: Lesson 3, Application; Chapter 3: Lessons 1, 5; Chapter 5, Lessons 1–9; Chapter 5: Application; Chapter 6: Lessons 1–8, Application; Chapter 7: Lessons 1–5, Application; Chapter 8: Lessons 1, 2, Application; Chapter 9: Lessons 1, 3–5; Chapter 10: Lessons 1–8; Chapter 11: Lessons 1–4, Application; Chapter 12: Lessons 1, 3, Application.** Coordinate geometry and graphic representation: **Chapter 4: Lessons 1–7; Chapter 6: Lessons 5, 7, 8; Chapter 7: Lessons 4, 5; Chapter 8: Lesson 6; Chapter 9: Lesson 3; Chapter 12: Lesson 2.** Transformations and symmetry: **Chapter 6: Lessons 5–8; Chapter 7: Lessons 4, 5.** Visualization, spatial reasoning, and geometric modeling: **Chapter 1: Lesson 4; Chapter 6: Lessons 5–8, Application; Chapter 9: Lessons 4, 5, Application; Chapter 10: Lesson 8; Chapter 11: Lessons 1–4, Application; Chapter 12: Lesson 1–3, Application.**

Standard 4: Measurement

Mathematics instructional programs should include attention to measurement to enable all students to—

- understand measurable attributes of objects and units, systems, and processes of measurement;
- apply appropriate techniques, tools, and formulas to determine measurements.

AGS GEOMETRY:

Measurement systems and units:
Chapter 1: Lesson 4; Chapter 7: Lesson 5; Chapter 9: Lesson 4; Chapter 11: Lesson 5.

Measurement tools, techniques, and formulas:
Chapter 1: Lessons 2, 4; Chapter 2: Lesson 4; Chapter 7: Lesson 3; Chapter 8: Lessons 5, 6; Chapter 9: Lessons 1, 4–6, Application; Chapter 10: Lessons 2, 3, 5; Chapter 11: Lessons 1–4.

Standard 5: Data Analysis and Probability

Mathematics instructional programs should include attention to data analysis and probability to enable all students to—

- formulate questions that can be addressed with data and collect, organize, and display relevant data to answer them;
- select and use appropriate statistical methods to analyze data;
- develop and evaluate inferences and predictions that are based on data;
- understand and apply basic concepts of probability.

AGS GEOMETRY:

Collect, organize, and represent data:
Chapter 10: Lesson 2.

Interpret data:
Chapter 4: Application; Chapter 10: Application.

Develop and evaluate conclusions based on data:
Chapter 6: Lesson 6; Chapter 10: Lesson 8.

Chance and probability:
Chapter 10: Lesson 4.

Standard 6: Problem Solving

Mathematics instructional programs should focus on solving problems as part of understanding mathematics to enable all students to—

- build new mathematical knowledge through problem solving;
- solve problems that arise in mathematics and and in other contexts;
- apply and adapt a variety of appropriate strategies to solve problems;
- monitor and reflect on the process of mathematical problem solving.

AGS GEOMETRY:

Problem-Solving exercises throughout provide students opportunities to apply problem-solving strategies. See Problem-Solving Strategies on **pages xii–xiii; pages 6, 10, 14, 43, 55, 72, 89, 100, 103, 118, 159, 183, 193, 199, 230, 239, 243, 265, 290, 299, 322, 353, 378.**

In addition, see Application exercises at the end of each chapter, **pages 28, 56, 90, 128, 172, 214, 244, 278, 310, 346, 370, 386.**

Standard 7: Reasoning and Proof

Mathematics instructional programs should focus on learning to reason and construct proofs as part of understanding mathematics to enable all students to—

- recognize reasoning and proof as fundamental aspects of mathematics;
- make and investigate mathematical conjectures;
- develop and evaluate mathematical arguments and proofs;
- select and use various types of reasoning and methods of proof.

AGS GEOMETRY:

Reasoning and proof:
Chapter 2: Lessons 1–5; Chapter 3: Lesson 2; Chapter 8: Lessons 2–4, 7.

Investigate conjectures, develop arguments and proofs:
Chapter 3: Lesson 5; Chapter 5: Lesson 9; Chapter 6: Lessons 1–4; Chapter 7: Lessons 1, 2; Chapter 8: Lessons 2, 3, 7.

Employ reasoning and proofs:
Chapter 2: Lessons 1–5; Chapter 3: Lesson 6; Chapter 5: Lessons 4, 5; Chapter 6: Lessons 1–4; Chapter 8: Lessons 2, 3; Chapter 10: Lesson 7.

Standard 8: Communication

Mathematics instructional programs should use communication to foster understanding of mathematics to enable all students to—

- organize and consolidate their mathematical thinking through communication;
- communicate their mathematical thinking coherently and clearly to peers, teachers, and others;
- analyze and evaluate the mathematical thinking and strategies of others;
- use the language of mathematics to express mathematical ideas precisely.

AGS GEOMETRY:

Oral explanation and discussion:
Problem-Solving exercises throughout provide opportunities for oral language. See Problem-Solving Strategies on **pages xii–xiii; pages 6, 10, 14, 43, 55, 72, 89, 100, 103, 118, 159, 183, 193, 199, 230, 239, 243, 265, 290, 299, 322, 353, 378.**

Writing About Mathematics exercises encourage students to express mathematical ideas in clear and logical form. See **pages 17, 23, 24, 35, 48, 63, 67, 86, 117, 122, 126, 135, 156, 187, 199, 209, 213, 228, 233, 274, 277, 290, 298, 322, 325, 335, 343, 377.**

Contributions of others: **Chapter 2: Lesson 2; Chapter 8: Lessons 1–5; Chapter 9: Lesson 6.**

Mathematical language: **Definitions of topic-relevant terms are included on the first page of each lesson throughout.**

Standard 9: Connections

Mathematics instructional programs should emphasize connections to foster understanding of mathematics to enable all students to—

- recognize and use connections among mathematical ideas;
- understand how mathematical ideas interconnect and build on one another to produce a coherent whole;
- recognize and apply mathematics in contexts outside of mathematics.

AGS GEOMETRY:

Relationships between geometric principles, as well as the progression of ideas from lines and planes, formal proofs, Euclid's Postulates, construction of geometric figures, and characteristics of specific groups of figures are explored throughout the text and underlying principles are presented as an integrated whole.

Algebra and geometry: Connections between geometry and algebra are presented in **Chapter 1: Lesson 6; Chapter 2, Lesson 4; Chapter 3: Lesson 4; Chapter 4: Lessons 1–7; Chapter 5: Lessons 5, 9; Chapter 6: Lesson 4; Chapter 7: Lesson 3; Chapter 8: Lessons 1, 3, 5; Chapter 9: Lessons 1–4; Chapter 10: Lesson 2; Chapter 11: Lessons 1, 4; Chapter 12: Lesson 2, Application.**

Outside contexts: The role of geometry in other disciplines is explored in the chapter openers on **pages 2, 32, 60, 94, 132, 176, 219, 248, 282, 314; in Application exercises on pages 28, 56, 90, 128, 172, 214, 244, 278, 310, 344, 368, 386 , 386; in Technology Connections on pages 10, 35, 64, 103, 151, 203, 234, 255, 308, 343, 357, 382; and in Geometry Around Us on pages 19, 45, 65, 79, 83, 102, 171, 200, 243, 277, 309, 327, 341, 358, 378.**

Standard 10: Representation

Mathematics instructional programs should emphasize mathematical representations to foster understanding of mathematics to enable all students to—

- create and use representations to organize, record, and communicate mathematical ideas;
- select, apply, and translate among mathematical representations to solve problems;
- use representations to model and interpret physical, social, and mathematical phenomena.

AGS GEOMETRY:

Representations: Throughout the text, construction examples and exercises allow students to build a repertoire of geometric representations. See **Chapter 1: Lessons 1–4; Chapter 2: Lesson 3; Chapter 3: Lessons 2, 6; Chapter 4: Lessons 5–7; Chapter 5: Lessons 1, 2, 7, 8; Chapter 6: Lessons 1, 2; Chapter 10: Lesson 7.**

Models: See the Experiencing Geometry activities on **pages 23, 27, 43, 48, 82, 106, 148, 167, 204, 213, 234, 262, 266, 296, 321, 340, 354, 377.**

How Do We Handle Proofs in AGS Geometry?

For more information, call 1-800-328-2560.

Visit our Web site at www.agsmath.com

or send e-mail to agsmail@agsnet.com

Reasoning and proofs are often new, but important, concepts to students studying geometry. For this reason, proofs in AGS Geometry are introduced very carefully. Three principles guide the inclusion of proofs in the Student Text:

Principle	Rationale	Examples
Theorems are introduced by using numerical examples or other demonstrations before presenting formal proofs.	Students use inductive thinking skills to go from the specific to the general, and then to the statement of the theorem. Students use familiar arithmetic facts to support geometric conclusions.	■ See Exercise C, page 67 for numerical examples and measurement, followed by Chapter 3, Lesson 3, Theorems About Parallel Lines (pages 68–72). ■ See numerical example on page 167, followed by proof of the Angle Sum Theorem on page 168. ■ The Pythagorean Theorem (Chapter 8) is motivated by the discussion of Pythagorean Triples (pages 250–252) and by geometric demonstrations on pages 254, 255, and 257.
The most important theorems are proven; the others are assumed as postulates.	Students are introduced to more formal, two-column proofs or demonstrations in cases where important theorems are not intuitively clear. Other theorems that appear more obvious to the learner are safely assumed as postulates. In either case, reasoning and thinking skills are emphasized. Guided proofs assist the student with structure and format.	■ The Angle Sum Theorem (page 168) and the Pythagorean Theorem (pages 255 and 262) are not intuitive and are proven. ■ The triangle congruency theorems SAS, SSS, and ASA (pages 179, 189, 190) are assumed as postulates and demonstrated by construction. ■ Area formulas are derived from the area of a rectangle (pages 295–296).
More than one proof or demonstration is presented for key theorems.	Students are allowed to form the concept of the theorem from several viewpoints, demonstrating various types of reasoning and methods of proof. Students understand the key theorems more completely by using a variety of problem-solving strategies and applications.	■ Angle sums for quadrilaterals are developed from square and rectangle examples on page 80, followed by Chapter 5, Triangles and Quadrilaterals, and the Angle Sum Theorem proof (page 168). ■ The Pythagorean Theorem is illustrated by Pythagorean Triples and demonstrated and proven in four different ways. (See Chapter 8.) ■ Similarity (Chapter 7) is discussed geometrically and is also illustrated using the coordinate system for transformations. Nested similar right triangles (page 228) are introduced in preparation for teaching the basic trigonometric functions (pages 340–341).

Geometry

by
Siegfried Haenisch

AGS®
American Guidance Service, Inc.
Circle Pines, Minnesota 55014-1796
1-800-328-2560

About the Author

Siegfried Haenisch, Ed.D., has taught mathematics at every level, from elementary to graduate school, most recently as Professor in the Department of Mathematics and Statistics at the College of New Jersey. The Mathematical Association of America granted him the 1995 Award for Distinguished Teaching of Mathematics. Dr. Haenisch was the site director for the training of teachers in the New Jersey Algebra Project. He was a member of the National Science Foundation Institutes in Mathematics at Rutgers University, Oberlin College, and Princeton University. At Yale University he was a member of the Seminar in the History of Mathematics, sponsored by the National Endowment in the Humanities. Dr. Haenisch currently serves as a mathematics curriculum consultant to school districts.

Photo Credits: pp. ix (top), xiv, EyeWire Images; p. ix (bottom), Myrleen Ferguson, PhotoEdit; p. x, PhotoDisc; p. xii, Skjold Photography; pp. xvi, 1, CORBIS/Roger Ressmeyer; pp. 6, 28, 62 (top), 118, Tony Freeman/PhotoEdit; p. 19, Paul Merideth/Tony Stone Images; p. 32, T. Stewart and R. Baxter/Tony Stone Images; p. 39, CORBIS/Bettmann; pp. 43, 370, Felicia Martinez/PhotoEdit; pp. 44, 100, 385, P. R. Productions/SuperStock; pp. 45, 378, Michael Newman/PhotoEdit; pp. 53, 277, Spencer Grant/PhotoEdit; p. 56, Giraudon/SuperStock; p. 60, CORBIS/Michael T. Sedam; p. 62 (middle), G. Outerbridge/SuperStock; p. 64 (top), Chigmaroff/Davison/SuperStock; pp. 64 (bottom), 89 (top), R. A. Lee/SuperStock; p. 65, BL Productions/SuperStock; p. 72, Cindy Charles/PhotoEdit; p. 79, Jonathan Nourok/PhotoEdit; p. 83, S. Vidler/SuperStock; p. 89 (bottom), Dennis O'Clair/Tony Stone Images; p. 90, John Elk/Tony Stone Images; pp. 94, 382, PhotoEdit; p. 102, R. Heinzen/SuperStock; pp. 103, 220 (right), 309, 359, 360, 374, SuperStock; p. 128, Bill Aron/PhotoEdit; p. 132, Koji Kitagawa/SuperStock; p. 134, Mark Joseph/Tony Stone Images; p. 159 (left), Kristian Hilsen/Tony Stone Images; p. 159 (right), Peter Langone/Tony Stone Images; p. 171 (left), Bruce Forster/Tony Stone Images; p. 171 (right), Mike McQueen/Tony Stone Images; p. 176, Tom and Dee Ann McCarthy/PhotoEdit; p. 200 (top), Art Wolfe/Tony Stone Images; p. 200 (bottom), Robert Daemrich/Tony Stone Images; p. 214, NASA Collection/SuperStock; p. 218, Museum of Roman Civilization, Rome, Italy/Canali PhotoBank, Milan/SuperStock; pp. 220 (left), 327, L. Prosor/SuperStock; p. 235, *I Saw the Figure 5 in Gold* by Charles Demuth/SuperStock; p. 243 (left), Jeff Greenberg/PhotoEdit; p. 243 (right), Forms and Proportions of Roman Capital Letters/Manuscripts/SuperStock; p. 248, Joseph Pobereskin/Tony Stone Images; p. 250, Pythagoras Mathematician Marble Bust/Roman Art/SuperStock; p. 278 (top), Peter Pearson/Tony Stone Images; p. 278 (bottom), G. Fritz/SuperStock; p. 282, Jason Hawkes/Tony Stone Images; p. 310, Leslye Borden/PhotoEdit; p. 314, Mark Wagner/Tony Stone Images; p. 343, ESA/K. Horgan/Tony Stone Images; p. 350, Tim Hursley/SuperStock; pp. 364, 386, Susan Van Etten/PhotoEdit

Printed in the United States of America

ISBN 0-7854-2221-8

Product Number 91200

A 0 9 8 7 6 5 4 3 2 1

Contents

Geometry *Contents* **iii**

iv *Contents* *Geometry*

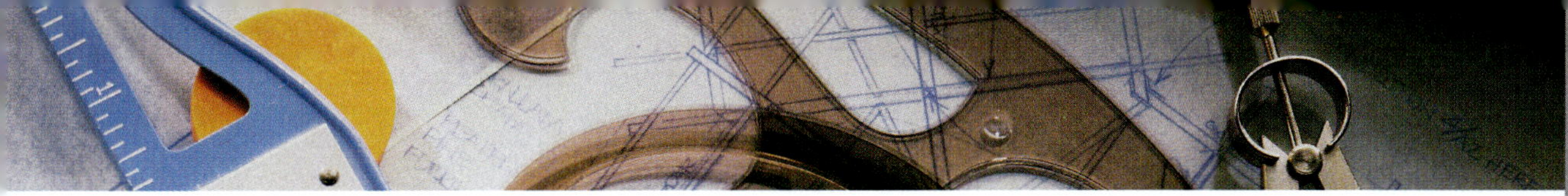

Geometry *Contents* v

vi *Contents* *Geometry*

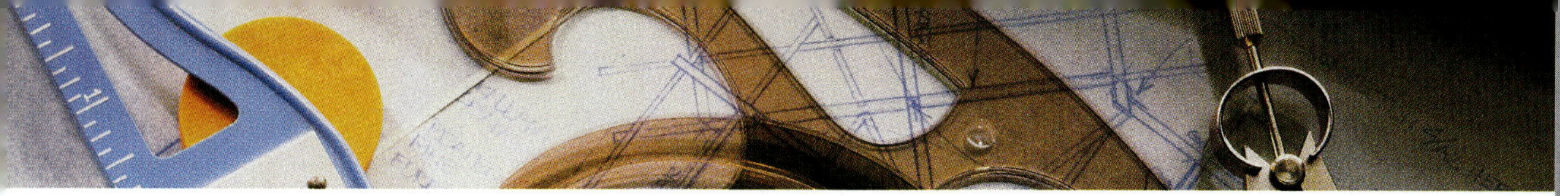

Chapter 9

Chapter 10

Geometry *Contents* **vii**

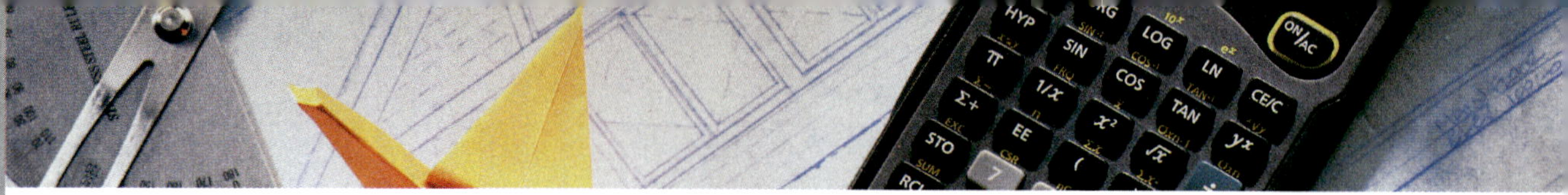

viii *Contents* *Geometry*

How to Study Geometry

Welcome to *Geometry*. Geometry is the study of how points, lines, and planes can be used to picture the space around us. In this book, the main attention will be placed on *plane* geometry. Plane geometry studies geometric figures in a plane, like squares, triangles, and circles. Later, we will consider *solid* geometry. In solid geometry, you will learn about solid figures like cubes, prisms, and spheres.

You may be asking yourself, "Why do I need to learn about geometry?"

Knowing about geometry will help you in many ways. Anytime you read a map, look at a floor plan for a house, or set up a baseball diamond, you are using the basic ideas of geometry. When you describe a shape or direction as rectangular, round, or at right angles, you are using terms from geometry. Everyday, carpenters, graphic designers, architects, engineers, and many others use geometry in their jobs. Knowing about geometric shapes and measurements for length, area, and volume will give you skills you can use throughout your life. Geometry will also teach you about thinking and problem solving.

How to Study

- Plan a regular time to study.
- Choose a quiet desk or table where you will not be distracted. Find a spot that has good lighting.
- Gather all the books, pencils, paper, and other equipment you will need to complete your assignments.
- Decide on a goal. For example: "I will finish reading and taking notes on Chapter 1, Lesson 1, by 8:00."
- Take a five- to ten-minute break every hour to keep alert.
- If you start to feel sleepy, take a short break and get some fresh air.

Name Date Period

page 1 — *How to Study Geometry* 1

Study Habits Survey

Directions Here are 25 statements about studying. If a statement describes what you really do—not what you think you should do—mark a *T* for True on the blank. If the statement is not how you study, mark an *F* for False on the blank.

Study Habits

_____ **1.** I have a regular study time.

_____ **2.** I listen to the radio or watch TV as I study.

_____ **3.** As I study, I try to express in my own words what the book is saying.

_____ **4.** My study time is often interrupted by phone calls or talking to people.

_____ **5.** I set goals for how much reading I will complete by a certain time.

_____ **6.** I usually sit in an easy chair or lie down to study.

_____ **7.** Before reading about a topic, I look over the material to see what it is about.

_____ **8.** If I start to feel sleepy, I take a nap.

_____ **9.** I try to figure out how what I am studying will be useful to me later.

Taking Notes

_____ **10.** I try to get the teacher's exact words down when I take notes.

_____ **11.** I put my notes in outline form as soon after taking them as possible.

_____ **12.** I rewrite all of my notes.

_____ **13.** When I rewrite my notes, I put them in outline form and fill in important details.

_____ **14.** I seldom take notes when I read my assignments.

_____ **15.** As I listen to teachers, I think about what they are saying.

_____ **16.** I write down everything the teacher says.

_____ **17.** During lectures, I look at the teacher when I am not taking notes.

 Geometry

TRL

Name Date Period

page 2 — *How to Study Geometry* 1

Study Habits Survey, *continued*

Taking Tests

_____ **18.** If I have trouble with a question, I don't move on until I can answer it.

_____ **19.** I think through the test questions before I begin to answer them.

_____ **20.** I only review by rereading or scanning the material on which I am to be tested.

_____ **21.** On math tests, I break down the problem into smaller parts.

_____ **22.** I think many test questions are tricky and are meant to fool me.

_____ **23.** I review my notes before a test and think about what questions will be asked.

_____ **24.** I stay up late the night before a test to study.

_____ **25.** I look over the entire test before I start answering questions.

The odd-numbered questions on this survey reflect good study habits. Give yourself a point for every odd-numbered question you answered with a True. Then give yourself a point for every even-numbered question you answered with a False. Decide on which study habits you would like to improve. Take the survey again after you have worked on improving your study skills. Compare your score on the odd-numbered items to see if you are making progress.

 Geometry

TRL

How to Study Geometry 1, pages 1 and 2

Using This Section

How to Study Geometry

Overview This section can be used to introduce the study of geometry, to preview the book's features, and to review effective study skills.

Objectives

- To introduce the study of geometry
- To preview the student textbook
- To review study skills

Student Pages ix–xv

Teacher's Resource Library TRL

How to Study Geometry 1–5

Introduction to the Book

Invite volunteers to read aloud the three paragraphs of the introduction. Discuss with students why studying geometry is important and what kinds of things people can learn from studying geometry.

How to Study

Read aloud each bulleted statement, pausing to discuss with students why the suggestion contributes to good study habits. Distribute copies of How to Study Geometry 1, pages 1 and 2. Read the directions with the class and then have students complete the survey. After they have scored their surveys, ask students to make a list of their study habits that need improving. After three or four weeks, have students complete the survey again to see whether they have improved their study habits. Suggest that they keep the survey handy and review it every month to see whether they are maintaining and improving their study habits.

Getting Ready to Study Geometry

Note that the text in the Teacher's Edition accompanying each chapter opener offers teaching suggestions and questions for the chapter introduction and photograph. The text also includes a list of the Teacher's Resource Library (TRL) materials designed for use with the chapter.

Point out to students that not all chapters will require the use of a straightedge, compass, protractor, or calculator. Even so, encourage students to keep these tools within reach whenever they are using the Geometry text.

Before Beginning Each Chapter

As students begin their study of Chapter 1, have a volunteer read aloud the Goals for Learning. Point out that the Goals for Learning appear at the beginning of each chapter. Stress to students that these goals describe what students are expected to learn in the chapter.

Using Chapter 1 as a model, ask students to find each of the features mentioned on page x. Discuss the importance of learning the vocabulary introduced in each chapter. Indicate to students that not all chapters have constructions; however, when students do encounter constructions, they must follow the step-by-step process presented.

Help students locate the *Experiencing Geometry* feature in Chapter 1. Point out that this feature, appearing in every chapter, often allows students to make physical models of difficult concepts.

Distribute How to Study Geometry 2, Writing About Mathematics to students, telling them that it can be used to record their responses to this feature.

Getting Ready to Study Geometry

What will you need to study geometry? Make sure you have enough paper, sharp pencils, and an eraser. You'll also need some basic tools for geometry.

- **Straightedge** A straightedge is a ruler without numbers. You'll use your straightedge to draw lines.
- **Compass** A compass is used to draw circles and arcs and to measure lengths.
- **Protractor** A transparent plastic protractor is used to measure angles.
- **Calculator** A scientific calculator will help with computation and some calculations.

Before you begin each chapter, read the chapter title and opening paragraph. Study the photograph. What does the photo tell you about the chapter? Read the Goals for Learning. They describe what you will learn in the chapter.

As you study each lesson, start with the lesson title. The title will tell you the main point of the lesson. Use the lesson features to help you understand what you read. You'll also discover how the lesson relates to what you already know.

Postulate
A statement about geometric figures accepted as true without proof

Vocabulary terms are the boldface words in your book. They are defined at the beginning of each lesson. You'll find these and other words in the **Glossary** at the back of the book. The Glossary also gives you examples. These words and definitions will help you to understand what you read and to think about geometry. It will be helpful if you memorize the definitions.

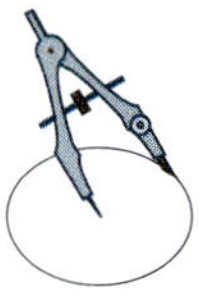

Constructions in some lessons show you step-by-step instructions to draw certain geometric figures. When you study the lesson, do your own construction on a separate sheet of paper. Be sure to follow each step carefully.

Experiencing Geometry gives you a chance to try out what you've learned about geometry. Some students find that they learn better if they can "do" or see something to illustrate a new idea.

Writing About Mathematics gives you a chance to think and write about what you've learned.

x *How to Study Geometry* *Geometry*

Name Date Period

How to Study Geometry 2

Writing About Mathematics

Topic

©AGS® American Guidance Service, Inc. Permission is granted to reproduce for classroom use only. Geometry

TRL

How to Study Geometry 2

Algebra Connection tells you where skills you learned in algebra are used to solve problems in geometry. If you need to review your algebra, turn to the **Algebra Review** section in the back of the book.

EXAMPLE **Examples** in each lesson show you how to use what you read to solve problems. Follow the examples when you do the **Exercises** and **Problem Solving.**

☞ **Selected Answers and Solutions** for some exercises are found in the back of the book. Use these answers to check your work after you've done the problems. A sample solution is also included for the first problem in each exercise set.

Calculator Practice exercises are included with each chapter. Use these exercises to practice using your calculator to solve problems.

There are many different kinds of calculators available. Some calculators perform only a few simple operations. Other calculators do many advanced calculations. It is important to know what your calculator can do and how to use it. To learn more about your calculator, read the instructions that come with it.

The diagram below shows an example of a scientific calculator. It describes the keys you will most likely use in your study of geometry.

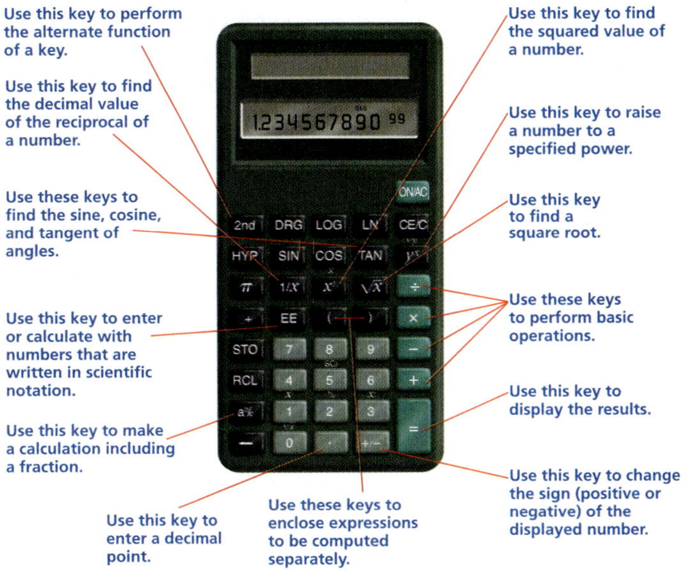

As You Study Each Lesson

Indicate to students that the Algebra Connection symbol appears whenever students need to recall the principles and concepts they learned in their study of algebra.

Make certain students recognize that each lesson's examples provide models for solving the problems found at the end of the lesson. Students can check some of their work by referring to the Selected Answers and Solutions, which are at the back of the book.

Choosing a Calculator

The geometry tool useful for the greatest number of situations may be the calculator. Distribute How to Study Geometry 3, Choosing a Classroom Calculator to students. Discuss the questions, pointing out to students that completing the list—and taking it with them when shopping —increases the likelihood they will choose a calculator that is capable of performing the functions necessary for geometry.

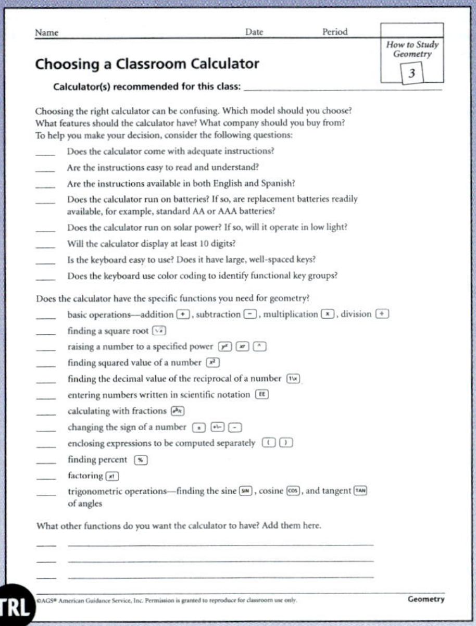

Name ______ Date ______ Period ______

How to Study Geometry 3

Choosing a Classroom Calculator

Calculator(s) recommended for this class: ______

Choosing the right calculator can be confusing. Which model should you choose? What features should the calculator have? What company should you buy from? To help you make your decision, consider the following questions:

____ Does the calculator come with adequate instructions?
____ Are the instructions easy to read and understand?
____ Are the instructions available in both English and Spanish?
____ Does the calculator run on batteries? If so, are replacement batteries readily available, for example, standard AA or AAA batteries?
____ Does the calculator run on solar power? If so, will it operate in low light?
____ Will the calculator display at least 10 digits?
____ Is the keyboard easy to use? Does it have large, well-spaced keys?
____ Does the keyboard use color coding to identify functional key groups?

Does the calculator have the specific functions you need for geometry?

____ basic operations—addition [+], subtraction [−], multiplication [×], division [÷]
____ finding a square root [√x]
____ raising a number to a specified power [y^x] [x^y] [^]
____ finding squared value of a number [x^2]
____ finding the decimal value of the reciprocal of a number [1\x]
____ entering numbers written in scientific notation [EE]
____ calculating with fractions [a^b/c]
____ changing the sign of a number [±] [+/−] [-]
____ enclosing expressions to be computed separately [(] [)]
____ finding percent [%]
____ factoring [x!]
____ trigonometric operations—finding the sine [SIN], cosine [COS], and tangent [TAN] of angles

What other functions do you want the calculator to have? Add them here.

____ ______
____ ______
____ ______

 Geometry

TRL

How to Study Geometry 3

Problem-Solving Strategies

Students will already realize that they will be asked to solve problems as they learn geometry. Help students recognize that a clear strategy for solving problems is the greatest asset they can have. Discourage students from jumping into problems, trying to find a formula or quick answer. Rather, stress that in the long run they will be better problem solvers if they follow four strategic steps: Read, Plan, Solve, and Reflect.

Read the sample problem with students. Then ask a volunteer to read the paragraph below that describes what the problem is asking, what information the problem provides that can be used in the solution, and what information the problem provides that is unnecessary. Poll students to ascertain their perceptions of what the problem asks and what information can be used to find the solution.

Problem-Solving Strategies

One reason for learning about geometry is to help you use mathematics to solve everyday problems. You will notice sets of problem-solving exercises throughout your *Geometry* book. When you learn a new skill, you will have a chance to apply this skill to a real problem.

Following these steps will help you to solve the problems.

1. Read

Read the problem to discover what information you are to gather. Study the problem to decide if you have all the information you need or if you need more data. Also study the problem to decide if it includes information you do not need to solve the problem. Begin thinking about the steps needed to solve the problem.

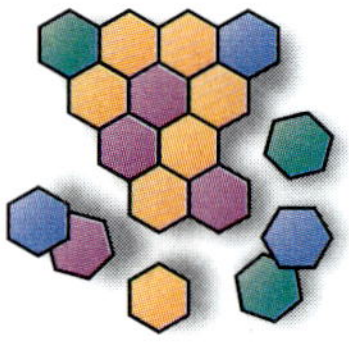

Ask yourself:
Am I looking for the measure of an angle?
Am I looking for the sum of all the angle measures?
Am I looking for a geometric shape?
Am I looking for a line or line segment?
Am I looking for a length, area, or volume?
Will solving the problem require multiple steps?

For example, read this problem ➧

Jackson wants to plant geraniums in a window box. The rectangular flower box is 2 feet long, 6 inches wide, and 6 inches high. The box must be filled with topsoil. Will one cubic foot of soil be enough to fill the box?

This problem asks you to decide if one cubic foot of soil is enough to fill the window box. First, you'll need to find the volume of the flower box. Then you will need to compare the volume of the box to one cubic foot. You'll decide if the volume is less than, equal to, or more than one cubic foot. The fact that geraniums are to be planted is unnecessary information.

xii *How to Study Geometry* *Geometry*

Name Date Period

How to Study Geometry 4

Problem-Solving Strategies

Step 1 Read

Read the problem to discover what information you are to gather.

Ask yourself:

1. What facts are given?
2. What does the problem tell me?
3. Do I know what all the words mean?
4. Did I read the whole problem carefully?
5. What is the problem asking me to find?

Read this problem: A library has 2,890 science books. The science books are classified into three categories: life, earth, and physical science. There are 190 more books in the earth science category than in each other category. How many books are in each category?

Step 2 Plan

Think about the steps you will need to do to solve the problem.

Ask yourself:

1. What am I trying to find out?
2. What facts do I need to know?
3. Should I make a chart, table, or diagram?
4. Can I do the problem in my head or will I need paper and pencil or a calculator?
5. What steps should I follow?

Make a plan to solve the problem.

Step 3 Solve

Review the steps you will need to solve the problem. Then do the calculations.

2,890 – 190 = 2,700. 2,700 ÷ 3 = 900   900 + 190 = 1,090

1, 090 earth science books, 900 life science books, 900 physical science books

Check your work: 1, 090 + 900 + 900 = 2,890.

Step 4 Reflect

Reread the problem. Does your answer make sense?

TRL

 Geometry

How to Study Geometry 4

Name Date Period

How to Study Geometry 5

Problem Solving

Problem-Solving Strategies

Step 1 Read

Step 2 Plan

Step 3 Solve

Step 4 Reflect

TRL

 Geometry

How to Study Geometry 5

2. Plan

Think about the steps you will need to do to solve the problem. Decide if you are going to calculate the answer in your head, on paper, or with a calculator. Will you need to draw a picture? Are the measurements in the same units? Will you need a formula or theorem? Will you need to do more than one step?

These strategies may help you to find a solution:

- ✔ Reword the problem
- ✔ Draw a picture
- ✔ Label the parts on the picture
- ✔ Make a chart or graph
- ✔ Estimate your answer
- ✔ Divide the problem into smaller parts
- ✔ Use a formula or write an equation
- ✔ Use a theorem
- ✔ Look for a pattern

For the example problem, you can use more than one strategy.

- ✔ Divide the problem into smaller parts.
- ✔ Draw a picture and label the parts.
- ✔ Change the dimensions of the box to feet, because the volume of the topsoil is in cubic feet. You know that 6 inches = $\frac{1}{2}$ foot so the dimensions of the box are 2 ft long by $\frac{1}{2}$ ft wide by $\frac{1}{2}$ ft high.
- ✔ Use a formula.
  $V = l \cdot w \cdot h = (\text{length})(\text{width})(\text{height})$

3. Solve

Follow your plan and do the calculations. Check your work. Make sure to label your answer correctly.

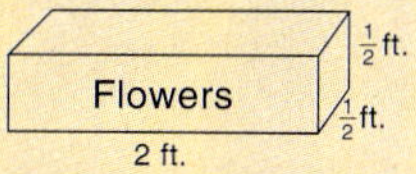

$V = l \cdot w \cdot h$

$V = 2(\frac{1}{2})(\frac{1}{2}) = \frac{2}{4} = \frac{1}{2}$ cubic foot

$\frac{1}{2}$ cu. ft < 1 cu. ft

$\frac{1}{2}$ cubic foot is less than one cubic foot, so one cubic foot of topsoil is more than enough to fill the flower box.

4. Reflect

Reread the problem. Does your answer make sense? Did you answer the question? You can also check your work to see if your answer is correct.

The volume of the flower box is $\frac{1}{2}$ cubic foot. You will need $\frac{1}{2}$ cubic foot of topsoil to fill the flower box. You have one cubic foot of topsoil. One cubic foot is more than $\frac{1}{2}$ cubic foot, so you have plenty of topsoil to fill the flower box. The answer is correct.

Indicate that once students have read the problem and have identified what the problem is asking, they can then plan how to solve the problem. Ask students to scan the list of possible problem-solving strategies. Ask volunteers to apply each of the strategies to the sample problem. Let students discover that they can—and may be required to—use more than one strategy to solve a problem. Open a discussion with students, allowing them to list what strategies they feel would be most useful in solving the problem.

Once students have developed a plan for problem solving, they can execute the plan. Remind them that part of a good solution to a problem is a check of the steps and calculations. Have them determine how much soil is needed for the box and then answer the question "Is 1 cubic foot of soil enough?"

Encourage students to develop the habit of rereading the problem and making certain that they have found the solution the problem requires. Remind them that estimating can be a useful way of checking to see if they performed any required arithmetic calculations correctly.

Preparing for Geometry and Mathematics Tests

Ask a volunteer to read the four bulleted items telling students some of the basic ways they can prepare for tests. Ask students if they think there are other ways they might prepare. Allow for discussion of student-offered preparations.

Point out that in geometry, as in other subjects, it pays to keep a journal or log of important terms, theorems, postulates, corollaries, and formulas.

Indicate to students that there are many things they can do once they have the test in their hands. One of the most important it to read the entire test before they start and to read each direction carefully.

Preparing for Geometry and Mathematics Tests

Taking tests can be stressful. You might wonder whether you've studied enough. You might worry that you studied the wrong material. You might feel like there is just too much to know! How can you prepare to take tests?

- Review the examples in the book and your completed homework assignments.
- Review the chapter vocabulary definitions. Review any rules or formulas or theorems in boxes. Read the notes in the side columns.
- Use the Chapter Reviews. The Chapter Reviews give you additional examples for solving problems.
- Read the Test-Taking Tip with each Chapter Review.

Research shows that when you have more test-taking skills and thinking strategies to use when you take a test, you are better prepared to solve the problems and more likely to be successful on the test. If you practice these test-preparation strategies each day, you will likely be an active, flexible thinker on test day. These strategies are helpful whether you are taking a chapter test or a standardized test.

Get organized and stay organized. Make your own geometry reference sheets or keep a journal of key concepts. Organize vocabulary terms, theorems, postulates, corollaries, and formulas under major headings for easy reference. Write your own examples of each term or idea.

Study your geometry reference sheets often so that your written notes become mental notes. Then when you see a test item that shows a *circle*, you can recall your mental notes for *circles*.

Ask what type of test it will be. For example: true/false, multiple choice, short answer, or long answer. Find out if you can use a reference sheet or a calculator.

Look over the entire test before you start.

Use your time wisely. Find out if the questions are all worth the same number of points. If they are, the easy questions will count the same as the hard ones. In this case, don't get stuck working too long on one question. Keep moving. Plan so that you will have time to complete the test.

Read each question twice. Read first to get the big idea. Take a moment to "hover." Then read the question again to zoom in and gather the specific numbers and other information you need to solve the problem.

Don't panic! When several geometry concepts are put together in one problem, it can look complicated. Stop and think about what you already know. Break the problem down into smaller parts.

Make a plan for solving the problem. Decide where to start, what information to use, and how to use it. Make notes about your plan so you don't forget or get lost in the middle of solving the problem.

Use the diagram or draw one yourself. If the question shows a diagram, you probably need information from the diagram to answer the question. If the question does not show a diagram, it might help to draw one yourself!

Use the answer choices. Sometimes the answer choices for multiple-choice questions contain hints. They give you a place to start finding the answer.

Stay focused. Most geometry questions are written to test your math reasoning—not your ability to perform basic calculations. The numbers used in geometry questions are often chosen to make calculations easier. If you find yourself bogged down in difficult calculations, you are probably on the wrong track. Start over.

Have an answer-checking strategy. Place a mark beside each test question when you finish it. You might put a ✔ beside questions you are sure you answered correctly. Place a **?** beside those you are still unsure about. Place an **X** beside those you think you missed. If you cannot write on the test booklet, use your scratch paper and write the numbers of the questions that you will need to recheck.

If time allows, read over all of your answers. Make sure you have answered the question that was asked.

Invite a volunteer to read each of the hints listed. Pause after each, allowing the class some discussion time. Ask students to describe in their own words each of the tips listed.

Graphically show students the merits of keeping organized, neat scratch paper. Present to students a scratch sheet on which the student has noted a plan and sketched a diagram. Present a second scratch paper where the information is sketchy and problems do not appear in any particular order. Point out that if their notes are neat and orderly, it is easier to go back and check their answers and much easier to find any error in calculations.

Chapter 1

Planning Guide

Exploring Geometry: Points, Lines, and Angles in the Plane

	Student Text Lesson						
	Student Pages	Vocabulary	Proof	Practice Exercises	Problem Solving	Calculator Practice	Solutions Key
Lesson 1 Points and Lines in the Plane	2–6	✔		✔	6		✔
Lesson 2 Ruler Postulates	7–10	✔		✔			✔
Lesson 3 Copying and Bisecting Angles	11–14	✔		✔	14		✔
Lesson 4 Angle Measurement	15–19	✔		✔	19		✔
Lesson 5 Complementary and Supplementary Angles	20–23	✔		✔		22	✔
Lesson 6 Algebra and Angles	24–27			✔	27		✔
Application Drawing Networks	28			✔			✔

Chapter Activities

Teacher's Resource Library
Geometry on the Job 1: Surveyors
Community Connection 1: Your Town

Teacher's Edition
Chapter 1 Project

Assessment Options

Student Text
Chapter 1 Review

Teacher's Resource Library
Chapter 1 Mastery Tests A and B
Chapter 1 Cumulative Mastery Test

Teacher's Edition
Chapter 1 Alternative Assessment

Student Text Features						Teaching Strategies								Learning Styles						Teacher's Resource Library		
Algebra Connection	Constructions	Experiencing Geometry	Technology Connection	Try This	Writing About Mathematics	Warm-Up Activity	Common Error	Algebra Connection	Mental Math	Manipulatives	Calculator	Applications	Group Problem Solving	Auditory/Verbal	Visual/Spatial	Body/Kinesthetic	Logical/Mathematical	Interpersonal/Group Learning	LEP/ESL	Workbook Activities	Additional Practice	Self-Study Guide
				6		2						4, 5			3, 5	3	6	4		1	1	✔
			10			7	8		9	8		10	10	9						2, 3	2	✔
	12–14					11								13		13		14	12	4	3	✔
	18, 19				17	15	16		19	18		18	17	19	17				17	5	4	✔
		23			23	20			23		22				22	23	23			6	5	✔
24		27			24	24	25	24		25	26		26						27	7	6	✔
						28						28								8		✔

Manipulatives

Factor Blocks

Angle Rule

Manipulatives Master 1 (Number Line)

Chapter at a Glance

Chapter 1: Exploring Geometry: Points, Lines, and Angles in the Plane

pages 1–31

Lessons

Teacher's Resource Library TRL

(Answer Keys for the Teacher's Resource Library begin on page 484 of this Teacher's Edition.)

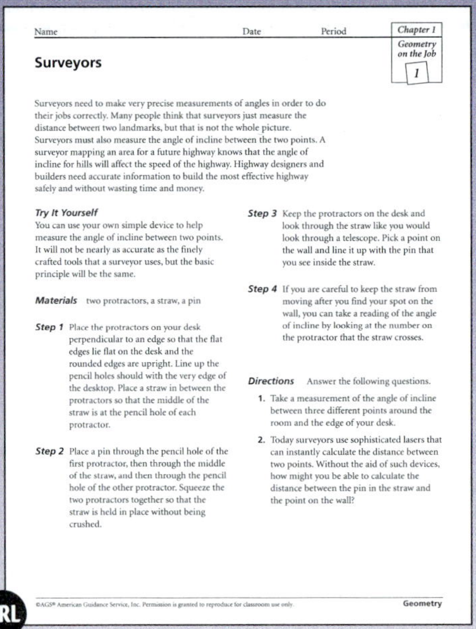

Name Date Period | Chapter 1 | Geometry on the Job 1

Surveyors

Surveyors need to make very precise measurements of angles in order to do their jobs correctly. Many people think that surveyors just measure the distance between two landmarks, but that is not the whole picture. Surveyors must also measure the angle of incline between the two points. A surveyor mapping an area for a future highway knows that the angle of incline for hills will affect the speed of the highway. Highway designers and builders need accurate information to build the most effective highway safely and without wasting time and money.

Try It Yourself
You can use your own simple device to help measure the angle of incline between two points. It will not be nearly as accurate as the finely crafted tools that a surveyor uses, but the basic principle will be the same.

Materials two protractors, a straw, a pin

Step 1 Place the protractors on your desk perpendicular to an edge so that the flat edges lie flat on the desk and the rounded edges are upright. Line up the pencil holes should with the very edge of the desktop. Place a straw in between the protractors so that the middle of the straw is at the pencil hole of each protractor.

Step 2 Place a pin through the pencil hole of the first protractor, then through the middle of the straw, and then through the pencil hole of the other protractor. Squeeze the two protractors together so that the straw is held in place without being crushed.

Step 3 Keep the protractors on the desk and look through the straw like you would look through a telescope. Pick a point on the wall and line it up with the pin that you see inside the straw.

Step 4 If you are careful to keep the straw from moving after you find your spot on the wall, you can take a reading of the angle of incline by looking at the number on the protractor that the straw crosses.

Directions Answer the following questions.

1. Take a measurement of the angle of incline between three different points around the room and the edge of your desk.
2. Today surveyors use sophisticated lasers that can instantly calculate the distance between two points. Without the aid of such devices, how might you be able to calculate the distance between the pin in the straw and the point on the wall?

 Geometry

TRL

Geometry on the Job 1

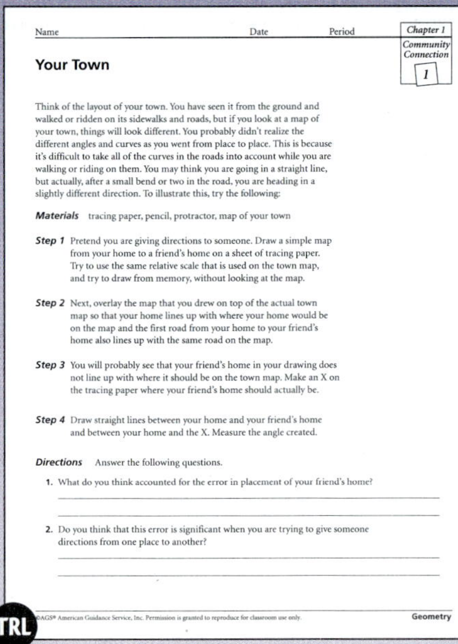

Name Date Period | Chapter 1 | Community Connection 1

Your Town

Think of the layout of your town. You have seen it from the ground and walked or ridden on its sidewalks and roads, but if you look at a map of your town, things will look different. You probably didn't realize the different angles and curves as you went from place to place. This is because it's difficult to take all of the curves in the roads into account while you are walking or riding on them. You may think you are going in a straight line, but actually, after a small bend or two in the road, you are heading in a slightly different direction. To illustrate this, try the following:

Materials tracing paper, pencil, protractor, map of your town

Step 1 Pretend you are giving directions to someone. Draw a simple map from your home to a friend's home on a sheet of tracing paper. Try to use the same relative scale that is used on the town map, and try to draw from memory, without looking at the map.

Step 2 Next, overlay the map that you drew on top of the actual town map so that your home lines up with where your home would be on the map and the first road from your home to your friend's home also lines up with the same road on the map.

Step 3 You will probably see that your friend's home in your drawing does not line up with where it should be on the town map. Make an X on the tracing paper where your friend's home should actually be.

Step 4 Draw straight lines between your home and your friend's home and between your home and the X. Measure the angle created.

Directions Answer the following questions.

1. What do you think accounted for the error in placement of your friend's home?
2. Do you think that this error is significant when you are trying to give someone directions from one place to another?

 Geometry

TRL

Community Connection 1

Chapter 1

Exploring Geometry: Points, Lines, and Angles in the Plane

Nighttime falls and you look up into a star-studded sky. Brilliant dots of light fill the sky as far as you can see. You, like millions of people before you, see patterns in the arrangement of stars. In the northern sky, you see seven stars that form the Big Dipper. These stars seem to line up to form a long-handled cup. The stars of the Big Dipper are points of light joined by imaginary lines. The points and lines appear to be on a plane, or a flat surface. Together the points, imaginary lines, and plane make up the Big Dipper.

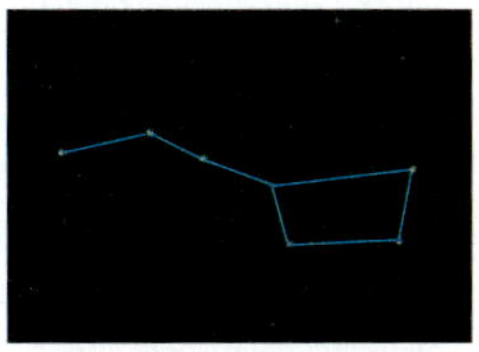

In Chapter 1, you will learn about points, lines, and angles in the plane.

Goals for Learning

- To recognize a point, a line, and a plane
- To identify line segments and rays
- To use postulates to determine how to use a ruler with geometric figures
- To construct angles—copying and bisecting
- To measure and classify angles
- To identify complementary and supplementary angles
- To use algebra to solve problems in geometry

Introducing the Chapter

Use the information on these pages to introduce the concepts of points and lines. Suggest that students move their fingers from bright star to bright star to trace imaginary lines for the Big Dipper. Then draw dots in the pattern of the Big Dipper on the board. Ask a volunteer to connect the dots, using the shortest line segments possible. What do students discover about the line segments? **(The shortest line segment between two points is a straight line.)** Then provide students with a definition of geometry: the branch of mathematics that studies points, lines, angles, surfaces, and solids. Help students identify the Big Dipper in terms of this definition. **(The Big Dipper appears to be seven points [stars] that form the shape of a water dipper when imaginary line segments are drawn from point to point on a plane.)** Introduce the Goals for Learning. Explain to students that the chapter will help them identify and use basic geometric figures.

Chapter Project

Assign students partners. Suggest that they produce one of the following for their chapter project.

- An illustrated dictionary—Students can identify important terms in the chapter, draw illustrations to reinforce the terms, and define the terms in their own words. Their completed project could be a booklet or poster of terms. Encourage students with limited English proficiency to include the names of the terms in their native language as well as English.
- Rules Handbook—Students can prepare a book that lists the postulates developed in the chapter. The book should present each postulate and provide one or more examples to illustrate it.

Encourage partners to agree on the information to include in their books or posters and determine the role each will perform in producing the product.

SELF-STUDY GUIDE

Name ____________

CHAPTER 1: Exploring Geometry: Points, Lines, and Angles in the Plane

Goals 1.1 and 1.2 *To recognize a point, a line, and a plane; to identify line segments and rays*

Date	Assignment	Score
____	1: Read pages 2–4. Complete Exercises A, B, C, D, and E on pages 4–6.	____
____	2: Complete Workbook Activity 1.	____

Comments:

Goal 1.3 *To use postulates to determine how to use a ruler with geometric figures*

Date	Assignment	Score
____	3: Read pages 7–9. Complete Exercises A, B, C, and D on pages 9–10.	____
____	4: Complete Workbook Activity 2.	____
____	5: Complete Exercises E, F, and G on page 10.	____
____	6: Complete Workbook Activity 3.	____

Comments:

Goal 1.4 *To construct angles—copying and bisecting*

Date	Assignment	Score
____	7: Read pages 11–13. Complete Exercises A, B, and C on page 14.	____
____	8: Complete Workbook Activity 4.	____

Comments:

TRL

 Geometry

SELF-STUDY GUIDE

Name ____________

CHAPTER 1: Exploring Geometry: Points, Lines, and Angles in the Plane, continued

Goal 1.5 *To measure and classify angles*

Date	Assignment	Score
____	9: Read pages 15–18. Complete Exercises A, B, C, and D on pages 18–19.	____
____	10: Complete Workbook Activity 5.	____

Comments:

Goal 1.6 *To identify complementary and supplementary angles*

Date	Assignment	Score
____	11: Read pages 20–21. Complete Exercise A on page 22.	____
____	12: Read the Calculator Practice and complete Exercise B on page 22.	____
____	13: Complete Exercises C and D on page 23.	____
____	14: Complete Workbook Activity 6.	____

Comments:

Goal 1.7 *To use algebra to solve problems in geometry*

Date	Assignment	Score
____	15: Read pages 24–25. Complete Exercises A, B, C, and D on pages 26–27.	____
____	16: Complete Workbook Activity 7.	____
____	17: Read and complete the Application Exercise on page 28.	____
____	18: Complete Workbook Activity 8.	____
____	19: Complete the Chapter 1 Review on pages 29–31.	____

Comments:

Student's Signature ____________ Date ____________

Instructor's Signature ____________ Date ____________

TRL

 Geometry

Chapter 1 Self-Study Guide

Lesson at a Glance

Chapter 1 Lesson 1

Overview This lesson introduces endpoints, line segments, lines, rays, and collinear points.

Objective

- To identify the symbols for points, lines, line segments, rays, and planes.

Student Pages 2–6

Teacher's Resource Library TRL

Workbook Activity 1

Additional Practice 1

Mathematics Vocabulary

collinear
endpoints
geometry
line
line segment
plane
point
ray

1. Warm-Up Activity

Duplicate a simple connect-the-dots pattern and distribute copies to students, or produce a numbered connect-the-dots pattern on the board. Have students connect the dots to form the pattern. Relate this activity to the concepts of points, lines, and planes.

2. Teaching the Lesson

Lesson 1 presents the most basic geometric figures that serve as building blocks for other geometric figures. As a result, understanding and recognizing the terms, representations, and symbols for the figures are essential to geometric knowledge. Present and reinforce the lesson's figures, vocabulary, and symbols with drawings and models. Introduce the concepts of points, lines, segments, rays, and planes with figures you draw on the board or on chart paper. Help students find examples or models to represent the concepts in the classroom. For example, a sheet of paper could represent a plane,

Lesson 1 Points and Lines in the Plane

Collinear
Points on the same line

Endpoints A B
A and B are the endpoints of $\overline{AB}$ *where* $\overline{AB}$ *is the set of points A, B, and all the points between A and B*

Geometry
The study of points, lines, angles, surfaces, and solids

Line $\overleftrightarrow{AB}$
A set of many points that extend in opposite directions without ending

Line segment $\overline{AB}$
Segment $\overline{AB}$ *is the set of points A, B, and all the points between A and B*

Plane □
A two-dimensional flat surface

In **geometry,** you study the size, shape, and position of objects. One way to make that study easier is to concentrate on the outline of an object. Look closely at the points and lines that make up geometric figures.

A **point** is the simplest of geometric figures and it has no formal definition. In geometry, a point is represented by a dot on a piece of paper. Capital letters are used to name points. A point has a location but no size or shape. You might think of a point as a star in the night sky. Here are some examples of points.

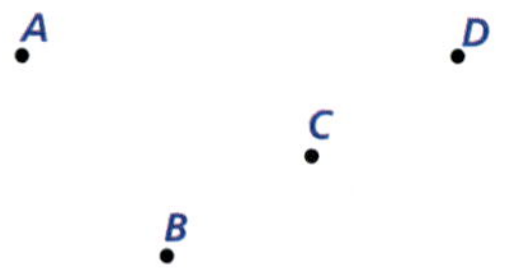

These points are named *A*, *B*, *C*, and *D*.

There are two other concepts in geometry that have no formal definition—a **line** and a **plane.** A straight two-headed arrow represents a line. A line is endless in two directions. A line can be named by naming any two points on it, or it can be named with a lowercase letter. You might think of a line as a straight, thin wire or thread, but a geometric line has no thickness. Here is an example of a line.

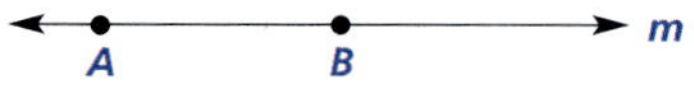

This line is named $\overleftrightarrow{AB}$ or $\overleftrightarrow{BA}$ or *m*.

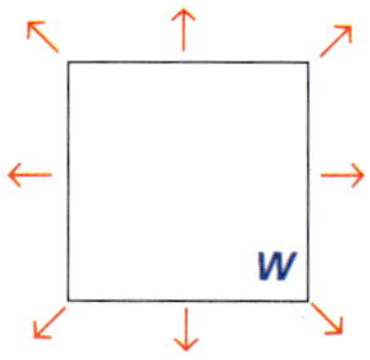

This is plane *W*.

The four-sided shape at the left represents a plane. A plane is a flat surface. It extends in all directions with no end and has no thickness. A plane is named using capital letters. You might think of a page from this book or a tabletop as part of a plane.

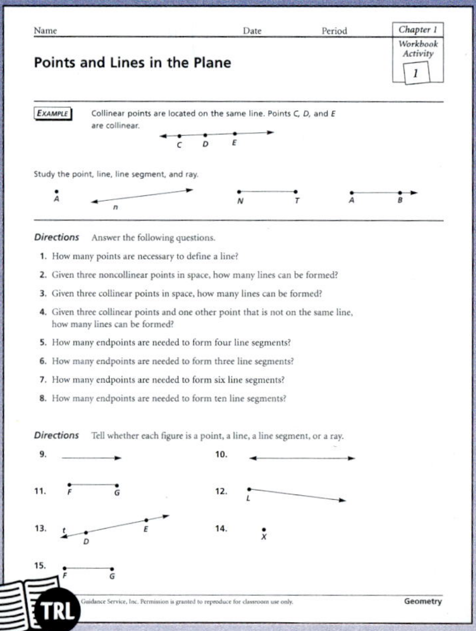

Name Date Period — Chapter 1 Workbook Activity 1

Points and Lines in the Plane

EXAMPLE Collinear points are located on the same line. Points *C*, *D*, and *E* are collinear.

Study the point, line, line segment, and ray.

Directions Answer the following questions.

1. How many points are necessary to define a line?
2. Given three noncollinear points in space, how many lines can be formed?
3. Given three collinear points in space, how many lines can be formed?
4. Given three collinear points and one other point that is not on the same line, how many lines can be formed?
5. How many endpoints are needed to form four line segments?
6. How many endpoints are needed to form three line segments?
7. How many endpoints are needed to form six line segments?
8. How many endpoints are needed to form ten line segments?

Directions Tell whether each figure is a point, a line, a line segment, or a ray.

9. 10. 11. 12. 13. 14. 15.

 Geometry

Workbook Activity 1

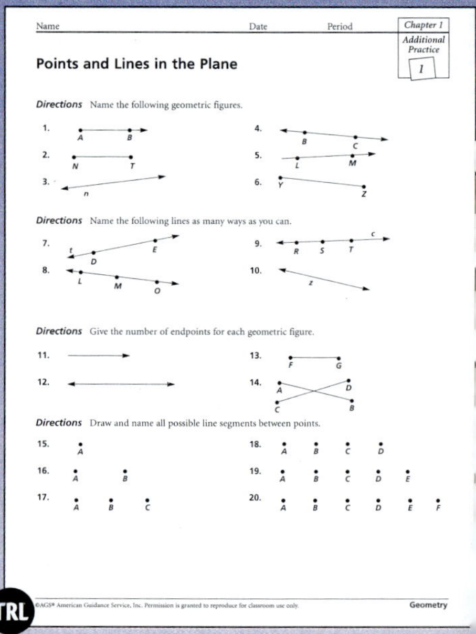

Name Date Period — Chapter 1 Additional Practice 1

Points and Lines in the Plane

Directions Name the following geometric figures.

1. 2. 3. 4. 5. 6.

Directions Name the following lines as many ways as you can.

7. 8. 9. 10.

Directions Give the number of endpoints for each geometric figure.

11. 12. 13. 14.

Directions Draw and name all possible line segments between points.

15. 16. 17. 18. 19. 20.

 Geometry

Additional Practice 1

EXAMPLE Name this line in nine different ways.

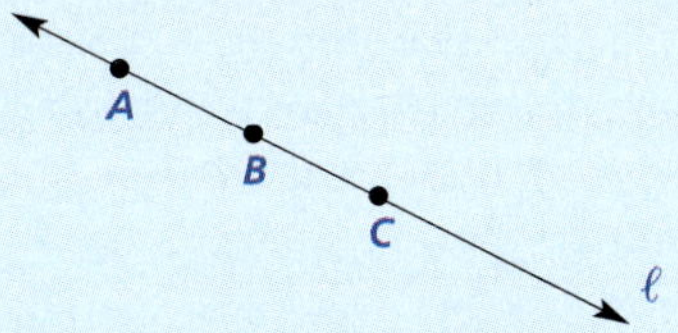

This line can be named $\overleftrightarrow{AB}$, $\overleftrightarrow{BA}$, $\overleftrightarrow{AC}$, $\overleftrightarrow{CA}$, $\overleftrightarrow{BC}$, $\overleftrightarrow{CB}$, $\overleftrightarrow{AC}$, $\overleftrightarrow{CA}$, or ℓ.

Point • *A*
A location in space represented by a dot

Ray $\overrightarrow{AB}$
A set of points that is part of a line; a ray has one endpoint and extends in one direction with no end

Points on the same line are called **collinear** points.
In the example above, *A*, *B*, and *C* are collinear points on line ℓ.

EXAMPLE Which three of these points are collinear?

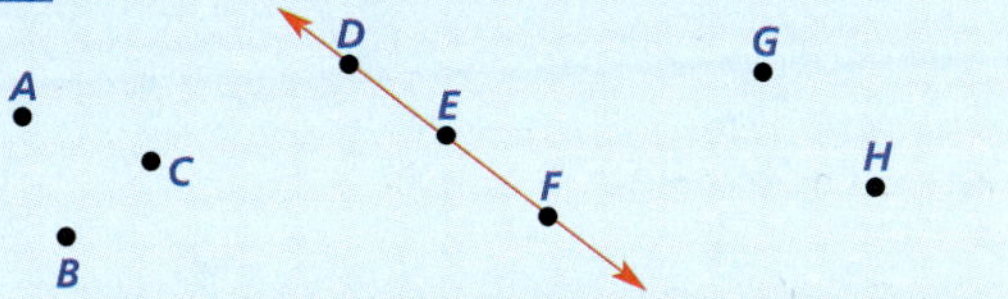

Points *D*, *E*, and *F* appear to be on the same line.

If you draw a line between any two points, such as *A* and *B*, you have drawn a **line segment.** A line segment is part of a line.

This is segment $\overline{AB}$.

Segment $\overline{AB}$ is the set of points *A*, *B*, and all points between *A* and *B*. *A* and *B* are called the **endpoints** of the segment. The symbol $\overline{AB}$ is read "segment *AB*."

Recall that a set is a collection of particular things, like the set of points on a line segment.

EXAMPLE Draw and name all the line segments between points *P*, *Q*, and *R*.

Remember that you need two endpoints for each segment. You can draw segments $\overline{PQ}$, $\overline{QR}$, and $\overline{RP}$.

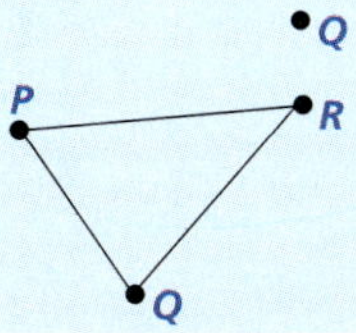

bookends could represent endpoints, and the book bindings lined up between the bookends could be collinear points along a line segment.

3. Reinforce and Extend

Learning Styles

Body/Kinesthetic Ask students to prepare one set of cards with symbols or illustrations for line, line segment, endpoints, ray, plane, and collinear points. Also have them make a set of cards with the terms *line, line segment, endpoints, ray, plane,* and *collinear points* and their definitions. Students can mix the two sets of cards and place them face down on a table. They can then play a game with the cards by matching definitions with illustrations.

Learning Styles

Visual/Spatial Using a color-coded system may help students more easily distinguish the illustrations or symbols used for points, endpoints, line segments, rays, lines, and planes. Prepare a color-coded system for the symbols and illustrations. Use different colored chalk to represent each symbol or figure. Write the code key on the board. Use the system as you teach the lesson, and encourage students to use the colors when writing the symbols. This system will provide visual reinforcement of the symbols.

Learning Styles

Interpersonal/Group Learning Have students work in groups of four to create a quiz and its answer key for Lesson 1. Review each group's work to make sure that the answers are correct and the questions reflect the lesson's topic. After a group makes any needed corrections, ask each member to copy the quiz. Each group can then exchange quizzes with another group whose members take the quiz independently. When everyone has completed the quiz, students should rejoin their groups to score and discuss their answers.

Application

Career Connection Who uses lines, line segments, and rays? Architects, engineers, carpenters, and sculptors are among the people whose careers require them to use geometry daily. Each has tools that enable him or her to precisely use lines and planes. Designate an area in the classroom where students can set up a career tool display. Each student should select a career in which geometry is used and then prepare an exhibit of tools and other materials that are used by people in the career. For example, an exhibit for an architect might include blueprints, T squares, and compasses. One for a carpenter might include a level, a tape measure, and a plumb. Have students explain the use and application of the tools they display.

Another part of a line is called a **ray.** A ray is a line with one endpoint. You might think of a ray as a ray of light leaving the sun and traveling into space forever. A straight arrow is used to represent a ray. The symbol $\overrightarrow{AB}$ is read "ray *AB*."

This ray is named $\overrightarrow{AB}$. *A* is its endpoint.

A ray extends from its endpoint in one direction with no end.

Selected Answers and Solutions begin on page 420.

Exercise A Which three points are collinear? There may be more than one correct answer.

1.
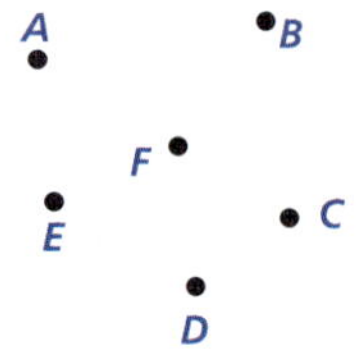

A, F, C

2.
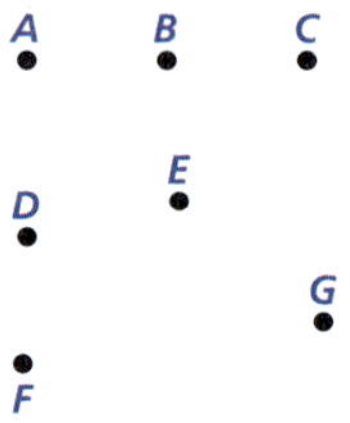

A, B, C; A, D, F; A, E, G; and C, E, F

3.
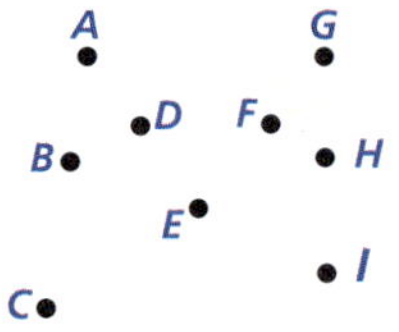

A, B, C; A, D, E; E, F, G; and G, H, I

Exercise B Tell whether each figure is a line, a line segment, or a ray.

4.  *ray*

5. 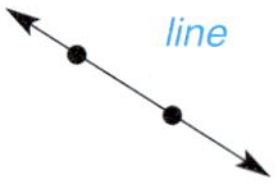 *line*

6.  *line segment*

7. 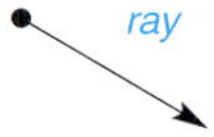 *ray*

8. *line*

9. 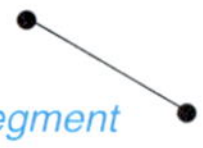 *line segment*

Exercise C Name the line as many ways as you can.

10.

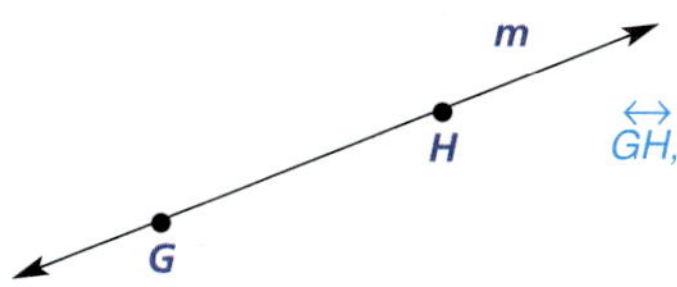

$\overleftrightarrow{GH}$, $\overleftrightarrow{HG}$, and *m*

Exercise D Answer each question.

11. How many endpoints does a line segment have? two

12. How many endpoints does a line have? none

13. How many endpoints does a ray have? one

14. Are two points always collinear? Explain why or why not.

14. Yes, they can each be the endpoint of a single line segment.

15. What geometric term(s) can you use to describe the lights in this picture? points

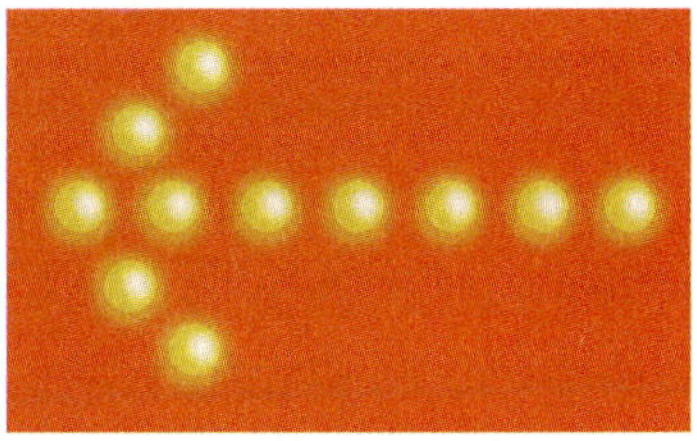

16. What geometric term(s) can you use to describe the lights on a straight string of holiday lights? points on a line segment

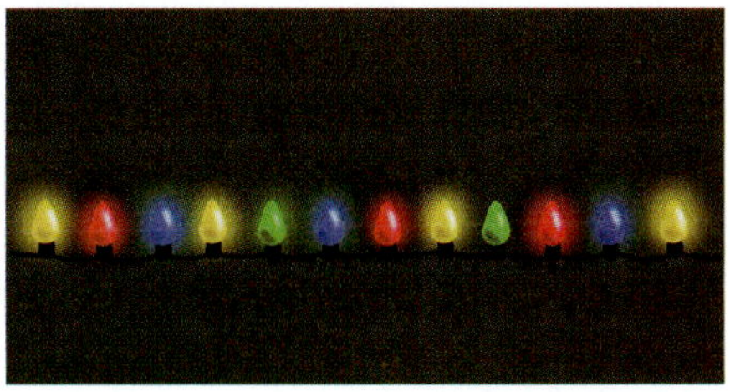

Learning Styles

Visual/Spatial Have groups of three students prepare a Basic Geometric Figure poster. Students should use string or yarn and paper dots and arrowheads to construct points, line segments, lines, and rays. The dots can represent endpoints and other collinear points on a line segment. Ask students to label their constructions and title their posters. Then display their work for reference.

Application

In the Community

Suggest that students look for and list examples of objects that could represent points, line segments, rays, lines, and planes in their community. For example, a left-turn arrow on a traffic signal could represent a ray, and a football field could be considered a plane through which line segments pass. Have students share their lists with the class.

Learning Styles

Logical/Mathematical Ask students to expand the chart shown in problem 20 to include the number of segments for figures with 7, 9, 10, 11, 12, 13, 14, and 15 points. **(21, 36, 45, 55, 66, 78, 91, and 105, respectively)** When students have completed their charts, ask them to identify the strategy they used.

17. What geometric term can you use to describe the light from a flashlight?

a ray

Problem Solving

Exercise E Answer the following questions.

18. How many line segments can you draw using 2 points as endpoints?

one

19. How many line segments can you draw using 3 points as endpoints?

three

20. Complete the following table to show how many segments can be drawn using the given number of points as endpoints. Hint: Draw a triangle (3 sides), square (4 sides), 5-sided figure, 6-sided figure, and 8-sided figure. Connect the corners (endpoints). Count the segments.

Number of points	2	3	4	5	6	8
Number of segments	1	3	6	10	15	28

Try This...

You experimented by counting segments. Can you discover a pattern or formula to give you the answer? Hint: Let n = the number of sides or points.

n = number of sides or points; $n \bullet \frac{n-1}{2}$ or count down from $n - 1$ to 1, adding each possibility: square = 4 sides → 3 + 2 + 1 = 6 segments; 5 points → 4 + 3 + 2 + 1 = 10 segments.

Lesson 2 Ruler Postulates

Measurement is an important part of geometry. You use measurement every day. The measuring tool you use most often is a ruler. You can think of a ruler as a line with numbers on it. Modern geometry has set some rules on how to use a ruler for geometric figures. The rules are called **postulates.** A postulate is a statement about geometric figures accepted as true without proof. Here are three postulates involving rulers.

> ***Postulate***
> *A statement about geometric figures accepted as true without proof*

RULER POSTULATE:
The points on a line can be placed in a one-to-one correspondence with the real numbers so that
1. for every point on the line, there is exactly one real number.
2. for every real number, there is exactly one point on the line.
3. the distance between any two points is the absolute value of the difference of the corresponding real numbers.

> Remember that the *absolute value* of a number is equivalent to its distance from zero on the number line. $|n|$ is the symbol for the absolute value of n. For example, $|-3| = 3$.

EXAMPLE Line

Line as Ruler

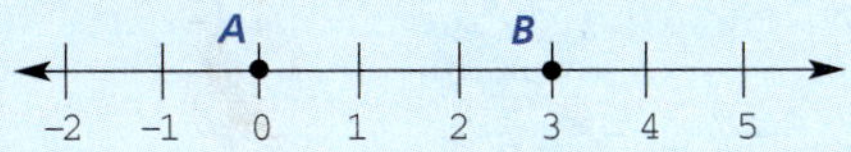

A corresponds to 0. *B* corresponds to 3.

The distance between *A* and *B* is 3.

Ruler Postulate $\overline{AB} = |3 - 0|$ or $|0 - 3| = 3$

Notice that *A* was placed at 0 and *B* at 3. This makes it easy to find the distance between *A* and *B*—you can read it directly from the number line. The following postulate allows you to do this for any two points.

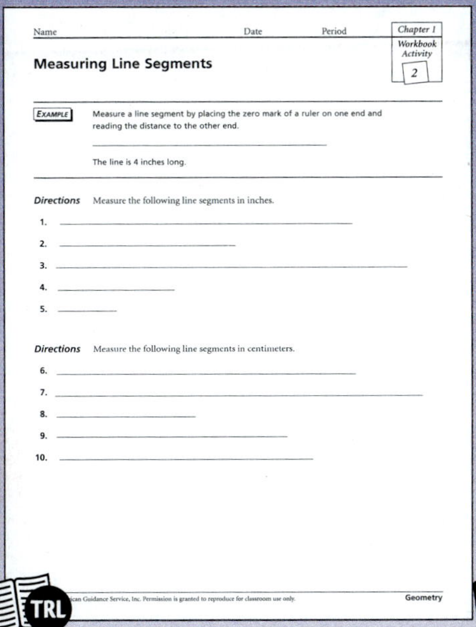

Name Date Period *Chapter 1 Workbook Activity* 2

Measuring Line Segments

EXAMPLE Measure a line segment by placing the zero mark of a ruler on one end and reading the distance to the other end.

The line is 4 inches long.

Directions Measure the following line segments in inches.

1.
2.
3.
4.
5.

Directions Measure the following line segments in centimeters.

6.
7.
8.
9.
10.

Geometry

Workbook Activity 2

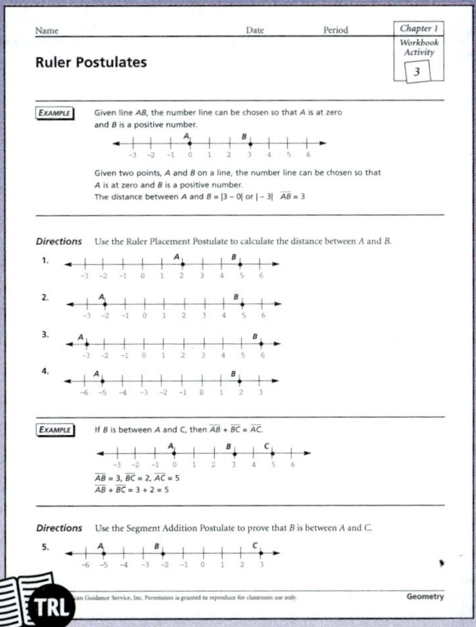

Name Date Period *Chapter 1 Workbook Activity* 3

Ruler Postulates

EXAMPLE Given line *AB*, the number line can be chosen so that *A* is at zero and *B* is a positive number.

Given two points, *A* and *B* on a line, the number line can be chosen so that *A* is at zero and *B* is a positive number.
The distance between *A* and $B = |3 - 0|$ or $|-3|$ $\overline{AB} = 3$

Directions Use the Ruler Placement Postulate to calculate the distance between *A* and *B*.

1.
2.
3.
4.

EXAMPLE If *B* is between *A* and *C*, then $\overline{AB} + \overline{BC} = \overline{AC}$.

$\overline{AB} = 3$, $\overline{BC} = 2$, $\overline{AC} = 5$
$\overline{AB} + \overline{BC} = 3 + 2 = 5$

Directions Use the Segment Addition Postulate to prove that *B* is between *A* and *C*.

5.

Geometry

Workbook Activity 3

Lesson at a Glance

Chapter 1 Lesson 2

Overview This lesson explains the concepts of postulates and the rules of measurement.

Objectives

- To explain and give examples for the ruler postulate, the ruler placement postulate, and the segment addition postulate.

Student Pages 7–10

Teacher's Resource Library TRL

Workbook Activities 2–3

Additional Practice 2

Mathematics Vocabulary

postulate

1. Warm-Up Activity

Draw a number line on the board, showing numbered segments –10 to 0 to 10. Point to different numbers on the line, and ask students to give the absolute values of the numbers. Continue reinforcing the concept of absolute value in this way and then ask students to give a definition for absolute value in terms of the number line. (Absolute value of a number is equivalent to its distance from 0 on the number line.)

2. Teaching the Lesson

Continue using the number line you constructed for the Warm-Up Activity. Use the number line to present and reinforce visually the ruler postulate, ruler placement postulate, and segment addition postulate. Identify line segments on the line, and ask students to determine the distances between the points of the segments. Encourage them to determine the distances both by counting segments and by using the arithmetic operation of subtraction.

Remind students that when they subtract integers in algebra, they add the opposite. Write these equations on the board, and have students use them to determine distances on the number line.

$a - (-b) = a + b$
$a - (+b) = a + (-b)$

Common Error

Students may confuse integers' positive and negative signs with the operational plus and minus signs. To help them distinguish the signs, have students read problems aloud. For example, they should read $\overline{AB} = 2 - (-2)$ as "line segment *AB* equals two minus negative two." The answer is then read "line segment *AB* equals four."

Manipulatives

On the Number Line

Materials: Factor Blocks, Manipulatives Master 1 (Number Line)

Group Practice: Review with students the Ruler Postulate and the Ruler Placement Postulate. Show them how to use the Number Line to determine the difference between the endpoints of a line segment and to shift the segment so that *A* is at zero.

Student Practice: Have students use the Factor Blocks and Number Line to reinforce visually and manipulatively the Ruler and Ruler Placement Postulates. Students should place Factor Blocks between the endpoints you name. Write the endpoints on the board and have students use their Factor Blocks to determine distance: $A = -2, B = 3$ **($\overline{AB} = 5$)**; $A = 3, B = 5$ **($\overline{AB} = 2$)**; $A = -3, B = 0$ **($\overline{AB} = 3$)**; $A = -6, B = -1$ **($\overline{AB} = 5$)**.

Students determine distance between the endpoints by counting and recording the number of blocks. Then have them shift the blocks so that endpoint *A* is at zero. They once again count and record the number of blocks. Have them use this experience to explain that position on the number line does not affect the distance between endpoints of a line segment.

Ruler Placement Postulate:
Given two points, *A* and *B* on a line, the number line can be chosen so that *A* is at zero and *B* is at a positive number.

Example Given $\overline{AB}$.

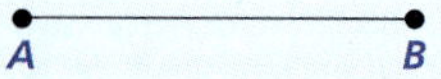

Ruler Postulate

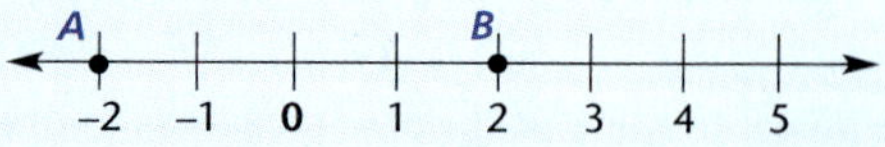

$\overline{AB} = |-2 - 2| = 4$ or $|2 - (-2)| = 4$

Ruler Placement Postulate

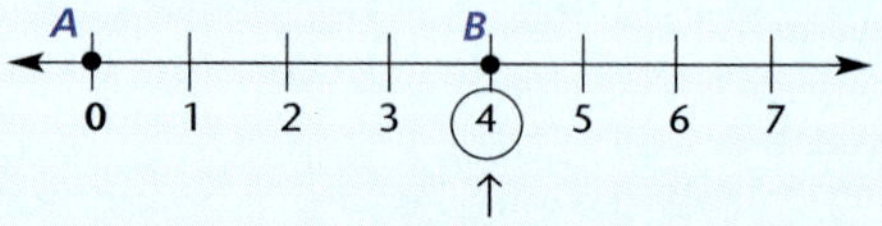

The distance between *A* and $B = |4 - 0|$ or $|0 - 4|$

$\overline{AB} = 4$

Using the Ruler Placement Postulate, you can read the distance between *A* and *B* directly from the number line "ruler."

Segment Addition Postulate:
If *B* is between *A* and *C*, then $\overline{AB} + \overline{BC} = \overline{AC}$.

Example Prove that *B* is between *A* and *C*.

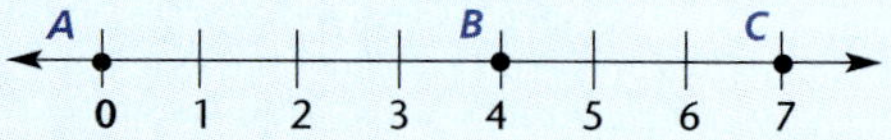

Segment Addition Postulate

$\overline{AB} = 4$, $\overline{BC} = 3$, $\overline{AC} = 7$

$\overline{AB} + \overline{BC} = 4 + 3 = 7$

B is between *A* and *C* because $4 + 3 = 7$

Also, 4 is between 0 and 7 on the number line.

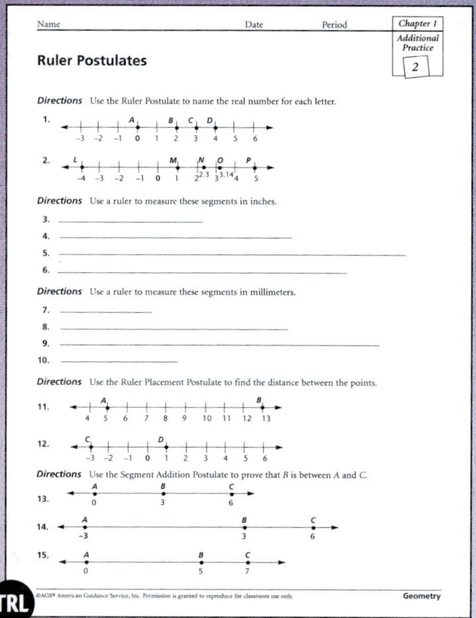

Name Date Period

Chapter 1 *Additional Practice* 2

Ruler Postulates

Directions Use the Ruler Postulate to name the real number for each letter.

1. (number line: −3 −2 −1 0 1 2 3 4 5 6; points A, B, C, D)

2. (number line: −4 −3 −2 −1 0 1 2 2.3 3 3.14 4 5; points L, M, N, O, P)

Directions Use a ruler to measure these segments in inches.

3. ______
4. ______
5. ______
6. ______

Directions Use a ruler to measure these segments in millimeters.

7. ______
8. ______
9. ______
10. ______

Directions Use the Ruler Placement Postulate to find the distance between the points.

11. (number line: 4 5 6 7 8 9 10 11 12 13; points A, B)

12. (number line: −3 −2 −1 0 1 2 3 4 5 6; points C, D)

Directions Use the Segment Addition Postulate to prove that *B* is between *A* and *C*.

13. A 0, B 3, C 6

14. A −3, B 3, C 6

15. A 0, B 5, C 7

 Geometry

TRL

Additional Practice 2

EXAMPLE Prove that *D* is between *C* and *E*.

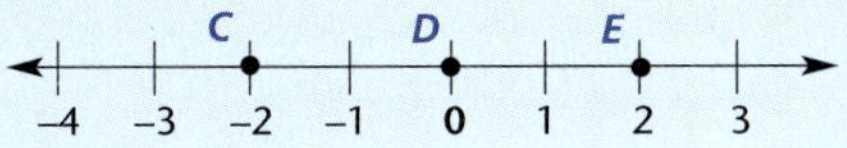

Segment Addition Postulate

$\overline{CD} = 2$, $\overline{DE} = 2$, $\overline{CE} = 4$
$\overline{CD} + \overline{DE} = 2 + 2 = 4$
D is between *C* and *E* because 2 + 2 = 4
Also, 0 is between –2 and 2 on the number line.

Exercise A Use the Ruler Postulate to name the real number corresponding to each letter.

1.

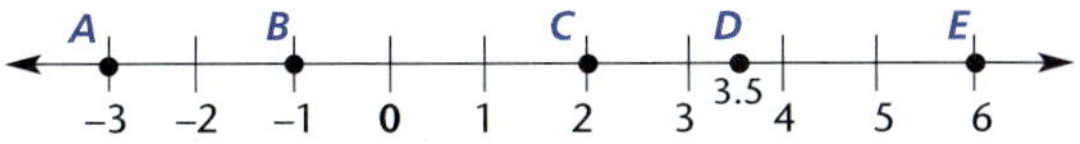

A –3, *B* –1, *C* 2, *D* 3.5, *E* 6

Exercise B Use the Ruler Postulate to name the letter corresponding to each real number.

2.

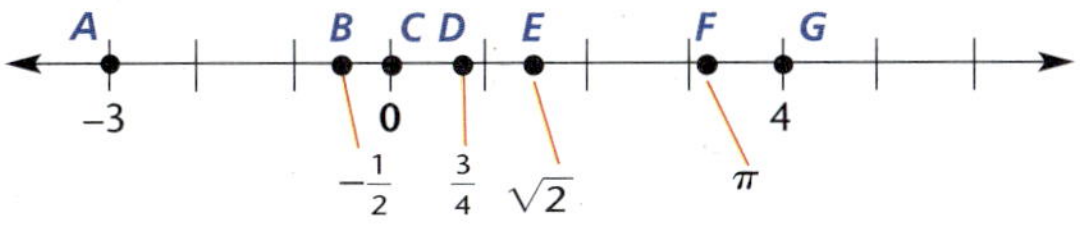

A –3, *B* $-\frac{1}{2}$, *C* 0, *D* $\frac{3}{4}$, *E* $\sqrt{2}$, *F* π, *G* 4

Exercise C Use the Ruler Placement Postulate to find the distance between the points. (Hint: Think 0 at the point at the left.)

3.

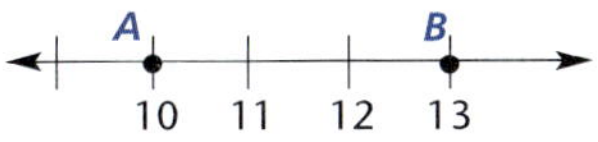

distance *A* to *B* = 3 units

4.

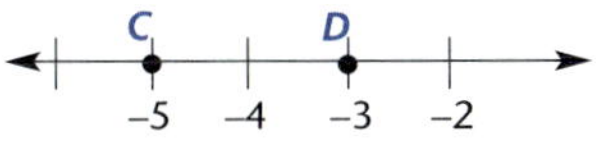

distance *C* to *D* = 2 units

3. Reinforce and Extend

MENTAL MATH

Tell students that on the number line $\overline{AB} + \overline{BC} = \overline{AC}$. Call out numbers for the lengths of $\overline{AB}$ and $\overline{BC}$. Have students use the Segment Addition Postulate to calculate mentally the length of segment *AC*.

- $\overline{AB} = 2$, $\overline{BC} = 6$ ($\overline{AC} = 8$)
- $\overline{AB} = 1$, $\overline{BC} = 4$ ($\overline{AC} = 5$)
- $\overline{AB} = 3$, $\overline{BC} = 7$ ($\overline{AC} = 10$)
- $\overline{AB} = 1$, $\overline{BC} = 2$ ($\overline{AC} = 3$)
- $\overline{AB} = 5$, $\overline{BC} = 1$ ($\overline{AC} = 6$)
- $\overline{AB} = 4$, $\overline{BC} = 2$ ($\overline{AC} = 6$)
- $\overline{AB} = 5$, $\overline{BC} = 11$ ($\overline{AC} = 16$)

LEARNING STYLES

Auditory/Verbal Read aloud each postulate and use a number line on the board for visual reinforcement as you give verbal examples based on the postulate. Then have students give oral answers to Exercises A, B, C, and F after they cite the applicable postulate for each problem aloud.

Group Problem Solving

Have students work in small groups to solve the following problem. When they have completed their work, ask volunteers to share their answers, problem-solving strategies, and applicable postulates.

The temperature at noon was –5°F. By 2 o'clock, the temperature had risen 4°; by 3 o'clock, it rose another 2°. How many degrees did the temperature go up from noon to 3 o'clock? What was the temperature at 3 o'clock? **(6°; 1°F)**

Application

In the Community

Have students work with the NASDAQ or Dow Jones Index in a daily newspaper. Allow students to choose a specific stock and follow its activity for a week. If necessary, help them identify the price and gain and/or loss in the stock's newspaper entry. Have them record the information on a number line. They should record the first day's price as point *A* on the line and the gains and/or losses in points over the week as points *B*, *C*, *D*, and *E*. Encourage each student to present his or her number line to the class and use it to explain the stock's activity.

Exercise D Use a ruler to measure these segments in inches. Round to the nearest inch.

5. ______________________ 2 in.

6. ________________________________ 3 in.

Exercise E Use a ruler to measure these segments in centimeters. Round to the nearest centimeter.

7. ____ 1 cm **8.** ________ 2 cm **9.** ____________ 3 cm

Exercise F Use the Segment Addition Postulate to prove that *B* is between *A* and *C*.

10.

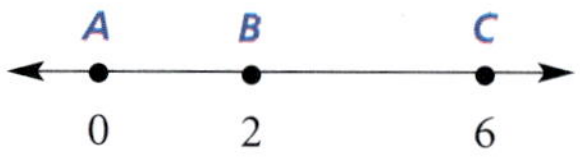

11.

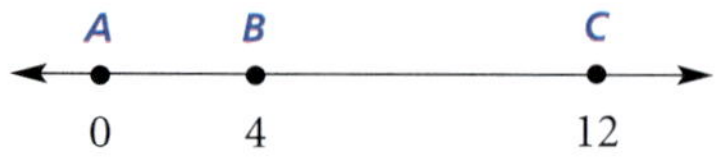

10. $\overline{AB} = 2$, $\overline{BC} = 4$, $\overline{AC} = 6$; $\overline{AB} + \overline{BC} = 2 + 4 = 6$, so *B* is between *A* and *C*

11. $\overline{AB} = 4$, $\overline{BC} = 8$, $\overline{AC} = 12$; $\overline{AB} + \overline{BC} = 4 + 8 = 12$, so *B* is between *A* and *C*

Exercise G Answer the questions.

12. Which postulate places numbers on a line?

13. Which postulate can you use to measure distances between points?

14. How did you use the ruler postulate in problems 5–9?

15. Do you think you can use the ruler postulate if the points *A*, *B*, and *C* are not collinear? Why or why not?

12. ruler postulate
13. ruler placement postulate
14. Sample: I placed the ruler, a number line, so that 0 was at one endpoint. Then I read the number at the other endpoint.
15. No. If the points are not in a straight line, $\overline{AB} + \overline{BC} \neq \overline{AC}$

Technology Connection

Accurate measurements of distances and angles, such as flight paths and re-entry angles, are important to space exploration. The Mars Climate Orbiter, which traveled 416,000,000 miles to Mars, may have failed because it came about $12\frac{1}{2}$ miles too close to the planet's surface. That's an error of about 0.00003%!

Lesson 3 Copying and Bisecting Angles

Geometric figures exist in a variety of shapes and sizes. As you look around, you will see one of the most important geometric figures, angles.

An **angle** is a geometric figure made up of two rays with a common endpoint called the **vertex.**

Examples of angles

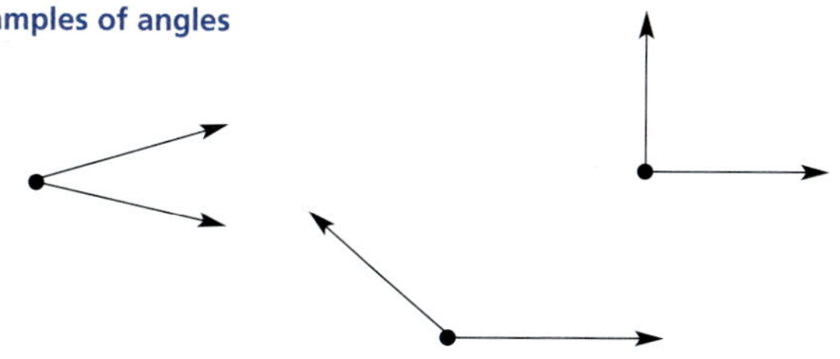

The symbol for an angle is ∠. Angles can be named in several ways.

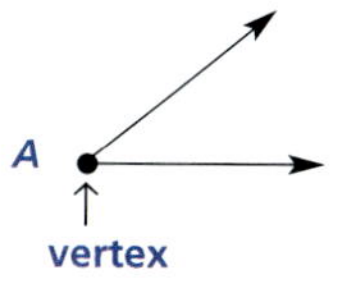

An angle can be named by its vertex.
Angle *A* or ∠*A*

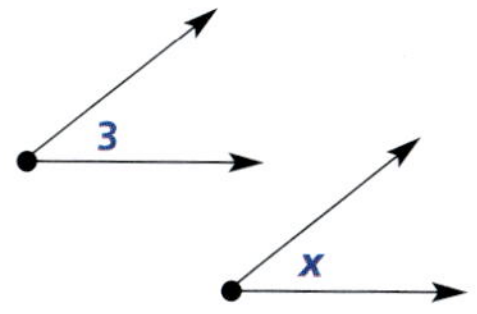

An angle can be named by a letter or number "inside" the angle.
Angle *x* or ∠*x*
Angle 3 or ∠3

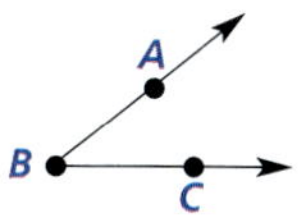

An angle can be named by the rays that form it. The vertex is always in the middle of the name.
∠*ABC* or ∠*CBA*

Angle ∠
A figure made up of two sides, or rays, with a common endpoint

Angle bisector
Ray that divides an angle into two equal parts

Arc
Part of a circle

Bisect
To divide into two equal parts

Compass
A tool used to draw circles and parts of circles called arcs

Construction
Process of making a line, angle, or figure according to specific requirements

Vertex
The point common to both sides of an angle

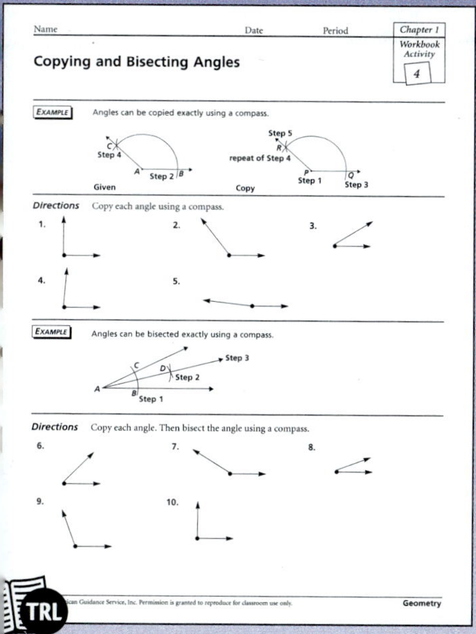
Name Date Period Chapter 1 Workbook Activity 4
Copying and Bisecting Angles
EXAMPLE Angles can be copied exactly using a compass.
Directions Copy each angle using a compass.
EXAMPLE Angles can be bisected exactly using a compass.
Directions Copy each angle. Then bisect the angle using a compass.
Geometry

Workbook Activity 4

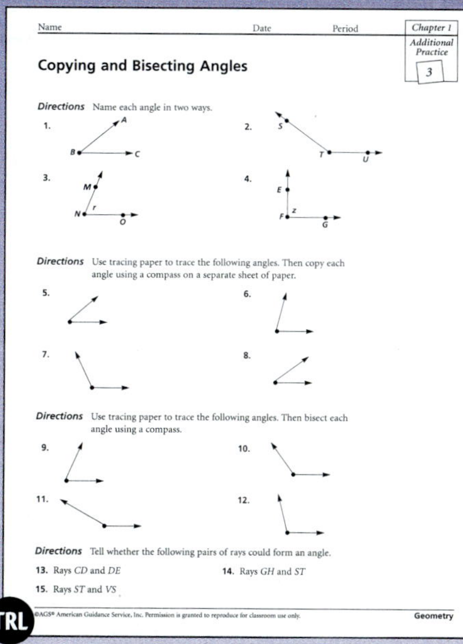
Name Date Period Chapter 1 Additional Practice 3
Copying and Bisecting Angles
Directions Name each angle in two ways.
Directions Use tracing paper to trace the following angles. Then copy each angle using a compass on a separate sheet of paper.
Directions Use tracing paper to trace the following angles. Then bisect each angle using a compass.
Directions Tell whether the following pairs of rays could form an angle.
13. Rays *CD* and *DE* 14. Rays *GH* and *ST*
15. Rays *ST* and *VS*
Geometry

Additional Practice 3

Lesson at a Glance

Chapter 1 Lesson 3

Overview This lesson defines angles and explains how to copy and bisect angles.

Objectives

- To introduce angles.
- To draw and bisect angles using a compass.

Student Pages 11–14

Teacher's Resource Library TRL

Workbook Activity 4

Additional Practice 3

Mathematics Vocabulary

angle
angle bisector
arc
bisect
compass
construction
vertex

1. Warm-Up Activity

Draw a line segment and a ray on the board. Ask students to distinguish between the two. Then form an angle by drawing another ray extending from the endpoint of the first. Tell students that this is an angle. Ask them to define an angle based on what they have observed. (An angle is a figure made up of two rays with the same endpoint.) Then introduce the text definitions of *angle* and *vertex.*

2. Teaching the Lesson

Draw several angles on the board and label each angle's vertex, inside, and a point on each ray. Be sure to use different letters and/or numbers for each angle drawn. Then give the name of one of the angles you drew. Have volunteers identify the other two names by which the angle could be known. Emphasize that to avoid confusion people usually use only one of the names when referring to a specific angle.

Draw an angle on the board or chart paper. Follow the steps given for copying the angle. As you follow the steps, explain what you are doing and why. Repeat by drawing and copying angles of different sizes so that students can generalize that the method for copying angles can be used with any size angle. Follow the same procedure for bisecting angles. Then ask volunteers to copy and bisect angles. Make sure that the volunteers copying and bisecting the angles on the board say the steps aloud as they do them so that the other students will have a chance to hear the steps repeated many times.

3. Reinforce and Extend

Learning Styles

LEP/ESL The vocabulary presented in the lesson will be used throughout the study of geometry. It is essential that students thoroughly understand the terms and their meanings as they apply to geometry. In some languages, the words for such words as *angle, bisect,* and *construction* are quite similar to the English terms. Students may wish to say the words in both their first and second languages to reinforce the concepts. Have students with limited English proficiency work with students whose first language is English. Together the students can produce cards or sheets that include the English term, its pronunciation, the term's translation in the ESL student's first language, a definition of the term, and perhaps an illustration. Students can use the cards or sheets to quiz one another on the terms and their meanings. Later the cards or sheets can be used for review.

Construction

Problem: Copy $\angle A$.

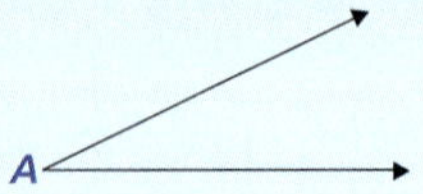

Step 1 Draw a ray. Label endpoint P.

Step 2 Place the needle of the **compass** at A, the vertex of $\angle A$. Draw any size **arc** that crosses each ray of $\angle A$. Once you draw the arc, do not change the opening of the compass. Label the points where the arc crosses the rays B and C.

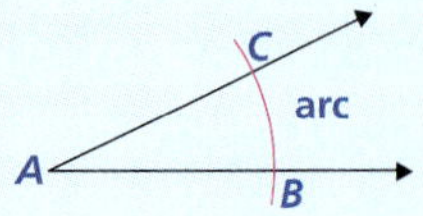

Step 3 Place the compass needle at point P. Keeping the compass opening the same, draw an arc that crosses the ray. Label that point Q.

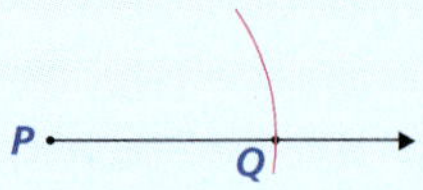

Step 4 Place the compass needle at point B. Fix the compass opening so that the pencil touches point C. Keeping the same opening, place the needle at point Q. Draw an arc that crosses the arc that passes through Q. Label the point where the arc crosses R.

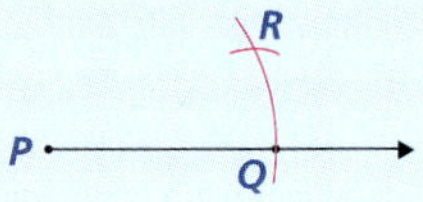

Step 5 Draw $\overrightarrow{PR}$. $\angle RPQ$ is a copy of $\angle A$.

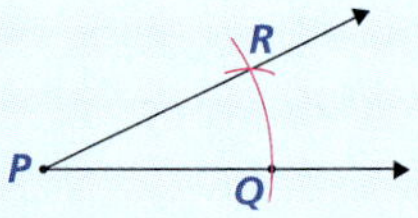

If you are asked to copy an angle, you will be expected to do all five steps on the given angle and its copy. The **construction** will look like this.

To **bisect** an angle means to divide an angle into two equal parts.

CONSTRUCTION

Problem: Bisect $\angle A$.

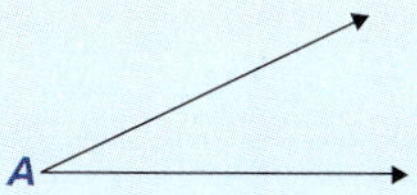

Step 1 Place the compass needle at *A*. Draw an arc of any size that crosses each ray of the angle. Label the points where the arc crosses the rays *B* and *C*.

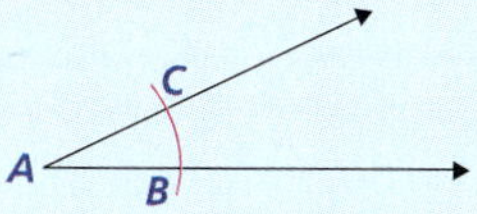

Step 2 Place the compass needle at *B*. Draw an arc. Once you draw the arc, do not change the opening of the compass. Place the needle at *C* and draw an arc that crosses the arc you just drew. Label the point where the arcs cross, *D*.

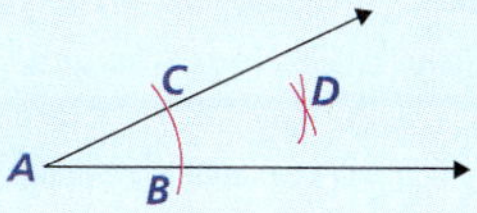

Step 3 Draw $\overrightarrow{AD}$. $\overrightarrow{AD}$ bisects $\angle A$.

Angle *A* has been bisected. $\overrightarrow{AD}$ is the **angle bisector.** Two angles are formed. $\angle CAD$ and $\angle DAB$ are equal.

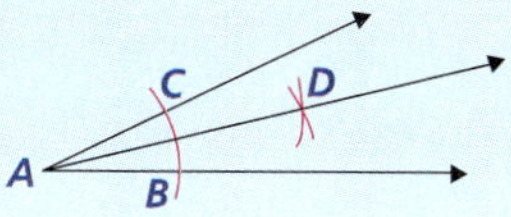

If you are asked to bisect an angle, you will be expected to do all three of the steps on the given angle. The construction will look like this.

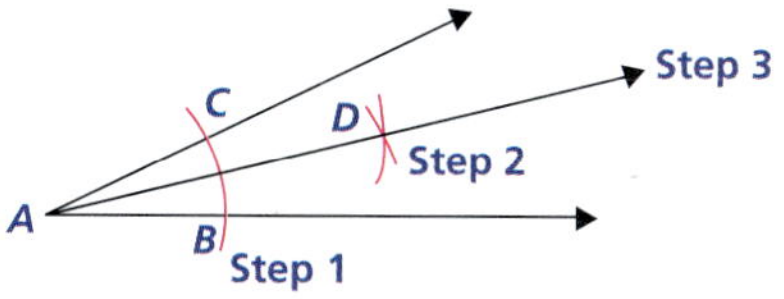

LEARNING STYLES

Auditory/Verbal Produce a step-by-step instructional audiotape for copying and bisecting angles. The tape should include the steps given in the book and any extra tips that may be helpful to students. Place the tape in an independent work center, and allow students to listen to it as they practice copying and bisecting angles.

LEARNING STYLES

Body/Kinesthetic Students can use their gross motor skills to draw, copy, and bisect angles on a school sidewalk or play area. Give pairs of students a meter stick, chalk, and string. Ask one student to use the meter stick to draw an angle on the ground. Then have the pair of students work together to copy and bisect the angle. Show the pair how they can work together to use the string and chalk as a large compass. When copying the angle, the partners draw a ray and arcs. They use the meter stick to draw the straight ray needed to copy the original angle.

Learning Styles

Interpersonal/Group Learning Have students work in pairs or small groups to produce, copy, and bisect angles. They should use compasses, colored pencils, tracing paper, and rulers in the activity. Every student draws an angle with rays that take up at least half of a sheet of tracing paper. The resulting angle should be labeled *ABC*. Students then exchange angles with partners, who reproduce the angle on a separate sheet of paper. Partners should use different colored pencils for their reproductions. Students can check their work by placing the copy over the original angle. If the vertices and rays line up, the angles are duplicated. Have students use the same procedure to bisect the duplicated angles.

Exercise A Name each angle in two other ways.

1. $\angle PQR$

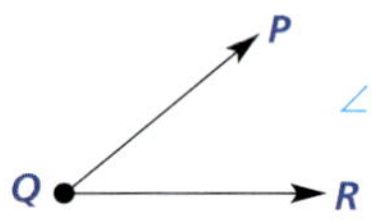

$\angle Q$, $\angle RQP$

2. $\angle y$

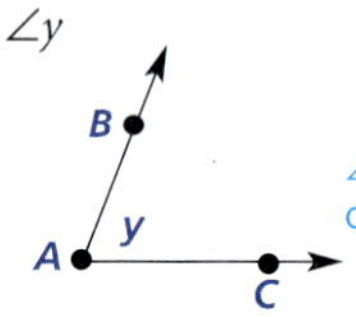

$\angle A$, $\angle BAC$ or $\angle CAB$

3. $\angle R$

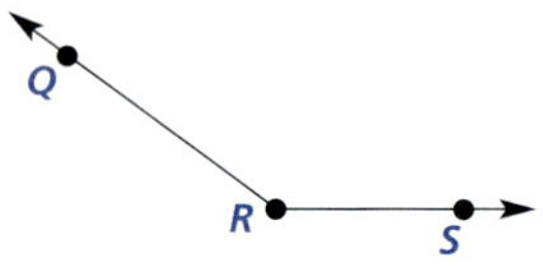

$\angle QRS$, $\angle SRQ$

Construction

Exercise B On a separate sheet of paper, draw an angle similar to each of the angles shown. Be sure to use a ruler or straightedge and be neat and accurate. Then copy each angle you have drawn.

4.

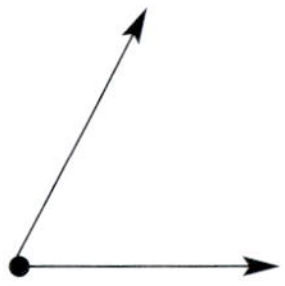

5.

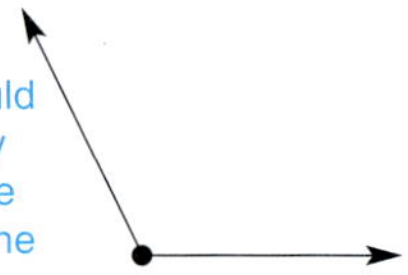

Students should draw and copy angles with the measures of the angles shown.

6.

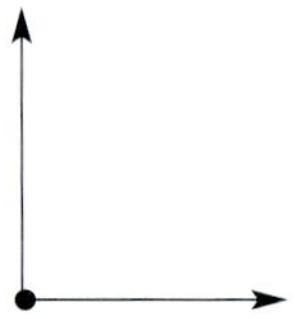

Exercise C On a sheet of paper, draw an angle similar to each of the angles 7–9. Remember to be accurate. Then bisect each angle you have drawn. You will need to use a compass and straightedge.

7.

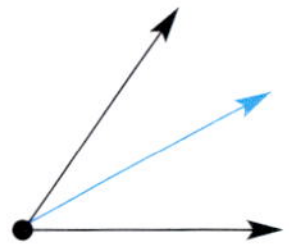

8.

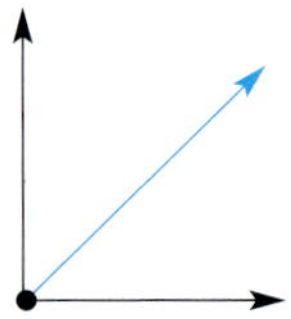

9.

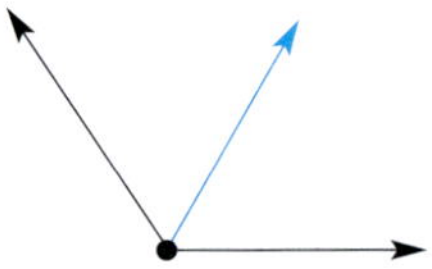

Problem Solving

Exercise D Answer the following question.

10. Can $\overrightarrow{AB}$ and $\overrightarrow{BC}$ form an angle? Why or why not?

Yes, common point *B* could be vertex.

Lesson 4 Angle Measurement

The ancient Babylonians created several systems of measurement. Historians believe that they were the first to divide a circle into 360 equal parts, called **degrees.** No matter who invented them, we all use degrees to measure angles.

Note that the numbers increase in a counter-clockwise direction—the direction opposite to the direction the hands of a clock move.

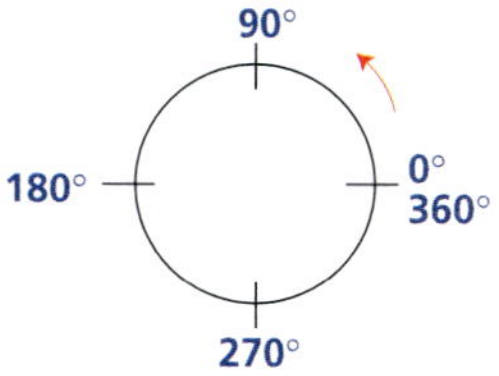

The symbol for a degree is °. An angle measurement of 10 degrees is written 10°. To measure an angle you can use a **protractor.** A protractor is a tool shaped like half a circle with degree markings from 0° to 180°.

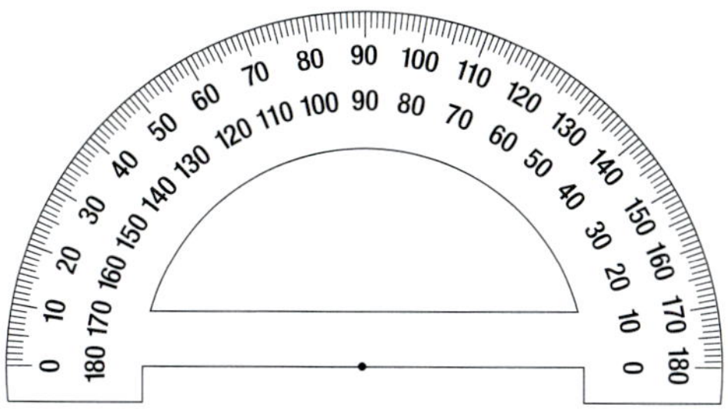

Acute angle
An angle whose measure is less than 90°

Degree
A unit of angle measure

Obtuse angle
An angle whose measure is greater than 90° but less than 180°

Perpendicular lines ⊥
Lines that form right angles

Protractor
A tool used to draw or measure angles

Right angle
An angle whose measure is 90°

Straight angle
An angle whose measure is 180°

EXAMPLE Find the measure of ∠*a*.

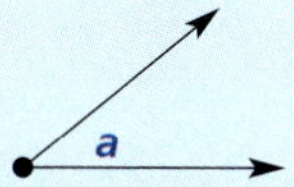

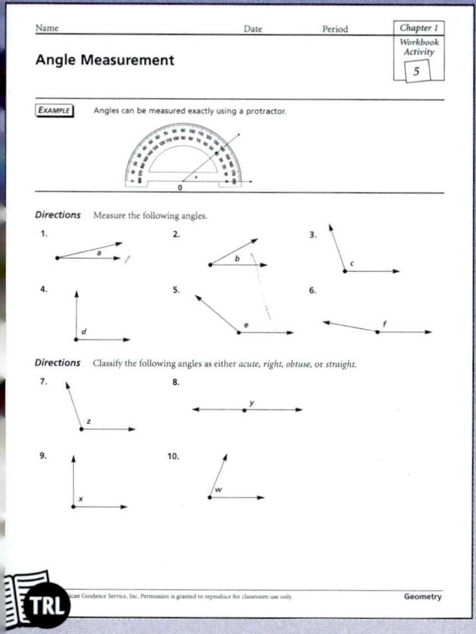

Workbook Activity 5

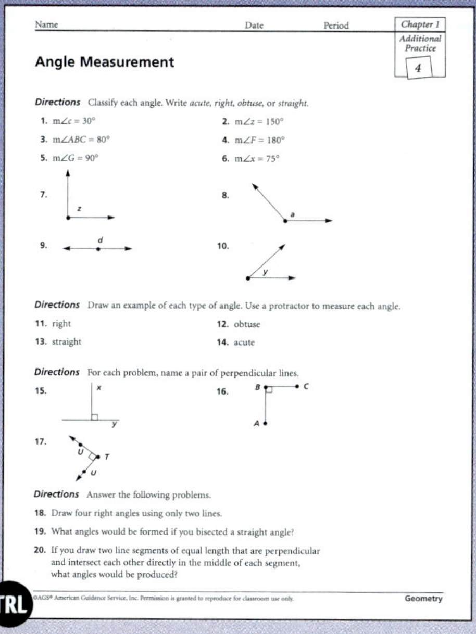

Additional Practice 4

Lesson at a Glance

Chapter 1 Lesson 4

Overview This lesson teaches how to measure angles and how to categorize the different kinds of angles.

Objectives

- To name angles based on their number of degrees.
- To draw and bisect angles using a compass.

Student Pages 15–19

Teacher's Resource Library TRL

Workbook Activity 5

Additional Practice 4

Mathematics Vocabulary

acute angle
degree
obtuse angle
perpendicular lines
protractor
right angle
straight angle

1. Warm-Up Activity

Draw an acute angle on the board. Place the needle of the compass at the vertex of the angle and draw a circle passing through the rays of the angle. Ask a volunteer to come to the board and trace the arc of the circle between the rays of the angle. Explain to students that angles are measured by the arc of the circle that lies between the rays. Use your illustration to emphasize that the arc is a part of the complete circle. Ask whether students know the unit of measurement used for arcs and angles. (degrees)

2. Teaching the Lesson

Demonstrate how to use a protractor to measure angles. If possible, draw a number of angles on acetate and use an overhead projector and a transparent protractor for your demonstration. Draw several angles with different measures on a sheet of paper. Be sure to include right angles and straight angles. Make and distribute photocopies. Have students measure the angles and record the measurements on their papers. Review the names used to classify angles by their measure—acute, right, obtuse, and straight. After defining the terms and providing examples on the board, ask students to label each angle they measured.

Demonstrate how to use the protractor to draw an angle with a specific measure. Then list the following ten angle measures on the board and have students draw the angles and label them as acute, right, obtuse, or straight angles.

- 30° (acute)
- 100° (obtuse)
- 90° (right)
- 45° (acute)
- 110° (obtuse)
- 125° (obtuse)
- 60° (acute)
- 170° (obtuse)
- 180° (straight)
- 80° (acute)

Common Error

Some students may read their angle measurements from the wrong set of degree numbers on the protractor. The numbers on the protractor are set up to be read from left to right or right to left. Explain to students that when they use the protractor, they place the center of the protractor at the vertex of the angle and align one ray with a zero on the protractor. If the ray lined up with zero is on the right side of the vertex, students should read the angle measurement from right to left. If the ray lined up with zero is on the left side of the vertex, they should read the angle measurement from left to right. You might use a memory device such as the following to help students remember which set of degree numbers to use: Ray on the right, use the inside; ray on the left, use the outer edge.

Example *(continued)*

Step 1 Place the 0 point on the base of the protractor at the vertex of the angle. Turn the protractor so that one ray of the angle passes through the 0° mark on the protractor.

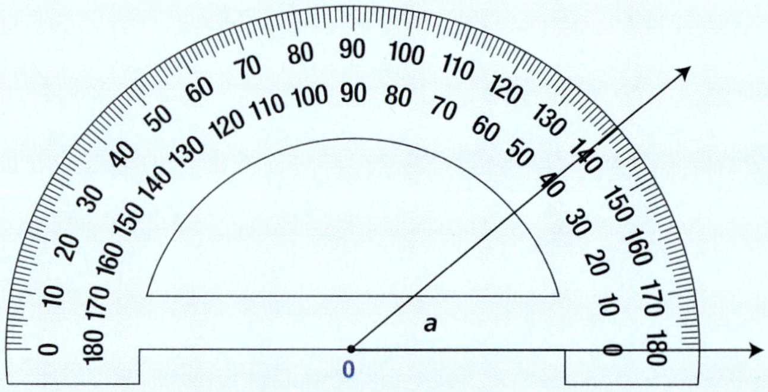

Step 2 Follow the second ray outward from the 0 on the protractor. Read the degree mark where the ray crosses the protractor.

The measure of angle *a* is 40°. You can write this as m∠*a* = 40°.

You can classify an angle using its measure.

An **acute angle** is an angle whose measure is greater than 0 and less than 90°.

Examples of acute angles

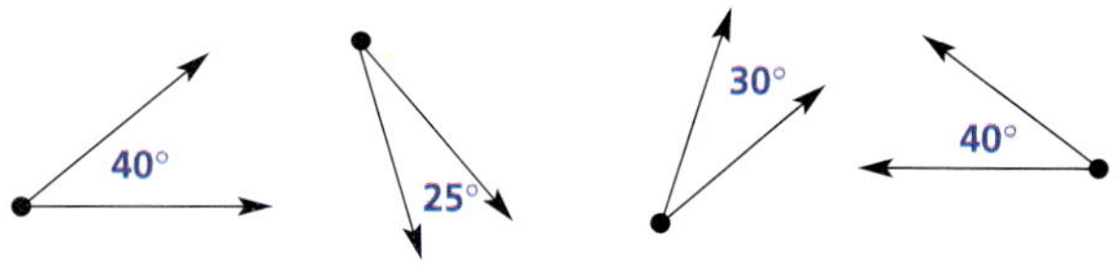

The symbol □ is sometimes used to show a right angle.

A **right angle** is an angle whose measure = 90°.

Examples of right angles

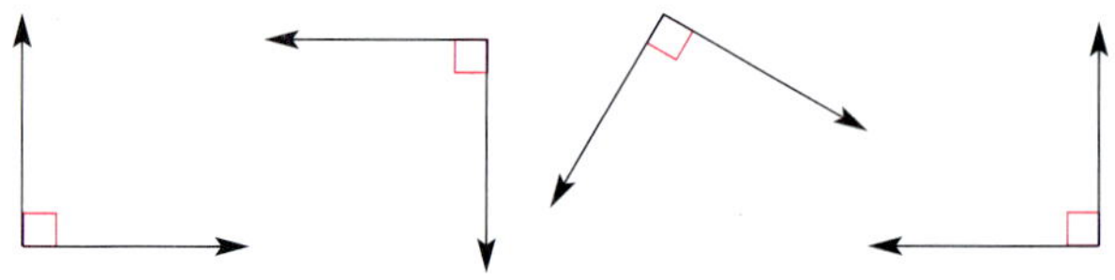

An **obtuse angle** is an angle whose measure is greater than 90° and less than 180°.

Examples of obtuse angles

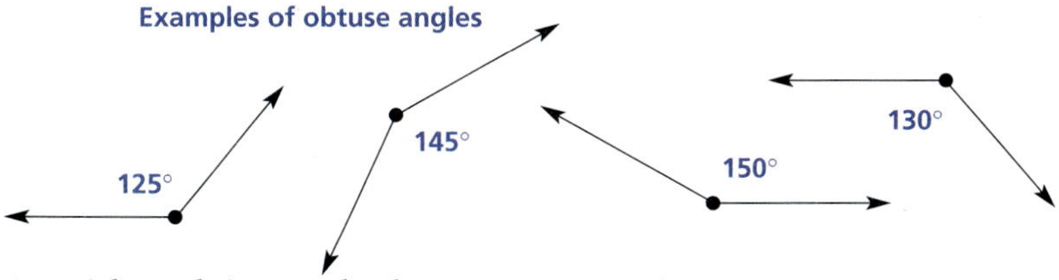

A **straight angle** is an angle whose measure = 180°.

Examples of straight angles

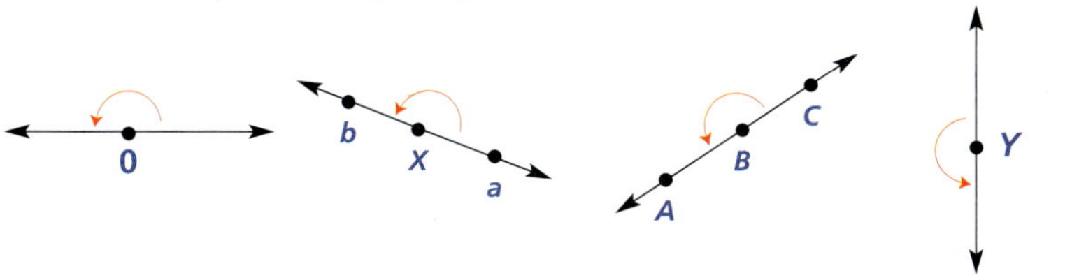

Look again at the right angles. Notice the lines that make up the angles.

Lines that form right angles are said to be **perpendicular lines.**

Examples of perpendicular lines

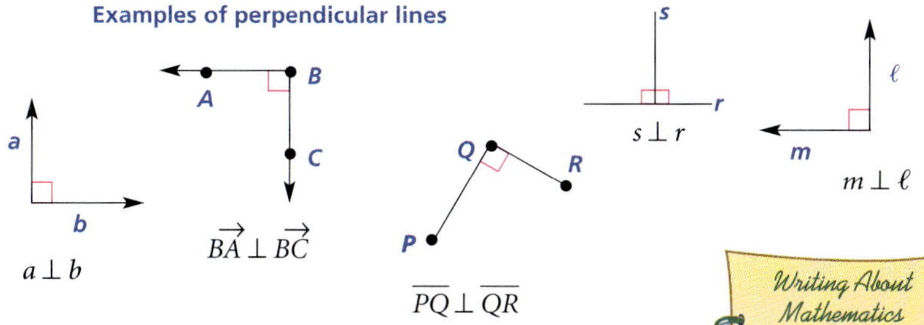

The symbol for "is perpendicular to" is ⊥. Here, a is perpendicular to b, $\overrightarrow{BA}$ is perpendicular to $\overrightarrow{BC}$, $\overline{PQ}$ is perpendicular to $\overline{QR}$, s is perpendicular to r, and m is perpendicular to ℓ. As part of your geometry study, you will need to know how to construct perpendicular lines.

Writing About Mathematics

Write a description of three objects. Tell whether the objects have acute, right, or obtuse angles. Point out any perpendiculars.

3. Reinforce and Extend

Learning Styles

LEP/ESL The terms *acute, obtuse,* and *right* used to identify angle measures are words with multiple meanings. Have students work together to reinforce the geometric meanings of these terms. Suggest that students prepare cards with illustrations and definitions of *acute, obtuse,* and *right* angles. They can use the cards to identify the angles shown and explain why the angle is an acute, obtuse, or right angle.

Learning Styles

Visual/Spatial Use a color-coded system when showing examples of angles on the board. Suggest that students use the color code to visually reinforce the classification of angles as acute, obtuse, right, and straight.

Group Problem Solving

Have each student name an angle measurement between 0° and 180°. List all measurements on the board. Divide the class into small groups. Have the groups add together all of the listed measurements. If necessary, remind students that a complete circle contains 360°. They then should determine how many complete circles the resulting answer would contain. How many degrees remain? What kind of angle would the remaining number of degrees form? Have groups compare answers and check their work if their answers differ.

Application

At Home

Have students find examples of angles in their homes. Ask them to conclude from their observations what kind of angle is most common in the home. (Examples may include room corners and the positions of the hands of a clock. Students should conclude that most angles they find in their homes are right angles.)

Manipulatives

Angle Draw

Materials: Angle Rule, colored pencils

Student Practice: Adjust the Angle Rule to show differing angle measures. Use the rule to identify the size of the angle and to classify the angle as acute, obtuse, right, or straight. Repeat several times.

Group Practice: Have students work in groups of three. Each member of the group should choose a colored pencil and use the same color throughout the activity. Every member of the group draws five angles of differing measures. Group members pass their papers to the left and use the Angle Rule to measure the angles on the page they now have. They should write the measure under each angle. They pass the pages to the left again and label each angle as *acute, obtuse, right,* or *straight.* Group members should verify each other's work.

Construction

Problem: Construct a line perpendicular to a given line and passing through a given point.

Step 1 Given line ℓ with point *P*.

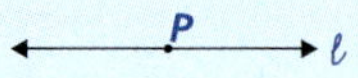

Step 2 Place the needle of the compass at *P*. Draw any size arc that crosses ℓ on both sides of point *P*. Label one place where the arc crosses *A*. Label the other place *B*.

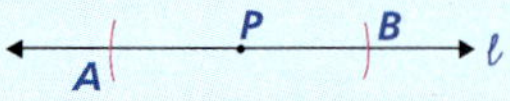

Step 3 Place the compass needle at *A*. Open the compass so that it reaches past point *P*. Keep the needle on *A* and draw an arc above *P*.

Step 4 Then, keeping the same compass opening, place the compass needle at point *B*. Draw a second arc above *P*. Be sure the arcs cross one another. Label the point where the arcs cross *C*.

Step 5 Draw $\overline{PC}$. $\overline{PC} \perp \ell$.

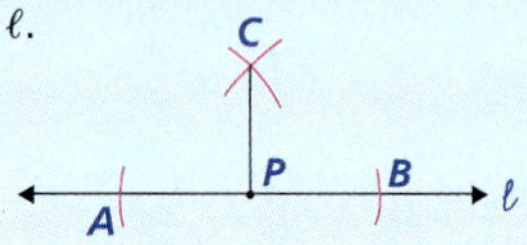

Exercise A Classify each angle. Write *acute, right, obtuse,* or *straight.*

1. $m\angle a = 15°$ acute
2. $m\angle ABC = 100°$ obtuse
3. $m\angle z = 180°$ straight
4. $m\angle D = 90°$ right

5. 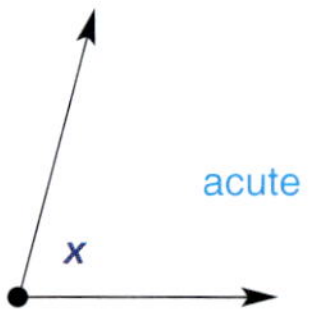

acute

6. 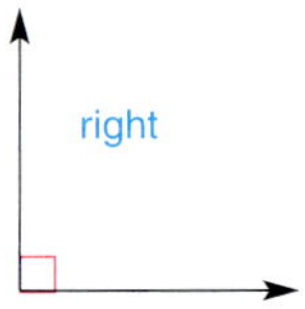
right

7. 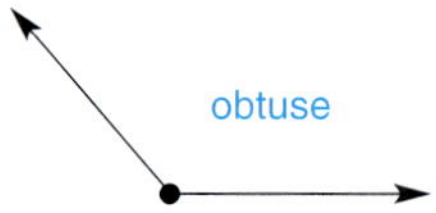
obtuse

8. 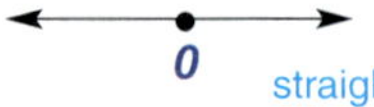

straight

Construction

Exercise B Draw an example of each type of angle. Then use a protractor to measure your angle.

9. acute 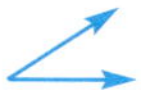 **10.** right  **11.** obtuse 

9. Students' angle measures will vary but should be less than 90°.
10. Students' angle measures should be 90°.
11. Students' angle measures will vary but should be greater than 90°.

Exercise C For each exercise, name a pair of perpendicular lines.

12.

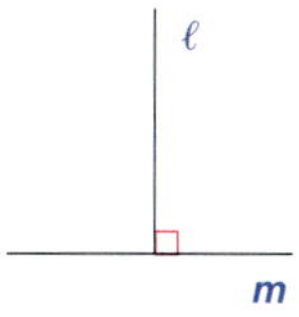

$\ell \perp m$

13.

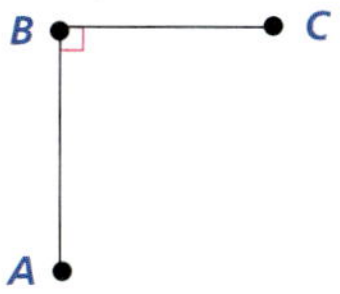

$\overline{AB} \perp \overline{BC}$

14.

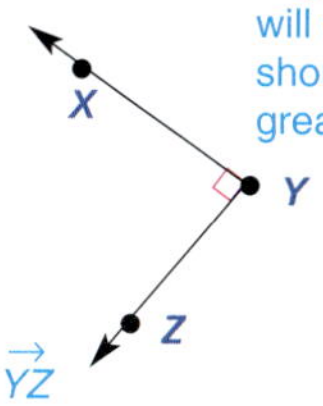

$\overrightarrow{YX} \perp \overrightarrow{YZ}$

Problem Solving

Exercise D Answer the following question.

15. Explain with a sentence or a construction how you can make four right angles using only two lines. When two perpendicular lines intersect, they form four right angles.

Geometry Around Us

Artists and architects use their knowledge of geometry and angles to produce special decorations. Frank Lloyd Wright used geometry to create this stained glass window.

Learning Styles

Auditory/Verbal Have partners draw perpendiculars. For auditory/verbal reinforcement, one partner should read the steps aloud as the other follows the steps to construct a perpendicular. Partners then switch roles and repeat the activity. Encourage them repeat the process several times.

Mental Math

Have students identify for each of the following angles the number of degrees in the angles formed by the angle bisector. Then ask students to explain what all of the angles formed by angle bisectors have in common, except for those of one angle.

- 30° (15°)
- 120° (60°)
- 180° (90°)
- 90° (45°)
- 72° (36°)
- 140° (70°)
- 116° (58°)
- 60° (30°)
- 158° (79°)
- 80° (40°)

(All of the angles formed by the angle bisectors are acute angles except for the right angles formed by bisecting the straight angle.)

Lesson at a Glance

Chapter 1 Lesson 5

Overview This lesson introduces the concepts of adjacent, complementary, supplementary, and vertical angles.

Objectives

- To identify and name adjacent, complementary, supplementary, and vertical angles.
- To determine missing angle measures based on known measures.

Student Pages 20–23

Teacher's Resource Library TRL

Workbook Activity 6

Additional Practice 5

Mathematics Vocabulary

adjacent angles
complementary angles
intersect
supplementary angles
vertical angles

1. Warm-Up Activity

Review the terms *acute angle, right angle, obtuse angle,* and *straight angle* introduced in the previous lesson. Draw each type of angle on the board and ask volunteers to write the name of each angle under it.

2. Teaching the Lesson

After presenting the concepts of adjacent angles and complementary angles, draw two right angles on the board. Bisect the angles, mark each bisected angle 45°, and label the bisected angles *a, b, c,* and *d*. Ask questions such as the following to help students understand the terms *adjacent angles* and *complementary angles.* Encourage students to explain their answers.

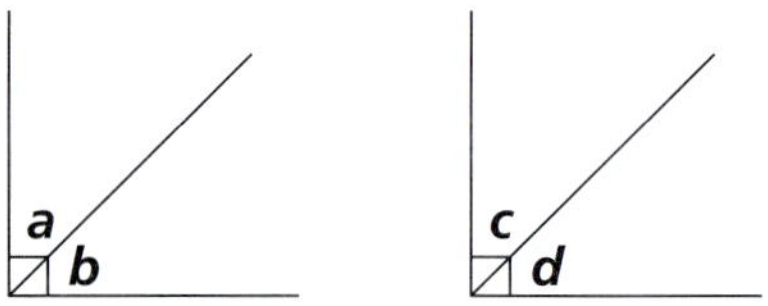

Lesson 5 Complementary and Supplementary Angles

Adjacent angles
Angles with a common vertex and one common side

Complementary angles
Two angles whose measures add to 90°

Intersect
Cross each other; point at which two or more lines cross each other

Supplementary angles
Two angles whose measures add to 180°

Vertical angles
Angles that have a common vertex and whose sides are formed by the same lines

You have seen that angles can be classified by their measure. Now you will see that pairs of angles can be classified by their position and their measure.

In the drawing at right, $\angle a$ and $\angle b$ are classified as **adjacent angles.** They have a common vertex, *C*. They also have a common side, ℓ.

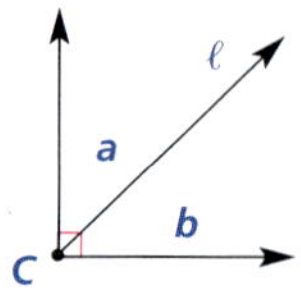

Adjacent angles have a common vertex and one common side. The □ symbol in the drawing tells you that $m\angle C = 90°$. So, $m\angle a + m\angle b = 90°$.

Two angles whose measures add to 90° are called **complementary angles.** In this case, *a* and *b* are also called complementary angles. Note: Two angles do not have to be adjacent in order to be complementary. For example, each of the following pairs of angles are complementary angles.

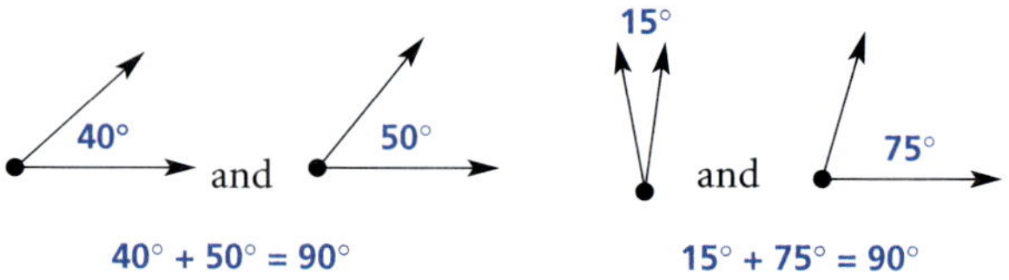

Suppose you have a straight angle such as the angle below.

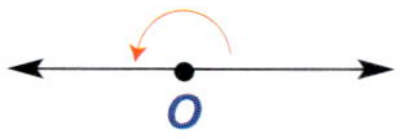

If you draw any ray from *O*, you will form two adjacent angles, *c* and *d*.

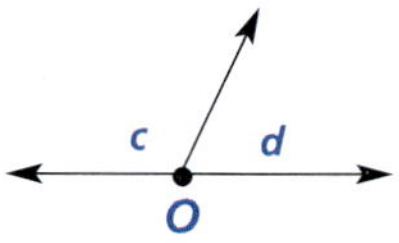

The drawing tells you that $m\angle O = 180°$.

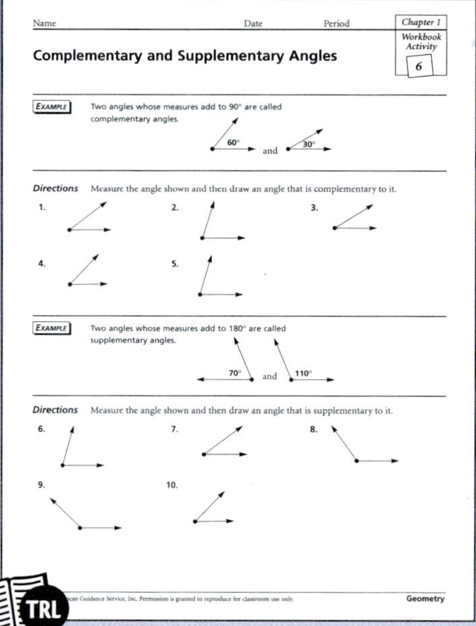
Name Date Period
Chapter 1 Workbook Activity 6

Complementary and Supplementary Angles

EXAMPLE Two angles whose measures add to 90° are called complementary angles.

60° and 30°

Directions Measure the angle shown and then draw an angle that is complementary to it.

1. 2. 3. 4. 5.

EXAMPLE Two angles whose measures add to 180° are called supplementary angles.

70° and 110°

Directions Measure the angle shown and then draw an angle that is supplementary to it.

6. 7. 8. 9. 10.

 Geometry

Workbook Activity 6

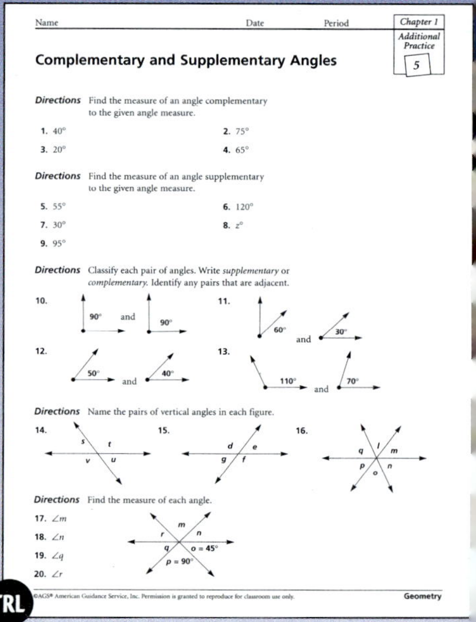
Name Date Period
Chapter 1 Additional Practice 5

Complementary and Supplementary Angles

Directions Find the measure of an angle complementary to the given angle measure.

1. 40°
2. 75°
3. 20°
4. 65°

Directions Find the measure of an angle supplementary to the given angle measure.

5. 55°
6. 120°
7. 30°
8. $z°$
9. 95°

Directions Classify each pair of angles. Write *supplementary* or *complementary.* Identify any pairs that are adjacent.

10. 90° and 90°
11. 60° and 30°
12. 50° and 40°
13. 110° and 70°

Directions Name the pairs of vertical angles in each figure.

14. s t v u
15. d e g f
16. q l m p o n

Directions Find the measure of each angle.

17. $\angle m$
18. $\angle n$
19. $\angle q$
20. $\angle r$

m r n q o = 45° p = 90°

 Geometry

Additional Practice 5

Two angles whose measures add to 180° are **supplementary angles.** So, $m\angle c + m\angle d = 180°$. In this case, *c* and *d* are also called supplementary angles. Note: Two angles do not have to be adjacent in order to be supplementary. For example, each of the following pairs of angles are supplementary angles.

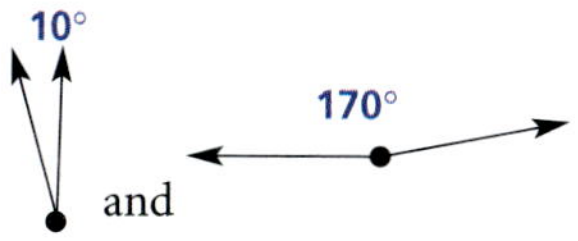

$10° + 170° = 180°$

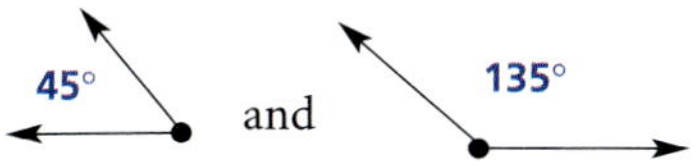

$45° + 135° = 180°$

When two lines **intersect,** four angles are formed.

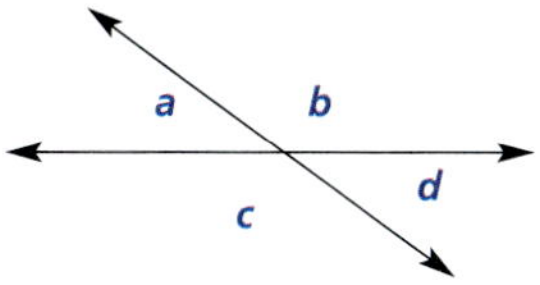

Vertical angles are angles that have a common vertex and whose sides are formed by the same lines. In the drawing above, $\angle a$ and $\angle d$ are vertical angles and $\angle b$ and $\angle c$ are vertical angles. Whenever two lines intersect, two pairs of equal vertical angles are formed. If $m\angle a = 45°$, then $m\angle d = 45°$. If $m\angle b = 135°$, then $m\angle c = 135°$. That is, $m\angle a = m\angle d$, and $m\angle b = m\angle c$.

Note that if the two lines are perpendicular, then four right angles are formed.

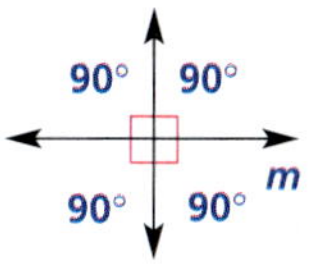

In this drawing, more than two lines intersect to create three pairs of vertical angles: $\angle n$ and $\angle q$, $\angle o$ and $\angle r$, and $\angle s$ and $\angle p$.

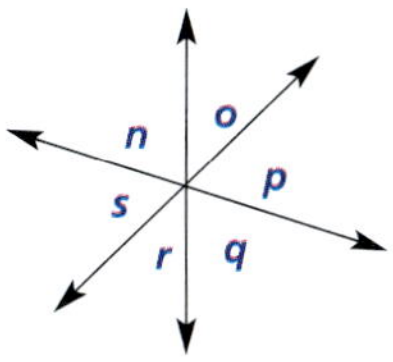

Are angles *a* and *b* complementary? **(yes, because the sum of their measures equals 90°)**

Are they adjacent? **(yes, because they have a common vertex and side)**

Are angles *c* and *d* complementary? **(yes, because the sum of their measures equals 90°)**

Are they adjacent? **(yes, because they have a common vertex and share a side)**

Extend the process to present and teach supplementary angles. Draw two straight angles on the board. Divide each into supplementary angles of 80° and 100°, mark the angle degrees, and label the angles *e*, *f*, *g*, and *h*. Then ask questions such as the following:

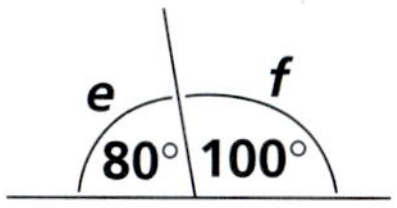

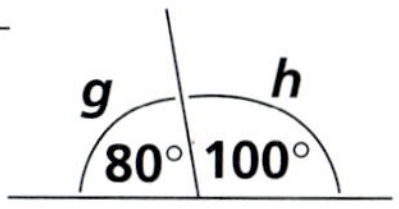

Are angles *e* and *h* supplementary? **(yes, because the sum of their measures equals 180°)**

Are they adjacent? **(no, because they do not have a common vertex or side)**

Then present the concepts of intersecting and vertical angles. Help students identify the pairs of vertical angles that are equal.

3. Reinforce and Extend

LEARNING STYLES

Visual/Spatial To help students organize and visualize angles that are complementary, have them prepare a chart that shows complementary angles in increments of 5°. The chart should show that the m∠A + m∠B = 90°. Students should set up the chart as in the example, or they may choose to illustrate each angle as well. Make sure students extend the chart for every combination of measures in increments of 5 that result in complementary angles. Students can then use the chart to quickly determine whether two angles are complementary.

Complementary Angles		
m∠A	+ m∠B	=
5°	85°	90°
10°	80°	90°
15°	75°	90°

Students can make similar charts for supplementary angles. They should use the formula m∠A + m∠B = 180° for this chart.

CALCULATOR

Have students use their calculators to determine whether the two angles given are supplementary. Students should press 47 [+] 123 [=] display 170. (Angles are not supplementary because they do not add up to 180.)

- 126°, 54° (180°, yes)
- 44°, 134° (178°, no)
- 91°, 89° (180°, yes)
- 65°, 115° (180°, yes)
- 122°, 68° (190°, no)

Exercise A Classify each pair of angles. Write *supplementary* or *complementary.* Identify any pairs that are *adjacent.*

1. 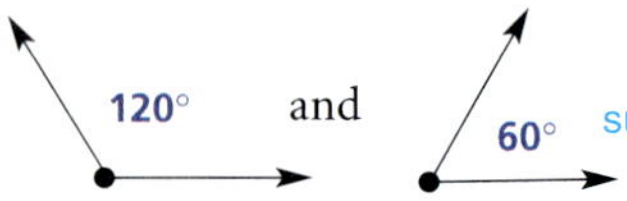

supplementary

2. 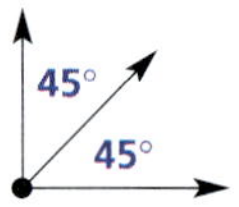

complementary and adjacent

3. 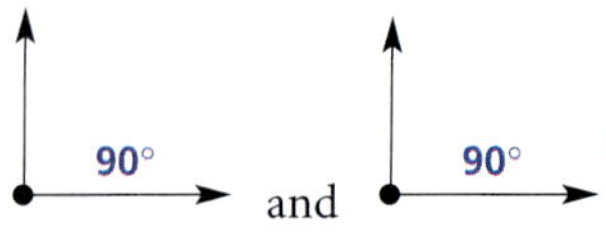

supplementary

4. 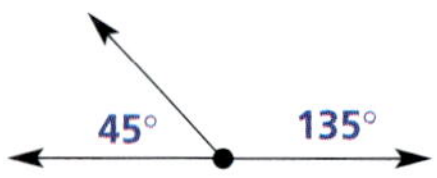

supplementary and adjacent

Calculator Practice

Use your calculator to find the measure of an angle.

EXAMPLE Find the measure of the angle.

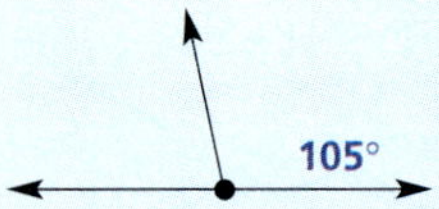

Use your calculator to find the answer.

Press 180 [–] 105, then press [=]. The answer is 75°.

Exercise B Use your calculator to find the measure of an angle supplementary to the given angle measure.

5. 100° 80°
6. 70° 110°
7. 62° 118°
8. 78° 102°
9. 53° 127°
10. 125° 55°
11. 81° 99°
12. 67° 113°

Exercise C Name the pairs of vertical angles in each figure.

13.

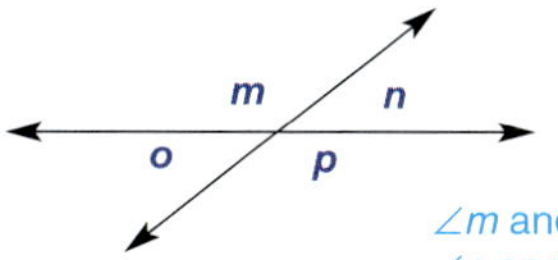

$\angle m$ and $\angle p$, $\angle o$ and $\angle n$

14.

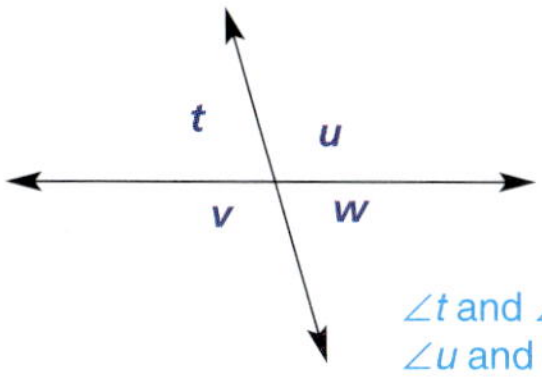

$\angle t$ and $\angle w$, $\angle u$ and $\angle v$

15.

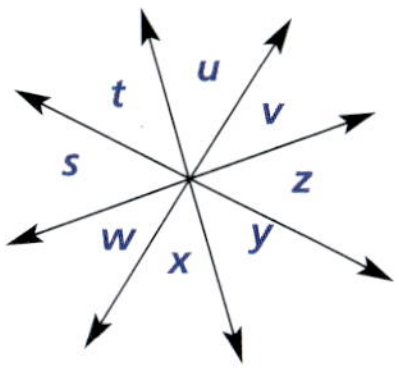

$\angle t$ and $\angle y$, $\angle u$ and $\angle x$, $\angle v$ and $\angle w$, $\angle s$ and $\angle z$

16.

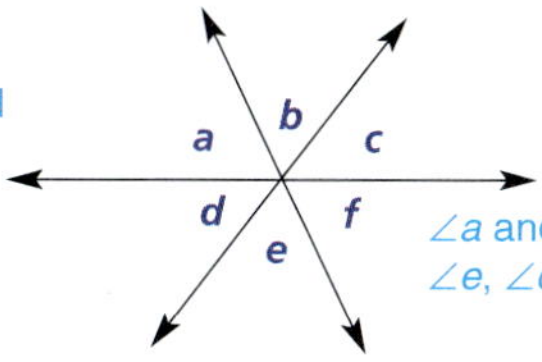

$\angle a$ and $\angle f$, $\angle b$ and $\angle e$, $\angle c$ and $\angle d$

Exercise D Find the measure of each angle.

17. $\angle f$ 90°

18. $\angle b$ 45°

19. $\angle d$ 45°

20. $\angle e$ 45°

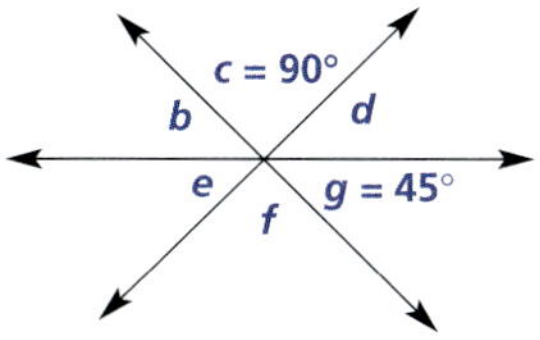

Writing About Mathematics

Think of a real-world example of vertical angles. Describe the example, telling about the angles.

Experiencing Geometry

Study a street map for your area. Look for streets that intersect to form angles and identify the angles that the streets form. Did you find complementary, supplementary, and vertical angles? Are there any adjacent angles?

Learning Styles

Body/Kinesthetic Have students make an Angle Poster. Using pipe cleaners, they can construct complementary angles; complementary, adjacent angles; supplementary angles; supplementary, adjacent angles; and vertical angles. Encourage students to use their protractors as they construct the angles. Ask them to glue the angles in place and label the angle measures and the angle identifications with different colored markers. The posters should be displayed in the classroom to provide visual and tactile reinforcement of lesson vocabulary and concepts.

Learning Styles

Logical/Mathematical Ask students to use intersecting lines to construct vertical angles of varying degrees. They should label each of the angles formed, identify the pairs of equal vertical angles by their labels, and measure the angles to "prove" that the pairs of equal vertical angles are indeed equal.

Mental Math

Have students add the two measures of angles together. They should then tell whether the two angles are supplementary, complementary, or neither.

- 40° + 50° = (90°, complementary)
- 130° + 50° = (180°, supplementary)
- 45° + 35° = (80°, neither)
- 65° + 25° = (90°, complementary)
- 123° + 62° = (185°, neither)
- 145° + 35° = (180°, supplementary)
- 12° + 108° = (120°, neither)
- 54° + 36° = (90°, complementary)

Lesson at a Glance

Chapter 1 Lesson 6

Overview This lesson explains how to use algebra and the definition of angles to solve problems in geometry.

Objectives

- To use algebra to solve for unknown angle measures.

Student Pages 24–27

Teacher's Resource Library TRL

Workbook Activity 7

Additional Practice 6

1. Warm-Up Activity

Ask students to define algebra. (the branch of mathematics that uses letters and numbers to show relationships between quantities) Review the four arithmetic operations—addition, subtraction, multiplication, and division. Explain that algebra uses these basic operations.

2. Teaching the Lesson

Algebra Connection

Write these four algebraic equations on the board, one equation for each basic arithmetic operation. Emphasize that algebraic equations contain at least one operation and a variable that is identified by a letter.

$45 + x = 90$

$90 - x = 45$

$2x = 90$

$90 \div x = 45$

Show students how to solve the equations for x.

Use the examples in the text to help students recognize how algebra and their understanding of complementary, supplementary, and vertical angles can be used to solve for unknown measures.

Lesson 6 Algebra and Angles

Algebra Connection

Basic principles of algebra can be used to find the measure of an unknown angle.

You can use the definitions of complementary, supplementary, and vertical angles along with algebra to solve problems in geometry.

Example The measure of one of two complementary angles is 20°. What is the measure of the second angle?

Step 1 Call the unknown angle measure x.

Step 2 $20° + x = 90°$ by definition of complementary angles

Step 3 $x = 90° - 20°$

$x = 70°$

The measure of the second angle is 70°.

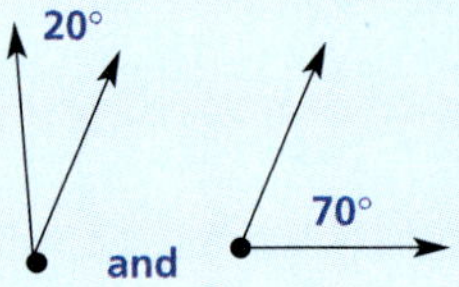

Use algebra and the definitions of angles to solve this problem.

Example The measure of one of two supplementary angles is 35°. What is the measure of the second angle?

Step 1 Call the unknown angle measure y.

Step 2 $35° + y = 180°$ by definition of complementary angles

Step 3 $y = 180° - 35°$

$y = 145°$

The measure of the second angle is 145°.

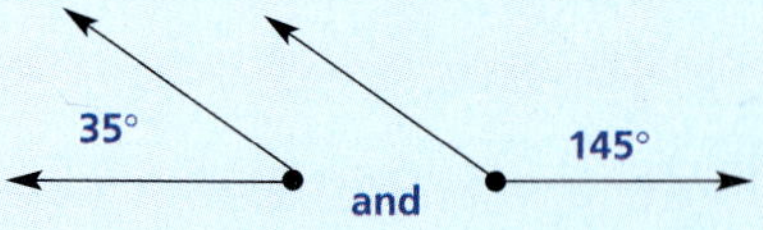

Writing About Mathematics

Two perpendicular lines always form four right angles. Write an explanation of why this is so.

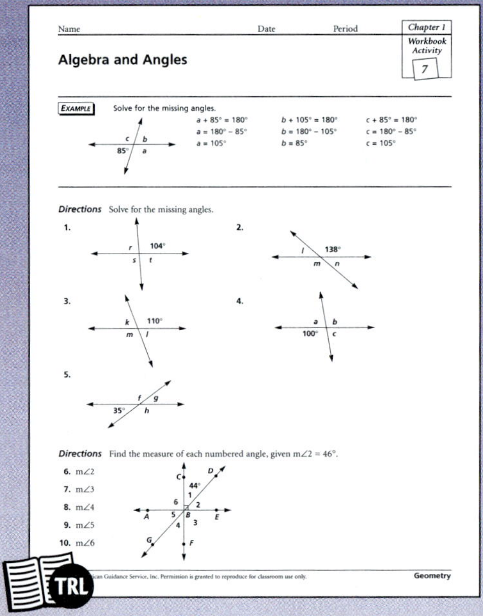

Name Date Period Chapter 1 Workbook Activity 7

Algebra and Angles

EXAMPLE Solve for the missing angles.

$a + 85° = 180°$, $a = 180° - 85°$, $a = 105°$; $b + 105° = 180°$, $b = 180° - 105°$, $b = 85°$; $c + 85° = 180°$, $c = 180° - 85°$, $c = 105°$

Directions Solve for the missing angles.

1. 2. 3. 4. 5.

Directions Find the measure of each numbered angle, given m∠2 = 46°.

6. m∠2
7. m∠3
8. m∠4
9. m∠5
10. m∠6

Geometry

Workbook Activity 7

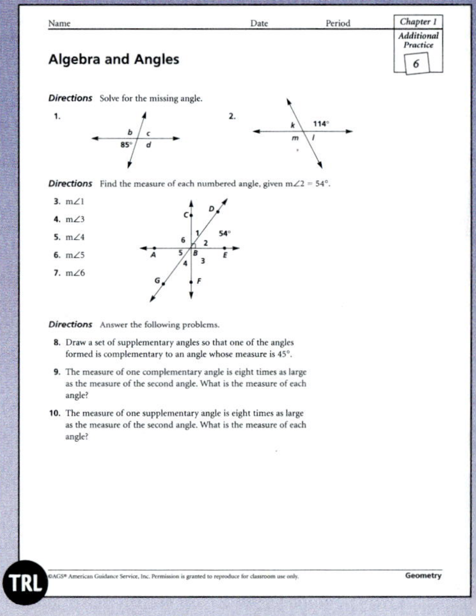

Name Date Period Chapter 1 Additional Practice 6

Algebra and Angles

Directions Solve for the missing angle.

1. 2.

Directions Find the measure of each numbered angle, given m∠2 = 54°.

3. m∠1
4. m∠3
5. m∠4
6. m∠5
7. m∠6

Directions Answer the following problems.

8. Draw a set of supplementary angles so that one of the angles formed is complementary to an angle whose measure is 45°.
9. The measure of one complementary angle is eight times as large as the measure of the second angle. What is the measure of each angle?
10. The measure of one supplementary angle is eight times as large as the measure of the second angle. What is the measure of each angle?

Geometry

Additional Practice 6

EXAMPLE The measure of one supplementary angle is twice the measure of the second supplementary angle. What is the measure of each angle?

Step 1 Let w be the measure of the smaller angle. Let $2w$ be the measure of the larger angle.

Step 2 $2w + w = 180°$ by definition of supplementary angles

Step 3 $3w = 180°$

$w = 60°$

The measure of one of the four angles is 60°.

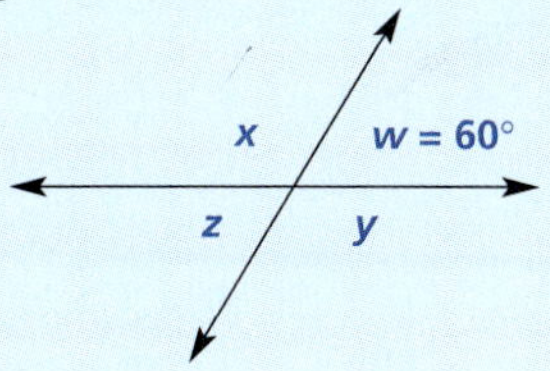

Now, find the measures of the other three angles.

Step 4 $x + 60° = 180°$ by definition of supplementary angles

$x = 180° - 60°$

$x = 120°$ so $m\angle x = 120°$

Step 5 $y + 60° = 180°$ by definition of supplementary angles

$y = 180° - 60°$

$y = 120°$ so $m\angle y = 120°$

OR $m\angle y = 120°$ because vertical angles are equal

Step 6 $z + 120° = 180°$ by definition of supplementary angles

$z = 180° - 120°$

$z = 60°$ so $m\angle z = 60°$

OR $m\angle z = 60°$ because vertical angles are equal

You may wish to supplement the examples by solving these problems on the board.

The measure of one of two complementary angles equals 30°. What is the measure of the second angle? **(60°)**

The measure of one of two supplementary angles equals 30°. What is the measure of the second angle? **(150°)**

The measure of one of two supplementary angles is 3 times larger than the other. What are the measures of the angles? **(135°, 45°)**

Common Error

Help students avoid assigning any significance to the letter *x*, which is commonly used as a placeholder for a number in algebraic equations. Emphasize that any letter from *a* to *z* may be used to represent a number in an algebraic equation.

3. Reinforce and Extend

Manipulatives

M **Angle Challenge**

Materials: Angle Rule, pencil, paper

Group Practice: Have students working in small groups or pairs use the Angle Rule to draw supplementary angles, complementary angles, and vertical angles of different measures. They should use the Angle Rule to measure one angle in each set of angles drawn and then label that angle measurement. Groups exchange papers and use algebra and the definitions of complementary, supplementary, and vertical angles to solve for the unknown angles. They can use the Angle Rule to verify their answers.

Calculator

Ask students to use their calculators to solve the following equations for the missing complementary or supplementary angle measures.

- $42° + x = 90°$ **(48°)**
- $5x + x = 180°$ **(150°, 30°)**
- $12° + x = 180°$ **(168°)**
- $2x + x = 90°$ **(60°, 30°)**
- $90° + x = 180°$ **(90°)**

Group Problem Solving

Divide the class into small groups. Have the group members draw intersecting lines and label the resulting angles from left to right *a*, *b*, *c*, *d*. Explain that *a* and *b* are supplementary angles as are *c* and *d*. Intersecting lines should be constructed so that the measure of angle *b* is 31°. Ask students to use their constructions, algebra, and their understanding of supplementary and vertical angles to identify the measures of angles *a*, *c*, and *d*. **(*a* = 149°, *b* = 31°, *c* = 31°, and *d* = 149°)**

Exercise A Solve for the missing angles.

1.

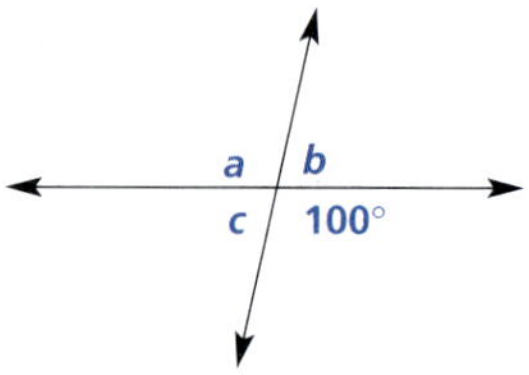

m∠a = 100°, m∠b = m∠c = 80°

2.

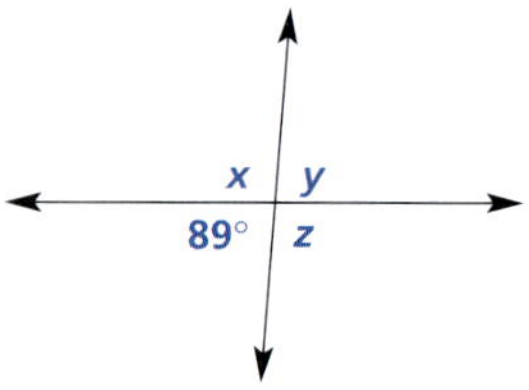

m∠y = 89°, m∠x = m∠z = 91°

3.

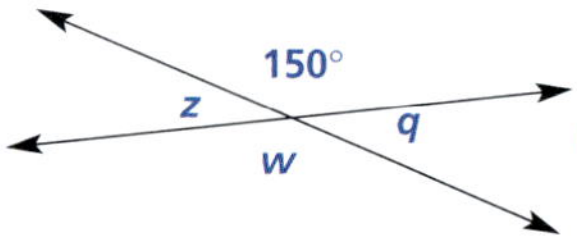

m∠w = 150°, m∠z = m∠q = 30°

Exercise B Find the measure of each numbered angle, given m∠2 = 42°.

4. m∠1 48°

5. m∠3 90°

6. m∠4 48°

7. m∠5 42°

8. m∠6 90°

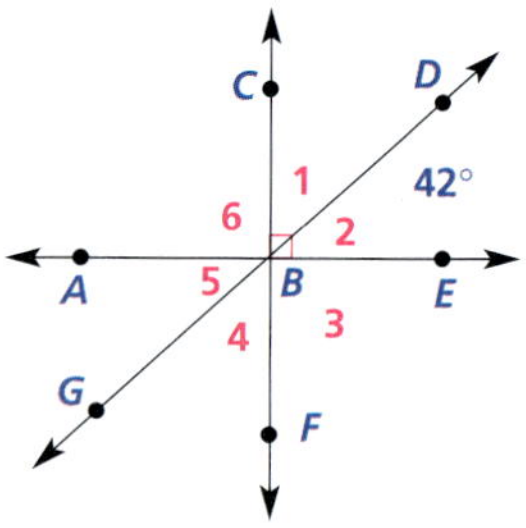

9. 180°
10. ∠ABC
11. ∠5, ∠2; ∠1, ∠4; or ∠6, ∠3
12. ∠1, ∠2; or ∠4, ∠5; also ∠2, ∠4 and ∠1, ∠5

Exercise C Use the figure above to do the following.

9. Find m∠6 and m∠3.

10. Give another name for ∠6.

11. Name a pair of vertical angles.

12. Name a pair of complementary angles.

Problem Solving

Exercise D Answer the following questions.

13. Draw a set of vertical angles. Compare the measures of the vertical angles. Write a statement that describes your observation about the measures of vertical angles. For any pair of vertical angles, the angles have equal measures.

14. The measure of one complementary angle is five times as large as the measure of the second angle. What is the measure of each angle? 15° and 75°

15. The measure of one supplementary angle is four times as large as the measure of the second angle. What is the measure of each angle? 36° and 144°

16. One complementary angle is half as large as the second angle. What is the measure of each angle? 30° and 60°

17. One supplementary angle is two-thirds as large as the second angle. What is the measure of each angle? 72° and 108°

18. The measure of one complementary angle is four times as large as the measure of the second angle. What is the measure of each angle? 18° and 72°

19. The measure of one supplementary angle is five times as large as the measure of the second angle. What is the measure of each angle? 30° and 150°

20. One complementary angle is two-thirds as large as the second angle. What is the measure of each angle? 36° and 54°

Experiencing Geometry

You will need construction paper and several drinking straws for this activity. Use the straws to make a model of each of the following angles: acute, obtuse, right, straight, vertical, supplementary, adjacent, and complementary. Once you have made your model, glue it on a sheet of construction paper. Label your model, telling what kind of angle(s) it shows.

Learning Styles

LEP/ESL To help students reinforce their understanding of complementary, supplementary, and vertical angles and the use of algebra to solve for unknown angle measures, ask them to draw a variety of such angles on index cards. Make sure the measure of at least one of the angles is labeled on each card. Suggest that they write the definition of the angles on the back of each card. Working with partners, they can use the cards to review the angles and their relationships. For example, when shown a pair of complementary angles with one angle marked 45°, a student should identify the angles as complementary and use an algebraic equation to determine that the measure of the unknown angle is 45°. This activity will reinforce vocabulary as well as process.

Lesson at a Glance

Chapter 1 Application

Overview This application explores the concepts of networks and traversable networks.

Objectives

- To explain the concepts of networks and traversable networks.
- To use understanding of geometric figures to identify traversable networks.

Student Page 28

Teacher's Resource Library TRL

Workbook Activity 8

1. Warm-Up Activity

Engage students in a discussion about the route they take for getting to school each day. Whether they walk, ride a bike, or take the bus, they are likely to follow the same route daily. Encourage students to discuss the route in terms of getting from point *a* (home) to point *b* (school) by a series of straight lines. They may wish to draw their route on the board or paper.

2. Teaching the Lesson

Give students a floor plan of the school. Together plan a route that would enable students to visit each classroom on the same floor as your classroom. Start the route from your classroom. As you develop routes, remind students that you want to avoid retracing steps.

3. Reinforce and Extend

Application

At Home

Have students discuss with family members the routes they take to and from various places, such as work, the mall, the grocery store, the health club, and so on. Together students and their family members might draw the routes presently taken, identify why the route is taken, and plan alternate routes. Have students share the findings with the class. Are family members using the shortest routes or taking routes that provide scenic interest or avoid congested areas?

Application

Drawing Networks Every morning, across the country, thousands of school buses pick up and deliver students to school. What do the buses have to do with geometry? Each bus route can be drawn as a series of points connected by lines.

EXAMPLE

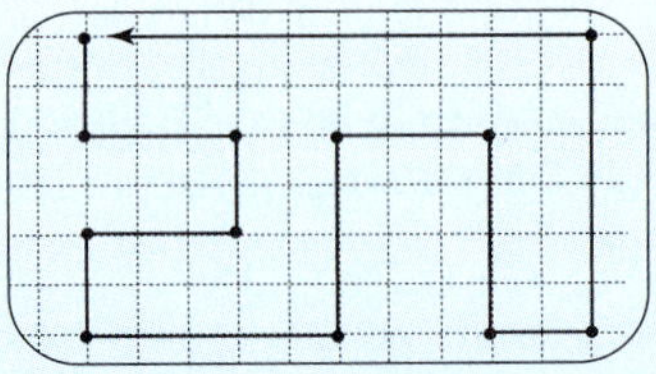

Notice that in this bus route, the bus travels over each street only once. It would be a waste of time and gasoline if a bus traveled a street more than once. So, you can see that geometry offers a way to plan a route.The point-and-line representation of the bus route is called a *network.* Think about tracing the bus route. If you can trace the bus route, without lifting your finger off the page and trace each section of the route only once, then the network is *traversable.*

Traversable networks are helpful in areas other than transportation. Architects use them to make sure people can walk through a museum, see all the exhibits, and not see the same exhibit twice. Telephone and electric companies use them so that repair crews can efficiently check existing lines to see where repair is needed.

Exercise Trace the following networks. Decide if they are traversable or not. Explain why they do not all work.

1. 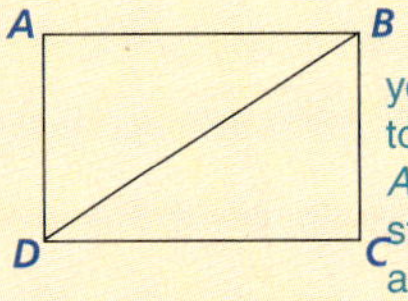

yes, *D* to *C,* to *B* to *D* to *A* to *B,* or start at *B* and reverse

2. 

no, because there are more than two odd vertices

3. 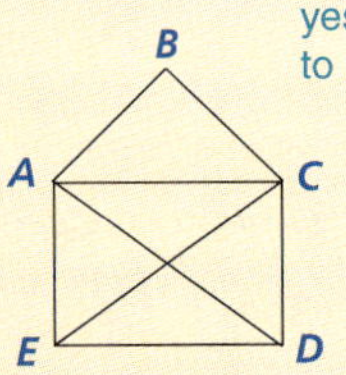

yes, *D* to *A* to *E* to *A* to *B* *C* to *D* to *E* or reverse

4. 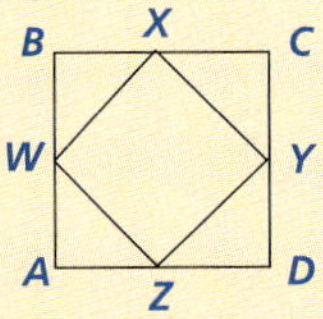

yes, *A* to *W* to *B* to *X* to *C* to *Y* to *D* to *Z* to *W* to *X* to *Y* to *Z* to *A,* or reverse

5. 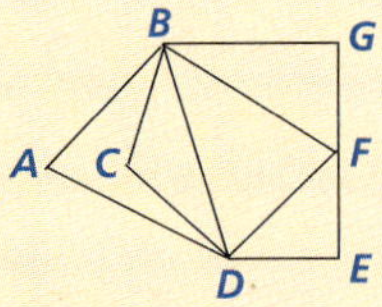

yes, *B* to *D* to *C* to *B* to *A* to *D* to *F* to *B* to *G* to *F* to *E* to *D,* or reverse

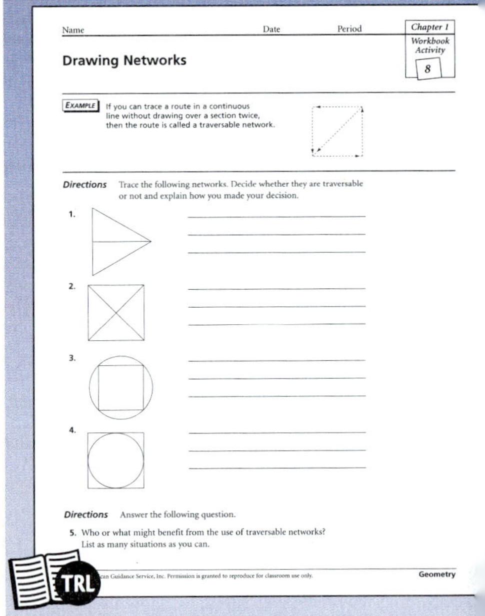
Name Date Period

Chapter 1 Workbook Activity 8

Drawing Networks

EXAMPLE If you can trace a route in a continuous line without drawing over a section twice, then the route is called a traversable network.

Directions Trace the following networks. Decide whether they are traversable or not and explain how you made your decision.

1.

2.

3.

4.

Directions Answer the following question.

5. Who or what might benefit from the use of traversable networks? List as many situations as you can.

 Geometry

TRL **Workbook Activity 8**

Tell whether each figure is a point, a line, a line segment, or a ray. Then use symbols to name each figure.

Example: Solution: line segment; $\overline{QW}$

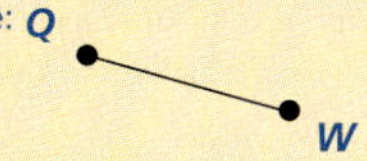

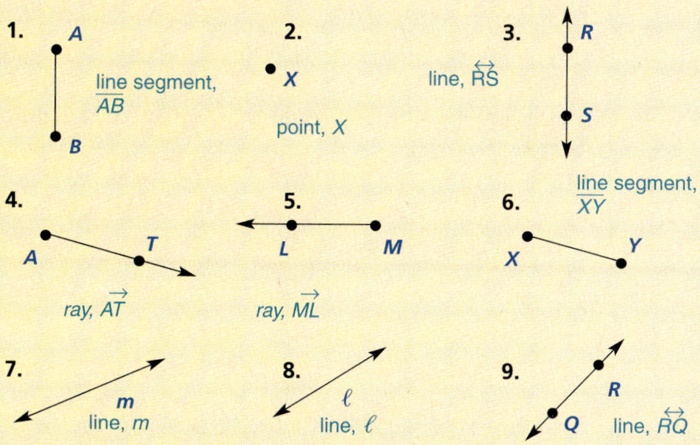

1. line segment, $\overline{AB}$
2. point, X
3. line, $\overleftrightarrow{RS}$
4. ray, $\overrightarrow{AT}$
5. ray, $\overrightarrow{ML}$
6. line segment, $\overline{XY}$
7. line, m
8. line, ℓ
9. line, $\overleftrightarrow{RQ}$

Write the letter of the answer to each question.

Example: How many endpoints does a line have?
Solution: A   A. none   B. one   C. two

10. How many endpoints does a line segment have?
A. none **B.** one **C.** two   C

11. How many endpoints does a ray have?
A. none **B.** one **C.** two   B

12. What geometric term(s) can you use to describe the light from the sun?
A. line **B.** ray **C.** segment   B

13. What geometric term(s) can you use to describe birds perched on a power line?
A. rays on a line **B.** segments on a line **C.** points on a line   C

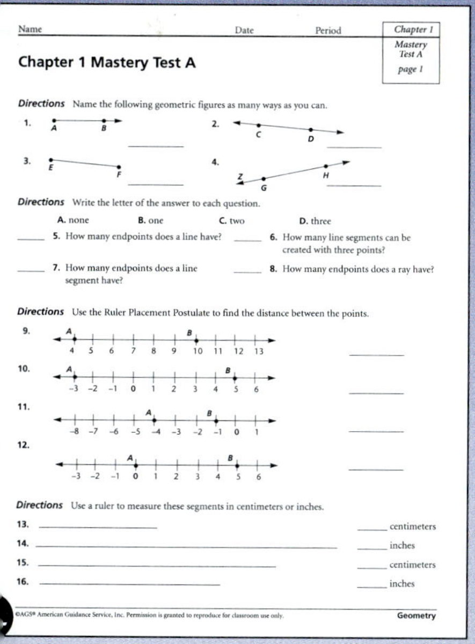
Name Date Period | *Chapter 1* *Mastery Test A* *page 1*

Chapter 1 Mastery Test A

Directions Name the following geometric figures as many ways as you can.

1. 2. 3. 4.

Directions Write the letter of the answer to each question.

A. none **B.** one **C.** two **D.** three

____ **5.** How many endpoints does a line have?
____ **6.** How many line segments can be created with three points?
____ **7.** How many endpoints does a line segment have?
____ **8.** How many endpoints does a ray have?

Directions Use the Ruler Placement Postulate to find the distance between the points.

9. 10. 11. 12.

Directions Use a ruler to measure these segments in centimeters or inches.

13. ____ centimeters
14. ____ inches
15. ____ centimeters
16. ____ inches

©AGS® American Guidance Service, Inc. Permission is granted to reproduce for classroom use only. Geometry

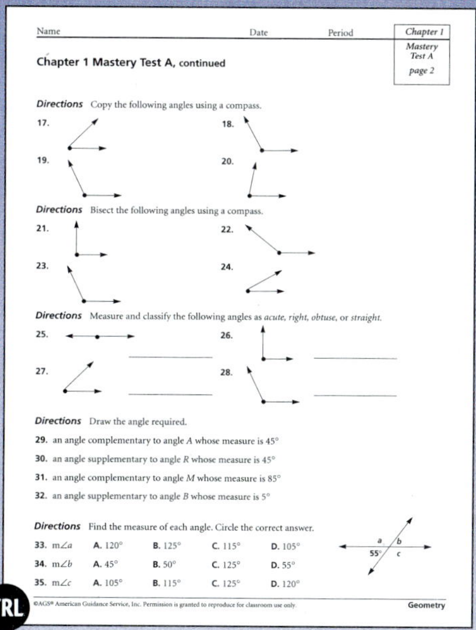
Name Date Period | *Chapter 1* *Mastery Test A* *page 2*

Chapter 1 Mastery Test A, continued

Directions Copy the following angles using a compass.

17. 18. 19. 20.

Directions Bisect the following angles using a compass.

21. 22. 23. 24.

Directions Measure and classify the following angles as *acute, right, obtuse,* or *straight.*

25. 26. 27. 28.

Directions Draw the angle required.

29. an angle complementary to angle *A* whose measure is 45°
30. an angle supplementary to angle *R* whose measure is 45°
31. an angle complementary to angle *M* whose measure is 85°
32. an angle supplementary to angle *B* whose measure is 5°

Directions Find the measure of each angle. Circle the correct answer.

33. m∠*a* **A.** 120° **B.** 125° **C.** 115° **D.** 105°
34. m∠*b* **A.** 45° **B.** 50° **C.** 125° **D.** 55°
35. m∠*c* **A.** 105° **B.** 115° **C.** 125° **D.** 120°

©AGS® American Guidance Service, Inc. Permission is granted to reproduce for classroom use only. Geometry

TRL

Chapter 1 Mastery Test A

Chapter 1 Review

Each set of problems in the Chapter Review includes an example and solution to illustrate the concept. Use the given examples for reteaching the material in Chapter 1. For additional practice, refer to the Supplementary Problems for Chapter 1 (pages 390–392) and Additional Practice 1–8 in the Teacher's Resource Library.

Chapter 1 Mastery Test TRL

The Teacher's Resource Library includes parallel forms of the Chapter 1 Mastery Test. The difficulty level of the two forms is equivalent. You may wish to use one form as a pretest and the other form as a posttest. In addition, the Chapter 1 Cumulative Mastery Test covers content from Chapter 1.

Alternative Assessment

Alternative Assessment items correlate with student Goals for Learning at the beginning of this chapter.

- Have students identify a point, line, and plane.

Draw and label each of the following geometric figures: plane, point, endpoints, line.

- Have students identify line segments and rays.

Write a description that compares and contrasts the following: line and line segment, collinear points and endpoints, line and ray, ray and line segment.

- Have students use postulates to determine how to use a ruler with geometric figures.

Draw a number of line segments. Exchange papers with a classmate, and measure each other's segments in both inches and centimeters.

Alternative Assessment

continued

- Have students construct angles—copying and bisecting.

Give a short oral presentation explaining how to construct and name an angle in three different ways. Explain and demonstrate how to copy and bisect each angle. Use visuals to enhance your presentation.

- Have students measure and classify angles.

Prepare a visual presentation that classifies angles as acute, obtuse, right, or straight. Define each term in the presentation. Make sure you define the angles based on the number of degrees in their measures.

- Have students identify complementary and supplementary angles.

Have students work with a partner to construct two drawings: one for complementary angles, one for supplementary angles.

- Have students write the plan to solve a given problem.

Construct drawings of complementary, supplementary, and vertical angles. Use your protractor to measure one angle and algebra to solve for the measure of the other angles.

More Review...

Use a ruler to measure these segments in inches. Round to the nearest inch.

Example: ______________________ Solution: 2 inches

14. ______________________________ 3 inches

15. __________ 1 inch

Use a ruler to measure these segments in centimeters. Round to the nearest centimeter.

Example: __________ Solution: 2 cm

16. ____________ 3 cm

17. ____________________ 5 cm

18. ________________ 4 cm

Name each angle in two other ways.

Example: angle *C* Solution: ∠*ACB*, ∠*BCA*

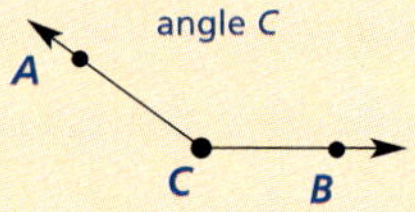

19. ∠*RST* ∠*S*, ∠*TSR*

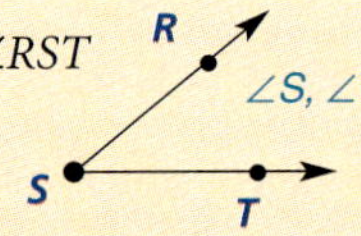

20. angle *y* ∠*B*, ∠*ABC*, or ∠*CBA*

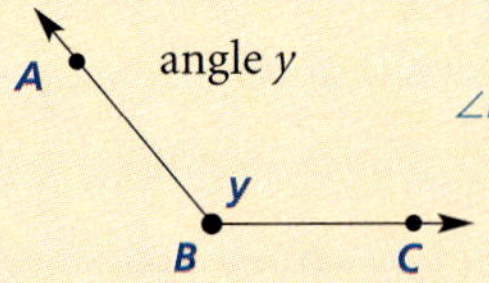

21. ∠*WXY* ∠*YXW*, ∠*X*, or ∠*a*

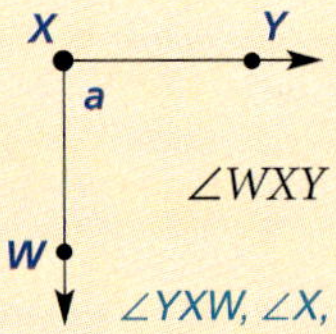

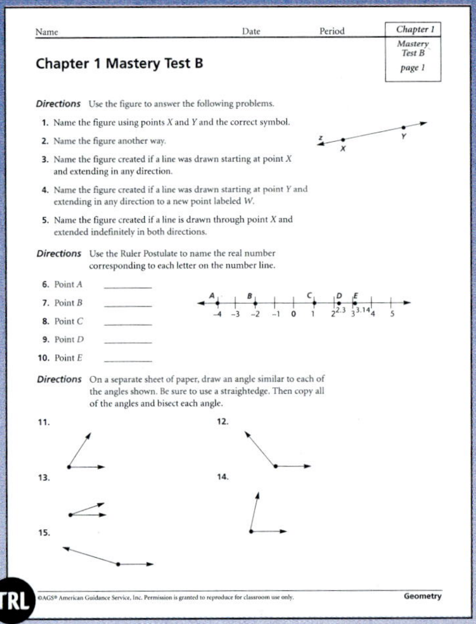

Name ______ Date ______ Period ______ *Chapter 1 Mastery Test B page 1*

Chapter 1 Mastery Test B

Directions Use the figure to answer the following problems.

1. Name the figure using points *X* and *Y* and the correct symbol.
2. Name the figure another way.
3. Name the figure created if a line was drawn starting at point *X* and extending in any direction.
4. Name the figure created if a line was drawn starting at point *Y* and extending in any direction to a new point labeled *W*.
5. Name the figure created if a line is drawn through point *X* and extended indefinitely in both directions.

Directions Use the Ruler Postulate to name the real number corresponding to each letter on the number line.

6. Point *A* ______
7. Point *B* ______
8. Point *C* ______
9. Point *D* ______
10. Point *E* ______

Directions On a separate sheet of paper, draw an angle similar to each of the angles shown. Be sure to use a straightedge. Then copy all of the angles and bisect each angle.

11. 12. 13. 14. 15.

 Geometry

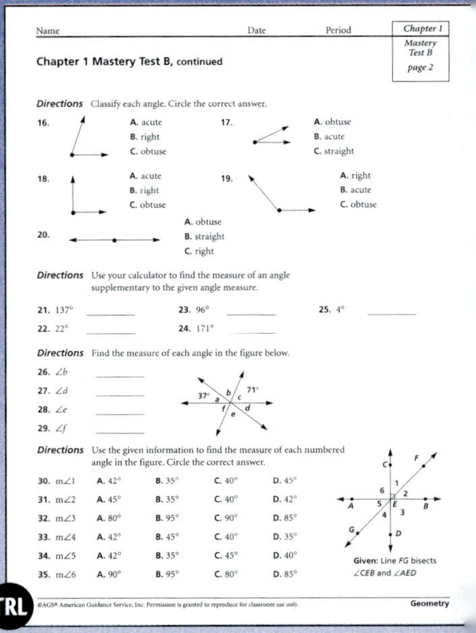

Name ______ Date ______ Period ______ *Chapter 1 Mastery Test B page 2*

Chapter 1 Mastery Test B, continued

Directions Classify each angle. Circle the correct answer.

16. A. acute B. right C. obtuse
17. A. obtuse B. acute C. straight
18. A. acute B. right C. obtuse
19. A. right B. acute C. obtuse
20. A. obtuse B. straight C. right

Directions Use your calculator to find the measure of an angle supplementary to the given angle measure.

21. 137° ______
22. 22° ______
23. 96° ______
24. 171° ______
25. 4° ______

Directions Find the measure of each angle in the figure below.

26. ∠*b* ______
27. ∠*d* ______
28. ∠*e* ______
29. ∠*f* ______

Directions Use the given information to find the measure of each numbered angle in the figure. Circle the correct answer.

30. m∠1	A. 42°	B. 35°	C. 40°	D. 45°
31. m∠2	A. 45°	B. 35°	C. 40°	D. 42°
32. m∠3	A. 80°	B. 95°	C. 90°	D. 85°
33. m∠4	A. 42°	B. 45°	C. 40°	D. 35°
34. m∠5	A. 42°	B. 35°	C. 45°	D. 40°
35. m∠6	A. 90°	B. 95°	C. 80°	D. 85°

Given: Line *FG* bisects ∠*CEB* and ∠*AED*

 Geometry

Chapter 1 Mastery Test B

Classify each angle. Write *acute, right, obtuse,* or *straight.*

Example: m∠C = 90° Solution: right

22. m∠x = 105° obtuse

23. m∠XYZ = 90° right

24. m∠a = 180° straight

25. m∠R = 45° acute

26.

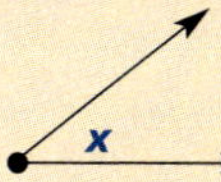

acute

27.

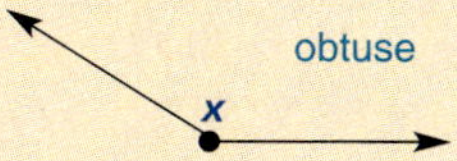

obtuse

Find the measure of an angle complementary to the given angle.

Example: m∠C = 35° Solution: 90° – 35° = 55°

28. m∠y = 15° 75°

29. m∠x = 30° 60°

30. m∠w = 45° 45°

Find the measure of an angle supplementary to the given angle.

Example: m∠C = 135° Solution: 180° – 135° = 45°

31. m∠m = 115° 65°

32. m∠n = 60° 120°

33. m∠o = 40° 140°

Draw an example of each type of angle, then use a protractor to measure your angle.

Example: right Solution:

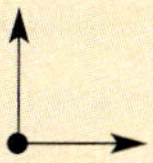

34. obtuse Student's angle measure will vary but should be greater than 90°.

35. straight Student's angle measure should be 180°.

TEST-TAKING TIP

When studying for a test, review the topics in the chapter. Then make up a practice test for yourself.

Name Date Period

Chapter 1 Cumulative Mastery Test page 1

Chapter 1 Cumulative Mastery Test

Directions Name each figure using symbols.

1. 2. 3. 4. 5.

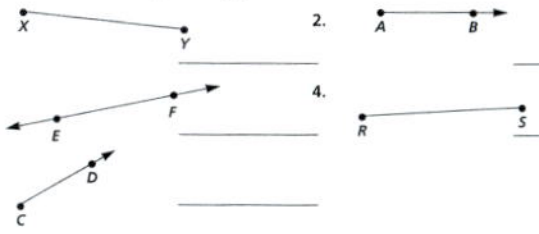

Directions Use the Ruler Placement Postulate to find the distance between the points.

6. 7. 8. 9. 10.

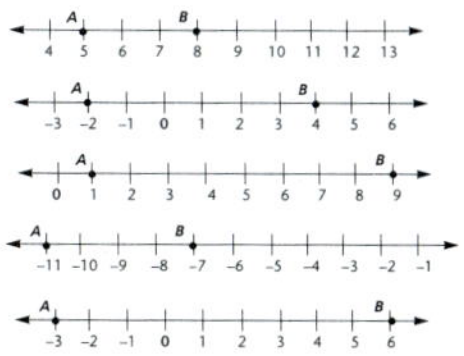

Directions On a sheet of paper, draw each angle. Then bisect each angle you have drawn. You will need to use a compass and a straightedge.

11. 12.

Geometry

Name Date Period

Chapter 1 Cumulative Mastery Test page 2

Chapter 1 Cumulative Mastery Test, continued

13. 14. 15.

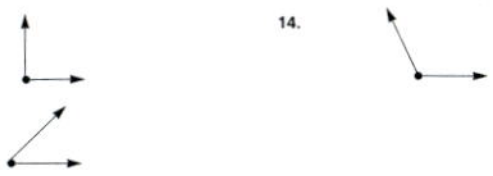

Directions Classify each angle. Circle the correct answer.

16. A. right B. straight C. obtuse

17. A. acute B. right C. obtuse

18. A. obtuse B. acute C. right

19. A. acute B. obtuse C. right

Directions Use your calculator to find the measure of an angle complementary to and an angle supplementary to the given angle measure.

20. 53° ______ 22. 42° ______ 24. 6° ______

21. 77° ______ 23. 33° ______

Directions Use the given information to find the measure of each numbered angle in the figure below. Circle the correct answer.

Given: The measure of ∠DBE is five times the measure of ∠CBD.

25. m∠1	A. 20°	B. 25°	C. 15°	D. 10°
26. m∠2	A. 75°	B. 85°	C. 90°	D. 70°
27. m∠3	A. 70°	B. 85°	C. 90°	D. 95°
28. m∠4	A. 10°	B. 15°	C. 25°	D. 20°
29. m∠5	A. 90°	B. 75°	C. 85°	D. 15°
30. m∠6	A. 80°	B. 95°	C. 90°	D. 85°

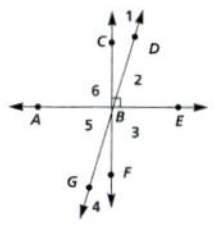

Geometry

Chapter 1 Cumulative Mastery Test

Chapter 2

Planning Guide

Thinking Geometrically: Using Proofs

	Student Text Lesson						
	Student Pages	Vocabulary	Proof	Practice Exercises	Problem Solving	Calculator Practice	Solutions Key
Lesson 1 If . . . Then Statements	34–38	✔		✔			✔
Lesson 2 Euclid's Five Postulates	39–43			✔	43		✔
Lesson 3 Using Euclid's Postulates	44–48			✔			✔
Lesson 4 Axioms or Common Notions	49–52	✔		✔		52	✔
Lesson 5 A Simple Proof	53–55	✔	✔	✔	55		✔
Application Art and Geometry	56			✔			✔

Chapter Activities

Teacher's Resource Library
Geometry on the Job 2: Detective Work

Community Connection 2: Sports and Geometry

Teacher's Edition
Chapter 2 Project

Assessment Options

Student Text
Chapter 2 Review

Teacher's Resource Library
Chapter 2 Mastery Tests A and B

Chapter 2 Cumulative Mastery Test

Teacher's Edition
Chapter 2 Alternative Assessment

Student Text Features						Teaching Strategies								Learning Styles						Teacher's Resource Library		
Algebra Connection	Constructions	Experiencing Geometry	Technology Connection	Try This	Writing About Mathematics	Warm-Up Activity	Common Error	Algebra Connection	Mental Math	Manipulatives	Calculator	Applications	Group Problem Solving	Auditory/Verbal	Visual/Spatial	Body/Kinesthetic	Logical/Mathematical	Interpersonal/Group Learning	LEP/ESL	Workbook Activities	Additional Practice	Self-Study Guide
			35	38	35	34	35		36			35, 37			37		36	36	38	9, 10	7	✔
		43				39			41			43			42	41			42	11, 12	8	✔
	47–48	48			48	44						46, 48	47	46, 48	47	45				13, 14	9	✔
49						49	52	50		50	52				51			52	51	15–17	10	✔
						53					55		55	55	54		55			18	11	✔
						56						56								19		✔

Manipulatives

Factor Blocks

Chapter at a Glance

Chapter 2: Thinking Geometrically: Using Proofs

pages 32–59

Lessons

Teacher's Resource Library TRL

(Answer Keys for the Teacher's Resource Library begin on page 484 of this Teacher's Edition.)

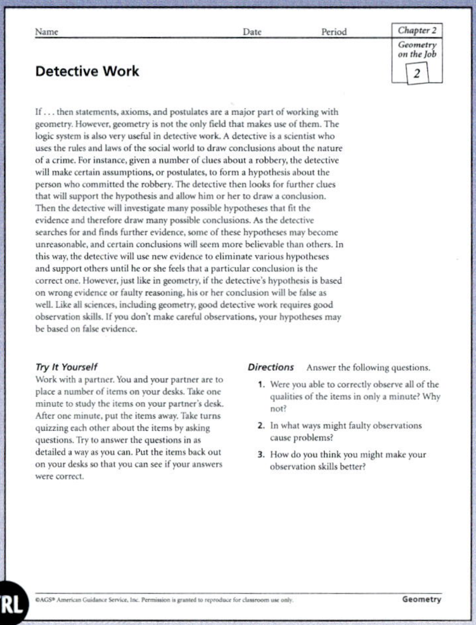

Name Date Period

Chapter 2
Geometry on the Job
2

Detective Work

If . . . then statements, axioms, and postulates are a major part of working with geometry. However, geometry is not the only field that makes use of them. The logic system is also very useful in detective work. A detective is a scientist who uses the rules and laws of the social world to draw conclusions about the nature of a crime. For instance, given a number of clues about a robbery, the detective will make certain assumptions, or postulates, to form a hypothesis about the person who committed the robbery. The detective then looks for further clues that will support the hypothesis and allow him or her to draw a conclusion. Then the detective will investigate many possible hypotheses that fit the evidence and therefore draw many possible conclusions. As the detective searches for and finds further evidence, some of these hypotheses may become unreasonable, and certain conclusions will seem more believable than others. In this way, the detective will use new evidence to eliminate various hypotheses and support others until he or she feels that a particular conclusion is the correct one. However, just like in geometry, if the detective's hypothesis is based on wrong evidence or faulty reasoning, his or her conclusion will be false as well. Like all sciences, including geometry, good detective work requires good observation skills. If you don't make careful observations, your hypotheses may be based on false evidence.

Try It Yourself
Work with a partner. You and your partner are to place a number of items on your desks. Take one minute to study the items on your partner's desk. After one minute, put the items away. Take turns quizzing each other about the items by asking questions. Try to answer the questions in as detailed a way as you can. Put the items back out on your desks so that you can see if your answers were correct.

Directions Answer the following questions.

1. Were you able to correctly observe all of the qualities of the items in only a minute? Why not?
2. In what ways might faulty observations cause problems?
3. How do you think you might make your observation skills better?

TRL

 Geometry

Geometry on the Job 2

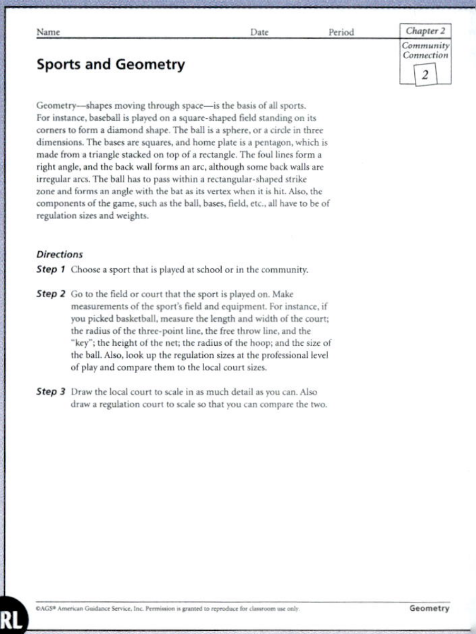

Name Date Period

Chapter 2
Community Connection
2

Sports and Geometry

Geometry—shapes moving through space—is the basis of all sports. For instance, baseball is played on a square-shaped field standing on its corners to form a diamond shape. The ball is a sphere, or a circle in three dimensions. The bases are squares, and home plate is a pentagon, which is made from a triangle stacked on top of a rectangle. The foul lines form a right angle, and the back wall forms an arc, although some back walls are irregular arcs. The ball has to pass within a rectangular-shaped strike zone and forms an angle with the bat as its vertex when it is hit. Also, the components of the game, such as the ball, bases, field, etc., all have to be of regulation sizes and weights.

Directions

Step 1 Choose a sport that is played at school or in the community.

Step 2 Go to the field or court that the sport is played on. Make measurements of the sport's field and equipment. For instance, if you picked basketball, measure the length and width of the court; the radius of the three-point line, the free throw line, and the "key"; the height of the net; the radius of the hoop; and the size of the ball. Also, look up the regulation sizes at the professional level of play and compare them to the local court sizes.

Step 3 Draw the local court to scale in as much detail as you can. Also draw a regulation court to scale so that you can compare the two.

TRL

 Geometry

Community Connection 2

Chapter 2

Thinking Geometrically: Using Proofs

You have just pressed the keys of your calculator to compute a sum. In the calculator, electric charges pulse through a system of circuits to perform the operations. The interconnected circuits, arranged in geometric shapes, work as a system to provide an answer.

In the same way, geometry consists of axioms and postulates. These are statements that are assumed to be true. If combined through logic, they lead to new conclusions. As on a circuit board, each does its part and makes the system work. The system of circuitry makes your calculator capable of performing complex calculations. Through practical application, the system of geometry makes the construction of buildings, spacecraft, and, yes, circuit boards possible.

In Chapter 2, you will learn about proofs as you apply some of the axioms and postulates of geometry.

Goals for Learning

- To evaluate and write conditionals
- To identify and apply the postulates that define lines and angles
- To identify postulates that apply to constructions
- To construct geometric lines and shapes based on postulates
- To use axioms in explanations
- To prove the vertical angle theorem and to use it to find measures of angles

33

SELF-STUDY GUIDE

Name ____________

CHAPTER 2: Thinking Geometrically: Using Proofs

Goal 2.1 *To evaluate and write conditionals*

Date	Assignment	Score
______	1: Read pages 34–36. Complete Exercises A and B on pages 37–38.	______
______	2: Complete Workbook Activity 9.	______
______	3: Complete Exercises C and D on page 38.	______
______	4: Complete Workbook Activity 10.	______

Comments:

Goals 2.2 and 2.3 *To identify and apply the postulates that define lines and angles; to identify postulates that apply to constructions*

Date	Assignment	Score
______	5: Read pages 39–41. Complete Exercises A–C on pages 42–43.	______
______	6: Complete Workbook Activities 11 and 12.	______

Comments:

Goal 2.4 *To construct geometric lines and shapes based on postulates*

Date	Assignment	Score
______	7: Read pages 44–45. Complete Exercise A on pages 45–47.	______
______	8: Complete Workbook Activity 13.	______
______	9: Complete Exercise B on pages 47–48.	______
______	10: Complete Workbook Activity 14.	______

Comments:

 Geometry

RL

SELF-STUDY GUIDE

Name ____________

CHAPTER 2: Thinking Geometrically: Using Proofs, continued

Goal 2.5 *To use axioms in explanations*

Date	Assignment	Score
______	11: Read pages 49–51. Complete Exercises A and B on pages 51–52.	______
______	12: Complete Workbook Activities 15 and 16.	______
______	13: Complete the Calculator Practice and Exercise C on page 52.	______
______	14: Complete Workbook Activity 17.	______

Comments:

Goal 2.6 *To prove the vertical angle theorem and to use it to find measures of angles*

Date	Assignment	Score
______	15: Read pages 53–54. Complete Exercises A–D on pages 54–55.	______
______	16: Complete Workbook Activity 18.	______
______	17: Read and complete the Application Exercise on page 56.	______
______	18: Complete Workbook Activity 19.	______
______	19: Complete the Chapter 2 Review on pages 57–59.	______

Comments:

Student's Signature ____________ Date ________

Instructor's Signature ____________ Date ________

 Geometry

TRL

Chapter 2 Self-Study Guide

Introducing the Chapter

Use the information on these pages to help students think of geometry as a system with many practical applications in the everyday world. Reinforce the comparison of the circuit board in a calculator as a system to geometry as a system. The circuitry enables students to perform simple and complex calculations by just pressing buttons. In the same way, the system of geometry enables people to design and construct many different things from highway signs to multileveled bridges. Have students look around the classroom and identify basic geometric shapes in objects. For example, they might identify a clock as a circle or a writing board as a rectangle. Point out that geometry provides a system for determining the sizes of these shapes and for constructing them. Introduce the Goals for Learning. Explain to students that in this chapter they will use axioms and postulates of geometry as they learn about proofs.

Chapter Project

Encourage students to work in groups to develop and produce a geometry game show. The show should concentrate on identifying the postulates, axioms, and theorems explained in the chapter. Remind students that they must produce a product. For example, they could develop storyboards, question-and-answer cards, game rules, and props for the show. Recommend that they stage their show either on videotape or in a live performance.

As an alternative project, students could develop a board game. For this game, their product will be the board and playing pieces as well as the game rules.

Lesson at a Glance

Chapter 2 Lesson 1

Overview This lesson explains conditionals and their converses and how to determine whether a conditional is true or false.

Objectives

- To identify the hypothesis and the conclusion of a conditional.
- To determine if a conditional is true or false.
- To write the converse of a conditional.

Student Pages 34–38

Teacher's Resource Library TRL

Workbook Activities 9–10

Additional Practice 7

Mathematics Vocabulary

conclusion
conditional
converse
hypothesis

1. Warm-Up Activity

Familiarize students with conditionals by giving them several examples of everyday If … then … statements, such as "If you study for a test, then you will get a better score." Then write If … statements on the board and ask students to provide then … statements. Ask them to make up their own true or false conditionals and write them on the board.

2. Teaching the Lesson

Discuss the terms *hypothesis* and *conclusion* as they relate to the If … then … statements of the conditionals used in the Warm-Up Activity. Explain the rule for determining whether a conditional is true or false. Then, using the conditionals in the examples, have students provide conclusions that would make the true conditional false and the false conditional true.

Lesson 1 If . . . Then Statements

Conclusion
The then . . . part of a conditional

Conditional
A statement in the If . . . then . . . form

Converse
A new conditional formed from a given conditional in which the hypothesis and conclusion are switched

Hypothesis
The given or If . . . part of a conditional

Every day you hear statements in the form—If . . . then . . . . "**If** you use Clean & Bright toothpaste, **then** your teeth will be sparkling." You know this may or may not be true depending on Clean & Bright toothpaste. If . . . then . . . statements are also used in geometry and mathematics. These statements describe what is necessary, the *conditions* necessary for a statement to be true. In geometry and mathematics there are strict rules for "If . . . then . . ." statements to be true.

EXAMPLE Study this example.

If two angles are complementary, →	**then** the sum of their measures equals 90°.
Given or Hypothesis "Two angles are complementary" is the given statement, also called the *hypothesis*.	**Conclusion** "The sum of their measures equals 90°" is the *conclusion*.

The entire If . . . then . . . statement is called a **conditional.** A statement of the form "If . . . then . . ." is a *conditional.*

The statement that follows the "If" in a conditional is the **hypothesis.**

The statement that follows the "then" in a conditional is the **conclusion.**

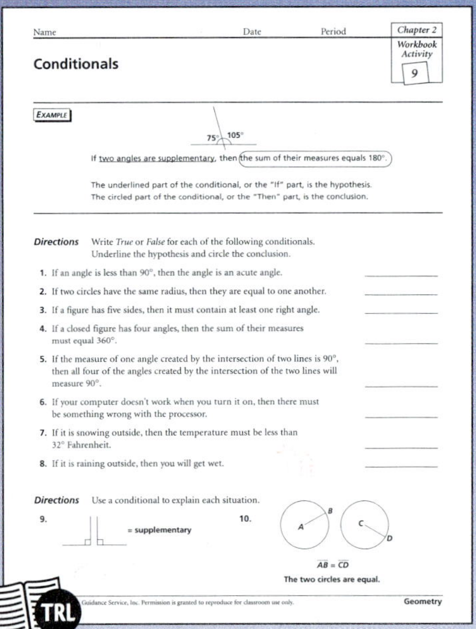

Name Date Period

Chapter 2 Workbook Activity 9

Conditionals

EXAMPLE

75° 105°

If two angles are supplementary, then the sum of their measures equals 180°.

The underlined part of the conditional, or the "If" part, is the hypothesis.
The circled part of the conditional, or the "Then" part, is the conclusion.

Directions Write *True* or *False* for each of the following conditionals. Underline the hypothesis and circle the conclusion.

1. If an angle is less than 90°, then the angle is an acute angle. ______
2. If two circles have the same radius, then they are equal to one another. ______
3. If a figure has five sides, then it must contain at least one right angle. ______
4. If a closed figure has four angles, then the sum of their measures must equal 360°. ______
5. If the measure of one angle created by the intersection of two lines is 90°, then all four of the angles created by the intersection of the two lines will measure 90°. ______
6. If your computer doesn't work when you turn it on, then there must be something wrong with the processor. ______
7. If it is snowing outside, then the temperature must be less than 32° Fahrenheit. ______
8. If it is raining outside, then you will get wet. ______

Directions Use a conditional to explain each situation.

9. = supplementary
10. A B C D
$\overline{AB} = \overline{CD}$
The two circles are equal.

 Geometry

TRL

Workbook Activity 9

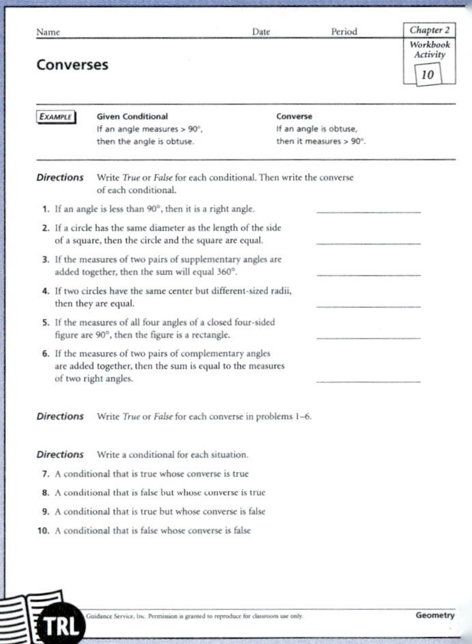

Name Date Period

Chapter 2 Workbook Activity 10

Converses

EXAMPLE	Given Conditional	Converse
	If an angle measures > 90°, then the angle is obtuse.	If an angle is obtuse, then it measures > 90°.

Directions Write *True* or *False* for each conditional. Then write the converse of each conditional.

1. If an angle is less than 90°, then it is a right angle. ______
2. If a circle has the same diameter as the length of the side of a square, then the circle and the square are equal. ______
3. If the measures of two pairs of supplementary angles are added together, then the sum will equal 360°. ______
4. If two circles have the same center but different-sized radii, then they are equal. ______
5. If the measures of all four angles of a closed four-sided figure are 90°, then the figure is a rectangle. ______
6. If the measures of two pairs of complementary angles are added together, then the sum is equal to the measures of two right angles. ______

Directions Write *True* or *False* for each converse in problems 1–6.

Directions Write a conditional for each situation.

7. A conditional that is true whose converse is true
8. A conditional that is false but whose converse is true
9. A conditional that is true but whose converse is false
10. A conditional that is false whose converse is false

 Geometry

TRL

Workbook Activity 10

The rules of logic tell you if a conditional statement is true or false.

Rule for Conditionals:
A conditional is always **true** except when a *true hypothesis* is followed by a *false conclusion.* A conditional in which a true hypothesis is followed by a false conclusion is **false.**

Examples

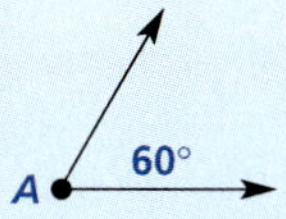

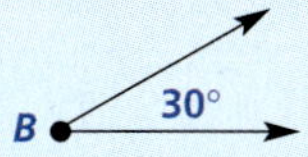

If $m\angle A + m\angle B = 90°$	then angles *A* and *B* are complementary.
true	true

The conditional is true because both the hypothesis and the conclusion are *true.*

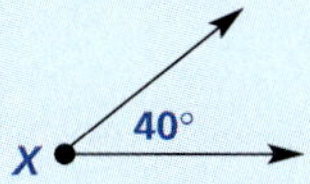

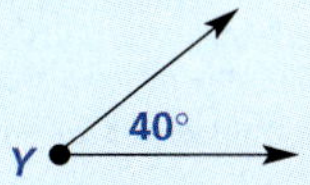

If $m\angle X = m\angle Y$	then angles *X* and *Y* are complementary.
true	false

The hypothesis is true but the conclusion is false so the conditional is *false.*

Writing About Mathematics

Copy an If… then … statement from an advertisement. Tell which part is the hypothesis and which is the conclusion. Tell whether you think the conditional is true or false. Explain why.

Technology Connection

If . . .then statements are at the heart of many computer programs. Here is an excerpt from a program designed to classify angles.

```
110   INPUT "ENTER MEASURE OF ∠A": A
120   IF (A > 0) AND (A < 90) THEN 170
130   IF (A = 90) THEN 180
```


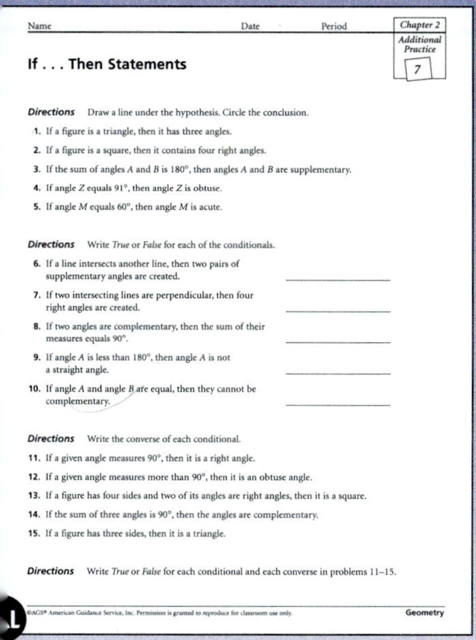

Name Date Period Chapter 2 Additional Practice 7

If . . . Then Statements

Directions Draw a line under the hypothesis. Circle the conclusion.

1. If a figure is a triangle, then it has three angles.
2. If a figure is a square, then it contains four right angles.
3. If the sum of angles *A* and *B* is 180°, then angles *A* and *B* are supplementary.
4. If angle *Z* equals 91°, then angle *Z* is obtuse.
5. If angle *M* equals 60°, then angle *M* is acute.

Directions Write *True* or *False* for each of the conditionals.

6. If a line intersects another line, then two pairs of supplementary angles are created. ________
7. If two intersecting lines are perpendicular, then four right angles are created. ________
8. If two angles are complementary, then the sum of their measures equals 90°. ________
9. If angle *A* is less than 180°, then angle *A* is not a straight angle. ________
10. If angle *A* and angle *B* are equal, then they cannot be complementary. ________

Directions Write the converse of each conditional.

11. If a given angle measures 90°, then it is a right angle.
12. If a given angle measures more than 90°, then it is an obtuse angle.
13. If a figure has four sides and two of its angles are right angles, then it is a square.
14. If the sum of three angles is 90°, then the angles are complementary.
15. If a figure has three sides, then it is a triangle.

Directions Write *True* or *False* for each conditional and each converse in problems 11–15.

 Geometry

Additional Practice 7

After introducing converses of conditionals, work with students to write converses of several conditionals. Show them how to use the truth table to determine whether converses are true.

3. Reinforce and Extend

Application

At Home

Have students listen for and list conditionals expressed at home. For example, they may hear such conditionals as "If you miss your curfew, you will be grounded" or "If I get home late, we will order out for dinner." Encourage students to share their lists with the class. You might suggest that they write the converses of their conditionals and state whether they are true or false.

Common Error

Some students may assume that the converse of a conditional is automatically true if the given conditional is true. To help students understand that converses may be true or false, put the following conditionals on the board and help volunteers write the converse. Then provide counterexamples to show why the converse may not be true:

- If $\angle W$ is a right angle, then its measure is 90°.
- If $\angle A$ and $\angle B$ are right angles, then they are supplementary. (counterexample: $\angle A = 50°$, $\angle B = 130°$)
- If two angles add up to 90°, then they are acute angles. (counterexample: one angle = 50° and the other angle = 50°)

Learning Styles

Logical/Mathematical Encourage students to use a truth table and the symbols p, q, and $\rightarrow$ to analyze conditionals. By arranging the symbols in a truth table, students can see the logic involved in determining the validity of each conditional.

Mental Math

Read the following conditionals to the class and have students decide mentally whether they are true or false. If they are false, have students change either the hypothesis or the conclusion to make the conditional true.

- If $1 + 2 = 3$, then $1 + 2 + 3 = 6$. (true)
- If $2 \cdot 3 = 6$, then $2 \cdot 3 \cdot 4 = 10$. (false)
- If $6 - 2 = 4$, then $2 - 6 = 4$. (false)
- If $x + 2 = 7$, then $x = 5$. (true)
- If $x - 1 = 10$, then $x - 10 = 1$. (true)
- If $4x = 20$, then $20x = 4$. (false)

Learning Styles

Interpersonal/Group Learning Have individuals write several conditionals. Tell students to write conditionals for a variety of situations, not just mathematical statements. Then in groups of 3 or 4, students share their conditionals and the group decides whether they are true or false. As a group, students should then write the converse of each conditional and decide whether it is true or false. Encourage students to use counterexamples to illustrate why a converse is false. Ask the groups to turn in one paper showing their work, including their reasoning for determining whether a conditional or its converse is true or false.

The Rule for Conditionals can also be written with symbols. The symbols are arranged in a table called a *truth table*. *T* stands for true, *F* stands for false.

In words—A conditional is always true except when a true hypothesis is followed by a false conclusion.

In a truth table—Let p stand for the hypothesis, let q stand for the conclusion, use $\rightarrow$ to mean "If . . . then . . ."

	p	$\rightarrow$	q
If $2 + 2 = 4$, then $4 - 2 = 2$.	T	T	T
If $3 > 0$, then $3 < 0$.	T	F	F
In mathematics, cases in which the hypothesis is false and the conclusion may be true or false are not useful.	F	T	T
	F	T	F

Conclusion may be true or false.

conditional

If you switch or reverse the hypothesis and conclusion in a conditional, you form a new conditional called the **converse** of the original conditional.

Given Conditional	Converse
If an angle measures $< 90°$, then it is acute.	If an angle is acute, then it measures $< 90°$.

Note that a given conditional can be true, and its converse can be false. Here is such an example.

Let p = The figure is a square.
Let q = The figure has four sides.

If a figure is a square, then it has four sides.

p	$\rightarrow$	q
T	T	T

If a figure has four sides, then it is a square.

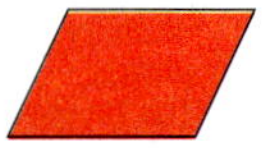

q	$\rightarrow$	p
T	F	F

Exercise A Copy each statement. Draw one line under the hypothesis. Circle the conclusion.

Selected Answers and Solutions begin on page 420.

1. If a figure is a triangle, then it has three sides.
   If a figure is a triangle, then (it has three sides.)
2. If an angle is obtuse, then it measures > 90°.
   If an angle is obtuse, then (it measures > 90°.)
3. If the sum of the measures of two angles is 90°, then the angles are complementary.
   If the sum of the measures of two angles is 90°, then (the angles are complementary.)
4. If two angles are supplementary, then the sum of their measures is 180°.
   If two angles are supplementary, then (the sum of their measures is 180°.)
5. If you use SuperBright detergent, then your clothes will be clean.
   If you use SuperBright detergent, then (your clothes will be clean.)

Exercise B Write *True* or *False* for each of the conditionals.

6. If an angle is a right angle, then it measures 90°. True

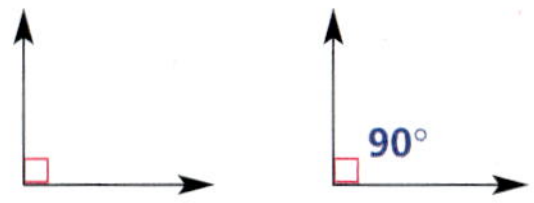

7. If ∠*A* and ∠*B* are supplementary, then their angle sum is 180°. True

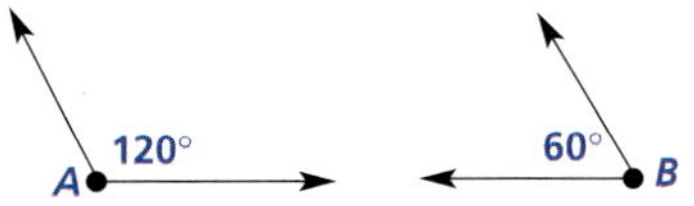

120° + 60° = 180°

8. If an angle is a straight angle, then it measures 180°. True

Application

In the Community

Have students visit an organization or a facility in the community, such as a recycling plant, the library, the police or fire department, a community service organization, etc. After learning about the purpose of the organization and how it serves the community, students should write three or four conditionals that apply to the organization. For example, after visiting a recycling plant, a student might write the conditional "If everyone recycles paper, then fewer trees will have to be cut down." Ask volunteers to share their statements with the class and look for similarities and differences among the conditionals as they apply to different community groups.

Learning Styles

LEP/ESL Students may have difficulty with the concepts presented in this lesson, particularly with identifying parts of a conditional and writing its converse. Pair students who are less proficient in English with those of greater ability. Suggest that partners prepare picture/word cards to reinforce vocabulary and concepts. For example, a card might show an obtuse angle with the words printed below the illustration. Partners can then use the cards to construct conditionals and converses through words and visuals.

Learning Styles

Visual/Spatial Have students draw pictures to illustrate the hypotheses and the conclusions of the conditionals in Exercise D. Their visual representations can help them determine whether the conditionals and converses are true.

False **9.** If $\angle Y$ and $\angle Z$ are equal, then they are supplementary.

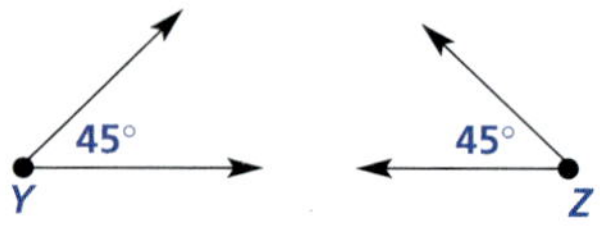

$45° + 45° = 90°$

True **10.** If *ABCD* is a square, then it is a rectangle.

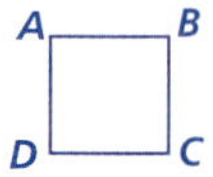

Exercise C Write the converse of each conditional.

True **11.** If an angle is obtuse, then its angle measure is $90° < m < 180°$.

False **12.** If a figure has four sides, then it is a rectangle.

True **13.** If the sum of the measures of two angles is 90°, then the angles are complementary.

True **14.** If a figure is a triangle, then it has three sides.

True **15.** If an animal is a dog, then it has four legs.

Exercise D Write *True* or *False* for each conditional and each converse in Exercise C.

11. If an angle measure is $90° < m < 180°$, the angle is obtuse; True
12. If a figure is a rectangle, then it has four sides; True
13. If two angles are complementary, then the sum of their measures is 90°; True
14. If a figure has three sides, then it is a triangle; True
15. If an animal has four legs, then it is a dog; False

Try This...

Collect six newspaper or magazine advertisements. See how many claims for the product you can write in the form of an "If . . . then . . ." statement.

Lesson 2 Euclid's Five Postulates

Euclid wrote the first complete geometry book. He lived in ancient Greece about 300 B.C. Euclid began his book with five "facts" or assumptions about geometry in a plane. He assumed these statements were true without proving them. These statements, called **Euclid's postulates,** form the building blocks for Plane Euclidean Geometry. They explain why certain constructions can be made. You will need to know these postulates as part of your study of geometry.

Euclid's Postulate 1
A straight line can be drawn from any point to any point.

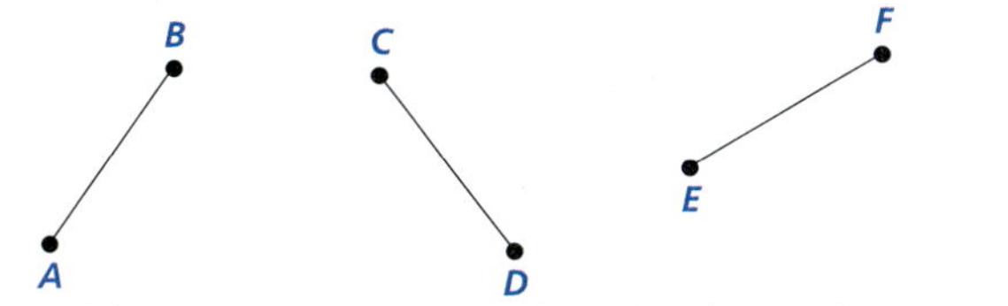

You could connect point *A*, not only with *B*, but with *C*, *D*, *E*, and *F*, and so on.

> Remember that a statement assumed to be true without proof is a postulate.

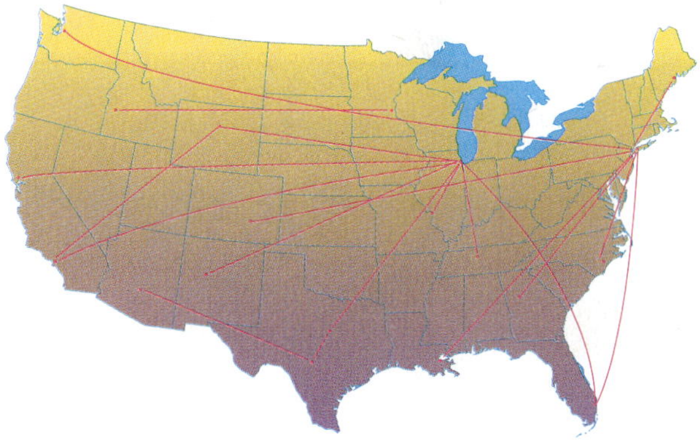

The lines on this map represent airline routes showing where planes travel from one point to another.

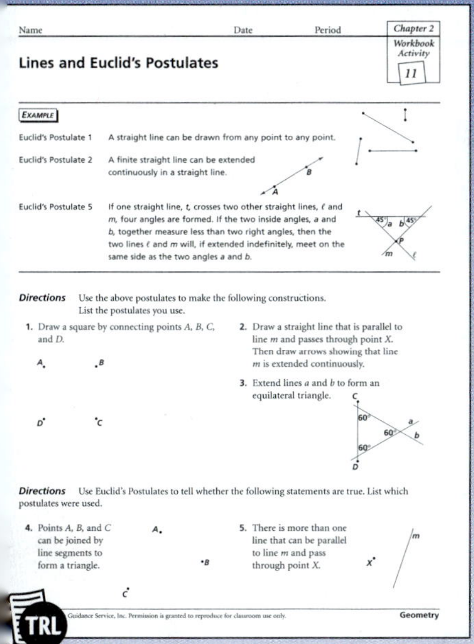

Name Date Period — Chapter 2 Workbook Activity 11

Lines and Euclid's Postulates

EXAMPLE

Euclid's Postulate 1 — A straight line can be drawn from any point to any point.

Euclid's Postulate 2 — A finite straight line can be extended continuously in a straight line.

Euclid's Postulate 5 — If one straight line, *t*, crosses two other straight lines, *ℓ* and *m*, four angles are formed. If the two inside angles, *a* and *b*, together measure less than two right angles, then the two lines *ℓ* and *m* will, if extended indefinitely, meet on the same side as the two angles *a* and *b*.

Directions Use the above postulates to make the following constructions. List the postulates you use.

1. Draw a square by connecting points *A*, *B*, *C*, and *D*.
2. Draw a straight line that is parallel to line *m* and passes through point *X*. Then draw arrows showing that line *m* is extended continuously.
3. Extend lines *a* and *b* to form an equilateral triangle.

Directions Use Euclid's Postulates to tell whether the following statements are true. List which postulates were used.

4. Points *A*, *B*, and *C* can be joined by line segments to form a triangle.
5. There is more than one line that can be parallel to line *m* and pass through point *X*.

 Geometry

TRL **Workbook Activity 11**

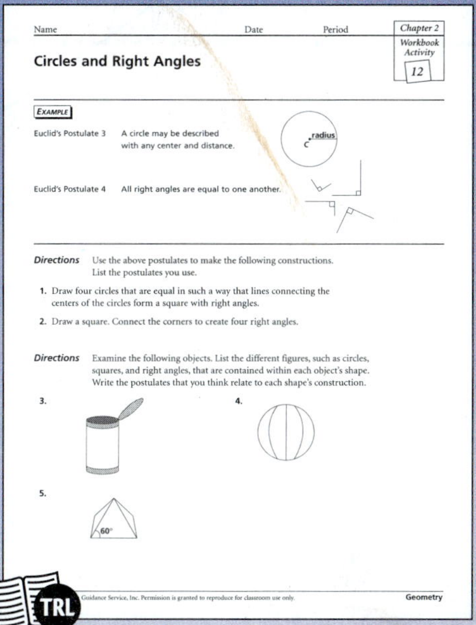

Name Date Period — Chapter 2 Workbook Activity 12

Circles and Right Angles

EXAMPLE

Euclid's Postulate 3 — A circle may be described with any center and distance.

Euclid's Postulate 4 — All right angles are equal to one another.

Directions Use the above postulates to make the following constructions. List the postulates you use.

1. Draw four circles that are equal in such a way that lines connecting the centers of the circles form a square with right angles.
2. Draw a square. Connect the corners to create four right angles.

Directions Examine the following objects. List the different figures, such as circles, squares, and right angles, that are contained within each object's shape. Write the postulates that you think relate to each shape's construction.

3.

4.

5.

 Geometry

TRL **Workbook Activity 12**

Lesson at a Glance

Chapter 2 Lesson 2

Overview This lessons introduces Euclid's five postulates as the building blocks of Plane Euclidean Geometry.

Objectives

- To identify Euclid's postulates.
- To use various constructions to illustrate Euclid's postulates.

Student Pages 39–43

Teacher's Resource Library TRL

Workbook Activities 11–12

Additional Practice 8

1. Warm-Up Activity

Begin by providing students with examples of statements that are accepted as true without proof, such as "My name is (your name)" or "Today is (day of the week)." Ask students to give their own examples of "accepted" statements. Then, using a calendar, show how the postulate "Today is (day of week)" can be used to prove the statement "Tomorrow is (next day)" or "In two days, it will be (day of week). Ask students to think of other examples in which statements accepted as true are used as the basis of proof for other statements.

2. Teaching the Lesson

Review the terms *point, line, line segment,* and *parallel.* After reading Postulates 1 and 2, provide students with examples that illustrate Postulate 1, such as a bridge, boundary lines drawn on a football field or tennis court, a net on a volleyball court or a table tennis table, a clothesline, etc. Ask students for other examples. Put an arrangement of three or four points on the board and ask students to draw as many line segments as possible using the points as endpoints. Then discuss what would happen if the line segments were extended in each direction. Illustrate how a line segment is part of a line.

To illustrate Postulate 3, instruct students to draw a circle around a given point freehand. Then, using a ruler, have students measure the radius in several different places to see how close to a true circle they drew. Ask students how they might use a compass to draw a more accurate circle. Describe how a compass is used to draw circles and discuss how Postulate 3 allows for this method of drawing circles.

After students read and discuss Postulate 4, ask them to rewrite the postulate as a conditional. **(If two angles are right angles, then they are supplementary.)** Then have students write the converse and tell whether it is true or false. **(If two angles are supplementary, then they are right angles. False.)** If they conclude it is false, ask that they provide a counterexample illustrating why.

After students read and discuss Postulate 5, have them draw a line and a point. Ask students to draw a line through the point that is parallel to the first line. Then tell them to draw more lines through the point until they can "see" and understand why Postulate 5 is true. Encourage students to describe in their own words why the postulate is true. Refer them to pages 458–459 for a list of the postulates, axioms, and theorems used in the program.

EUCLID'S POSTULATE 2
A finite straight line can be extended continuously in a straight line. (A finite straight line is a line segment.)

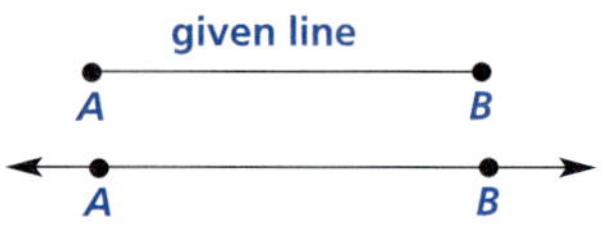

Postulate 2 allows you to extend the line segment in each direction.

EUCLID'S POSTULATE 3
A circle may be described with any center and distance.

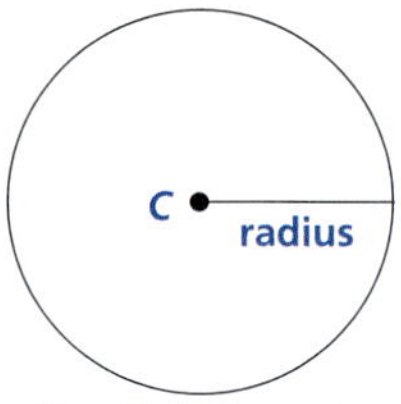

Postulate 3 means that if you are given a point *C*, and a distance, you can draw a circle with the center at *C* and a radius equal to the given distance. [distance = radius]

EUCLID'S POSTULATE 4
All right angles are equal to one another.

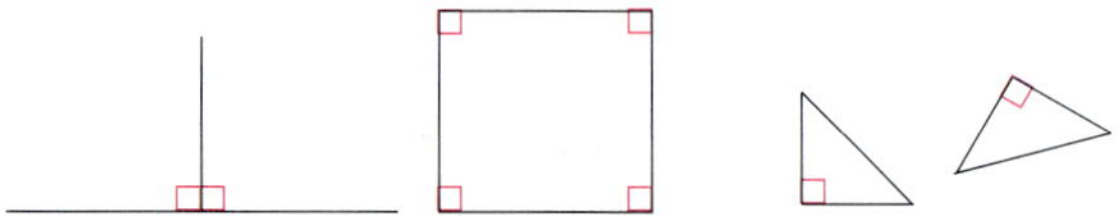

This postulate assumes that the measures of all right angles are equal to each other and to 90°.

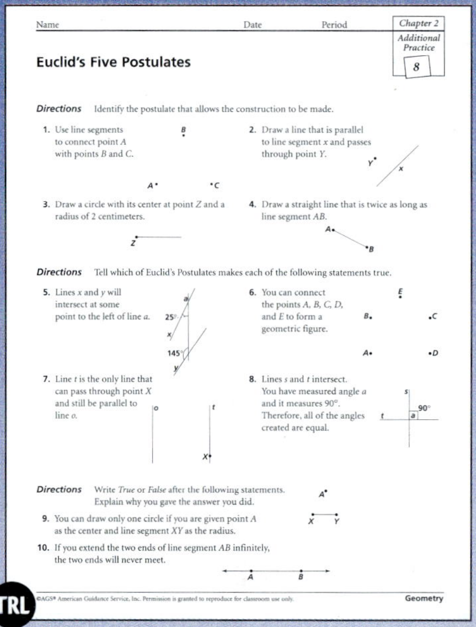

Name Date Period

Chapter 2 Additional Practice 8

Euclid's Five Postulates

Directions Identify the postulate that allows the construction to be made.

1. Use line segments to connect point *A* with points *B* and *C*.
2. Draw a line that is parallel to line segment *x* and passes through point *Y*.
3. Draw a circle with its center at point *Z* and a radius of 2 centimeters.
4. Draw a straight line that is twice as long as line segment *AB*.

Directions Tell which of Euclid's Postulates makes each of the following statements true.

5. Lines *x* and *y* will intersect at some point to the left of line *a*.
6. You can connect the points *A, B, C, D,* and *E* to form a geometric figure.
7. Line *t* is the only line that can pass through point *X* and still be parallel to line *o*.
8. Lines *s* and *t* intersect. You have measured angle *a* and it measures 90°. Therefore, all of the angles created are equal.

Directions Write *True* or *False* after the following statements. Explain why you gave the answer you did.

9. You can draw only one circle if you are given point *A* as the center and line segment *XY* as the radius.
10. If you extend the two ends of line segment *AB* infinitely, the two ends will never meet.

 Geometry

TRL

Additional Practice 8

EUCLID'S POSTULATE 5
If two lines ℓ and m are cut by a third line t, and the two inside angles, a and b, together measure less than two right angles, then the two lines ℓ and m, if extended, will meet on the same side as the two angles a and b.

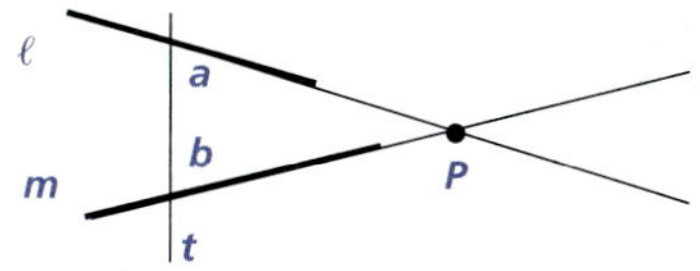

This postulate actually defines the conditions necessary for two lines to meet or *intersect.* The illustration of the postulate shows that if $m\angle a + m\angle b < 2 \cdot 90°$, then ℓ intersects m at P.

There is a modern equivalent to Euclid's Postulate 5.

When two lines intersect, they cross each other.

PARALLEL POSTULATE 5
If there is a line ℓ and a point P not on ℓ, then there is only one line that passes through P and is parallel to ℓ.

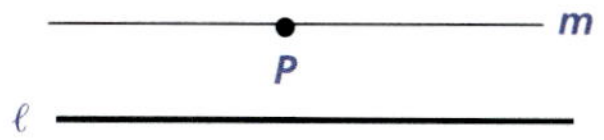

Line m is the only line through P that is parallel to the given line ℓ.

This statement has the same meaning as Euclid's Postulate 5. Therefore, Euclid's Postulate 5 is often called the "Parallel Postulate."

This is an example of Euclid's Postulate 5. Notice where the blades of the shears cross the stem.

3. Reinforce and Extend

MENTAL MATH

Draw figures on the board that illustrate Euclid's five postulates. Ask questions about each figure. Have students name the postulate that answers your questions about the figure. For example, draw two endpoints connected by a line and ask, "What postulate allows this figure to be drawn?" **(Postulate 1)** Draw a line and a point; then draw a line through the point parallel to the other line. Ask: "Is there another line that can be drawn through this point and parallel to the first line?" **(no, Parallel Postulate 5)** Also, draw circle *P* with radius 2 inches, and ask, "What postulate allows me to draw circle *P* with radius 2 inches?" **(Postulate 3)**

LEARNING STYLES

Body/Kinesthetic Tie a piece of chalk to a string. Draw a dot on the board. Give a student a specific measurement for the length of the string. Holding the string at the specified length on the dot, the student draws a circle around the dot. Repeat with varying lengths. This activity will provide kinesthetic reinforcement of Euclid's Postulate 3.

Learning Styles

Visual/Spatial To help students better understand Euclid's Postulate 5, draw several different representations similar to the figure in the text (line *t* intersecting lines *l* and *m*), and in each representation, draw lines *l* and *m* closer and closer to parallel, until finally they are parallel. (You may also want to draw the figure so that lines *l* and *m* intersect on the left side of line *t*.) Using a protractor, measure the interior angles in each of the figures, find their sum, and compare the sums as the two lines get closer and closer to being parallel. Help students see that as the lines get closer to being parallel, the sum of the measures of the interior angles gets larger but is always less than 180° until the lines become parallel.

Learning Styles

LEP/ESL Have English language learners work with students who are more proficient in English. Together they can prepare flashcards with illustrations depicting the postulate on one side and the written postulate on the reverse side. Students can use the cards to help one another review the postulates.

Exercise A Identify the postulate that allows each construction to be made.

Post. 3

1. Draw a circle with center *A* and distance *d*.

Post. 1

2. Use line segments to connect point *P* to points *A*, *B*, and *C*.

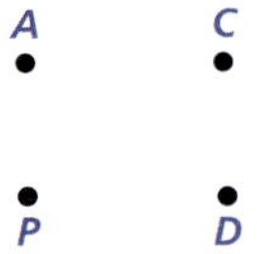

Post. 5, Parallel Post.

3. Draw a line that is parallel to line ℓ and passes through point *X*.

Post. 1, Post. 2

4. Draw a straight line between points *P* and *Q*.

Exercise B Tell which of Euclid's postulates makes each of the following statements true.

Post. 4

5. Angles *A*, *B*, and *C* are all right angles; therefore, their measures are equal.

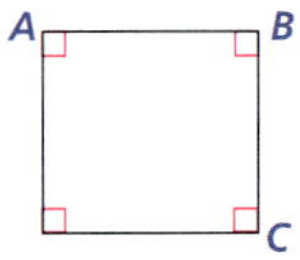

Post. 5, Parallel Post.

6. There is no line other than line *s* that can be parallel to ℓ and pass through *Q*.

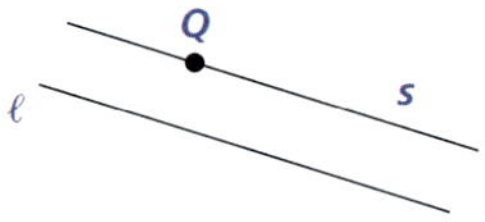

Post. 1

7. You can connect the points *W*, *X*, *Y*, and *Z* to form a geometric figure.

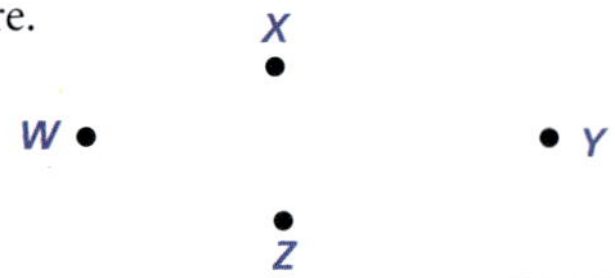

8. Lines s and t will intersect at some point to the right of line p.

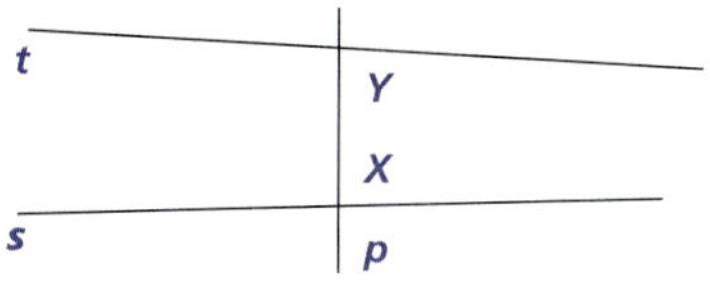

$m\angle X + m\angle Y < 2 \cdot 90°$

Post. 5

9. You can use a point Q and a line segment $\overline{AB}$ to draw a circle.

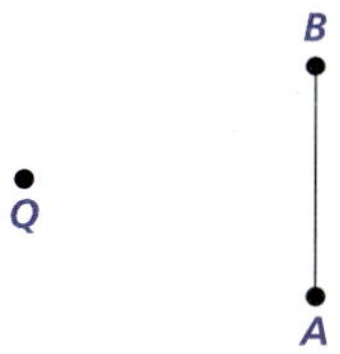

Post. 3

Problem Solving

Exercise C Solve the following problem.

10. Line ℓ forms two right angles when it crosses lines m and n. Are lines m and n parallel? Why do you think so?

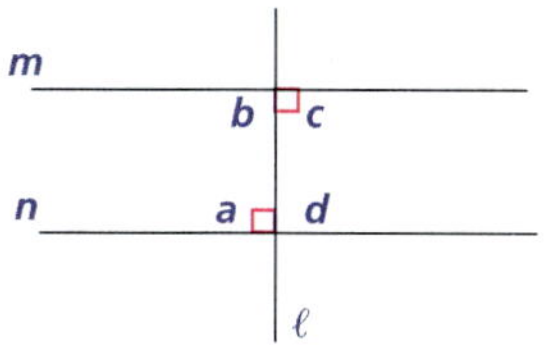

Students should recognize that the $m\angle d = 90°$ because $\angle a$ and $\angle d$ are supplementary; and $m\angle b = 90°$ because $\angle b$ and $\angle c$ are supplementary. Then, since $m\angle c + m\angle d = 180°$, the lines do not meet the requirement for intersection and so they must be parallel.

Experiencing Geometry

Fold a sheet of lined paper so that the horizontal lines make right angles with the fold. What is the measure of each of the four angles that that fold makes with one of the lines? How do you know?

All four are right angles and they are all equal to one another; Post. 4

Application

Career Connection

Have students describe a job or career in which some or all of the work is based on a set of postulates or assumptions. For example, the work of research scientists is almost entirely based on scientific postulates and/or theorems. Computer scientists, medical practitioners, and even musicians base much of their work on a given set of postulates and theorems. Instruct students to identify individuals in such a career and discuss with them the use of postulates in their work. Encourage students to find out how old the postulate or theorem is, who is responsible for it, and how it is applied in the given field of work.

Lesson at a Glance

Chapter 2 Lesson 3

Overview This lessons demonstrates how Euclid's postulates form the basis of some of geometry's basic constructions.

Objectives

- To identify the postulates used in given constructions.
- To make constructions using Euclid's postulates.

Student Pages 44–48

Teacher's Resource Library TRL

Workbook Activities 13–14

Additional Practice 9

1. Warm-Up Activity

Review Euclid's five postulates from Lesson 2. Draw simple constructions on the board that illustrate each of Euclid's postulates and ask students to identify and recite the postulate that allows the construction to be drawn.

Lesson 3 Using Euclid's Postulates

In geometry, you must be able to explain why you can take certain steps. Euclid's five postulates form the basis of constructions in geometry. They provide the reasons for some of geometry's basic constructions.

EXAMPLE **Given:** point P and segment $\overline{AB}$

Draw a circle with P as the center and $\overline{AB}$ as the radius.

Remember, $\overline{AB}$ *is the symbol for line segment AB.* $\overleftrightarrow{XY}$ *is the symbol for line XY.*

Euclid's Postulate 3 allows you to make this construction.

Euclid's Postulate 3 allows the greens keeper to make a green with a radius of any size.

EXAMPLE **Given:** $\overline{AC} \perp \overline{BD}$

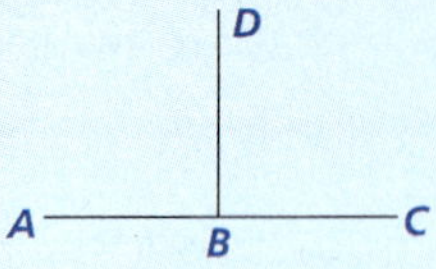

Then $m\angle ABD = m\angle CBD$ is true by Euclid's Postulate 4.

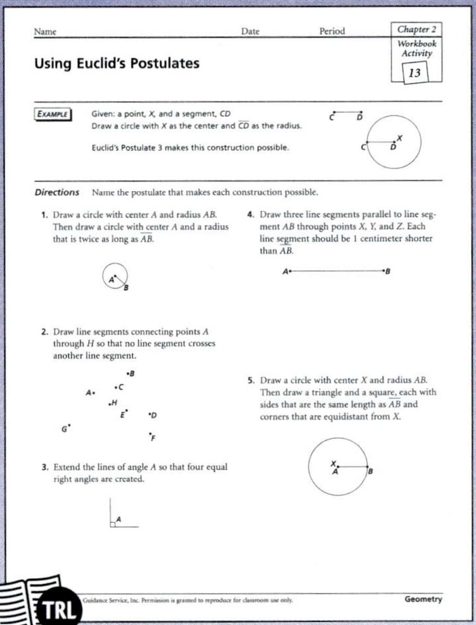

Name Date Period | Chapter 2 Workbook Activity 13

Using Euclid's Postulates

EXAMPLE Given: a point, X, and a segment, CD
Draw a circle with X as the center and $\overline{CD}$ as the radius.

Euclid's Postulate 3 makes this construction possible.

Directions Name the postulate that makes each construction possible.

1. Draw a circle with center A and radius AB. Then draw a circle with center A and a radius that is twice as long as $\overline{AB}$.
2. Draw line segments connecting points A through H so that no line segment crosses another line segment.
3. Extend the lines of angle A so that four equal right angles are created.
4. Draw three line segments parallel to line segment AB through points X, Y, and Z. Each line segment should be 1 centimeter shorter than $\overline{AB}$.
5. Draw a circle with center X and radius AB. Then draw a triangle and a square, each with sides that are the same length as $\overline{AB}$ and corners that are equidistant from X.

 Geometry

Workbook Activity 13

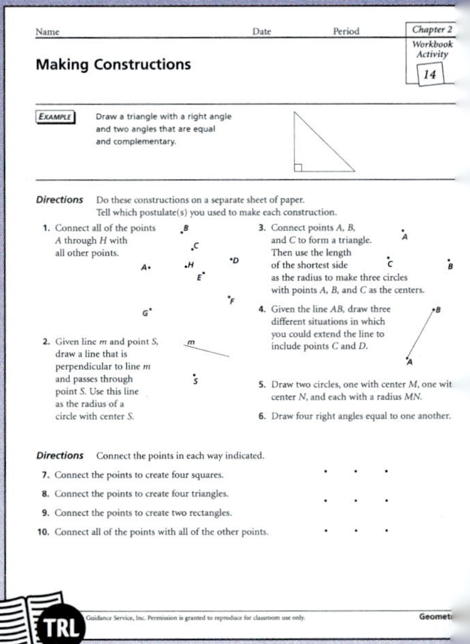

Name Date Period | Chapter 2 Workbook Activity 14

Making Constructions

EXAMPLE Draw a triangle with a right angle and two angles that are equal and complementary.

Directions Do these constructions on a separate sheet of paper. Tell which postulate(s) you used to make each construction.

1. Connect all of the points A through H with all other points.
2. Given line m and point S, draw a line that is perpendicular to line m and passes through point S. Use this line as the radius of a circle with center S.
3. Connect points A, B, and C to form a triangle. Then use the length of the shortest side as the radius to make three circles with points A, B, and C as the centers.
4. Given the line AB, draw three different situations in which you could extend the line to include points C and D.
5. Draw two circles, one with center M, one wit[h] center N, and each with a radius MN.
6. Draw four right angles equal to one another.

Directions Connect the points in each way indicated.

7. Connect the points to create four squares.
8. Connect the points to create four triangles.
9. Connect the points to create two rectangles.
10. Connect all of the points with all of the other points.

 Geomet[ry]

Workbook Activity 14

EXAMPLE Draw all the lines that pass through three points that are not on a straight line.

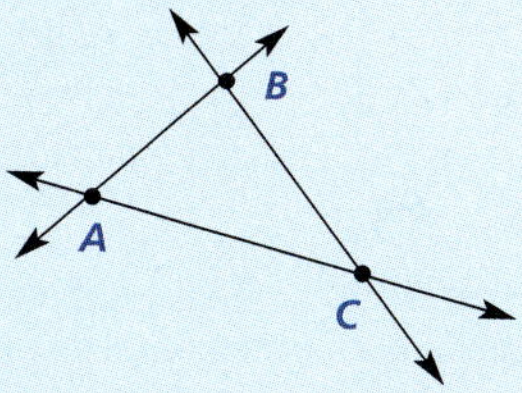

$\overleftrightarrow{AB}$, $\overleftrightarrow{AC}$, and $\overleftrightarrow{BC}$ can be drawn using Euclid's Postulate 1.

Geometry Around Us

This draftsperson uses Euclid's postulates to be able to draw geometric figures.

Exercise A Name the postulate that makes each construction possible.

1. Draw lines passing through point *A* and points *B* and *C*.

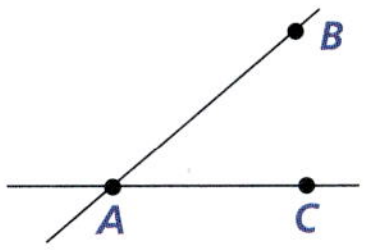

Post. 1

2. Draw two circles, one with center *A*, one with center *B*, and each with a radius *AB*.

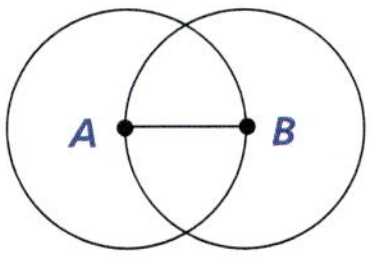

Post. 3

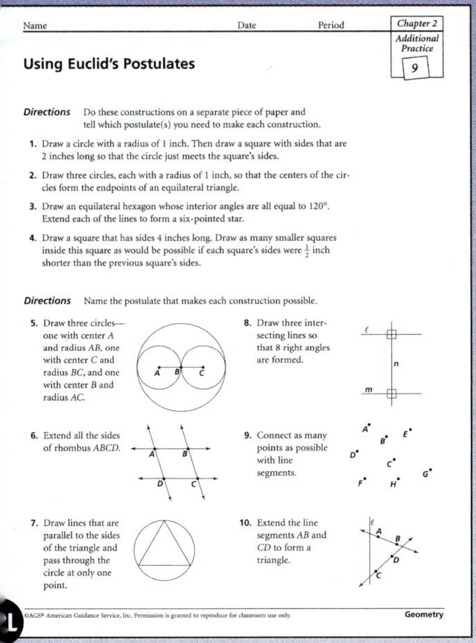

Name Date Period

Chapter 2 *Additional Practice* 9

Using Euclid's Postulates

Directions Do these constructions on a separate piece of paper and tell which postulate(s) you need to make each construction.

1. Draw a circle with a radius of 1 inch. Then draw a square with sides that are 2 inches long so that the circle just meets the square's sides.
2. Draw three circles, each with a radius of 1 inch, so that the centers of the circles form the endpoints of an equilateral triangle.
3. Draw an equilateral hexagon whose interior angles are all equal to 120°. Extend each of the lines to form a six-pointed star.
4. Draw a square that has sides 4 inches long. Draw as many smaller squares inside this square as would be possible if each square's sides were $\frac{1}{2}$ inch shorter than the previous square's sides.

Directions Name the postulate that makes each construction possible.

5. Draw three circles—one with center *A* and radius *AB*, one with center *C* and radius *BC*, and one with center *B* and radius *AC*.
6. Extend all the sides of rhombus *ABCD*.
7. Draw lines that are parallel to the sides of the triangle and pass through the circle at only one point.
8. Draw three intersecting lines so that 8 right angles are formed.
9. Connect as many points as possible with line segments.
10. Extend the line segments *AB* and *CD* to form a triangle.

 Geometry

Additional Practice 9

2. Teaching the Lesson

After reading the lesson and discussing the constructions in the examples, ask students to create their own constructions using each of the postulates. Encourage them to make constructions that incorporate two or more of the postulates in one drawing. Ask for volunteers to share their drawings with the class and have other students identify and/or recite the postulate(s) illustrated by the constructions.

Note: If you have physically challenged students who cannot draw the constructions, you might prepare the constructions as transparencies and show them to students as you review the steps in making the constructions. You might also have the students work with a partner or small group to draw the constructions.

LEARNING STYLES

Body/Kinesthetic Provide students with string and straws or spaghetti. Have them use the materials to build the constructions shown in the examples and Exercise A and described in Exercise B.

3. Reinforce and Extend

Learning Styles

Auditory/Verbal Both auditory and verbal learners may benefit from hearing and/or giving verbal descriptions of constructions. Have pairs of students work together. One student can give a verbal description of the constructions in Exercise A on pages 45–47 to help the other student name the postulates that make the constructions possible.

Application

Career Connection
Have students investigate how a draftsperson, graphic artist, or landscape architect makes use of constructions that are based on Euclid's five postulates. Ask them to find specific constructions for each postulate and describe what the figures may represent in that given line of work. Encourage students to use sources such as encyclopedias, how-to manuals, the Internet, or interviews with someone working in the field the student has chosen. If possible, display graphic or landscape architectural drawings.

Post. 2 **3.** Extend all the sides of triangle *ABC*.

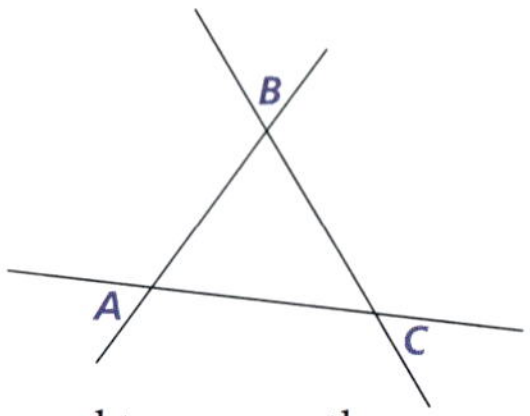

Post. 4 **4.** Draw four right angles equal to one another.

Post. 2 **5.** Extend all the sides of pentagon *ABCDE*.

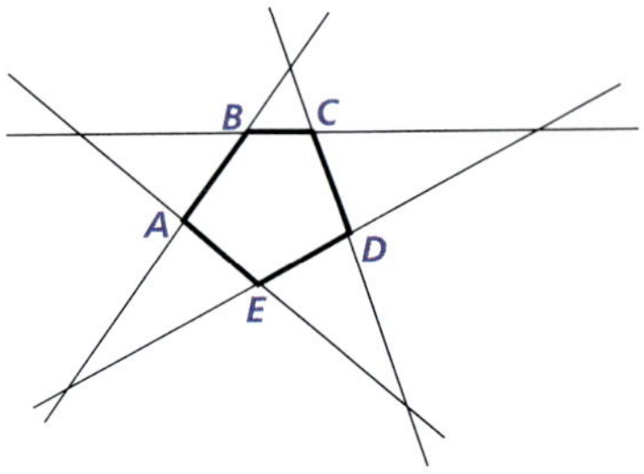

Post. 1 **6.** Draw a line passing through point *P* and points *A, B, C, D, E,* and *F.*

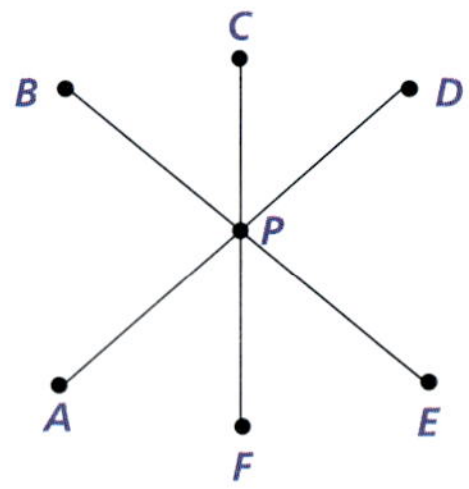

Post. 3 **7.** Draw three circles, each with radius *AB* and centers *A, B,* and *C.*

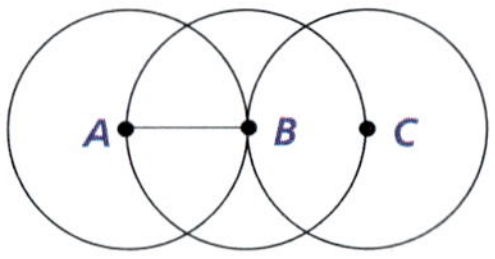

8. Draw two lines parallel to line ℓ, one through point P and one through point Q.

Post. 5

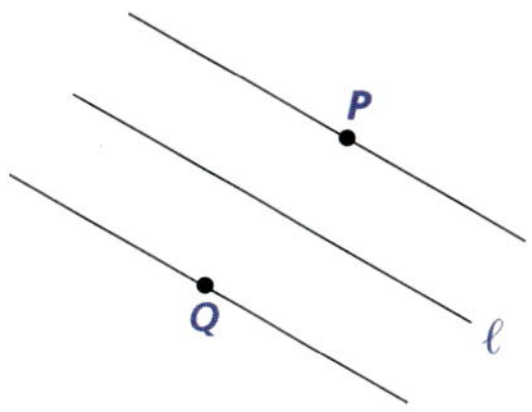

9. Draw a circle with center P and with a radius equal to $\overline{PQ}$.

Post. 1 and 3

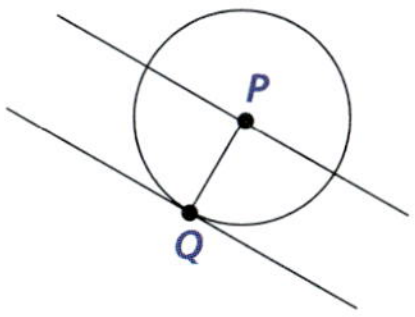

10. Draw three perpendicular lines with four equal angles.

Post. 4

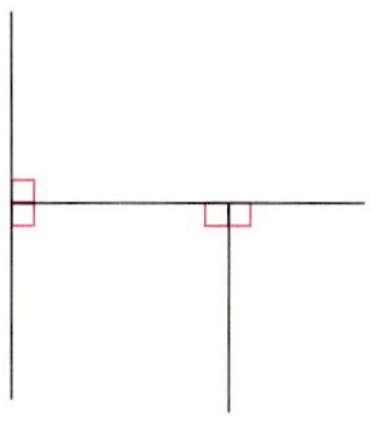

Construction

Exercise B Do these constructions on a separate sheet of paper. Tell which postulate(s) you used to make each construction.

11. Draw a line between points K and L.

Post. 1

12. Draw a line segment MN; then extend each end.

Post. 2

13. Draw a rectangle; state which angles are equal.

Post. 4

All four angles are equal right angles.

Group Problem Solving

Ask each group of three to four students to create a drawing that combines figures based on all of Euclid's postulates. Explain that their final product should be a construction that incorporates all of the postulates. However, they should also produce a progressive "slide show" of the stages leading up to their final drawing. The "slide show" would start with a single construction using one postulate, then show a second construction based on a different postulate added on to the first construction, and so on. The end product should show a series of six drawings, each labeled with the postulate that was used to make each new construction in the series of drawings. Encourage students to be creative in their design of constructions, use of color, and layout, but remind them to limit their constructions to those that are based on the postulates in Lesson 2. Display the "slide shows" in your room or school.

Learning Styles

Visual/Spatial Instruct students to copy the figures in Exercise A on pages 45–47 on index cards, writing the postulate that allows the construction to be made on the back. Students can then look at the constructions and practice reciting the postulates that they are based on. Students may prefer to make constructions of their own design for each of the postulates. If so, check their drawings before they begin quizzing themselves to make sure each construction accurately portrays the given postulate.

Learning Styles

Auditory/Verbal To complete Exercise B, pair up students and have one student read aloud the description of the construction to be drawn, while the other student listens and draws the construction. They can then use the constructions to quiz one another. One student should read the postulate and the other choose the appropriate construction illustrating it.

Application

At Home

Ask students to look around their homes for objects whose designs reflect, at least partially, one or more of Euclid's five postulates. For example, Postulate 4 allows the corners of walls or ceilings to be equal, and Postulate 1 allows a fence to be built or a flag to be raised up a flagpole.

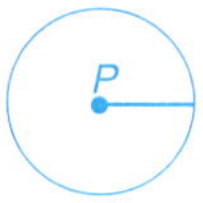

14. Given point P and a segment of 2 inches; draw a circle.
Post. 3

15. Given line ℓ and point Q; draw a line through Q and parallel to ℓ.
Post. 5

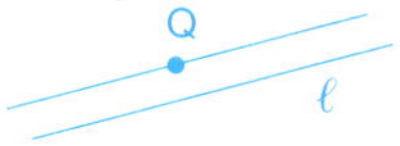

Writing About Mathematics

Which version of the parallel postulate do you use to draw parallel lines? Why? Which version do you use to draw intersecting lines? Why?

16. Given $\overline{FG}$; extend the line to include points H and J.
Post. 2

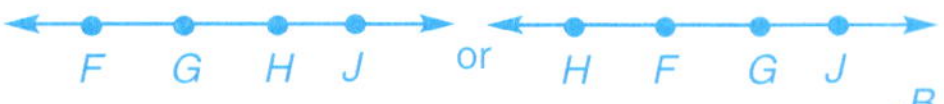

17. Connect points A, B, and C to form a triangle.
Post. 1

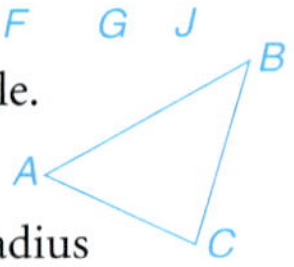

18. Draw a circle with center at point P, and a radius twice the length of $\overline{CD}$. Let $CD = 3$ cm.
Post. 1 and 3

19. Given line ℓ and points A and B not on line ℓ; draw two lines parallel to ℓ, one passing through A, one passing through B. Post. 5

20. Given $\overline{QR}$, and point P; draw a circle with radius equal to three times $\overline{QR}$ and center at point P.
Let $QR = 4$ cm. Post. 2 and 3

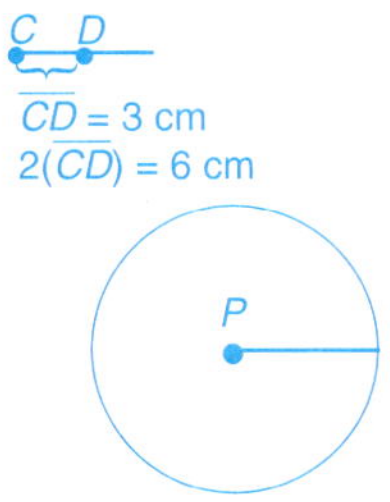

Experiencing Geometry

Copy the points at the right on a sheet of a paper. Then connect the points to form a figure. You may use a point more than once, but may not draw over any line you have already made. What postulate allows you to complete this puzzle?

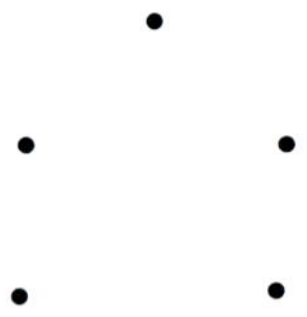

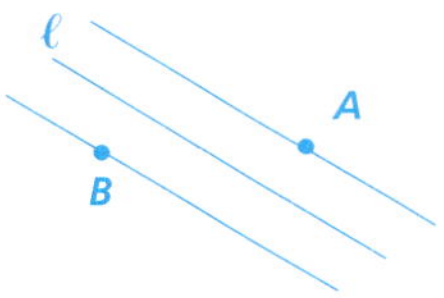

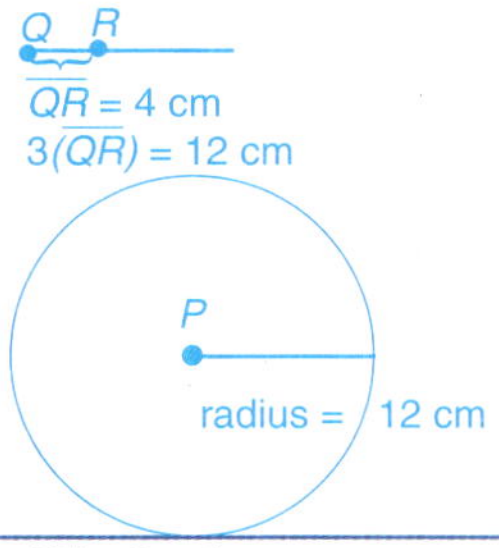

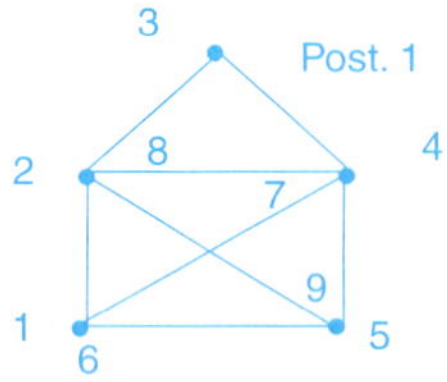

An **axiom**, like a postulate, is a statement taken to be true without proof. Axioms are statements about properties of equality. Axioms give us a formal way to refer to facts that seem obviously true. You may remember some axioms from your study of algebra. You used them to solve equations. In addition to his postulates, Euclid included five axioms or common notions (ideas) in his geometry.

> **Axiom**
> *A common notion taken to be true without proof*

	Euclid	Algebra Connection
Axiom 1	Things that are equal to the same thing are equal to each other.	If $a = b$ and $b = c$, then $a = c$. You might think of this as substituting c for b.
Axiom 2	If equals are added to equals, the sums are equal.	If $a = b$ $c = d$ then $a + c = b + d$
Axiom 3	If equals are subtracted from equals, the differences are equal.	If $a = b$ $c = d$ then $a - c = b - d$
Axiom 4	Things that are alike or coincide with one another are equal to one another.	Two figures that can be made to match one another, point for point, are equal.
Axiom 5	The whole, or sum, is greater than the parts.	If $a + b = c$ then $c > a$ and $c > b$.

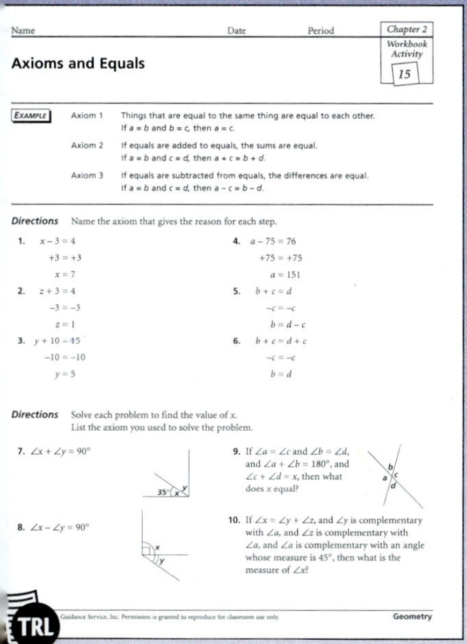
Name Date Period

Chapter 2 Workbook Activity 15

Axioms and Equals

EXAMPLE

Axiom 1 Things that are equal to the same thing are equal to each other. If $a = b$ and $b = c$, then $a = c$.

Axiom 2 If equals are added to equals, the sums are equal. If $a = b$ and $c = d$, then $a + c = b + d$.

Axiom 3 If equals are subtracted from equals, the differences are equal. If $a = b$ and $c = d$, then $a - c = b - d$.

Directions Name the axiom that gives the reason for each step.

1. $x - 3 = 4$; $+3 = +3$; $x = 7$
2. $z + 3 = 4$; $-3 = -3$; $z = 1$
3. $y + 10 = 15$; $-10 = -10$; $y = 5$
4. $a - 75 = 76$; $+75 = +75$; $a = 151$
5. $b + c = d$; $-c = -c$; $b = d - c$
6. $b + c = d + c$; $-c = -c$; $b = d$

Directions Solve each problem to find the value of x. List the axiom you used to solve the problem.

7. $\angle x + \angle y = 90°$
8. $\angle x - \angle y = 90°$
9. If $\angle a = \angle c$ and $\angle b = \angle d$, and $\angle a + \angle b = 180°$, and $\angle c + \angle d = x$, then what does x equal?
10. If $\angle x = \angle y + \angle z$, and $\angle y$ is complementary with $\angle a$, and $\angle z$ is complementary with $\angle a$, and $\angle a$ is complementary with an angle whose measure is 45°, then what is the measure of $\angle x$?

 Geometry

TRL

Workbook Activity 15

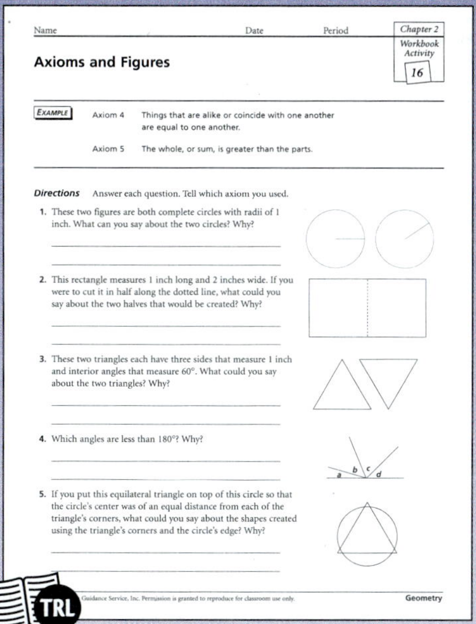
Name Date Period

Chapter 2 Workbook Activity 16

Axioms and Figures

EXAMPLE

Axiom 4 Things that are alike or coincide with one another are equal to one another.

Axiom 5 The whole, or sum, is greater than the parts.

Directions Answer each question. Tell which axiom you used.

1. These two figures are both complete circles with radii of 1 inch. What can you say about the two circles? Why?
2. This rectangle measures 1 inch long and 2 inches wide. If you were to cut it in half along the dotted line, what could you say about the two halves that would be created? Why?
3. These two triangles each have three sides that measure 1 inch and interior angles that measure 60°. What could you say about the two triangles? Why?
4. Which angles are less than 180°? Why?
5. If you put this equilateral triangle on top of this circle so that the circle's center was of an equal distance from each of the triangle's corners, what could you say about the shapes created using the triangle's corners and the circle's edge? Why?

 Geometry

TRL

Workbook Activity 16

Lesson at a Glance

Chapter 2 Lesson 4

Overview This lesson describes how axioms are used in algebra and geometry to solve problems.

Objectives

- To identify the axioms used to solve algebraic equations.
- To use axioms to solve geometric problems.

Student Pages 49–52

Teacher's Resource Library TRL

Workbook Activities 15–17

Additional Practice 10

Mathematics Vocabulary

axiom

1. Warm-Up Activity

Discuss the role of properties in mathematics. Review the following properties from algebra: commutative properties of addition and multiplication, associative properties of addition and multiplication, and the distributive property of multiplication over addition and subtraction. Put numeric examples of each on the board and have students identify the property being illustrated. Discuss the similarity between properties and axioms. Remind students that they can find a list of the axioms presented in the program on page 458.

2. Teaching the Lesson

Review solving equations of the form $x + a = b$ and $ax + b = c$, and $a(x + b) = c$. Also, review measurement equivalents for linear measurements (inches, feet, etc.), liquid measurements (cup, pint, etc.), and time measurements (minute, hour, day, etc.)

Algebra Connection

After reading the lesson and discussing the axioms used in the examples, present the following equations and statements to the class. Ask students to identify which axioms from this lesson are used to find a solution or justify the statement:

- $x + 5 = 9$ (**$x = 4$; Axiom 3**)
- $x - 13 = 27$ (**$x = 40$; Axiom 2**)
- $2x - 3 = 5$ (**$x = 4$; Axiom 2**)
- $3x + 1 = 13$ (**$x = 4$; Axiom 3**)
- $2(x - 3) = 20$ (**$x = 13$; Axiom 2**)
- If $y = 4$, and $4 = x$, then $y = x$. (**Axiom 1**)
- $5(x + 2) - 4 = 21$ (**$x = 3$; Axiom 3**)
- If $y = 3 + x$, and $3 + x = 9$, then $y = 9$. (**Axiom 1**)
- If $x + y = 10$, then $x < 10$ and $y < 10$. (**Axiom 5**)
- If $y = 3x + 9$, then $y > 3x$. (**Axiom 5**)

If square *ABCD* has six-inch sides and square *WXYZ* has $\frac{1}{2}$-foot sides, then square *ABCD* and square *WXYZ* are equal. (**Axiom 4**)

Manipulatives

Equations and Axioms

Materials: Factor Blocks

Group Practice: Model $x - 3 = 10$. To solve for x, add +3 to both sides of the equation. Ask students what axiom is demonstrated by this step. Repeat the process with the equation $x + 3 = 10$.

Student Practice: Instruct students to use manipulatives in this same manner to model the equations and the steps of the solutions for problems 1–5 in Exercise A.

EXAMPLES

		Reason
Given:	$x - 3 = 10$ $+3 = +3$ $x = 13$	Axiom 2 If equals are added to equals, the sums are equal.

		Reason
Given:	$x + 3 = 10$ $-3 = -3$ $x = 7$	Axiom 3 If equals are subtracted from equals, the differences are equal.

		Reason
Given:	1 yard = 3 feet 1 foot = 12 inches 1 yard = 3(12 inches) = 36 inches	Axiom 1 Equals can be substituted for equals.

		Reason
Given:	The upper half circle, called a *semicircle*, completely matches the lower half. (Think of folding them together along the center line or diameter.) Therefore, the semicircles are equal.	Axiom 4 Things that coincide, point for point, are equal.

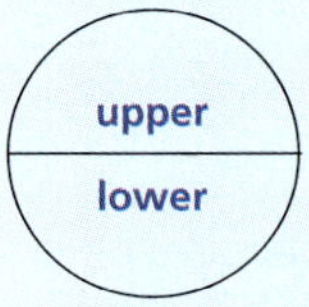

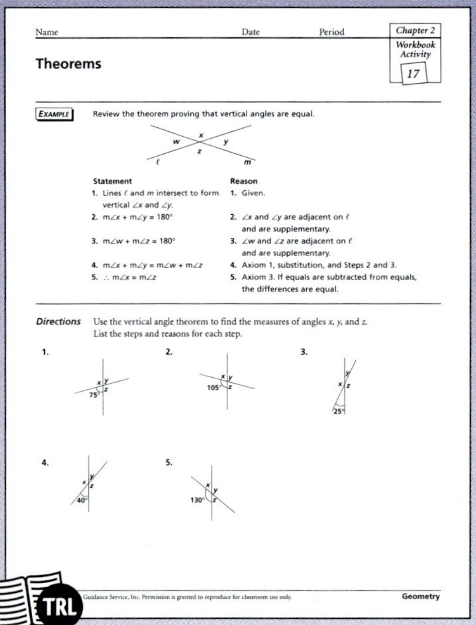

Name Date Period

Chapter 2 Workbook Activity 17

Theorems

EXAMPLE Review the theorem proving that vertical angles are equal.

Statement	Reason
1. Lines ℓ and *m* intersect to form vertical ∠*x* and ∠*y*.	1. Given.
2. m∠*x* + m∠*y* = 180°	2. ∠*x* and ∠*y* are adjacent on ℓ and are supplementary.
3. m∠*w* + m∠*z* = 180°	3. ∠*w* and ∠*z* are adjacent on ℓ and are supplementary.
4. m∠*x* + m∠*y* = m∠*w* + m∠*z*	4. Axiom 1, substitution, and Steps 2 and 3.
5. ∴ m∠*x* = m∠*z*	5. Axiom 3. If equals are subtracted from equals, the differences are equal.

Directions Use the vertical angle theorem to find the measures of angles *x*, *y*, and *z*. List the steps and reasons for each step.

1. 75°
2. 105°
3. 25°
4. 40°
5. 130°

 Geometry

Workbook Activity 17

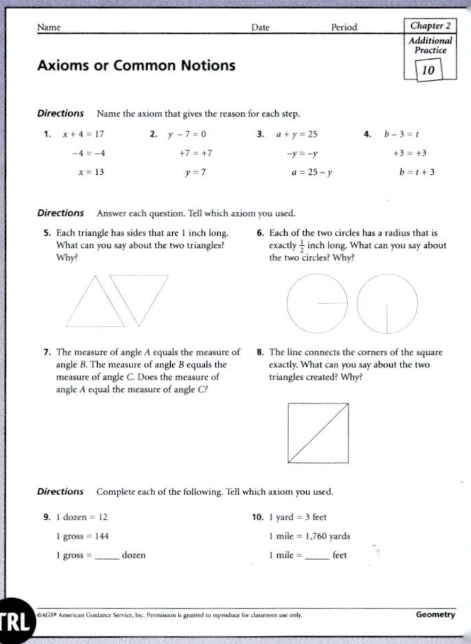

Name Date Period

Chapter 2 Additional Practice 10

Axioms or Common Notions

Directions Name the axiom that gives the reason for each step.

1. $x + 4 = 17$; $-4 = -4$; $x = 13$
2. $y - 7 = 0$; $+7 = +7$; $y = 7$
3. $a + y = 25$; $-y = -y$; $a = 25 - y$
4. $b - 3 = t$; $+3 = +3$; $b = t + 3$

Directions Answer each question. Tell which axiom you used.

5. Each triangle has sides that are 1 inch long. What can you say about the two triangles? Why?
6. Each of the two circles has a radius that is exactly $\frac{1}{2}$ inch long. What can you say about the two circles? Why?
7. The measure of angle *A* equals the measure of angle *B*. The measure of angle *B* equals the measure of angle *C*. Does the measure of angle *A* equal the measure of angle *C*?
8. The line connects the corners of the square exactly. What can you say about the two triangles created? Why?

Directions Complete each of the following. Tell which axiom you used.

9. 1 dozen = 12
   1 gross = 144
   1 gross = _____ dozen
10. 1 yard = 3 feet
    1 mile = 1,760 yards
    1 mile = _____ feet

 Geometry

Additional Practice 10

EXAMPLES *(continued)*

		Reason
Given:	m∠*ABD* + m∠*DBC* = 180°	Axiom 5
	Therefore, 180° > m∠*ABD* and 180° > m∠*DBC*	The whole is greater than the parts.

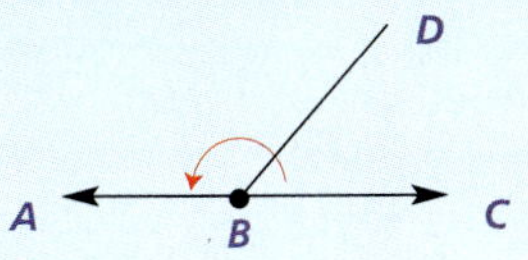

Exercise A Name the axiom that gives the reason for each step.

1. $y - 5 = 10$, $+5 = +5$, $y = 15$ — Axiom 2, equals added to equals
2. $w + 5 = 10$, $-5 = -5$, $w = 5$ — Axiom 3, equals subtracted from equals
3. $z + 7 = 30$, $-7 = -7$, $z = 23$ — Axiom 3, equals subtracted from equals
4. $q + 6 = 12$, $-6 = -6$, $q = 6$ — Axiom 3, equals subtracted from equals
5. $m - 4 = 10$, $+4 = +4$, $m = 14$ — Axiom 2, equals added to equals

Exercise B Answer each question. Tell which axiom you used.

6. Which angle(s) are less than 90°? Why?

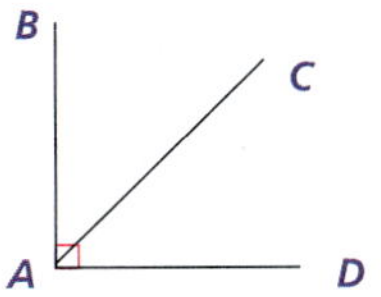

∠*BAC* < 90°, ∠*CAD* < 90°; Axiom 5, whole is greater than parts

7. The m∠*a* = m∠*b*, and m∠*b* = m∠*c*. Does m∠*a* = m∠*c*? Why?

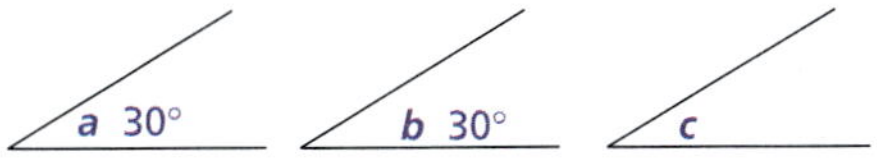

Yes; Axiom 1, substitution for equals

3. Reinforce and Extend

LEARNING STYLES

Visual/Spatial Have students write all the axioms on one sheet of paper with an algebraic and a geometric example illustrating each axiom. Students can then use this page as a reference to help them choose the appropriate axiom for the problems in Exercises A and B.

LEARNING STYLES

LEP/ESL Students whose second language is English may be from countries where the metric system of measurement is used rather than the customary system. Suggest that pairs of students—English proficient and students with limited proficiency—work together to make a table of metric equivalents to illustrate various axioms. Students familiar with the customary system of measurement can use rulers, yardsticks, and measuring containers to visually present and reinforce customary system equivalents as students more familiar with the metric system use meter sticks and measuring containers to visually present and reinforce metric equivalents.

Learning Styles

Interpersonal/Group Learning Have students working in pairs practice writing and solving algebraic equations and geometric problems in which the axioms in this lesson are used to find the solutions. Students should take turns writing problems for each of the axioms and naming and reciting the axioms used to solve the problems.

Common Error

Some students may not identify all of the angles in problem 9. To help them identify the combined angles 1 and 2 and 2 and 3, draw the construction on the board. Assign letters to each endpoint and identify every angle by letters. Have volunteers trace each angle on the board and tell whether it is less than 180°.

Calculator

As students use their calculators, ask them to both multiply and divide to verify their answers to problems 11–15. For problem 11, they can

press 4 [×] 2 [=] 8 and

press [÷] 2 [=] 4.

In this way, they verify their calculations twice—once when they arrive at the answer to the problem and once when they arrive at one of the factors used in the problem.

8. The $m\angle 1 = m\angle 3$ and $m\angle 2 = m\angle 4$. What do you know about $m\angle 1 + m\angle 3$ and $m\angle 2 + m\angle 4$? Why?

$m\angle 1 + m\angle 2 = m\angle 3 + m\angle 4$; Axiom 2, sums of equals are equal

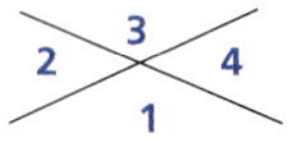

9. Which angles are less than 180°? Why?

Angles 1, 2, 3, and 1 + 2, 1 + 3, and 2 + 3 are all less than 180°; Axiom 5, whole is greater than parts

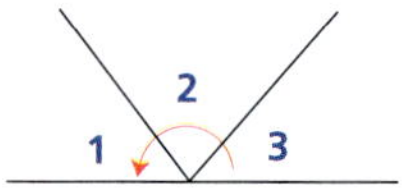

10. Every point on square *ABCD* can be matched to a point on *EFGH*. What can you say about the two squares? Why?

The squares are equal; Axiom 4, if all points match, items are equal

Calculator Practice

Exercise C Complete each of the following. Tell which axiom you used. Use your calculator to check your computations.

11. 1 gal = 4 qt
1 qt = 2 pt
1 gal = ____ pt

8 pt; Axiom 1, substitution

12. 1 qt = 2 pt
1 pt = 16 oz
1 qt = _____ oz

32 oz; Axiom 1, substitution

13. 1 pt = 2 c
1 c = 8 oz
1 pt = ___ oz

16 oz; Axiom 1 substitution

14. 1 day = 24 hr
1 hr = 60 min.
1 day = ___ min.

1,440 min.; Axiom 1, substitution

15. 1 decade = 10 years
1 year = 365 days
1 decade = ____ days

3,650 days; Axiom 1, substitution

Lesson 5 A Simple Proof

In your study of geometry in this book, you have learned about points, lines, and planes. You have also learned that axioms and postulates are assumed to be true without proof. Definitions, too, are true statements. **Theorems,** however, are statements that need to be proven true. A **proof** is a series of true statements. The last statement in a proof is the desired conclusion.

> ***Proof***
> *A series of true statements leading to a desired conclusion*
>
> ***Theorem***
> *A statement that can be proven true*

EXAMPLE **Theorem:** If angles are vertical angles, then their measures are equal.

To start a proof, clearly state what is *given* and what is *to prove.* In this case the *given,* or *hypothesis,* is "Angles are vertical." The *conclusion* to be proven true is "their measures are equal."

Next, use a construction to draw a picture of the given. You may also write out the given information.

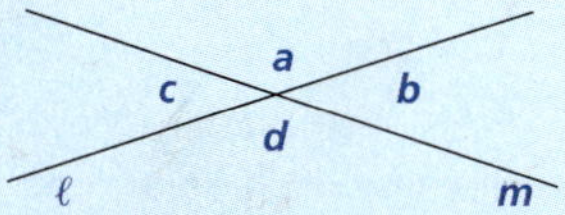

Recall that vertical angles have a common vertex and the same sides.

$\angle a$ and $\angle d$ are vertical angles.

$\angle c$ and $\angle b$ are vertical angles.

You may use symbols to write what you need to prove.

To prove: $m\angle a = m\angle d$ and $m\angle c = m\angle b$

Now you are ready to start your proof. You will need to work in two columns. On the left, list the true statements you need to reach the conclusion. On the right, list the definition, axiom, or postulate that makes the statement true.

Like Auguste Rodin's "The Thinker" shown here, you may need to think about what you already know to prove something new.

> *A*djacent angles have a common vertex and one common side.

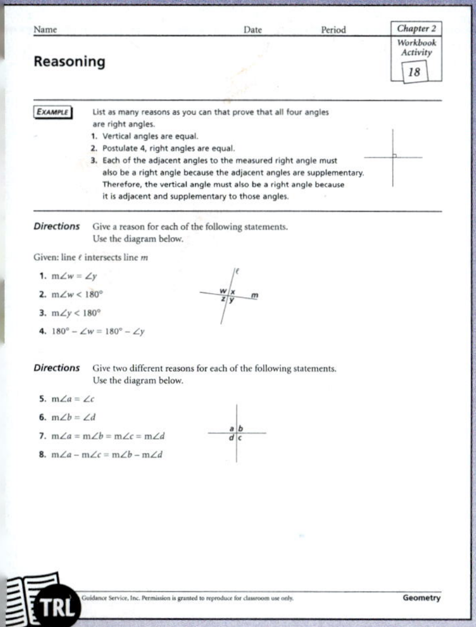

Name Date Period | *Chapter 2* *Workbook Activity* *18*

Reasoning

EXAMPLE List as many reasons as you can that prove that all four angles are right angles.

1. Vertical angles are equal.
2. Postulate 4, right angles are equal.
3. Each of the adjacent angles to the measured right angle must also be a right angle because the adjacent angles are supplementary. Therefore, the vertical angle must also be a right angle because it is adjacent and supplementary to those angles.

Directions Give a reason for each of the following statements. Use the diagram below.

Given: line ℓ intersects line *m*

1. $m\angle w = \angle y$
2. $m\angle w < 180°$
3. $m\angle y < 180°$
4. $180° - \angle w = 180° - \angle y$

Directions Give two different reasons for each of the following statements. Use the diagram below.

5. $m\angle a = \angle c$
6. $m\angle b = \angle d$
7. $m\angle a = m\angle b = m\angle c = m\angle d$
8. $m\angle a - m\angle c = m\angle b - m\angle d$

TRL Guidance Service, Inc. Permission is granted to reproduce for classroom use only. **Geometry**

Workbook Activity 18

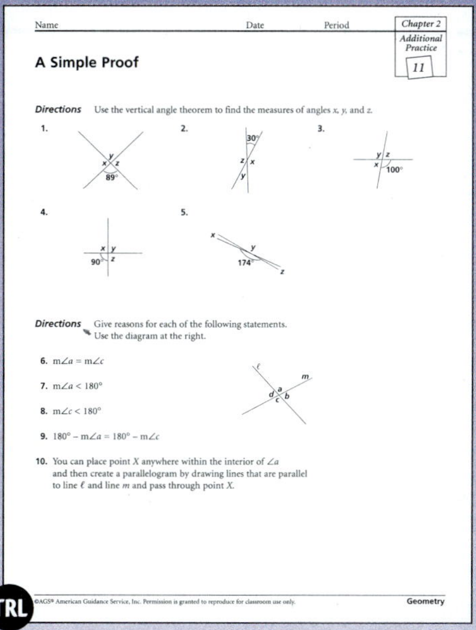

Name Date Period | *Chapter 2* *Additional Practice* *11*

A Simple Proof

Directions Use the vertical angle theorem to find the measures of angles *x, y,* and *z.*

1. 2. 3. 4. 5.

Directions Give reasons for each of the following statements. Use the diagram at the right.

6. $m\angle a = m\angle c$
7. $m\angle a < 180°$
8. $m\angle c < 180°$
9. $180° - m\angle a = 180° - m\angle c$
10. You can place point *X* anywhere within the interior of $\angle a$ and then create a parallelogram by drawing lines that are parallel to line ℓ and line *m* and pass through point *X.*

TRL ©AGS® American Guidance Service, Inc. Permission is granted to reproduce for classroom use only. **Geometry**

Additional Practice 11

Lesson at a Glance

Chapter 2 Lesson 5

Overview This lesson defines what a proof is and shows how to write a proof for a given hypothesis and conclusion.

Objectives

- To identify the postulates, axioms, and theorems that justify the statements in a proof.
- To use a theorem to solve problems.

Student Pages 53–55

Teacher's Resource Library TRL

Workbook Activity 18

Additional Practice 11

Mathematics Vocabulary

proof
theorem

1. Warm-Up Activity

Review the terms *vertical angles, supplementary angles,* and *adjacent angles.* Put the following figure on the board and ask, "What is the sum of $m\angle x$ and $m\angle y$?" (180°)

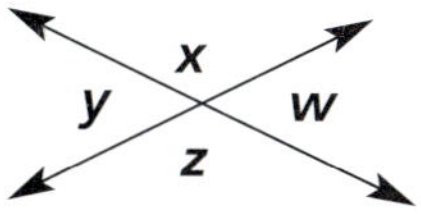

Have students explain how they know that. (because they are adjacent and supplementary) Then ask students what kind of angles $\angle y$ and $\angle w$ are and have them explain how they know that. (They are vertical angles because they have a common vertex and the same sides.) Also ask, "If $m\angle w = 130°$, what is $m\angle z$?" and have students explain how they got their answer. ($130° + m\angle z = 180°$, $m\angle z = 50°$)

2. Teaching the Lesson

Read the theorem in the example on page 53 and ask students what kind of statement it is. **(conditional)** Ask them to identify the hypothesis and conclusion. Relate these terms to the terms *given* and *to prove* as they are used to write a proof. Then put the following statements on the board and ask students for a postulate, theorem, or definition that justifies each statement:

- If $x = 7$ and $y = 7$, then $x = y$. **(Axiom 1)**
- If $x + 9 = 9 + 4$, then $x = 4$. **(Axiom 3)**
- If $\angle x$ and $\angle y$ are supplementary, what is the sum of their measures? **(180°; definition of supplementary angles)**

Once students seem familiar with the process of stating reasons that justify statements, work through the proof on page 54 together. Make sure students understand that the statements and reasons in the proof are the same as those they have been working with in this and previous lessons. Refer students to pages 459–461 for a list of theorems for easy reference.

3. Reinforce and Extend

Learning Styles

Visual/Spatial Encourage students to prepare a notebook in which they write and illustrate postulates, axioms, definitions, and theorems. They can continue to add to the notebook in subsequent lessons. Students can use their notebooks as review and reference guides.

Refer to the figure in the example on page 53. Prove m$\angle a$ = m$\angle d$.

Proof

Statement	Reason
1. Lines ℓ and m intersect to form vertical angles a and d.	1. Given.
2. m$\angle a$ + m$\angle b$ = 180°	2. $\angle a$ and $\angle b$ are adjacent on m and are supplementary.
3. m$\angle b$ + m$\angle d$ = 180°	3. $\angle b$ and $\angle d$ are adjacent on ℓ and are supplementary.
4. m$\angle a$ + m$\angle b$ = m$\angle b$ + m$\angle d$	4. Axiom 1, substitution, and steps 2 and 3.
5. ∴ m$\angle a$ = m$\angle d$	5. Axiom 3. If equals are subtracted from equals, the differences are equal.

The vertical angles a and d have been proven equal. Notice that angles c and b also form vertical angles, so they too must be equal. For practice, you can work out the proof that shows angles c and b are equal in the following exercises.

Remember that ∴ means "therefore" and is used to signal the last step of a proof.

Exercise A Copy the proof. Write the reason(s) for each step.

To Prove: m$\angle c$ = m$\angle b$

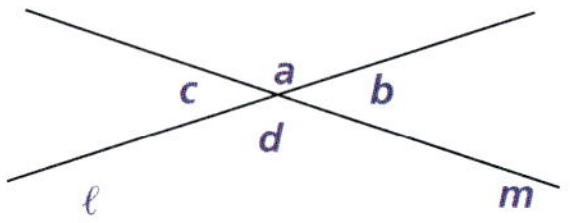

Proof

Statement	Reason
1. Lines ℓ and m intersect to form vertical angles b and c.	1. Given. 2. $\angle a$ and $\angle c$ are adjacent on ℓ and are supplementary. 3. $\angle a$ and $\angle b$ are adjacent on m and are supplementary. 4. Axiom 1, substitution, and steps 2 and 3. 5. Axiom 3. If equals are subtracted from equals, the differences are equal.
2. m$\angle a$ + m$\angle c$ = 180°	2.
3. m$\angle a$ + m$\angle b$ = 180°	3.
4. m$\angle a$ + m$\angle c$ = m$\angle a$ + m$\angle b$	4.
5. ∴ m$\angle c$ = m$\angle b$	5.

Exercise B Use the vertical angle theorem to find the measures of angles x, y, and z.

6. $x = 60°$, $y = 120°$, $z = 60°$

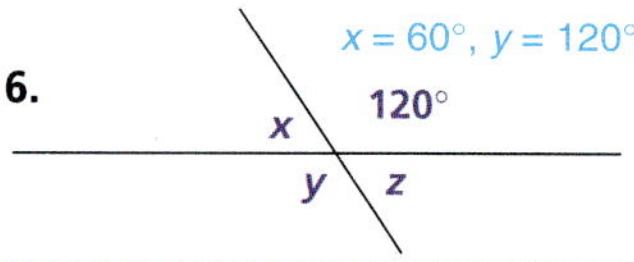

7. $x = 160°$, $y = 20°$, $z = 160°$

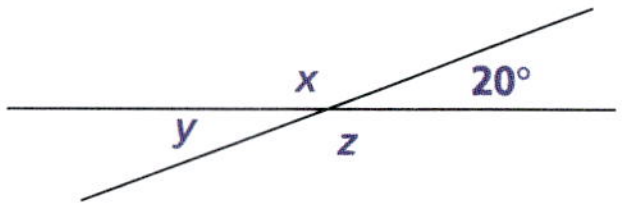

8.

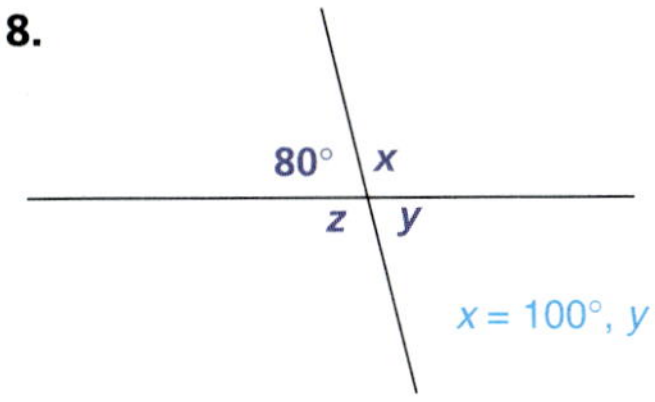

$x = 100°$, $y = 80°$, $z = 100°$

9.

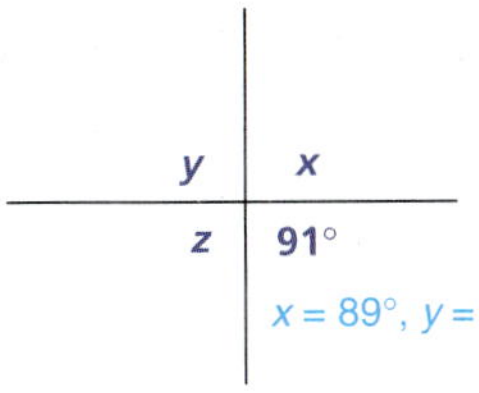

$x = 89°$, $y = 91°$, $z = 89°$

10.

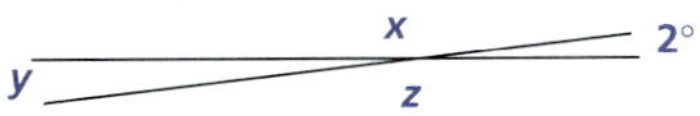

$x = 178°$, $y = 2°$, $z = 178°$

Exercise C Give a reason for each of the following statements. Use the diagram at the right.

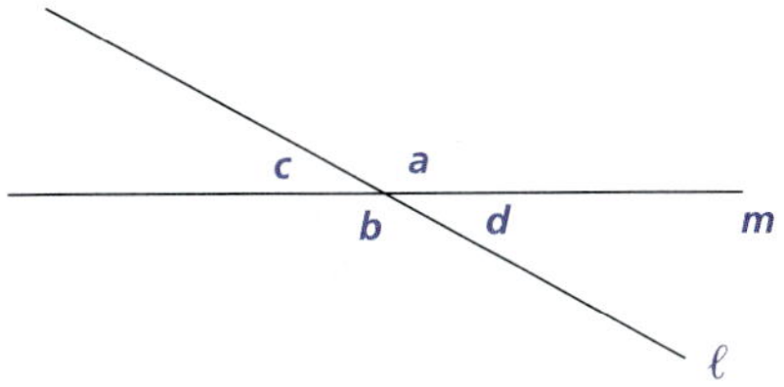

Given ℓ intersects m

11. $m\angle a = m\angle b$

12. $m\angle a < 180°$

13. $m\angle b < 180°$

14. $180° - m\angle a = 180° - m\angle b$

11. Theorem: vertical angles are equal
12. $a + c = 180°$, so $a < 180°$, Axiom 5, whole is greater than parts
13. $b + d = 180°$, so $b < 180°$, Axiom 5, whole is greater than parts
14. Axiom 3, equals subtracted from equals

Problem Solving

Exercise D Follow the directions.

15. Give two different reasons for each of the following statements. Use the diagram at the right.

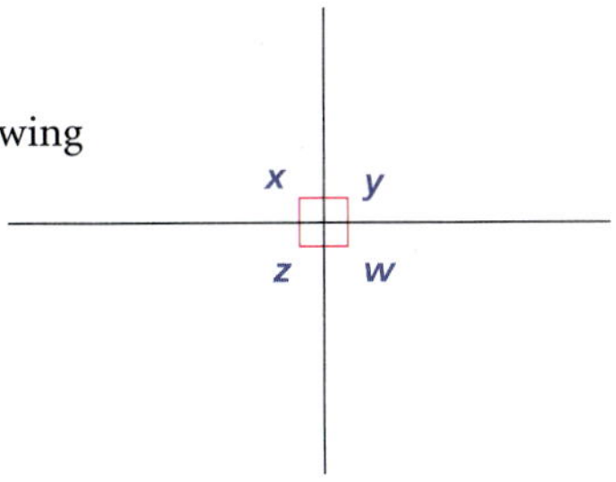

a. $m\angle x = m\angle w$

b. $m\angle y = m\angle z$

c. $m\angle x = m\angle y = m\angle w = m\angle z$

a. vertical angles equal, Post. 4: right angles equal
b. vertical angles equal, Post. 4: right angles equal
c. Post. 4: right angles equal, Axiom 1, substitution

Learning Styles

Auditory/Verbal Have students restate proofs and theorems in their own words. Ask them to write their restatements. Check to make sure that they are correct, and then have students record their restatements. Students may use the recordings to aurally review the proofs and theorems.

Learning Styles

Logical/Mathematical Have students begin Exercise B by writing equations relating the measures of the angles they are trying to find. For example, in problem 6, some of the equations that relate angles x, y, and z are $m\angle x = m\angle z$; $m\angle x + 120° = 180°$; $m\angle z = m\angle y$; $m\angle x + m\angle y = 180°$. Students may find such equations helpful as they determine the measures of the angles in problems 6–10 and in similar problems.

Group Problem Solving

Have students in groups of four or five write a proof for the theorem that states all right angles are congruent. The groups should write up a formal proof that includes statements and reasons columns, and all statements and reasons should be clearly, completely, and accurately stated. The groups should prepare posters as final products for display in the classroom.

Calculator

Encourage students to use calculators to verify their answers in Exercise B.

Lesson at a Glance

Chapter 2 Application

Overview This lesson explores how geometric shapes are used to produce works of art.

Objectives

- To create an artistic pattern using geometric shapes.
- To investigate how geometric shapes are used in the environment.

Student Page 56

Teacher's Resource Library TRL

Workbook Activity 19

1. Warm-Up Activity

Display prints of works of art or art books to show to students. Help students identify geometric shapes in the art.

2. Teaching the Lesson

Discuss the use of the different geometric shapes in the painting. Point out to students that Piet Mondrian was well-known for his use of simple geometric shapes in his paintings. He called his style of abstract painting *neoplasticism.* Particularly in his later works, Mondrian avoided techniques that give a sense of depth to a painting. Ask for students' interpretations of the painting.

3. Reinforce and Extend

Application

In the Community

Have students visit art museums, galleries, libraries, or stores where artwork, prints, or cards with reprints of artwork are displayed. Have them choose a work that makes use of geometric shapes. Ask students to describe the work of art, tell where they saw it, and identify the artist. Invite them to share their findings with the class. Encourage them to show a print or a sketch of the artwork if possible.

Application

Art and Geometry A knowledge of geometry can do more than build bridges, buildings, and highways. Geometry can also produce patterns that people find pleasing and restful.

Example In this painting, the artist Piet Mondrian has chosen simple geometric shapes, including lines, squares, and rectangles, to express his views. The artist's choice of shape and color gives us another way to view the world.

Exercise Use construction paper or colored acetate to create several squares, rectangles, and lines of various colors. Arrange the cut pieces of paper or acetate into a pattern you find pleasing.

1. Compare your construction with those of your classmates. Decide who used the greatest number of geometric shapes and who used the least.
2. Do you find some arrangements or constructions more pleasing than others? If so, describe them.
3. Search newspapers and periodicals for advertisements that rely on simple geometric shapes. Trace or copy the advertisements and present them to the class for discussion. Why do you think advertisers use geometric shapes?
4. Look for buildings and landscapes in your neighborhood. Identify the geometric shapes portrayed.
5. Find other examples of geometric shapes in objects such as windows, paintings, or ceramics. Describe the shapes.

Students' answers will vary.

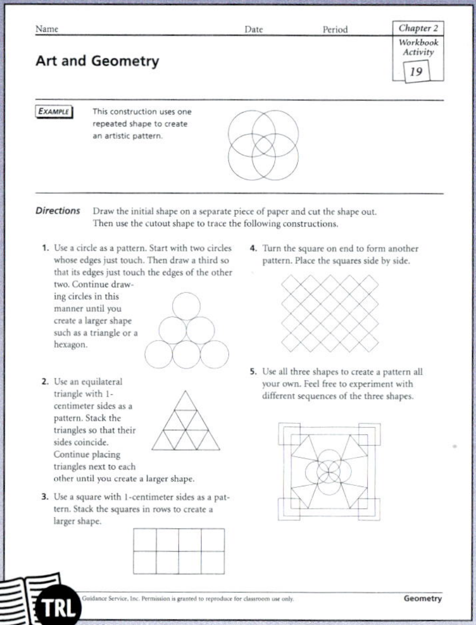

Name Date Period

Chapter 2 *Workbook Activity* *19*

Art and Geometry

Example This construction uses one repeated shape to create an artistic pattern.

Directions Draw the initial shape on a separate piece of paper and cut the shape out. Then use the cutout shape to trace the following constructions.

1. Use a circle as a pattern. Start with two circles whose edges just touch. Then draw a third so that its edges just touch the edges of the other two. Continue drawing circles in this manner until you create a larger shape such as a triangle or a hexagon.
2. Use an equilateral triangle with 1-centimeter sides as a pattern. Stack the triangles so that their sides coincide. Continue placing triangles next to each other until you create a larger shape.
3. Use a square with 1-centimeter sides as a pattern. Stack the squares in rows to create a larger shape.
4. Turn the square on end to form another pattern. Place the squares side by side.
5. Use all three shapes to create a pattern all your own. Feel free to experiment with different sequences of the three shapes.

 Geometry

Workbook Activity 19

Chapter 2 Review

Write the hypothesis for statements 1–4.

Example: If an angle is a straight angle, then it measures 180°.

Solution: An angle is a straight angle.

1. If an angle is acute, then it measures less than 90°. — An angle is acute.
2. If the sum of the measures of two angles is 90°, then the angles are complementary. — The sum of the measures of two angles is 90°.
3. If a figure has three sides, then it is a triangle. — A figure has three sides.
4. If two angles are supplementary, then the sum of their measures is 180°. — Two angles are supplementary.

Write the conclusion for the items listed.

Example: If an angle is a straight angle, then it measures 180°.

Solution: It measures 180°.

5. Write the conclusion for statement 1. — It measures less than 90°.
6. Write the conclusion for statement 2. — The angles are complementary.
7. Write the conclusion for statement 3. — It is a triangle.

Write the converse for the items listed.

Example: If an angle is a straight angle, then it measures 180°.

Solution: If an angle measures 180°, then it is a straight angle.

8. Write the converse of statement 1. — If an angle measures less than 90°, then the angle is acute.
9. Write the converse of statement 2. — If two angles are complementary, then the sum of their measures is 90°.
10. Write the converse of statement 4. — If the sum of the measures of two angles is 180°, then the angles are supplementary.

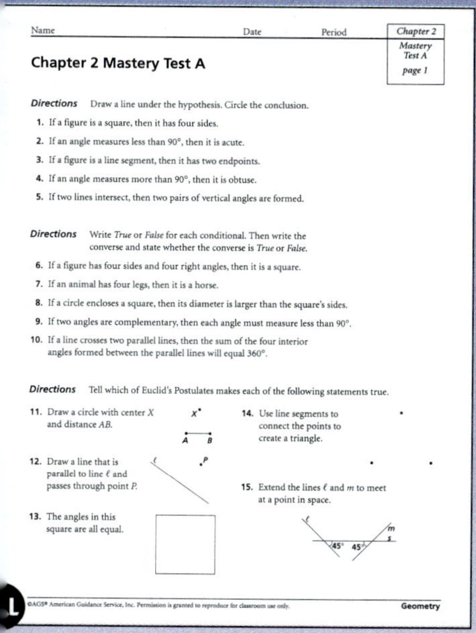

Name Date Period — Chapter 2 Mastery Test A page 1

Chapter 2 Mastery Test A

Directions Draw a line under the hypothesis. Circle the conclusion.

1. If a figure is a square, then it has four sides.
2. If an angle measures less than 90°, then it is acute.
3. If a figure is a line segment, then it has two endpoints.
4. If an angle measures more than 90°, then it is obtuse.
5. If two lines intersect, then two pairs of vertical angles are formed.

Directions Write *True* or *False* for each conditional. Then write the converse and state whether the converse is *True* or *False*.

6. If a figure has four sides and four right angles, then it is a square.
7. If an animal has four legs, then it is a horse.
8. If a circle encloses a square, then its diameter is larger than the square's sides.
9. If two angles are complementary, then each angle must measure less than 90°.
10. If a line crosses two parallel lines, then the sum of the four interior angles formed between the parallel lines will equal 360°.

Directions Tell which of Euclid's Postulates makes each of the following statements true.

11. Draw a circle with center X and distance AB.
12. Draw a line that is parallel to line ℓ and passes through point P.
13. The angles in this square are all equal.
14. Use line segments to connect the points to create a triangle.
15. Extend the lines ℓ and m to meet at a point in space.

 Geometry

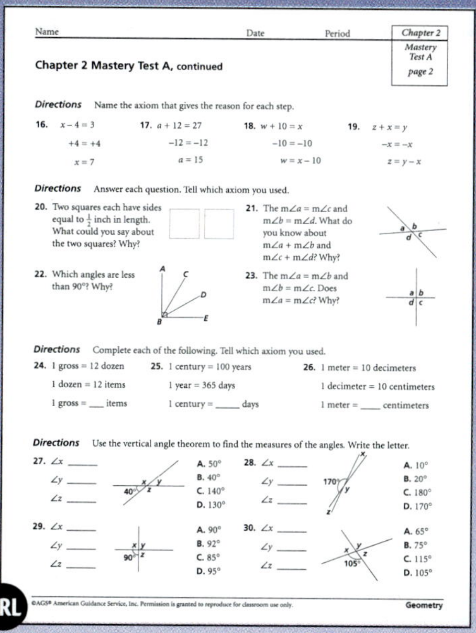

Name Date Period — Chapter 2 Mastery Test A page 2

Chapter 2 Mastery Test A, continued

Directions Name the axiom that gives the reason for each step.

16. $x - 4 = 3$; $+4 = +4$; $x = 7$
17. $a + 12 = 27$; $-12 = -12$; $a = 15$
18. $w + 10 = x$; $-10 = -10$; $w = x - 10$
19. $z + x = y$; $-x = -x$; $z = y - x$

Directions Answer each question. Tell which axiom you used.

20. Two squares each have sides equal to $\frac{1}{2}$ inch in length. What could you say about the two squares? Why?
21. The $m\angle a = m\angle c$ and $m\angle b = m\angle d$. What do you know about $m\angle a + m\angle b$ and $m\angle c + m\angle d$? Why?
22. Which angles are less than 90°? Why?
23. The $m\angle a = m\angle b$ and $m\angle b = m\angle c$. Does $m\angle a = m\angle c$? Why?

Directions Complete each of the following. Tell which axiom you used.

24. 1 gross = 12 dozen; 1 dozen = 12 items; 1 gross = ___ items
25. 1 century = 100 years; 1 year = 365 days; 1 century = ___ days
26. 1 meter = 10 decimeters; 1 decimeter = 10 centimeters; 1 meter = ___ centimeters

Directions Use the vertical angle theorem to find the measures of the angles. Write the letter.

27. $\angle x$ ___ $\angle y$ ___ $\angle z$ ___ — A. 50° B. 40° C. 140° D. 130°
28. $\angle x$ ___ $\angle y$ ___ $\angle z$ ___ — A. 10° B. 20° C. 180° D. 170°
29. $\angle x$ ___ $\angle y$ ___ $\angle z$ ___ — A. 90° B. 92° C. 85° D. 95°
30. $\angle x$ ___ $\angle y$ ___ $\angle z$ ___ — A. 65° B. 75° C. 115° D. 105°

 Geometry

TRL

Chapter 2 Mastery Test A

Chapter 2 Review

Each set of problems in the Chapter Review includes an example and solution to illustrate the concept. Use the given examples for reteaching the material in Chapter 2. For additional practice, refer to the Supplementary Problems for Chapter 2 (pages 393–395) and Additional Practice 7–11 in the Teacher's Resource Library.

Chapter 2 Mastery Test TRL

The Teacher's Resource Library includes parallel forms of the Chapter 2 Mastery Test. The difficulty level of the two forms is equivalent. You may wish to use one form as a pretest and the other form as a posttest. In addition, the Chapter 2 Cumulative Mastery Test covers content from Chapters 1–2.

Alternative Assessment

Alternative Assessment items correlate with student Goals for Learning at the beginning of this chapter.

- Have students evaluate and write conditionals.

Write three conditionals. For each conditional, write its converse. Use a truth table to determine whether each conditional and its converse are true or false.

- Have students identify and apply the postulates that define lines and angles.

Write each of Euclid's postulates and make a drawing to show the meaning of the postulate.

- Have students identify postulates that apply to constructions.

Draw on a sheet of paper figures representing each of Euclid's postulates in random order. Label each illustration with the name of the appropriate postulate.

Alternative Assessment

continued

- Have students construct geometric lines and shapes based on postulates.

Reproduce the drawings you made in item 2. This time do not label the illustrations. Exchange papers with a partner. Identify the postulate each construction represents.

- Have students use axioms in explanations.

Use axioms to explain why each of the following is true.

If triangle *ABC* is a right triangle with legs of 3 and 4 units and a hypotenuse of 5, and triangle *DEF* is a right triangle with one leg 3 units long and one leg 4 units long, then triangles *ABC* and *DEF* are equal.

If $d = e$ and $e = f$, then $f = d$.

If $d = e$ and $f = g$, then $d + f = e + g$.

If $d = e$ and $f = g$, then $d - f = e - g$.

- Have students prove the vertical angle theorem and use it to find measures of angles.

Draw intersecting lines and label the angles with letters. Use the vertical angle theorem to identify angles with the same measurement and explain why they are the same.

More Review...

Euclid's Postulates

1 A straight line can be drawn from any point to any point.
2 A finite straight line can be extended continuously in a straight line.
3 A circle may be described with any center and distance.
4 All right angles are equal to one another.
5 If two lines ℓ and m are cut by a third line t, and the two inside angles, a and b, together measure less than two right angles, then the two lines will meet.

Parallel Postulate 5

If there is a line ℓ and a point P not on ℓ, then there is only one line that passes through P and is parallel to ℓ.

Axioms

1 Things that are equal to the same thing are equal to each other.
2 If equals are added to equals, the sums are equal.
3 If equals are subtracted from equals, the differences are equal.
4 Things that are alike or coincide with one another are equal to one another.
5 The whole, or sum, is greater than the parts.

Decide which postulate allows each construction to be made. Write the number of the postulate.

Example:

Solution: Postulate 2, A finite straight line can be repeated continuously in a straight line.

11. Connect point P to points A, B, C, and D.
Postulate 1

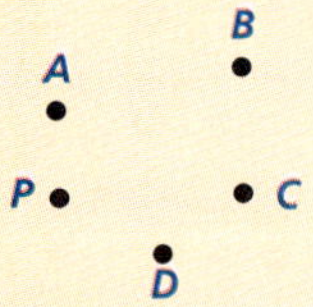

12. Draw a line that is parallel to line m and passes through point X.
Parallel Postulate 5

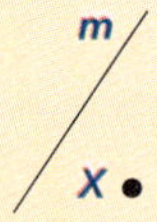

13. Draw a circle with center A and distance 3 cm.
Postulate 3

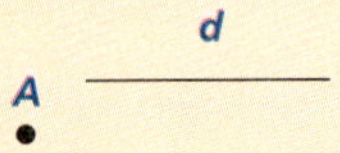

Decide which axiom allows each operation. Write the letter of the axiom.

Example: $m + 13 = 17$
$-13 = -13$
$m = 4$

Solution: Axiom 3, If equals are subtracted from equals, the differences are equal.

14. $m - 9 = 20$
$+9 = +9$
$m = 29$
Axiom 2

15. Angle $x + y$ is less than 180°. Axiom 5

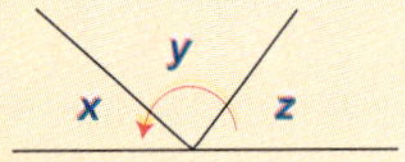

Copy the proof at the top of the next page. Write the missing steps to prove $m\angle c = m\angle b$.

Example: Lines ℓ and m intersect.
Solution: Given.

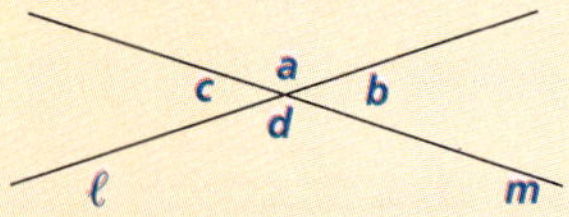

58 *Chapter 2 Thinking Geometrically: Using Proofs*

Name Date Period | *Chapter 2 Mastery Test B page 1*

Chapter 2 Mastery Test B

Directions Draw a line under the hypothesis. Circle the conclusion.

1. If a figure is a ray, then it has one endpoint.
2. If a figure is a triangle, then it has three sides.
3. If a figure is an angle, then it has two rays.
4. If an angle is straight, then it measures 180°.
5. If an angle measures 90°, then it is a right angle.

Directions Write *True* or *False* for each of the conditionals. Then write the converse and state whether the converse is *True* or *False*.

6. If a figure has no line segments, then it is a circle.
7. If a figure has four sides and four right angles, then it is a rectangle.
8. If an angle is complementary to an angle measuring 42°, then the angle measures 48°.
9. If two angles are supplementary, then each angle must measure less than 180°.
10. If all angles equaled 90°, then all angles would be right angles.

Directions Tell which of Euclid's Postulates makes each statement true.

11. Use line segments to connect the points to create a hexagon.

12. Draw a circle with center U and distance a.

13. Draw a line that is parallel to line d and passes through point X.

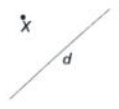

14. All of the angles in this rectangle are equal.

15. Extend the lines r and s to meet at a point in space.

16. Extend the line segment to three times its length.

 Geometry

Name Date Period | *Chapter 2 Mastery Test B page 2*

Chapter 2 Mastery Test B, continued

Directions Name the axiom that gives the reason for each step.

17. $w + 7 = 3$; $-7 = -7$; $w = -4$
18. $t - 2 = 14$; $+2 = +2$; $t = 16$
19. $g - 14 = x$; $+14 = +14$; $g = x + 14$
20. $a + b = c$; $-b = -b$; $a = c - b$

Directions Answer each question. Tell which axiom you used.

21. The $m\angle t = m\angle u$ and $m\angle v = m\angle w$. What do you know about $m\angle t + m\angle v$ and $m\angle u + m\angle w$? Why?

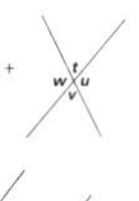

22. Which angles are less than 90°? Why?

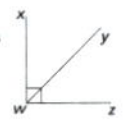

23. The $m\angle e = m\angle f$ and $m\angle f = m\angle g$. Does $m\angle e = m\angle g$? Why?

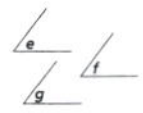

24. Two triangles each have sides $\frac{1}{2}$ inch in length. What could you say about the two triangles? Why?

Directions Use the vertical angle theorem to find the measures of angles. Write the letter.

25. $\angle m$ ____ $\angle n$ ____ $\angle o$ ____

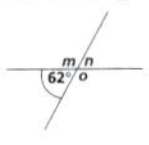

A. 62° B. 52° C. 128° D. 118°

26. $\angle m$ ____ $\angle n$ ____ $\angle o$ ____

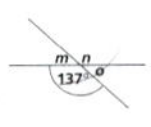

A. 137° B. 147° C. 43° D. 33°

27. $\angle m$ ____ $\angle n$ ____ $\angle o$ ____

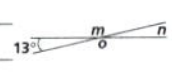

A. 18° B. 13° C. 147° D. 167°

28. $\angle m$ ____ $\angle n$ ____ $\angle o$ ____

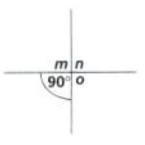

A. 80° B. 90° C. 95° D. 85°

29. $\angle m$ ____ $\angle n$ ____ $\angle o$ ____

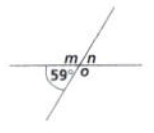

A. 111° B. 121° C. 59° D. 69°

30. $\angle m$ ____ $\angle n$ ____ $\angle o$ ____

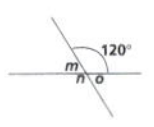

A. 120° B. 50° C. 130° D. 60°

 Geometry

TRL

Chapter 2 Mastery Test B

PROOF

Statement	Reason
16. Lines ℓ and m intersect. ∠b and ∠c are vertical angles.	16. Given.
17. m∠a + m∠c = 180°	17. ∠a and ∠b are adjacent on line m and are supplementary.
18. m∠b + m∠d = 180°	18. ∠d and ∠b are adjacent on ℓ and are supplementary.
19. m∠a + m∠c = m∠a + m∠b	19. Axiom 1, substitution, and steps 2 and 3.
20. ∴ m∠c = m∠b	20. Axiom 3. If equals are subtracted from equals, the differences are equal.

Find the measure of each angle. Write the letter of the correct measure.

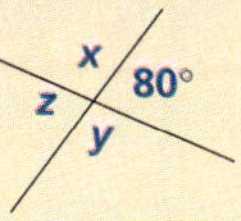

Example: ∠z A. 80° B. 100° C. 180° D. 50°

Solution: A, the 80° angle and z are vertical angles, vertical angles are equal, so z = 80°

Use the illustration at the right for problems 21–23.

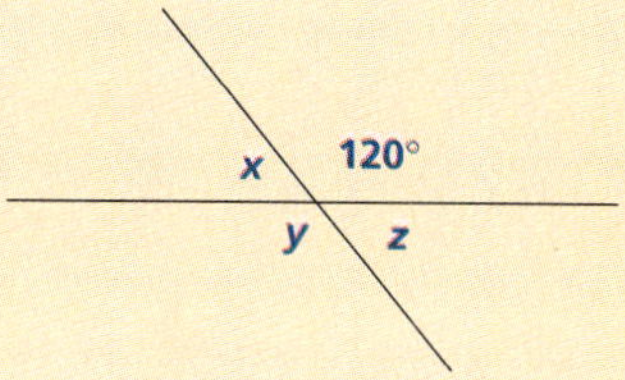

21. ∠x A

A. 60° **B.** 120° **C.** 180° **D.** 30°

22. ∠y B

A. 60° **B.** 120° **C.** 180° **D.** 30°

23. ∠z A

A. 60° **B.** 120° **C.** 180° **D.** 30°

Use the illustration at the right for problems 24–25.

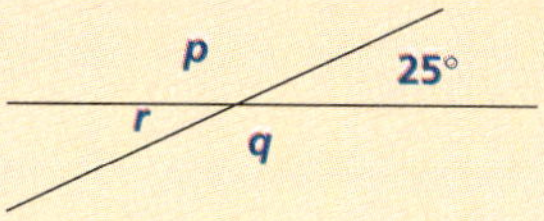

24. ∠p D

A. 25° **B.** 65° **C.** 180° **D.** 155°

25. ∠r A

A. 25° **B.** 65° **C.** 180° **D.** 155°

TEST-TAKING TIP

When answering multiple-choice questions, first identify the answers you know are incorrect.

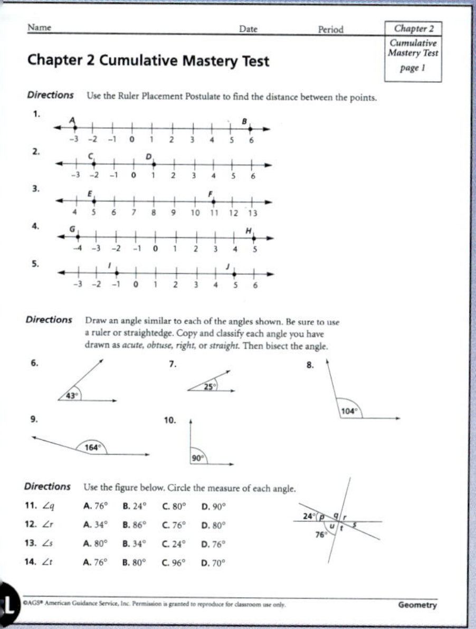

Name Date Period

Chapter 2 Cumulative Mastery Test page 1

Chapter 2 Cumulative Mastery Test

Directions Use the Ruler Placement Postulate to find the distance between the points.

1. A at −3, B at 6
2. C at −2, D at 1
3. E at 5, F at 11
4. G at −4, H at 5
5. I at −1, J at 5

Directions Draw an angle similar to each of the angles shown. Be sure to use a ruler or straightedge. Copy and classify each angle you have drawn as *acute, obtuse, right,* or *straight.* Then bisect the angle.

6. 43°
7. 25°
8. 104°
9. 164°
10. 90°

Directions Use the figure below. Circle the measure of each angle.

11. ∠q A. 76° B. 24° C. 80° D. 90°
12. ∠r A. 34° B. 86° C. 76° D. 80°
13. ∠s A. 80° B. 34° C. 24° D. 76°
14. ∠t A. 76° B. 80° C. 96° D. 70°

©AGS® American Guidance Service, Inc. Permission is granted to reproduce for classroom use only. Geometry

TRL

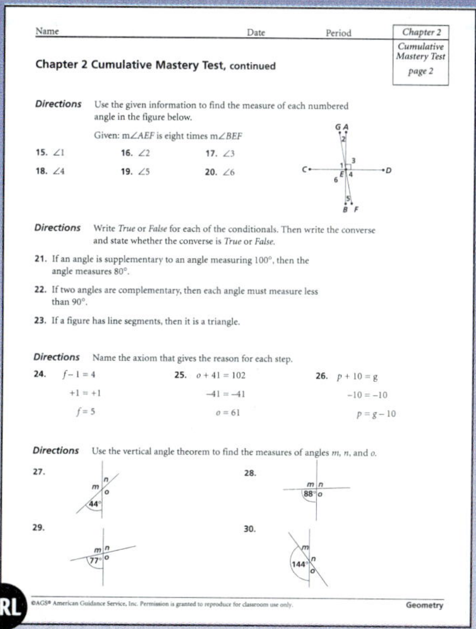

Name Date Period

Chapter 2 Cumulative Mastery Test page 2

Chapter 2 Cumulative Mastery Test, continued

Directions Use the given information to find the measure of each numbered angle in the figure below.

Given: m∠AEF is eight times m∠BEF

15. ∠1 16. ∠2 17. ∠3
18. ∠4 19. ∠5 20. ∠6

Directions Write *True* or *False* for each of the conditionals. Then write the converse and state whether the converse is *True* or *False.*

21. If an angle is supplementary to an angle measuring 100°, then the angle measures 80°.
22. If two angles are complementary, then each angle must measure less than 90°.
23. If a figure has line segments, then it is a triangle.

Directions Name the axiom that gives the reason for each step.

24. $f - 1 = 4$, $+1 = +1$, $f = 5$
25. $o + 41 = 102$, $-41 = -41$, $o = 61$
26. $p + 10 = g$, $-10 = -10$, $p = g - 10$

Directions Use the vertical angle theorem to find the measures of angles *m, n,* and *o.*

27. 44°
28. 88°
29. 77°
30. 144°

©AGS® American Guidance Service, Inc. Permission is granted to reproduce for classroom use only. Geometry

TRL

Chapter 2 Cumulative Mastery Test

Chapter 3

Planning Guide

Parallel Lines and Transversals

	Student Text Lesson						
	Student Pages	Vocabulary	Proof	Practice Exercises	Problem Solving	Calculator Practice	Solutions Key
Lesson 1 Parallel Lines	62–64	✔		✔			✔
Lesson 2 Transversals	65–67	✔		✔			✔
Lesson 3 Theorems About Parallel Lines	68–72	✔	✔	✔	72	72	✔
Lesson 4 Constructions and Problem Solving	73–76			✔			✔
Lesson 5 Quadrilaterals and Parallels	77–82	✔	✔	✔			✔
Lesson 6 Proving Lines Parallel	83–89		✔	✔	89		✔
Application Geometric Quilt Shapes	90			✔			✔

Chapter Activities

Teacher's Resource Library
Geometry on the Job 3: Drafting
Community Connection 3: Construction Sites

Teacher's Edition
Chapter 3 Project

Assessment Options

Student Text
Chapter 3 Review

Teacher's Resource Library
Chapter 3 Mastery Tests A and B
Chapter 3 Cumulative Mastery Test

Teacher's Edition
Chapter 3 Alternative Assessment

Student Text Features						Teaching Strategies								Learning Styles						Teacher's Resource Library		
Algebra Connection	Constructions	Experiencing Geometry	Technology Connection	Try This	Writing About Mathematics	Warm-Up Activity	Common Error	Algebra Connection	Mental Math	Manipulatives	Calculator	Applications	Group Problem Solving	Auditory/Verbal	Visual/Spatial	Body/Kinesthetic	Logical/Mathematical	Interpersonal/Group Learning	LEP/ESL	Workbook Activities	Additional Practice	Self-Study Guide
			64	64	63	62						63				63	64		64	20, 21	12	✔
	67				67	65	67					66	67		67	66	67		66	22, 23	13	✔
72						68	71	72			72	70	71		70	69			71	24, 25	14	✔
74	73, 76					73		74	75			75	76	75	76			76		26	15	✔
		82				77	80			78		81	82	79, 81	78	80, 82	82		79	27	16	✔
	83–89				86	83	88		85			84, 85	89	89	88		86		87	28	17	✔
						90						90			90					29		✔

Manipulatives

Overhead Pattern Blocks

Manipulite Pattern Blocks

Chapter at a Glance

Chapter 3: Parallel Lines and Transversals
pages 60–93

(Answer Keys for the Teacher's Resource Library begin on page 484 of this Teacher's Edition.)

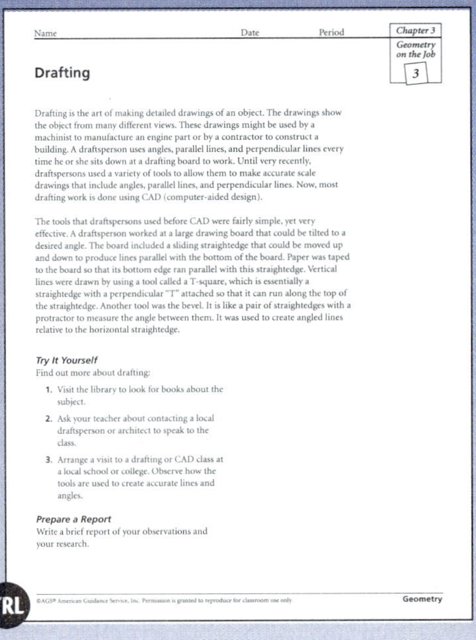

Name Date Period — Chapter 3 *Geometry on the Job* 3

Drafting

Drafting is the art of making detailed drawings of an object. The drawings show the object from many different views. These drawings might be used by a machinist to manufacture an engine part or by a contractor to construct a building. A draftsperson uses angles, parallel lines, and perpendicular lines every time he or she sits down at a drafting board to work. Until very recently, draftspersons used a variety of tools to allow them to make accurate scale drawings that include angles, parallel lines, and perpendicular lines. Now, most drafting work is done using CAD (computer-aided design).

The tools that draftspersons used before CAD were fairly simple, yet very effective. A draftsperson worked at a large drawing board that could be tilted to a desired angle. The board included a sliding straightedge that could be moved up and down to produce lines parallel with the bottom of the board. Paper was taped to the board so that its bottom edge ran parallel with this straightedge. Vertical lines were drawn by using a tool called a T-square, which is essentially a straightedge with a perpendicular "T" attached so that it can run along the top of the straightedge. Another tool was the bevel. It is like a pair of straightedges with a protractor to measure the angle between them. It was used to create angled lines relative to the horizontal straightedge.

Try It Yourself
Find out more about drafting:

1. Visit the library to look for books about the subject.
2. Ask your teacher about contacting a local draftsperson or architect to speak to the class.
3. Arrange a visit to a drafting or CAD class at a local school or college. Observe how the tools are used to create accurate lines and angles.

Prepare a Report
Write a brief report of your observations and your research.

TRL

 Geometry

Geometry on the Job 3

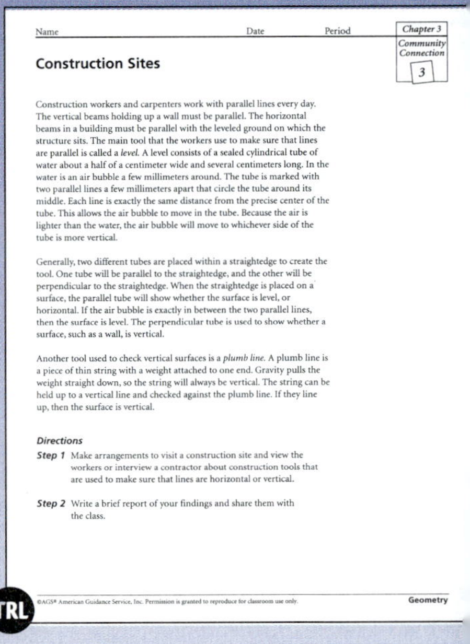

Name Date Period — Chapter 3 *Community Connection* 3

Construction Sites

Construction workers and carpenters work with parallel lines every day. The vertical beams holding up a wall must be parallel. The horizontal beams in a building must be parallel with the leveled ground on which the structure sits. The main tool that the workers use to make sure that lines are parallel is called a *level.* A level consists of a sealed cylindrical tube of water about a half of a centimeter wide and several centimeters long. In the water is an air bubble a few millimeters around. The tube is marked with two parallel lines a few millimeters apart that circle the tube around its middle. Each line is exactly the same distance from the precise center of the tube. This allows the air bubble to move in the tube. Because the air is lighter than the water, the air bubble will move to whichever side of the tube is more vertical.

Generally, two different tubes are placed within a straightedge to create the tool. One tube will be parallel to the straightedge, and the other will be perpendicular to the straightedge. When the straightedge is placed on a surface, the parallel tube will show whether the surface is level, or horizontal. If the air bubble is exactly in between the two parallel lines, then the surface is level. The perpendicular tube is used to show whether a surface, such as a wall, is vertical.

Another tool used to check vertical surfaces is a *plumb line.* A plumb line is a piece of thin string with a weight attached to one end. Gravity pulls the weight straight down, so the string will always be vertical. The string can be held up to a vertical line and checked against the plumb line. If they line up, then the surface is vertical.

Directions

Step 1 Make arrangements to visit a construction site and view the workers or interview a contractor about construction tools that are used to make sure that lines are horizontal or vertical.

Step 2 Write a brief report of your findings and share them with the class.

TRL

 Geometry

Community Connection 3

Chapter 3

Parallel Lines and Transversals

Rows upon rows of tulips color a quiet country scene. Only roads paralleling and crossing one another break the colorful fields of flowers. Many farmers grow crops in rectangular fields. The shape of the fields are defined by parallel lines.

In the United States, the Land Ordinance of 1785 set up a system for settling and developing land in square parcels. Today, as a result, people in airplanes flying over much of the United States will see farm fields and cities divided into grids. These grids are marked by parallel lines and transversals, or lines that cross over two or more parallel lines. Geometry definitely has played a role in the development and use of land.

In Chapter 3, you will learn how to identify and measure angles in parallels and quadrilaterals.

Goals for Learning

- To distinguish parallel, intersecting, and skew lines
- To identify and name angles formed by transversals crossing parallel lines
- To use theorems to compare and find the measure of angles and identify parallel lines
- To use postulates and theorems to construct parallel lines
- To define quadrilaterals and parallels and apply theorems to them

61

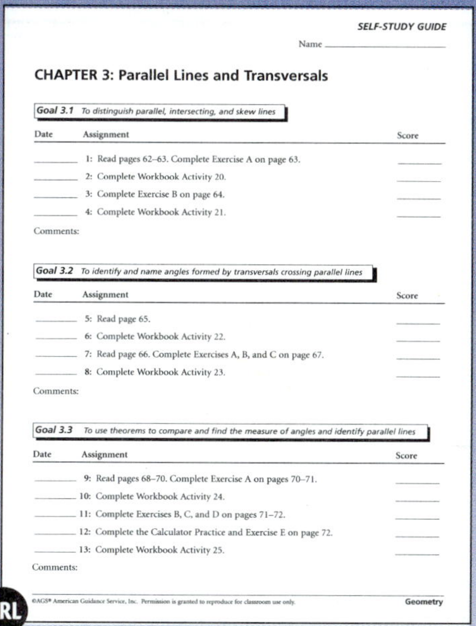

SELF-STUDY GUIDE

Name ____________

CHAPTER 3: Parallel Lines and Transversals

Goal 3.1 To distinguish parallel, intersecting, and skew lines

Date	Assignment	Score
______	1: Read pages 62–63. Complete Exercise A on page 63.	______
______	2: Complete Workbook Activity 20.	______
______	3: Complete Exercise B on page 64.	______
______	4: Complete Workbook Activity 21.	______

Comments:

Goal 3.2 To identify and name angles formed by transversals crossing parallel lines

Date	Assignment	Score
______	5: Read page 65.	______
______	6: Complete Workbook Activity 22.	______
______	7: Read page 66. Complete Exercises A, B, and C on page 67.	______
______	8: Complete Workbook Activity 23.	______

Comments:

Goal 3.3 To use theorems to compare and find the measure of angles and identify parallel lines

Date	Assignment	Score
______	9: Read pages 68–70. Complete Exercise A on pages 70–71.	______
______	10: Complete Workbook Activity 24.	______
______	11: Complete Exercises B, C, and D on pages 71–72.	______
______	12: Complete the Calculator Practice and Exercise E on page 72.	______
______	13: Complete Workbook Activity 25.	______

Comments:

 Geometry

TRL

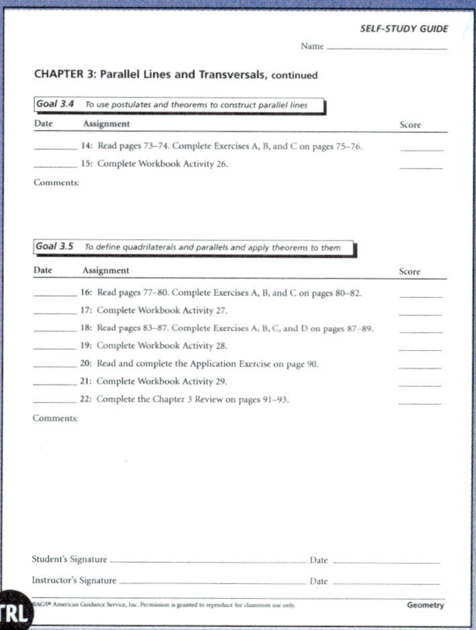

SELF-STUDY GUIDE

Name ____________

CHAPTER 3: Parallel Lines and Transversals, continued

Goal 3.4 To use postulates and theorems to construct parallel lines

Date	Assignment	Score
______	14: Read pages 73–74. Complete Exercises A, B, and C on pages 75–76.	______
______	15: Complete Workbook Activity 26.	______

Comments:

Goal 3.5 To define quadrilaterals and parallels and apply theorems to them

Date	Assignment	Score
______	16: Read pages 77–80. Complete Exercises A, B, and C on pages 80–82.	______
______	17: Complete Workbook Activity 27.	______
______	18: Read pages 83–87. Complete Exercises A, B, C, and D on pages 87–89.	______
______	19: Complete Workbook Activity 28.	______
______	20: Read and complete the Application Exercise on page 90.	______
______	21: Complete Workbook Activity 29.	______
______	22: Complete the Chapter 3 Review on pages 91–93.	______

Comments:

Student's Signature ____________ Date ____________

Instructor's Signature ____________ Date ____________

 Geometry

TRL

Chapter 3 Self-Study Guide

Introducing the Chapter

Use the information on these pages to help students think about ways parallels and transversals are practically applied and seen in everyday constructions and landscapes. Reinforce the use of parallels in platting and dividing land parcels. If possible, display a plat map of the school property or a blueprint of the school building. Help students identify parallel lines and transversals on the map or blueprint. Introduce the Goals for Learning, defining terms as necessary.

Chapter Project

Ask students to work in groups to plan and construct a model business complex or downtown area for a community. Suggest that students use parallel lines in their designs for building exteriors and parallel lines and transversals for roadways and grounds around the complex or downtown area. Encourage them to add details to their models such as landscaping vehicles on the roads, roadway signs, and so on. In addition to the model, students should provide a list that identifies the materials used on the project and notes the buildings, roadways, and details that include parallel lines and transversals.

As an alternative project, students could prepare a plat drawing of a farm. The drawings should include the boundaries of the farmland, the fields, farmhouse, driveways, garden areas, and other structures on the land. With the drawing, they should provide a list of buildings, land areas, and details that include parallel lines and transversals.

Lesson at a Glance

Chapter 3 Lesson 1

Overview This lesson defines parallel, skew, and coplanar lines and explains how to distinguish parallel and skew lines.

Objective

- To identify parallel lines.
- To use symbols to describe pairs of lines.

Student Pages 62–64

Teacher's Resource Library TRL

Workbook Activities 20–21

Additional Practice 12

Mathematics Vocabulary

coplanar
parallel lines
skew lines

1. Warm-Up Activity

Review what students know about parallel lines and planes. Ask them to identify examples of parallel lines and planes in the classroom. Remind them that the objects identified in the classroom are actually line segments, which represent a portion of a line.

2. Teaching the Lesson

Be sure students can distinguish coplanar and non-coplanar lines. Use a box as a reference and have volunteers identify various coplanar and non-coplanar lines as parallel or skew lines. Help students see that there are infinitely many lines in any plane. Also, be certain that students understand that the lines and planes represented by the sides and edges of the box extend infinitely. Ask, "How are parallel and skew lines similar? How are they different?" (They are similar in that they never intersect. They are different because parallel lines are coplanar and skew lines are not coplanar.)

Lesson 1 Parallel Lines

Coplanar
Lines or points in the same plane

Parallel lines ||
Lines in the same plane that never meet

Skew lines
Lines in space that are not coplanar and never meet

Look for parallel lines. You'll see them everywhere. Parallel lines are in the same plane and never meet.

Parallel lines are all around us. Look at a window and you will see that the sides are parallel to one another. The top and the sill are parallel to one another as well.

Parallel lines are lines that do not meet or intersect. Another condition for parallel lines is that the lines must be in the same plane. Lines in the same plane are called **coplanar.** Look at the picture of the shoe box to the right. Notice that the bottom edge and side edge do not meet. These lines are not in the same plane. Lines in space that never meet and are not in the same plane are **skew lines.**

Remember, a *plane* is one of geometry's undefined terms. You can think of a page from this book or a desktop as part of a plane.

These bottom and side edges do not meet. The lines are not in the same plane. The lines are skew. What other lines or edges on the shoe box are skew?

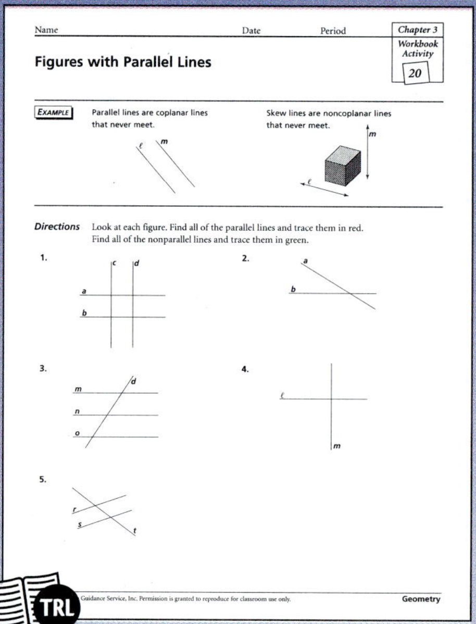

Name Date Period | *Chapter 3 Workbook Activity* 20

Figures with Parallel Lines

EXAMPLE Parallel lines are coplanar lines that never meet. Skew lines are noncoplanar lines that never meet.

Directions Look at each figure. Find all of the parallel lines and trace them in red. Find all of the nonparallel lines and trace them in green.

1.
2.
3.
4.
5.

Guidance Service, Inc. Permission is granted to reproduce for classroom use only. **Geometry**

Workbook Activity 20

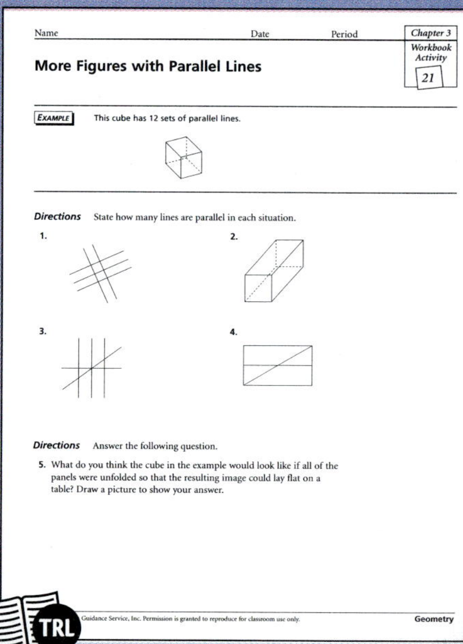

Name Date Period | *Chapter 3 Workbook Activity* 21

More Figures with Parallel Lines

EXAMPLE This cube has 12 sets of parallel lines.

Directions State how many lines are parallel in each situation.

1.
2.
3.
4.

Directions Answer the following question.

5. What do you think the cube in the example would look like if all of the panels were unfolded so that the resulting image could lay flat on a table? Draw a picture to show your answer.

Guidance Service, Inc. Permission is granted to reproduce for classroom use only. **Geometry**

Workbook Activity 21

Parallel lines are coplanar lines that never meet. The symbol for parallel lines is ||.

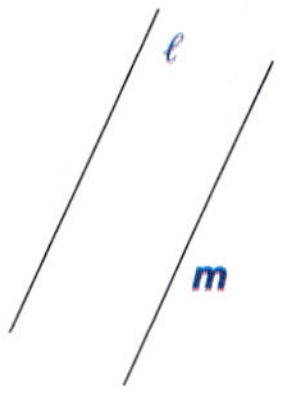

ℓ is parallel to m
$\ell \parallel m$

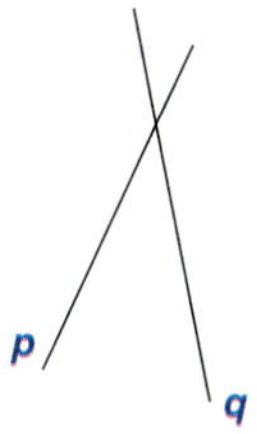

p is not parallel to q
$p \nparallel q$

Parallel

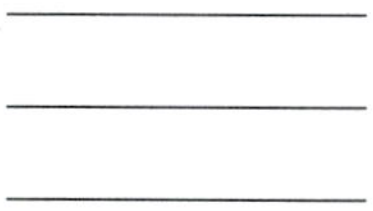

in the same plane

Not Parallel

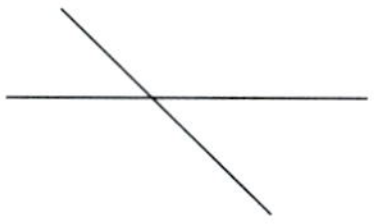

intersect in the same plane

Skew

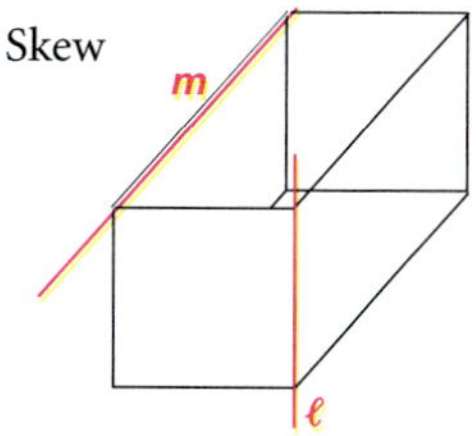

ℓ and m are skew lines not in the same plane

Exercise A Use the figure below. Write *T* if you think the statement is *true*. Write *F* if you think the statement is *false*.

Selected Answers and Solutions begin on page 420.

1. $\overline{AE} \parallel \overline{BD}$ T
2. $\overline{BE} \parallel \overline{ED}$ F
3. $\overline{FC} \nparallel \overline{BE}$ T
4. $\overline{AB} \parallel \overline{ED}$ T
5. $\overline{FC} \parallel \overline{ED}$ T
6. $\overline{AB} \parallel \overline{FC}$ T
7. $\overline{AB} \nparallel \overline{BE}$ T
8. $\overline{AF} \nparallel \overline{FE}$ T
9. $\overline{AF} \parallel \overline{CD}$ T
10. $\overline{AB} \parallel \overline{FC} \parallel \overline{ED}$ T

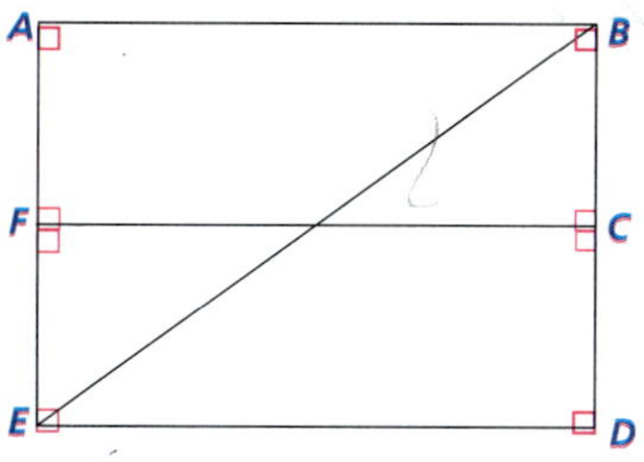

Writing About Mathematics

Describe a pair of skew lines you find in the classroom.

3. Reinforce and Extend

Application

At Home

Have students think of games that they play with family members, either at home or elsewhere, that make use of parallel lines. For example, most court and field games, such as tennis, basketball, and soccer, use parallel lines to form the boundaries of the playing field. Also, many board games use parallel lines, as do playing cards. Tell students to think of as many games as they can and provide a brief written description or sketch to describe how parallel lines are used in each game.

Learning Styles

Body/Kinesthetic Ask pairs of students to model parallel, intersecting, and skew lines in various ways around the room. For example, the pair could stand against the same wall parallel to each other to model parallel lines, or one could stand leaning slightly to model intersecting lines. Ask for volunteers to display their line models to the class.

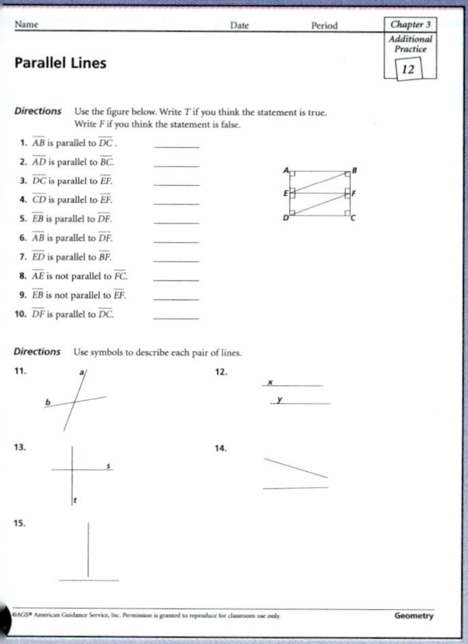

Name Date Period

Chapter 3 Additional Practice 12

Parallel Lines

Directions Use the figure below. Write *T* if you think the statement is true. Write *F* if you think the statement is false.

1. $\overline{AB}$ is parallel to $\overline{DC}$. ______
2. $\overline{AD}$ is parallel to $\overline{BC}$. ______
3. $\overline{DC}$ is parallel to $\overline{EF}$. ______
4. $\overline{CD}$ is parallel to $\overline{EF}$. ______
5. $\overline{EB}$ is parallel to $\overline{DF}$. ______
6. $\overline{AB}$ is parallel to $\overline{DF}$. ______
7. $\overline{ED}$ is parallel to $\overline{BF}$. ______
8. $\overline{AE}$ is not parallel to $\overline{FC}$. ______
9. $\overline{EB}$ is not parallel to $\overline{EF}$. ______
10. $\overline{DF}$ is parallel to $\overline{DC}$. ______

Directions Use symbols to describe each pair of lines.

11. a b

12. x y

13. s t

14.

15.

 Geometry

Additional Practice 12

Try This...

Remind students that to distinguish between parallel and skew lines they must first determine whether the lines are coplanar or not. Many lines in the photo do not intersect and are not coplanar; therefore, they are skew lines. Lines that intersect in the same plane are not parallel.

Learning Styles

LEP/ESL With a student more proficient in English, LEP/ESL students can make a set of nine cards. On three of the cards, they should write the terms *parallel lines, skew lines,* and *intersecting lines.* On three other cards, they should draw a sketch that illustrates each of the terms. On the remaining three cards, they should write the definitions of the three terms. Have students mix the cards and place them face down in a pile on a desk. They take turns turning over a card and placing it in one of three piles, so that the three cards for each term are together in a single pile. They can repeat the process as needed to review the terms.

Learning Styles

Logical/Mathematical It may help some students to make an organizational chart or hierarchy for the following terms used in this lesson: *lines, parallel lines, intersecting lines, skew lines, coplanar, non-coplanar.* A sample chart is shown here:

lines

coplanar → parallel lines; coplanar ↘ intersecting lines

non-coplanar → skew lines

coplanar		non-coplanar
↓	↘	↓
parallel lines	intersecting lines	skew lines

Invite volunteers to share their charts with the class.

Exercise B Use symbols $\parallel$ or $\nparallel$ to describe each pair of lines.

11.

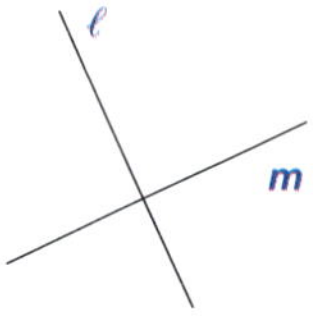

$\ell \nparallel m$

12.

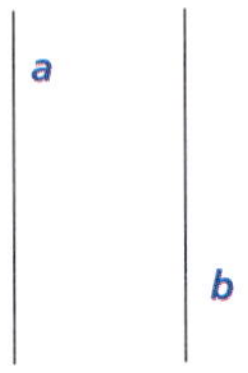

$a \parallel b$

13.

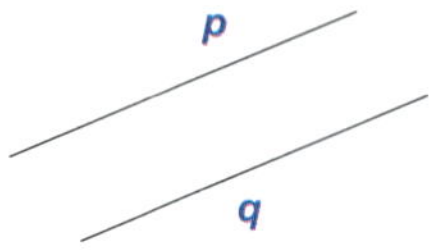

$p \parallel q$

14.

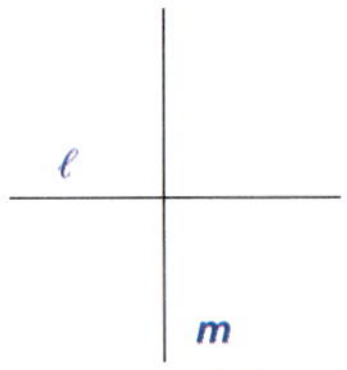

$\ell \nparallel m$

15.

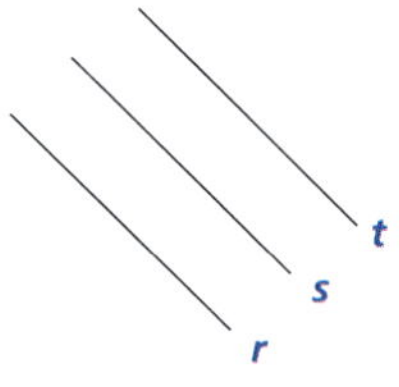

$r \parallel s \parallel t$

Technology Connection

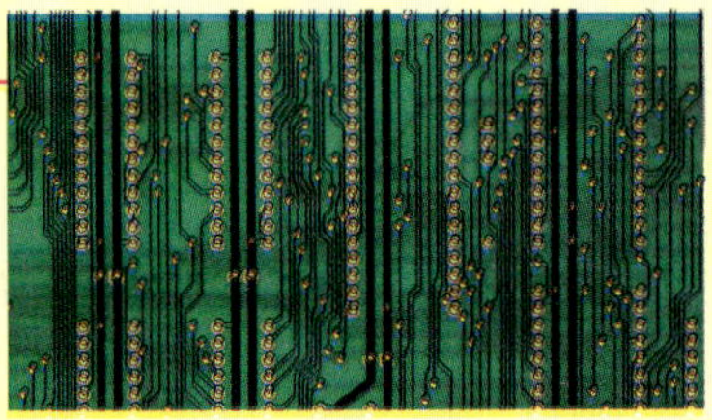

You can see many parallel lines on this circuit board. Electricity travels along these lines. People and machines making this board must be careful that the lines stay straight and parallel so that the circuits are accurate. One way to do this is to make a grid that covers the entire circuit board. By keeping the angle at which the circuits cross the grid equal, the manufacturer can be sure the lines are parallel. What other way might the manufacturer use to be sure the lines are parallel?

Try This...

Are these lines parallel or skew? Why?

Lesson 2 Transversals

When one line is crossed by a second line, four angles are formed.

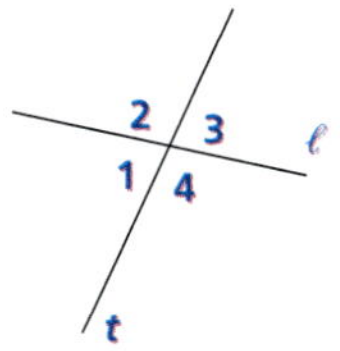

If two parallel lines are crossed by a third line, eight angles are formed. The line that crosses the parallel lines is a **transversal.** The eight angles formed have names that identify their location.

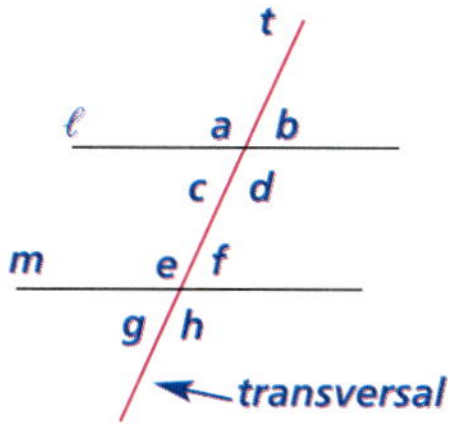

▶ **Exterior Angles**—angles outside the parallel lines ℓ and m

There are four exterior angles—a, b, g, and h

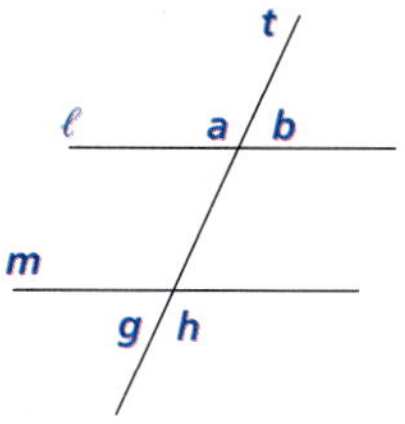

Alternate exterior angles
Pairs of exterior angles on opposite sides of the transversal

Alternate interior angles
Pairs of interior angles on opposite sides of the transversal

Corresponding angles
Pairs of angles in similar positions

Exterior angle
Angle outside the parallel lines

Interior angle
Angle inside the parallel lines

Transversal
A line that crosses parallel lines

Geometry Around Us

Here the fence pickets are parallel to each other and the crossbar is a transversal.

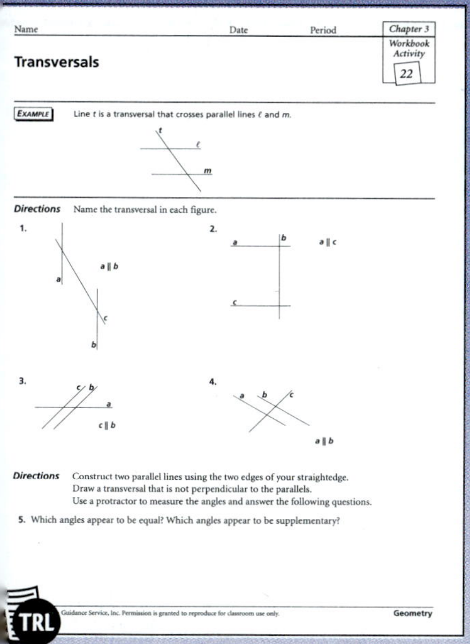

Name Date Period | *Chapter 3 Workbook Activity* 22

Transversals

EXAMPLE Line t is a transversal that crosses parallel lines ℓ and m.

Directions Name the transversal in each figure.

1. $a \parallel b$
2. $a \parallel c$
3. $c \parallel b$
4. $a \parallel b$

Directions Construct two parallel lines using the two edges of your straightedge.
Draw a transversal that is not perpendicular to the parallels.
Use a protractor to measure the angles and answer the following questions.

5. Which angles appear to be equal? Which angles appear to be supplementary?

 Geometry

Workbook Activity 22

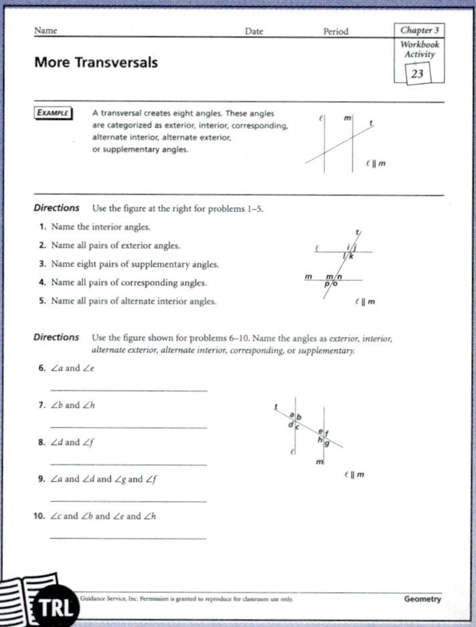

Name Date Period | *Chapter 3 Workbook Activity* 23

More Transversals

EXAMPLE A transversal creates eight angles. These angles are categorized as exterior, interior, corresponding, alternate interior, alternate exterior, or supplementary angles. $\ell \parallel m$

Directions Use the figure at the right for problems 1–5.

1. Name the interior angles.
2. Name all pairs of exterior angles.
3. Name eight pairs of supplementary angles.
4. Name all pairs of corresponding angles.
5. Name all pairs of alternate interior angles.

$\ell \parallel m$

Directions Use the figure shown for problems 6–10. Name the angles as *exterior, interior, alternate exterior, alternate interior, corresponding,* or *supplementary.*

6. $\angle a$ and $\angle e$
7. $\angle b$ and $\angle h$
8. $\angle d$ and $\angle f$
9. $\angle a$ and $\angle d$ and $\angle g$ and $\angle f$
10. $\angle c$ and $\angle b$ and $\angle e$ and $\angle h$

$\ell \parallel m$

 Geometry

Workbook Activity 23

Lesson at a Glance

Chapter 3 Lesson 2

Overview This lesson defines transversals and discusses the angles that are formed by a transversal and the lines it intersects.

Objectives

- To identify exterior, interior, corresponding, alternate interior, and supplementary angles.
- To use a protractor to measure and classify angles.

Student Pages 65–67

Teacher's Resource Library TRL

Workbook Activities 22–23

Additional Practice 13

Mathematics Vocabulary

alternate exterior angles
alternate interior angles
corresponding angles
exterior angle
interior angle
transversal

1. Warm-Up Activity

Review *acute, right, obtuse, straight, vertical, complementary,* and *supplementary angles.* Draw sketches of different angles, and have students identify them, using the terms above. Also, write statements, such as $m\angle a = 78°$ or $m\angle x + m\angle z = 180°$ and have students identify angles by their measures.

2. Teaching the Lesson

Be sure students can distinguish interior and exterior angles and alternate-side and same-side angles. Draw sketches of parallel lines and transversals with one pair of each type of angle shaded on separate sketches. Compare and contrast the position of the different angles formed.

3. Reinforce and Extend

Learning Styles

LEP/ESL Success in this and subsequent lessons depends on students' ability to distinguish the different pairs of angles formed by a transversal. Encourage a student proficient in English to help a student with limited English proficiency make flashcards with the name and definition of a pair of angles on one side and a sketch of the angles shaded on the other side. One student can look at the sketch and name the angles, and the other student can read the name and definition.

Learning Styles

Body/Kinesthetic Suggest that students make three-dimensional models of parallel lines with transversals. They can paste their models on posterboard and label the angles. Suggest that students use such materials as drinking straws, craft sticks, and pipe cleaners to make their models.

Application

At Home
Have students find an object at home that models a transversal intersecting two or more lines, such as a rung and the legs of a chair. Ask students to make a simple sketch of the object, highlighting the transversal and the lines it crosses, and then name the various pairs of angles that are formed. Also, have students estimate the measures of the different angles both in the actual object and in their drawing to check the accuracy of their sketch. Students should also identify pairs of angles that should be supplementary.

▶ **Interior Angles**—angles inside the parallel lines ℓ and m

There are four interior angles—c, d, e, and f.

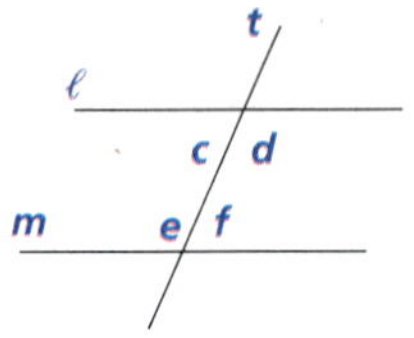

▶ **Corresponding Angles**—pairs of angles that are in similar positions

There are four pairs of corresponding angles.

a and e c and g
b and f d and h

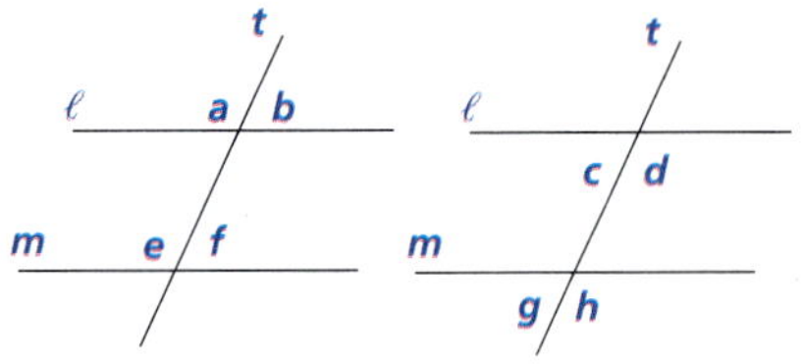

▶ **Alternate Interior Angles**—pairs of interior angles on opposite sides of the transversal

Angles c and f are alternate interior angles, so are angles d and e.

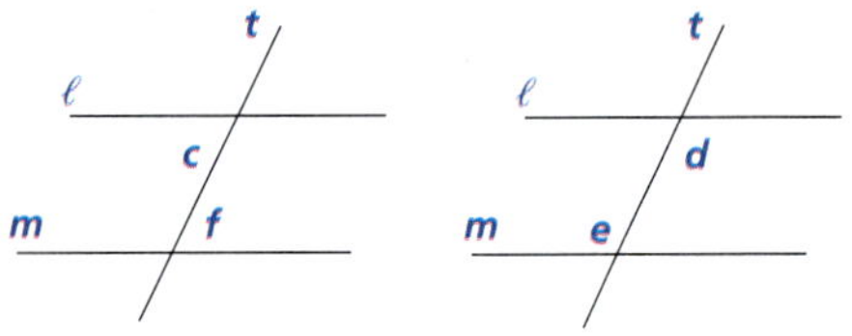

▶ **Alternate Exterior Angles**—pairs of exterior angles on opposite sides of the transversal

Angles a and h are alternate exterior angles, so are angles b and g.

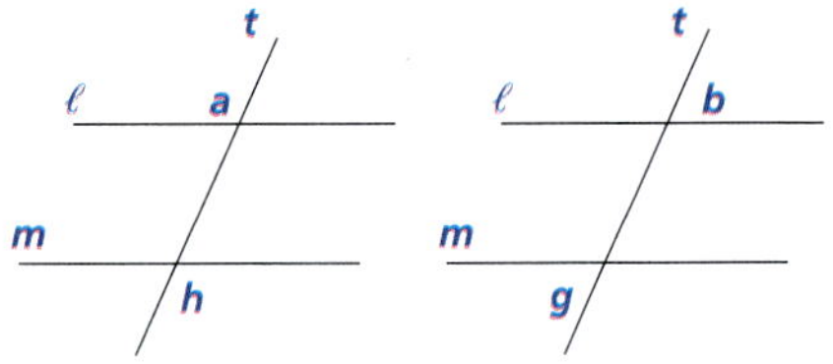

Example

Look closely at the diagram and you can see that adjacent angles whose exterior sides are in a straight line are supplementary angles.

There are eight pairs of supplementary angles: *a* and *b*, *b* and *d*, *a* and *c*, *c* and *d*, *e* and *f*, *f* and *h*, *e* and *g*, *g* and *h*.

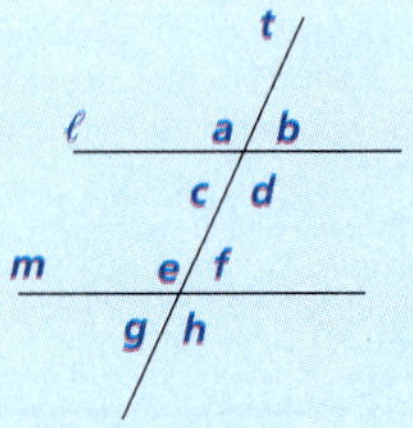

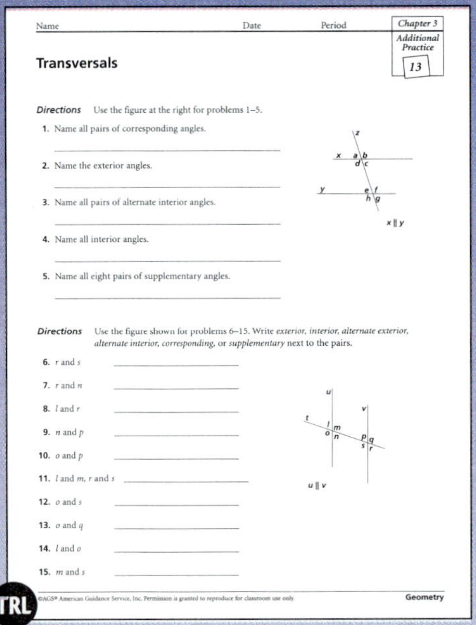

Name Date Period

Chapter 3 Additional Practice 13

Transversals

Directions Use the figure at the right for problems 1–5.

1. Name all pairs of corresponding angles.
2. Name the exterior angles.
3. Name all pairs of alternate interior angles.
4. Name all interior angles.
5. Name all eight pairs of supplementary angles.

Directions Use the figure shown for problems 6–15. Write *exterior, interior, alternate exterior, alternate interior, corresponding,* or *supplementary* next to the pairs.

6. r and s
7. r and n
8. l and r
9. n and p
10. o and p
11. l and m, r and s
12. o and s
13. o and q
14. l and o
15. m and s

 Geometry

TRL

Additional Practice 13

Exercise A Use the figure at the right for problems 1–5.

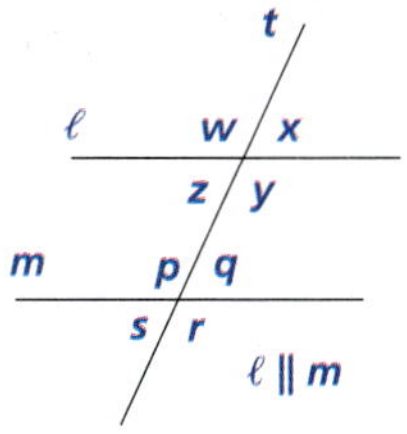

1. Name the exterior angles.
   w, x, s, r
2. Name the interior angles.
   z, y, p, q
3. Name all pairs of corresponding angles.
   w and p, x and q, z and s, y and r
4. Name all pairs of alternate interior angles.
   z and q, y and p
5. Name eight pairs of supplementary angles.
   w and x, z and y, w and z, x and y, p and q, s and r, p and s, q and r

Exercise B Use the figure shown for problems 6–15.
Name the angles as *exterior, interior, alternate exterior, alternate interior, corresponding,* or *supplementary.*

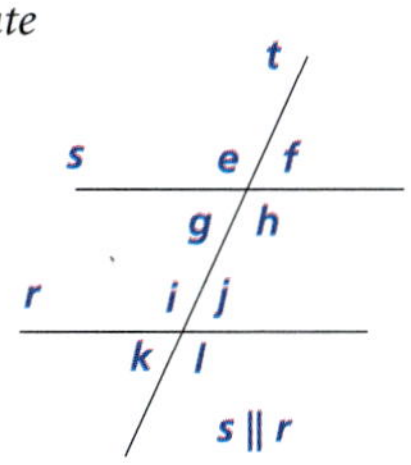

6. f and k
   alternate exterior
7. k and ℓ, e and f
   exterior
8. g and j
   alternate interior
9. g and k
   corresponding
10. h and i, j and g
    alternate interior
11. e and f
    supplementary
12. j and ℓ
    supplementary
13. h and ℓ
    corresponding
14. i and h
    alternate interior
15. i and j
    supplementary

Construction

Exercise C Construct two parallel lines using the two edges of your straightedge. Draw a transversal that is not perpendicular to the parallels. Use a protractor to measure the angles. Which of the following angles appear to be equal? Write *yes* or *no.*

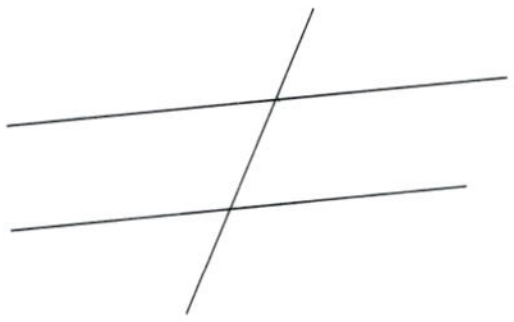

16. Corresponding angles Yes
17. Supplementary angles No
18. Alternate interior angles Yes
19. Alternate exterior angles Yes
20. Interior angles on the same side of the transversal No

Writing About Mathematics

Discuss what you think will happen to the eight angles if the transversal is perpendicular to the two lines?

Learning Styles

Logical/Mathematical Some students may benefit from writing the descriptions of the different pairs of angles as conditionals. Show students an example of how to describe alternate interior angles using the If . . . then . . . format. (If a pair of interior angles are not adjacent and are on opposite sides of the transversal, then they are alternate interior angles.) Encourage students to write descriptions in this format for other angles described in this lesson.

Learning Styles

Visual/Spatial Have students draw several sets of parallel lines cut by a transversal. Tell them to label and shade a different pair of angles on each set of lines and write the name of the angles below the picture. Encourage students to use this visual reference to complete Exercises A and B.

Common Error

When measuring angles with a protractor, students may read the measure from the wrong side of the scale. For example, they may read 150° on one scale when measuring a 30° angle. Remind students to always find 0° on the scale to start measuring and then to follow that scale up to the other side of the angle to find the measure.

Group Problem Solving

Ask groups of three or four students to draw five sets of lines being cut by a transversal. Instruct them to draw the first four sets with nonparallel lines crossed by a transversal, with each set getting closer to being parallel. The fifth set should be parallel lines crossed by a transversal. Ask them to number the sets 1–5. Have students measure the same pair of alternate interior angles to find the sum of the same-side exterior angles in each set of lines and to record their data in an organized chart. After reviewing and discussing the data in their chart, students should draw conclusions about the measures of alternate interior angles and the sum of same-side exterior angles for lines that are not parallel and lines that are parallel.

Lesson at a Glance

Chapter 3 Lesson 3

Overview This lesson illustrates how to use theorems, postulates, and definitions to support statements in a proof and to solve problems.

Objectives

- To use theorems and definitions to find the measures of angles.
- To use theorems and definitions to write a formal proof.

Student Pages 68–72

Teacher's Resource Library TRL

Workbook Activities 24–25

Additional Practice 14

Mathematics Vocabulary

indirect proof

1. Warm-Up Activity

Review the definition of supplementary angles, Euclid's Postulate 5, the definition of adjacent angles, Axiom 1, and Axiom 3 with students. Write examples on the board to illustrate each of the items and ask students to draw a conclusion about each statement and to identify the postulate, axiom, or definition that supports their conclusion. For example, to illustrate Axiom 1, write m∠a = 90°, m∠c = 90°. Students can conclude that m∠a = m∠c and identify Axiom 1 as support for their conclusion.

2. Teaching the Lesson

Define *theorem* for students. (a rule that can be established or proved by using previously established or proved premises) Explain that a theorem uses known facts and a step-by-step logical process to prove a mathematical truth. Explain that a proof for a theorem does not have a set number of steps. It flows from logical reasoning and uses whatever number of steps are needed to arrive at the conclusion.

Lesson 3 Theorems About Parallel Lines

Indirect proof

A proof that shows that the conclusion cannot be false because accepted facts would be contradicted

At the end of the last lesson you measured certain angles. In your drawing, you may have found that the corresponding angles were equal. In geometry, measurement is not enough. We need a formal proof about corresponding angles before we can say that all corresponding angles are equal. There are several theorems about which angles are equal. The first, and most important, depends on Euclid's original Parallel Postulate. (Now would be a good time to turn back to page 41 and review Euclid's Postulate 5, the Parallel Postulate.)

A statement can be either *true* or *false.* An argument, or proof, can be either *valid* or *invalid.* A proof is valid if the conclusion is true whenever the given (the hypothesis) is true.

THEOREM: If two lines are parallel, then the interior angles on the same side of the transversal are supplementary.

For this theorem, we will use a method of proof called *Proof by Contradiction,* also known as an **indirect proof.** In an indirect proof, you begin with the *opposite* of the desired conclusion. Then you work, in logical steps, to find a contradiction. When the same statement is shown to be true as well as false, it is a contradiction. If you reach a contradiction, then you know it was incorrect, or *false,* to assume the opposite of the conclusion. Therefore the conclusion must be true.

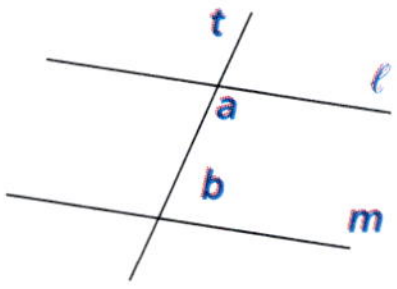

Proof by Contradiction (Indirect Proof)

To Prove: ∠a and ∠b are supplementary

PROOF

Statement	Reason
1. ℓ ∥ m, transversal t	1. Given.
2. ∠a and ∠b are not supplementary	2. Indirect proof assumption (something taken to be true without proof); the opposite of "To prove."
3. m∠a + m∠b ≠ 180°	3. Definition of not supplementary angles.
4. m∠a + m∠b < 180° or m∠a + m∠b > 180°	4. Definition of not equal means greater than or less than.
5. ℓ must intersect m	5. Euclid's Postulate 5.

This contradicts the given: ℓ ∥ m

∴ Theorem is valid and proof is complete.

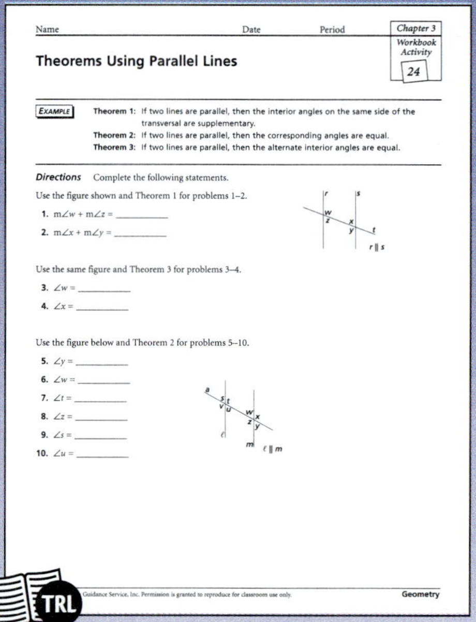

Name Date Period — Chapter 3 Workbook Activity 24

Theorems Using Parallel Lines

EXAMPLE **Theorem 1:** If two lines are parallel, then the interior angles on the same side of the transversal are supplementary.
Theorem 2: If two lines are parallel, then the corresponding angles are equal.
Theorem 3: If two lines are parallel, then the alternate interior angles are equal.

Directions Complete the following statements.

Use the figure shown and Theorem 1 for problems 1–2.

1. m∠w + m∠z = ________
2. m∠x + m∠y = ________

Use the same figure and Theorem 3 for problems 3–4.

3. ∠w = ________
4. ∠x = ________

Use the figure below and Theorem 2 for problems 5–10.

5. ∠y = ________
6. ∠w = ________
7. ∠t = ________
8. ∠z = ________
9. ∠s = ________
10. ∠u = ________

 Geometry

Workbook Activity 24

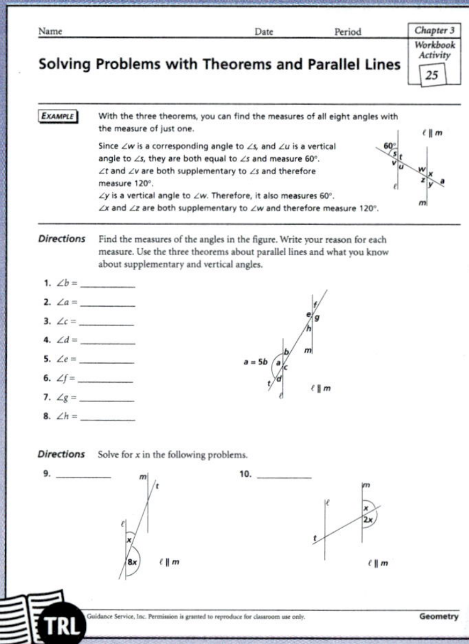

Name Date Period — Chapter 3 Workbook Activity 25

Solving Problems with Theorems and Parallel Lines

EXAMPLE With the three theorems, you can find the measures of all eight angles with the measure of just one.

Since ∠w is a corresponding angle to ∠s, and ∠u is a vertical angle to ∠s, they are both equal to ∠s and measure 60°.
∠t and ∠v are both supplementary to ∠s and therefore measure 120°.
∠y is a vertical angle to ∠w. Therefore, it also measures 60°.
∠x and ∠z are both supplementary to ∠w and therefore measure 120°.

Directions Find the measures of the angles in the figure. Write your reason for each measure. Use the three theorems about parallel lines and what you know about supplementary and vertical angles.

1. ∠b = ________
2. ∠a = ________
3. ∠c = ________
4. ∠d = ________
5. ∠e = ________
6. ∠f = ________
7. ∠g = ________
8. ∠h = ________

Directions Solve for x in the following problems.

9. ________
10. ________

 Geometry

Workbook Activity 25

What does this theorem mean? It means every time you have two parallel lines, ℓ and *m*, then you know that

$$m\angle a + m\angle c = 180°$$

and

$$m\angle b + m\angle d = 180°$$

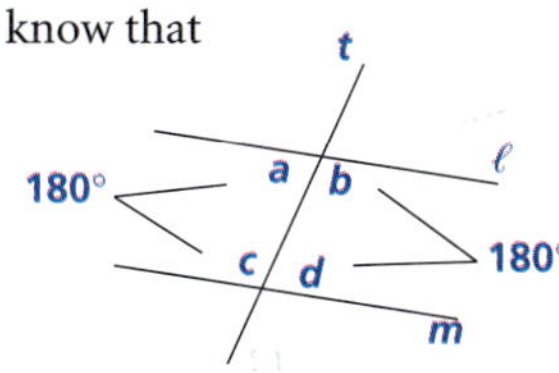

> *Remember*, ∴ is the symbol for therefore, and notes the last statement of a proof.

This theorem about supplementary angles forms the foundation for the proofs of other theorems about parallel lines.

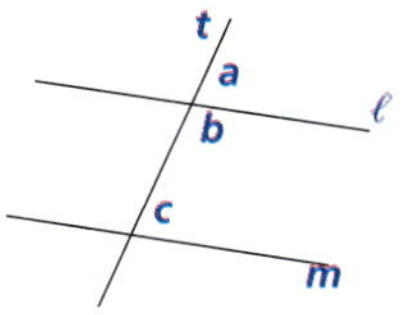

THEOREM: If two lines are parallel, then the corresponding angles are equal.

To Prove: m∠a = m∠c

PROOF

Statement	Reason
1. ℓ ∥ *m*, transversal *t*	1. Given.
2. m∠b + m∠c = 180°	2. Theorem, interior angles on the same side of the transversal are supplementary.
3. m∠a + m∠b = 180°	3. Adjacent angles whose exterior sides are a straight line are supplementary. Definition of adjacent and supplementary.
4. m∠b + m∠c = m∠a + m∠b	4. Axiom 1, things equal to the same thing are equal to each other, and steps 2 and 3.
5. ∴ m∠c = m∠a	5. Axiom 3, equals subtracted from equals give equal differences.

What does this theorem mean? It means every time you have two parallel lines, ℓ and *m*, then you know that the angles in each pair of corresponding angles are equal.

$$m\angle x = m\angle y$$

So, if you know the m∠*x*, you also know the m∠*y*.

$$m\angle x = 80° = m\angle y$$

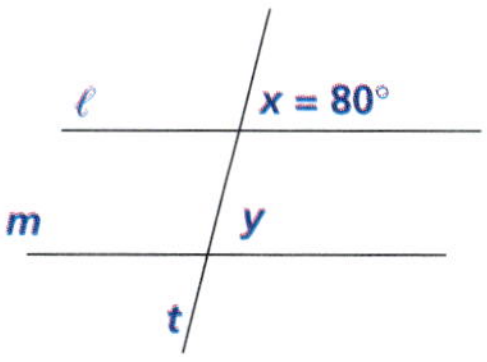

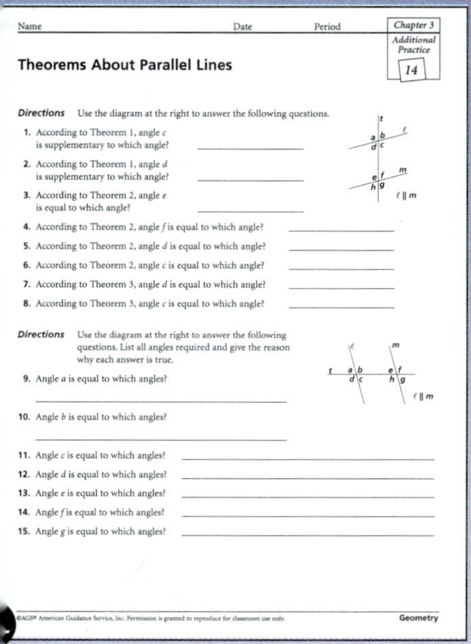

Name Date Period

Chapter 3 Additional Practice 14

Theorems About Parallel Lines

Directions Use the diagram at the right to answer the following questions.

1. According to Theorem 1, angle *c* is supplementary to which angle? ______
2. According to Theorem 1, angle *d* is supplementary to which angle? ______
3. According to Theorem 2, angle *e* is equal to which angle? ______
4. According to Theorem 2, angle *f* is equal to which angle? ______
5. According to Theorem 2, angle *d* is equal to which angle? ______
6. According to Theorem 2, angle *c* is equal to which angle? ______
7. According to Theorem 3, angle *d* is equal to which angle? ______
8. According to Theorem 3, angle *c* is equal to which angle? ______

t, a, b, ℓ, d, c, e, f, m, h, g, ℓ ∥ m

Directions Use the diagram at the right to answer the following questions. List all angles required and give the reason why each answer is true.

9. Angle *a* is equal to which angles? ______
10. Angle *b* is equal to which angles? ______
11. Angle *c* is equal to which angles? ______
12. Angle *d* is equal to which angles? ______
13. Angle *e* is equal to which angles? ______
14. Angle *f* is equal to which angles? ______
15. Angle *g* is equal to which angles? ______

ℓ, m, t, a, b, d, c, e, f, h, g, ℓ ∥ m

 Geometry

Additional Practice 14

On the board, draw the figure for the proof of the first theorem in the lesson and write the theorem. Discuss how the "given" and "to prove" statements for the proof are derived from the If … then … parts of the theorem. Review the information on page 68 about an indirect proof or proof by contradiction. Explain that an indirect proof shows that it is impossible for the thing to be proved to be false because if it is false some accepted facts are contradicted.

Before going through the formal proof, ask students for their own informal reasoning to prove the theorem. Use their responses as a basis for introducing and understanding how the formal proof is developed. After reviewing the proof of the second theorem, discuss whether each of the following statements is derived from the "given," from a statement that precedes it, or from the figure for the proof. Repeat for the proof of the third theorem in the lesson.

3. Reinforce and Extend

LEARNING STYLES

Body/Kinesthetic Tell students to draw two identical sets of parallel lines cut by a transversal (not perpendicular to the parallel lines) large enough so that they can cut the angles out, and then label each set of eight angles *a, b, c, d, e, f, g, h*. Have students cut the angles out of one drawing and find and record all the pairs of angles that are congruent or supplementary. They can compare their findings to the angles in the uncut set of lines to verify the theorems presented in the lesson.

Learning Styles

Visual/Spatial Help students draw sketches and write statements on index cards to identify the theorems and other axioms, definitions, and properties they need to review. For example, they might draw parallel lines with alternate interior angles labeled x and y and write the statement $m\angle x = m\angle y$. Working with a partner, one student holds up a card and the other recites the theorem, property, etc., that justifies the statement on the card. Students can take turns holding the cards and reciting the theorems.

Application

Career Connection Have students find and research a specific job or occupation in which proofs and deductive reasoning are used either directly or indirectly by persons working in the field. Brainstorm a list of occupations with the group—scientists such as chemists, physicists, or geologists and medical personnel such as doctors, paramedics, and lab technicians. Tell students to describe the job or occupation and explain how proofs are used and by whom.

There is one more parallel line theorem left to prove.

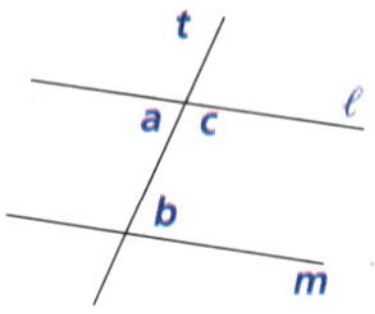

THEOREM: If two lines are parallel, then the alternate interior angles are equal.

To Prove: $m\angle a = m\angle b$

PROOF

Statement	Reason
1. $\ell \parallel m$, transversal t	1. Given.
2. Label $\angle c$ supplementary to $\angle a$. $m\angle a + m\angle c = 180°$	2. Definition of adjacent and supplementary.
3. $m\angle b + m\angle c = 180°$	3. Theorem, interior angles on the same side of the transversal are supplementary.
4. $m\angle a + m\angle c = m\angle b + m\angle c$	4. Axiom 1, things equal to the same things are equal to each other, and steps 2 and 3.
5. $\therefore m\angle a = m\angle b$	5. Axiom 3, equals subtracted from equals give equal differences.

Let's look at the meaning of this theorem. It means that if ℓ and m are parallel, then any pair of alternate interior angles are equal. You can prove this to yourself by writing a proof of the theorem for angles c and d.

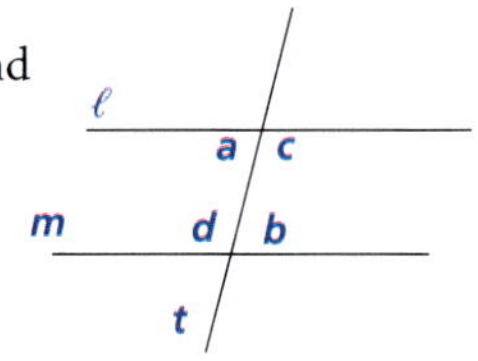

Exercise A Complete the following statements. For problems 1 and 2, use the figure shown and the theorem, if $\ell \parallel m$, then interior angles on the same side of the transversal are supplementary.

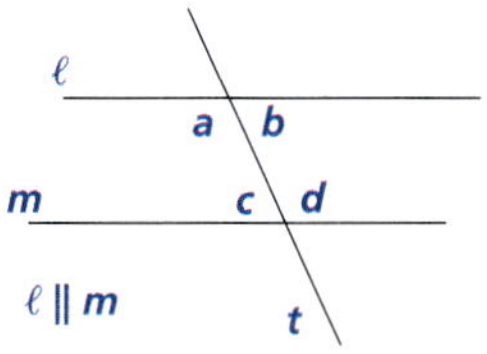

1. $m\angle a + m\angle c =$ _____ 180°

2. $m\angle b + m\angle d =$ _____ 180°

For problems 3 and 4, use the figure shown and the theorem, if $\ell \parallel m$, then alternate interior angles are equal.

3. $m\angle a =$ _____ $m\angle d$

4. $m\angle b =$ _____ $m\angle c$

For problems 5–12, use the figure shown and the theorem, if $\ell \parallel m$, then corresponding angles are equal.

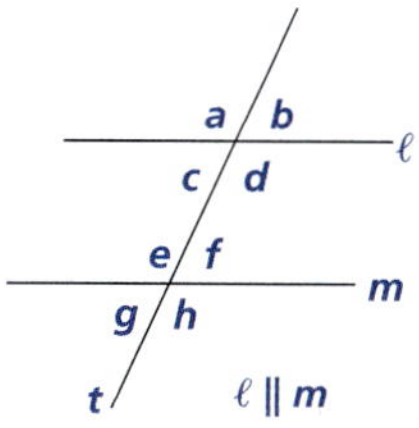

5. $m\angle a =$ ____ $m\angle e$		**9.** $m\angle e =$ ____ $m\angle a$	
6. $m\angle b =$ ____ $m\angle f$		**10.** $m\angle f =$ ____ $m\angle b$	
7. $m\angle c =$ ____ $m\angle g$		**11.** $m\angle g =$ ____ $m\angle c$	
8. $m\angle d =$ ____ $m\angle h$		**12.** $m\angle h =$ ____ $m\angle d$	

Exercise B Find the measures of the angles in the figure below. Write your reason for each measure. Use the three theorems about parallel lines and what you know about supplementary and vertical angles.

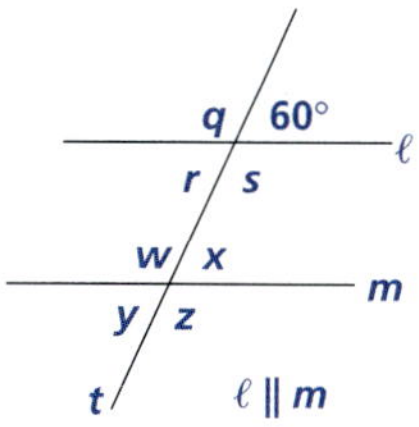

13. $\angle q$ **16.** $\angle w$ **19.** $\angle z$

14. $\angle r$ **17.** $\angle x$

15. $\angle s$ **18.** $\angle y$

See Teacher's Edition page for answers.

Exercise C Copy and then complete the following Indirect Proof (Proof by Contradiction). Use the proof of the theorem on page 68 as a guide.

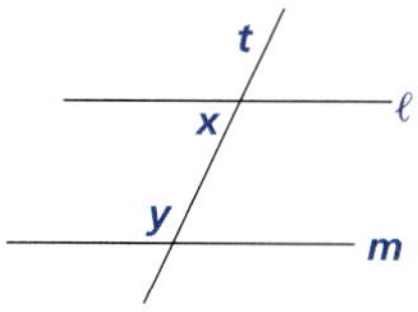

If $\ell \parallel m$ with transversal t, then $m\angle x + m\angle y = 180°$.

To Prove: $\angle x$ and $\angle y$ are supplementary

PROOF

Statement	Reason
1. $\ell \parallel m$, transversal t	1. Given.
2. **20.** ____ $\angle x$ and $\angle y$ are not supplementary	2. Indirect proof assumption; the opposite of "To prove."
3. $m\angle x + m\angle y \neq 180°$	3. **21.** ____ Definition of not supplementary angles.
4. **22.** ____ $m\angle x + m\angle y < 180°$ or $m\angle x + m\angle y > 180°$	4. Definition of not equal means greater than or less than.
5. ℓ must intersect m This contradicts the given: $\ell \parallel m$ $\therefore m\angle x + m\angle y = 180°$	5. **23.** ____ Euclid's Postulate 5.

Answers to Exercise B 13. 120°; $\angle q$ is supplementary with the angle whose measure is 60°; 180° – 60° = 120° **14.** 60°; adjacent angles with common side a straight line are supplementary angles; 180° – 120° = 60°; also $\angle r$ or the 60° angles are vertical angles and so are equal **15.** 120°; $\angle s$ and $\angle q$ are vertical angles and therefore equal—$\angle s = \angle q = 120°$ **16.** 120°; $\angle w = \angle q = \angle s = 120°$; alternate interior angles or corresponding angles **17.** 60°; $\angle x =$ and the 60° angle are corresponding angles and therefore equal **18.** 60°; $\angle y = \angle r = 60°$; corresponding angles or $\angle x$ and $\angle y$ are vertical angles so $m\angle x = m\angle y = 60°$ **19.** 120°; $\angle z$ and $\angle x$ are supplementary; $m\angle x = 60°$; 180° – 60° = 120°

Common Error

Developing a proof may be challenging for even the best students. They may have difficulty organizing their thoughts in a logical way. If students are struggling with the development of the statements for their proof, you may want to provide one or more of the statements and/or reasons to help guide them through the steps.

Learning Styles

LEP/ESL Show students how to write theorems using symbols and "shorthand." For example, the shorthand version of the third theorem in the lesson can be written "If 2 lines ||, then alternate interior $\angle s$ =." Let students write their own "shorthand" versions of theorems and definitions in this lesson. Then pair a student with limited English with a student more proficient in English. Students can take turns reciting the theorems aloud, using their shorthand versions as cues. Encourage students to help one another with vocabulary as they review the theorems and definitions.

Group Problem Solving

Ask small groups of students to write a complete and formal proof of the following statement: If $m \parallel p$, then $\angle a = \angle h$.

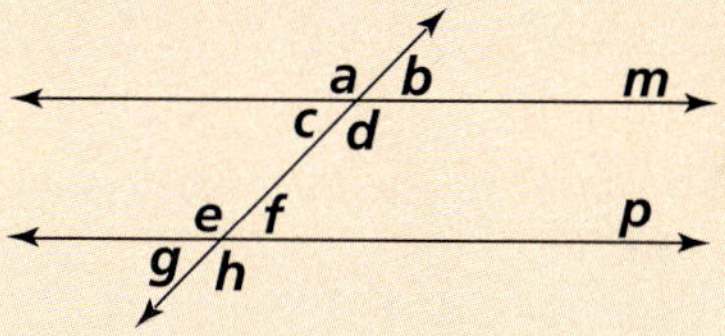

Have each group write its proof on a large sheet of paper that can be displayed in the classroom. Also, instruct students to write a brief summary of the steps they used to set up, develop, and check the proof. Ask for volunteers to present and defend their proof to the class.

Possible solution:

Statements

1. Line m is parallel to line p.
2. $\angle a = \angle d$; $\angle h = \angle e$
3. $\angle d = \angle e$
4. $\angle a = \angle h$

Reasons

1. Given.
2. Vertical angles are equal.
3. Theorem 3
4. Axiom 1 (quantities equal to the same, or equal, quantities are equal to each other; Steps 2 and 3).

Algebra Connection

Students write and solve one- and two-step equations using theorems, postulates, definitions, and properties in this and other lessons. Use the following equations to review with students how to solve equations:

$4x = 92$

$3x + 4 = 13$

$180 - x = 50$

If $x + y = 180$ and $x = 40$, find y.

Before students begin solving for the measures of angles in problem 25, have them identify all pairs of equal and supplementary angles. You may want to tell students that they should start by using the statement $w = 3x$ and the fact that $\angle w$ and $\angle x$ are supplementary angles to find x. Remind students that their equations must have only one unknown quantity.

Calculator

Have students use a calculator to solve the equations they write for Exercise E. For example, to solve $4x = 180°$;

Press 180 [÷] 4 [=].

The display shows 45.

The measure of $\angle x$ is 45°.

For an equation such as $180° - z = 45°$;

Press 180 [−] 45 [=].

The display shows 135.

The measure of $\angle z$ is 135°.

Problem Solving

Exercise D Solve the following problems.

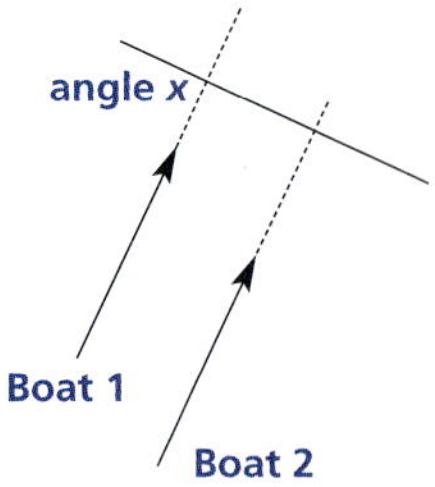

90°, corresponding angles are equal

24. Each sailboat will cross under the bridge during a race. Boat 2 will cross at a right angle to the bridge. If the boats are following parallel courses, what will be the measure of angle x? Why?

Algebra Connection

25. Find the measures of angles w, x, y, z, h, i, j, and k. Use the figure below and any theorems or information about supplementary angles.

$m\angle x = m\angle y = m\angle i = m\angle j = 45°$;
$m\angle w = m\angle h = m\angle k = m\angle z = 135°$

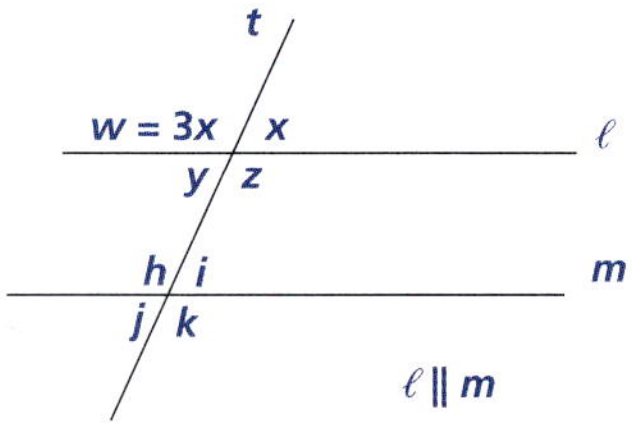

Calculator Practice

Exercise E Use a calculator to solve the following problems.

26. supplementary angles, one angle = 52° 128°

27. supplementary angles, one angle = 28° 152°

28. supplementary angles, one angle = 36° 144°

29. supplementary angles, one angle = 63° 117°

30. supplementary angles, one angle = 74° 106°

Lesson 4 Constructions and Problem Solving

You now know three useful theorems about parallel lines and their angles. You can use the information from these theorems to develop other information about parallel lines. You can also use the information to construct parallel lines.

> *R*emember, if $p \rightarrow q$ is any conditional, then $q \rightarrow p$ is its *converse.*

Recall the theorem, if two lines are parallel, then alternate interior angles are equal. You will need the *converse* of this theorem: if alternate interior angles are equal, then the lines that are not common to the angles are parallel. This statement will be taken to be true as a postulate. It is called the Alternate Interior Angles Postulate. You can use this postulate to construct parallel lines.

Alternate Interior Angles Postulate: If a transversal intersects two lines so that the alternate interior angles are equal, then the lines are parallel.

Construction

Suppose you wish to construct a line that passes through point *P* and is parallel to line ℓ

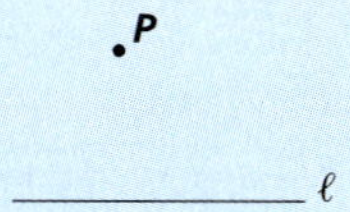

Step 1 Draw a transversal through point *P*.

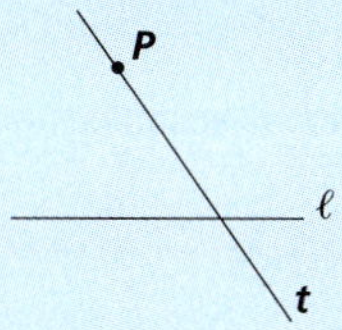

Step 2 Copy angle *x* at point *P*. This will produce alternate interior angles that are equal.

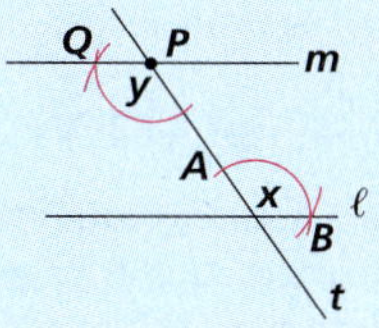

> *R*emember, to copy an angle:
>
> Draw arc through rays of angle *x*.
>
> Draw same size arc at vertex *P*.
>
> Then complete the construction for copying the angle.
>
> See page 12 for more review.

You have now constructed line *m* parallel to line ℓ using only a straightedge and a compass.

Lesson at a Glance

Chapter 3 Lesson 4

Overview This lesson explains how to use theorems about parallel lines to develop additional information and to construct parallel lines.

Objectives

- To use the converse of a theorem to construct parallel lines.
- To use theorems to find the measures of angles formed by parallel lines and transversals.

Student Pages 73–76

Teacher's Resource Library TRL

Workbook Activity 26

Additional Practice 15

1. Warm-Up Activity

Review parallel-line theorems, substitution property (Axiom 3), the vertical angle theorem, and the definitions of supplementary and adjacent angles with the class. Say part of a theorem or definition and ask students to complete the statement. For example, say, "If two lines are parallel and cut by a transversal, then… ," and ask students to complete the statement in as many ways as possible to make a true statement. (the interior angles on the same side of the transversal are supplementary; the corresponding angles are equal; the alternate interior angles are equal.)

2. Teaching the Lesson

Write this theorem on the board: if two lines are parallel, then alternate interior angles are equal. Ask for its converse. (If the alternate interior angles of two lines are equal, then the two lines are parallel.) Discuss how the theorem and converse are both true and how they are different. Ask students if they think the converses of other parallel-line theorems would be true statements also. Encourage them to explain their reasoning. Read the first example on page 74 and before going through the steps, ask several different questions such as: "If $m\angle a = 40$, what is $m\angle b$?" Write down all

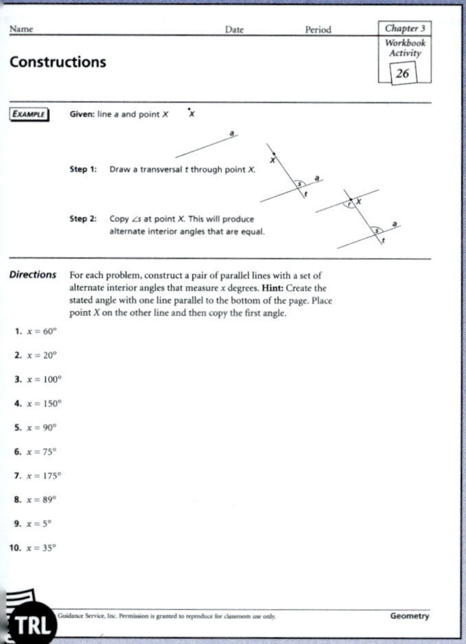

Name Date Period | Chapter 3 Workbook Activity 26

Constructions

EXAMPLE **Given:** line *a* and point *X*

Step 1: Draw a transversal *t* through point *X*.

Step 2: Copy ∠*s* at point *X*. This will produce alternate interior angles that are equal.

Directions For each problem, construct a pair of parallel lines with a set of alternate interior angles that measure *x* degrees. **Hint:** Create the stated angle with one line parallel to the bottom of the page. Place point *X* on the other line and then copy the first angle.

1. $x = 60°$
2. $x = 20°$
3. $x = 100°$
4. $x = 150°$
5. $x = 90°$
6. $x = 75°$
7. $x = 175°$
8. $x = 89°$
9. $x = 5°$
10. $x = 35°$

 Geometry

Workbook Activity 26

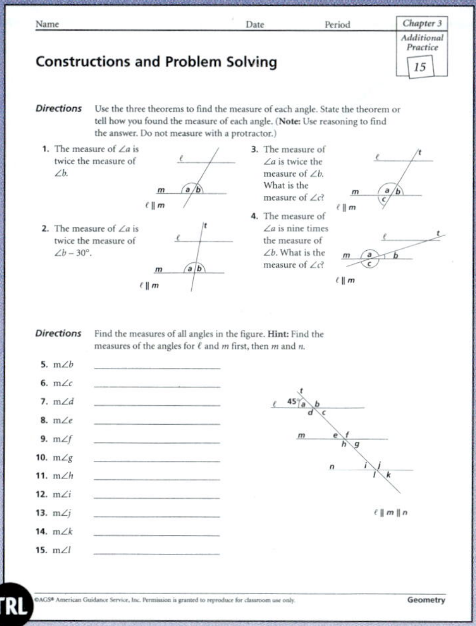

Name Date Period | Chapter 3 Additional Practice 15

Constructions and Problem Solving

Directions Use the three theorems to find the measure of each angle. State the theorem or tell how you found the measure of each angle. (**Note:** Use reasoning to find the answer. Do not measure with a protractor.)

1. The measure of ∠*a* is twice the measure of ∠*b*. ($\ell \parallel m$)
2. The measure of ∠*a* is twice the measure of ∠*b* – 30°. ($\ell \parallel m$)
3. The measure of ∠*a* is twice the measure of ∠*b*. What is the measure of ∠*c*? ($\ell \parallel m$)
4. The measure of ∠*a* is nine times the measure of ∠*b*. What is the measure of ∠*c*? ($\ell \parallel m$)

Directions Find the measures of all angles in the figure. **Hint:** Find the measures of the angles for ℓ and *m* first, then *m* and *n*.

5. m∠*b* ________
6. m∠*c* ________
7. m∠*d* ________
8. m∠*e* ________
9. m∠*f* ________
10. m∠*g* ________
11. m∠*h* ________
12. m∠*i* ________
13. m∠*j* ________
14. m∠*k* ________
15. m∠*l* ________

$\ell \parallel m \parallel n$

 Geometry

Additional Practice 15

the pairs of answers and ask students if any can be the correct answer for the example. Have them explain why or why not.

Algebra Connection

Emphasize that students must write equations with only one unknown. As you review the examples, help students identify the letter used for the unknown.

After going through the steps in the first example, ask: "How did they find m∠b = 120°?"(m∠b = $2x$. x = m∠a = 60°, so $2x$ = 120° = m∠b, or subtract 60° from 180°.)

After presenting the steps in the second example, ask students to describe two methods for finding m∠a. (Subtract 30° from 150°, or multiply 30° by 5.)

Before reviewing the steps for the third example, ask students to write equations to describe the relationship between ∠x and ∠z, ∠x and ∠y, and ∠z and ∠y.

Algebra Connection

Using the theorems, you can solve for the measure of any angle formed by a transversal and a set of parallel lines. Then you can use algebra to determine the angle measure. Here are some examples.

EXAMPLES The measure of ∠b is twice the measure of ∠a. What is the measure of each angle?

Let x = m∠a and $2x$ = m∠b

m∠a + m∠b = 180° by theorem, interior angles are supplementary

so $x + 2x$ = 180° by substitution

$3x$ = 180°

x = 60°

m∠a = 60° and m∠b = 120°

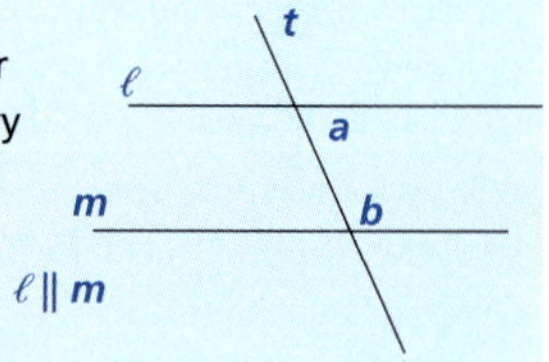

The measure of ∠a is five times the measure of ∠b. What is the measure of ∠y?

By theorem, you know that corresponding angles are equal and m∠b = m∠y. So you need only to solve for m∠b.

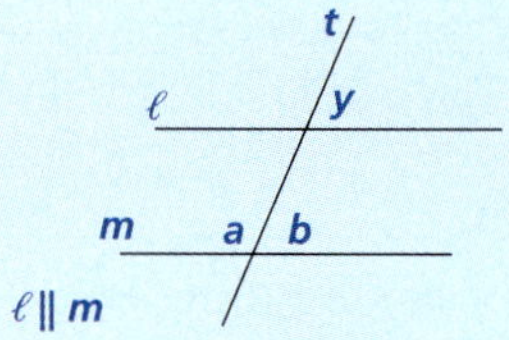

Let x = m∠b and $5x$ = m∠a

m∠a + m∠b = 180°, because they are supplementary angles

so $5x + x$ = 180° by substitution

$6x$ = 180°

x = 30°

m∠b = 30° = m∠y

Using the figure shown, give two ways to find m∠y.

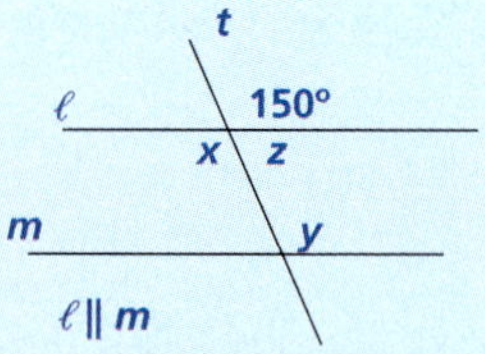

First, the angle of 150° and ∠y are corresponding angles. By theorem, corresponding angles are equal, so m∠y = 150°.

Second, ∠z and the angle of 150° are supplementary angles, so m∠z = 180° − 150° = 30°. Since ∠x and ∠z are supplementary, m∠x + 30° = 180°, so m∠x = 150°. Note that ∠x and ∠y are alternate interior angles, so by theorem, they are equal.

m∠x = m∠y = 150°

Exercise A Use the three theorems to find the measure of each angle. State the theorem or tell how you found the measure of each angle. (Note: Use reasoning to find the answer and do not measure with a protractor.)

1. The measure of $\angle z$ is three times that of $\angle y$.

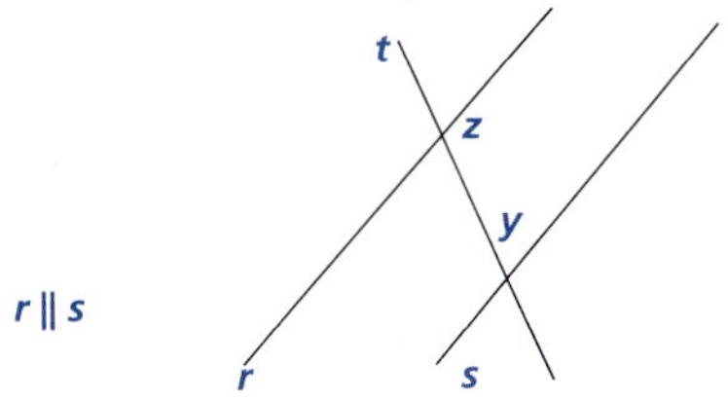

m$\angle z$ =135°, m$\angle y$ = 45°, interior angles are supplementary.

2. Eight times the m$\angle p$ = m$\angle q$.

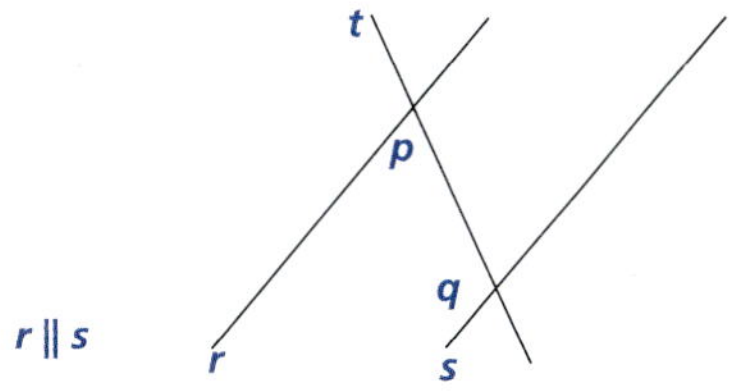

m$\angle p$ = 20°, m$\angle q$ = 160°, interior angles are supplementary.

3. Nine times the m$\angle w$ = m$\angle z$. What is m$\angle a$?

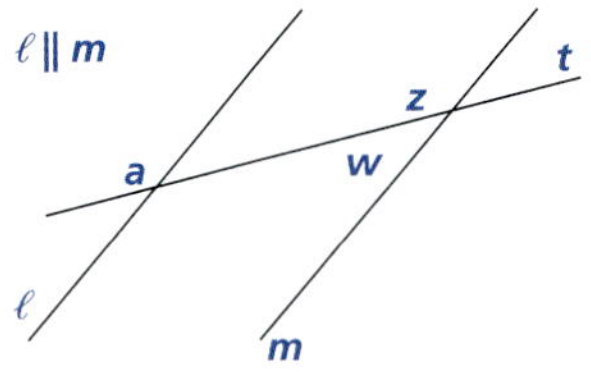

m$\angle a$ = m$\angle z$,corresponding angles are equal; m$\angle z$ = m$\angle a$ = 162° (m$\angle w$ = 18°)

4. Three times the m$\angle a$ equals two times the m$\angle b$.

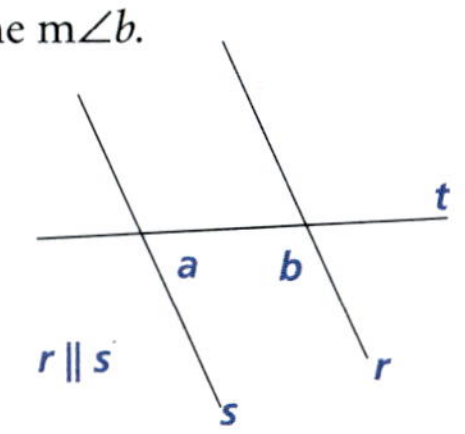

m$\angle a$ = 72°, m$\angle b$ = 108°, interior angles are supplementary.

5. m$\angle b$ is $\frac{1}{3}$ m$\angle a$. What is m$\angle x$?

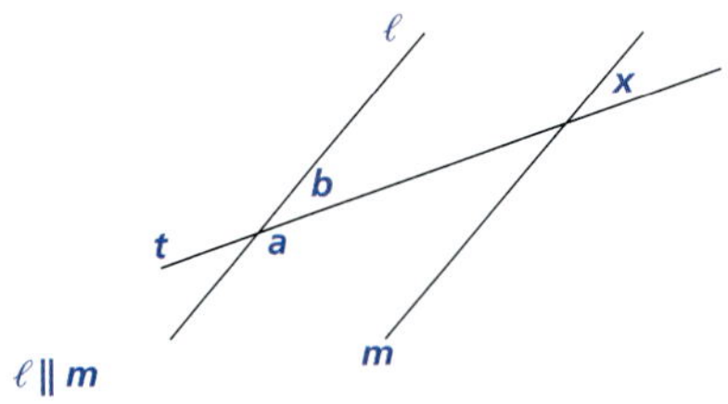

m$\angle b$ = m$\angle x$, corresponding angles are equal; m$\angle x$ = m$\angle b$ = 45° (m$\angle a$ = 135°)

6.

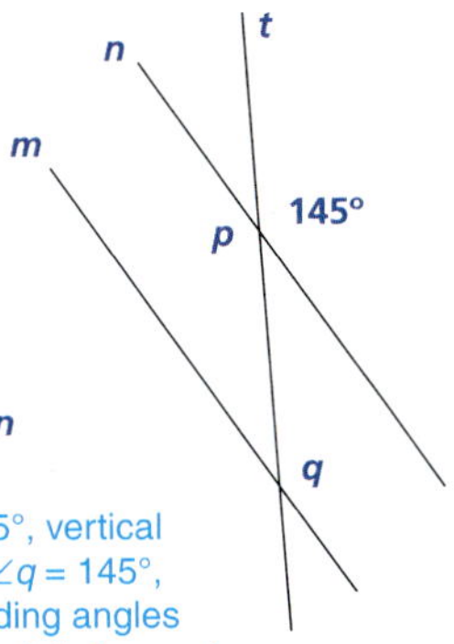

m || n

m$\angle p$ = 145°, vertical angles; m$\angle q$ = 145°, corresponding angles or alternate interior angles

3. Reinforce and Extend

Mental Math

Because so many calculations in this lesson involve supplementary angles, have students practice finding pairs of addends whose sum is 180. Have one student say a number less than 180 and another student say the addend that will make a sum of 180. You may wish to introduce a strategy in which students subtract the first addend from 200, then subtract 20 from that number.

Application

Career Connection

Have students contact a carpenter or stonemason to discuss something that he or she built that required verifying that two lines or planes were parallel or two angles were congruent. Ask students to provide a description of the project, identification of the builder, and the method the builder used to verify parallelism or congruence. Invite students to share their findings with the class.

Learning Styles

Auditory/Verbal For Exercise A, have students work in pairs. Encourage them to take turns orally describing each figure and its related information. For example, for problem 1, one student would say, "Lines *r* and *s* are parallel, $\angle z$ and $\angle y$ are same-side interior angles and are supplementary, and $\angle z$ is three times larger than $\angle y$." Then students can solve the problem together.

Learning Styles

Visual/Spatial For Exercise B, suggest that students make an enlarged drawing of the figure and label it accordingly. Then, using two different colored markers, they should label each angle's measure and color congruent angles the same color. Remind students not to cover the labels when they color the angles.

Learning Styles

Interpersonal/Group Learning Have students complete Exercise C in groups of two or three. In addition to doing the constructions, instruct groups to write a clear and detailed description of the steps they used to make each construction. Ask for volunteers to share their methods with the class, then compare and contrast the different methods used by the groups.

Group Problem Solving

Give students working in groups of three or four the following problem to solve. In addition to solving the problem, groups should prepare a presentation describing how they solved the problem. When they present their solution process to the class, they should be prepared to answer questions.

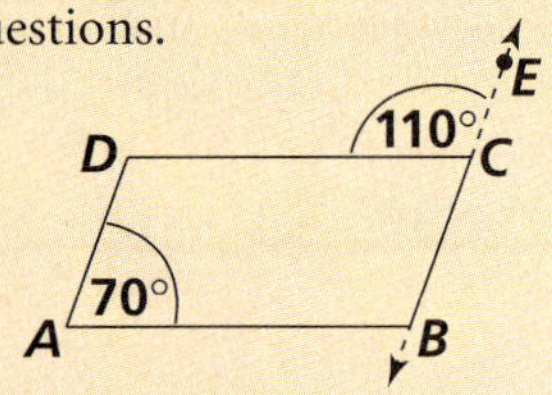

A parallelogram is defined as a four-sided figure with two pairs of parallel sides. In the figure above, ∠A and ∠D are supplementary, with angle measures as indicated. Determine whether figure *ABCD* is a parallelogram and provide a description of how you arrived at your answer, including any theorems or postulates you used.

Exercise B Find the measures of all angles in the figure. Hint: find the measures of the angles for ℓ || *m* first, then *m* || *n*.

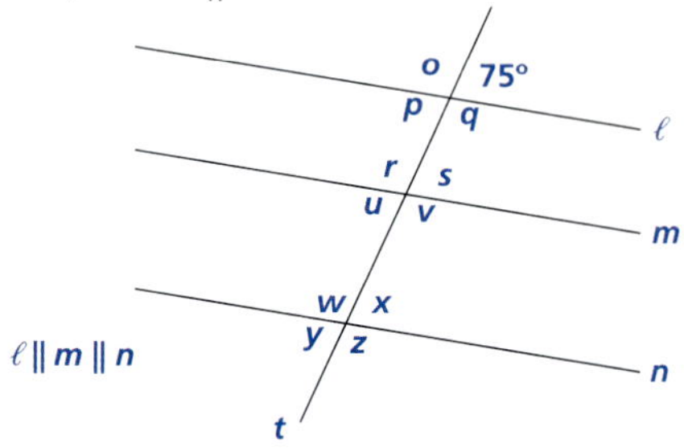

7. m∠*o* 105°

8. m∠*p* 75°

9. m∠*q* 105°

10. m∠*r* 105°

11. m∠*s* 75°

12. m∠*u* 75°

13. m∠*v* 105°

14. m∠*w* 105°

15. m∠*x* 75°

16. m∠*y* 75°

17. m∠*z* 105°

Construction

Exercise C Copy and complete the following constructions on a separate sheet of paper. Use the Alternate Interior Angles Postulate.

18. Construct line *m* parallel to line ℓ and through point *P*.

Student constructions should have appropriate arcs and meet conditions set by exercise.

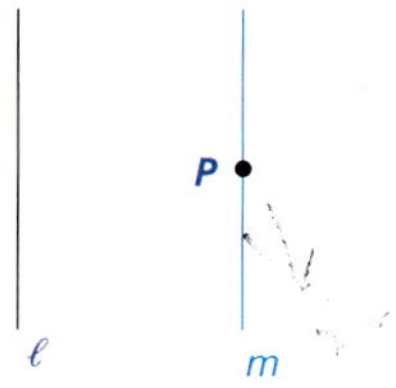

19. Construct line *t*, parallel to line *m* and through point *X*.

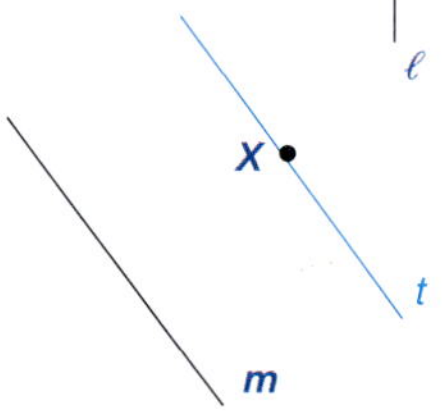

20. Construct *m*, parallel to ℓ and through *X*. Also construct *n*, parallel to *m* and through *Y*.

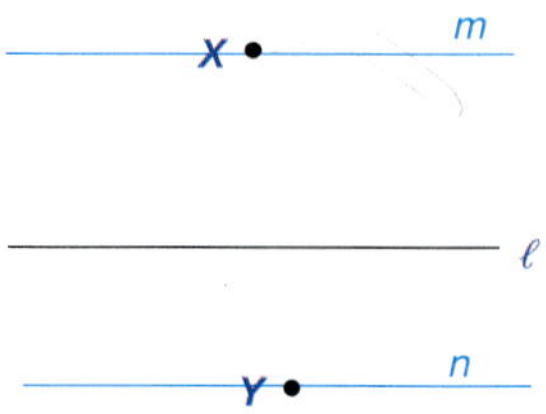

Lesson 5 Quadrilaterals and Parallels

Look around you. Concentrate on the shapes and outlines of buildings, parking lots, and houses. Notice that many of the things you see could be drawn using a four-sided geometric figure called a **quadrilateral.** In fact, many of the things you see can be drawn using specific kinds of quadrilaterals such as **rectangles** and **squares.**

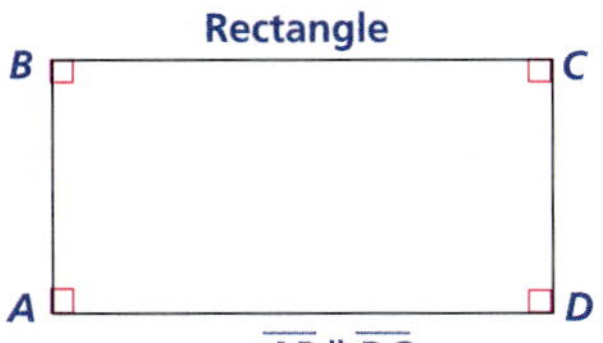

$\overline{AB} \parallel \overline{DC}$
$\overline{AD} \parallel \overline{BC}$

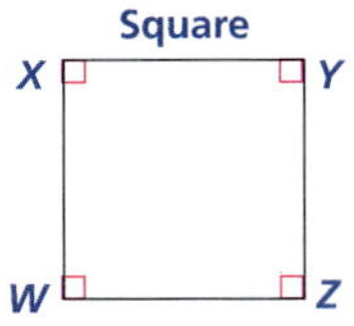

$\overline{WX} \parallel \overline{ZY}$
$\overline{WZ} \parallel \overline{XY}$
$\overline{WX} = \overline{XY} = \overline{YZ} = \overline{ZW}$

The **parallelogram** and **rhombus** are two other figures whose opposite sides are parallel.

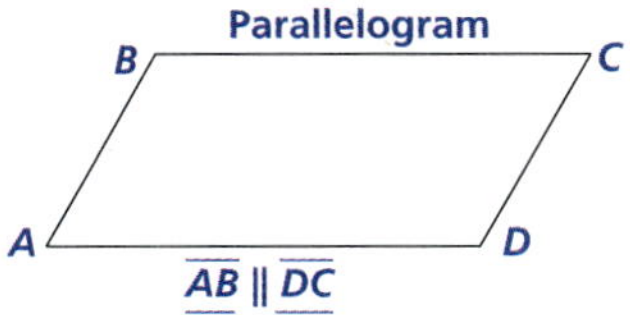

$\overline{AB} \parallel \overline{DC}$
$\overline{BC} \parallel \overline{AD}$

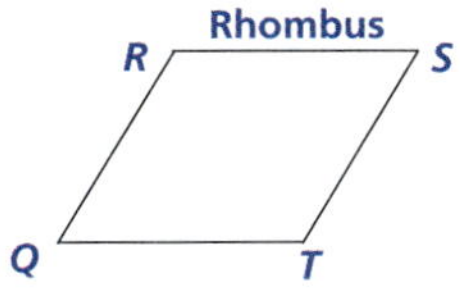

$\overline{QR} \parallel \overline{TS}$
$\overline{QT} \parallel \overline{RS}$
$\overline{QR} = \overline{RS} = \overline{ST} = \overline{TQ}$

These figures have some characteristics in common. They are made up of two pairs of parallel lines. Another way to describe these figures is to say that their opposite sides are parallel.

A parallelogram is a quadrilateral whose opposite sides are parallel. A rectangle is also a parallelogram, but its angles are all right angles. A rectangle is a parallelogram with four right angles.

Isosceles
A geometric figure with two sides of equal length

Parallelogram
A quadrilateral whose opposite sides are parallel

Quadrilateral
A geometric figure with four sides

Rectangle
A parallelogram with four right angles

Rhombus
A parallelogram with four equal sides

Scalene
A geometric figure with sides of unequal lengths

Square
A rectangle with sides of equal length

Trapezoid
A quadrilateral with exactly one pair of parallel sides

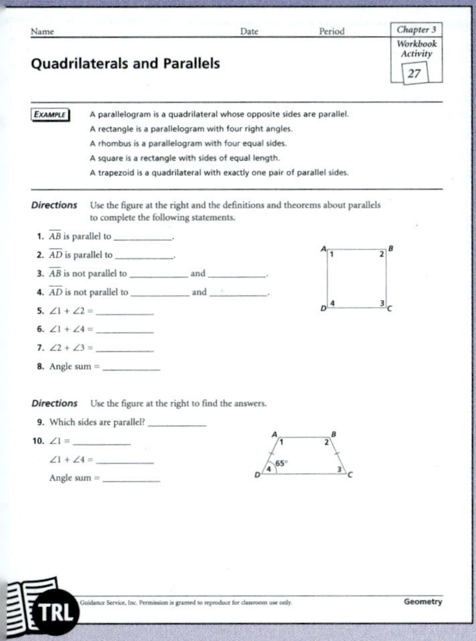

Name Date Period

Chapter 3 Workbook Activity 27

Quadrilaterals and Parallels

EXAMPLE A parallelogram is a quadrilateral whose opposite sides are parallel.
A rectangle is a parallelogram with four right angles.
A rhombus is a parallelogram with four equal sides.
A square is a rectangle with sides of equal length.
A trapezoid is a quadrilateral with exactly one pair of parallel sides.

Directions Use the figure at the right and the definitions and theorems about parallels to complete the following statements.

1. $\overline{AB}$ is parallel to ________.
2. $\overline{AD}$ is parallel to ________.
3. $\overline{AB}$ is not parallel to ________ and ________.
4. $\overline{AD}$ is not parallel to ________ and ________.
5. $\angle 1 + \angle 2 =$ ________
6. $\angle 1 + \angle 4 =$ ________
7. $\angle 2 + \angle 3 =$ ________
8. Angle sum = ________

Directions Use the figure at the right to find the answers.

9. Which sides are parallel? ________
10. $\angle 1 =$ ________
$\angle 1 + \angle 4 =$ ________
Angle sum = ________

 Geometry

Workbook Activity 27

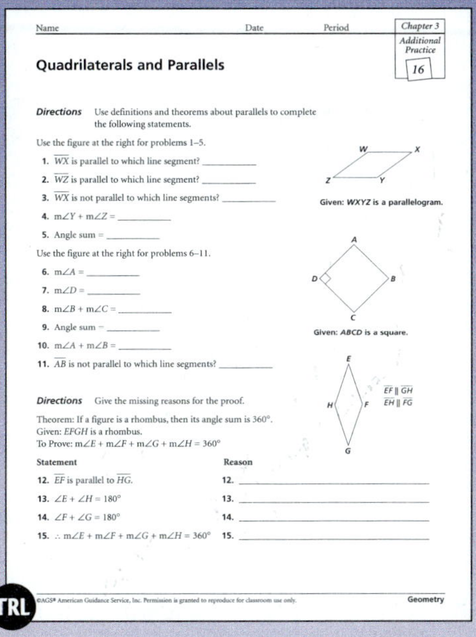

Name Date Period

Chapter 3 Additional Practice 16

Quadrilaterals and Parallels

Directions Use definitions and theorems about parallels to complete the following statements.

Use the figure at the right for problems 1–5.

1. $\overline{WX}$ is parallel to which line segment? ________
2. $\overline{WZ}$ is parallel to which line segment? ________
3. $\overline{WX}$ is not parallel to which line segments? ________
4. $m\angle Y + m\angle Z =$ ________
5. Angle sum = ________

Given: *WXYZ* is a parallelogram.

Use the figure at the right for problems 6–11.

6. $m\angle A =$ ________
7. $m\angle D =$ ________
8. $m\angle B + m\angle C =$ ________
9. Angle sum = ________
10. $m\angle A + m\angle B =$ ________
11. $\overline{AB}$ is not parallel to which line segments? ________

Given: *ABCD* is a square.

$\overline{EF} \parallel \overline{GH}$
$\overline{EH} \parallel \overline{FG}$

Directions Give the missing reasons for the proof.

Theorem: If a figure is a rhombus, then its angle sum is 360°.
Given: *EFGH* is a rhombus.
To Prove: $m\angle E + m\angle F + m\angle G + m\angle H = 360°$

Statement	Reason
12. $\overline{EF}$ is parallel to $\overline{HG}$.	12. ________
13. $\angle E + \angle H = 180°$	13. ________
14. $\angle F + \angle G = 180°$	14. ________
15. $\therefore m\angle E + m\angle F + m\angle G + m\angle H = 360°$	15. ________

 Geometry

Additional Practice 16

Lesson at a Glance

Chapter 3 Lesson 5

Overview This lesson defines various quadrilaterals and parallelograms and discusses their characteristics as they relate to parallel lines.

Objectives

- To learn about the characteristics of parallelograms.
- To use knowledge of the characteristics of parallelograms to solve problems.

Student Pages 77–82

Teacher's Resource Library TRL

Workbook Activity 27

Additional Practice 16

Mathematics Vocabulary

isosceles
parallelogram
quadrilateral
rectangle
rhombus
scalene
square
trapezoid

1. Warm-Up Activity

Ask students to name as many specific types of four-sided figures as possible. Ask for volunteers to draw sketches on the board. Have students share what they know about the characteristics of each figure and encourage them to find similarities and differences among the quadrilaterals. Ask: "How many of the figures have two pairs of parallel sides? equal sides? one pair of parallel sides?"

2. Teaching the Lesson

Discuss with students the characteristics of a quadrilateral that make it a parallelogram. On the board, draw several quadrilaterals that are parallelograms and several that are not. Ask for a volunteer to identify the quadrilaterals that are parallelograms and to tell why they are parallelograms.

Manipulatives

M Examples of Quadrilaterals

Materials: Overhead Pattern Blocks, Manipulite Pattern Blocks, overhead projector

Group Practice: Show examples of a square, rectangle (put two squares together), rhombus (blue and/or gold pieces), and an isosceles trapezoid. Discuss characteristics of each figure. Point out differences between the red isosceles trapezoid model and a right and/or scalene trapezoid. Place two red trapezoids together to form a parallelogram. Ask, "What does this show about the relationship between the upper and lower base angles of a trapezoid?" If necessary, help students conclude that the angles are supplementary.

Student Practice: Provide small groups of students with a collection of Pattern Blocks and have them sort them in any way they choose. Have them describe, in writing, the rule(s) they used for sorting. Then have them draw conclusions about the consecutive angles in a rhombus, square, and rectangle by placing shapes together as in the group presentation with the trapezoids.

3. Reinforce and Extend

Learning Styles

Visual/Spatial Have students draw each of the quadrilaterals introduced in this lesson and then extend each of the sides in both directions. Instruct students to label all right angles and supplementary angles and write the theorem that supports each pair of angle labels.

If you wished, you could shorten this definition to read "A rectangle is a parallelogram with one right angle." Let's see why this is possible.

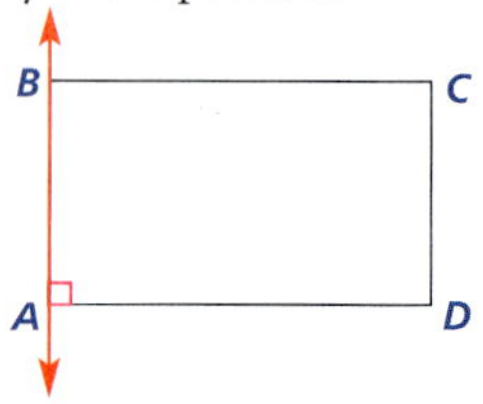

$\overline{AB} \parallel \overline{DC}$
$\overline{AD} \parallel \overline{BC}$
$m\angle A = 90°$

Example

Suppose $ABCD$ is a rectangle with $\angle A$ a right angle. What is true about $\angle B$? It must also be a right angle! You know this because one theorem states that interior angles on the same side of a transversal are supplementary.

If $\overline{AD} \parallel \overline{BC}$, with transversal $\overleftrightarrow{AB}$, then $m\angle A + m\angle B = 180°$,

$m\angle A = 90°$, so

$90° + m\angle B = 180°$

$m\angle B = 90°$

*Short lines, called *ticks,* can be used to mark equal sides of a figure. A single line represents one length. If the figure has all its sides marked with a single line, then all the sides are of equal length. More than one tick may be used in the same manner.*

You can make the same argument to show that $m\angle C$ and $m\angle D$ are also $90°$. Therefore, all four angles are right angles.

Now look at the next figure. This figure is a rhombus. It has four sides of equal length. A rhombus is a parallelogram with four equal sides.

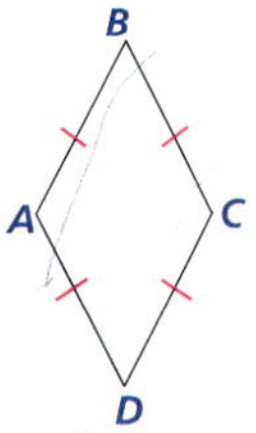

$\overline{AB} \parallel \overline{DC}$
$\overline{BC} \parallel \overline{AD}$
$\overline{AB} = \overline{BC} = \overline{CD} = \overline{DA}$

It is possible to have a quadrilateral with *exactly one pair* of parallel lines. A **trapezoid** is a quadrilateral with exactly one pair of parallel sides. Here are some examples.

EXAMPLES

Examples of Trapezoids
$\overline{AB} \parallel \overline{CD}$ for each figure

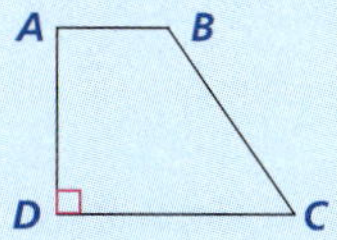

Right-Angled Trapezoid

$m\angle D = 90°$ makes $ABCD$ a *right* trapezoid

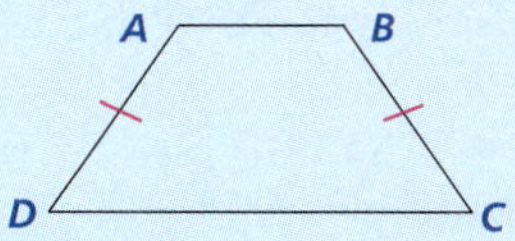

Isosceles Trapezoid

$AD = BC$ makes it **isosceles.** Figures with two sides of equal length are called isosceles.

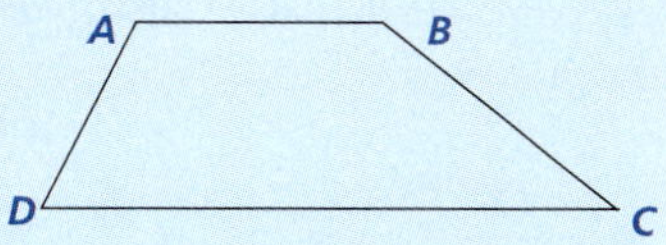

Scalene Trapezoid

Figures with no sides of equal length are called **scalene.**

$\overline{AB} \parallel \overline{CD}$ and $\overline{AD} \nparallel \overline{BC}$ make $ABCD$ a trapezoid in each case.

What do you know about the measures of angles A, B, C, and D in each trapezoid?

$m\angle A + m\angle D = ?$ (180°, supplementary, because they are on the same side between parallels)

$m\angle B + m\angle C = ?$ (180°, supplementary)

From this we can conclude that for any trapezoid $ABCD$, $m\angle A + m\angle B + m\angle C + m\angle D = 360°$.

Now you've worked with angles in parallelograms and trapezoids. What statement do you think you can make about the sum of the four angle measures in any parallelogram or trapezoid?

Geometry Around Us

You can think of the edges of a crystal as line segments. In some crystals the edges form quadrilaterals.

LEARNING STYLES

Auditory/Verbal Review terms by describing each of the quadrilaterals defined in this lesson and having students name the quadrilateral verbally and/or by drawing a sketch on the board.

LEARNING STYLES

LEP/ESL Have students create a "companion sheet" with a sketch of each of the different quadrilaterals discussed in this lesson. Pair a student whose second language is English with a student more proficient in English. Ask students to take turns pointing to the figures and describing the quadrilateral, using as many of the terms from this lesson as possible.

Learning Styles

Body/Kinesthetic Have large groups of students take turns making a formation to represent one of the quadrilaterals in this lesson while the other students verify the accuracy of each formation by counting and comparing the number of students that are supposed to represent pairs of congruent and noncongruent sides.

Common Error

Students may have difficulty keeping track of the various characteristics that distinguish one quadrilateral from another. Help them make an organizational chart or a Venn diagram that illustrates the characteristics that quadrilaterals have in common and those that make each unique.

Here is an example of inductive thinking. With inductive thinking, you use the similarities of particular cases to make a general statement. Look at the examples on this and previous pages. We can conclude that all parallelograms and trapezoids will behave in the same way.

Parallelograms

Square

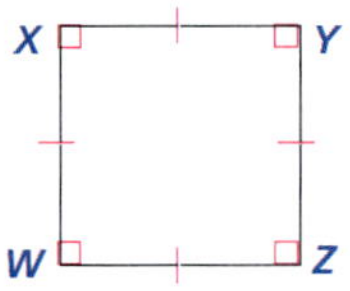

angle sum = 4 • 90° = 360°

Rectangle

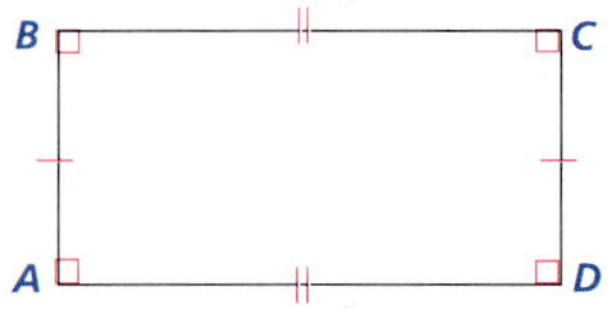

angle sum = 4 • 90° = 360°

Rhombus

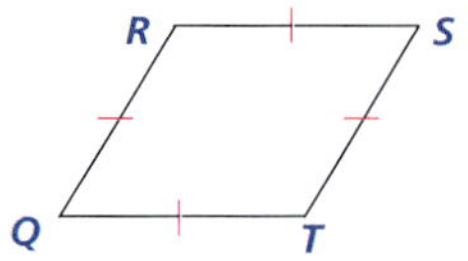

m∠Q + m∠R = 180°, m∠S + m∠T = 180°

angle sum = 2 • 180° = 360°

Trapezoids

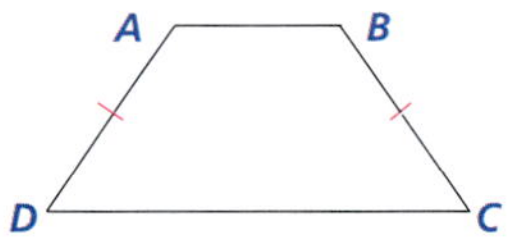

m∠A + m∠D = 180°, m∠B + m∠C = 180°

angle sum = 2 • 180° = 360°

Conclusion: The sum of the angle measures inside a parallelogram or trapezoid is 360° for the examples shown. The formal proofs follow in Exercise B.

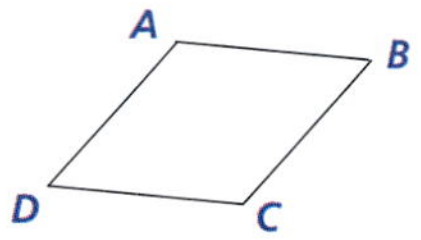

Exercise A Use definitions and theorems on parallels to complete the following statements.

Use the figure at right for problems 1–8.

Given: *ABCD* is a parallelogram.

1. $\overline{AB} \parallel$ ___ $\overline{DC}$
2. $\overline{AD} \parallel$ ___ $\overline{BC}$
3. $\overline{AB} \nparallel$ ___ and $\overline{AB} \nparallel$ ___ $\overline{AD}, \overline{BC}$
4. m∠A + m∠D = ___ 180°
5. m∠B + m∠C = ___ 180°
6. m∠A + m∠B = ___ 180°
7. m∠C + m∠D = ___ 180°
8. angle sum = ___ 360°

Use the figure below for problems 9–14.

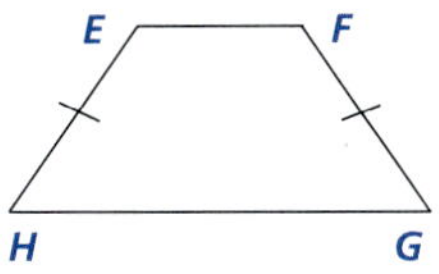

Given: *EFGH* is an isosceles trapezoid.

9. $\overline{EF} \parallel$ ___ $\overline{HG}$

10. $\overline{EH} \nparallel$ ___, $\overline{EH} \nparallel$ ___, $\overline{EH} \nparallel$ ___ $\overline{EF}, \overline{FG}, \overline{GH}$

11. $\overline{EH} =$ ___ $\overline{FG}$

12. $m\angle E + m\angle H =$ ___ 180°

13. $m\angle F + m\angle G =$ ___ 180°

14. $m\angle E + m\angle F + m\angle G + m\angle H =$ ___ 360°

Exercise B Give the missing reasons in each proof. Use the figure below for problems 15–18.

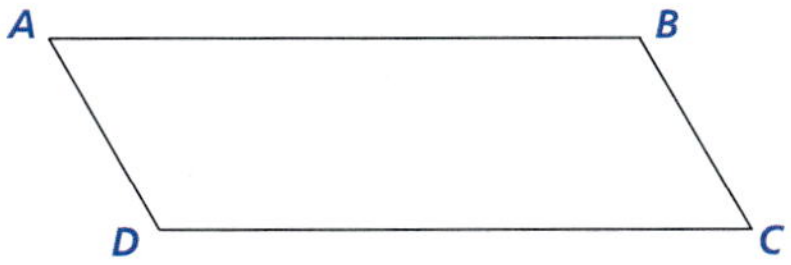

Theorem: If a figure is a parallelogram, then its angle sum is 360°.
Given: *ABCD* is a parallelogram.
To Prove: $m\angle A + m\angle B + m\angle C + m\angle D = 360°$

PROOF

Statement	Reason
1. $\overline{AB} \parallel \overline{CD}$ and $\overline{AD} \parallel \overline{BC}$	1. Given: **15.** ___ Definition of parallelogram.
2. $m\angle A + m\angle D = 180°$	2. (Hint: Think of $\overline{AD}$ as a transversal) **16.** ___ Theorem, interior angles are supplementary.
3. $m\angle B + m\angle C = 180°$	3. **17.** ___ Theorem, interior angles are supplementary.
4. $\therefore m\angle A + m\angle B + m\angle C + m\angle D = 360°$	4. **18.** ___ Axiom 2, equals added to equals are equal.

LEARNING STYLES

Auditory/Verbal Have students in pairs or small groups quiz each other on the theorems and definitions used in this lesson. Students can prompt each other with hints such as "Tell me a theorem about corresponding angles" or "Tell me a postulate about adding equal amounts to both sides of an equation."

APPLICATION

In the Community

Ask students to find examples of quadrilaterals in the community. They should list and perhaps draw several of the quadrilaterals they find. Examples include road signs, the faces of buildings such as malls and office buildings, gates, fences, the sides of columns on some buildings, windows, and plazas.

Group Problem Solving

Ask small groups of students to draw a trapezoid and label it. Have groups write a complete and formal proof to show that the sum of the measures of the angles in their trapezoid is 360°. Ask groups to volunteer to present their proofs to the class and explain their reasoning.

Learning Styles

Body/Kinesthetic Tell students to make each of the quadrilaterals in this lesson by combining two or more Pattern Blocks. They may use several of the same shapes or a combination of shapes to make the quadrilaterals. Encourage students to describe their designs using terms from this lesson.

Learning Styles

Logical/Mathematical To help students sort out the similarities and differences between the quadrilaterals, have them provide written responses to the following questions:

- Why aren't trapezoids parallelograms? (They have only two parallel sides.)
- Why are all squares rectangles? (They are parallelograms with four right angles.)
- Why are all parallelograms quadrilaterals? (They have four sides.)
- Why are all squares rhombuses? (They have four equal sides.)
- Why aren't all rectangles squares? (They have four right angles, but their lengths and widths are not equal.)
- Why aren't all quadrilaterals parallelograms? (Some do not have two pairs of parallel sides.)

Use the figure shown for problems 19–22.

Theorem: If a figure is a trapezoid, then its angle sum is 360°.
Given: $WXYZ$ is a trapezoid.
To Prove: $m\angle W + m\angle X + m\angle Y + m\angle Z = 360°$

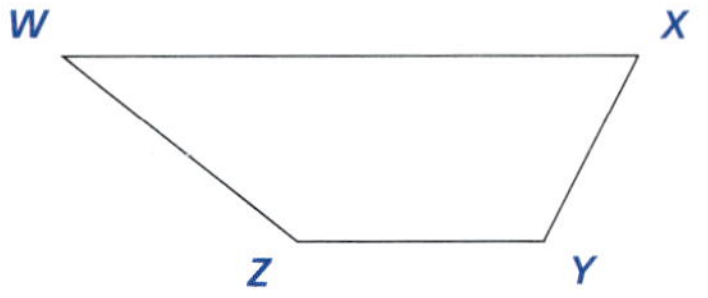

Proof

Statement	Reason
1. $\overline{WX} \parallel \overline{ZY}$	1. Given: **19.** Definition of trapezoid.
2. $m\angle W + m\angle Z = 180°$	2. **20.** Theorem, interior angles are supplementary.
3. $m\angle X + m\angle Y = 180°$	3. **21.** Theorem, interior angles are supplementary.
4. $\therefore m\angle W + m\angle X + m\angle Y + m\angle Z = 360°$	4. **22.** Axiom 2, equals added to equals are equal.

Exercise C Tell whether each statement is *True* or *False*. Give reasons for your answer.

23. A square is a rhombus with one right angle. True. Rhombus has four equal sides, by theorem, interior angles on same side of transversal, sum of each pair of angles is 180°. If one of the angles is 90°, then the other angle in the pair must be 90°.

24. A trapezoid with one right angle has at least two right angles. True. The alternate interior angles on the same side of the transversal are supplementary if one angle is 90°, then the other angle must also be 90°.

25. Every square is also a rectangle. True. Rectangle has four right angles, square has four right angles.

Experiencing Geometry

Gather a variety of cardboard boxes, such as shoe boxes, moving boxes, and cereal boxes. Remove the tops and bottoms of each box. Reshape the boxes by folding or pulling them so that the angles of the box change. Tell what shapes you can make with the edges of the boxes, e.g., square, rectangle, rhombus, parallelogram, and so on.

Lesson 6 Proving Lines Parallel

Look closely at this picture of the Golden Gate Bridge in San Francisco, California. Try to find at least four sets of parallel lines. Then look for transversals to the parallel lines. How could you prove that some of the transversal lines are parallel too?

You have learned how to construct parallel lines using alternate interior angles. Here are the steps you use. Notice that in Step 2 you construct an angle equal to the existing angle and that the pair of angles are alternate and interior.

CONSTRUCTION **Given:** line ℓ and point P not on ℓ

Problem: Construct a line $\parallel$ to ℓ, through point P.

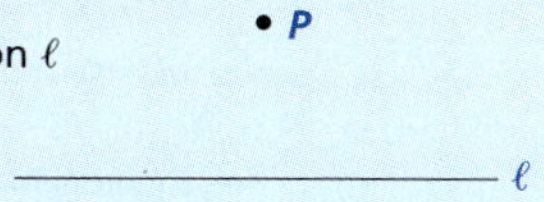

Solution:

Step 1 Draw a transversal t through P.

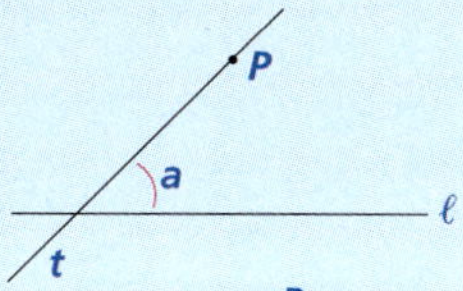

Step 2 Copy $\angle a$ at P. $\ell \parallel m$

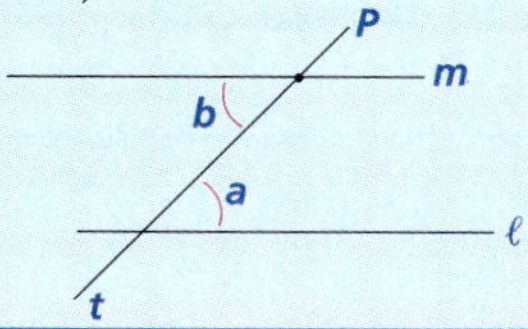

As with all constructions in geometry, there must be an axiom, postulate, or theorem that makes the construction *legal,* or correct. In this case, the construction is based on a postulate about alternate interior angles.

Lesson at a Glance

Chapter 3 Lesson 6

Overview This lesson explores how to construct parallel lines using theorems about angles formed by transversals and how to determine if two lines are parallel.

Objectives

- To construct parallel lines.
- To use converses of theorems and postulates to determine if lines are parallel.

Student Pages 83–89

Teacher's Resource Library TRL

- **Workbook Activity** 28
- **Additional Practice** 17

1. Warm-Up Activity

Have pairs of students draw two figures: a transversal across two intersecting lines and a transversal across almost parallel lines. They then should measure a pair of corresponding, alternate interior, or same-side interior angles. Ask: "What appears to be true about the lines as the corresponding or alternate interior angles get closer to being equal, or as same-side interior angles get closer to being supplementary." (They appear to become parallel.)

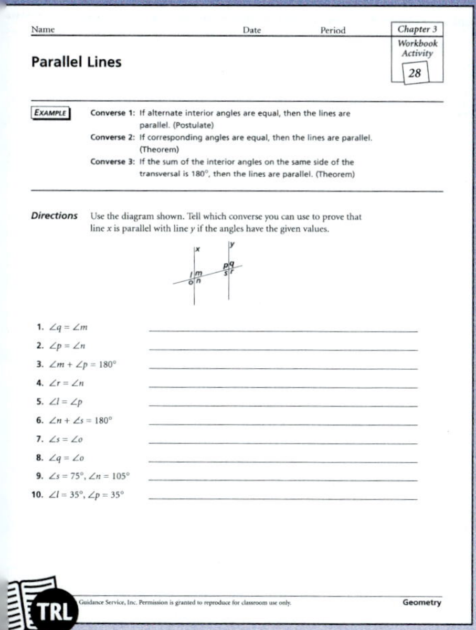

Name Date Period — *Chapter 3 Workbook Activity 28*

Parallel Lines

EXAMPLE
Converse 1: If alternate interior angles are equal, then the lines are parallel. (Postulate)
Converse 2: If corresponding angles are equal, then the lines are parallel. (Theorem)
Converse 3: If the sum of the interior angles on the same side of the transversal is 180°, then the lines are parallel. (Theorem)

Directions Use the diagram shown. Tell which converse you can use to prove that line x is parallel with line y if the angles have the given values.

1. $\angle q = \angle m$ ______
2. $\angle p = \angle n$ ______
3. $\angle m + \angle p = 180°$ ______
4. $\angle r = \angle n$ ______
5. $\angle l = \angle p$ ______
6. $\angle n + \angle s = 180°$ ______
7. $\angle s = \angle o$ ______
8. $\angle q = \angle o$ ______
9. $\angle s = 75°, \angle n = 105°$ ______
10. $\angle l = 35°, \angle p = 35°$ ______

Guidance Service, Inc. Permission is granted to reproduce for classroom use only. **Geometry**

Workbook Activity 28

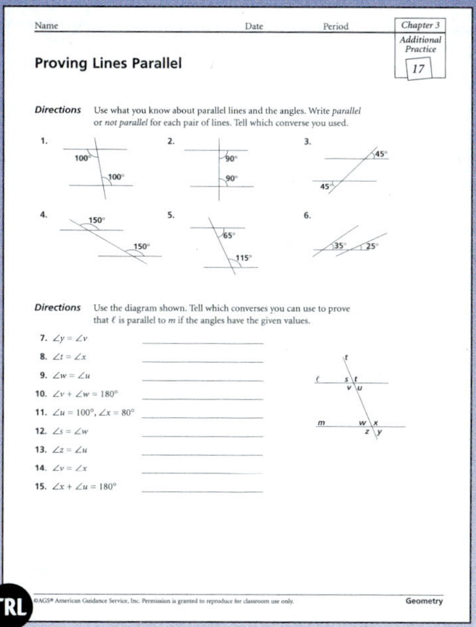

Name Date Period — *Chapter 3 Additional Practice 17*

Proving Lines Parallel

Directions Use what you know about parallel lines and the angles. Write *parallel* or *not parallel* for each pair of lines. Tell which converse you used.

1. 100° 100°
2. 90° 90°
3. 45° 45°
4. 150° 150°
5. 65° 115°
6. 35° 25°

Directions Use the diagram shown. Tell which converses you can use to prove that ℓ is parallel to m if the angles have the given values.

7. $\angle y = \angle v$ ______
8. $\angle t = \angle x$ ______
9. $\angle w = \angle u$ ______
10. $\angle v + \angle w = 180°$ ______
11. $\angle u = 100°, \angle x = 80°$ ______
12. $\angle s = \angle w$ ______
13. $\angle z = \angle u$ ______
14. $\angle v = \angle x$ ______
15. $\angle x + \angle u = 180°$ ______

©AGS® American Guidance Service, Inc. Permission is granted to reproduce for classroom use only. **Geometry**

Additional Practice 17

2. Teaching the Lesson

Summarize for students the three conditions to look for to determine if two lines are parallel: alternate interior angles equal, corresponding angles equal, same-side interior angles supplementary. Explain that the measures may not be stated explicitly and sometimes they must use the measures of other angles to determine if the necessary angle measures are equal or supplementary. Show students examples such as the following figure to help them recall how to find the measures of unknown angles.

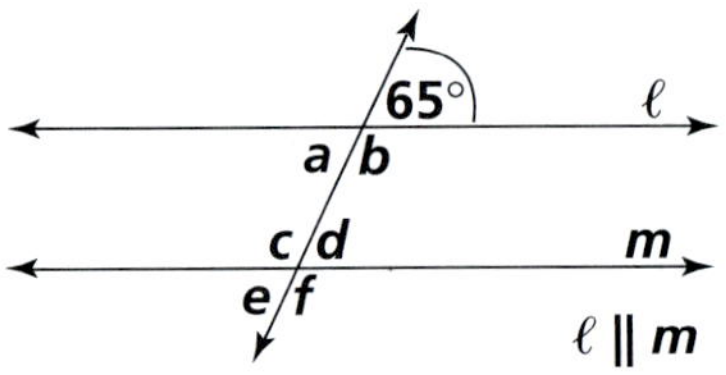

3. Reinforce and Extend

APPLICATION

At Home

Tell students to find places in or around their homes where it is important to know that two lines are parallel, for example, opposite sides of a door frame, fence posts, etc. Tell them to report on where the parallel lines are found and why it is important that they are parallel.

APPLICATION

In the Community

Instruct students to visit or call their local county surveyor or similar office and report on the methods used by employees to construct parallel lines and why it is necessary in their work to know for certain that lines are parallel. Tell students to include a description of some of the objects that this office uses parallel lines to represent.

Look closely—this is the second time you've met a statement about alternate interior angles and parallel lines. This postulate is the converse of a theorem you've proved earlier: If two lines are parallel, then their alternate interior angles are equal. You used this postulate to construct parallel lines in Lesson 4.

Arcs are used like ticks to show angles. If a pair of angles is marked with a single arc, those angles are of equal measure. More than one arc may be used in the same manner.

ALTERNATE INTERIOR ANGLES POSTULATE: If a transversal intersects two lines so that the alternate interior angles are equal, then the lines are parallel.

If $m\angle a = m\angle b$, then $m \parallel \ell$.

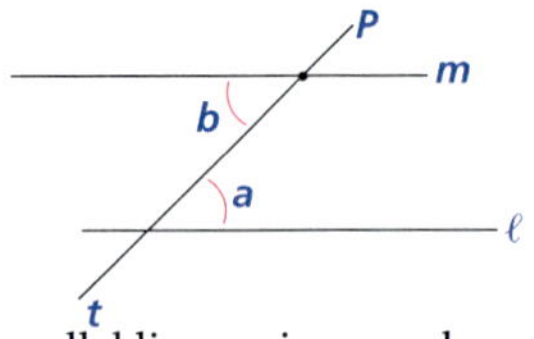

Suppose you wanted to construct parallel lines using equal corresponding angles. You would want to know that if corresponding angles are equal, then lines are parallel. Let's prove that theorem now.

THEOREM: If corresponding angles are equal, then the lines are parallel.

Given: t is a transversal, $m\angle a = m\angle b$

To Prove: $\ell \parallel m$

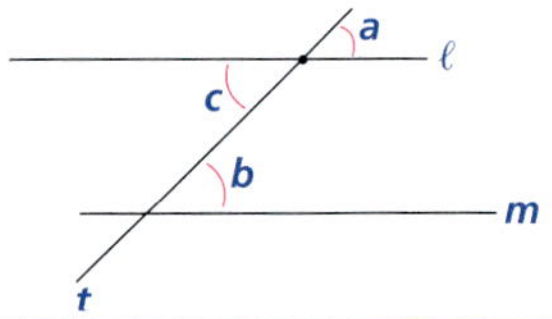

PROOF

Statement	Reason
1. t is a transversal, $m\angle a = m\angle b$	1. Given.
2. $m\angle a = m\angle c$	2. Vertical angles are equal.
3. $m\angle b = m\angle c$	3. Substitution of equals and steps 1 and 2.
4. $\therefore \ell \parallel m$	4. Alternate Interior Angles Postulate.

This theorem allows you to construct parallel lines using equal corresponding angles. Study the following construction.

CONSTRUCTION

Given: line *m* and point *R* not on *m*

Problem: Construct a line ∥ to *m*, through point *R*, using equal corresponding angles.

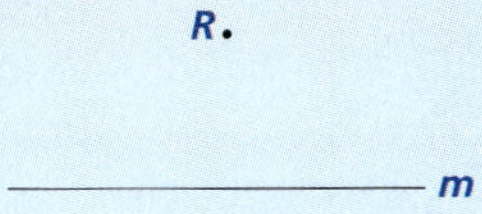

Solution:

Step 1 Draw a transversal *t* through *R*. Mark ∠*x* at *m*.

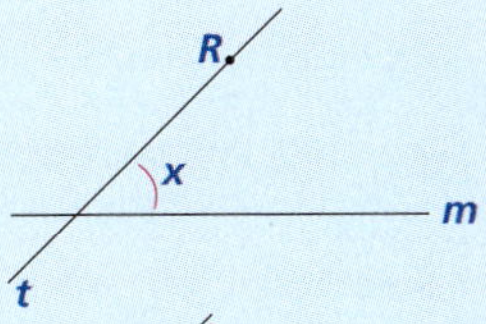

Step 2 Copy ∠*x* at *R*, corresponding to ∠*x* at *m*. Label the new angle *y*.

Conclusion: *m* ∥ *n* based on the corresponding angles theorem.

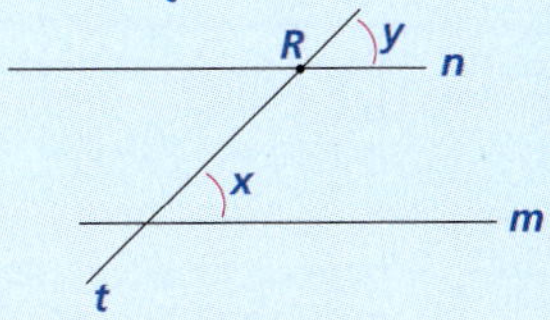

Recall the first theorem you learned about parallels. If parallel lines are cut by a transversal, then the interior angles on the same side of the transversal are supplementary. Here is the converse of that theorem.

THEOREM: If two lines are cut by a transversal so that the angles on the same side of the transversal are supplementary, then the lines are parallel.

Let us prove this new theorem.

Given: *t*, a transversal for ℓ and *m*

$m\angle a + m\angle b = 180°$

To Prove: ℓ ∥ m

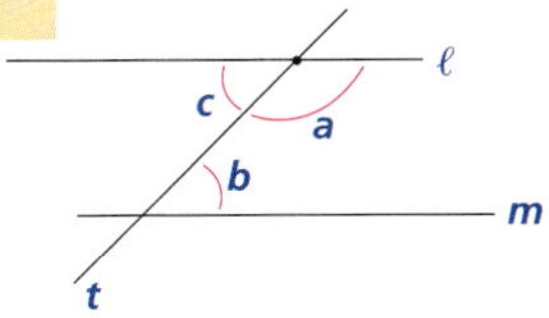

PROOF

Statement	Reason
1. *t* is a transversal for ℓ and *m*; $m\angle a + m\angle b = 180°$	1. Given.
2. $m\angle a + m\angle c = 180°$	2. If exterior sides of adjacent angles form a straight angle, the angles are supplementary.
3. $m\angle a + m\angle b = m\angle a + m\angle c$	3. Axiom 1, things equal to the same things are equal to each other.
4. $m\angle b = m\angle c$	4. Axiom 3, equals subtracted from equals give equal remainders.
5. ∴ ℓ ∥ *m*	5. Alternate Interior Angles Postulate.

MENTAL MATH

Review the theorems in this lesson by prompting students with a figure and asking them to recite the theorem that could be used to prove that the lines are parallel, based on the angle measures given in the figure.

APPLICATION

Career Connection

Have students choose and research a job or career, such as architect, surveyor, or builder, that involves constructing parallel lines. Students should find out about the training or education required.

Learning Styles

Logical/Mathematical Show students figures similar to the following one and ask them what the measures of the missing angles need to be in order to conclude that the lines are parallel.
($y = 70°$; $x = 110°$)

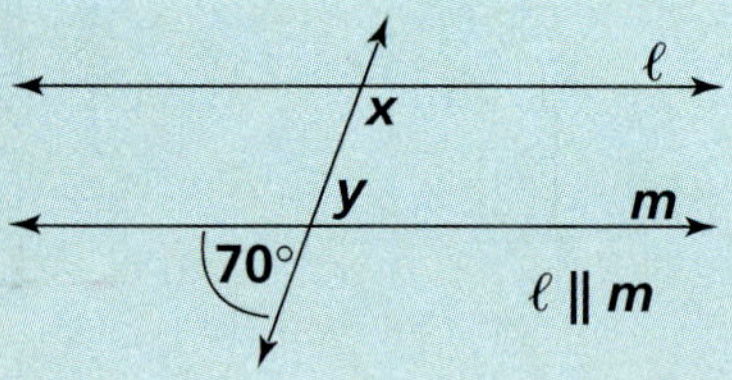

Now you have a third method for constructing parallel lines.

Construction **Given:** line s and point X not on s

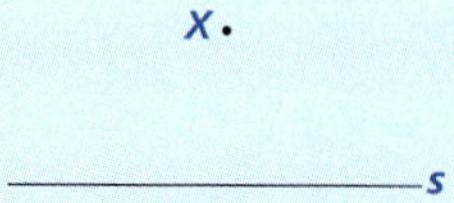

Problem: Construct a line || to s, through point X, using the theorem that if angles on the same side of the transversal are supplementary, then the lines are parallel.

Solution:

Step 1 Draw a transversal t through X. Mark $\angle h$ and $\angle j$ at s.
$m\angle h + m\angle j = 180°$

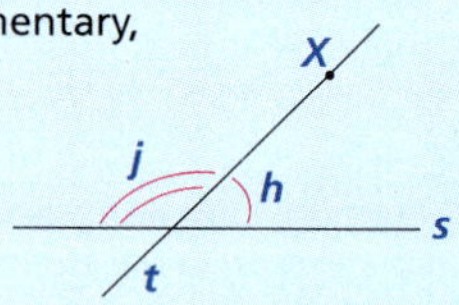

Step 2 Copy $\angle j$ at X and label it $\angle k$.
$m\angle h + m\angle k = 180°$

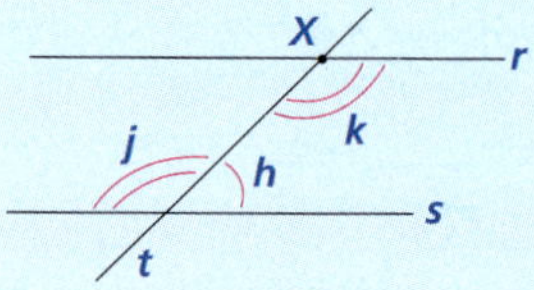

Conclusion: So $s \parallel r$ by the theorem just proved.

The following table summarizes the three theorems and their converses in this lesson.

Writing About Mathematics

Which construction of parallel lines do you think is the easiest for you? Why?

Theorems	Converses
Given Parallels: If lines are parallel, then the alternate interior angles are equal.	Given Angles: If alternate interior angles are equal, then the lines are parallel. (Postulate)
If lines are parallel, then the corresponding angles are equal.	If corresponding angles are equal, then the lines are parallel. (Theorem)
If lines are parallel, then the sum of the interior angles on the same side of the transversal is 180°.	If the sum of the interior angles on the same side of the transversal is 180°, then the lines are parallel. (Theorem)

You will find the converses helpful when you need to prove lines are parallel. Notice the converses in the following examples.

EXAMPLES Given the illustration, prove that lines ℓ and m are parallel.

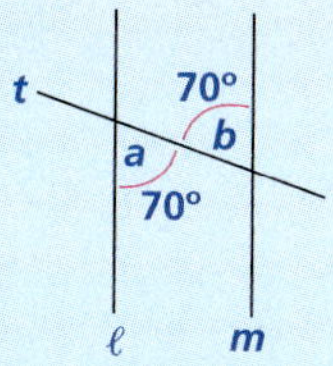

Alternate interior angles are equal.
$\therefore \ell \parallel m$

Given the illustration, prove that lines s and t are parallel.

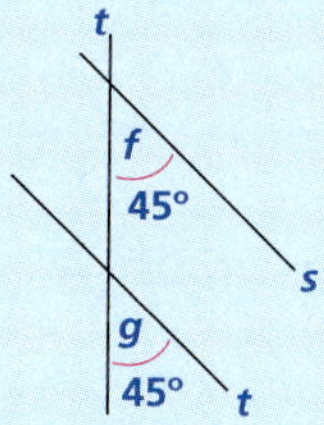

Corresponding angles are equal.
$\therefore s \parallel t$

Given the illustration, prove that lines m and n are parallel.

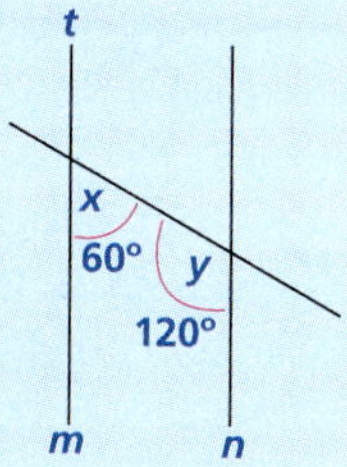

$60° + 120° = 180°$

The angles are supplementary and they are on the same side of the transversal.
$\therefore m \parallel n$

1. parallel; equal alternate interior angles
2. parallel; equal alternate interior angles
3. parallel; equal corresponding angles

Exercise A Use what you know about parallel lines and the angles. Write *parallel* or *not parallel* for each pair of lines. Tell which converse you used.

1.

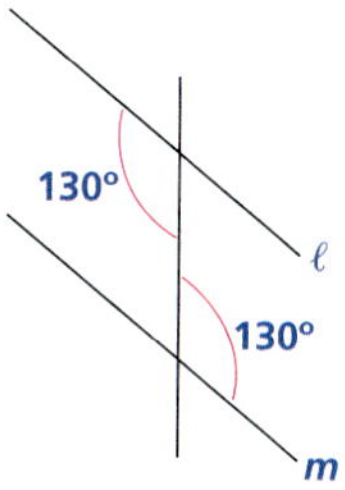

2.

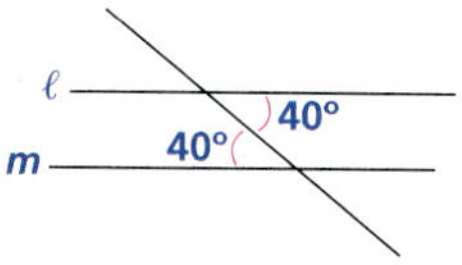

3.

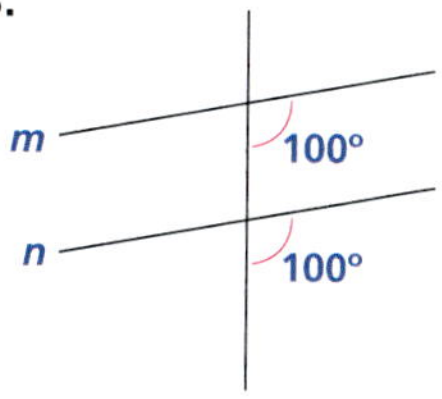

LEARNING STYLES

LEP/ESL Pair students with different levels of English proficiency. Instruct one student to recite a parallel-line theorem or postulate and the other student to state its converse. Students should switch roles and repeat the process until both can successfully state each theorem and its converse fluently.

COMMON ERROR

Students may look at problem 5 in Exercise A and conclude that the lines are not parallel because the angle measures given do not support this conclusion. Remind them that often it is necessary to use other theorems, postulates, or definitions to find the measures of other angles before they can determine if the lines are parallel.

LEARNING STYLES

Visual/Spatial Encourage students to copy the figure in Exercise B for each problem. Tell them to label, then color only the equal or supplementary angles named in the problem before deciding which converse can be used to prove that the lines are parallel.

4.

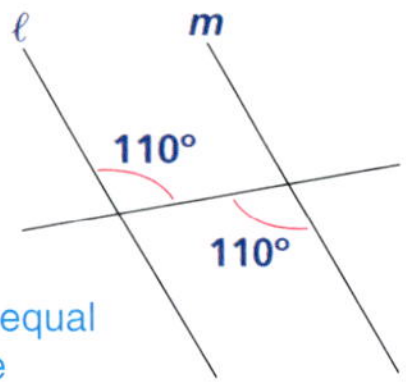

parallel; equal alternate interior angles

5.

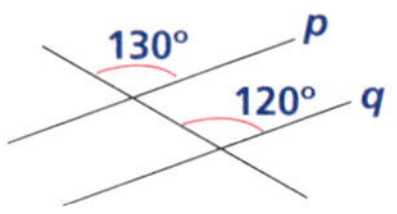

not parallel; unequal corresponding angles and not alternate interior angles

6.

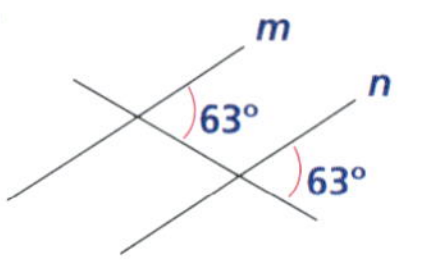

parallel; equal corresponding angles

7.

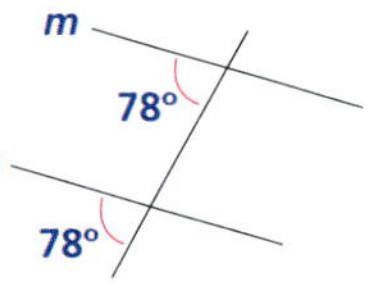

parallel; corresponding angles equal

8.

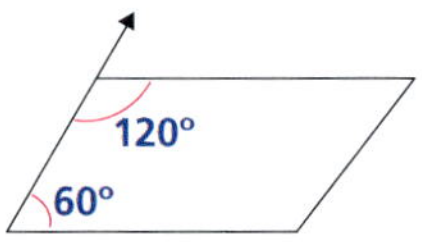

parallel, sum of interior angles on same side of transversal equal 180°

9.

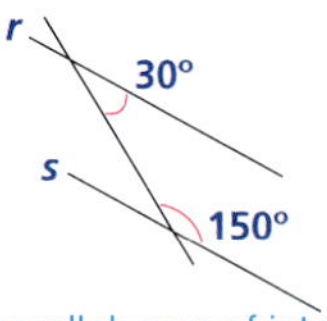

parallel; sum of interior angles on same side of transversal equal 180°

10.

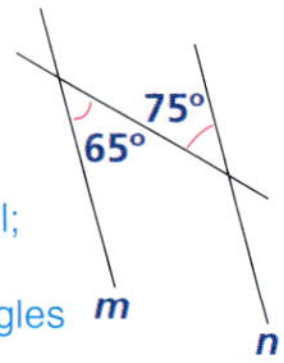

not parallel; alternate interior angles not equal

11.

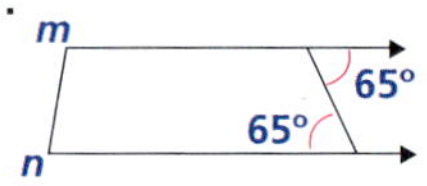

parallel, alternate interior angles equal

12.

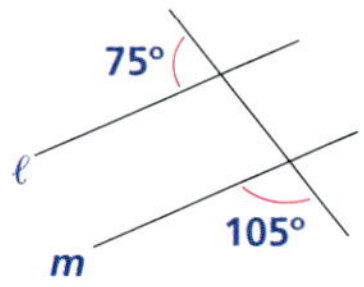

parallel; sum of interior angles on same side of transversal equal 180° and corresponding angles

Exercise B Use the diagram shown. Tell which converse you can use to prove $\ell \parallel m$ if the angles have the given values.

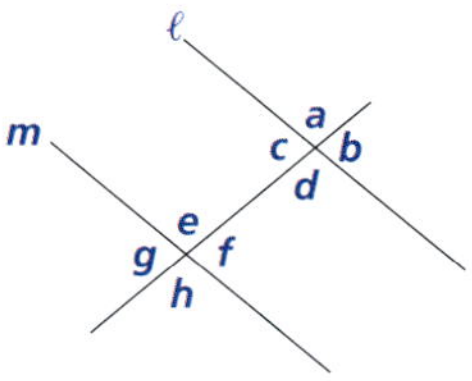

13. $m\angle d = m\angle e$

14. $m\angle b = m\angle f$

15. $m\angle c = m\angle f$

16. $m\angle a = m\angle e$

17. $m\angle g = m\angle c$

18. $m\angle d + m\angle f = 180°$

19. $m\angle c + m\angle e = 180°$

20. $m\angle d = m\angle h$

21. $m\angle g = 120°, m\angle c = 120°$

22. $m\angle d = 105°, m\angle f = 75°$

23. $m\angle d = 75°, m\angle e = 75°$

24. $m\angle c = 120°, m\angle e = 60°$

25. $m\angle h = 110°, m\angle d = 110°$

13. alternate interior angles eq
14. corresponding angles equa
15. alternate interior angles eq
16. corresponding angles equa
17. corresponding angles equa
18. interior angles on same sid transversal = 180°
19. interior angles on same sid transversal = 180°
20. corresponding angles equa
21. corresponding angles equa
22. interior angles on same sic transversal = 180°
23. alternate interior angles eq
24. interior angles on same sic transversal = 180°
25. corresponding angles equa

Construction

Exercise C Complete each of the following constructions.

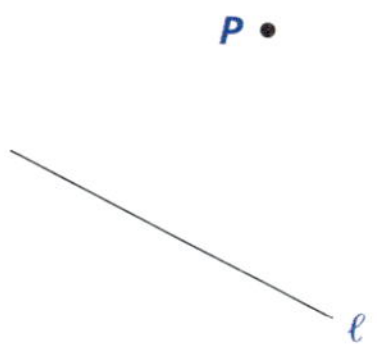

26. Draw line ℓ and point P, not on ℓ, on a separate sheet of paper. Construct m parallel to ℓ using equal alternate interior angles.

27. Draw line ℓ and point Q, not on ℓ, on a separate sheet of paper. Construct k parallel to ℓ using equal corresponding angles.

28. Draw line ℓ and point R, not on ℓ, on a separate sheet of paper. Construct j parallel to ℓ using angles that are interior angles, on the same side of the transversal, and supplementary.

Student constructions should have appropriate arcs and meet conditions set by exercise. See Teacher's Edition page for answers.

Problem Solving

Exercise D Solve the following problems.

29. The tool with the green handle is a carpenter's bevel. It is used to draw parallel line segments. Jane, the carpenter, has set the tool to a 45° angle. Then she draws several lines using the bevel. How can she be sure the lines will be parallel?

The corresponding angles always = 45° so the lines are always parallel.

30. Two bicyclists have crossed the 10 m mark. Each crossed at the angle marked in the diagram. If each cyclist stays on the same course, will they cross each other's path?

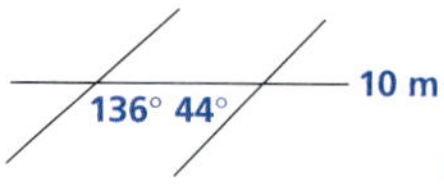

No. The given angles are interior, on the same side of the transversal, and supplementary. ∴ paths are parallel.

Note: For students who are physically challenged, complete the constructions on a transparency as a group or have students work with partners to draw the constructions.

Learning Styles

Auditory/Verbal Have students complete the constructions in Exercise C in pairs. Instruct one student to read the problem aloud after which the other student describes verbally the steps involved in the construction. Have students switch roles so that both have a chance to complete a construction.

Group Problem Solving

Have groups of students write five word problems similar to problems 29 and 30 on page 89. Tell students that their problems should be based on realistic situations involving parallel lines. Explain to groups this may require them to research a particular area to find a realistic situation or application of parallel lines. Groups then exchange and solve one another's problems.

Answers to Exercise C

26.

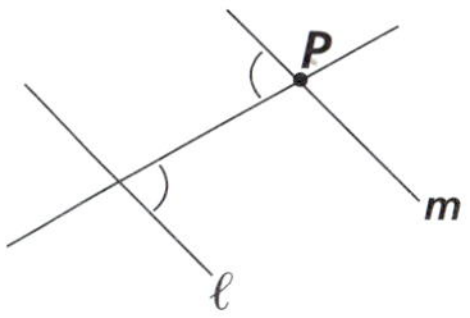

27.

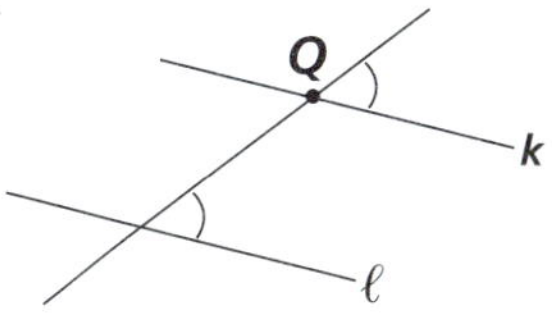

28.

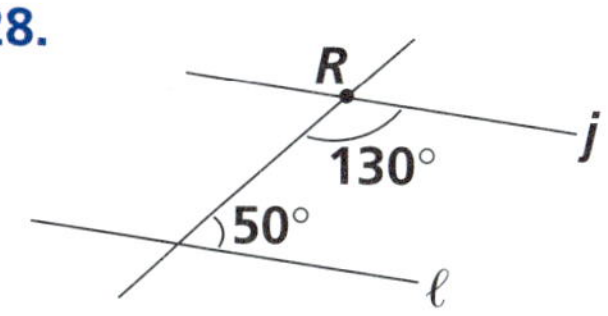

Lesson at a Glance

Chapter 3 Application

Overview This lesson defines polygons and explains how to distinguish between polygons and nonpolygons.

Objectives

- To determine if a shape is a polygon or not.
- To explain why a shape is or is not a polygon.

Student Page 90

Teacher's Resource Library TRL

Workbook Activity 29

1. Warm-Up Activity

Have each student choose a number between 3 and 8. Ask them to use a straightedge to draw a shape of any kind that has the same number of sides as they number they chose. Have students share their drawings with the class and save them for later in the lesson.

2. Teaching the Lesson

After reviewing the definition of a polygon and the examples, ask students whether the shape they drew in the Warm-Up Activity is a polygon and have them explain to the class why it is or is not a polygon.

3. Reinforce and Extend

Application

In the Community Ask students to find examples of polygons in quilts or other objects made of fabric. Tell them to choose one and provide a written description of the work, including how polygons are used, who the artist or craftsperson is, and when the work was created. Craft books often include pictures of quilts and other artwork that uses geometric shapes.

Application

Geometric Quilt Shapes The rectangle, parallelogram, square, and rhombus are only a few of the many geometric shapes you might find in a quilt pattern. All of these shapes can be classified as *polygons.* Remember, a polygon is a closed, many-sided figure that is made up of line segments. Quilters sew one or more shapes together in a pattern that tells a story or creates a pleasing design.

A polygon consists of three or more line segments, in the same plane, so that each segment intersects with others, one at each endpoint.

Examples *ABCDE* is a polygon. Each segment intersects two others, one at each endpoint.

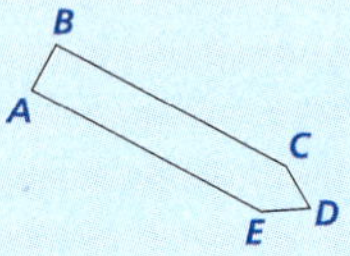

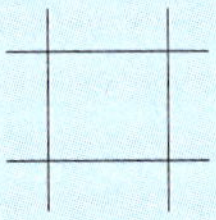

The figure at the left is not a polygon. Its segments intersect at points other than their endpoints.

The figure at the right is not a polygon. Its segments intersect more than two other segments.

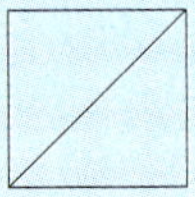

Exercise Write *polygon* or *not a polygon* for each of the figures. Give reasons why you think any figure is not a polygon.

1. 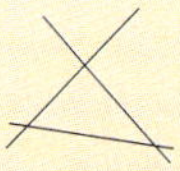
2. 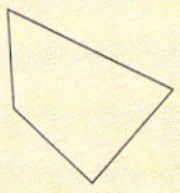
3. 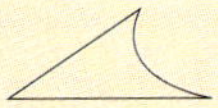
4. 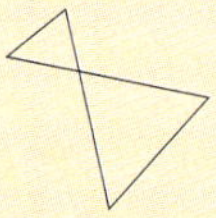
5. 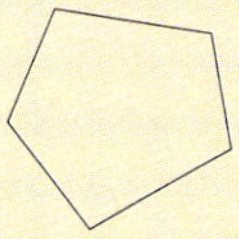
6. 
7. 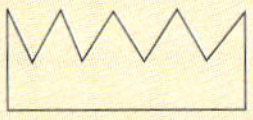
8. 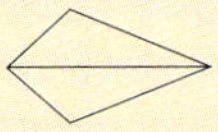
9. 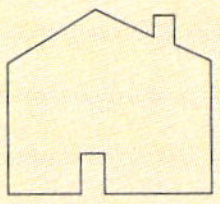
10. 

2, 5, 6, 7, 9, 10 are polygons. 1. not a polygon; segments do not intersect at end points 3. not a polygon, not all segments 4. not a polygon, intersects more than 2 other segments 8. not a polygon, intersects more than 2 other segments

Learning Styles

Visual/Spatial Have pairs of students make a set of 3-by-5 cards with various shapes on them, some polygons, some not polygons. They can take turns identifying shapes that are polygons and those that are not and explaining why the shapes are or are not polygons.

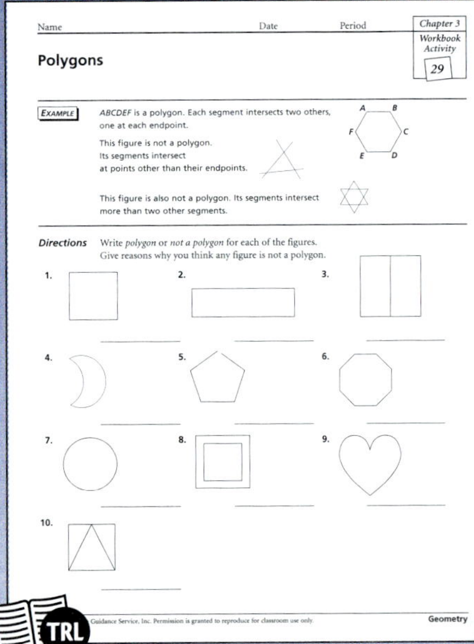

Name Date Period — Chapter 3 Workbook Activity 29

Polygons

Example *ABCDEF* is a polygon. Each segment intersects two others, one at each endpoint.

This figure is not a polygon. Its segments intersect at points other than their endpoints.

This figure is also not a polygon. Its segments intersect more than two other segments.

Directions Write *polygon* or *not a polygon* for each of the figures. Give reasons why you think any figure is not a polygon.

1. 2. 3. 4. 5. 6. 7. 8. 9. 10.

Guidance Service, Inc. Permission is granted to reproduce for classroom use only. Geometry

Workbook Activity 29

Chapter 3 Review

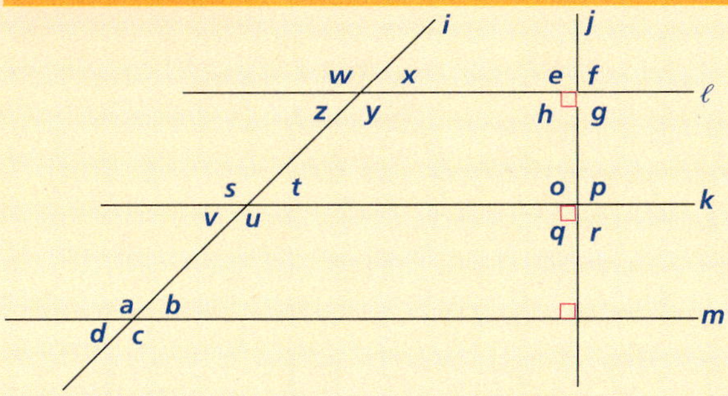

Classify each line or pair of lines in the above figure. Write the letter of the answer. There may be more than one answer.

Example: k and ℓ A. parallel B. not parallel C. transversal D. skew

Solution: A angles h and q are equal and are corresponding angles, so k and ℓ are parallel

1. ℓ and m A
A. parallel **B.** not parallel **C.** transversal **D.** skew

2. i and j B, C
A. parallel **B.** not parallel **C.** transversal **D.** skew

3. j and ℓ B
A. parallel **B.** not parallel **C.** transversal **D.** skew

---

Classify each angle or pair of angles in the figure at the top of the page. Write *acute, right, obtuse, straight, adjacent, complementary, supplementary, interior, exterior, alternate interior,* or *corresponding.* More than one name may apply to an angle or pair of angles.

Example: $\angle w$
Solution: obtuse, exterior

4. $\angle z$
5. $\angle t$
6. $\angle h$ and $\angle g$
7. $\angle a$ and $\angle b$
8. $\angle y$ and $\angle s$
9. $\angle x$ and $\angle b$
10. $\angle y$ and $\angle u$

4. acute, interior
5. acute, interior
6. right, interior, supplementary, adjacent
7. adjacent, supplementary
8. obtuse, alternate interior
9. acute, corresponding
10. obtuse and corresponding

Chapter 3 Review

Each set of problems in the Chapter Review includes an example and solution to illustrate the concept. Use the given examples for reteaching the material in Chapter 3. For additional practice, refer to the Supplementary Problems for Chapter 3 (pages 396–398) and Additional Practice 12–17 in the Teacher's Resource Library.

Chapter 3 Mastery Test TRL

The Teacher's Resource Library includes parallel forms of the Chapter 3 Mastery Test. The difficulty level of the two forms is equivalent. You may wish to use one form as a pretest and the other form as a posttest. In addition, the Chapter 3 Cumulative Mastery Test covers content from Chapters 1–3.

Alternative Assessment

Alternative Assessment items correlate with student Goals for Learning at the beginning of this chapter.

- Have students distinguish parallel, intersecting, and skew lines.

Draw and label three sets of lines: parallel, intersecting, and skew lines.

- Have students identify and name angles formed by transversals crossing parallel lines.

Draw two parallel lines crossed by a transversal and label all angles. Identify these angles by their letters on your drawing: alternate interior angles, alternate exterior angles, corresponding angles, supplementary angles.

- Have students use theorems to compare and find the measures of angles and identify parallel lines.

Study the figure for Exercise B on page 71. Identify the theorems that prove $\angle x = 60°$, $\angle r = 60°$, $\angle w = 120°$, $\angle s = 120°$, and $\angle q = 120°$.

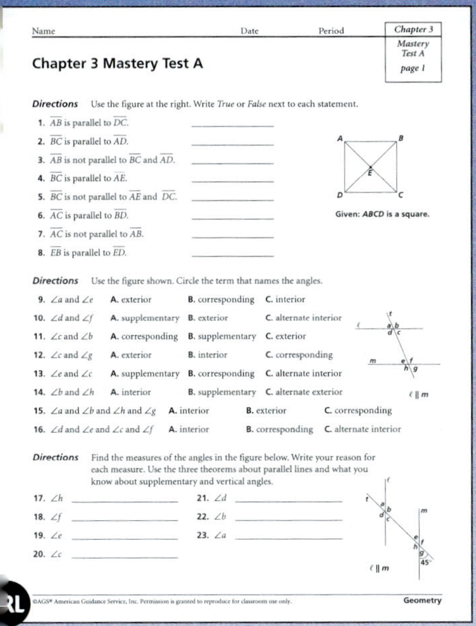

Name Date Period | Chapter 3 Mastery Test A page 1

Chapter 3 Mastery Test A

Directions Use the figure at the right. Write *True* or *False* next to each statement.

1. $\overline{AB}$ is parallel to $\overline{DC}$. ________
2. $\overline{BC}$ is parallel to $\overline{AD}$. ________
3. $\overline{AB}$ is not parallel to $\overline{BC}$ and $\overline{AD}$. ________
4. $\overline{BC}$ is parallel to $\overline{AE}$. ________
5. $\overline{BC}$ is not parallel to $\overline{AE}$ and $\overline{DC}$. ________
6. $\overline{AC}$ is parallel to $\overline{BD}$. ________
7. $\overline{AC}$ is not parallel to $\overline{AB}$. ________
8. $\overline{EB}$ is parallel to $\overline{ED}$. ________

Given: *ABCD* is a square.

Directions Use the figure shown. Circle the term that names the angles.

9. $\angle a$ and $\angle e$ A. exterior B. corresponding C. interior
10. $\angle d$ and $\angle f$ A. supplementary B. exterior C. alternate interior
11. $\angle c$ and $\angle b$ A. corresponding B. supplementary C. exterior
12. $\angle c$ and $\angle g$ A. exterior B. interior C. corresponding
13. $\angle e$ and $\angle c$ A. supplementary B. corresponding C. alternate interior
14. $\angle b$ and $\angle h$ A. interior B. supplementary C. alternate exterior
15. $\angle a$ and $\angle b$ and $\angle h$ and $\angle g$ A. interior B. exterior C. corresponding
16. $\angle d$ and $\angle e$ and $\angle c$ and $\angle f$ A. interior B. corresponding C. alternate interior

$\ell \parallel m$

Directions Find the measures of the angles in the figure below. Write your reason for each measure. Use the three theorems about parallel lines and what you know about supplementary and vertical angles.

17. $\angle h$ ________ 21. $\angle d$ ________
18. $\angle f$ ________ 22. $\angle b$ ________
19. $\angle e$ ________ 23. $\angle a$ ________
20. $\angle c$ ________

$\ell \parallel m$ 45°

 Geometry

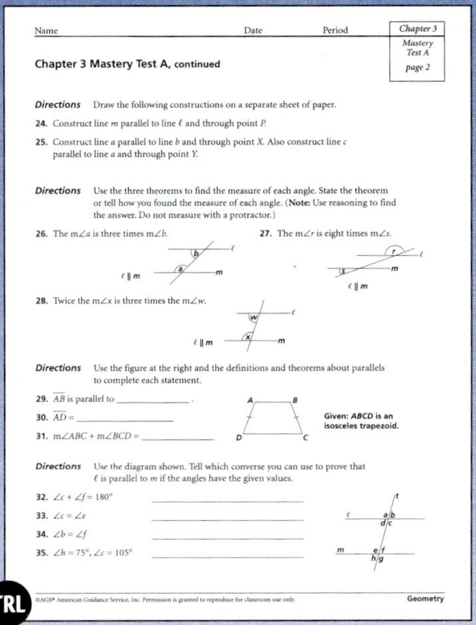

Name Date Period | Chapter 3 Mastery Test A page 2

Chapter 3 Mastery Test A, continued

Directions Draw the following constructions on a separate sheet of paper.

24. Construct line m parallel to line ℓ and through point P.
25. Construct line a parallel to line b and through point X. Also construct line c parallel to line a and through point Y.

Directions Use the three theorems to find the measure of each angle. State the theorem or tell how you found the measure of each angle. (**Note:** Use reasoning to find the answer. Do not measure with a protractor.)

26. The $m\angle a$ is three times $m\angle b$. ($\ell \parallel m$)
27. The $m\angle r$ is eight times $m\angle s$. ($\ell \parallel m$)
28. Twice the $m\angle x$ is three times the $m\angle w$. ($\ell \parallel m$)

Directions Use the figure at the right and the definitions and theorems about parallels to complete each statement.

29. $\overline{AB}$ is parallel to ________.
30. $\overline{AD}$ = ________
31. $m\angle ABC + m\angle BCD$ = ________

Given: *ABCD* is an isosceles trapezoid.

Directions Use the diagram shown. Tell which converse you can use to prove that ℓ is parallel to m if the angles have the given values.

32. $\angle c + \angle f = 180°$ ________
33. $\angle c = \angle e$ ________
34. $\angle b = \angle f$ ________
35. $\angle h = 75°$, $\angle c = 105°$ ________

 Geometry

TRL

Chapter 3 Mastery Test A

Alternative Assessment

continued

- Have students use postulates and theorems to construct parallel lines.

Draw a line and label it *Y*. Draw a point not on the line and draw a transversal through the point and your line. Choose a theorem, identify it, and use it to help you draw a line parallel to *Y*.

- Have students define quadrilaterals and parallels and apply theorems to them.

Write definitions for quadrilaterals and parallelograms and identify examples of each. Tell whether all quadrilaterals are parallelograms and why or why not. Use a theorem to explain why there are 360° in a parallelogram.

Use the vertical angle theorem to identify angles with the same measurement and explain why they are the same.

More Review...

11. $\angle w = \angle q = 135°$ corresponding angles
12. $\angle r + \angle q = 180°$, angles with common side a straight line are supp angles; $180° - 135° = 45°$; OR $\angle r$ and $\angle p$ are vertical angles, so $\angle r = 45°$
13. $\angle y = \angle r = 45°$; corresponding angles
14. $\angle x = 45°$ corresponding angles
15. $\angle z = \angle s = 135°$ corresponding angles OR $\angle w$ and $\angle z$ are vertical angles and so $m\angle w = m\angle z = 135°$
16. Indirect proof assumption is opposite of "To prove."
17. $m\angle a + m\angle b \neq 180°$
18. $m\angle a + m\angle b < 180°$ or $m\angle a + m\angle b > 180°$
19. Euclid's Postulate 5
20. This contradicts the given: $\ell \parallel m$

In this figure, $m \parallel \ell$, t is a transversal, and $m\angle q = 135°$. Find the measure (m) of each of the angles listed.

Example: $\angle p$ Solution: $\angle p$ and $\angle q$ are supplementary because they form a straight line. $m\angle q = 135°$, $m\angle p = 180° - 135° = 45°$

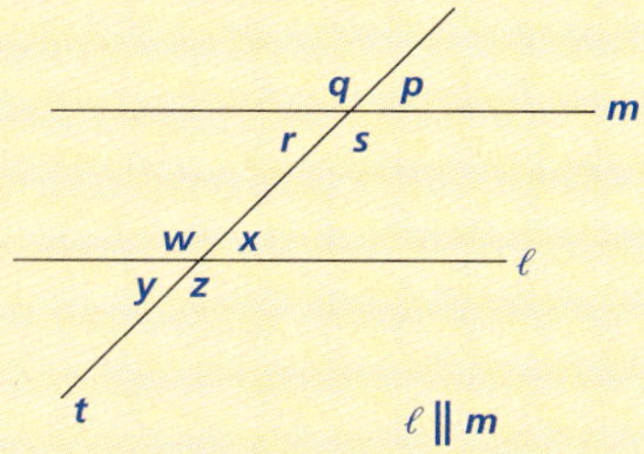

11. $\angle w$ **12.** $\angle r$ **13.** $\angle y$ **14.** $\angle x$ **15.** $\angle z$

Complete the following indirect proof (proof by contradiction).

Given: $\ell \parallel m$, transversal t

To Prove: $\angle a$ and $\angle b$ are supplementary

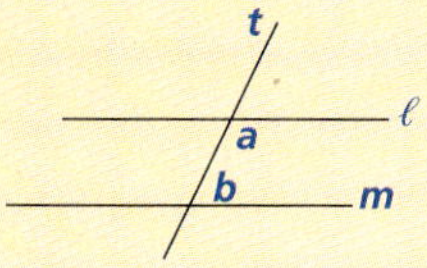

Proof

Statement	Reason
1. $\ell \parallel m$, transversal t	1. Given.
2. $\angle a$ and $\angle b$ are not supplementary.	2. **16.** ____
3. **17.** ____	3. Definition of not supplementary angles.
4. **18.** ____	4. Definition of not equal means greater than or less than.
5. ℓ must intersect m	5. **19.** ____
6. **20.** ____	6. Alternate Interior Angles Postulate.

∴ Theorem is valid, and proof is complete.

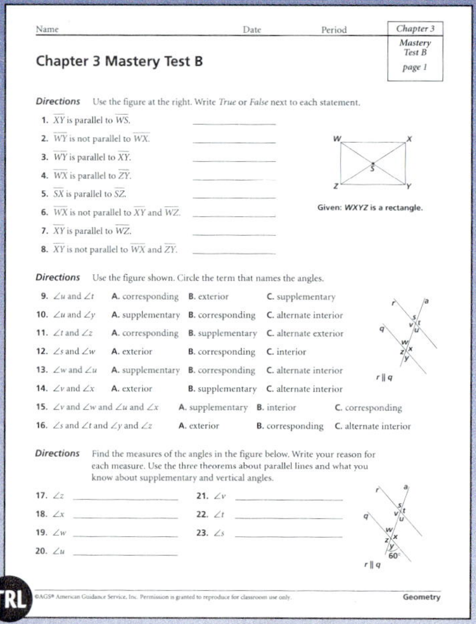

Name Date Period | *Chapter 3 Mastery Test B page 1*

Chapter 3 Mastery Test B

Directions Use the figure at the right. Write *True* or *False* next to each statement.

1. $\overline{XY}$ is parallel to $\overline{WS}$. ____
2. $\overline{WY}$ is not parallel to $\overline{WX}$. ____
3. $\overline{WY}$ is parallel to $\overline{XY}$. ____
4. $\overline{WX}$ is parallel to $\overline{ZY}$. ____
5. $\overline{SX}$ is parallel to $\overline{SZ}$. ____
6. $\overline{WX}$ is not parallel to $\overline{XY}$ and $\overline{WZ}$. ____
7. $\overline{XY}$ is parallel to $\overline{WZ}$. ____
8. $\overline{XY}$ is not parallel to $\overline{WX}$ and $\overline{ZY}$. ____

Given: *WXYZ* is a rectangle.

Directions Use the figure shown. Circle the term that names the angles.

9. $\angle u$ and $\angle t$ — A. corresponding B. exterior C. supplementary
10. $\angle u$ and $\angle y$ — A. supplementary B. corresponding C. alternate interior
11. $\angle t$ and $\angle z$ — A. corresponding B. supplementary C. alternate exterior
12. $\angle s$ and $\angle w$ — A. exterior B. corresponding C. interior
13. $\angle w$ and $\angle u$ — A. supplementary B. corresponding C. alternate interior
14. $\angle v$ and $\angle x$ — A. exterior B. supplementary C. alternate interior
15. $\angle v$ and $\angle w$ and $\angle u$ and $\angle x$ — A. supplementary B. interior C. corresponding
16. $\angle s$ and $\angle t$ and $\angle y$ and $\angle z$ — A. exterior B. corresponding C. alternate interior

Directions Find the measures of the angles in the figure below. Write your reason for each measure. Use the three theorems about parallel lines and what you know about supplementary and vertical angles.

17. $\angle z$ ____ 21. $\angle v$ ____
18. $\angle x$ ____ 22. $\angle t$ ____
19. $\angle w$ ____ 23. $\angle s$ ____
20. $\angle u$ ____

TRL

 Geometry

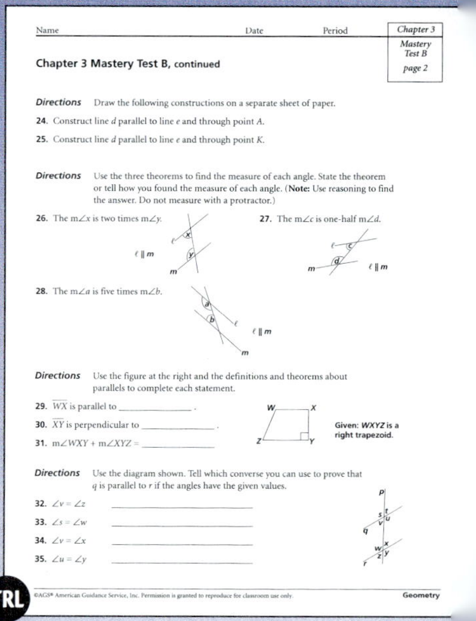

Name Date Period | *Chapter 3 Mastery Test B page 2*

Chapter 3 Mastery Test B, continued

Directions Draw the following constructions on a separate sheet of paper.

24. Construct line *d* parallel to line *e* and through point *A*.
25. Construct line *d* parallel to line *e* and through point *K*.

Directions Use the three theorems to find the measure of each angle. State the theorem or tell how you found the measure of each angle. (**Note:** Use reasoning to find the answer. Do not measure with a protractor.)

26. The $m\angle x$ is two times $m\angle y$.

27. The $m\angle c$ is one-half $m\angle d$.

28. The $m\angle a$ is five times $m\angle b$.

Directions Use the figure at the right and the definitions and theorems about parallels to complete each statement.

29. $\overline{WX}$ is parallel to ____.
30. $\overline{XY}$ is perpendicular to ____.
31. $m\angle WXY + m\angle XYZ =$ ____

Given: *WXYZ* is a right trapezoid.

Directions Use the diagram shown. Tell which converse you can use to prove that *q* is parallel to *r* if the angles have the given values.

32. $\angle v = \angle z$ ____
33. $\angle s = \angle w$ ____
34. $\angle v = \angle x$ ____
35. $\angle u = \angle y$ ____

TRL

 Geometry

Chapter 3 Mastery Test B

Find the measure of each angle. Use reasoning to determine the answer. Write the letter of the correct answer.

Example:

m∠x = 2m∠y

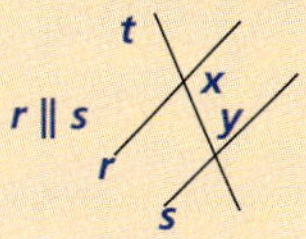

Solution: Since $r \parallel s$, $m\angle x + m\angle y = 180°$. Let $2m\angle y = m\angle x$; then $m\angle y + 2m\angle y = 180°$, $3m\angle y = 180°$, $m\angle y = 60°$, $m\angle x = 120°$ B

Find m∠x.

A. 60° B. 120° C. 180° D. 80°

Eight times m∠a = m∠b

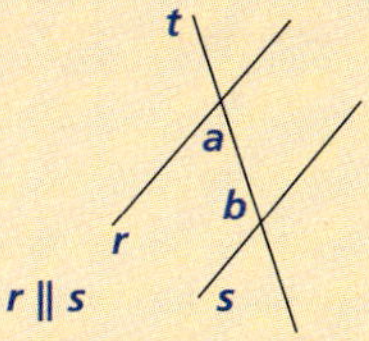

21. m∠*a*	**A.** 180°	**B.** 22.5°	**C.** 20°	**D.** 160°	C
22. m∠*b*	**A.** 180°	**B.** 220°	**C.** 160°	**D.** 20°	C

Name these geometric figures.

Example:

$\overline{AB} \parallel \overline{DC}$, $\overline{AD} \parallel \overline{BC}$ Solution: rectangle

23.

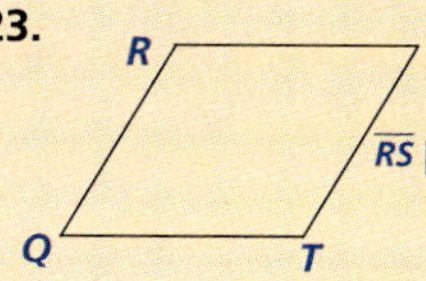

$\overline{RS} \parallel \overline{QT}$, $\overline{RQ} \parallel \overline{ST}$

24.

$\overline{AB} \parallel \overline{DC}$

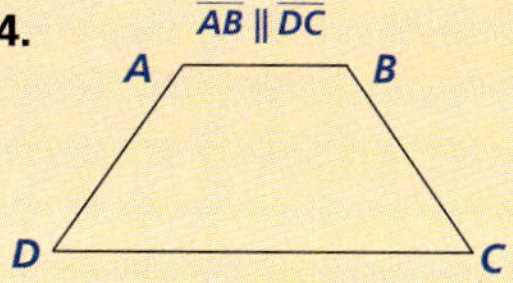

23. parallelogram

24. trapezoid

Follow the directions.

25. Draw line ℓ and point *X*, not on ℓ. Construct *m* parallel to ℓ using equal alternate interior angles.

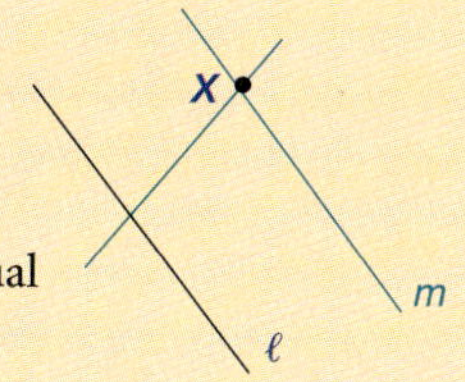

TEST-TAKING TIP

It is a good idea to double-check the location of numbers or letters that you identify in a diagram or figure.

Name Date Period

Chapter 3
Cumulative Mastery Test
page 1

Chapter 3 Cumulative Mastery Test

Directions Draw and copy each angle, identify it as *acute, obtuse, right,* or *straight,* and bisect the angle. Be sure to use a ruler or straightedge and a compass.

1.

______

2.

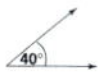

______

3.

______

4.

______

Directions Circle the correct measure of each angle in the figure.

5. ∠*b*	**A.** 77°	**B.** 57°	**C.** 103°	**D.** 20°
6. ∠*c*	**A.** 77°	**B.** 57°	**C.** 103°	**D.** 20°
7. ∠*d*	**A.** 77°	**B.** 57°	**C.** 103°	**D.** 20°
8. ∠*f*	**A.** 77°	**B.** 57°	**C.** 103°	**D.** 20°

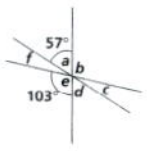

Directions Do these constructions on a separate sheet of paper. Tell which postulate(s) you used to make each construction.

9. Draw a circle using point *U* as the center and line *LP* as a 1-inch radius.

10. Draw two rectangles and state which angles are equal.

11. Draw a line between two points. Label the points *P* and *Q*.

12. Draw two intersecting lines to form four equal angles.

Directions Use the theorems to find the measures of angles *p*, *q*, and *r*.

13.

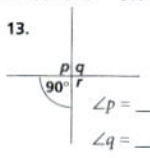

∠*p* = ______
∠*q* = ______
∠*r* = ______

14.

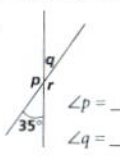

∠*p* = ______
∠*q* = ______
∠*r* = ______

15.

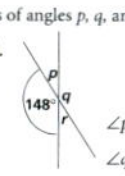

∠*p* = ______
∠*q* = ______
∠*r* = ______

16.

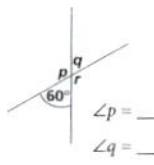

∠*p* = ______
∠*q* = ______
∠*r* = ______

TRL

 Geometry

Name Date Period

Chapter 3
Cumulative Mastery Test
page 2

Chapter 3 Cumulative Mastery Test, continued

Directions Use the figure shown. Name the following angles as *exterior, interior, alternate interior, alternate exterior, corresponding,* or *supplementary.*

17. ∠*c* and ∠*e* ______

18. ∠*a* and ∠*b* and ∠*g* and ∠*h* ______

19. ∠*c* and ∠*d* and ∠*e* and ∠*f* ______

20. ∠*e* and ∠*f* ______

21. ∠*b* and ∠*h* ______

22. ∠*b* and ∠*f* ______

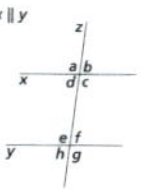

Directions Find the measures of the angles in the figure shown. Write your reason for each measure. Use the three theorems about parallel lines and what you know about supplementary and vertical angles.

23. ∠*g* ______ **26.** ∠*b* ______

24. ∠*f* ______ **27.** ∠*c* ______

25. ∠*d* ______

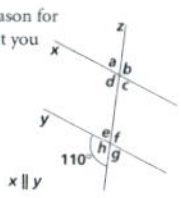

Directions Use the figure at the right and the definitions and theorems about parallels to complete each statement.

28. ∠*OLM* + ∠*NOL* = ______

29. ∠*LMN* + ∠*MNO* = ______

30. Angle sum = ______

31. $\overline{LM}$ is parallel to ______.

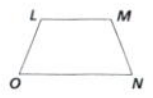

Given: *LMNO* is a trapezoid.

Directions Use the diagram shown. Tell which converse you can use to prove that *f* is parallel to *g* if the angles have the given values.

32. ∠*m* = ∠*q* **34.** ∠*s* = 85°, ∠*n* = 95°

33. ∠*n* = ∠*q* **35.** ∠*s* = ∠*o*

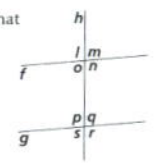

TRL

 Geometry

Chapter 3 Cumulative Mastery Test

Chapter 4

Planning Guide

Using Algebra: Lines in the Coordinate Plane

	Student Text Lesson						
	Student Pages	Vocabulary	Proof	Practice Exercises	Problem Solving	Calculator Practice	Solutions Key
Lesson 1 Naming Points	96–100	✔		✔	100		✔
Lesson 2 Horizontal Lines	101–103			✔	103		✔
Lesson 3 Vertical Lines	104–106			✔	106		✔
Lesson 4 Slope of a Line	107–112	✔		✔		111	✔
Lesson 5 Using Algebra: $y = mx + b$	113–118	✔		✔	118		✔
Lesson 6 The Point-Slope Formula	119–122			✔			✔
Lesson 7 Finding the Midpoint of a Segment	123–127	✔		✔		126	✔
Application Screaming Slopes	128			✔			✔

Chapter Activities

Teacher's Resource Library
Geometry on the Job 4: Coordinate Planes and Radar

Community Connection 4: Using a Coordinate Plane at Home

Teacher's Edition
Chapter 4 Project

Assessment Options

Student Text
Chapter 4 Review

Teacher's Resource Library
Chapter 4 Mastery Tests A and B

Chapter 4 Cumulative Mastery Test

Teacher's Edition
Chapter 4 Alternative Assessment

Student Text Features						Teaching Strategies								Learning Styles						Teacher's Resource Library		
Algebra Connection	Constructions	Experiencing Geometry	Technology Connection	Try This	Writing About Mathematics	Warm-Up Activity	Common Error	Algebra Connection	Mental Math	Manipulatives	Calculator	Applications	Group Problem Solving	Auditory/Verbal	Visual/Spatial	Body/Kinesthetic	Logical/Mathematical	Interpersonal/Group Learning	LEP/ESL	Workbook Activities	Additional Practice	Self-Study Guide
						96	98		99			100	100	99	97	97			98	30, 31	18	✔
01	103		103			101	102	102						103	102					32, 33	19	✔
04		106				104	105	104	106					105	106					34, 35	20	✔
07						107	109	107			111	108, 110		111	112		109			36, 37	21	✔
14	117			118	117	113	115	114	116						117		115		118	38, 39	22	✔
20	119–122				122	119	121	120			121			122	122		120, 121	121		40, 41	23	✔
24	123, 126				126	123		124	125	124	126			127	126	126	123			42, 43	24	✔
						128									128		128			44		✔

Manipulatives

Factor Blocks

Manipulatives Master 2 (Coordinate Grid)

Chapter at a Glance

Chapter 4: Using Algebra: Lines in the Coordinate Plane

pages 94–131

Lessons

Teacher's Resource Library TRL

- Workbook Activities 30–44
- Additional Practice 18–24
- Geometry on the Job 4
- Community Connection 4
- Chapter 4 Self-Study Guide
- Chapter 4 Mastery Tests A and B
- Chapter 4 Cumulative Mastery Test

(Answer Keys for the Teacher's Resource Library begin on page 484 of this Teacher's Edition.)

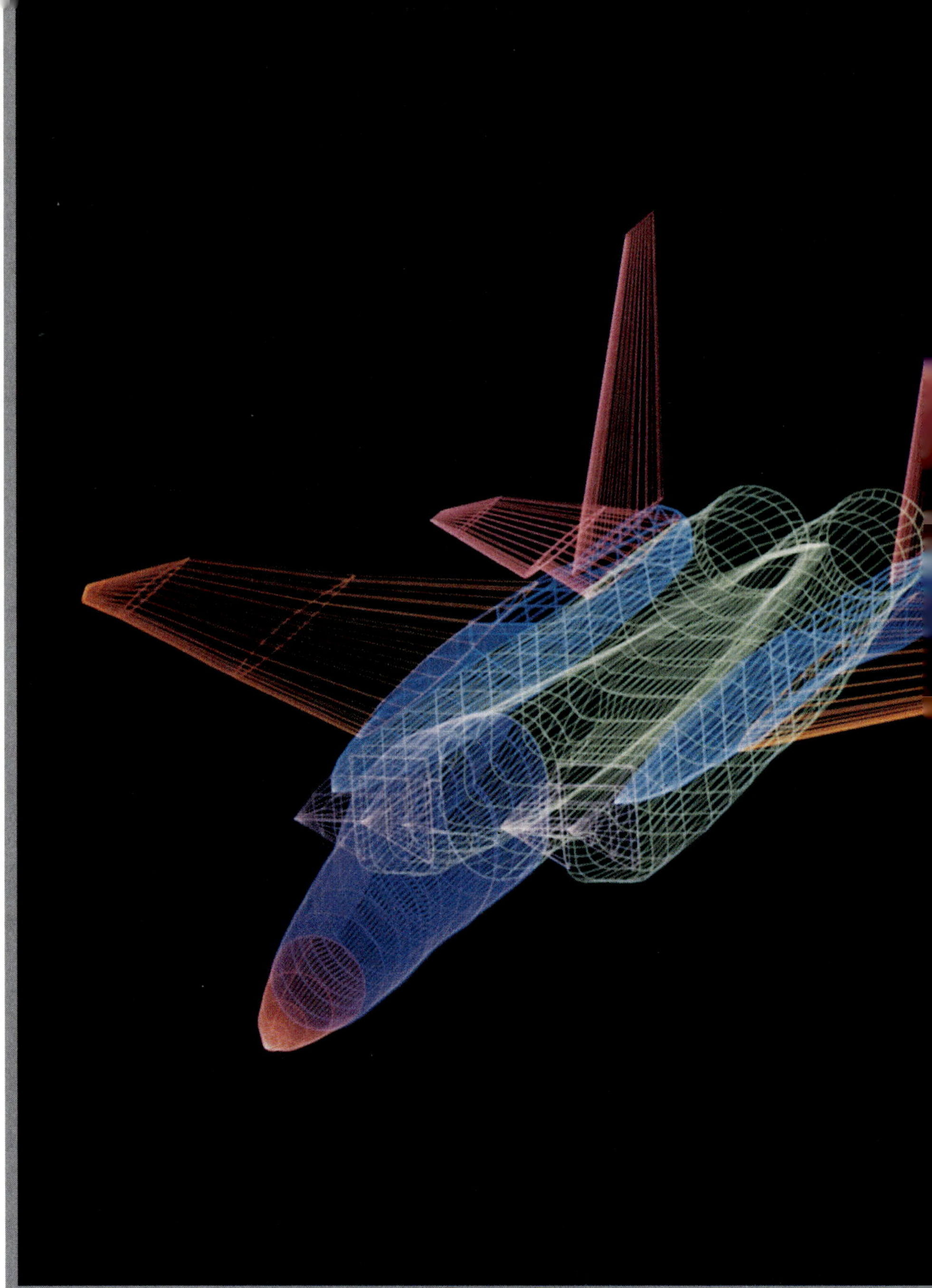

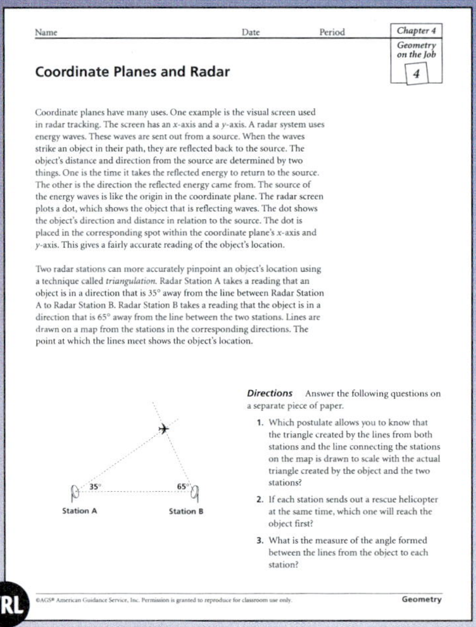

Name Date Period | Chapter 4 *Geometry on the Job* 4

Coordinate Planes and Radar

Coordinate planes have many uses. One example is the visual screen used in radar tracking. The screen has an *x*-axis and a *y*-axis. A radar system uses energy waves. These waves are sent out from a source. When the waves strike an object in their path, they are reflected back to the source. The object's distance and direction from the source are determined by two things. One is the time it takes the reflected energy to return to the source. The other is the direction the reflected energy came from. The source of the energy waves is like the origin in the coordinate plane. The radar screen plots a dot, which shows the object that is reflecting waves. The dot shows the object's direction and distance in relation to the source. The dot is placed in the corresponding spot within the coordinate plane's *x*-axis and *y*-axis. This gives a fairly accurate reading of the object's location.

Two radar stations can more accurately pinpoint an object's location using a technique called *triangulation*. Radar Station A takes a reading that an object is in a direction that is 35° away from the line between Radar Station A to Radar Station B. Radar Station B takes a reading that the object is in a direction that is 65° away from the line between the two stations. Lines are drawn on a map from the stations in the corresponding directions. The point at which the lines meet shows the object's location.

Directions Answer the following questions on a separate piece of paper.

1. Which postulate allows you to know that the triangle created by the lines from both stations and the line connecting the stations on the map is drawn to scale with the actual triangle created by the object and the two stations?
2. If each station sends out a rescue helicopter at the same time, which one will reach the object first?
3. What is the measure of the angle formed between the lines from the object to each station?

 Geometry

TRL

Geometry on the Job 4

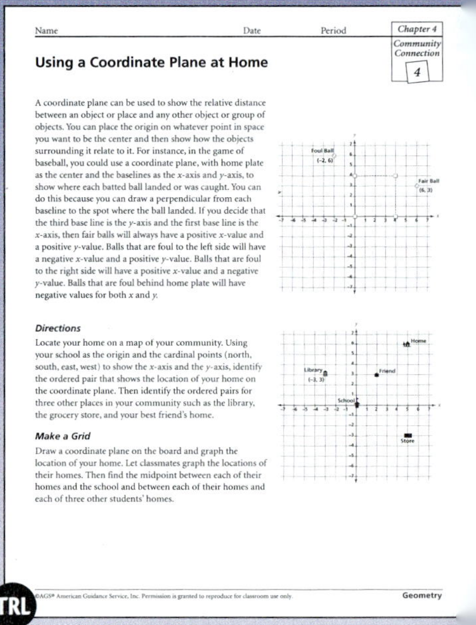

Name Date Period | Chapter 4 *Community Connection* 4

Using a Coordinate Plane at Home

A coordinate plane can be used to show the relative distance between an object or place and any other object or group of objects. You can place the origin on whatever point in space you want to be the center and then show how the objects surrounding it relate to it. For instance, in the game of baseball, you could use a coordinate plane, with home plate as the center and the baselines as the *x*-axis and *y*-axis, to show where each batted ball landed or was caught. You can do this because you can draw a perpendicular from each baseline to the spot where the ball landed. If you decide that the third base line is the *y*-axis and the first base line is the *x*-axis, then fair balls will always have a positive *x*-value and a positive *y*-value. Balls that are foul to the left side will have a negative *x*-value and a positive *y*-value. Balls that are foul to the right side will have a positive *x*-value and a negative *y*-value. Balls that are foul behind home plate will have negative values for both *x* and *y*.

Directions

Locate your home on a map of your community. Using your school as the origin and the cardinal points (north, south, east, west) to show the *x*-axis and the *y*-axis, identify the ordered pair that shows the location of your home on the coordinate plane. Then identify the ordered pairs for three other places in your community such as the library, the grocery store, and your best friend's home.

Make a Grid

Draw a coordinate plane on the board and graph the location of your home. Let classmates graph the locations of their homes. Then find the midpoint between each of their homes and the school and between each of their homes and each of three other students' homes.

 Geometry

TRL

Community Connection 4

Chapter 4

Using Algebra: Lines in the Coordinate Plane

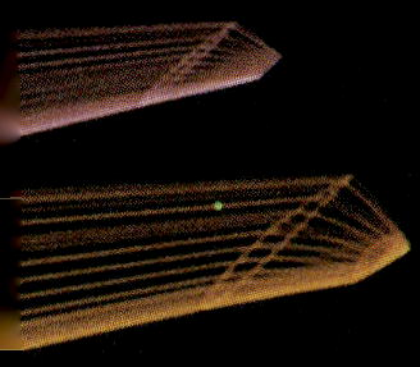

Leonardo da Vinci, scientist and artist, drew his plans for aircraft during the Renaissance. Hundreds of years later in the 1900s, the Wright Brothers developed plans for airplanes that could actually fly. Today, aerospace engineers draw computer models of aircraft. These increasingly sophisticated drawings are the first steps in designing and building jets. All the drawings, from those of da Vinci to those of today, have something in common. The people who drew them applied mathematics and science. Engineers of today have put geometry to practical use. The result is their aircraft designs for today and tomorrow.

In Chapter 4, you will learn about points, lines, and angles in coordinate planes.

Goals for Learning

- To identify the algebraic equation for a line
- To use ordered pairs to graph lines on a coordinate plane
- To define and find the slope of a line
- To write and calculate the equations for lines
- To graph lines given slope and ordered pairs
- To find the midpoint of a line segment

SELF-STUDY GUIDE

Name ____________

CHAPTER 4: Using Algebra: Lines in the Coordinate Plane

Goal 4.1 *To identify the algebraic equation for a line*

Date	Assignment	Score
______	1: Read pages 96–98. Complete Exercises A–E on pages 99–100.	______
______	2: Complete Workbook Activities 30–31.	______

Comments:

Goal 4.2 *To use ordered pairs to graph lines on a coordinate plane*

Date	Assignment	Score
______	3: Read pages 101–102. Complete Exercises A–E on pages 102–103.	______
______	4: Complete Workbook Activities 32–33.	______
______	5: Read pages 104–105. Complete Exercises A–E on page 106.	______
______	6: Complete Workbook Activities 34–35.	______

Comments:

Goal 4.3 *To define and find the slope of a line*

Date	Assignment	Score
______	7: Read pages 107–110.	______
______	8: Read and complete the Calculator Practice and Exercise A on page 111.	______
______	9: Complete Exercises B–D on pages 111–112.	______
______	10: Complete Workbook Activities 36–37.	______

Comments:

 Geometry

TRL

SELF-STUDY GUIDE

Name ____________

CHAPTER 4: Using Algebra: Lines in the Coordinate Plane, continued

Goal 4.4 *To write and calculate the equations for lines*

Date	Assignment	Score
______	11: Read pages 113–117. Complete Exercises A–D on pages 117–118.	______
______	12: Complete Workbook Activities 38–39.	______

Comments:

Goal 4.5 *To graph lines given slope and ordered pairs*

Date	Assignment	Score
______	13: Read pages 119–120. Complete Exercises A and B on page 122.	______
______	14: Complete Workbook Activities 40–41.	______

Comments:

Goal 4.6 *To find the midpoint of a line segment*

Date	Assignment	Score
______	15: Read pages 123–125.	______
______	16: Read and complete the Calculator Practice and Exercise A on page 126.	______
______	17: Complete Exercises B and C on pages 126–127.	______
______	18: Complete Workbook Activities 42–43.	______
______	19: Read and complete the Application Exercise on page 128.	______
______	20: Complete Workbook Activity 44.	______
______	21: Complete the Chapter 4 Review on pages 129–131.	______

Comments:

Student's Signature ____________ Date ____________

Instructor's Signature ____________ Date ____________

 Geometry

TRL

Chapter 4 Self-Study Guide

Introducing the Chapter

Use the information on these pages to help students think about how geometry is used to design and build airplanes and other constructions, such as buildings and cars. Discuss the use of lines in scale drawings. If possible, display several scale drawings with grids. Present the Goals for Learning, briefly discussing each one.

Chapter Project

Encourage students to work in groups of three or four. Suggest that they produce one of the following for their chapter project.

- A model ski resort—Students can create a resort complete with beginners, intermediate, and advanced slopes. In addition to the slopes, students should plan and construct resort chalets, lifts, hotels, shops, and roadways, using a grid pattern. Encourage students to use the slope formula to determine the grades of the various ski slopes. They may wish to use the Internet to find out about the slopes used at actual resorts.
- Formula Notebook—Throughout the text, algebraic formulas are used to solve geometric problems. Suggest that students prepare a notebook containing these algebraic formulas. They should list each formula, identify the elements in the formula, and provide step-by-step instructions for using it or examples showing its use. Students can add to their notebook throughout their study of geometry and use the notebook as a study tool.

Lesson at a Glance

Chapter 4 Lesson 1

Overview This lesson introduces the coordinate plane and explains how to name and locate points using ordered pairs.

Objective

- To identify the parts of a coordinate plane.
- To locate and plot points on a coordinate plane.

Student Pages 96–100

Teacher's Resource Library TRL

Workbook Activities 30–31

Additional Practice 18

Mathematics Vocabulary

coordinate plane
graph
horizontal
ordered pair
origin
quadrant
vertical

1. Warm-Up Activity

With the class, create a map for a fictional town where all the streets are straight and perpendicular to one another. Encourage students to name the streets, include some landmarks, and mark two specific locations. Discuss ways to describe how to get from one of the marked locations to the other. Lead students to realize that their instructions will generally include two components: a direction and a distance (for example, 5 units up, down, right, or left.) Discuss other uses for this type of grid system.

2. Teaching the Lesson

After using the board to demonstrate the example on page 96, Point *P* located at (2, 3), discuss how a point's location is based on its position relative to the origin. Emphasize that points are always plotted by moving left or right, then up or down.

Lesson 1 Naming Points

Coordinate plane
Plane formed by placing two real number lines at right angles

Graph
A diagram showing how one quantity depends on another

Horizontal
Parallel to the horizon

Ordered pair
A set of two real numbers that locate a point on a plane

Origin
The point at which the x-axis and y-axis in the coordinate system intersect

Quadrant
One-fourth of the coordinate plane

Vertical
Straight up and down

A plane, as Euclid saw it, has no named points except those that you place in it. Another way to think about a plane is to imagine that every point in the plane has a specific name. The Ruler Postulate allows you to do this. First assign real numbers on a **horizontal** line and on a **vertical** line.

Real Number Lines

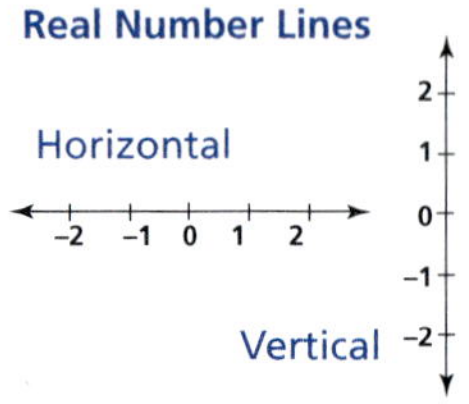

Then place the two number lines at right angles to each other, allowing the zero points of each to match. You have now formed a **coordinate plane.**

Coordinate Plane

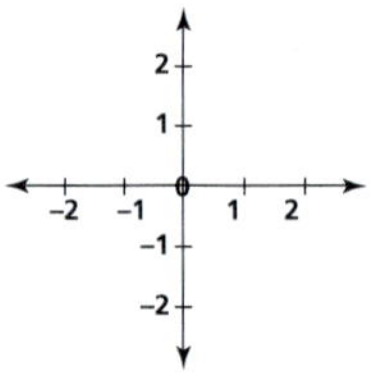

Points in the coordinate plane are named using their position relative to the *x*- and *y*-axis. The points are placed on a **graph.**

EXAMPLE To name point *P*

Step 1 Draw a perpendicular to the *x*-axis. Locate the point where the perpendicular meets the *x*-axis. This number is the *x*-value of point *P*.

Step 2 Draw a perpendicular to the *y*-axis. Locate the point where the perpendicular meets the *y*-axis. This number is the *y*-value of point *P*.

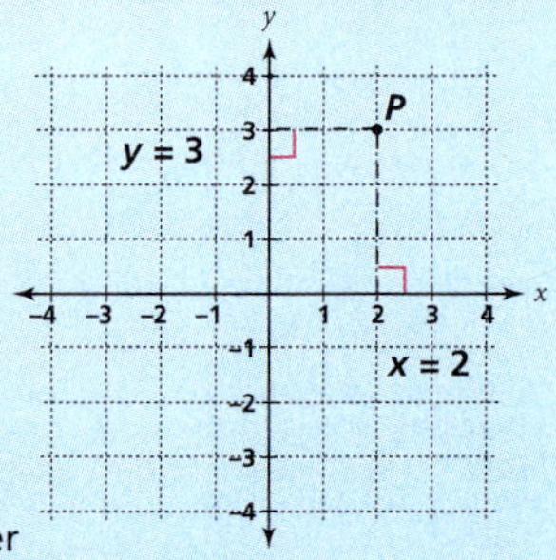

Point *P* can be named by using both the *x*- and *y*-values together. In this case, $x = 2$ and $y = 3$. Therefore you can name the point $P = (2, 3)$.

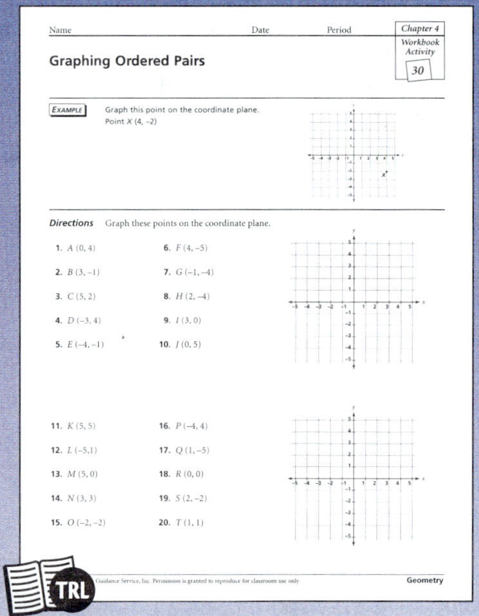
Name Date Period Chapter 4 Workbook Activity 30

Graphing Ordered Pairs

EXAMPLE Graph this point on the coordinate plane. Point X (4, –2)

Directions Graph these points on the coordinate plane.

1. A (0, 4) 6. F (4, –5)
2. B (3, –1) 7. G (–1, –4)
3. C (5, 2) 8. H (2, –4)
4. D (–3, 4) 9. I (3, 0)
5. E (–4, –1) 10. J (0, 5)

11. K (5, 5) 16. P (–4, 4)
12. L (–5, 1) 17. Q (1, –5)
13. M (5, 0) 18. R (0, 0)
14. N (3, 3) 19. S (2, –2)
15. O (–2, –2) 20. T (1, 1)

Geometry

Workbook Activity 30

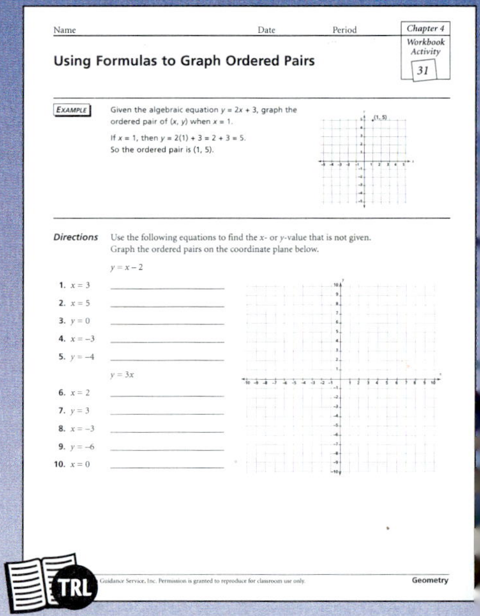
Name Date Period Chapter 4 Workbook Activity 31

Using Formulas to Graph Ordered Pairs

EXAMPLE Given the algebraic equation $y = 2x + 3$, graph the ordered pair of (x, y) when $x = 1$.
If $x = 1$, then $y = 2(1) + 3 = 2 + 3 = 5$.
So the ordered pair is (1, 5).

Directions Use the following equations to find the x- or y-value that is not given. Graph the ordered pairs on the coordinate plane below.

$y = x - 2$
1. $x = 3$
2. $x = 5$
3. $y = 0$
4. $x = -3$
5. $y = -4$

$y = 3x$
6. $x = 2$
7. $y = 3$
8. $x = -3$
9. $y = -6$
10. $x = 0$

Geometry

Workbook Activity 31

An **ordered pair** of real numbers (x, y) names each and every point on the coordinate plane. The pair is called an ordered pair because the x-value is always given first and the y-value is always given second.

First Second
(x, y) ordered pair of real numbers

EXAMPLES Name the points *A, B,* and *C* as ordered pairs.

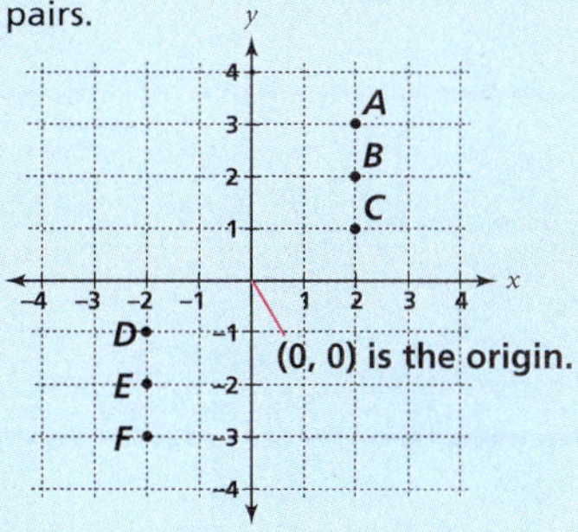

$A = (2, 3)$ x-value is 2, y-value is 3

$B = (2, 2)$ x-value is 2, y-value is 2

$C = (2, 1)$ x-value is 2, y-value is 1

Notice that the x-values for these three points are on the same vertical line, $x = 2$.

Name the points *D, E,* and *F* as ordered pairs.

$D = (-2, -1)$ x-value is -2, y-value is -1

$E = (-2, -2)$ x-value is -2, y-value is -2

$F = (-2, -3)$ x-value is -2, y-value is -3

The x-values for these three points are on the same vertical line, $x = -2$.

The ordered pair (0, 0) is called the **origin.**

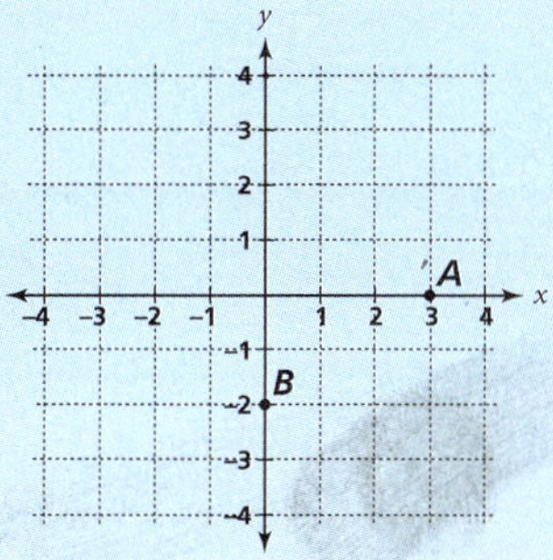

Points on the axes can be named in two ways. *A* can be named as (3, 0). The x-value is 3 and the y-value is 0. *A* can also be named $x = 3$.

B can be named as (0, −2). The x-value is 0, the y-value is −2. *B* can also be named $y = -2$.

It is usually better to use the ordered pair name even if a point is on the x- or y-axis. The ordered pair name specifies a single point and excludes all other points. This eliminates any confusion about the point's location.

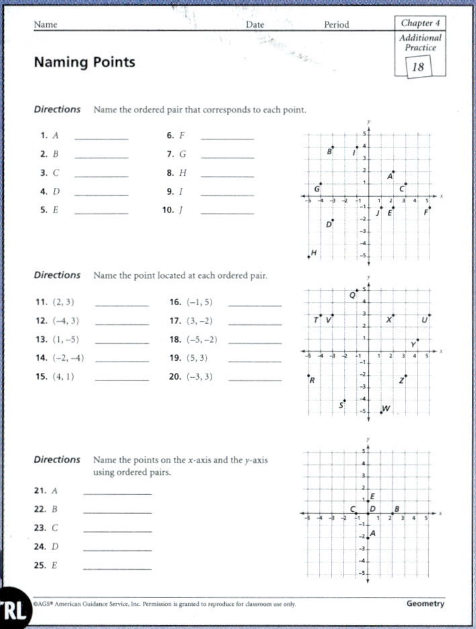

Name Date Period

Chapter 4 *Additional Practice* *18*

Naming Points

Directions Name the ordered pair that corresponds to each point.

1. *A* ________
2. *B* ________
3. *C* ________
4. *D* ________
5. *E* ________
6. *F* ________
7. *G* ________
8. *H* ________
9. *I* ________
10. *J* ________

Directions Name the point located at each ordered pair.

11. (2, 3) ________
12. (−4, 3) ________
13. (1, −5) ________
14. (−2, −4) ________
15. (4, 1) ________
16. (−1, 5) ________
17. (3, −2) ________
18. (−5, −2) ________
19. (5, 3) ________
20. (−3, 3) ________

Directions Name the points on the x-axis and the y-axis using ordered pairs.

21. *A* ________
22. *B* ________
23. *C* ________
24. *D* ________
25. *E* ________

 Geometry

RL

Additional Practice 18

After reviewing the last example on page 98, ask for volunteers to explain in their own words why all the points in Quadrants I, II, III, and IV have the x- and y-values as described in the example. (A sample answer may include both x and y are positive in Quadrant I because all the points are to the right of and above the origin.)

3. Reinforce and Extend

LEARNING STYLES

Visual/Spatial Suggest that students work in pairs to prepare a study sheet that they can use to reinforce vocabulary. Tell them to list vocabulary words and their definitions on the top half of a page and draw a graph on the bottom half of the page, labeling its quadrants I, II, III, and IV. On the graph, they should draw a coordinate plane, horizontal and vertical lines, and a point labeled with its ordered pair. Students can then use the study sheet to help one another learn and understand the lesson's vocabulary. For example, one student points to a word or something on the graph. The other student reads and defines the word and locates its illustration on the graph or identifies the part of the graph by name and definition. Then students reverse roles.

LEARNING STYLES

Body/Kinesthetic Have students become a living graph. Ask them to form the axes of a coordinate "plane" by rearranging their desks into two lines at right angles. They should label each desk with appropriate numbers for the x- and y-axes. Call on a student by name, give him or her an ordered pair, and ask the student to stand at the approximate location of the point on the coordinate plane. Continue until all the students are standing at points on the graph.

Learning Styles

LEP/ESL Some students may not be familiar with the Roman numerals used to label quadrants on a graph. To help them equate the numerals to those in the more common Arabic numbering system, write I = 1, II = 2, III = 3, IV = 4 on the board and have students copy the statements on paper. Ask pairs of students of varying English proficiency to help one another reinforce the equivalents in the numbering systems. Encourage them to assess one another's understanding of the Roman numerals by preparing a short quiz, using any format they wish, such as multiple choice, matching, or fill in the blank. For example, on their quizzes, they might write the Roman numerals in random order and ask their partners to fill in the equivalent Arabic numerals.

Common Error

Students sometimes plot points in the wrong quadrant. Encourage them to determine the two directions they will move from the origin to plot a point (as indicated by the signs of the coordinates) before they graph the point. Once they know the quadrant a point is in, they can then focus on the specific location of the point within that quadrant. Ask students to draw a grid on a sheet of paper and label the quadrants by their Roman numerals. Explain that the quadrants are numbered counter-clockwise, beginning with the upper right quadrant. In Quadrant I, students should write $x = +, y = +$; in Quadrant II, they write $x = +, y = -$; in Quadrant III, $x = -, y = -$; and in Quadrant IV, $x = -, y = +$. Suggest that students refer to their quadrants as they plot points on a graph.

Examples Name each of the points as an ordered pair.

$A = (-2, 0)$, $B = (-1, 0)$, $C = (0, 0)$,
$D = (1, 0)$, $E = (2, 0)$, $F = (3, 0)$

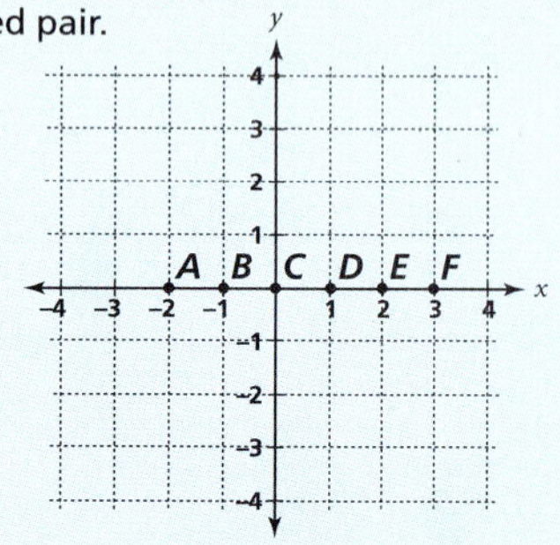

Name each of the points as an ordered pair.

$U = (0, -2)$, $V = (0, -1)$, $W = (0, 0)$,
$X = (0, 1)$, $Y = (0, 2)$, $Z = (0, 3)$

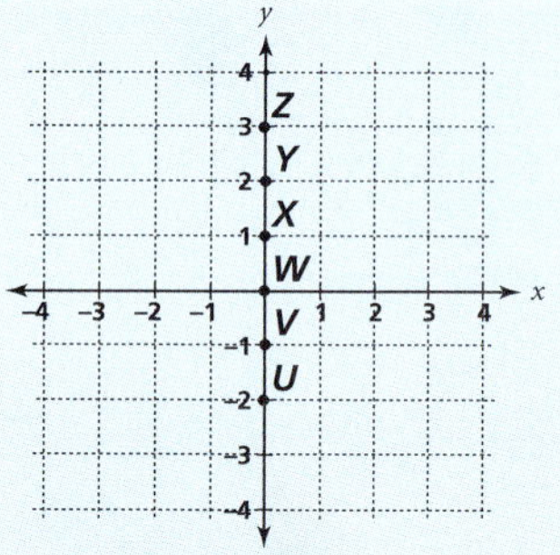

Example The x-axis and the y-axis divide the coordinate system into four regions called **quadrants.** Name the quadrant of each point's location.

Point $A = (2, 2)$ is in Quadrant I.

Point $B = (-3, 1)$ is in Quadrant II.

Point $C = (-1, -3)$ is in Quadrant III.

Point $D = (3, -3)$ is in Quadrant IV.

In Quadrant I, all points have x- and y-values greater than 0. That is $x > 0$ and $y > 0$.

In Quadrant II, all points have $x < 0$ and $y > 0$.

In Quadrant III, all points have $x < 0$ and $y < 0$.

In Quadrant IV, all points have $x > 0$ and $y < 0$.

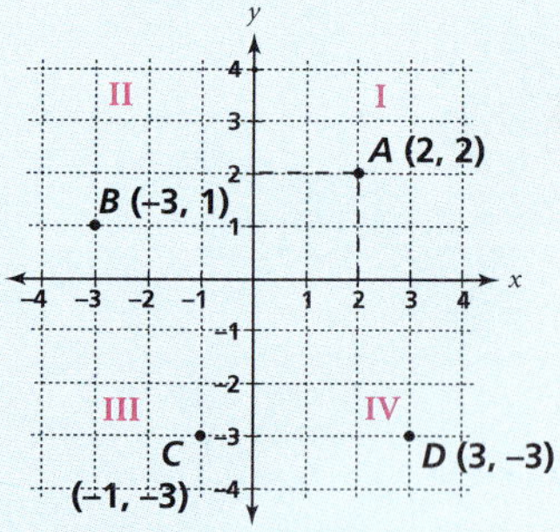

Exercise A Name the ordered pair that corresponds to each point.

1. *A*	(1, 1)	**6.** *F*	(−4, 1)
2. *B*	(3, 2)	**7.** *G*	(−1, −3)
3. *C*	(2, 3)	**8.** *H*	(−3, −2)
4. *D*	(4, 3)	**9.** *I*	(2, −2)
5. *E*	(−2, 1)	**10.** *J*	(4, −4)

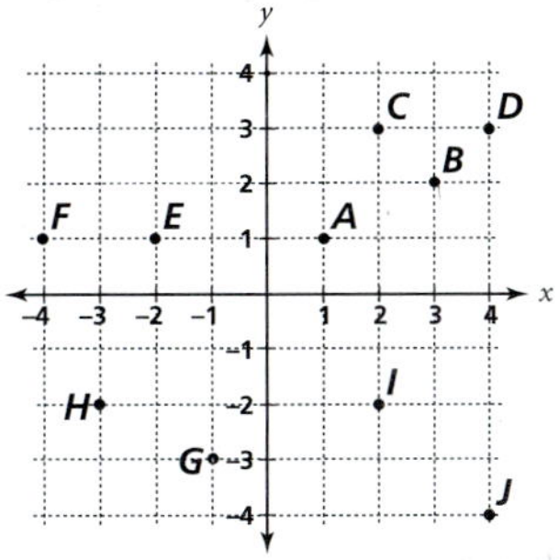

Exercise B Name the point located at each ordered pair.

11. (−2, 2)	Q	**16.** (−2, −2)	Y
12. (3, −3)	X	**17.** (3, 5)	R
13. (−3, −3)	T	**18.** (−4, 3)	Z
14. (2, 2)	S	**19.** (−4, 0)	W
15. (1, −1)	U	**20.** (0, 0)	V

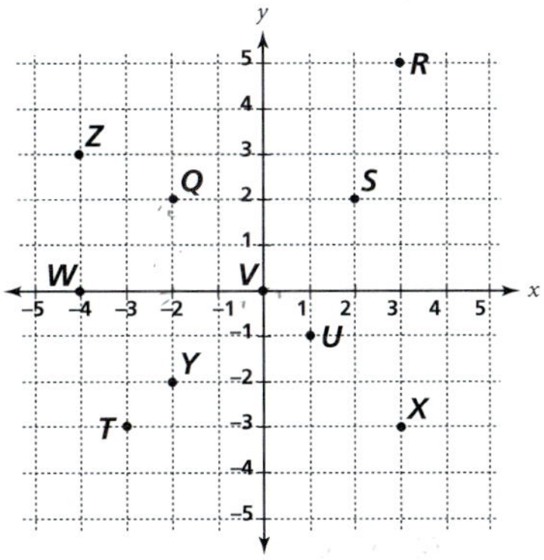

Exercise C Name the points on the *x*-axis using ordered pairs.

21. *A*	(−2, 0)
22. *B*	(−1, 0)
23. *C*	(0, 0)
24. *D*	(1, 0)
25. *E*	(2, 0)

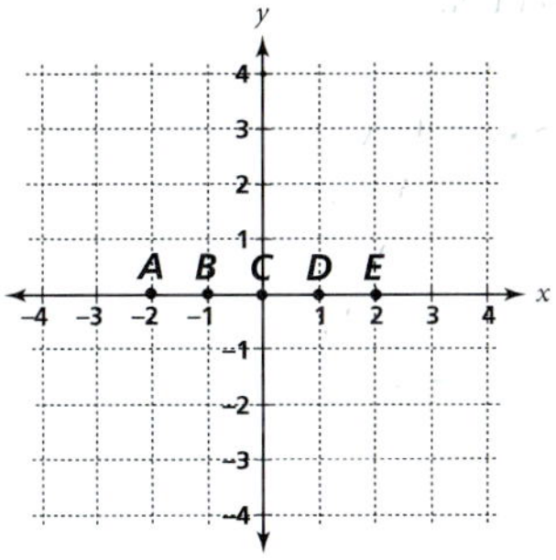

Selected Answers and Solutions begin on page 420.

Mental Math

Have one student name an ordered pair and ask another student to name the quadrant that pair is in on the coordinate plane. Reverse the activity by having one student name a quadrant and the other give an ordered pair for a point in that quadrant.

Learning Styles

Auditory/Verbal Have students work in pairs to complete Exercise A. Tell one student to describe the location of the points in the problems while the other student writes the ordered pair for that point.

Group Problem Solving

Ask groups of students to imagine they work at a travel agency and are responsible for coordinating a sightseeing tour. Have them make a tour map, using a coordinate plane. They may choose an actual geographical location or a fictional one. Explain that if they use an actual location, they will have to superimpose their grid over a map of the location. They should pinpoint several places of interest on the map and provide clear and accurate directions for how to get from one point to the next. Along with their map, students should prepare a formal travel route plan that can be displayed in the classroom.

Application

In the Community

There are many government departments, such as the County Surveyor's Office and the Municipal Public Works Department, and private businesses that use coordinate grids. Encourage students to make an appointment to visit one of these departments or businesses to talk with an employee who uses coordinate graphs. Ask students to find out how the workers use the grids and why. Provide time for students to offer a brief summary of their discussion to the class.

31. $x = 3$; $-3 \le y \le 2$
32. $-2 \le x \le 3$; $y = 2$
33. $x \le 0$
34. Each system uses a grid, the globe grid has lines of latitude roughly similar to lines || the x-axis; the lines of longitude are roughly similar to lines || the y-axis. They rely on two scales of real numbers. The global grid names points on the surface of a sphere, the coordinate system names points on a plane.

Exercise D Name the points on the y-axis using ordered pairs.

26. V (0, 2)

27. W (0, 1)

28. X (0, 0)

29. Y (0, –1)

30. Z (0, –2)

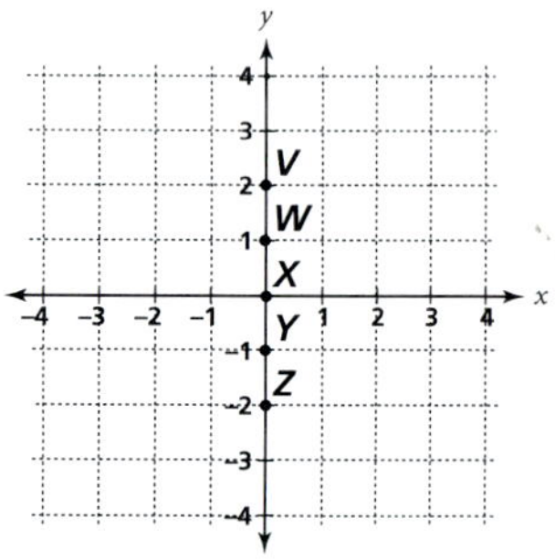

Problem Solving

Exercise E Answer the following questions.

31. What can you say about the locations of these points in the coordinate plane: (3, 1), (3, 2), (3, 0), (3, –1), (3, –2) and (3, –3)?

32. What generalization can you make about the locations of these points in the coordinate plane: (–2, 2) (–1, 2), (0, 2), (1, 2), (2, 2), and (3, 2)?

33. Use algebraic terms to describe the points on the x-axis in quadrant II.

34. You can locate any city in the world using latitude and longitude. For example, Denver, Colorado, is located at 40°N, 105°W; Belo Horizonte, Brazil, is located at 20°S and 45°W. How are the lines of latitude and longitude on a map like a coordinate plane? How are they different?

35. In chess, each square has a name consisting of a letter and number, such as *b6* or *f8*. Why might this be useful to people playing chess by E-mail? How is the identification system for a chessboard like a coordinate plane? How is it different?

35. Each square can be specified using a letter-number combination. Unlike the coordinate system, which names intersections or points, the chess system names specific areas.

Lesson 2 Horizontal Lines

Horizontal lines are lines parallel to the x-axis. To help yourself remember this idea, think of the "horizon." The horizon appears as a straight line at the edge of your vision as you look across a field, lake, or ocean.

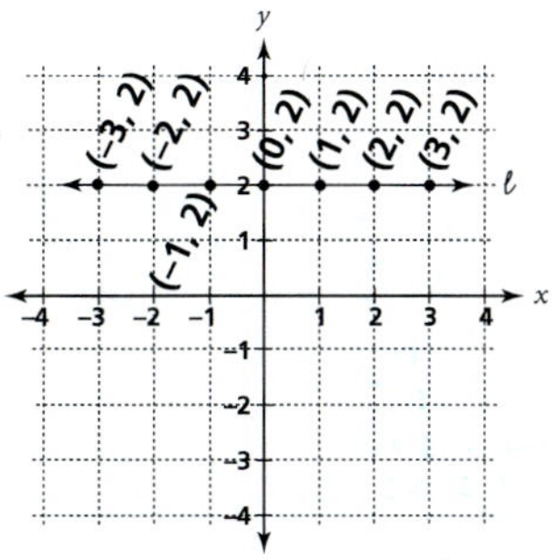

On the graph, line ℓ passes through the point (0, 2). Each point on ℓ has a different x-value, but all points have the same y-value, 2. If you wanted to describe the line using ordered pairs, it would be impossible to name all the points. For this reason, ℓ is often described as $y = 2$. A more useful way to describe ℓ is to say $y = 2$ and x is any real number. Or, you can describe the line using ordered pairs $(x, 2)$.

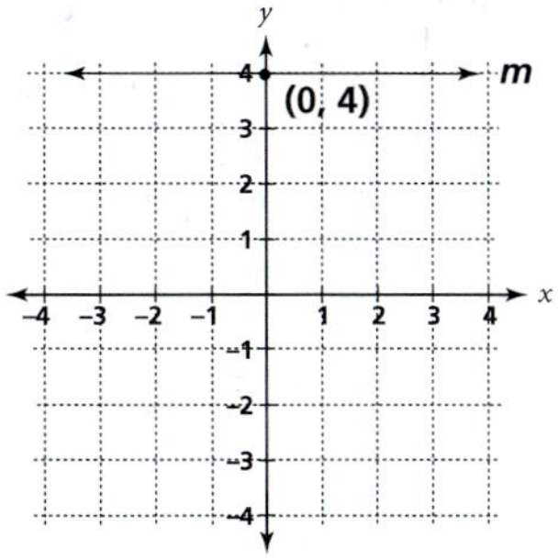

Algebra Connection What is the algebraic expression, or equation, for line m? Since m passes through (0, 4), the algebraic expression is $y = 4$ and x is any real number. The ordered pairs are all $(x, 4)$.

EXAMPLES What are the algebraic expressions for these two lines, each parallel to the x-axis and exactly 5 units from the x-axis?

Solution:
Line ℓ passes through (0, 5) and the line m passes through (0, −5). Each of the lines is parallel to the x-axis so each has an x-value equal to any real number. Therefore the algebraic expression for ℓ is $y = 5$ and x is any real number. The algebraic expression for m is $y = -5$ and x is any real number.

The ordered pairs for ℓ are $(x, 5)$.
The ordered pairs for m are $(x, -5)$.

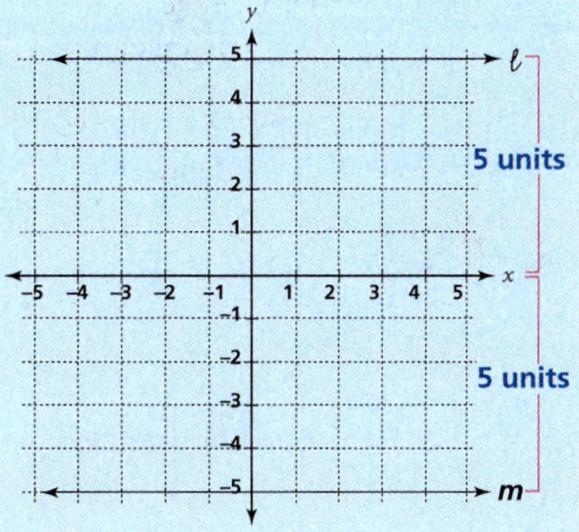

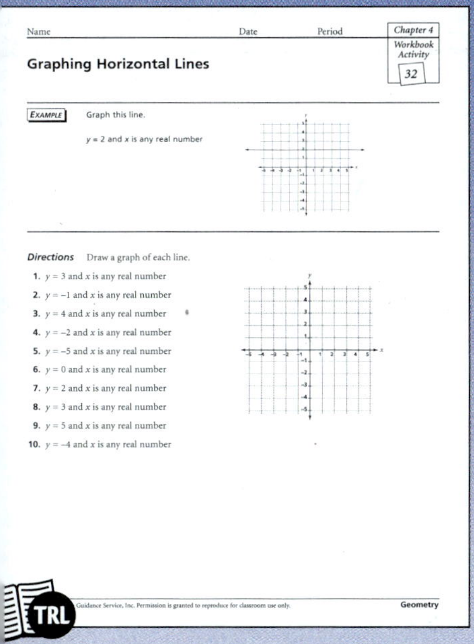

Name Date Period | *Chapter 4 Workbook Activity* 32

Graphing Horizontal Lines

EXAMPLE Graph this line.

$y = 2$ and x is any real number

Directions Draw a graph of each line.

1. $y = 3$ and x is any real number
2. $y = -1$ and x is any real number
3. $y = 4$ and x is any real number
4. $y = -2$ and x is any real number
5. $y = -5$ and x is any real number
6. $y = 0$ and x is any real number
7. $y = 2$ and x is any real number
8. $y = 3$ and x is any real number
9. $y = 5$ and x is any real number
10. $y = -4$ and x is any real number

Geometry

Workbook Activity 32

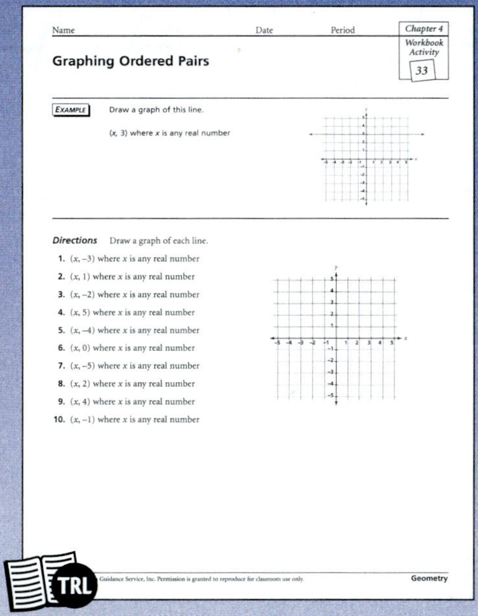

Name Date Period | *Chapter 4 Workbook Activity* 33

Graphing Ordered Pairs

EXAMPLE Draw a graph of this line.

$(x, 3)$ where x is any real number

Directions Draw a graph of each line.

1. $(x, -3)$ where x is any real number
2. $(x, 1)$ where x is any real number
3. $(x, -2)$ where x is any real number
4. $(x, 5)$ where x is any real number
5. $(x, -4)$ where x is any real number
6. $(x, 0)$ where x is any real number
7. $(x, -5)$ where x is any real number
8. $(x, 2)$ where x is any real number
9. $(x, 4)$ where x is any real number
10. $(x, -1)$ where x is any real number

Geometry

Workbook Activity 33

Lesson at a Glance

Chapter 4 Lesson 2

Overview This lesson discusses horizontal lines and demonstrates how to write the algebraic expression, or equation, for a horizontal line.

Objectives

- To write the algebraic expression for a horizontal line.
- To graph a horizontal line, given its equation or an ordered pair on the line.

Student Pages 101–103

Teacher's Resource Library TRL

Workbook Activities 32–33

Additional Practice 19

1. Warm-Up Activity

Draw a coordinate plane on the board or overhead projector, and ask students to name the axis that is a horizontal line. Then ask students to name coordinates of points that lie on the x-axis. Encourage them to describe points that do not appear on the given coordinate graph, such as (100, 0). Write the ordered pairs on the board or projector and ask: "What do you notice about the ordered pairs that lie on the horizontal axis, or x-axis?" **(The x-coordinates are all different, and the y-coordinate is always 0.)** Draw another horizontal line on the graph, and ask students to make predictions about the ordered pairs that lie on this new line. Then ask for their predictions about a line whose coordinates all have a y-coordinate of −2.

2. Teaching the Lesson

Make sure students understand that the y-value of a point on a horizontal line does not change and does not depend on the corresponding x-value. Write the coordinates for several points on a horizontal line in a table of values to help illustrate this point. Demonstrate how an equation such as $y = 5$ describes *any* point that has a y-value of 5, regardless of its location to the right or left of the origin.

Common Error

Because the x-axis is the horizontal axis, students often get confused and think that it is the x-value that is the same in each ordered pair. Remind them that any time you move right or left on the coordinate plane, you change the location on the x-axis and thus change the x-value.

Algebra Connection

Have students graph several horizontal lines on a coordinate plane and then practice writing equations for them. Instruct them to list the coordinates of several points on each line in a table of values before they write the equation for the line.

3. Reinforce and Extend

Learning Styles

Visual/Spatial Have students copy the coordinate plane for Exercise A and write the ordered pairs for three or four points on each line before they write the equation. It may be easier for students to understand the equation of the line if they see the ordered pairs with the same y-coordinates written on the the line.

Geometry Around Us

In order to be sure that a line is absolutely horizontal, a carpenter uses a level. If the surface is level, the air bubble is centered between the two lines on the level.

EXAMPLE What is the algebraic expression, or equation, for a line passing through (6, 4) and parallel to the line passing through (4, 2) and (6, 2)?

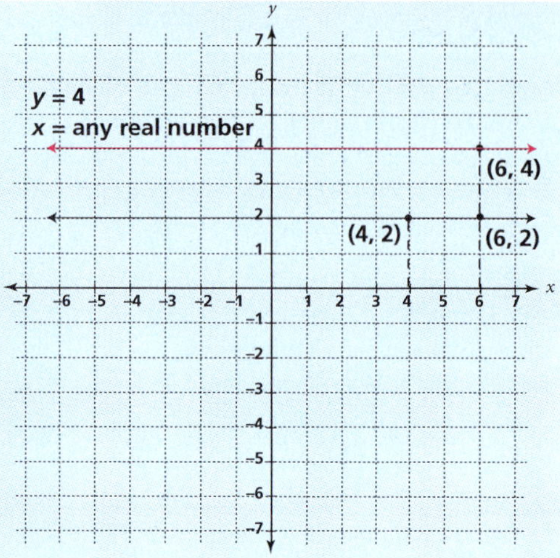

Solution:
The line passing through (4, 2) and (6, 2) is parallel to the x-axis. So the line passing through (6, 4) must also be parallel to the x-axis. The algebraic expression is $y = 4$ and x is any real number. The ordered pairs are $(x, 4)$.

Exercise A Write the algebraic expression for each line.

1. ℓ $y = 3$, x is any real number
2. m $y = 2$, x is any real number
3. n $y = 1$, x is any real number
4. o $y = -1$, x is any real number
5. p $y = -2$, x is any real number

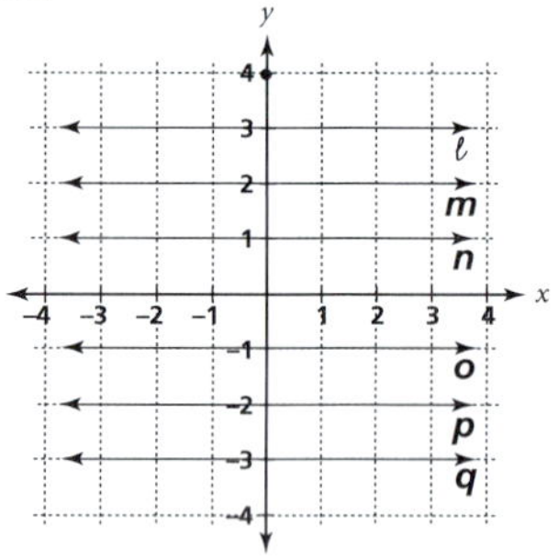

Exercise B Write the ordered pairs for each line.

6. ℓ $(x, 3)$
7. m $(x, 2)$
8. n $(x, 1)$
9. o $(x, -1)$
10. p $(x, -2)$

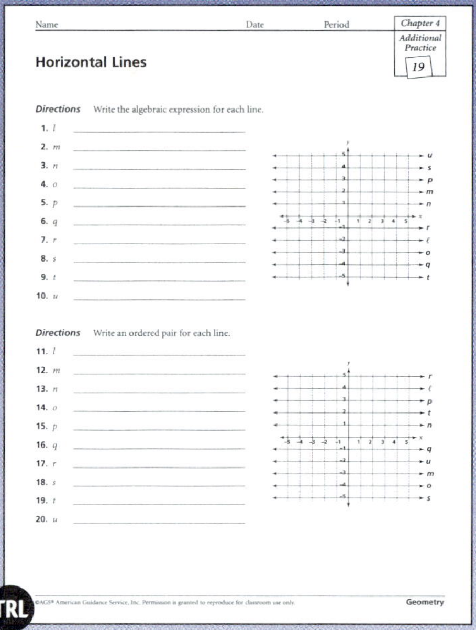

Name Date Period

Chapter 4 Additional Practice 19

Horizontal Lines

Directions Write the algebraic expression for each line.

1. l ______
2. m ______
3. n ______
4. o ______
5. p ______
6. q ______
7. r ______
8. s ______
9. t ______
10. u ______

Directions Write an ordered pair for each line.

11. l ______
12. m ______
13. n ______
14. o ______
15. p ______
16. q ______
17. r ______
18. s ______
19. t ______
20. u ______

 Geometry

TRL

Additional Practice 19

Construction

Exercise C Draw a graph of each line.

11. $y = 2$ and x is any real number

12. $y = -2$ and x is any real number

13. $y = 1\frac{1}{2}$ and x is any real number

14. $y = -3$ and x is any real number

See Teacher's Edition page for answers.

Exercise D Graph all the ordered pairs.

15. $(x, 1)$ where x is any real number

16. $(x, -3)$ where x is any real number

17. $(x, -5)$ where x is any real number

18. $(x, 0)$ where x is any real number

> **Technology Connection**
>
> We can plot many different equations on the coordinate grid. For example, we can plot the path of a projectile such as a rocket. We can then use the curve to predict where the rocket will be at a specific time. This allows us to adjust the speed and other rocket variables to make sure that the rocket reaches its target.

Problem Solving

Exercise E Answer the following questions.

19. What is the equation for the x-axis?

$y = 0$, x is any real number

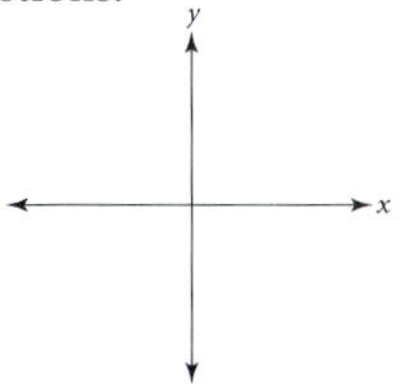

20. Downhill skiers often talk of "slopes." Beginning skiers look for gentle slopes, while more advanced skiers look for steeper slopes. Using your knowledge of slopes and the diagram below, what do you think is the slope of a level surface? zero

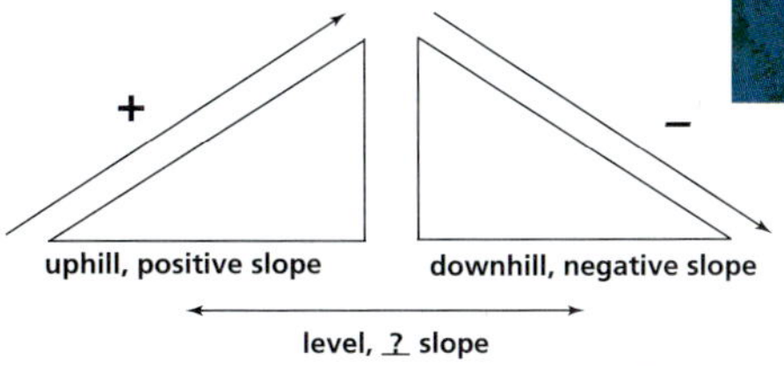

Learning Styles

Auditory/Verbal Suggest that students work in pairs to complete Exercises C and D. One student should name ordered pairs for points on the lines described in Exercises C and D and the other student plots the points and draws the lines on the graph.

Answers to Exercise C 11–14

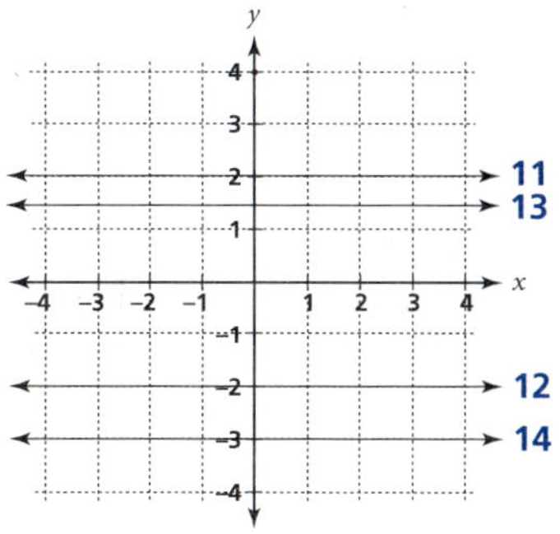

Answers to Exercise D 15–18

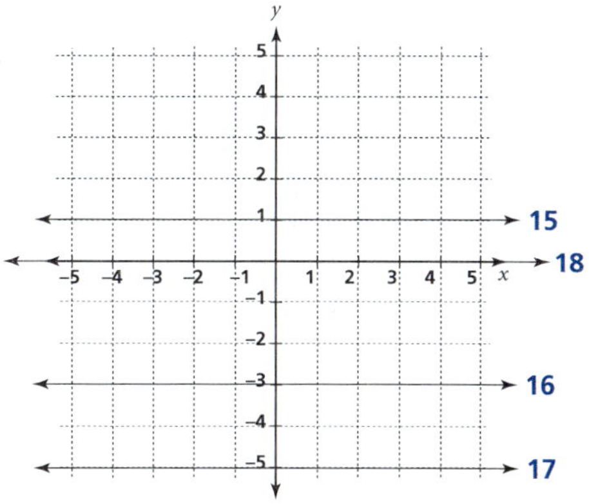

Lesson at a Glance

Chapter 4 Lesson 3

Overview This lesson discusses vertical lines and demonstrates how to write an algebraic expression, or equation, to describe them.

Objectives

- To write the algebraic expression for a vertical line.
- To graph a horizontal line, given its equation or an ordered pair on the line.

Student Pages 104–106

Teacher's Resource Library TRL

Workbook Activities 34–35

Additional Practice 20

1. Warm-Up Activity

Draw a coordinate plane on the board or overhead projector, and ask students to name the axis that is a vertical line. Then ask students to name the coordinates of points that lie on the *y*-axis. Encourage them to describe points that do not appear on the given coordinate graph, such as (0, –50). Write the ordered pairs on the board or projector and ask: "How do the ordered pairs that lie on the vertical axis, or *y*-axis compare to those that lie on a horizontal line?" (The *y*-coordinates are all different, and the *x*-coordinate is always 0.) Draw another vertical line on the graph and ask students to make predictions about the ordered pairs that lie on this new line. Then ask for their predictions about a line whose coordinates all have an *x*-coordinate of –10.

2. Teaching the Lesson

Compare the ordered pairs on a vertical line to those on a horizontal line. Use the comparison to help students understand why a vertical line is described by an equation of the form $x = a$. Point out how one value of x can be paired with infinitely many y-values.

Lesson 3 Vertical Lines

Remember that the *y*-axis is the vertical axis. Therefore, lines parallel to the *y*-axis are vertical lines. These lines pass through a single point on the *x*-axis.

Line m is parallel to the *y*-axis and passes through the point $x = 3$ or $(3, 0)$.

Line m passes through the point $(3, 0)$. Each point on m has a different y-value, but all points have the same x-value, 3. If you tried to describe the line using ordered pairs, you would find it impossible to name all the points. However, you can say that line m consists of all ordered pairs where $x = 3$ and y is any real number. Notice that $m \perp x$-axis.

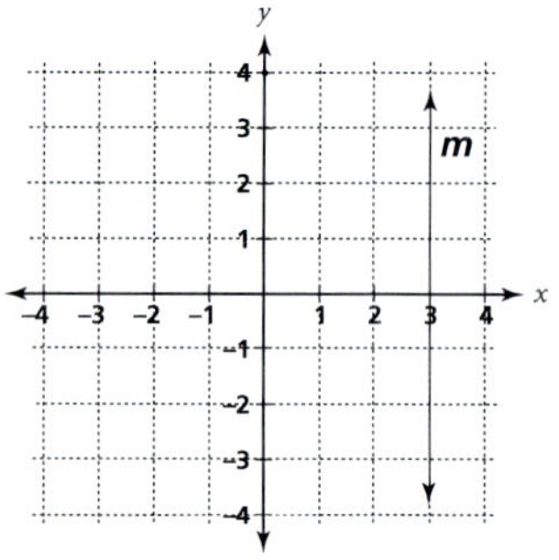

EXAMPLE Here, ℓ is parallel to the *y*-axis. What are the ordered pairs for points on ℓ?

$A = (5, 4)$

$B = (5, 3)$

$C = (5, 2)$

and so on to $G = (5, -2)$

Algebra Connection Notice that although all the *x*-values are 5, the *y*-values change. The algebraic expression for line ℓ is $x = 5$ and y is any real number.

Since ℓ passes through (5, 0), the algebraic expression is $x = 5$ and y is any real number. The ordered pairs are all $(5, y)$.

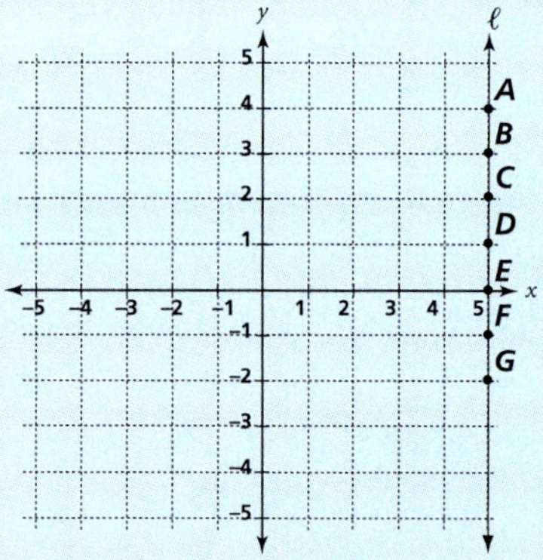

Algebra Connection

Each of the following points lies on a different vertical line: $(3, y)$; $(-4, y)$; $(0, y)$; $(25, y)$; $(-55, y)$; $(\frac{1}{2}, y)$; $(-5.4, y)$. Have students write an ordered pair for another point on each line. They then should write the equation for each line.

EXAMPLES What are the algebraic expressions for these two lines, each parallel to the y-axis at a distance of 3 units?

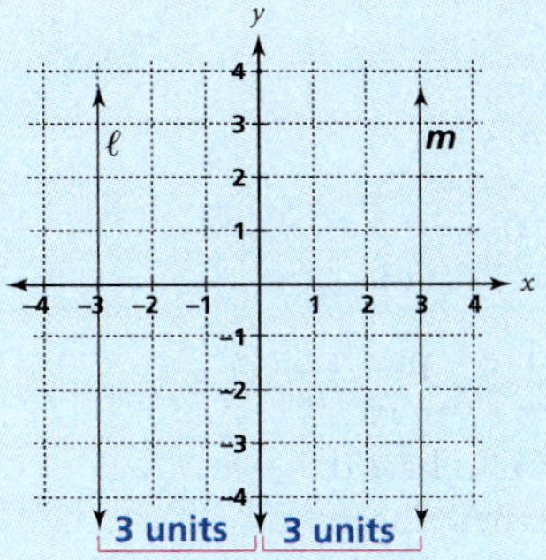

Solution:
Line ℓ passes through $(-3, 0)$ and line m passes through $(3, 0)$. Each of the lines is parallel to the y-axis so each has a y-value equal to any real number. Therefore the algebraic expression for ℓ is $x = -3$ and y is any real number. The algebraic expression for m is $x = 3$ and y is any real number.

The ordered pairs for ℓ are $(-3, y)$.
The ordered pairs for m are $(3, y)$.

If you draw $\overline{BC}$, is it parallel to the y-axis? How do you know?

Solution:
$\overline{BC}$ passes through $x = 4$ because both B and C have the same x-value, 4. $\overline{BC}$ is a portion of the line that has the equation $x = 4$ and y is any real number. Therefore $\overline{BC}$ must be parallel to the y-axis.

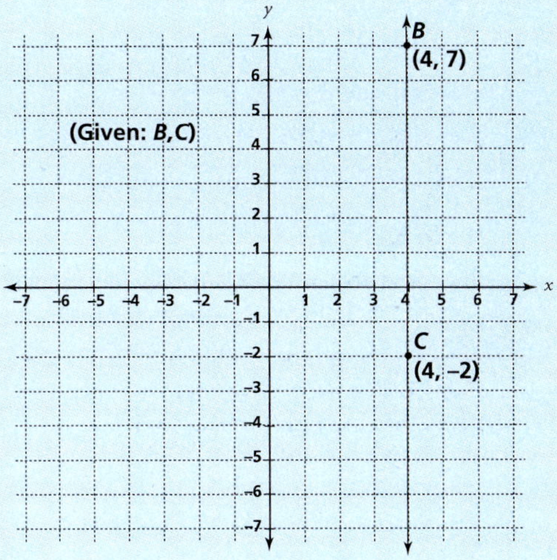

COMMON ERROR

Students are likely to confuse the coordinate that has a constant value in a horizontal or a vertical line. Encourage them to always check their equations by writing and plotting ordered pairs derived from their equations.

LEARNING STYLES

Auditory/Verbal Ask students to express in their own words why all the points on a vertical line have the same x-value. Remind them to express themselves clearly, using mathematical terminology.

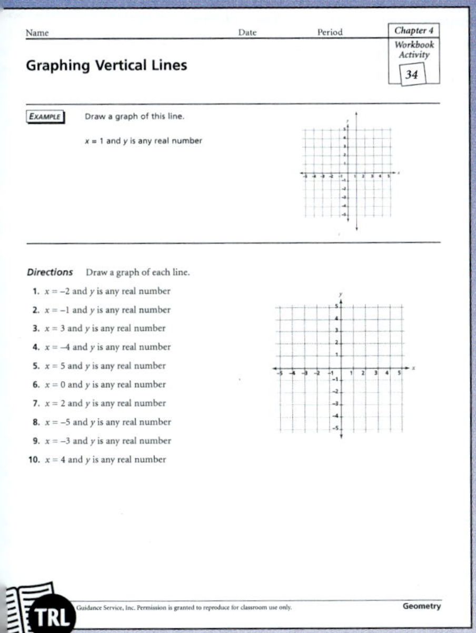

Name Date Period | *Chapter 4 Workbook Activity 34*

Graphing Vertical Lines

EXAMPLE Draw a graph of this line.

$x = 1$ and y is any real number

Directions Draw a graph of each line.

1. $x = -2$ and y is any real number
2. $x = -1$ and y is any real number
3. $x = 3$ and y is any real number
4. $x = -4$ and y is any real number
5. $x = 5$ and y is any real number
6. $x = 0$ and y is any real number
7. $x = 2$ and y is any real number
8. $x = -5$ and y is any real number
9. $x = -3$ and y is any real number
10. $x = 4$ and y is any real number

 Geometry

TRL **Workbook Activity 34**

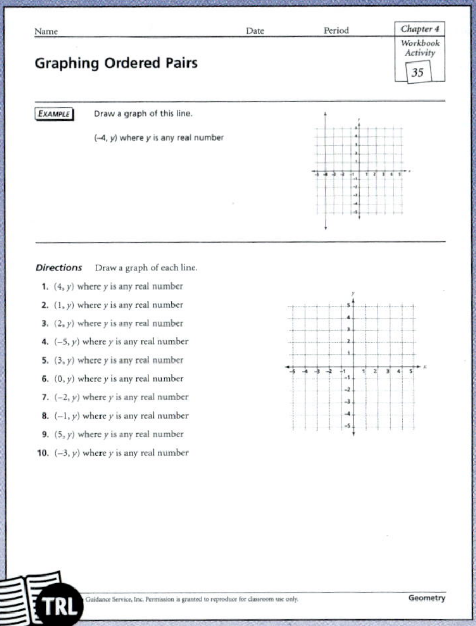

Name Date Period | *Chapter 4 Workbook Activity 35*

Graphing Ordered Pairs

EXAMPLE Draw a graph of this line.

$(-4, y)$ where y is any real number

Directions Draw a graph of each line.

1. $(4, y)$ where y is any real number
2. $(1, y)$ where y is any real number
3. $(2, y)$ where y is any real number
4. $(-5, y)$ where y is any real number
5. $(3, y)$ where y is any real number
6. $(0, y)$ where y is any real number
7. $(-2, y)$ where y is any real number
8. $(-1, y)$ where y is any real number
9. $(5, y)$ where y is any real number
10. $(-3, y)$ where y is any real number

 Geometry

TRL **Workbook Activity 35**

3. Reinforce and Extend

Learning Styles

Visual/Spatial Have students copy the coordinate plane for Exercise A and write the ordered pairs for three or four points on each line before they write the equation.

Mental Math

Ask students to state whether the ordered pair or equation you name describes a horizontal or vertical line.

$(x, 5)$ (horizontal)

$(-3, y)$ (vertical)

$x = 2$, y is any real number (vertical)

$y = -2$, x is any real number (horizontal)

Answers to Exercise C 11–14

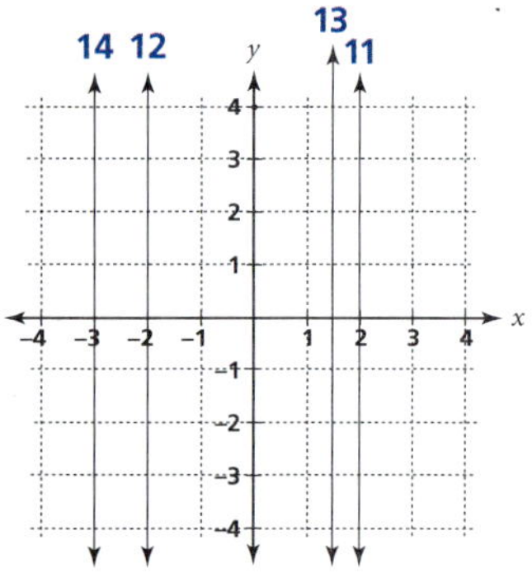

Answers to Exercise D 15–18

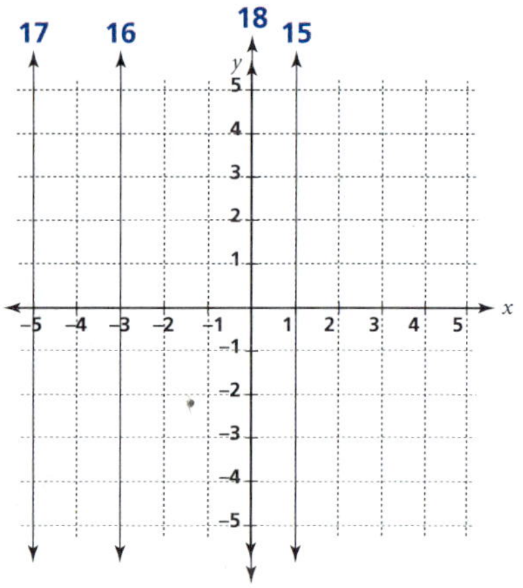

Exercise A Write the equation for each line.

1. ℓ
2. m
3. n
4. p
5. q

Exercise B Write the ordered pairs for each line.

6. ℓ
7. m
8. n
9. r
10. s

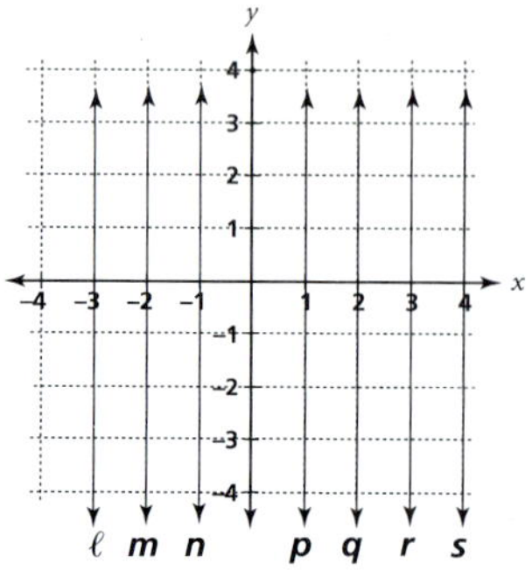

1. $x = -3$, y is any real number
2. $x = -2$, y is any real number
3. $x = -1$, y is any real number
4. $x = 1$, y is any real number
5. $x = 2$, y is any real number
6. $(-3, y)$
7. $(-2, y)$
8. $(-1, y)$
9. $(3, y)$
10. $(4, y)$

Exercise C Graph the line described by the equations.

11. $x = 2$ and y is any real number
12. $x = -2$ and y is any real number
13. $x = 1\frac{1}{2}$ and y is any real number
14. $x = -3$ and y is any real number

See Teacher's Edition page for answers.

Exercise D Graph all the ordered pairs.

15. $(1, y)$ where y is any real number
16. $(-3, y)$ where y is any real number
17. $(-5, y)$ where y is any real number
18. $(0, y)$ where y is any real number

See Teacher's Edition page for answers.

Problem Solving

Exercise E Answer the following questions.

19. What is the equation for the y-axis?
    $x = 0$, y is any real number
20. Explain why any line parallel to the y-axis is perpendicular to the x-axis. (Hint: Review the definition of the coordinate plane and theorems from Chapter 3).

The interior angles on the same side of the transversal (the line ∥ to the y-axis) will be equal to each other and equal to 90°. Therefore the line ∥ to the y-axis must be perpendicular to the x-axis.

Experiencing Geometry

Set up a coordinate grid on your classroom floor. Play a game of Simon Says in which "Simon" tells a player to move from one location to another using ordered pairs.

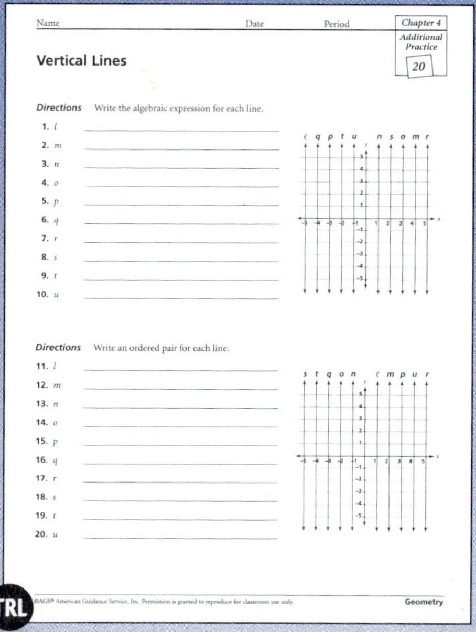

Name Date Period Chapter 4 Additional Practice 20

Vertical Lines

Directions Write the algebraic expression for each line.

1. l
2. m
3. n
4. o
5. p
6. q
7. r
8. s
9. t
10. u

Directions Write an ordered pair for each line.

11. l
12. m
13. n
14. o
15. p
16. q
17. r
18. s
19. t
20. u

 Geometry

TRL

Additional Practice 20

Slope of a Line

Carpenters and other craftspeople interested in the **slope of a line** define slope as rise divided by run.

slope $= \frac{\text{rise}}{\text{run}}$

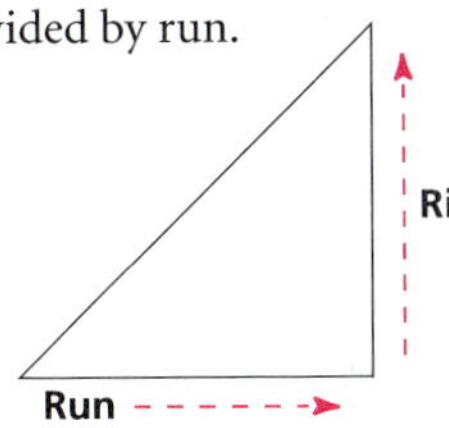

Slope of a line
Ratio of the difference in y-values to the difference in x-values

In the coordinate plane, slope has an algebraic definition.

Algebra Connection

Here, line ℓ passes through points (2, 3) and (5, 4). Remember, Euclid's postulate states that any two points determine a line.

slope of line $\ell = \frac{\text{rise}}{\text{run}} = \frac{\text{difference in } y\text{-values}}{\text{difference in } x\text{-values}}$

slope of $\ell = \frac{4-3}{5-2} = \frac{1}{3}$

The slope is positive; the line is moving uphill.

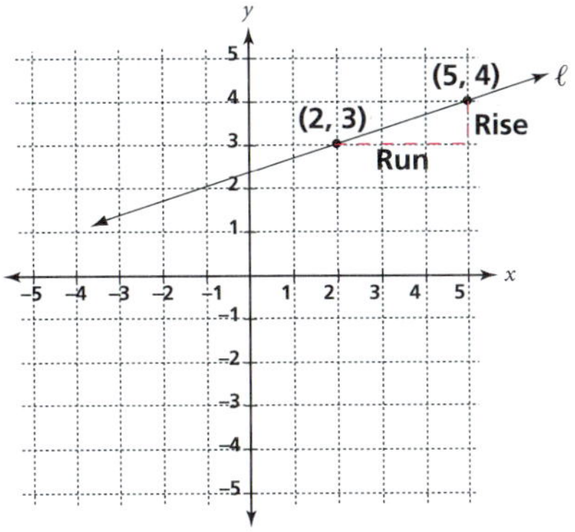

Line m passes through points (–1, 3) and (4, –1).

slope of line $m = \frac{\text{rise}}{\text{run}} = \frac{\text{difference in } y\text{-values}}{\text{difference in } x\text{-values}}$

$= \frac{3-(-1)}{-1-4} = -\frac{4}{5}$

slope $= -\frac{4}{5}$

The slope is negative; the line is moving downhill.

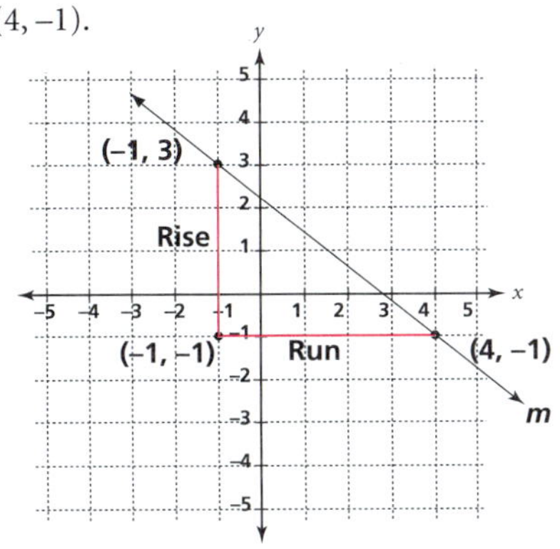

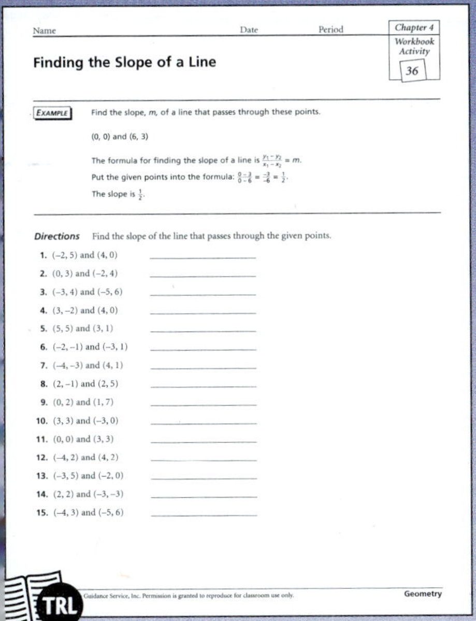

Name Date Period

Chapter 4 Workbook Activity 36

Finding the Slope of a Line

EXAMPLE Find the slope, m, of a line that passes through these points.

(0, 0) and (6, 3)

The formula for finding the slope of a line is $\frac{y_1 - y_2}{x_1 - x_2} = m$.

Put the given points into the formula: $\frac{0-3}{0-6} = \frac{-3}{-6} = \frac{1}{2}$.

The slope is $\frac{1}{2}$.

Directions Find the slope of the line that passes through the given points.

1. (–2, 5) and (4, 0)
2. (0, 3) and (–2, 4)
3. (–3, 4) and (–5, 6)
4. (3, –2) and (4, 0)
5. (5, 5) and (3, 1)
6. (–2, –1) and (–3, 1)
7. (–4, –3) and (4, 1)
8. (2, –1) and (2, 5)
9. (0, 2) and (1, 7)
10. (3, 3) and (–3, 0)
11. (0, 0) and (3, 3)
12. (–4, 2) and (4, 2)
13. (–3, 5) and (–2, 0)
14. (2, 2) and (–3, –3)
15. (–4, 3) and (–5, 6)

Guidance Service, Inc. Permission is granted to reproduce for classroom use only. Geometry

Workbook Activity 36

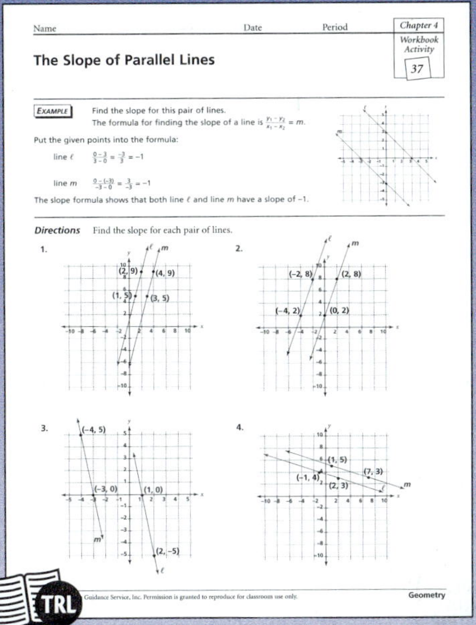

Name Date Period

Chapter 4 Workbook Activity 37

The Slope of Parallel Lines

EXAMPLE Find the slope for this pair of lines.
The formula for finding the slope of a line is $\frac{y_1 - y_2}{x_1 - x_2} = m$.

Put the given points into the formula:

line ℓ $\frac{0-3}{3-0} = \frac{-3}{3} = -1$

line m $\frac{0-(-3)}{-3-0} = \frac{3}{-3} = -1$

The slope formula shows that both line ℓ and line m have a slope of –1.

Directions Find the slope for each pair of lines.

1. (2, 9) (4, 9) (1, 5) (3, 5)
2. (–2, 8) (2, 8) (–4, 2) (0, 2)
3. (–4, 5) (–3, 0) (1, 0) (2, –5)
4. (1, 5) (7, 3) (–1, 4) (2, 3)

Guidance Service, Inc. Permission is granted to reproduce for classroom use only. Geometry

Workbook Activity 37

Lesson at a Glance

Chapter 4 Lesson 4

Overview This lesson discusses the slope of a line and demonstrates how to use a formula to calculate the slope.

Objectives

- To understand what the slope of a line is.
- To find the slope of a line.

Student Pages 107–112

Teacher's Resource Library TRL

Workbook Activities 36–37

Additional Practice 21

Mathematics Vocabulary

slope of a line

1. Warm-Up Activity

Engage students in a discussion about slope. Ask them for examples of slopes, such as ski or sledding hills, roofs, escalators, and the path of a plane after liftoff. Have students suggest ways to compare slopes.

2. Teaching the Lesson

On a coordinate grid, draw several lines with varying positive slopes. Outline the rise and run between two points on each line to form a triangle with the line itself. Ask students to rank the lines from least to most steep and make observations about the triangles formed by the line and the rise and run for the slopes.

Algebra Connection

Let students simplify these expressions to review the rules for subtracting integers:

6 + –9 = **(–3)**	–3 + 8 = **(5)**
–10 + 4 = **(–6)**	–7 + –2 = **(–9)**
–5 + 1 = **(–4)**	–2 + –2 = **(–4)**
7 – 9 = **(–2)**	–4 – 3 = **(–7)**
–3 – 2 = **(–5)**	–10 – (–4) = **(–6)**
–5 – (–5) = **(0)**	8 – (–8) = **(16)**
–9 – 9 = **(–18)**	1 – (–5) = **(6)**

3. Reinforce and Extend

Application

At Home

Have students find and list examples of positive, negative, and zero slopes in their homes. Suggest that they may find such slopes in the architecture of their homes or in pictures or objects around the home. Ask students to share their lists with the class.

The difference in y-values can be written as $y_1 - y_2$. The difference in x-values can be written as $x_1 - x_2$. We can use these expressions to write the slope of a line in symbol form.

The letter m is generally used to represent the value of the slope of a line. This leads to the following formula.

When you use the formula for the slope of a line, it does not matter which point you label (x_1, y_1) and which you label (x_2, y_2); the results are the same.

Formula for Slope of a Line:

$$m = \frac{y_1 - y_2}{x_1 - x_2}$$

where m is the slope of the line, and (x_1, y_1) and (x_2, y_2) are any two points on the line.

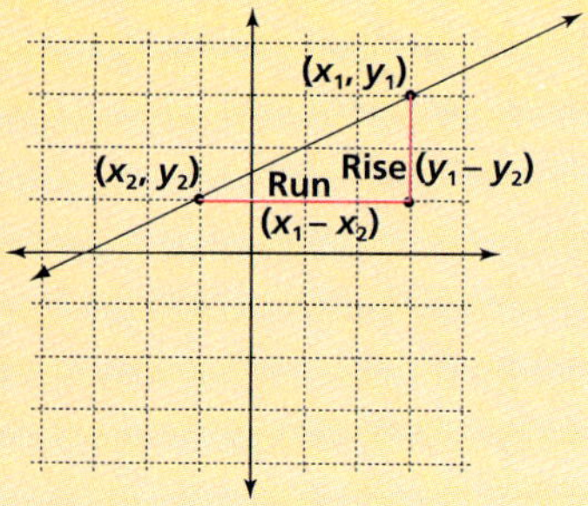

Let $(x_1, y_1) = (-2, 1)$

$(x_2, y_2) = (6, 3)$

$$\text{slope} = \frac{y_1 - y_2}{x_1 - x_2} = \frac{1 - 3}{-2 - 6} = \frac{-2}{-8} = \frac{1}{4}$$

OR

Let $(x_1, y_1) = (6, 3)$

$(x_2, y_2) = (-2, 1)$

$$\text{slope} = \frac{y_1 - y_2}{x_1 - x_2} = \frac{3 - 1}{6 - (-2)} = \frac{2}{8} = \frac{1}{4}$$

The slope is $\frac{1}{4}$ no matter which way you assign the points to the formula.

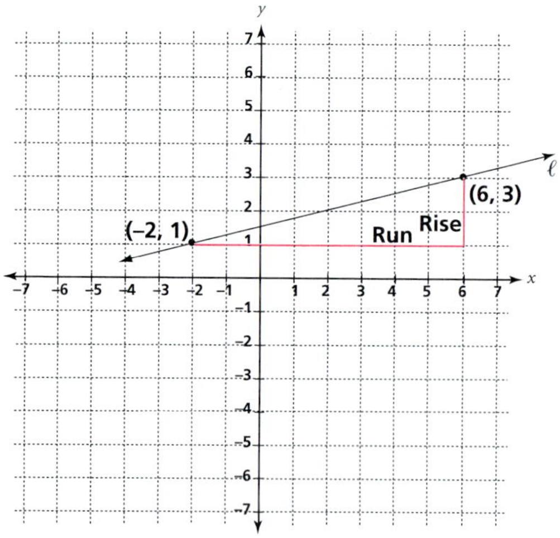

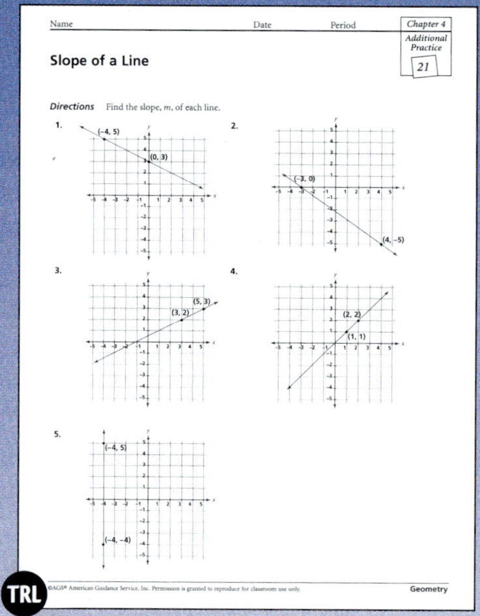
Name Date Period

Chapter 4 Additional Practice 21

Slope of a Line

Directions Find the slope, *m*, of each line.

1. 2. 3. 4. 5.

Geometry

TRL

Additional Practice 21

EXAMPLES

Calculate the slope of the line passing through (–4, 2) and (3, 1).

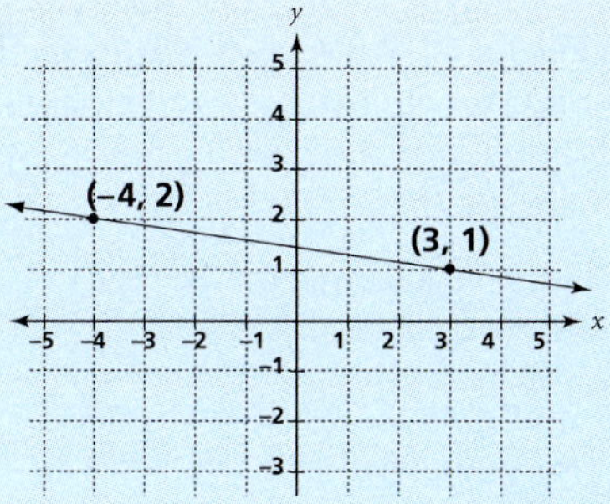

Solution:

$$m = \frac{y_1 - y_2}{x_1 - x_2}$$

Let $(x_1, y_1) = (-4, 2)$

$(x_2, y_2) = (3, 1)$

then $m = \frac{2-1}{-4-3} = \frac{1}{-7} = -\frac{1}{7}$

$m = -\frac{1}{7}$

Calculate the slope of the horizontal line passing through (1, 4) and (3, 4).

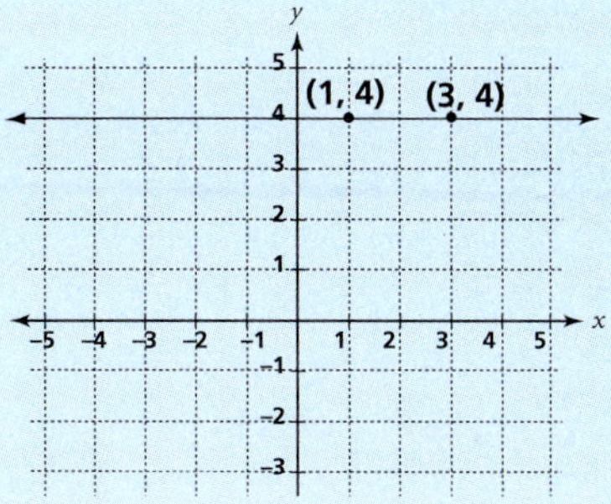

Solution:

$$m = \frac{y_1 - y_2}{x_1 - x_2}$$

Let $(x_1, y_1) = (1, 4)$

$(x_2, y_2) = (3, 4)$

then $m = \frac{4-4}{1-3} = \frac{0}{-2} = 0$

$m = 0$

Notice that the y-values of all lines parallel to the x-axis are the same; and their difference will always be zero. Therefore, the slope of any line parallel to the x-axis is zero. However, for vertical lines—lines parallel to the y-axis—the slope formula fails. Here is why.

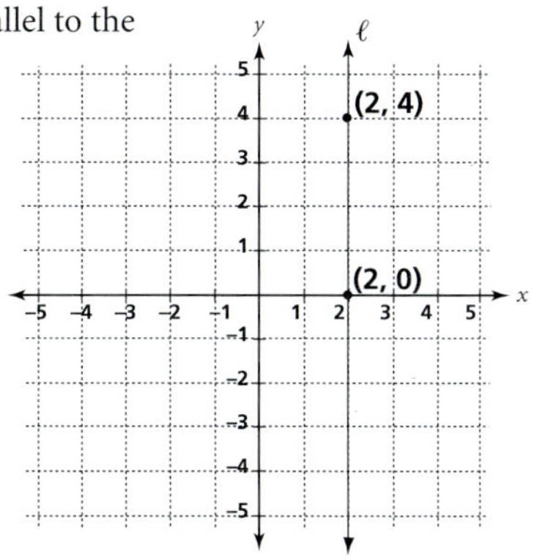

Suppose ℓ, parallel to the y-axis, passes through (2, 0) and (2, 4).

Let $(x_1, y_1) = (2, 0)$

$(x_2, y_2) = (2, 4)$

then $m = \frac{y_1 - y_2}{x_1 - x_2} = \frac{0-4}{2-2} = \frac{-4}{0}$

But division by zero is *undefined.* Therefore the slope of a line ℓ parallel to the y-axis is also *undefined.*

COMMON ERROR

When using the slope formula, students may write the ratio as $\frac{x}{y}$ or as $\frac{y_1 - y_2}{x_2 - x_1}$.

To help prevent this, tell students to write the ordered pairs one above the other. Remind them to work from right to left and from top to bottom when choosing x- and y-values to place in the formula.

LEARNING STYLES

Logical/Mathematical Have students find the slopes of the lines that pass through the following pairs of points: (1, 1) and (5, 3); (2, 4) and (0, 0); and (–2, 2) and (1, –4) **($m = \frac{1}{2}$; $m = 2$; $m = -2$).** Suggest that students serve as tutors for others who may be having difficulty using the formula for determining slope.

Application

In the Community

Encourage small groups of students to contact the state or county Department of Transportation to find out how the slopes, or grades, of roads are measured. They might also inquire about whether the slope measurements are used to regulate the speed limits on or usage of roadways and whether there are limits placed on the slopes of various types of roads, such as freeways and residential streets. Encourage groups to share their findings with the class.

Any line parallel to the y-axis has equal x-values for all of its points. Therefore, the denominator of the slope formula, $(x_1 - x_2)$, will always equal zero and the fraction will be undefined.

Suppose you construct $p \parallel q$ on the coordinate plane as shown.

slope of $p = \frac{0-1}{-2-0} = \frac{-1}{-2} = \frac{1}{2}$

slope of $q = \frac{-2-(-1)}{0-2} = \frac{-1}{-2} = \frac{1}{2}$

Parallel lines have equal slope.

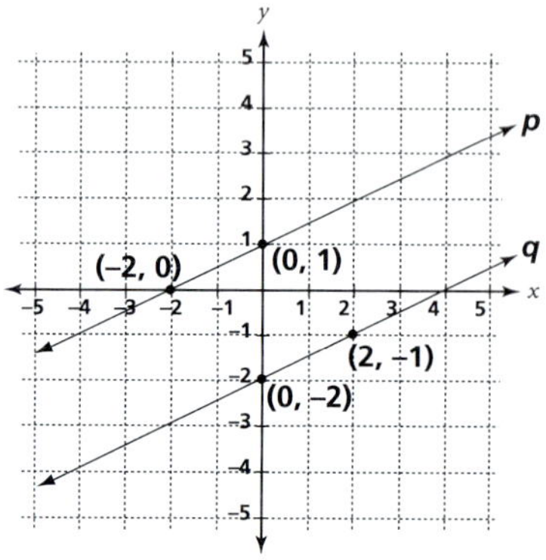

Summary of Information about Slope

Negative Slope

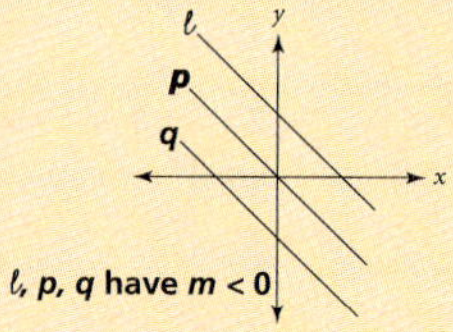

Positive Slope

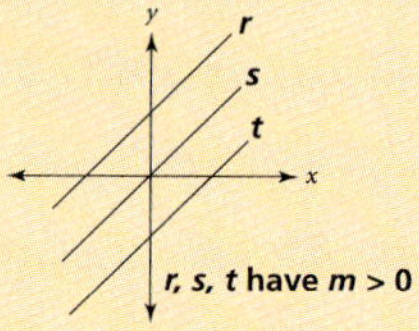

Undefined Slope

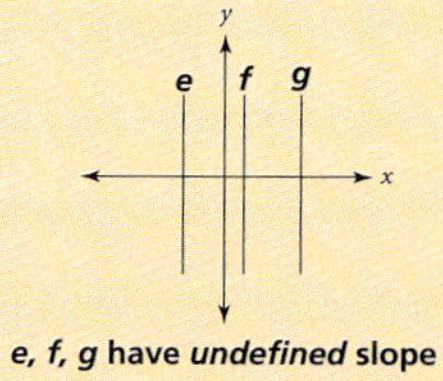

Zero Slope

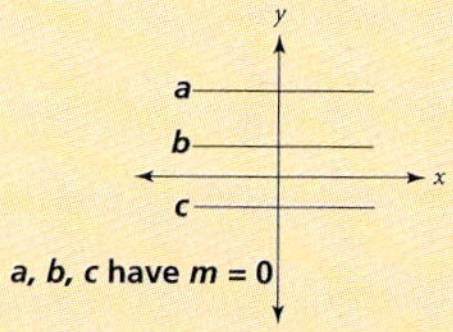

Parallel lines have equal slopes

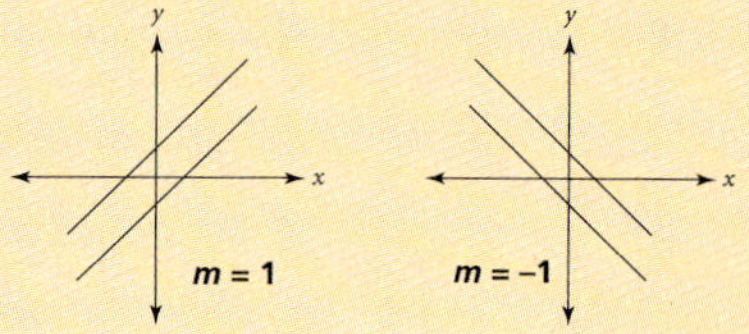

Calculator Practice

Using the parentheses keys on your calculator can help you find the slope of a line easily.

EXAMPLES Let $(x_1, y_1) = (8, 6)$

$(x_2, y_2) = (4, 2)$

then $m = \frac{y_1 - y_2}{x_1 - x_2} = \frac{6-2}{8-4}$

Use your calculator to find the answer.

Press [( ] 6 [−] 2 [ )], then press [÷] [( ] 8 [−] 4 [ )] [=].
The answer is 1.

Let $(x_1, y_1) = (8, 6)$

$(x_2, y_2) = (4, -2)$

then $m = \frac{y_1 - y_2}{x_1 - x_2} = \frac{6-(-2)}{8-4}$

Use your calculator to find the answer.

Press [( ] 6 [−] 2 [±] [ )], then press [÷] [( ] 8 [−] 4 [ )] [=].
The answer is 2.

Exercise A Use a calculator to find the following quotients.

1. $\frac{6-3}{4-(-2)}$ 0.5
2. $\frac{20-15}{7-(-3)}$ 0.5
3. $\frac{4-(-3)}{12-(-2)}$ 0.5
4. $\frac{9-3}{-15-(-3)}$ −0.5
5. $\frac{(-9)-(-3)}{15-3}$ −0.5

Exercise B Find the slope, m, of each line.

6.
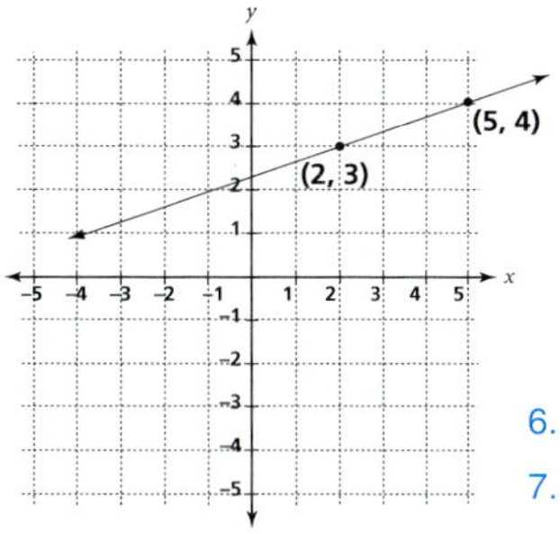

7.
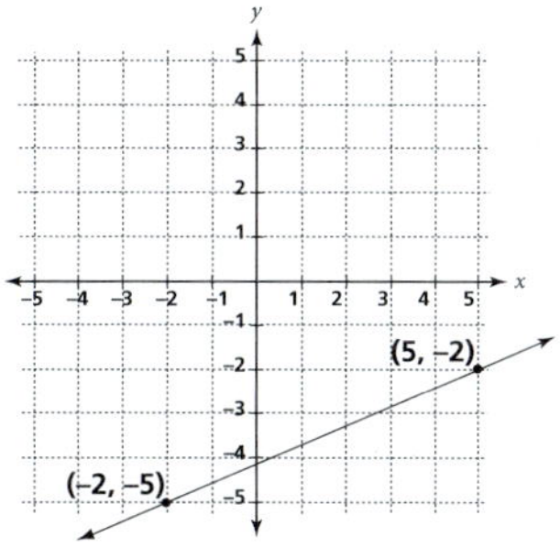

6. $m = \frac{1}{3}$
7. $m = \frac{3}{7}$

CALCULATOR

Provide calculator practice in computing slope. Have students find the slopes of lines that pass through the following pairs of points. Then have them check their calculations by graphing each line and drawing the rise and run to find the slope.

(2, 3), (4, 6) **($\frac{3}{2}$)**

(5, −3), (1, −7) **(1)**

(1, 3), (−6, −2) **($\frac{5}{7}$)**

(−4, 8), (5, 10) **($\frac{2}{9}$)**

(3, 1), (8, 11) **(2)**

(−2, −2), (−3, −6) **(4)**

(−6, 1), (3, −3) **($-\frac{4}{9}$)**

LEARNING STYLES

Auditory/Verbal Allow students to work in pairs to complete Exercises B and D. Students can take turns providing a verbal description of the line, including the points labeled on the graph. They might say, for example, "The rise and run between the two points are 1 and 3, in that order." The other student uses the information to determine the slope.

Learning Styles

Visual/Spatial For Exercise C, students calculate the slopes of lines, given points on the lines. Encourage students to graph each line and draw the rise and run lines between the given points for the line. They can use their graphs to check their calculations.

8.

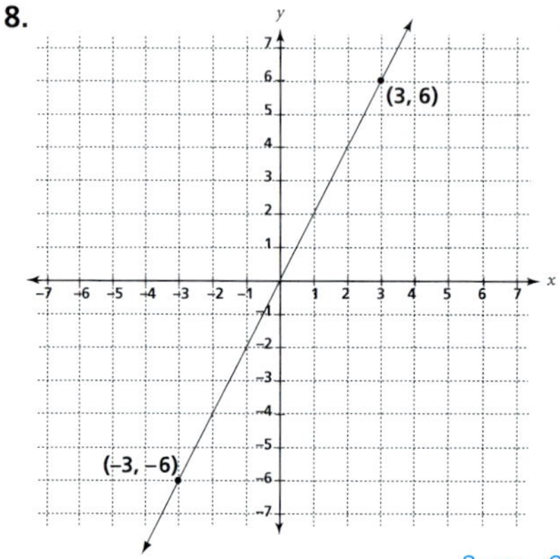

9.

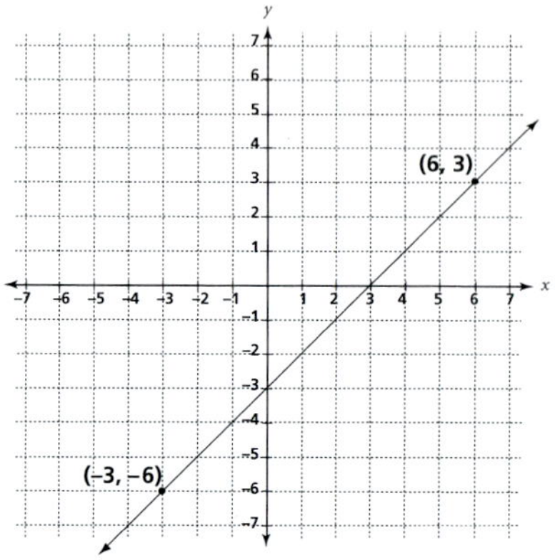

8. $m = 2$ 9. $m = 1$

Exercise C Find the slope, m, of a line that passes through the given points.

10. (3, 4) and (7, 16) $m = 3$

11. (−3, 4) and (7, 16) $m = \frac{6}{5}$

12. (4, 6) and (8, 12) $m = \frac{3}{2}$

13. (−3, −7) and (−7, −3) $m = -1$

Exercise D Find the slope for each pair of lines.

14.

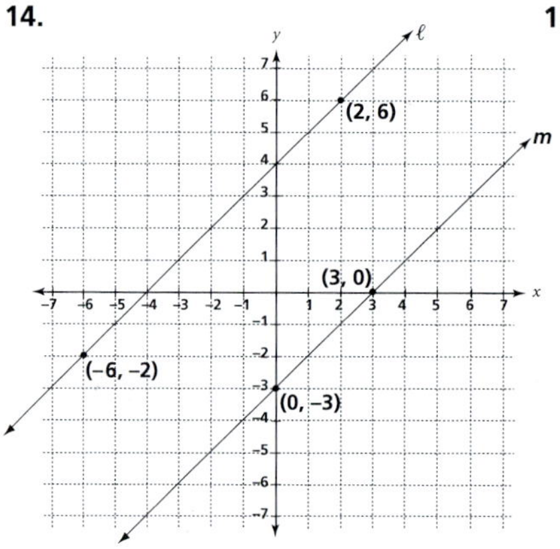

15.

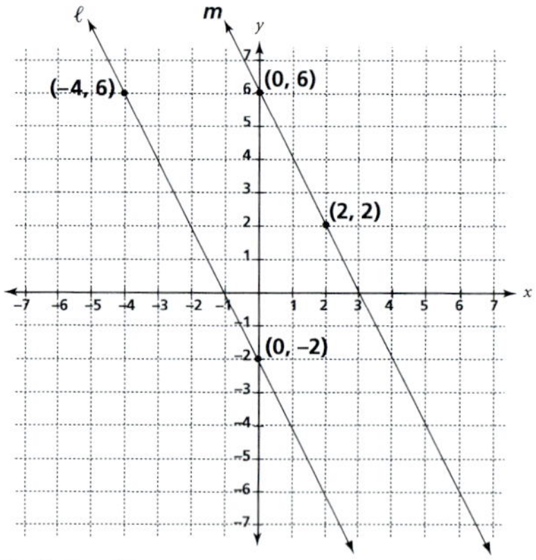

14. slope of ℓ = slope of $m = 1$
15. slope of ℓ = slope of $m = -2$

Lesson 5 Using Algebra: $y = mx + b$

In geometry, Euclid's first postulate states that for any two points there exists one, and only one, line passing through those points. If the points are A and B, then line $\overleftrightarrow{AB}$ or segment $\overline{AB}$ is determined.

Line $\overleftrightarrow{AB}$

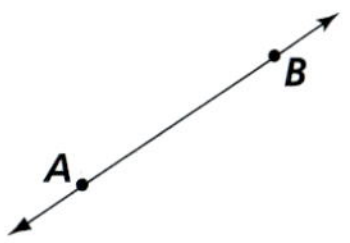

Segment $\overline{AB}$

You can draw the line or line segment on the coordinate plane as well. All you need are the ordered pairs that locate points A and B.

> ***Domain***
> *The x-values of a function*
>
> ***Function***
> *A rule that pairs every x-value with one and only one y-value*
>
> ***Range***
> *The y-values of a function*
>
> ***Y-intercept***
> *The y-value of the point where a line crosses the y-axis (when x = 0)*

Let $A = (-1, 1)$ and $B = (5, 3)$.

Place A and B in the plane.

Draw the line through A and B.

(Use a straightedge.)

You have now graphed line $\overleftrightarrow{AB}$.

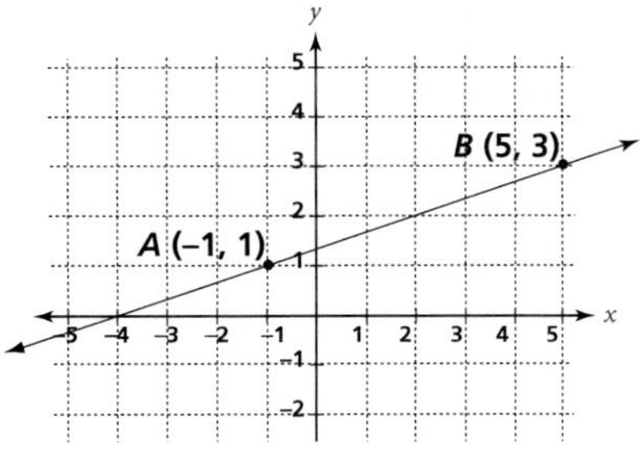

Let $A = (-1, 1)$ and $B = (5, 3)$.

Place A and B in the plane.

Draw a line between A and B.

(Use a straightedge.)

You have now graphed segment $\overline{AB}$.

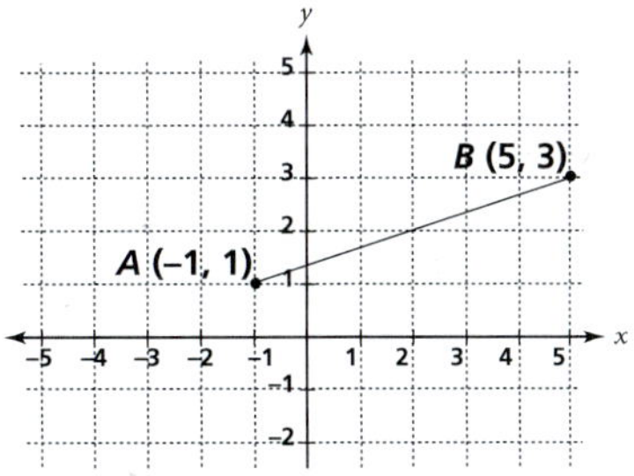

Name Date Period | *Chapter 4* *Workbook Activity* *38*

$y = mx + b$

EXAMPLE Write the equation of line ℓ. Use the form $y = mx + b$.

The slope formula shows that $\frac{y_1 - y_2}{x_1 - x_2} = \frac{4-2}{4-3} = \frac{2}{1} = 2 = m$.

To solve for b, put one point's x- and y-values plus the value for m into the formula $y = mx + b$.

Using the first point, $2 = 2(3) + b$; $2 = 6 + b$; $b = -4$.

The equation is written as $y = 2x - 4$.

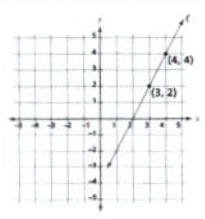

Directions Write the equation of line ℓ. Use the form $y = mx + b$.

1. 2. 3.

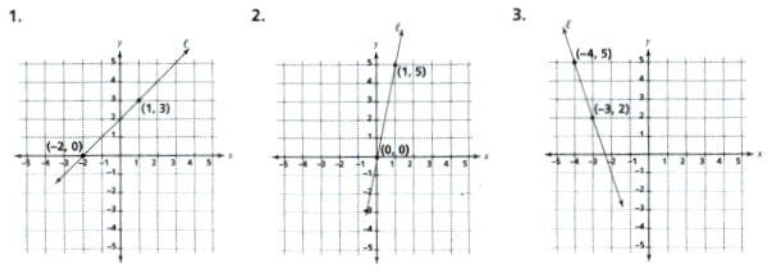

4. 5.

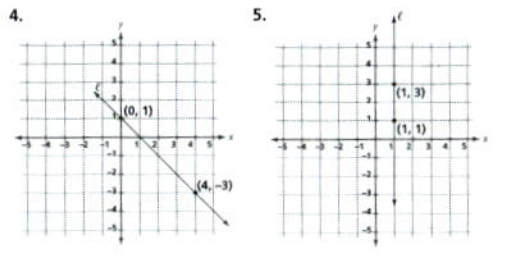

 Geometry

Workbook Activity 38

Name Date Period | *Chapter 4* *Workbook Activity* *39*

Using Slope in Real-Life Examples

EXAMPLE A staircase has 10 stairs that are 8 inches high and 12 inches deep. If the base of the staircase is at (0, 0), what are the staircase's domain, range, and slope? What would the ordered pair for the top of the staircase be if the numbers were counted in inches?

The domain of the staircase is 108 inches. Because the depth of the last stair is actually part of the second floor, the domain is the depth of each stair, or 12 times the nine stairs that have depth, giving 108 inches.

The range of the staircase is 80 inches—the number of stairs (10) times the height of each stair (8).

The slope of the staircase is $\frac{2}{3}$.

The ordered pair for the top of the staircase is (108, 80).

Directions Solve the following problems.

A surveyor finds that a nearby hill's crest has a horizontal distance of 300 feet from the spot where she is standing. The elevation of the hill's crest is 100 feet.

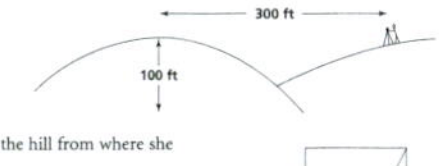

1. What are the domain, range, and slope of the hill from where she is standing? ____________

A diagonal is drawn from opposite corners of a rectangle. The rectangle is standing on one of its shorter sides, which is 4 inches, and the slope of the diagonal is 2.

2. What is the length of the two longer sides? ____________
3. What are the domain and range of the diagonal? ____________

A seesaw has a horizontal distance of 10 feet from one seat to the other. The vertical distance of the seat not resting on the ground is 4 feet.

4. What are the domain, range, and slope of the seesaw?
5. Write an equation for each problem.

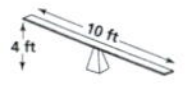

hill ____________

diagonal ____________

seesaw ____________

 Geometry

Workbook Activity 39

Lesson at a Glance

Chapter 4 Lesson 5

Overview This lesson introduces the concepts of domain, range, and function and demonstrates how to use algebra to represent a function.

Objectives

- To graph a function and find its domain and range.
- To write the equation of a line using the form $y = mx + b$.

Student Pages 113–118

Teacher's Resource Library TRL

Workbook Activities 38–39

Additional Practice 22

Mathematics Vocabulary

domain
function
range
***y*-intercept**

1. Warm-Up Activity

Review the concept of a function by using a "function box." Draw the function box below on the board and write an expression in the box using the variable x. Explain to students that the "function box" takes an input, or x-value, performs the given function on the value, then produces a corresponding output, or y-value. Choose different input values and ask students to find the corresponding output value. Emphasize the point that each value of x will produce one and only one corresponding value for y.

Function Box

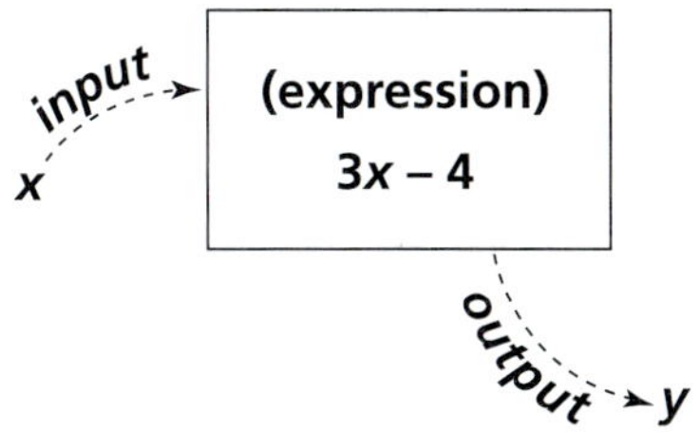

Change the expression in the box and repeat the activity.

2. Teaching the Lesson

After reviewing the example on page 114, demonstrate how the only solutions to the equation are the ordered pairs that lie on the line. Substitute x and y in the equation with ordered pairs for points that lie on the line and those that do not. Simplify the equations together and emphasize how the coordinates of points not on the line do not make the equation true.

When discussing domain and range, use a given line's equation to illustrate that its domain and range have infinitely many values. Demonstrate that no matter what value is chosen for x, there is always a corresponding y-value. Compare these results with the limited range of a line segment as a result of the limited values of x in the domain of the line segment.

ALGEBRA CONNECTION

Provide practice in working with the algebraic equation for a line, $y = mx + b$, by giving students values for three of the variables and asking them to solve for the fourth. For example, give students these values $x = 2$, $y = 4$, and $b = 6$. Ask them to solve the equation for m.

$$4 = m(2) + 6$$

$$-2 = m(2)$$

$$-1 = m$$

Continue the activity, substituting values for all variables except one and asking students to solve for the missing variable. Emphasize that the equation can be used to solve for any variable in the equation when the three other values are known.

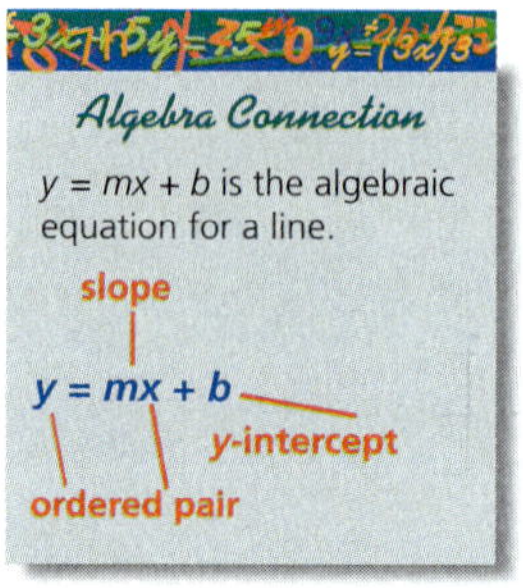

To represent $\overleftrightarrow{AB}$ and $\overline{AB}$ using only algebra, you need to use the equation for the line, $y = mx + b$. The equation $y = mx + b$ has three parts: an ordered pair, (x, y); the slope, m; and the y-intercept, b. The **y-intercept** is the y-value of a **function** when $x = 0$; it is the point at which the line crosses the y-axis.

$y = mx + b$ is a rule that pairs every x-value with one and only one y-value, therefore $y = mx + b$ is an example of a function. The graph of $y = mx + b$ is a straight line, so the equation is sometimes called the straight line function. Sometimes $y = mx + b$ is written as $f(x) = mx + b$ and read as "the function of x equals mx plus b."

The x-values of a function are called the **domain** of the function. The y-values of a function are called the **range** of the function. Let's see how we can represent $\overleftrightarrow{AB}$ using $y = mx + b$.

EXAMPLE

Step 1 Solve for m.

$$m = \frac{1-3}{-1-5} = \frac{-2}{-6} = \frac{1}{3}$$

Step 2 The equation $y = \frac{1}{3}x + b$ is true for all ordered pairs on $\overleftrightarrow{AB}$. This means we can substitute the coordinates of any one point on the line for the x- and y-values in the equation. This will allow us to solve for b.

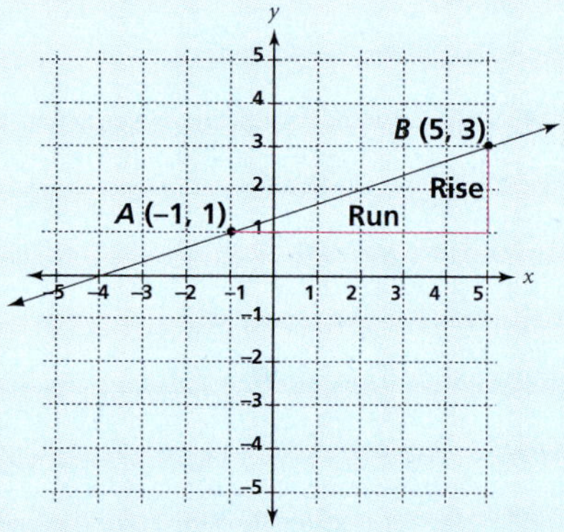

Take $A = (-1, 1)$
$x = -1$, $y = 1$
so $y = mx + b$ becomes
$1 = \frac{1}{3}(-1) + b$
$\therefore \frac{4}{3} = b$
Conclusion: $y = \frac{1}{3}x + \frac{4}{3}$

Take $B = (5, 3)$
$x = 5$, $y = 3$
so $y = mx + b$ becomes
$3 = \frac{5}{3} + b$
$\therefore \frac{4}{3} = b$
Conclusion: $y = \frac{1}{3}x + \frac{4}{3}$

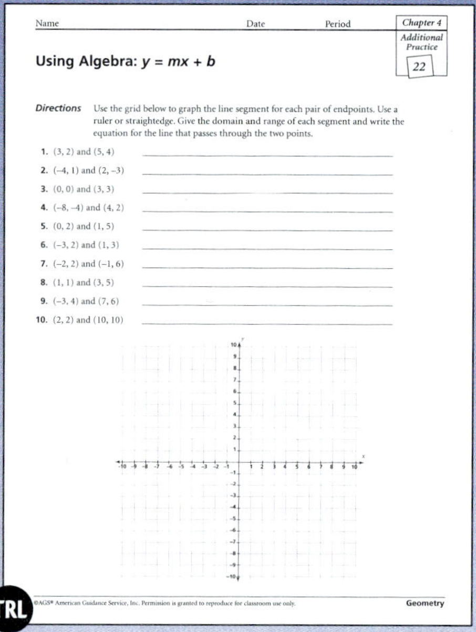

Name Date Period

Chapter 4 Additional Practice 22

Using Algebra: $y = mx + b$

Directions Use the grid below to graph the line segment for each pair of endpoints. Use a ruler or straightedge. Give the domain and range of each segment and write the equation for the line that passes through the two points.

1. (3, 2) and (5, 4) ______
2. (–4, 1) and (2, –3) ______
3. (0, 0) and (3, 3) ______
4. (–8, –4) and (4, 2) ______
5. (0, 2) and (1, 5) ______
6. (–3, 2) and (1, 3) ______
7. (–2, 2) and (–1, 6) ______
8. (1, 1) and (3, 5) ______
9. (–3, 4) and (7, 6) ______
10. (2, 2) and (10, 10) ______

 Geometry

TRL

Additional Practice 22

The substitution of either A or B in the equation leads to the same conclusion.

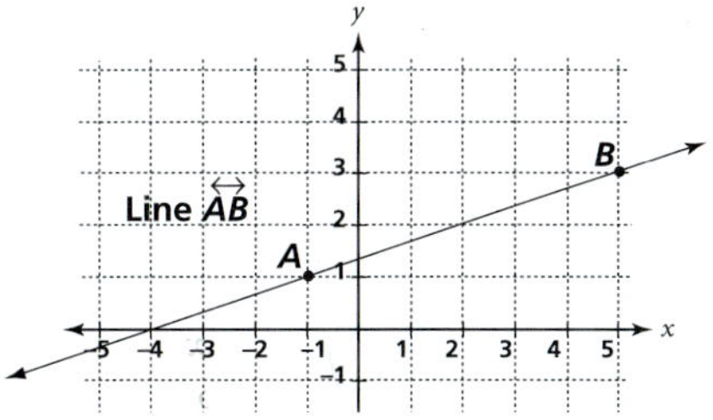

$y = \frac{1}{3}x + \frac{4}{3}$ represents $\overleftrightarrow{AB}$

Domain of x = all real numbers
Range of y = all real numbers

Here, the x-values can be any real number. The y-values can also be any real number for the ordered pairs on $\overleftrightarrow{AB}$.

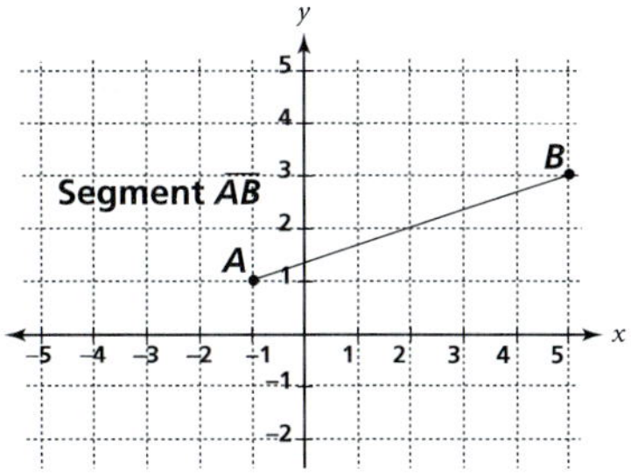

$y = \frac{1}{3}x + \frac{4}{3}$ represents the segment $\overline{AB}$

Domain of $-1 \leq x \leq 5$
Range of $1 \leq y \leq 3$

EXAMPLE Write the equation for segment $\overline{CD}$ and line $\overleftrightarrow{CD}$ when $C = (2, 1)$ and $D = (6, 5)$.

Step 1 Calculate $m = \frac{1-5}{2-6} = \frac{-4}{-4} = 1$

If we substitute 1 for m in the equation, the equation for the line becomes $y = x + b$.

Step 2 Substitute the coordinates of any one point on the line for the x- and y-values in the equation. Then solve for b.

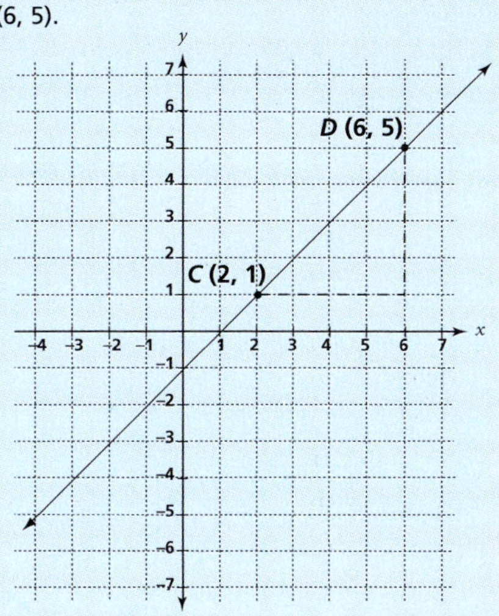

3. Reinforce and Extend

COMMON ERROR

Because of the many steps and number of variables involved in writing and solving the equation of a line, students may have difficulty mastering the process. Provide them with the following summary of the process and encourage them to refer to it when necessary.

1. Find the slope. Use $m = \frac{y_2 - y_1}{x_2 - x_1}$.
2. Replace x and y in the equation with the coordinates of any point on the line. Find b.
3. Write the equation. Substitute values for m and b only; the variables x and y appear in the final form of the equation.

LEARNING STYLES

Logical/Mathematical Ask students to describe in simple terms how to write the equation of a line. The steps should be numbered or otherwise organized in a logical way and described using terminology from this lesson. You might also suggest that students provide examples or some other kind of visual aid to accompany their written instructions. After reviewing their work, ask for volunteers to share their instructions with the class or small groups of students.

Mental Math

Write an expression such as $y = 3x + 2$ on the board. Below the equation, write five to seven values for x. Ask students to mentally calculate the corresponding y-value for each x-value and to record their answers on paper. After all students have recorded their answers, write the answers on the board and give students time to check their mental math skills.

Example *(continued)*

Take $C = (2, 1)$	Take $D = (6, 5)$
$x = 2, y = 1$	$x = 6, y = 5$
so $y = mx + b$ becomes	so $y = mx + b$ becomes
$1 = 2 + b$	$5 = 6 + b$
$\therefore -1 = b$	$\therefore -1 = b$
Conclusion: $y = x - 1$	**Conclusion:** $y = x - 1$

For line $\overleftrightarrow{CD}$, $y = mx + b$ becomes $y = x - 1$
or $f(x) = x - 1$; domain equals all real numbers
and range equals all real numbers

For segment $\overline{CD}$, $y = mx + b$ becomes $y = x - 1$
or $f(x) = x - 1$; domain $2 \le x \le 6$ and range $1 \le y \le 5$

You can generally read the domain and range of a function from the graph of that function.

Domain is all points on the x-axis needed to form the ordered pairs on $\overline{AB}$, and the range is the y-values.

Examples Name the domain and range for each graph.

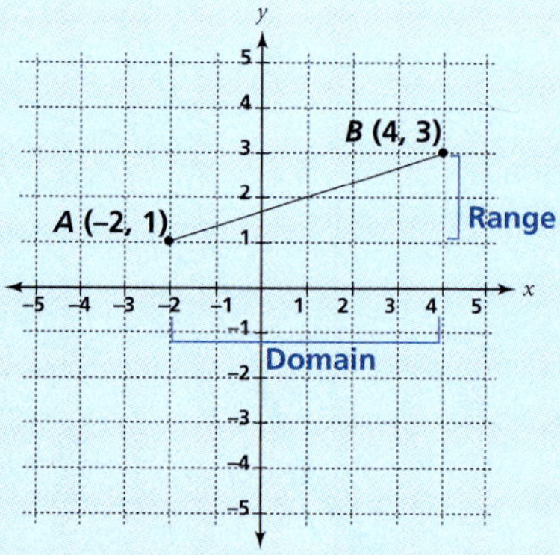

Solution:

Domain is $-2 \le x \le 4$

Range is $1 \le y \le 3$

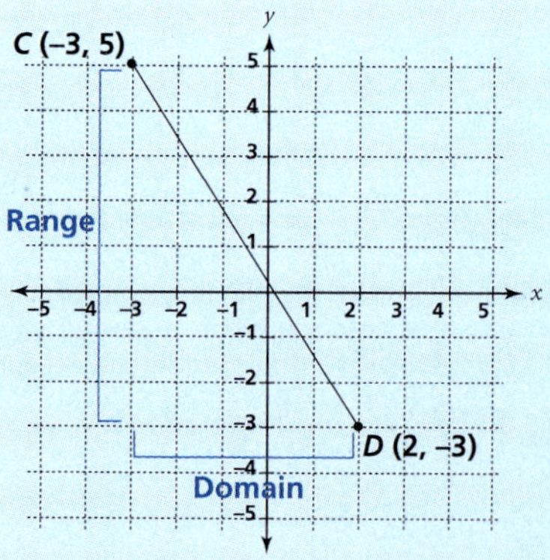

Solution:

Domain is $-3 \le x \le 2$

Range is $-3 \le y \le 5$

EXAMPLES *(continued)*

Special Case: Remember, $m = 0$ for lines parallel to the x-axis. In these cases $y = mx + b$ becomes $y = (0)x + b$ or $y = b$. In the example, the line passes through (1, 1) and (5, 1) and

$$m = \frac{1 - 1}{1 - 5} = \frac{0}{-4} = 0.$$

So, $y = mx + b$ becomes $y = b$, but $y = 1$.

Solution:

The domain for this function is any real number and the range is 1.

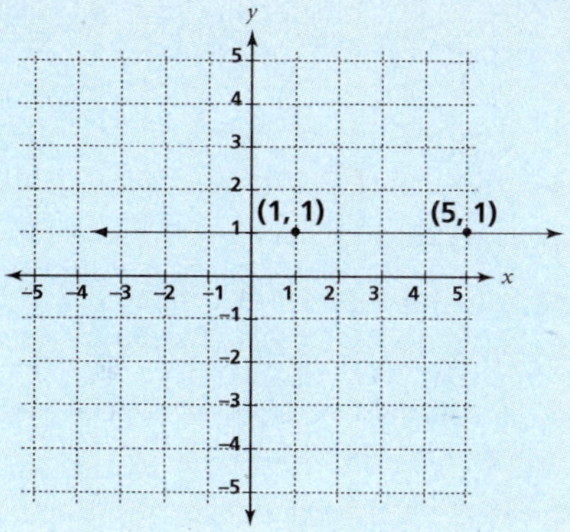

Construction

Exercise A Graph the line that passes through each pair of points. Use graph paper and a straightedge.

1. (2, 1) and (1, 2)
2. (3, 4) and (–3, –4)
3. (5, 1) and (10, 1)
4. (3, 2) and (2, 3)
5. (2, 2) and (5, 5)
6. (–3, –3) and (4, –3)
7. (– 2, 3) and (4, –3)

See Teacher's Edition page for answers.

Exercise B Graph the line segments for the following pairs of endpoints. Use graph paper and a straightedge. Give the domain and range of each.

8. (2, 1) and (1, 2)
9. (–2, 6) and (1, 1)
10. (3, 1) and (2, –5)
11. (4, 2) and (–1, 2)
12. (6, 3) and (–3, 4)

See Teacher's Edition page for answers.

Writing About Mathematics

Why are lines at right angles to the x-axis not functions?

Learning Styles

Visual/Spatial To help students find the domains and ranges in Exercise B, tell them to draw vertical lines from each endpoint to the x-axis and horizontal lines from each endpoint to the y-axis. Suggest that they shade or darken the portion of the axes between each set of lines to help highlight the domain and range of each segment.

Answers to Exercise A 1–4

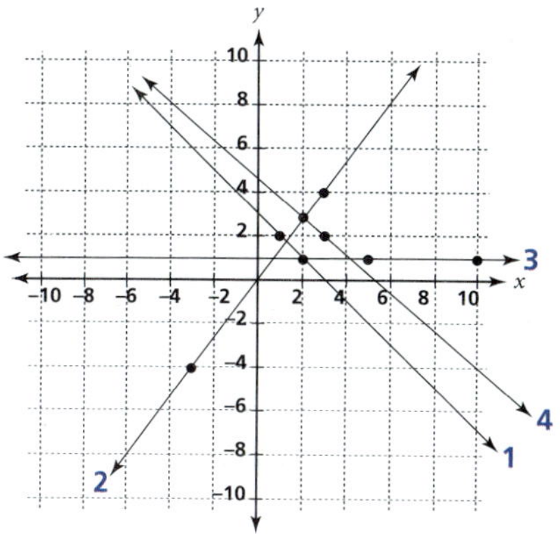

Answers to Exercise A 5–7

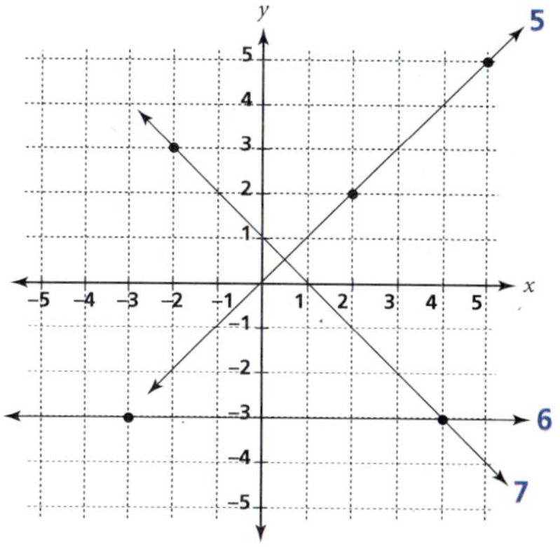

Answers to Exercise B 8–12

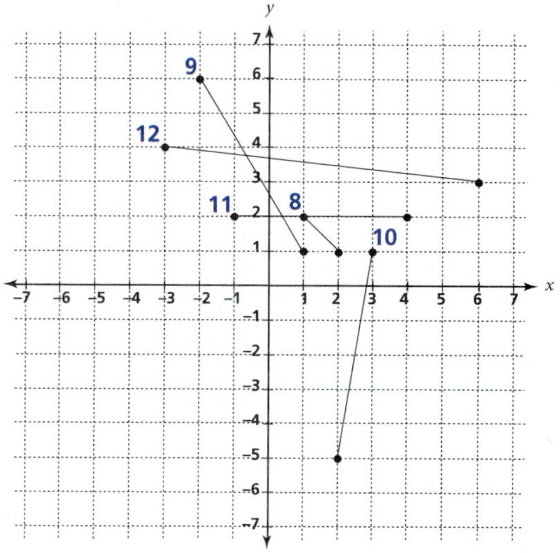

8. Domain: $1 \le x \le 2$;
Range: $1 \le y \le 2$

9. Domain: $-2 \le x \le 1$;
Range: $1 \le y \le 6$

10. Domain: $2 \le x \le 3$;
Range: $-5 \le y \le 1$

11. Domain: $-1 \le x \le 4$;
Range: $y = 2$

12. Domain: $-3 \le x \le 6$;
Range: $3 \le y \le 4$

Learning Styles

LEP/ESL To reinforce their understanding of the vocabulary in this lesson, students with limited English proficiency can describe each line in Exercise C by copying and completing the following sentences. Encourage students to work with partners who are more proficient in English. The following example answer is for problem 13.

1. The function that describes each point on this line is ______. ($y = -x + 3$)
2. The y-intercept of this line is ______. (+3)
3. ______ are in the domain of this function and ______ are the corresponding values in the range. ($1 \le x \le 2$) ($1 \le y \le 2$)

Try This...

Answers will vary but should reflect an understanding that a vertical line does not represent a function nor does a graph in which only one value of x is paired with more than one y-value (not a one-to-one correspondence).

Exercise C Write the equation of the line that passes through each pair of points. Use the form $y = mx + b$.

13. (2, 1) and (1, 2) $y = -x + 3$

14. (3, 4) and (−3, −4) $y = \frac{4}{3}x$

15. (5, 1) and (10, 1) $y = 1$

16. (3, 2) and (2, 3) $y = -x + 5$

17. (2, 2) and (5, 5) $y = x$

Try This...

Look in a magazine or newspaper for a line graph. Explain whether the graph is or is not an example of a function. If it is, what is its domain and range?

Problem Solving

Exercise D Solve the following problems.

18. At the right is a design for a playground slide. What is the slope of the slide?

slope = $\frac{7}{10}$ or $-\frac{7}{10}$

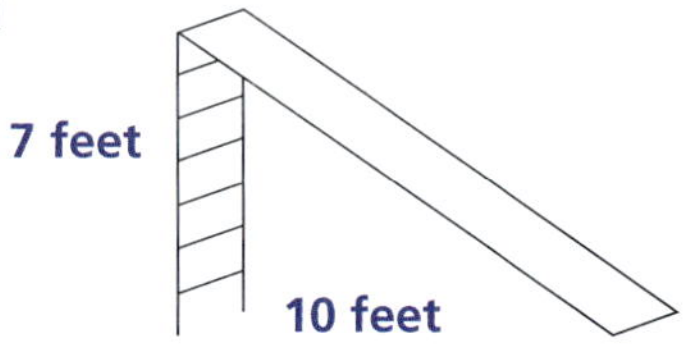

19. A ski lift carries skiers from an elevation of 5,200 ft to an elevation of 7,700 ft. The horizontal distance from the start to the end of the lift is 5,000 ft. Draw a diagram showing this information and find the slope of the lift line.

slope = $\frac{1}{2}$

20. The distance between the first and second floor of a building is 10 ft. The horizontal distance between the bottom step and the top step is 72 in. What is the slope of the stairway?

slope = $\frac{5}{3}$

Lesson 6 The Point-Slope Formula

Suppose you are given a point A at (1, 3) and a slope $m = \frac{1}{2}$. How would you graph the line and then write the equation for that line?

CONSTRUCTION

Given: Point A (1, 3) and slope $m = \frac{1}{2}$

Graph: The line with slope $m = \frac{1}{2}$ that passes through A (1, 3)

Graph $\overleftrightarrow{AB}$.

Step 1 Locate (1, 3) on the plane, in Quadrant I.

Step 2 Use $m = \frac{1}{2}$ to locate a second point on the line. (Remember Euclid's Postulate 1.)

$m = \frac{1}{2} = \frac{\text{change in } y}{\text{change in } x}$ count UP 1 from point A (1, 3) to reach point B (3, 4) count 2 to the RIGHT

Draw $\overleftrightarrow{AB}$. This is your graph.

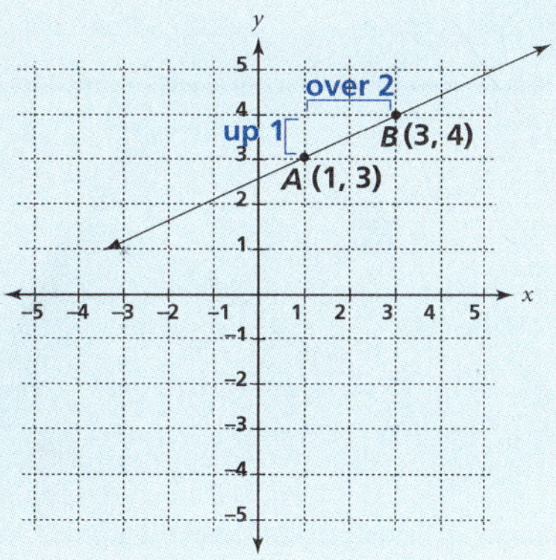

You can also calculate the coordinates for point B. The given slope, $m = \frac{1}{2}$, indicates that as the x-value increases by 2, the y-value increases by 1. If point A is (x, y), then the ordered pair for point B is $(x + 2, y + 1)$.

Point $A = (1, 3)$, so point $B = (1 + 2, 3 + 1) = (3, 4)$.

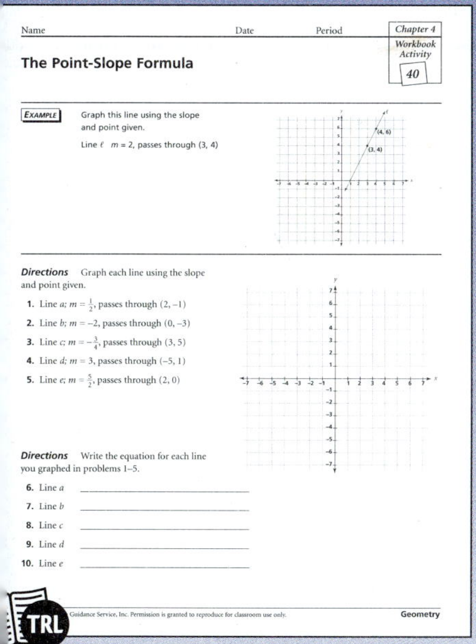

Name Date Period

Chapter 4 Workbook Activity 40

The Point-Slope Formula

EXAMPLE Graph this line using the slope and point given.

Line ℓ $m = 2$, passes through (3, 4)

(4, 6)

(3, 4)

Directions Graph each line using the slope and point given.

1. Line a; $m = \frac{1}{2}$, passes through (2, –1)
2. Line b; $m = -2$, passes through (0, –3)
3. Line c; $m = -\frac{3}{4}$, passes through (3, 5)
4. Line d; $m = 3$, passes through (–5, 1)
5. Line e; $m = \frac{5}{2}$, passes through (2, 0)

Directions Write the equation for each line you graphed in problems 1–5.

6. Line a _____
7. Line b _____
8. Line c _____
9. Line d _____
10. Line e _____

TRL

Geometry

Workbook Activity 40

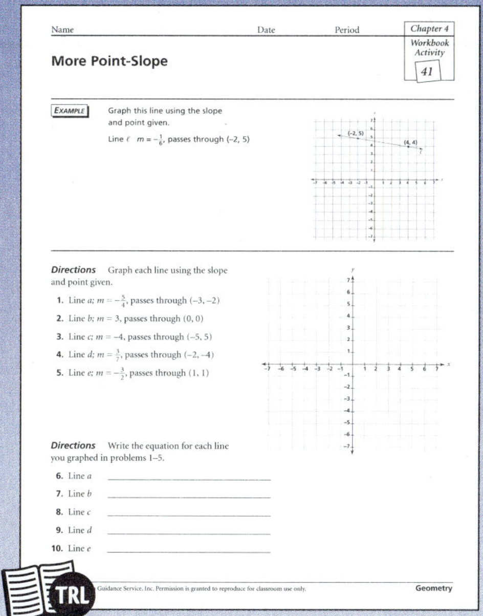

Name Date Period

Chapter 4 Workbook Activity 41

More Point-Slope

EXAMPLE Graph this line using the slope and point given.

Line ℓ $m = -\frac{1}{6}$, passes through (–2, 5)

(–2, 5)

(4, 4)

Directions Graph each line using the slope and point given.

1. Line a; $m = -\frac{5}{4}$, passes through (–3, –2)
2. Line b; $m = 3$, passes through (0, 0)
3. Line c; $m = -4$, passes through (–5, 5)
4. Line d; $m = \frac{3}{7}$, passes through (–2, –4)
5. Line e; $m = -\frac{3}{2}$, passes through (1, 1)

Directions Write the equation for each line you graphed in problems 1–5.

6. Line a _____
7. Line b _____
8. Line c _____
9. Line d _____
10. Line e _____

TRL

Geometry

Workbook Activity 41

Lesson at a Glance

Chapter 4 Lesson 6

Overview This lesson explains how to graph a line and write its equation using the point-slope formula.

Objectives

- To graph a line given its slope and a point on the line.
- To write the equation of a line using the point-slope formula.

Student Pages 119–122

Teacher's Resource Library TRL

Workbook Activities 40–41

Additional Practice 23

1. Warm-Up Activity

Review with students that slope is a ratio between the rise and the run of a line. Draw some lines on a coordinate graph and ask students to find the slope by finding the rise-to-run ratio. Then draw the line $y = x$ and ask, "If you begin at the origin and move 4 units up and 4 units to the right, what point would you be at on the line?" **(4, 4)** Ask the same question but change the starting point to (1, 1), then (2, 2), then (20, 20), etc., until students begin to see a pattern. **(5, 5), (6, 6), (24, 24)**

2. Teaching the Lesson

After reviewing the examples on pages 119–121, discuss the similarities and differences between using the point-slope formula to write the equation of a line and using the method in Lesson 5.

Algebra Connection

Have students find the coordinates of another point on each line with the slope and point indicated. Remind students that the slope is a ratio of the change in y to the change in x. Ask students to show their work.

(Possible answers are given.)

- $(3, 6); m = \frac{1}{3}$ **(6, 7)**
- $(-2, 4); m = \frac{3}{2}$ **(0, 7)**
- $(8, -5); m = 1$ **(9, –4)**
- $(1, 2); m = -\frac{1}{2}$ **(3, 1) or (–1, 3)**
- $(10, -3); m = -5$ **(11, –8) or (9, 2)**
- $(\frac{1}{2}, -6); m = \frac{1}{4}$ **$(4\frac{1}{2}, -5)$**

3. Reinforce and Extend

Learning Styles

Logical/Mathematical Give students the slope of a line and a point on the line. Ask them to work with partners to calculate other points on that line and to draw the line on a graph.

Algebra Connection

Example Write the equation for $\overleftrightarrow{AB}$.

Step 1 $y = mx + b$ and $m = \frac{1}{2}$, so $y = \frac{x}{2} + b$.

Step 2 The equation $y = \frac{1}{2}x + b$ is true for all ordered pairs on $\overleftrightarrow{AB}$. Because (1, 3) is on the line, we can substitute $x = 1$ and $y = 3$ and solve for b.

$y = \frac{x}{2} + b$ becomes $3 = \frac{1}{2} + b$

$2\frac{1}{2} = b$

Step 3 Using this value for b, the equation becomes

$y = \frac{x}{2} + 2\frac{1}{2}$

$\therefore y = \frac{x}{2} + 2\frac{1}{2}$ is the equation for $\overleftrightarrow{AB}$.

Construction

Given: Point $C(-1, -2)$ and slope $m = -2$

Graph: The line with slope $m = -2$ that passes through $C(-1, -2)$

Graph $\overleftrightarrow{CD}$.

Step 1 Locate $C(-1, -2)$ on the plane, in Quadrant III.

Step 2 Use $m = -2$ to locate a second point on the line. (Remember Euclid's Postulate 1.)

$m = -\frac{2}{1} = \frac{\text{change in } y}{\text{change in } x}$

count DOWN 2 / count 1 to the RIGHT — from point $C(-1, -2)$ to reach point $D(0, -4)$

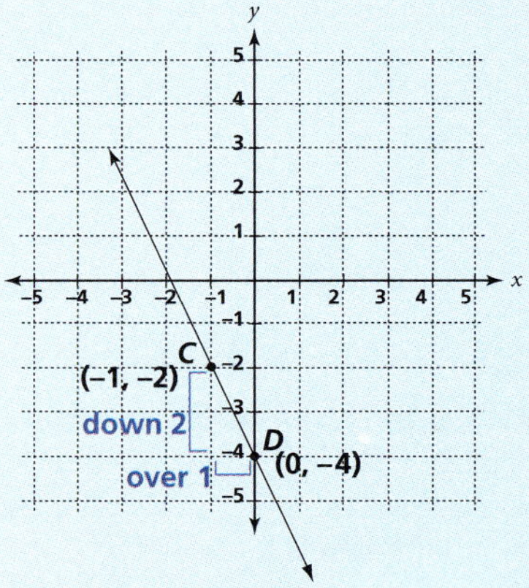

Draw $\overleftrightarrow{CD}$. This is your graph.

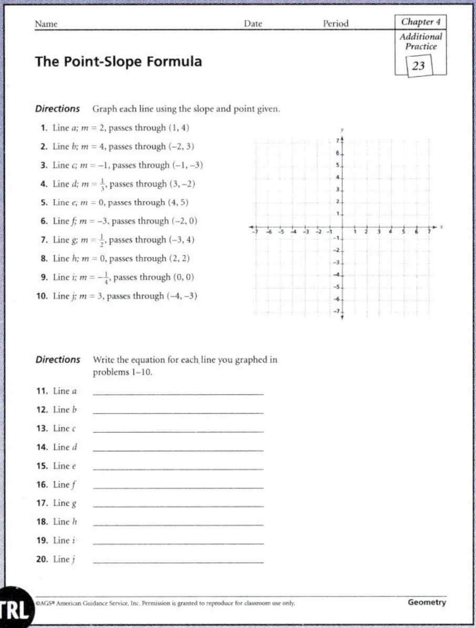

Name ____ Date ____ Period ____ *Chapter 4 Additional Practice 23*

The Point-Slope Formula

Directions Graph each line using the slope and point given.

1. Line a; $m = 2$, passes through (1, 4)
2. Line b; $m = 4$, passes through (–2, 3)
3. Line c; $m = -1$, passes through (–1, –3)
4. Line d; $m = \frac{1}{3}$, passes through (3, –2)
5. Line e; $m = 0$, passes through (4, 5)
6. Line f; $m = -3$, passes through (–2, 0)
7. Line g; $m = \frac{1}{2}$, passes through (–3, 4)
8. Line h; $m = 0$, passes through (2, 2)
9. Line i; $m = -\frac{1}{4}$, passes through (0, 0)
10. Line j; $m = 3$, passes through (–4, –3)

Directions Write the equation for each line you graphed in problems 1–10.

11. Line a ____
12. Line b ____
13. Line c ____
14. Line d ____
15. Line e ____
16. Line f ____
17. Line g ____
18. Line h ____
19. Line i ____
20. Line j ____

TRL

 Geometry

Additional Practice 23

Check by calculating the coordinates for point D. The given slope, $m = -2 = -\frac{2}{1}$, indicates that as the x-value increases by 1, the y-value decreases by 2. If point C is (x, y), then the ordered pair for point D is $(x + 1, y - 2)$.

Point $C = (-1, -2)$, so point $D = (-1 + 1, -2 - 2) = (0, -4)$.

EXAMPLE Write the equation for $\overleftrightarrow{CD}$.

Step 1 $y = mx + b$ and $m = -2$, so $y = -2x + b$.

Step 2 The equation $y = -2x + b$ is true for all ordered pairs on $\overleftrightarrow{CD}$. Because $(-1, -2)$ is on the line, we can substitute $x = -1$ and $y = -2$ and solve for b.

$y = -2x + b$ becomes $-2 = (-2)(-1) + b$

$-4 = b$

Step 3 Using this value for b, the equation becomes $y = -2x - 4$.

$\therefore$ $y = -2x - 4$ is the equation for $\overleftrightarrow{CD}$.

Special Case: Suppose the point is any point, but the slope is $m = 0$. You know that all lines with $m = 0$ are parallel to the x-axis. In these cases, you need only plot the point and draw a line through that point parallel to the x-axis.

CONSTRUCTION

Given: Point P (3, 4) and slope $m = 0$

Graph: The line ℓ, with slope $m = 0$ that passes through P (3, 4).

Graph line ℓ.

Step 1 Locate P (3, 4) on the plane.

Step 2 Draw a line parallel to the x-axis passing through (3, 4).

This is your graph.

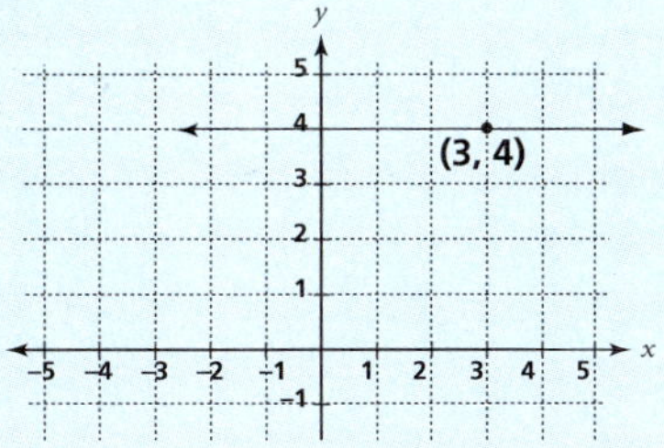

CALCULATOR

Have students use their calculators to find b for each line with the given point and slope:

- $(2, 3)$; $m = \frac{1}{2}$ ($b = 2$)
- $(-5, 4)$; $m = -2$ ($b = -6$)
- $(1.2, -3.4)$; $m = -8$ ($b = 6.2$)
- $(-\frac{2}{3}, 1.3)$; $m = 6$ ($b = 4.3$)
- $(6.8, 2.1)$; $m = -\frac{3}{4}$ ($b = 7.2$)
- $(2\frac{1}{2}, -4\frac{1}{8})$; $m = 4$ ($b = -14.125$)

LEARNING STYLES

Logical/Mathematical Have students explain in their own words the two methods for finding points on a line when you know the slope and another point on the line. (You can write an equation and use it to find ordered pairs, or you can count units on the graph using the slope.) Ask students to write the steps of each method in a clear and logical manner and to include examples and illustrations as necessary.

COMMON ERROR

When graphing a line by using the slope to count units, students may start counting from the origin rather than from a point on the line. Remind students that the first step in this method of graphing a line is to plot a point. Then, without lifting their pencil, they can use the slope to count units to find another point on the line.

LEARNING STYLES

Interpersonal/Group Learning Have small groups of students draw a triangle with vertices on points with integer coordinate values. Ask each group to draw its triangle so that it lies in at least two quadrants. Then on a separate sheet of paper, the group writes the equation for each of the segments that form the sides of the triangle. Collect and display the triangle drawings, and have groups exchange equations. Each group matches the given equation to a triangle on display.

Learning Styles

Visual/Spatial Encourage students to graph the lines in Exercise A on grid paper and to extend each line across the y-axis. Then have them use their graphs to compare the y-intercept on the graph with the value of b in the equation they wrote for each line in Exercise B. The comparison can help them check their work.

Learning Styles

Auditory/Verbal Have pairs of students complete Exercise B together. Instruct one student to describe the lines to be graphed. He or she should include a verbal description of the slope of the line and the numerical description. For example, a student may say, "Line l in problem 1 slopes up to the right at an equal distance from the x- and y-axes." The other student can use the information to write the equation.

Example Write the equation for line ℓ.

Step 1 $y = mx + b$ and $m = 0$, so $y = 0x + b$, or $y = b$

Step 2 The equation $y = b$ is true for all ordered pairs on ℓ. Because (3, 4) is on the line, we can substitute $y = b = 4$ and solve for b.

$y = mx + b$ becomes $y = 4$

$\therefore$ $y = 4$ is the equation for ℓ. The domain is all real numbers. Specifying the domain makes sure that $y = 4$ will not be mistaken for a single point.

Construction

Exercise A Graph each line using the slope and point given.

See Teacher's Edition page for answers.

1. ℓ, $m = 1$, passes through (2, 3)
2. k, $m = -1$, passes through (−2, −3)
3. n, $m = -1$, passes through (3, 4)
4. o, $m = \frac{1}{3}$, passes through (4, 5)
5. p, $m = 0$, passes through (3, −2)
6. q, $m = 5$, passes through (1, −1)
7. r, $m = 3$, passes through (−6, 2)
8. s, $m = 2$, passes through (4, 3)
9. t, $m = -2$, passes through (4, 3)
10. u, $m = 0$, passes through (3, 6)

Exercise B Write the equation for each of the lines you graphed in problems 1–10 above.

11. line ℓ
 $y = x + 1$
12. line k
 $y = -x - 5$
13. line n
 $y = -x + 7$
14. line o
 $3y = x + 11$
15. line p
 $y = -2$, x = any real number
16. line q
 $y = 5x - 6$
17. line r
 $y = 3x + 20$
18. line s
 $y = 2x - 5$
19. line t
 $y = -2x + 11$
20. line u
 $y = 6$, x = any real number

Writing About Mathematics

Which method do you prefer for finding a second point on a line—counting on a graph or calculating using the slope and the ordered pair? Tell why.

Answers to Exercise A 1–5

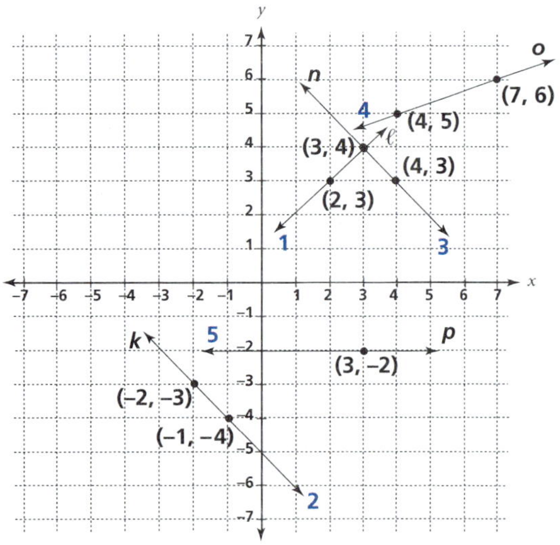

Answers to Exercise A 6–10

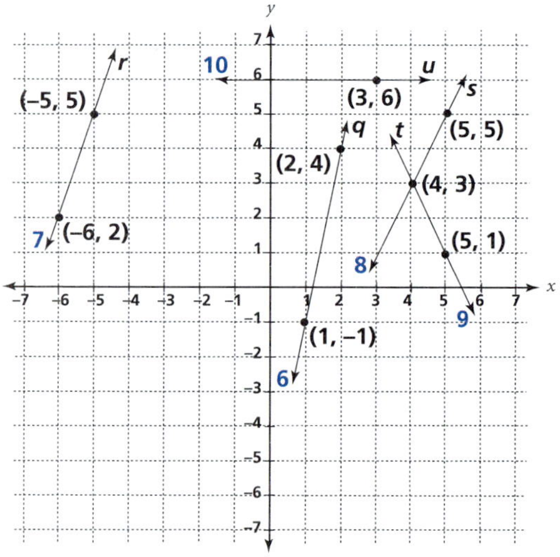

Lesson 7 Finding the Midpoint of a Segment

When you do constructions in geometry, you must be careful to use only a straightedge and a compass. Remember that Euclid's Postulate 1—two points determine a line—allows you to use a straightedge. Euclid's Postulate 3—a circle may be described by any center and distance—allows you to use a compass.

You have used a straightedge and compass in your earlier constructions. In this lesson, you will learn how to use them to find the **midpoint** of any line segment.

Bisect
To divide into two equal parts

Midpoint
Point that divides a line segment into two equal parts

Remember, a postulate is a statement accepted as true without proof. For a review of Euclid's postulates, turn back to Chapter 2.

CONSTRUCTION

Given: $\overline{AB}$

Construct: Find midpoint of $\overline{AB}$

Step 1 Open your compass to a distance somewhat larger than $\frac{1}{2}$ of $\overline{AB}$.

Step 2 Using *A* as the center of a circle, draw an arc on each side of $\overline{AB}$. Do not change your compass opening.

Step 3 Using *B* as a center, draw an arc on each side of $\overline{AB}$. Make sure that these arcs intersect the arcs you have already drawn.

Step 4 Draw a line connecting the two points where the arcs intersect. This line is perpendicular to $\overline{AB}$ and bisects $\overline{AB}$. The point *M*, at which this line crosses $\overline{AB}$, is the midpoint of $\overline{AB}$. $\overline{AM} = \overline{MB}$.

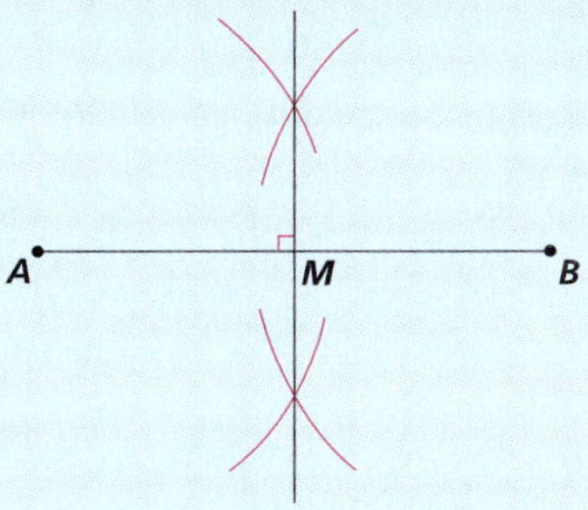

Lesson at a Glance

Chapter 4 Lesson 7

Overview This lesson demonstrates how to construct the bisector of a line segment and to use the midpoint formula to find the midpoint of a segment.

Objectives

- To bisect a line segment.
- To find the midpoint of a line segment.

Student Pages 123–127

Teacher's Resource Library TRL

Workbook Activities 42–43

Additional Practice 24

Mathematics Vocabulary

bisect
midpoint

1. Warm-Up Activity

Ask pairs of students to take linear measurements of any surface in the classroom and to round their measurements to the nearest $\frac{1}{4}$ inch. Discuss ideas for how to find the middle of their measured lines and why some midpoint measurements may be easier to calculate than others.

2. Teaching the Lesson

Introduce the lesson vocabulary and demonstrate how to use a compass to find the midpoint of a line segment. Help students draw horizontal and vertical lines on a grid and find the midpoints of these lines, using algebra, coordinate geometry, and arithmetic. Then work through the example for finding the midpoint of a line that is not parallel to either axis (see page 125). Make sure students understand that the coordinates of the midpoint have no relation to the length of the segments formed by the midpoint. Point out to students, however, that the length of the two segments formed by the midpoint of a horizontal or vertical line can be calculated easily.

Learning Styles

Logical/Mathematical Have students examine the construction example on page 123. Ask them to explain why the point where the two arcs intersect is equidistant from points *A* and *B*. If necessary, provide a hint by telling them to draw complete circles with centers *A* and *B*, rather than just arcs. (Student answers should reflect the understanding that the segments are radii of congruent circles and are therefore the same length.)

Manipulatives

M At the Midpoint

Materials: Factor Blocks, Manipulatives Master 2 (Coordinate Grid)

Group Practice: Review with students the methods for finding the midpoint of horizontal and vertical lines described on student page 124. Show them how to use the Factor Blocks to represent the endpoints on the Coordinate Grid. They can then count the number of points between the two blocks to find the midpoint.

Student Practice: Have students use the Factor Blocks and Coordinate Grid to reinforce visually and manipulatively the midpoints of horizontal and vertical lines. Write ordered pairs on the board and have students place their Factor Blocks on the pairs. They can count the total number of grid points between the blocks and then count half that total to find the midpoint. Have them identify the coordinate pair of the midpoint. Use these ordered pairs: (–3, 3); (5, 3) [midpoint = (2, 3)]; (3, 3) (3, 7) [midpoint = (3, 5)]. Continue with other ordered pairs for vertical and horizontal lines if necessary.

After the manipulative experience, have students draw vertical and horizontal lines on the grid and find their midpoints.

Algebra Connection

To **bisect** a line or an angle means to divide that line or angle into two equal parts.

You can also find the midpoint of a line using algebra, coordinate geometry, and arithmetic.

Examples **Given:** $\overline{AB}$

Construct: Find midpoint of $\overline{AB}$

The midpoint of $\overline{AB}$ at right is the point where the perpendicular bisector of the segment $1 \le x \le 7$ meets $\overline{AB}$.

The x-value of the midpoint can be found by dividing the sum of the x-values of the endpoints by 2.

$x\text{-value} = \frac{7+1}{2} = \frac{8}{2} = 4$

You already know the y-value of the midpoint of $\overline{AB}$; it is 3.

That means the midpoint M of $\overline{AB}$ is (4, 3).

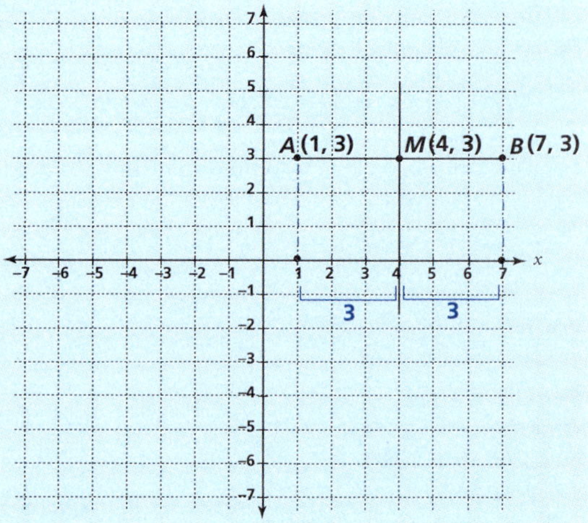

Given: $\overline{CD}$

Construct: Find midpoint of $\overline{CD}$

The midpoint of $\overline{CD}$ at right is the point where the perpendicular bisector of the segment $-2 \le y \le 4$ meets $\overline{CD}$.

The y-value of the midpoint can be found by dividing the sum of the y-values of the endpoints by 2.

$y\text{-value} = \frac{4+(-2)}{2} = \frac{2}{2} = 1$

You already know the x-value of the midpoint of $\overline{CD}$; it is 5.

That means the midpoint M of $\overline{CD}$ is (5, 1).

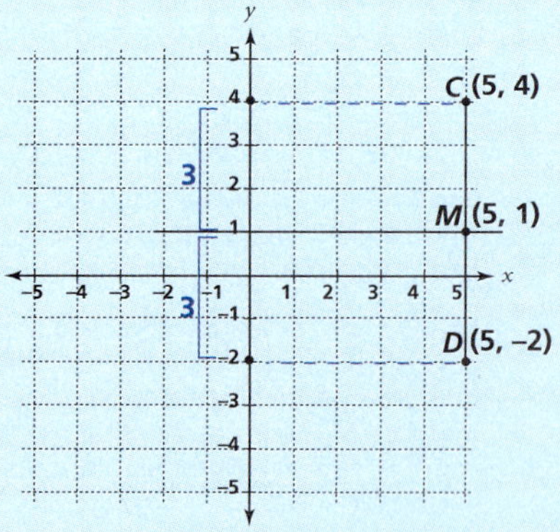

3. Reinforce and Extend

Algebra Connection

Review the first example on student page 124. Help students develop an algebraic expression for finding the x-value of the midpoint of a line segment.

(midpoint $= \frac{x_1 + x_2}{2}$)

Then have them name two x-values and use their expression to determine the midpoint of the two values. Follow the same procedure for developing and using an algebraic expression for finding the y-value of the midpoint of a line segment.

(midpoint $= \frac{y_1 + y_2}{2}$)

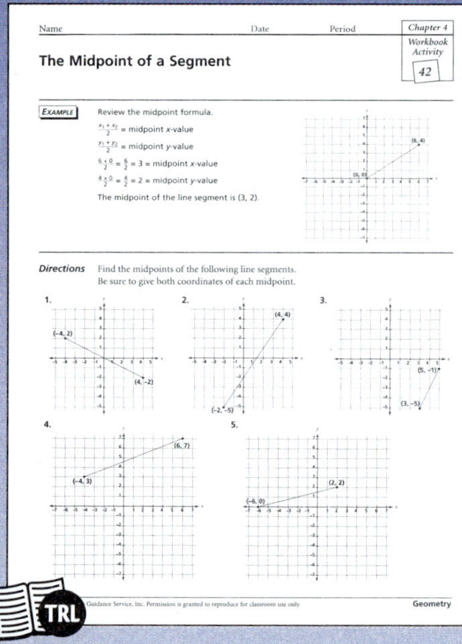

Workbook Activity 42

EXAMPLE **Given:** $\overline{PQ}$

Construct: Find midpoint of $\overline{PQ}$

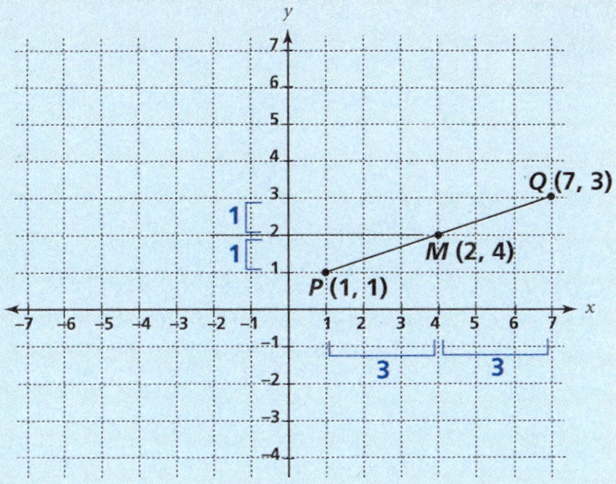

The midpoint of $\overline{PQ}$, which is not parallel to either the x-axis or the y-axis, can be calculated in three steps.

Step 1 Calculate the midpoint of the x-values of P and Q. x-value of midpoint $= \frac{1+7}{2} = \frac{8}{2} = 4$

Step 2 Calculate the midpoint of the y-values of P and Q. y-value of midpoint $= \frac{1+3}{2} = \frac{4}{2} = 2$

Step 3 Use the x- and y-values you calculated to form the ordered pair for the midpoint M.

$M = (4, 2)$

Note that when you calculated the midpoint, you calculated the midpoint of the domain $(1 \le x \le 7)$ and the midpoint of the range $(1 \le y \le 3)$ for segment $\overline{PQ}$.

These examples lead to the following formula.

MIDPOINT FORMULA: The midpoint of a segment between points (x_1, y_1) and (x_2, y_2) is

$$\left(\frac{x_1 + x_2}{2}, \frac{y_1 + y_2}{2}\right)$$

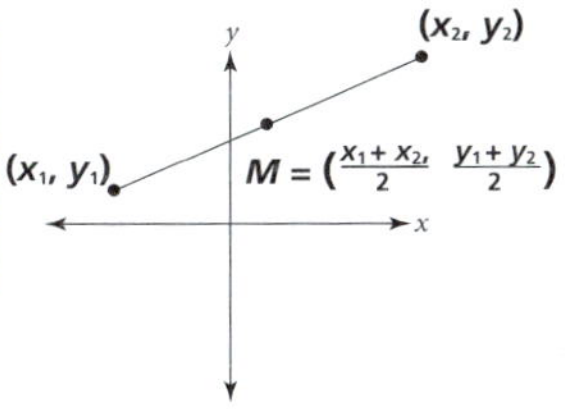

Use the midpoint formula to find the midpoint between $A = (3, 5)$ and $B = (4, -1)$.

Let $x_1 = 3, x_2 = 4, y_1 = 5, y_2 = -1$

$$\text{Midpoint} = \left(\frac{x_1 + x_2}{2}, \frac{y_1 + y_2}{2}\right)$$

$$= \left(\frac{3+4}{2}, \frac{5+(-1)}{2}\right) = \left(\frac{7}{2}, 2\right)$$

$$\therefore \text{ Midpoint } M = \left(\frac{7}{2}, 2\right)$$

You can graph the segment to check if your answer is reasonable.

MENTAL MATH

Provide students with practice in calculating the x- and y-coordinates of a midpoint. Call out the numbers below and have students mentally calculate their sum divided by 2.

- 3 and 7 (5)
- 10 and 16 (13)
- –4 and 12 (4)
- 0 and –14 (–7)
- –1 and –5 (–3)
- 6 and 7 $(6\frac{1}{2})$
- –5 and 8 $(1\frac{1}{2})$
- –9 and 9 (0)
- –21 and 14 $(-3\frac{1}{2})$
- $8\frac{1}{2}$ and $-6\frac{1}{2}$ (1)

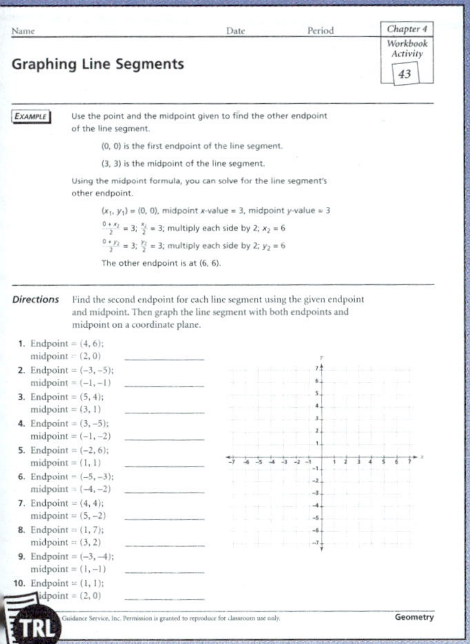

Name Date Period | Chapter 4 Workbook Activity 43

Graphing Line Segments

EXAMPLE Use the point and the midpoint given to find the other endpoint of the line segment.

(0, 0) is the first endpoint of the line segment.

(3, 3) is the midpoint of the line segment.

Using the midpoint formula, you can solve for the line segment's other endpoint.

$(x_1, y_1) = (0, 0)$, midpoint x-value = 3, midpoint y-value = 3

$\frac{0 + x_2}{2} = 3$; $\frac{x_2}{2} = 3$; multiply each side by 2; $x_2 = 6$

$\frac{0 + y_2}{2} = 3$; $\frac{y_2}{2} = 3$; multiply each side by 2; $y_2 = 6$

The other endpoint is at (6, 6).

Directions Find the second endpoint for each line segment using the given endpoint and midpoint. Then graph the line segment with both endpoints and midpoint on a coordinate plane.

1. Endpoint = (4, 6); midpoint = (2, 0) ______
2. Endpoint = (–3, –5); midpoint = (–1, –1) ______
3. Endpoint = (5, 4); midpoint = (3, 1) ______
4. Endpoint = (3, –5); midpoint = (–1, –2) ______
5. Endpoint = (–2, 6); midpoint = (1, 1) ______
6. Endpoint = (–5, –3); midpoint = (–4, –2) ______
7. Endpoint = (4, 4); midpoint = (5, –2) ______
8. Endpoint = (1, 7); midpoint = (3, 2) ______
9. Endpoint = (–3, –4); midpoint = (1, –1) ______
10. Endpoint = (1, 1); midpoint = (2, 0) ______

Geometry

Workbook Activity 43

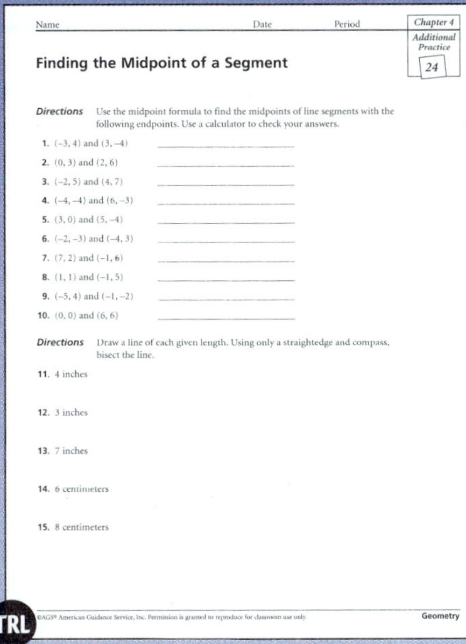

Name Date Period | Chapter 4 Additional Practice 24

Finding the Midpoint of a Segment

Directions Use the midpoint formula to find the midpoints of line segments with the following endpoints. Use a calculator to check your answers.

1. (–3, 4) and (3, –4) ______
2. (0, 3) and (2, 6) ______
3. (–2, 5) and (4, 7) ______
4. (–4, –4) and (6, –3) ______
5. (3, 0) and (5, –4) ______
6. (–2, –3) and (–4, 3) ______
7. (7, 2) and (–1, 6) ______
8. (1, 1) and (–1, 5) ______
9. (–5, 4) and (–1, –2) ______
10. (0, 0) and (6, 6) ______

Directions Draw a line of each given length. Using only a straightedge and compass, bisect the line.

11. 4 inches
12. 3 inches
13. 7 inches
14. 6 centimeters
15. 8 centimeters

Geometry

Additional Practice 24

Learning Styles

Visual/Spatial Ask students to graph the line segments described in Exercise A. They should then draw lines from each endpoint perpendicular to each axis and identify the midpoint of the segment formed on the axis by the two lines. Suggest that they compare these coordinates to those calculated using the midpoint formula to check their work. It may also be helpful for students to draw a line from the midpoints marked on the axes to the graph of the line segment to ensure that they are the coordinates of the midpoint.

Learning Styles

Body/Kinesthetic Have students check their work in Exercise A by using a compass to graph each line segment and find its midpoint.

Calculator

Suggest that students verify their answers for Exercise B with their calculators. Have them use their calculators to divide the given length by two. Then have them measure the length of the bisected line segment from the midpoint to an endpoint. The measurement and the calculator display should match.

Calculator Practice

Using the parentheses keys on your calculator can help you find the midpoint of a line easily.

EXAMPLE Let $(x_1, y_1) = (8, 6)$

$(x_2, y_2) = (4, 2)$

Then the *x*-value of $M = \frac{x_1 + x_2}{2}$ and the *y*-value of $M = \frac{y_1 + y_2}{2}$.

Use your calculator to find the answer.

Press [( ] 8 [+] 4 [ )], then press [÷] 2 [=] 6 (the *x*-value)

Press [( ] 6 [+] 2 [ )], then press [÷] 2 [=] 4 (the *y*-value)

$M = (6, 4)$

Exercise A Use the midpoint formula to find the midpoints of line segments having the following endpoints. Use a calculator to check your answers.

1. (6, 9) and (12, 18)
$(9, 13\frac{1}{2})$; calculator (9, 13.5)

2. (2, 1) and (3, 7)
$(2\frac{1}{2}, 4)$; calculator (2.5, 4)

3. (–3, 1) and (–1, 4)
$(-2, \frac{5}{2})$; calculator (–2, 2.5)

4. (4, –5) and (6, 3)
(5, –1); calculator same

5. (–3, 5) and (–7, 10)
$(-5, 7\frac{1}{2})$; calculator (–5, 7.5)

6. (2, –6) and (4, –12)
(3, –9); calculator same

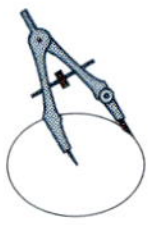

Construction

Exercise B Draw lines having each of the given lengths on a separate sheet of paper. Using only a straightedge and compass, bisect each line.

7. 3 in.

8. 5 in.

9. 10 cm

10. 20 cm

11. a length of your choice

Student constructions should include the arcs necessary to find the midpoint.

Writing About Mathematics

What happens to your construction of a bisector if your compass opening is less than $\frac{1}{2}\overline{AB}$?

Exercise C Find the midpoints of the following line segments. Be sure to give both coordinates of each midpoint.

12.

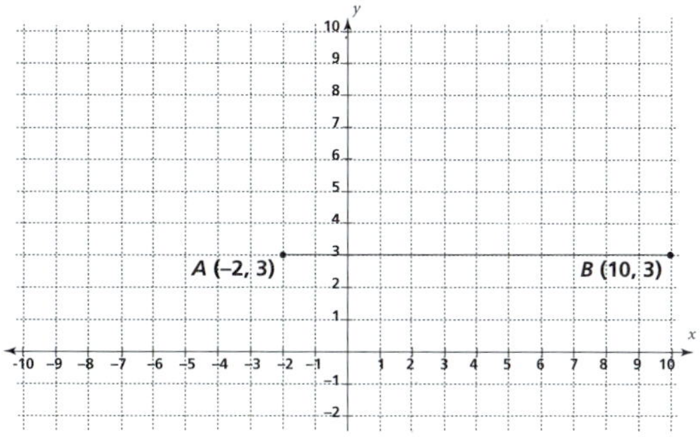

(4, 3)

13.

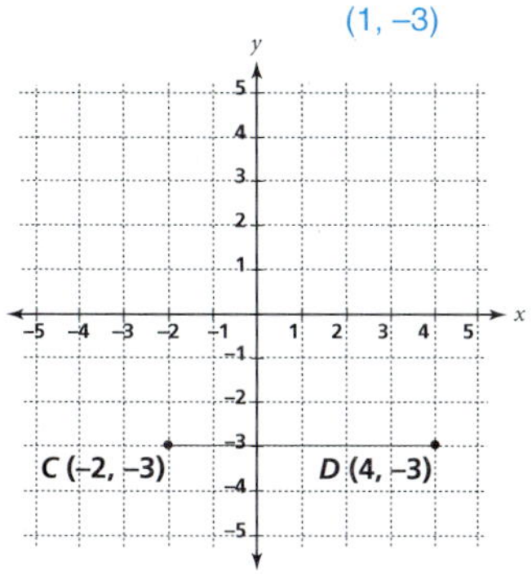

(1, –3)

14.

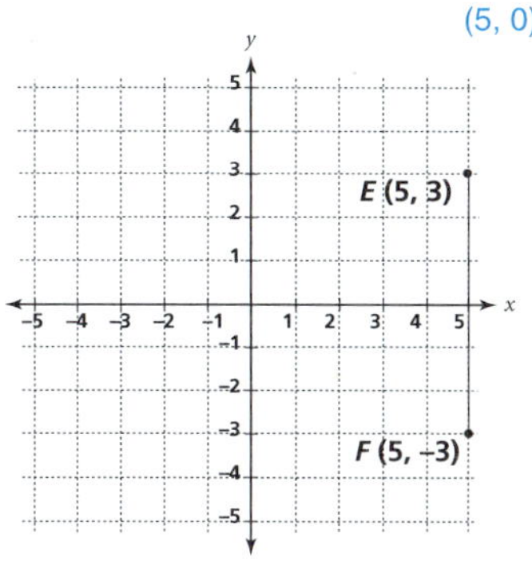

(5, 0)

15.

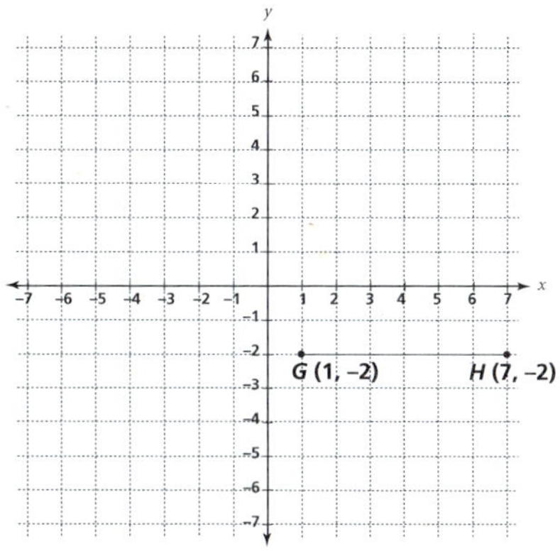

(4, –2)

Learning Styles

Auditory/Verbal Have students compare the two methods for finding the midpoint of a segment. Ask them to describe different situations in which one of the methods seems easier to use than the other and to explain why.

Lesson at a Glance

Chapter 4 Application

Overview This lesson relates the concept of slope to the design and function of roller coasters.

Objectives

- To find the slope of various roller coaster rides.
- To compare the slopes of various roller coaster rides.

Student Page 90

Teacher's Resource Library TRL

Workbook Activity 44

1. Warm-Up Activity

Invite students to share their experiences with roller coasters. Discuss other geometric applications such as parallel lines, transversals, and circles, as well as slope, found in the design of roller coasters or other amusement park rides.

2. Teaching the Lesson

Make a simple drawing of a roller coaster on a coordinate grid to illustrate how the height, horizontal distance, and roller coaster track form a right triangle. Remind students that the height represents the rise and the horizontal distance represents the run. It is the ratio between these two measurements that determines the slope.

3. Reinforce and Extend

Learning Styles

Visual/Spatial Have students draw and label sketches of each of the roller coasters before computing the slope. They might also find it helpful to label the rise and run in the sketch.

Application

Screaming Slopes Clickity-clack, clickity-clack, clickity-clack, you grip the bar tightly. Clickity-clack, clickity-clack, clickity-clack, you're at the top of the first hill. The coaster pauses for a moment—just long enough for you to see the drop in front of you and then EEEEEEE! Down the slope you rattle and roll.

Roller coasters are surely one of the most loved examples of applied geometry and physics. The golden rule of coaster builders seems to be steeper, faster, scarier. All important to the coaster ride is the first hill. It must be high enough to give the coaster enough energy to run its course. And it must be steep enough to give the riders a thrill. (Of course, not so steep that the cars leave the track!) In other words, roller coaster designers need to know what slope will give the best ride.

Exercise The table below gives the height of the first hill for several coasters. It also gives the distance, along the ground, covered during the first drop. Copy and complete the table, finding the slope for each drop. Then answer questions 8–10. Remember, slope = rise/run.

Coaster	Height of first hill (ft)	Horizontal distance covered (ft)	Slope
1. Cyclone	135	101	1.34
2. Tornado	78	59	1.32
3. Screamer	85	49	1.73
4. Shaker	166	92	1.80
5. Rattler	137	103	1.33
6. Terror	155	121	1.28
7. Beast	135	135	1.0

8. Which coaster has the greatest slope? Shaker
9. Which coaster offers the gentlest slope? Beast
10. What seems to be the "middle value" or *median* slope of these coasters? about 1.33

128 *Chapter 4 Using Algebra: Lines in the Coordinate Plane*

Learning Styles

Logical/Mathematical Ask students to explain how to find the slope of an escalator. They should identify the measurements needed to calculate the slope. Then ask them to estimate what the slope of an average escalator might be and to compare its slope to that of an elevator.

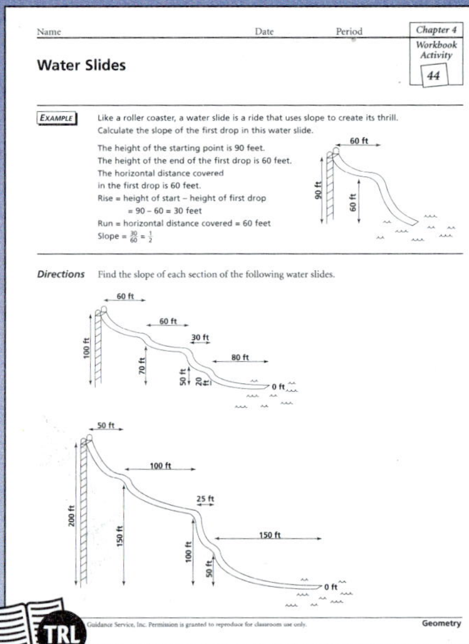

Name Date Period *Chapter 4 Workbook Activity 44*

Water Slides

EXAMPLE Like a roller coaster, a water slide is a ride that uses slope to create its thrill. Calculate the slope of the first drop in this water slide.

The height of the starting point is 90 feet.
The height of the end of the first drop is 60 feet.
The horizontal distance covered in the first drop is 60 feet.
Rise = height of start – height of first drop = 90 – 60 = 30 feet
Run = horizontal distance covered = 60 feet
Slope = $\frac{30}{60} = \frac{1}{2}$

Directions Find the slope of each section of the following water slides.

Geometry

TRL

Workbook Activity 44

Find the slope m for each line. Write the letter of the answer for each problem.

Example:

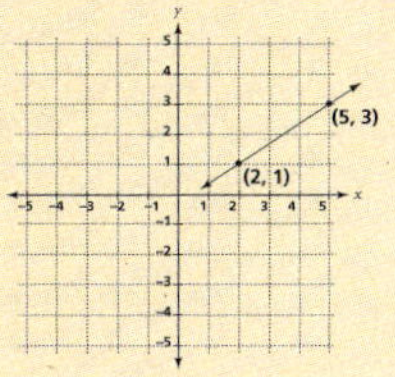

Solution: A

$m = \frac{y_1 - y_2}{x_1 - x_2}$

$m = \frac{3-1}{5-2} = \frac{2}{3}$

A. $\frac{2}{3}$ B. $\frac{3}{2}$ C. 2 D. 3

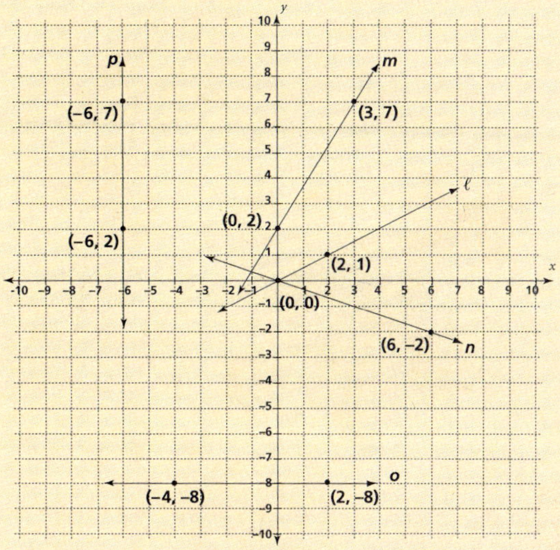

1.	ℓ	**A.** $\frac{1}{2}$	**B.** 2	**C.** 4	**D.** 0	A
2.	m	**A.** $\frac{3}{5}$	**B.** $\frac{5}{3}$	**C.** 3	**D.** 5	B
3.	n	**A.** 1	**B.** −3	**C.** $-\frac{1}{3}$	**D.** $\frac{1}{3}$	C
4.	o	**A.** 4	**B.** −4	**C.** 0	**D.** undefined	C
5.	p	**A.** 1	**B.** 0	**C.** −2	**D.** undefined	D

Chapter 4 Review

Each set of problems in the Chapter Review includes an example and solution to illustrate the concept. Use the given examples for reteaching the material in Chapter 4. For additional practice, refer to the Supplementary Problems for Chapter 4 (pages 399–401) and Additional Practice 18–24 in the Teacher's Resource Library.

Chapter 4 Mastery Test TRL

The Teacher's Resource Library includes parallel forms of the Chapter 4 Mastery Test. The difficulty level of the two forms is equivalent. You may wish to use one form as a pretest and the other form as a posttest. In addition, the Chapter 4 Cumulative Mastery Test covers content from Chapters 1–4.

Alternative Assessment

Alternative Assessment items correlate with student Goals for Learning at the beginning of this chapter.

- Have students identify the algebraic equation for a line.

Write the algebraic equation for a line and identify each of its elements. Then draw a line on a grid and label each of the elements from the equation on the line.

- Have students use ordered pairs to graph lines on a coordinate plane.

Draw a line segment on a coordinate plane, using these ordered numbers: (−2, 6), (6, −2).

- Have students define and find the slope of a line.

Write a definition of "slope of a line" and use the formula for slope of a line to find the slope of a line with these ordered pairs: (1, 8), (3, −1).

- Have students write and calculate the equations for lines.

Write the equation for a line, $y = mx + b$. Then use the equation to solve for the slope, y-axis intercept, and another ordered pair on the line with these ordered pairs: (5, 7), (−3, 10).

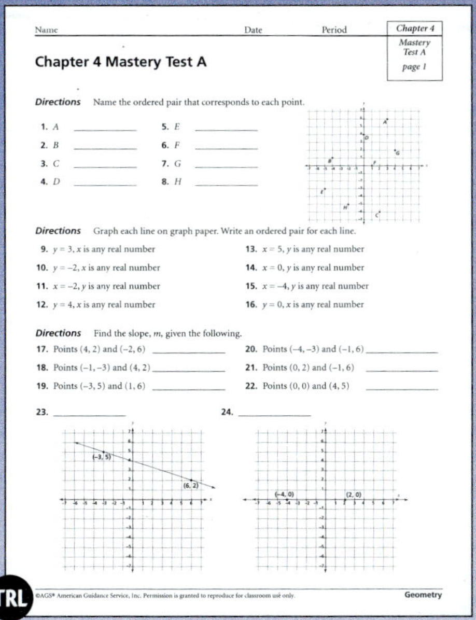

Name Date Period | *Chapter 4* *Mastery Test A* *page 1*

Chapter 4 Mastery Test A

Directions Name the ordered pair that corresponds to each point.

1. A ________ 5. E ________
2. B ________ 6. F ________
3. C ________ 7. G ________
4. D ________ 8. H ________

Directions Graph each line on graph paper. Write an ordered pair for each line.

9. $y = 3$, x is any real number
10. $y = -2$, x is any real number
11. $x = -2$, y is any real number
12. $y = 4$, x is any real number
13. $x = 5$, y is any real number
14. $x = 0$, y is any real number
15. $x = -4$, y is any real number
16. $y = 0$, x is any real number

Directions Find the slope, m, given the following.

17. Points (4, 2) and (−2, 6) ________
18. Points (−1, −3) and (4, 2) ________
19. Points (−3, 5) and (1, 6) ________
20. Points (−4, −3) and (−1, 6) ________
21. Points (0, 2) and (−1, 6) ________
22. Points (0, 0) and (4, 5) ________

23. ________ 24. ________

(−3, 5) (6, 2) (−4, 0) (2, 0)

TRL ©AGS® American Guidance Service, Inc. Permission is granted to reproduce for classroom use only. Geometry

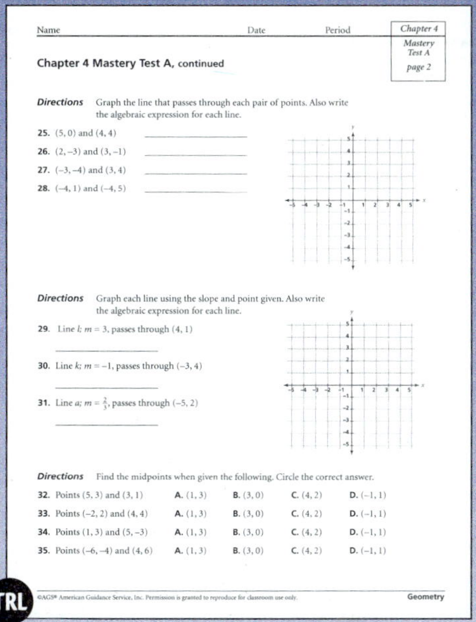

Name Date Period | *Chapter 4* *Mastery Test A* *page 2*

Chapter 4 Mastery Test A, continued

Directions Graph the line that passes through each pair of points. Also write the algebraic expression for each line.

25. (5, 0) and (4, 4) ________
26. (2, −3) and (3, −1) ________
27. (−3, −4) and (3, 4) ________
28. (−4, 1) and (−4, 5) ________

Directions Graph each line using the slope and point given. Also write the algebraic expression for each line.

29. Line l; $m = 3$, passes through (4, 1) ________
30. Line k; $m = -1$, passes through (−3, 4) ________
31. Line a; $m = \frac{2}{3}$, passes through (−5, 2) ________

Directions Find the midpoints when given the following. Circle the correct answer.

32.	Points (5, 3) and (3, 1)	A. (1, 3)	B. (3, 0)	C. (4, 2)	D. (−1, 1)
33.	Points (−2, 2) and (4, 4)	A. (1, 3)	B. (3, 0)	C. (4, 2)	D. (−1, 1)
34.	Points (1, 3) and (5, −3)	A. (1, 3)	B. (3, 0)	C. (4, 2)	D. (−1, 1)
35.	Points (−6, −4) and (4, 6)	A. (1, 3)	B. (3, 0)	C. (4, 2)	D. (−1, 1)

TRL ©AGS® American Guidance Service, Inc. Permission is granted to reproduce for classroom use only. Geometry

Chapter 4 Mastery Test A

ALTERNATIVE ASSESSMENT

continued

- ▶ Have students graph lines given slope and ordered pairs.

Graph the line for which you solved the equation in the immediately preceding Alternative Assessment item.

- ▶ Have students find the midpoint of a line segment.

Draw a line segment on a graph. Bisect the segment and use the midpoint formula to verify that you have bisected the line correctly.

More Review...

Find the slope m of a line that passes through the given points. Write the letter of the slope.

Example: (3, 4) and (–7, –16) Solution: C $m = \frac{y_1 - y_2}{x_1 - x_2}$

A. $\frac{2}{3}$ B. $\frac{3}{2}$ C. 2 D. 3

$m = \frac{4-(-16)}{3-(-7)} = \frac{4+16}{3+7} = \frac{20}{10} = 2$

6. (14, 2) and (2, 14) B

A. $\frac{1}{2}$ **B.** –1 **C.** 2 **D.** 3

7. (2, 2) and (8, 8) D

A. 6 **B.** $\frac{3}{4}$ **C.** 8 **D.** 1

8. (–2, –2) and (4, 12) A

A. $\frac{7}{3}$ **B.** $\frac{5}{7}$ **C.** $\frac{5}{3}$ **D.** $\frac{5}{7}$

9. (0, –2) and (3, –12) C

A. $\frac{3}{14}$ **B.** $\frac{14}{3}$ **C.** $-\frac{10}{3}$ **D.** $-\frac{14}{3}$

10. (4, 3) and (–12, 4) C

A. $\frac{1}{8}$ **B.** 16 **C.** $-\frac{1}{16}$ **D.** 8

Write the equation of the line that passes through each pair of points. Use the form $y = mx + b$.

Example: (4, 3) and (8, 9)

Solution: $m = \frac{y_1 - y_2}{x_1 - x_2} = \frac{3-9}{4-8} = \frac{-6}{-4} = \frac{3}{2}$ $\therefore y = \frac{3}{2}x + b$

Substitute x- and y-values from one of the ordered pairs.

Solve for b. $9 = (\frac{3}{2})(8) + b$ $9 = (\frac{24}{2}) + b$

$9 = 12 + b$ $-3 = b$ $\therefore y = \frac{3}{2}x - 3$

11. (3, 2) and (2, 3)

12. (8, 1) and (17, 1)

13. (3, 5) and (–6, 8)

14. (–6, –3) and (4, 2)

15. (–7, –4) and (–5, –2)

11. $y = -x + 5$

12. $y = 1$

13. $y = -\frac{1}{3}x + 6$

14. $y = \frac{1}{2}x$

15. $y = x + 3$

130 *Chapter 4 Using Algebra: Lines in the Coordinate Plane*

Name Date Period | *Chapter 4 Mastery Test B page 1*

Chapter 4 Mastery Test B

Directions Name the ordered pair that corresponds to each point.

1. *A* ______ **5.** *E* ______

2. *B* ______ **6.** *F* ______

3. *C* ______ **7.** *G* ______

4. *D* ______ **8.** *H* ______

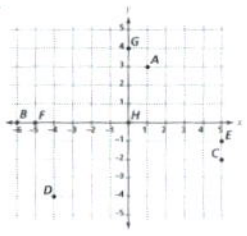

Directions Graph each line on graph paper. Write an ordered pair for each line.

9. $x = 5$, y is any real number

10. $y = 4$, x is any real number

11. $y = -1$, x is any real number

12. $x = 2$, y is any real number

13. $y = -5$, x is any real number

14. $x = -3$, y is any real number

15. $y = 1$, x is any real number

16. $y = -3$, x is any real number

Directions Find the slope, m, given the following.

17. Points (5, 1) and (2, 6) ______

18. Points (3, 0) and (5, 3) ______

19. Points (–4, 2) and (0, 4) ______

20. Points (2, 2) and (5, 5) ______

21. Points (–6, 9) and (–3, 6) ______

22. Points (0, 0) and (3, 6) ______

23. ______

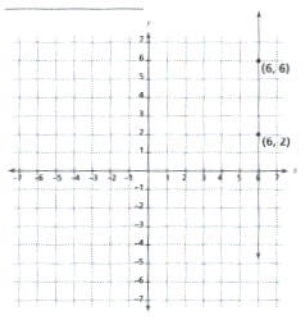

24. ______

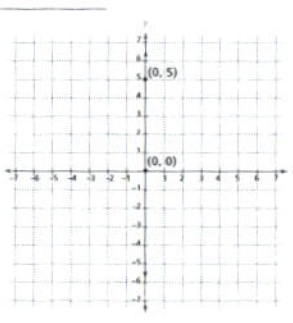

TRL Geometry

Name Date Period | *Chapter 4 Mastery Test B page 2*

Chapter 4 Mastery Test B, continued

Directions Graph the line that passes through each pair of points. Also write the algebraic expression for each line.

25. (2, 3) and (4, 5) ______

26. (0, 2) and (1, 4) ______

27. (–2, 4) and (3, 5) ______

28. (5, 5) and (–4, 5) ______

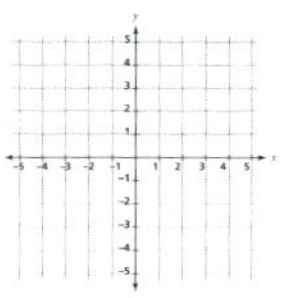

Directions Graph each line using the slope and point given. Also write the algebraic expression for each line.

29. Line *a*; $m = 1$, passes through (3, 3) ______

30. Line *b*; $m = 2$, passes through (2, 4) ______

31. Line *c*; $m = \frac{1}{3}$, passes through (3, 2) ______

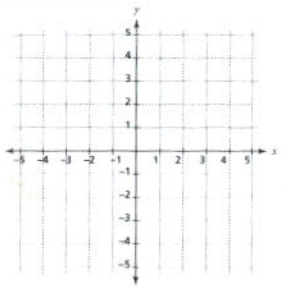

Directions Find the midpoints when given the following. Circle the correct answer.

32. Points (4, 2) and (2, 0) **A.** (0, 0) **B.** (2, 1) **C.** (1, 2) **D.** (–1, 1)

33. Points (5, 3) and (–1, –1) **A.** (3, 1) **B.** (–1, 1) **C.** (1, 2) **D.** (2, 1)

34. Points (–3, 0) and (5, 4) **A.** (0, 0) **B.** (2, 1) **C.** (1, 2) **D.** (–1, 1)

35. Points (–4, –6) and (4, 6) **A.** (–1, 1) **B.** (2, 1) **C.** (1, 2) **D.** (0, 0)

TRL Geometry

Chapter 4 Mastery Test B

Use the midpoint formula to find the midpoints of line segments having the following endpoints.

Example: (2, 1) and (6, 4)

Solution: Midpoint $M = (\frac{x_1 + x_2}{2}, \frac{y_1 + y_2}{2})$

$M = (\frac{2+6}{2}, \frac{1+4}{2}) = (\frac{8}{2}, \frac{5}{2}) = (4, \frac{5}{2})$

16. (4, 3) and (–2, 3) (1, 3)

17. (–6, 9) and (–2, 7) (–4, 8)

18. (–1, –1) and (7, 7) (3, 3)

19. (–3, 4) and (–3, 10) (–3, 7)

20. (3, 6) and (12, 4) $(\frac{15}{2}, 5)$

Graph each line using the slope and point given. Use a separate sheet of graph paper and a straightedge.

Example: $m = -1$; (4, 2)

Solution: Mark (4, 2). Count down 1 and over 1 to (5, 1). Connect the points.

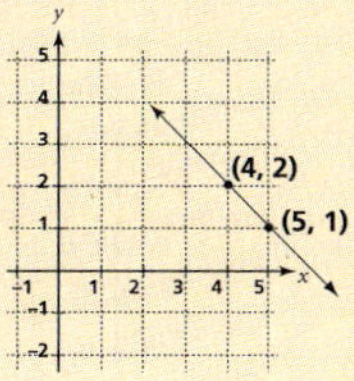

21. $m = -1$; (–2, –3)

22. $m = \frac{2}{3}$; (0, 3)

23. $m = \frac{1}{4}$; (1, 1)

24. $m = -2$; (–2, –3)

25. $m = 2$; (1, 1)

See Teacher's Edition page for answers.

> ***Test-Taking Tip***
>
> To prepare for a geometry test, study in short sessions for several days rather than one long session the night before the test. Spend some time writing an example for each of the vocabulary words. In the week before the test, spend time each day reviewing your notes.

Answers to problems 21–25

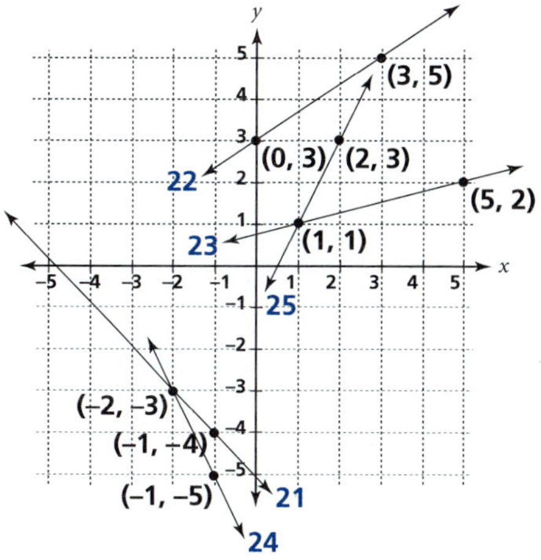

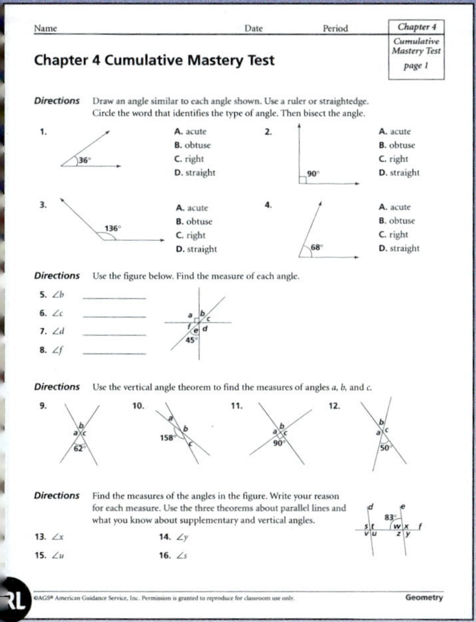

Name Date Period

Chapter 4 *Cumulative Mastery Test* *page 1*

Chapter 4 Cumulative Mastery Test

Directions Draw an angle similar to each angle shown. Use a ruler or straightedge. Circle the word that identifies the type of angle. Then bisect the angle.

1. 36° **A.** acute **B.** obtuse **C.** right **D.** straight

2. 90° **A.** acute **B.** obtuse **C.** right **D.** straight

3. 136° **A.** acute **B.** obtuse **C.** right **D.** straight

4. 68° **A.** acute **B.** obtuse **C.** right **D.** straight

Directions Use the figure below. Find the measure of each angle.

5. ∠b ______

6. ∠c ______

7. ∠d ______

8. ∠f ______

Directions Use the vertical angle theorem to find the measures of angles *a*, *b*, and *c*.

9. 62° **10.** 158° **11.** 90° **12.** 50°

Directions Find the measures of the angles in the figure. Write your reason for each measure. Use the three theorems about parallel lines and what you know about supplementary and vertical angles.

13. ∠x **14.** ∠y

15. ∠u **16.** ∠s

©AGS® American Guidance Service, Inc. Permission is granted to reproduce for classroom use only. **Geometry**

TRL

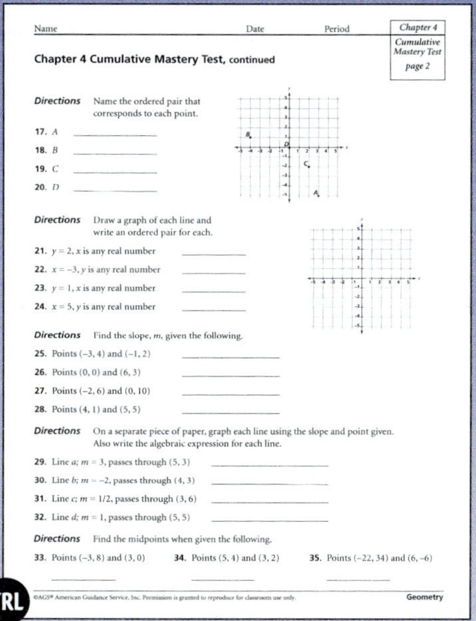

Name Date Period

Chapter 4 *Cumulative Mastery Test* *page 2*

Chapter 4 Cumulative Mastery Test, continued

Directions Name the ordered pair that corresponds to each point.

17. *A* ______

18. *B* ______

19. *C* ______

20. *D* ______

Directions Draw a graph of each line and write an ordered pair for each.

21. $y = 2$, x is any real number ______

22. $x = -3$, y is any real number ______

23. $y = 1$, x is any real number ______

24. $x = 5$, y is any real number ______

Directions Find the slope, *m*, given the following.

25. Points (–3, 4) and (–1, 2) ______

26. Points (0, 0) and (6, 3) ______

27. Points (–2, 6) and (0, 10) ______

28. Points (4, 1) and (5, 5) ______

Directions On a separate piece of paper, graph each line using the slope and point given. Also write the algebraic expression for each line.

29. Line *a*; $m = 3$, passes through (5, 3) ______

30. Line *b*; $m = -2$, passes through (4, 3) ______

31. Line *c*; $m = 1/2$, passes through (3, 6) ______

32. Line *d*; $m = 1$, passes through (5, 5) ______

Directions Find the midpoints when given the following.

33. Points (–3, 8) and (3, 0) **34.** Points (5, 4) and (3, 2) **35.** Points (–22, 34) and (6, –6)

©AGS® American Guidance Service, Inc. Permission is granted to reproduce for classroom use only. **Geometry**

TRL

Chapter 4 Cumulative Mastery Test

Chapter 5

Planning Guide

Triangles and Quadrilaterals

	Student Text Lesson						
	Student Pages	Vocabulary	Proof	Practice Exercises	Problem Solving	Calculator Practice	Solutions Key
Lesson 1 Naming Triangles Using Their Sides	134–138	✔		✔			✔
Lesson 2 Naming Triangles Using Their Angles	139–142	✔		✔			✔
Lesson 3 Special Quadrilaterals	143–145	✔		✔			✔
Lesson 4 Quadrilaterals and Their Diagonals	146–148	✔	✔	✔			✔
Lesson 5 Angle Sum of Any Triangle	149–153		✔	✔			✔
Lesson 6 Concave and Convex Polygons	154–159	✔		✔	159		✔
Lesson 7 Constructing Perpendiculars	160–162	✔		✔			✔
Lesson 8 Altitudes, Angle Bisectors, and Medians	163–166	✔		✔			✔
Lesson 9 Proof of the Angle Sum Theorem	167–171	✔	✔	✔		171	✔
Application Geometric Puzzles	172			✔			✔

Chapter Activities

Teacher's Resource Library
Geometry on the Job 5: Architects and Designers

Community Connection 5: Building Shapes

Teacher's Edition
Chapter 5 Project

Assessment Options

Student Text
Chapter 5 Review

Teacher's Resource Library
Chapter 5 Mastery Tests A and B

Chapter 5 Cumulative Mastery Test

Teacher's Edition
Chapter 5 Alternative Assessment

Student Text Features						Teaching Strategies								Learning Styles						Teacher's Resource Library		
Algebra Connection	Constructions	Experiencing Geometry	Technology Connection	Try This	Writing About Mathematics	Warm-Up Activity	Common Error	Algebra Connection	Mental Math	Manipulatives	Calculator	Applications	Group Problem Solving	Auditory/Verbal	Visual/Spatial	Body/Kinesthetic	Logical/Mathematical	Interpersonal/Group Learning	LEP/ESL	Workbook Activities	Additional Practice	Self-Study Guide
	135–138				135	134						136	137	136, 138	135	138			136	45	25	✔
	141–142			142		139	140			140				141, 142	141	142			141	46, 47	26	✔
					145	143	144										145		145	48	27	✔
		148				146				147						147		148	147	49, 50	28	✔
149			151			149		152	153	151		151					153	153	151	51, 52	29	✔
					156	154						157	159	156	158		156, 159	158	156	53, 54	30	✔
	160–162			162		160							161						162	55, 56	31	✔
	165–166					163								166				166	165	57, 58	32	✔
70		167				167		170	171		171			169				169		59, 60	33	✔
						172										172				61		✔

Manipulatives

Manipulite Triangles
Polygon Power Pack
Manipulite Polygons

Chapter at a Glance

Chapter 5: Triangles and Quadrilaterals

pages 132–175

Lessons

Teacher's Resource Library TRL

(Answer Keys for the Teacher's Resource Library begin on page 484 of this Teacher's Edition.)

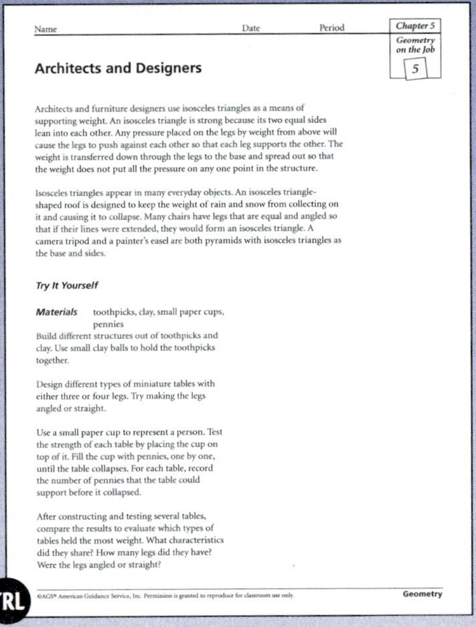

Name Date Period | Chapter 5 Geometry on the Job 5

Architects and Designers

Architects and furniture designers use isosceles triangles as a means of supporting weight. An isosceles triangle is strong because its two equal sides lean into each other. Any pressure placed on the legs by weight from above will cause the legs to push against each other so that each leg supports the other. The weight is transferred down through the legs to the base and spread out so that the weight does not put all the pressure on any one point in the structure.

Isosceles triangles appear in many everyday objects. An isosceles triangle-shaped roof is designed to keep the weight of rain and snow from collecting on it and causing it to collapse. Many chairs have legs that are equal and angled so that if their lines were extended, they would form an isosceles triangle. A camera tripod and a painter's easel are both pyramids with isosceles triangles as the base and sides.

Try It Yourself

Materials toothpicks, clay, small paper cups, pennies

Build different structures out of toothpicks and clay. Use small clay balls to hold the toothpicks together.

Design different types of miniature tables with either three or four legs. Try making the legs angled or straight.

Use a small paper cup to represent a person. Test the strength of each table by placing the cup on top of it. Fill the cup with pennies, one by one, until the table collapses. For each table, record the number of pennies that the table could support before it collapsed.

After constructing and testing several tables, compare the results to evaluate which types of tables held the most weight. What characteristics did they share? How many legs did they have? Were the legs angled or straight?

 Geometry

TRL

Geometry on the Job 5

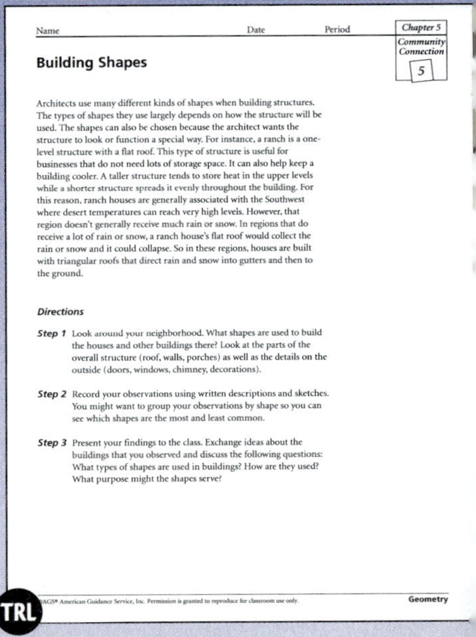

Name Date Period | Chapter 5 Community Connection 5

Building Shapes

Architects use many different kinds of shapes when building structures. The types of shapes they use largely depends on how the structure will be used. The shapes can also be chosen because the architect wants the structure to look or function a special way. For instance, a ranch is a one-level structure with a flat roof. This type of structure is useful for businesses that do not need lots of storage space. It can also help keep a building cooler. A taller structure tends to store heat in the upper levels while a shorter structure spreads it evenly throughout the building. For this reason, ranch houses are generally associated with the Southwest where desert temperatures can reach very high levels. However, that region doesn't generally receive much rain or snow. In regions that do receive a lot of rain or snow, a ranch house's flat roof would collect the rain or snow and it could collapse. So in these regions, houses are built with triangular roofs that direct rain and snow into gutters and then to the ground.

Directions

Step 1 Look around your neighborhood. What shapes are used to build the houses and other buildings there? Look at the parts of the overall structure (roof, walls, porches) as well as the details on the outside (doors, windows, chimney, decorations).

Step 2 Record your observations using written descriptions and sketches. You might want to group your observations by shape so you can see which shapes are the most and least common.

Step 3 Present your findings to the class. Exchange ideas about the buildings that you observed and discuss the following questions: What types of shapes are used in buildings? How are they used? What purpose might the shapes serve?

 Geometry

TRL

Community Connection 5

Chapter 5

Triangles and Quadrilaterals

Colors and shapes changing before your eyes—this is the kaleidoscope! In the early 1800s, David Brewster from Scotland invented the kaleidoscope. Since then, children and adults have delighted in the color-and-shape show kaleidoscopes provide. With a gentle turn of a tube, you can see an ever-changing pattern of triangles, quadrilaterals, and other shapes.

The richly colored patterns change with each turn. Yet the shapes remain the same, and you see that basic geometric forms can be found everywhere!

In Chapter 5, you will learn how to identify and measure sides and angles in triangles and quadrilaterals.

Goals for Learning

- To construct triangles using sides
- To name triangles using their angles
- To name and distinguish quadrilaterals
- To find the measures of angles in quadrilaterals and triangles
- To define quadrilaterals and parallels and apply theorems to them
- To construct perpendiculars
- To construct triangles based on attributes
- To use the angle sum theorem to define triangles

Introducing the Chapter

Use the kaleidoscope photograph to engage students in a discussion of geometric shapes and forms. Invite volunteers to describe the image in the photograph. Reinforce descriptions that include named geometric figures such as rectangles, squares, and so on. Encourage students to name other geometric shapes they might be familiar with. Indicate to students that geometry provides a method for classifying these shapes and allows us to make comparisons as well as generalizations about the shapes. Read the Goals for Learning with students and point out that this chapter will help them identify and measure sides and angles in triangles and quadrilaterals.

Chapter Project

Encourage students to work in small groups to research and construct a small kaleidoscope. The contents of the scope should be cutout polygons in a variety of shapes and sizes. Some shapes could be made of tissue paper; others could be poster board.

As an alternative project, student groups could develop a connect-the-dots game in which players connect the dots to form different polygons. The completed project would include a game board, goal of the game, number of players, and game rules.

SELF-STUDY GUIDE

Name ______

CHAPTER 5: Triangles and Quadrilaterals

Goal 5.1 *To construct triangles using sides*

Date	Assignment	Score
______	1: Read pages 134–137. Complete Exercises A and B on page 138.	______
______	2: Complete Workbook Activity 45.	______

Comments:

Goal 5.2 *To name triangles using their angles*

Date	Assignment	Score
______	3: Read pages 139–141. Complete Exercises A and B on page 142.	______
______	4: Complete Workbook Activities 46–47.	______

Comments:

Goal 5.3 *To name and distinguish quadrilaterals*

Date	Assignment	Score
______	5: Read pages 143–144. Complete Exercises A–C on page 145.	______
______	6: Complete Workbook Activity 48.	______

Comments:

Goal 5.4 *To find the measures of angles in quadrilaterals and triangles*

Date	Assignment	Score
______	7: Read pages 146–147. Complete Exercises A–C on page 148.	______
______	8: Complete Workbook Activities 49–50.	______
______	9: Read pages 149–152. Complete Exercises A–C on page 153.	______
______	10: Complete Workbook Activities 51–52.	______

Comments:

 Geometry

TRL

SELF-STUDY GUIDE

Name ______

CHAPTER 5: Triangles and Quadrilaterals, continued

Goal 5.5 *To define quadrilaterals and parallels and apply theorems to them*

Date	Assignment	Score
______	11: Read pages 154–158. Complete Exercises A and B on pages 158–159.	______
______	12: Complete Workbook Activities 53–54.	______

Comments:

Goals 5.6 and 5.7 *To construct perpendiculars and to construct triangles based on attributes*

Date	Assignment	Score
______	13: Read pages 160–161. Complete Exercise A on page 162.	______
______	14: Complete Workbook Activities 55–56.	______
______	15: Read pages 163–164. Complete Exercises A and B on pages 165–166.	______
______	16: Complete Workbook Activities 57–58.	______

Comments:

Goal 5.8 *To use the angle sum theorem to define triangles*

Date	Assignment	Score
______	17: Read pages 167–170. Complete Exercises A–C on pages 170–171.	______
______	18: Complete Workbook Activities 59–60.	______
______	19: Read and complete the Application Exercise on page 172.	______
______	20: Complete Workbook Activity 61.	______
______	21: Complete the Chapter 5 Review on pages 173–175.	______

Comments:

Student's Signature ______ Date ______

Instructor's Signature ______ Date ______

 Geometry

TRL

Chapter 5 Self-Study Guide

Lesson at a Glance

Chapter 5 Lesson 1

Overview This lesson introduces different-shaped triangles and classifies these triangles based on the relative lengths of their sides.

Objectives

- To identify triangles as equilateral, isosceles, or scalene.
- To construct equilateral, isosceles, and scalene triangles whose sides match given line segments for side and base lengths.

Student Pages 134–138

Teacher's Resource Library TRL

Workbook Activity 45

Additional Practice 25

Mathematics Vocabulary

equilateral triangle
isosceles triangle
scalene triangle

1. Warm-Up Activity

Familiarize students with triangles by asking them to identify three-sided figures in the classroom. List these figures on chart paper. Ask students to describe how the triangles differ from one another. Have them devise their own classification system for triangles. Record their classification systems.

Lesson 1 Naming Triangles Using Their Sides

Equilateral triangle
A triangle with three equal sides

Isosceles triangle
A triangle with two equal sides

Scalene triangle
A triangle with no equal sides

Remember, a tick mark (|) is used to represent sides of the same length.

Look for triangles around you. You can find them in houses, bridges, and other structures. For example, this tower is built in the shape of a triangle with triangle supports. As you look for triangles, you will see that they have different shapes. One way to name and identify triangles is to use their sides.

Different-Shaped Triangles

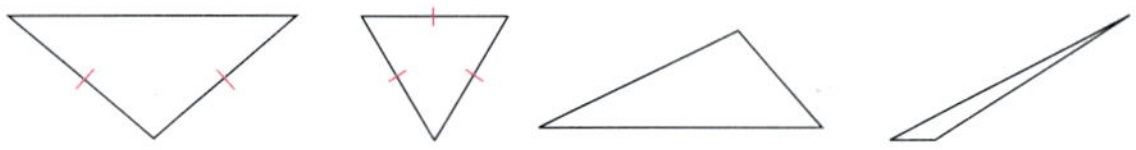

You may find it easy to recognize a triangle that has three equal sides. An **equilateral triangle** is a triangle with three sides of equal length. Below are some examples.

Equilateral Triangles

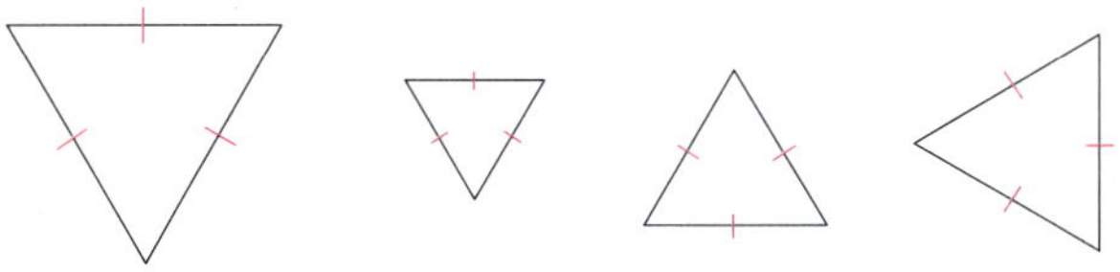

Another group of triangles have two equal sides. In an **isosceles triangle** the equal sides are called *legs,* and the remaining side is called the *base.* Below are some examples of isosceles triangles.

Isosceles Triangles

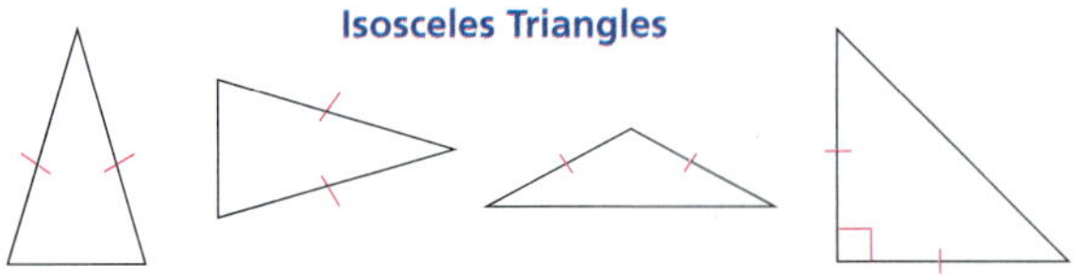

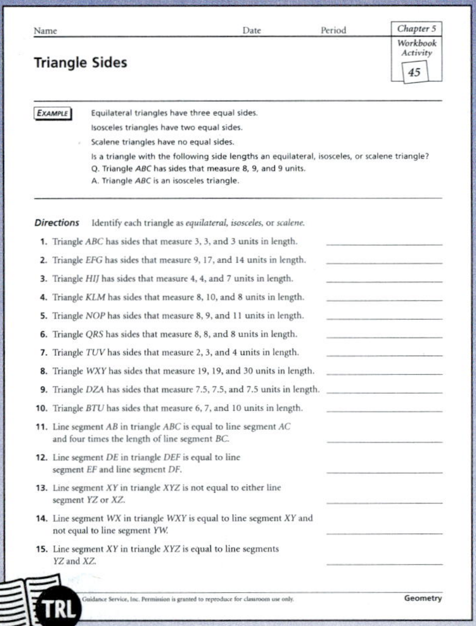

Name Date Period

Chapter 5 Workbook Activity 45

Triangle Sides

EXAMPLE Equilateral triangles have three equal sides.
Isosceles triangles have two equal sides.
Scalene triangles have no equal sides.
Is a triangle with the following side lengths an equilateral, isosceles, or scalene triangle?
Q. Triangle *ABC* has sides that measure 8, 9, and 9 units.
A. Triangle *ABC* is an isosceles triangle.

Directions Identify each triangle as *equilateral, isosceles,* or *scalene.*

1. Triangle *ABC* has sides that measure 3, 3, and 3 units in length. ______
2. Triangle *EFG* has sides that measure 9, 17, and 14 units in length. ______
3. Triangle *HIJ* has sides that measure 4, 4, and 7 units in length. ______
4. Triangle *KLM* has sides that measure 8, 10, and 8 units in length. ______
5. Triangle *NOP* has sides that measure 8, 9, and 11 units in length. ______
6. Triangle *QRS* has sides that measure 8, 8, and 8 units in length. ______
7. Triangle *TUV* has sides that measure 2, 3, and 4 units in length. ______
8. Triangle *WXY* has sides that measure 19, 19, and 30 units in length. ______
9. Triangle *DZA* has sides that measure 7.5, 7.5, and 7.5 units in length. ______
10. Triangle *BTU* has sides that measure 6, 7, and 10 units in length. ______
11. Line segment *AB* in triangle *ABC* is equal to line segment *AC* and four times the length of line segment *BC.* ______
12. Line segment *DE* in triangle *DEF* is equal to line segment *EF* and line segment *DF.* ______
13. Line segment *XY* in triangle *XYZ* is not equal to either line segment *YZ* or *XZ.* ______
14. Line segment *WX* in triangle *WXY* is equal to line segment *XY* and not equal to line segment *YW.* ______
15. Line segment *XY* in triangle *XYZ* is equal to line segments *YZ* and *XZ.* ______

 Geometry

TRL **Workbook Activity 45**

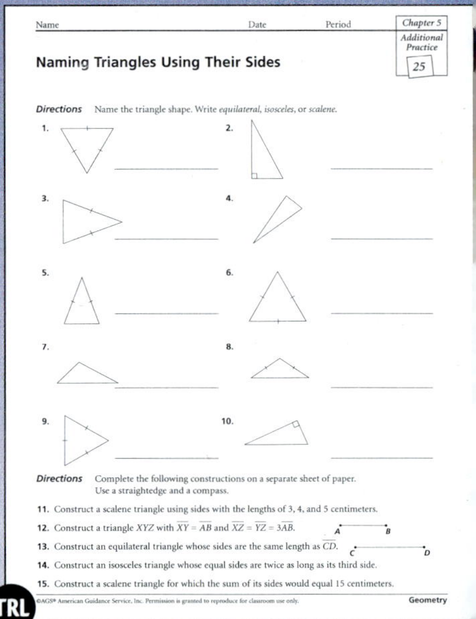

Name Date Period

Chapter 5 Additional Practice 25

Naming Triangles Using Their Sides

Directions Name the triangle shape. Write *equilateral, isosceles,* or *scalene.*

1. ______ 2. ______
3. ______ 4. ______
5. ______ 6. ______
7. ______ 8. ______
9. ______ 10. ______

Directions Complete the following constructions on a separate sheet of paper. Use a straightedge and a compass.

11. Construct a scalene triangle using sides with the lengths of 3, 4, and 5 centimeters.
12. Construct a triangle *XYZ* with $\overline{XY} = \overline{AB}$ and $\overline{XZ} = \overline{YZ} = 3\overline{AB}$. A B
13. Construct an equilateral triangle whose sides are the same length as $\overline{CD}$. C D
14. Construct an isosceles triangle whose equal sides are twice as long as its third side.
15. Construct a scalene triangle for which the sum of its sides would equal 15 centimeters.

 Geometry

TRL **Additional Practice 25**

Many triangles have no equal sides. A **scalene triangle** is a triangle with no sides of equal length. Below are some examples.

Scalene Triangles

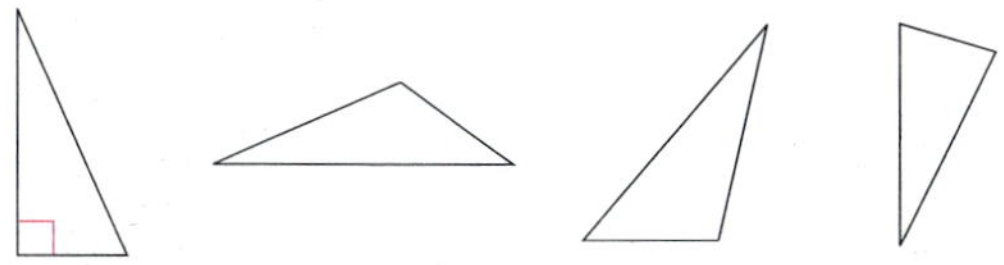

If you are given the length of a side, you can construct a triangle using only a straightedge and a compass.

Writing About Mathematics

When constructing a scalene triangle, what must be true about the lengths of sides *b* and *c* in relation to the length of side *a*, the base?

CONSTRUCTION

Given: $\overline{AB}$

Construct: an equilateral triangle with each side equal to the given length $\overline{AB}$

Step 1 Use your compass to copy $\overline{AB}$ on ℓ.

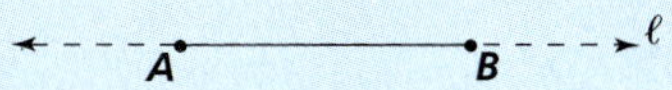

Step 2 Set your compass opening to match $\overline{AB}$. Draw an arc with its center at *A*. Draw a second arc with its center at *B*. Make sure the arcs intersect. (Note: The arcs may intersect both above and below ℓ. Choose one intersection to use.)

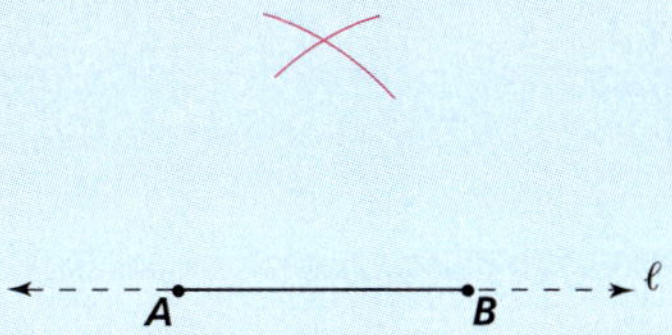

Step 3 Connect the point of intersection with point *A* and then point *B*. You have now constructed an equilateral triangle with each side equal to $\overline{AB}$.

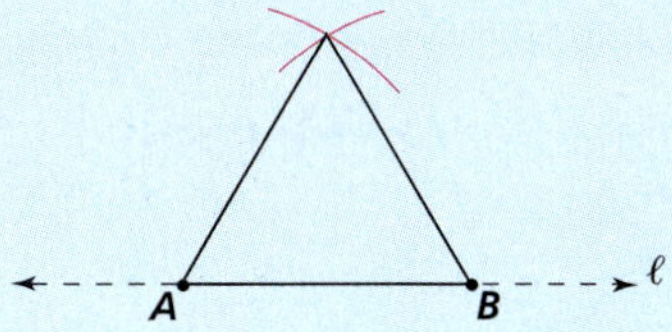

2. Teaching the Lesson

Discuss the terms *equilateral, isosceles,* and *scalene* using the examples given in the text. Explain that using side length is one useful way to classify triangles. Ask students to use the terms as they classify the triangles they observed during the Warm-Up Activity.

Draw students' attention to the diagrams of equilateral and isosceles triangles. Point out the small tick marks on the sides of the triangles. Ask students what the tick marks indicate. Have them explain how they can classify a triangle using only a diagram with tick marks. (Sides of the same length will have the same number of tick marks. If three sides are marked with the same number of ticks, then the sides are of equal length, and so on.)

Remind students of Euclid's Postulates. Indicate how the postulates can be applied to triangle construction. Work with students as they follow the steps for constructing an equilateral triangle. Ask them to explain or justify each step in the construction.

3. Reinforce and Extend

LEARNING STYLES

Visual/Spatial Have student pairs work together to prepare flashcards. Each card could have an illustration and name of one kind of triangle on the front of the card. Students can write the description of the triangle on the back of the card. They can use the cards to help reinforce the images and names of the different triangles. Encourage students to draw some of the triangles "upside-down" or "sideways." This will further aid visual identification and generalization. Encourage students to keep the cards for the next lesson.

To construct an isosceles triangle, you need the length of one of the legs. You will also need the length of the base, which is not equal to the legs.

Construction

Given: base $\overline{CD}$ and one of the equal sides, s

Construct: an isosceles triangle using the given side and base

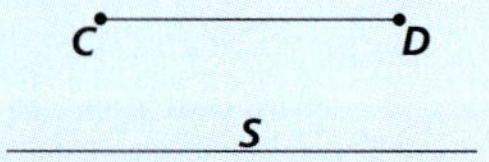

Step 1 Use your compass to copy $\overline{CD}$ on ℓ.

Step 2 Using s as a radius, and C and D as centers, draw intersecting arcs above $\overline{CD}$.

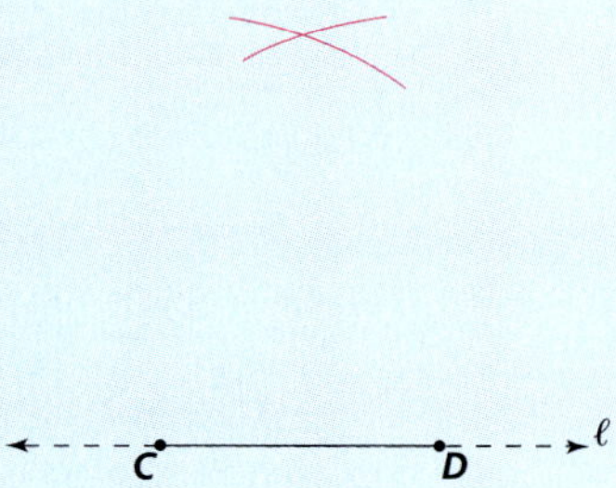

Step 3 Connect the point of intersection with point C and then point D. You have now constructed an isosceles triangle with base equal to $\overline{CD}$ and two legs each equal to s.

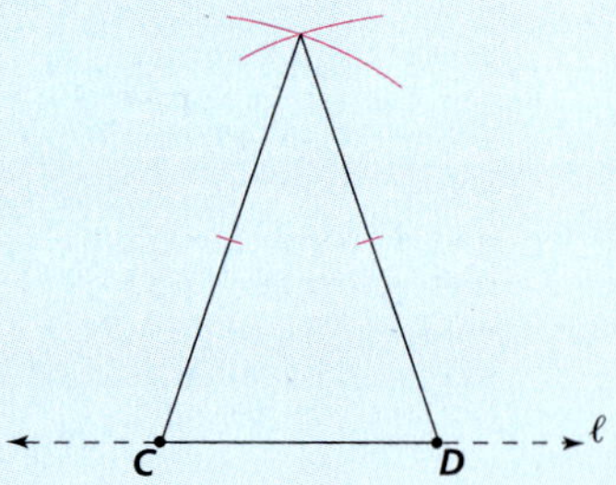

Learning Styles

Auditory/Verbal Have a volunteer demonstrate—on the board or overhead projector—the steps necessary to construct an isosceles triangle. As the volunteer performs each step, ask a second volunteer to describe the step. Invite a third student to justify each step. Repeat the procedure for a scalene triangle.

Learning Styles

LEP/ESL Point out to students that the word *equilateral* is formed from two root words: *equi*—which means "equal or equally," and *lateral*—which means "side." Ask students to read and pronounce several other words that have the prefix *equi-*. (For example, *equidistant, equiangular, equation,* and *equivalent*) Invite students to describe how all the words are related. Have them offer their own definition for each term. Encourage students to compare their definitions with the definitions found in a dictionary.

Application

At Home
Direct students to spend ten minutes on a home triangle hunt. Students can list all the triangles they find at home within ten minutes. Encourage them to use a ruler (either metric or English units) to determine whether the triangles are equilateral, isosceles, or scalene. Allow for creativity in describing triangles such "pie slice" or something similar. Students can share their lists and discuss which items best fit the definition of a triangle.

To construct a scalene triangle, you need to be given the lengths of all three sides.

CONSTRUCTION

Given: three lines of unequal lengths; *a, b, c*

Construct: a scalene triangle using the given lines as side lengths

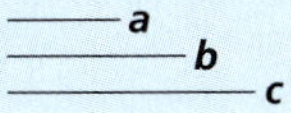

Step 1 Choose one side to be the base, *XY*, of the triangle. For this example, we'll use line *a* as the base. Use your compass to copy *a* on ℓ. Label the segment $\overline{XY}$.

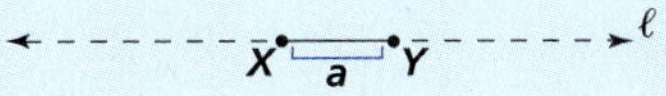

Step 2 Using *b* as a radius, and *X* as the center, draw an arc above $\overline{XY}$.

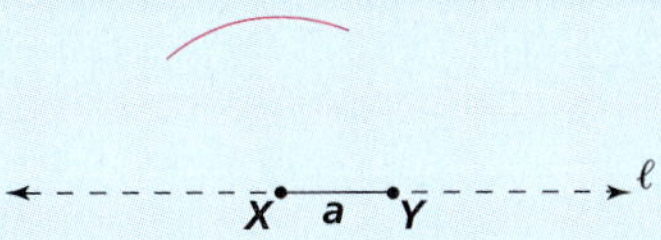

Step 3 Using *c* as a radius, and *Y* as the center, draw a second arc above $\overline{XY}$. Make sure the arcs intersect.

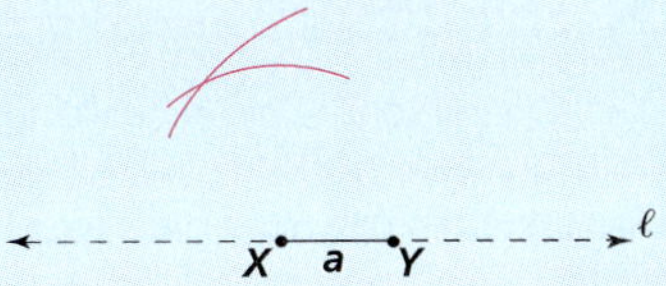

Step 4 Connect the point of intersection with point *X* and then point *Y*. You have now constructed a scalene triangle with sides equal to line segments *a, b,* and *c*. Because the line segments are different lengths, the triangle is scalene.

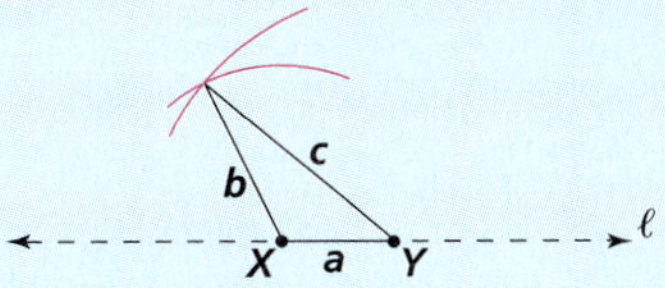

GROUP PROBLEM SOLVING

Ask each group of four students to create a comic strip or computer animation that shows how to construct a scalene triangle given three line segments of different lengths. Explain that the final strip may contain a variety of pictures, illustrations, and images but cannot contain any dialogue or written directions.

Students can assume different roles, such as director, artist, colorist, and production coordinator. Encourage students to be creative in their designs, approach, and layout. Arrange for them to show their solutions to the class and have the class critique the products. Allow students to make any revisions or improvements they wish. Completed strips or animations can be shown to other classes.

Learning Styles

Auditory/Verbal To complete Exercise A, allow students to work in pairs. For each item, one student should "think out loud"—describing how he or she uses the information in the diagram to determine the classification of each triangle. Partners should exchange speaking and listening roles after each item.

Learning Styles

Body/Kinesthetic Direct students to find at least one Manipulite Triangle that matches each triangle in Exercise A. Encourage students to use both the text illustration as well as the actual triangle to determine how to classify each triangle.

Selected Answers and Solutions begin on page 420.

Exercise A Name the triangle shape. Write *equilateral, isosceles,* or *scalene.*

1. 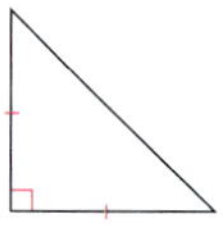 isosceles

2.  scalene

3.  scalene

4. 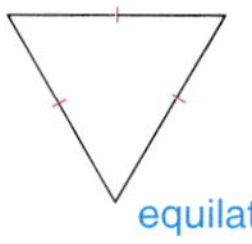 equilateral

5. 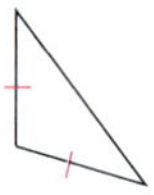 isosceles

6. 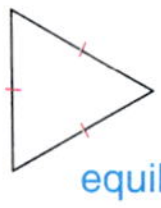 equilateral

7. 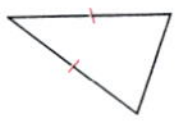 isosceles

8. 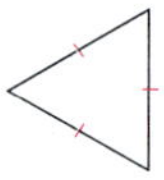 equilateral

9. 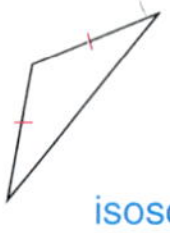 isosceles

10.  scalene

Construction

Exercise B Complete the following constructions. Use a separate sheet of paper. You will need a straightedge and a compass.

See Teacher's Edition page for answers.

11. Construct an equilateral triangle whose sides are the same length as $\overline{AB}$.

12. Construct an isosceles triangle using the length of $\overline{CD}$ as the base and with two sides equal in length to $\overline{EF}$.

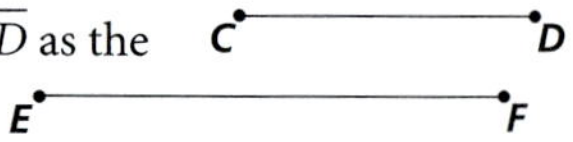

13. Construct a scalene triangle using the lengths of segments *q*, *r*, and *s* as the lengths of the sides.

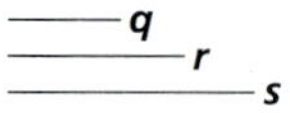

14. Construct triangle *XYZ* with $\overline{XY} = \overline{RS}$ and $\overline{XZ} = \overline{YZ} = 2\overline{RS}$.

15. Construct triangle *ABC* with $\overline{AB} = 5\overline{MN}$, $\overline{BC} = 4\overline{MN}$, and $\overline{CA} = 3\overline{MN}$.

Answers to Exercise B

11.

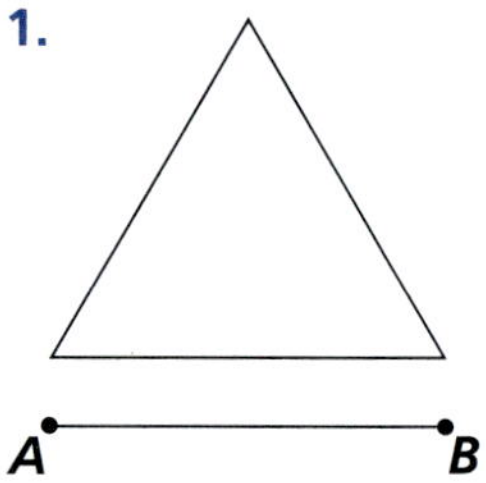

12.

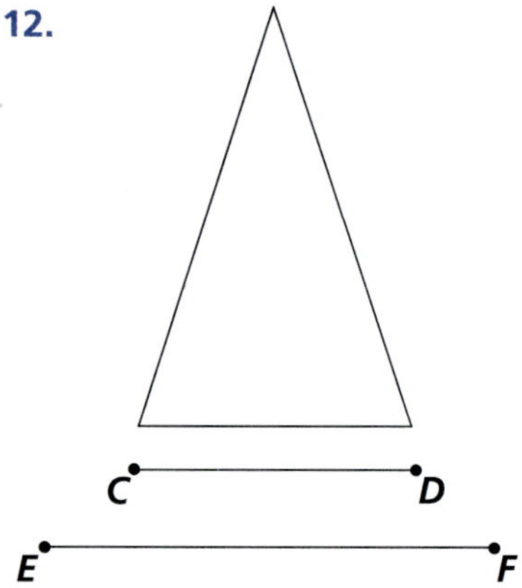

13.

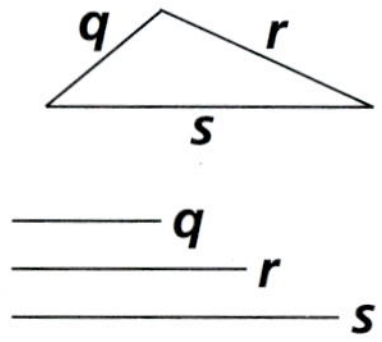

14.

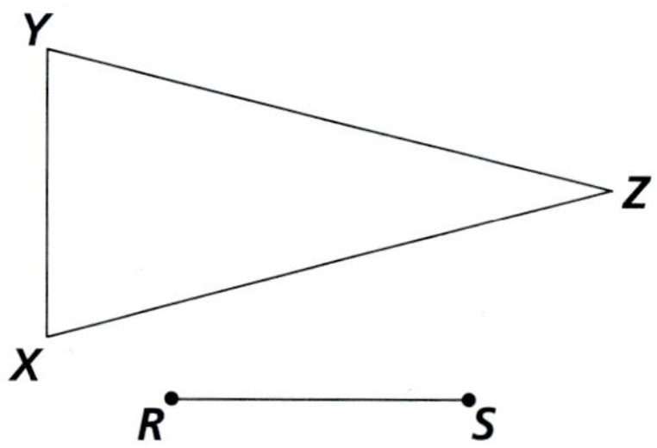

15.

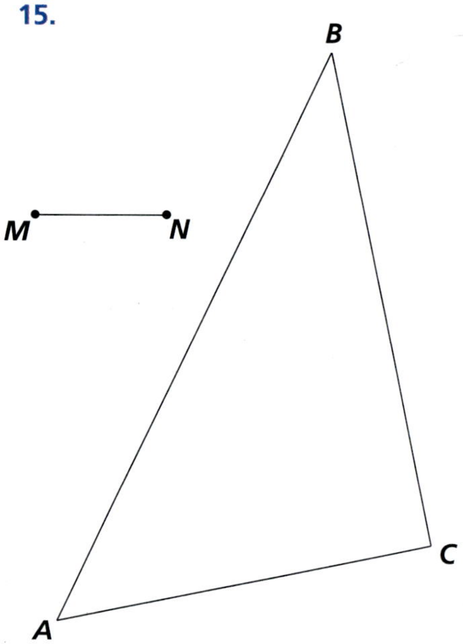

Lesson 2 Naming Triangles Using Their Angles

A second way to name and identify triangles is to use their angles.

In some triangles, all the angles are less than 90°.

Acute Triangles

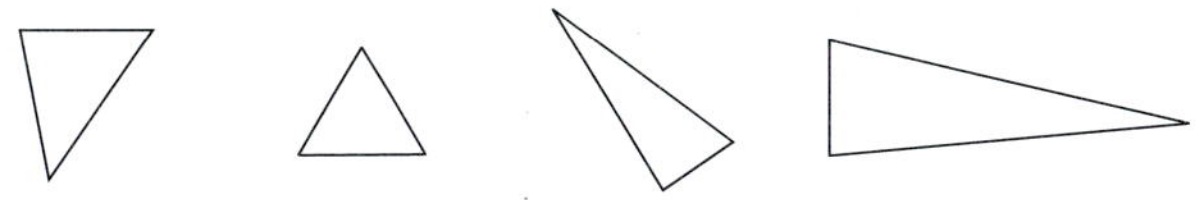

An **acute triangle** is a triangle with three angles less than 90°.

Some triangles have one 90° angle.

Right Triangles

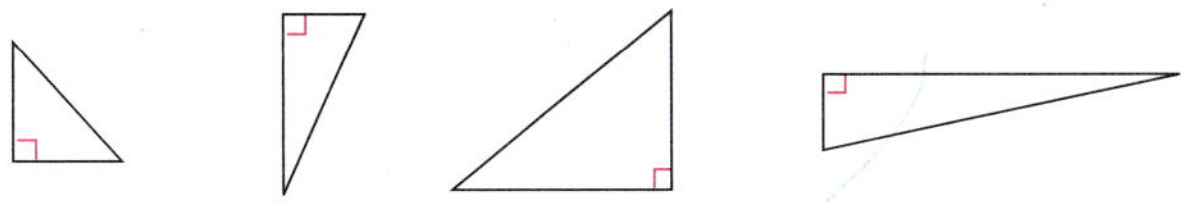

A **right triangle** is a triangle with a right angle. The sides adjacent to the right angle are often called the *legs* of a right triangle. The side opposite the right angle is called the *hypotenuse.*

The third group of triangles has one obtuse angle.

Obtuse Triangles

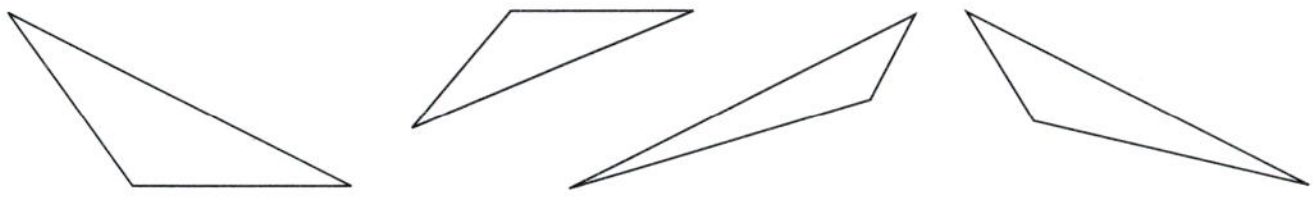

An **obtuse triangle** is a triangle with one angle greater than 90°.

Acute triangle
A triangle whose angles are all less than 90°

Obtuse triangle
A triangle having one angle greater than 90°

Right triangle
A triangle having one right angle

Lesson at a Glance

Chapter 5 Lesson 2

Overview This lesson presents a classification of triangles based on their angles.

Objectives

- To identify triangles as acute, right, or obtuse.
- To construct right and obtuse triangles.

Student Pages 139–142

Teacher's Resource Library TRL

Workbook Activities 46–47

Additional Practice 26

Mathematics Vocabulary

acute triangle
obtuse triangle
right triangle

1. Warm-Up Activity

Revisit the list of triangles students generated in Lesson 1. Have them reconsider their classification schemes. Reinforce any schemes that use angle measure as a way to distinguish one triangle from another.

2. Teaching the Lesson

Be sure to bring the use of the arc to students' attention. Indicate that like the tick mark, an arc can help them read a diagram and recognize which angles have the same measure.

Name ____ Date ____ Period ____ *Chapter 5 Workbook Activity 46*

Naming Triangles by Their Angles

EXAMPLE Acute triangles have three angles less than 90°.
Obtuse triangles have one obtuse angle.
Right triangles have one right angle.
Equilateral triangles have three equal angles.
Isosceles triangles have two equal angles.
Scalene triangles have no equal angles.

Directions Name each of these triangles using its angles.

1. Triangle *ABC* has angles that measure 100°, 40°, and 40°. ____
2. Triangle *DEF* has angles that measure 60°, 70°, and 50°. ____
3. Triangle *GHI* has angles that measure 60°, 60°, and 60°. ____
4. Triangle *JKL* has angles that measure 110°, 30°, and 40°. ____
5. Triangle *MNO* has angles that measure 90°, 45°, and 45°. ____
6. Triangle *PQR* has angles that measure 90°, 40°, and 50°. ____
7. Triangle *STU* has angles that measure 80°, 50°, and 50°. ____
8. Triangle *VWX* has angles that measure 130°, 30°, and 20°. ____
9. $\angle BAC$ in triangle *ABC* measures 60° and is equal to $\angle ABC$. ____
10. $\angle BAC$ in triangle *ABC* measures 90°, and $\angle ABC$ and $\angle BCA$ are not equal to each other. ____

 Geometry

Workbook Activity 46

Name ____ Date ____ Period ____ *Chapter 5 Workbook Activity 47*

Constructing Triangles Using Their Angles

EXAMPLE Construct $\triangle ABC$ with $\angle CAB = 2\angle BCA$ and $m\angle B = 90°$.

Step 1 Draw a right angle. Label it $\angle B$.

Step 2 Pick a point along one of the rays of $\angle B$. Label this point *A*. Using your protractor, draw an angle that measures 30° using point *A* as the vertex and one ray of $\angle B$ as one side. Extend the new ray so that it crosses the other ray of $\angle B$. Label the point where it crosses, point *C*.

Step 3 Measure $\angle ACB$. It should measure 60°. You have now constructed a right triangle with acute angles that have a ratio of 2 to 1. This triangle is a type of right scalene triangle.

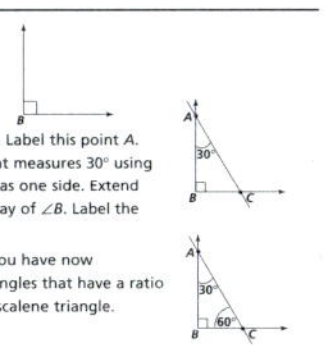

Directions Complete the following constructions. Use a separate sheet of paper. You will need a straightedge and a protractor.

1. Construct an isosceles triangle in which the sum of its two equal angles equals the measure of its third angle.
2. Construct an isosceles triangle in which the sum of its two equal angles is equal to $\frac{1}{2}$ the measure of its third angle.
3. Construct a scalene triangle that has a right angle and two other angles with a ratio of 8 to 1.
4. Construct an equilateral triangle by drawing one 60° angle and then another 60° angle off one of the first angle's rays.
5. Construct an isosceles triangle in which the sum of its two equal angles is equal to twice the measure of its third angle.

 Geometry

Workbook Activity 47

MANIPULATIVES

M Exploring Triangles

Materials: Manipulite Triangles

Group Practice: Allow pairs of students to work with an assortment of Manipulite Triangles. Have one student call out the name of a type of triangle such as "right isosceles" or "obtuse." The partner then must retrieve such a triangle from the assortment and record its description in a list. Alternately, one student can choose a triangle and his or her partner can classify the triangle as precisely as possible, recording the classification in the list. Student pairs should continue until no triangles remain or until their list reflects all triangles mentioned in the lesson.

Student Practice: Individual students work independently to sort and label the triangles.

COMMON ERROR

Students may incorrectly assume that any triangle classified by its angles has three of the same angles. For example, a student might think that an *obtuse triangle* has three obtuse angles. Work with students as they explore the following possibilities:

acute triangle: three acute angles

right triangle: one right, two acute angles

obtuse triangle: obtuse, two acute angles

Indicate to students that in Lesson 9 they will prove that the following combinations of angles cannot describe triangles:

two right, one acute

two obtuse

three obtuse

one obtuse, one right, one acute

Triangles can be named by the characteristics of their sides, the characteristics of their angles, or the characteristics of both. For example, an equilateral triangle is also acute.

> Remember, an arc mark (\ ) is used to represent angles of the same size.

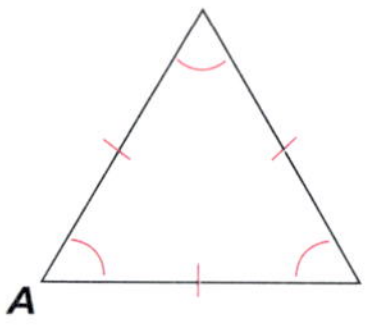

Later you will prove that an equilateral triangle has three equal angles that each measure 60°.

A right triangle, ΔABC, can be isosceles.

$m\angle A = 90°$

$\overline{AB} = \overline{AC}$

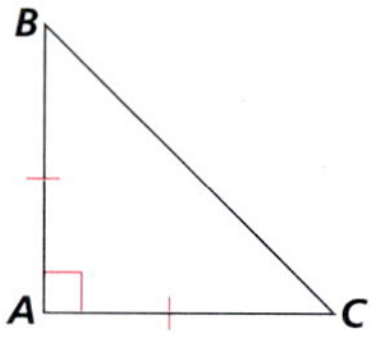

> The symbol Δ is used to represent a triangle. ΔABC is read "triangle ABC."

An obtuse triangle, ΔABC, can be scalene.

$m\angle A > 90°$

$\overline{AB} \neq \overline{BC}$, $\overline{BC} \neq \overline{AC}$, and $\overline{AC} \neq \overline{AB}$

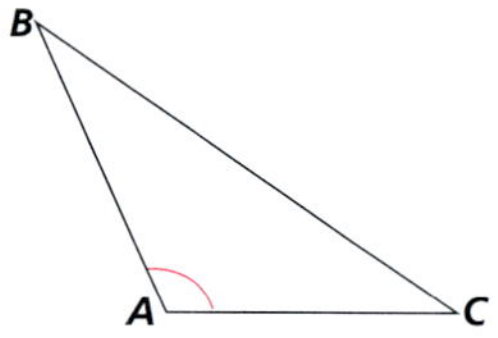

An obtuse triangle, ΔABC, can be isosceles.

$m\angle B > 90°$, $\overline{AB} = \overline{BC}$

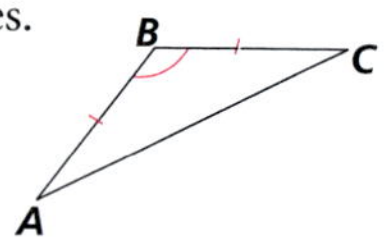

If you are given information about the sides and angles of a triangle, you can construct the triangle. Remember, you need to do the construction using only a straightedge and a compass.

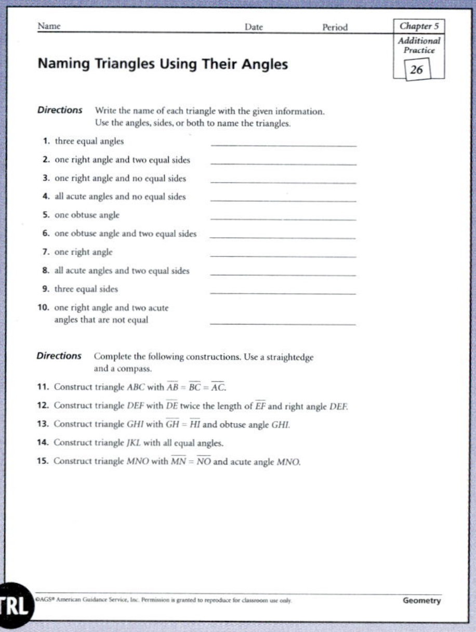

Name Date Period

Chapter 5 *Additional Practice* 26

Naming Triangles Using Their Angles

Directions Write the name of each triangle with the given information. Use the angles, sides, or both to name the triangles.

1. three equal angles ____________
2. one right angle and two equal sides ____________
3. one right angle and no equal sides ____________
4. all acute angles and no equal sides ____________
5. one obtuse angle ____________
6. one obtuse angle and two equal sides ____________
7. one right angle ____________
8. all acute angles and two equal sides ____________
9. three equal sides ____________
10. one right angle and two acute angles that are not equal ____________

Directions Complete the following constructions. Use a straightedge and a compass.

11. Construct triangle ABC with $\overline{AB} = \overline{BC} = \overline{AC}$.
12. Construct triangle DEF with $\overline{DE}$ twice the length of $\overline{EF}$ and right angle DEF.
13. Construct triangle GHI with $\overline{GH} = \overline{HI}$ and obtuse angle GHI.
14. Construct triangle JKL with all equal angles.
15. Construct triangle MNO with $\overline{MN} = \overline{NO}$ and acute angle MNO.

TRL

 Geometry

Additional Practice 26

CONSTRUCTION

Given: $\overline{AB} = \overline{BC}$ and $m\angle B = 90°$
Construct: $\triangle ABC$ with $\overline{AB} = \overline{BC}$ and $m\angle B = 90°$

Step 1 Draw a right angle. Label the angle *B*.

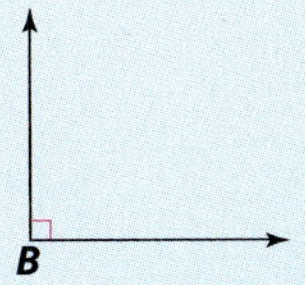

Step 2 Using your compass and *B* as the center, draw an arc that crosses each ray of angle *B*. Label the points where the arc crosses the rays *A* and *C*.

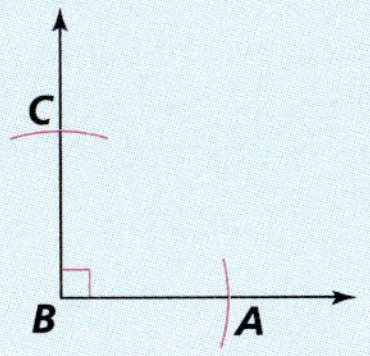

Step 3 Connect points *A* and *C*. You have now constructed a right triangle with two equal sides, or an isosceles right triangle.

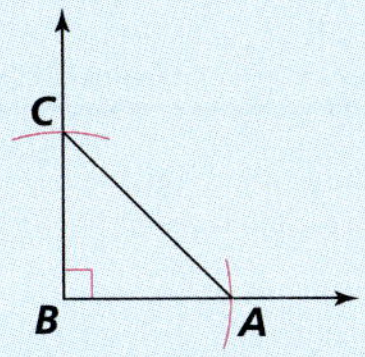

Given: $\overline{CD} = \overline{DE}$ and $m\angle D > 90°$
Construct: $\triangle CDE$ with $\overline{CD} = \overline{DE}$ and $m\angle D > 90°$

Step 1 Draw any obtuse angle. Label the angle *D*.

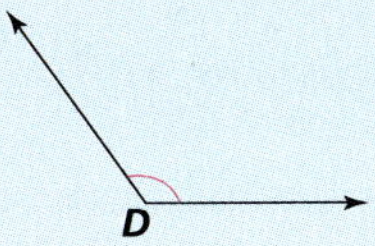

Step 2 Using your compass and point *D* as the center, draw an arc that crosses each ray of the obtuse angle. Label the points where the arc crosses the rays *C* and *E*.

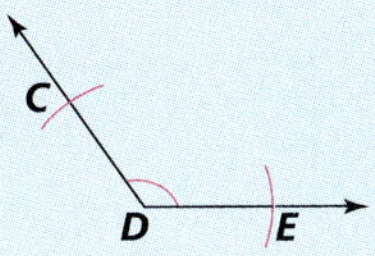

Step 3 Draw $\overline{EC}$. You have now constructed an obtuse isosceles triangle.

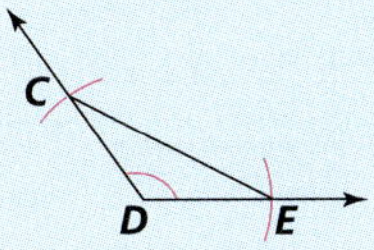

3. Reinforce and Extend

Note: For students who are physically challenged, complete the constructions on a transparency as a group or have students work with partners to draw the constructions.

LEARNING STYLES

Visual/Spatial Draw a selection of acute, right, and obtuse triangles on the board or overhead projector. For each triangle, ask a volunteer to place an *X* on the angle responsible for the triangle's classification as *right, obtuse,* or *acute.*

LEARNING STYLES

LEP/ESL Some students may need a review of the terms *obtuse, right,* and *acute.* Ask pairs of students to work together to develop an illustration and a written description of these terms as well as the terms *acute triangle, right triangle,* and *obtuse triangle.* Encourage students to keep their illustrations and descriptions as part of a pictionary of geometric terms.

LEARNING STYLES

Auditory/Verbal Have a volunteer describe the steps needed to construct a right triangle. Invite a second volunteer to demonstrate—on the board or overhead projector—each step as it is described. Be sure the student follows directions precisely. Once each step is completed, ask a third student to justify each step.

Learning Styles

Auditory/Verbal To complete Exercise A, allow students to work in pairs. For each item, one student should draw or construct a triangle that matches the description. His or her partner can determine whether or not the drawing accurately matches the description. Partners should exchange drawing and evaluating roles after each item.

Learning Styles

Body/Kinesthetic Direct student pairs to find at least one Manipulite Triangle that matches each description in Exercise A. Then have them determine if a single Manipulite Triangle can match more than one definition. Partners should be able to explain to one another why or why not the triangle matches more than one description.

Exercise A Give the name of each triangle. Use the angles, sides, or both to name the triangles.

1. one right angle — right triangle
2. one obtuse angle — obtuse triangle
3. three acute angles — acute triangle
4. one right angle and two equal sides — right isosceles triangle
5. three equal angles — equilateral triangle
6. two equal sides and one obtuse angle — obtuse isosceles triangle
7. all acute angles and three equal sides — acute equilateral triangle
8. all acute angles and no equal sides — acute scalene triangle
9. one right angle and no equal sides — right scalene triangle
10. two acute angles, one right angle, no equal sides — right scalene triangle

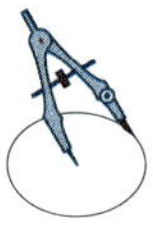

Construction

Exercise B Complete the following constructions. Use a separate sheet of paper. You will need a straightedge and a compass.

See Teacher's Edition page for answers.

11. Construct ΔABC with $\overline{AC} = \overline{BC}$ and $m\angle C = 90°$.
12. Construct ΔDEF with $\overline{DE} = \overline{DF}$ and $m\angle D < 90°$.
13. Construct ΔGHI with $\overline{GH} = \overline{HI}$ and $m\angle H > 90°$.
14. Construct ΔJKL with $\overline{JK} = 3s$, $\overline{JL} = 4s$, and $m\angle J = 90°$.

s

15. Construct ΔMNO with $m\angle M = 90°$, $\overline{MO} = r$, and $\overline{MN} = 2r$.

r

Try This Try to construct a right triangle with three equal sides. Explain your observations.

Answers to Exercise B

11.

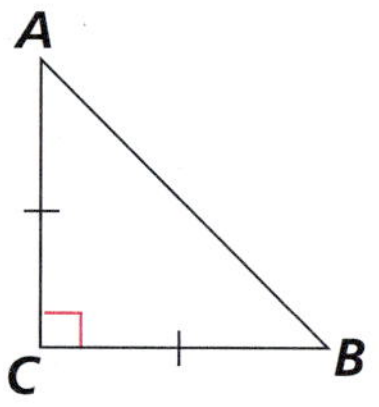

12.

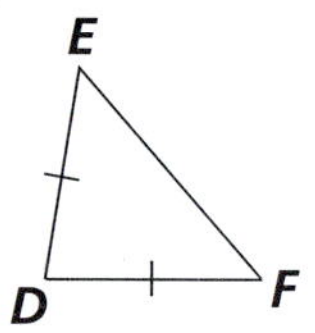

13.

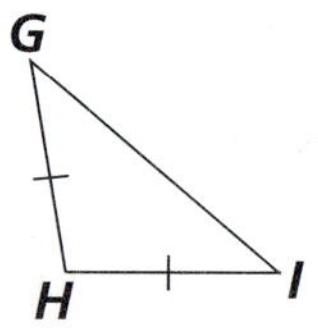

14.

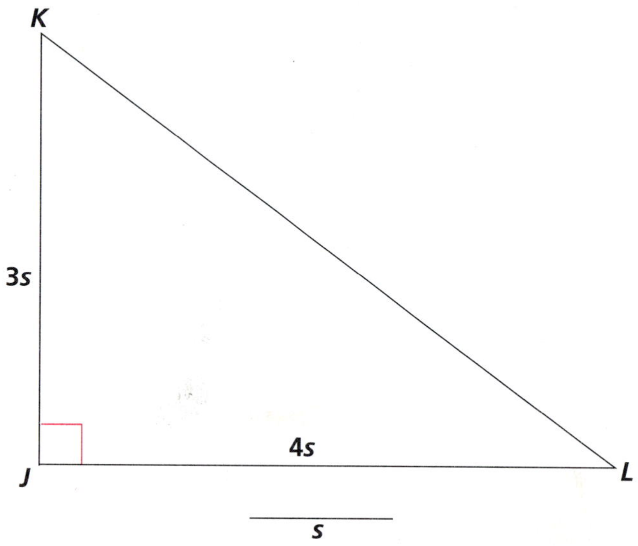

15.

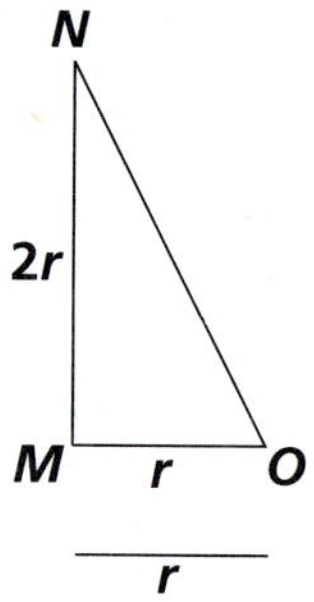

Try This...

Constructions will vary. A right triangle cannot be equilateral because it does not have three 60° angles.

Special Quadrilaterals

In Chapter 3, you were introduced to geometric figures called quadrilaterals—figures with four sides. Here is some more information about quadrilaterals.

Examples of Quadrilaterals

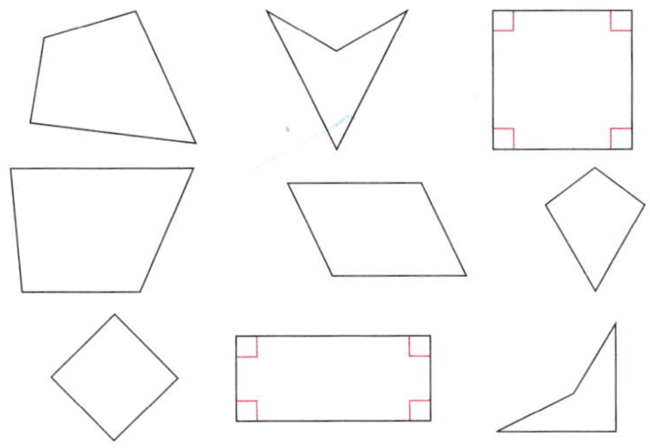

If we take a closer look at quadrilaterals, we will find that some of them have "special" characteristics. Some have parallel sides. Some have sides of equal length. During your study of parallel lines, you learned that **parallelograms** are quadrilaterals formed by two pairs of parallel sides.

Rectangles and squares are parallelograms with special properties. **Rectangles** have four equal angles; squares have four equal angles and equal sides.

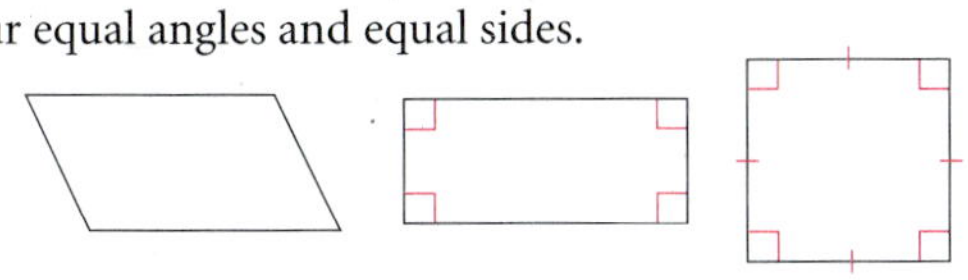

A **rhombus** is a parallelogram with four equal sides. Note that this means you can call a square a rhombus with four equal angles.

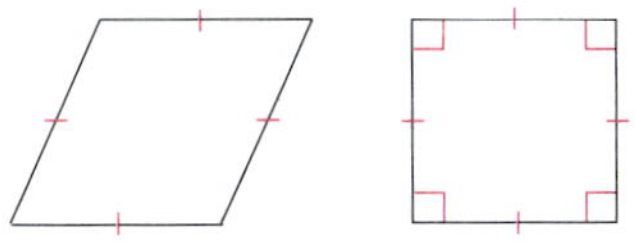

Counterexample
A single example that proves a statement false

Parallelogram
A quadrilateral whose opposite sides are parallel

Rectangle
A parallelogram with four right angles

Rhombus
A parallelogram with four equal sides

Trapezoid
A quadrilateral with exactly one pair of parallel sides

The word *quadrilateral* means "four sides."

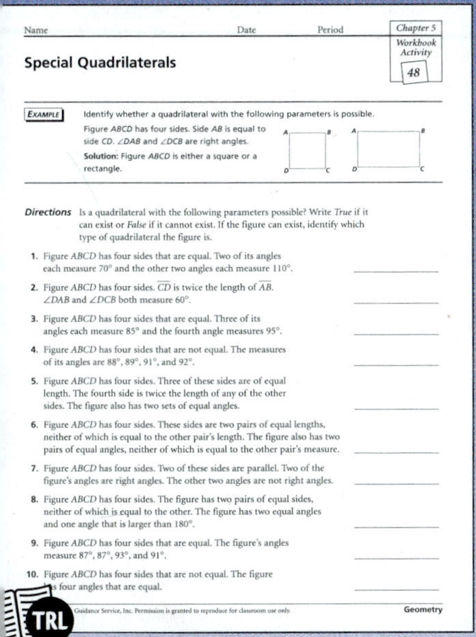

Name Date Period | *Chapter 5 Workbook Activity* 48

Special Quadrilaterals

EXAMPLE Identify whether a quadrilateral with the following parameters is possible.
Figure *ABCD* has four sides. Side *AB* is equal to side *CD*. ∠*DAB* and ∠*DCB* are right angles.
Solution: Figure *ABCD* is either a square or a rectangle.

Directions Is a quadrilateral with the following parameters possible? Write *True* if it can exist or *False* if it cannot exist. If the figure can exist, identify which type of quadrilateral the figure is.

1. Figure *ABCD* has four sides that are equal. Two of its angles each measure 70° and the other two angles each measure 110°. ______
2. Figure *ABCD* has four sides. $\overline{CD}$ is twice the length of $\overline{AB}$. ∠*DAB* and ∠*DCB* both measure 60°. ______
3. Figure *ABCD* has four sides that are equal. Three of its angles each measure 85° and the fourth angle measures 95°. ______
4. Figure *ABCD* has four sides that are not equal. The measures of its angles are 88°, 89°, 91°, and 92°. ______
5. Figure *ABCD* has four sides. Three of these sides are of equal length. The fourth side is twice the length of any of the other sides. The figure also has two sets of equal angles. ______
6. Figure *ABCD* has four sides. These sides are two pairs of equal lengths, neither of which is equal to the other pair's length. The figure also has two pairs of equal angles, neither of which is equal to the other pair's measure. ______
7. Figure *ABCD* has four sides. Two of these sides are parallel. Two of the figure's angles are right angles. The other two angles are not right angles. ______
8. Figure *ABCD* has four sides. The figure has two pairs of equal sides, neither of which is equal to the other. The figure has two equal angles and one angle that is larger than 180°. ______
9. Figure *ABCD* has four sides that are equal. The figure's angles measure 87°, 87°, 93°, and 91°. ______
10. Figure *ABCD* has four sides that are not equal. The figure has four angles that are equal. ______

 Geometry

Workbook Activity 48

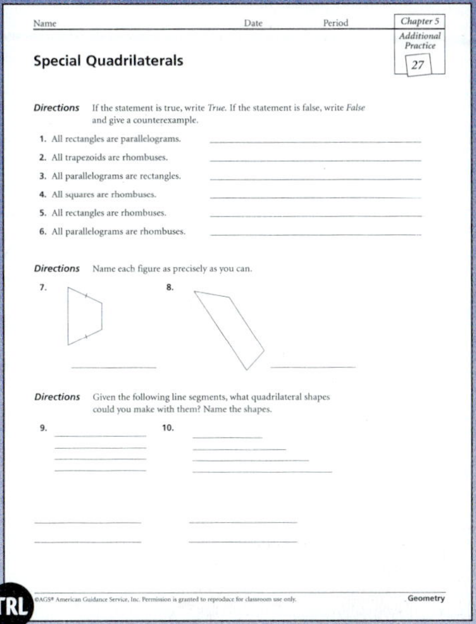

Name Date Period | *Chapter 5 Additional Practice* 27

Special Quadrilaterals

Directions If the statement is true, write *True*. If the statement is false, write *False* and give a counterexample.

1. All rectangles are parallelograms. ______
2. All trapezoids are rhombuses. ______
3. All parallelograms are rectangles. ______
4. All squares are rhombuses. ______
5. All rectangles are rhombuses. ______
6. All parallelograms are rhombuses. ______

Directions Name each figure as precisely as you can.

7. ______
8. ______

Directions Given the following line segments, what quadrilateral shapes could you make with them? Name the shapes.

9. ______
10. ______

 Geometry

Additional Practice 27

Lesson at a Glance

Chapter 5 Lesson 3

Overview This lesson introduces quadrilaterals—geometric figures with four sides.

Objectives

- To distinguish quadrilaterals from other polygons.
- To identify specific quadrilaterals, including parallelograms, rectangles, rhombuses, trapezoids, and squares.

Student Pages 143–145

Teacher's Resource Library TRL

Workbook Activity 48

Additional Practice 27

Mathematics Vocabulary

counterexample
parallelogram
rectangle
rhombus
trapezoid

1. Warm-Up Activity

Remind students that they worked with quadrilaterals in Chapter 3. Ask volunteers to draw each of the following quadrilaterals on the board or overhead projector: quadrilateral, parallelogram, rectangle, and rhombus. Indicate to students that in this lesson they will learn more about quadrilaterals with special characteristics.

2. Teaching the Lesson

Remind students that a diagram can have tick marks indicating that some segments are the same length. A diagram may also indicate which angles, if any, are right angles.

Discuss with students the idea of a *counterexample* using a nonmathematical context. Point out that all that is needed to prove that a statement such as "All boys play football" is false, is one boy who does *not* play football. Indicate to students that the *number* of examples that do fit the statement (millions of boys play football) is irrelevant. One boy who does not play football makes the statement false.

Common Error

In their earlier exposure to quadrilaterals, students may have concentrated on distinguishing one figure from another, for instance, rectangles from squares. This may have left the students with the impression that quadrilateral figures have no common characteristics. Indicate to students that quadrilaterals in this lesson all share at least one characteristic in common (four sides) and many share additional characteristics.

Trapezoids are also quadrilaterals. All trapezoids share one characteristic—they have exactly one pair of parallel sides. However, some trapezoids have additional characteristics. You can use the same method you use to name triangles to name trapezoids.

Examples of Trapezoids

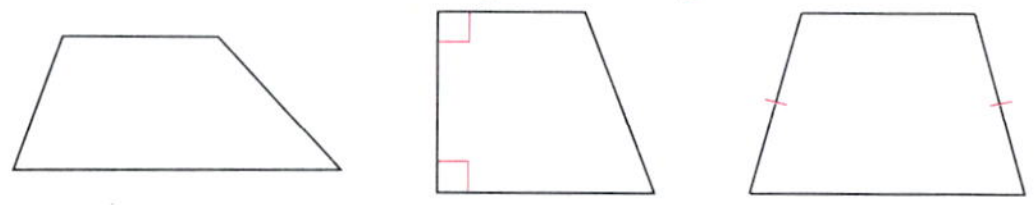

Scalene Trapezoid | Right-Angled Trapezoid | Isosceles Trapezoid

When you think about quadrilaterals and parallelograms, you will find there are some general statements that are always true—such as all squares are rectangles. But some statements are not always true. The following is an example.

Example

Statement All trapezoids are isosceles.

Question Is this statement true?

Answer No. It is possible to construct a trapezoid that does not have a pair of equal sides and so it is not isosceles.

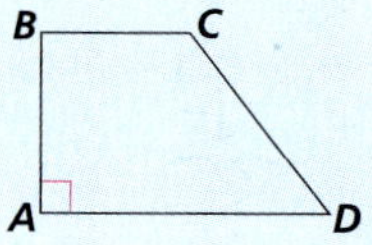

In this case, you were able to provide an example for which the statement was not true. Such an example is called a **counterexample.** One counterexample makes an *all*-statement false.

Example

Statement All parallelograms are rectangles.

Question Is this statement true?

Answer No. You can construct a parallelogram that is not a rectangle. Therefore, you can show a counterexample. The counterexample proves the statement false.

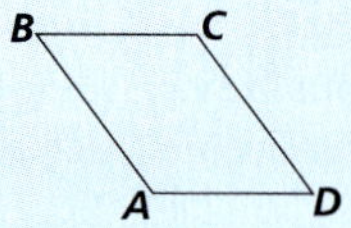

144 *Chapter 5 Triangles and Quadrilaterals*

Exercise A If the statement is true, write *True.* If the statement is false, write *False* and give a counterexample.

1. All parallelograms are squares. False, rectangle
2. All squares are parallelograms. True
3. All trapezoids are scalene. False, isosceles trapezoid
4. All squares are rhombuses. True
5. All rhombuses are squares. False, not all rhombuses have equal angles
6. All rectangles are squares. False, $\ell > w$
7. All squares are rectangles. True

Writing About Mathematics

You know that "All rectangles are squares" is false because you can give a counterexample. Why is "Some rectangles are squares" a true statement?

Exercise B Copy the chart. Fill in the missing information.

Quadrilateral	Parallelogram	Rectangle	Rhombus	Square
Figure with four sides	Figure with four sides	Figure with four sides	**8.** ______	Figure with four sides
	Opposite sides are parallel and equal	Opposite sides are parallel and equal	Opposite sides are parallel and equal	Opposite sides are **9.** ______
		Four **10.** ______		Four equal angles

8. Figure with four sides
9. parallel and equal
10. equal angles

Exercise C Name each figure as precisely as you can.

11.

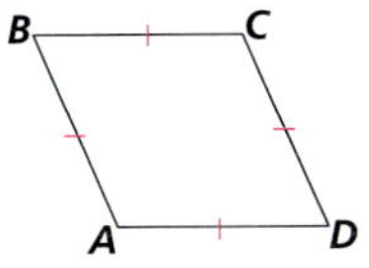

rhombus

12.

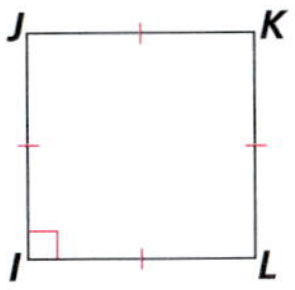

square

13.

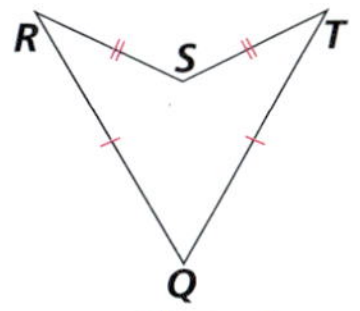

quadrilateral

14.

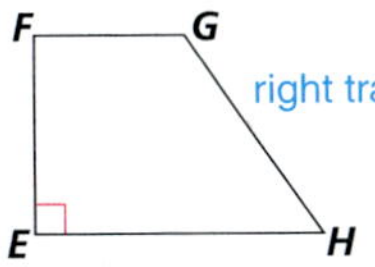

right trapezoid

15.

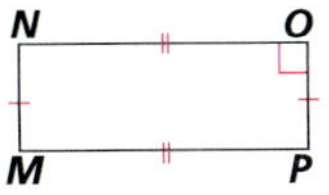

rectangle

3. Reinforce and Extend

Learning Styles

Logical/Mathematical Challenge students adept at organization to create a word web or other logical scheme incorporating the following figures: quadrilateral, trapezoid, parallelogram, isosceles trapezoid, rhombus, rectangle, square. Students should be able to explain their scheme to the class. For example, in the following scheme, every property held by a figure is also held by each of the figures listed below it.

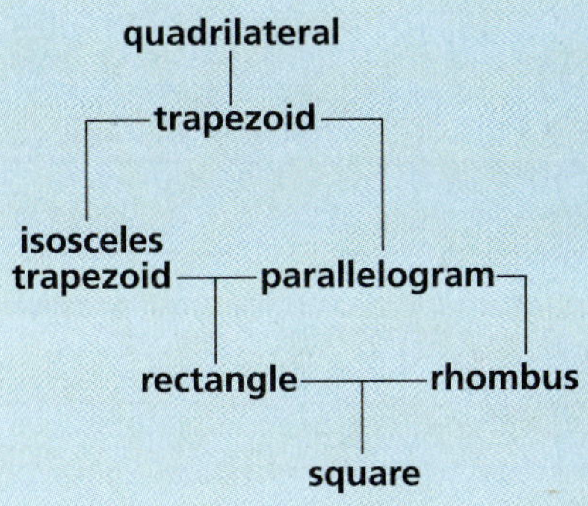

Learning Styles

LEP/ESL Pair students less proficient in English with students who have more advanced language skills. Direct student pairs to work together to prepare pairs of cards to use in a memory game. One card in the pair could have an illustration of a quadrilateral; the second card could have the name of the quadrilateral. All cards are placed face down and one student turns over cards one at a time, trying to match the illustration and its identification. Students gain points for each match they make. Extra points can be gained by describing the characteristics of the matched quadrilateral.

Lesson at a Glance

Chapter 5 Lesson 4

Overview This lesson discusses the characteristics of quadrilateral diagonals—line segments that connect nonadjacent vertices of quadrilaterals.

Objectives

- To identify the diagonals of a quadrilateral.
- To use parallel line postulates to determine the relationships of angles formed by a quadrilateral's diagonals.

Student Pages 146–148

Teacher's Resource Library TRL

Workbook Activities 49–50

Additional Practice 28

Mathematics Vocabulary

diagonal

1. Warm-Up Activity

Students are likely to be familiar with the term *diagonal*. Ask for a volunteer to describe, or demonstrate, a common meaning for *diagonal*. Allow the class to agree or disagree with the description, folding all views into the discussion. At the close of the discussion, indicate that in this lesson, students will learn about diagonals in geometry.

2. Teaching the Lesson

Capitalize on students' notion of *diagonal* to lead them to an understanding of the diagonals in quadrilaterals.

List each of the statements of the proof on page 147 on the board or an overhead projector. Invite students to provide the reason for each step. You may wish to have a volunteer circle or otherwise mark the elements central to each step of the proof. For example, in Step 1, a volunteer would highlight the part of the illustration that resembles parallel lines and a transversal, isolating it from the rest of the illustration.

Lesson 4 Quadrilaterals and Their Diagonals

Diagonal
A line segment connecting any two nonadjacent vertices of a polygon

Quadrilaterals not only have four sides, they have four angles and four vertices. In quadrilateral *ABCD* below, *A*, *B*, *C*, and *D* are vertices.

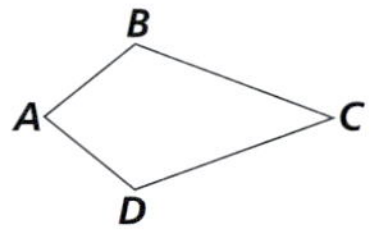

You can draw a line segment from each vertex to a vertex that does not share a common side. This is called a *nonadjacent* vertex. For example, you can draw a segment from *A* to *C*. $\overline{AC}$ is an example of a **diagonal.** The diagonal from *C* is also $\overline{AC}$ (or $\overline{CA}$).

Remember, a vertex is the common endpoint of two rays that form an angle. The plural of *vertex* is *vertices*. *Adjacent* vertices share a common side. *Nonadjacent* vertices do not share a common side.

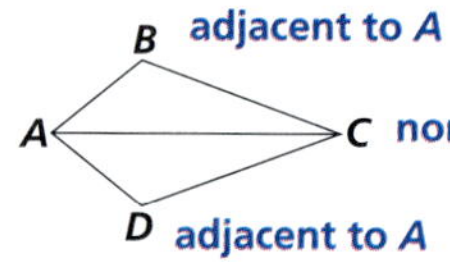

A diagonal is a line segment connecting any two nonadjacent vertices of a geometric figure. There is a second diagonal that can be drawn for *ABCD*. It is the segment from *B* to *D*, $\overline{DB}$. There are no other diagonals possible for quadrilateral *ABCD*.

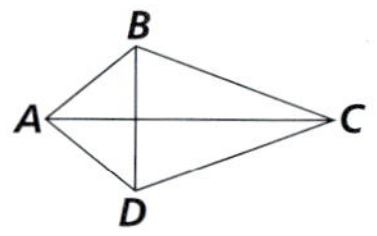

Suppose *ABCD* is a rectangle. Using Euclid's Postulate 1, you can draw diagonal $\overline{AC}$. This will form angles 1, 2, 3, and 4. Which, if any, of these angles are equal?

THEOREM: In rectangle *ABCD* with diagonal $\overline{AC}$, $m\angle 1 = m\angle 4$ and $m\angle 2 = m\angle 3$.

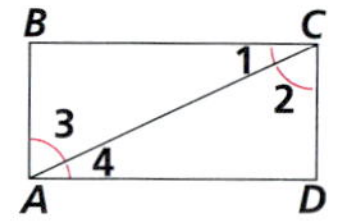

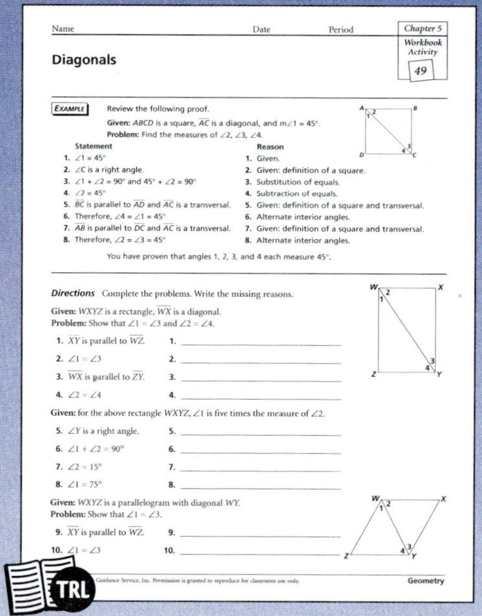

Name Date Period | *Chapter 5 Workbook Activity 49*

Diagonals

EXAMPLE Review the following proof.

Given: *ABCD* is a square, $\overline{AC}$ is a diagonal, and $m\angle 1 = 45°$.
Problem: Find the measures of $\angle 2$, $\angle 3$, $\angle 4$.

Statement	Reason
1. $\angle 1 = 45°$	1. Given.
2. $\angle C$ is a right angle.	2. Given: definition of a square.
3. $\angle 1 + \angle 2 = 90°$ and $45° + \angle 2 = 90°$	3. Substitution of equals.
4. $\angle 2 = 45°$	4. Subtraction of equals.
5. $\overline{BC}$ is parallel to $\overline{AD}$ and $\overline{AC}$ is a transversal.	5. Given: definition of a square and transversal.
6. Therefore, $\angle 4 = \angle 1 = 45°$	6. Alternate interior angles.
7. $\overline{AB}$ is parallel to $\overline{DC}$ and $\overline{AC}$ is a transversal.	7. Given: definition of a square and transversal.
8. Therefore, $\angle 2 = \angle 3 = 45°$	8. Alternate interior angles.

You have proven that angles 1, 2, 3, and 4 each measure 45°.

Directions Complete the problems. Write the missing reasons.

Given: *WXYZ* is a rectangle, $\overline{WX}$ is a diagonal.
Problem: Show that $\angle 1 = \angle 3$ and $\angle 2 = \angle 4$.

1. $\overline{XY}$ is parallel to $\overline{WZ}$. 1. ______
2. $\angle 1 = \angle 3$ 2. ______
3. $\overline{WX}$ is parallel to $\overline{ZY}$. 3. ______
4. $\angle 2 = \angle 4$ 4. ______

Given: for the above rectangle *WXYZ*, $\angle 1$ is five times the measure of $\angle 2$.

5. $\angle Y$ is a right angle. 5. ______
6. $\angle 1 + \angle 2 = 90°$ 6. ______
7. $\angle 2 = 15°$ 7. ______
8. $\angle 1 = 75°$ 8. ______

Given: *WXYZ* is a parallelogram with diagonal *WY*.
Problem: Show that $\angle 1 = \angle 3$.

9. $\overline{XY}$ is parallel to $\overline{WZ}$. 9. ______
10. $\angle 1 = \angle 3$ 10. ______

Guidance Service, Inc. Permission is granted to reproduce for classroom use only. Geometry

TRL **Workbook Activity 49**

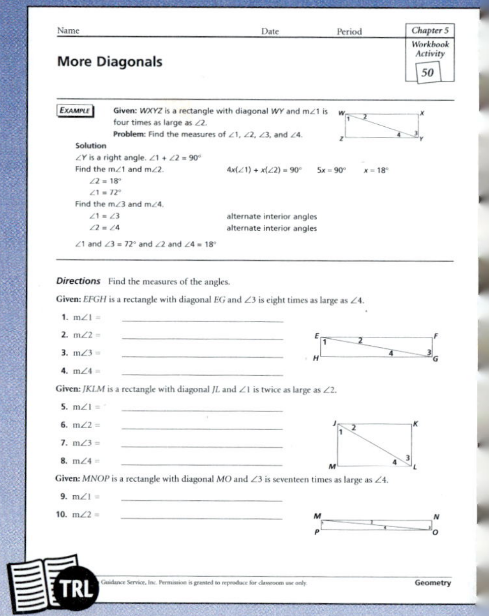

Name Date Period | *Chapter 5 Workbook Activity 50*

More Diagonals

EXAMPLE **Given:** *WXYZ* is a rectangle with diagonal *WY* and $m\angle 1$ is four times as large as $\angle 2$.
Problem: Find the measures of $\angle 1$, $\angle 2$, $\angle 3$, and $\angle 4$.

Solution
$\angle Y$ is a right angle. $\angle 1 + \angle 2 = 90°$
Find the $m\angle 1$ and $m\angle 2$. $4x(\angle 1) + x(\angle 2) = 90°$ $5x = 90°$ $x = 18°$
$\angle 2 = 18°$
$\angle 1 = 72°$
Find the $m\angle 3$ and $m\angle 4$.
$\angle 1 = \angle 3$ alternate interior angles
$\angle 2 = \angle 4$ alternate interior angles
$\angle 1$ and $\angle 3 = 72°$ and $\angle 2$ and $\angle 4 = 18°$

Directions Find the measures of the angles.

Given: *EFGH* is a rectangle with diagonal *EG* and $\angle 3$ is eight times as large as $\angle 4$.

1. $m\angle 1 =$ ______
2. $m\angle 2 =$ ______
3. $m\angle 3 =$ ______
4. $m\angle 4 =$ ______

Given: *JKLM* is a rectangle with diagonal *JL* and $\angle 1$ is twice as large as $\angle 2$.

5. $m\angle 1 =$ ______
6. $m\angle 2 =$ ______
7. $m\angle 3 =$ ______
8. $m\angle 4 =$ ______

Given: *MNOP* is a rectangle with diagonal *MO* and $\angle 3$ is seventeen times as large as $\angle 4$.

9. $m\angle 1 =$ ______
10. $m\angle 2 =$ ______

Guidance Service, Inc. Permission is granted to reproduce for classroom use only. Geometry

TRL **Workbook Activity 50**

PROOF

Statement	Reason
1. $\overline{BC} \parallel \overline{AD}$ and $\overline{AC}$ is a transversal	1. Given; by definition of a rectangle and transversal.
2. $\therefore m\angle 4 = m\angle 1$	2. By theorem, alternate interior angles are equal.
3. $\overline{AB} \parallel \overline{DC}$ and $\overline{AC}$ is a transversal	3. Given; by definition of a rectangle and transversal.
4. $\therefore m\angle 2 = m\angle 3$	4. By theorem, alternate interior angles are equal.

You can use this information to solve problems like the following.

EXAMPLE **Given:** *ABCD* is a rectangle, $\overline{AC}$ is a diagonal, and $m\angle 1 = 40°$

Problem: find the measures of $\angle 2$, $\angle 3$, and $\angle 4$

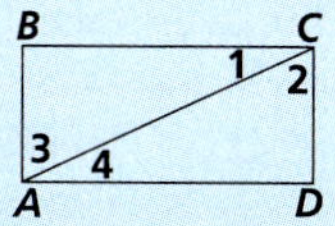

Solution:	Reason:
$m\angle 1 = m\angle 4$, so $m\angle 4 = 40°$	alternate interior angles
$\angle C$ is a right angle	definition of rectangle
so $m\angle 1 + m\angle 2 = 90°$, and $40° + m\angle 2 = 90°$	substitution of equals
$m\angle 2 = 50°$	subtraction of equals
$m\angle 2 = m\angle 3 = 50°$	alternate interior angles
OR you can argue that $\angle A = 90°$	definition of rectangle
so $m\angle 3 + m\angle 4 = 90°$, and $m\angle 3 + 40° = 90°$	substitution of equals
$m\angle 3 = 50°$	subtraction of equals

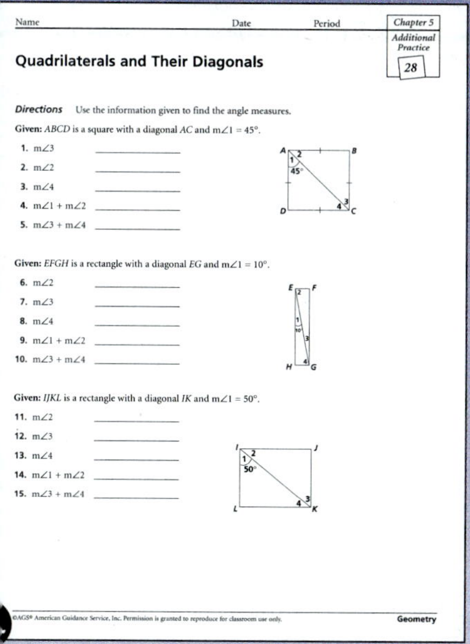

Name Date Period

Chapter 5 *Additional Practice* 28

Quadrilaterals and Their Diagonals

Directions Use the information given to find the angle measures.

Given: *ABCD* is a square with a diagonal *AC* and $m\angle 1 = 45°$.

1. $m\angle 3$ ________
2. $m\angle 2$ ________
3. $m\angle 4$ ________
4. $m\angle 1 + m\angle 2$ ________
5. $m\angle 3 + m\angle 4$ ________

Given: *EFGH* is a rectangle with a diagonal *EG* and $m\angle 1 = 10°$.

6. $m\angle 2$ ________
7. $m\angle 3$ ________
8. $m\angle 4$ ________
9. $m\angle 1 + m\angle 2$ ________
10. $m\angle 3 + m\angle 4$ ________

Given: *IJKL* is a rectangle with a diagonal *IK* and $m\angle 1 = 50°$.

11. $m\angle 2$ ________
12. $m\angle 3$ ________
13. $m\angle 4$ ________
14. $m\angle 1 + m\angle 2$ ________
15. $m\angle 3 + m\angle 4$ ________

 Geometry

Additional Practice 28

Learning Styles

LEP/ESL Certain students might benefit from a review session stressing the meanings of the terms *vertex, vertices, adjacent,* and *nonadjacent.* Pair students learning English as a second language with those more proficient in English. Direct one student to call out one of the terms. His or her partner should then draw an illustration and give a description of the term. Students can reverse roles after each term.

Lead students through the example on page 147 by duplicating the diagram on the board or overhead projector. Be certain to draw their attention to the two different ways to determine the measure of $\angle 3$.

Manipulatives

M **Polygon Practice**

Materials: Polygon Power Pack

Student Practice: Provide students with a selection of polygons from the Polygon Power Pack. Ask them to use the polygons, a china marker or grease pencil, and a straightedge to determines what happens when they attempt to draw a diagonal between adjacent vertices. Invite volunteers to describe their findings. **(The diagonal will rest along the edge of the polygon.)**

3. Reinforce and Extend

Learning Styles

Body/Kinesthetic Use masking tape or yarn to mark out a variety of quadrilaterals on the floor. Label each quadrilateral *ABCD.* Have students form groups of four and give each group a length of yarn or string. Ask students to arrange themselves at the vertices of the quadrilaterals. Then pursue a game of "Simon Says." For example: "Simon Says form a diagonal from *A* to *C*." In response students at *A* and *C* should stretch a length of yarn between them. Other directions might include identification of adjacent or nonadjacent angles. Once diagonals are formed, directions might include identification of the transversal and of alternate interior angles.

Learning Styles

Interpersonal/Group Learning Direct students to work in groups of four to complete Exercise C. Members of the group can work together to produce a "Math Minute" news segment that gives the measures for problems 7–11 and 12–15 as well as the reasoning supporting each answer. Students should include alternate reasoning wherever appropriate.

Experiencing Geometry

Students should recognize that angles 1, 2, 3, and 4 are equal to one another. When the square is folded along the diagonal, the triangles formed, ΔABC and ΔADC, exactly match one another so they must be congruent. That would mean that $\angle B = \angle D$ and $\frac{1}{2}\angle A = \frac{1}{2}\angle C$ so $\angle A = \angle C$.

Exercise A Copy the problem below. Fill in the missing reasons using the information in this lesson about rectangles.

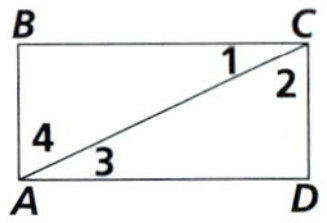

Given: *ABCD* is a parallelogram, $\overline{AC}$ is a diagonal

Problem: show m∠1 = m∠3 and m∠2 = m∠4

Solution:	Reason:
1. $\overline{BC} \parallel \overline{AD}$	definition of parallelogram
2. $m\angle 1 = m\angle 3$	alternate interior angles
3. $\overline{AB} \parallel \overline{DC}$	definition of parallelogram
4. $m\angle 2 = m\angle 4$	alternate interior angles

Exercise B Use the information given to answer the questions.

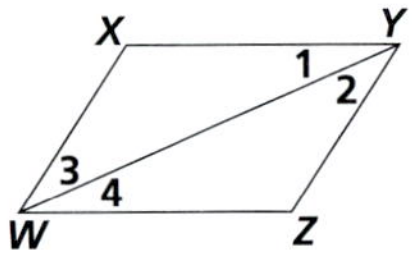

Given: *WXYZ* is a parallelogram with diagonal $\overline{WY}$

5. If $m\angle 1 = 30°$, which remaining angle is also 30°? 4

6. If $m\angle 2 = 40°$, which remaining angle is also 40°? 3

Experiencing Geometry

Cut a square from a sheet of paper and label the square *ABCD*. Draw diagonal $\overline{AC}$. Now fold the square in half along $\overline{AC}$, matching vertex *B* with vertex *D*. Label angles 1, 2, 3, and 4. Notice that ΔABC matches ΔADC exactly. What can you say about the measures of angles 1, 2, 3, and 4? Give reasons for your answer.

Exercise C Use the information given to supply the angle measures.

Given: *EFGH* is a rectangle with a diagonal $\overline{EG}$ and m∠1 = 25°

7. $m\angle 4$ 25°

8. $m\angle 1 + m\angle 2$ 90°

9. $m\angle 2$ 65°

10. $m\angle 3 + m\angle 4$ 90°

11. $m\angle 3$ 65°

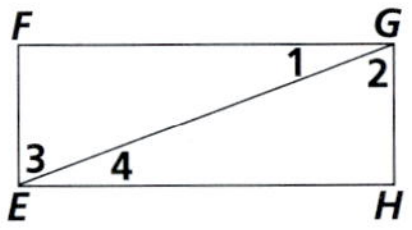

Given: *IJKL* is a rectangle with a diagonal $\overline{JL}$ and m∠3 = 35°

12. $m\angle 4$ 55°

13. $m\angle 1$ 55°

14. $m\angle 2 + m\angle 1$ 90°

15. $m\angle 2$ 35°

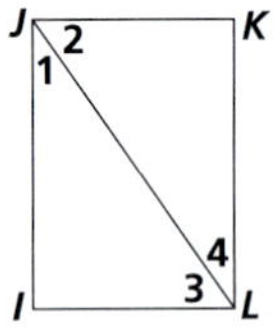

148 *Chapter 5 Triangles and Quadrilaterals*

Lesson 5 Angle Sum of Any Triangle

In Chapter 3, Lesson 5, you used inductive thinking to conclude that the sum of the angle measures inside a parallelogram or trapezoid is 360°.

Algebra Connection

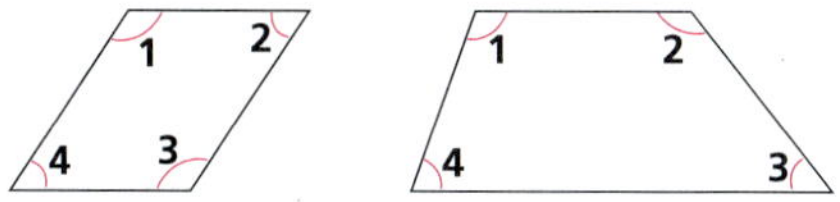

$m\angle 1 + m\angle 4 = 180°$
$m\angle 2 + m\angle 3 = 180°$ — angles on the same side of a transversal between parallels are supplementary

$m\angle 1 + m\angle 2 + m\angle 3 + m\angle 4 = 180° + 180° = 360°$ by addition of equals

> *W*ith inductive thinking, you use the similarities of particular cases to make a general statement.

If we draw a diagonal in each figure, we can conclude that the sum of the angles in two triangles is 360°.

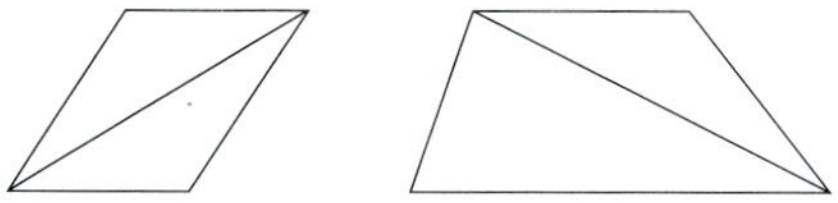

How can we find the sum of the angles in *one* triangle?

EXAMPLES

Given: *ABCD*, a square

$4 \bullet 90° = 360°$, $360° \div 2 = 180°$

If we make the two triangles alike or coincide by matching the two right triangles, we can see that the angle sum of right $\triangle ABC = 180°$.

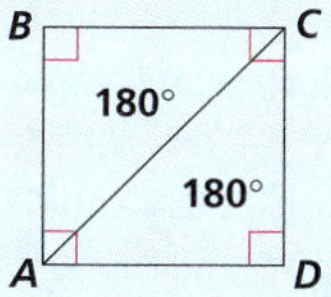

Conclusion: The sum of the angles in an isosceles right triangle is 180°.

Given: *ABCD*, a rectangle, and diagonal $\overline{AC}$

$4 \bullet 90° = 360°$

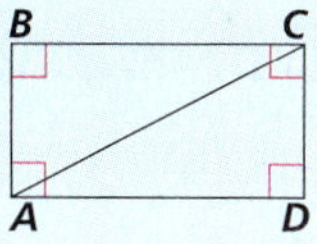

Lesson at a Glance

Chapter 5 Lesson 5

Overview This lesson presents an informal proof of the angle sum theorem: the sum of the angles of a triangle equals 180°.

Objectives

- To identify the sum of the angles of any triangle as 180°.
- To use the angle sum theorem to find the measure of unknown angles.

Student Pages 149–153

Teacher's Resource Library TRL

Workbook Activities 51–52

Additional Practice 29

1. Warm-Up Activity

Present the following data to students: On Thursday and Friday of last week and on Tuesday and Wednesday of this week, it rained and Selene was late for school. On all the other days it did not rain and Selene was on time for school. What generalization might you make about the weather and Selene's timeliness?

Students are likely to assume that if it rains, Selene will be tardy. Indicate to students that to reach this conclusion, they used inductive reasoning. They took the information from specific examples and generalized it to a new set of circumstances. Point out that much of this lesson depends on inductive reasoning as well.

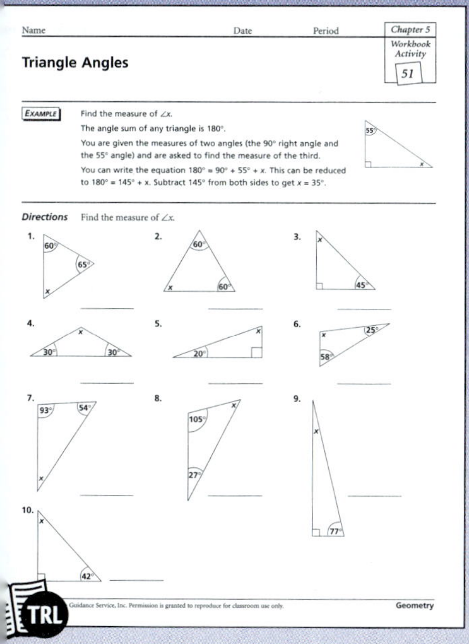

Name Date Period — Chapter 5 Workbook Activity 51

Triangle Angles

EXAMPLE Find the measure of $\angle x$.

The angle sum of any triangle is 180°.

You are given the measures of two angles (the 90° right angle and the 55° angle) and are asked to find the measure of the third.

You can write the equation $180° = 90° + 55° + x$. This can be reduced to $180° = 145° + x$. Subtract 145° from both sides to get $x = 35°$.

Directions Find the measure of $\angle x$.

1. 60°, 65°, x
2. 60°, 60°, x
3. x, 45°
4. 30°, x, 30°
5. 20°, x
6. 25°, 58°, x
7. 93°, 54°, x
8. 105°, 27°, x
9. x, 77°
10. x, 42°

Guidance Service, Inc. Permission is granted to reproduce for classroom use only. Geometry

Workbook Activity 51

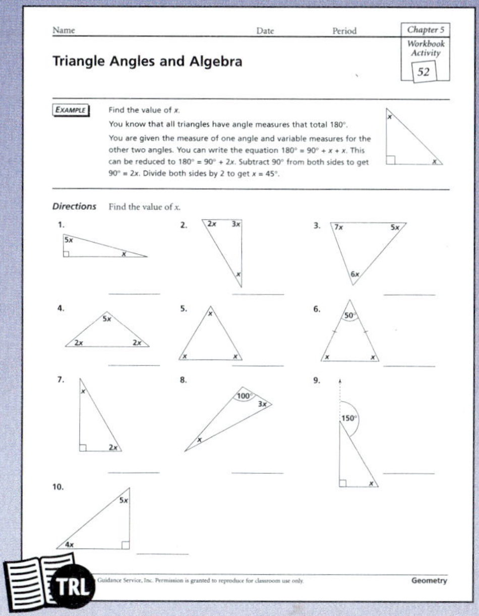

Name Date Period — Chapter 5 Workbook Activity 52

Triangle Angles and Algebra

EXAMPLE Find the value of x.

You know that all triangles have angle measures that total 180°.

You are given the measure of one angle and variable measures for the other two angles. You can write the equation $180° = 90° + x + x$. This can be reduced to $180° = 90° + 2x$. Subtract 90° from both sides to get $90° = 2x$. Divide both sides by 2 to get $x = 45°$.

Directions Find the value of x.

1. 5x, x
2. 2x, 3x, x
3. 7x, 5x, 6x
4. 5x, 2x, 2x
5. x, x
6. 50°, x, x
7. x, 2x
8. 100°, 3x, x
9. 150°, x
10. 5x, 4x

Guidance Service, Inc. Permission is granted to reproduce for classroom use only. Geometry

Workbook Activity 52

2. Teaching the Lesson

Reproduce the examples on page 149 and at the top of page 150 on the board or overhead projector. Invite a volunteer to use a large protractor or an angle ruler to measure each angle in the triangles. Have the volunteer list the measures and ask the class to confirm that the measures do, indeed, add to 180°.

Reproduce the diagram for the example at the bottom of page 150. Invite a volunteer to explain and confirm that $m\angle a = (m\angle 1 + m\angle 2) + m\angle b$ = the sum of the angles in ΔABC. Invite other volunteers to explain and confirm each of the remaining steps in the example.

EXAMPLES *(continued)*

Separate the rectangle, along its diagonal, forming two right triangles. Turn ΔADC so that ΔABC can be placed on ΔADC, matching the angles and sides of the two triangles. From this we can see that the angles and sides of the triangles are equal.

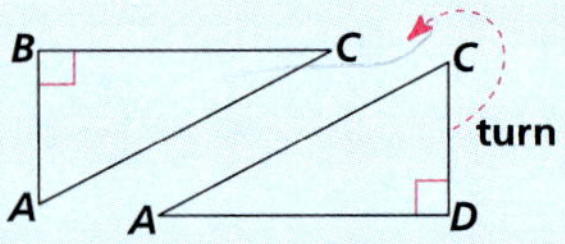

$360° \div 2 = 180°$ each

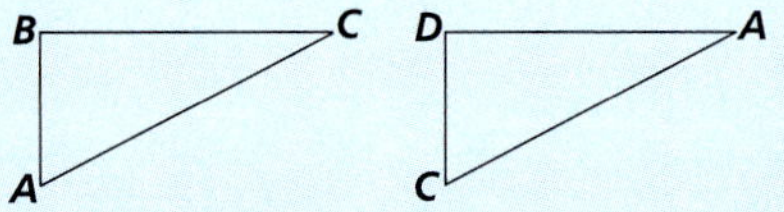

Conclusion: The sum of the angles in each right triangle is 180°.

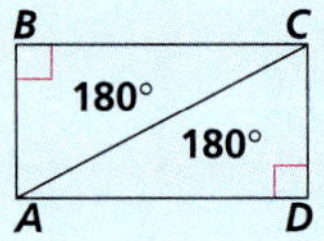

Given: ΔABC, and draw $\overline{BD} \perp \overline{AC}$ at D

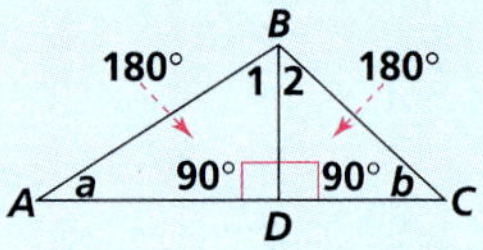

$m\angle a + (m\angle 1 + m\angle 2) + m\angle b =$ angle sum of ΔABC	Definition of sum of angles.
$m\angle a + m\angle 1 + 90° = 180°$, and $m\angle b + m\angle 2 + 90° = 180°$	The sum of the angles in every right Δ is 180°.
$m\angle a + m\angle 1 + m\angle 2 + m\angle b + 90° + 90° = 360°$	Addition of equals.
$m\angle a + (m\angle 1 + m\angle 2) + m\angle b = 180°$	Subtract 180° from each side.

Conclusion: The sum of the angles in any Δ is 180°.

150 *Chapter 5 Triangles and Quadrilaterals*

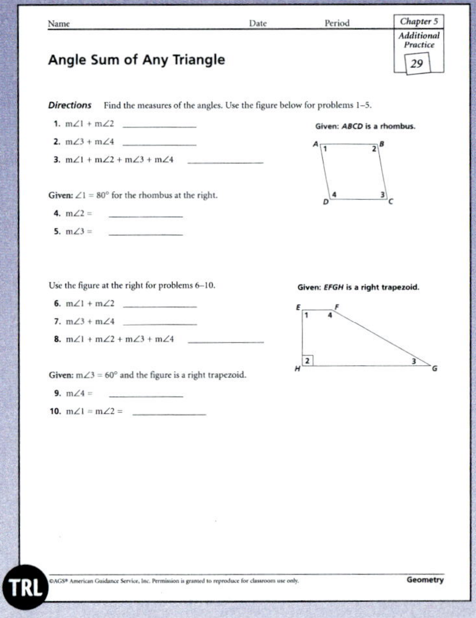

Name Date Period

Chapter 5 *Additional Practice* 29

Angle Sum of Any Triangle

Directions Find the measures of the angles. Use the figure below for problems 1–5.

1. $m\angle 1 + m\angle 2$ ________
2. $m\angle 3 + m\angle 4$ ________
3. $m\angle 1 + m\angle 2 + m\angle 3 + m\angle 4$ ________

Given: *ABCD* **is a rhombus.**

Given: $\angle 1 = 80°$ for the rhombus at the right.

4. $m\angle 2 =$ ________
5. $m\angle 3 =$ ________

Use the figure at the right for problems 6–10.

Given: *EFGH* **is a right trapezoid.**

6. $m\angle 1 + m\angle 2$ ________
7. $m\angle 3 + m\angle 4$ ________
8. $m\angle 1 + m\angle 2 + m\angle 3 + m\angle 4$ ________

Given: $m\angle 3 = 60°$ and the figure is a right trapezoid.

9. $m\angle 4 =$ ________
10. $m\angle 1 = m\angle 2 =$ ________

 Geometry

TRL

Additional Practice 29

Summary of Inductive/Informal Argument

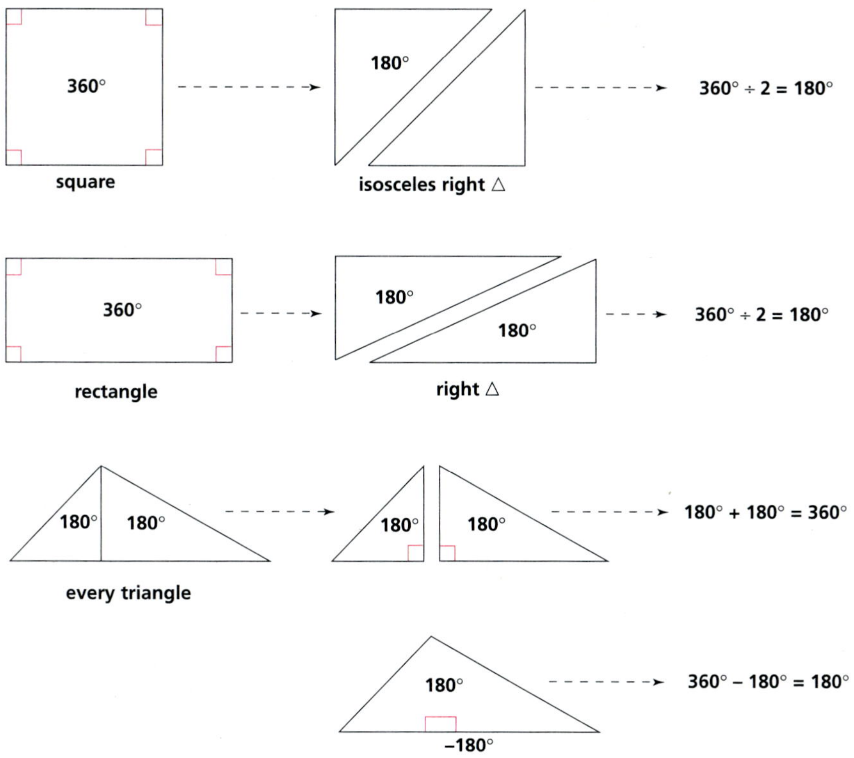

Conclusion: The sum of the angles in any triangle is 180°.

You will prove this theorem more formally later. Using this theorem, you can solve many kinds of angle measure problems involving triangles.

> **TECHNOLOGY CONNECTION**
>
> Scientists and researchers rely on basic geometric terms to describe the physical world. For example, thousands of solid chemical compounds contain a regular arrangement of particles. These arrangements—called crystals—are described using geometric terms such as *square, rectangle,* and *rhombus.*

Technology Connection

Cubic system, each face is a square

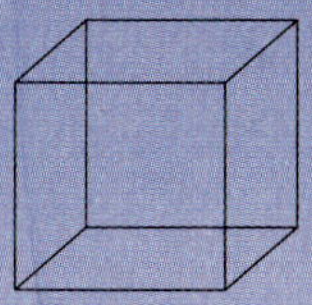

Tetragonal system, four faces are rectangles, two faces are squares

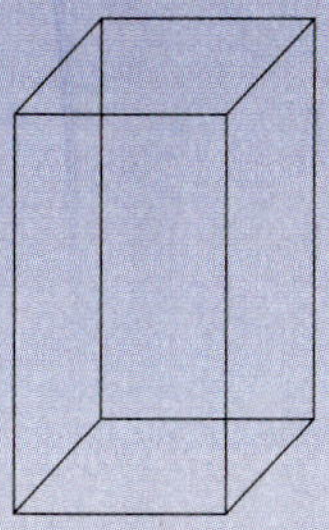

Orthorhombic system, three pairs of rectangular faces

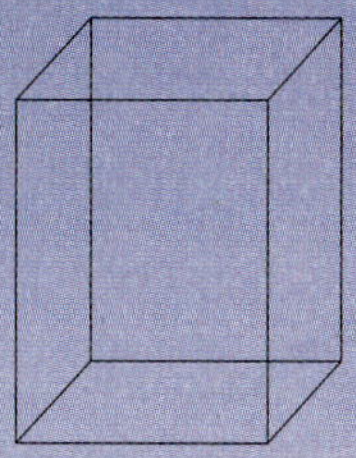

Rhombohedral or trigonal system, each face is a rhombus

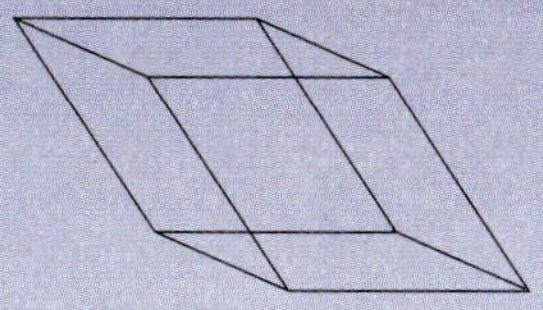

MANIPULATIVES

M Polygons

Materials: Manipulite Polygons or Polygon Power Pack

Student Practice: Encourage students to use either the Manipulite Polygons or the polygons in the Polygon Power Pack to find triangles and rectangles with which they can model each of the possibilities shown on page 151.

3. Reinforce and Extend

LEARNING STYLES

LEP/ESL All students, but particularly those acquiring English as a second language, can benefit from "reading" the illustrations on page 151. Have students explain each illustration. After a discussion, students can compose captions for each illustration. Ask volunteers to read their captions aloud. Then have the class determine if the caption accurately describes the illustration. Allow caption writers to revise or rewrite captions as needed.

Note that the formal proof of this theorem is given in Chapter 9.

APPLICATION

Career Connection

Encourage students to learn what kinds of careers rely on a knowledge of the geometry and structure of crystals. Encourage them to examine areas such as crystallography, mineralogy, and the study and use of liquid crystals. Invite them to present their findings as a two-minute commercial for the "Crystal Career Academy."

Algebra Connection

After discussing the examples on page 152 with students, present the following problems. Direct students to write an equation for the unknown, then solve the equation.

The three angles of a triangle are equal. ($3x = 180°$, $x = 60°$; all three angles = 60°)

$\angle A$ is twice $\angle B$, $\angle B$ is twice $\angle C$. (Let $\angle B = x$, then $2x + x + \frac{1}{2}x = 180°$, $x = 51.4°$; $\angle A = 102.8°$, $\angle B = 51.4°$, $\angle C = 25.8°$)

$\angle A = \angle B$, and $\angle B$ is four times greater than $\angle C$. (Let $x = \angle A$, $x + x + 4x = 180°$, $x = 30°$; $\angle A = \angle B = 30°$, $\angle C = 120°$)

Examples

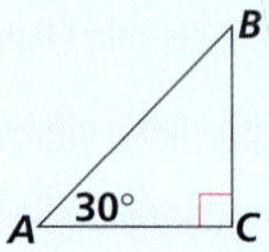

Given: ΔABC a right triangle, with $m\angle A = 30°$

Problem: find the measure of $\angle B$

Solution:	Reason:
$m\angle A + m\angle B + m\angle C = 180°$	angle sum theorem
$30° + m\angle B + 90° = 180°$	substitution
$120° + m\angle B = 180°$	simplification
$m\angle B = 60°$	subtraction of equals

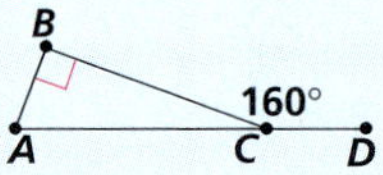

Given: ΔABC is a right triangle, with $m\angle BCD = 160°$

Problem: find the measure of $\angle A$

Solution:	Reason:
$m\angle ACB + 160° = 180°$	*ACD* forms a straight angle
$m\angle ACB = 20°$	subtraction of equals
$m\angle A + m\angle ACB + m\angle B = 180°$	angle sum theorem
$m\angle A + 20° + 90° = 180°$	subtraction of equals
$m\angle A = 70°$	subtraction of equals

Given: ΔABC is an isosceles triangle, with $m\angle ABC = 15°$

Problem: find the measures of the base angles *A* and *C*

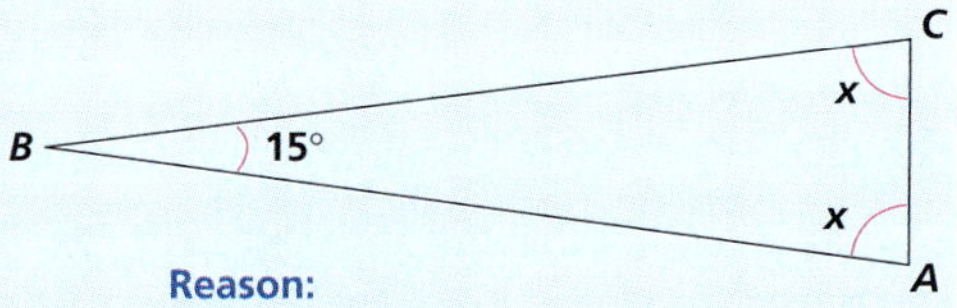

Solution:	Reason:
Let $x = m\angle A$ (or *C*)	$m\angle A = m\angle C$
$15° + x + x = 180°$	angle sum theorem
$2x = 165°$	simplification
$x = 82.5°$	division of equals

Conclusion: Each base angle, $\angle A$ and $\angle C$, $= 82.5°$.

Exercise A Find the measure of the angles.

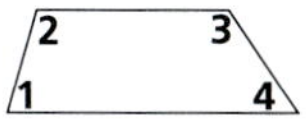

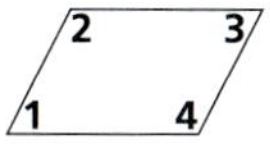

1. m∠1 + m∠2 180°

2. m∠3 + m∠4 180°

3. m∠1 + m∠2 + m∠3 + m∠4 360°

4. m∠1 + m∠2 180°

5. m∠3 + m∠4 180°

6. m∠1 + m∠2 + m∠3 + m∠4 360°

Exercise B Find the measure of ∠*x*.

7.

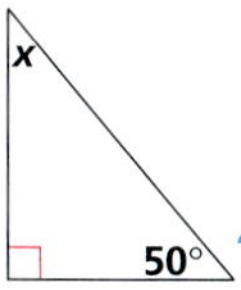

40°

8.

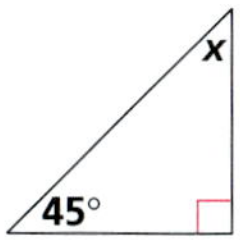

45°

9.

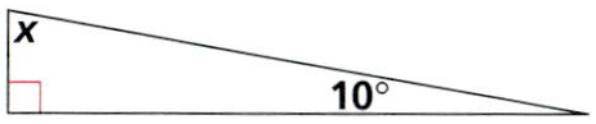

80°

10.

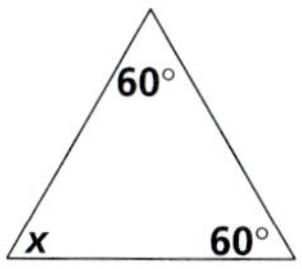

60°

11.

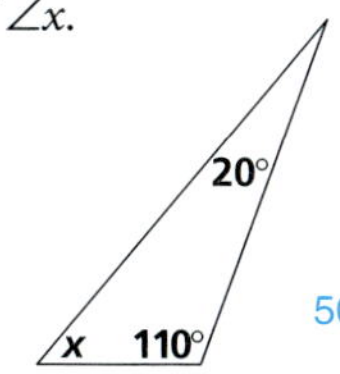

50°

12.

10°

13.

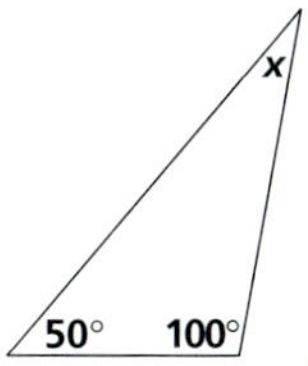

30°

14.

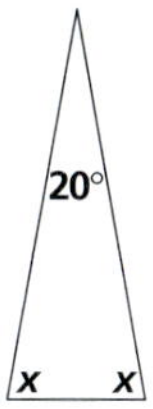

80°

15.

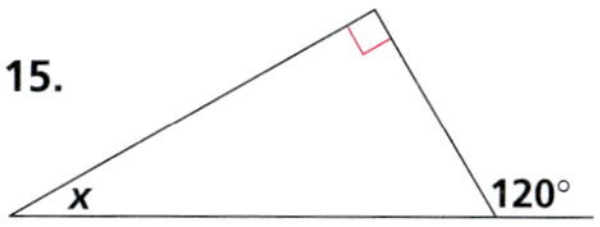

30°

16.

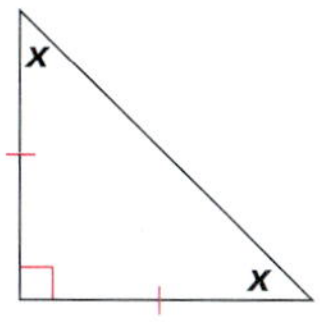

45°

17.

20°

Exercise C Tell whether or not the following triangles are possible. Give reasons for your answer.

18. A right, obtuse triangle

no, would give an angle sum > 180°

19. An isosceles, obtuse triangle

yes, one angle > 90° and two equal sides

20. An equilateral, right triangle

no, equilateral is also equiangular

LEARNING STYLES

Interpersonal/Group Learning Have students working in pairs practice writing and solving algebraic equations to find the measures of the angles in a triangle. Students should take turns writing the equations and have their partners solve the equation.

MENTAL MATH

Present the following equations for students. Ask them to find the solutions without using paper and pencil.

The sum of two angles is 135°. What is the measure of the third angle? **(45°)**

The sum of two angles is 110°. What is the measure of the third angle? **(70°)**

Two angles of a triangle are equal. Each is half the measure of the third angle. What are the measures of the angles? **(45°, 45°, 90°)**

One angle of a triangle is twice the smallest angle. The third angle is three times the smallest angle. What are the measures of the angles? **(30°, 60°, 90°)**

LEARNING STYLES

Logical/Mathematical Have students extend Exercise A by writing equations relating the measures of the angles they are trying to find. For example, in problem 3, some of the equations that relate angles 1, 2, 3, and 4 are m∠1 + m∠2 + m∠3 = 360° – m∠4; m∠1 + m∠2 = 360° – (m∠3 + m∠4); m∠1 = 360° – (m∠2 + m∠3 + m∠4). Manipulating equations in this manner will help students gain facility with equations.

Lesson at a Glance

Chapter 5 Lesson 6

Overview This lesson introduces concave, convex, and regular polygons.

Objectives

- To identify a geometric figure as a polygon.
- To distinguish between concave and convex polygons.
- To describe the characteristics of regular polygons.
- To classify or name polygons according to the number of sides.

Student Pages 154–159

Teacher's Resource Library TRL

Workbook Activities 53–54

Additional Practice 30

Mathematics Vocabulary

concave polygon
convex polygon
polygon
regular polygon

1. Warm-Up Activity

Present students with a selection of polygons from the Polygon Power Pack by placing them on an overhead projector. Ask students to indicate which polygons they might consider "regular." Ask the class to come up with an operational definition based on the classification.

You might also ask students to indicate which polygons they might categorize as "elbows out" and which they would consider as "elbows in." Point out that in this lesson students will learn two more ways to classify polygons based on characteristics in addition to the number of sides.

Lesson 6 Concave and Convex Polygons

Concave polygon
A polygon for which a segment joining any two points in the interior does not lie entirely within the figure

Convex polygon
A polygon for which the line segment joining any two points in the interior lies entirely within the figure

Polygon
A closed, many-sided geometric figure

Regular polygon
A polygon with sides of equal length and angles of equal measure

The word **polygon** means "many sided" and is used to describe geometric figures. A **regular polygon** has angles of equal measure and sides of equal length.

Most of the polygons you have studied had either three sides (triangle) or four sides (quadrilateral). In addition, most of the quadrilaterals you have studied have been **convex polygons.**

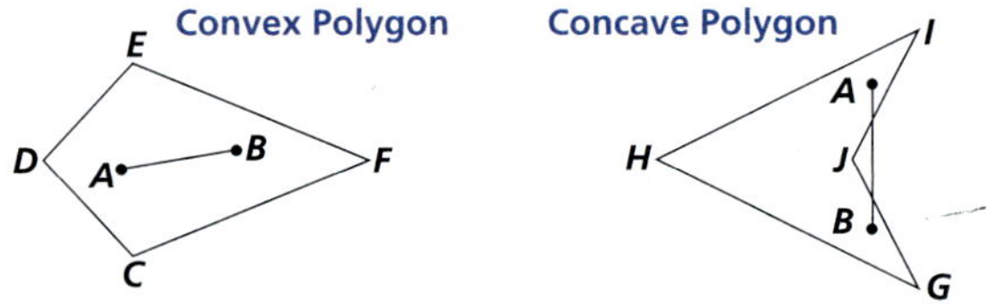

A plane figure is convex if, given any two points in the interior (inside), the segment joining the points lies wholly within the figure. The figure is nonconvex, or **concave,** if the segment does not lie wholly within the figure.

Another way to tell if a figure is concave or convex is to study the polygon's diagonals.

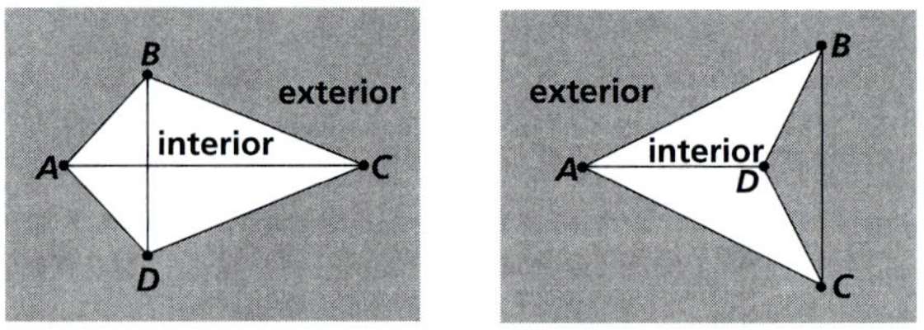

Polygons are named by the number of sides that form the figure.

Polygon	Number of Sides	Regular Polygon (= sides, = angles)
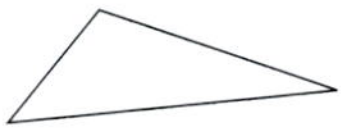 Triangle	3	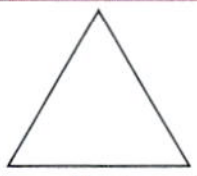 Equilateral Triangle

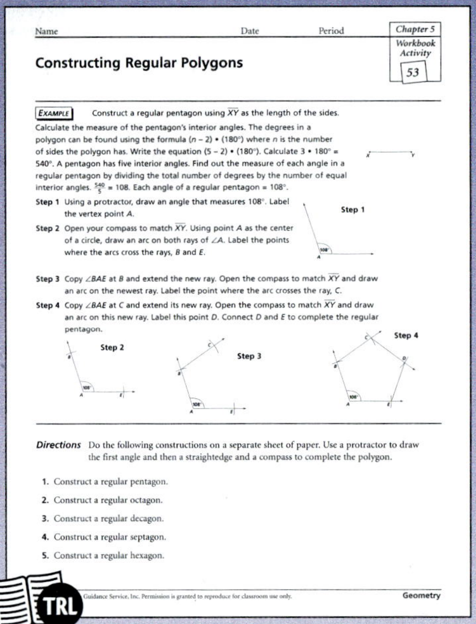

Name Date Period

Chapter 5 Workbook Activity 53

Constructing Regular Polygons

EXAMPLE Construct a regular pentagon using $\overline{XY}$ as the length of the sides.

Calculate the measure of the pentagon's interior angles. The degrees in a polygon can be found using the formula $(n - 2) \cdot (180°)$ where n is the number of sides the polygon has. Write the equation $(5 - 2) \cdot (180°)$. Calculate $3 \cdot 180° = 540°$. A pentagon has five interior angles. Find out the measure of each angle in a regular pentagon by dividing the total number of degrees by the number of equal interior angles. $\frac{540}{5} = 108$. Each angle of a regular pentagon = 108°.

Step 1 Using a protractor, draw an angle that measures 108°. Label the vertex point A.

Step 2 Open your compass to match $\overline{XY}$. Using point A as the center of a circle, draw an arc on both rays of $\angle A$. Label the points where the arcs cross the rays, B and E.

Step 3 Copy $\angle BAE$ at B and extend the new ray. Open the compass to match $\overline{XY}$ and draw an arc on the newest ray. Label the point where the arc crosses the ray, C.

Step 4 Copy $\angle BAE$ at C and extend its new ray. Open the compass to match $\overline{XY}$ and draw an arc on this new ray. Label this point D. Connect D and E to complete the regular pentagon.

Directions Do the following constructions on a separate sheet of paper. Use a protractor to draw the first angle and then a straightedge and a compass to complete the polygon.

1. Construct a regular pentagon.
2. Construct a regular octagon.
3. Construct a regular decagon.
4. Construct a regular septagon.
5. Construct a regular hexagon.

 Geometry

Workbook Activity 53

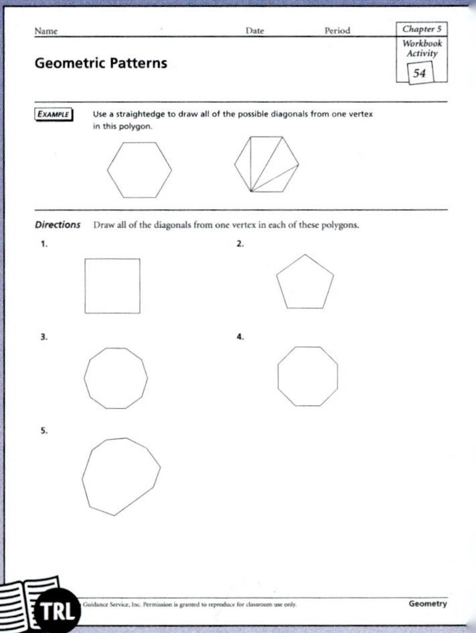

Name Date Period

Chapter 5 Workbook Activity 54

Geometric Patterns

EXAMPLE Use a straightedge to draw all of the possible diagonals from one vertex in this polygon.

Directions Draw all of the diagonals from one vertex in each of these polygons.

1.
2.
3.
4.
5.

 Geometry

Workbook Activity 54

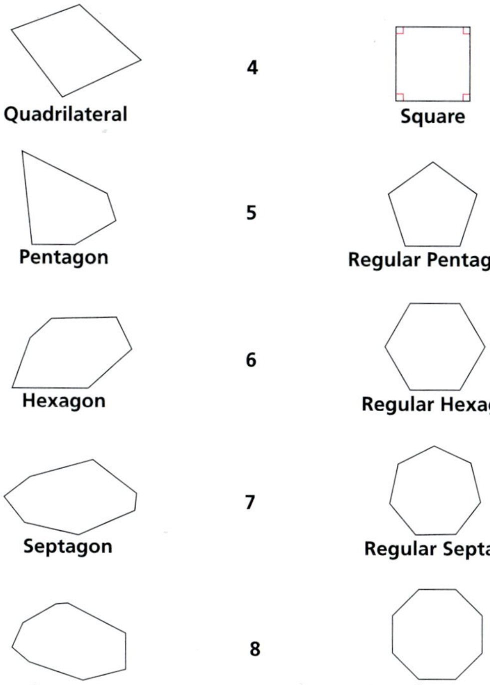

Polygon	Number of Sides	Regular Polygon (= sides, = angles)
Quadrilateral	4	Square
Pentagon	5	Regular Pentagon
Hexagon	6	Regular Hexagon
Septagon	7	Regular Septagon
Octagon	8	Regular Octagon

As the number of sides, n, of a polygon increases, the number of diagonals also increases. Let's look for a pattern in how the numbers increase.

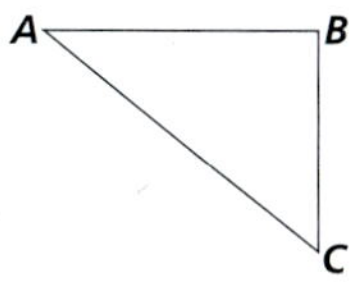

number of vertices = 3,

number of diagonals = 0

2. Teaching the Lesson

Read and discuss the chart. Have students identify the differences between the polygons and the regular polygons, paying attention to the number of sides and how they are arranged in each figure.

Learning Styles

LEP/ESL Pair students proficient in English with those having difficulty. Direct students' attention to the figures on page 155. Ask students to take turns describing each pair of illustrations to their partner. Descriptions should match each illustration as well as indicate how the illustrations are different.

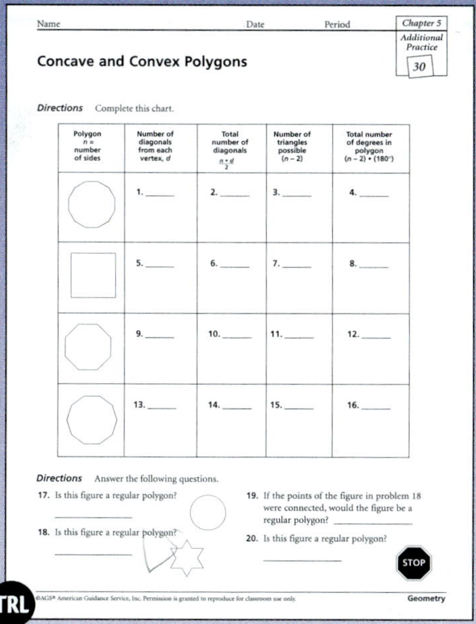

Name Date Period

Chapter 5 *Additional Practice* *30*

Concave and Convex Polygons

Directions Complete this chart.

Polygon n = number of sides	Number of diagonals from each vertex, d	Total number of diagonals $\frac{n \cdot d}{2}$	Number of triangles possible $(n - 2)$	Total number of degrees in polygon $(n - 2) \cdot (180°)$
	1. ______	2. ______	3. ______	4. ______
	5. ______	6. ______	7. ______	8. ______
	9. ______	10. ______	11. ______	12. ______
	13. ______	14. ______	15. ______	16. ______

Directions Answer the following questions.

17. Is this figure a regular polygon? ______________

18. Is this figure a regular polygon? ______________

19. If the points of the figure in problem 18 were connected, would the figure be a regular polygon? ______________

20. Is this figure a regular polygon? ______________

STOP

TRL

 Geometry

Additional Practice 30

Discuss the example at the bottom of page 156 with students. To determine if they grasp the relationship between the number of vertices and the number of diagonals, have them predict how many diagonals can be drawn in a 5-, 6-, 7-, 8-, and 10-sided polygon. Students can check their predictions using the formula on page 156.

Ask students to describe the difference, if any, in the number of diagonals that can be drawn in regular and nonregular polygons with the same number of sides. **(no difference)**

3. Reinforce and Extend

Learning Styles

Logical/Mathematical Ask students if the following statement is (1) always true; (2) always false; (3) sometimes true, sometimes false. Students must defend their position. "Every rectangle has all angles equal and every rhombus has all sides equal; therefore, a figure that is both a rectangle and a rhombus is a regular polygon." **(always true)**

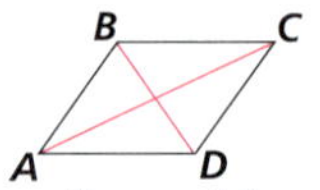

number of vertices = 4,

number of diagonals = 2

There is one diagonal from each vertex. But each diagonal can be drawn twice:

A to $C = C$ to A and B to $D = D$ to B

We can write this information as a formula:

Formula:

$$\frac{\text{(\# vertices) times (\# diagonals from each vertex)}}{2} = \text{total number of diagonals}$$

For the above quadrilateral this would be

$$\frac{4(1)}{2} = 2 \text{ diagonals}$$

Does the formula work for a pentagon?

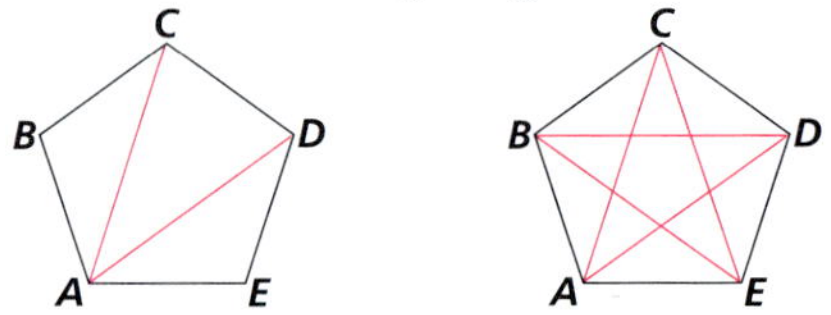

$$\frac{\text{(\# vertices) times (\# diagonals from each)}}{2} = \text{number of diagonals}$$

$$\frac{5(2)}{2} = 5$$

There is a pattern for finding the number of diagonals in a polygon.

Writing About Mathematics

What is the sum of the interior angle measures for a 4-sided convex polygon? Explain.

Example

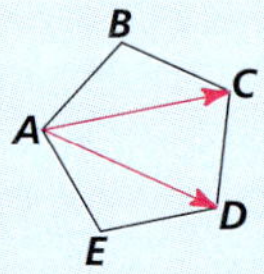

Step 1 Determine the number of diagonals that can be drawn from a single vertex.

Step 2 Multiply the number of diagonals for each vertex by *n*, the number of sides of the polygon.

Step 3 Divide the product by 2 to avoid counting any diagonal twice.

Remember, you can check your answer by drawing the polygon and all its diagonals.

As you draw all the possible diagonals from one vertex, you are dividing the polygon into triangles. If there is no overlap of areas or angles, you will have divided the polygon into triangles.

Every n-sided polygon can be divided into $(n - 2)$ triangles.

Quadrilateral	Pentagon	Hexagon

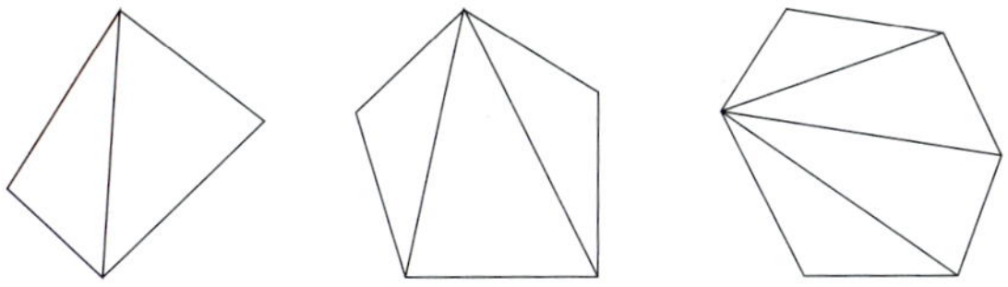

2 triangles	3 triangles	4 triangles

There is a pattern here as well.

number of sides, n, for the polygon	number of triangles formed by diagonals from one vertex	difference between number of sides and number of triangles
3	**1**	$3 - 1 = 2$
4	2	$4 - 2 = 2$
5	3	$5 - 3 = 2$
6	4	$6 - 4 = 2$
....		
n	$(n - 2)$	2

Remember that the sum of the angles in any triangle is 180°.

Therefore, we can say that (the number of triangles) • (180°) = sum of the angles in the polygon.

The sum of the angles in a polygon, with number of sides = n, is $(n - 2)(180°)$.

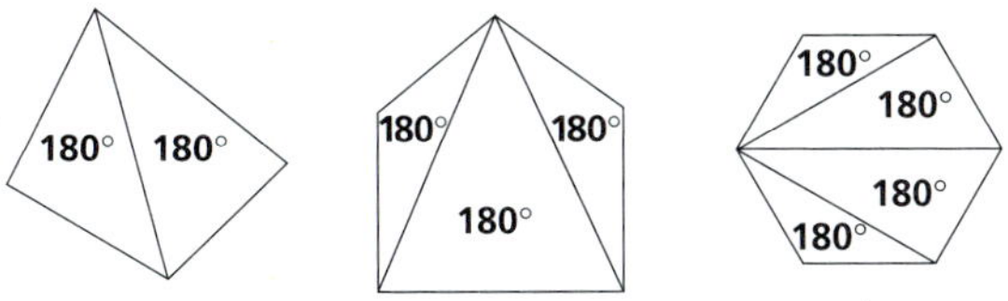

Learning Styles

Auditory/Verbal Direct students to apply the formula in the note at the top of page 157 to confirm the number of triangles in the quadrilateral, pentagon, and hexagon shown in the illustrations.

Application

At Home

Encourage interested students to use library and electronic reference sources to determine what relationship, if any, exists between the names of some polygons and names assigned to sports (triathlon, decathlon, etc.) or hydrocarbon compounds (pentane, heptane, octane, etc). Students can present their findings in a two-minute news brief entitled "Know Your Number Prefixes."

Learning Styles

Interpersonal/Group Learning Direct students to form groups of four. Have the group work together to complete Exercise A. Students should take turns explaining each of the entries in the chart to the group. Encourage group members to evaluate each explanation and allow for revision if needed.

Learning Styles

Visual/Spatial Encourage students to work in pairs to produce illustrations for Exercise A showing (a) the number of diagonals from each vertex, (b) the total number of diagonals, and (c) the number of triangles possible for each polygon. Pairs can work together to make an illustrated chart. Display students' charts in the classroom and allow students to choose the three charts they feel would be most helpful as a review tool.

Because regular polygons have angles of equal measure, you can use this information to determine the measures of the angles in a regular polygon.

Examples

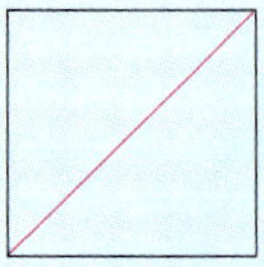

Square, $n = 4$
$(n - 2)(180°)$ = number of degrees in polygon
$2 \bullet 180° = 360°$
4 equal angles means each angle equals $360° \div 4 = 90°$
Each angle in a square measures 90°.

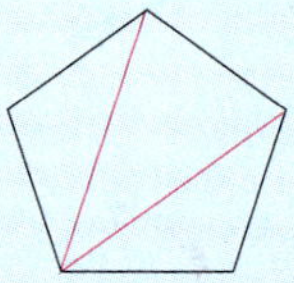

Pentagon, $n = 5$
$(n - 2)(180°)$ = number of degrees in polygon
$3 \bullet 180° = 540°$
5 equal angles means each angle equals $540° \div 5 = 108°$
Each angle in a pentagon measures 108°.

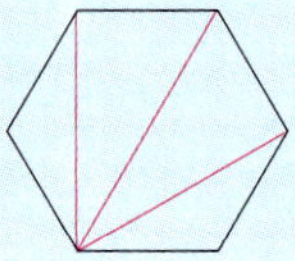

Hexagon, $n = 6$
$(n - 2)(180°)$ = number of degrees in polygon
$4 \bullet 180° = 720°$
6 equal angles means each angle equals $720° \div 6 = 120°$
Each angle in a hexagon measures 120°.

Exercise A Copy and complete this chart on a separate sheet of paper.

Polygon, n = number of sides	Number of diagonals from each vertex, d	Total number of diagonals $\frac{n \bullet d}{2}$	Number of triangles possible $(n - 2)$	Total number of degrees in polygon $(n - 2) \bullet (180°)$
 $n = 8$	1. 5	2. 20	6	3. 1,080°
 $n = 9$	6	4. 27	5. 7	6. 1,260°

Polygon, n = number of sides	Number of diagonals from each vertex, d	Total number of diagonals $\frac{n \cdot d}{2}$	Number of triangles possible $(n - 2)$	Total number of degrees in polygon $(n - 2) \cdot (180°)$
 $n = 10$	7. 7	8. 35	9. 8	10. 1,440°
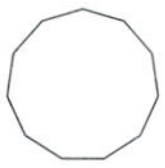 $n = 11$	8	11. 44	12. 9	13. 1,620°
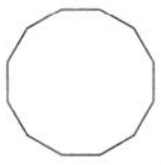 $n = 12$	14. 9	15. 54	16. 10	1,800°

Problem Solving

Exercise B Answer the following questions.

17. What polygon provides the outside shape for the snowflake shown? Is it regular? How many diagonals are shown? Explain your answer.

It's a regular hexagon, all the angles are equal; only 3 diagonals are shown

18. Is the figure below a regular dodecagon (12-sided polygon)? Give reasons for your answer.

no, it is concave (also not all the angles are equal)

19. The object pictured below is called a *hex nut.* What is the sum of the interior angles for this nut?

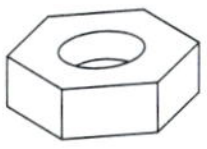

720°

20. This umbrella has eight equal sides, and eight equal angles. What is the measure of angle *a*? Explain.

$\angle 1 + \angle 2 = 135°$, but the triangles are isosceles, so $\angle 2 + \angle 3 = 135°$; $\angle a + \angle 2 + \angle 3 = 180°$, so $\angle a = 45°$

Learning Styles

Logical/Mathematical Ask students to consider an additional column for the chart in Exercise A. The new column heading would be "Number of degrees in each angle." Ask students to develop an explanation and, if possible, a formula to determine entries in this column.

Group Problem Solving

Ask students to find a way to use a circle to test whether a polygon is a square or a rectangle. Students can display their findings as part of a Geometric Side Show. Encourage them to try their test on several other polygons—some regular and some not regular. Suggest students use existing manipulatives, or manipulatives of their own making, to prepare models for the side show. Direct students to share the responsibilities of the show, some gathering materials, some preparing the models, and others preparing the script for the side show barker. (A circle drawn around the outside of a square will touch each vertex; a circle drawn inside the square will touch each side. No such circle can be drawn for a rectangle.)

Triangles and Quadrilaterals Chapter 5 **159**

Lesson at a Glance

Chapter 5 Lesson 7

Overview This lesson demonstrates two new constructions: constructing a perpendicular to ℓ from a point not on ℓ and constructing a perpendicular to ℓ at point P on ℓ.

Objectives

- To construct a perpendicular to ℓ from a point not on ℓ.
- To construct a perpendicular to ℓ at point P on ℓ.
- To construct a perpendicular bisector to ℓ.

Student Pages 160–162

Teacher's Resource Library TRL

Workbook Activity 55–56

Additional Practice 31

Mathematics Vocabulary

perpendicular bisector

1. Warm-Up Activity

Review with students how to use a compass and a straightedge to find the midpoint of a line. Ask students to identify which of Euclid's postulates can be used to justify the construction.
(Post 1)

Lesson 7 Constructing Perpendiculars

> ***Perpendicular bisector***
> *Line that bisects a given line segment and forms a right angle with it*

Remember that in an earlier chapter you learned how to use a compass and a straightedge to find the midpoint of a line.

To find the midpoint of $\overline{AB}$, draw an arc above and below $\overline{AB}$. Keeping your compass opening fixed, use A as a center and then B as a center. Connect the points where the arcs intersect by drawing line ℓ. The point, M, located where ℓ crosses $\overline{AB}$, is the midpoint of $\overline{AB}$.

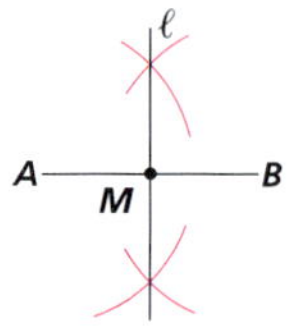

Line ℓ is also perpendicular to $\overline{AB}$ so we can say that ℓ is the **perpendicular bisector** of $\overline{AB}$. We will prove this in a later lesson.

Now we will learn two new constructions.

CONSTRUCTION A

Given: point P, not on ℓ

Construct: a line perpendicular to ℓ from point P, not on ℓ

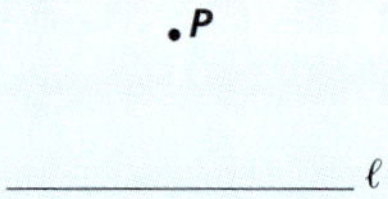

Step 1 Open your compass to a distance somewhat greater than the distance from P to ℓ. Using P as the center of a circle, draw an arc that intersects ℓ in two places. Label the points A and B.

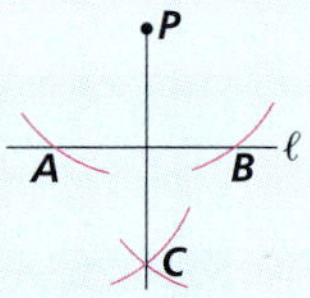

Step 2 Keeping your compass opening constant, and using A as the center of a circle, draw an arc below ℓ. Then, with the same compass opening, draw a second arc below ℓ using B as the center. Label the point where the arcs intersect C.

Step 3 Connect P and C. $\overline{PC} \perp \ell$

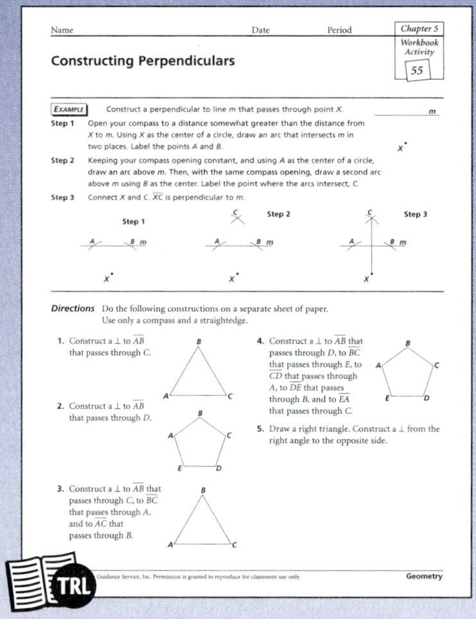

Name Date Period | Chapter 5 Workbook Activity 55

Constructing Perpendiculars

EXAMPLE Construct a perpendicular to line m that passes through point X.

Step 1 Open your compass to a distance somewhat greater than the distance from X to m. Using X as the center of a circle, draw an arc that intersects m in two places. Label the points A and B.

Step 2 Keeping your compass opening constant, and using A as the center of a circle, draw an arc above m. Then, with the same compass opening, draw a second arc above m using B as the center. Label the point where the arcs intersect, C.

Step 3 Connect X and C. $\overline{XC}$ is perpendicular to m.

Step 1 Step 2 Step 3

Directions Do the following constructions on a separate sheet of paper. Use only a compass and a straightedge.

1. Construct a $\perp$ to $\overline{AB}$ that passes through C.
2. Construct a $\perp$ to $\overline{AB}$ that passes through D.
3. Construct a $\perp$ to $\overline{AB}$ that passes through C, to $\overline{BC}$ that passes through A, and to $\overline{AC}$ that passes through B.
4. Construct a $\perp$ to $\overline{AB}$ that passes through D, to $\overline{BC}$ that passes through E, to $\overline{CD}$ that passes through A, to $\overline{DE}$ that passes through B, and to $\overline{EA}$ that passes through C.
5. Draw a right triangle. Construct a $\perp$ from the right angle to the opposite side.

Guidance Service, Inc. Permission is granted to reproduce for classroom use only. Geometry

Workbook Activity 55

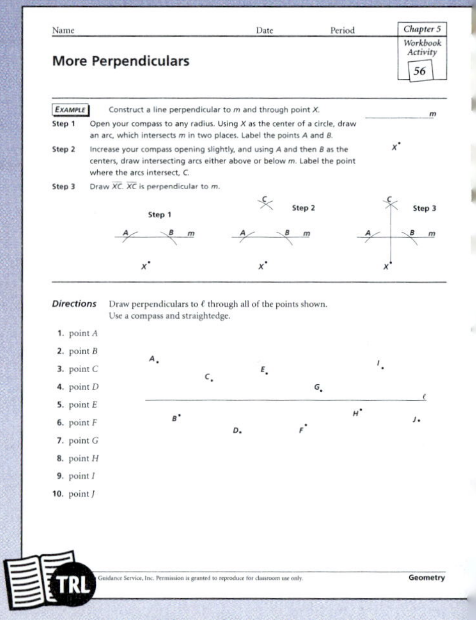

Name Date Period | Chapter 5 Workbook Activity 56

More Perpendiculars

EXAMPLE Construct a line perpendicular to m and through point X.

Step 1 Open your compass to any radius. Using X as the center of a circle, draw an arc, which intersects m in two places. Label the points A and B.

Step 2 Increase your compass opening slightly, and using A and then B as the centers, draw intersecting arcs either above or below m. Label the point where the arcs intersect, C.

Step 3 Draw $\overline{XC}$. $\overline{XC}$ is perpendicular to m.

Step 1 Step 2 Step 3

Directions Draw perpendiculars to ℓ through all of the points shown. Use a compass and straightedge.

1. point A
2. point B
3. point C
4. point D
5. point E
6. point F
7. point G
8. point H
9. point I
10. point J

Guidance Service, Inc. Permission is granted to reproduce for classroom use only. Geometry

Workbook Activity 56

CONSTRUCTION B

Given: point P, on ℓ

Construct: a line perpendicular to ℓ and at point P

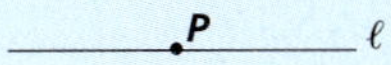

Step 1 Open your compass to any radius. Using P as the center of a circle, draw an arc that intersects ℓ in two places. Label the points A and B.

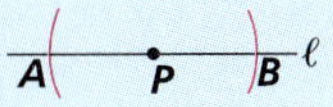

Step 2 Increase your compass opening, and using A and then B as the centers, draw intersecting arcs either above or below ℓ. Label the point where the arcs intersect C.

Step 3 Draw $\overline{PC}$. $\overline{PC} \perp \ell$ at P. P is called the *foot* of the perpendicular.

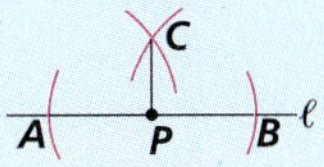

You can use these constructions in the following ways.

EXAMPLES

Given: $\triangle ABC$

Construct: a line $\perp$ from B to $\overline{AC}$

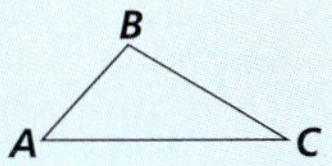

B is a point not on $\overleftrightarrow{AC}$.
Use the construction of a perpendicular from a point not on the line. (Construction A)

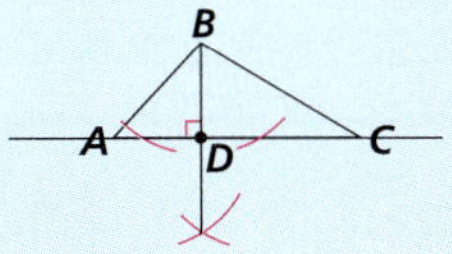

You may have to extend $\overline{AC}$ to $\overleftrightarrow{AC}$.
Note that $\overleftrightarrow{BD} \perp \overleftrightarrow{AC}$ and $\overline{BD} \perp \overline{AC}$.

Given: $\triangle DEF$

Construct: a line $\perp$ to $\overline{DF}$ at point D

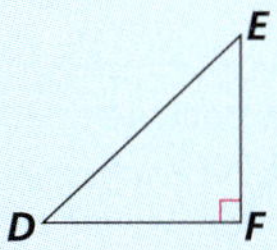

D is a point on $\overleftrightarrow{DF}$.
Use the construction of a perpendicular to a line through a point on the line. (Construction B)

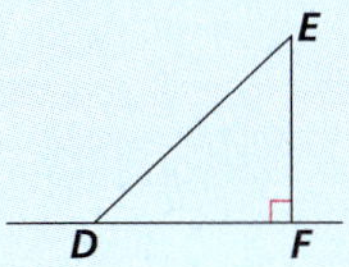

You may need to extend $\overline{DF}$.
Note that $\overleftrightarrow{GD} \perp \overleftrightarrow{DF}$ and $\overline{GD} \perp \overline{DF}$.

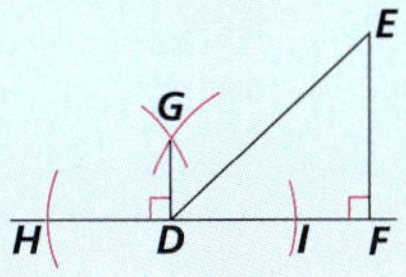

2. Teaching the Lesson

Demonstrate Construction A for students either on the board or an overhead projector. Draw students' attention to the opening in the drawing compass and the placement of the compass needle for Steps 1 and 2. Be sure to indicate that to be able to draw the line in Step 3, the arcs drawn in Step 2 must meet. You may want to duplicate the procedure drawing the arcs from point B in such a way that they will not intersect the arcs drawn from point A. Show students how to remedy this problem by repositioning the compass needle at point B and drawing larger arcs.

Demonstrate Construction B in a similar manner.

3. Reinforce and Extend

LEARNING STYLES

Auditory/Verbal Encourage pairs of students to work together to demonstrate the examples on page 161 for themselves. Indicate that one student in each pair can read the directions while his or her partner executes the directions. Circulate as students work through the examples, making certain students have correctly applied the construction techniques from Constructions A and B.

Name Date Period | *Chapter 5* *Additional Practice* *31*

Constructing Perpendiculars

Directions Complete the following constructions. Use a straightedge and a compass.

1. Draw a line m and a point P not on m. Construct a perpendicular to m through P. Label this perpendicular n. Then construct a line that is perpendicular to n and that also passes through P. Label this perpendicular o.

2. In problem 1, what is the relationship between lines m and o? State your reasons.

____

____

____

3. Construct a scalene triangle as shown. Draw a perpendicular to $\overline{BC}$ that passes through point A.

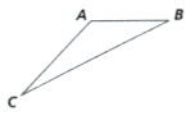

4. Construct a right isosceles triangle as shown. Draw a perpendicular to $\overline{AB}$ that passes through point C.

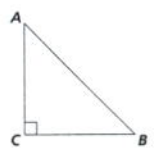

5. Draw a square with diagonal AC as shown. Draw a perpendicular to $\overline{AC}$ that passes through both point B and point D.

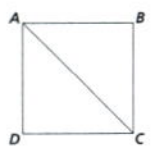

 Geometry

RL

Additional Practice 31

Learning Styles

LEP/ESL Pair students with advanced language skills with students less advanced. Suggest they complete the constructions in Exercise A together. Encourage each pair to read the directions, allowing them to sketch what they believe will be the final product. Direct students to discuss whether they will be using Construction A or Construction B to complete the item. Then have one student execute the construction while the other provides a narration for the activity.

Try This...

Students should produce a variety of triangles, each with perpendicular bisectors. They should note that all perpendicular bisectors are perpendicular to the sides of the triangle and pass through the midpoints of the respective sides. As an additional observation, students who construct the perpendicular bisectors carefully may observe that the bisectors of all three sides meet at a single point.

Construction

Exercise A Complete the following constructions on a separate sheet of paper. Use only a straightedge and a compass.

1. Draw a line ℓ and a point P, not on ℓ. Construct a $\perp$ to ℓ through P.

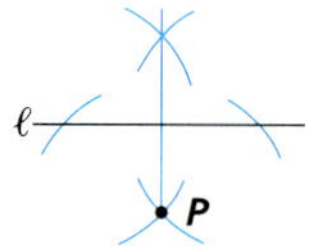

2. Draw a line m and point Q, on m. Construct a $\perp$ to m at Q.

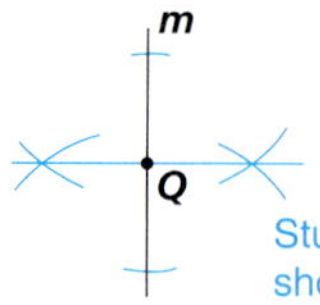

Student constructions should reflect correct use of construction steps for constructing a line perpendicular to a given line either from a point not on the line or through a point on the line.

3. Construct any right triangle. Label the right angle R and the other two angles S and T. Construct a perpendicular from the right angle vertex, R, to the side opposite the right angle, $\overline{ST}$.

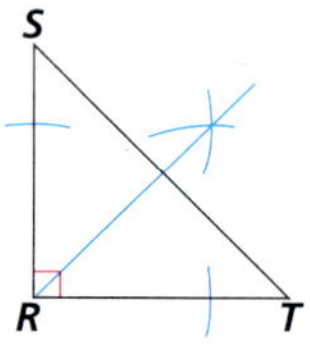

4. Draw a trapezoid $TRAP$ as shown. Construct a line $\perp$ to $\overline{TP}$ at point A.

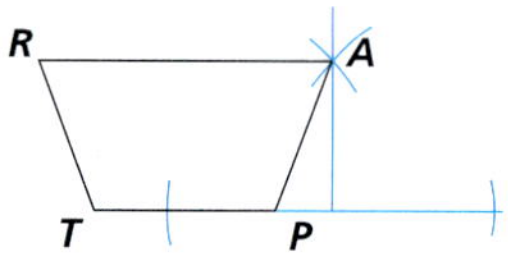

5. In $TRAP$ at right, $\overline{XT} \perp \overline{TP}$ by construction. Is $\overline{XT} \perp \overline{RA}$? Explain your answer. Yes, because $\overline{TP} \parallel \overline{RA}$, alternate interior angles are equal. $\therefore$ $RXT = XTP = 90°$ and $\overline{XT}$ is perpendicular to both $\overline{RA}$ and $\overline{TP}$.

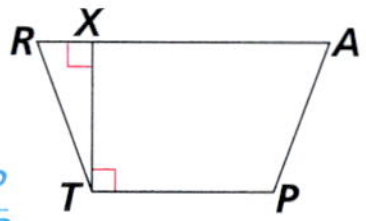

Try This...

You can find triangles all around you—for example, a yield sign, a roof line, and a recycle symbol. Choose one and copy it on paper. Name the triangle as an acute, a right, or an obtuse triangle. Then draw a perpendicular bisector through it. Compare your triangles with those of your classmates. How are all perpendicular bisectors alike?

162 *Chapter 5 Triangles and Quadrilaterals*

Lesson 8 Altitudes, Angle Bisectors, and Medians

Because triangles have three vertices, or angles, they have three **altitudes** and three angle bisectors.

A line from a vertex of a triangle and perpendicular to the opposite side is the altitude of the triangle.

Acute Case

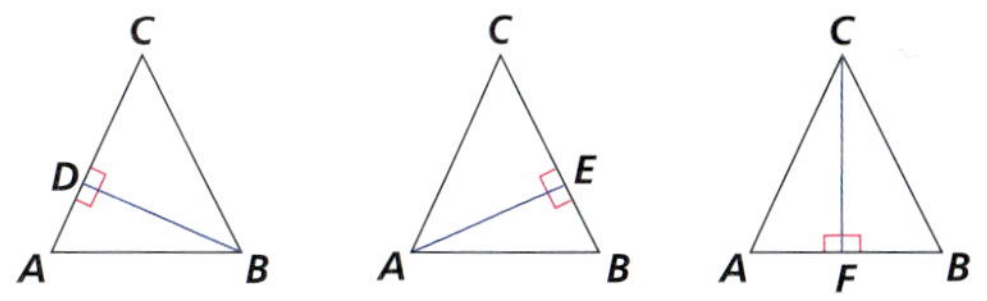

In an acute triangle, all the altitudes are internal. That is, they lie wholly within the triangle. The altitudes meet at one point called the **orthocenter.**

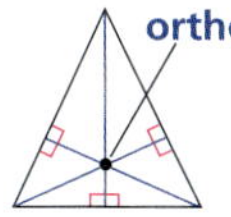

To construct an altitude, you can use the construction for constructing a perpendicular to a line from a point (vertex) not on the line. (See Construction A, Lesson 7)

Obtuse Case

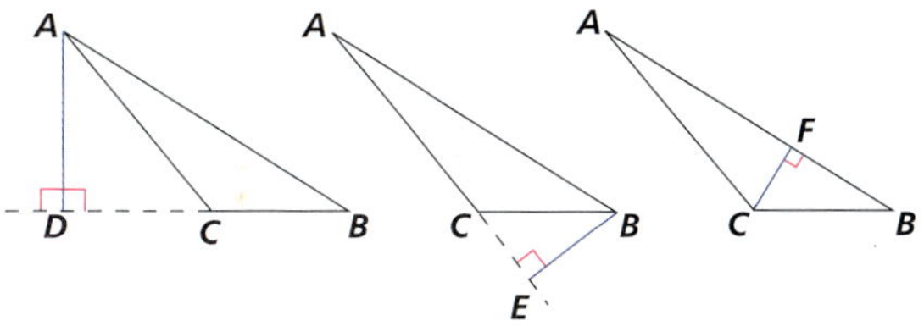

In the obtuse triangle ABC above, altitudes $\overline{AD}$ and $\overline{BE}$ are external while altitude $\overline{CF}$ is internal. Obtuse triangles have two external altitudes and one internal altitude.

Triangles have three altitudes and they have three angle bisectors. This is a good time to review how to bisect an angle.

Altitude
Line from the vertex of a triangle and perpendicular to the opposite side

Centroid
Point where the three medians of a triangle intersect

In-center
The point where the angle bisectors of the angles of a triangle intersect

Median
Line from a vertex of a triangle to the midpoint of the opposite side

Orthocenter
The point where the altitudes of a triangle meet

Review acute and obtuse triangles on page 139.

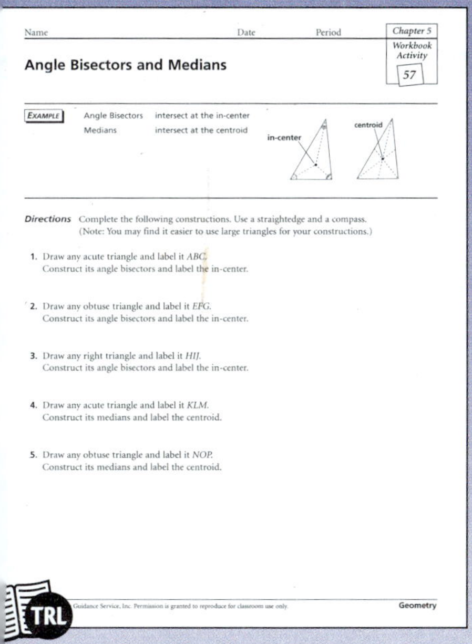

Name Date Period

Chapter 5 Workbook Activity 57

Angle Bisectors and Medians

EXAMPLE Angle Bisectors intersect at the in-center
Medians intersect at the centroid

in-center centroid

Directions Complete the following constructions. Use a straightedge and a compass. (Note: You may find it easier to use large triangles for your constructions.)

1. Draw any acute triangle and label it *ABC*. Construct its angle bisectors and label the in-center.
2. Draw any obtuse triangle and label it *EFG*. Construct its angle bisectors and label the in-center.
3. Draw any right triangle and label it *HIJ*. Construct its angle bisectors and label the in-center.
4. Draw any acute triangle and label it *KLM*. Construct its medians and label the centroid.
5. Draw any obtuse triangle and label it *NOP*. Construct its medians and label the centroid.

 Geometry

TRL

Workbook Activity 57

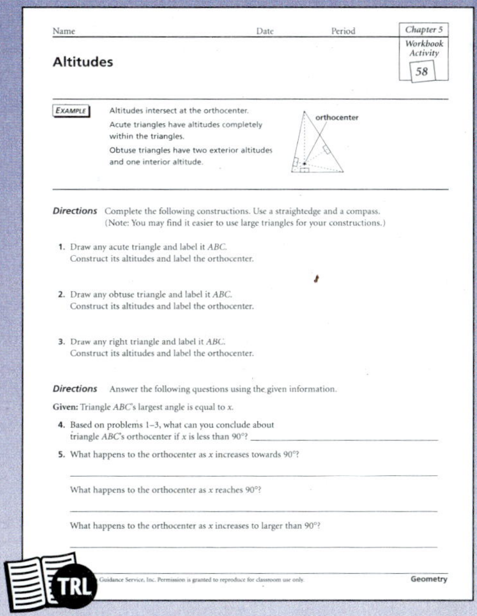

Name Date Period

Chapter 5 Workbook Activity 58

Altitudes

EXAMPLE Altitudes intersect at the orthocenter.
Acute triangles have altitudes completely within the triangles.
Obtuse triangles have two exterior altitudes and one interior altitude.

orthocenter

Directions Complete the following constructions. Use a straightedge and a compass. (Note: You may find it easier to use large triangles for your constructions.)

1. Draw any acute triangle and label it *ABC*. Construct its altitudes and label the orthocenter.
2. Draw any obtuse triangle and label it *ABC*. Construct its altitudes and label the orthocenter.
3. Draw any right triangle and label it *ABC*. Construct its altitudes and label the orthocenter.

Directions Answer the following questions using the given information.

Given: Triangle *ABC*'s largest angle is equal to x.

4. Based on problems 1–3, what can you conclude about triangle *ABC*'s orthocenter if x is less than 90°? ______
5. What happens to the orthocenter as x increases towards 90°? ______

   What happens to the orthocenter as x reaches 90°? ______

   What happens to the orthocenter as x increases to larger than 90°? ______

 Geometry

TRL

Workbook Activity 58

Lesson at a Glance

Chapter 5 Lesson 8

Overview This lesson presents several characteristics and properties of triangles including altitudes, medians, angle bisectors, in-centers, and orthocenters.

Objectives

- To locate and identify the point where a triangle's altitudes meet as its orthocenter.
- To locate and identify the point at which the angle bisectors of a triangle meet as its in-center.
- To locate and identify the point at which the medians of a triangle meet as its centroid.

Student Pages 163–166

Teacher's Resource Library TRL

Workbook Activities 57–58

Additional Practice 32

Mathematics Vocabulary

altitude
centroid
in-center
median
orthocenter

1. Warm-Up Activity

Review with students the differences between acute, right, and obtuse triangles. You might also wish to review how to construct a perpendicular from a point to a line, how to bisect an angle, and how to find the midpoint of a line. Indicate to students that they will be using these constructions to draw a triangle's altitudes, angle bisectors, and medians, which they can use to find the triangle's orthocenter, in-center, and centroid, respectively.

Triangles and Quadrilaterals Chapter 5 **163**

2. Teaching the Lesson

Demonstrate the constructions in the lesson for students using either the board or an overhead projector. You may want to produce a series of constructions—showing altitude, angle bisector, and median construction for acute triangles.

Arrange for a pair of volunteers to demonstrate the same constructions—altitude, angle bisector, and median—for an obtuse triangle. One student can execute the construction while the second student explains the steps involved. You may wish to repeat this for right triangles. For each series of constructions, have students identify the in-center, orthocenter, and centroid.

To bisect an angle, use A as a center, and draw an arc that intersects both rays of the angle. Mark as points B and C. Then, with the compass opening the same, use B and then C as centers and draw intersecting arcs, labeling the intersection D. Finally, draw $\overline{AD}$, the bisector of $\angle A$.

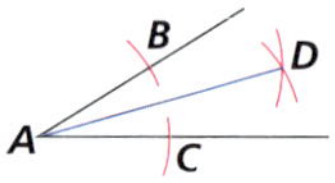

The angle bisectors of the angles of a triangle intersect at a point called the **in-center** of the triangle.

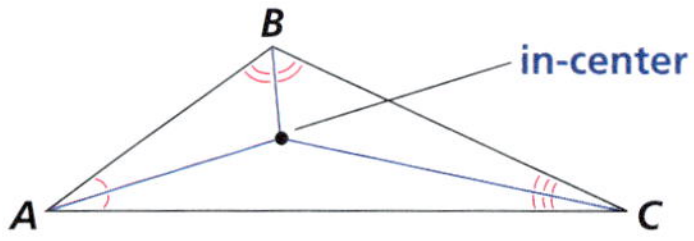

Because triangles have three sides, they have three **medians.** A line from a vertex of a triangle to the midpoint of the opposite side is a median of the triangle.

To construct the median from vertex A, find the midpoint, M_1, of $\overline{BC}$. To find the midpoint, construct the $\perp$ bisector of $\overline{BC}$. Draw $\overline{AM_1}$, a median of ΔABC.

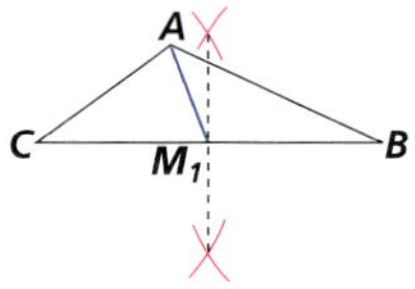

Repeat the steps to draw the other two medians, $\overline{BM_2}$ and $\overline{CM_3}$.

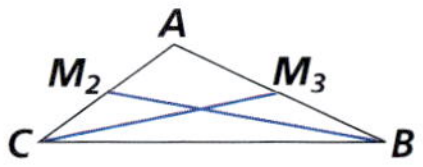

The point where the three medians of a triangle intersect is the **centroid.**

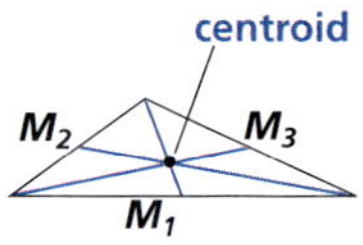

Altitudes	intersect at the orthocenter
Angle bisectors	intersect at the in-center
Medians	intersect at the centroid

Acute triangles have altitudes completely within the triangles.

Obtuse triangles have two exterior altitudes and one interior altitude.

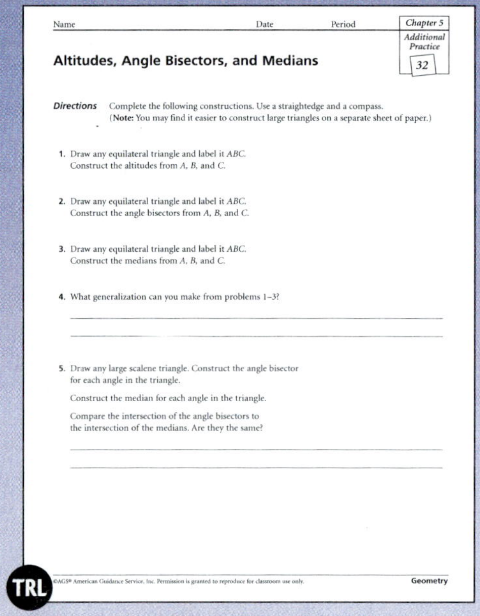
Name Date Period

Chapter 5 *Additional Practice* 32

Altitudes, Angle Bisectors, and Medians

Directions Complete the following constructions. Use a straightedge and a compass. (**Note:** You may find it easier to construct large triangles on a separate sheet of paper.)

1. Draw any equilateral triangle and label it ABC. Construct the altitudes from A, B, and C.
2. Draw any equilateral triangle and label it ABC. Construct the angle bisectors from A, B, and C.
3. Draw any equilateral triangle and label it ABC. Construct the medians from A, B, and C.
4. What generalization can you make from problems 1–3?
5. Draw any large scalene triangle. Construct the angle bisector for each angle in the triangle.

   Construct the median for each angle in the triangle.

   Compare the intersection of the angle bisectors to the intersection of the medians. Are they the same?

TRL

Geometry

Additional Practice 32

Construction

Exercise A Complete the following constructions on a separate sheet of paper. Use only a straightedge and a compass. (Note: You may find it easier to use large triangles for your constructions.)

1. Draw any acute triangle and label it ABC. Construct altitudes from A and C.

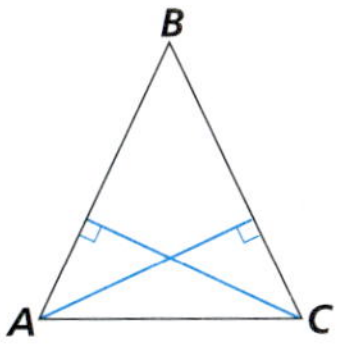

2. Draw any obtuse triangle and label the obtuse angle D and the triangle DEF. Construct the altitude from E to $\overline{DF}$.

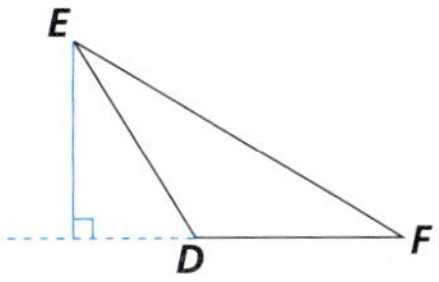

3. Draw any right triangle and label the right angle G and the triangle GHI. Construct the altitude from G to $\overline{HI}$.

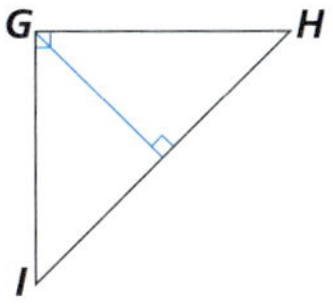

4. Draw an obtuse triangle like ΔJKL. Label the obtuse angle J. Construct all the medians of the triangle.

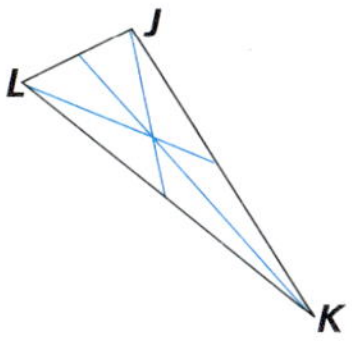

5. Draw any large scalene triangle. Construct the angle bisectors for each angle in the triangle.

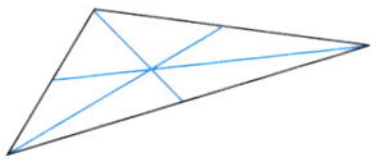

6. Draw any isosceles triangle ABC with $\overline{AB} = \overline{BC}$. (You may need to review how to construct an isosceles triangle on page 136.) Construct the bisector of the vertex angle B.

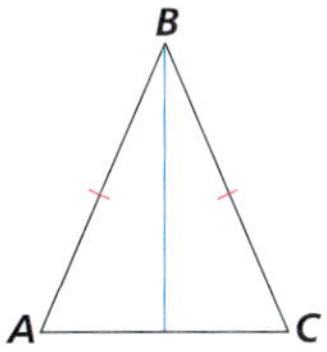

3. Reinforce and Extend

Learning Styles

LEP/ESL Many students may benefit from making their own flashcards for medians, altitudes, perpendicular bisectors, and angle bisectors. One side of the card should be an illustration with labels such as *medians* and *centroid.* The other side of the card should have the definitions of any terms used on the front. In the example, the back of the card would contain the definitions for *median* and *centroid.*

Have students collect all flashcards and place them, illustrations up, in a single stack. Use the cards for a geometry term contest in which each student contestant draws a card and describes the illustration including the definitions of any labels. Students who correctly identify the illustrations and definitions on any card they choose are winners.

Learning Styles

Auditory/Verbal Using the questions in Exercise B as a guide, have pairs of students work together to write as many generalizations about altitudes, medians, and angle bisectors, and perpendicular bisectors as they can. For example, "The medians and the angle bisectors of an equilateral triangle are the same" and "The median from the vertex of an isosceles triangle to the base of the triangle is the same as the angle bisector of the vertex angle." Encourage students to group their generalizations according to what type of triangle—acute, right, obtuse, isosceles, or equilateral—is involved.

Learning Styles

Interpersonal/Group Learning Direct students to form groups of four. Give each group a line segment length $\overline{AB}$. Challenge students to use only the line length, a drawing compass, and a straightedge to draw a triangle with angles of 120°, 15°, and 45°. Students should work together to develop a plan, writing out sequential steps as appropriate.

Once the group has reached a consensus on how to solve the problem, group members can execute the plan with each student performing a different step. Have groups present their findings to the class. Allow time for a class discussion, including alternate plans for accomplishing the task. (Sample plan: Construct equilateral triangle using $\overline{AB}$. Bisect one angle to get an angle of 30°, bisect a second time to get an angle of 15°. Construct second triangle, copying 15° angle for one vertex, copying angle of 15° + 30° for second angle, and completing the triangle. The third angle will, of necessity, be 120°.)

7. Using an isosceles triangle like the one in Problem 6, draw the median to the base, $\overline{AC}$, of the triangle. The median is the same as $\angle B$ bisector.

8. Construct a large equilateral triangle. (You may need to review the directions on page 135.) Draw the perpendicular bisector from each vertex to the opposite side.

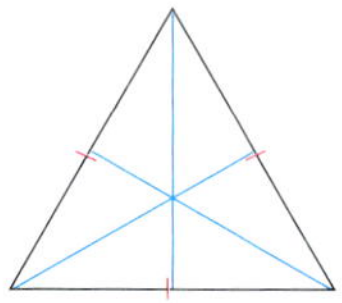

9. Using an equilateral triangle like the one in Problem 8, draw the angle bisector for each angle of the triangle.

10. Using an equilateral triangle like the one in Problem 8, draw the median to each side. The medians are the same as the angle bisectors.

Exercise B Answer the following questions about your constructions.

11. For a right triangle like the triangle in Problem 3, what can you say about the altitudes from H to $\overline{GI}$ and from I to $\overline{HG}$? They are the triangle's sides or legs.

12. At what point do the altitudes in a right triangle meet? at the right angle

13. In the isosceles triangle used in Problems 6 and 7, what seems to be true about the angle bisector of the vertex angle and the median to the base? They seem to be the same.

14. For the equilateral triangle used in Problems 8, 9, and 10, what seems to be true about the perpendicular bisectors of the sides, the angle bisectors, and the medians to each side? They seem to be the same.

15. Look again at the bisectors for the angles in Problem 5. Label the point where the bisectors meet P. Draw a perpendicular line from P to each side of the triangle. Measure the length of each perpendicular line segment. What appears to be true about the segment lengths? What statement can you make about the point P where the angle bisectors meet and its distance from the sides of the triangle?

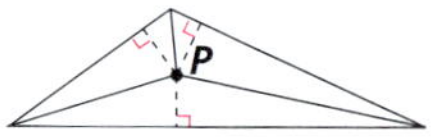

Point P, the intersection of the angle bisectors, appears to be equidistant from each side of the triangle.

Lesson 9 Proof of the Angle Sum Theorem

In Lesson 5, you found that the sum of the measures of the angles in a triangle is 180°. Before we complete the formal proof of the theorem for all triangles, let's look at one more example.

> ***Corollary***
> *A conclusion that follows easily from a given theorem*
>
> ***Exterior angle***
> *Angle formed by extending one side of a triangle through any vertex*

Given: $\triangle ABC$, with $m\angle A = 30°$ and $m\angle C = 50°$

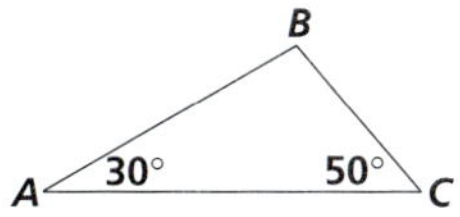

By the equivalent of Euclid's Parallel Postulate (Postulate 5), there is a unique line $\ell \parallel \overline{AC}$ passing through vertex B.

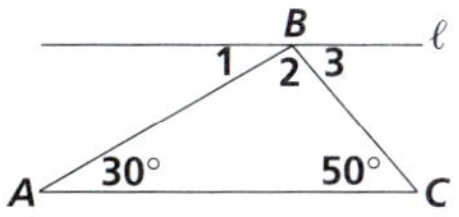

Think of $\overline{AB}$ as a transversal, then $m\angle 1 = 30°$ by the alternate interior angles postulate.

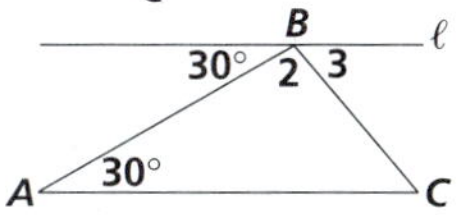

Think of $\overline{BC}$ as a transversal, then $m\angle 3 = 50°$ by the alternate interior angles postulate.

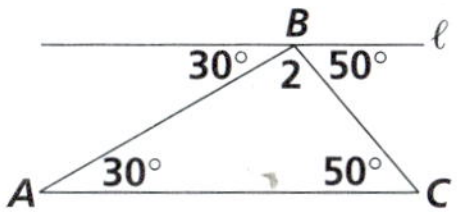

30°	+	$m\angle 2$	+	50°	=	180°	ℓ is a straight line
$m\angle A$	+	$m\angle 2$	+	$m\angle C$	=	180°	substitution
		$\therefore m\angle A + m\angle B + m\angle C$			=	180°	

Experiencing Geometry

Draw any $\triangle ABC$ on a sheet of paper. Cut out the triangle. Fold your triangle so that point B lies on $\overline{AC}$; the fold should be parallel to $\overline{AC}$. Then fold your triangle so that A and C are at B. Notice that $\angle A + \angle B + \angle C$ seem to equal a straight angle. Try it again with another triangle, one that is not equilateral. Does $\angle A + \angle B + \angle C$ still seem to equal a straight angle? What does this tell you about the sum of the angles in any triangle?

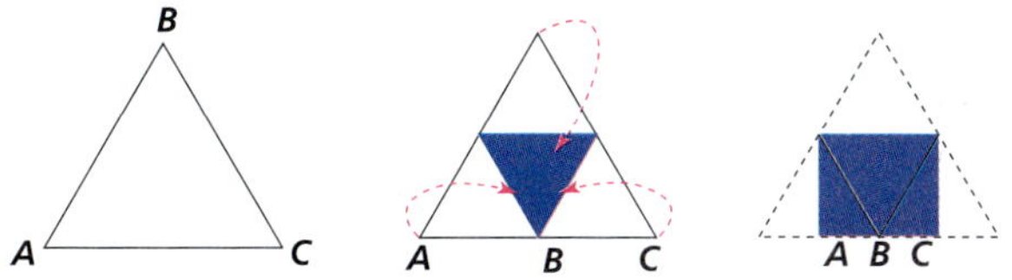

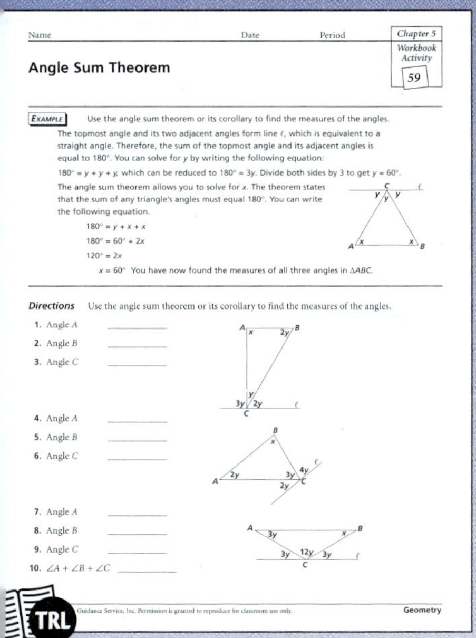
Name Date Period

Chapter 5 Workbook Activity 59

Angle Sum Theorem

EXAMPLE Use the angle sum theorem or its corollary to find the measures of the angles.

The topmost angle and its two adjacent angles form line ℓ, which is equivalent to a straight angle. Therefore, the sum of the topmost angle and its adjacent angles is equal to 180°. You can solve for y by writing the following equation:

$180° = y + y + y$, which can be reduced to $180° = 3y$. Divide both sides by 3 to get $y = 60°$.

The angle sum theorem allows you to solve for x. The theorem states that the sum of any triangle's angles must equal 180°. You can write the following equation.

$180° = y + x + x$

$180° = 60° + 2x$

$120° = 2x$

$x = 60°$ You have now found the measures of all three angles in $\triangle ABC$.

Directions Use the angle sum theorem or its corollary to find the measures of the angles.

1. Angle A ______
2. Angle B ______
3. Angle C ______
4. Angle A ______
5. Angle B ______
6. Angle C ______
7. Angle A ______
8. Angle B ______
9. Angle C ______
10. $\angle A + \angle B + \angle C$ ______

Guidance Service, Inc. Permission is granted to reproduce for classroom use only. Geometry

Workbook Activity 59

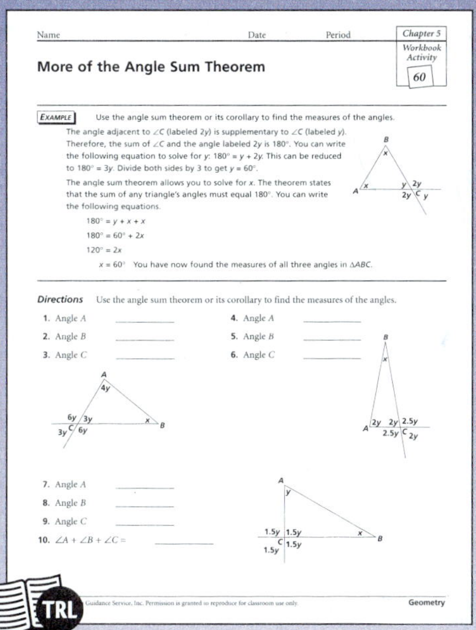
Name Date Period

Chapter 5 Workbook Activity 60

More of the Angle Sum Theorem

EXAMPLE Use the angle sum theorem or its corollary to find the measures of the angles.

The angle adjacent to $\angle C$ (labeled $2y$) is supplementary to $\angle C$ (labeled y). Therefore, the sum of $\angle C$ and the angle labeled $2y$ is 180°. You can write the following equation to solve for y: $180° = y + 2y$. This can be reduced to $180° = 3y$. Divide both sides by 3 to get $y = 60°$.

The angle sum theorem allows you to solve for x. The theorem states that the sum of any triangle's angles must equal 180°. You can write the following equations.

$180° = y + x + x$

$180° = 60° + 2x$

$120° = 2x$

$x = 60°$ You have now found the measures of all three angles in $\triangle ABC$.

Directions Use the angle sum theorem or its corollary to find the measures of the angles.

1. Angle A ______
2. Angle B ______
3. Angle C ______
4. Angle A ______
5. Angle B ______
6. Angle C ______
7. Angle A ______
8. Angle B ______
9. Angle C ______
10. $\angle A + \angle B + \angle C =$ ______

Guidance Service, Inc. Permission is granted to reproduce for classroom use only. Geometry

Workbook Activity 60

Lesson at a Glance

Chapter 5 Lesson 9

Overview This lesson presents a formal proof of the angle sum theorem.

Objectives

- To prove that the sum of the angles in a triangle equals 180°.
- To prove the corollary to the angle sum theorem.
- To use the angle sum theorem and its corollary to determine the measures of unknown angles in a triangle.

Student Pages 167–171

Teacher's Resource Library TRL

Workbook Activities 59–60

Additional Practice 33

Mathematics Vocabulary

corollary
exterior angle

1. Warm-Up Activity

Remind students that they have already determined that the sum of the angle measures in a triangle is 180°. Prompt students' memories by reproducing and discussing the following diagrams from Lesson 5.

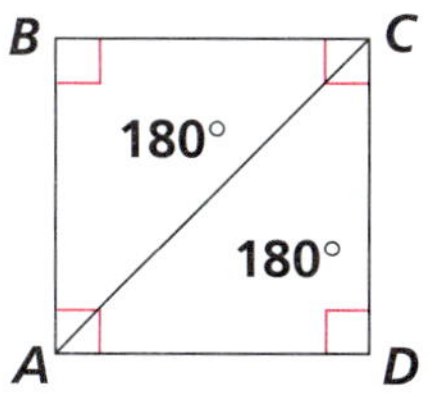

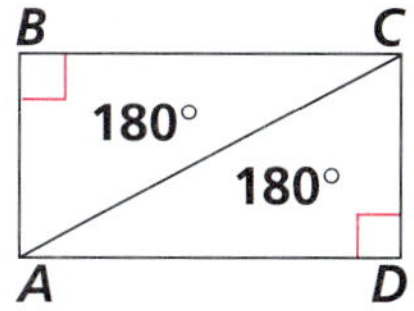

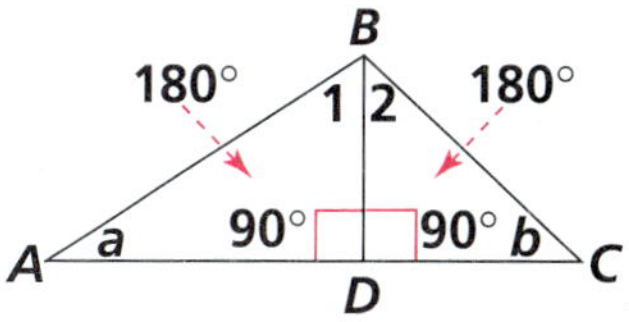

2. Teaching the Lesson

To help students recognize the parallel lines and transversals in the triangle illustrations on page 167, you may wish to reconstruct the illustrations on the board or overhead projector. Use one color to draw line ℓ and $\overline{AC}$ parallel to one another. Use a second color to draw $\overline{AB}$, identifying it as a transversal. Finally, with a third color, draw $\overline{BC}$ identifying it too as a transversal.

Work through the proof on page 168 with students. As you make each statement, ask students to provide the reason. Invite a volunteer to identify the parts cited in each reason. For example, in Step 3, ask the student to locate on the drawing the adjacent angles that form the straight angle. Have the volunteer explain why the angles are adjacent and why they form a straight angle. In Step 4, have a volunteer identify the parallel lines as well as the alternate interior angles. Continue the process for Steps 5 and 6.

Repeat the process for the proof of the corollary.

You can write a formal proof of the theorem in nearly the same manner. The only difference is that you may not specify the measure of any angle—you must prove the theorem is true for any triangle and its angles.

ANGLE SUM THEOREM: The sum of the angle measures in any triangle is 180°.

Given: any $\triangle ABC$

To Prove: $m\angle A + m\angle B + m\angle C = 180°$

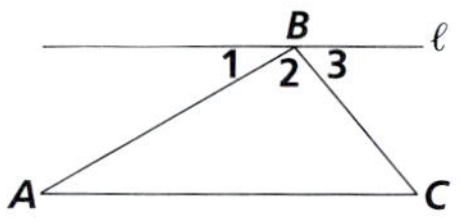

PROOF

Statement	Reason
1. $\triangle ABC$	1. Given.
2. Construct $\ell \parallel \overline{AC}$ through B	2. Equivalent of Euclid's Parallel Postulate.
3. $m\angle 1 + m\angle 2 + m\angle 3 = 180°$	3. Adjacent angles with exterior sides in a straight line form a straight angle; straight angle measures 180°.
4. $m\angle 1 = m\angle A$, $m\angle 3 = m\angle C$	4. Alternate interior angles between parallel lines are equal.
5. $m\angle 2 = m\angle B$	5. Same angle.
6. $\therefore m\angle 1 + m\angle 2 + m\angle 3 = 180°$ becomes $m\angle A + m\angle B + m\angle C = 180°$	6. Substitution of equals for equals.

With this theorem, we can prove another, related theorem. A conclusion that follows easily from a given theorem is often called a **corollary** to the original theorem.

COROLLARY TO THE ANGLE SUM THEOREM: If $\triangle ABC$ is any triangle, then an exterior angle of ABC has the same measure as the sum of the measures of the two nonadjacent interior angles.

An **exterior angle** of a triangle is formed by extending one side of the triangle through any vertex. The extension forms an exterior angle with the adjacent side of the triangle.

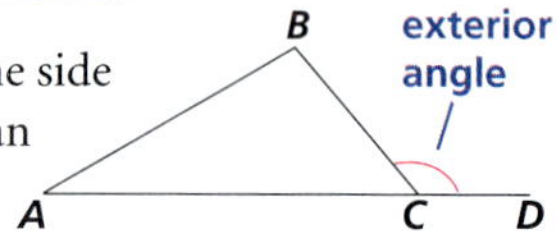

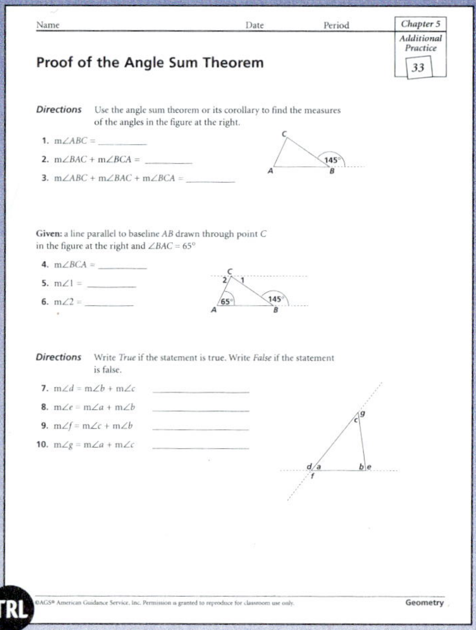

Name Date Period

Chapter 5 Additional Practice 33

Proof of the Angle Sum Theorem

Directions Use the angle sum theorem or its corollary to find the measures of the angles in the figure at the right.

1. $m\angle ABC =$ ______
2. $m\angle BAC + m\angle BCA =$ ______
3. $m\angle ABC + m\angle BAC + m\angle BCA =$ ______

Given: a line parallel to baseline AB drawn through point C in the figure at the right and $\angle BAC = 65°$

4. $m\angle BCA =$ ______
5. $m\angle 1 =$ ______
6. $m\angle 2 =$ ______

Directions Write *True* if the statement is true. Write *False* if the statement is false.

7. $m\angle d = m\angle b + m\angle c$ ______
8. $m\angle e = m\angle a + m\angle b$ ______
9. $m\angle f = m\angle c + m\angle b$ ______
10. $m\angle g = m\angle a + m\angle c$ ______

 Geometry

TRL

Additional Practice 33

Let's prove the corollary.

Given: any $\triangle ABC$, with exterior angle BCD

To Prove: $m\angle BCD = m\angle A + m\angle B$

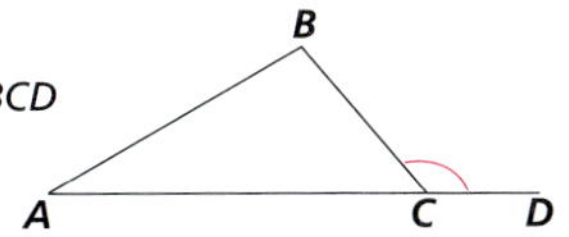

PROOF

Statement	Reason
1. $\triangle ABC$ with exterior angle BCD	1. Given.
2. $\angle C$ and $\angle BCD$ are adjacent angles	2. Definition of adjacent angles.
3. $m\angle BCD + m\angle C = 180°$	3. Adjacent angles with exterior sides in a straight line form a straight angle; straight angle measures 180°.
4. $m\angle A + m\angle B + m\angle C = 180°$	4. Angle sum theorem.
5. $m\angle BCD + m\angle C = m\angle A + m\angle B + m\angle C$	5. Steps 3 and 4; quantities equal to the same quantity are equal to each other.
6. $\therefore m\angle BCD = m\angle A + m\angle B$	6. Subtracting equals from equals gives equal differences.

This corollary guarantees the following equalities.

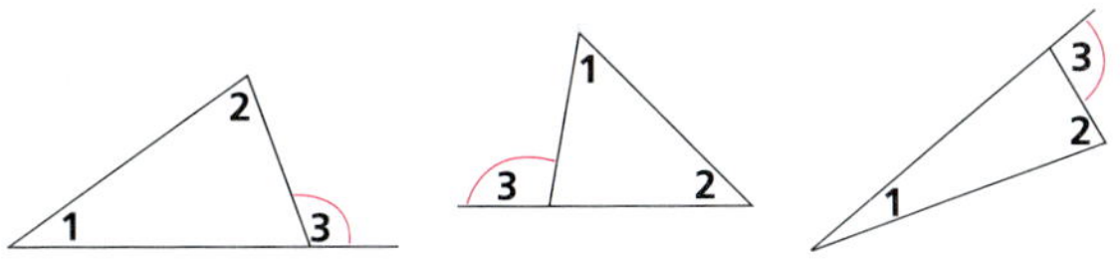

$m\angle 3 = m\angle 1 + m\angle 2$ in each triangle

Measure of an exterior angle equals the sum of the two nonadjacent interior angles.

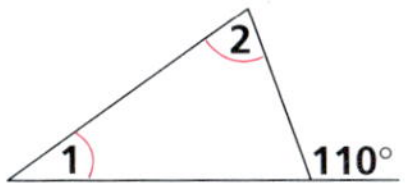

$m\angle 1 + m\angle 2 = 110°$

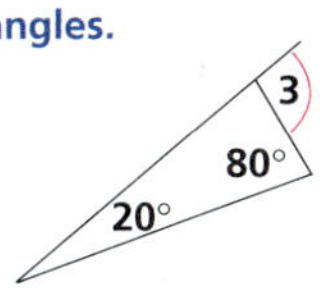

$m\angle 3 = 20° + 80° = 100°$

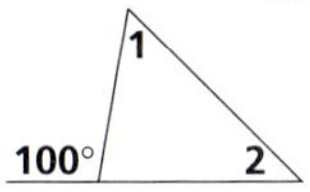

$m\angle 1 + m\angle 2 = 100°$

3. Reinforce and Extend

LEARNING STYLES

Auditory/Verbal Have students work in pairs. Draw students' attention to the relationship guaranteed by the corollary. Have one partner explain to the other why the equalities given at the bottom of page 169 are true. Then ask each partner to take a turn matching the written description "Measure of an exterior angle equals the sum of the two nonadjacent interior angles" to the appropriate parts of the triangle.

LEARNING STYLES

Interpersonal/Group Learning Ask students to work in groups of four. Challenge each group to make up four diagrams, each similar to the diagrams at the bottom of page 169 and each specifying an exterior angle measure and two nonadjacent interior angles. Each diagram can be on a 3×5 card. Direct groups to exchange cards with one another. Once cards are exchanged, have the groups write an equality using the information in the diagram.

Triangles and Quadrilaterals Chapter 5 **169**

Algebra Connection

After discussing the examples on page 170 with students, present the following problems to them. Direct students to use their knowledge of the angle sum theorem and its corollary to draw a diagram reflecting the equation and then solve the equation.

$x + 75° = 140°$ [$x = 65°$]

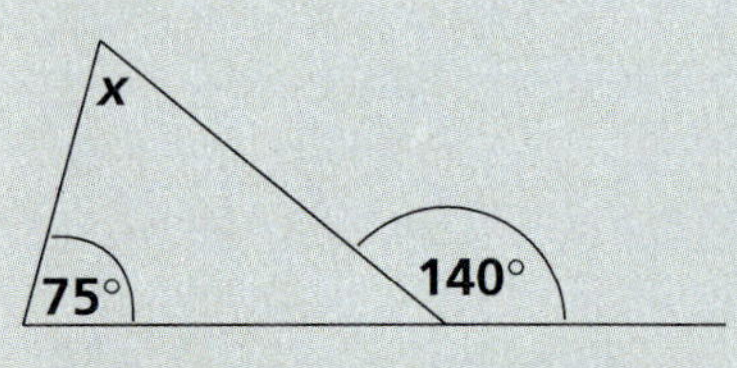

$55° + x = 180°$; [$x = 125°$]

$55° + y = 125°$; [$y = 70°$]

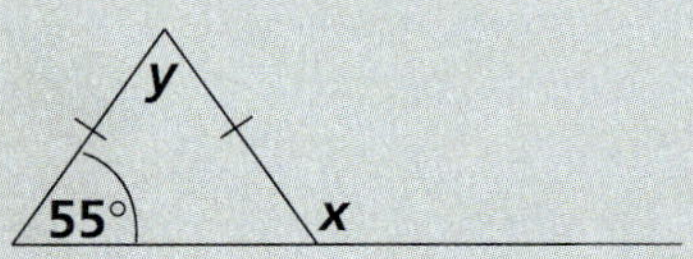

$x + 43° = 97°$ [$x = 54°$]

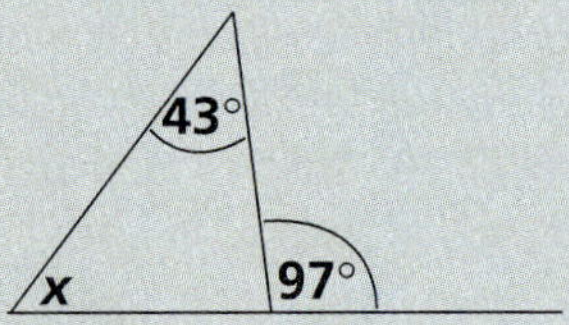

Examples Solve for x.

$x + 50° = 150°$ (by corollary)

$x = 100°$ (subtraction)

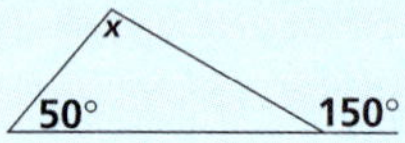

Solve for x.

$70° + 70° + x = 180°$ (angle sum theorem)

$x = 40°$ (subtraction)

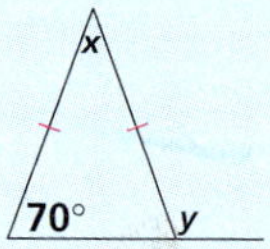

Solve for y.

$70° + 40° = y$ (by corollary) OR $70° + y = 180°$

$y = 110°$ (addition) $y = 110°$ (subtraction)

Algebra Connection

Exercise A Use the angle sum theorem or its corollary to find the measures of the angles.

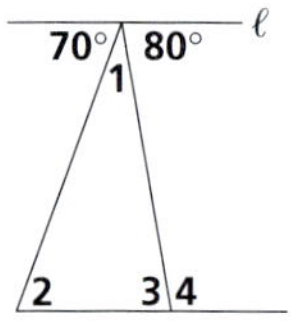

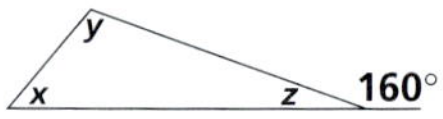

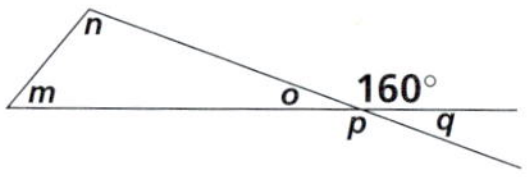

1. $\angle 1$ 30°
2. $\angle 2$ 70°
3. $\angle 3$ 80°
4. $\angle 4$ 100
5. $m\angle x + m\angle y$ 160°
6. $m\angle z$ 20°
7. $m\angle x + m\angle y + m\angle z$ 180°
8. o 20°
9. p 160°
10. q 20°
11. $m + n$ 160°
12. $m + n + o$ 180°

Exercise B Write *True* if the statement is true. Write *False* if the statement is false. Use the figure at the right.

13. $m\angle d = m\angle a + m\angle b$ True
14. $m\angle d = m\angle b + m\angle c$ False
15. $m\angle e = m\angle b + m\angle c$ True
16. $m\angle f = m\angle b + m\angle c$ True
17. $m\angle f = m\angle d$ False
18. $m\angle d + m\angle c = 180°$ True

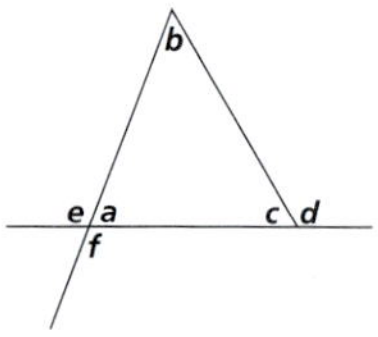

Exercise C Answer the following questions. Give reasons for each answer.

19. Suppose a triangle is equiangular. How many degrees are in each angle? $60°$; $180° \div 3 = 60°$

20. How many degrees are in each angle of an isosceles right triangle? $90°, 45°, 45°$; $180° = 90° + 2x, x = 45°$

Calculator Practice

You can use your calculator to help find missing angle measures.

EXAMPLES $60° + 20° = y$

Press 60 [+] 20 [=] 80.

$80° + y = 180°$

Press 180 [−] 80 [=] 100.

Exercise D Find each unknown using a calculator.

21. $43° + 27° = y$ $70°$

22. $92° + y = 180°$ $88°$

23. $34° + 63° = y$ $97°$

24. $21.5° + 34.5° = y$ $56°$

25. $78° + y = 180°$ $102°$

Geometry Around Us

Give yourself three days to find examples of at least twenty different polygons. Once you spot a polygon, write its name, a description, and its location. For example, you might include your observation of a regular octagon as a stop sign on the corner of Green and Pine.

CALCULATOR

Remind students that they can also use a calculator to check their answers to numerical problems. Indicate that they can use subtraction to check the answers to addition problems. For example, to check problem 21, the key sequence is

Press 70 [−] 27 [=] 43.

Suggest that students use their calculator to check the remaining answers for Exercise D.

MENTAL MATH

Students can use the angle sum theorem to find the value of the third angle in these sequences.

$45°, 45°, x$ [90°]

$30°, 60°, x$ [90°]

$100°, 35°, x$ [45°]

$70°, 30°, x$ [80°]

$15°, 140°, x$ [25°]

$60°, 60°, x$ [60°]

Lesson at a Glance

Chapter 5 Application

Overview This lesson explores how a specific set of polygon shapes can be used to produce a variety of geometric figures.

Objectives

- To use tangrams to produce specified polygons.
- To arrange tangrams in fanciful shapes.

Student Page 172

Teacher's Resource Library TRL

Workbook Activity 61

1. Warm-Up Activity

Ask students to identify each of the tangram shapes. Encourage them to be as specific as possible.

2. Teaching the Lesson

Make certain students understand that the shapes in a tangram are specific and related to one another by size as well as shape.

Students can use the square in the example as a model or template to form their own tangrams.

3. Reinforce and Extend

Learning Styles

Body/Kinesthetic Allow students to work with complete tangram sets. Have them close their eyes and try to identify each piece. Encourage students to use the pieces to make squares, parallelograms, and larger triangles.

Application

Geometric Puzzles

There are dozens of puzzles that rely on geometric figures and shapes. One of the most versatile puzzles is the tangram. The tangram puzzle consists of these seven shapes—five triangles, a square, and a parallelogram.

Example

What makes tangrams interesting is combining the seven shapes to produce a variety of figures. For example, it is possible to combine the seven shapes to form a solid square.

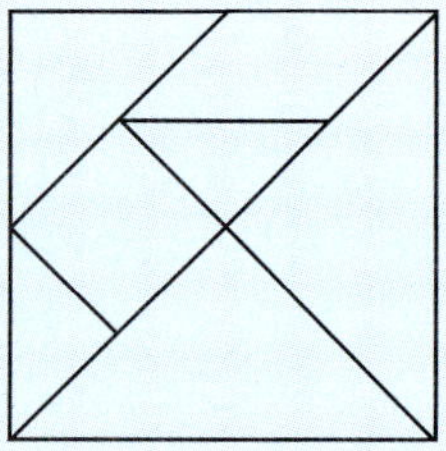

You can make your own set of tangrams by making the shapes shown above.

Exercise

Form each of the following polygons using the tangram pieces. Sketch how you placed the shapes together. Compare your sketches with your classmates'. Find out if there was more than one way to form the polygon.

1. a rectangle using five shapes
2. a trapezoid using five shapes
3. a hexagon using six shapes
4. a pentagon using four shapes
5. Have some fun with tangrams! See if you can use all seven shapes to form the cat, rabbit, and ship shown. Then see what other interesting figures you can form using all seven tangram shapes.

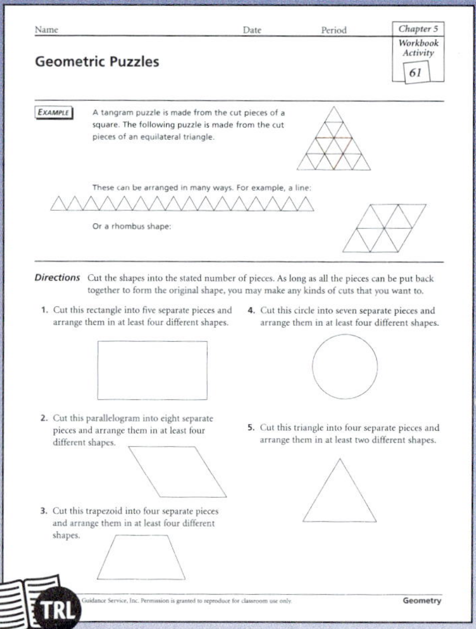

Name Date Period | Chapter 5 Workbook Activity 61

Geometric Puzzles

Example A tangram puzzle is made from the cut pieces of a square. The following puzzle is made from the cut pieces of an equilateral triangle.

These can be arranged in many ways. For example, a line:

Or a rhombus shape:

Directions Cut the shapes into the stated number of pieces. As long as all the pieces can be put back together to form the original shape, you may make any kinds of cuts that you want to.

1. Cut this rectangle into five separate pieces and arrange them in at least four different shapes.
2. Cut this parallelogram into eight separate pieces and arrange them in at least four different shapes.
3. Cut this trapezoid into four separate pieces and arrange them in at least four different shapes.
4. Cut this circle into seven separate pieces and arrange them in at least four different shapes.
5. Cut this triangle into four separate pieces and arrange them in at least two different shapes.

 Geometry

TRL

Workbook Activity 61

Chapter 5 Review

If the statement is true, write *True.* If the statement is false, write *False* and give a counterexample or explain why.

Example: All trapezoids are scalene.

Solution: False, a trapezoid can have two sides of equal length, an isosceles trapezoid.

1. A rhombus, a square, and a trapezoid each have two diagonals.
2. A scalene triangle and a scalene trapezoid each have no sides of equal length.
3. A right triangle always has two sides of equal length.
4. The sum of the angles in a parallelogram equals the sum of the angles in a pentagon.
5. The sum of the angles in an obtuse right triangle is 180°.
6. Each angle in an octagon measures 135°.

1. True
2. True
3. False, a right triangle can also be scalene.
4. False, sum in a parallelogram = 360°, sum in a pentagon = $(5 - 2) \cdot (180°) = 540°$.
5. False, there cannot be an obtuse right triangle.
6. False, an octagon that is not regular can have angles of different measures.

Use the information given to supply the angle measures.

Example: $m\angle z$

Solution: $z + 170° = 180°$; $z = 10°$

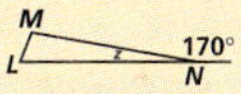

Given: *ABCD*, a rectangle with diagonal $\overline{BD}$ and $m\angle 1 = 35°$

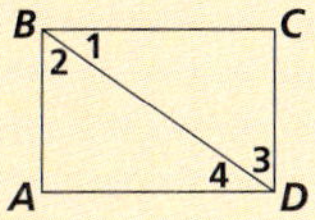

7.	$m\angle 4$	35°
8.	$m\angle 1 + m\angle 2$	90°
9.	$m\angle 2$	55°
10.	$m\angle 3 + m\angle 4$	90°
11.	$m\angle 3$	55°
12.	$m\angle 4 + m\angle 2 + m\angle BAD$	180°
13.	$m\angle ABC + m\angle BCA + m\angle CAB$	180°

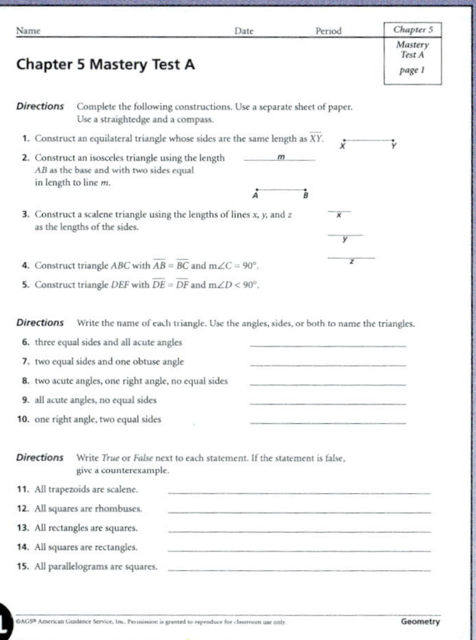

Name Date Period | *Chapter 5* *Mastery Test A* *page 1*

Chapter 5 Mastery Test A

Directions Complete the following constructions. Use a separate sheet of paper. Use a straightedge and a compass.

1. Construct an equilateral triangle whose sides are the same length as $\overline{XY}$.
2. Construct an isosceles triangle using the length *AB* as the base and with two sides equal in length to line *m*.
3. Construct a scalene triangle using the lengths of lines *x*, *y*, and *z* as the lengths of the sides.
4. Construct triangle *ABC* with $\overline{AB} = \overline{BC}$ and $m\angle C = 90°$.
5. Construct triangle *DEF* with $\overline{DE} = \overline{DF}$ and $m\angle D < 90°$.

Directions Write the name of each triangle. Use the angles, sides, or both to name the triangles.

6. three equal sides and all acute angles ______
7. two equal sides and one obtuse angle ______
8. two acute angles, one right angle, no equal sides ______
9. all acute angles, no equal sides ______
10. one right angle, two equal sides ______

Directions Write *True* or *False* next to each statement. If the statement is false, give a counterexample.

11. All trapezoids are scalene. ______
12. All squares are rhombuses. ______
13. All rectangles are squares. ______
14. All squares are rectangles. ______
15. All parallelograms are squares. ______

TRL

 Geometry

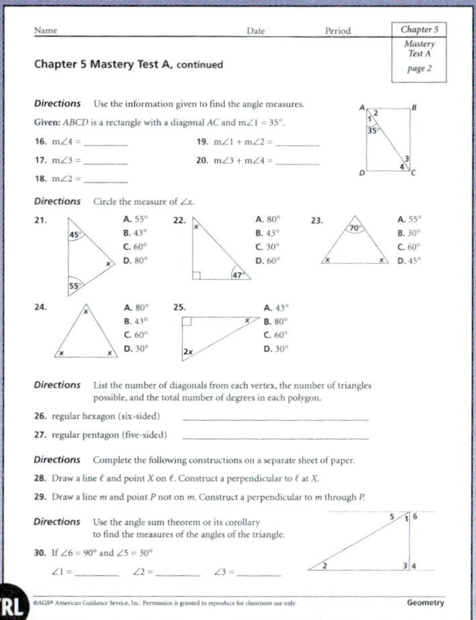

Name Date Period | *Chapter 5* *Mastery Test A* *page 2*

Chapter 5 Mastery Test A, continued

Directions Use the information given to find the angle measures.

Given: *ABCD* is a rectangle with a diagonal *AC* and $m\angle 1 = 35°$.

16. $m\angle 4 =$ ______
17. $m\angle 3 =$ ______
18. $m\angle 2 =$ ______
19. $m\angle 1 + m\angle 2 =$ ______
20. $m\angle 3 + m\angle 4 =$ ______

Directions Circle the measure of $\angle x$.

21. A. 55° B. 43° C. 60° D. 80°
22. A. 80° B. 43° C. 30° D. 60°
23. A. 55° B. 30° C. 60° D. 45°
24. A. 80° B. 43° C. 60° D. 30°
25. A. 43° B. 80° C. 60° D. 30°

Directions List the number of diagonals from each vertex, the number of triangles possible, and the total number of degrees in each polygon.

26. regular hexagon (six-sided) ______
27. regular pentagon (five-sided) ______

Directions Complete the following constructions on a separate sheet of paper.

28. Draw a line ℓ and point *X* on ℓ. Construct a perpendicular to ℓ at *X*.
29. Draw a line *m* and point *P* not on *m*. Construct a perpendicular to *m* through *P*.

Directions Use the angle sum theorem or its corollary to find the measures of the angles of the triangle.

30. If $\angle 6 = 90°$ and $\angle 5 = 30°$

$\angle 1 =$ ______ $\angle 2 =$ ______ $\angle 3 =$ ______

TRL

 Geometry

Chapter 5 Mastery Test A

Chapter 5 Review

Each set of problems in the Chapter Review includes an example and solution to illustrate the concept. Use the given examples for reteaching the materials in Chapter 5. For additional practice, refer to the Supplementary Problems for Chapter 5 (pages 402–404) and Additional Practice 25–33 in the Teacher's Resource Library.

Chapter 5 Mastery Test TRL

The Teacher's Resource Library includes parallel forms of the Chapter 5 Mastery Test. The difficulty level of the two forms is equivalent. You may wish to use one form as a pretest and the other form as a posttest. In addition, the Chapter 5 Cumulative Mastery Test covers content from Chapters 1–5.

Alternative Assessment

Alternative Assessment items correlate with student Goals for Learning at the beginning of this chapter.

- Have students construct triangles using sides.

Provide students with a compass, a straightedge, and line segments $\overline{XY}$, $\overline{ZW}$, and $\overline{UV}$. Direct students to construct three triangles, one with all sides equal to $\overline{XY}$, one with two sides equal to $\overline{ZW}$, and one with side lengths XY, ZW, and UV.

- Have students name triangles using their angles.

Provide students with a variety of triangles from the Manipulite set as well as the Polygon Power Pack. Direct students to group and identify the triangles by comparing their angles. Students should produce three groups—acute, right, and obtuse.

- Have students name and distinguish quadrilaterals.

Provide students with a selection of polygons from the Polygon Power Pack. Direct students to identify and name the quadrilaterals.

Alternative Assessment

continued

- Have students find the measures of angles in quadrilaterals and triangles.
- Have students define quadrilaterals and parallels and apply theorems to them.

Direct students to write the descriptions, and provide illustrations for, a series of "Wanted" posters for quadrilaterals and triangles. Each poster or bulletin should show sketches of a quadrilateral and a triangle and include information on side length, parallel sides, and angle measures. Additional identifying information such as diagonals may be included.

- Have students construct perpendiculars.

Provide students with a compass and a straightedge, a line ℓ, and a line *m* with point *P* not on the line. Direct students to construct a perpendicular to ℓ. Students should also construct a perpendicular to *m* that passes through *P*.

- Have students construct triangles based on attributes.

Ask students to create a construction problem for a triangle using several attributes. Then have them exchange problems and solve them.

- Have students use the angle sum theorem to define triangles.

Ask students how they could use angle measure to determine whether an unseen polygon is a triangle. Have them prepare their answer as part of a mystery story.

More Review...

Given: triangle *XYZ*, line ℓ ∥ *m*

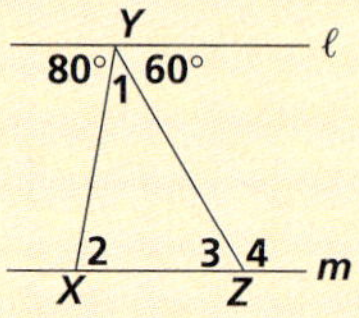

14. m∠1 40°

15. m∠2 80°

16. m∠3 60°

17. m∠4 120°

Describe each line. Write the letter of the correct description.

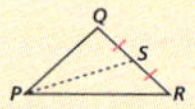

Example: $\overline{PS}$

A. altitude B. median C. angle bisector D. perpendicular bisector

Solution: B; $\overline{QS} = \overline{SR}$, so $\overline{PS}$ is the median from *P*.

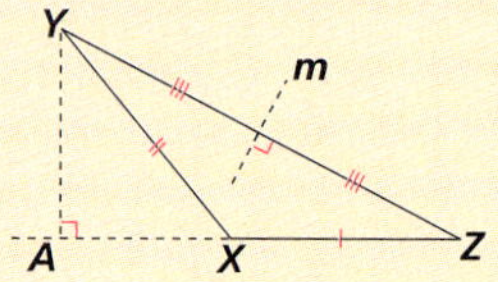

18. $\overline{YA}$ A

A. altitude **B.** median
C. angle bisector **D.** perpendicular bisector

19. *m* C

A. altitude **B.** median
C. perpendicular bisector **D.** *A* and *B*

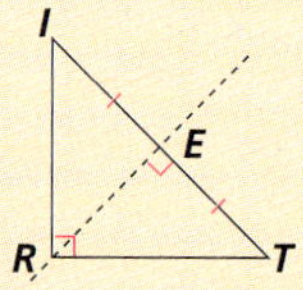

20. $\overline{RE}$ D

A. altitude **B.** diagonal
C. perpendicular bisector **D.** *A* and *C*

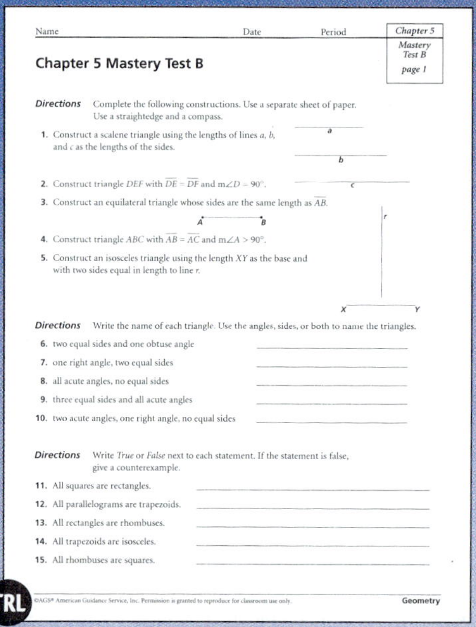

Name Date Period

Chapter 5 Mastery Test B page 1

Chapter 5 Mastery Test B

Directions Complete the following constructions. Use a separate sheet of paper. Use a straightedge and a compass.

1. Construct a scalene triangle using the lengths of lines *a*, *b*, and *c* as the lengths of the sides.
2. Construct triangle *DEF* with $\overline{DE} = \overline{DF}$ and m∠*D* = 90°.
3. Construct an equilateral triangle whose sides are the same length as $\overline{AB}$.
4. Construct triangle *ABC* with $\overline{AB} = \overline{AC}$ and m∠*A* > 90°.
5. Construct an isosceles triangle using the length *XY* as the base and with two sides equal in length to line *r*.

Directions Write the name of each triangle. Use the angles, sides, or both to name the triangles.

6. two equal sides and one obtuse angle ______
7. one right angle, two equal sides ______
8. all acute angles, no equal sides ______
9. three equal sides and all acute angles ______
10. two acute angles, one right angle, no equal sides ______

Directions Write *True* or *False* next to each statement. If the statement is false, give a counterexample.

11. All squares are rectangles. ______
12. All parallelograms are trapezoids. ______
13. All rectangles are rhombuses. ______
14. All trapezoids are isosceles. ______
15. All rhombuses are squares. ______

TRL

 Geometry

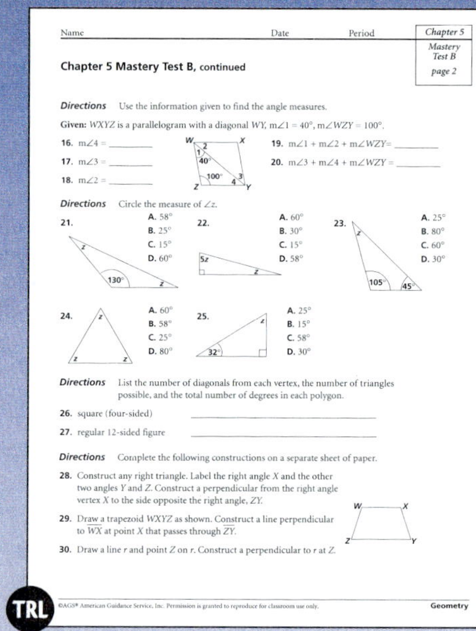

Name Date Period

Chapter 5 Mastery Test B page 2

Chapter 5 Mastery Test B, continued

Directions Use the information given to find the angle measures.

Given: *WXYZ* is a parallelogram with a diagonal *WY*, m∠1 = 40°, m∠*WZY* = 100°.

16. m∠4 = ______
17. m∠3 = ______
18. m∠2 = ______
19. m∠1 + m∠2 + m∠*WZY* = ______
20. m∠3 + m∠4 + m∠*WZY* = ______

Directions Circle the measure of ∠*z*.

21. A. 58° B. 25° C. 15° D. 60°
22. A. 60° B. 30° C. 15° D. 58°
23. A. 25° B. 80° C. 60° D. 30°
24. A. 60° B. 58° C. 25° D. 80°
25. A. 25° B. 15° C. 58° D. 30°

Directions List the number of diagonals from each vertex, the number of triangles possible, and the total number of degrees in each polygon.

26. square (four-sided) ______
27. regular 12-sided figure ______

Directions Complete the following constructions on a separate sheet of paper.

28. Construct any right triangle. Label the right angle *X* and the other two angles *Y* and *Z*. Construct a perpendicular from the right angle vertex *X* to the side opposite the right angle, *ZY*.
29. Draw a trapezoid *WXYZ* as shown. Construct a line perpendicular to $\overline{WX}$ at point *X* that passes through $\overline{ZY}$.
30. Draw a line *r* and point *Z* on *r*. Construct a perpendicular to *r* at *Z*.

TRL

 Geometry

Chapter 5 Mastery Test B

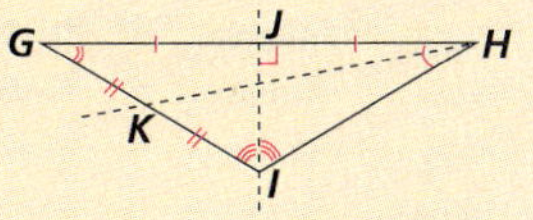

21. $\overline{KH}$ B

A. altitude　　**B.** median
C. angle bisector　　**D.** *A* and *C*

22. $\overline{JI}$ D

A. altitude　　**B.** median
C. angle bisector　　**D.** *A*, *B*, and *C*

Student constructions should reflect correct use of steps for constructing an angle bisector, a line perpendicular to a given line from a point not on the line, and finding the midpoint of a line.

Complete the following constructions on a separate sheet of paper. Use only a straightedge and a compass.

23. Construct any right triangle. Label the right angle *R* and the other two angles *S* and *T*. Construct an angle bisector for angle *R*.

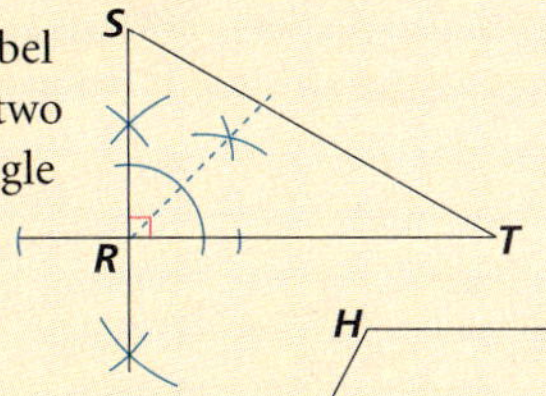

24. Draw a rhombus *RHOM* as shown. Construct an altitude from *O*.

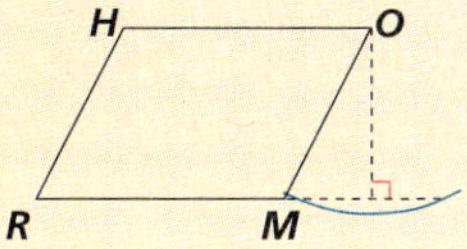

25. Draw any obtuse triangle and label the obtuse angle *D* and the triangle *DEF*. Construct the three medians of the triangle.

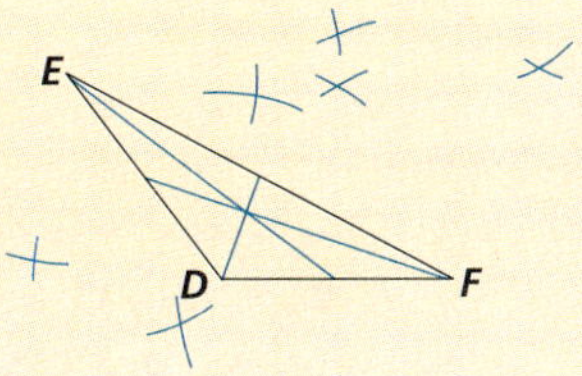

TEST-TAKING TIP

When taking a multiple-choice test, read every choice before you answer a question. Put a line through choices you know are wrong. Then choose the best answer from the remaining choices.

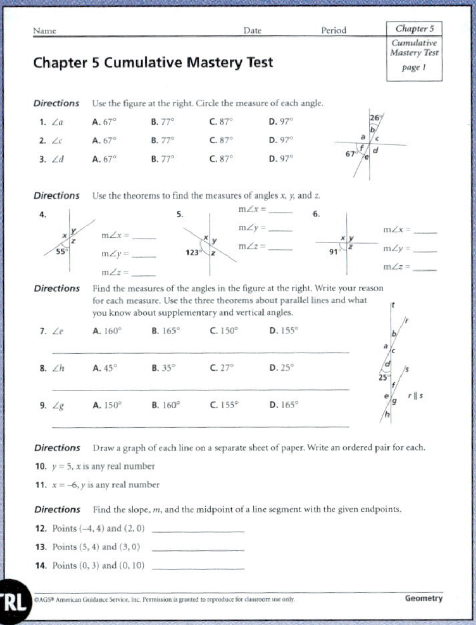

Name　Date　Period　*Chapter 5 Cumulative Mastery Test page 1*

Chapter 5 Cumulative Mastery Test

Directions Use the figure at the right. Circle the measure of each angle.

1. ∠*a*　**A.** 67°　**B.** 77°　**C.** 87°　**D.** 97°
2. ∠*c*　**A.** 67°　**B.** 77°　**C.** 87°　**D.** 97°
3. ∠*d*　**A.** 67°　**B.** 77°　**C.** 87°　**D.** 97°

Directions Use the theorems to find the measures of angles *x*, *y*, and *z*.

4. m∠*x* = ____ m∠*y* = ____ m∠*z* = ____
5. m∠*x* = ____ m∠*y* = ____ m∠*z* = ____
6. m∠*x* = ____ m∠*y* = ____ m∠*z* = ____

Directions Find the measures of the angles in the figure at the right. Write your reason for each measure. Use the three theorems about parallel lines and what you know about supplementary and vertical angles.

7. ∠*e*　**A.** 160°　**B.** 165°　**C.** 150°　**D.** 155°
8. ∠*h*　**A.** 45°　**B.** 35°　**C.** 27°　**D.** 25°
9. ∠*g*　**A.** 150°　**B.** 160°　**C.** 155°　**D.** 165°

Directions Draw a graph of each line on a separate sheet of paper. Write an ordered pair for each.

10. $y = 5$, x is any real number
11. $x = -6$, y is any real number

Directions Find the slope, *m*, and the midpoint of a line segment with the given endpoints.

12. Points (–4, 4) and (2, 0) ________
13. Points (5, 4) and (3, 0) ________
14. Points (0, 3) and (0, 10) ________

 Geometry

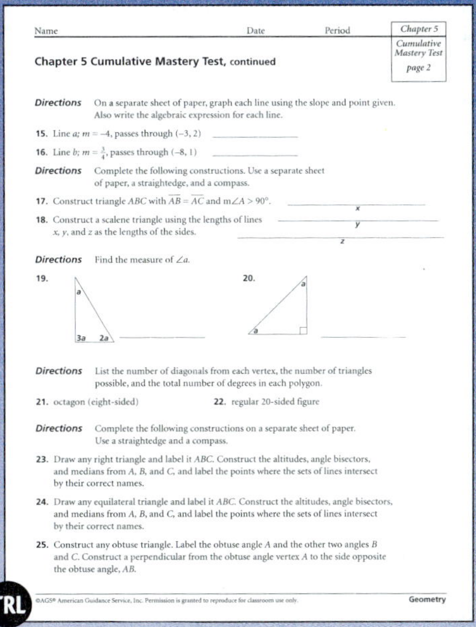

Name　Date　Period　*Chapter 5 Cumulative Mastery Test page 2*

Chapter 5 Cumulative Mastery Test, continued

Directions On a separate sheet of paper, graph each line using the slope and point given. Also write the algebraic expression for each line.

15. Line *a*; $m = -4$, passes through (–3, 2) ________
16. Line *b*; $m = \frac{3}{4}$, passes through (–8, 1) ________

Directions Complete the following constructions. Use a separate sheet of paper, a straightedge, and a compass.

17. Construct triangle *ABC* with $\overline{AB} = \overline{AC}$ and m∠*A* > 90°.
18. Construct a scalene triangle using the lengths of lines *x*, *y*, and *z* as the lengths of the sides.

Directions Find the measure of ∠*a*.

19. ________
20. ________

Directions List the number of diagonals from each vertex, the number of triangles possible, and the total number of degrees in each polygon.

21. octagon (eight-sided)
22. regular 20-sided figure

Directions Complete the following constructions on a separate sheet of paper. Use a straightedge and a compass.

23. Draw any right triangle and label it *ABC*. Construct the altitudes, angle bisectors, and medians from *A*, *B*, and *C*, and label the points where the sets of lines intersect by their correct names.
24. Draw any equilateral triangle and label it *ABC*. Construct the altitudes, angle bisectors, and medians from *A*, *B*, and *C*, and label the points where the sets of lines intersect by their correct names.
25. Construct any obtuse triangle. Label the obtuse angle *A* and the other two angles *B* and *C*. Construct a perpendicular from the obtuse angle vertex *A* to the side opposite the obtuse angle, *AB*.

 Geometry

Chapter 5 Cumulative Mastery Test

Triangles and Quadrilaterals　*Chapter 5*　**175**

Chapter 6

Planning Guide

Congruent Triangles and Transformations

	Student Text Lesson						
	Student Pages	Vocabulary	Proof	Practice Exercises	Problem Solving	Calculator Practice	Solutions Key
Lesson 1 Corresponding Parts: SAS Postulate	178–183	✔	✔	✔	183		✔
Lesson 2 SSS Construction and Triangle Inequality	184–188	✔	✔	✔			✔
Lesson 3 SSS and ASA Congruencies	189–193		✔	✔	193		✔
Lesson 4 Special Congruencies: Right Triangles	194–199		✔	✔	199	199	✔
Lesson 5 Reflections in the Coordinate Plane	200–204	✔		✔			✔
Lesson 6 Special Reflections: Symmetries	205–206	✔		✔			✔
Lesson 7 Slides and Translations	207–209	✔		✔			✔
Lesson 8 Rotations	210–213	✔		✔			✔
Application Geometry and Computer Drawings	214			✔			✔

Chapter Activities

Teacher's Resource Library
Geometry on the Job 6: Computer-Generated Special Effects

Community Connection 6: Views of History

Teacher's Edition
Chapter 6 Project

Assessment Options

Student Text
Chapter 6 Review

Teacher's Resource Library
Chapter 6 Mastery Tests A and B

Chapter 6 Cumulative Mastery Test

Chapters 1–6 Midterm Mastery Test

Teacher's Edition
Chapter 6 Alternative Assessment

Student Text Features						Teaching Strategies								Learning Styles						Teacher's Resource Library		
Algebra Connection	Constructions	Experiencing Geometry	Technology Connection	Try This	Writing About Mathematics	Warm-Up Activity	Common Error	Algebra Connection	Mental Math	Manipulatives	Calculator	Applications	Group Problem Solving	Auditory/Verbal	Visual/Spatial	Body/Kinesthetic	Logical/Mathematical	Interpersonal/Group Learning	LEP/ESL	Workbook Activities	Additional Practice	Self-Study Guide
	179					178	180					182	183		180	181	183			62	34	✔
	184, 188				187	184	188		187						185	186	187	187		63	35	✔
	189, 193					189	190		193					191		191			192	64	36	✔
197					199	194	195	197	195		199	198			196, 198			199		65	37	✔
		204	203			200				201		203			204	203	204	202		66, 67	38	✔
				206		205				205		206			206					68	39	✔
					209	207			209						209	209			207	69	40	✔
		213			213	210			213	210		211	213	211	212		212			70	41	✔
						214						214								71		✔

Manipulatives

GeoReflector

Manipulite Pattern Blocks

Polygons Plus Power Pack

Manipulatives Master 2 (Coordinate Grid)

Chapter at a Glance

Chapter 6: Congruent Triangles and Transformations pages 176–217

(Answer Keys for the Teacher's Resource Library begin on page 484 of this Teacher's Edition.)

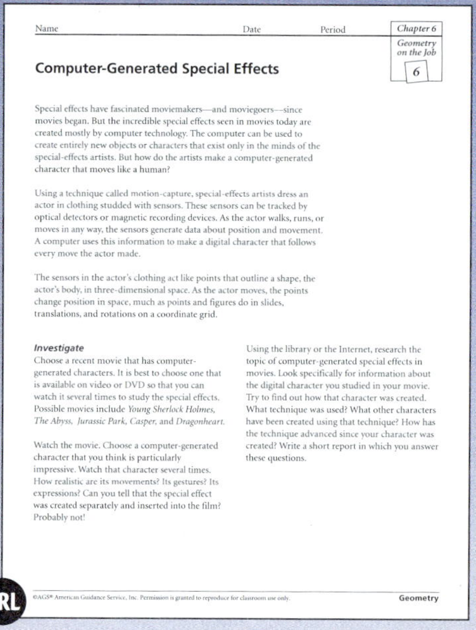

Name Date Period | Chapter 6 | *Geometry on the Job* | 6

Computer-Generated Special Effects

Special effects have fascinated moviemakers—and moviegoers—since movies began. But the incredible special effects seen in movies today are created mostly by computer technology. The computer can be used to create entirely new objects or characters that exist only in the minds of the special-effects artists. But how do the artists make a computer-generated character that moves like a human?

Using a technique called motion-capture, special-effects artists dress an actor in clothing studded with sensors. These sensors can be tracked by optical detectors or magnetic recording devices. As the actor walks, runs, or moves in any way, the sensors generate data about position and movement. A computer uses this information to make a digital character that follows every move the actor made.

The sensors in the actor's clothing act like points that outline a shape, the actor's body, in three-dimensional space. As the actor moves, the points change position in space, much as points and figures do in slides, translations, and rotations on a coordinate grid.

Investigate
Choose a recent movie that has computer-generated characters. It is best to choose one that is available on video or DVD so that you can watch it several times to study the special effects. Possible movies include *Young Sherlock Holmes, The Abyss, Jurassic Park, Casper,* and *Dragonheart.*

Watch the movie. Choose a computer-generated character that you think is particularly impressive. Watch that character several times. How realistic are its movements? Its gestures? Its expressions? Can you tell that the special effect was created separately and inserted into the film? Probably not!

Using the library or the Internet, research the topic of computer-generated special effects in movies. Look specifically for information about the digital character you studied in your movie. Try to find out how that character was created. What technique was used? What other characters have been created using that technique? How has the technique advanced since your character was created? Write a short report in which you answer these questions.

TRL

 Geometry

Geometry on the Job 6

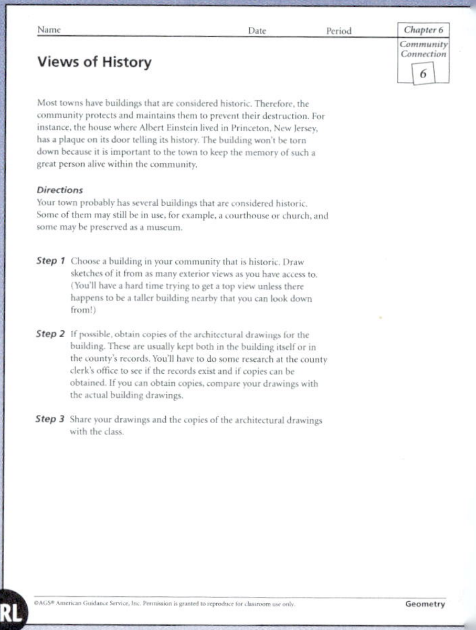

Name Date Period | Chapter 6 | *Community Connection* | 6

Views of History

Most towns have buildings that are considered historic. Therefore, the community protects and maintains them to prevent their destruction. For instance, the house where Albert Einstein lived in Princeton, New Jersey, has a plaque on its door telling its history. The building won't be torn down because it is important to the town to keep the memory of such a great person alive within the community.

Directions
Your town probably has several buildings that are considered historic. Some of them may still be in use, for example, a courthouse or church, and some may be preserved as a museum.

Step 1 Choose a building in your community that is historic. Draw sketches of it from as many exterior views as you have access to. (You'll have a hard time trying to get a top view unless there happens to be a taller building nearby that you can look down from!)

Step 2 If possible, obtain copies of the architectural drawings for the building. These are usually kept both in the building itself or in the county's records. You'll have to do some research at the county clerk's office to see if the records exist and if copies can be obtained. If you can obtain copies, compare your drawings with the actual building drawings.

Step 3 Share your drawings and the copies of the architectural drawings with the class.

TRL

 Geometry

Community Connection 6

Chapter 6

Congruent Triangles and Transformations

Air and water meet. You see a sparkling reflection of buildings in the water. Take a photo and you can note characteristics of reflection. The object reflected and its image are congruent, or the same size and shape. Every point on the object matches a point on its image. Fold the photo in half. What will you find? The image and the object match in size, shape, and position. They form one picture. This symmetry, or correspondence of size, shape, and position, can be found throughout nature, science, and geometry.

In Chapter 6, you will learn about congruent triangles and transformations.

Goals for Learning

- To use the SAS postulate to identify corresponding parts in triangles
- To construct triangles using SSS information
- To identify congruent triangles based on SSS and ASA structures
- To identify angle and side characteristics of congruent triangles
- To make generalizations about reflections and lines of reflection
- To identify lines of symmetry in reflections
- To map translations in a coordinate plane
- To describe and graph rotations

RL

SELF-STUDY GUIDE

Name ____________

CHAPTER 6: Congruent Triangles and Transformations

Goal 6.1 *To use the SAS postulate to identify corresponding parts in triangles*

Date	Assignment	Score
______	1: Read pages 178–181. Complete Exercises A–D on pages 182–183.	______
______	2: Complete Workbook Activity 62.	______

Comments:

Goal 6.2 *To construct triangles using SSS information*

Date	Assignment	Score
______	3: Read pages 184–187. Complete Exercises A–C on page 188.	______
______	4: Complete Workbook Activity 63.	______

Comments:

Goal 6.3 *To identify congruent triangles based on SSS and ASA structures*

Date	Assignment	Score
______	5: Read pages 189–191. Complete Exercises A–D on pages 192–193.	______
______	6: Complete Workbook Activity 64.	______

Comments:

Goal 6.4 *To identify angle and side characteristics of congruent right triangles*

Date	Assignment	Score
______	7: Read pages 194–197. Complete Exercises A–C on pages 198–199.	______
______	8: Read and complete the Calculator Practice and Exercise D on page 199.	______
______	9: Complete Workbook Activity 65.	______

Comments:

 Geometry

TRL

SELF-STUDY GUIDE

Name ____________

CHAPTER 6: Congruent Triangles and Transformations, continued

Goals 6.5 and 6.6 *To make generalizations about reflections and lines of reflection; to identify lines of symmetry in reflections*

Date	Assignment	Score
______	10: Read pages 200–203. Complete Exercises A–C on page 204.	______
______	11: Complete Workbook Activities 66–67.	______
______	12: Read page 205. Complete Exercises A–B on page 206.	______
______	13: Complete Workbook Activity 68.	______

Comments:

Goal 6.7 *To map translations in a coordinate plane*

Date	Assignment	Score
______	14: Read pages 207–208. Complete Exercises A–D on page 209.	______
______	15: Complete Workbook Activity 69.	______

Comments:

Goal 6.8 *To describe and graph rotations*

Date	Assignment	Score
______	16: Read pages 210–211. Complete Exercises A–C on pages 212–213.	______
______	17: Complete Workbook Activity 70.	______
______	18: Read and complete the Application Exercise on page 214.	______
______	19: Complete Workbook Activity 71.	______
______	20: Complete the Chapter 6 Review on pages 215–217.	______

Comments:

Student's Signature ____________ Date ____________

Instructor's Signature ____________ Date ____________

 Geometry

Chapter 6 Self-Study Guide

Introducing the Chapter

Use the information and photograph on these pages to help students think about triangles, congruencies, and reflections. Help them identify the triangular shapes of some of the roof lines. Discuss how the buildings and their reflections are alike and different. Ask students to read the Goals for Learning, encouraging them to use their understanding of triangles to hypothesize what SAS and SSS mean.

Chapter Project

Encourage students to work in pairs or groups of three or four. Suggest that they produce one of the following for their chapter project.

- A Photo Album of Reflections—Students interested in photography may work in pairs or small groups to produce a Photo Album of Reflections. Encourage students to carry their cameras with them and take pictures as they encounter reflections in mirrors, windows, bodies of water, and so on. Students develop their photos and mount them in books. Ask students to include a one-page discussion of lines of reflection or lines of symmetry and how they relate to the album photos.

  Alternatively, students could prepare a Photo or Sketch Album of Triangles in the Real World. They should take or draw pictures of triangles they find in their community and prepare an album with information similar to that in the Album of Reflections. Their one-page discussion would focus on the chapter's characteristics of triangles.

- Geometric Poetry—Pairs or small groups of students choose three concepts discussed in the chapter and write poems to relate each concept to their understanding of geometry. For example, students could write a poem about the SAS postulate, translations, and symmetry lines. Remind them that poems need not rhyme. Have them present each poem on a colorful poster.

Lesson at a Glance

Chapter 6 Lesson 1

Overview This lesson introduces congruent triangles, segments, and angles and demonstrates how to use the Side-Angle-Side postulate to determine if two triangles are congruent.

Objectives

- To identify congruent segments and congruent triangles.
- To identify the parts of triangles that must be congruent to satisfy the SAS Postulate.
- To comprehend that triangles are congruent if they have two sides and an included angle that are congruent.

Student Pages 178–183

Teacher's Resource Library TRL

Workbook Activity 62

Additional Practice 34

Mathematics Vocabulary

congruent angles
congruent segments
congruent triangles
corresponding angles

1. Warm-Up Activity

Tell students to fold a square sheet of paper diagonally from corner to corner to form two triangles. Ask them how the triangles are alike. Discuss which sides are the same length and whether the angles are alike. Ask them to tell whether the triangles are adjacent. Then explain that *congruent figures* are figures with the same size and shape. Ask students if their triangles are congruent.

Lesson 1 Corresponding Parts: SAS Postulate

Congruent angles
Angles that have equal sides and equal measures

Congruent segments
Segments that have the same length

Congruent triangles
Triangles whose corresponding angles and sides are congruent

Corresponding angles
Pairs of angles in similar positions

In geometry, it is possible to know that one figure is equal to another without measuring either one! The word *congruent* is used to describe angles whose measures are equal, sides with equal lengths, and polygons whose angles and sides are equal to one another. The symbol for congruent is ≅.

Congruent angles have the same measure.

$\angle A$ and $\angle A'$ are congruent angles.

$\angle A \cong \angle A'$ and $m\angle A = m\angle A'$

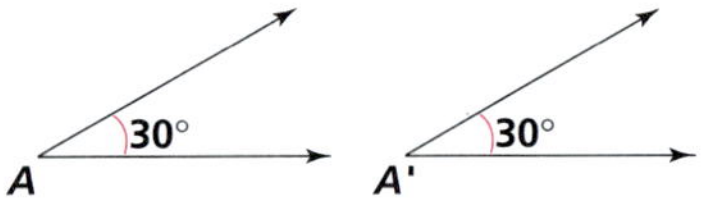

Congruent segments have the same length.

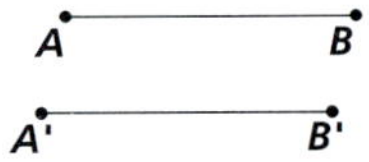

Segments $\overline{AB}$ and $\overline{A'B'}$ are congruent; the length of $\overline{AB}$ equals the length of $\overline{A'B'}$. We can write this as $\overline{AB} \cong \overline{A'B'}$ and $AB = A'B'$.

Congruent triangles are triangles whose **corresponding angles** and sides are congruent. The symbol ↔ means "corresponds to."

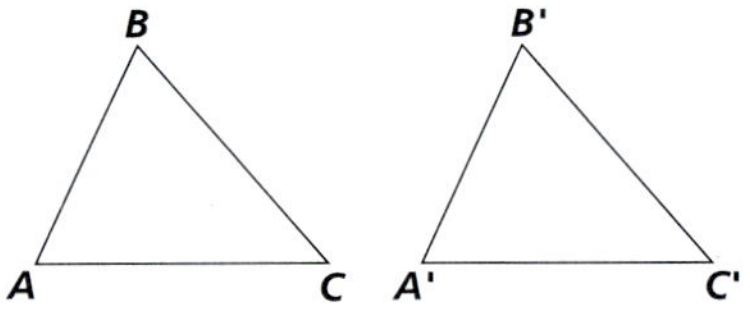

The ' mark is used to show that two items are related. A' is read "A prime" and $\overline{A'B'}$ is read "segment A prime B prime."

Corresponding Sides	Corresponding Angles
$\overline{AB} \leftrightarrow \overline{A'B'}$	$\angle A \leftrightarrow \angle A'$
$\overline{BC} \leftrightarrow \overline{B'C'}$	$\angle B \leftrightarrow \angle B'$
$\overline{AC} \leftrightarrow \overline{A'C'}$	$\angle C \leftrightarrow \angle C'$

$$\Delta ABC \cong \Delta A'B'C'$$

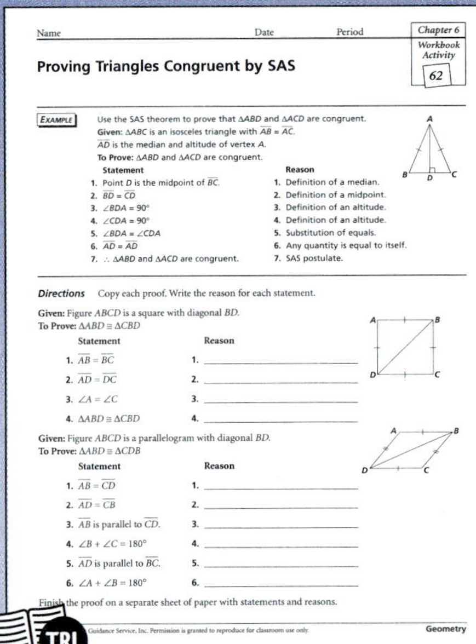

Name Date Period — Chapter 6 Workbook Activity 62

Proving Triangles Congruent by SAS

EXAMPLE Use the SAS theorem to prove that ΔABD and ΔACD are congruent.
Given: ΔABC is an isosceles triangle with $\overline{AB} = \overline{AC}$.
$\overline{AD}$ is the median and altitude of vertex A.
To Prove: ΔABD and ΔACD are congruent.

Statement	Reason
1. Point D is the midpoint of $\overline{BC}$.	1. Definition of a median.
2. $\overline{BD} = \overline{CD}$	2. Definition of a midpoint.
3. $\angle BDA = 90°$	3. Definition of an altitude.
4. $\angle CDA = 90°$	4. Definition of an altitude.
5. $\angle BDA = \angle CDA$	5. Substitution of equals.
6. $\overline{AD} = \overline{AD}$	6. Any quantity is equal to itself.
7. ∴ ΔABD and ΔACD are congruent.	7. SAS postulate.

Directions Copy each proof. Write the reason for each statement.

Given: Figure $ABCD$ is a square with diagonal BD.
To Prove: $\Delta ABD \cong \Delta CBD$

Statement	Reason
1. $\overline{AB} = \overline{BC}$	1. ______
2. $\overline{AD} = \overline{DC}$	2. ______
3. $\angle A = \angle C$	3. ______
4. $\Delta ABD \cong \Delta CBD$	4. ______

Given: Figure $ABCD$ is a parallelogram with diagonal BD.
To Prove: $\Delta ABD \cong \Delta CDB$

Statement	Reason
1. $\overline{AB} = \overline{CD}$	1. ______
2. $\overline{AD} = \overline{CB}$	2. ______
3. $\overline{AB}$ is parallel to $\overline{CD}$.	3. ______
4. $\angle B + \angle C = 180°$	4. ______
5. $\overline{AD}$ is parallel to $\overline{BC}$.	5. ______
6. $\angle A + \angle B = 180°$	6. ______

Finish the proof on a separate sheet of paper with statements and reasons.

 Geometry

Workbook Activity 62

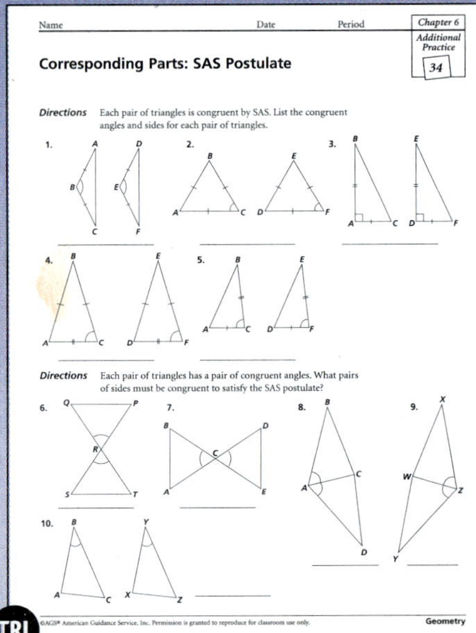

Name Date Period — Chapter 6 Additional Practice 34

Corresponding Parts: SAS Postulate

Directions Each pair of triangles is congruent by SAS. List the congruent angles and sides for each pair of triangles.

1. 2. 3. 4. 5.

Directions Each pair of triangles has a pair of congruent angles. What pairs of sides must be congruent to satisfy the SAS postulate?

6. 7. 8. 9. 10.

 Geometry

Additional Practice 34

In congruent triangles, the corresponding angles have equal measures, and the corresponding sides have equal lengths.

CONSTRUCTION

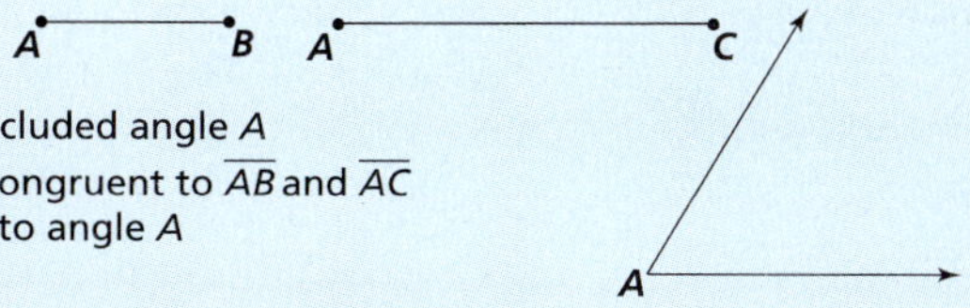

Given: segments $\overline{AB}$ and $\overline{AC}$, included angle A

Construct: triangle with sides congruent to $\overline{AB}$ and $\overline{AC}$ and included angle congruent to angle A

Step 1 Draw any line ℓ. Using a compass, copy $\overline{AB}$ onto ℓ.

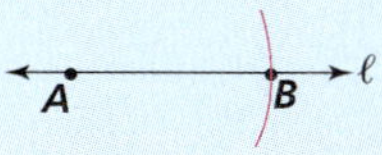

Step 2 Copy angle A, using A as the vertex.

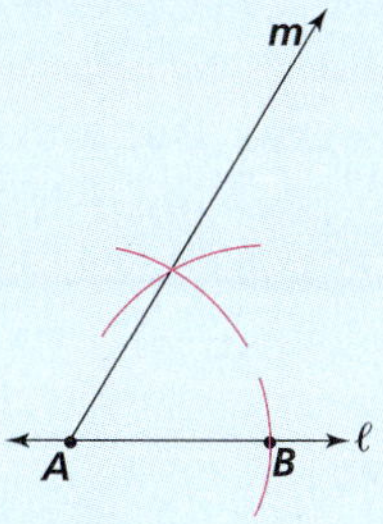

Step 3 Copy $\overline{AC}$ onto m. Then draw $\overline{BC}$.

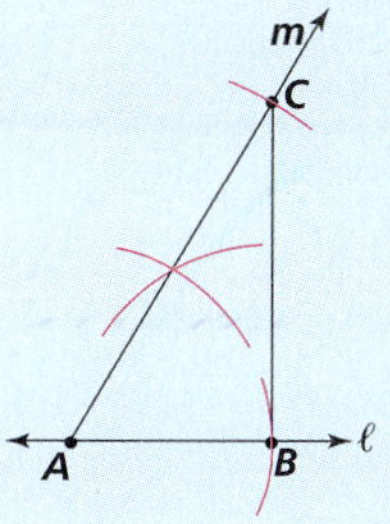

ΔABC has the given sides $\overline{AB}$ and $\overline{AC}$ and the given angle A.

Note that this is the only triangle you can construct using the given sides and the angle they form. If you are given a side of an angle, the angle, and the second side of the angle (abbreviated SAS) of a triangle, then there is only one possible construction for ΔABC.

Because there is only one possible construction of a triangle given the SAS, all triangles using the same SAS will be congruent. This idea is expressed by the following postulate.

SAS (SIDE-ANGLE-SIDE) POSTULATE: If two sides and the included angle of one triangle are congruent to the corresponding two sides and included angle of a second triangle, the triangles are congruent.

2. Teaching the Lesson

To reinforce the concept of congruence, have students cut their sheet of paper from the Warm-Up Activity along the fold line and compare the size, shape, sides, and angles of the two triangles. Tell them to label the corresponding sides and angles using the prime symbol for the second triangle (A, A'; B, B', and so on).

After demonstrating and reviewing the construction on page 179, give students time to practice constructing the same triangle. You may wish to assign partners so that all students have help in constructing congruent angles and sides. Ask pairs of students to compare their triangles, using the SAS postulate to determine whether their triangles are congruent.

3. Reinforce and Extend

Learning Styles

Visual/Spatial To reinforce the SAS postulate as proof of congruency, students can draw pairs of triangles—some that are congruent by SAS and others that are not congruent by SAS. Tell students to use ticks and arcs to indicate congruent and noncongruent sides and angles. Have each student make a key, indicating which triangles can and cannot be proven congruent by SAS on a separate sheet of paper. Students can exchange papers and identify each pair of triangles as congruent by SAS or not. (Emphasize that students should not conclude that triangles that are not congruent by SAS are noncongruent. They are simply determining whether or not each pair of triangles can be proven congruent using the SAS postulate.)

Common Error

Students sometimes forget that in order to use SAS to prove triangles congruent, the congruent angle must be between, or formed by, the pair of congruent sides. To help prevent this misconception, tell students to always mark pairs of congruent sides and angles before testing for SAS. They might find it helpful to use a colored pencil to highlight the two pairs of congruent sides. The highlighting can help them double-check that the sides of the congruent angle are the same as the two congruent sides.

Note that using ticks (|) and arcs (\ ) indicates which sides and angles are equal to one another.

If you know that $\Delta ABC \cong \Delta DEF$ by the SAS postulate, then you can use the SAS postulate, as well as the markings on the triangles, to identify which sides and angles are congruent.

$\angle A \cong \angle D$ $\quad \overline{AB} \cong \overline{DE}$

$\angle B \cong \angle E$ $\quad \overline{AC} \cong \overline{DF}$

$\angle C \cong \angle F$ $\quad \overline{BC} \cong \overline{EF}$

Given: isosceles ΔABC with $\overline{AB} \cong \overline{BC}$, and $\overline{BD}$ bisects $\angle ABC$

To Prove: $\Delta ABD \cong \Delta CBD$

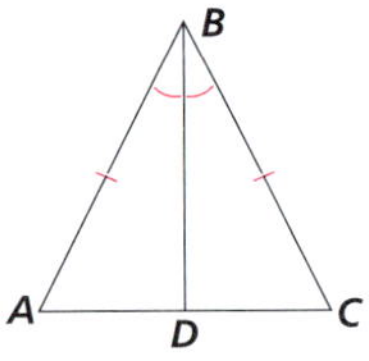

Proof

Statement	Reason
1. $\overline{AB} \cong \overline{BC}$	1. Given.
2. $\overline{BD}$ bisects $\angle ABC$	2. Given.
3. $\angle ABD \cong \angle DBC$	3. Definition of angle bisector.
4. $\overline{BD} \cong \overline{BD}$	4. Property of equality, any quantity is equal to itself.
5. $\therefore \Delta ABD \cong \Delta CBD$	5. SAS Postulate; side, angle, and side of $\Delta ABD \cong$ side, angle, and side of ΔCBD.

Given: G is the midpoint of $\overline{FI}$ and $\overline{EH}$

To Prove: $\triangle EFG \cong \triangle HIG$

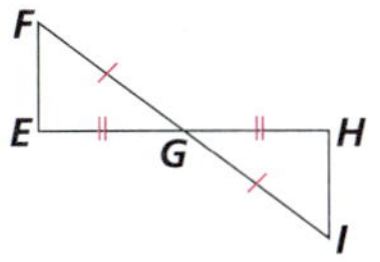

PROOF

Statement	Reason
1. G is the midpoint of $\overline{FI}$ and $\overline{EH}$	**1.** Given.
2. $\overline{EG} \cong \overline{GH}$ and $\overline{FG} \cong \overline{GI}$	**2.** Definition of midpoint.
3. $\angle EGF \cong \angle IGH$	**3.** Vertical angles are equal.
4. $\therefore \triangle EFG \cong \triangle HIG$	**4.** SAS Postulate; side, angle, and side of $\triangle EFG \cong$ side, angle, and side of $\triangle HIG$.

EXAMPLES $\triangle UVW \cong \triangle XYZ$ by SAS
Which sides of the triangles are *given* congruent?

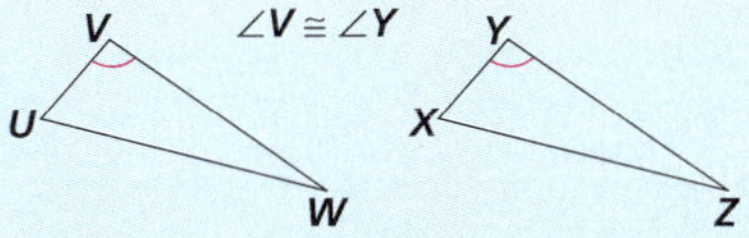

The segments that form the congruent angles are congruent.
$\overline{VU} \cong \overline{YX}$ and $\overline{VW} \cong \overline{YZ}$

$\triangle KLM \cong \triangle QRP$ by SAS
The markings show the congruent angles and one pair of congruent sides. What remaining sides must be congruent to satisfy the SAS postulate?

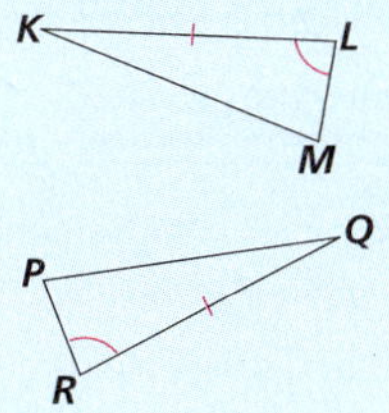

The congruent sides must be the second side of the congruent angles. Because $\angle L \cong \angle R$ and $\overline{KL} \cong \overline{QR}$, then $\overline{LM} \cong \overline{RP}$.

LEARNING STYLES

Body/Kinesthetic Encourage students to construct large congruent triangles similar to those on pages 180 and 181. Students can use string or ribbon, poster board, and markers for their constructions. Ask them to write the proof on their posters. Suggest that they expand on the reasons in the proof by defining terms and concepts. For example, they could define *midpoint* as the point that divides a line into two equal parts. Display the posters in the classroom.

APPLICATION

At Home

To reinforce their understanding of congruent triangles, segments, and angles, students can look around their homes for products or designs that contain these figures. Ask them to describe where they found the figures and to draw a simple sketch of each congruent figure within the context of its location.

Selected Answers and Solutions begin on page 420.

Exercise A Each pair of triangles is congruent by SAS. List the given congruent angles and sides for each pair of triangles.

1.

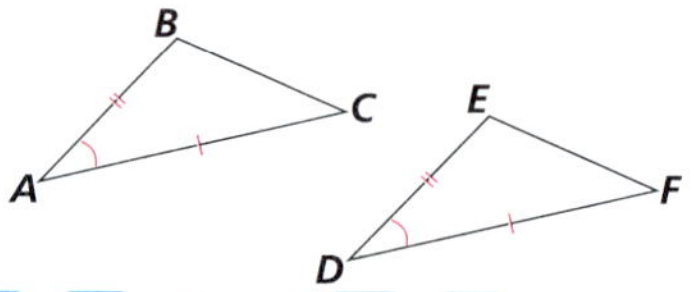

$\overline{AB} \cong \overline{DE}$, $\angle A \cong \angle D$, $\overline{AC} \cong \overline{DF}$

2.

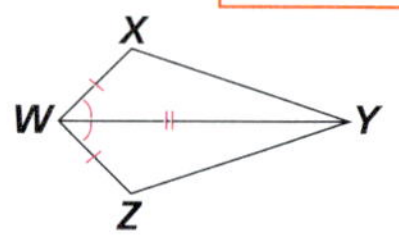

$\overline{WX} \cong \overline{WZ}$, $\angle YWX \cong \angle YWZ$, $\overline{WY} \cong \overline{WY}$

Exercise B Each pair of triangles has a pair of congruent angles. What pairs of sides must be congruent to satisfy the SAS postulate?

3.

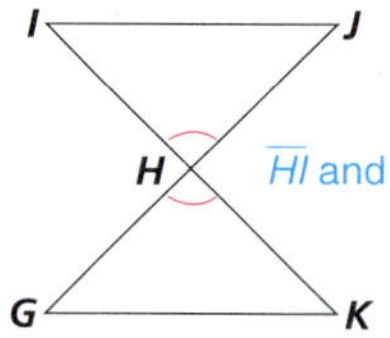

$\overline{HI}$ and $\overline{HK}$; $\overline{HJ}$ and $\overline{HG}$

4.

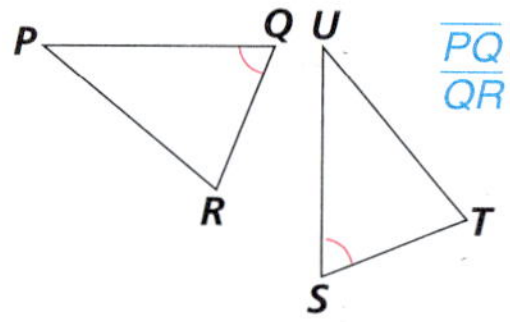

$\overline{PQ}$ and $\overline{US}$; $\overline{QR}$ and $\overline{ST}$

5.

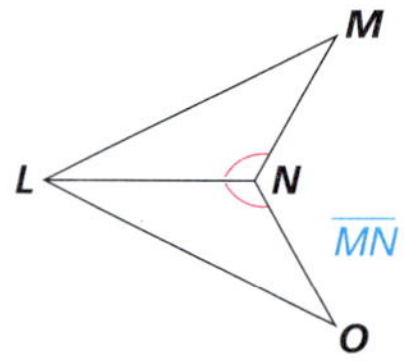

$\overline{MN}$ and $\overline{ON}$; $\overline{NL}$ and $\overline{NL}$

6.

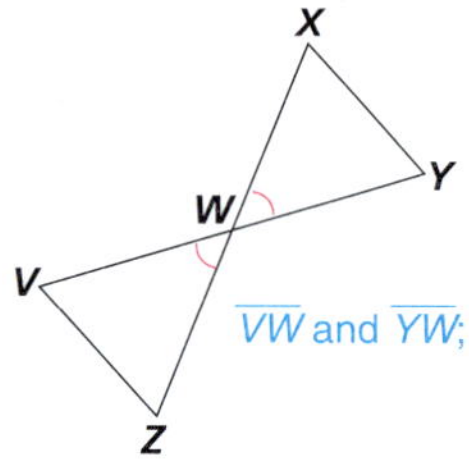

$\overline{VW}$ and $\overline{YW}$; $\overline{WZ}$ and $\overline{WX}$

Exercise C Copy each proof. Fill in the missing steps.

Given: $\triangle ABC$, $\triangle ADC$, $\overline{AC}$ bisects $\angle BCD$ and $\overline{CD} \cong \overline{CB}$

To Prove: $\triangle ABC \cong \triangle ADC$

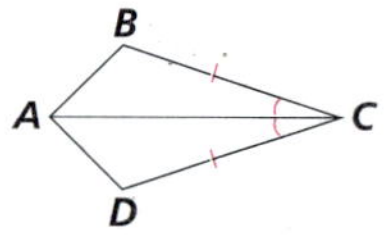

PROOF

Statement	Reason
1. $\overline{CD} \cong \overline{CB}$	1. __7.__ Given.
2. $\overline{AC}$ bisects $\angle BCD$	2. __8.__ Given.
3. $\angle ACD \cong$ __9.__ $\angle ACB$	3. Definition of angle bisector.
4. $\overline{AC} \cong \overline{AC}$	4. __10.__ Any quantity is equal to itself.
5. $\therefore \triangle ABC \cong \triangle ADC$	5. SAS Postulate; side, angle, and side of $\triangle ABC \cong$ side, angle, and side of $\triangle ADC$.

Given: right triangles ΔEFG and ΔHIK,
$\overline{EF} \cong \overline{HI}$ and $\overline{FG} \cong \overline{IK}$

To Prove: $\Delta EFG \cong \Delta HIK$

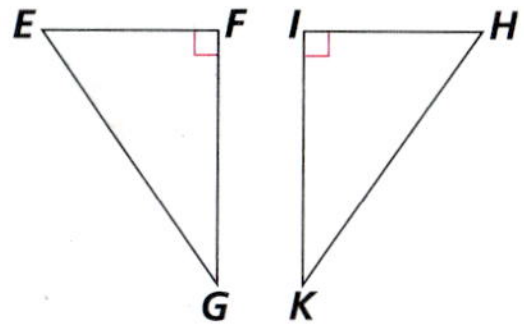

PROOF

Statement	Reason
1. $\overline{EF} \cong \overline{HI}$ and $\overline{FG} \cong \overline{IK}$	1. **11.** Given.
2. $m\angle EFG = 90°$, $m\angle HIK = 90°$	2. **12.** Given.
3. **13.** $\angle EFG \cong \angle HIK$	3. Things equal to the same thing are equal to each other.
4. $\therefore \Delta EFG \cong \Delta HIK$	4. **14.** SAS Postulate; side, angle, and side of $\Delta EFG \cong$ side, angle, and side of ΔHIK.

Problem Solving

Exercise D Find the answer to the following problem.

15. Tamisha wants to swim across the widest part of the pond, but first she wants to know the distance she must swim. She places a marker at point *X*. She moves to point *A* and walks 25 paces in a straight line to *X*. At *X*, she continues in a straight line for another 25 paces. She marks that point *A*'.

Tamisha then moves to point *B*. Walking in a straight line, she takes 35 paces to reach point *X*. She walks an additional 35 paces in a straight line and marks the endpoint *B*'. She walks the distance from *B*' to *A*' and finds it is 40 paces. What is the distance across the pond from point *A* to point *B*? How do you know?

40 paces, SAS congruency

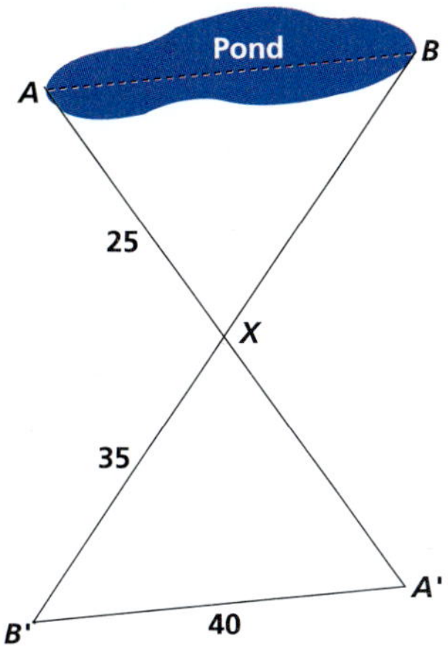

Group Problem Solving

Have groups of students write problems, modeled on problem 15, in which similar triangles need to be drawn to find a measurement. Before organizing students into groups, brainstorm scenarios for problems. Groups should present their problems in a clear and organized way and include good illustrations so that the problems may be shared with and solved by other groups. They should also prepare solution keys for their problems on separate sheets of paper. Groups exchange and solve one another's problems.

Learning Styles

Logical/Mathematical To help them understand the importance of the position of congruent parts when using SAS to prove congruency, have students draw a pair of triangles to illustrate an SSA congruence of parts; that is, triangles with two sides and a nonincluded angle that are congruent. Students should use a compass and straightedge to draw their triangles as accurately as possible. Display the drawings. Ask students to compare the different examples of SSA pairs and to explain why SSA cannot be used to establish that triangles are congruent.

Lesson at a Glance

Chapter 6 Lesson 2

Overview This lesson demonstrates the Side-Side-Side construction of a triangle and explains why the sum of the lengths of any two sides of a triangle must be greater than the length of the third side.

Objectives

- To determine if three given segments can be used to construct a triangle.
- To construct a triangle using the SSS construction technique.

Student Pages 184–188

Teacher's Resource Library TRL

Workbook Activity 63

Additional Practice 35

Mathematics Vocabulary

hypotenuse

1. Warm-Up Activity

Have each student cut a straw into three pieces of any length and try to form a triangle with the three pieces. Have students glue the pieces on a sheet of paper in the shape of a triangle or a figure as close to a triangle as possible. Some students may create a figure in which the three sides do not connect to one another or the sides overhang the triangle they formed. Then ask students to measure the length of each piece. Discuss whether all students were able to form triangles with their three pieces and ask for their thoughts on why some pieces could be used to form triangles and some could not. Accept all reasonable comments and ideas.

Lesson 2 **SSS Construction and Triangle Inequality**

> ***Hypotenuse***
> *The side opposite the right angle in a right triangle*

Suppose you are given three sides of a triangle, *a*, *b*, and *c*. You can use a compass and a straightedge to construct the triangle. Because you are given three sides of the triangle, this construction is called the SSS construction.

CONSTRUCTION

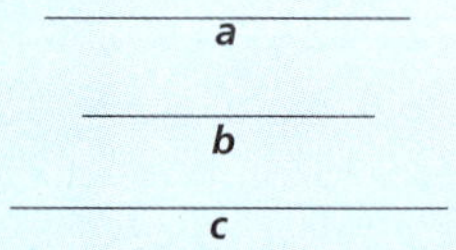

Given: segments *a*, *b*, and *c*

Construct: triangle with sides congruent to *a*, *b*, and *c*

Step 1 Draw any line ℓ. Using your compass, copy side *c* onto ℓ. Label the endpoints *A* and *B*.

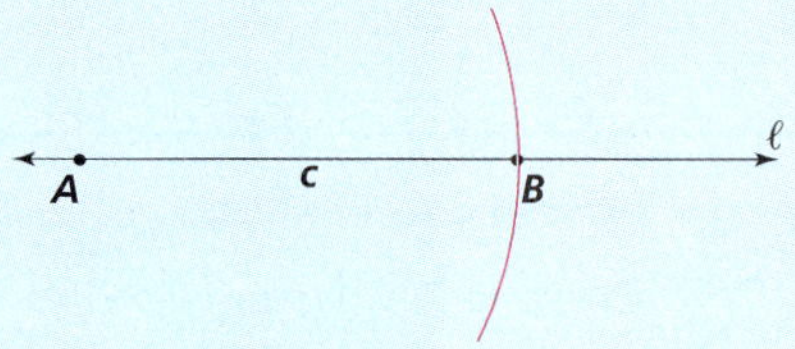

Step 2 Draw an arc using *A* as the center and *b* as the radius.

Step 3 Draw an arc using *B* as the center and *a* as the radius. Label the point where the arcs intersect *C*.

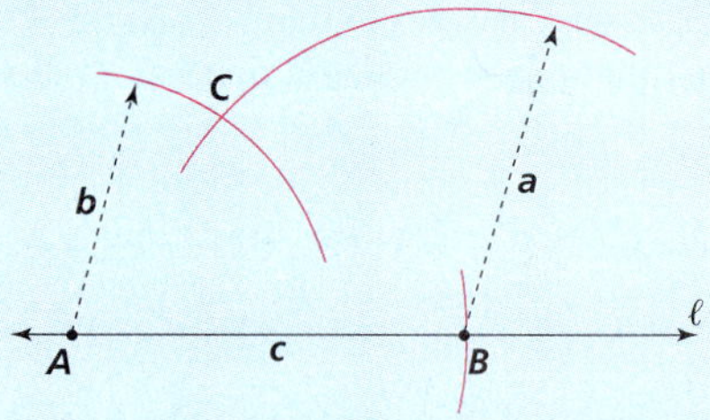

Step 4 Draw segments $\overline{AC}$ and $\overline{BC}$.

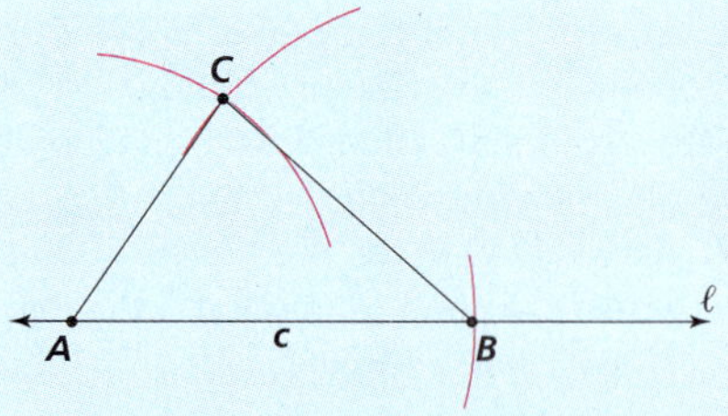

Triangle *ABC* has sides *a*, *b*, and *c* equal to the given segment lengths.

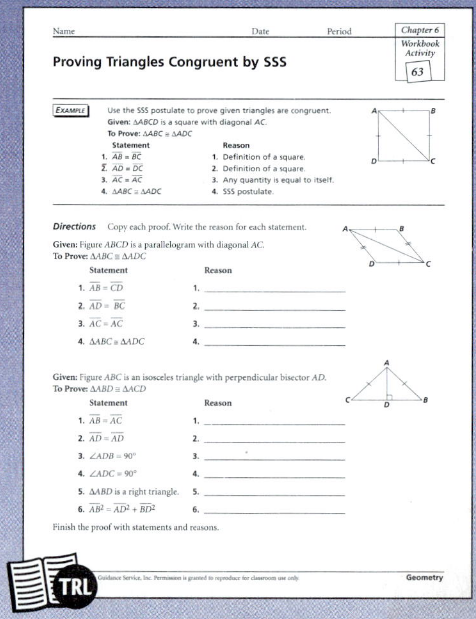

Name Date Period

Chapter 6 Workbook Activity 63

Proving Triangles Congruent by SSS

EXAMPLE Use the SSS postulate to prove given triangles are congruent.
Given: $\triangle ABCD$ is a square with diagonal *AC*.
To Prove: $\triangle ABC \cong \triangle ADC$

Statement	Reason
1. $\overline{AB} = \overline{BC}$	1. Definition of a square.
2. $\overline{AD} = \overline{DC}$	2. Definition of a square.
3. $\overline{AC} = \overline{AC}$	3. Any quantity is equal to itself.
4. $\triangle ABC \cong \triangle ADC$	4. SSS postulate.

Directions Copy each proof. Write the reason for each statement.

Given: Figure *ABCD* is a parallelogram with diagonal *AC*.
To Prove: $\triangle ABC \cong \triangle ADC$

Statement	Reason
1. $\overline{AB} = \overline{CD}$	1. ______
2. $\overline{AD} = \overline{BC}$	2. ______
3. $\overline{AC} = \overline{AC}$	3. ______
4. $\triangle ABC \cong \triangle ADC$	4. ______

Given: Figure *ABC* is an isosceles triangle with perpendicular bisector *AD*.
To Prove: $\triangle ABD \cong \triangle ACD$

Statement	Reason
1. $\overline{AB} = \overline{AC}$	1. ______
2. $\overline{AD} = \overline{AD}$	2. ______
3. $\angle ADB = 90°$	3. ______
4. $\angle ADC = 90°$	4. ______
5. $\triangle ABD$ is a right triangle.	5. ______
6. $\overline{AB}^2 = \overline{AD}^2 + \overline{BD}^2$	6. ______

Finish the proof with statements and reasons.

 Geometry

Workbook Activity 63

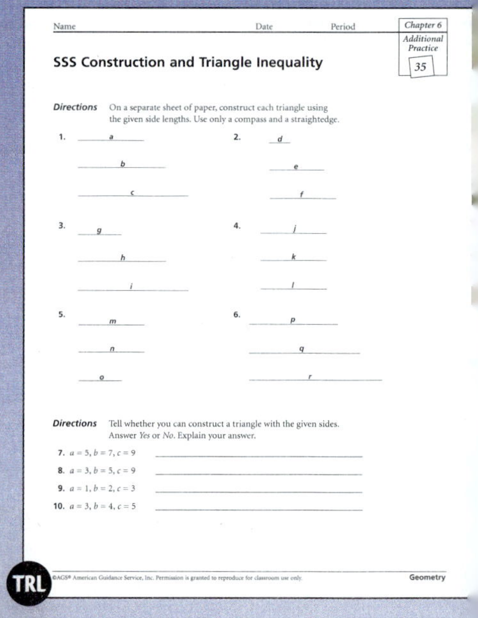

Name Date Period

Chapter 6 Additional Practice 35

SSS Construction and Triangle Inequality

Directions On a separate sheet of paper, construct each triangle using the given side lengths. Use only a compass and a straightedge.

1. a b c
2. d e f
3. g h i
4. j k l
5. m n o
6. p q r

Directions Tell whether you can construct a triangle with the given sides. Answer *Yes* or *No*. Explain your answer.

7. $a = 5, b = 7, c = 9$ ______
8. $a = 3, b = 5, c = 9$ ______
9. $a = 1, b = 2, c = 3$ ______
10. $a = 3, b = 4, c = 5$ ______

 Geometry

Additional Practice 35

Note that there is a general rule for naming the sides and angles of a triangle: side a is opposite angle A, side b is opposite angle B, and side c is opposite angle C.

Although you were able to use SSS to construct triangle ABC in the example, you cannot use any three segments to construct a triangle! For example:

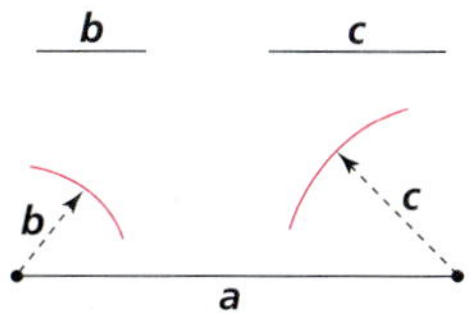

Given: segments a, b, and c

Construct: triangle with sides congruent to a, b, and c

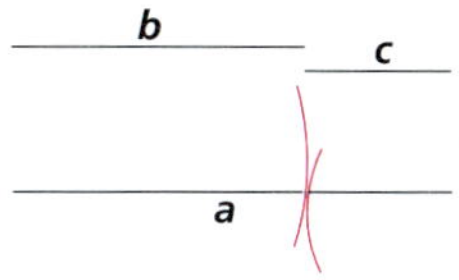

The arcs drawn from A and B do not intersect because the length of segment $b + c$ is less than a. Therefore, no triangle can be constructed.

Suppose that $b + c = a$. In this case, the arcs intersect on a, so again no triangle can be constructed.

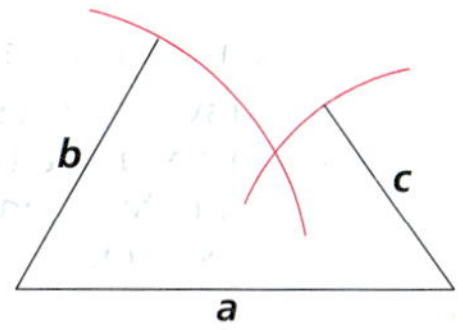

Therefore, we can conclude that $b + c$ must be greater than a in order to construct a triangle.

Conclusion: The sum of the lengths of any two sides of a triangle must be greater than the length of the third side.

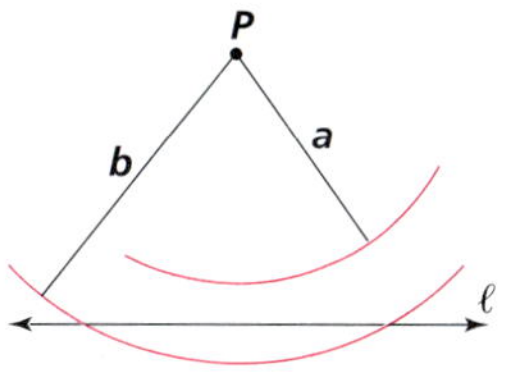

In order to formally prove this conclusion, we first need to learn about the shortest distance to a line.

Given a point, P, not on line ℓ, what is the shortest distance from P to ℓ?

Radius a is too short, radius b is too long. The shortest distance is a perpendicular from P to ℓ. $\overline{PQ} \perp \ell$ and $\overline{PQ}$ is the shortest distance from P to ℓ. This leads us to the *shortest distance theorem,* which will be taken as true without proof.

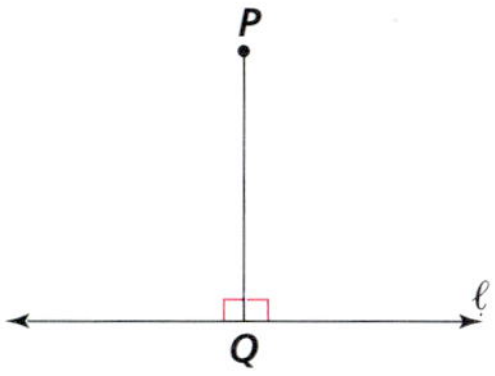

Remember, the radius is the distance from the center of a circle to the circle.

2. Teaching the Lesson

After they complete the Warm-Up activity, have students read the lesson to learn about the relationship between the lengths of the sides of a triangle. Allow them to experiment with SSS constructions similar to those shown in the lesson. Ask, "What is significant about the point where the arcs intersect?" (That is the point where the two sides will meet and form the third vertex of the triangle.)

After students conclude that the sum of the lengths of any two sides of the triangle must be greater than the length of the third side, introduce the Triangle Inequality Theorem. Apply the theorem to the following lengths and, with the class, determine whether they form the sides of a triangle. Draw the lengths on the board, using inches or centimeters as the unit of measure.

5, 8, 10	(yes)
26, 20, 5	(no)
2, 3, 1	(no)
9, 5, 6	(yes)
40, 50, 15	(yes)

3. Reinforce and Extend

Learning Styles

Visual/Spatial Following the lesson construction techniques, have students try to construct triangles with the following numbers of units in inches or centimeters. Ask students to paste each construction on construction paper and explain why the constructions do or do not result in triangles.

2, 8, 9	(yes)
4, 5, 6	(yes)
10, 10, 1	(yes)
15, 35, 35	(no)
19, 2, 6	(no)

LEARNING STYLES

Body/Kinesthetic To help students explore the Shortest Distance Theorem on their own, tell them to glue string on paper to form a line segment two inches or longer and glue a button as a point above the segment. Have them attach strings to form several line segments from the point to the line, including one that is perpendicular to the line. Then ask students to use a straightedge to measure the length of each segment to verify that the perpendicular segment is indeed the shortest. Tell students to make generalizations about the lengths of the line segments as they get farther away from the perpendicular segment.

SHORTEST DISTANCE THEOREM: The perpendicular segment from a point to a line is the shortest segment from that point to the line. The length of the perpendicular segment is called the *distance* from the point to the line.

$\overline{PQ}$ = shortest distance from P to ℓ.

$\overline{PR} > \overline{PQ}$, $\overline{PR}$ is **not** the shortest distance.

Likewise, $\overline{RQ}$ = shortest distance from R to $\overline{PQ}$.

$\overline{PR} > \overline{RQ}$, $\overline{PR}$ is **not** the shortest distance.

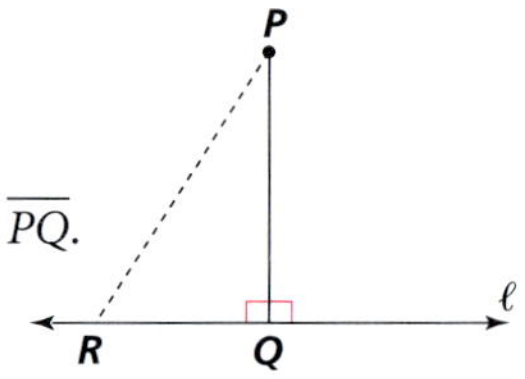

Be sure you recognize that ΔPQR is a right triangle and that $\overline{PR}$, opposite the right angle, is the longest side of the triangle.

In a right triangle, the side opposite the right angle is the longest side of the triangle and is called the **hypotenuse.**

Now we have the tools we need to prove the Triangle Inequality Theorem.

TRIANGLE INEQUALITY THEOREM: The sum of the lengths of any two sides of a triangle must be greater than the length of the third side.

Given: ΔABC, any triangle

To Prove: $\overline{AC} + \overline{BC} > \overline{BA}$
$\overline{AB} + \overline{AC} > \overline{BC}$
$\overline{AB} + \overline{BC} > \overline{AC}$

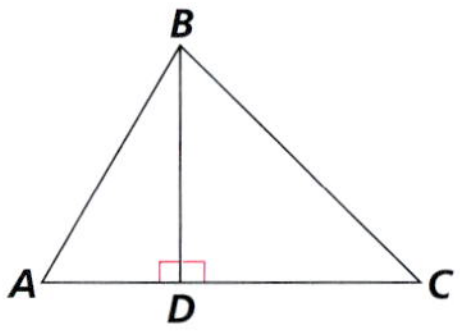

PROOF

Statement	Reason
1. $\triangle ABC$	1. Given.
2. Construct $\overline{BD}$, a segment perpendicular to $\overline{AC}$ from B	2. Construction of a perpendicular to a line.
3. $\overline{AD}$ is the shortest distance from A to $\overline{BD}$, which means $\overline{AB} > \overline{AD}$ and $\overline{CD}$ is the shortest distance from C to $\overline{BD}$, which means $\overline{BC} > \overline{DC}$	3. Shortest distance theorem.
4. $\overline{AB} > \overline{AD}$ $\overline{BC} > \overline{DC}$ $\overline{AB} + \overline{BC} > \overline{AD} + \overline{DC}$	4. Addition.
5. $\therefore \overline{AB} + \overline{BC} > \overline{AC}$	5. Substitution.

Note that it is enough to prove $\overline{AB} + \overline{BC} > \overline{AC}$. You can prove the other two cases by relabeling the vertices.

You can use this theorem to help you determine if any three given sides can form a triangle.

EXAMPLES Can three segments of lengths 3 inches, 4 inches, and 5 inches form a triangle?

$3 + 4 >? 5$ true

$3 + 5 >? 4$ true

$4 + 5 >? 3$ true

The segments satisfy the triangle inequality theorem and so such a triangle exists.

Can three segments of lengths 10 ft, 15 ft, and 4 ft form a triangle?

$10 + 4 >? 15$ false

The segments do not satisfy the triangle inequality theorem and so no such triangle exists.

Writing About Mathematics

Line segment BD is an altitude from angle B to side AC of a triangle. What can you say about $\overline{BD}$ and the distance to $\overline{AC}$? (Hint: Draw a picture of the triangle and its altitude.)

LEARNING STYLES

Interpersonal/Group Learning Have each student write the lengths of two sides of a triangle on one index card and the length of the third side on another card. Ask students to leave their card with the two lengths on the top of their desks and to stand up, holding their second card. Collect the cards students are holding, mix them, and randomly redistribute them to the students. Invite students to move from desk to desk to find a card that has lengths on it that could form a triangle with the length on the card they are holding. There will be more than one way to match up the cards, so some cards may be unmatched. Help students work together to find a way to make matches for all of the cards.

LEARNING STYLES

Logical/Mathematical Have students make a flowchart or other form of organizational chart that can be used to guide a person through the process of determining if three given segments can form the sides of a triangle. A sample flowchart is shown below.

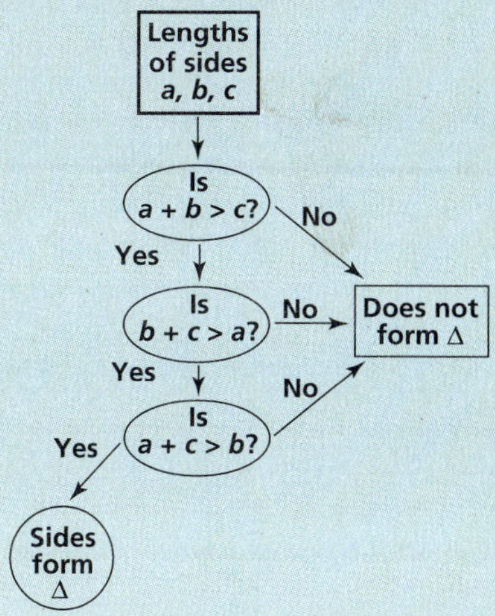

MENTAL MATH

Write the lengths below on the board and ask students to determine mentally if the lengths can be the sides of a triangle.

- 4, 5, 6 (yes)
- 2, 9, 10 (yes)
- 1, 9, 10 (no)
- 20, 10, 5 (no)
- 4, 4, 10 (no)
- 3, 17, 20 (no)
- 4, 17, 20 (yes)
- 6, 6, 13 (yes)
- 6, 6, 12 (no)
- 1, 1, 1 (yes)

Common Error

When adding the lengths of sides to see whether they form a triangle, students sometimes pair the same two numbers twice but in a different order (for example, $a + b > c$ and $b + a > c$). To help them avoid this, tell students to draw and label a triangle with the given lengths and the letters *a, b, c.* They can start with *a* and move clockwise around the triangle to find the lengths to substitute into the inequalities $a + b > c$. Next, they start with *b* and solve for the inequality $b + c > a$, then *c* and solve for the inequality $c + a > b$.

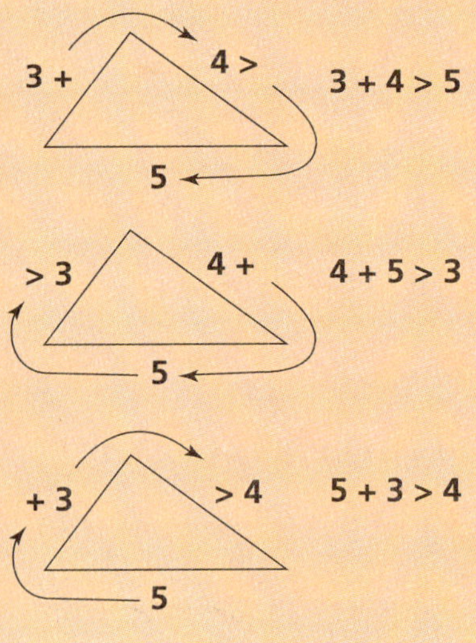

Exercise A Tell whether you can construct a triangle with the given sides. Answer *yes* or *no*. Explain your answer.

1. 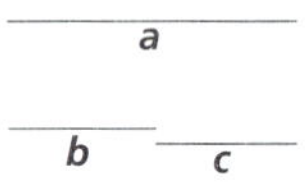

no, $b + c = a$

2. 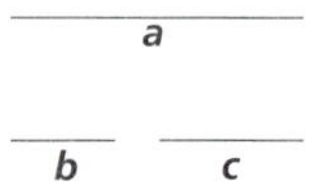

no, $b + c < a$

3. 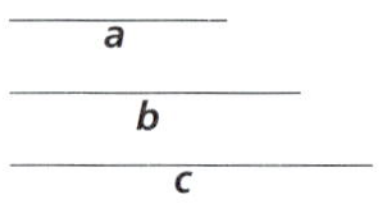

yes, $a + b > c$, $b + c > a$, and $c + a > b$

4. 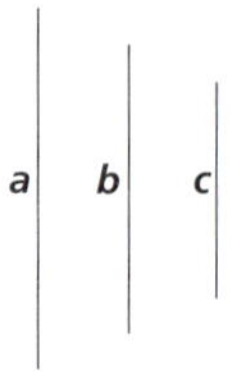

yes, $a + b > c$, $b + c > a$, and $c + a > b$

5. $a = 4, b = 7, c = 10$
   yes, $14 > 7$, $11 > 10$, and $17 > 4$
6. $a = 8, b = 2, c = 5$
   no, $7 < 8$
7. $a = 9, b = 11, c = 12$
   yes, $20 > 12$, $23 > 9$, and $21 > 11$
8. $a = 9, b = 12, c = 15$
   yes, $21 > 15$, $27 > 9$, and $24 > 12$
9. $a = 3, b = 3, c = 5$
   yes, $6 > 5$ and $8 > 3$
10. $a = 6, b = 7, c = 14$
    no, $13 < 14$

Exercise B Write >, <, or = to make each statement true.

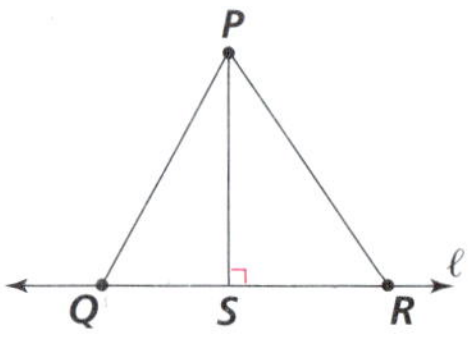

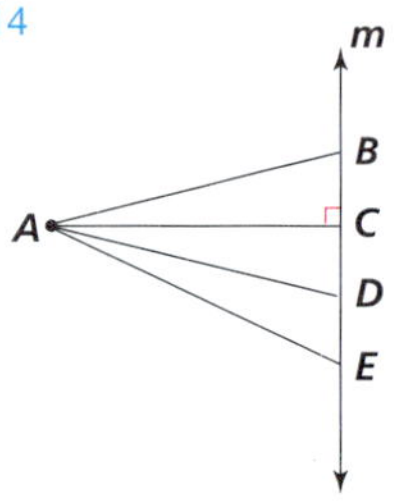

11. $\overline{PQ}$ __>__ shortest distance from P to ℓ
12. $\overline{PR}$ __>__ shortest distance from P to ℓ
13. $\overline{PS}$ __=__ shortest distance from P to ℓ
14. $\overline{AB}$ __>__ shortest distance from A to m
15. $\overline{AC}$ __=__ shortest distance from A to m
16. $\overline{AD}$ __>__ shortest distance from A to m
17. $\overline{AE}$ __>__ shortest distance from A to m

Student constructions should include all the steps necessary for constructing SSS congruent triangles.

Construction

Exercise C Construct each triangle using the given side lengths. Use only a compass and a straightedge.

18. 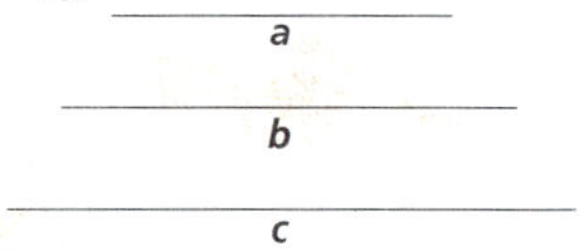

19. 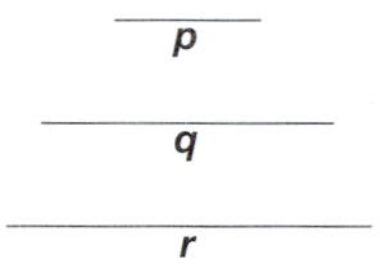

20. 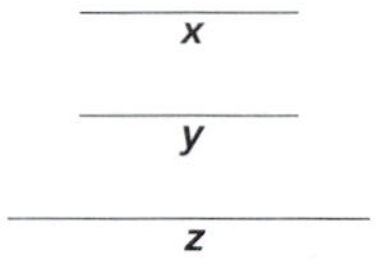

SSS and ASA Congruencies

You've shown that if you are given any three segments that satisfy the triangle inequality theorem, you can construct a triangle whose sides are congruent to the segments. In addition, any other triangles that have those same sides are congruent to each other.

SSS (Side-Side-Side) Postulate: If three sides of a triangle are congruent to three sides of a second triangle, then the triangles are congruent.

This means you now have two ways—SAS and SSS—to show that one triangle is congruent to another. There is a third way—angle-side-angle or ASA. The following construction is an example of ASA congruency.

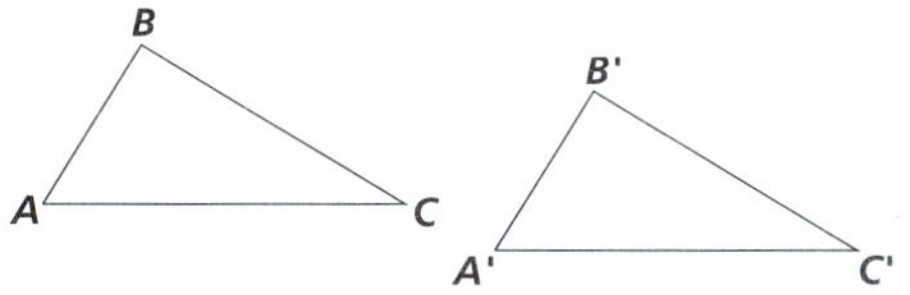

Construction

Given: angle A, angle B, and $\overline{AB}$. $\overline{AB}$ is called the included side because it is between the vertices of the given angles.

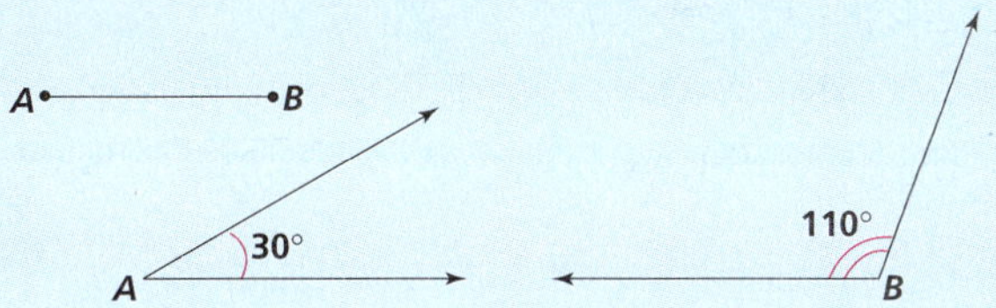

Construct: a triangle with angles congruent to $\angle A$ and $\angle B$ and included side congruent to $\overline{AB}$

Step 1 Use a compass to copy $\overline{AB}$ onto line ℓ.

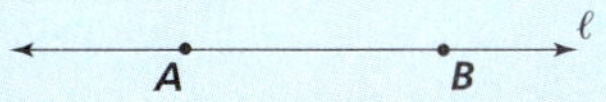

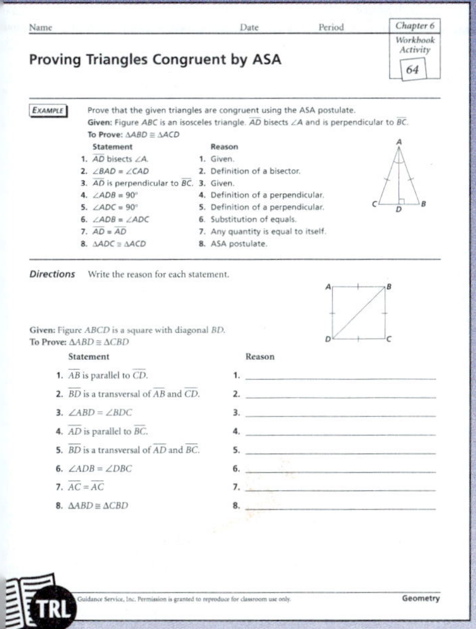
Name Date Period

Chapter 6 Workbook Activity 64

Proving Triangles Congruent by ASA

Example Prove that the given triangles are congruent using the ASA postulate.
Given: Figure ABC is an isosceles triangle. $\overline{AD}$ bisects $\angle A$ and is perpendicular to $\overline{BC}$.
To Prove: $\Delta ABD \cong \Delta ACD$

Statement	Reason
1. $\overline{AD}$ bisects $\angle A$.	1. Given.
2. $\angle BAD = \angle CAD$	2. Definition of a bisector.
3. $\overline{AD}$ is perpendicular to $\overline{BC}$.	3. Given.
4. $\angle ADB = 90°$	4. Definition of a perpendicular.
5. $\angle ADC = 90°$	5. Definition of a perpendicular.
6. $\angle ADB = \angle ADC$	6. Substitution of equals.
7. $\overline{AD} = \overline{AD}$	7. Any quantity is equal to itself.
8. $\Delta ADC \cong \Delta ACD$	8. ASA postulate.

Directions Write the reason for each statement.

Given: Figure $ABCD$ is a square with diagonal BD.
To Prove: $\Delta ABD \cong \Delta CBD$

Statement	Reason
1. $\overline{AB}$ is parallel to $\overline{CD}$.	1. ______
2. $\overline{BD}$ is a transversal of $\overline{AB}$ and $\overline{CD}$.	2. ______
3. $\angle ABD = \angle BDC$	3. ______
4. $\overline{AD}$ is parallel to $\overline{BC}$.	4. ______
5. $\overline{BD}$ is a transversal of $\overline{AD}$ and $\overline{BC}$.	5. ______
6. $\angle ADB = \angle DBC$	6. ______
7. $\overline{AC} = \overline{AC}$	7. ______
8. $\Delta ABD \cong \Delta CBD$	8. ______

 Geometry

TRL **Workbook Activity 64**

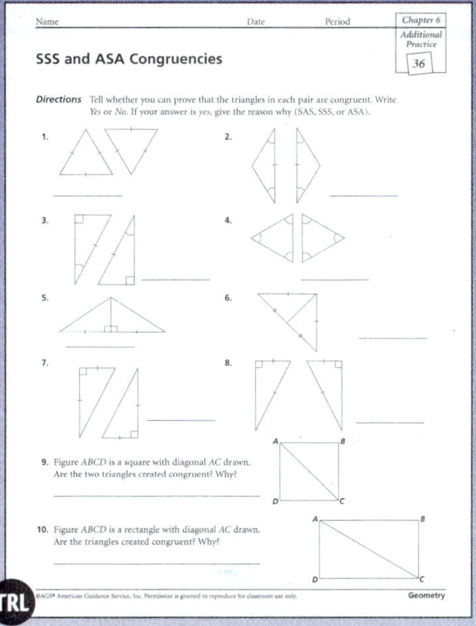
Name Date Period

Chapter 6 Additional Practice 36

SSS and ASA Congruencies

Directions Tell whether you can prove that the triangles in each pair are congruent. Write *Yes* or *No*. If your answer is *yes*, give the reason why (SAS, SSS, or ASA).

1. ______
2. ______
3. ______
4. ______
5. ______
6. ______
7. ______
8. ______
9. Figure $ABCD$ is a square with diagonal AC drawn. Are the two triangles created congruent? Why? ______
10. Figure $ABCD$ is a rectangle with diagonal AC drawn. Are the triangles created congruent? Why? ______

 Geometry

TRL **Additional Practice 36**

Lesson at a Glance

Chapter 6 Lesson 3

Overview This lesson introduces the Side-Side-Side and Angle-Side-Angle postulates for proving triangles congruent and demonstrates how to construct a triangle using the ASA congruency.

Objectives

- To identify two triangles as congruent when their three sides are congruent.
- To use the ASA postulate to solve problems.

Student Pages 189–193

Teacher's Resource Library TRL

Workbook Activity 64

Additional Practice 36

1. Warm-Up Activity

Tell students to cut one straw into three lengths that will form the sides of a triangle. (It may be helpful to review the Triangle Inequality Theorem from Lesson 2 to ensure that students cut their straws into pieces that will form a triangle). Then have students cut another straw into three pieces equal in length to the first three. Ask them to make two *noncongruent* triangles from each set of straws (this is, of course, impossible). After students have had some time to experiment with their pieces of straw, ask for their hypotheses as to why the task is impossible.

2. Teaching the Lesson

Introduce and discuss the SSS and ASA postulates and review the SAS postulate. Then have students consider the similarities and differences between the SSS, ASA, and SAS congruency postulates. After a thorough discussion, ask, "If you know that three angles of one triangle are congruent to three angles of another triangle, does this establish that the two triangles are congruent?" (no) Ask students to support their responses to this question with drawings or concrete examples.

When referring to the three pairs of congruent triangles at the bottom of page 190, review the property of equality and make sure students understand that the side the triangles have in common belongs to both triangles. Drawing the triangles separately may also help illustrate this concept.

Common Error

As students learn new postulates, they may have difficulty identifying the correct postulate for proving a given pair of triangles congruent. Suggest that students label congruent sides and angles with an *S* or an *A*. Seeing the letters on the triangles can help students easily identify the sides and/or angles that are known to be congruent. If a pair of triangles shares a common side, you might suggest that students draw the triangles separately before labeling the congruent sides and angles.

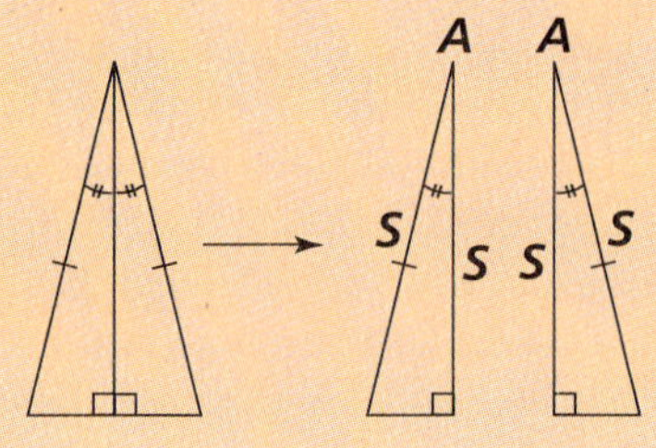

Construction *(continued)*

Step 2 Copy ∠*A* at *A*.

Step 3 Copy ∠*B* at *B*. Label the point where the rays intersect *C*.

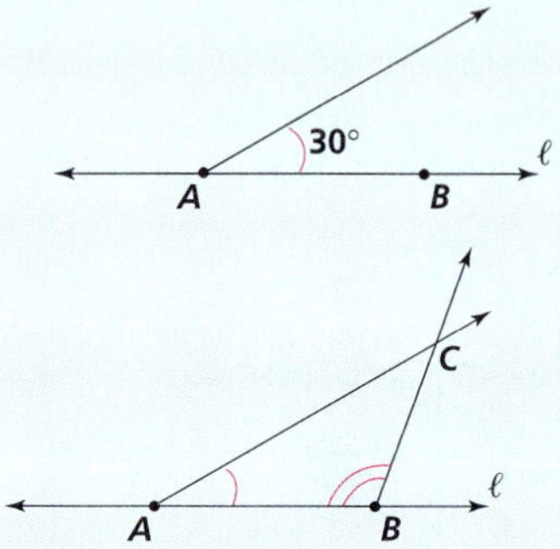

Δ*ABC* has ∠*A* and ∠*B* as given and $\overline{AB}$, the included side, as its base. Any triangle using the same angle, side, and angle—ASA—will be congruent to Δ*ABC*. This leads to the following postulate.

ASA (Angle-Side-Angle) Postulate: If two angles and the included side of a triangle are congruent to the corresponding angles and included side of a second triangle, then the triangles are congruent.

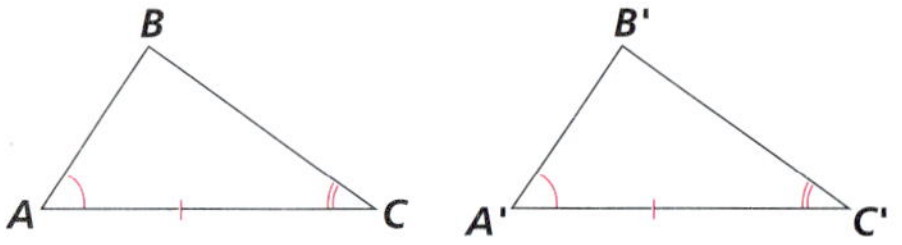

For each diagram below, one postulate—SAS, SSS, or ASA—applies and indicates the two triangles are congruent.

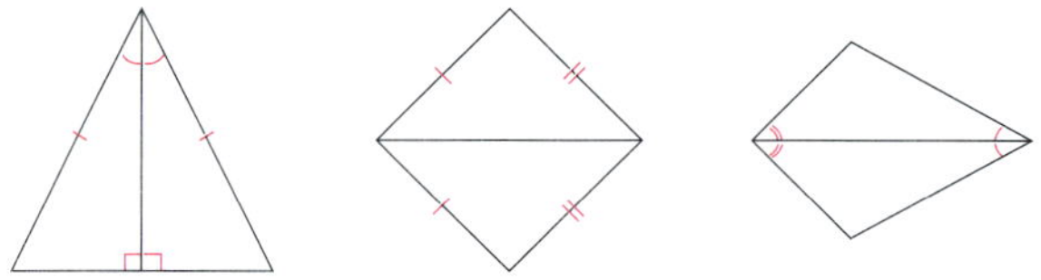

SAS **SSS** **ASA**

Given: quadrilateral *ABCD* with diagonal *BD* that bisects angles *B* and *D*

To Prove: $\triangle ABD \cong \triangle CBD$

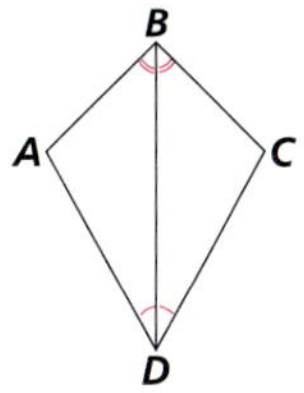

PROOF

Statement	Reason
1. $\overline{BD}$ bisects $\angle D$; $\overline{BD}$ bisects $\angle B$	1. Given.
2. $\angle ABD \cong \angle CBD$; $\angle ADB \cong \angle CDB$	2. Definition of angle bisector.
3. $\overline{BD} \cong \overline{BD}$	3. Property of equality, any quantity is equal to itself.
4. $\therefore \triangle ABD \cong \triangle CBD$	4. ASA Postulate; angle, side, and angle of $\triangle ABD$ is congruent to an angle, side, and angle of $\triangle CBD$.

Given: quadrilateral *WXYZ* with diagonal *WY*, $\overline{WZ} = \overline{WX}$, and $\overline{YX} = \overline{YZ}$

To Prove: $\triangle WXY \cong \triangle WZY$

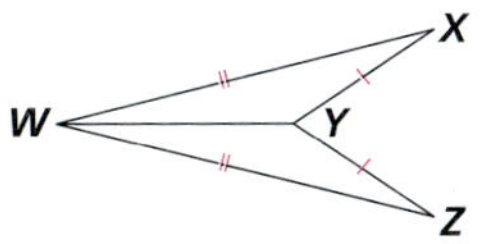

PROOF

Statement	Reason
1. $\overline{WZ} = \overline{WX}$ and $\overline{YX} = \overline{YZ}$	1. Given.
2. $\overline{WY} = \overline{WY}$	2. Property of equality, any quantity is equal to itself.
3. $\therefore \triangle WXY \cong \triangle WZY$	3. SSS Postulate.

In summary, you have three ways to show that one triangle is congruent to another.

Method 1 SAS

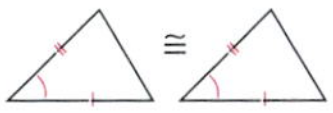

Method 2 SSS

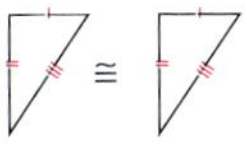

Method 3 ASA

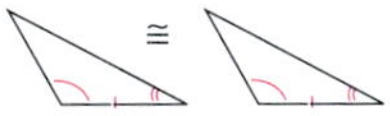

3. Reinforce and Extend

Learning Styles

Body/Kinesthetic Have students construct three pairs of congruent triangles using a compass and protractor. One pair should be made using the SSS construction, one pair with the SAS construction, and the last pair with the ASA construction.

Learning Styles

Auditory/Verbal Ask students to write in their own words the steps involved in constructing triangles using the SSS, ASA, and SAS constructions. Ask for volunteers to demonstrate their methods for the entire class or small groups of students.

Learning Styles

LEP/ESL Make examples of pairs of triangles that are congruent by each of the postulates on one side of a large index card and write the postulate abbreviation on the other side. Students work in pairs, with one partner describing the pair of triangles and naming the postulate that is illustrated while the other holds the card and verifies the answer. Then partners switch roles and repeat. The students should use both sides of the cards as cues.

Exercise A Tell whether the triangles in each pair are congruent. Write *yes* or *no*. If your answer is *yes*, give the reason why (SAS, SSS, or ASA).

1.

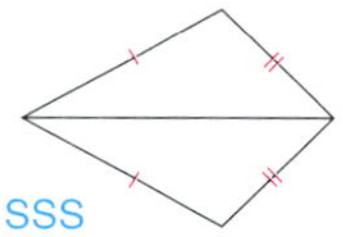

yes, SSS

2.

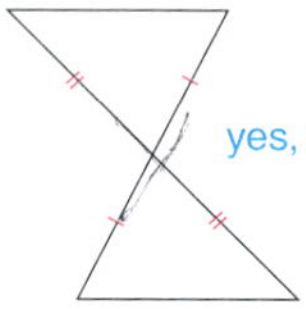

yes, SAS

3.

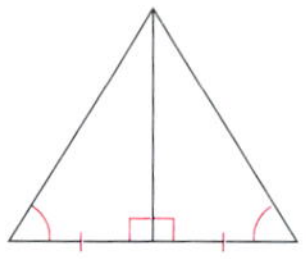

yes, ASA

4.

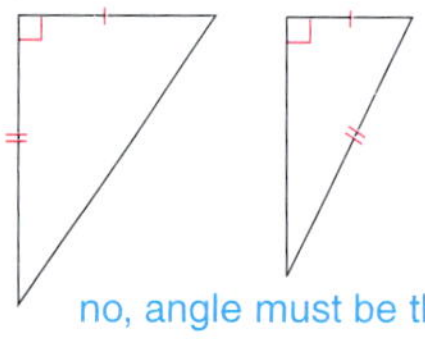

no, angle must be the included angle

5.

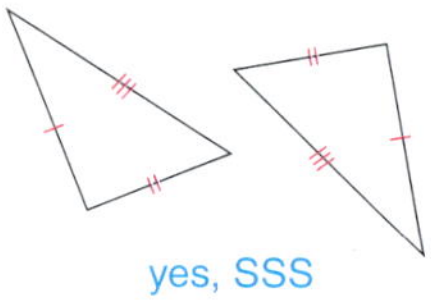

yes, SSS

Exercise B Copy each proof. Fill in the missing steps.

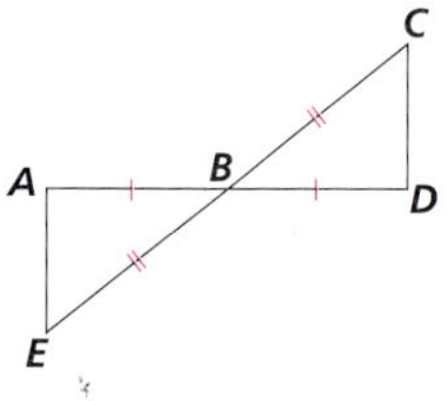

Given: $\triangle ABE$ and $\triangle DBC$, $\overline{AD}$ and $\overline{CE}$ bisect each other at B

To Prove: $\triangle ABE \cong \triangle DBC$

PROOF

Statement	Reason
1. $\triangle ABE$ and $\triangle DBC$	1. **6.** Given.
2. $\overline{AD}$ bisects $\overline{EC}$ at B and $\overline{EC}$ bisects $\overline{AD}$ at B	2. **7.** Given.
3. **8.** $\overline{AB} = \overline{DB}$ and $\overline{EB} = \overline{CB}$	3. Definition of bisector.
4. **9.** $\angle ABE = \angle DBC$	4. Vertical angles are equal.
5. $\therefore \triangle ABE \cong \triangle DBC$	5. **10.** SAS Postulate.

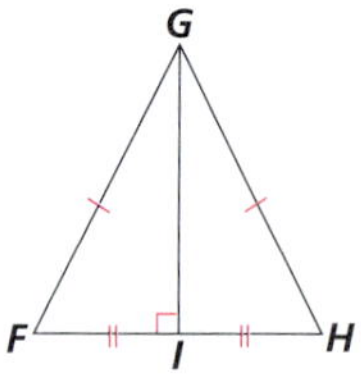

Given: isosceles $\triangle FGH$ with perpendicular bisector GI

To Prove: $\triangle FIG \cong \triangle HIG$

PROOF

	Statement		Reason
1.	isosceles ΔFGH	1.	**11.** Given.
2.	**12.** $\overline{FG} = \overline{GH}$	2.	Definition of isosceles.
3.	$\overline{GI}$ bisects $\overline{FH}$	3.	**13.** Given.
4.	**14.** $\overline{FI} = \overline{IH}$	4.	Definition of bisector.
5.	**15.** $\overline{GI} = \overline{GI}$	5.	Property of equality, any quantity is equal to itself.
6.	$\therefore \Delta FIG \cong \Delta HIG$	6.	**16.** SSS Postulate.

Construction

Student constructions should follow the appropriate steps for copying side lengths and angles.

Exercise C Complete the following constructions.

17. Draw any acute scalene triangle *ABC*. Draw a second triangle *A'B'C'* congruent to ΔABC using the SSS Postulate.

18. Draw any obtuse triangle. Label the obtuse angle *D* and the triangle *DEF*. Draw a second triangle *D'E'F'* congruent to ΔDEF using SAS where *D* is the included angle.

Problem Solving

Exercise D Solve the following problems.

19. Triangle *ABC* has three equal sides and three equal angles. Triangle *DEF* has three equal sides and three equal angles, each equal to the angles of *ABC*. Is $\Delta ABC \cong \Delta DEF$? If yes, explain why. If no, give a counterexample.

See Teacher's Edition page for answer.

20. At ten in the morning, a 30-foot flagpole casts a 10-foot shadow across the ground. A nearby tree casts a shadow of equal length. Rodney, a forester, drew the diagram at right. He assumed the m$\angle P$ = m$\angle T$ because both are formed by the sun's rays as they reached the ground. How tall is the tree? How do you know?

30 ft, the triangles are congruent by ASA

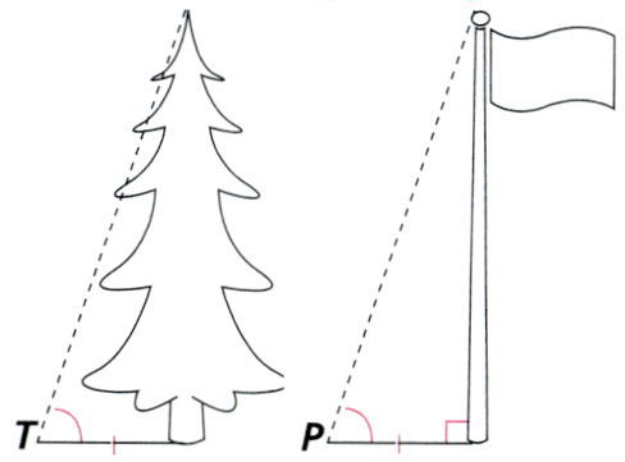

MENTAL MATH

Have groups create three word problems similar to problem 20. Each problem should require a different postulate to solve it. Each group should prepare a solution key on a separate sheet of paper for its problems. Groups then exchange and solve another group's problems. Ask for volunteers to share one or more of their problems and solutions with the class.

Answer to Problem 19

No. Two triangles can have three equal angles but sides of different lengths. In this case, the triangles are similar but not congruent.

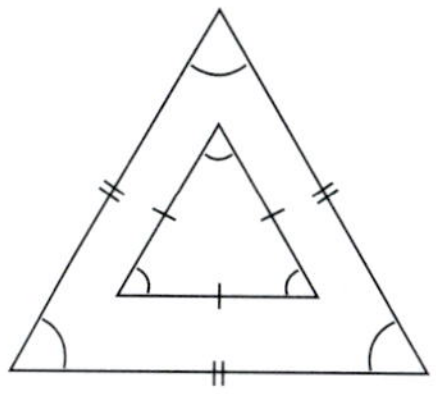

Lesson at a Glance

Chapter 6 Lesson 4

Overview This lesson introduces the Angle-Angle-Side Theorem and two theorems that can be used to prove right triangles are congruent. The lesson also illustrates how to use the special characteristics of right triangles to find angle measures.

Objectives

- To determine that two triangles are congruent based on AAS.
- To use the Hypotenuse-Leg theorem to prove that two right triangles are congruent.
- To calculate unknown angle measures in right triangles.

Student Pages 194–199

Teacher's Resource Library TRL

Workbook Activity 65

Additional Practice 37

1. Warm-Up Activity

Discuss the special characteristics of right triangles and review the terms *hypotenuse, leg, acute angle, right angle,* and *perpendicular.* Make sure students understand that the two nonright angles in a right triangle are both acute angles. Challenge students to explain why the nonright angles are always acute. (A triangle has three angles whose sum are 180°. The right angle is 90°, so the other two angles must add up to a sum of 90°, which means both must be less than 90°, or acute angles.)

2. Teaching the Lesson

Draw and label the vertices of several right triangles and have students identify the hypotenuse and legs and explain how to distinguish between the two. (The hypotenuse is the side opposite the right angle; the legs are the sides that form the right angle.)

Lesson 4 **Special Congruencies: Right Triangles**

You already know that the right angles in any triangle are congruent to one another because the measure of all right angles equals 90°. There are other special conditions that apply only to right angles.

> ***AAS (Angle-Angle-Side) Theorem:***
> If two angles and one side not included between those angles are congruent to two angles and the corresponding side of a second triangle, then the two triangles are congruent.

You can prove this theorem using the Angle Sum Theorem for triangles.

Proof

Given: ΔABC and $\Delta A'B'C'$ where $\overline{AB} \cong \overline{A'B'}$, $\angle A \cong \angle A'$, and $\angle C \cong \angle C'$

To Prove: $\Delta ABC \cong \Delta A'B'C'$

You already know that the sum of the measures of the three angles in any triangle is 180°. So you can write

$(m\angle A + m\angle C) + m\angle B = 180°$ and

$(m\angle A' + m\angle C') + m\angle B' = 180°$.

You can substitute equals for equals, and write

$(m\angle A + m\angle C) + m\angle B = (m\angle A' + m\angle C') + m\angle B'$.

By subtraction you get $m\angle B = m\angle B'$.

Therefore, by ASA, $\Delta ABC \cong \Delta A'B'C'$.

In a right triangle, you know that one angle must be 90°. Next we will prove that the sum of the remaining angles must also be 90°.

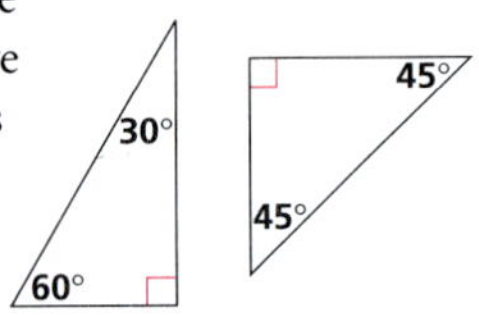

194 *Chapter 6 Congruent Triangles and Transformations*

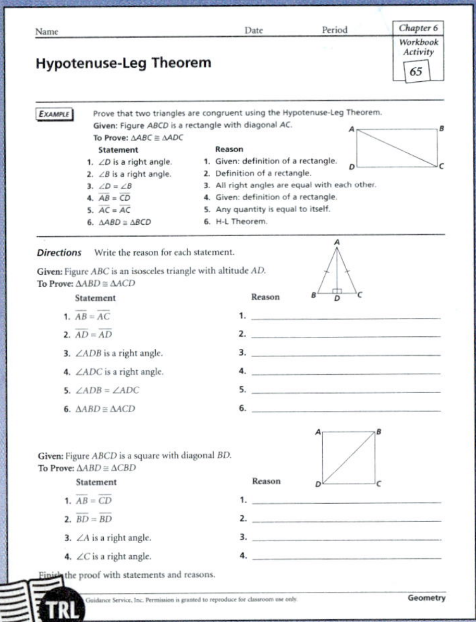

Name Date Period | Chapter 6 Workbook Activity 65

Hypotenuse-Leg Theorem

EXAMPLE Prove that two triangles are congruent using the Hypotenuse-Leg Theorem.

Given: Figure *ABCD* is a rectangle with diagonal *AC*.

To Prove: $\Delta ABC \cong \Delta ADC$

Statement	Reason
1. $\angle D$ is a right angle.	1. Given: definition of a rectangle.
2. $\angle B$ is a right angle.	2. Definition of a rectangle.
3. $\angle D = \angle B$	3. All right angles are equal with each other.
4. $\overline{AB} = \overline{CD}$	4. Given: definition of a rectangle.
5. $\overline{AC} = \overline{AC}$	5. Any quantity is equal to itself.
6. $\Delta ABD \cong \Delta BCD$	6. H-L Theorem.

Directions Write the reason for each statement.

Given: Figure *ABC* is an isosceles triangle with altitude *AD*.

To Prove: $\Delta ABD \cong \Delta ACD$

Statement	Reason
1. $\overline{AB} = \overline{AC}$	1. ________
2. $\overline{AD} = \overline{AD}$	2. ________
3. $\angle ADB$ is a right angle.	3. ________
4. $\angle ADC$ is a right angle.	4. ________
5. $\angle ADB = \angle ADC$	5. ________
6. $\Delta ABD \cong \Delta ACD$	6. ________

Given: Figure *ABCD* is a square with diagonal *BD*.

To Prove: $\Delta ABD \cong \Delta CBD$

Statement	Reason
1. $\overline{AB} = \overline{CD}$	1. ________
2. $\overline{BD} = \overline{BD}$	2. ________
3. $\angle A$ is a right angle.	3. ________
4. $\angle C$ is a right angle.	4. ________

Finish the proof with statements and reasons.

 Geometry

TRL **Workbook Activity 65**

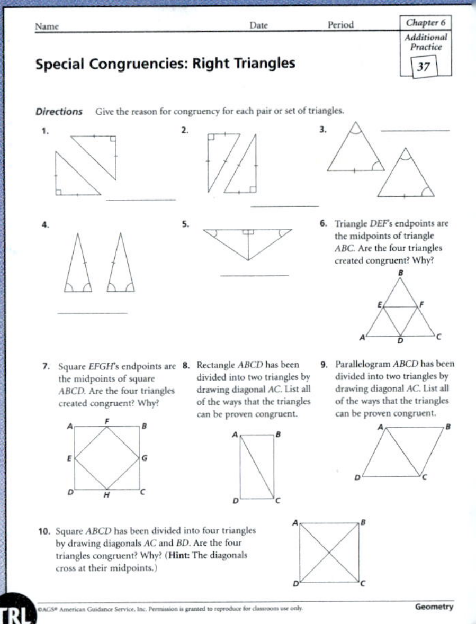

Name Date Period | Chapter 6 Additional Practice 37

Special Congruencies: Right Triangles

Directions Give the reason for congruency for each pair or set of triangles.

1. ________
2. ________
3. ________
4. ________
5. ________
6. Triangle *DEF*'s endpoints are the midpoints of triangle *ABC*. Are the four triangles created congruent? Why?
7. Square *EFGH*'s endpoints are the midpoints of square *ABCD*. Are the four triangles created congruent? Why?
8. Rectangle *ABCD* has been divided into two triangles by drawing diagonal *AC*. List all of the ways that the triangles can be proven congruent.
9. Parallelogram *ABCD* has been divided into two triangles by drawing diagonal *AC*. List all of the ways that the triangles can be proven congruent.
10. Square *ABCD* has been divided into four triangles by drawing diagonals *AC* and *BD*. Are the four triangles congruent? Why? (**Hint:** The diagonals cross at their midpoints.)

 Geometry

TRL **Additional Practice 37**

THEOREM: The acute angles in any right triangle are complementary.

PROOF

Given: right triangle ABC with $m\angle A = 90°$

To Prove: $m\angle B + m\angle C = 90°$

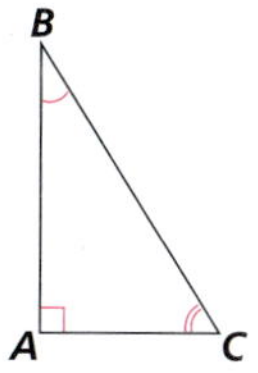

$m\angle A + m\angle B + m\angle C = 180°$	by the Angle Sum Theorem
$90° + m\angle B + m\angle C = 180°$	given $m\angle A = 90°$
$\therefore\ m\angle B + m\angle C = 90°$	subtraction

This theorem makes it possible for you to prove that right triangles are congruent using the congruency postulates for the following cases:

Case 1

Given: hypotenuse and one acute angle

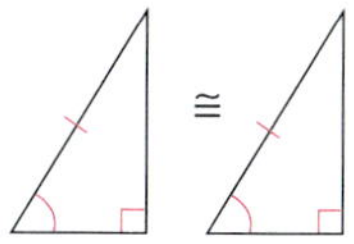

Congruent by ASA or AAS.

Case 2

Given: one leg and one acute angle

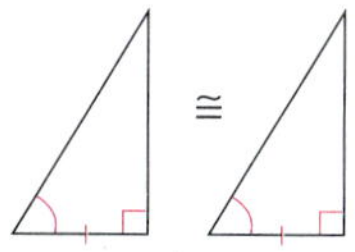

or

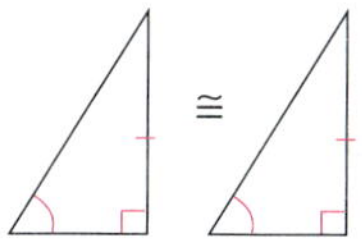

Congruent by ASA or AAS.

After introducing the Hypotenuse-Leg Theorem on page 196, ask: "What postulate could you use to prove two right triangles are congruent if the legs of one right triangle were congruent to the legs of another right triangle?" **(SAS)** Ask students to explain their reasoning.

3. Reinforce and Extend

MENTAL MATH

Say the measure of one acute angle of a right triangle and have students mentally calculate the measure of the other acute angle in the triangle. For example, say, "45°." Students should respond, "45°," since 45° plus 45° equals 90°.

COMMON ERROR

Students sometimes have difficulty distinguishing between the hypotenuse and the legs of a right triangle. To help them keep the terms straight, tell students that the legs of a right triangle form the letter *L* (for "leg") and the hypotenuse is the side that makes a triangle out of the letter *L*.

Learning Styles

Visual/Spatial Have students draw on the board pairs of congruent triangles with tick marks and arcs showing congruent sides and/or angles and pairs of noncongruent triangles. They can then take turns pointing to a pair of triangles and identifying the triangles as congruent or not congruent. If the triangles are congruent, students should write the name of the postulate or theorem that proves congruency.

Case 3

Given: two legs

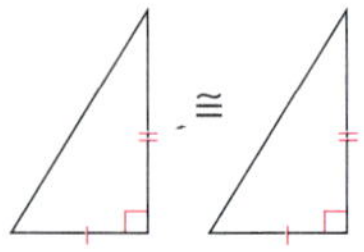

Congruent by SAS.

There is another way to prove the congruency of right triangles.

> The leg of a right triangle is either one of the sides next to the right angle.

Case 4

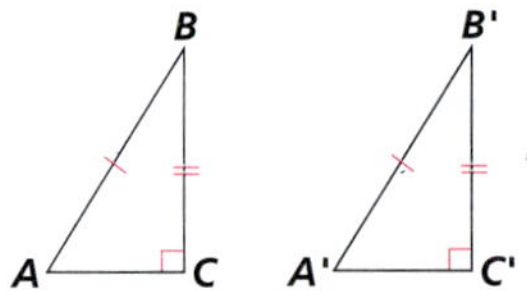

In this case, you are given that the hypotenuse and leg of one triangle are congruent to the hypotenuse and corresponding leg of a second triangle.

Hypotenuse-Leg (H-L) Theorem: Two right triangles are congruent if the hypotenuse and leg of one triangle are congruent to the hypotenuse and corresponding leg of a second triangle.

This theorem is easily proven after our discussion of another theorem called the Pythagorean Theorem in Chapter 8. For now, we will take the theorem as true without proof.

Now you have all the information you need to prove right triangles congruent. What's more, you can use the fact that the acute angles in a right triangle are complementary to calculate angle measure.

EXAMPLES

Given: two right triangles with congruent parts as marked

Are these triangles congruent? Why?

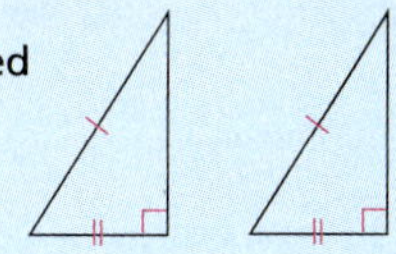

The hypotenuses and corresponding legs are congruent so the triangles are congruent by the H-L Theorem.

Given: right triangles *ABC* and *DEC*,
C is the midpoint of $\overline{BE}$

Are these triangles congruent? Why?

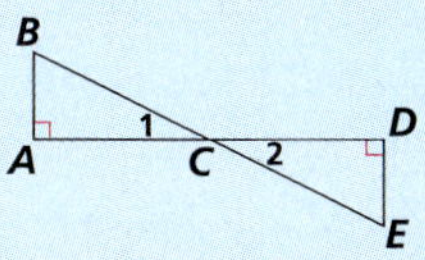

$\angle 1 \cong \angle 2$, vertical angles

$\overline{BC} \cong \overline{CE}$, definition of midpoint

$\therefore \triangle ABC \cong \triangle DEC$ by AAS of right triangles

---

Algebra Connection

m∠1 = 80°, what is m∠2?

$m\angle 1 + m\angle 2 = 90°$	by theorem
$80° + m\angle 2 = 90°$	substitution
$m\angle 2 = 10°$	subtraction

Given: m∠1 is three times m∠2.

What is the measure of each angle?

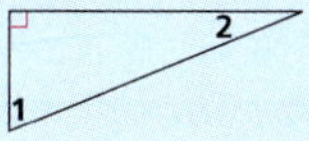

Let $m\angle 2 = x$, $m\angle 1 = 3x$

$m\angle 1 + m\angle 2 = 90°$	by theorem
$3x + x = 90°$	by substitution
$4x = 90°$	addition
$x = 22.5°$	divide
$m\angle 2 = 22.5°$, $m\angle 1 = 67.5°$	substitution for x and $3x$

ALGEBRA CONNECTION

Give students practice writing expressions with variables to show the relationship between two quantities. Use the following examples.

- One number is three more than another number. (x = one number; $x + 3$ = the other number)
- A number is four times larger than another number. (x = one number; $4x$ = the larger number)
- Jamie is five years younger than her brother. (x = Jamie's age in years, $x + 5$ = Jamie's brother's age; or x = Jamie's brother's age, $x - 5$ = Jamie's age)
- m∠*A* is two times the measure of m∠*B*. ($m\angle A = 2x$; $m\angle B = x$)
- ∠1 and ∠2 are supplementary. ($m\angle 1 = x$; $m\angle 2 = 180 - x$)

Learning Styles

Visual/Spatial Have pairs of students take turns orally describing the congruent parts of the pairs of triangles in Exercise A and providing reasons for the congruency of each pair.

Application

At Home

Ask students to look around their homes for triangles. Suggest that they look at tools, product labels and packaging, or magazine and newspaper pictures for examples. Have them choose a triangle and construct a congruent one. Encourage students to display in the classroom the labels, tools, or pictures that include triangles as well as their constructions of congruent triangles.

Exercise A Give the reason for congruency for each pair of triangles.

1.  SAS

2. 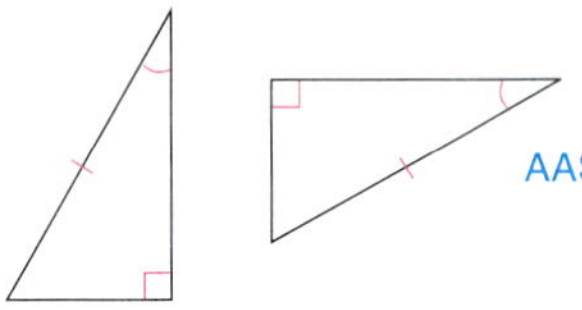 AAS

3. 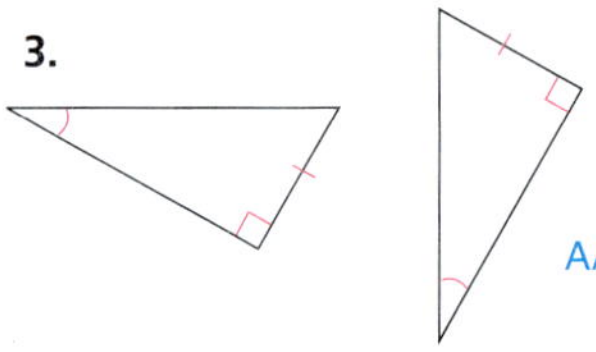 AAS

4. 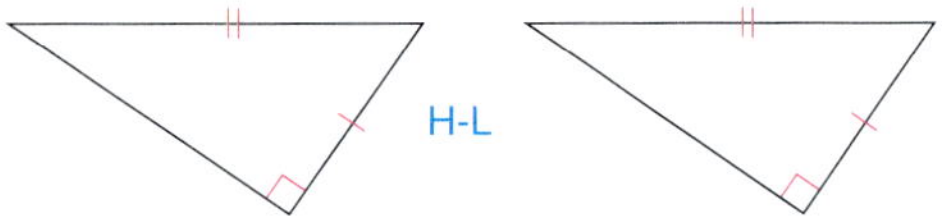 H-L

Exercise B Copy each proof. Fill in the missing steps.

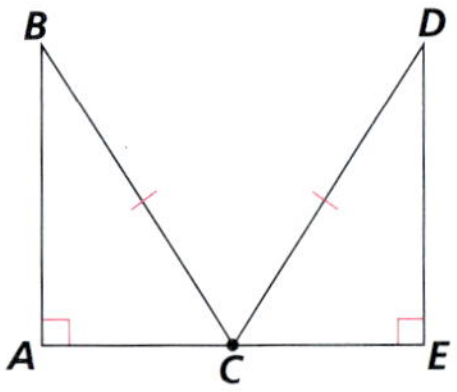

Given: $\triangle ABC$, $\triangle EDC$, C midpoint $\overline{AE}$
$\overline{AB}$ and $\overline{DE} \perp \overline{AE}$; $\overline{BC} \cong \overline{DC}$

To Prove: $\triangle ABC \cong \triangle EDC$

PROOF

Statement	Reason
1. $m\angle A = 90°$, $m\angle E = 90°$	1. **5.** Given/Def of perpendicular lines.
2. $\overline{AC} \cong \overline{CE}$	2. **6.** Given/Def of midpoint.
3. $\overline{BC} \cong \overline{DC}$	3. Given.
4. $\therefore \triangle ABC \cong \triangle EDC$	4. **7.** H-L Theorem.

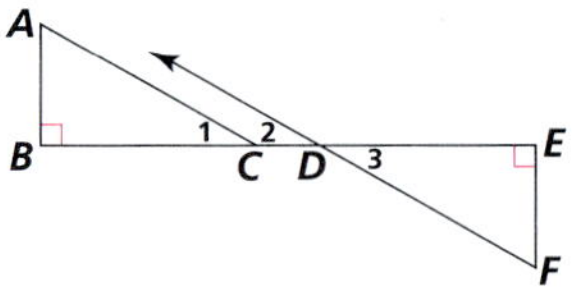

Given: $\triangle ABC$, $\triangle FED$, $\overline{AC} \parallel \overleftrightarrow{DF}$
$\overline{AB}$ and $\overline{EF} \perp \overline{BE}$; $\overline{AB} \cong \overline{EF}$

To Prove: $\triangle ABC \cong \triangle FED$

PROOF

Statement	Reason
1. $m\angle B = 90°$, $m\angle E = 90°$	1. **8.** Given/Def of perpendicular lines.
2. $m\angle 1 = m\angle 2$	2. **9.** Given/parallel lines have equal corresponding angles.
3. **10.** $m\angle 2 = m\angle 3$	3. Vertical angles are equal.
4. **11.** $m\angle 1 = m\angle 3$	4. Substitution of equals.
5. $\therefore \triangle ABC \cong \triangle FED$	5. **12.** AAS Theorem.

Problem Solving

Exercise C Solve the following problems.

13. If m∠2 is twice m∠1, what is the measure of ∠1? ∠2?

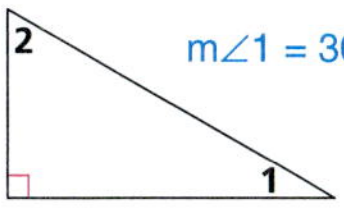

m∠1 = 30°, m∠2 = 60°

15. If m∠3 is ten more than m∠4, what is the measure of ∠3? ∠4?

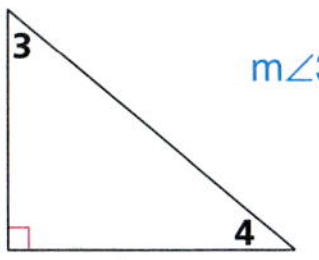

m∠3 = 50°, m∠4 = 40°

14. You plan to use two same-size brackets to support a printer shelf. You plan to attach one bracket 10 inches below the shelf on the left side. How far below the shelf will you attach the bracket on the right? Explain why.

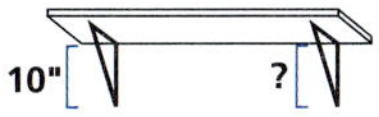

10 inches. For the shelf to hang straight, the triangles formed by the shelf, brackets, and wall must be congruent. By the H-L Theorem, the second legs must be equal.

Calculator Practice

You can use your calculator to help find missing angle measures.

EXAMPLE ∠1and ∠2 are complementary angles.

m∠1 is twice m∠2. Let x = m∠2, $2x$ = m∠1.

$2x + x = 90°$

$3x = 90°$

Press 90 [÷] 3 [=] 30.

m∠2 = 30°, m∠1 = 60°

Exercise D Find each unknown using a calculator.

16. $5x + x = 120°$ $x = 20°$

17. $3x + 2x = 100°$ $x = 20°$

18. $2x + 3x = 80°$ $x = 16°$

19. $2x + 3x = 120°$ $x = 24°$

20. $3x + 4x = 160°$ $x = 22.9°$

Writing About Mathematics

Suppose you have an isosceles triangle such as the one in the drawing below. What congruency theorem would you use to show △ADB ≅ △CDB? Explain your choice.

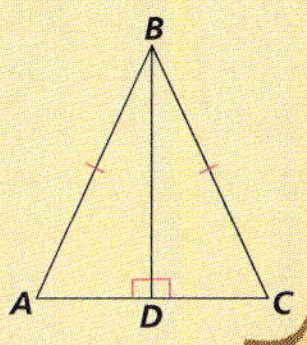

LEARNING STYLES

Interpersonal/Group Learning Set up stations at which you place pairs of triangles congruent by SSS, SAS, AAS, ASA, or H-L and pairs of noncongruent triangles. At each station, provide information that students can use to determine congruency. Invite small groups of students to move from station to station, identifying which triangles are congruent and stating the postulate or theorem that supports their conclusions.

CALCULATOR

Suggest that students use a calculator to determine the unknown acute angle measurements of right triangles. Remind them that by theorem (page 195) measurements of the two acute angles in a right triangle must add up to 90°. If necessary, demonstrate how to use the calculator for basic subtraction. For example, to determine the measurement of the unknown acute angle of a right triangle when the known acute angle is 42°:

Press 90 [−] 42 [=] 48.

Ask students to find the unknown measurements for right triangles with the following known acute angle measurements.

- 63° (27°)
- 31° (59°)
- 49° (41°)
- 56° (34°)
- 18° (72°)

Lesson at a Glance

Chapter 6 Lesson 5

Overview This lesson defines what an image and a line of reflection are and explains how to reflect a geometric figure over a given line of reflection on a coordinate plane.

Objectives

- To understand that an image is an object's reflection over a specific line of reflection.
- To reflect an image on the coordinate plane over a given line of reflection.

Student Pages 200–204

Teacher's Resource Library TRL

Workbook Activities 66–67

Additional Practice 38

Mathematics Vocabulary

image

1. Warm-Up Activity

Hang a framed mirror in the classroom and allow each student to examine his or her reflection. Then lead a discussion about how the images are alike and different from the people reflected.

2. Teaching the Lesson

Tell students to draw a straight line across the middle of a blank sheet of paper and then draw a simple geometric figure on the top half of their paper. Tell them to imagine that the line is a mirror and then draw the reflection of their figure as it would appear in a mirror. Observe students' drawings and discuss the methods they used to draw the figure's reflection and whether or not they did indeed draw the image of the original figure. Then have them read the lesson to learn how to use their knowledge of coordinate planes to draw an object's image.

Lesson 5 Reflections in the Coordinate Plane

Image

Reflection of an object

You've probably noticed reflections of trees in a pond or lake or even reflections of buildings in puddles of water. In fact, you probably see a reflection of yourself every day when you look in a mirror! As you may already know, the reflection of an object is called its **image.**

In the picture, the object is the flamingo. The image is the reflection of the flamingo in the water.

As you work with objects and their reflections, or images, you will also work with the *line of reflection.* The object and its image are at equal distances from the line of reflection.

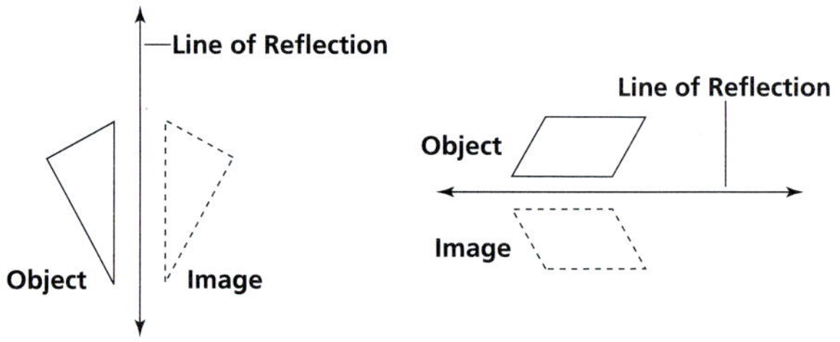

Geometry Around Us

An object and its image are at equal distances from the line of reflection. Look into a mirror. As you move toward the mirror, your image appears to move as well. Move away from the mirror and your image will move away. Explain how this "distance" changes if you are looking at your reflection in a second mirror.

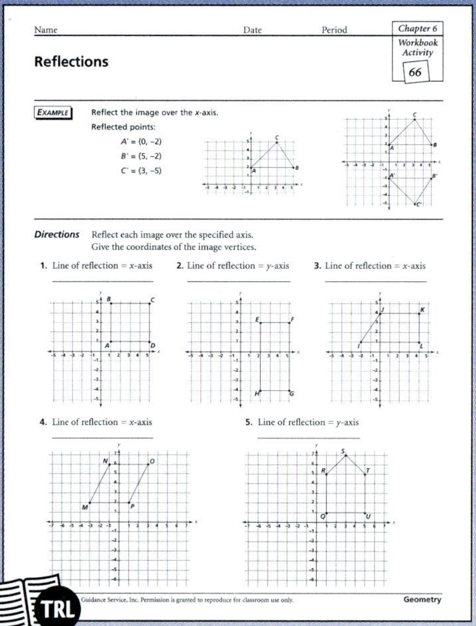
Name Date Period

Chapter 6 Workbook Activity 66

Reflections

EXAMPLE Reflect the image over the *x*-axis.

Reflected points:

$A' = (0, -2)$

$B' = (5, -2)$

$C' = (3, -5)$

Directions Reflect each image over the specified axis. Give the coordinates of the image vertices.

1. Line of reflection = *x*-axis
2. Line of reflection = *y*-axis
3. Line of reflection = *x*-axis
4. Line of reflection = *x*-axis
5. Line of reflection = *y*-axis

Guidance Service, Inc. Permission is granted to reproduce for classroom use only. Geometry

Workbook Activity 66

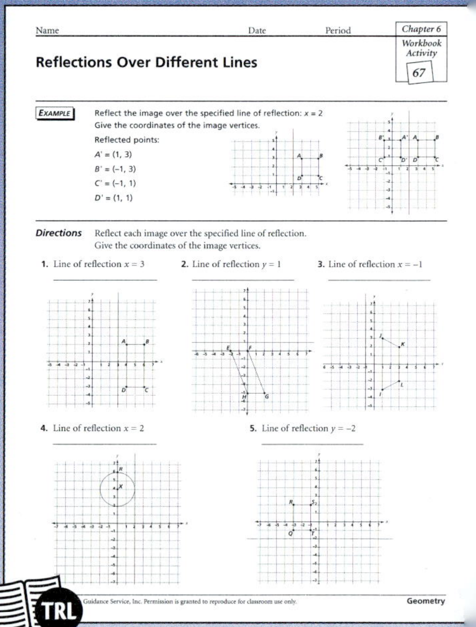
Name Date Period

Chapter 6 Workbook Activity 67

Reflections Over Different Lines

EXAMPLE Reflect the image over the specified line of reflection: $x = 2$

Give the coordinates of the image vertices.

Reflected points:

$A' = (1, 3)$

$B' = (-1, 3)$

$C' = (-1, 1)$

$D' = (1, 1)$

Directions Reflect each image over the specified line of reflection. Give the coordinates of the image vertices.

1. Line of reflection $x = 3$
2. Line of reflection $y = 1$
3. Line of reflection $x = -1$
4. Line of reflection $x = 2$
5. Line of reflection $y = -2$

Guidance Service, Inc. Permission is granted to reproduce for classroom use only. Geometry

Workbook Activity 67

The distance between the object and its image can be shown and calculated if both are placed in a coordinate plane.

The point (3, 2) reflected over the x-axis has as its image point (3, –2).

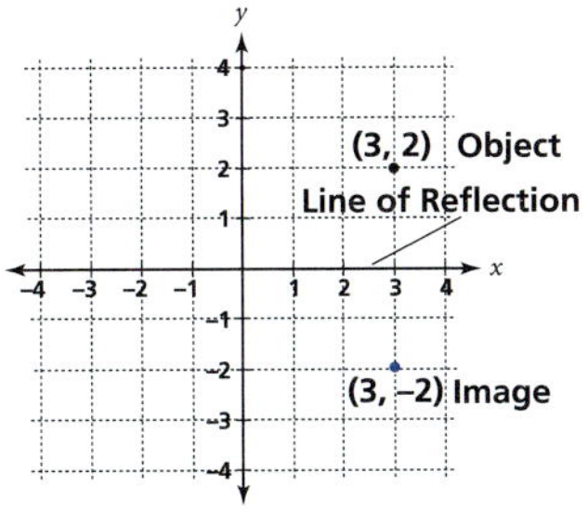

The same point, (3, 2), reflected over the y-axis has as its image point (–3, 2).

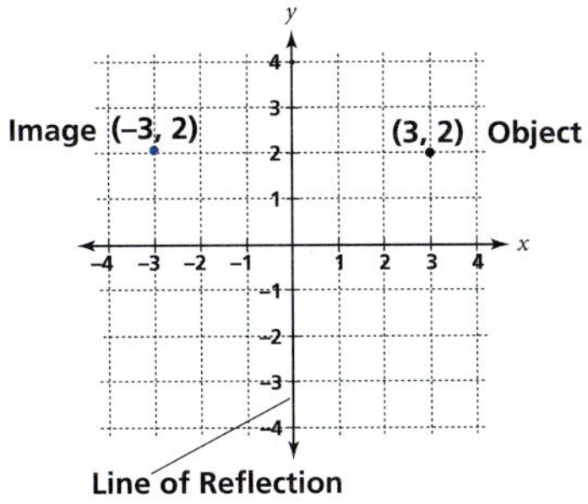

In fact, you can make the following generalizations or statements.

Reflection over *x*-axis

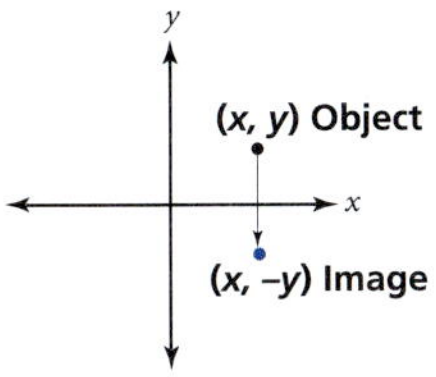

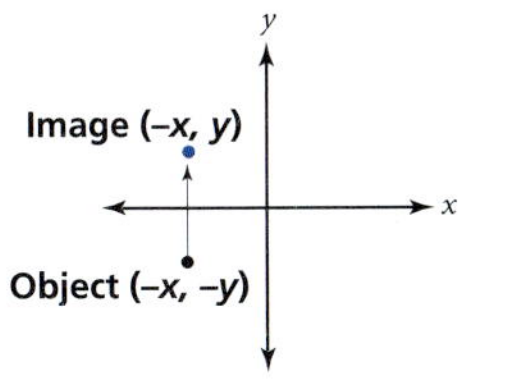

Image: has constant *x*-value
***y*-value has opposite sign**

Reflection over *y*-axis

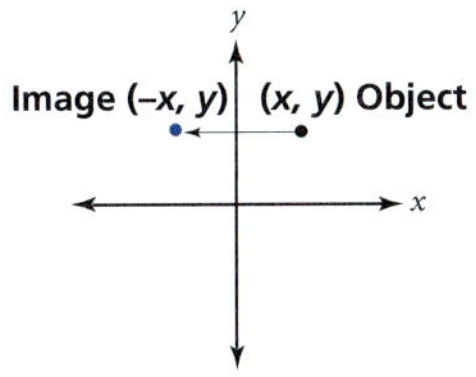

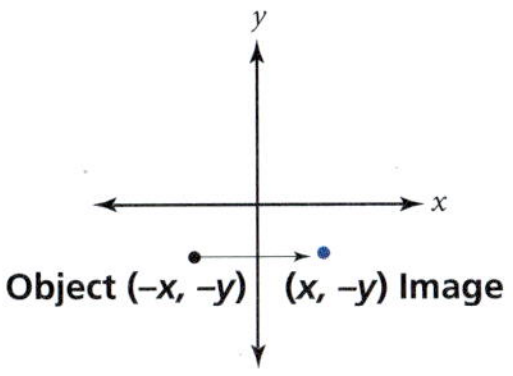

Image: has constant *y*-value
***x*-value has opposite sign**

Manipulatives

M Reflections

Materials: GeoReflector, Manipulite Pattern Blocks, Manipulatives Master 2 (Coordinate Grid)

Group Practice: Demonstrate how to use the GeoReflector by placing an object in front of it. Explain that the reflection on the face of the GeoReflector is the image of the object and that the line that the GeoReflector lies on is the line of reflection.

Ask students to make a design with the pattern blocks and then study the reflection on the face of the GeoReflector. Invite students to share their observations, comments, and questions about the images. Demonstrate how to trace the image of a figure and draw the line of reflection along the beveled edge of the GeoReflector.

Next, demonstrate how to use the GeoReflector on the coordinate grid. Suggest that students experiment with graphing points and their images on the grid as well as with simple geometric figures such as line segments and squares. Encourage students to share their observations and questions about tracing images on a coordinate plane.

Student Practice: Ask students to use the GeoReflector and the coordinate grid to reflect these points over the x-axis and then give the coordinates of the image: (2, –4) **(2, 4)**; (5, 3) **(5, –3)**; (–1, –7) **(–1, 7)**.

Name ______ Date ______ Period ______

Chapter 6 *Additional Practice* 38

Reflections in the Coordinate Plane

Directions Reflect each point over the x-axis. Give the coordinates of the image point.

1. $A = (4, 1)$ ______
2. $B = (1, -6)$ ______
3. $C = (-5, 0)$ ______
4. $D = (7, -3)$ ______
5. $E = (-2, 1)$ ______
6. $F = (-4, -5)$ ______
7. $G = (-3, 2)$ ______
8. $H = (6, 0)$ ______
9. $I = (-9, -4)$ ______
10. $J = (3, 3)$ ______

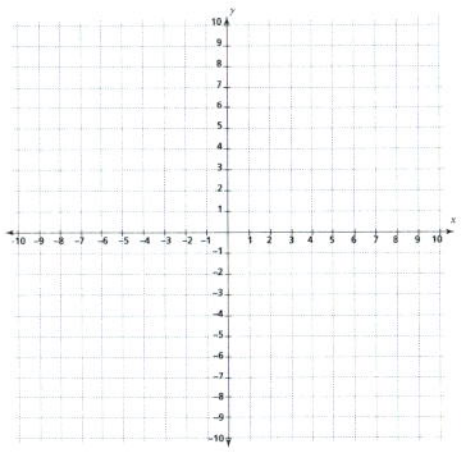

Directions Reflect each point over the y-axis. Give the coordinates of the image point.

11. $K = (4, 1)$ ______
12. $L = (5, 3)$ ______
13. $M = (-2, 0)$ ______
14. $N = (-7, 4)$ ______
15. $O = (-1, 1)$ ______
16. $P = (-3, 3)$ ______
17. $Q = (4, -5)$ ______
18. $R = (0, -5)$ ______
19. $S = (3, -2)$ ______
20. $T = (-7, 0)$ ______

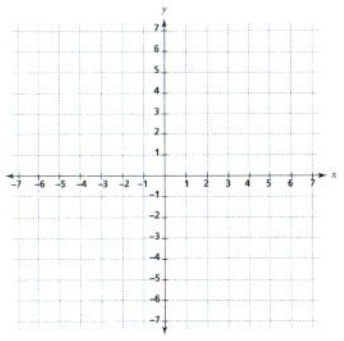

 Geometry

Additional Practice 38

3. Reinforce and Extend

Learning Styles

Interpersonal/Group Learning Provide groups of students with a GeoReflector and two identical Manipulite Pattern Blocks. Have them place the two geometric figures anywhere on a sheet of white paper. Challenge them to position the GeoReflector so that the reflection of one block fits on top of the second block. When they find that position, students should draw a line on the paper along the beveled edge of the GeoReflector. Encourage them to experiment to determine if they can find another position for the GeoReflector to match up the shapes. Help students conclude that there is only one position that causes the reflection of one block to fit over the other block.

You can use this information to reflect entire geometric figures over either the x- or y-axis. You need only reflect the vertices of the figures and then draw the line segments. Here are some examples.

Examples Reflect $\triangle ABC$ over the x-axis.
$A = (2, 5)$, $B = (2, 2)$, $C = (5, 2)$

Object vertices	Image vertices
$A = (2, 5)$	$A' = (2, -5)$
$B = (2, 2)$	$B' = (2, -2)$
$C = (5, 2)$	$C' = (5, -2)$

Notice that the y-values of the image are opposite the y-values of the object.

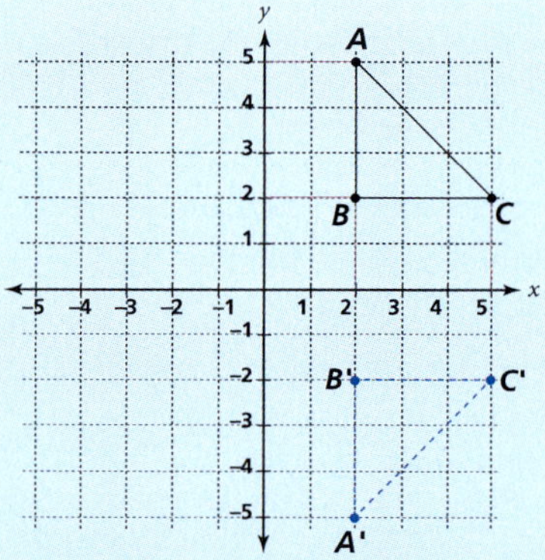

Reflect $\triangle DEF$ over the y-axis.
$D = (3, 2)$, $E = (5, 2)$, $F = (3, -1)$

Object vertices	Image vertices
$D = (3, 2)$	$D' = (-3, 2)$
$E = (5, 2)$	$E' = (-5, 2)$
$F = (3, -1)$	$F' = (-3, -1)$

Notice that the x-values of the image are opposite the x-values of the object.

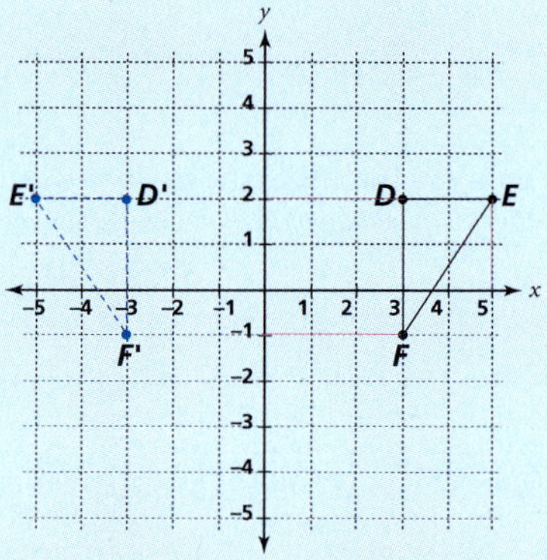

In each example above, you were able to find the coordinates of the image by finding the distance from the object points to the line of reflection and then using that same distance to place the image points.

You may, of course, use a line other than the x- or y-axis as the line of reflection. The geometry is still the same—the object points and their images are the same distance from the line of reflection.

EXAMPLES

Reflect $\triangle LMN$ over the line $y = 2$.
$L = (1, 5)$, $M = (5, 5)$, $N = (4, 3)$

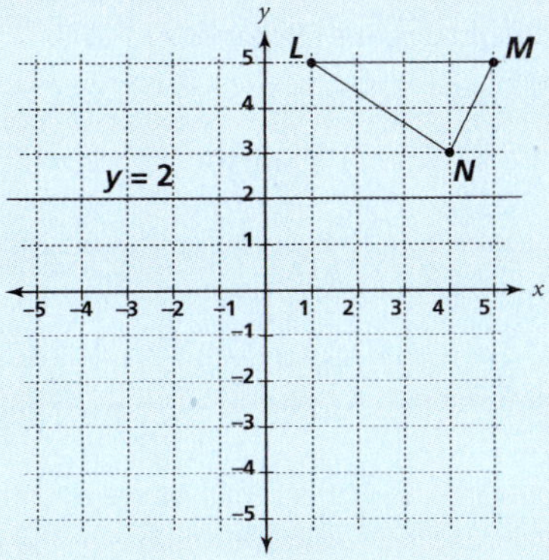

To locate the image, you need to find the distance of each object point to the line of reflection, $y = 2$. Then apply this same distance in the opposite direction from $y = 2$.

Object Point	Distance to Line of Reflection ($y = 2$)	Image Point
$L = (1, 5)$	$5 - 2 = 3$	$L' = (1, 2 - 3)$ $L' = (1, -1)$
$M = (5, 5)$	$5 - 2 = 3$	$M' = (5, 2 - 3)$ $M' = (5, -1)$
$N = (4, 3)$	$3 - 2 = 1$	$N' = (4, 2 - 1)$ $N' = (4, 1)$
Vertices for object $\triangle LMN$ (1, 5), (5, 5), (4, 3)	Vertices for image $\triangle L'M'N'$ (1, −1), (5, −1), (4, 1)	

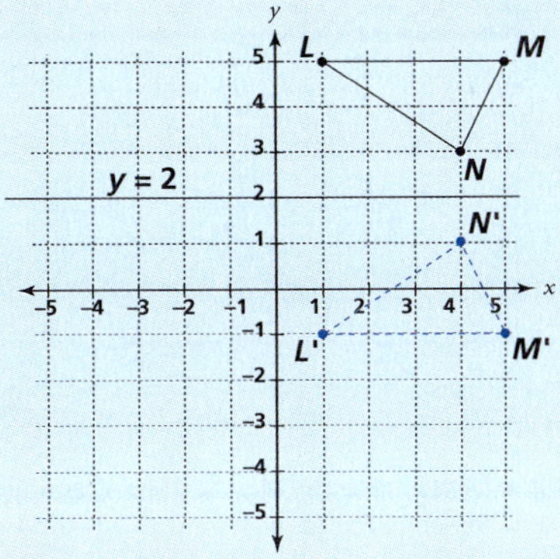

TECHNOLOGY CONNECTION

Reflections play an important role in billiards and miniature golf. The angle at which the ball strikes a barrier and the angle at which it is reflected back onto the playing surface are related—in a specific and predictable way. This relationship holds true not just for games but also for sound, radio, and light waves. Knowing these reflection relationships have helped researchers determine what's inside an atom.

LEARNING STYLES

Body/Kinesthetic Have students work with the GeoReflector, Pattern Blocks, and copies of Manipulatives Master 2. Allow them to explore independently by tracing figures made with Pattern Blocks and shapes drawn on the coordinate grid. When they are tracing images on the coordinate grid, encourage students to investigate how the positions of images change with different lines of reflection.

APPLICATION

At Home

Tell students to look in and around their homes for surfaces that reflect images. Have them describe the surface and the quality of the image produced (for example, clear, distorted, and larger or smaller than the original object). Ask students to make generalizations about which kinds of surfaces reflect the most accurate images and why.

Learning Styles

Visual/Spatial Have students work with the GeoReflector and copies of Manipulatives Master 2. Ask them to label the coordinate grid with axis names (x and y) and one unit interval values. Students should draw and label a polygon whose vertices all lie on points with integer coordinate values. Tell them to place the GeoReflector on the x-axis and then write the ordered pairs of the vertices of the image, using the prime symbol (') to name the reflected vertices. Then have students draw the image on the grid by graphing the points of the vertices. Repeat, placing the GeoReflector along the y-axis.

Learning Styles

Logical/Mathematical Have students use dotted grid paper to draw and label a polygon.

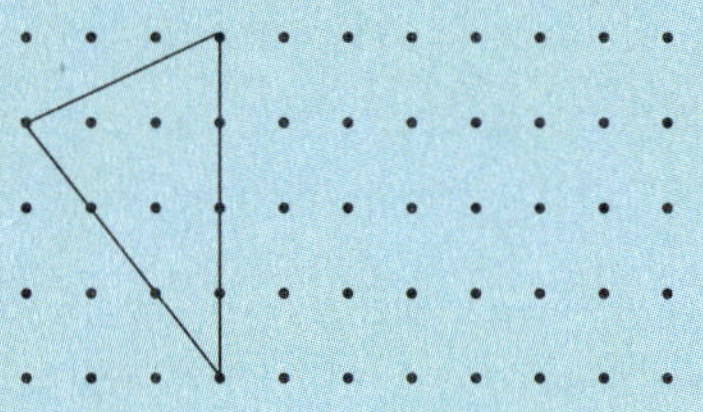

Tell students to add the x-axis and y-axis and write the ordered pairs for the image's vertices. Have them make a horizontal or vertical reflection of their polygon on the grid and use the prime symbol (') to name the vertices of the reflection.

Exercise A Reflect each point over the x-axis. Give the coordinates of the image.

1. $A = (3, 5)$ $A' = (3, -5)$
2. $B = (3, -5)$ $B' = (3, 5)$
3. $C = (-3, 5)$ $C' = (-3, -5)$
4. $D = (-3, -5)$ $D' = (-3, 5)$
5. $E = (2, 0)$ $E' = (2, 0)$
6. $F = (0, 4)$ $F' = (0, -4)$

Exercise B Reflect each point over the y-axis. Give the coordinates of the image.

7. $G = (4, 2)$ $G' = (-4, 2)$
8. $H = (4, -2)$ $H' = (-4, -2)$
9. $I = (-4, 2)$ $I' = (4, 2)$
10. $J = (-4, -4)$ $J' = (4, -4)$
11. $K = (3, 0)$ $K' = (-3, 0)$
12. $L = (0, 5)$ $L' = (0, 5)$

Exercise C Reflect each image over the specified line of reflection. Give the coordinates of the image vertices.

See Teacher's Edition page for answers.

13. $x = 2$

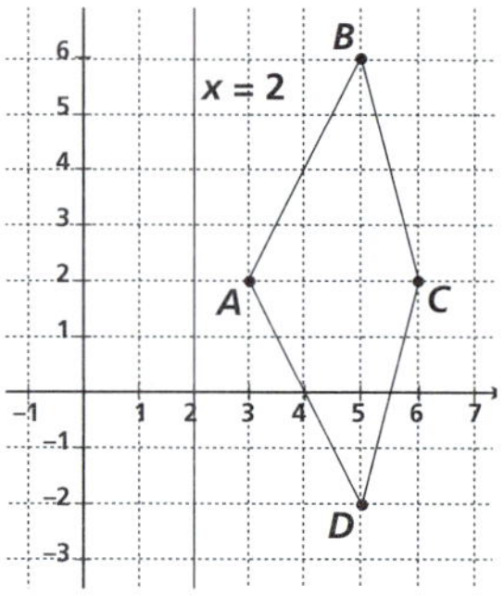

$A = (3, 2)$ $B = (5, 6)$
$C = (6, 2)$ $D = (5, -2)$

14. $y = -3$

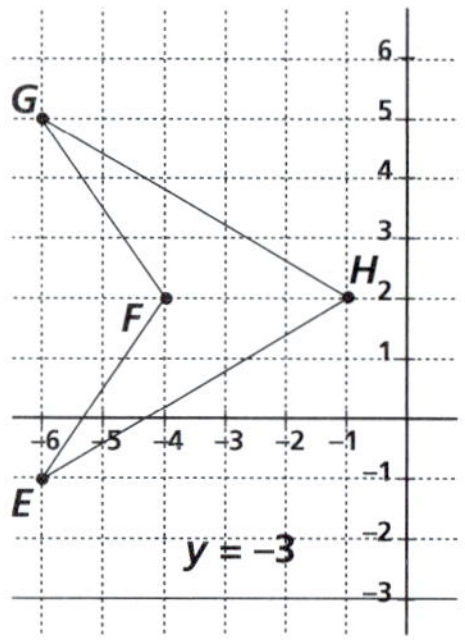

$E = (-6, -1)$ $F = (-4, 2)$
$G = (-6, 5)$ $H = (-1, 2)$

15. $x = 1$

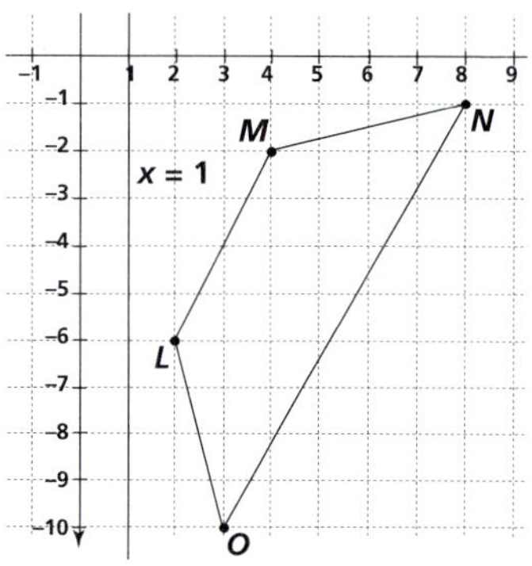

$L = (2, -6)$ $M = (4, -2)$
$N = (8, -1)$ $O = (3, -10)$

Experiencing Geometry

Choose a small photograph from a magazine. Trace the object onto a sheet of paper. Then fold the paper and trace the outline of the object from the image. Unfold the paper. How are the object and image pictures alike? How are they different? Where is the line of reflection? How can you prove the object and image are each the same distance from the line of reflection?

Answers to Exercise C

13.

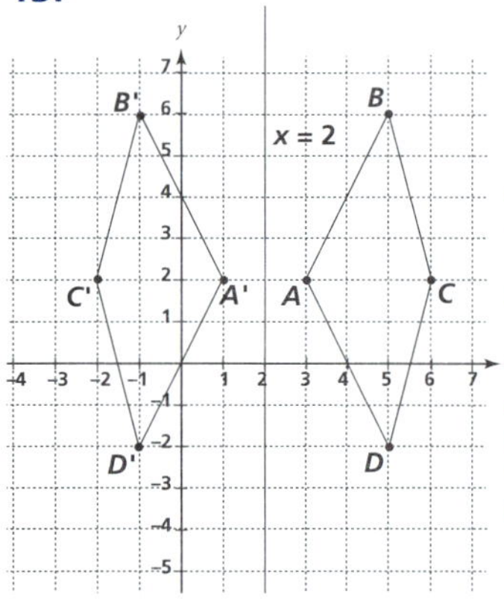

$A = (3, 2)$
$B = (5, 6)$
$C = (6, 2)$
$D = (5, -2)$

$A' = (1, 2)$
$B' = (-1, 6)$
$C' = (-2, 2)$
$D' = (-1, -2)$

14.

$E = (-6, -1)$
$F = (-4, 2)$
$G = (-6, 5)$
$H = (-1, 2)$

$E' = (-6, -5)$
$F' = (-4, -8)$
$G' = (-6, -11)$
$H' = (-1, -8)$

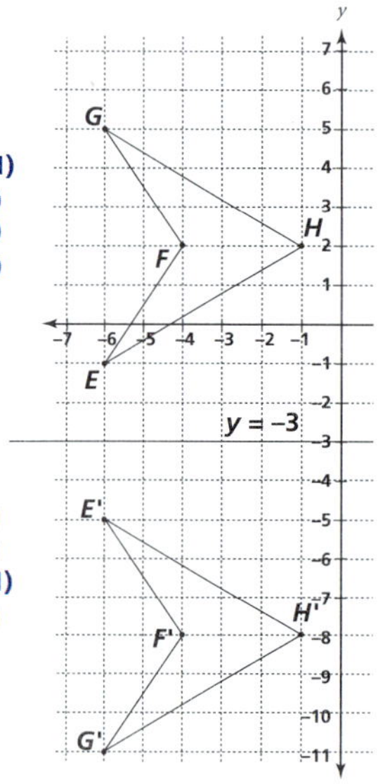

15.

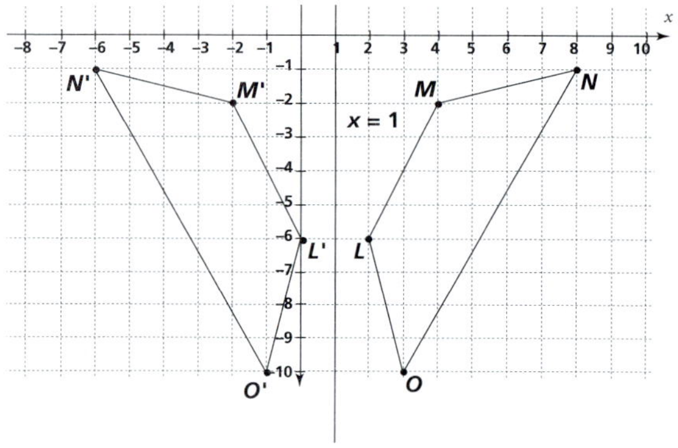

$L' = (0, -6)$
$M' = (-2, -2)$
$N' = (-6, -1)$
$O' = (-1, -10)$

$L = (2, -6)$
$M = (4, -2)$
$N = (8, -1)$
$O = (3, -10)$

Lesson 6 Special Reflections: Symmetries

A special class of geometric reflections leaves the object and image in the same location. That is, the object and its image appear as one. If you reflect a square over one of its diagonals, you get a triangle in which the object and image meet. The reflection line is called a **line of symmetry.** A figure with at least one line of symmetry is called *symmetric.*

> ***Line of symmetry***
> *Reflection line of an object for which the object coincides with the image*

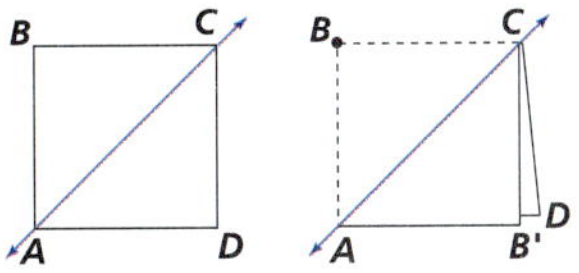

A square has four lines of symmetry. Each diagonal is a line of symmetry. Each line connecting the midpoints of opposite sides of the square are also lines of symmetry.

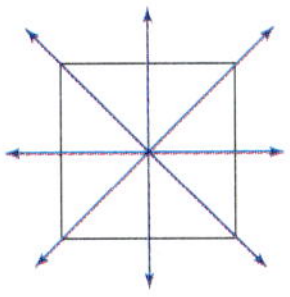

You can see that for rectangles, the lines connecting the midpoints of opposite sides are lines of symmetry.

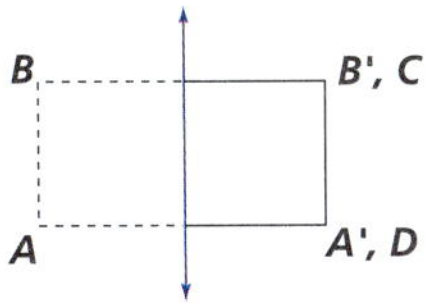

However, unlike the square, the diagonals of a rectangle are not lines of symmetry.

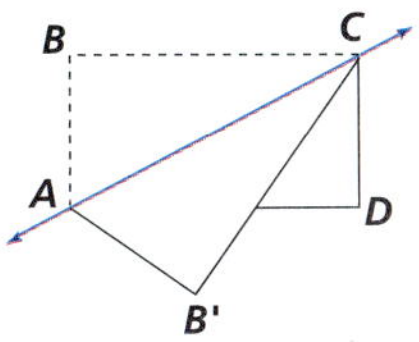

An isosceles triangle has one line of symmetry. The line of symmetry is both the bisector of the vertex angle and the perpendicular bisector of the base.

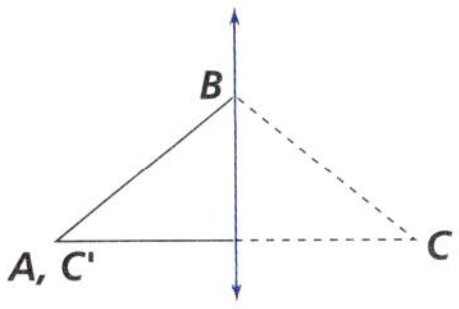

An equilateral triangle has three lines of symmetry. Each line of symmetry is the median of a side. It is also a bisector of the opposite angle.

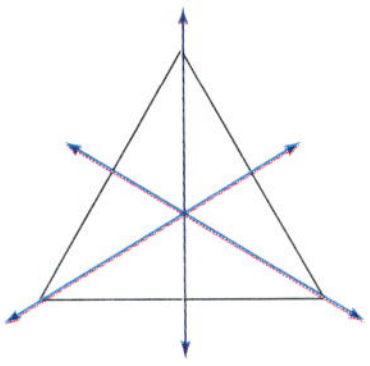

Sometimes you may meet the terms *vertical symmetry* or *horizontal symmetry.* The letter A, as printed below, has vertical symmetry. The letter E, as printed below, has horizontal symmetry.

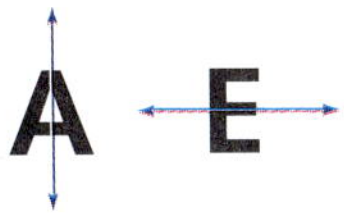

Lesson at a Glance

Chapter 6 Lesson 6

Overview This lesson introduces the concept of symmetry and explains how to determine whether a geometric figure is symmetrical.

Objectives

- To determine if a line of reflection is also a line of symmetry.
- To determine the total number of lines of symmetry of a polygon.
- To draw conclusions about lines of symmetry in relation to a given symmetrical figure.

Student Pages 205–206

Teacher's Resource Library TRL

Workbook Activity 68

Additional Practice 39

Mathematics Vocabulary

line of symmetry

1. Warm-Up Activity

Have students draw a figure that contains at least two line segments. Tell them to position the GeoReflector so that the beveled edge lies on the line segment. Ask, "How is the image in the GeoReflector the same as the figure you drew? How is it different?" (It has the same size and shape. It differs in the direction it faces.)

2. Teaching the Lesson

Manipulatives

M Symmetry

Materials: GeoReflector

Group Practice: Provide students with a GeoReflector and blank white paper and ask that they experiment with you during the following demonstration. Draw a square and ask for suggestions for positioning the mirror so that the image matches up with a part of the figure on the other side of the GeoReflector. Each time you find a line of symmetry, draw the line on the figure. Repeat until all four lines of symmetry have been drawn on the square. Repeat this procedure with other polygons, both symmetrical and nonsymmetrical. Be sure to include regular and nonregular polygons, such as a regular pentagon and a nonregular pentagon. Help students draw conclusions about the characteristics of symmetrical and nonsymmetrical figures.

3. Reinforce and Extend

Application

Career Connection Many artists have used symmetry in their work. Ask students to choose an artist whose work includes symmetry. Suggest that they research information about the artist and his or her work. Ask them to write a report that discusses the artist's work, specifically identifying three or four pieces that include symmetry. The report should describe each piece and identify its artistic genre. If possible, students should provide sketches or prints of the artwork, highlighting its symmetry. Ask volunteers to share what they discovered about symmetry in the artist's work.

Learning Styles

Visual/Spatial Have students draw and cut out large copies of each Exercise B figure. Ask them to verify the lines of symmetry shown in Exercise B by folding their figures so that the sides of the fold match up. They should draw a line on each fold to represent a line of symmetry. They can then use their figures to complete the exercise.

Exercise A Decide if the given line of reflection is also a line of symmetry. Answer *yes* or *no*.

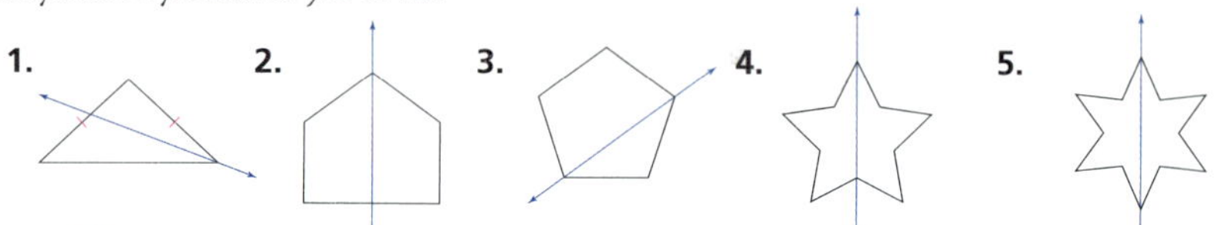

1. no
2. yes
3. no
4. yes
5. yes

Exercise B Complete the following table. Then answer the questions.

	Polygon	Number of Sides	Number of Diagonals Acting as Lines of Symmetry	Total Number of Lines of Symmetry
6. 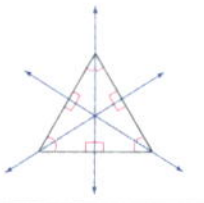	equilateral triangle	3	none	3
7. 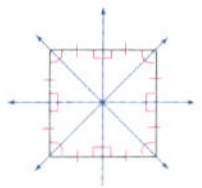	square	4	2	4
8. 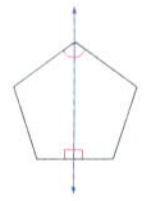	regular pentagon	5	none	5

9. What appears to be necessary for a diagonal to be a line of symmetry?
10. What appears to be true about the perpendicular bisector of a side of a polygon and the polygon's lines of symmetry?

9. It must bisect an interior angle.
10. To be a line of symmetry, a perpendicular bisector of a side must also bisect a second side or bisect an angle.

Try This...

What do you think is the relationship between a diameter (distance across a circle through the center) of a circle and a circle's symmetry? How many lines of symmetry do you think a circle has?

Any diameter appears to be a line of symmetry, there are an infinite number of diameters so there must be an infinite number of lines of symmetry.

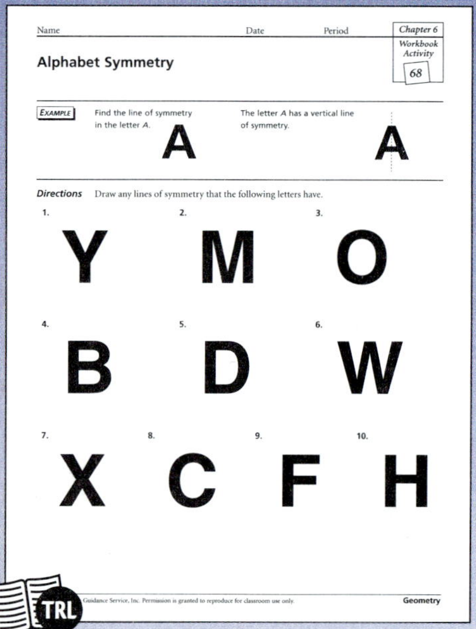
Name Date Period Chapter 6 Workbook Activity 68

Alphabet Symmetry

EXAMPLE Find the line of symmetry in the letter *A*. A The letter *A* has a vertical line of symmetry. A

Directions Draw any lines of symmetry that the following letters have.

1. Y 2. M 3. O
4. B 5. D 6. W
7. X 8. C 9. F 10. H

TRL Geometry

Workbook Activity 68

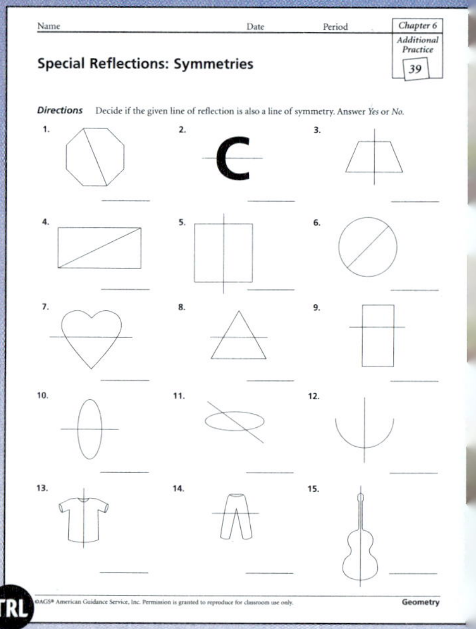
Name Date Period Chapter 6 Additional Practice 39

Special Reflections: Symmetries

Directions Decide if the given line of reflection is also a line of symmetry. Answer *Yes* or *No*.

1. 2. 3. 4. 5. 6. 7. 8. 9. 10. 11. 12. 13. 14. 15.

TRL Geometry

Additional Practice 39

Lesson 7 Slides and Translations

A **transformation** is the movement of a figure from one place to another. Reflections are one group of geometric transformations. A *slide,* or **translation,** is another example of a transformation. A slide is a transformation in which you "slide" the figure from one location to another. You do this without changing its size or shape. Sometimes this transformation is called *mapping;* you map point *A* onto point *A'*. A slide, or translation, leaves the object in one location and its image in another.

Transformation
Movement of a geometric figure from one location to another

Translation
Transformation in which a geometric figure slides from one location to another without affecting its size or shape

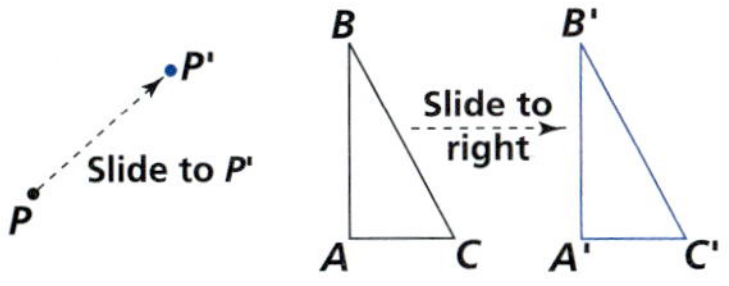

One way to study this kind of transformation is to look at slides or translations that occur in the coordinate plane.

EXAMPLE Slide point *P* (−1, −1) to location *P'*.

Looking at the grid you can see that to reach *P'* you need to move 4 units to the right (*x* + 4) and 4 units up (*y* + 4).

Here is how you can use the coordinates of *P* to find the coordinates of *P'*.

P = (−1, −1)

(*x* + 4) (*y* + 4)

(−1 + 4) (−1 + 4) = (3, 3) = *P'*

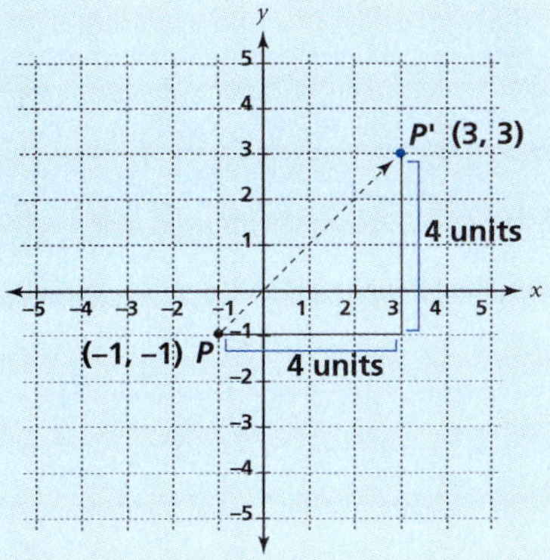

FORMULA: To translate or slide point *P* *a* units horizontally (change in *x*) and *b* units vertically (change in *y*), you must change the *x*- and *y*-values of the coordinates of *P*.

P = (*x*, *y*) translates to *P'* = (*x* + *a*, *y* + *b*)

Lesson at a Glance

Chapter 6 Lesson 7

Overview This lesson describes what a translation is and explains how to map a given translation on a coordinate plane.

Objectives

- To graph the image of a geometric figure according to a given translation.
- To identify an image point when an object point is mapped by a given translation.
- To find the image point of any object point under a given translation.

Student Pages 207–209

Teacher's Resource Library TRL

Workbook Activity 69

Additional Practice 40

Mathematics Vocabulary

transformation
translation

1. Warm-Up Activity

Ask students to observe as you slide any object on your desk from one place to another. Encourage students to discuss what they observed. Ask questions, such as "What changed about the object?" (its position) "What remained the same?" (its size, shape, and composition). Explain that Lesson 7 discusses such movement on a coordinate grid.

LEARNING STYLES

LEP/ESL Students may be familiar with the meaning of *translation* as it applies to language but not to geometry. Suggest that students with limited English proficiency work with students with greater proficiency. Encourage them to find *translation*'s etymology in a dictionary. (*trans-*, Latin prefix meaning "across" or "through" and *latus,* Latin verb meaning "carry.") Help them consider how the meanings of the Latin words relate to the meaning of the term *translation* as it applies to language and geometry.

2. Teaching the Lesson

To provide practice moving, or translating, points on a coordinate grid, tell students to make a coordinate grid on graph paper, marking both axes in one-unit intervals from –5 to 5. Then have students plot the point (–5, –5). As you read the directions below, tell students to follow the directions to move from point to point, plotting each new point as they go. After the last step, ask students where they ended up. **(the origin)** Tell them that what they were doing was making a translation of each point to a different location on the coordinate grid.

Directions to move from point to point
Start at (–5, –5).

1. 3 units right; 2 units up
2. 4 units right; 1 unit down
3. 1 unit left; 3 units up
4. 1 unit right; 2 units up
5. 2 units left; 1 unit down

After reviewing the first example on page 208, refer students to the drawing and ask: "How would the location of the image change if it were mapped four units to the left of the object instead of four units to the right?" **(It would be the same distance away from and below the object, but it would be to the left of the object rather than to the right.)**

Discuss the similarities and differences between reflections and translations. Guide students to realize that they are similar in that each pair of object and image points in a translation is also a reflection (that is, there is a line of reflection for each pair). They are also similar in that both reflected images and translated images are congruent to the original object. They are different because in a translation the image faces the same way as the object, and therefore there is no line of reflection for the image as a whole.

Note that if $a > 0$, the image is to the right, if $a < 0$, the image is to the left; if $b > 0$, the image is above the object, if $b < 0$, the image is below the object.

EXAMPLES Slide ΔABC 4 units to the right and 3 units down.

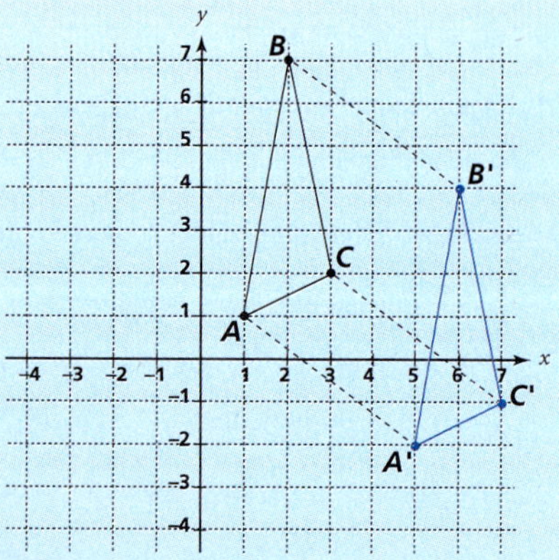

First slide each vertex.
$A = (1, 1)$, $B = (2, 7)$, and $C = (3, 2)$
$a = 4$ (4 units to the right)
$b = -3$ (3 units down)

$P = (x, y)$ translates to $P' = (x + a, y + b)$
$A = (1, 1)$ translates to $A' = (1 + 4, 1 + -3) = (5, -2)$
$B = (2, 7)$ translates to $B' = (2 + 4, 7 + -3) = (6, 4)$
$C = (3, 2)$ translates to $C' = (3 + 4, 2 + -3) = (7, -1)$
Once you have located each image vertex, draw $\Delta A'B'C'$.

A translation maps $(x, y) \rightarrow (x - 5, y + 3)$.
What are the coordinates of the image of (–3, –7)?
$(x, y) \rightarrow (x - 5, y + 3)$
$(-3, -7) \rightarrow (-3 - 5, -7 + 3) = (-8, -4)$

A translation maps $(4, -7) \rightarrow (0, 3)$.
What is the image of (x, y) under this translation?
$(4, -7) \rightarrow (0, 3)$
$(0, 3) \rightarrow (x + a, y + b) = (4 + a, -7 + b)$
$0 = 4 + a$ $3 = -7 + b$
$a = -4$ $b = 10$
The image of (x, y) is $(a - 4, y + 10)$.

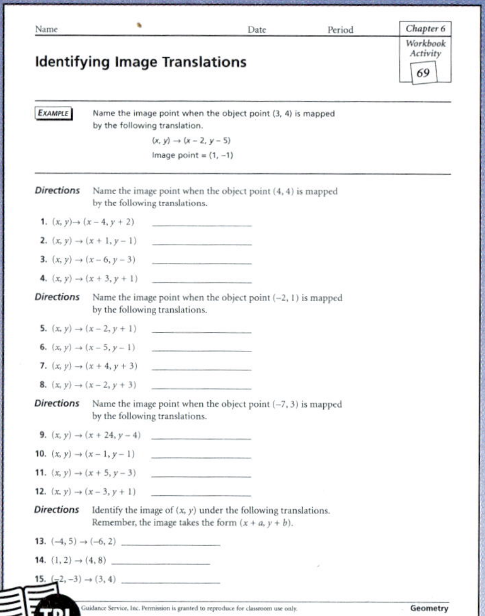
Name Date Period
Chapter 6 Workbook Activity 69

Identifying Image Translations

EXAMPLE Name the image point when the object point (3, 4) is mapped by the following translation.
$(x, y) \rightarrow (x - 2, y - 5)$
Image point = (1, –1)

Directions Name the image point when the object point (4, 4) is mapped by the following translations.

1. $(x, y) \rightarrow (x - 4, y + 2)$ ____
2. $(x, y) \rightarrow (x + 1, y - 1)$ ____
3. $(x, y) \rightarrow (x - 6, y - 3)$ ____
4. $(x, y) \rightarrow (x + 3, y + 1)$ ____

Directions Name the image point when the object point (–2, 1) is mapped by the following translations.

5. $(x, y) \rightarrow (x - 2, y + 1)$ ____
6. $(x, y) \rightarrow (x - 5, y - 1)$ ____
7. $(x, y) \rightarrow (x + 4, y + 3)$ ____
8. $(x, y) \rightarrow (x - 2, y + 3)$ ____

Directions Name the image point when the object point (–7, 3) is mapped by the following translations.

9. $(x, y) \rightarrow (x + 24, y - 4)$ ____
10. $(x, y) \rightarrow (x - 1, y - 1)$ ____
11. $(x, y) \rightarrow (x + 5, y - 3)$ ____
12. $(x, y) \rightarrow (x - 3, y + 1)$ ____

Directions Identify the image of (x, y) under the following translations. Remember, the image takes the form $(x + a, y + b)$.

13. $(-4, 5) \rightarrow (-6, 2)$ ____
14. $(1, 2) \rightarrow (4, 8)$ ____
15. $(-2, -3) \rightarrow (3, 4)$ ____

 Geometry

TRL **Workbook Activity 69**

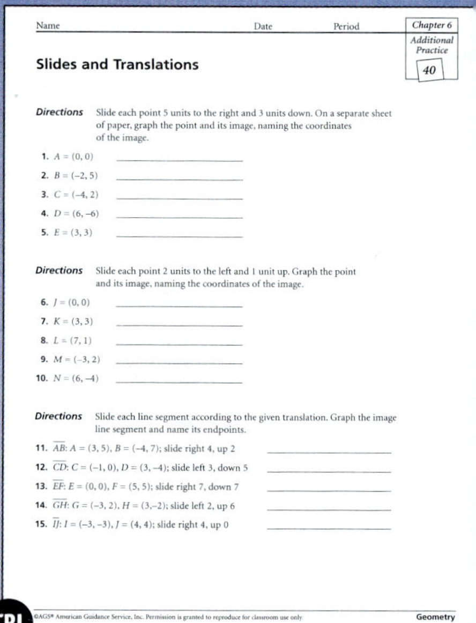
Name Date Period
Chapter 6 Additional Practice 40

Slides and Translations

Directions Slide each point 5 units to the right and 3 units down. On a separate sheet of paper, graph the point and its image, naming the coordinates of the image.

1. $A = (0, 0)$ ____
2. $B = (-2, 5)$ ____
3. $C = (-4, 2)$ ____
4. $D = (6, -6)$ ____
5. $E = (3, 3)$ ____

Directions Slide each point 2 units to the left and 1 unit up. Graph the point and its image, naming the coordinates of the image.

6. $J = (0, 0)$ ____
7. $K = (3, 3)$ ____
8. $L = (7, 1)$ ____
9. $M = (-3, 2)$ ____
10. $N = (6, -4)$ ____

Directions Slide each line segment according to the given translation. Graph the image line segment and name its endpoints.

11. $\overline{AB}$: $A = (3, 5)$, $B = (-4, 7)$; slide right 4, up 2 ____
12. $\overline{CD}$: $C = (-1, 0)$, $D = (3, -4)$; slide left 3, down 5 ____
13. $\overline{EF}$: $E = (0, 0)$, $F = (5, 5)$; slide right 7, down 7 ____
14. $\overline{GH}$: $G = (-3, 2)$, $H = (3, -2)$; slide left 2, up 6 ____
15. $\overline{IJ}$: $I = (-3, -3)$, $J = (4, 4)$; slide right 4, up 0 ____

 Geometry

TRL **Additional Practice 40**

Exercise A Slide each point 3 units to the left and 2 units up. Graph the point and its image, naming the coordinates of the image.

1. $A = (3, 6)$ $A' = (0, 8)$
2. $B = (-3, 6)$ $B' = (-6, 8)$
3. $C = (3, -6)$ $C' = (0, -4)$
4. $D = (-3, -6)$ $D' = (-6, -4)$
5. $E = (0, 5)$ $E' = (-3, 7)$
6. $F = (3, -2)$ $F' = (0, 0)$
7. $G = (0, 0)$ $G' = (-3, 2)$

Exercise B Slide each triangle according to the given translation. Graph the image triangle and name the vertices of the image triangle.

8. ΔABC: $A = (-7, 3)$, $B = (-2, 1)$, $C = (4, 6)$; slide 4 right, 2 up
9. ΔEFG: $E = (-3, -2)$, $F = (2, 2)$, $G = (2, -3)$; slide 2 left, 3 up
10. ΔHIJ: $H = (2, 1)$, $I = (1, 4)$, $J = (5, 4)$; slide 6 left, 6 down

8. $A' = (-3, 5)$, $B' = (2, 3)$, $C' = (8, 8)$
9. $E' = (-5, 1)$, $F' = (0, 5)$, $G' = (0, 0)$
10. $H' = (-4, -5)$, $I' = (-5, -2)$, $J' = (-1, -2)$

Exercise C Name the image point when the object point $(3, -1)$ is mapped by the following translations.

11. $(x, y) \rightarrow (x + 1, y + 5)$ $(4, 4)$
12. $(x, y) \rightarrow (x + 3, y - 2)$ $(6, -3)$
13. $(x, y) \rightarrow (x - 2, y - 4)$ $(1, -5)$
14. $(x, y) \rightarrow (x, y - 3)$ $(3, -4)$
15. $(x, y) \rightarrow (x + 4, y)$ $(7, -1)$

Exercise D Identify the image of (x, y) under the following translations. Remember, the image takes the form $(x + a, y + b)$.

16. $(4, -1) \rightarrow (0, 3)$ $(x - 4, y + 4)$
17. $(6, 3) \rightarrow (9, 6)$ $(x + 3, y + 3)$
18. $(6, 3) \rightarrow (-3, -6)$ $(x - 9, y - 9)$
19. $(4, 2) \rightarrow (6, -2)$ $(x + 2, y - 4)$
20. $(1, 3) \rightarrow (-3, 5)$ $(x - 4, y + 2)$

Writing About Mathematics

Describe how you would slide a circle with a radius of 5 and center at (0, 0) 5 units to the left and 3 units up. What is the center of the image?

(–5, 3)

3. Reinforce and Extend

Learning Styles

Auditory/Verbal Provide auditory reinforcement of where translated points on the coordinate plane lie. Orally read and expand on, if necessary, the paragraph at the top of page 208. Repeat the paragraph several times as you demonstrate location on a grid you draw on the board. Ask students to explain in their own words why this statement is true.

Mental Math

Write the coordinates of the points below on the board, as well as the translation to be mapped, and have students mentally find the image points when they are mapped by the translation $(x, y) \rightarrow (x - 1, y + 1)$.

$A = (4, 8)$	(3, 9)
$B = (-1, 0)$	(–2, 1)
$C = (2, -3)$	(1, –2)
$D = (1, -1)$	(0, 0)
$E = (10, 10)$	(9, 11)
$F = (0, 0)$	(–1, 1)
$G = (-12, 12)$	(–13, 13)

Learning Styles

Visual/Spatial Encourage students to check their answers to Exercise C by graphing the object and image points and comparing their graphs to the ordered pairs they calculated.

Graphing the points for problems in Exercise D before trying to find the images of (x, y) will provide students with a visual reference. In addition to graphing the points, they can also draw the mapping lines from the object points to the image points to get a clearer picture of the relative position of each point.

Lesson at a Glance

Chapter 6 Lesson 8

Overview This lesson defines what a rotation is and demonstrates how to transform an object using a rotation.

Objectives

- To draw an object image using a 90° rotation.
- To draw an object image using a 180° rotation.

Student Pages 210–213

Teacher's Resource Library TRL

Workbook Activity 70

Additional Practice 41

Mathematics Vocabulary

rotation

1. Warm-Up Activity

Briefly discuss clockwise and counterclockwise directions of movement. Then have students (in multiples of four) form a circle around one student. The student directly facing the student in the center holds a sign labeled 0°, and the students standing at each 90° position and the 180° position hold signs indicating their positions. The center student tells the circle of students to move either 90° or 180° clockwise or counterclockwise. The circle of students must then move to reflect this rotation and when done, the center student determines if the circle responded correctly to his or her instructions. Repeat the activity several times, allowing students to take turns being in the center.

Lesson 8 Rotations

Rotation
Transformation in which a geometric figure turns around a center without affecting its size or shape

The final group of transformations you will study are called **rotations.** You see and use rotations every day. Wheels on a bus, door knobs, and swivel chairs are examples of objects that turn, or rotate.

Below are diagrams of the rotations of some common geometric figures. Notice that each diagram shows an *O*, the center of rotation. This is the point that remains in place as the geometric figure turns around it. It is rather like the needle of a drawing compass.

Remember, an object that moves *clockwise* moves in the same direction as the hands on a clock move. An object that moves *counterclockwise* moves in the direction opposite to the direction the hands on a clock move.

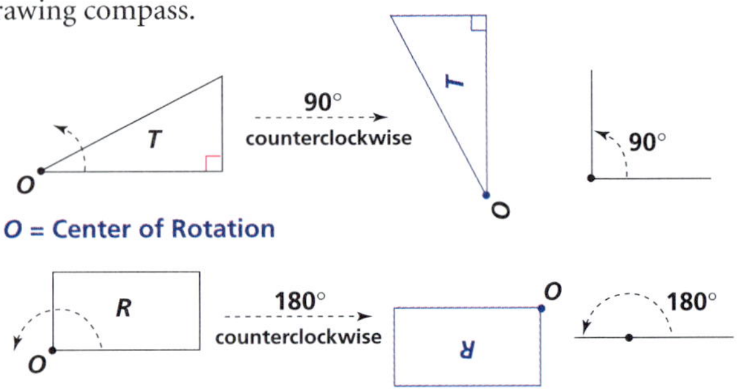

As with other transformations, you can specify a rotation by placing the figure on a coordinate plane.

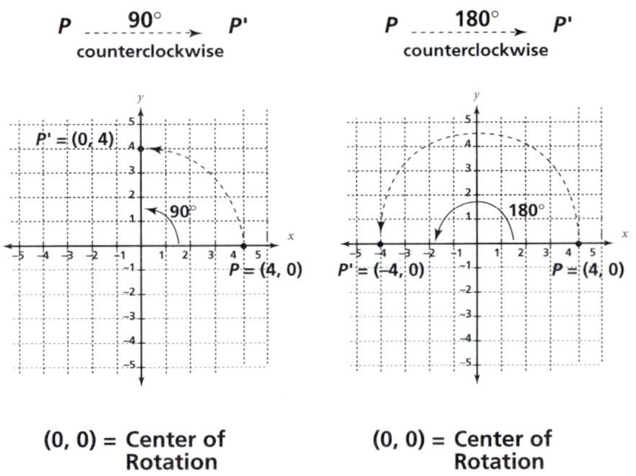

2. Teaching the Lesson

MANIPULATIVES

M Using Rotations

Materials: Polygons Plus Power Pack, Manipulite Pattern Blocks, GeoReflector, Manipulatives Master 2 (Coordinate Grid)

Group Practice: After students read page 210, demonstrate rotations using the polygons in the Polygons Plus Power Pack and an overhead projector. Change the center of rotation for any given rotation and ask students to describe how the image changes. To demonstrate how a rotation is different from a reflection, use the GeoReflector to trace the reflection image of a pattern block. Then rotate the pattern block and try to match it with the reflected image. Ask: "How is the image of a 180° rotation different from the image of a reflection?" (The direction it faces is different.)

Student Practice: Allow students time to explore independently with rotations of various shapes. Encourage them to use the GeoReflector and Pattern Blocks to compare reflection and rotation images.

These examples show that knowing the location of the center of rotation, *O*, is key to making the transformation. Without knowing the center of rotation, you have no way of knowing how the object figure is transformed.

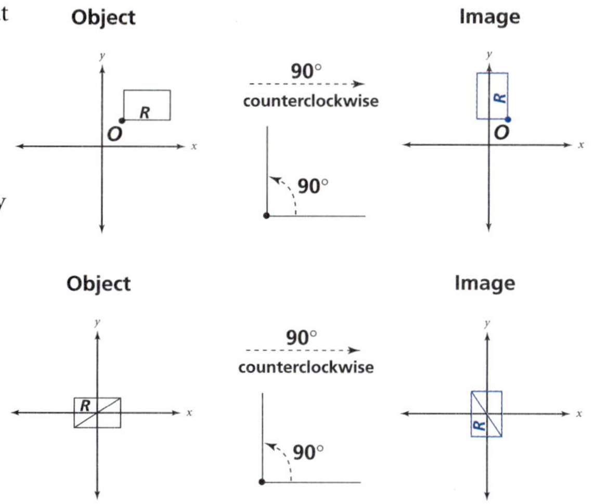

Center of Rotation
(0, 0)

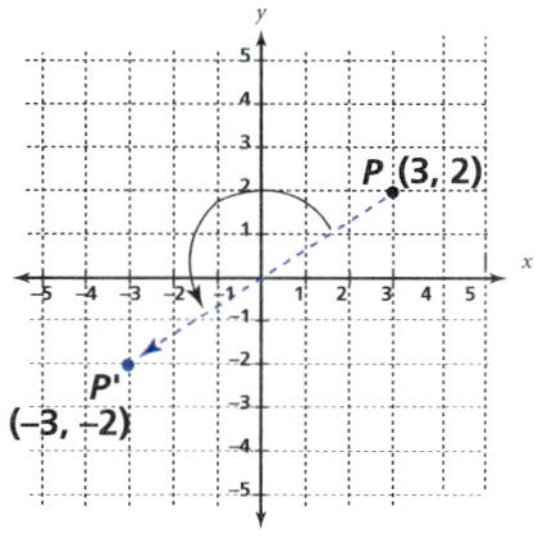

(3, 2) ---180°---> (−3, −2)
counterclockwise

(*x*, *y*) ---180°---> (−*x*, −*y*)
counterclockwise

Center of Rotation
(0, 0)

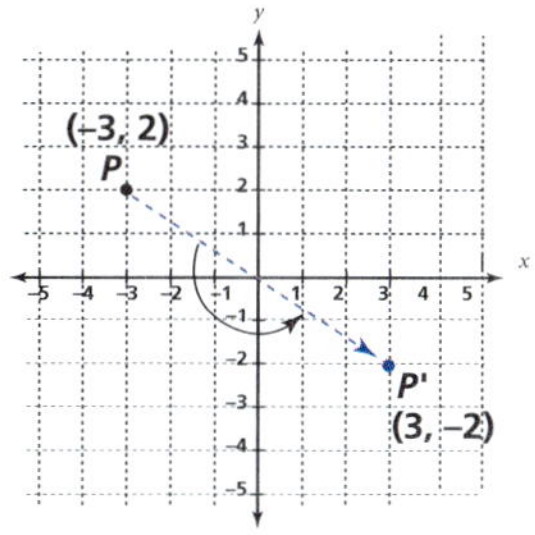

(−3, 2) ---180°---> (3, −2)
counterclockwise

(−*x*, *y*) ---180°---> (*x*, −*y*)
counterclockwise

Remember, always check the location of the center of rotation before finding the image of the rotated object.

3. Reinforce and Extend

Learning Styles

Auditory/Verbal Have students examine the examples of rotations and read the paragraph at the top of page 211 aloud. Ask them to explain in their own words why it is necessary to know the center of rotation in order to know how an object figure is transformed.

Application

Career Connection

Invite a graphic artist to speak to your class about the use of transformations in his or her work. Ask students to prepare questions ahead of time to ask the speaker about his or her work and how he or she uses transformations. Have students write a brief summary of the visitor's presentation.

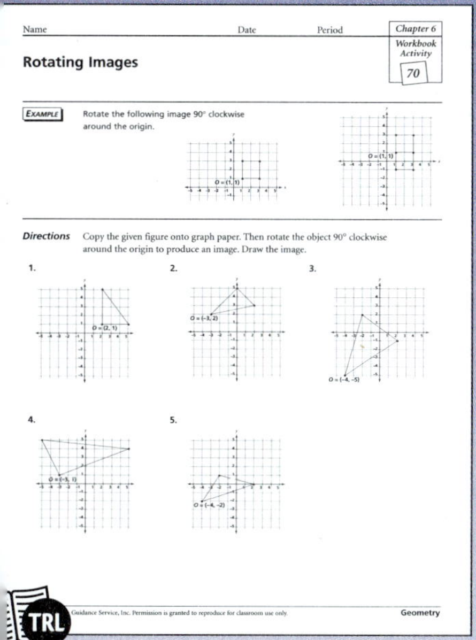

Name Date Period | Chapter 6 Workbook Activity 70

Rotating Images

EXAMPLE Rotate the following image 90° clockwise around the origin.

Directions Copy the given figure onto graph paper. Then rotate the object 90° clockwise around the origin to produce an image. Draw the image.

1. 2. 3. 4. 5.

Guidance Service, Inc. Permission is granted to reproduce for classroom use only. Geometry

Workbook Activity 70

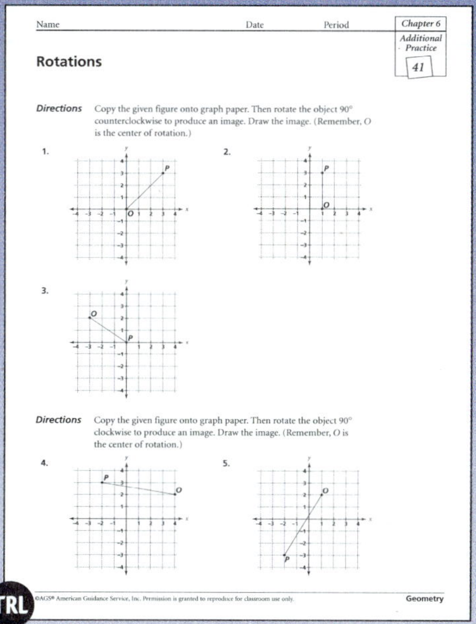

Name Date Period | Chapter 6 Additional Practice 41

Rotations

Directions Copy the given figure onto graph paper. Then rotate the object 90° counterclockwise to produce an image. Draw the image. (Remember, O is the center of rotation.)

1. 2. 3.

Directions Copy the given figure onto graph paper. Then rotate the object 90° clockwise to produce an image. Draw the image. (Remember, O is the center of rotation.)

4. 5.

©AGS® American Guidance Service, Inc. Permission is granted to reproduce for classroom use only. Geometry

Additional Practice 41

Learning Styles

Logical/Mathematical Have students place the GeoReflector on a sheet of white paper and draw the line of reflection along the base of the mirror. Then tell them to place a pattern block so it touches the face of the mirror and trace the reflection image behind the mirror. After removing the mirror, have students respond to this question: "Is there any way to check if the reflection is also a rotation of the object image?" **(Rotate the pattern block around different centers of rotation to see if the two images match up.)** Have students repeat this process with polygons they draw on the paper. Afterwards, challenge students to make a generalization about the category of polygons whose reflection images can be matched up with a rotation image. **(A possible answer is regular polygons.)** Ask students to provide a explanation to support their generalization.

Learning Styles

Visual/Spatial Make copies of Manipulatives Master 2 and the pattern blocks available. Allow students to use them to complete Exercises A and B or to check their work.

Exercise A Copy the given figure onto graph paper. Then rotate the object 90° counterclockwise to produce an image. Draw the image. (Remember, *O* is the center of rotation.)

1.

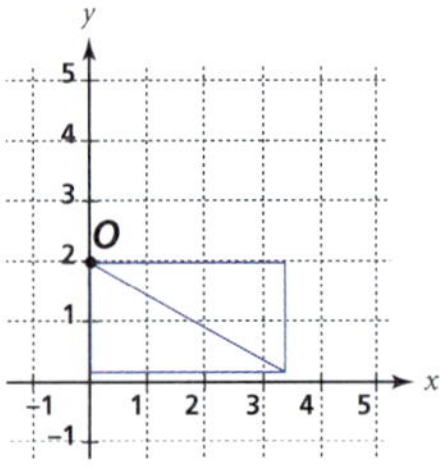

2.

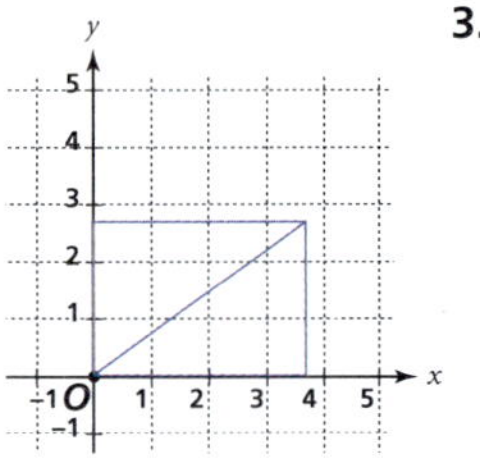

3.

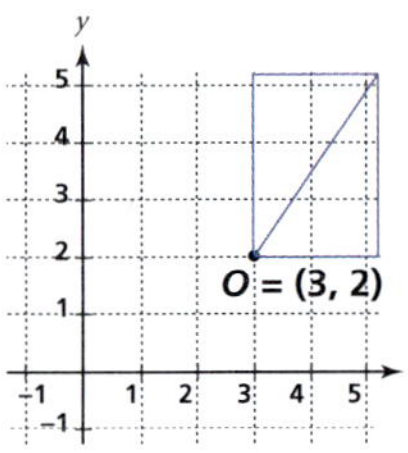

4.

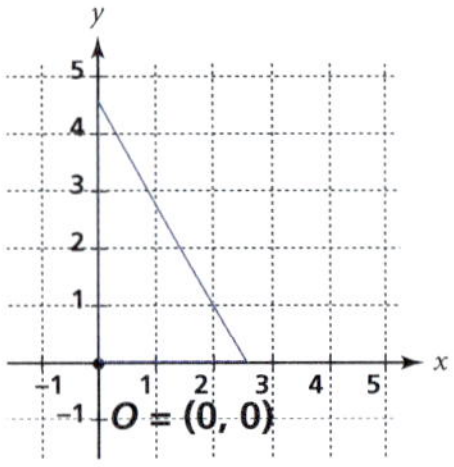

5.

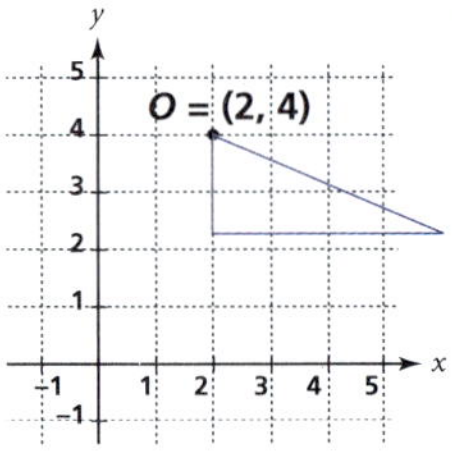

6.

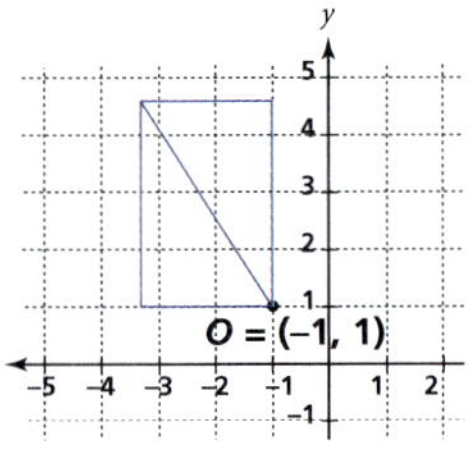

7.

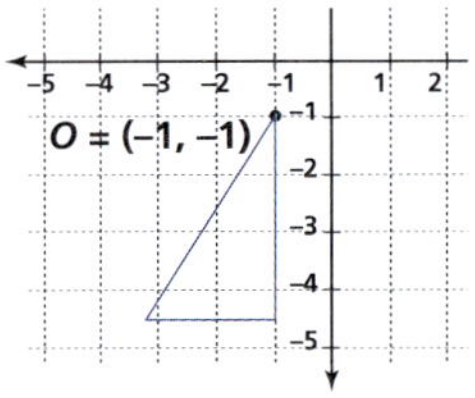

See Teacher's Edition page for answers.

Exercise B Copy the given figure onto graph paper. Then rotate the object 180° counterclockwise to produce an image. Draw the image. (Remember, *O* is the center of rotation.)

See Teacher's Edition page for answers.

8.

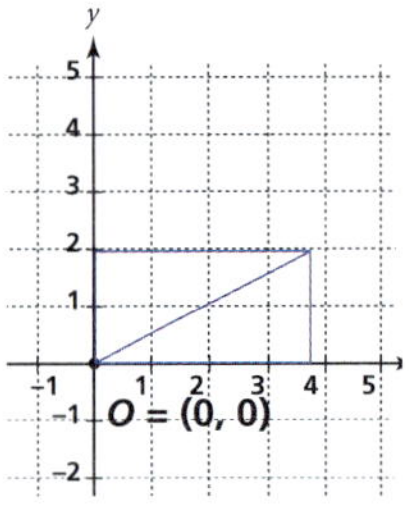

9.

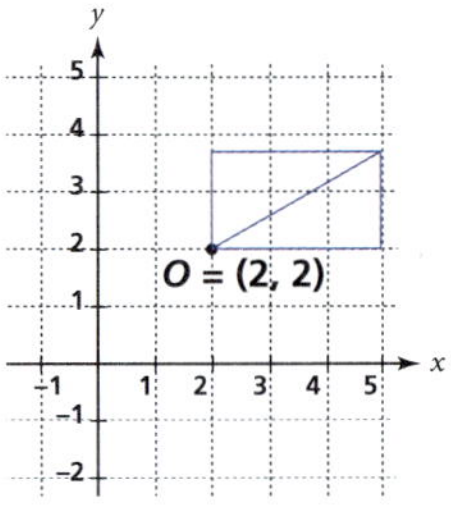

10.

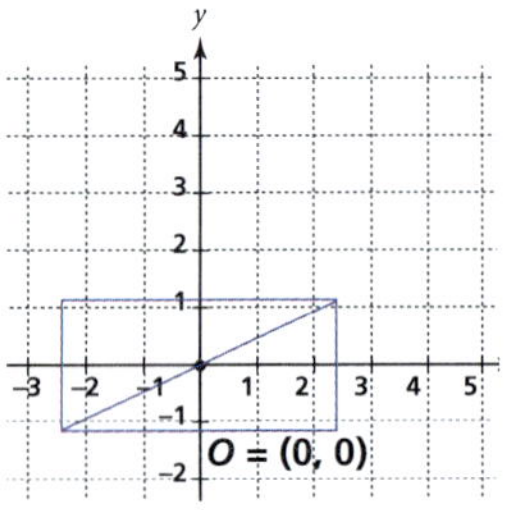

Answers to Exercises A and B

1.

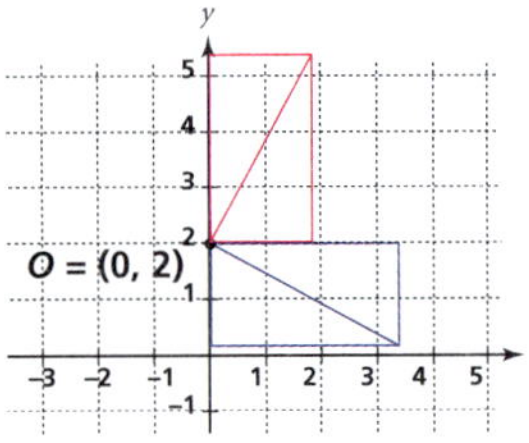

2.

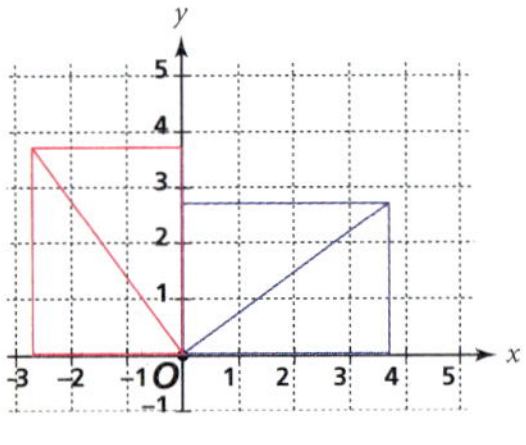

3.

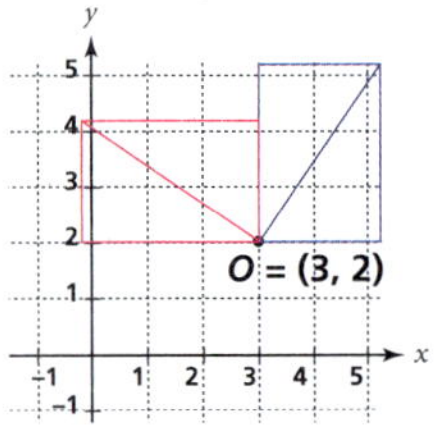

4.

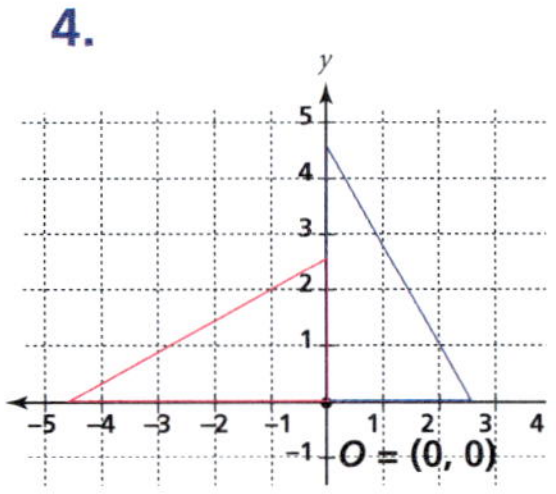

5.

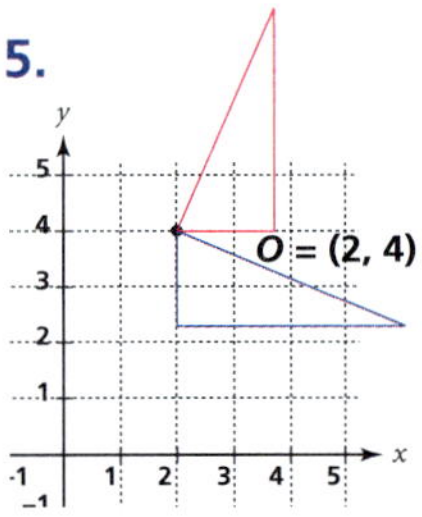

6.

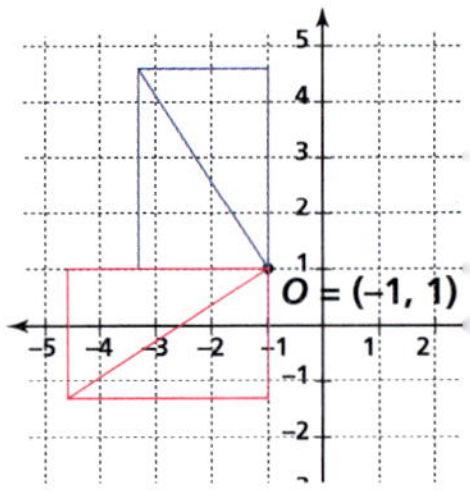

7.

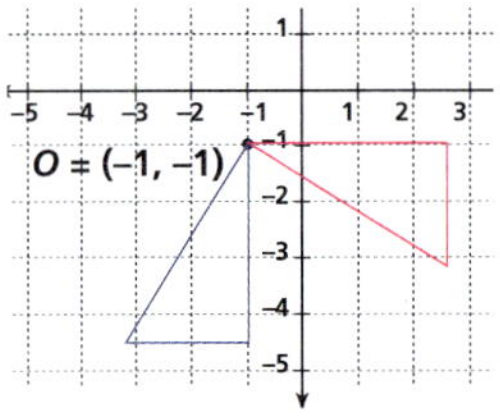

8.

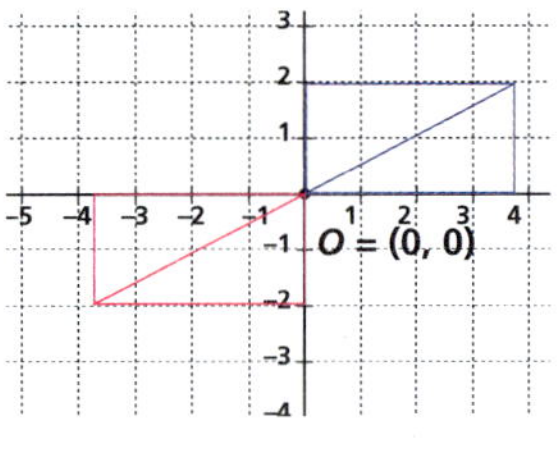

9.

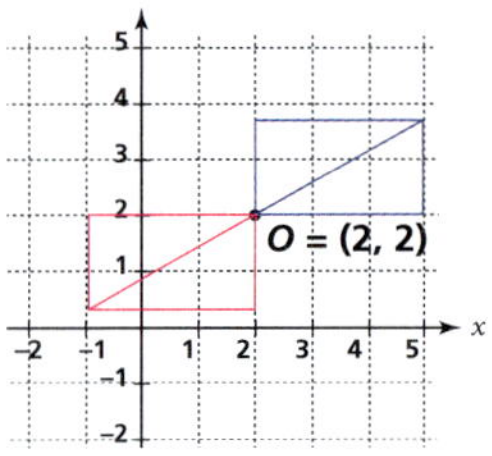

10.

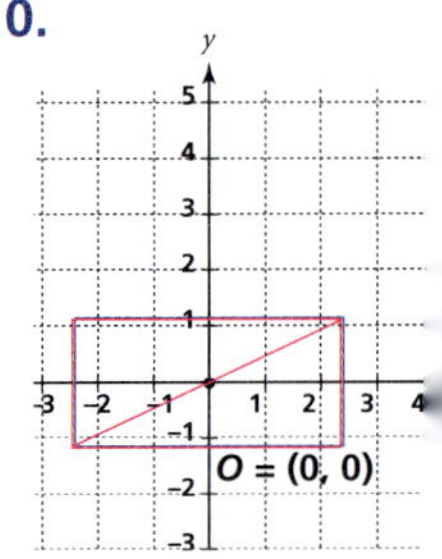

11.

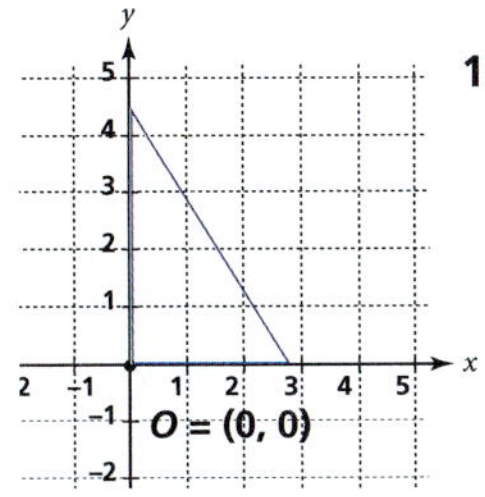

12.

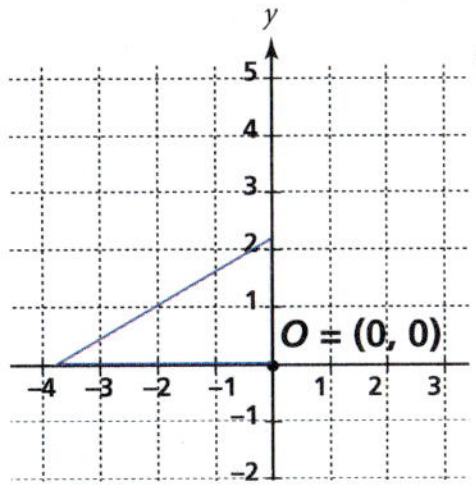

13.

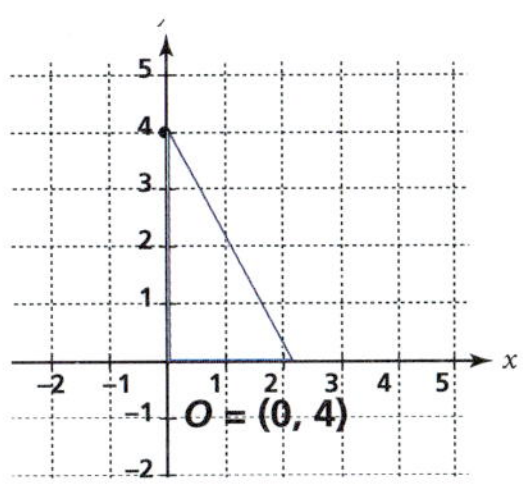

14.

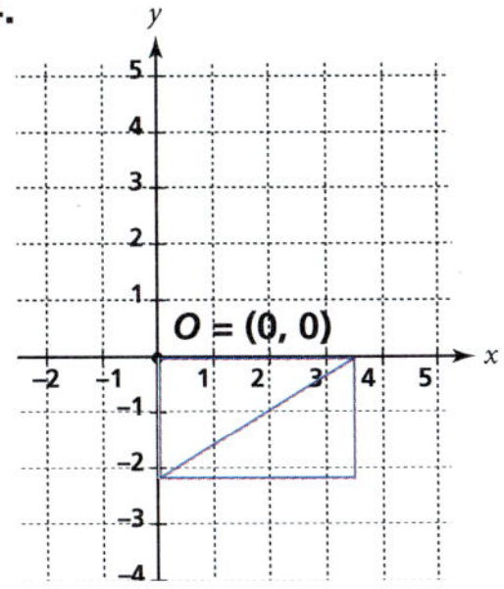

Writing About Mathematics

Describe the image of the circle in the diagram if the circle is rotated 90° or rotated 180° with a center of rotation at (0, 0).

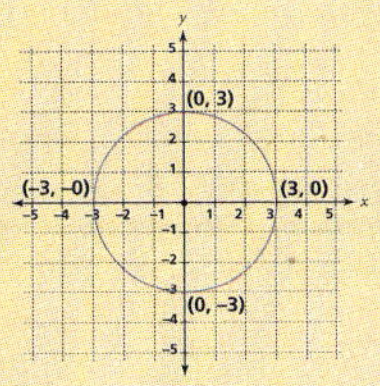

Exercise C Rotate each of the following points 180° counterclockwise around the origin. Give the coordinates of the image point.

15. (4, −5) (−4, 5)

16. (5, 5) (−5, −5)

17. (7, 1) (−7, −1)

18. (−8, 3) (8, −3)

19. (8, −3) (−8, 3)

20. (4, −6) (−4, 6)

Experiencing Geometry

Cut a triangle, square, and rectangle from heavy construction paper. Mark a center of rotation similar to the one in each of the diagrams. Then, using your compass needle to anchor the center of rotation, turn your triangle, square, and rectangle to match the diagrams in this lesson.

Mental Math

Write the following points on the board, and ask students to mentally calculate the image point as indicated.

(90° counterclockwise)

(1, 0)	(0, 1)
(−3, 0)	(0, −3)
(0, 1)	(−1, 0)
(0, −2)	(2, 0)

(180° clockwise)

(4, −1)	(−4, 1)
(−5, 3)	(5, −3)
(−2, −1)	(2, 1)
(6, 8)	(−6, −8)

Group Problem Solving

Discuss briefly with the class how animated cartoons can be made by translating and rotating the figures in a series of pictures. Then have groups of three or four students make "flip books." They can cut and staple 10 to 15 pieces of grid paper together to form a book and draw a series of pictures that involve gradual and sequential transformations of one or more aspects of the picture. When the book is flipped through quickly, the figures in the picture should appear to be moving. Each group should pick a theme or idea for its flip book. Encourage students to focus on animating one or two simple aspects of a given scene and to make the pictures colorful and clearly drawn. Invite groups to share their flip books with the class.

11.

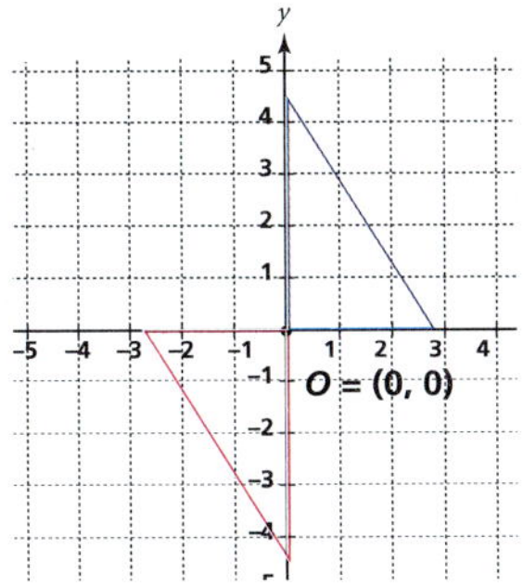

12.

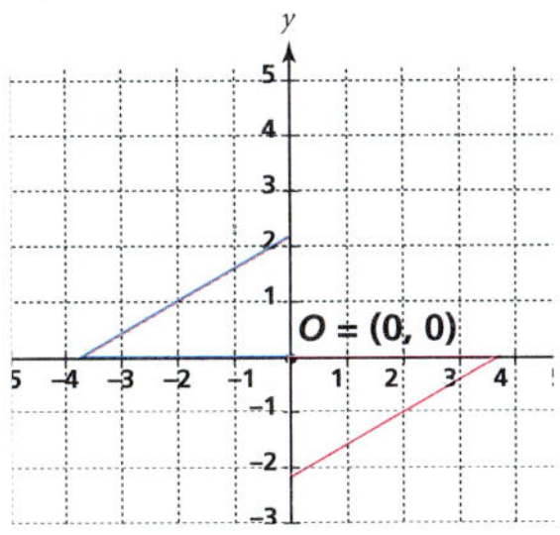

13.

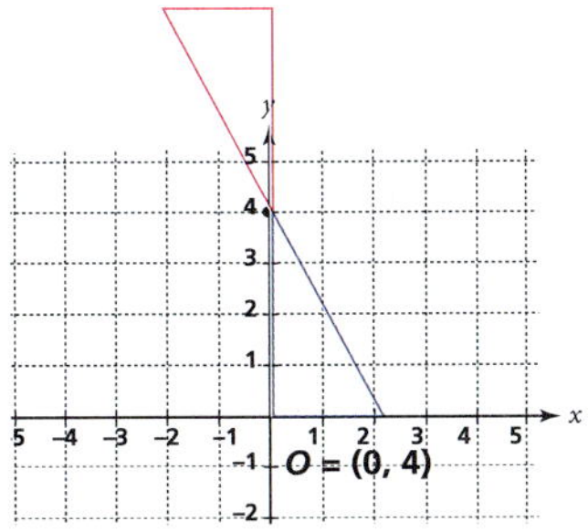

14.

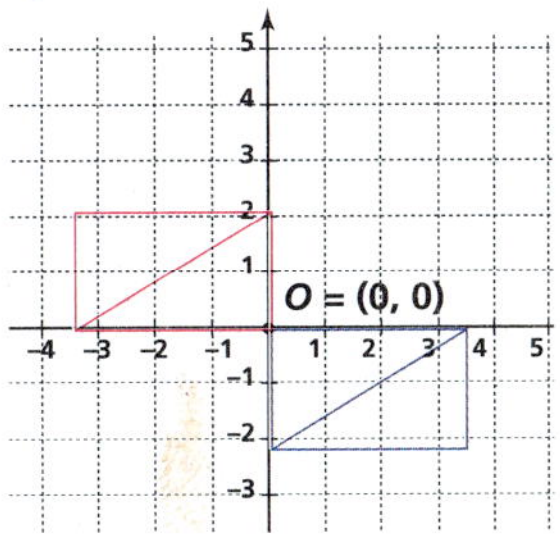

Lesson at a Glance

Chapter 6 Application

Overview This lesson introduces students to how computers are used to draw different views of an object.

Objectives

- To draw several different views of a three-dimensional object.
- To identify an object from its drawing.

Student Page 214

Teacher's Resource Library TRL

Workbook Activity 71

1. Warm-Up Activity

Place a large box where all students can see it and ask them to draw a sketch of it. Tell them to try to draw a three-dimensional representation of the box, rather than just the face of it. Ask students to share their drawings with the class and then discuss the methods they used to add depth to their drawings.

2. Teaching the Lesson

Discuss the computer drawing of the automobile manifold in the text. Then demonstrate techniques to make three-dimensional drawings of objects. For example, when drawing a cube, you can draw two squares, one behind the other, to represent the faces, then connect the corners with line segments to complete the cube. Ask students to share examples of their drawing techniques.

3. Reinforce and Extend

Application

In the Community

Computer-aided design programs (CAD) are used in many different industries. Ask students to find a company or organization in their community that uses some kind of computer-aided design program and to find out what kind of program or programs are used and what they are used for. For example, an electronics firm may use CAD programs to design circuit boards, or the plumbing industry may use CAD to design plumbing parts or plumbing systems. You might suggest some places to start such as engineering firms, electronics firms, architecture and drawing firms, civil engineers, and city planning and development offices.

Ask students to write a brief report on their discoveries.

Application

Geometry and Computer Drawings

Rotations and symmetry play an important part in preparing drawings of objects to be constructed. The computer drawing of the car part shown here presents one view of the part. With sophisticated drawing programs, engineers and others can rotate the drawing. The computer automatically adjusts the drawing to show the desired view of the part. These kinds of computer drawing programs provide numerous—and accurate—views of an object within seconds, saving both time and money.

EXAMPLE

Here are four views of a simple object—a top view, a side view, a reflected view, and a front view. As you can see, each view provides different information about the object as it rotates in front of the viewer.

Exercise

1. Choose a small three-dimensional object. It might be easiest if the sides and faces of the object are regular polygons.
2. Draw several different views of your object, rotating it in space.
3. Look for any symmetry that your object may have. Be sure your drawing shows the symmetry.
4. Place the object you chose in a group with other objects chosen by your classmates.
5. Share your drawings with a partner. See if your partner can identify your object using your drawings. Ask your partner which drawing(s) provided the greatest clue to the identity of the object. Answers will vary.

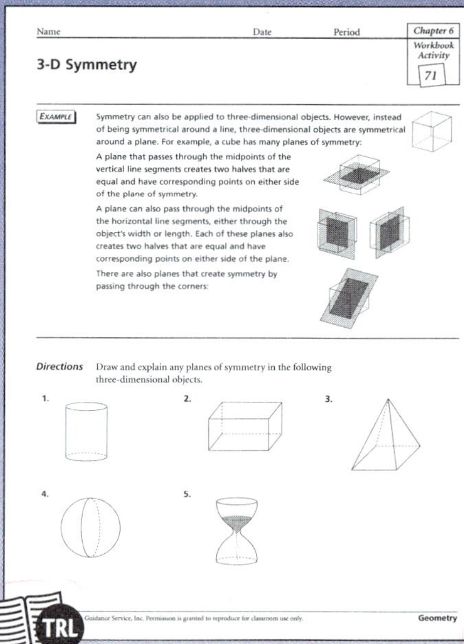

Name Date Period

Chapter 6 Workbook Activity 71

3-D Symmetry

EXAMPLE Symmetry can also be applied to three-dimensional objects. However, instead of being symmetrical around a line, three-dimensional objects are symmetrical around a plane. For example, a cube has many planes of symmetry:

A plane that passes through the midpoints of the vertical line segments creates two halves that are equal and have corresponding points on either side of the plane of symmetry.

A plane can also pass through the midpoints of the horizontal line segments, either through the object's width or length. Each of these planes also creates two halves that are equal and have corresponding points on either side of the plane.

There are also planes that create symmetry by passing through the corners:

Directions Draw and explain any planes of symmetry in the following three-dimensional objects.

1.

2.

3.

4.

5.

 Geometry

TRL

Workbook Activity 71

Chapter 6 Review

For each pair of congruent triangles, write the letter of the postulate or theorem.

Example:

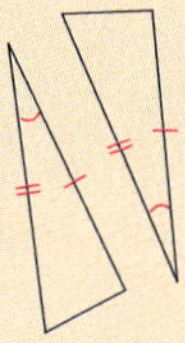

Solution: A; The corresponding sides and the included angle of the triangle are marked as congruent.

A. SAS
B. SSS
C. ASA
D. H-L

1. 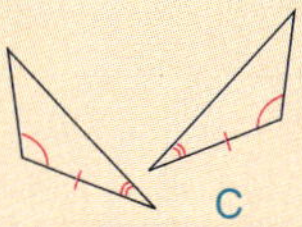

C

A. SAS
B. SSS
C. ASA
D. AAS

2. 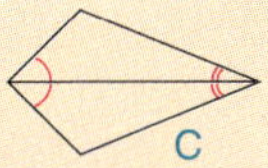

C

A. SAS
B. SSS
C. ASA
D. H-L

3. 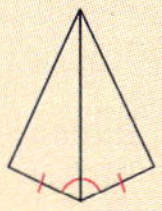

A

A. SAS
B. SSS
C. AAS
D. H-L

4. 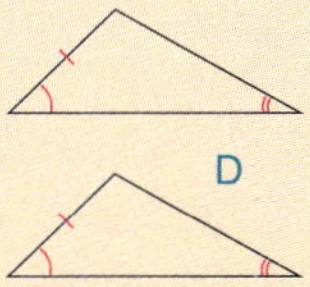

D

A. SAS
B. SSS
C. ASA
D. AAS

5. 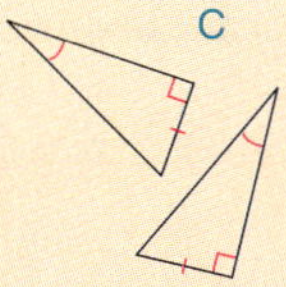

C

A. SAS
B. SSS
C. AAS
D. H-L

6. 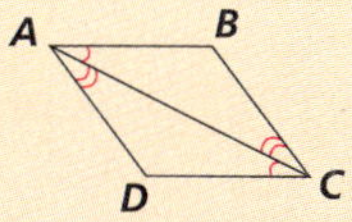

D

A. SAS
B. SSS
C. AAS
D. ASA

7. 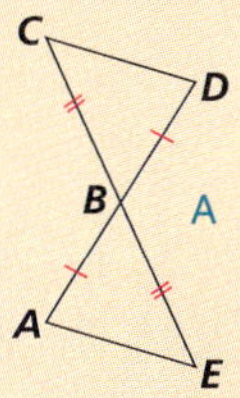

A

A. SAS
B. SSS
C. ASA
D. not congruent

8. 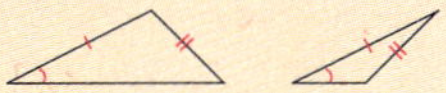

D

A. SAS
B. SSS
C. ASA
D. not congruent

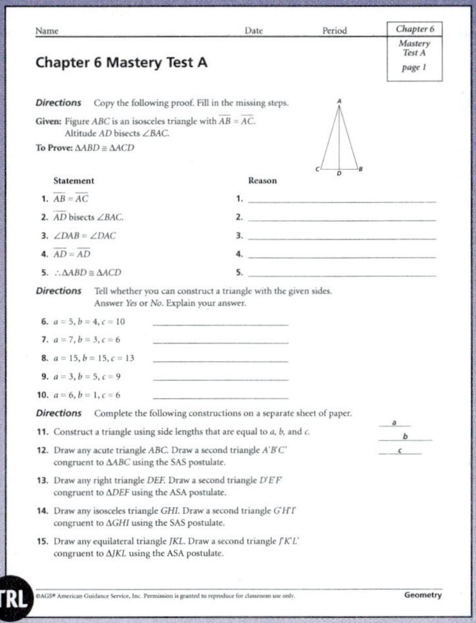

Name Date Period | *Chapter 6* *Mastery Test A* *page 1*

Chapter 6 Mastery Test A

Directions Copy the following proof. Fill in the missing steps.

Given: Figure *ABC* is an isosceles triangle with $\overline{AB} = \overline{AC}$.
Altitude *AD* bisects $\angle BAC$.

To Prove: $\Delta ABD \cong \Delta ACD$

Statement	Reason
1. $\overline{AB} = \overline{AC}$	1. ______
2. $\overline{AD}$ bisects $\angle BAC$.	2. ______
3. $\angle DAB = \angle DAC$	3. ______
4. $\overline{AD} = \overline{AD}$	4. ______
5. $\therefore \Delta ABD \cong \Delta ACD$	5. ______

Directions Tell whether you can construct a triangle with the given sides. Answer *Yes* or *No*. Explain your answer.

6. $a = 5, b = 4, c = 10$ ______
7. $a = 7, b = 3, c = 6$ ______
8. $a = 15, b = 15, c = 13$ ______
9. $a = 3, b = 5, c = 9$ ______
10. $a = 6, b = 1, c = 6$ ______

Directions Complete the following constructions on a separate sheet of paper.

11. Construct a triangle using side lengths that are equal to *a*, *b*, and *c*.

12. Draw any acute triangle *ABC*. Draw a second triangle *A'B'C'* congruent to ΔABC using the SAS postulate.

13. Draw any right triangle *DEF*. Draw a second triangle *D'E'F'* congruent to ΔDEF using the ASA postulate.

14. Draw any isosceles triangle *GHI*. Draw a second triangle *G'H'I'* congruent to ΔGHI using the SAS postulate.

15. Draw any equilateral triangle *JKL*. Draw a second triangle *J'K'L'* congruent to ΔJKL using the ASA postulate.

TRL ©AGS® American Guidance Service, Inc. Permission is granted to reproduce for classroom use only. **Geometry**

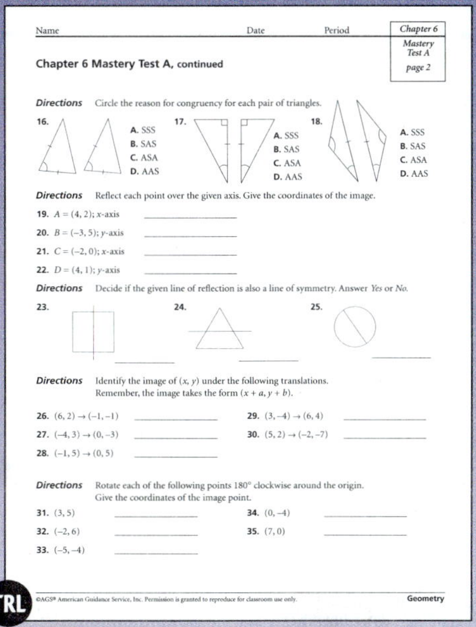

Name Date Period | *Chapter 6* *Mastery Test A* *page 2*

Chapter 6 Mastery Test A, continued

Directions Circle the reason for congruency for each pair of triangles.

16. A. SSS B. SAS C. ASA D. AAS

17. A. SSS B. SAS C. ASA D. AAS

18. A. SSS B. SAS C. ASA D. AAS

Directions Reflect each point over the given axis. Give the coordinates of the image.

19. $A = (4, 2)$; *x*-axis ______
20. $B = (-3, 5)$; *y*-axis ______
21. $C = (-2, 0)$; *x*-axis ______
22. $D = (4, 1)$; *y*-axis ______

Directions Decide if the given line of reflection is also a line of symmetry. Answer *Yes* or *No*.

23. ______ **24.** ______ **25.** ______

Directions Identify the image of (x, y) under the following translations. Remember, the image takes the form $(x + a, y + b)$.

26. $(6, 2) \to (-1, -1)$ ______
27. $(-4, 3) \to (0, -3)$ ______
28. $(-1, 5) \to (0, 5)$ ______
29. $(3, -4) \to (6, 4)$ ______
30. $(5, 2) \to (-2, -7)$ ______

Directions Rotate each of the following points 180° clockwise around the origin. Give the coordinates of the image point.

31. $(3, 5)$ ______
32. $(-2, 6)$ ______
33. $(-5, -4)$ ______
34. $(0, -4)$ ______
35. $(7, 0)$ ______

TRL ©AGS® American Guidance Service, Inc. Permission is granted to reproduce for classroom use only. **Geometry**

Chapter 6 Mastery Test A

Chapter 6 Review

Each set of problems in the Chapter Review includes an example and solution to illustrate the concept. Use the given examples for reteaching the material in Chapter 6. For additional practice, refer to the Supplementary Problems for Chapter 6 (pages 405–407) and Additional Practice 34–41 in the Teacher's Resource Library.

Chapter 6 Mastery Test TRL

The Teacher's Resource Library includes parallel forms of the Chapter 6 Mastery Test. The difficulty level of the two forms is equivalent. You may wish to use one form as a pretest and the other form as a posttest. In addition, the Chapter 6 Cumulative Mastery Test covers content from Chapters 1–6.

Alternative Assessment

Alternative Assessment items correlate with student Goals for Learning at the beginning of this chapter.

- Have students use the SAS postulate to identify corresponding parts in triangles.

Draw five triangles including two that are congruent. Exchange papers with a classmate. Your classmate can use a straightedge and a protractor to match the congruent triangles by SAS.

- Have students construct triangles using SSS information.

Use SSS information to construct a triangle with these side unit measurements: 5, 4, 3 centimeters.

- Have students identify congruent triangles based on SSS and ASA structures.

Write an explanation of how to identify congruent triangles by SSS and ASA structures. Include illustrations with your explanations.

Alternative Assessment

continued

- Have students identify angle and side characteristics of congruent right triangles.

Make a numbered list of the side and angle characteristics for a right triangle. Use the numbers to label the characteristics on a right triangle you draw.

- Have students make generalizations about reflections and lines of reflection.

Write your generalizations about reflections and lines of reflection and use the GeoReflector and other manipulatives to verify your generalizations.

- Have students identify lines of symmetry in reflections.

Draw a regular polygon and its reflection. Find their lines of symmetry.

- Have students map translations in a coordinate plane.

Map this translation by sliding point A at coordinates $(-2, -2)$. A' is $(x + 3)$ and $(y + 3)$. Identify the coordinate pair of A'.

- Have students describe and graph rotations.

Draw a regular polygon on a grid and rotate it 90° counterclockwise.

Chapters 1–6 Midterm Mastery Test (TRL)

The Teacher's Resource Library includes the Midterm Mastery Test. The test is pictured on page 481 of this Teacher's Edition. The Midterm Mastery Test assesses the major learning objectives for Chapters 1–6.

More Review...

Find the measurement for each angle listed. Write the measurement.

Example:

Given: m∠1 is twice m∠2

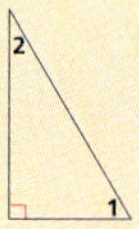

Solution: m∠2 = 30°; m∠1 = 60°
let x = m∠2, $2x$ = m∠1
$x + 2x + 90° = 180°$ sum of angles in triangle
$3x = 180° - 90° = 90°$
$x = 30°$ m∠2 = 30° m∠1 = 2(30°) = 60°

9.

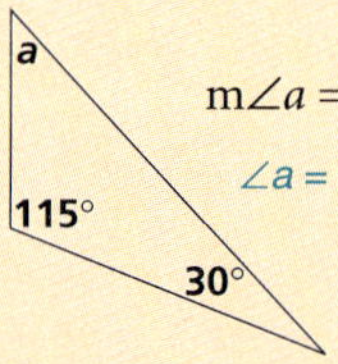

m∠a = __

∠a = 35°

10.

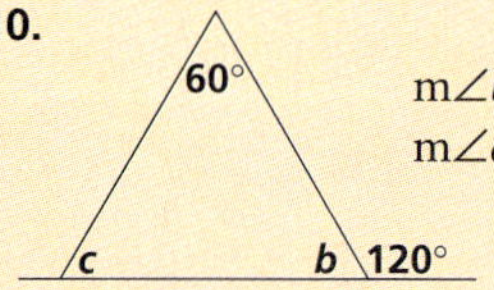

m∠b = __
m∠c = __

∠b = 60°, ∠c = 60°

11.

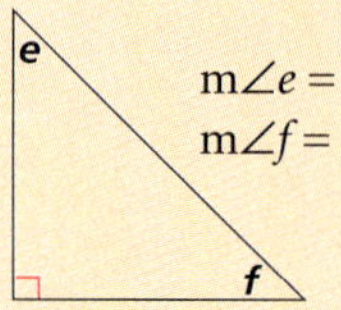

m∠e = m∠f
m∠f = __

∠f = 45°

12. m∠g is 10° greater than m∠h

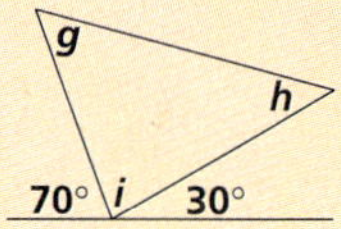

m∠g = __
m∠h = __
m∠i = __

∠g = 55°, ∠h = 45°, ∠i = 80°

Tell whether the given line of reflection is also a line of symmetry. Write *yes* or *no*.

Example:

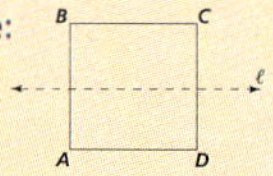

Solution: Yes, the image and object will coincide.

13. yes
14. yes

13.

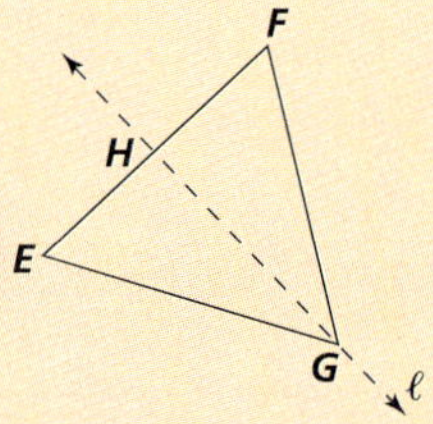

14.

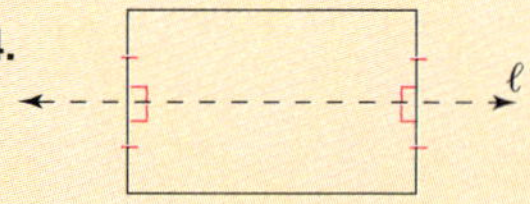

216 *Chapter 6 Congruent Triangles and Transformations*

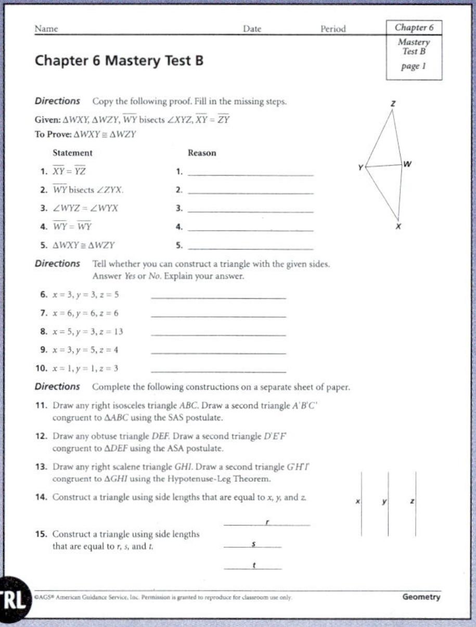

Name Date Period

Chapter 6 Mastery Test B page 1

Chapter 6 Mastery Test B

Directions Copy the following proof. Fill in the missing steps.

Given: ΔWXY, ΔWZY, $\overline{WY}$ bisects $\angle XYZ$, $\overline{XY} = \overline{ZY}$

To Prove: $\Delta WXY \cong \Delta WZY$

Statement	Reason
1. $\overline{XY} = \overline{YZ}$	1. ______
2. $\overline{WY}$ bisects $\angle ZYX$.	2. ______
3. $\angle WYZ = \angle WYX$	3. ______
4. $\overline{WY} = \overline{WY}$	4. ______
5. $\Delta WXY \cong \Delta WZY$	5. ______

Directions Tell whether you can construct a triangle with the given sides. Answer *Yes* or *No*. Explain your answer.

6. $x = 3, y = 3, z = 5$ ______
7. $x = 6, y = 6, z = 6$ ______
8. $x = 5, y = 3, z = 13$ ______
9. $x = 3, y = 5, z = 4$ ______
10. $x = 1, y = 1, z = 3$ ______

Directions Complete the following constructions on a separate sheet of paper.

11. Draw any right isosceles triangle ABC. Draw a second triangle $A'B'C'$ congruent to ΔABC using the SAS postulate.
12. Draw any obtuse triangle DEF. Draw a second triangle $D'E'F'$ congruent to ΔDEF using the ASA postulate.
13. Draw any right scalene triangle GHI. Draw a second triangle $G'H'I'$ congruent to ΔGHI using the Hypotenuse-Leg Theorem.
14. Construct a triangle using side lengths that are equal to x, y, and z.
15. Construct a triangle using side lengths that are equal to r, s, and t.

TRL

 Geometry

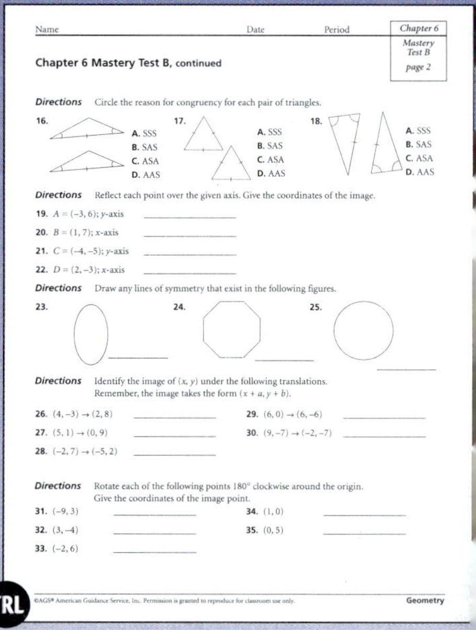

Name Date Period

Chapter 6 Mastery Test B page 2

Chapter 6 Mastery Test B, continued

Directions Circle the reason for congruency for each pair of triangles.

16. A. SSS B. SAS C. ASA D. AAS
17. A. SSS B. SAS C. ASA D. AAS
18. A. SSS B. SAS C. ASA D. AAS

Directions Reflect each point over the given axis. Give the coordinates of the image.

19. $A = (-3, 6)$; y-axis ______
20. $B = (1, 7)$; x-axis ______
21. $C = (-4, -5)$; y-axis ______
22. $D = (2, -3)$; x-axis ______

Directions Draw any lines of symmetry that exist in the following figures.

23. ______ 24. ______ 25. ______

Directions Identify the image of (x, y) under the following translations. Remember, the image takes the form $(x + a, y + b)$.

26. $(4, -3) \rightarrow (2, 8)$ ______
27. $(5, 1) \rightarrow (0, 9)$ ______
28. $(-2, 7) \rightarrow (-5, 2)$ ______
29. $(6, 0) \rightarrow (6, -6)$ ______
30. $(9, -7) \rightarrow (-2, -7)$ ______

Directions Rotate each of the following points 180° clockwise around the origin. Give the coordinates of the image point.

31. $(-9, 3)$ ______
32. $(3, -4)$ ______
33. $(-2, 6)$ ______
34. $(1, 0)$ ______
35. $(0, 5)$ ______

TRL

 Geometry

Chapter 6 Mastery Test B

15.

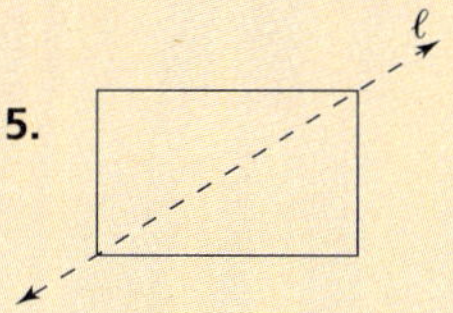

16.

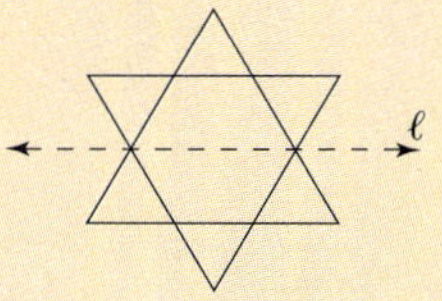

15. no

16. yes

Tell whether the object is a *slide, reflection,* or *rotation.*

Example:

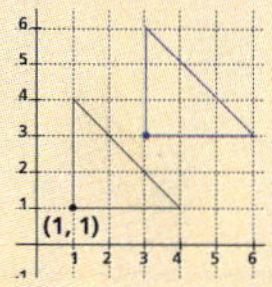

Solution: Slide; the image is above and to the right of the object.

17.

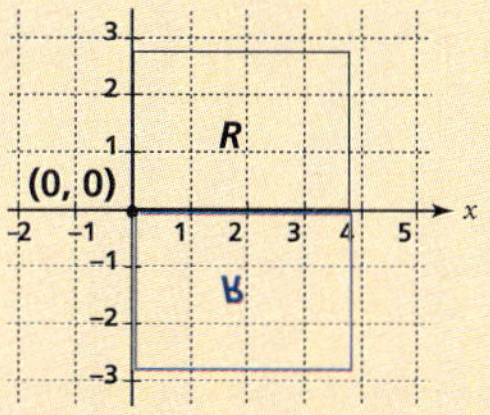

18.

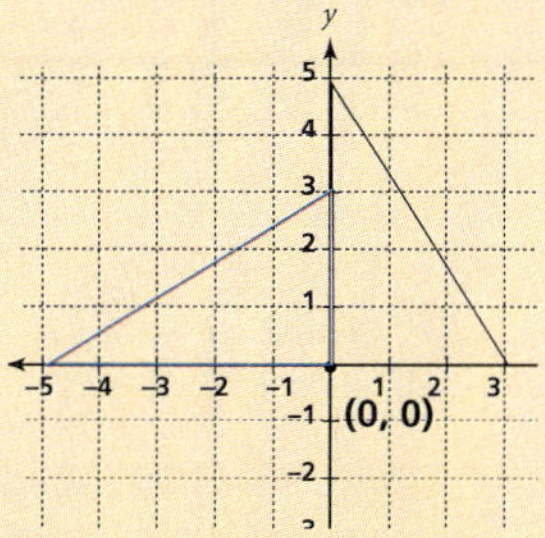

17. reflection

18. rotation through 90° counterclockwise

19. rotation through 180° counterclockwise

20. slide

19.

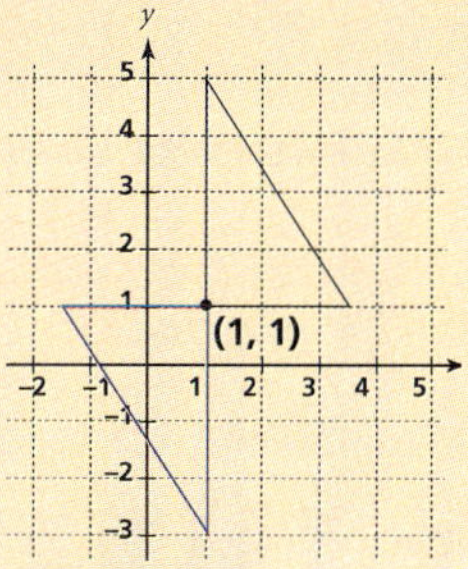

20.

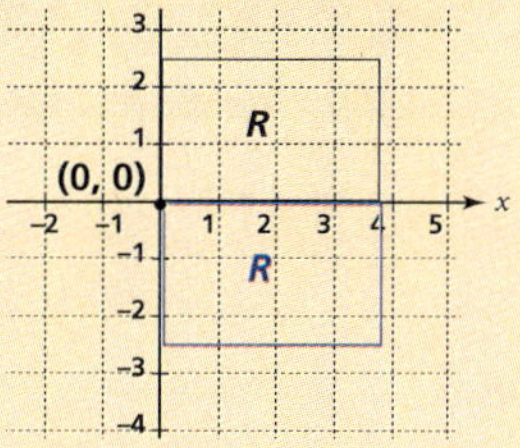

TEST-TAKING TIP

When studying for a test, review any tests or quizzes you took earlier that cover the same information. Make sure you have the correct answers for any items you missed.

Name Date Period | Chapter 6 Cumulative Mastery Test page 1

Chapter 6 Cumulative Mastery Test

Directions Use the vertical angle theorem to find the measures of angles *x*, *y*, and *z*.

1.

$\angle x =$ ______
$\angle y =$ ______
$\angle z =$ ______

A. 53° **B.** 37° **C.** 47° **D.** 143°

2.

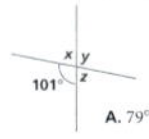

$\angle x =$ ______
$\angle y =$ ______
$\angle z =$ ______

A. 79° **B.** 91° **C.** 69° **D.** 101°

Directions Find the measures of the angles in the figure below. Write your reason for each measure. Use the three theorems about parallel lines and what you know about supplementary and vertical angles.

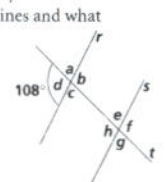

3. $\angle h$ ______________

4. $\angle a$ ______________

5. $\angle g$ ______________

Directions Find the slope, *m*, and the midpoint of a line segment with the given endpoints.

6. Points (–3, 6) and (0, 5) ______________

7. Points (2, 8) and (–4, 2) ______________

8. Points (4, –5) and (5, 1) ______________

Directions On a separate sheet of paper, graph each line using the slope and point given. Also write the algebraic expression for each line.

9. Line *a*; $m = 2$, passes through (5, 1)

10. Line *b*; $m = \frac{1}{4}$, passes through (4, 0)

Directions Complete the following constructions. Use a separate sheet of paper. You will need a straightedge and a compass.

11. Construct an equilateral triangle using the length of line *x*. ______ *x* ______

12. Construct triangle *XYZ* with $\overline{XY} = \overline{XZ}$ and $m\angle X < 90°$.

Directions Find the measure of $\angle x$.

13.

____

14.

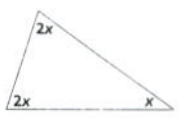

____

RL

 Geometry

Name Date Period | Chapter 6 Cumulative Mastery Test page 2

Chapter 6 Cumulative Mastery Test, continued

Directions Complete the following constructions on a separate sheet of paper. Use only a straightedge and a compass.

15. Draw any obtuse triangle and label it *XYZ*. Construct the altitudes, angle bisectors, and medians from *X*, *Y*, and *Z*, and label the points where each set of lines cross by their correct names.

16. Draw any acute triangle and label it *RST*. Construct a second triangle *R'S'T'* that is congruent to ΔRST using the ASA postulate.

17. Draw any obtuse triangle and label it *ABC*. Construct a second triangle *A'B'C'* that is congruent to ΔABC using the AAS postulate.

18. Draw any right scalene triangle and label it *DEF*. Construct a second triangle *D'E'F'* that is congruent to ΔDEF using the H-L Theorem.

19. Draw any equilateral triangle and label it *JKL*. Construct a second triangle *J'K'L'* that is congruent to ΔJKL using the SSS postulate.

20. Draw any isosceles triangle and label it *MNO*. Construct a second triangle *M'N'O'* that is congruent to ΔMNO using the SAS postulate.

Directions Reflect each figure over the given line of reflection. Give the coordinates of the image.

21. $x = -1$ ______

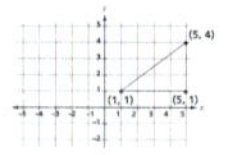

22. $y = 2$ ______

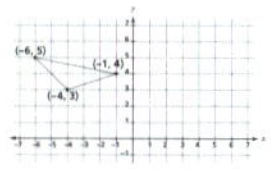

Directions Identify the image of (*x*, *y*) under the following translations. Remember, the image takes the form $(x + a, y + b)$.

23. $(-3, -5) \rightarrow (2, 4)$ ______________

24. $(4, 7) \rightarrow (1, 8)$ ______________

Directions Copy the given figure onto graph paper. Rotate the figure 180° clockwise around the origin. Give the coordinates of the image's endpoints.

25. ______

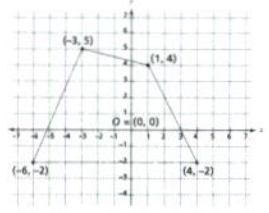

TRL

 Geometry

Chapter 6 Cumulative Mastery Test

Chapter 7

Planning Guide

Proportion and Similarity

	Student Text Lesson						
	Student Pages	Vocabulary	Proof	Practice Exercises	Problem Solving	Calculator Practice	Solutions Key
Lesson 1 Similar Figures and Proportions	220–225	✔	✔	✔		224	✔
Lesson 2 Similar Triangles	226–230	✔	✔	✔	230		✔
Lesson 3 Angle Measure in Regular Polygons	231–234	✔		✔			✔
Lesson 4 Enlarging Geometric Figures	235–239	✔		✔	239		✔
Lesson 5 Shrinking Geometric Figures	240–243			✔	243		✔
Application Maps and Globes	244			✔			✔

Chapter Activities

Teacher's Resource Library
Geometry on the Job 7: Cartography

Community Connection 7: Models

Teacher's Edition
Chapter 7 Project

Assessment Options

Student Text
Chapter 7 Review

Teacher's Resource Library
Chapter 7 Mastery Tests A and B

Chapter 7 Cumulative Mastery Test

Teacher's Edition
Chapter 7 Alternative Assessment

Student Text Features						Teaching Strategies								Learning Styles						Teacher's Resource Library		
Algebra Connection	Constructions	Experiencing Geometry	Technology Connection	Try This	Writing About Mathematics	Warm-Up Activity	Common Error	Algebra Connection	Mental Math	Manipulatives	Calculator	Applications	Group Problem Solving	Auditory/Verbal	Visual/Spatial	Body/Kinesthetic	Logical/Mathematical	Interpersonal/Group Learning	LEP/ESL	Workbook Activities	Additional Practice	Self-Study Guide
				225		220	223		222	223	224					224	225		225	72, 73	42	✔
					228	226						228	230	229	230				229	74–76	43	✔
231, 232		234	234		233	231	232	232										234	233	77, 78	44	✔
	238–239					235									238	237				79, 80	45	✔
	242					240						242						241		81, 82	46	✔
						244						244								83		✔

Manipulatives

Polygon Power Pack

Pattern Blocks

Pattern Blocks Geoboard

Chapter at a Glance

Chapter 7: Proportion and Similarity

pages 218–247

(Answer Keys for the Teacher's Resource Library begin on page 484 of this Teacher's Edition.)

Name Date Period | *Chapter 7* *Geometry on the Job* 7

Cartography

Cartography is the making and studying of maps. The oldest map was found on a clay tablet in Babylonia around 2000 B.C. The map shows waterways, mountains, and settlements. The Babylonians developed our system of latitude and longitude by dividing a circle into 360 degrees.

The Egyptians used geometric techniques to make property boundaries after the Nile River flooded its bank each year. The Greeks developed map projections and used geometry and surveying techniques.

Today, mapmaking is aided by technology such as high-powered satellites and computers, and there is little need for a cartographer to set foot in the actual places being mapped.

Try It Yourself

Step 1 Decide what kind of map you will make. Put a title on your map. Draw the area or use an outline map.

- general reference map—one that shows roads and streets or transportation routes
- thematic map—one that shows population, precipitation (rainfall/snow), natural resources
- inventory map—one that shows a specific feature—campsites in your state, pollution in your town, schools in your district

Step 2 Include a scale and legend. Add a north arrow to show direction.

Step 3 Create an index to help people locate items on your map.

TRL

 Geometry

Geometry on the Job 7

Name Date Period | *Chapter 7* *Community Connection* 7

Models

Architects often make scale models of their creations.The models show the people who want to construct the building what the building will look like when it is finished. Models are also used for most commercial products in the design phase to save time and cost. The model can be used to test the design or the product with consumers to help the builder or manufacturer determine whether the product will meet the needs of the market. For every product that you use, a model was constructed first and tested with consumer groups. If the model is well received, then a company will consider producing the product. Models for buildings are often on display in city or corporate offices long before the final decisions to build are made.

Directions

Step 1 Work in groups to make a model. Choose a local building of interest and work together to make measurements of the building. Use a tape measure to obtain the outer dimensions. Line of sight measurements can be made by standing in one place and holding a ruler at arm's length to get fairly accurate measurements of the relative sizes of the building's parts.

Step 2 After making measurements of the building from all different views, work together to build a simple model of the building using colored construction paper, cardboard, markers, and tape.

Step 3 Display the models and decide the accuracy for the scale used in each model.

TRL

 Geometry

Community Connection 7

Chapter 7

Proportion and Similarity

Rome was the capital city and heart of the ancient Roman empire. It was also a major center of trade and culture. The architects and engineers of Rome planned and built a grand and elegant city. Arched bridges, huge arenas, and open-air theaters helped make ancient Rome a busy and inviting place.

This reconstruction, or model, shows how part of the ancient city may have looked. The model is drawn to scale. Its structures are similar to, or have the same shape as, the actual ancient buildings. The model is also made in proportion, so that every model building is an exact miniature of the real building. The model gives you an idea of what ancient Rome may have looked like. Thanks to geometry, a reconstruction of the city is possible.

In Chapter 7, you will learn about similarities and proportions in geometric shapes.

Goals for Learning

- To identify equal proportions
- To use the theorem that the product of the extremes equals the product of the means in solutions to problems
- To solve for the missing sides of similar triangles
- To give coordinates of dilations

Introducing the Chapter

Use the information on these pages to help students recognize the value of proportion and similarity in reconstructing ancient cities. Direct students to look closely at the scale model of Rome. Have them indicate which buildings were likely public buildings and which were likely residences.

Ask students how a similar model of the city they live in might be helpful to city planners. Prompt students to think about what kinds of information the model provides. For example, how could a model help locate where traffic might be a problem? How might a model be useful in determining where fire or police stations should be located? Extend the discussion to everyday life, helping students realize that similarity and proportion can play an important role in their lives. Present the Goals for Learning, briefly discussing each one.

Chapter Project

Encourage students to work in groups to develop and produce a scale model of their classroom, school lunchroom, library, or other location of their choosing. Arrange for students to take measurements of the location they choose. Indicate that they will need to choose a scale for their model and draw a to-scale floor plan of the location. Students can present their plans as part of a fire safety drill, showing the best way to exit their location in case of an emergency.

As an alternative, interested students may construct a three-dimensional model of their chosen location. Advise students that in this case, they may use a scale different from that used for the floor plan. Students should indicate emergency escape routes in their model.

SELF-STUDY GUIDE

Name ______________

CHAPTER 7: Proportion and Similarity

Goal 7.1 *To identify equal proportions*

Date	Assignment	Score
________	1: Read pages 220–224.	________
________	2: Read the Calculator Practice on page 224.	________
________	3: Complete Exercises A–C on pages 224–225.	________
________	4: Complete Workbook Activities 72–73.	________

Comments:

Goal 7.2 *To use the theorem that the product of the extremes equals the product of the means in solutions to problems*

Date	Assignment	Score
________	5: Read pages 226–229. Complete Exercises A–D on pages 229–230.	________
________	6: Complete Workbook Activities 74–76.	________

Comments:

Goal 7.3 *To solve for the missing sides of similar triangles*

Date	Assignment	Score
________	7: Read pages 231–233. Complete Exercises A and B on pages 233–234.	________
________	8: Complete Workbook Activities 77–78.	________

Comments:

 Geometry

TRL

SELF-STUDY GUIDE

Name ______________

CHAPTER 7: Proportion and Similarity, continued

Goal 7.4 *To give coordinates of dilations*

Date	Assignment	Score
________	9: Read pages 235–238. Complete Exercises A–C on pages 238–239.	________
________	10: Complete Workbook Activities 79–80.	________
________	11: Read pages 240–241. Complete Exercises A–C on pages 242–243.	________
________	12: Complete Workbook Activities 81–82.	________
________	13: Read and complete the Application Exercise on page 244.	________
________	14: Complete Workbook Activity 83.	________
________	15: Complete the Chapter 7 Review on pages 245–247.	________

Comments:

Student's Signature ______________ Date ______________

Instructor's Signature ______________ Date ______________

 Geometry

TRL

Chapter 7 Self-Study Guide

Lesson at a Glance

Chapter 7 Lesson 1

Overview This lesson introduces the concept of similar figures—figures having the same shape but different sizes.

Objectives

- To identify similar figures.
- To determine the ratio of similarity for similar figures.
- To use the ratio of similarity to form proportions.
- To find the value of an unknown using the product of the means and the product of the extremes.

Student Pages 220–225

Teacher's Resource Library TRL

Workbook Activities 72–73

Additional Practice 42

Mathematics Vocabulary

diameter
proportion
ratio
similar

1. Warm-Up Activity

Initiate a discussion with students to help them focus on similar figures they meet in everyday life. If feasible, display a selection of similar objects such as two pizzas of different sizes, two same-style sport shoes of different sizes, and so on. Prompt students' thinking by mentioning dolls, model trains, and the sets of figures from science-fiction and space-fantasy movies. Help students conclude that the difference between similar objects is size only; all other details are identical.

Lesson 1 Similar Figures and Proportions

Diameter
Distance across a circle through the center

Proportion
The equality of two ratios
$\frac{a}{b} = \frac{c}{d}$

Ratio
Comparison of two like quantities
$\frac{a}{b}$

Similar
Having the same shape but a different size

Look around and you may notice that many things are the same shape but are very different sizes. For example, a bicycle and a Ferris wheel both have wheels. The wheels have the same circle shapes but are very different sizes.

In geometry, too, **similar** figures are figures that have the same shape. They do not need to be the same size. The different sizes can be compared using a **ratio**—a fraction that represents the relative sizes.

For the circles at the right, the ratio of **diameters** is 4 to 2. We can write this as 4:2 or $\frac{4}{2}$.

This can also be stated as *the ratio of similarity* is $\frac{4}{2}$ or $\frac{2}{1}$.

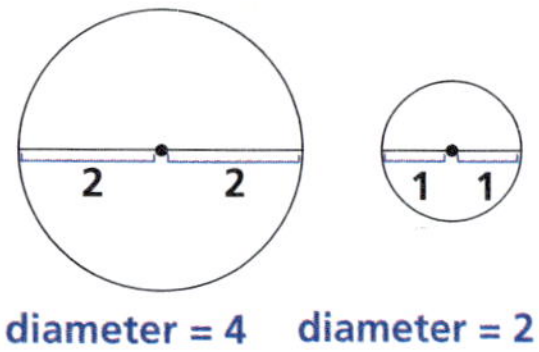

You already know that all squares are similar. All squares have the same shape and all have four right angles. The squares below have sides measuring 2 to 6 units long.

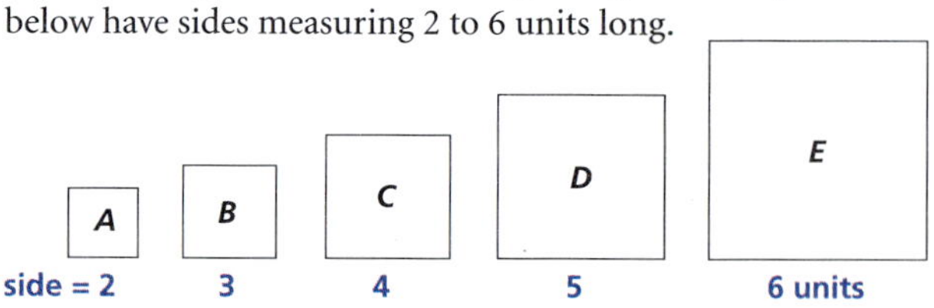

The ratio of similarity between the sides of A and B is 2 to 3 or $\frac{2}{3}$. The ratio of similarity between the sides of A and E is 2 to 6 or $\frac{2}{6} = \frac{1}{3}$.

Name Date Period | Chapter 7 Workbook Activity 72

Proportions

Example Find the missing value in the following proportion.

$\frac{10}{15} = \frac{20}{x}$

The outside elements, the 10 and the x, are called the *extremes*. The inside elements, the 15 and the 20, are called the *means*.

The theorem states that the product of the extremes = the product of the means. Write the following equation.

10x = 300 Divide both sides by 10 to get $x = 30$.

Directions Find the missing value in each proportion.

1. $\frac{3}{4} = \frac{x}{12}$ ______
2. $\frac{5}{8} = \frac{10}{x}$ ______
3. $\frac{11}{23} = \frac{44}{x}$ ______
4. $\frac{x}{3} = \frac{3}{9}$ ______
5. $\frac{7}{x} = \frac{12}{24}$ ______
6. $\frac{5}{x} = \frac{20}{24}$ ______
7. $\frac{2}{5} = \frac{8}{x}$ ______
8. $\frac{8}{13} = \frac{24}{x}$ ______
9. $\frac{x}{34} = \frac{11}{33}$ ______
10. $\frac{5}{9} = \frac{x}{27}$ ______
11. $\frac{19}{x} = \frac{20}{25}$ ______
12. $\frac{33}{33} = \frac{x}{24}$ ______
13. $\frac{8}{15} = \frac{x}{30}$ ______
14. $\frac{x}{60} = \frac{10}{24}$ ______
15. $\frac{4}{x} = \frac{12}{26}$ ______

TRL

 Geometry

Workbook Activity 72

Name Date Period | Chapter 7 Workbook Activity 73

Determining Equal Ratios

Example Determine if the following ratios are equal to each other. Use the theorem that the product of the extremes equals the product of the means.

$\frac{5}{3}$ and $\frac{25}{9}$

The product of the extremes, 5 and 9, is 45.

The product of the means, 3 and 25, is 75.

Since the product of the means must equal the product of the extremes in order for two ratios to be equal, these two ratios are not equal.

Directions Are the following ratios equal? Write *Yes* or *No*. Use the theorem that the product of the extremes equals the product of the means.

1. $\frac{5}{8}$ and $\frac{7}{10}$ ______
2. $\frac{6}{10}$ and $\frac{12}{20}$ ______
3. $\frac{8}{3}$ and $\frac{24}{9}$ ______
4. $\frac{2}{5}$ and $\frac{10}{25}$ ______
5. $\frac{9}{4}$ and $\frac{17}{8}$ ______
6. $\frac{35}{40}$ and $\frac{7}{8}$ ______
7. $\frac{10}{1}$ and $\frac{30}{1}$ ______
8. $\frac{15}{16}$ and $\frac{17}{18}$ ______
9. $\frac{21}{24}$ and $\frac{14}{16}$ ______
10. $\frac{60}{66}$ and $\frac{10}{11}$ ______
11. $\frac{99}{100}$ and $\frac{9}{10}$ ______
12. $\frac{85}{50}$ and $\frac{17}{10}$ ______
13. $\frac{6}{3}$ and $\frac{7}{4}$ ______
14. $\frac{9}{16}$ and $\frac{27}{48}$ ______
15. $\frac{36}{60}$ and $\frac{6}{10}$ ______

TRL

 Geometry

Workbook Activity 73

In the same way, you can see that all equilateral triangles have the same shape. Each angle measures 60°. The sides of the triangles below are 1, 2, or 3 units long.

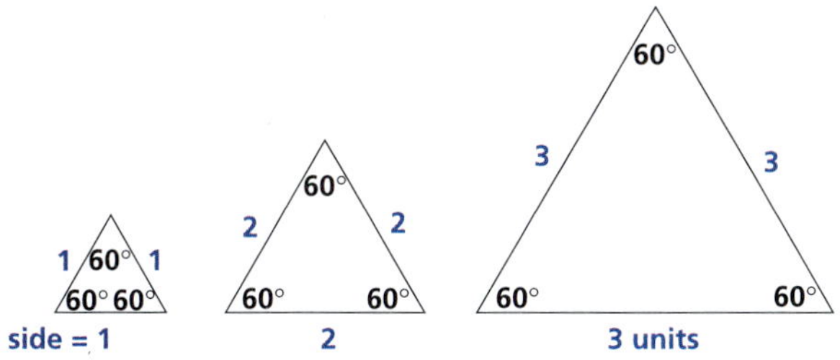

The ratio of similarity between the smallest and largest triangles is 1 to 3 or $\frac{1}{3}$.

The ratio of similarity between the largest and the middle triangles is 3 to 2 or $\frac{3}{2}$.

Triangles need not be equilateral to be similar. Any two triangles can be similar if certain conditions are present.

ΔABC is similar to ΔDEF—the corresponding angles of the triangles are equal; and the corresponding pairs of sides have the same ratio of similarity.

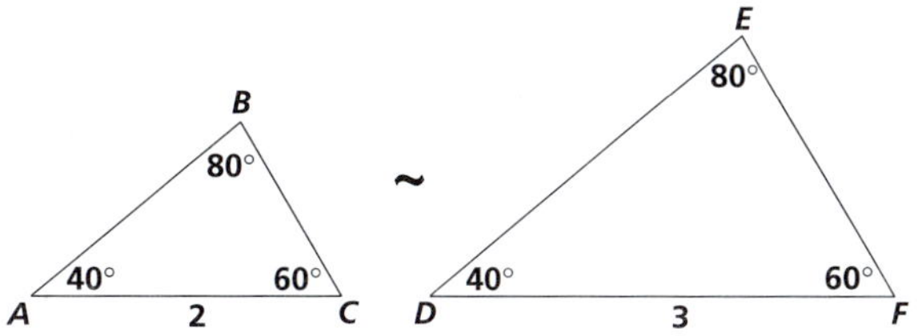

Because the corresponding sides of the triangles have the same ratio of similarity, we know that

$$\frac{AB}{DE} = \frac{BC}{EF} = \frac{AC}{DF}$$

The ratio $\frac{AC}{DF}$ is 2 to 3 or $\frac{2}{3}$ and is given.

Therefore, the ratios $\frac{AB}{DE}$ and $\frac{BC}{EF}$ are also equal to $\frac{2}{3}$.

The symbol ~ is used to show that two geometric figures are similar. Read it as "is similar to."

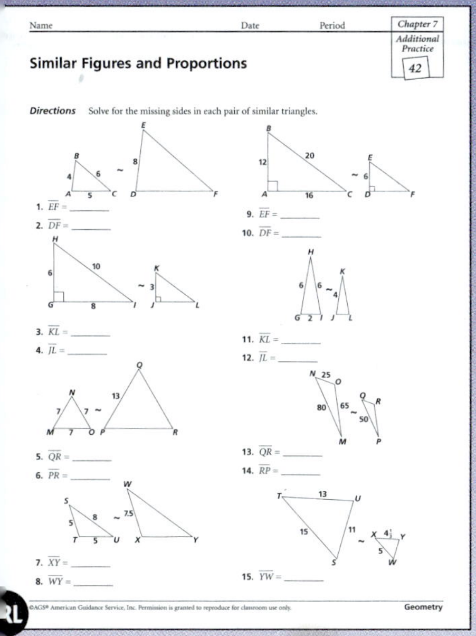

Name Date Period

Chapter 7 Additional Practice 42

Similar Figures and Proportions

Directions Solve for the missing sides in each pair of similar triangles.

1. $\overline{EF}$ = ______
2. $\overline{DF}$ = ______
3. $\overline{KL}$ = ______
4. $\overline{JL}$ = ______
5. $\overline{QR}$ = ______
6. $\overline{PR}$ = ______
7. $\overline{XY}$ = ______
8. $\overline{WY}$ = ______
9. $\overline{EF}$ = ______
10. $\overline{DF}$ = ______
11. $\overline{KL}$ = ______
12. $\overline{JL}$ = ______
13. $\overline{QR}$ = ______
14. $\overline{RP}$ = ______
15. $\overline{YW}$ = ______

Geometry

Additional Practice 42

2. Teaching the Lesson

Build on students' observations that all circles and all squares are similar to one another. Ask volunteers to explain the ratio of similarity for the squares on page 220. On the board or overhead projector, draw squares *F*, *G*, and *H* with side lengths 8, 10, and 20 units, respectively. Ask students to determine the ratio of similarity between squares *A* and *G*

(2 to 10 or $\frac{2}{10}$ or $\frac{1}{5}$)

and between squares *A* and *H*

(2 to 20 or $\frac{2}{20}$ or $\frac{1}{10}$).

Continue the activity with equilateral triangles 4, 5, and 10 with side lengths of 4, 5, and 10 units, respectively. Ask students to determine the ratio of similarity between triangle 1 and 5

(1 to 5 or $\frac{1}{5}$)

and between triangle 10 and 4

(10 to 4 or $\frac{10}{4}$ or $\frac{5}{2}$).

Work through the proof on page 222, inviting a volunteer to complete the multiplication in Step 2. Invite a second volunteer to simplify the product.

Stress to students the equivalence of the forms $\frac{a}{b} = \frac{c}{d}$ and $a{:}b{::}c{:}d$. Indicate the means and the extremes for each form:

means

$a{:}b{::}c{:}d$

extremes

means

$\frac{a}{b} = \frac{c}{d}$

extremes

Ask a volunteer to explain each of the examples on page 223. Be certain students understand that $\frac{15}{20}$ can be reduced to its lowest terms by dividing by the numerator and denominator by 5, showing that $\frac{15}{20} = \frac{3}{4}$.

Mental Math

Provide students with an opportunity to use ratios and proportions to find equivalent fractions. Ask:

How many twelfths equal $\frac{3}{4}$? (9)

How many ninths equal $\frac{1}{3}$? (3)

How many sixteenths equal $\frac{5}{8}$? (10)

Then have students complete the following:

$\frac{6}{21} = \frac{x}{7}$ $(x = 2)$

$\frac{45}{90} = \frac{x}{2}$ $(x = 1)$

$\frac{22}{66} = \frac{x}{3}$ $(x = 1)$

Invite students to explain how they found the value for x.

Remember, the hypotenuse is the longest side in a right triangle.

An example of a ratio you can form with right triangles is the ratio of the hypotenuse of one triangle to the hypotenuse of a second triangle. Another ratio is the shortest side of a triangle to the shortest side of a second triangle, and so on. You can write these ratios as fractions such as

$$\frac{\text{hypotenuse}}{\text{hypotenuse}} \qquad \frac{\text{base}}{\text{base}} \qquad \frac{\text{shortest side of } \Delta}{\text{shortest side of second } \Delta}$$

If two ratios are equal, we can write them as a **proportion.** A proportion, written $\frac{a}{b} = \frac{c}{d}$, is the equality of two ratios.

In a proportion, terms a and d are called the *extremes* while terms b and c are called the *means.*

Note that $\frac{a}{b} = \frac{c}{d}$ can also be written as $a{:}b{::}c{:}d$ (read "a is to b as c is to d"). This form may help you to know the means from the extremes.

means

$a{:}b{::}c{:}d$

extremes

Theorem: In any proportion, the product of the extremes equals the product of the means.

Here's the proof of this theorem.

Given: $\frac{a}{b} = \frac{c}{d}$, any proportion

To Prove: $ad = bc$

Proof

Statement	Reason
1. $\frac{a}{b} = \frac{c}{d}$	1. Given.
2. $\frac{a}{b}(b)(d) = \frac{c}{d}(b)(d)$	2. Multiply by equals.
3. $ad = cb$	3. Simplification.

EXAMPLES **Given:** ΔABC is similar to ΔDEF
Find: length of $\overline{DE}$

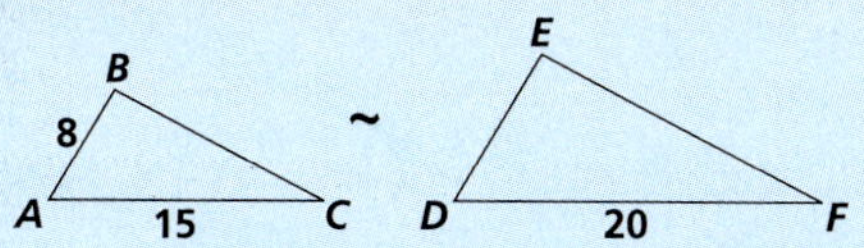

Because the triangles are similar, their corresponding sides are in proportion. This means

$\frac{AC}{DF} = \frac{15}{20}$, which can be reduced to $\frac{3}{4}$.

Therefore, $\frac{3}{4} = \frac{AB}{DE}$

or $\frac{3}{4} = \frac{8}{DE}$

Using product of the means = product of the extremes,

$\frac{8}{DE} \times \frac{3}{4}$

so $3(\overline{DE}) = 8 \bullet 4$

and $3(\overline{DE}) = 32$

so $\overline{DE} = \frac{32}{3} = 10\frac{2}{3}$

The length of $\overline{DE}$ is $10\frac{2}{3}$.

Given: ΔABC is similar to ΔDEF
Find: length of leg DE
and length of hypotenuse EF

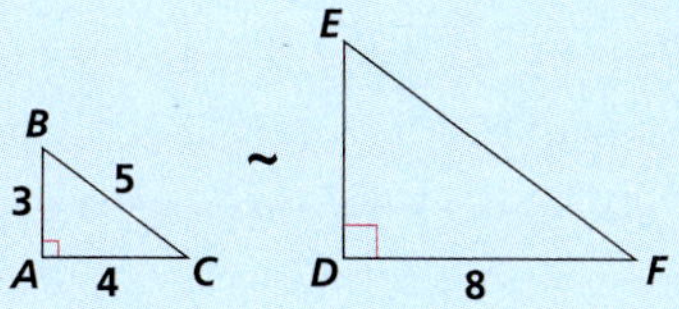

Because the triangles are similar, their corresponding sides are in proportion. This means

$\frac{AC}{DF} = \frac{4}{8} = \frac{1}{2}$

Therefore, $\frac{1}{2} = \frac{AB}{DE}$

or $\frac{1}{2} = \frac{3}{DE}$

Using product of the extremes = product of the means,

$\overline{DE} = 2 \bullet 3 = 6$ The length of $\overline{DE}$ is 6.

Also, $\frac{1}{2} = \frac{BC}{EF}$

or $\frac{1}{2} = \frac{5}{EF}$

Using product of the extremes = product of the means,

$\overline{EF} = 2 \bullet 5 = 10$ The length of $\overline{EF}$ is 10.

3. Reinforce and Extend

MANIPULATIVES

M Polygons

Materials: Polygon Power Pack, Pattern Blocks

Group Practice: Provide students with a variety of polygons from the Polygon Power Pack or Pattern Blocks. Suggest they work in pairs to identify pairs or groups of similar polygons. For each group they form, students should be able to explain how they decided the figures were similar.

Student Practice: Students may also work independently for additional practice.

COMMON ERROR

Some students may become confused when similar triangles are not positioned in the same manner. Draw several similar triangles in different orientations on the board or overhead projector.

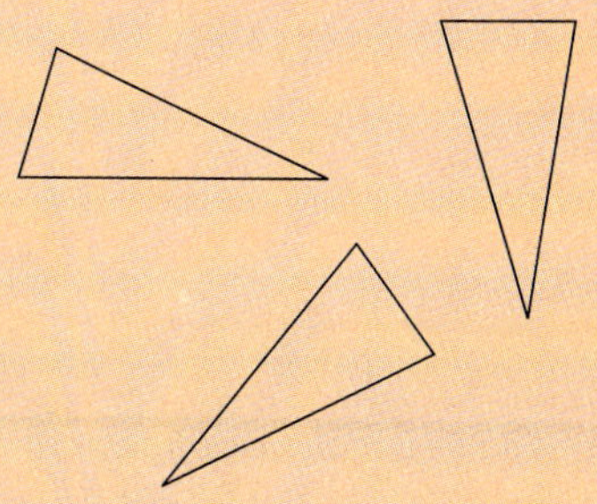

Indicate to students that they need to compare similar parts of the triangles—longest sides, shortest sides, and so on. Be sure they understand that the base of a triangle can be drawn at the bottom, the top, the left, or even the right of the triangle.

Learning Styles

Body/Kinesthetic Provide groups of students with a Pattern Blocks Geoboard as well as several similar pattern blocks. Have students use the geoboard to make outlines of the pattern blocks. Once outlines have been made, students should explain to one another how they can determine which outlines are similar.

Calculator

Take students through the steps for finding the missing value in a proportion. Work through Exercise A with the class. Offer students the opportunity to complete the following proportions:

$\frac{11}{33} = \frac{x}{3}$ (x = 1)

$\frac{45}{135} = \frac{1}{x}$ (x = 3)

$\frac{26}{78} = \frac{x}{3}$ (x = 2)

$\frac{36}{60} = \frac{x}{5}$ (x = 3)

$\frac{27}{135} = \frac{3}{x}$ (x = 5)

In order to find out if $\Delta ABC \sim \Delta DEF$, we need to look at the ratios $\frac{3}{6}$ and $\frac{5}{9}$ to find out if they are equal.

Example Are the ratios $\frac{3}{6}$ and $\frac{5}{9}$ equal?

If the ratios are equal, then they form a proportion. In a proportion, the product of the extremes equals the product of the means.

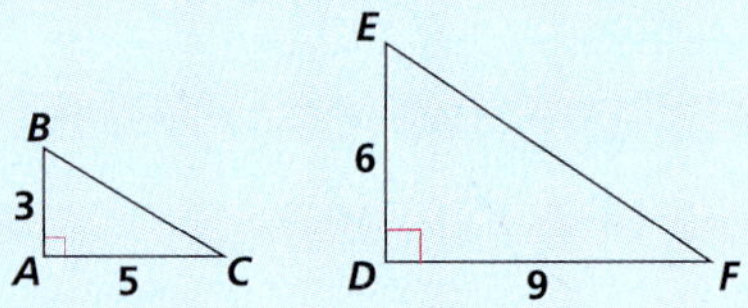

$\frac{3}{6} \stackrel{?}{=} \frac{5}{9}$

(extremes) $3 \cdot 9 \stackrel{?}{=} 5 \cdot 6$ (means)

$27 = 30$ False, the ratios are not equal and the triangles are not similar.

Calculator Practice

You can use your calculator to find the missing value in a proportion.

Example $\frac{3}{5} = \frac{6}{x}$

(extremes) $3 \cdot x = 5 \cdot 6$ (means)

$x = \frac{(5 \cdot 6)}{3}$

Press 5 [×] 6 [÷] 3 [=] 10.

Exercise A Find the missing value in each proportion.

1. $\frac{2}{3} = \frac{x}{9}$ 6
2. $\frac{3}{5} = \frac{6}{y}$ 10
3. $\frac{7}{8} = \frac{z}{4}$ 3.5
4. $\frac{3}{4} = \frac{a}{12}$ 9
5. $\frac{5}{4} = \frac{b}{32}$ 40

Exercise B Solve for the missing sides in each pair of similar triangles.

☞ *Selected Answers and Solutions begin on page 420.*

6. $\overline{DE}$ 10
7. $\overline{EF}$ 24
8. $\overline{JK}$ 3
9. $\overline{KL}$ 4.2

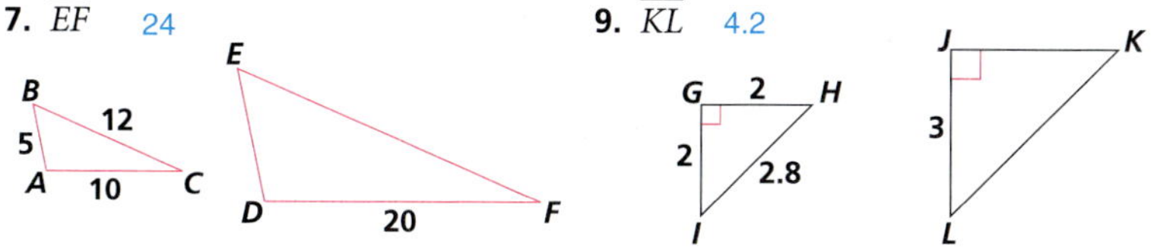

10. $\overline{PR}$ 6.7
11. $\overline{RQ}$ 8
12. $\overline{VW}$ 19.6
13. $\overline{WU}$ 15.4
14. $\overline{AB}$ 11.7
15. $\overline{CA}$ 13.33

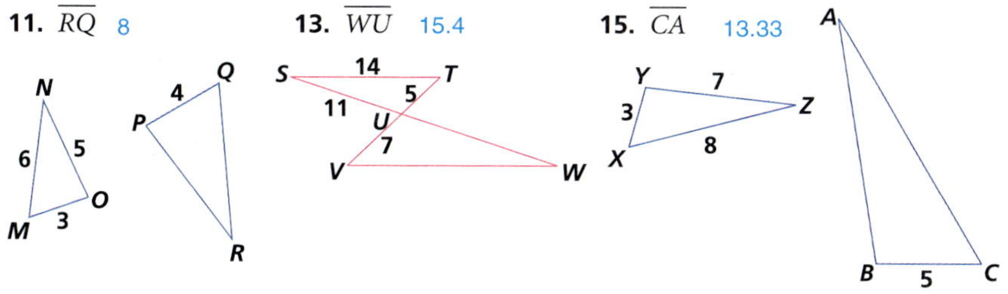

Exercise C Are the following ratios equal? Write *yes* or *no*. Use the theorem that the product of the extremes equals the product of the means.

16. $\frac{3}{5}$ and $\frac{10}{16}$ no
17. $\frac{3}{7}$ and $\frac{5}{8}$ no
18. $\frac{1}{3}$ and $\frac{7}{21}$ yes
19. $\frac{4}{9}$ and $\frac{28}{63}$ yes
20. $\frac{6}{13}$ and $\frac{12}{39}$ no
21. $\frac{7}{9}$ and $\frac{49}{81}$ no
22. $\frac{6}{17}$ and $\frac{18}{51}$ yes
23. $\frac{3}{23}$ and $\frac{27}{21}$ no
24. $\frac{6}{11}$ and $\frac{16}{61}$ no
25. $\frac{7}{21}$ and $\frac{12}{36}$ yes

TRY THIS...

You know all equilateral triangles are similar. Is the same true for all isosceles triangles? Give reasons or an example to support your answer.

Students can give a single counterexample such as ΔABC with sides 2, 2, and 3 and ΔDEF with sides of 4, 4, and 5.

LEARNING STYLES

Logical/Mathematical Invite students to draw a third triangle for each pair of triangles in Exercise B. The third triangle should be similar to the two given in the Exercise. Direct students to write an explanation telling how they determined the side lengths of the third similar triangle.

LEARNING STYLES

LEP/ESL Pair students proficient in English with those less proficient. Have each student pair work through the items in Exercise C. For each item, one student should give the answer and explain how he or she used the product of the extremes equals the product of the means to determine the answer. Students should exchange roles for each item.

Proportion and Similarity Chapter 7 **225**

Lesson at a Glance

Chapter 7 Lesson 2

Overview This lessons focuses on similar triangles—triangles with proportional sides or equal angles.

Objectives

- To identify similar triangles using the Angle-Angle Similarity Postulate.
- To analyze the ratio of corresponding sides to distinguish similar triangles.
- To solve for the missing sides of similar triangles.

Student Pages 226–230

Teacher's Resource Library TRL

Workbook Activities 74–76

Additional Practice 43

Mathematics Vocabulary

similar triangles

1. Warm-Up Activity

Remind students that in the previous lesson they relied on triangle side length to determine similar triangles. Present several similar triangles to students either on the board or on the overhead projector. Ask them to describe how they think the angles of the similar triangles compare. Allow students to measure the angles to confirm their thoughts.

Lesson 2 Similar Triangles

Similar triangles
Triangles whose corresponding angles are equal and whose corresponding sides are proportional

In the last lesson, you looked at several groups of similar geometric figures such as squares, circles, and equilateral triangles. You even worked with triangles that were not equilateral but were similar.

Similar triangles are triangles whose corresponding angles are equal and whose corresponding sides are proportional.

The following two triangles illustrate the definition of similar.

Corresponding angles are equal:
$m\angle A = m\angle D$
$m\angle B = m\angle E$
$m\angle C = m\angle F$

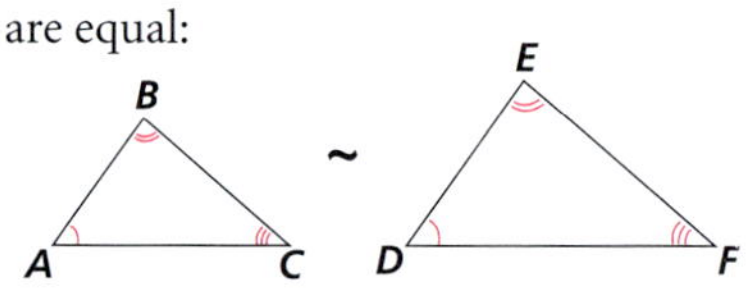

Opposite each pair of equal angles is a pair of corresponding sides that are proportional:

$\frac{BC}{EF}$	=	$\frac{AC}{DF}$	=	$\frac{AB}{DE}$
sides opposite $\angle A$ and $\angle D$		sides opposite $\angle B$ and $\angle E$		sides opposite $\angle C$ and $\angle F$

If triangles are similar, you can calculate the lengths of unknown sides of the triangles using the lengths of corresponding sides. Given any three sides in a proportion, you can always solve for the fourth side.

EXAMPLES **Given:** $\triangle ABC \sim \triangle DFE$, $\overline{AB} = 10$, $\overline{DF} = 5$, $\overline{AC} = 8$, $\overline{EF} = 7$

Find: values for x and y

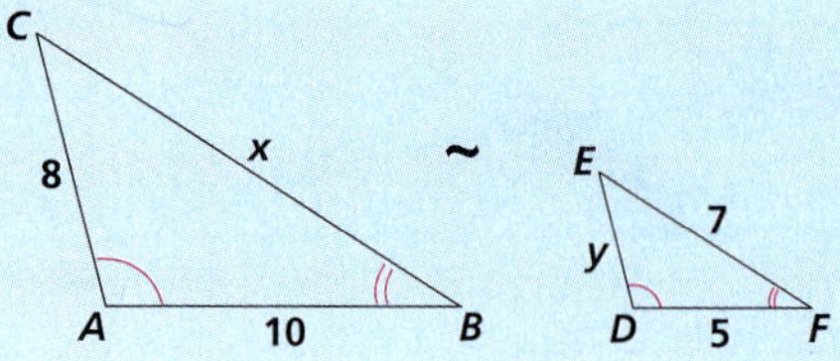

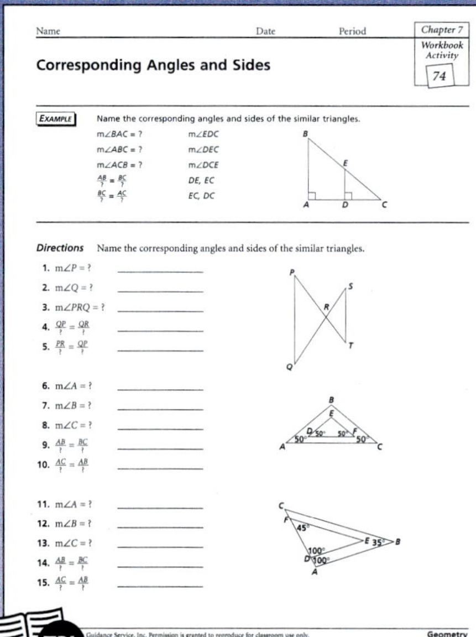

Name Date Period | Chapter 7 Workbook Activity 74

Corresponding Angles and Sides

EXAMPLE Name the corresponding angles and sides of the similar triangles.

$m\angle BAC = ?$ $m\angle EDC$
$m\angle ABC = ?$ $m\angle DEC$
$m\angle ACB = ?$ $m\angle DCE$
$\frac{AB}{?} = \frac{BC}{?}$ DE, EC
$\frac{BC}{?} = \frac{AC}{?}$ EC, DC

Directions Name the corresponding angles and sides of the similar triangles.

1. $m\angle P = ?$ ______
2. $m\angle Q = ?$ ______
3. $m\angle PRQ = ?$ ______
4. $\frac{QP}{?} = \frac{QR}{?}$ ______
5. $\frac{PR}{?} = \frac{QP}{?}$ ______
6. $m\angle A = ?$ ______
7. $m\angle B = ?$ ______
8. $m\angle C = ?$ ______
9. $\frac{AB}{?} = \frac{BC}{?}$ ______
10. $\frac{AC}{?} = \frac{AB}{?}$ ______
11. $m\angle A = ?$ ______
12. $m\angle B = ?$ ______
13. $m\angle C = ?$ ______
14. $\frac{AB}{?} = \frac{BC}{?}$ ______
15. $\frac{AC}{?} = \frac{AB}{?}$ ______

 Geometry

TRL **Workbook Activity 74**

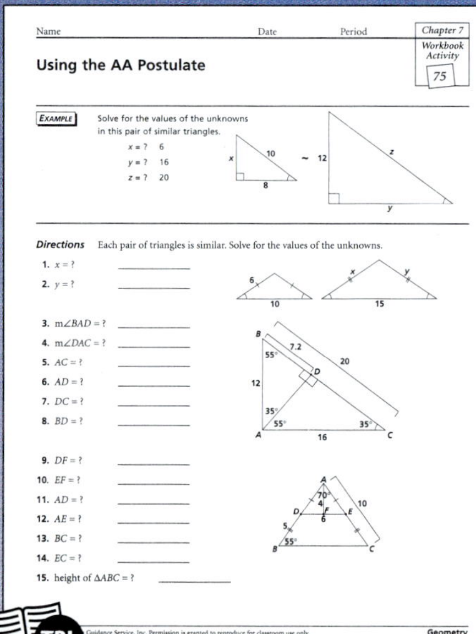

Name Date Period | Chapter 7 Workbook Activity 75

Using the AA Postulate

EXAMPLE Solve for the values of the unknowns in this pair of similar triangles.

$x = ?$ 6
$y = ?$ 16
$z = ?$ 20

Directions Each pair of triangles is similar. Solve for the values of the unknowns.

1. $x = ?$ ______
2. $y = ?$ ______
3. $m\angle BAD = ?$ ______
4. $m\angle DAC = ?$ ______
5. $AC = ?$ ______
6. $AD = ?$ ______
7. $DC = ?$ ______
8. $BD = ?$ ______
9. $DF = ?$ ______
10. $EF = ?$ ______
11. $AD = ?$ ______
12. $AE = ?$ ______
13. $BC = ?$ ______
14. $EC = ?$ ______
15. height of $\triangle ABC = ?$ ______

 Geometry

TRL **Workbook Activity 75**

EXAMPLES *(continued)*

For a similar triangle, pairs of corresponding sides have the same ratio of similarity.

Because $\frac{AB}{DF} = \frac{10}{5} = \frac{2}{1}$ we know all other pairs of corresponding sides are in the ratio 2:1.

We can use the products of the means and the extremes to find the values for x and y.

$\frac{x}{7} = \frac{2}{1}$ becomes $(x \bullet 1) = (7 \bullet 2)$; $x = 14$

Check: $\frac{14}{7} = \frac{2}{1}$ True

$\frac{8}{y} = \frac{2}{1}$ becomes $(8 \bullet 1) = (y \bullet 2)$; $y = 4$

Check: $\frac{8}{4} = \frac{2}{1}$ True

Given: right $\triangle ABC \sim$ right $\triangle DEC$

Find: the ratio of $\overline{AB}$ to $\overline{DE}$ or $\frac{AB}{DE}$

$\frac{AB}{DE} = \frac{\text{larger altitude}}{\text{smaller altitude}} = \frac{\text{larger base}}{\text{smaller base}} = \frac{15}{5} = \frac{3}{1}$

Therefore, $\frac{AB}{DE} = \frac{3}{1}$

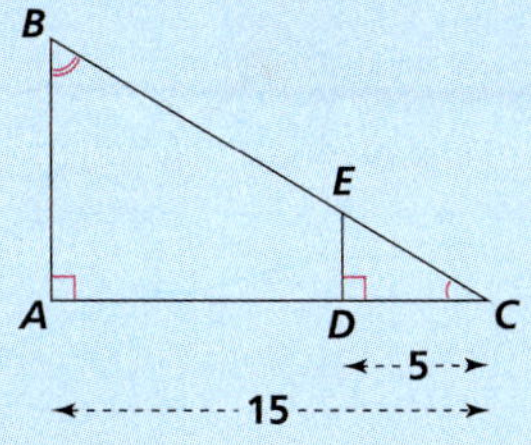

One way to show that two triangles are similar is to show that two angles of one triangle are equal to two angles of a second triangle.

Because every triangle has 180° as its angle sum, if two angles in one triangle equal two angles in the second triangle, then the third angles must also be equal. If the triangles have three pairs of corresponding equal angles, the triangles are similar. Examples of similar triangles follow.

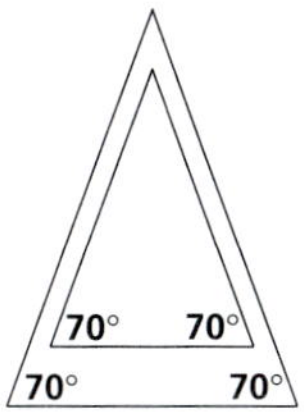

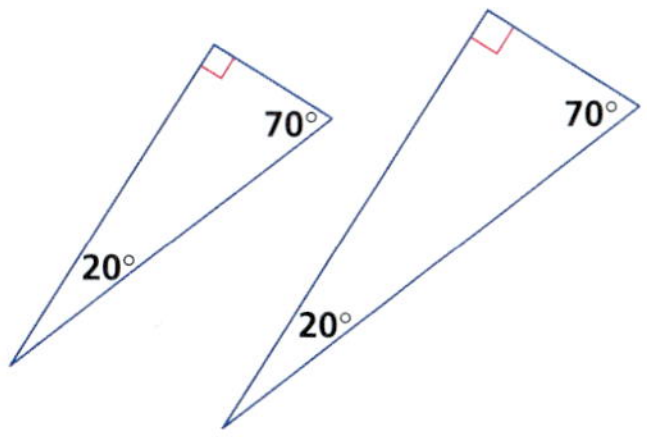

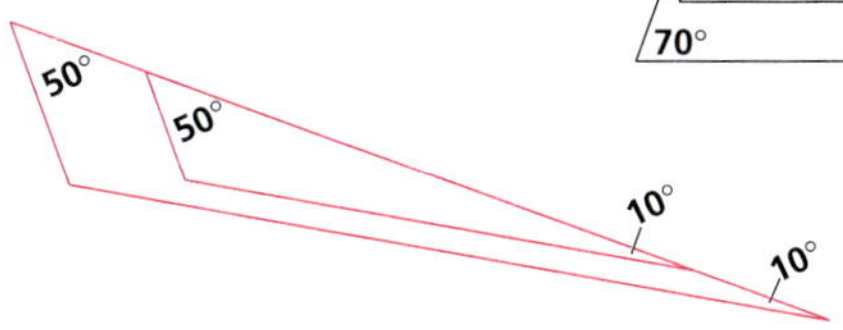

2. Teaching the Lesson

Work through the first example on page 226 with students. Have them generate ratios involving all three sides of the triangles, $\frac{AB}{DF}$, $\frac{AC}{DE}$, and $\frac{CB}{EF}$. Point out that all of these ratios are equal. Ask students to explain why. **(similar triangles)** Be sure they understand how to use the products of the means and the extremes to find the values for the missing sides.

Invite a volunteer to explain each of the steps used in the second example on page 227.

Use the measures of the angles at the bottom of page 227 to show that if two angles of two triangles are equal, then the third angles must be equal.

90° + 70° + 20° = 180°

50° + 10° = 60°, 180° – 60° = 120°

70° + 70° = 140°, 180° – 140° = 40°

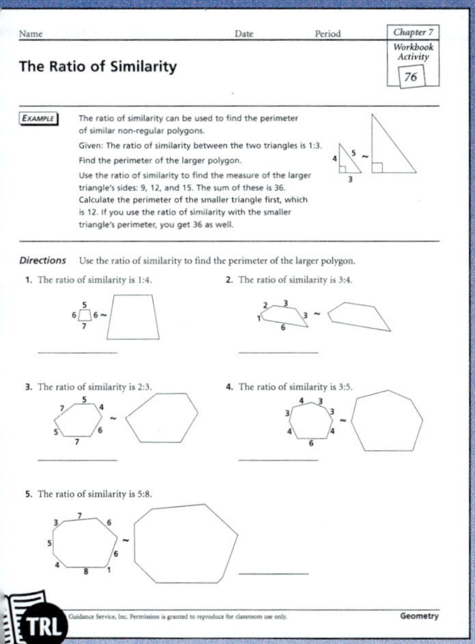

Name Date Period — Chapter 7 Workbook Activity 76

The Ratio of Similarity

EXAMPLE The ratio of similarity can be used to find the perimeter of similar non-regular polygons. Given: The ratio of similarity between the two triangles is 1:3. Find the perimeter of the larger polygon. Use the ratio of similarity to find the measure of the larger triangle's sides: 9, 12, and 15. The sum of these is 36. Calculate the perimeter of the smaller triangle first, which is 12. If you use the ratio of similarity with the smaller triangle's perimeter, you get 36 as well.

Directions Use the ratio of similarity to find the perimeter of the larger polygon.

1. The ratio of similarity is 1:4.
2. The ratio of similarity is 3:4.
3. The ratio of similarity is 2:3.
4. The ratio of similarity is 3:5.
5. The ratio of similarity is 5:8.

Geometry

Workbook Activity 76

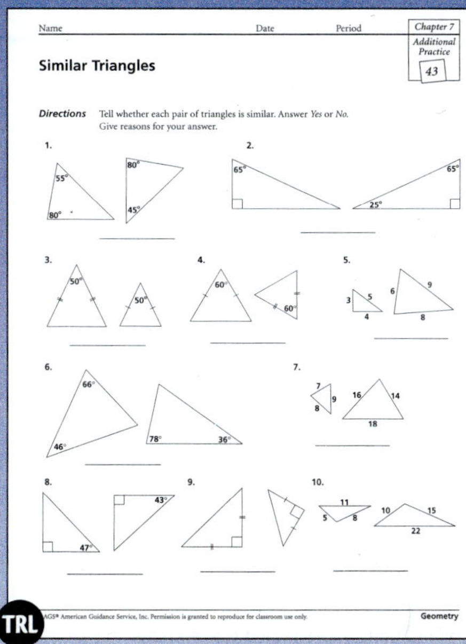

Name Date Period — Chapter 7 Additional Practice 43

Similar Triangles

Directions Tell whether each pair of triangles is similar. Answer *Yes* or *No*. Give reasons for your answer.

Geometry

Additional Practice 43

3. Reinforce and Extend

Writing About Mathematics

Students should recognize that the triangles are similar. All the triangles have a right angle and all share vertex *C*. Therefore, by the AA Similarity Postulate, the triangles are similar.

APPLICATION

At Home

Encourage students to look for pairs or groups of similar objects in their school or at home. Encourage them to consider items such as wallet- and portrait-sized photographs, graduated measuring spoons, soft-drink cups, and so on. Allow students to keep a running list in the classroom. If space allows, have them bring in some of the similar objects they find. Remind them that they should be able to explain their choices.

Writing About Mathematics

Are the nested triangles shown in the diagram below similar? Why or why not?

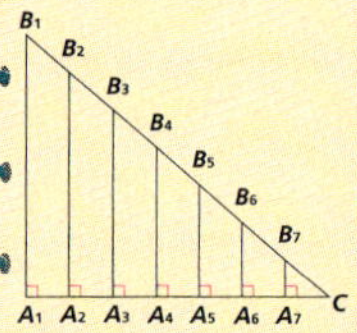

The following statement is assumed to be true without proof.

AA SIMILARITY POSTULATE: If two angles of a triangle have the same measures as two angles of a second triangle, then the two triangles are similar.

You can use this postulate to prove the following theorem.

THEOREM: If one acute angle of a right triangle has the same measure as one acute angle of a second right triangle, then the two right triangles are similar.

Given: right ΔABC and right ΔDEF with $m\angle C = m\angle F$

To Prove: $\Delta ABC \sim \Delta DEF$

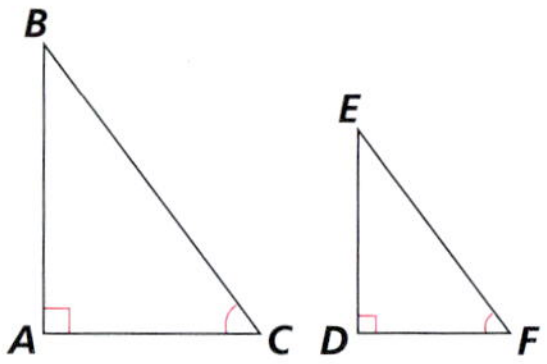

PROOF

Statement	Reason
1. $m\angle A = m\angle D$	1. All right angles are equal.
2. $m\angle C = m\angle F$	2. Given.
3. $\therefore$ right $\Delta ABC \sim$ right ΔDEF	3. AA Similarity Postulate.

Using the AA Similarity Postulate, we can show $\Delta ABC \sim \Delta EDC$.

EXAMPLE

Given: $\overline{AB} \parallel \overline{DE}$, $AB = 10$, $DE = 5$, $BC = 7$

Find: the value of x

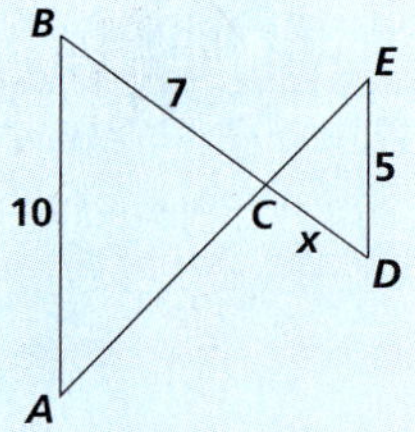

$m\angle ACB = m\angle DCE$

Vertical angles are equal.

$m\angle A = m\angle E$

$\overline{AB} \parallel \overline{DE}$ (given) and alternate interior angles are equal.

$\therefore \Delta ABC \sim \Delta EDC$ by AA Similarity Postulate.

$\Delta ABC \sim \Delta EDC$ so $\frac{AB}{DE} = \frac{BC}{DC}$

EXAMPLE *(continued)*

and $\frac{10}{5} = \frac{7}{x}$

$10x = 35$

$x = 3.5$ The value of x is 3.5.

Check: $\frac{10}{5} \stackrel{?}{=} \frac{7}{3.5}$

$35 = 35$ True

Exercise A Tell whether each pair of triangles is similar. Answer *yes* or *no*. Give reasons for your answer.

1.

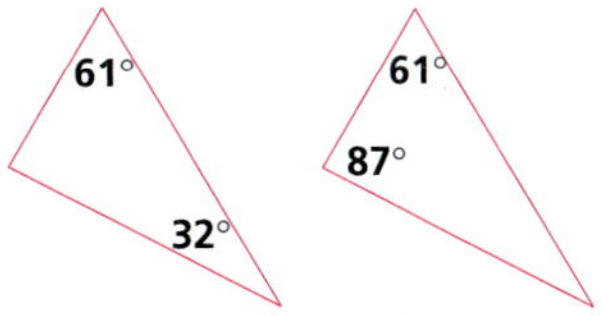

yes, 3 equal angles

2.

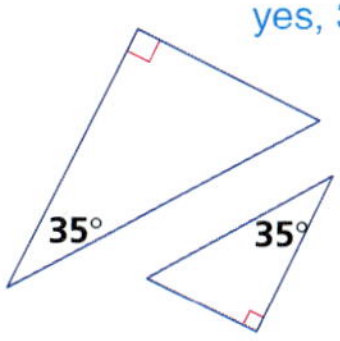

yes,
180° – (35° + 90°) =
180° – (35° + 90°)

3.

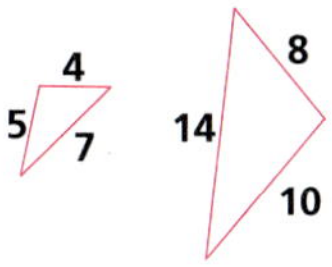

yes, sides are in proportion

4.

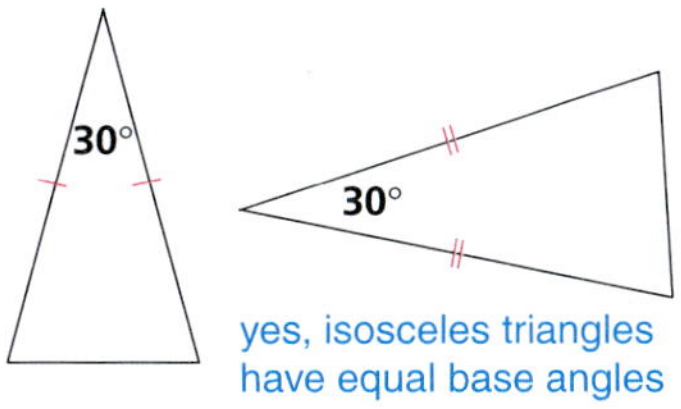

yes, isosceles triangles have equal base angles

5.

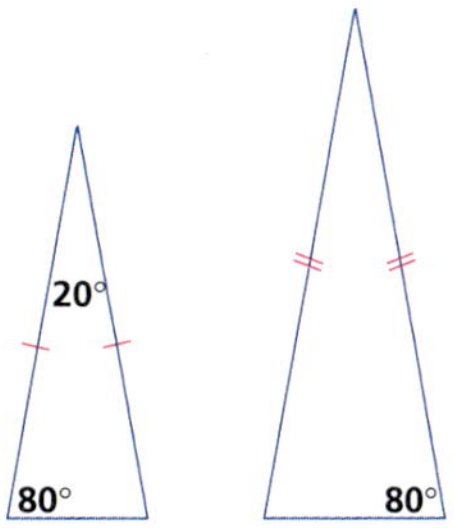

yes, angle sum and equal base angles

Exercise B Name the corresponding angles and sides of the similar triangles.

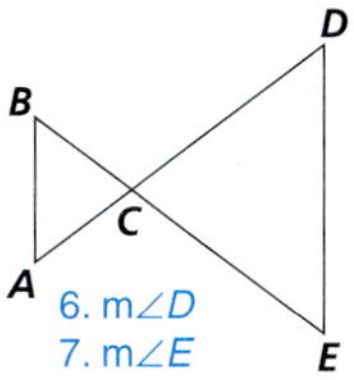

6. $m\angle A = ?$

7. $m\angle B = ?$

8. $m\angle ACB = ?$

9. $\frac{AB}{?} = \frac{AC}{?}$

10. $\frac{AC}{?} = \frac{BC}{?}$

6. $m\angle D$
7. $m\angle E$
8. $m\angle DCE$
9. $\overline{ED}$, $\overline{DC}$
10. $\overline{DC}$, $\overline{EC}$

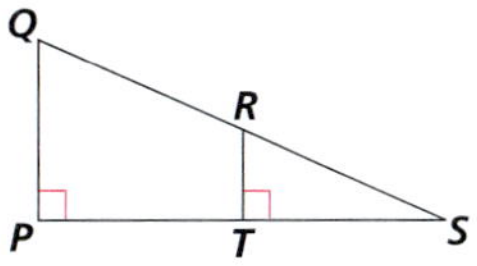

11. $m\angle QPS = ?$

12. $m\angle PQR = ?$

13. $m\angle PSQ = ?$

14. $\frac{PQ}{?} = \frac{QS}{?}$

15. $\frac{QS}{?} = \frac{PS}{?}$

11. $m\angle RTS$
12. $m\angle TRS$
13. $m\angle TSR$
14. $\overline{TR}$, $\overline{RS}$
15. $\overline{RS}$, $\overline{TS}$

LEARNING STYLES

Auditory/Verbal Direct students to work in pairs to find the measure of each angle for problems 1, 2, 4, and 5 in Exercise A. Encourage students to record each step of their thought process as they try to determine whether or not the triangles are similar. Students should take turns explaining and listening to one another's thought process.

LEARNING STYLES

LEP/ESL Have students work in pairs to cast the answers for Exercise B in sentence form. Provide the following model: "Angle *A* and angle *D* are corresponding angles, so the measure of angle *A* equals the measure of angle *D*." Encourage students to form similar sentences for the angles and sides given in Exercise B.

Proportion and Similarity *Chapter 7* **229**

Learning Styles

Visual/Spatial Encourage students to work in pairs to "read" each pair of triangles in Exercise C. Students should use the arc marks on the diagrams to determine which angles in a pair of triangles have the same measure and use the tick marks to determine which sides have the same length. Direct students to form each thought as a complete sentence. You may wish to provide the following model: "In the triangle with sides *x*, *y*, and 22, each angle has a different number of tick marks, so each angle must be a different measure."

Group Problem Solving

Assign students to groups of four. Tell each group that it will be opening a pizza kitchen. Have students decide the price for a regular 12-inch round cheese pizza with one topping. Then have them use that price to determine the prices of large and extra-large pizzas based on ingredients only.

Cheese, one topping		
regular 12 in.	large 14 in.	extra-large 16 in.
$_____	$_____	$_____

Once students complete their pricing, have them list other factors to consider as they determine the prices. Groups can present their findings as a mini-business plan in which they describe their product, identify why their product is better than the competition's, explain their pricing policy, and list factors in addition to ingredients that may affect pricing.

(Students should recognize that the change in price should be proportional to the areas of the pizzas, not just the diameters. Rent, utilities, labor, and packaging are examples of factors that may affect pricing.)

Exercise C Each pair of triangles is similar. Solve for the values of the unknowns.

16. x — $x = 16$

17. y — $y = 12$

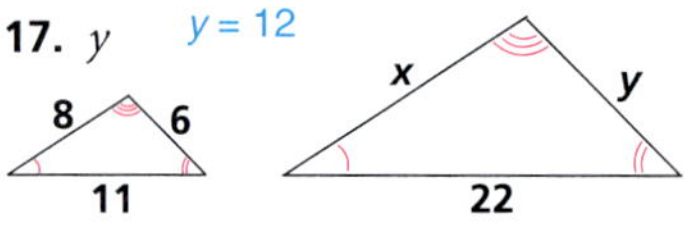

20. s — $s = 7.5$

21. t — $t = 7.5$

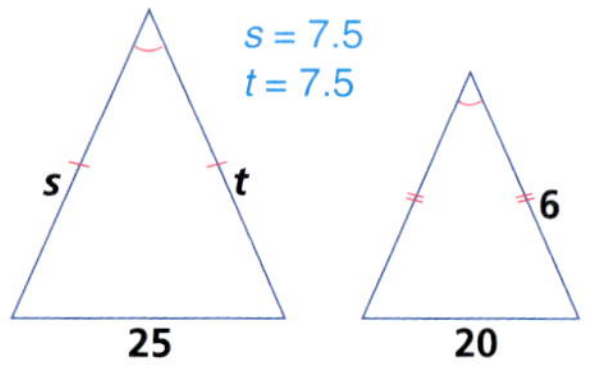

18. a — $a = 8$

19. b — $b = 12$

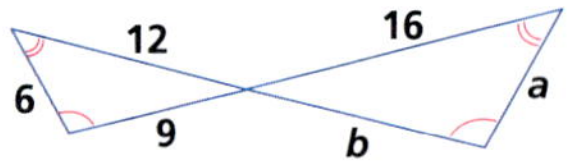

22. m — $m = 6.25$

23. n — $n = 16$

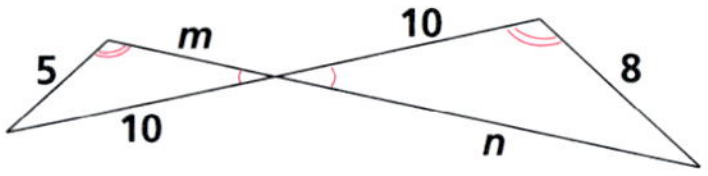

Problem Solving

Exercise D Find the answer to each of the following problems.

24. A landscape planner used similar triangles and the shadow of a tree to draw the diagram shown. What is the height of the tree?

343 inches

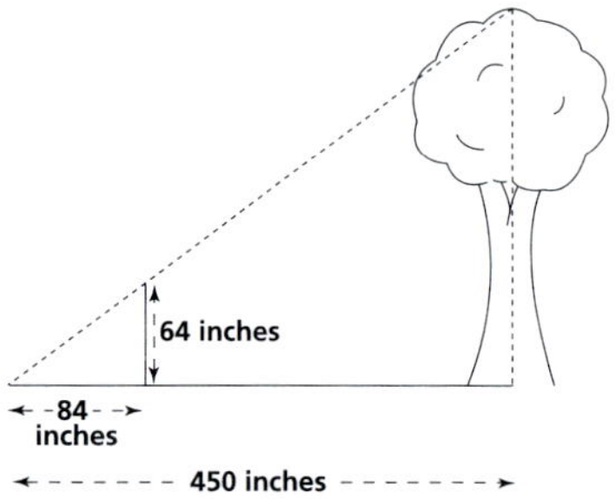

25. A surveyor needs to estimate the width of a pond. She has made the measurements shown in the diagram. How wide is the pond?

24 feet

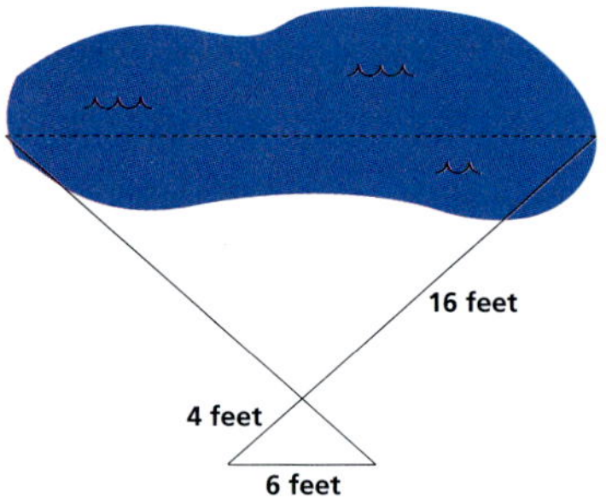

Lesson 3 Angle Measure in Regular Polygons

From your work with polygons you already know that regular polygons are polygons that have angles of equal measure and sides of equal length. Because regular polygons have equal angles, regular polygons with the same number of sides are similar. You are already familiar with some regular polygons.

All equilateral triangles are similar.

n = number of sides

$n = 3$

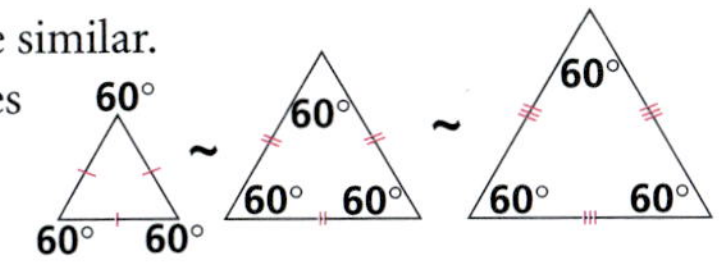

All squares are similar.

$n = 4$

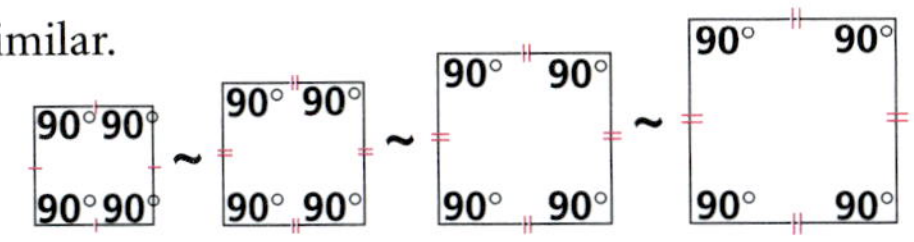

A five-sided regular polygon is called a pentagon. All regular pentagons are similar.

$n = 5$

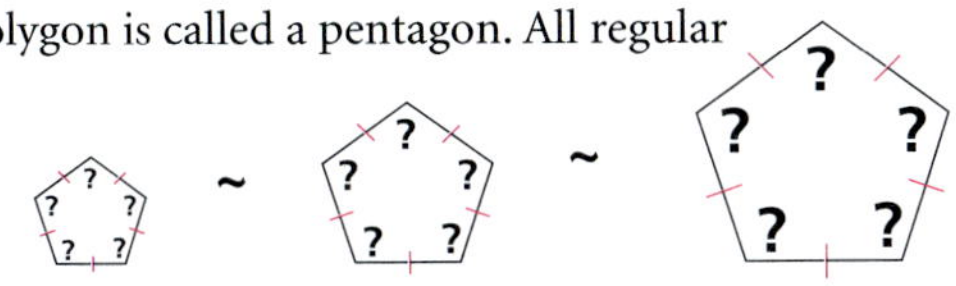

What is the angle measure of each of the five equal angles? We can reason out the answer for one pentagon and know that it will apply to all pentagons because they are similar figures.

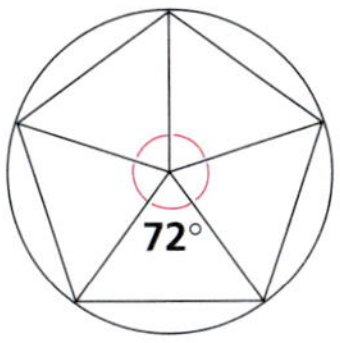

Draw a circle around a regular pentagon. Each angle at the center—each central angle—measures $\frac{360°}{5} = 72°$.

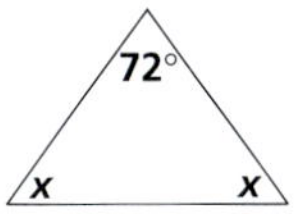

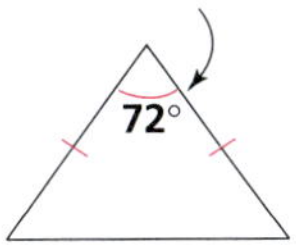

The triangles forming the pentagon are each isosceles. Can you tell why? Each is formed using a **radius** of the circle. Each radius of a circle is equal to the others.

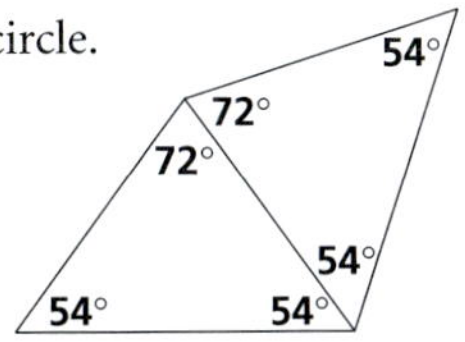

To find the measure of each base angle (x),

$72° + x + x = 180°$	angle sum theorem
$2x = 108°$	simplify
$x = 54°$	divide

> ***Perimeter***
> *The distance around the outside of a shape*
>
> ***Radius***
> *The distance from the center of a circle to the circle*

Algebra Connection

Lesson at a Glance

Chapter 7 Lesson 3

Overview This lesson relates angle measure in regular polygons to the number of polygon sides.

Objectives

- To determine the measure of each angle in a regular polygon.
- To find perimeters of similar polygons.

Student Pages 231–234

Teacher's Resource Library TRL

Workbook Activities 77–78

Additional Practice 44

Mathematics Vocabulary

perimeter
radius

1. Warm-Up Activity

Ask students to use their prior knowledge to describe or define a regular polygon. Ask a volunteer how he or she would distinguish between a polygon and a regular polygon. Point out that this lesson focuses on the characteristics of the angles of a regular polygon.

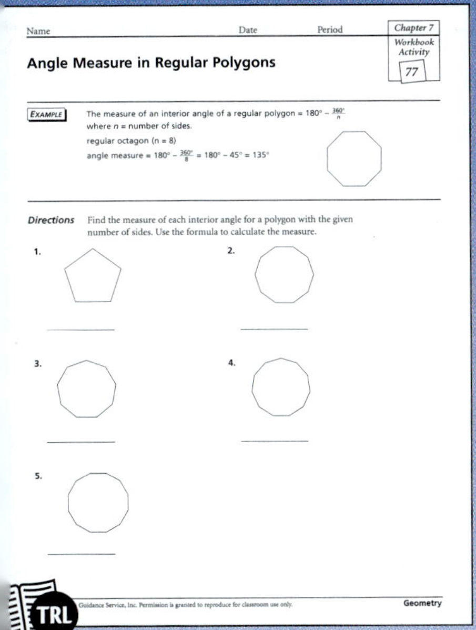

Name Date Period | Chapter 7 Workbook Activity 77

Angle Measure in Regular Polygons

EXAMPLE The measure of an interior angle of a regular polygon = $180° - \frac{360°}{n}$ where n = number of sides.
regular octagon ($n = 8$)
angle measure = $180° - \frac{360°}{8} = 180° - 45° = 135°$

Directions Find the measure of each interior angle for a polygon with the given number of sides. Use the formula to calculate the measure.

1. 2. 3. 4. 5.

Guidance Service, Inc. Permission is granted to reproduce for classroom use only. Geometry

Workbook Activity 77

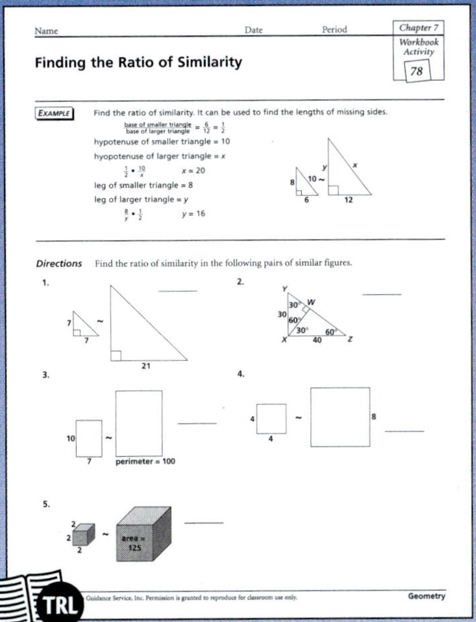

Name Date Period | Chapter 7 Workbook Activity 78

Finding the Ratio of Similarity

EXAMPLE Find the ratio of similarity. It can be used to find the lengths of missing sides.
$\frac{\text{base of smaller triangle}}{\text{base of larger triangle}} = \frac{6}{12} = \frac{1}{2}$
hypotenuse of smaller triangle = 10
hyopotenuse of larger triangle = x
$\frac{1}{2} \cdot \frac{10}{x}$ $x = 20$
leg of smaller triangle = 8
leg of larger triangle = y
$\frac{8}{y} \cdot \frac{1}{2}$ $y = 16$

Directions Find the ratio of similarity in the following pairs of similar figures.

1. 2. 3. 4. 5.

Guidance Service, Inc. Permission is granted to reproduce for classroom use only. Geometry

Workbook Activity 78

2. Teaching the Lesson

Help students follow the logical steps—from $n = 3$ and $n = 4$ to $n = 5$—in computing angle measure for regular polygons.

Common Error

Some students may be quick to make a generalization based on the number of degrees in a circle and the number of angles in a polygon. Upon seeing that each angle in a square is $360° \div 4$; and that a pentagon can be inscribed in a circle, students might generalize that each angle in a pentagon is $360° \div 5 = 72°$. Allow volunteers to explain that dividing 360° by the number of sides gives a central angle, not the measure of an interior angle.

Discuss the hexagon example with students, first modeling how to use the formula for interior angles, then modeling how to use reasoning to determine the interior angle measure.

Algebra Connection

Allow students to practice using the formula for interior angles by asking them to find the measures of the base interior angles for regular polygons with 7, 10, 20, 25, 50, and 100 sides. **(51°, 36°, 18°, 145.4°, and 3.4°)**

Have students use the information to predict what will happen to the measure of the interior angle as the number of sides increases. **(The angle measure approaches zero).**

But two angles of 54° make up one interior angle of a regular pentagon.

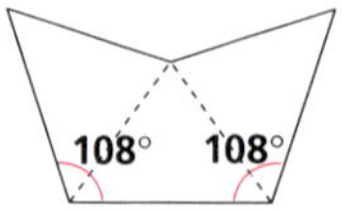

$\therefore 2(x) =$ interior angle

$2(54°) = 108°$ the measure of each interior angle of a regular pentagon

Algebra Connection

We can also use algebra to find the angle measure.
Let n = number of equal sides, x = the base angle of the triangle

then $\frac{360°}{n} + 2x = 180°$

$2x = 180° - \frac{360°}{n}$

Formula: The measure of an interior angle of a regular polygon $= 180° - \frac{360°}{n}$ where n = number of sides.

Example **Given:** regular hexagon ($n = 6$)
Find: the size of the interior angles

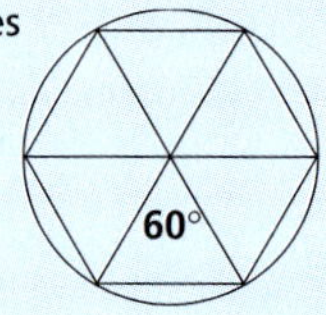

Using the formula:
$n = 6$ for hexagon

angle measure =
$180° - \frac{360°}{6} = 180° - 60° = 120°$

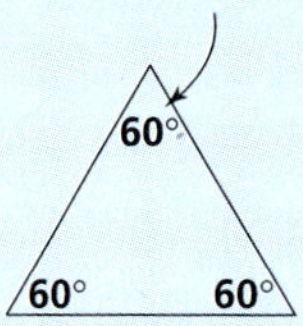

Using reasoning:
central angle measures $\frac{360°}{60°}$

$180° - 60° = 120°$, $120° \div 2 = 60°$

But the triangle is isosceles.
(In fact, it's equilateral for $n = 6$.)

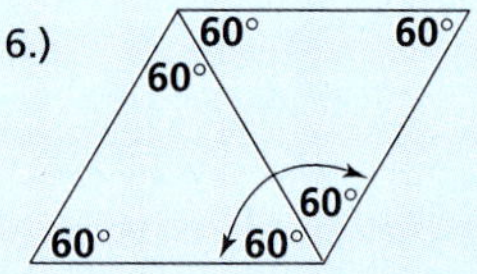

measure of the angle of a regular hexagon =
$2 \bullet 60° = 120°$

Of course, all regular hexagons are similar.

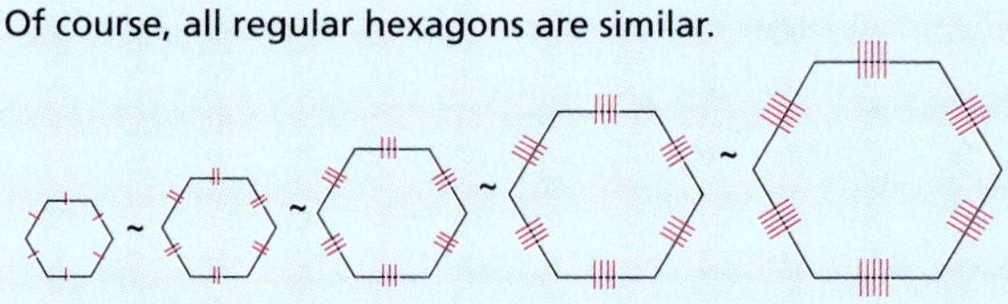

232 *Chapter 7 Proportion and Similarity*

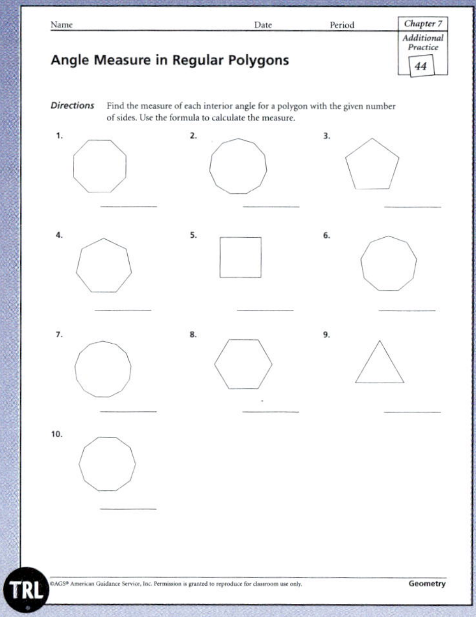

Name Date Period *Chapter 7 Additional Practice 44*

Angle Measure in Regular Polygons

Directions Find the measure of each interior angle for a polygon with the given number of sides. Use the formula to calculate the measure.

1. 2. 3. 4. 5. 6. 7. 8. 9. 10.

 Geometry

TRL

Additional Practice 44

EXAMPLE **Given:** two regular hexagons, $\overline{AB} = 5$ and the ratio of similarity is 5:8

Find: the perimeter of the larger hexagon

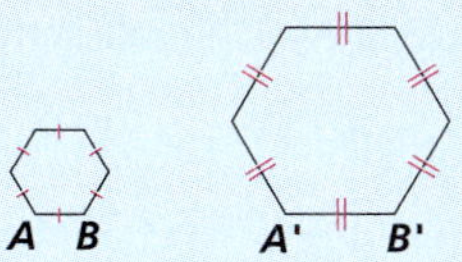

Perimeter of (distance around) smaller hexagon is 5 + 5 + 5 + 5 + 5 + 5 = 5 • 6 = 30.

The ratio of similarity is $\frac{AB}{A'B'} = \frac{5}{8}$.

$\frac{\text{smaller}}{\text{larger}} = \frac{5}{8} = \frac{30}{x} = \frac{\text{perimeter of smaller}}{\text{perimeter of larger}}$

$5x = 8(30)$ product of extremes = product of means

$5x = 240$ simplify

$x = 48$

Exercise A Find the measure of each interior angle for a polygon with the given number of sides. Use the formula to calculate the measure.

1. $n = 3$ 60°

2. $n = 4$ 90°

3. $n = 5$ 108°

4. $n = 6$ 120°

5. $n = 8$ 135°

6. $n = 10$ 144°

7. $n = 12$ 150°

8. $n = 20$ 162°

9. $n = 30$ 168°

10. $n = 40$ 171°

Writing About Mathematics

Use reasoning, rather than the formula, to find the measure of each interior angle of a regular octagon ($n = 8$).

Exercise B Use the ratio of similarity to find the perimeter of the larger polygon.

11. The perimeter of the smaller triangle is 6; the ratio of similarity is 3:5.

10

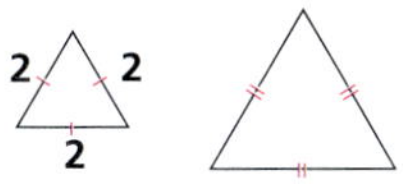

Confirm that students understand perimeter by asking these questions: If the side of an equilateral triangle is 10, what is the perimeter of the triangle? **(30)** If the side of a square is a, what is the perimeter of the square? **($4a$)** If a polygon has n sides, and the length of each side is ℓ, what is the perimeter of the polygon? **($n\ell$)**

3. Reinforce and Extend

LEARNING STYLES

LEP/ESL Some students might benefit from a short review of the names of polygons with regular sides. Sketch regular polygons with 4, 5, 6, 8, and 10 sides on the board or overhead projector. Direct students to follow your example and sketch each of the polygons. Then as the class names each figure you have drawn, have students name the polygons they have drawn. Indicate that they should include these pictures in their picture dictionary of geometric terms.

Writing About Mathematics

Students' reasoning should resemble reasoning on page 232: Using diagonals, we can draw a circle around the octagon and draw eight triangles in the octagon. The central angle or vertex angle of each triangle is the same measure: $\frac{360°}{8} = 45°$. Each of the triangles is an isosceles triangle (the legs are radii of the circle), so the base angles are equal. Therefore, 180° – 45° = 135°, 135° ÷ 2 = 67.5°. One interior angle of the octagon is made up of two base angles of the triangles, so one base angle = 2 • 67.5° = 135°.

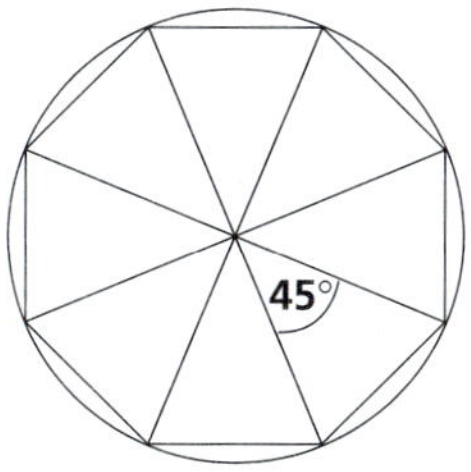

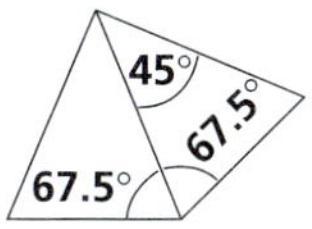

Learning Styles

Interpersonal/Group Learning Direct students to form groups of four. Each group can work together to complete problems 11–15 in Exercise B. Indicate to students that, in addition to finding the perimeter of the second figure, they should determine a way to find the side length. Once they have the side length for each figure, groups can determine the ratio of similarity of the sides of each figure. Ask students to explain their results. (The ratio of similarity of the sides equals that of the perimeters because each side length is multiplied by the same number to get the perimeter. Equals multiplied by equals are equal.)

Experiencing Geometry

Students should be able to generate ratios such as 1:2 or $\frac{1}{2}$ and 1:3 or $\frac{1}{3}$.

Technology Connection

Some retail outlets may be willing to lend a hand-held GPS for a class demonstration. Allow students to move about the classroom or school, noting how their position change is reflected in the reading of the GPS unit.

12. The perimeter of the smaller square is 16; the ratio of similarity is 2:3.
24

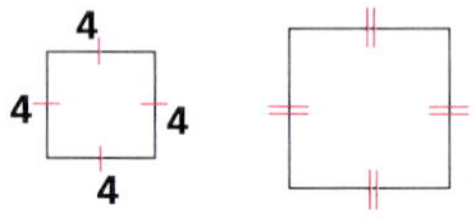

13. One side of the smaller pentagon measures 5; the ratio of similarity is 1:3.
75

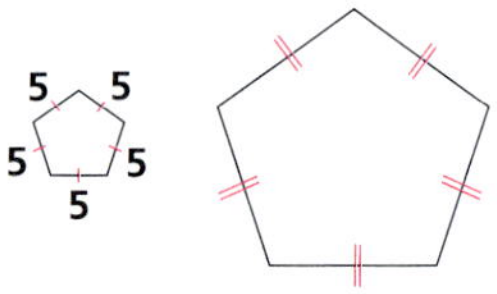

14. One side of the smaller hexagon measures 4; the ratio of similarity is 4:7.
42

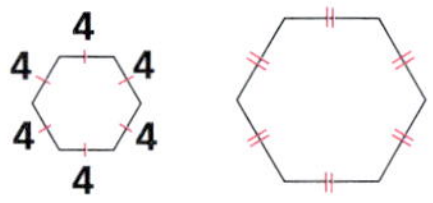

15. One side of the smaller octagon measures 3; the ratio of similarity is 3:7.
56

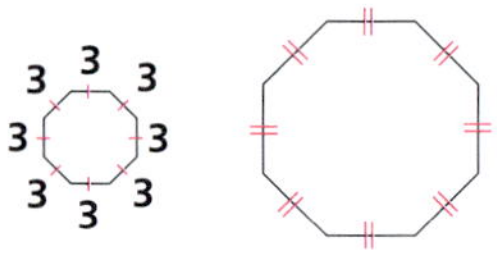

Experiencing Geometry

Look for ratios and proportions in your kitchen! Packaged foods such as rice, instant cereals, or dry soups often have directions such as "use $\frac{1}{2}$ cup of cereal for every cup of water" or "mix 1 cup rice with 2 cups water." Rewrite any ratios you find in the form $x{:}y$ or $\frac{x}{y}$.

Technology Connection

The orbits of some satellites keep them constantly above a single location on Earth. Global Positioning Systems use the known location of a standard object, the time, specially coded satellite signals, and similar triangles to pinpoint the location of a second object. Once used exclusively by the armed forces, these systems are now installed on some automobiles to make sure the driver does not get lost.

234 *Chapter 7 Proportion and Similarity*

Lesson 4 Enlarging Geometric Figures

Look at the sizes of the figures in this painting, "I Saw the Figure Five in Gold," by Charles Demuth. His later works are very geometric and precise in form.

We can use the fact that similar geometric figures have the same shape but different sizes to help us understand how to make geometric figures larger (or smaller).

> ***Dilation***
> *Process by which a geometric figure is enlarged or shrunk*

In the diagram you can see that $\Delta A'B'C'$ is a larger version of ΔABC. Each of the sides of the smaller triangle has been multiplied by a factor of 2 to produce the larger triangle. The triangles are similar; the ratio of similarity of ΔABC to $\Delta A'B'C'$ is 1 to 2.

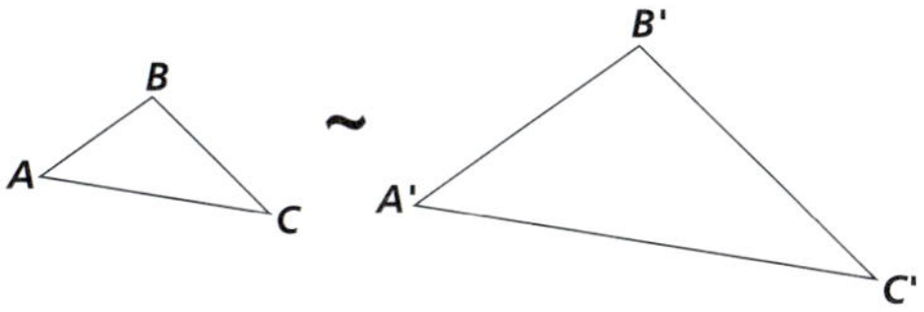

RULE: If the ratio of similarity is greater than one, then the resulting figure is enlarged.

The number used to produce the second figure—the multiplier—is equal to the ratio of similarity. Multiplication by a number greater than one will produce an image that is larger than the original figure. The process is called **dilation.**

One way to illustrate enlargements, or dilations, is to place the figures in the coordinate plane. The ordered pairs of the vertices provide a way to read or record the original lengths as well as the dilations.

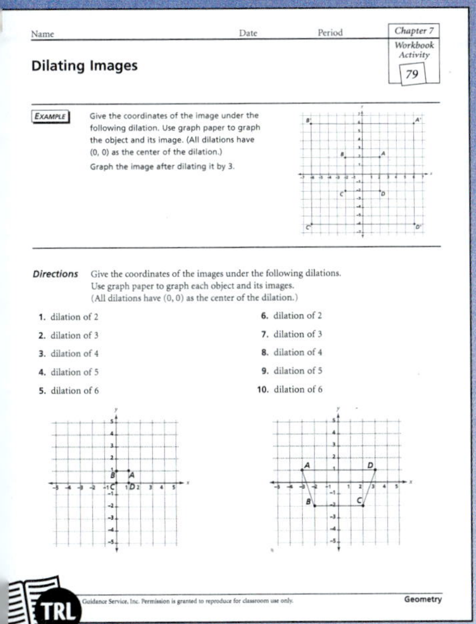
Name Date Period Chapter 7 Workbook Activity 79

Dilating Images

EXAMPLE Give the coordinates of the image under the following dilation. Use graph paper to graph the object and its image. (All dilations have (0, 0) as the center of the dilation.)
Graph the image after dilating it by 3.

Directions Give the coordinates of the images under the following dilations. Use graph paper to graph each object and its images. (All dilations have (0, 0) as the center of the dilation.)

1. dilation of 2
2. dilation of 3
3. dilation of 4
4. dilation of 5
5. dilation of 6
6. dilation of 2
7. dilation of 3
8. dilation of 4
9. dilation of 5
10. dilation of 6

Guidance Service, Inc. Permission is granted to reproduce for classroom use only. Geometry

Workbook Activity 79

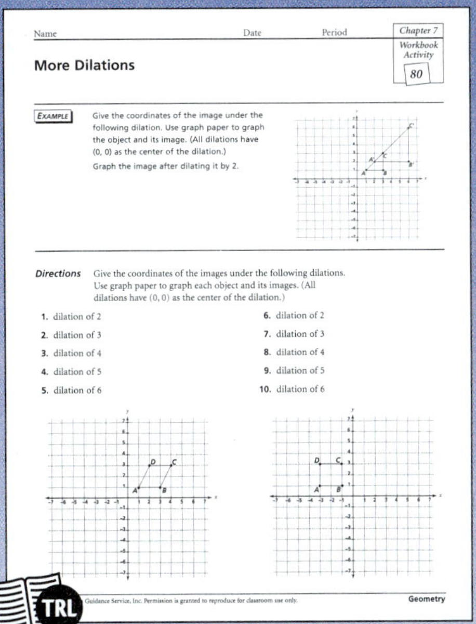
Name Date Period Chapter 7 Workbook Activity 80

More Dilations

EXAMPLE Give the coordinates of the image under the following dilation. Use graph paper to graph the object and its image. (All dilations have (0, 0) as the center of the dilation.)
Graph the image after dilating it by 2.

Directions Give the coordinates of the images under the following dilations. Use graph paper to graph each object and its images. (All dilations have (0, 0) as the center of the dilation.)

1. dilation of 2
2. dilation of 3
3. dilation of 4
4. dilation of 5
5. dilation of 6
6. dilation of 2
7. dilation of 3
8. dilation of 4
9. dilation of 5
10. dilation of 6

Guidance Service, Inc. Permission is granted to reproduce for classroom use only. Geometry

Workbook Activity 80

Lesson at a Glance

Chapter 7 Lesson 4

Overview This lesson describes how to enlarge geometric figures using a ratio of similarity greater than one.

Objectives

- To enlarge geometric figures using a ratio of similarity.
- To give the coordinates of enlarged figures.

Student Pages 235–239

Teacher's Resource Library TRL

Workbook Activities 79–80

Additional Practice 45

Mathematics Vocabulary

dilation

1. Warm-Up Activity

Introduce this lesson and Lesson 5 with a short description of *Gulliver's Travels*. Tell students that in this eighteenth-century classic by Jonathan Swift, Lemuel Gulliver travels first to a land where all are smaller than he, then to a land where he is smaller than all others. Invite students to speculate about what it might be like to live in a land of giants. Ask them how houses, castles, and even dogs and cats might be different if they were in a land of giants. Point out that this chapter shows a way to make figures larger by using the coordinate plane.

2. Teaching the Lesson

Point out to students that the arrows used in dilations on page 236 can be read as "changes to" or "transforms to." Help students recognize that placing objects on the coordinate plane allows them to enlarge the figures by multiplying the coordinates in different parts of the figures.

Draw students' attention to the transformation at the bottom of page 236. Point out that the original black triangle has coordinates (1, 1), (2, 4) and (2, 1). To make a triangle twice the size of the first, each coordinate was multiplied by 2, giving the smaller blue triangle these coordinates (2, 2), (4, 8) and(4, 2). Ask students to give the coordinates for the larger blue triangle. Then ask them what factor was used to produce this triangle. Have them justify their answer. **(5; Each coordinate is five times that of the original triangle.)**

EXAMPLE **Given:** ΔABC with $A = (1, 1)$, $B = (2, 4)$, and $C = (2, 1)$

Enlarge: ΔABC by a factor of 3

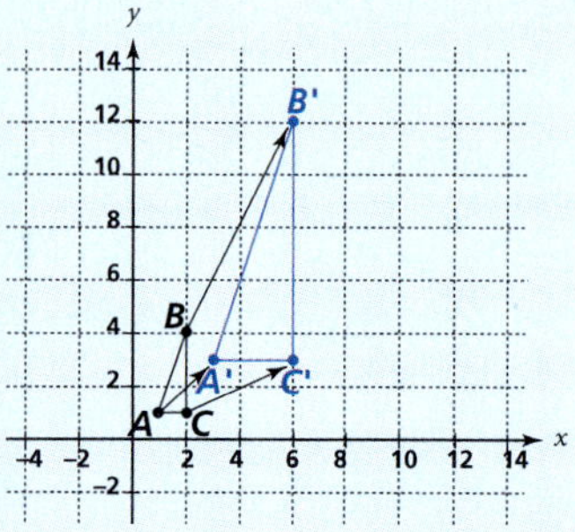

To enlarge ΔABC using a ratio of similarity of $\frac{3}{1}$, multiply each coordinate by 3 or $\frac{3}{1}$.

$A = (1, 1) \rightarrow A' = (1 \bullet 3, 1 \bullet 3) = (3, 3)$

$B = (2, 4) \rightarrow B' = (2 \bullet 3, 4 \bullet 3) = (6, 12)$

$C = (2, 1) \rightarrow C' = (2 \bullet 3, 1 \bullet 3) = (6, 3)$

This will give the coordinates of the image $\Delta A'B'C'$.

> An arrow $\rightarrow$ can be used to show the movement of points to new coordinates.

Below is a diagram of two dilations. ΔABC has been *transformed* into two larger triangles.

Note that all the triangles are similar to ΔABC and that each triangle has its vertices on $\overrightarrow{OA}$, $\overrightarrow{OB}$, and $\overrightarrow{OC}$. The point O, (0, 0), is called the *center of the dilation*. Each of these triangles was formed by multiplying the coordinates by the ratio of similarity.

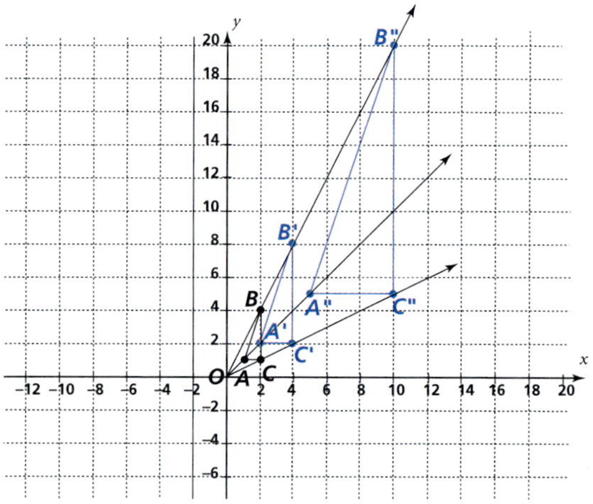

236 *Chapter 7 Proportion and Similarity*

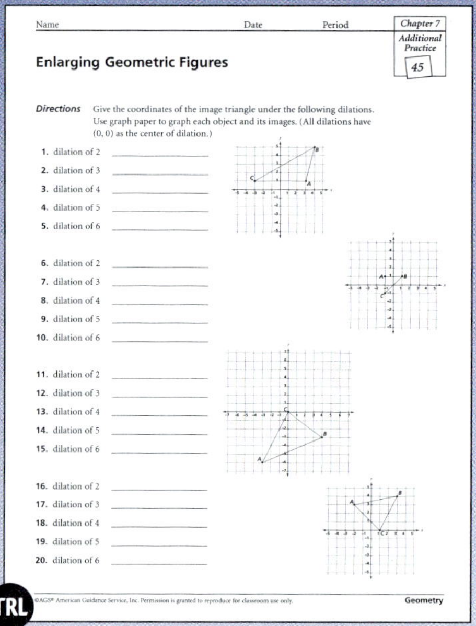
Name Date Period Chapter 7 Additional Practice 45

Enlarging Geometric Figures

Directions Give the coordinates of the image triangle under the following dilations. Use graph paper to graph each object and its images. (All dilations have (0, 0) as the center of dilation.)

1. dilation of 2
2. dilation of 3
3. dilation of 4
4. dilation of 5
5. dilation of 6
6. dilation of 2
7. dilation of 3
8. dilation of 4
9. dilation of 5
10. dilation of 6
11. dilation of 2
12. dilation of 3
13. dilation of 4
14. dilation of 5
15. dilation of 6
16. dilation of 2
17. dilation of 3
18. dilation of 4
19. dilation of 5
20. dilation of 6

Geometry

TRL

Additional Practice 45

EXAMPLES **Given:** $D = (-2, -1)$, $E = (6, 5)$, and $F = (6, 1)$

Find: the coordinates of the image of ΔDEF, $\Delta D'E'F'$, under a dilation of 4 or $\frac{4}{1}$. Use (0, 0) as the center of the dilation. Graph both triangles.

To enlarge ΔDEF using a ratio of similarity of $\frac{4}{1}$, multiply each coordinate by $\frac{4}{1}$.

$D = (-2, -1) \rightarrow D' = (-2 \bullet 4, -1 \bullet 4) = (-8, -4)$

$E = (6, 5) \rightarrow E' = (6 \bullet 4, 5 \bullet 4) = (24, 20)$

$F = (6, 1) \rightarrow F' = (6 \bullet 4, 1 \bullet 4) = (24, 4)$

Again, note that $\Delta DEF \sim \Delta D'E'F'$.

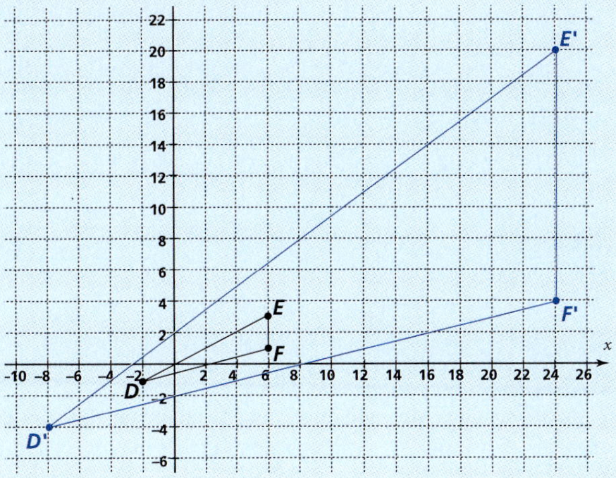

Given: square *KLMN* with coordinates (−1, −1), (−1, 1), (1, 1), and (1, −1)

Find: the coordinates of a similar square under a dilation of $\frac{5}{1}$, with (0, 0) as the center of the dilation. Graph both squares.

To enlarge the square, using a ratio of similarity of $\frac{5}{1}$, multiply each coordinate by $\frac{5}{1}$. The squares—like all squares—are similar to one another.

$K = (-1, -1) \rightarrow (-1 \bullet 5, -1 \bullet 5) = (-5, -5) = K'$

$L = (-1, 1) \rightarrow (-1 \bullet 5, 1 \bullet 5) = (-5, 5) = L'$

$M = (1, 1) \rightarrow (1 \bullet 5, 1 \bullet 5) = (5, 5) = M'$

$N = (1, -1) \rightarrow (1 \bullet 5, -1 \bullet 5) = (5, -5) = N'$

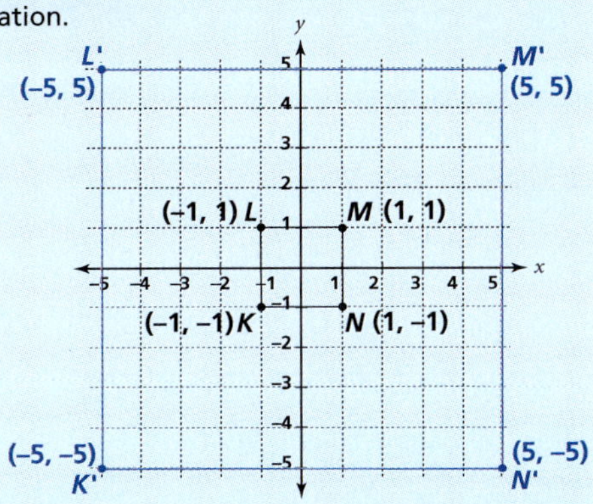

3. Reinforce and Extend

LEARNING STYLES

Body/Kinesthetic Using classroom floor tiles as a guide, mark out Quadrant I of a coordinate grid. Be sure the *x*- and *y*-axes extend to 20, and reproduce the dilations shown on page 236. Direct three students so that each one stands at a coordinate of the smallest triangle: (1, 1), (2, 4), (2, 1). Ask the students to hold a colored string so that it forms a triangle with each student at a vertex. Repeat the exercise for the second set of coordinates: (2, 2), (4, 8), (4, 2). And again for the third set of coordinates: (5, 5), (10, 20), (10, 5).

Ask students how they think the perimeters of the triangles compare. Have students not forming the triangles walk around—pace out—the perimeter of each triangle. Record students' results in a chart. Help students recognize that, taking the variation of pace length into account, the perimeters of the triangles have the ratios 1:2 and 1:5.

Learning Styles

Visual/Spatial Invite interested students to draw relatively simple sketches of everyday items such as a car, house, or flower on a coordinate grid. They may even choose to duplicate a simple cartoon character. Have students mark important points such as corners or turns. Then ask them to produce dilations greater than 1 for each object. Display completed dilations for the class to enjoy.

Answers to Exercise A

1.

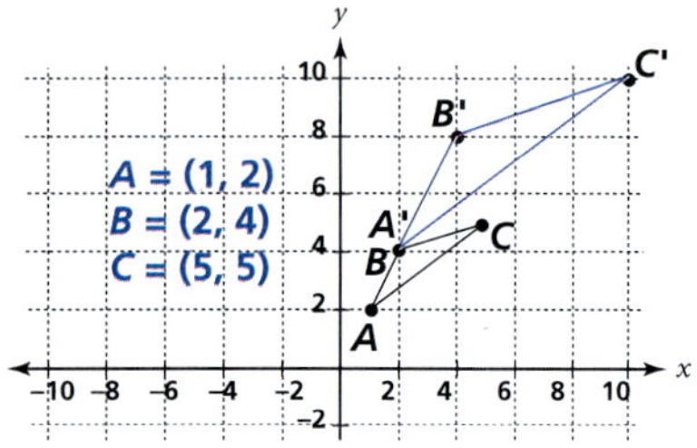

2.

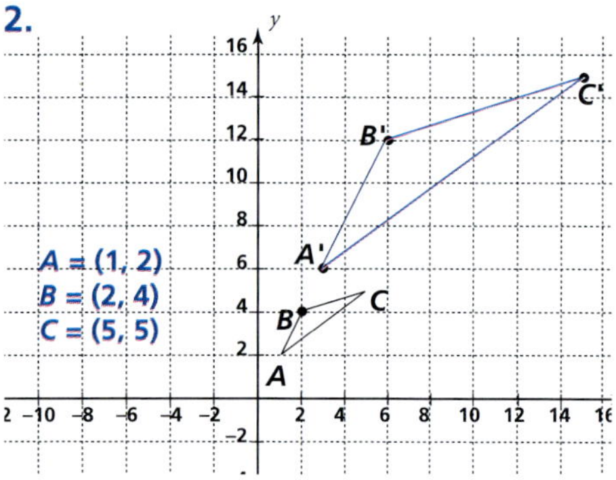

3.

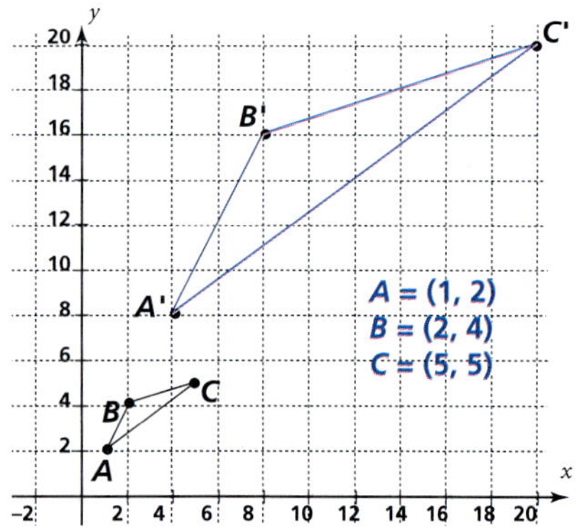

4.

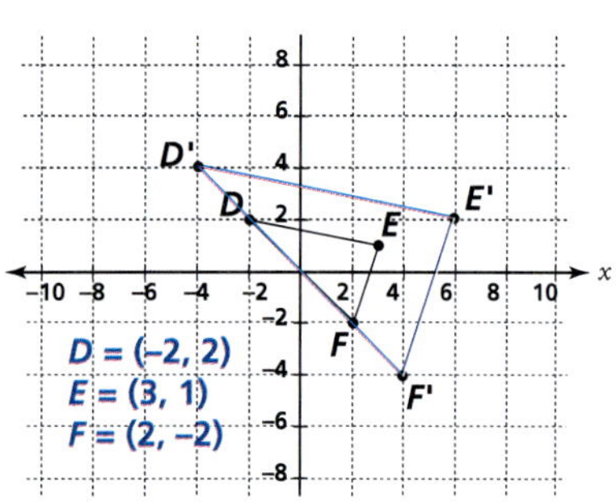

EXAMPLE **Given:** rectangle *PQRS* with coordinates (–2, 0), (–2, 2), (3, 2), and (3, 0)

Find: the coordinates of a similar rectangle under a dilation of 2 or $\frac{2}{1}$, with (0, 0) as the center of the dilation. Graph both rectangles.

To enlarge the rectangle, using a ratio of similarity of $\frac{2}{1}$, multiply each coordinate by $\frac{2}{1}$.

$P = (-2, 0) \rightarrow (-2 \bullet 2, 0 \bullet 2) = (-4, 0) = P'$

$Q = (-2, 2) \rightarrow (-2 \bullet 2, 2 \bullet 2) = (-4, 4) = Q'$

$R = (3, 2) \rightarrow (3 \bullet 2, 2 \bullet 2) = (6, 4) = R'$

$S = (3, 0) \rightarrow (3 \bullet 2, 0 \bullet 2) = (6, 0) = S'$

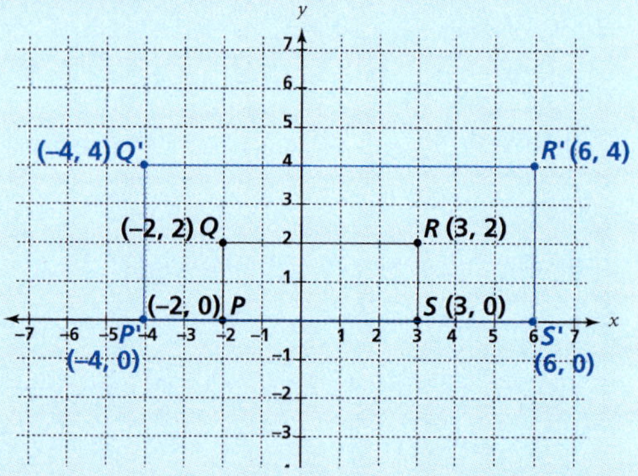

Exercise A Give the coordinates of the image triangle under the following dilations. Use graph paper to graph the object and its images. (All dilations have (0, 0) as the center of the dilation.)

1. $A' = (2, 4)$, $B' = (4, 8)$, $C' = (10, 10)$
2. $A' = (3, 6)$, $B' = (6, 12)$, $C' = (15, 15)$
3. $A' = (4, 8)$, $B' = (8, 16)$, $C' = (20, 20)$

1. dilation of 2

2. dilation of 3

3. dilation of 4

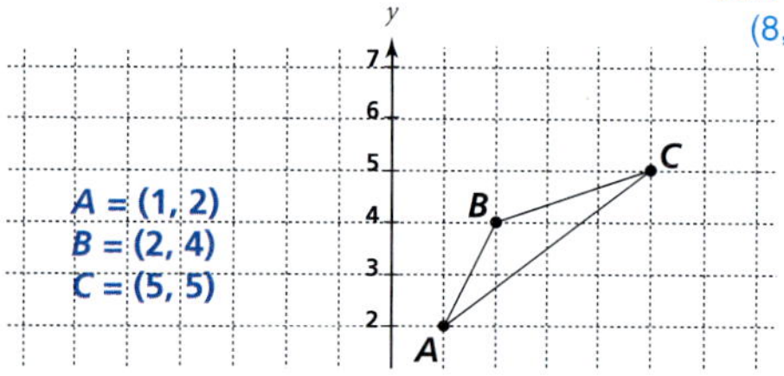

4. $D' = (-4, 4)$, $E' = (6, 2)$, $F' = (4, -4)$
5. $D' = (-6, 6)$, $E' = (9, 3)$, $F' = (6, -6)$

See Teacher's Edition page for graph answers.

4. dilation of 2

5. dilation of 3

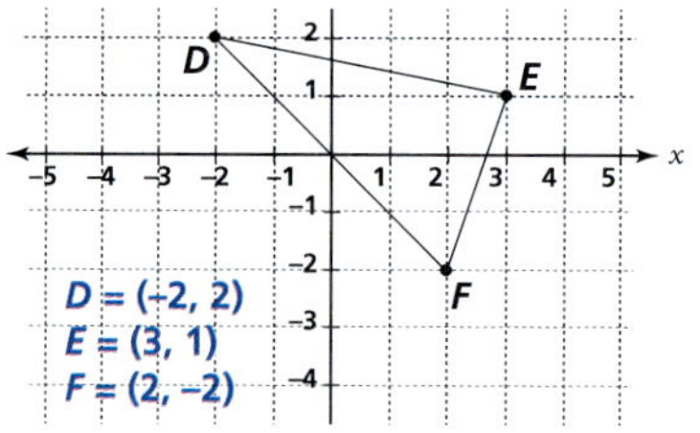

6. $J' = (-12, 4)$, $K' = (-8, 16)$, $L' = (0, 0)$
7. $J' = (-9, 3)$, $K' = (-6, 12)$, $L' = (0, 0)$

6. dilation of 4

7. dilation of 3

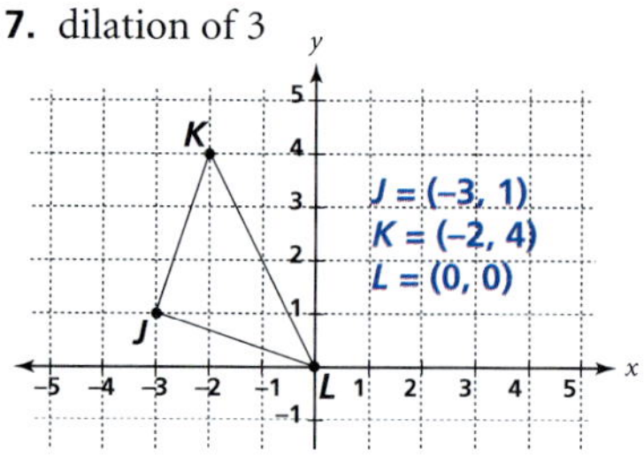

5.

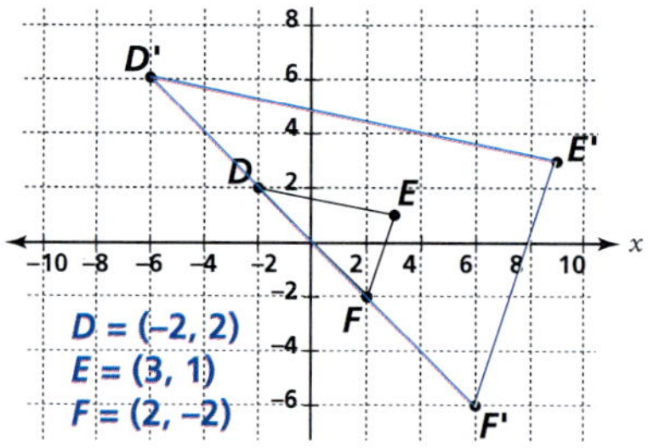

6.

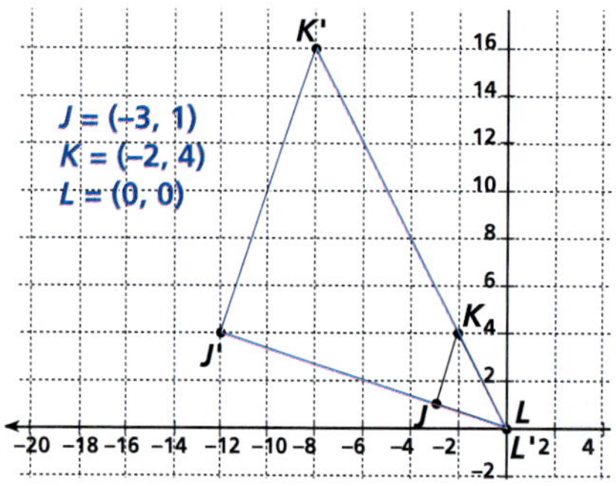

7.

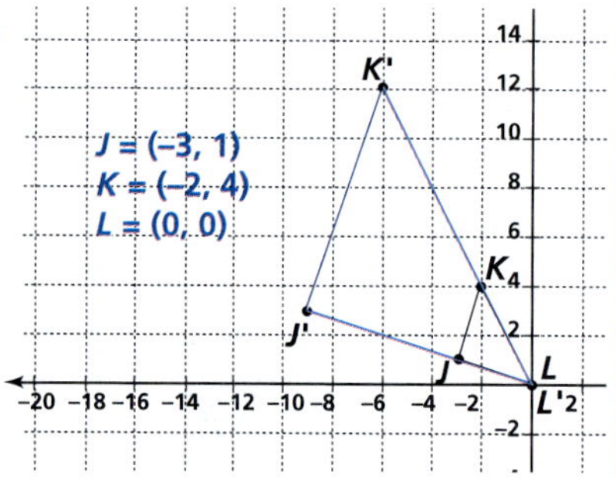

Exercise B Give the coordinates of the image under the following dilations. Use graph paper to graph the object and its images. (All dilations have (0, 0) as the center of the dilation.)

8. dilation of 2

9. dilation of 3

10. dilation of 4

See Teacher's Edition page for graph answers.

11. dilation of 2

12. dilation of 3

11. $A' = (-4, -4)$, $B' = (-8, 2)$, $C' = (2, 2)$, $D' = (6, -4)$
12. $A' = (-6, -6)$, $B' = (-12, 3)$, $C' = (3, 3)$, $D' = (9, -6)$

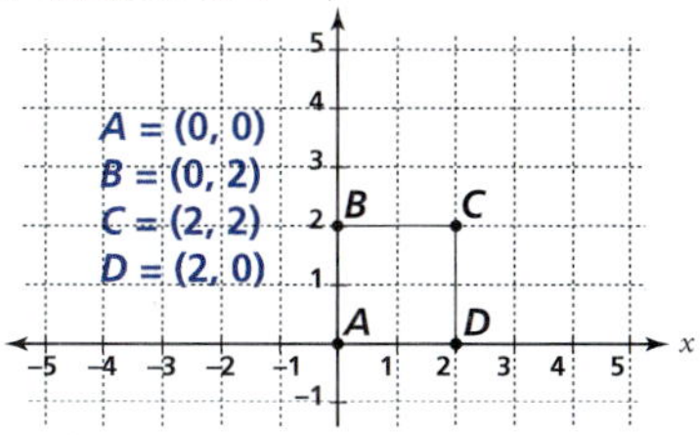

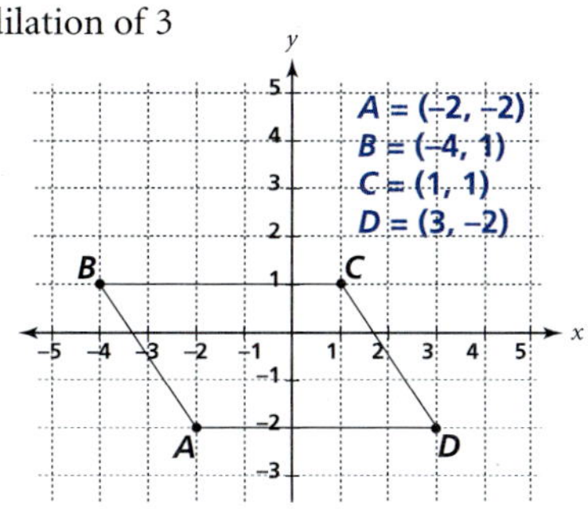

8. $A' = (0, 0)$, $B' = (0, 4)$, $C' = (4, 4)$, $D' = (4, 0)$
9. $A' = (0, 0)$, $B' = (0, 6)$, $C' = (6, 6)$, $D' = (6, 0)$
10. $A' = (0, 0)$, $B' = (0, 8)$, $C' = (8, 8)$, $D' = (8, 0)$

Problem Solving

Exercise C Solve the following problems.

13. Haley is in charge of making flyers and posters announcing a swim meet at her school. She wants the posters to be three times as large as the flyers. She drew the size of the flyers, 6 inches by 4 inches, on a grid. What will be the size of the posters?
18 inches x 12 inches

14. Bailey is planning the swim meet with Haley. Bailey's drawing of the flyers, and the enlargement for the posters, is shown at the right. Will Bailey's posters be the same size as Haley's? How can you tell?
yes, count number of squares or read the dimensions

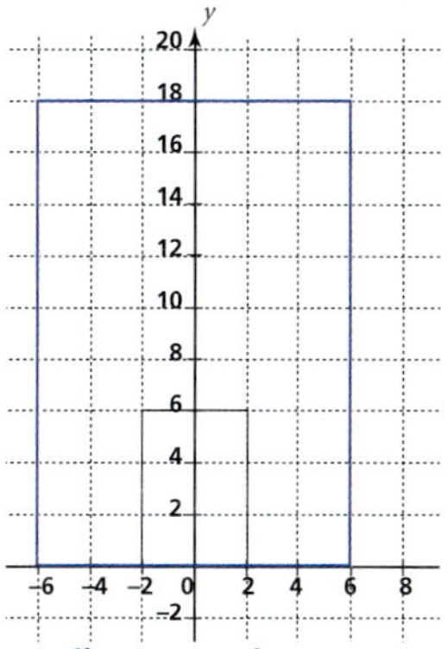

Bailey's Drawing

15. A photo processing machine is programmed to make enlargements. The basic, wallet-size photo is 2 inches $\times 2\frac{1}{2}$ inches. Customers can request 4×5 or 8×10 enlargements. Use what you know about dilations to graph the original photo and the enlargements. What is the size of the dilation? How do you know?

The magnitude of the dilation to get 4 x 5's is 2; the magnitude to get 10 x 8's is 4. Students should indicate that the coordinates of the vertices have been multiplied by 2.

Answers to Exercise A continued

8.

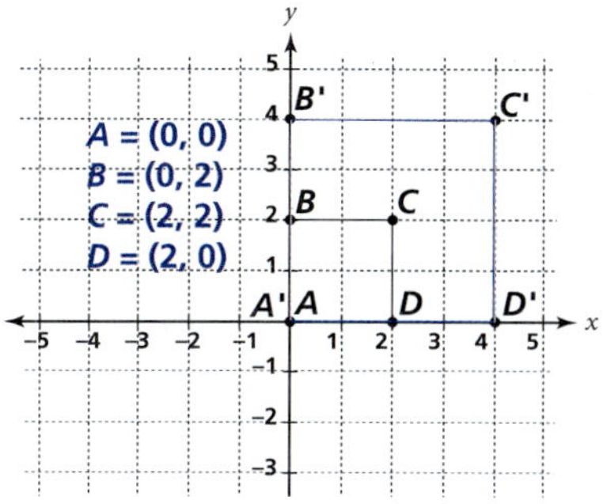

9.

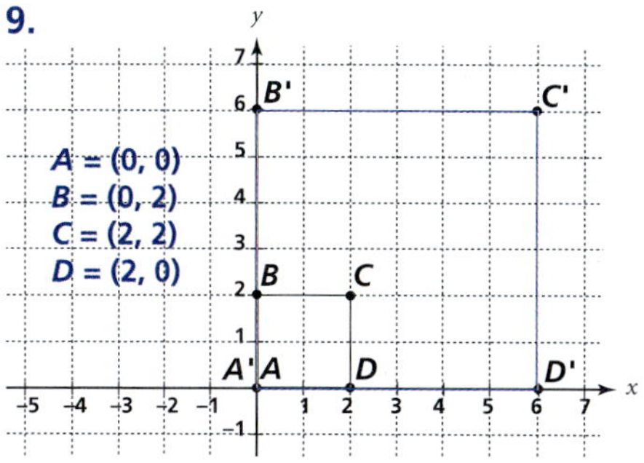

10.

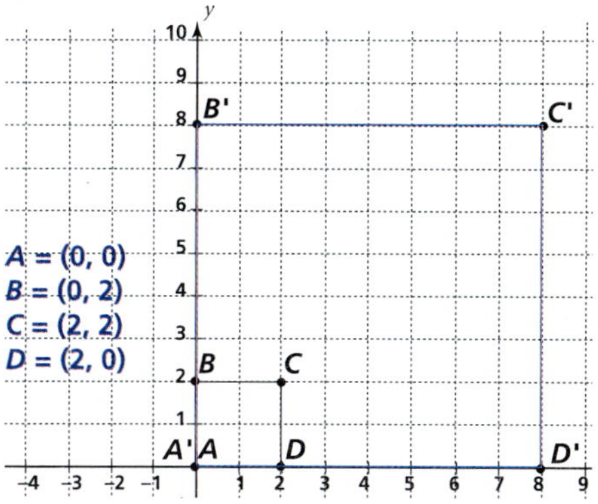

11.

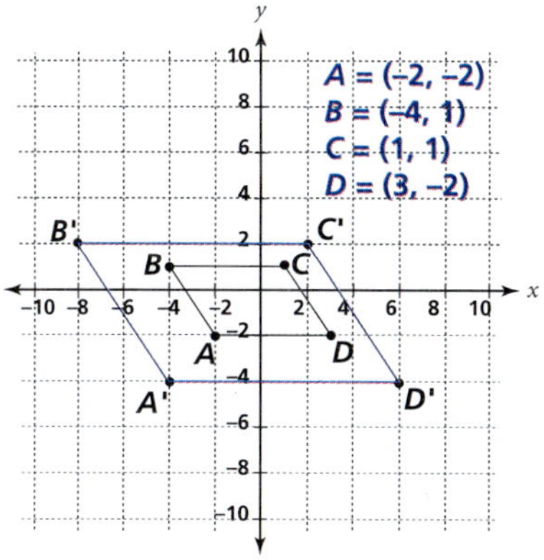

12.

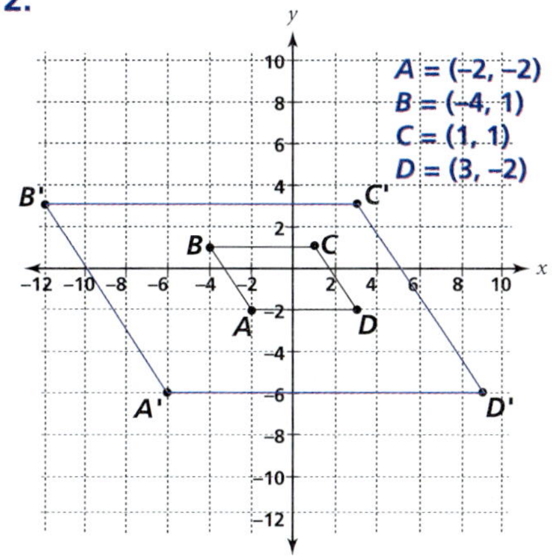

Lesson at a Glance

Chapter 7 Lesson 5

Overview This lesson describes how to shrink geometric figures using a ratio of similarity less than one.

Objectives

- To shrink geometric figures using a ratio of similarity.
- To give the coordinates of shrunken figures.

Student Pages 240–243

Teacher's Resource Library TRL

Workbook Activities 81–82

Additional Practice 46

1. Warm-Up Activity

Ask students to describe some special effects they may have seen in a recent space-fantasy or action movie. Point out that in some cases, the images they see on the screen appear much larger than the actual objects. Indicate that many special effects are created using miniature models of cities or spacecraft. Have students speculate on how easy or difficult it might be to build an accurate miniature version of their town or city so that it could be demolished by a monster in a movie.

Lesson 5 Shrinking Geometric Figures

In the last lesson, you enlarged geometric figures while keeping their shapes. To enlarge the figures you multiplied by the ratio of similarity, which was greater than one. In this lesson, you will learn how to shrink geometric figures by multiplying by a ratio of similarity *less than one.*

RULE: If the ratio of similarity is less than one, then the resulting figure is shrunk.

EXAMPLE **Given:** ΔABC with $A = (4, 2)$, $B = (10, 8)$ and $C = (10, 2)$

Find: the coordinates of the image of ΔABC under a dilation of $\frac{1}{2}$. Use (0, 0) as the center of the dilation.

Shrinking a geometric figure is similar to enlarging the figure—you multiply each coordinate by the ratio of similarity, in this case, $\frac{1}{2}$. Your work will be easier if you use graph paper.

$A = (4, 2) \rightarrow (4 \bullet \frac{1}{2}, 2 \bullet \frac{1}{2}) = (2, 1) = A'$

$B = (10, 8) \rightarrow (10 \bullet \frac{1}{2}, 8 \bullet \frac{1}{2}) = (5, 4) = B'$

$C = (10, 2) \rightarrow (10 \bullet \frac{1}{2}, 2 \bullet \frac{1}{2}) = (5, 1) = C'$

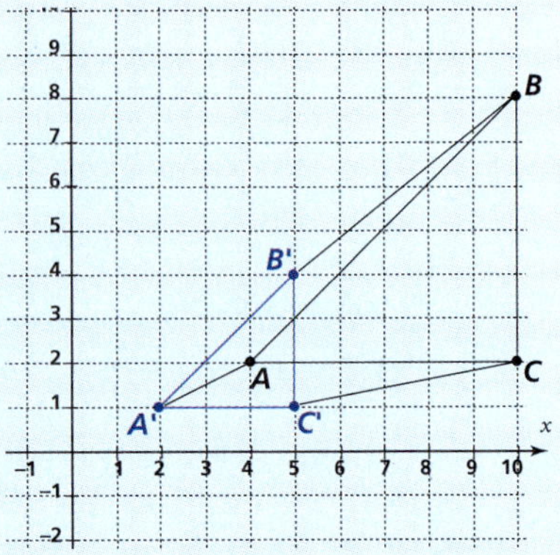

You now have the coordinates of the image $\Delta A'B'C'$. Note that ΔABC and $\Delta A'B'C'$ are similar.

As you can see, when the ratio of similarity is less than one, the figure shrinks. However, like the enlargement, the figure's vertices remain in the rays $\overrightarrow{OA}$, $\overrightarrow{OB}$, and $\overrightarrow{OC}$. All the triangles similar to ΔABC, whether larger or smaller, have their vertices on these three rays.

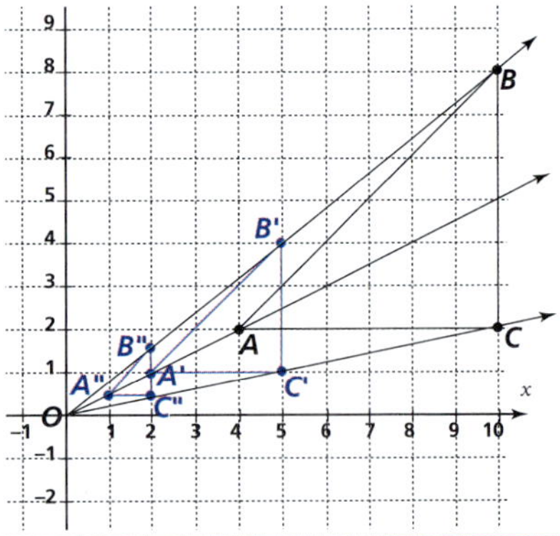

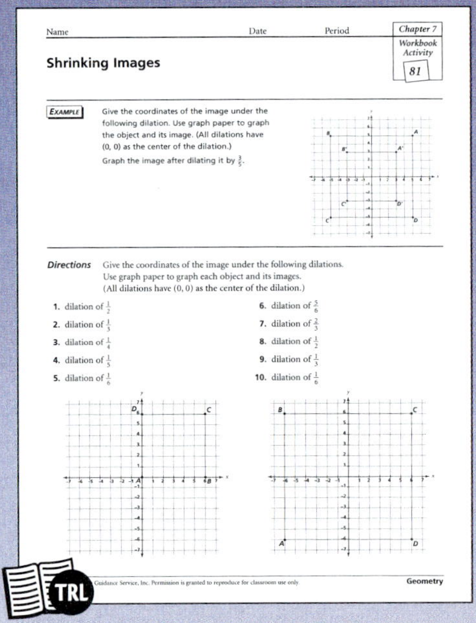

Name Date Period

Chapter 7 Workbook Activity 81

Shrinking Images

EXAMPLE Give the coordinates of the image under the following dilation. Use graph paper to graph the object and its image. (All dilations have (0, 0) as the center of the dilation.)

Graph the image after dilating it by $\frac{3}{5}$.

Directions Give the coordinates of the image under the following dilations. Use graph paper to graph each object and its images. (All dilations have (0, 0) as the center of the dilation.)

1. dilation of $\frac{1}{2}$
2. dilation of $\frac{1}{3}$
3. dilation of $\frac{1}{4}$
4. dilation of $\frac{1}{5}$
5. dilation of $\frac{1}{6}$
6. dilation of $\frac{5}{6}$
7. dilation of $\frac{2}{3}$
8. dilation of $\frac{1}{2}$
9. dilation of $\frac{1}{3}$
10. dilation of $\frac{1}{6}$

Geometry

Workbook Activity 81

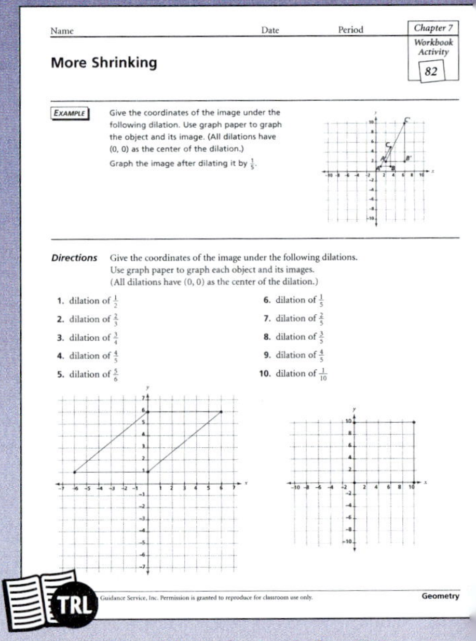

Name Date Period

Chapter 7 Workbook Activity 82

More Shrinking

EXAMPLE Give the coordinates of the image under the following dilation. Use graph paper to graph the object and its image. (All dilations have (0, 0) as the center of the dilation.)

Graph the image after dilating it by $\frac{1}{5}$.

Directions Give the coordinates of the image under the following dilations. Use graph paper to graph each object and its images. (All dilations have (0, 0) as the center of the dilation.)

1. dilation of $\frac{1}{2}$
2. dilation of $\frac{2}{3}$
3. dilation of $\frac{3}{4}$
4. dilation of $\frac{4}{5}$
5. dilation of $\frac{5}{6}$
6. dilation of $\frac{1}{5}$
7. dilation of $\frac{2}{5}$
8. dilation of $\frac{3}{5}$
9. dilation of $\frac{4}{5}$
10. dilation of $\frac{1}{10}$

Geometry

Workbook Activity 82

EXAMPLES **Given:** $D = (-2, -2)$, $E = (6, 12)$, and $F = (6, -2)$

Find: the coordinates of the image of ΔDEF, $\Delta D'E'F'$, under a dilation of $\frac{1}{3}$

Use (0, 0) as the center of the dilation. Graph both triangles.

To shrink ΔDEF using a ratio of similarity of $\frac{1}{3}$, multiply each coordinate by $\frac{1}{3}$.

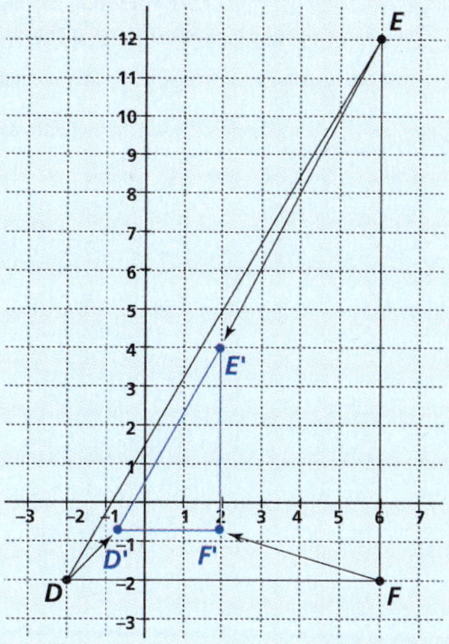

$D = (-2, -2) \rightarrow (-2 \bullet \frac{1}{3}, -2 \bullet \frac{1}{3}) = (-\frac{2}{3}, -\frac{2}{3}) = D'$

$E = (6, 12) \rightarrow (6 \bullet \frac{1}{3}, 12 \bullet \frac{1}{3}) = (2, 4) = E'$

$F = (6, -2) \rightarrow (6 \bullet \frac{1}{3}, -2 \bullet \frac{1}{3}) = (2, -\frac{2}{3}) = F'$

Again, note that $\Delta DEF \sim \Delta D'E'F'$.

Given: square $SQAR$ with coordinates (–5, –5), (–5, 5), (5, 5), and (5, –5)

Find: the coordinates of a similar square under a dilation of $\frac{1}{5}$, with (0, 0) as the center of the dilation. Graph both squares.

To shrink the square, using a ratio of similarity of $\frac{1}{5}$, multiply each coordinate by $\frac{1}{5}$. The squares—like all squares—are similar to one another.

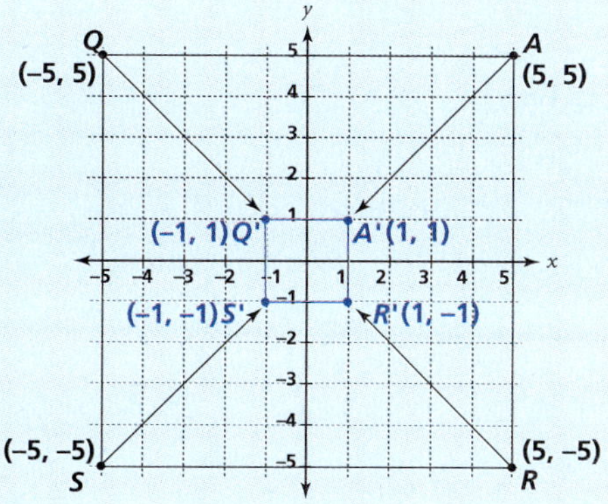

$S = (-5, -5) \rightarrow (-5 \bullet \frac{1}{5}, -5 \bullet \frac{1}{5}) = (-1, -1) = S'$

$Q = (-5, 5) \rightarrow (-5 \bullet \frac{1}{5}, 5 \bullet \frac{1}{5}) = (-1, 1) = Q'$

$A = (5, 5) \rightarrow (5 \bullet \frac{1}{5}, 5 \bullet \frac{1}{5}) = (1, 1) = A'$

$R = (5, -5) \rightarrow (5 \bullet \frac{1}{5}, -5 \bullet \frac{1}{5}) = (1, -1) = R'$

2. Teaching the Lesson

Bring students to the realization that shrinking geometric figures requires an operation opposite to that of enlarging figures. Lead students through the thought process that if they enlarge figures by multiplying by a factor greater than one, they can shrink figures by dividing by a factor greater than one (or multiplying by a factor less than one.)

As you work through the example at the top of page 241, ask a volunteer to show that multiplying by $\frac{1}{3}$ is the same as dividing by 3.

3. Reinforce and Extend

LEARNING STYLES

Interpersonal/Group Learning Direct groups of four to perform another dilation on square *SQAR* in the second example. Indicate that the dilation should be $\frac{1}{2}$ with (0, 0) as the center. Student groups should produce a sequence similar to the example, writing a *Given* and a *Find* as well as the directions and answers to the problem. Remind groups that they should also include an illustration of the original *SQAR* as well as $S'Q'A'R'$. Post student efforts and allow the class to choose one that seems the clearest and best suited to use as a review example.

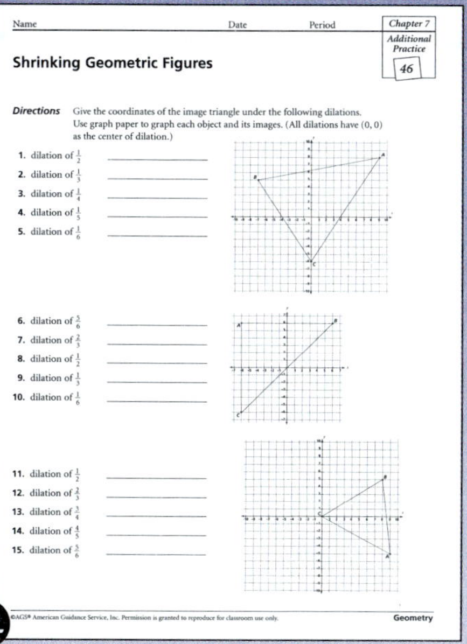

Name ____ Date ____ Period ____ *Chapter 7* *Additional Practice* *46*

Shrinking Geometric Figures

Directions Give the coordinates of the image triangle under the following dilations. Use graph paper to graph each object and its images. (All dilations have (0, 0) as the center of dilation.)

1. dilation of $\frac{1}{2}$ ____
2. dilation of $\frac{1}{3}$ ____
3. dilation of $\frac{1}{4}$ ____
4. dilation of $\frac{1}{5}$ ____
5. dilation of $\frac{1}{6}$ ____

6. dilation of $\frac{5}{6}$ ____
7. dilation of $\frac{2}{3}$ ____
8. dilation of $\frac{1}{2}$ ____
9. dilation of $\frac{1}{3}$ ____
10. dilation of $\frac{1}{6}$ ____

11. dilation of $\frac{1}{2}$ ____
12. dilation of $\frac{2}{3}$ ____
13. dilation of $\frac{3}{4}$ ____
14. dilation of $\frac{4}{5}$ ____
15. dilation of $\frac{5}{6}$ ____

 Geometry

Additional Practice 46

Proportion and Similarity *Chapter 7* **241**

APPLICATION

Career Connection

Encourage students to investigate model making as a career. Point out that model makers are employed by museums to prepare exhibits, by animators and puppeteers to prepare figures, and by special effects companies to produce model cities, spacecraft, and other locations to be used—and often destroyed—in motion pictures. Students can present their results in a two-minute recruiting speech in which they role-play a company representative describing the necessary qualifications and likely responsibilities for a model maker at the company.

Exercise A Give the coordinates of the image triangle under the following dilations. Use graph paper to graph the object and its images. (All dilations have (0, 0) as the center of the dilation.)

1. dilation of $\frac{1}{2}$
   $A' = (-2, -2)$, $B' = (-2, 2)$, $C' = (2, 0)$
2. dilation of $\frac{1}{4}$
   $A' = (-1, -1)$, $B' = (-1, 1)$, $C' = (1, 0)$
3. dilation of $\frac{1}{3}$
   $D' = (1, -2)$, $E' = (0, 2)$, $F' = (2, 1)$
4. dilation of $\frac{2}{3}$
   $D' = (2, -4)$, $E' = (0, 4)$, $F' = (4, 2)$

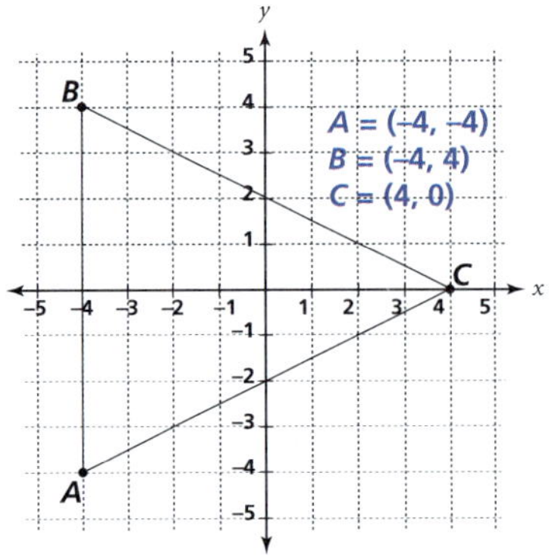

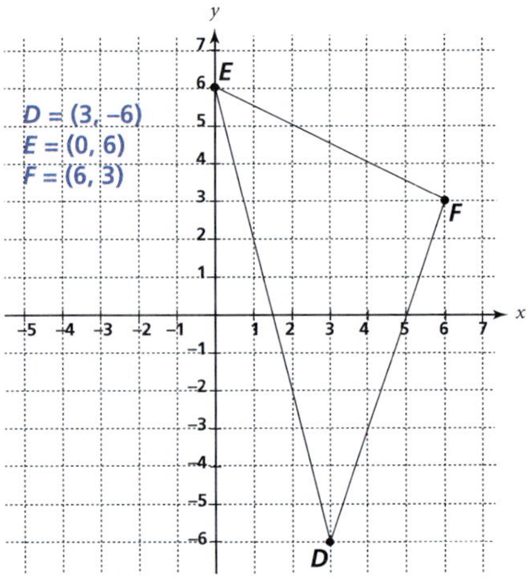

See Teacher's Edition page for graph answers.

Exercise B Give the coordinates of the image under the following dilations. Use graph paper to graph the object and its images. (All dilations have (0, 0) as the center of the dilation.)

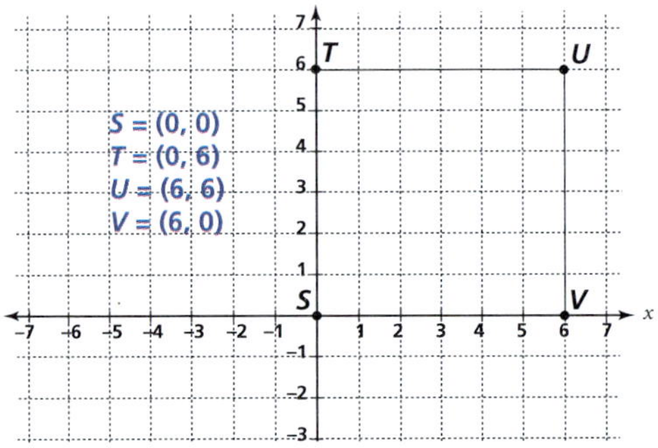

5. dilation of $\frac{1}{6}$ $S' = (0, 0)$, $T' = (0, 1)$, $U' = (1, 1)$, $V' = (1, 0)$
6. dilation of $\frac{1}{3}$ $S' = (0, 0)$, $T' = (0, 2)$, $U' = (2, 2)$, $V' = (2, 0)$
7. dilation of $\frac{2}{3}$ $S' = (0, 0)$, $T' = (0, 4)$, $U' = (4, 4)$, $V' = (4, 0)$
8. dilation of $\frac{5}{6}$ $S' = (0, 0)$, $T' = (0, 5)$, $U' = (5, 5)$, $V' = (5, 0)$
9. dilation of $\frac{1}{2}$ $S' = (0, 0)$, $T' = (0, 3)$, $U' = (3, 3)$, $V' = (3, 0)$
10. dilation of $\frac{1}{4}$ $S' = (0, 0)$, $T' = (0, 1.5)$, $U' = (1.5, 1.5)$, $V' = (1.5, 0)$

Answers to Exercise A

1. blue **2. red**

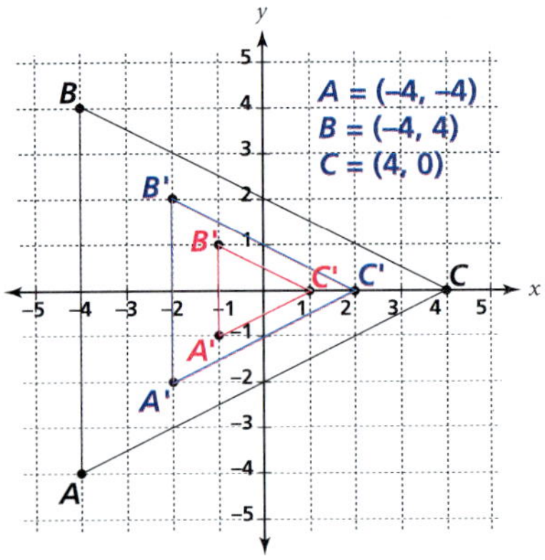

3. blue **4. red**

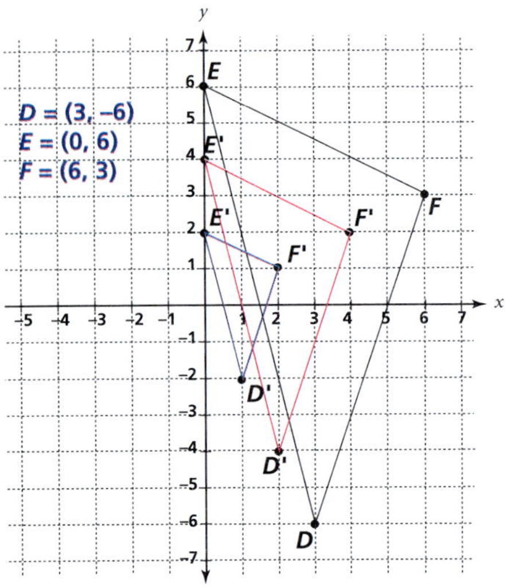

5. blue **6. red**

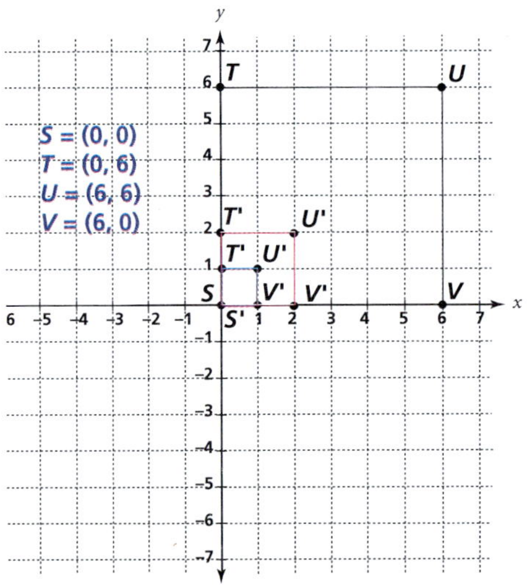

Problem Solving

Exercise C Use what you know about proportions and ratios of similarity to solve the following problems.

11. The cars for a model railroad are built in proportion to actual railroad cars. HO models represent actual dimensions with a ratio of similarity 1:80. If a HO-scale freight car is 14 cm long, what is the approximate length of the actual rail car? 1,120 cm or about 11 m

12. In N-scale model trains, the ratio of similarity is 1:160. If an N-scale freight car is 7 cm, what is the approximate length of the actual rail car? 1,120 cm or about 11 m

13. 45 in., about $3\frac{3}{4}$ ft

13. Shenequa wants to make a model of her favorite dinosaur. She will use a ratio of similarity of 1 in.:1 ft—1 inch of the model represents 1 foot (12 in.) of dinosaur. She knows that the dinosaur was about 45 feet from snout to tail. What will her model measure from snout to tail?

14. Raphael's miniature racer collection has several cars made with a ratio of similarity of 1:60. If the Red Menace car measures $1\frac{1}{4}$ inches, what is the measurement of the actual Red Menace racer? 75 in. or $6\frac{1}{4}$ ft

15. A working model of a pump delivers 1 gallon of water every 10 minutes. The model has a ratio of similarity with the actual pump of $\frac{1}{2750}$. How much water does the actual pump deliver in 10 minutes? in an hour?

2,750 gal in 10 min, 16,500 gal in an hour

Geometry Around Us

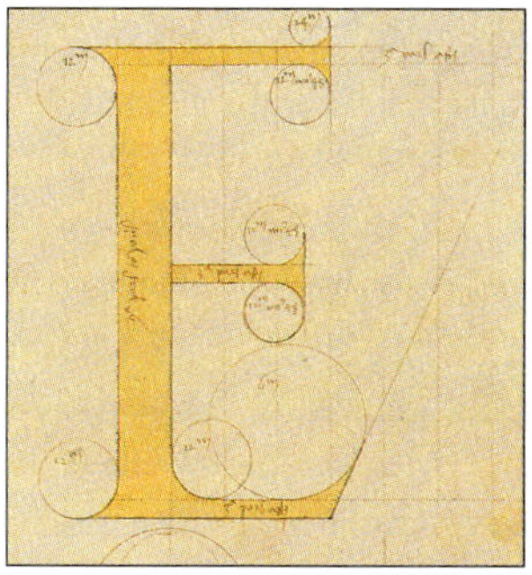

You read different sizes of letters every day—letters in newspaper headlines are large; letters at the bottom of milk cartons are small. Designers work to make sure that the small and large letters are similar so that the only difference is their size.

Answers to Exercise A continued

7. blue **8. red**

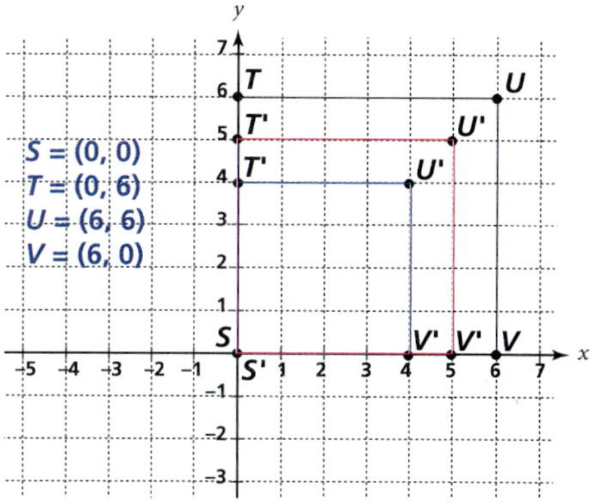

9. blue **10. red**

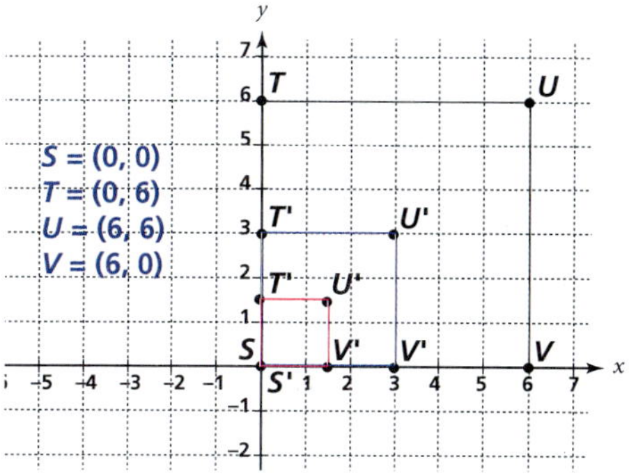

Proportion and Similarity *Chapter 7* **243**

Lesson at a Glance

Chapter 7 Application

Overview This lesson explores how ratios of similarities are applied to produce maps and globes.

Objectives

- To recognize a variety of map scales.
- To compute actual distance given a map scale and distance.

Student Page 244

Teacher's Resource Library TRL

Workbook Activity 83

1. Warm-Up Activity

Provide students with a variety of maps including ones with a variety of scales. Ask students to find the map distance between two given locations on the maps. Record these distances.

2. Teaching the Lesson

Point out to students how they can use the map scale and the on-map distance from the Warm-Up Activity to compute the actual distance between locations. Make certain students understand that maps with the smallest scale show the greatest amount of detail and maps with large scales show the greatest areas.

3. Reinforce and Extend

Application

In the Community

Direct students to chose four locations within a mile of the school, for example, a library, a grocery store, a church, and a playground. Have them prepare a map of the neighborhood that includes the locations in their relative positions. Indicate that students can use paces rather than miles or kilometers as the basis for their distance measurements. Share completed maps with other classes that are studying maps, for example, social studies.

Application

Maps and Globes On maps and globes the ratio of similarity is usually called a *scale*. You can see in the map at the right that the scale gives you a clear indication of the ratio the map-makers used. This map indicates that 1 inch→600 miles and 1 cm→400 kilometers. You can use this ratio to form a proportion and calculate the distances between cities on the map.

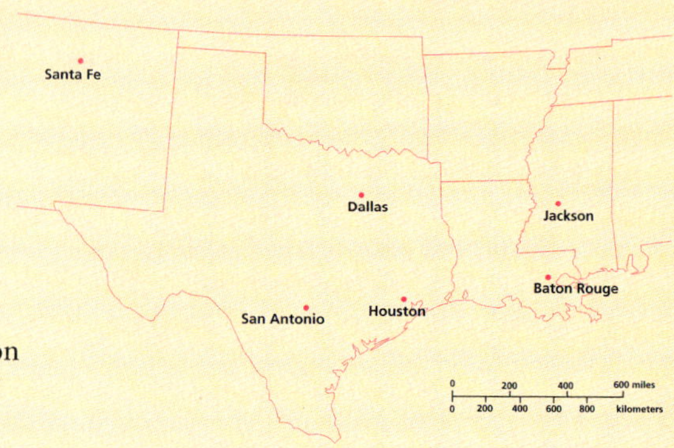

EXAMPLE The cities of Santa Fe, NM, and San Antonio, TX, are about 2 inches apart on the map. Let x = the actual distance between the cities. Using the scale, you can write

$$\frac{1}{600} = \frac{2}{x} \qquad x = 1{,}200$$

Santa Fe and San Antonio are about 1,200 miles apart.

Exercise Answer the questions based on these scales that you might find on a map or globe.

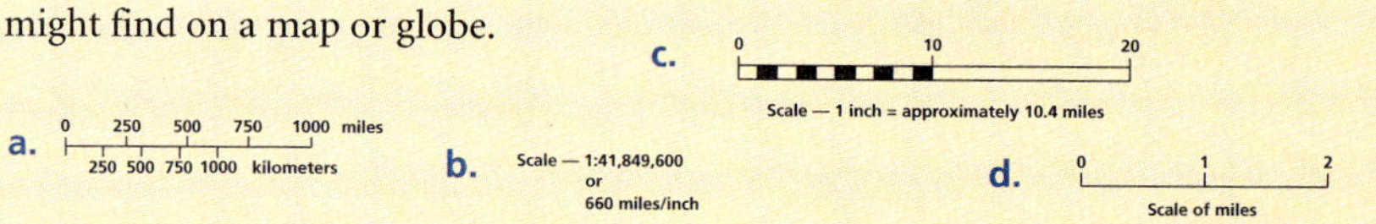

1. Which of the scales allows a map-maker to show the greatest amount of area? Explain why. a, it has the largest scale
2. Which of the scales allows a map-maker to show the greatest amount of detail? d, it has the smallest scale
3. On a globe with scale *b*, two cities are $3\frac{1}{2}$ inches apart. What is the actual distance between the cities? 2,310 miles
4. A city map using scale *d* shows that the city hall is $\frac{3}{4}$ inch from the art museum. How far will you have to walk to reach the art museum from city hall? 1.5 miles
5. A frequent flyer knows that Memphis, TN, and Chicago, IL, are about 500 miles apart. If she is using a map with scale *a*, what distance separates the cities on the map? $\frac{1}{2}$ inch

---

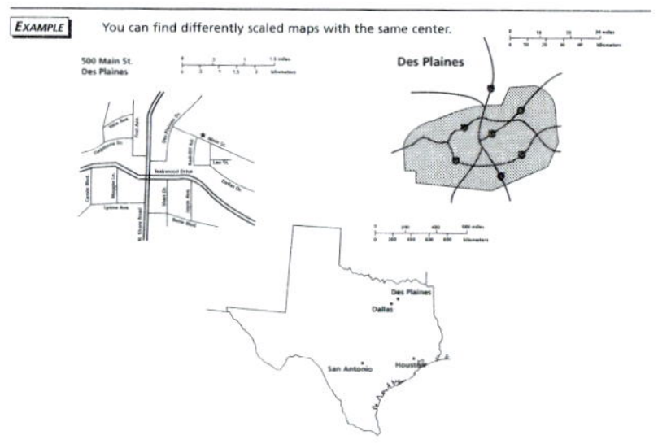

Name Date Period | Chapter 7 Workbook Activity 83

Zooming In

EXAMPLE You can find differently scaled maps with the same center.

Directions Use encyclopedias or atlases to choose a specific landmark or special landform. Then follow the directions to show each of the following.

1. Immediate field of vision—at landmark or landform
2. Town size
3. State size
4. Country size
5. World size

For example, if you choose Niagara Falls, you can map its location in Niagara and Niagara's location in the state of New York. Then you can map the state in relation to the country and then the world. Draw a sketched map of the places according to the given scale. Place a star on the center for the place that you chose.

 Geometry

Workbook Activity 83

Find the missing value in each proportion. Write the letter of the correct answer.

Example: $\frac{2}{3} = \frac{y}{18}$ Solution: D

A. 36 B. 54 C. 16 D. 12

product of extremes = product of means
so $2 \cdot 18 = 3y$ $36 = 3y$ $12 = y$

	A	B	C	D	Answer	
1. $\frac{2}{3} = \frac{x}{9}$	A. 27	B. 18	C. 6	D. 13.5	$x = 6$	C
2. $\frac{3}{5} = \frac{12}{y}$	A. 60	B. 20	C. 15	D. 36	$y = 20$	B
3. $\frac{8}{9} = \frac{z}{18}$	A. 72	B. 18	C. 56	D. 16	$z = 16$	D
4. $\frac{3}{4} = \frac{a}{12}$	A. 9	B. 48	C. 36	D. 16	$a = 9$	A
5. $\frac{3}{5} = \frac{b}{40}$	A. 24	B. 15	C. 120	D. 200	$b = 24$	A
6. $\frac{3}{4} = \frac{m}{32}$	A. 96	B. 128	C. 24	D. 8	$m = 24$	C
7. $\frac{3}{8} = \frac{b}{64}$	A. 24	B. 170	C. 32	D. 8	$b = 24$	A

Solve for the missing side(s) in each pair of similar triangles.

Example: Find x.

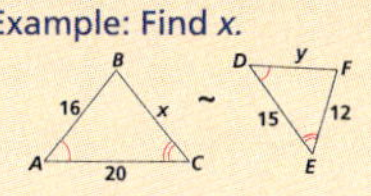

Solution: $\frac{AB}{DF} = \frac{BC}{FE} = \frac{AC}{DE}$

corresponding sides of similar triangles have the same ratio

$\frac{AC}{DE} = \frac{20}{15} = \frac{4}{3}$ so $\frac{BC}{FE} = \frac{4}{3}$

Let $x = \overline{BC}$ and substitute $\frac{x}{12} = \frac{4}{3}$

$3x = 48$ product of extremes = product of means

$x = 16$ divide

8. $\overline{BC}$ $\overline{BC} = 15$, $\overline{EF} = 8$

9. $\overline{EF}$

10. $\overline{JL}$ $\overline{JL} = 18$, $\overline{KJ} = 24$

11. $\overline{KJ}$

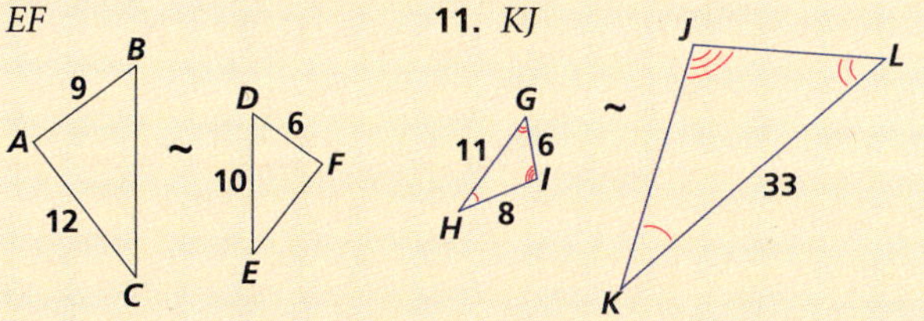

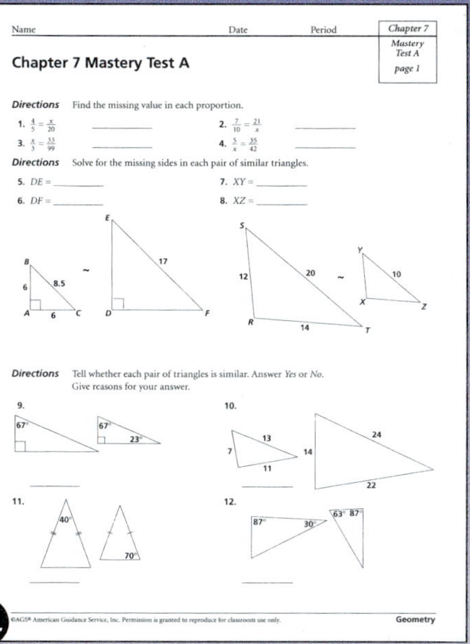
Name Date Period

Chapter 7 Mastery Test A page 1

Chapter 7 Mastery Test A

Directions Find the missing value in each proportion.

1. $\frac{4}{5} = \frac{x}{20}$ ______ 2. $\frac{7}{10} = \frac{21}{x}$ ______
3. $\frac{x}{3} = \frac{33}{99}$ ______ 4. $\frac{5}{x} = \frac{35}{42}$ ______

Directions Solve for the missing sides in each pair of similar triangles.

5. DE = ______ 7. XY = ______
6. DF = ______ 8. XZ = ______

Directions Tell whether each pair of triangles is similar. Answer *Yes* or *No*. Give reasons for your answer.

9. 10. 11. 12.

 Geometry

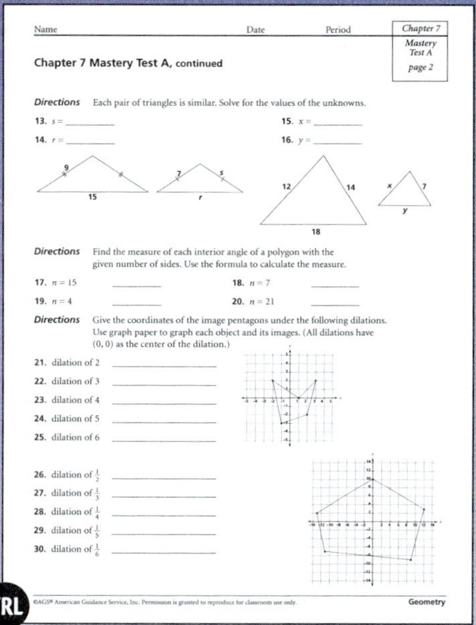
Name Date Period

Chapter 7 Mastery Test A page 2

Chapter 7 Mastery Test A, continued

Directions Each pair of triangles is similar. Solve for the values of the unknowns.

13. s = ______ 15. x = ______
14. r = ______ 16. y = ______

Directions Find the measure of each interior angle of a polygon with the given number of sides. Use the formula to calculate the measure.

17. $n = 15$ ______ 18. $n = 7$ ______
19. $n = 4$ ______ 20. $n = 21$ ______

Directions Give the coordinates of the image pentagons under the following dilations. Use graph paper to graph each object and its images. (All dilations have (0, 0) as the center of the dilation.)

21. dilation of 2 ______
22. dilation of 3 ______
23. dilation of 4 ______
24. dilation of 5 ______
25. dilation of 6 ______

26. dilation of $\frac{1}{2}$ ______
27. dilation of $\frac{1}{3}$ ______
28. dilation of $\frac{1}{4}$ ______
29. dilation of $\frac{1}{5}$ ______
30. dilation of $\frac{1}{6}$ ______

 Geometry

Chapter 7 Mastery Test A

Chapter 7 Review

Each set of problems in the Chapter Review includes an example and solution to illustrate the concept. Use the given examples for reteaching the material in Chapter 7. For additional practice, refer to the Supplementary Problems for Chapter 7 (pages 408–409) and Additional Practice 42–46 in the Teacher's Resource Library.

Chapter 7 Mastery Test TRL

The Teacher's Resource Library includes parallel forms of the Chapter 7 Mastery Test. The difficulty level of the two forms is equivalent. You may wish to use one form as a pretest and the other form as a posttest. In addition, the Chapter 7 Cumulative Mastery Test covers content from Chapters 1–7.

Alternative Assessment

Alternative Assessment items correlate with student Goals for Learning at the beginning of this chapter.

- Have students identify equal proportions.

Using the ratio $\frac{5}{6}$, write three equal proportions.

- Have students use the theorem that the product of the extremes equals the product of the means in solutions to problems.

Given $\frac{a}{b} = \frac{c}{d}$, prove that $ad = bc$. State your results as a theorem.

- Have students solve for the missing sides of similar triangles.

Draw three pairs of similar triangles. For each pair, give the measurements of all three sides of the first triangle and the measurement of one side of the second triangle. Exchange sets of triangles with a partner. Find the measurements of the missing sides for each pair of triangles.

ALTERNATIVE ASSESSMENT

continued

- ▶ Have students give coordinates of dilations.

Draw a five-sided polygon on a coordinate plane. Specify a dilation greater than one and a dilation less than one. Indicate that the center of both dilations is (0, 0). Exchange polygons with a partner. Find the coordinates of each dilation and draw both the enlarged and shrunken polygons.

More Review...

12. $\overline{AB}$ $\overline{AB} = 12$

13. $\overline{CD}$ $\overline{CD} = 7\frac{1}{3}$

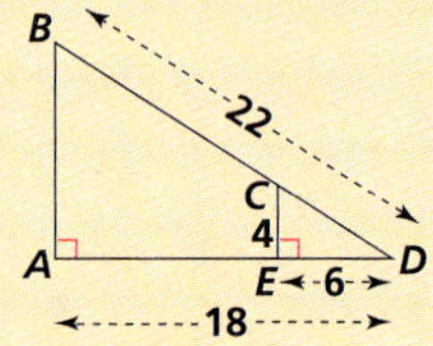

14. $\overline{RS}$ $\overline{RS} = 20$

15. $\overline{TS}$ $\overline{TS} = 11$

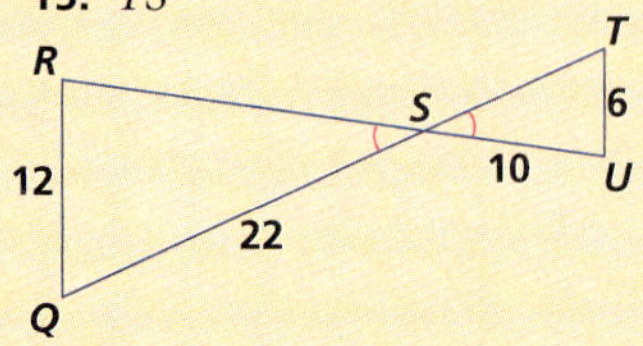

16. $\overline{HJ}$

17. $\overline{HI}$

$\overline{HJ} = 42$
$\overline{HI} = 56$

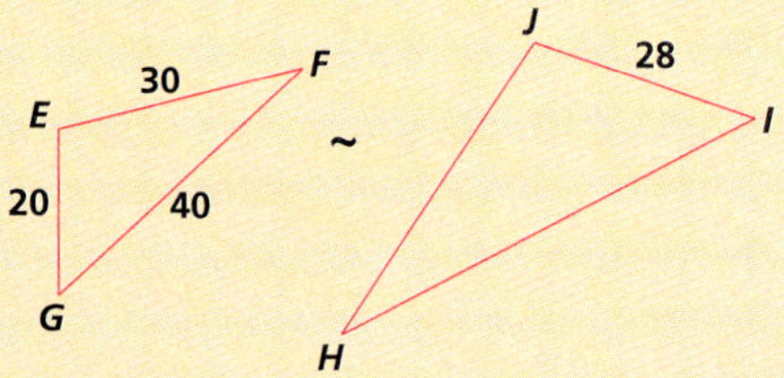

Give the coordinates of the image under the following dilations. Use graph paper to graph the object and its image(s). (All dilations have (0, 0) as the center of the dilation.)

Example: enlarge $\triangle ABC$ using a dilation of $\frac{2}{1}$

Solution: to enlarge $\triangle ABC$, multiply the coordinates of each vertex by the ratio of similarity $\frac{2}{1}$

$A = (-4, -2) \rightarrow (-4 \bullet \frac{2}{1}, -2 \bullet \frac{2}{1}) = (-8, -4) = A'$

$B = (0, 2) \rightarrow (0 \bullet \frac{2}{1}, 2 \bullet \frac{2}{1}) = (0, 4) = B'$

$C = (4 -2) \rightarrow (4 \bullet \frac{2}{1}, -2 \bullet \frac{2}{1}) = (8, -4) = C'$

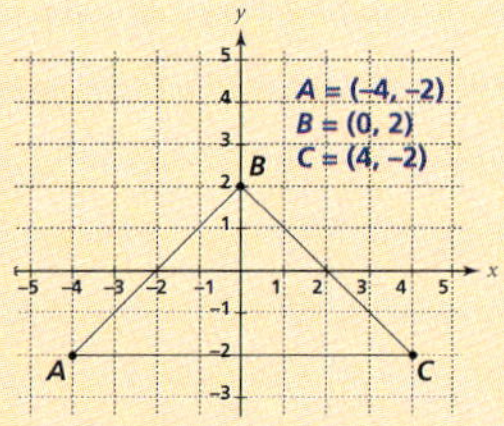

18. dilation of 2

$D' = (-4, -4)$, $E' = (-2, 4)$, $F' = (8, -6)$

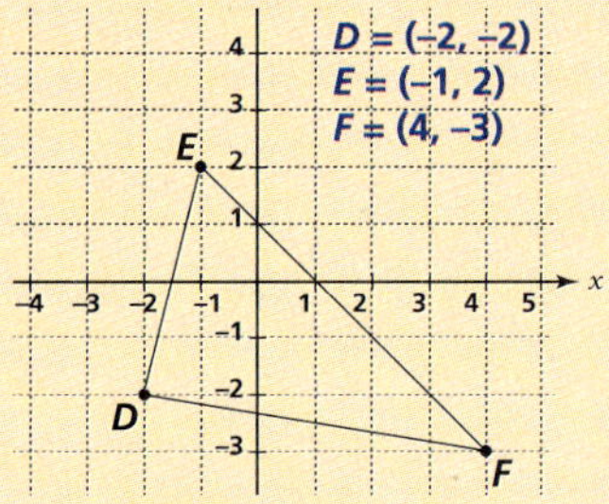

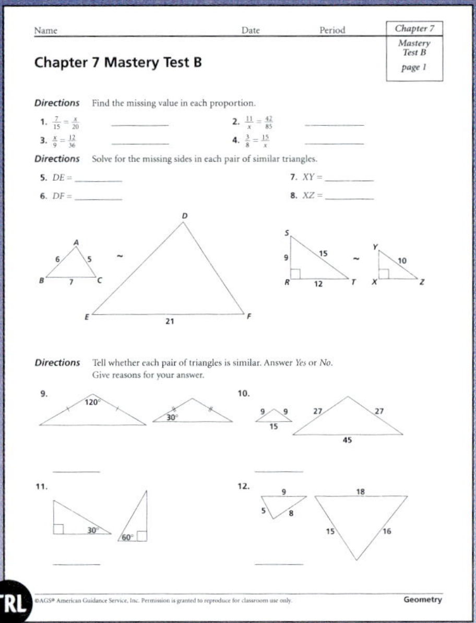

Name Date Period | *Chapter 7 Mastery Test B page 1*

Chapter 7 Mastery Test B

Directions Find the missing value in each proportion.

1. $\frac{7}{15} = \frac{x}{20}$ ______ 2. $\frac{11}{x} = \frac{42}{85}$ ______
3. $\frac{x}{9} = \frac{12}{36}$ ______ 4. $\frac{3}{8} = \frac{15}{x}$ ______

Directions Solve for the missing sides in each pair of similar triangles.

5. $DE =$ ______ 7. $XY =$ ______
6. $DF =$ ______ 8. $XZ =$ ______

Directions Tell whether each pair of triangles is similar. Answer *Yes* or *No*. Give reasons for your answer.

9. ______ 10. ______
11. ______ 12. ______

©AGS® American Guidance Service, Inc. Permission is granted to reproduce for classroom use only. Geometry

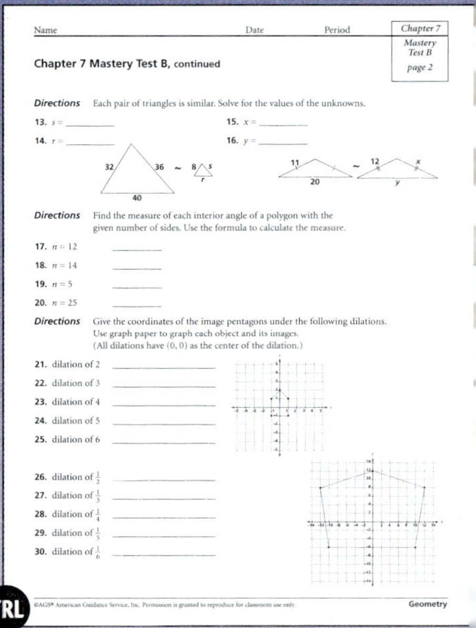

Name Date Period | *Chapter 7 Mastery Test B page 2*

Chapter 7 Mastery Test B, continued

Directions Each pair of triangles is similar. Solve for the values of the unknowns.

13. $s =$ ______ 15. $x =$ ______
14. $r =$ ______ 16. $y =$ ______

Directions Find the measure of each interior angle of a polygon with the given number of sides. Use the formula to calculate the measure.

17. $n = 12$ ______
18. $n = 14$ ______
19. $n = 5$ ______
20. $n = 25$ ______

Directions Give the coordinates of the image pentagons under the following dilations. Use graph paper to graph each object and its images. (All dilations have (0, 0) as the center of the dilation.)

21. dilation of 2 ______
22. dilation of 3 ______
23. dilation of 4 ______
24. dilation of 5 ______
25. dilation of 6 ______

26. dilation of $\frac{1}{2}$ ______
27. dilation of $\frac{1}{3}$ ______
28. dilation of $\frac{1}{4}$ ______
29. dilation of $\frac{1}{5}$ ______
30. dilation of $\frac{1}{6}$ ______

©AGS® American Guidance Service, Inc. Permission is granted to reproduce for classroom use only. Geometry

Chapter 7 Mastery Test B

19. dilation of 3

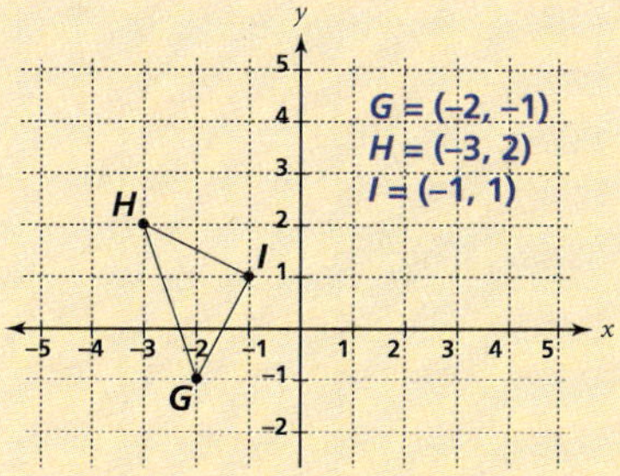

19. $G' = (-6, -3)$, $H' = (-9, 6)$, $I' = (-3, 3)$

20. dilation of 2

21. dilation of $\frac{1}{3}$

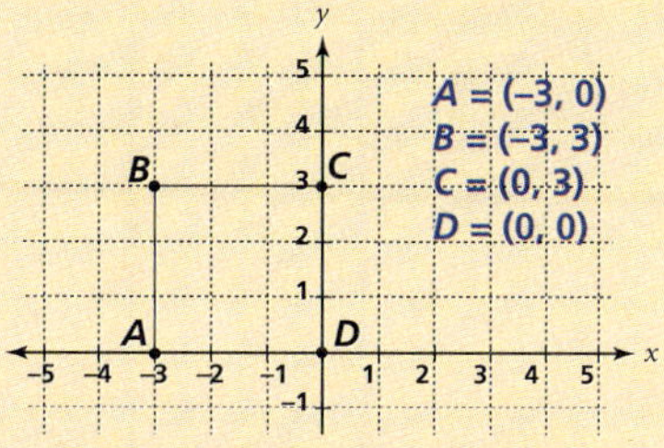

20. $A' = (-6, 0)$, $B' = (-6, 6)$, $C' = (0, 6)$, $D' = (0, 0)$

21. $A' = (-1, 0)$, $B' = (-1, 1)$, $C' = (0, 1)$, $D' = (0, 0)$

22. dilation of $\frac{1}{2}$

23. dilation of 2

24. dilation of $\frac{1}{4}$

25. dilation of 3

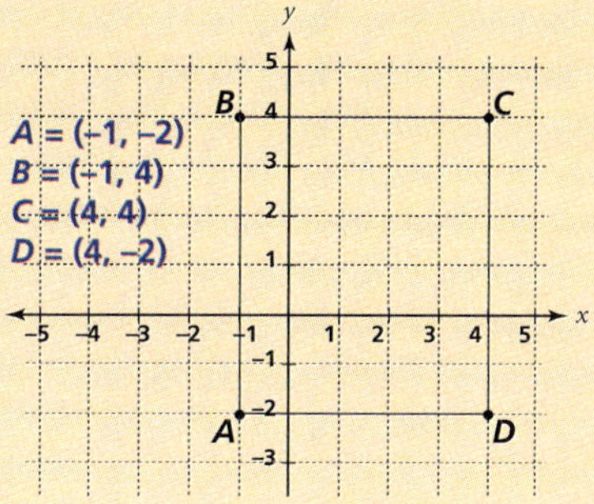

22. $A' = (-\frac{1}{2}, -1)$, $B' = (-\frac{1}{2}, 2,)$, $C' = (2, 2)$, $D' = (2, -1)$

23. $A' = (-2, -4)$, $B' = (-2, 8)$, $C' = (8, 8)$, $D' = (8, -4)$

24. $A' = (-\frac{1}{4}, -\frac{1}{2})$, $B' = (-\frac{1}{4}, 1)$, $C' = (1, 1)$, $D' = (1, -\frac{1}{2})$

25. $A' = (-3, -6)$, $B' = (-3, 12)$, $C' = (12, 12)$, $D' = (12, -6)$

Test-Taking Tip

Studying together in small groups and asking questions of one another is one way to review material for tests.

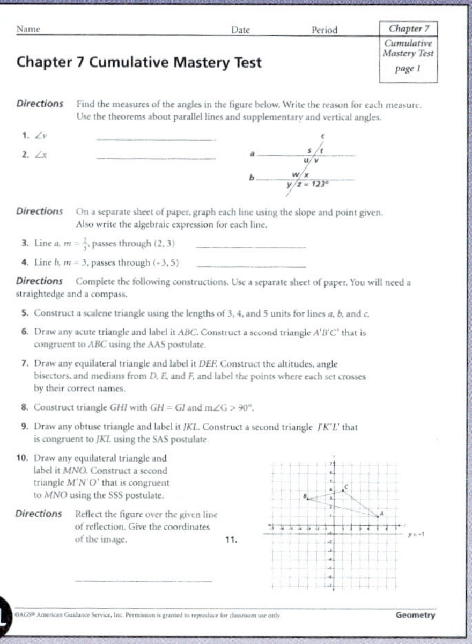

Name Date Period

Chapter 7 Cumulative Mastery Test page 1

Chapter 7 Cumulative Mastery Test

Directions Find the measures of the angles in the figure below. Write the reason for each measure. Use the theorems about parallel lines and supplementary and vertical angles.

1. ∠v ______
2. ∠x ______

Directions On a separate sheet of paper, graph each line using the slope and point given. Also write the algebraic expression for each line.

3. Line a, $m = \frac{2}{3}$, passes through (2, 3) ______
4. Line b, $m = 3$, passes through (−3, 5) ______

Directions Complete the following constructions. Use a separate sheet of paper. You will need a straightedge and a compass.

5. Construct a scalene triangle using the lengths of 3, 4, and 5 units for lines a, b, and c.
6. Draw any acute triangle and label it ABC. Construct a second triangle $A'B'C'$ that is congruent to ABC using the AAS postulate.
7. Draw any equilateral triangle and label it DEF. Construct the altitudes, angle bisectors, and medians from D, E, and F, and label the points where each set crosses by their correct names.
8. Construct triangle GHI with $GH = GI$ and $m\angle G > 90°$.
9. Draw any obtuse triangle and label it JKL. Construct a second triangle $J'K'L'$ that is congruent to JKL using the SAS postulate.
10. Draw any equilateral triangle and label it MNO. Construct a second triangle $M'N'O'$ that is congruent to MNO using the SSS postulate.

Directions Reflect the figure over the given line of reflection. Give the coordinates of the image.

11. ______

 Geometry

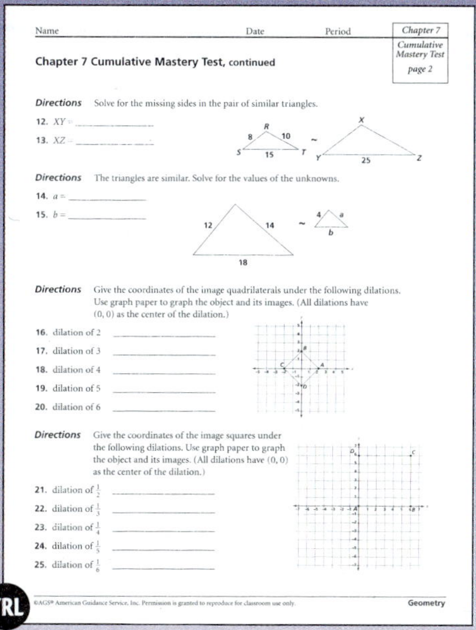

Name Date Period

Chapter 7 Cumulative Mastery Test page 2

Chapter 7 Cumulative Mastery Test, continued

Directions Solve for the missing sides in the pair of similar triangles.

12. XY = ______
13. XZ = ______

Directions The triangles are similar. Solve for the values of the unknowns.

14. a = ______
15. b = ______

Directions Give the coordinates of the image quadrilaterals under the following dilations. Use graph paper to graph the object and its images. (All dilations have (0, 0) as the center of the dilation.)

16. dilation of 2 ______
17. dilation of 3 ______
18. dilation of 4 ______
19. dilation of 5 ______
20. dilation of 6 ______

Directions Give the coordinates of the image squares under the following dilations. Use graph paper to graph the object and its images. (All dilations have (0, 0) as the center of the dilation.)

21. dilation of $\frac{1}{2}$ ______
22. dilation of $\frac{1}{3}$ ______
23. dilation of $\frac{1}{4}$ ______
24. dilation of $\frac{1}{5}$ ______
25. dilation of $\frac{1}{6}$ ______

 Geometry

TRL

Chapter 7 Cumulative Mastery Test

Proportion and Similarity *Chapter 7* **247**

Chapter 8

Planning Guide

The Pythagorean Theorem

	Student Text Lesson						
	Student Pages	Vocabulary	Proof	Practice Exercises	Problem Solving	Calculator Practice	Solutions Key
Lesson 1 Pythagorean Triples	250–253	✔		✔		253	✔
Lesson 2 Pythagorean Triples and a Proof	254–256	✔	✔	✔			✔
Lesson 3 Pythagorean Demonstration	257–261		✔	✔		260	✔
Lesson 4 Pythagorean Theorem and Similar Triangles	262–265		✔	✔	265		✔
Lesson 5 Pythagorean Proof and Trapezoids	266–269	✔	✔	✔			✔
Lesson 6 Distance Formula: Pythagorean Theorem	270–274			✔		274	✔
Lesson 7 Converse of the Pythagorean Theorem	275–277		✔	✔			✔
Application Architecture and Geometric Shapes	278			✔			✔

Chapter Activities

Teacher's Resource Library
Geometry on the Job 8: Ramps at Work

Community Connection 8: Supporting Triangles

Teacher's Edition
Chapter 8 Project

Assessment Options

Student Text
Chapter 8 Review

Teacher's Resource Library
Chapter 8 Mastery Tests A and B

Chapter 8 Cumulative Mastery Test

Teacher's Edition
Chapter 8 Alternative Assessment

Student Text Features						Teaching Strategies								Learning Styles						Teacher's Resource Library		
Algebra Connection	Constructions	Experiencing Geometry	Technology Connection	Try This	Writing About Mathematics	Warm-Up Activity	Common Error	Algebra Connection	Mental Math	Manipulatives	Calculator	Applications	Group Problem Solving	Auditory/Verbal	Visual/Spatial	Body/Kinesthetic	Logical/Mathematical	Interpersonal/Group Learning	LEP/ESL	Workbook Activities	Additional Practice	Self-Study Guide
251				252		250		251			253						252	252	253	84, 85	47	✔
			255			254									256	255			256	86	48	✔
		262				257	259				260	261	260				261		261	87	49	✔
						262				264		265		263						88, 89	50	✔
268		266				266		268						268	267		269	269	268	90	51	✔
					274	270	272		273		274			271		272				91, 92	52	✔
268		266			277	275											276		277	93	53	✔
						278						278								94		✔

Manipulatives

Proofs of Pythagoras

Manipulatives Master 3 (Pythagorean Model)

Chapter at a Glance

Chapter 8: The Pythagorean Theorem

pages 248–281

Lessons

Teacher's Resource Library TRL

Workbook Activities 84–94

Additional Practice 47–53

Geometry on the Job 8

Community Connection 8

Chapter 8 Self-Study Guide

Chapter 8 Mastery Tests A and B

Chapter 8 Cumulative Mastery Test

(Answer Keys for the Teacher's Resource Library begin on page 484 of this Teacher's Edition.)

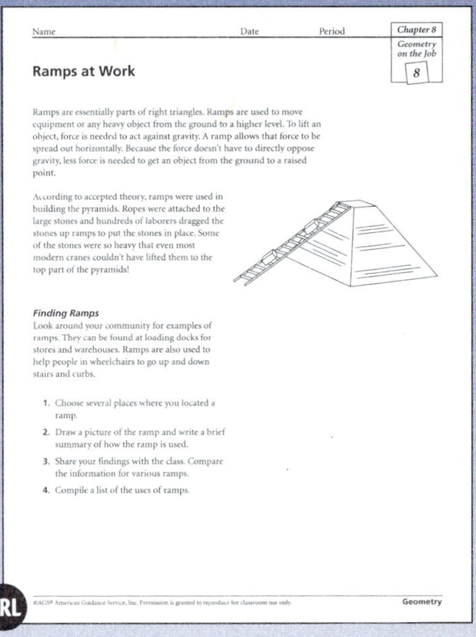

Name Date Period | Chapter 8 Geometry on the Job 8

Ramps at Work

Ramps are essentially parts of right triangles. Ramps are used to move equipment or any heavy object from the ground to a higher level. To lift an object, force is needed to act against gravity. A ramp allows that force to be spread out horizontally. Because the force doesn't have to directly oppose gravity, less force is needed to get an object from the ground to a raised point.

According to accepted theory, ramps were used in building the pyramids. Ropes were attached to the large stones and hundreds of laborers dragged the stones up ramps to put the stones in place. Some of the stones were so heavy that even most modern cranes couldn't have lifted them to the top part of the pyramids!

Finding Ramps

Look around your community for examples of ramps. They can be found at loading docks for stores and warehouses. Ramps are also used to help people in wheelchairs to go up and down stairs and curbs.

1. Choose several places where you located a ramp.
2. Draw a picture of the ramp and write a brief summary of how the ramp is used.
3. Share your findings with the class. Compare the information for various ramps.
4. Compile a list of the uses of ramps.

TRL

 Geometry

Geometry on the Job 8

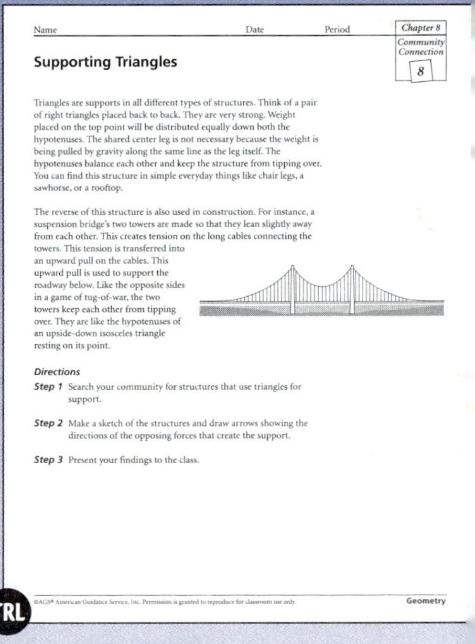

Name Date Period | Chapter 8 Community Connection 8

Supporting Triangles

Triangles are supports in all different types of structures. Think of a pair of right triangles placed back to back. They are very strong. Weight placed on the top point will be distributed equally down both the hypotenuses. The shared center leg is not necessary because the weight is being pulled by gravity along the same line as the leg itself. The hypotenuses balance each other and keep the structure from tipping over. You can find this structure in simple everyday things like chair legs, a sawhorse, or a rooftop.

The reverse of this structure is also used in construction. For instance, a suspension bridge's two towers are made so that they lean slightly away from each other. This creates tension on the long cables connecting the towers. This tension is transferred into an upward pull on the cables. This upward pull is used to support the roadway below. Like the opposite sides in a game of tug-of-war, the two towers keep each other from tipping over. They are like the hypotenuses of an upside-down isosceles triangle resting on its point.

Directions

Step 1 Search your community for structures that use triangles for support.

Step 2 Make a sketch of the structures and draw arrows showing the directions of the opposing forces that create the support.

Step 3 Present your findings to the class.

TRL

 Geometry

Community Connection 8

Chapter 8 The Pythagorean Theorem

The Manhattan Bridge spans New York City's East River. You can see the Empire State Building in the background. The bridge was built in 1909. It's one of the three suspension bridges that connect Manhattan Island to Brooklyn. Every day cars, trucks, and even the subway cross back and forth over its double decks.

Notice the right triangles in the design of the bridge. These are trusses, or beams, that attach the lower deck to the upper deck. The trusses also strengthen the bridge and support the lower level. With the help of the Pythagorean Theorem, engineers could make sure that the steel trusses were all the right size!

In Chapter 8, you will learn how to prove and apply the Pythagorean Theorem.

Goals for Learning

- To use the Pythagorean Theorem to determine the length of an unknown side of a right triangle
- To determine the area of a trapezoid
- To find the distance between points, using the distance formula
- To identify Pythagorean Triples

SELF-STUDY GUIDE

Name ____________

CHAPTER 8: The Pythagorean Theorem

Goal 8.1 *To use the Pythagorean Theorem to determine the length of an unknown side of a triangle*

Date	Assignment	Score
______	1: Read pages 250–253.	______
______	2: Read the Calculator Practice and complete Exercise A on page 253.	______
______	3: Complete Exercises B–D on page 253.	______
______	4: Complete Workbook Activity 84.	______
______	5: Read pages 254–256. Complete Exercises A and B on page 256.	______
______	6: Complete Workbook Activity 85.	______
______	7: Read pages 257–260.	______
______	8: Read the Calculator Practice and complete Exercise A on page 260.	______
______	9: Complete Exercises B–C on page 261.	______
______	10: Complete Workbook Activity 86.	______
______	11: Read pages 262–264. Complete Exercises A–B on pages 264–265.	______
______	12: Complete Workbook Activity 87.	______

Comments:

Goal 8.2 *To determine the area of a trapezoid*

Date	Assignment	Score
______	13: Read pages 266–268. Complete Exercises A–B on pages 268–269.	______
______	14: Complete Workbook Activity 88.	______

Comments:

 Geometry

SELF-STUDY GUIDE

Name ____________

CHAPTER 8: The Pythagorean Theorem, continued

Date

Goal 8.3 *To find the distance between points, using the distance formula*

Assignment Score

______	15: Read pages 270–272. Complete Exercises A and B on pages 273–274.	______
______	16: Read the Calculator Practice and complete Exercise C on page 274.	______
______	17: Complete Workbook Activity 89.	______

Comments:

Goal 8.4 *To identify Pythagorean Triples*

Date	Assignment	Score
______	18: Read pages 275–277. Complete Exercises A and B on page 277.	______
______	19: Complete Workbook Activity 90.	______
______	20: Read and complete the Application Exercise on page 278.	______
______	21: Complete Workbook Activity 91.	______
______	22: Complete the Chapter 8 Review on pages 279–281.	______

Comments:

Student's Signature ____________ Date ____________

Instructor's Signature ____________ Date ____________

 Geometry

TRL

Chapter 8 Self-Study Guide

Introducing the Chapter

Use the photograph on these pages to help students recognize the prevalence of right triangles in their surroundings. Reinforce the practical application of geometry by pointing out that engineers for this bridge needed to order each piece of steel reinforcement. For the bridge to be completed and functional, the steel beams must be made to accurate measurements, because pieces cannot be lengthened or shortened on site. Point out that geometry, and in this case the Pythagorean Theorem, provides a system for determining the sizes of these shapes and for constructing them. Read the Goals for Learning and discuss each one with students.

Chapter Project

Encourage students to work in groups to develop and produce a model bridge. Point out that like the bridge in the photograph, their bridge should be supported by triangular trusses. Indicate that students may make their bridges from craft sticks, toothpicks, or rolled paper strips. They can "weld" pieces together with glue. Challenge students to make a bridge that is at least 18 inches long and includes at least four triangular trusses.

As an enhancement to the chapter project, students may choose to enter their bridge in a competition. Students can decide that to enter the competition, bridges must be able to support a specific weight, perhaps a one-pound bag of beans or rice. Once bridges pass the entry test, they can compete to see which will support the greatest weight. Ask students to relate the bridge's ability to support weight to the number of trusses.

Lesson at a Glance

Chapter 8 Lesson 1

Overview This lesson introduces Pythagorean triples, a number pattern that is a precursor to the Pythagorean theorem.

Objectives

- To identify numbers that form Pythagorean triples.
- To use a formula to generate Pythagorean triples.

Student Pages 250–253

Teacher's Resource Library TRL

Workbook Activities 84–85

Additional Practice 47

Mathematics Vocabulary

integer
Pythagorean Triple

1. Warm-Up Activity

Remind students that they are already familiar with several number patterns. For example, the pattern 2, 4, 6, 8, 10, 12, ... or 1, 2, 3, 5, 8, 13, 21, 34, ... Ask students to explain each pattern and how to find the next number in each. **(add 2; add the last two numbers in the pattern)** Point out that in this lesson students will learn about number patterns that will be useful in their work with right triangles.

Lesson 1 Pythagorean Triples

Integer
One of the set of positive and negative whole numbers including zero

Pythagorean Triple
Integers a, b, c, such that $a^2 + b^2 = c^2$

An ancient Greek mathematician, Pythagoras, became famous for his theorems and thought that all things were based in numbers. We use numbers to describe geometric figures. Sometimes the numbers occur in patterns. These patterns can help us learn even more about the figures they describe. One of the most useful number patterns includes positive **integers** and their squares.

N:	1	2	3	4	5	6	7	8	9	10	11	12	13	14	15...
N^2:	1	4	9	16	25	36	49	64	81	100	121	144	169	196	225...

Let's look at three squares: 9, 16, and 25. Note that 9 + 16 = 25 or $3^2 + 4^2 = 5^2$.

Can we find other numbers among the squared integers so that the sum of the two smaller squares is equal to the largest square?

You know that $3^2 + 4^2 = 5^2$ is an equation. So you know that you can multiply each side of the equation by the same number and you will have another equation.

EXAMPLES

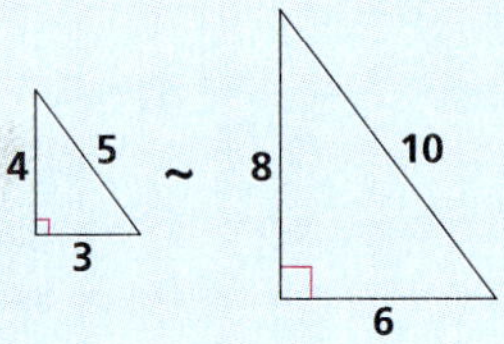

$3^2 + 4^2 = 5^2$ (3, 4, 5)

$(3 \bullet 2)^2 + (4 \bullet 2)^2 = (5 \bullet 2)^2$ ↓

$6^2 + 8^2 = 10^2$ (6, 8, 10)

$36 + 64 = 100$

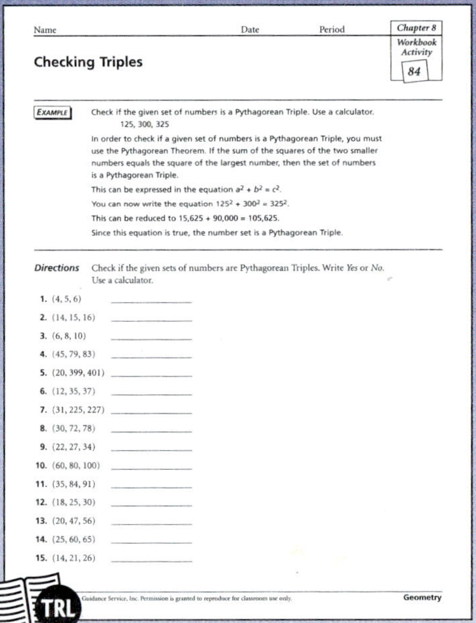

Name Date Period | Chapter 8 Workbook Activity 84

Checking Triples

EXAMPLE Check if the given set of numbers is a Pythagorean Triple. Use a calculator.
125, 300, 325

In order to check if a given set of numbers is a Pythagorean Triple, you must use the Pythagorean Theorem. If the sum of the squares of the two smaller numbers equals the square of the largest number, then the set of numbers is a Pythagorean Triple.

This can be expressed in the equation $a^2 + b^2 = c^2$.

You can now write the equation $125^2 + 300^2 = 325^2$.

This can be reduced to 15,625 + 90,000 = 105,625.

Since this equation is true, the number set is a Pythagorean Triple.

Directions Check if the given sets of numbers are Pythagorean Triples. Write *Yes* or *No*. Use a calculator.

1. (4, 5, 6) ________
2. (14, 15, 16) ________
3. (6, 8, 10) ________
4. (45, 79, 83) ________
5. (20, 399, 401) ________
6. (12, 35, 37) ________
7. (31, 225, 227) ________
8. (30, 72, 78) ________
9. (22, 27, 34) ________
10. (60, 80, 100) ________
11. (35, 84, 91) ________
12. (18, 25, 30) ________
13. (20, 47, 56) ________
14. (25, 60, 65) ________
15. (14, 21, 26) ________

 Geometry

Workbook Activity 84

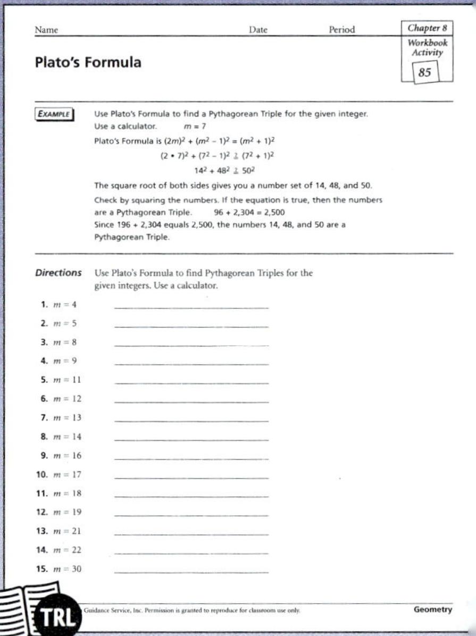

Name Date Period | Chapter 8 Workbook Activity 85

Plato's Formula

EXAMPLE Use Plato's Formula to find a Pythagorean Triple for the given integer.
Use a calculator. $m = 7$

Plato's Formula is $(2m)^2 + (m^2 - 1)^2 = (m^2 + 1)^2$

$(2 \bullet 7)^2 + (7^2 - 1)^2 \stackrel{?}{=} (7^2 + 1)^2$

$14^2 + 48^2 \stackrel{?}{=} 50^2$

The square root of both sides gives you a number set of 14, 48, and 50.

Check by squaring the numbers. If the equation is true, then the numbers are a Pythagorean Triple. 96 + 2,304 = 2,500

Since 196 + 2,304 equals 2,500, the numbers 14, 48, and 50 are a Pythagorean Triple.

Directions Use Plato's Formula to find Pythagorean Triples for the given integers. Use a calculator.

1. $m = 4$ ________
2. $m = 5$ ________
3. $m = 8$ ________
4. $m = 9$ ________
5. $m = 11$ ________
6. $m = 12$ ________
7. $m = 13$ ________
8. $m = 14$ ________
9. $m = 16$ ________
10. $m = 17$ ________
11. $m = 18$ ________
12. $m = 19$ ________
13. $m = 21$ ________
14. $m = 22$ ________
15. $m = 30$ ________

 Geometry

Workbook Activity 85

EXAMPLES *(continued)*

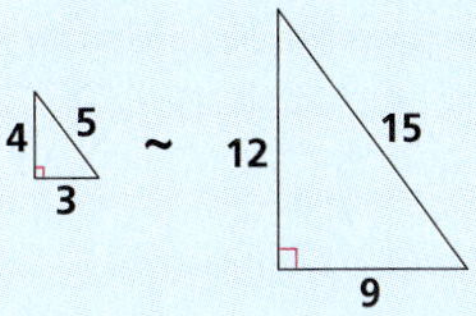

$3^2 + 4^2 = 5^2$ (3, 4, 5)

$(3 \cdot 3)^2 + (4 \cdot 3)^2 = (5 \cdot 3)^2$ ↓

$9^2 + 12^2 = 15^2$ (9, 12, 15)

$81 + 144 = 225$

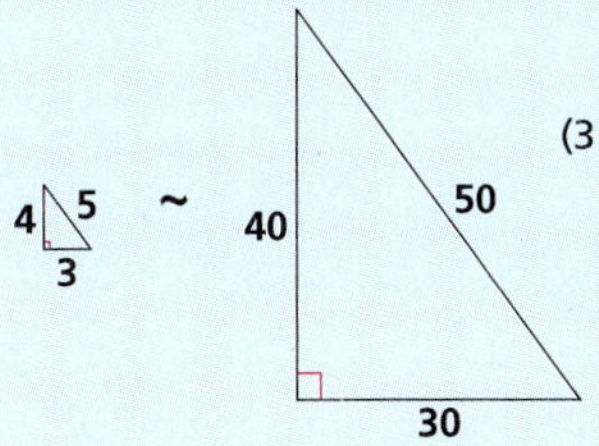

$3^2 + 4^2 = 5^2$ (3, 4, 5)

$(3 \cdot 10)^2 + (4 \cdot 10)^2 = (5 \cdot 10)^2$ ↓

$30^2 + 40^2 = 50^2$ (30, 40, 50)

$900 + 1{,}600 = 2{,}500$

The integers *a, b, c* form a **Pythagorean Triple** if $a^2 + b^2 = c^2$.

Algebra Connection

Another Pythagorean Triple is 5, 12, 13:

$$5^2 + 12^2 = 13^2$$
$$25 + 144 = 169$$

Again, you can create more triples by multiplying 5, 12, and 13 by the same factor. (Use a calculator.)

EXAMPLES

$5^2 + 12^2 = 13^2$ (5, 12, 13)

$(5 \cdot 2)^2 + (12 \cdot 2)^2 = (13 \cdot 2)^2$ ↓

$10^2 + 24^2 + = 26^2$ (10, 24, 26)

$100 + 576 = 676$

$5^2 + 12^2 = 13^2$ (5, 12, 13)

$(5 \cdot 3)^2 + (12 \cdot 3)^2 = (13 \cdot 3)^2$ ↓

$15^2 + 36^2 + = 39^2$ (15, 36, 39)

$225 + 1{,}296 = 1{,}521$

2. Teaching the Lesson

Duplicate the information from the first example on the board or overhead projector. Ask students to explain the second step in generating another triple:

$$(3 \cdot 2)^2 + (4 \cdot 2)^2 = (5 \cdot 2)^2$$

Have them explain why they can be certain the result is an equal relationship. **(Equals multiplied by equals are equal.)**

ALGEBRA CONNECTION

Ask students to generate at least four additional sets of triples using 3, 4, and 5 as a base. Suggest they use multiples such as 4 **(12, 16, 20)**, 5 **(15, 20, 25)**, 25 **(75, 100, 125)**, and 100 **(300, 400, 500)**. Record their responses. Ask students how they could confirm that each set is a triple. Then ask a volunteer to confirm that the sets are indeed triples.

Ask students to estimate how many sets of triples can be made from the triple 3, 4, 5. **(infinite)** Have students give an equation for finding triples based on 3, 4, 5. **(3*n*, 4*n*, 5*n*)**

Name ______ Date ______ Period ______

Chapter 8 *Additional Practice* *47*

Pythagorean Triples

Directions Find additional triples using the given triples and multiplier. Use a calculator.

1. triple (3, 4, 5) by 8 ______
2. triple (3, 4, 5) by 9 ______
3. triple (3, 4, 5) by 10 ______
4. triple (3, 4, 5) by 11 ______
5. triple (3, 4, 5) by 12 ______
6. triple (3, 4, 5) by 13 ______
7. triple (3, 4, 5) by 14 ______
8. triple (3, 4, 5) by 15 ______
9. triple (3, 4, 5) by 16 ______
10. triple (3, 4, 5) by 25 ______
11. triple (5, 12, 13) by 8 ______
12. triple (5, 12, 13) by 9 ______
13. triple (5, 12, 13) by 10 ______
14. triple (5, 12, 13) by 11 ______
15. triple (5, 12, 13) by 12 ______
16. triple (5, 12, 13) by 13 ______
17. triple (5, 12, 13) by 14 ______
18. triple (5, 12, 13) by 15 ______
19. triple (5, 12, 13) by 16 ______
20. triple (5, 12, 13) by 25 ______

RL

 Geometry

Additional Practice 47

3. Reinforce and Extend

Learning Styles

Logical/Mathematical Invite a volunteer to use Plato's Formula to generate a triple using $m = 1$. Have students check the results using $a^2 + b^2 = c^2$. Ask them whether the triple can represent a right triangle. (No, because one side length is zero.)

Learning Styles

Interpersonal/Group Learning Invite students to work in groups of four. Encourage them to write an explanation of why the following sets of numbers are (or are not) Pythagorean Triples: 32, 225, 287; 50, 576, 676. Have students explain their findings using Plato's Formula for triples. (The numbers are not triples, the first number is $2(m^2)$ rather than $(2m)^2$.)

Try This...

Students should recognize that they can produce similar right triangles by multiplying each side of a 3, 4, 5 triangle by the same number.

Following this model, you can show that 10, 24, and 26 and 15, 36, and 39 are also Pythagorean Triples.

Of course, not all triples of positive integers are Pythagorean Triples. To determine if three numbers are a triple, add the squares of the two smaller numbers and see if the sum equals the square of the largest number.

Example Is (4, 5, 6) a Pythagorean Triple?

Check: 4, 5 are the smaller, 6 is the largest

$4^2 + 5^2 \stackrel{?}{=} 6^2$

$16 + 25 \stackrel{?}{=} 36$ False

$\therefore$ (4, 5, 6) is not a Pythagorean Triple.

There are algebraic formulas for creating some, but not all, Pythagorean Triples. Here is a formula credited to the Greek philosopher Plato:

$$(2m)^2 + (m^2 - 1)^2 = (m^2 + 1)^2 \quad \text{for } m\text{, any positive integer}$$

Example Use Plato's Formula for $m = 2$. Check if the results are a Pythagorean Triple.

for $m = 2$: $2m = 4$, $m^2 - 1 = 3$ and $m^2 + 1 = 5$

$(2m)^2 + (m^2 - 1)^2 \stackrel{?}{=} (m^2 + 1)^2$

$4^2 + 3^2 \stackrel{?}{=} 5^2$

$16 + 9 = 25$

$\therefore$ Plato's Formula shows that (3, 4, 5) is a Pythagorean Triple.

Try This...

Suggest a formula to create triples that would produce right triangles similar to ΔABC shown at the right with leg 3, leg 4, and hypotenuse 5.

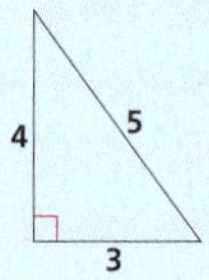

252 Chapter 8 The Pythagorean Theorem

Calculator Practice

You can use x^2 on your calculator to find the sum of the squares of two numbers.

EXAMPLE **Find:** $3^2 + 4^2$

Press 3 x^2 + 4 x^2 = 25.

Exercise A Find the sum of the squares of the following numbers.

1. 5 and 12 169
2. 9 and 12 225
3. 12 and 16 400
4. 6 and 7 85
5. 6 and 8 100

Selected Answers and Solutions begin on page 420.

Exercise B Find additional triples using the given triples and multiplier. Use a calculator.

6. triple (3, 4, 5) by 4 12, 16, 20
7. triple (3, 4, 5) by 5 15, 20, 25
8. triple (3, 4, 5) by 6 18, 24, 30
9. triple (3, 4, 5) by 7 21, 28, 35
10. triple (3, 4, 5) by 20 60, 80, 100
11. triple (5, 12, 13) by 4 20, 48, 52
12. triple (5, 12, 13) by 5 25, 60, 65
13. triple (5, 12, 13) by 6 30, 72, 78
14. triple (5, 12, 13) by 7 35, 84, 91
15. triple (5, 12, 13) by 20 100, 240, 260

Exercise C Check if the given sets of numbers are Pythagorean Triples. Use a calculator.

16. (3, 6, 9) no
17. (65, 72, 97) yes
18. (119, 120, 169) yes
19. (10, 20, 30) no
20. (56, 90, 106) yes

Exercise D Use Plato's Formula to find Pythagorean Triples for the given integers. Use a calculator.

21. $m = 3$ 6, 8, 10
22. $m = 6$ 12, 35, 37
23. $m = 10$ 20, 99,101
24. $m = 15$ 30, 224, 226
25. $m = 20$ 40, 399, 401

CALCULATOR

Encourage students to work in pairs to complete Exercise A. Once they have found the sum of the squares of the numbers, direct them to find the square root of the sum. On many calculators the square root is found using a "shift" key and a square root key. Model the process for students by finding the square root of 25:

Press 25 2nd √. The display reads 5. For many calculators it is not necessary to press =.

Have students find the following square roots: $\sqrt{169}$ (13), $\sqrt{225}$ (15), $\sqrt{400}$ (20), $\sqrt{85}$ (98.2), $\sqrt{100}$ (10).

LEARNING STYLES

LEP/ESL Pair students proficient in English with less proficient students. Have the pair work together to complete Exercise C. For each problem, one student should explain the steps needed to show that the sets are triples. The second student should execute the steps. Students can exchange roles after each problem.

The Pythagorean Theorem *Chapter 8* **253**

Lesson at a Glance

Chapter 8 Lesson 2

Overview This lesson relates the number pattern of the Pythagorean Triple to the Pythagorean Theorem.

Objective

- To use a diagram developed by Bhāskara and Pythagorean Triples to prove the Pythagorean Theorem.

Student Pages 254–256

Teacher's Resource Library TRL

Workbook Activity 86

Additional Practice 48

Mathematics Vocabulary

area

1. Warm-Up Activity

Invite students to share what they may already know about the Pythagorean Theorem. Often students know that $a^2 + b^2 = c^2$ is some sort of formula but may not recognize the name. Point out that in this lesson, and in the following lessons, students will learn several ways to prove the theorem.

Lesson 2 Pythagorean Triples and a Proof

> ***Area***
> *The number of square units inside a closed region*

We can use a Pythagorean Triple to illustrate one of the more famous theorems in history. First, let's look at a specific case.

We can form a square that is 5 units on a side by placing four 3, 4, 5 right triangles in the arrangement shown at the right. Notice that the arrangement leaves a 1×1 "window" at the center of a square with 5 units on each side.

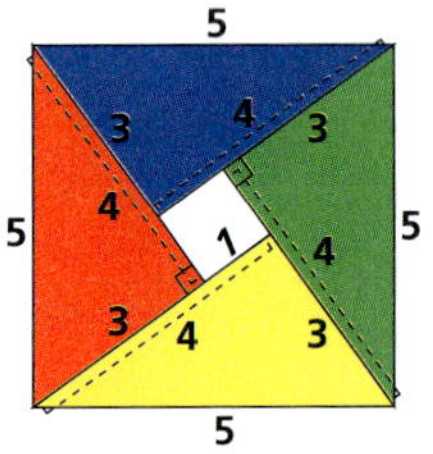

> Remember, the area of a triangle is $\frac{1}{2}$(base • height).

We know that the **area** of the larger square is $5 \times 5 = 25$. But we can think of the area in another way. The areas of the four right triangles plus the area of the center square equal the area of the larger square. Let's add the areas inside the larger square.

$1 \times 1 = 1$	area of center square
$\frac{1}{2}(3 \cdot 4) = 6$	area of red Δ
$\frac{1}{2}(3 \cdot 4) = 6$	area of blue Δ
$\frac{1}{2}(3 \cdot 4) = 6$	area of green Δ
$\frac{1}{2}(3 \cdot 4) = 6$	area of yellow Δ

Total area = 25 square units = area of large square

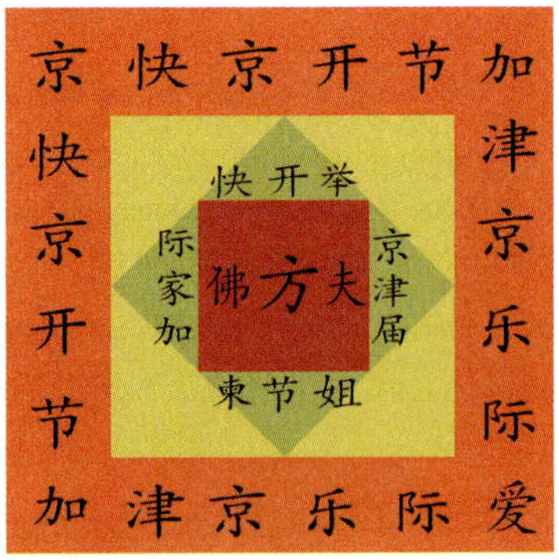

This illustration of a Chinese block print shows a series of triangles within a larger square. How many squares and triangle sets can you find?

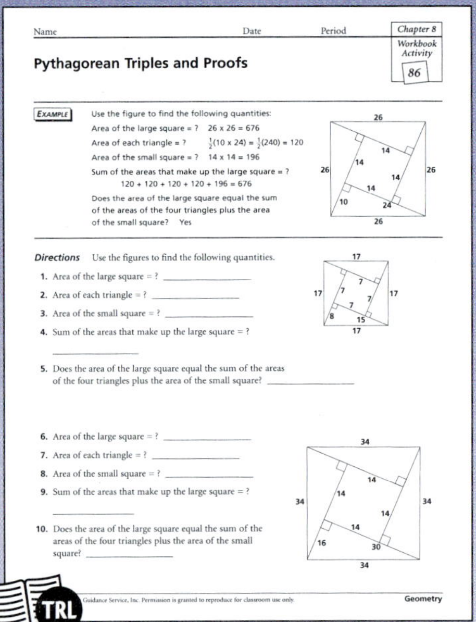

Name Date Period | Chapter 8 Workbook Activity 86

Pythagorean Triples and Proofs

EXAMPLE Use the figure to find the following quantities:

Area of the large square = ? 26 x 26 = 676

Area of each triangle = ? $\frac{1}{2}(10 \times 24) = \frac{1}{2}(240) = 120$

Area of the small square = ? 14 x 14 = 196

Sum of the areas that make up the large square = ?

120 + 120 + 120 + 120 + 196 = 676

Does the area of the large square equal the sum of the areas of the four triangles plus the area of the small square? Yes

Directions Use the figures to find the following quantities.

1. Area of the large square = ? ______
2. Area of each triangle = ? ______
3. Area of the small square = ? ______
4. Sum of the areas that make up the large square = ? ______
5. Does the area of the large square equal the sum of the areas of the four triangles plus the area of the small square? ______
6. Area of the large square = ? ______
7. Area of each triangle = ? ______
8. Area of the small square = ? ______
9. Sum of the areas that make up the large square = ? ______
10. Does the area of the large square equal the sum of the areas of the four triangles plus the area of the small square? ______

 Geometry

TRL

Workbook Activity 86

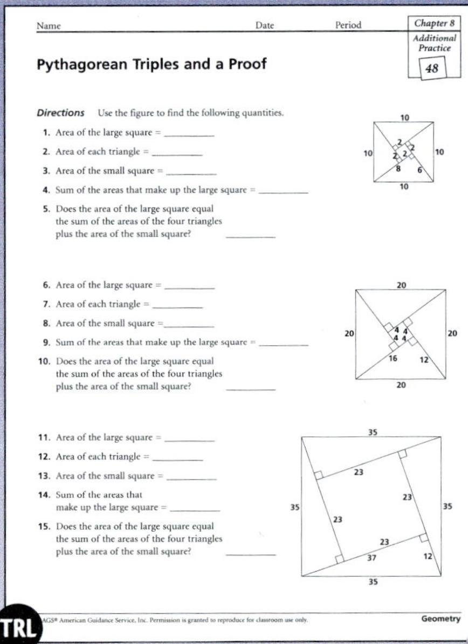

Name Date Period | Chapter 8 Additional Practice 48

Pythagorean Triples and a Proof

Directions Use the figure to find the following quantities.

1. Area of the large square = ______
2. Area of each triangle = ______
3. Area of the small square = ______
4. Sum of the areas that make up the large square = ______
5. Does the area of the large square equal the sum of the areas of the four triangles plus the area of the small square? ______
6. Area of the large square = ______
7. Area of each triangle = ______
8. Area of the small square = ______
9. Sum of the areas that make up the large square = ______
10. Does the area of the large square equal the sum of the areas of the four triangles plus the area of the small square? ______
11. Area of the large square = ______
12. Area of each triangle = ______
13. Area of the small square = ______
14. Sum of the areas that make up the large square = ______
15. Does the area of the large square equal the sum of the areas of the four triangles plus the area of the small square? ______

 Geometry

TRL

Additional Practice 48

Now suppose the right triangles have legs a and b and a hypotenuse, c. We can use this arrangement to prove the theorem $a^2 + b^2 = c^2$.

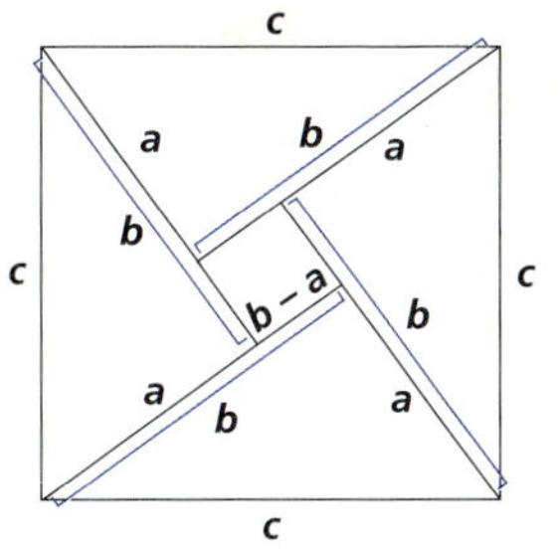

PYTHAGOREAN THEOREM: If a triangle is a right triangle with legs a and b and hypotenuse c, then $a^2 + b^2 = c^2$.

Given: area of large square = c^2
area of inner square = $(b - a)^2$
area of each right triangle = $\frac{1}{2}ab$

To Prove: $a^2 + b^2 = c^2$

PROOF

Statement	Reason
1. Area of 4 right Δs + area of inner square = area of larger square	**1.** By construction.
2. $4(\frac{1}{2}ab) + (b - a)^2 = c^2$	**2.** Whole is equal to sum of its parts.
3. $2ab + b^2 - 2ab + a^2 = c^2$	**3.** Simplification.
4. $2ab - 2ab + a^2 + b^2 = c^2$	**4.** Rearrange terms.
5. $\therefore a^2 + b^2 = c^2$	**5.** Simplification.

This equality is known as the Pythagorean Theorem. The method of this proof is based on Bhāskara, a leading Indian mathematician who lived in the 12th century. Note that in the diagram, the right triangles can be shown to be congruent.

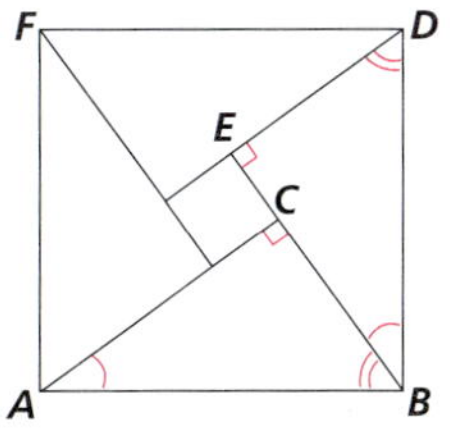

TECHNOLOGY CONNECTION

The ancient Egyptians developed a simple technology in which they used the Pythagorean Theorem to survey land. They positioned 12 equally spaced knots in a rope. They then bent the rope to form a triangle with side lengths 3, 4, and 5, forming a right angle. The justification for this tool depends on the converse of the Pythagorean Theorem, which you will prove at the end of this chapter.

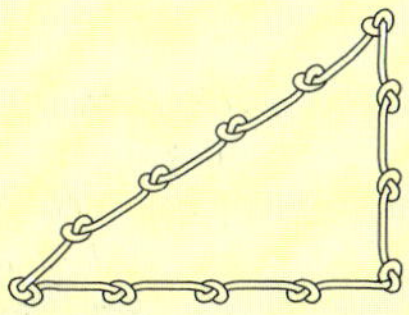

2. Teaching the Lesson

Invite a volunteer to give the formula for finding the area of a triangle. ($\frac{1}{2}(b \bullet h)$) Draw a single 3, 4, 5 right triangle on the board or overhead projector. Ask students to determine the area of the triangle. ($\frac{1}{2}(b \bullet h) = \frac{1}{2}(3 \bullet 4) = 6$)

Use a set of plastic 3, 4, 5 triangles (such as those included in the Proofs of Pythagoras) to model the argument on page 254. First, place the central square on the overhead projector and indicate that its area = 1. Then add the triangles, one at a time, to the square gradually building a figure that matches the one at the top of page 255. Keep a running total of the area as each triangle is added to the figure.

Use the plastic triangles to model each step of the proof on page 255. Invite a volunteer to provide a narrative for each step as you use the plastic triangles to model each step of the proof.

LEARNING STYLES

Body/Kinesthetic Prepare a knotted rope similar to the one described in the *Technology Connection.* Invite groups of four students to use the rope to form a right triangle. (Students can follow the seams on a tile floor to ensure they have formed a right angle.) Ask students to try to form a right triangle using only 11 equally spaced knots. Have them explain what happens.

3. Reinforce and Extend

Learning Styles

Visual/Spatial Have pairs of students prepare index cards with the statements and reasons for the proof on page 256. Each statement or reason, without its identifying number, should be on a separate card. Direct students to exchange cards with another student pair. Then have students arrange cards in the correct Statement and Reason order.

Learning Styles

LEP/ESL Direct pairs of students to work together to complete Exercise A. Students can take turns explaining to one another how to find the quantities for each item. One student can explain the process; the second student can execute each step.

Given: square *ABDF* (on page 255)
To Prove: $\triangle ABC \cong \triangle BDE$

Proof

Statement	Reason
1. *ABDF* is a square.	1. Given.
2. $\overline{AB} \cong \overline{BD}$	2. Definition of square; sides of a square are equal.
3. $m\angle ABC + m\angle DBE = m\angle ABD = 90^\circ$	3. Given; definition of square.
4. $m\angle BAC + m\angle ABC = 90^\circ$	4. Acute angles of a right triangle are complementary.
5. $m\angle ABC + m\angle DBE = m\angle ABC + m\angle BAC$	5. Quantities equal to the same quality are equal to each other.
6. $m\angle DBE = m\angle BAC$	6. Subtraction.
7. $\therefore m\angle BDE \cong m\angle ABC$	7. Complements of equal angles are equal.
8. $\therefore \triangle ABC \cong \triangle BDE$	8. Angle-Side-Angle.

Exercise A Use the figure to find the following quantities.

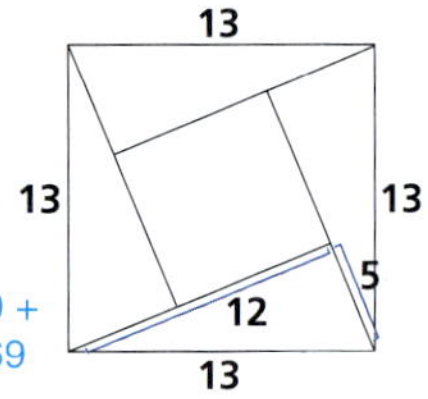

1. Area of large square $13 \cdot 13 = 169$ sq units
2. Area of each right triangle $\frac{1}{2}bh = \frac{1}{2}(5)(12) = \frac{1}{2}(60) = 30$
3. Side length and area of the small square 7, 49
4. Sum of the areas that make up the large square $30 + 30 + 30 + 30 + 49 = 169$
5. Does the area of the large square equal the sum of the areas of the four triangles plus the area of the small square? yes

Exercise B Use the figure to answer the questions.

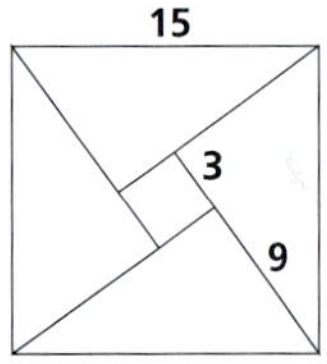

6. What are the side lengths of the right triangles? 9, 12, 15
7. What is the area of each right triangle? $\frac{1}{2}bh = \frac{1}{2}(9 \cdot 12) = \frac{1}{2}(108) = 54$
8. What is the area of the center square? 9 square units
9. How can you use the area of the large square to find the area of one of the right triangles? Find the area. Subtract the area of the small square from the area of the large square and then divide by 4; $(225 - 9) \div 4 = 216 \div 4 = 54$ sq units
10. Describe another way to find the areas of the four triangles. $4(\frac{1}{2}bh) = 4 \cdot 54 = 216$ sq ur

Lesson 3 Pythagorean Demonstration

Pythagoras lived in ancient Greece about 540 B.C. He developed a "demonstration," or proof, of what we now call the Pythagorean Theorem. The following is believed to be the demonstration he developed.

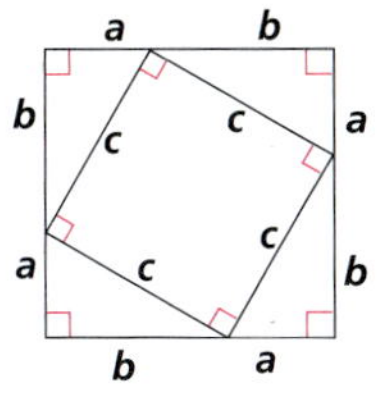

$\cong$

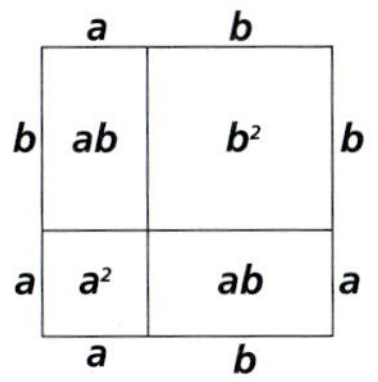

Area of square	=	Area of square	
$c^2 + \frac{1}{2}ab + \frac{1}{2}ab + \frac{1}{2}ab + \frac{1}{2}ab$	=	$a^2 + b^2 + ab + ab$	
$c^2 + 2ab$	=	$a^2 + b^2 + 2ab$	simplify
c^2	=	$a^2 + b^2$	subtract

Subtract four right triangles

Subtract two rectangles

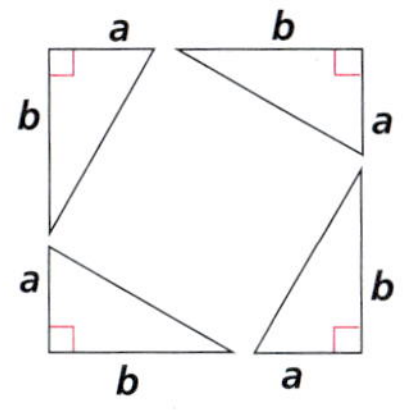

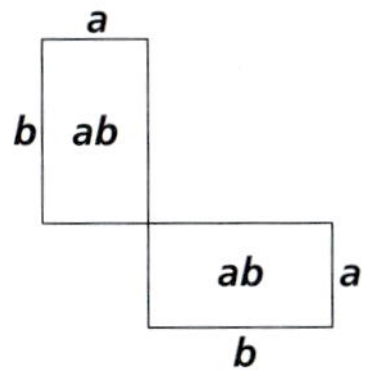

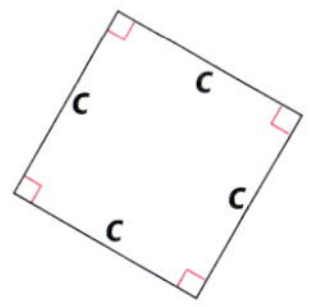

$\cong$

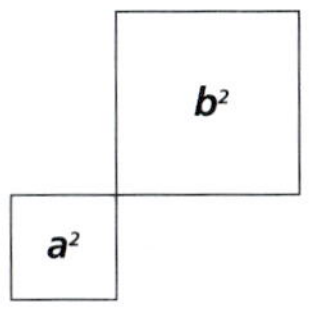

For the above demonstration to be valid, the center portion of the original diagram must be a square. Here is that proof.

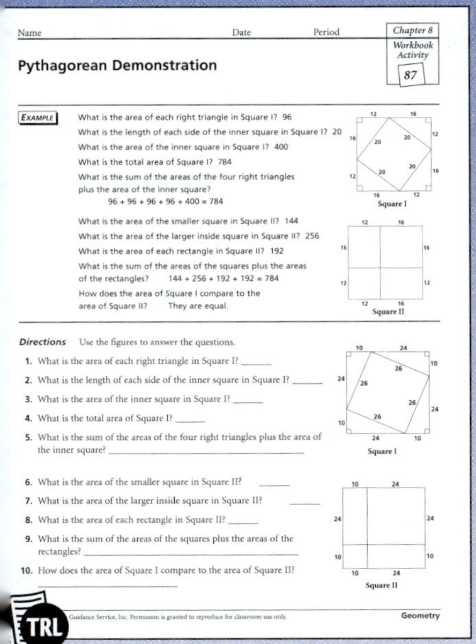

Name Date Period

Chapter 8 Workbook Activity 87

Pythagorean Demonstration

EXAMPLE

What is the area of each right triangle in Square I? 96
What is the length of each side of the inner square in Square I? 20
What is the area of the inner square in Square I? 400
What is the total area of Square I? 784
What is the sum of the areas of the four right triangles plus the area of the inner square?
96 + 96 + 96 + 96 + 400 = 784

What is the area of the smaller square in Square II? 144
What is the area of the larger inside square in Square II? 256
What is the area of each rectangle in Square II? 192
What is the sum of the areas of the squares plus the areas of the rectangles? 144 + 256 + 192 + 192 = 784
How does the area of Square I compare to the area of Square II? They are equal.

Directions Use the figures to answer the questions.

1. What is the area of each right triangle in Square I? ______
2. What is the length of each side of the inner square in Square I? ______
3. What is the area of the inner square in Square I? ______
4. What is the total area of Square I? ______
5. What is the sum of the areas of the four right triangles plus the area of the inner square? ______
6. What is the area of the smaller square in Square II? ______
7. What is the area of the larger inside square in Square II? ______
8. What is the area of each rectangle in Square II? ______
9. What is the sum of the areas of the squares plus the areas of the rectangles? ______
10. How does the area of Square I compare to the area of Square II? ______

Square I

Square II

Geometry

TRL **Workbook Activity 87**

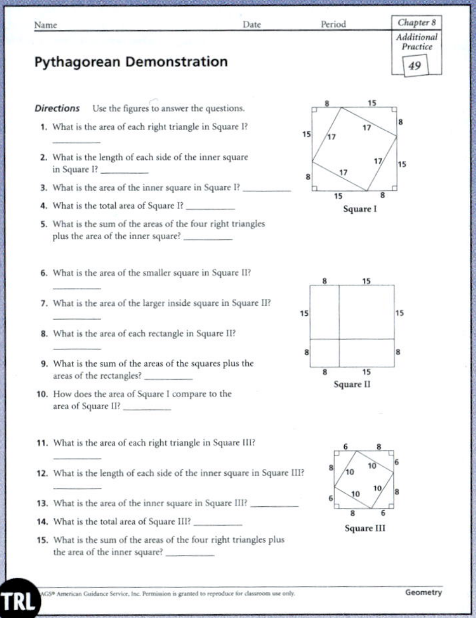

Name Date Period

Chapter 8 Additional Practice 49

Pythagorean Demonstration

Directions Use the figures to answer the questions.

1. What is the area of each right triangle in Square I? ______
2. What is the length of each side of the inner square in Square I? ______
3. What is the area of the inner square in Square I? ______
4. What is the total area of Square I? ______
5. What is the sum of the areas of the four right triangles plus the area of the inner square? ______
6. What is the area of the smaller square in Square II? ______
7. What is the area of the larger inside square in Square II? ______
8. What is the area of each rectangle in Square II? ______
9. What is the sum of the areas of the squares plus the areas of the rectangles? ______
10. How does the area of Square I compare to the area of Square II? ______
11. What is the area of each right triangle in Square III? ______
12. What is the length of each side of the inner square in Square III? ______
13. What is the area of the inner square in Square III? ______
14. What is the total area of Square III? ______
15. What is the sum of the areas of the four right triangles plus the area of the inner square? ______

Square I

Square II

Square III

Geometry

TRL **Additional Practice 49**

Lesson at a Glance

Chapter 8 Lesson 3

Overview This lesson presents another demonstration or proof of the Pythagorean Theorem.

Objectives

- To emulate Pythagoras' demonstration of the Pythagorean Theorem.
- To use the Pythagorean Theorem to find the length of the unknown side of a right triangle.

Student Pages 257–261

Teacher's Resource Library TRL

Workbook Activity 87

Additional Practice 49

1. Warm-Up Activity

Tell students that there are some theorems in mathematics, algebra, and geometry that can be proved in a number of ways. Point out to students that there are several ways to think through, as well as to formally prove, the Pythagorean Theorem. Indicate to students that in this lesson they will learn a second way to prove $a^2 + b^2 = c^2$.

2. Teaching the Lesson

Work through the demonstration on page 257 with students. Point out that the demonstration begins with two squares of equal side length. The square on the left is divided into a square and four triangles; the second square—same size as the first—is divided into two squares and two rectangles. Complete the subtractions as shown on page 257.

If feasible, emulate the demonstration using manipulatives from among the Proofs of Pythagoras materials.

Given: *ACEG* a square with equal segments as marked

To Prove: *BDFH* is a square

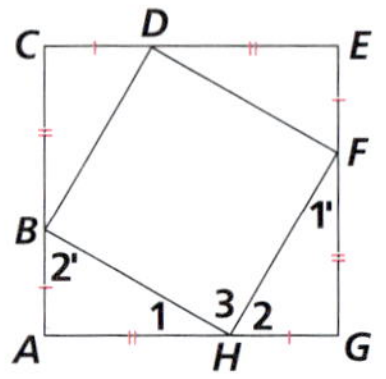

PROOF

Statement	Reason
1. *ACEG* is a square	1. Given.
2. ∠*A*, ∠*C*, ∠*E*, and ∠*G* are right angles	2. Definition of square.
3. m∠1 + m∠2 + m∠3 = 180°	3. Adjacent angles with exterior sides in a straight line are supplementary.
4. right △*ABH* ≅ right △*CDB* ≅ △*EFD* ≅ △*GHF*	4. SAS, sides are given as equal.
5. m∠2 = m∠2' and $\overline{HB} = \overline{BD} = \overline{DF} = \overline{FH}$	5. Corresponding parts of congruent triangles are equal.
6. m∠1 + m∠2' = 90°	6. Acute angles of right triangles are complementary.
7. m∠1 + m∠2 = 90°	7. Substitution using Step 5.
8. 90° + m∠3 = 180°	8. Substitution using Steps 3 and 7.
9. ∴ m∠3 = 90°	9. Subtraction.
10. Argument holds for angles at *D*, *F*, and *H*	10. Same argument applies.
11. ∴ *BDFH* is a square	11. Definition of a square is satisfied.

We can use specific dimensions to illustrate the proof.

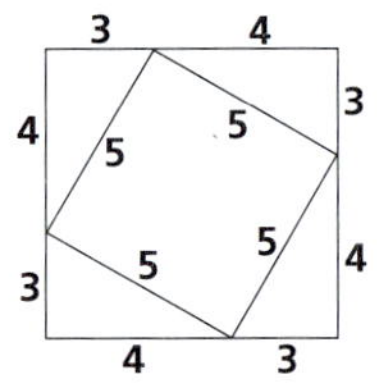

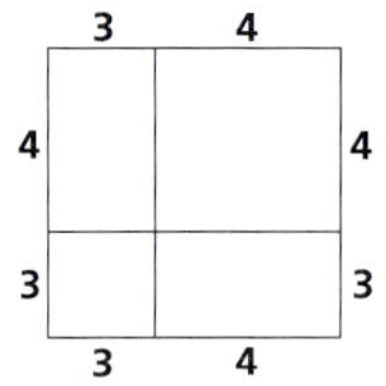

$$25 + 4[\tfrac{1}{2}(3 \cdot 4)] \stackrel{?}{=} 9 + 16 + 12 + 12$$

$$49 = 49 \quad \text{True}$$

You can use the $a^2 + b^2 = c^2$ equation to check if the lengths you are given satisfy the formula. This is similar to checking for Pythagorean Triples—except in this case, a, b, and c can be any real numbers, not just integers. You can also use the $a^2 + b^2 = c^2$ equation if you are given two side lengths of a right triangle and you wish to find the length of the third side. Use a calculator.

EXAMPLES

Do 3.1, 7.2, and 8.4 satisfy the formula $a^2 + b^2 = c^2$? Round to the nearest tenth.

Let $a = 3.1$, $b = 7.2$, and $c = 8.4$

$a^2 + b^2 = c^2$

$(3.1)^2 + (7.2)^2 \stackrel{?}{=} (8.4)^2$

$9.6 + 51.8 \stackrel{?}{=} 70.6$ False

$\therefore$ The given sides do not satisfy the formula.

The legs of a right triangle are 8.1 and 6.7.

What is the length, to the nearest tenth, of the hypotenuse?

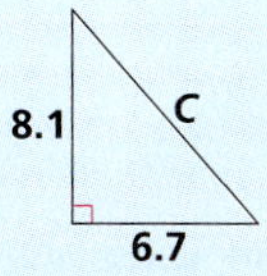

Let $a = 8.1$, $b = 6.7$

$a^2 + b^2 = c^2$

$(8.1)^2 + (6.7)^2 = c^2$

$65.6 + 44.9 = c^2$

$110.5 = c^2$ simplify

$10.5 = c$ take square root of both sides

3. Reinforce and Extend

COMMON ERROR

Remind students to write their equations and their solutions neatly. Often as students work through a problem, they may forget to write the exponents associated with a variable. One common mistake is to miswrite the equation, writing $a^2 + b^2 = c$. Students feel they have completed the problem without finding the square root of c because they have not kept the "2" exponent with the equation.

Group Problem Solving

Provide a group of four students with the Pythagorean Theorem Model in the Proofs of Pythagoras manipulatives. Direct students to arrange the manipulatives so that their edges form a triangle:

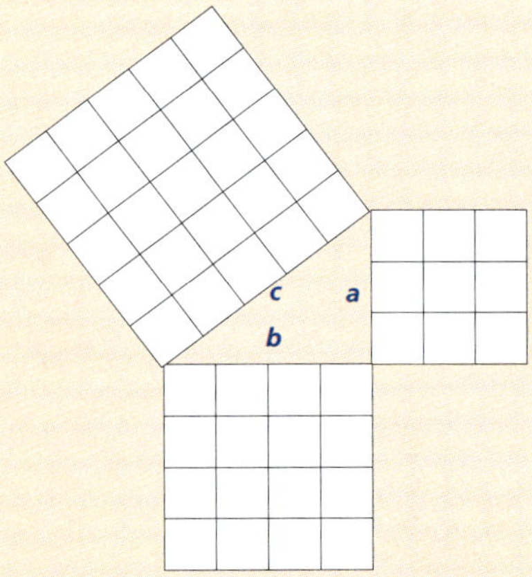

Have students use this configuration to illustrate the Pythagorean Theorem. See Manipulatives Master 3 for a copy of the model. Students should work together to produce a step-by-step proof, complete with illustrations. They can first use the illustration to prove that $9 + 16 = 25$ and then prove the general statement, $a^2 + b^2 = c^2$. Students can present their proofs as part of a "Do-It-Yourself" mathematics show in which audience members are invited to try their hand at producing proofs at home.

Calculator

Offer students additional practice in finding square roots by having them first estimate the square root of each of the following numbers to the nearest whole number and then find the actual square root using a calculator.

230 (15, 15.1), 312 (18, 17.7), 63 (8, 7.9), 112 (10, 10.6), 220 (15, 14.8), 320 (18, 17.9), 820 (28, 28.6)

EXAMPLES **Given:** one leg = 31, hypotenuse = 70
Find: the length of the second leg

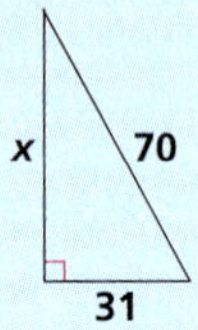

Let x = length of the second leg

$x^2 + 31^2 = 70^2$

$x^2 + 961 = 4{,}900$

$x^2 = 3{,}939$ simplify

$x = 62.8$ take square root of both sides, round to the nearest tenth

Given: isosceles right Δ with a hypotenuse of 10
Find: the length of the legs

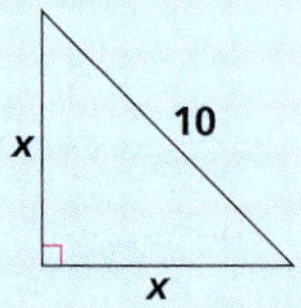

Let x = length of the legs

$x^2 + x^2 = 10^2$

$2x^2 = 100$

$x^2 = 50$ simplify

$x = 7.1$ take square root of both sides, round to the nearest tenth

Calculator Practice

You can use [√] on your calculator to find the square root of a number. On some calculators you need to press [2nd] [x^2] to find the square root of x.

EXAMPLE Find the square root of 225.

Press 225 [√] 15.

OR Press 225 [2nd] [x^2] [=] 15.

Exercise A Find the square root of each number.

1. 625 25

2. 169 13

3. 3,969 63

4. 2,209 47

260 *Chapter 8 The Pythagorean Theorem*

Exercise B Use the figures to answer the questions.

5. What is the area of each right triangle in Square I?
$\frac{1}{2}bh = \frac{1}{2}(12 \cdot 5) = 30$

6. What is the area of the inner square in Square I?
$13^2 = 169$

7. What is the total area of Square I?
$17^2 = 289$

8. What is the sum of the areas of the four right triangles plus the area of the inner square?
$30 + 30 + 30 + 30 + 169 = 289$

9. What is the area of the smaller square in Square II?
25

10. What is the area of the larger inside square?
144

11. What is the area of each rectangle in Square II?
60

12. What is the sum of the areas of the squares plus the areas of the rectangles?
$25 + 144 + 60 + 60 = 289$

13. How does the area of Square I compare to the area of Square II?
the same

Square I

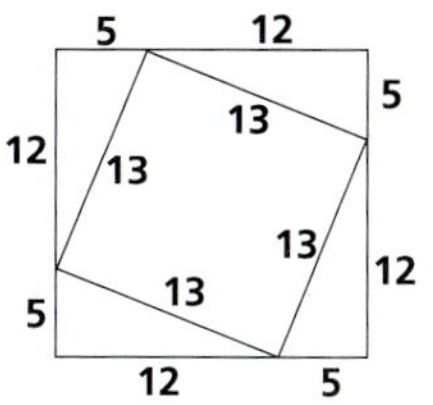

Square II

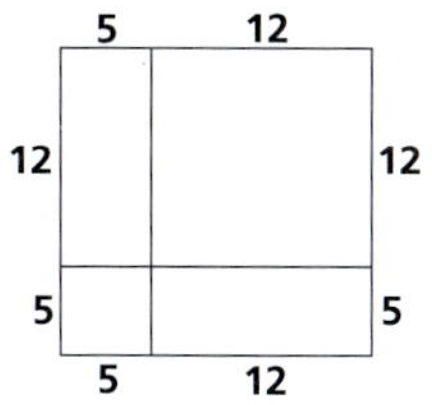

Exercise C Find the length of the unknown side of each right triangle.

14.

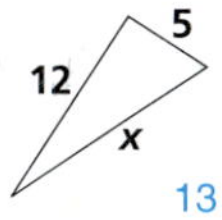

13

15.

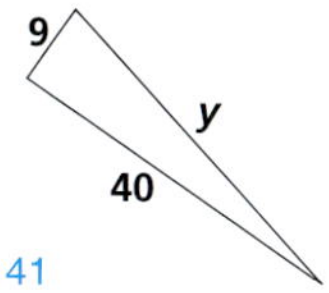

41

16.

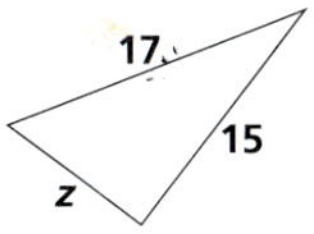

8

17.

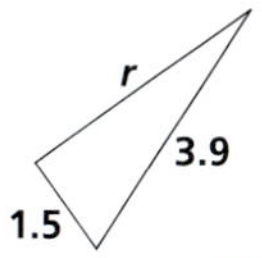

3.6

18.

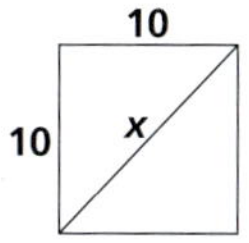

14.1

19.

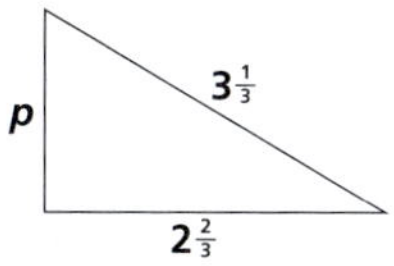

2

20.

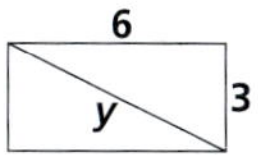

6.7

Learning Styles

LEP/ESL Students will benefit from using complete sentences to answer the questions in Exercise B. Have pairs of students take turns formulating the answer as well as an explanation for each problem in Exercise B. Encourage listeners to identify the subject and verb for each sentence.

Learning Styles

Logical/Mathematical Ask students to explain what must be true about each triangle in Exercise C in order for students to be able to find the length of the unknown side. (Each triangle must be a right triangle as given.)

Application

At Home

Invite students to collect several advertisements for television sets. Suggest they keep a record of the screen sizes for the advertised sets. Point out to students advertisements that specify that the size of the television screen is measured diagonally across the screen. Have students discuss why this measurement is more informative than a measurement that gives only the height or width of a set. Direct students to use the given diagonals and calculators to determine possible height and width measurements for the advertised televisions. Point out that diagonal measurements are also used to describe the sizes of computer screens.

The Pythagorean Theorem *Chapter 8* **261**

Lesson at a Glance

Chapter 8 Lesson 4

Overview This lesson describes how to use similar triangles to prove the Pythagorean Theorem.

Objectives

- To use similar triangles to prove the Pythagorean Theorem.
- To use the Pythagorean Theorem to solve geometric problems.

Student Pages 262–265

Teacher's Resource Library TRL

Workbook Activities 88–89

Additional Practice 50

1. Warm-Up Activity

Review the characteristics of similar triangles with students. Remind them that for similar triangles the ratios of similar parts are equal. For example, if ΔABC is similar to ΔDEF, then the corresponding sides form the same ratio:

$$\frac{EF}{BC} = \frac{FD}{BA} = \frac{DE}{AC}$$

Lesson 4

Pythagorean Theorem and Similar Triangles

One of the interesting facts about the Pythagorean Theorem, as well as other theorems, is that there are a number of ways to prove the theorem true. In this lesson, you will "make the case" using similar triangles.

You already know that the corresponding sides of similar triangles are equal. One way to produce similar triangles is to take any right triangle and draw a perpendicular line to the hypotenuse from the right angle vertex.

Given: ΔABC, a right Δ, $\overline{BD} \perp \overline{AC}$

To Prove: $\Delta ADB \sim \Delta BDC \sim \Delta ABC$

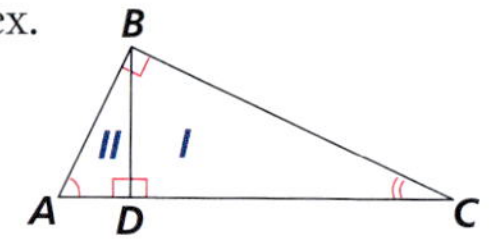

Split the given figure into two separate triangles.

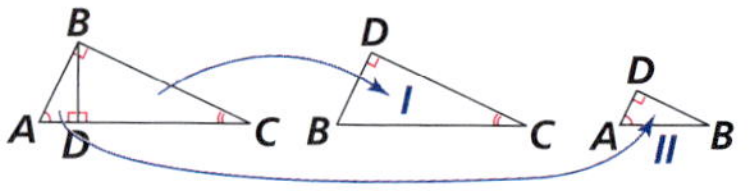

Because the acute angles of a right triangle are complementary, the corresponding angles of all three triangles are equal.

right angle:	$m\angle B$	=	$m\angle BDC$	=	$m\angle BDA$
$m\angle A$:	$m\angle A$	=	$m\angle DBC$	=	$m\angle DAB$
$m\angle C$:	$m\angle C$	=	$m\angle DCB$	=	$m\angle ABD$

Pythagorean Theorem: In any right triangle ABC, with $\overline{AC}$ the hypotenuse, $AC^2 = AB^2 + BC^2$.

Experiencing Geometry

Prove the similarity of the triangles to yourself. Cut two large congruent right angle triangles from construction paper. Label each triangle *ABC*. On one, draw a perpendicular line from the right angle to the hypotenuse. Label the point of intersection *D* to match the illustration above. Now cut along the perpendicular line to make two triangles. Compare all three triangles. Hold the triangles so that the right angles and the hypotenuses match. See which of the remaining angles correspond.

Name ______ Date ______ Period ______ Chapter 8 Workbook Activity 88

Pythagorean Theorem and Similar Triangles

Example ΔABC is similar to both ΔDBA and ΔDAC. ΔDBA is similar to ΔDAC.

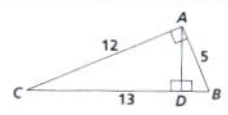

Directions Use the figure to answer the following questions.

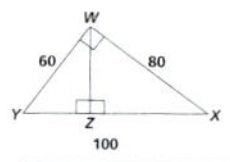

1. What is the length of $\overline{WZ}$? ______
2. What is the length of $\overline{XZ}$? ______
3. What is the length of $\overline{YZ}$? ______
4. What segment makes up the hypotenuse of the largest triangle? ______
5. What segment makes up the hypotenuse of the middle-sized triangle? ______
6. What segment makes up the hypotenuse of the smallest triangle? ______
7. Is ΔWXZ similar to ΔWYZ? Why? ______
8. What is the ratio of similarity between ΔWXY and ΔZXW? ______
9. What is the ratio of similarity between ΔWXY and ΔZWY? ______
10. What is the ratio of similarity between ΔZXW and ΔZWY? ______
11. What is the perimeter of ΔWXY? ______
12. What is the perimeter of ΔZWY? ______
13. What is the perimeter of ΔZXW? ______
14. What is the ratio between the three triangles' perimeters? ______
15. Is the ratio between the three triangles' perimeters and the sides of the largest triangle the same? **Hint:** Divide each number of the perimeter ratio by each corresponding number in the largest triangle's side ratio. The results should be equal.

______

TRL

Geometry

Workbook Activity 88

Name ______ Date ______ Period ______ Chapter 8 Workbook Activity 89

More with Similar Triangles

Example ΔABC is similar to both ΔDAB and ΔDAC. $\Delta ABC \sim \Delta DAB \sim \Delta DAC$

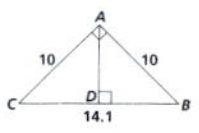

Directions Use the figure to answer the following questions.

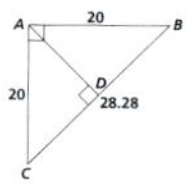

1. What is the length of $\overline{AD}$? ______
2. What is the length of $\overline{BD}$? ______
3. What is the length of $\overline{CD}$? ______
4. What segment makes up the hypotenuse of the largest triangle? ______
5. What segment makes up the hypotenuse of the left smaller triangle? ______
6. What segment makes up the hypotenuse of the right smaller triangle? ______
7. Is ΔDAB similar to ΔDAC? Why? ______
8. What is the ratio of similarity between ΔABC and ΔDAB? ______
9. What is the ratio of similarity between ΔABC and ΔDAC? ______
10. What is the ratio of similarity between ΔDAB and ΔDAC? ______
11. What is the perimeter of ΔABC? ______
12. What is the perimeter of ΔDAB? ______
13. What is the perimeter of ΔDAC? ______
14. What is the ratio between the three triangles' perimeters? ______
15. Is the ratio between the three triangles' perimeters and the sides of the largest triangle the same? **Hint:** Divide each number of the perimeter ratio by each corresponding number in the largest triangle's side ratio. The results should be equal.

______

TRL

Geometry

Workbook Activity 89

Given: right ΔABC, right angle at B, and $\overline{BD} \perp \overline{AC}$

To Prove: $\overline{AC}^2 = \overline{AB}^2 + \overline{BC}^2$

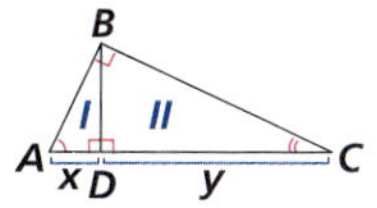

PROOF

Statement	Reason
1. right ΔABC, right angle at B, and $\overline{BD} \perp \overline{AC}$	1. Given.
2. $\Delta ABC \sim \Delta I \sim \Delta II$	2. By above discussion.
3. Let $x = \overline{AD}$ and $y = \overline{CD}$	3. Substitution.
4. $\frac{x}{\overline{AB}} = \frac{\overline{AB}}{\overline{AC}}$	4. $\Delta I \sim \Delta ABC$; $\frac{\text{small leg}}{\text{small leg}} = \frac{\text{hypotenuse}}{\text{hypotenuse}}$
5. $x \bullet \overline{AC} = \overline{AB}^2$	5. Product of means = product of extremes.
6. $x = \frac{\overline{AB}^2}{\overline{AC}}$	6. Division by equals.
7. $\frac{y}{\overline{BC}} = \frac{\overline{BC}}{\overline{AC}}$	7. $\Delta II \sim \Delta ABC$; $\frac{\text{large leg}}{\text{large leg}} = \frac{\text{hypotenuse}}{\text{hypotenuse}}$
8. $y \bullet \overline{AC} = \overline{BC}^2$	8. Product of means = product of extremes.
9. $y = \frac{\overline{BC}^2}{\overline{AC}}$	9. Division by equals.
10. $x + y = \frac{\overline{AB}^2 + \overline{BC}^2}{\overline{AC}}$	10. Addition of equals, and Steps 6 and 9.
11. $x + y = \overline{AD} + \overline{DC} = \overline{AC}$	11. Substitution, Step 3.
12. $\overline{AC} = \frac{\overline{AB}^2 + \overline{BC}^2}{\overline{AC}}$	12. Substitution for $x + y$ from Step 10.
13. $\therefore \overline{AC}^2 = \overline{AB}^2 + \overline{BC}^2$	13. Multiply by equals.

The Pythagorean Theorem is one of the most useful theorems in geometry. The formula in the theorem offers a way to solve many practical problems.

EXAMPLE **Given:** a 40-ft flagpole casts a 60-ft shadow
Find: length AB to the nearest hundredth
(Use a calculator.)

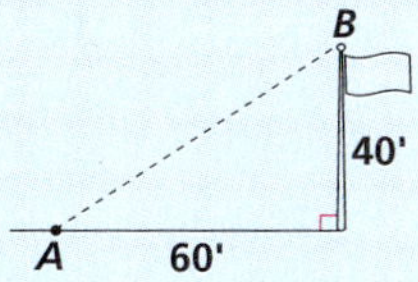

$40^2 + 60^2 = \overline{AB}^2$	substitute in formula
$1{,}600 + 3{,}600 = \overline{AB}^2$	square
$5{,}200 = \overline{AB}^2$	simplify
$72.11 = \overline{AB}$	take square root of each side

Name Date Period

Chapter 8 *Additional Practice* 50

Pythagorean Theorem and Similar Triangles

Directions Use the figure to answer the following questions.

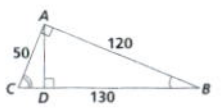

1. What is the length of $\overline{AD}$? ________
2. What is the length of $\overline{BD}$? ________
3. What is the length of $\overline{CD}$? ________
4. Name the three similar triangles. ________
5. Which angles in ΔABC correspond to which angles in ΔDAB? ________
6. Which angles in ΔDAC correspond to which angles in ΔDAB? ________
7. Which angles in ΔDAC correspond to which angles in ΔABC? ________
8. What segment makes up the hypotenuse of the largest triangle? ________
9. What segment makes up the hypotenuse of the middle-sized triangle? ________
10. What segment makes up the hypotenuse of the smallest triangle? ________
11. Is ΔDAB similar to ΔDAC? Why? ________
12. What is the ratio of similarity between ΔABC and ΔDAB? ________
13. What is the ratio of similarity between ΔABC and ΔDAC? ________
14. What is the ratio of similarity between ΔDAB and ΔDAC? ________
15. What is the perimeter of ΔABC? ________
16. What is the perimeter of ΔDAC? ________
17. What is the perimeter of ΔDAB? ________
18. What is the ratio between the three triangles' perimeters? ________
19. What is the ratio between the sides of the largest triangle? ________
20. Is the ratio between the three triangles' perimeters and the sides of the largest triangle the same? **Hint:** Divide each number of the perimeter ratio by each corresponding number in the largest triangle's side ratio. The results should be approximately equal because some of the lengths of the sides were rounded when calculated.

____

 Geometry

Additional Practice 50

2. Teaching the Lesson

Before addressing the proof on page 263, prompt students' memories about solving proportions for an unknown. Write a simple proportion such as $\frac{3}{5} = \frac{x}{15}$ on the board or overhead projector. Ask students what process they can use to find the value for x. Reinforce processes that specify multiplying the means and the extremes and then dividing.

Point out that although the proof on page 263 involves 13 steps, students are already familiar with the tools they will need to complete the proof.

LEARNING STYLES

Auditory/Verbal Suggest students form two competing teams for a geometry bee. Write the first statement of the proof on page 263 on the board or overhead projector. Ask the first student on Team A to supply the reason for the statement. If the student cannot answer, he or she moves to the last place on Team A. Ask the first student on Team B to supply the answer. Continue asking alternate team members until an accurate reason is supplied. Students supplying correct answers leave the team. Proceed, writing statements and asking for reasons until the proof is complete. Declare the team with the greater number of "retired" players the winner.

3. Reinforce and Extend

MANIPULATIVES

M Pythagoras' Proofs

Materials: Proofs of Pythagoras set

Group Practice: Provide or display a set of similar triangles from the Proofs of Pythagoras materials. Challenge students to duplicate the figure from Exercise A—finding two similar triangles that can be placed together to form a third triangle, similar to the other two.

Student Practice: Invite students to complete Exercise A using the plastic triangles as models for each problem.

EXAMPLES

Given: a 100-ft-long ramp is to be constructed from the top of a 20-ft wall to ground level

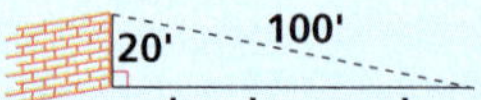

Find: the distance along the ground from the wall to the end of the ramp (Use a calculator. Compute to the nearest hundredth.)

Let x equal distance from wall.

$20^2 + x^2 = 100^2$	substitute in formula
$400 + x^2 = 10{,}000$	square
$x^2 = 9{,}600$	subtract
$x = 97.98$	take square root of each side

Given: $\overline{AB} = 100$ km, $\overline{AC} = 300$ km, $\overline{AC} \perp \overline{AB}$

Find: the shortest distance between B and C (Use a calculator. Compute to the nearest hundredth.)

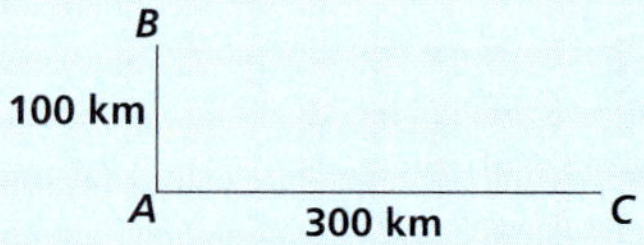

$\overline{BC}$ is the shortest distance, it is also the hypotenuse of right ΔABC.

$\overline{AB}^2 + \overline{AC}^2 = \overline{BC}^2$	theorem
$100^2 + 300^2 = \overline{BC}^2$	substitution
$10{,}000 + 90{,}000 = \overline{BC}^2$	square
$100{,}000 = \overline{BC}^2$	simplify
$316.23 \text{ km} = \overline{BC}$	take square root of each side

Exercise A Answer the following questions. Use the figure.

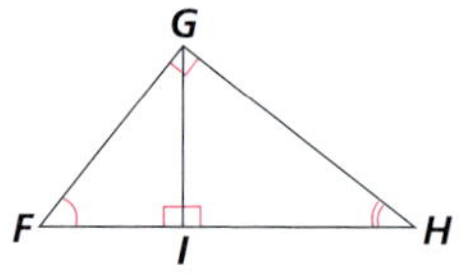

1. Name the three similar triangles. ΔFGH, ΔFIG, ΔGIH
2. Which angles in ΔFIG correspond to which angles in ΔFGH? $\angle F = ?$ $\angle I = ?$ $\angle FGI = ?$ F, G, H
3. Which angles in ΔGIH correspond to which angles in ΔFGH? $\angle I = ?$ $\angle H = ?$ $\angle IGH = ?$ G, H, F
4. What segment makes up the hypotenuse of the largest triangle? The middle-sized triangle? The smallest triangle? $\overline{FH}$, $\overline{HG}$, $\overline{FG}$
5. Is $\Delta FGI \sim \Delta GHI$? Why? yes, AA Similary Postulate

Problem Solving

Exercise B Solve the following problems. Use a calculator. Round answers to the nearest hundredth.

6. The minute hand of a clock is 14 cm long, the hour hand is 9 cm long. What is the distance between the hands when it is 9 A.M.?

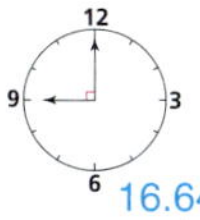

16.64 cm

7. A corner of a park has been set aside for dogs. The park fencing along the shorter leg is 27 ft. The length along the longer leg is 36 ft. How much more fencing will be needed to close in the doggie corner?

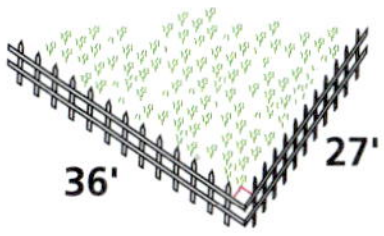

45 ft

8. The base of a 10-foot ladder is placed 2 feet away from a wall. Will the ladder reach a window that is 12 feet above ground? Why or why not?

No, the ladder will reach only 9.8 ft up the wall.

9. Janice uses a shortcut to go home from school. She knows that she walks 100 paces across the playground. She also knows that her house is 30 paces from the corner. What is the distance, in paces, from the corner to the school?

95.39 paces

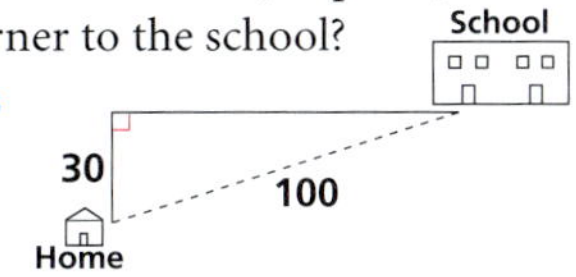

10. A wooden hanger forms a right triangle with legs equal to 12 inches each. How wide is the bottom?

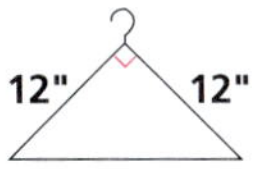

16.97 in.

11. Ephraim is making a corner shelf. One edge will be 8 inches, the second edge will be 12 inches. How long will the front edge of the shelf be?

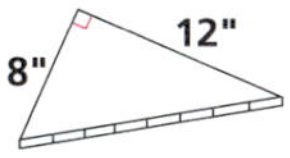

14.42 in.

12. Edna has a wheelchair ramp. Her front door is 6 feet above ground level and the ramp is 18 feet long. How far from the house is the start of the ramp?

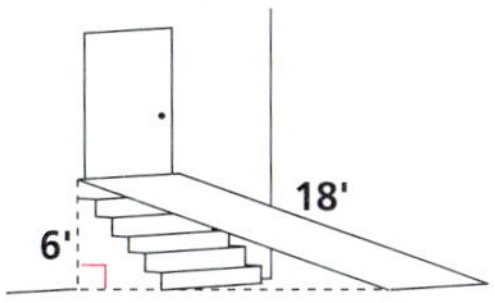

16.97 ft

13. A piece of wallboard is 4 feet wide and 6 feet long. The opening in Ken's van is 3 feet tall by 4 feet wide. Will the wallboard fit in Ken's van? Why or why not?

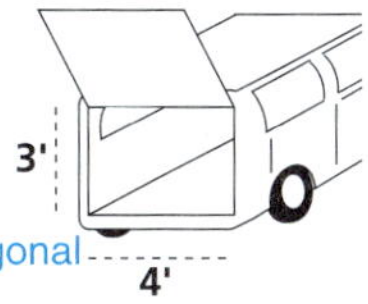

Yes, the opening has a diagonal of 5 ft, wide enough.

14. A transmission tower stands 22 feet tall. It will be secured by a wire attached 4 feet from the base of the tower. How long will the wire need to be? 22.36 ft

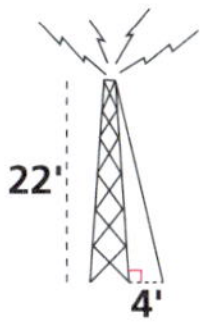

15. The sailboat is 400 m from a lighthouse. The cruise ship is 600 m from it. How far apart are the boats? 721.11 m

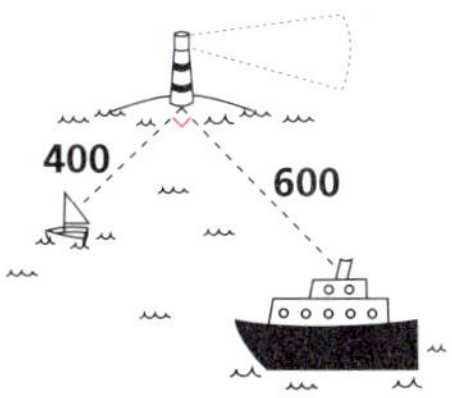

Application

In the Community

Direct students to survey the selection of straight, or extension, ladders at a local hardware store or hardware superstore. Tell them they will need to find out and record the minimum and maximum lengths for each extension ladder available at the store. Next, suggest students contact the ladder manufacturer or local fire department to determine the recommended and safest wall-to-ladder distance for an extension ladder.

Armed with the length of the ladder and the safest wall-to-ladder distance, students can compute which ladder would be the best to use to reach a window 12 feet above ground. Have them repeat the calculations for a window 18 feet, 24 feet, and 32 feet above ground.

The Pythagorean Theorem Chapter 8 265

Lesson at a Glance

Chapter 8 Lesson 5

Overview This lesson offers yet another proof of the Pythagorean Theorem—this one dependent on trapezoids.

Objectives

- To emulate Garfield's proof of the Pythagorean Theorem using trapezoids.
- To find the area of a trapezoid.

Student Pages 266–269

Teacher's Resource Library TRL

Workbook Activity 90

Additional Practice 51

Mathematics Vocabulary

average

1. Warm-Up Activity

Present students with a selection of quadrilaterals including squares, rectangles, rhombuses, and trapezoids. Ask them to identify the characteristics of a trapezoid. (four sides, one set of parallel sides)

Lesson 5 Pythagorean Proof and Trapezoids

> ***Average***
> *A number found by dividing the sum of all the quantities by the number of quantities*

There is yet another way to prove the Pythagorean Theorem. This proof, which relies on the characteristics of a trapezoid, was developed by James Garfield, the 20th president of the United States!

Before we begin the proof, we need to review the steps used to find the area of a trapezoid. Recall that the area of a rectangle is the product of its base times its height.

Area = base × height

$A = b \bullet h$

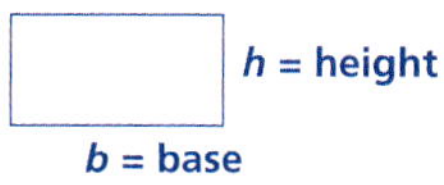

While a rectangle has two pairs of parallel sides, a trapezoid has only one pair of parallel sides. Usually the parallel sides of a trapezoid are called the *bases*. In this diagram the bases are *a* and *b*.

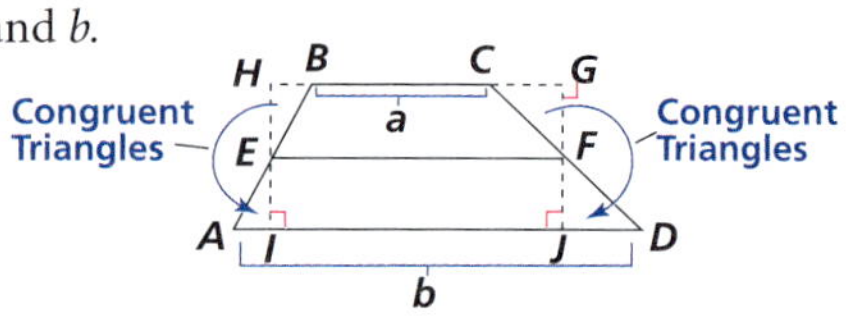

We can find a segment *EF* whose length is the **average** of the lengths of *a* and *b*.

$\overline{EF} = \frac{(a + b)}{2}$ = average of two bases

Note that $\overline{EF}$ joins the midpoints of the nonparallel sides. Also note that the pair of triangles formed at either edge of the trapezoid are congruent—here $\Delta BHE \sim \Delta AIE$ and $\Delta CGF \sim \Delta DJF$. Therefore, the rectangle *GHIJ* has the same area as the trapezoid.

Area of trapezoid = (average of bases)(height) = $\frac{(a + b)}{2}h$

Experiencing Geometry

You can prove this for yourself using a construction paper trapezoid. Mark the trapezoid bases *a* and *b*. Then cut the trapezoid into two triangles. Write an algebraic expression for the area of each triangle. Then write an algebraic expression for the area of the trapezoid. Simplify the expression.

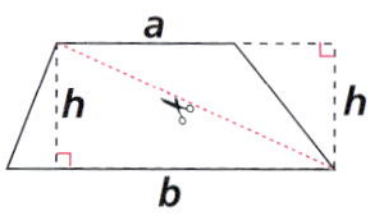

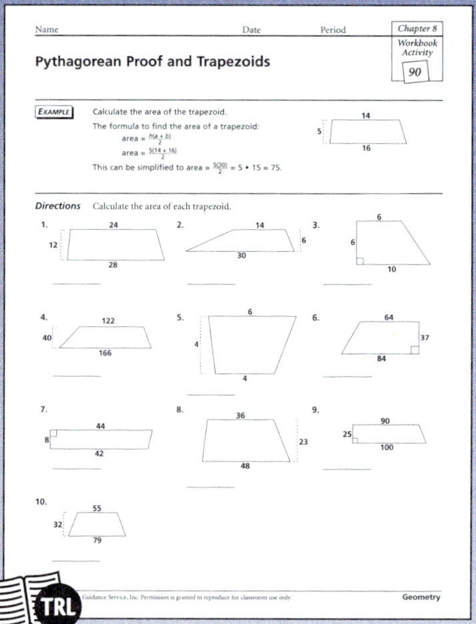

Name Date Period Chapter 8 Workbook Activity 90

Pythagorean Proof and Trapezoids

EXAMPLE Calculate the area of the trapezoid.

The formula to find the area of a trapezoid: area = $\frac{h(a+b)}{2}$

area = $\frac{5(14+16)}{2}$

This can be simplified to area = $\frac{5(30)}{2}$ = 5 • 15 = 75.

Directions Calculate the area of each trapezoid.

Geometry

Workbook Activity 90

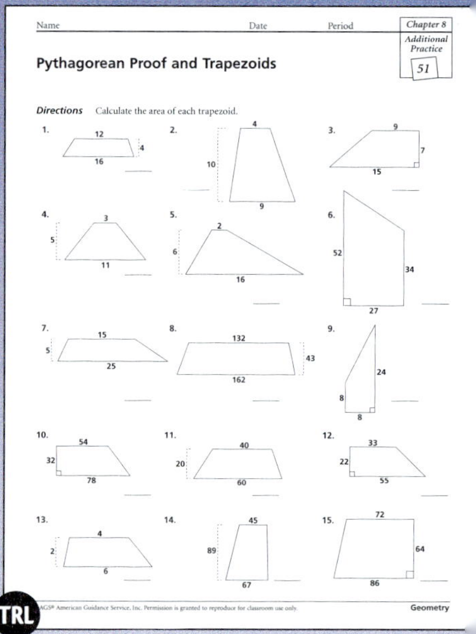

Name Date Period Chapter 8 Additional Practice 51

Pythagorean Proof and Trapezoids

Directions Calculate the area of each trapezoid.

Geometry

Additional Practice 51

Once you know how to find the area of a trapezoid, you can solve problems similar to these.

EXAMPLES **Given:** trapezoid with dimensions as marked

Find: area of the trapezoid

Area of trapezoid $= (\frac{a+b}{2})h$

$= (\frac{10+20}{2})8$

$= (\frac{30}{2})8$

$= 15 \bullet 8 = 120$

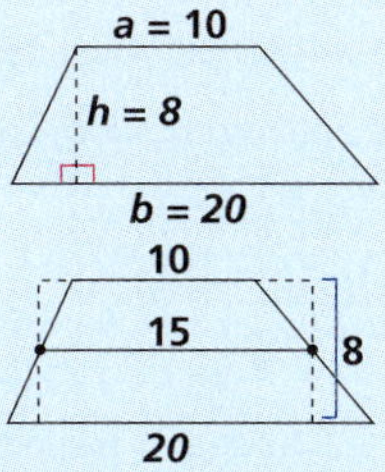

Note that the formula actually calculates the area of the rectangle, $15 \bullet 8 = 120$.

Given: trapezoid with dimensions as marked

Find: area of the trapezoid

Area of trapezoid $= (\frac{a+b}{2})h$

$= (\frac{22+32}{2})10$

$= (\frac{54}{2})10$

$= 27 \bullet 10 = 270$

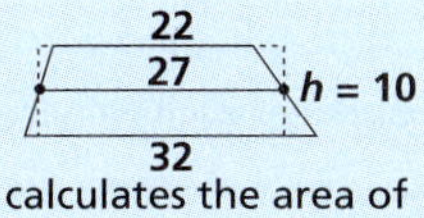

Again, the formula actually calculates the area of the rectangle, $27 \bullet 10 = 270$.

Now that we can compute the area of a trapezoid, we can continue with Garfield's proof of the Pythagorean Theorem.

Given: right $\triangle ABC$, $m\angle C = 90°$, legs a and b, hypotenuse c

To Prove: $a^2 + b^2 = c^2$

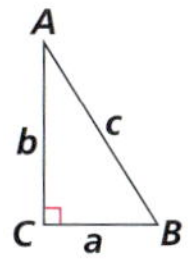

PROOF Copy the given triangle so that a and b form a straight line as shown at the right.

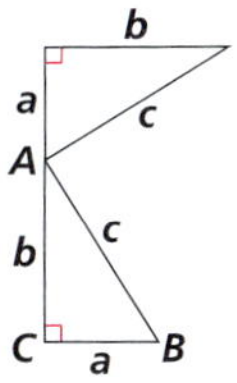

2. Teaching the Lesson

Work through the general case for finding the area of a trapezoid on page 266. Then have a volunteer explain the steps in each of the examples on page 267.

Experiencing Geometry (p.266)

Model this exercise for the class. Use colored acetate or tracing paper and an overhead projector. Draw the trapezoid, the diagonal, and the altitudes for the two resulting triangles on the model. Lead a discussion in which students indicate that the area of the trapezoid is $\frac{ah}{2} + \frac{bh}{2}$. Invite a volunteer to simplify the expression.

LEARNING STYLES

Visual/Spatial Invite students to make a flip book showing how to determine the area of a trapezoid. Each consecutive page of the book should contain a diagram that shows the next step in finding the area. For example:

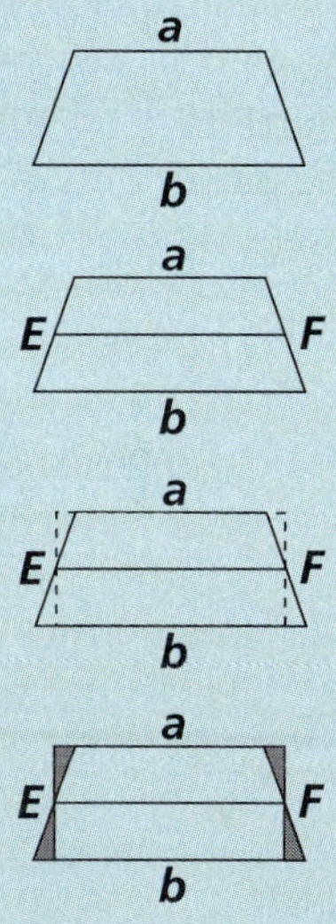

Once students have the diagrams in the correct sequence, they can write an explanatory caption for each step.

Alternative Proof

Here is another proof of the Pythagorean Theorem that relies only on areas of rectangles and right triangles.

Given: any right triangle with legs a and b and hypotenuse c.

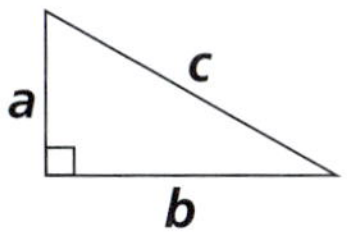

Make two of triangle abc as shown and connect the endpoints. Redraw the trapezoid with the same dimensions.

2 triangles + $\frac{1}{2}$ square = rectangle + triangle

$$2(\tfrac{1}{2}ab) + \tfrac{1}{2}c^2 = a(a+b) + \tfrac{1}{2}(b-a)(a+b)$$

$$2ab + c^2 = 2a(a+b) + (b-a)(a+b)$$

$$2ab + c^2 = 2a^2 + 2ab + ab + b^2 - a^2 - ab$$

$$\therefore c^2 = a^2 + b^2$$

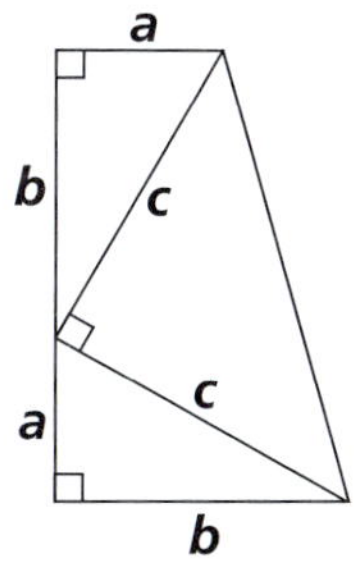

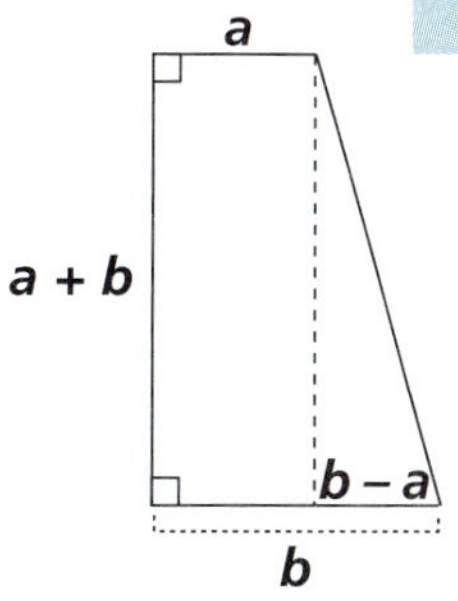

Learning Styles

Auditory/Verbal Invite a volunteer to draw two copies of right ΔABC on the board as shown at the bottom of page 267. Ask a second volunteer to connect the endpoints to form a trapezoid. Encourage that volunteer to explain why drawing the line is an acceptable step in the proof. (Euclid's Postulate 1)

Point out that in the proof, ΔI and ΔII must have the same area because they are the same triangle drawn in two different positions.

3. Reinforce and Extend

Learning Styles

LEP/ESL Students will benefit from a close examination of the proof on pages 267–268. Direct pairs of students to explain each step in the proof to one another, starting with the statement that the angles at A ($\angle 1$, $\angle 2$, and $\angle 3$) must equal 180° and the conclusion that $\angle 3 = 90°$. Students can work together to transcribe the proof in the traditional two-column style, one column for statements and the other for reasons. Students can exchange completed transcriptions with one another for evaluation. Proofs should be returned to the writers for any revisions or additions that might be necessary.

Algebra Connection

In addition to the product of $(a + b)^2$, challenge students to find the products of the following:

$(a - b)^2$	$(a^2 - 2ab + b^2)$
$a(ab + c)$	$(a^2b + ac)$
$(a + b)(a - b)$	$(a^2 - b^2)$

Proof *(continued)*

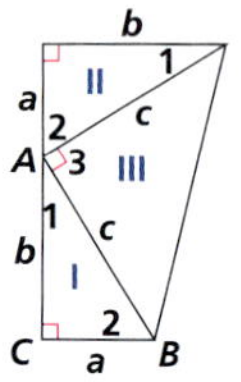

Connect the endpoints to form a trapezoid as shown. The angles at A must equal 180°. The acute angles ($\angle 2$ and $\angle 1$) are complementary because the given triangle is a right triangle. So the middle angle, $\angle 3$, must be 90°.

The height (h) of this trapezoid is $a + b$, so

Area of trapezoid = $(\frac{a+b}{2})h$ becomes

$(\frac{a+b}{2})(a + b)$ by substitution

Area of three triangles = Area of trapezoid

ΔI + ΔII + ΔIII = Area of trapezoid

$\frac{1}{2}ab + \frac{1}{2}ab + \frac{1}{2}c^2 = (\frac{a+b}{2})(a + b)$

$ab + (\frac{1}{2})c^2 = (\frac{a+b}{2})(a + b)$ simplify

$2ab + c^2 = (a + b)^2$ multiply by 2

$2ab + c^2 = a^2 + 2ab + b^2$ square of quantity $(a + b)$

$c^2 = a^2 + b^2$ subtract $2ab$

This completes the fourth proof of the Pythagorean Theorem.

Algebra Connection

Remember from algebra that $(a + b)^2 = (a + b)(a + b) = a^2 + 2ab + b^2$

Exercise A Calculate the area of each trapezoid.

1.

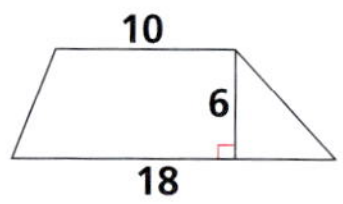

84

2.

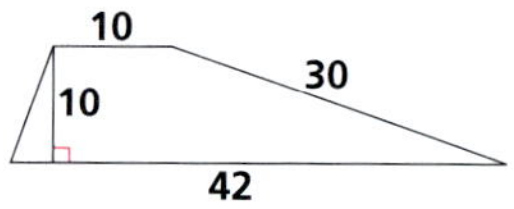

260

3.

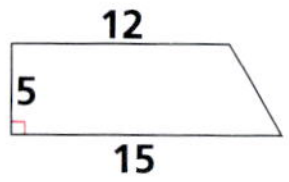

67.5

4.

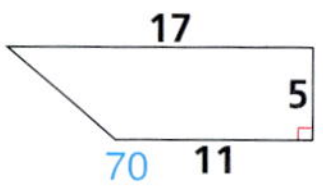

70

5.

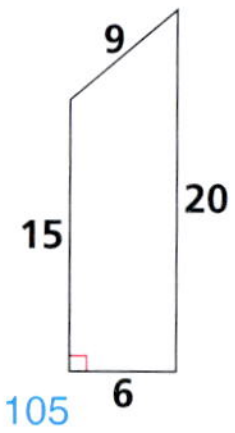

105

6.

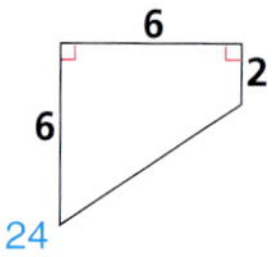

24

7. 21

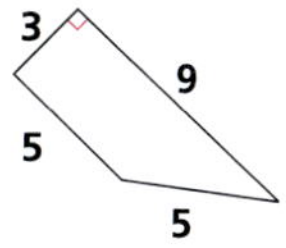

8. 58

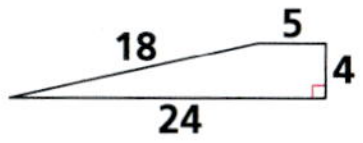

9. 19

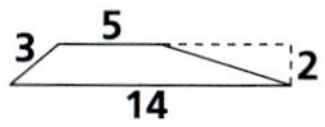

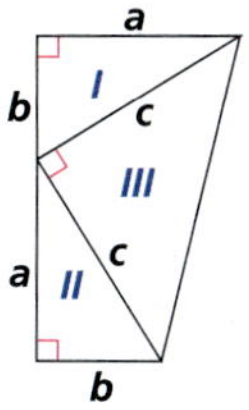

Exercise B Copy and complete the chart by calculating the areas for the given side lengths.

	Area Δ I	Area Δ II	Area Δ III	Trapezoid Area
$a = 4, b = 3, c = 5$	6 **10.** ____	6 **11.** ____	12.5 **12.** ____	24.5 **13.** ____
$a = 4, b = 12, c = 13$	24 **14.** ____	24 **15.** ____	84.5 **16.** ____	132.5 **17.** ____
$a = 8, b = 6, c = 10$	24 **18.** ____	24 **19.** ____	50 **20.** ____	98 **21.** ____
$a = 20, b = 48, c = 52$	480 **22.** ____	480 **23.** ____	1,352 **24.** ____	2,312 **25.** ____

Learning Styles

Interpersonal/Group Learning Direct groups of four to work together to complete Exercise A. Each student in the group can demonstrate and explain each step necessary to find the area. Students listening to the explanation can ask for clarification of any steps or points they do not understand.

Learning Styles

Logical/Mathematical Direct students' attention to the chart in Exercise B. Ask students which column of the chart is not necessary. Have them explain why. (Because ΔI and ΔII are the same (congruent), the chart need only ask for the area of one or the other. Twice the area of ΔI plus the area of ΔIII will equal the area of the trapezoid.)

The Pythagorean Theorem Chapter 8 **269**

Lesson at a Glance

Chapter 8 Lesson 6

Overview This lesson demonstrates how to use the Pythagorean Theorem to find a distance on the coordinate plane.

Objective

- To use the distance formula to find the distance between points on a coordinate plane.

Student Pages 270–274

Teacher's Resource Library TRL

Workbook Activities 91–92

Additional Practice 52

1. Warm-Up Activity

Review with students the conventions for naming a point on the coordinate plane. Emphasize to students that they must be careful to always name the *x*-coordinate first, *y*-coordinate second. Also review with students the symbol $|x|$ for the absolute value of *x*. Indicate to students that in this lesson they will learn how to use the coordinates of the endpoints of a line to determine the length of the line.

Lesson 6 Distance Formula: Pythagorean Theorem

One of the most useful applications of the Pythagorean Theorem is finding the distance between any two points in the coordinate plane.

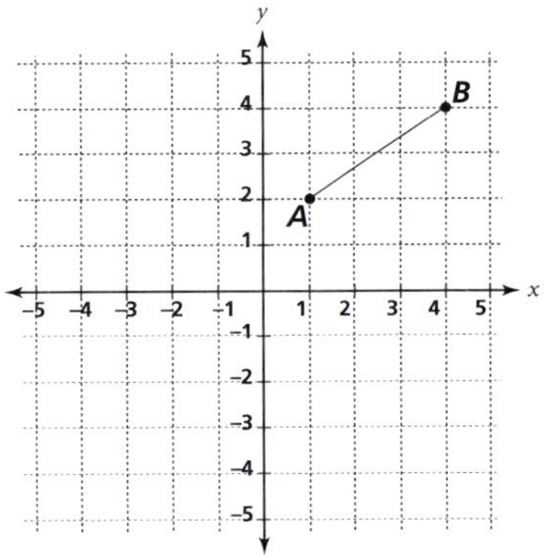

Find the distance between points *A* and *B* in the coordinate plane.

The axes form a right angle at their intersection, so it is possible to form a right triangle with hypotenuse $\overline{AB}$. If you know the coordinates of *A* and *B*, then you can use the Pythagorean Theorem to find the distance that separates them.

EXAMPLE **Given:** $A = (2, 3)$ and $B = (6, 6)$

Find: the distance $\overline{AB}$

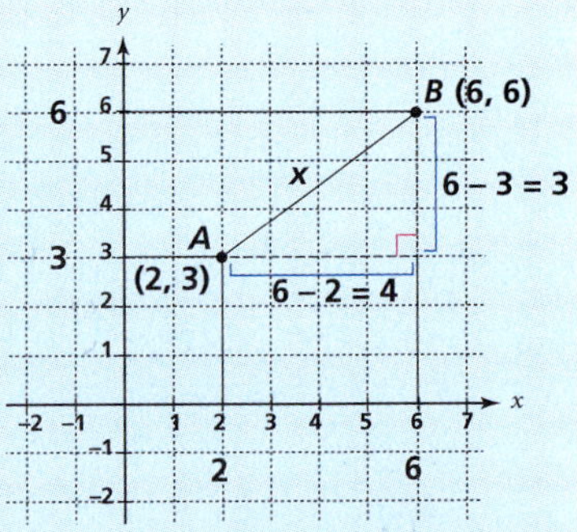

Complete the right triangle. The legs' lengths are 6 – 2 = 4 for the leg || to the *x*-axis and 6 – 3 = 3 for the leg || to the *y*-axis.

$x^2 = 3^2 + 4^2$ by Pythagorean Theorem

$x^2 = 9 + 16$

$x^2 = 25$

$x = 5$

The distance between (2, 3) and (6, 6) is 5.

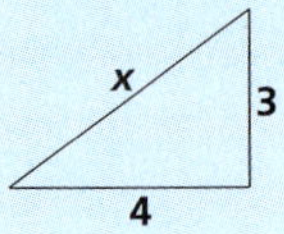

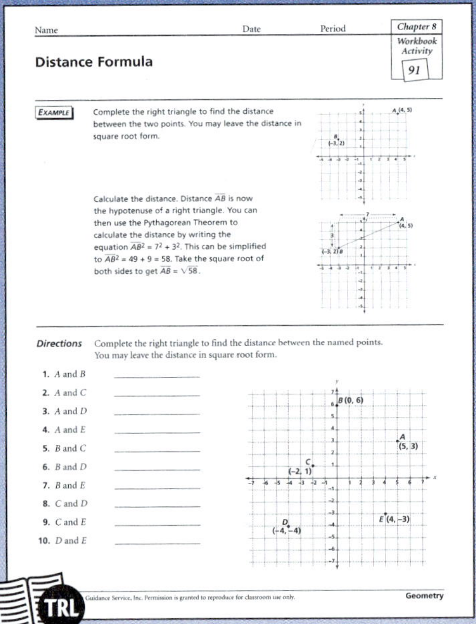

Name Date Period

Chapter 8 Workbook Activity 91

Distance Formula

EXAMPLE Complete the right triangle to find the distance between the two points. You may leave the distance in square root form.

Calculate the distance. Distance $\overline{AB}$ is now the hypotenuse of a right triangle. You can then use the Pythagorean Theorem to calculate the distance by writing the equation $\overline{AB}^2 = 7^2 + 3^2$. This can be simplified to $\overline{AB}^2 = 49 + 9 = 58$. Take the square root of both sides to get $\overline{AB} = \sqrt{58}$.

Directions Complete the right triangle to find the distance between the named points. You may leave the distance in square root form.

1. *A* and *B* ______
2. *A* and *C* ______
3. *A* and *D* ______
4. *A* and *E* ______
5. *B* and *C* ______
6. *B* and *D* ______
7. *B* and *E* ______
8. *C* and *D* ______
9. *C* and *E* ______
10. *D* and *E* ______

Geometry

TRL

Workbook Activity 91

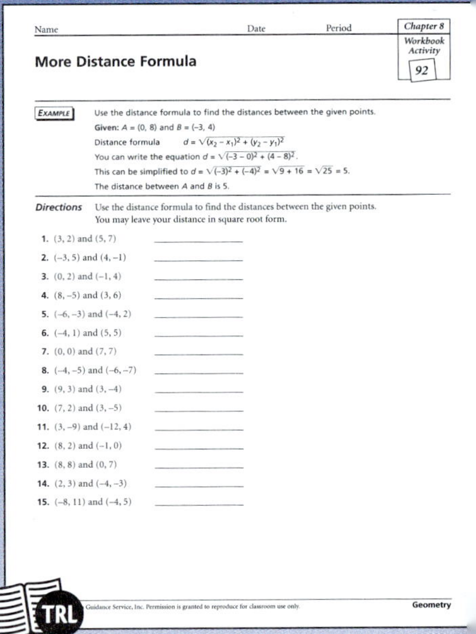

Name Date Period

Chapter 8 Workbook Activity 92

More Distance Formula

EXAMPLE Use the distance formula to find the distances between the given points.

Given: $A = (0, 8)$ and $B = (-3, 4)$

Distance formula $d = \sqrt{(x_2 - x_1)^2 + (y_2 - y_1)^2}$

You can write the equation $d = \sqrt{(-3 - 0)^2 + (4 - 8)^2}$.

This can be simplified to $d = \sqrt{(-3)^2 + (-4)^2} = \sqrt{9 + 16} = \sqrt{25} = 5$.

The distance between *A* and *B* is 5.

Directions Use the distance formula to find the distances between the given points. You may leave your distance in square root form.

1. (3, 2) and (5, 7) ______
2. (–3, 5) and (4, –1) ______
3. (0, 2) and (–1, 4) ______
4. (8, –5) and (3, 6) ______
5. (–6, –3) and (–4, 2) ______
6. (–4, 1) and (5, 5) ______
7. (0, 0) and (7, 7) ______
8. (–4, –5) and (–6, –7) ______
9. (9, 3) and (3, –4) ______
10. (7, 2) and (3, –5) ______
11. (3, –9) and (–12, 4) ______
12. (8, 2) and (–1, 0) ______
13. (8, 8) and (0, 7) ______
14. (2, 3) and (–4, –3) ______
15. (–8, 11) and (–4, 5) ______

Geometry

TRL

Workbook Activity 92

EXAMPLES **Given:** $D = (-2, -3)$ and $C = (1, 4)$

Find: the distance $\overline{CD}$

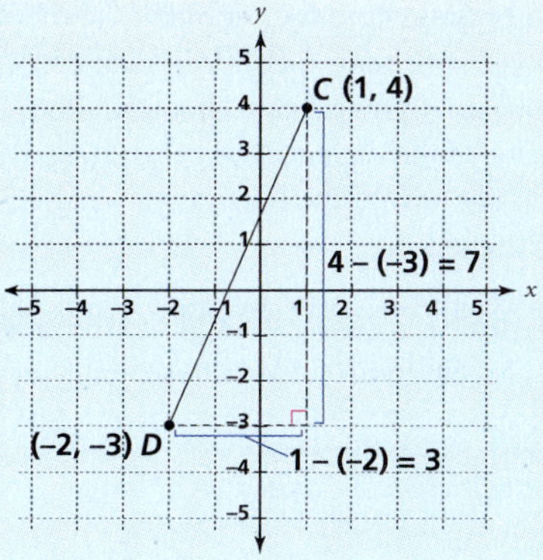

Complete the right triangle.
Let x = the distance $\overline{CD}$

$x^2 = 7^2 + 3^2$ by Pythagorean Theorem

$x^2 = 49 + 9$

$x^2 = 58$

$x = \sqrt{58}$

The distance between (–2, –3) and (1, 4) is $\sqrt{58}$.

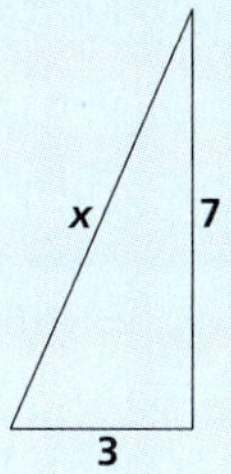

Given: $E = (-2, 5)$ and $F = (3, -4)$

Find: the distance $\overline{EF}$

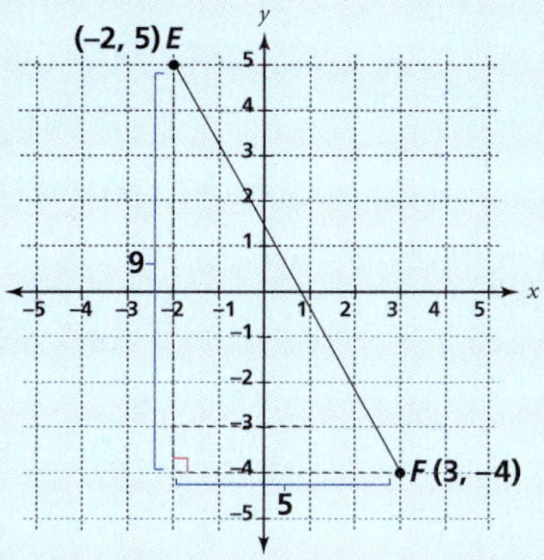

Complete the right triangle and use the Pythagorean Theorem. Let x = the distance $\overline{EF}$.

The length of the leg parallel to the x-axis is $-2 - 3 = -5$. You will need to use the absolute value of –5 or $|-5| = 5$.

$x^2 = 9^2 + 5^2$

$x^2 = 81 + 25$

$x^2 = 106$

$x = \sqrt{106}$

The distance between (–2, 5) and (3, –4) is $\sqrt{106}$.

2. Teaching the Lesson

Demonstrate each step in the example on page 270. Point out to students that one way to check that they have the correct length for each leg of the triangle is to count off the x-distance from point A to the right angle. They can also count off the y-distance from point B to the right angle. To check accuracy, they can compare that number to the number they get when they subtract the coordinates.

Invite a volunteer lead the class through the example at the top of page 271. Make certain that students recognize that the triangle with legs x, 7, and 3 is a copy of the triangle on the coordinate grid.

LEARNING STYLES

Auditory/Verbal Suggest students work in pairs and explain the example at the bottom of page 271 to one another. Direct students to draw the triangle on a sheet of paper separate from the coordinate grid. Circulate, offering reinforcement to students whose explanations include finding the absolute value of –5.

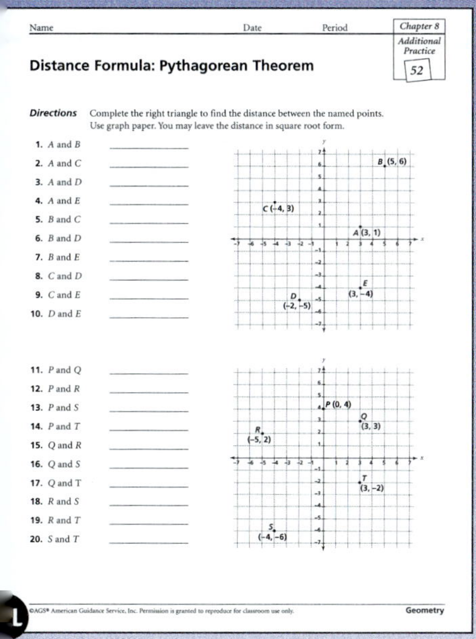

Name Date Period — *Chapter 8 Additional Practice* 52

Distance Formula: Pythagorean Theorem

Directions Complete the right triangle to find the distance between the named points. Use graph paper. You may leave the distance in square root form.

1. *A* and *B* ________
2. *A* and *C* ________
3. *A* and *D* ________
4. *A* and *E* ________
5. *B* and *C* ________
6. *B* and *D* ________
7. *B* and *E* ________
8. *C* and *D* ________
9. *C* and *E* ________
10. *D* and *E* ________

11. *P* and *Q* ________
12. *P* and *R* ________
13. *P* and *S* ________
14. *P* and *T* ________
15. *Q* and *R* ________
16. *Q* and *S* ________
17. *Q* and T ________
18. *R* and *S* ________
19. *R* and *T* ________
20. *S* and *T* ________

 Geometry

Additional Practice 52

3. Reinforce and Extend

LEARNING STYLES

Body/Kinesthetic Produce an 8 x 8 coordinate grid on the classroom floor, allowing a distance of one foot between grid lines. Direct one volunteer to stand at A (3, 4). This student should hold a card giving the coordinates of the point. Have a second student stand at B (6, 8). Ask a third student to stand at C (6, 4). Ask the students at A and C to estimate and then compute the distances between their locations and point B. Have them use the information to find the distance between A and B. Students will recognize this as a 3, 4, 5 triangle. Encourage them to confirm the length of $\overline{AB}$ using a tape measure.

COMMON ERROR

Students may be quick to find the distance by subtracting the coordinates of a single point. That is, they might try to subtract y_1 from x_1 and y_2 from x_2. Advise students to think carefully when they apply the distance formula, making sure they subtract x-coordinates from x-coordinates and y-coordinates from y-coordinates.

All of these kinds of problems can be solved using this formula, which comes from the Pythagorean Theorem. Because we need the distances to be positive, we will use absolute values for the lengths of the legs.

$\overline{AC} = |\text{difference in } x\text{-values}| = |x_2 - x_1|$

$\overline{BC} = |\text{difference in } y\text{-values}| = |y_2 - y_1|$

ΔABC is a right triangle because the coordinate axes form a right angle.

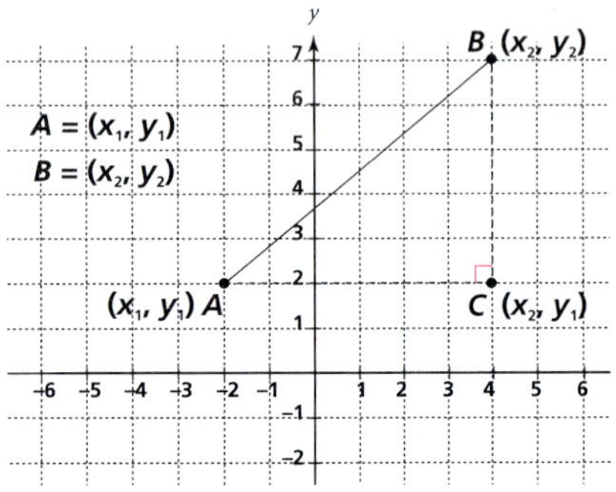

Let $d = \overline{AB}$, then by the Pythagorean Theorem,

$$d^2 = \overline{AC}^2 + \overline{BC}^2$$

$$d^2 = (x_2 - x_1)^2 + (y_2 - y_1)^2$$

DISTANCE FORMULA: $d = \sqrt{(x_2 - x_1)^2 + (y_2 - y_1)^2}$

EXAMPLES Use the distance formula to find the distance between (3, 4) and (6, 10). You may leave the answer in the square root form.

$d = \sqrt{(x_2 - x_1)^2 + (y_2 - y_1)^2}$ let $(x_1, y_1) = (3, 4)$, $(x_2, y_2) = (6, 10)$

$d = \sqrt{(6 - 3)^2 + (10 - 4)^2}$

$d = \sqrt{9 + 36} = \sqrt{45}$

Use the distance formula to find the distance between (–2, 3) and (–5, –7). You may leave the answer in the square root form.

$d = \sqrt{(x_2 - x_1)^2 + (y_2 - y_1)^2}$ let $(x_1, y_1) = (-2, 3)$, $(x_2, y_2) = (-5, -7)$

$d = \sqrt{(-5 - (-2))^2 + (-7 - 3)^2}$

$d = \sqrt{(-3)^2 + (-10)^2}$

$d = \sqrt{9 + 100} = \sqrt{109}$

272 Chapter 8 The Pythagorean Theorem

Exercise A Complete the right triangle to find the distance between the named points. Use graph paper. You may leave the distance in square root form.

1. *A* and *B* $\sqrt{26}$

2. *B* and *C* $\sqrt{58}$

3. *D* and *E* $\sqrt{52}$

4. *C* and *E* 9

5. *B* and *D* 11

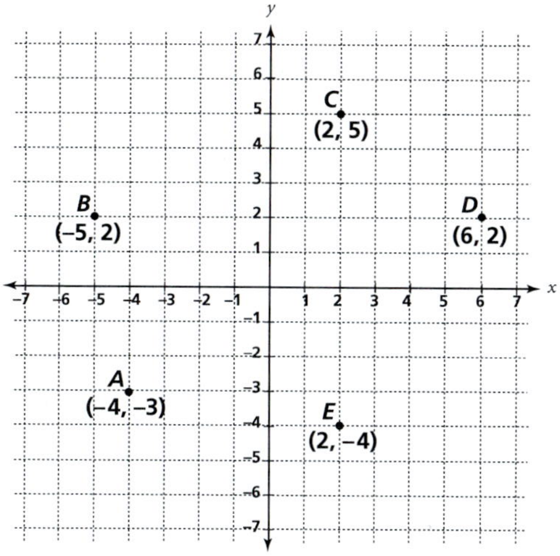

6. (8, 9) and (5, 6) $\sqrt{18} = 3\sqrt{2}$

7. (–12, 1) and (6, 9) $\sqrt{388} = 2\sqrt{97}$

8. (8, 9) and (5, 6) $\sqrt{18} = 3\sqrt{2}$

9. (–13, 7) and (5, 6) $\sqrt{325} = 5\sqrt{13}$

10. (–30, –25) and (–20, –15)
$\sqrt{200} = 10\sqrt{2}$

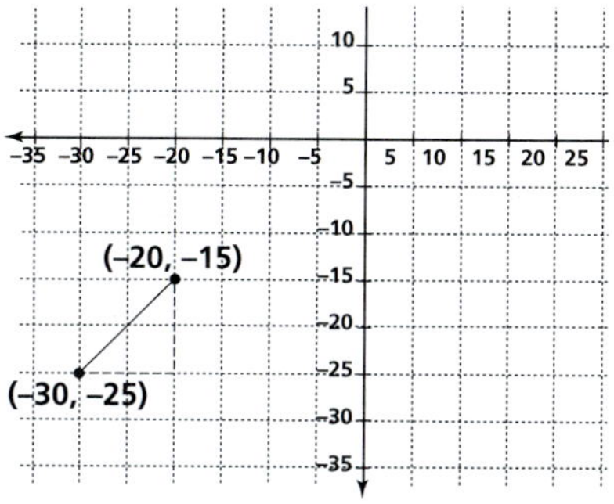

Mental Math

Review with students the process of presenting a radical in simplest form. Indicate that students should factor the number, then take the square root of any factors that are themselves squares. For example, $\sqrt{20} = \sqrt{4 \cdot 5} = 2\sqrt{5}$. Have students use mental math to present the following radicals in the simplest form.

$\sqrt{200}$ $(\sqrt{100 \bullet 2} = 10\sqrt{2})$

$\sqrt{98}$ $(\sqrt{49 \bullet 2} = 7\sqrt{2})$

$\sqrt{242}$ $(\sqrt{121 \bullet 2} = 11\sqrt{2})$

$\sqrt{27}$ $(\sqrt{3 \bullet 9} = 3\sqrt{3})$

$\sqrt{99}$ $(\sqrt{11 \bullet 9} = 3\sqrt{11})$

$\sqrt{80}$ $(\sqrt{20 \bullet 4} = \sqrt{4 \bullet 4 \bullet 5} = 4\sqrt{5})$

Calculator

Use the parentheses keys on the calculator to point out to students the necessity for order in certain computations. Have students perform the following calculations with correct and incorrect key use. Emphasize the importance of the location of the parentheses in terms of performing the calculations. Then lead a discussion on the results, reviewing with students the terms *associative* and *noncommutative.*

$4(6)^2$ becomes 46^2 or 2,116.

$(4 \cdot 6)^2$ becomes 24^2 or 576.

$(4)(6^2)$ becomes $4 \cdot 36$ or 144.

Exercise B Use the distance formula to find the distances between the given points.

11. (–2, –3) and (4, 5) 10

12. (2, 3) and (0, 0) $\sqrt{13}$

13. (6, 4) and (4, 9) $\sqrt{29}$

14. (2, 5) and (6, –3) $\sqrt{80} = 4\sqrt{5}$

15. (–6, –2) and (–7, –9) $\sqrt{50} = 5\sqrt{2}$

Calculator Practice

Exercise C Use the distance formula and your calculator to find the distances between the points.

If your calculator has parentheses keys, you can find the distance between points in one step. You may keep the answer in square root form.

Example Find the distance between (4, 6) and (2, 3).

$d = \sqrt{(x_2 - x_1)^2 + (y_2 - y_1)^2}$ let $(x_1, y_1) = (4, 6)$, $(x_2, y_2) = (2, 3)$

$d^2 = (2 - 4)^2 + (3 - 6)^2$

To find d^2, press (2 − 4) x^2 + (3 − 6) x^2 = 13

$d^2 = 13$; $d = \sqrt{13}$

Writing About Mathematics

Which method of finding distance do you prefer—completing the triangle or using the distance formula? Explain why.

16. (–8, 50) and (30, –11) $\sqrt{5{,}165}$

17. (–10, 15) and (–20, 20) $\sqrt{125} = 5\sqrt{5}$

18. (9, 16) and (25, 36) $\sqrt{656} = 4\sqrt{41}$

19. (17, 16) and (21, 19) 5

20. (21, 20) and (20, 12) $\sqrt{65}$

274 Chapter 8 The Pythagorean Theorem

Lesson 7 Converse of the Pythagorean Theorem

The Pythagorean Theorem tells us that $a^2 + b^2 = c^2$ for right triangles. Does that also mean that if you have a triangle such that $a^2 + b^2 = c^2$ you can be sure it is a right triangle? Let's look at a specific case: $3^2 + 4^2 = 5^2$ because $9 + 16 = 25$ is a true statement. You might think you have a right triangle, but can you be sure? To be sure you have to prove the following.

> Remember, in a converse, the If . . . then . . . terms of a conditional are reversed.

CONVERSE OF THE PYTHAGOREAN THEOREM: If for ΔABC, $\overline{BC}^2 = \overline{AB}^2 + \overline{AC}^2$, then ΔABC is a right triangle.

Given: ΔABC, $\overline{BC}^2 = \overline{AB}^2 + \overline{AC}^2$

To Prove: ΔABC is a right triangle

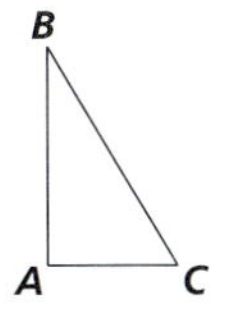

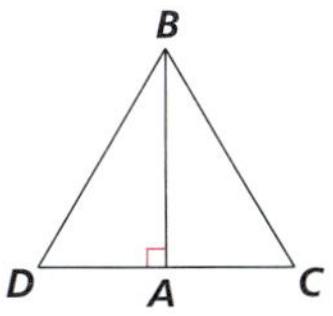

PROOF

Statement	Reason
1. ΔABC with $\overline{BC}^2 = \overline{AB}^2 + \overline{AC}^2$	1. Given.
2. Draw $\overline{AD}$ at right angles to $\overline{BA}$ so that $\overline{AD} = \overline{AC}$.	2. Construction of perpendiculars and copy line segments.
3. Draw $\overline{DB}$.	3. Postulate 1, two points determine a line.
4. $\overline{AD}^2 = \overline{AC}^2$	4. Square both sides from Step 2.
5. $\overline{AD}^2 + \overline{AB}^2 = \overline{AC}^2 + \overline{AB}^2$	5. Equals added to equals.
6. $\overline{DB}^2 = \overline{AD}^2 + \overline{AB}^2$	6. Pythagorean Theorem for ΔBAD.
7. $\therefore \overline{DB}^2 = \overline{BC}^2$	7. Steps 1, 4, and 6 and substitution.
8. $\overline{DB} = \overline{BC}$	8. Take square root of each side of Step 7.
9. $\overline{AB} = \overline{AB}$	9. A quantity is equal to itself.
10. $\therefore \Delta BAD \cong \Delta BAC$	10. SSS.
11. $m\angle BAD \cong m\angle BAC$	11. Corresponding parts of congruent triangles are equal.
12. $m\angle BAC$ is a right angle and ΔABC is a right triangle.	12. All right angles are equal.

Lesson at a Glance

Chapter 8 Lesson 7

Overview This lesson takes students through the proof of the converse of the Pythagorean Theorem.

Objective

- To use the converse of the Pythagorean Theorem to identify right triangles.

Student Pages 275–277

Teacher's Resource Library TRL

Workbook Activity 93

Additional Practice 53

1. Warm-Up Activity

Review with students the concept of a converse and the idea that a converse of a true theorem may not be true. For example, if an animal is a horse, then it has four legs. The converse, "If an animal has four legs, then it is a horse", is not true. Tell students that in this lesson they will prove that the converse of the Pythagorean Theorem is true.

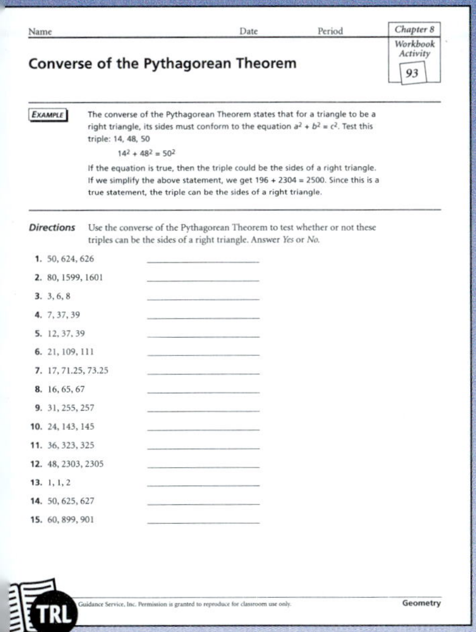

Name Date Period

Chapter 8 Workbook Activity 93

Converse of the Pythagorean Theorem

EXAMPLE The converse of the Pythagorean Theorem states that for a triangle to be a right triangle, its sides must conform to the equation $a^2 + b^2 = c^2$. Test this triple: 14, 48, 50

$$14^2 + 48^2 = 50^2$$

If the equation is true, then the triple could be the sides of a right triangle. If we simplify the above statement, we get $196 + 2304 = 2500$. Since this is a true statement, the triple can be the sides of a right triangle.

Directions Use the converse of the Pythagorean Theorem to test whether or not these triples can be the sides of a right triangle. Answer *Yes* or *No.*

1. 50, 624, 626 ________
2. 80, 1599, 1601 ________
3. 3, 6, 8 ________
4. 7, 37, 39 ________
5. 12, 37, 39 ________
6. 21, 109, 111 ________
7. 17, 71.25, 73.25 ________
8. 16, 65, 67 ________
9. 31, 255, 257 ________
10. 24, 143, 145 ________
11. 36, 323, 325 ________
12. 48, 2303, 2305 ________
13. 1, 1, 2 ________
14. 50, 625, 627 ________
15. 60, 899, 901 ________

 Geometry

Workbook Activity 93

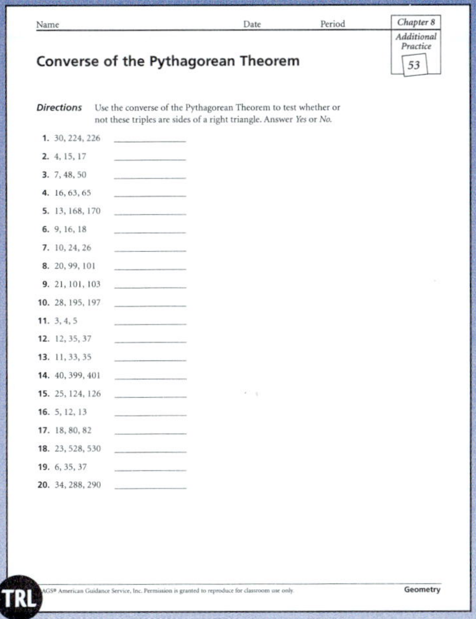

Name Date Period

Chapter 8 Additional Practice 53

Converse of the Pythagorean Theorem

Directions Use the converse of the Pythagorean Theorem to test whether or not these triples are sides of a right triangle. Answer *Yes* or *No.*

1. 30, 224, 226 ________
2. 4, 15, 17 ________
3. 7, 48, 50 ________
4. 16, 63, 65 ________
5. 13, 168, 170 ________
6. 9, 16, 18 ________
7. 10, 24, 26 ________
8. 20, 99, 101 ________
9. 21, 101, 103 ________
10. 28, 195, 197 ________
11. 3, 4, 5 ________
12. 12, 35, 37 ________
13. 11, 33, 35 ________
14. 40, 399, 401 ________
15. 25, 124, 126 ________
16. 5, 12, 13 ________
17. 18, 80, 82 ________
18. 23, 528, 530 ________
19. 6, 35, 37 ________
20. 34, 288, 290 ________

 Geometry

Additional Practice 53

2. Teaching the Lesson

Work through the proof on page 275 with students, reminding them that each reason listed is a postulate, theorem, or operation with which they are already familiar. At Step 10, make certain students can identify the steps in which the sides of the triangle were proved congruent. **(2, 8, and 9)** At Step 12, ask students how they know $\angle BAC$ is a right angle. **(Step 11)** Then ask how they know $\angle BAD$ is a right angle. **(Step 2, construction)**

Learning Styles

Logical/Mathematical Suggest students work in pairs to solve the example problems. Ask students to explain how the examples could also be used to prove whether the numbers form a Pythagorean Triple.

Now that you have proved this converse of the Pythagorean Theorem, you know that anytime the sum of the squares of two sides of a triangle equals the square of the third side of the triangle, the triangle must be a right triangle.

Pythagorean Theorem

Given: ΔABC is a right triangle, then $\overline{BC}^2 = \overline{AC}^2 + \overline{AB}^2$

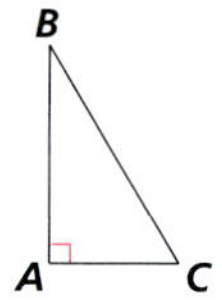

Converse of Pythagorean Theorem

Given: where $\overline{BC}^2 = \overline{AC}^2 + \overline{AB}^2$ then ΔABC is a right triangle.

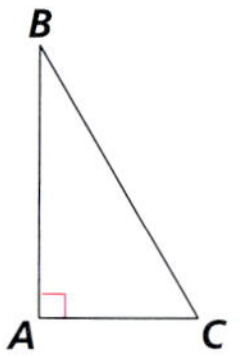

EXAMPLES The sides of a triangle are 6, 8, and 12. Is the triangle a right triangle?

If it is a right triangle, then $a^2 + b^2 = c^2$. Substitute the triangle lengths for a, b, and c.

$a^2 + b^2 \stackrel{?}{=} c^2$

$6^2 + 8^2 \stackrel{?}{=} 12^2$

$36 + 64 \stackrel{?}{=} 144$

$100 \neq 144$, the triangle is not a right triangle.

The sides of a triangle are 5, 12, and 13. Is the triangle a right triangle?

Substitute the triangle lengths for a, b, and c.

$a^2 + b^2 \stackrel{?}{=} c^2$

$5^2 + 12^2 \stackrel{?}{=} 13^2$

$25 + 144 \stackrel{?}{=} 169$

$169 = 169$, the triangle is a right triangle.

Geometry Around Us

You can find several examples of triangles in nature—the arms of a starfish or the points in a star fruit (called a carambola). How would you determine if these triangles were right triangles?

Exercise A Use the converse of the Pythagorean Theorem to test whether or not these triples are sides of a right triangle. Answer *yes* or *no*.

1. 3, 4, 5	yes	**6.** 9, 12, 15	yes
2. 9, 10, 11	no	**7.** 10, 20, 30	no
3. 7, 8, 10	no	**8.** 7, 6, 5	no
4. 6, 8, 10	yes	**9.** 11, 12, 13	no
5. 10, 25, 36	no	**10.** 17, 60, 70	no

Writing About Mathematics

Use algebra to show why multiples of a triple such as 3, 4, 5 correspond to similar right triangles.

Exercise B Answer the following questions about isosceles ΔABC.

11. What is the length of $\overline{BD}$?

12. How long is $\overline{AB}$?

13. Why must $\overline{AD} = \overline{CD}$?

14. What do you know about $\angle DAB$ and $\angle DCB$?

15. Why must $\angle ADB \cong \angle CDB$?

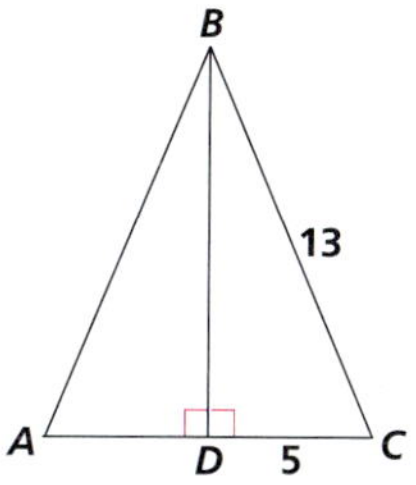

11. 12
12. 13, isosceles triangle
13. For isosceles triangle, line from vertex perpendicular to base, bisects base.
14. equal; base angles in isosceles triangle
15. both right angles

3. Reinforce and Extend

Learning Styles

LEP/ESL Suggest students work in pairs to complete Exercise A. Students can explain to their partners how to use the converse of the Pythagorean Theorem to test the triples. They can also explain the results of the test in complete sentences, for example, "The numbers 9, 10, and 11 cannot form the sides of a right triangle because they do not satisfy $a^2 + b^2 = c^2$." Students should take turns explaining and listening.

Geometry Around Us

Students should recognize that they can measure the sides of the triangle and then test to see if $a^2 + b^2 = c^2$. If so, then the triangles are right triangles according to the converse of the Pythagorean Theorem.

Writing About Mathematics

The following should be at the heart of students' explanations:

If $a^2 + b^2 = c^2$ is true for $a = 3$, $b = 4$, and $c = 5$, then $(na)^2 + (nb)^2 + (nc)^2$ is true for $n =$ any number because equals multiplied by equals are equal.

Lesson at a Glance

Chapter 8 Application

Overview This lesson explores how geometric shapes are used in art and architecture.

Objective

- To investigate how geometric shapes are used in architecture.

Student Page 278

Teacher's Resource Library TRL

Workbook Activity 94

1. Warm-Up Activity

Open the lesson with a discussion of any large exposed structural elements found in your community. Encourage students to look for triangles within the structure.

2. Teaching the Lesson

Draw students' attention to the cross bracing shown in the Hancock building. Have students identify which triangles are congruent and which are similar.

3. Reinforce and Extend

APPLICATION

Career Connection

Encourage students to contact an area architecture association to find out which buildings or structures in their area or another location were designed by well-known architects. Students can then investigate the architect's life including his or her schooling, style, and work. Students can present their findings as an Architecture Bio-Minute for the evening news.

Alternately, students can pool their research, writing an architectural tour of their area, including information about the architect's schooling and style as they discuss each building.

Application

Architecture and Geometric Shapes

Triangles play an important part in large structures such as buildings and bridges. Most skyscrapers are supported by steel frameworks. Usually these frameworks are hidden inside the buildings. But in the 100-story John Hancock tower in Chicago, the builder let the triangles formed by the bracing be seen. These triangles, each congruent to another, help stabilize the building against the wind and its own weight.

EXAMPLE The Air Force Academy Chapel is a series of tetrahedrons. (A tetrahedron is a solid shape with four faces, each a triangle.) As you can see, the triangles are not only similar, they are congruent.

Exercise

1. Find several photographs of large structures such as the buildings pictured here or the bridge shown at the beginning of this chapter.
2. Examine each photograph and determine how triangles were used in the structure. Identify any triangles that may be similar to one another and any that may be congruent.
3. List three examples in which a triangle shape (perhaps in the form of a cross brace) is used to strengthen an object.
4. Review the pictures. Look for other geometric shapes. Identify each shape and tell how you identified it.
5. Make a sketch of a building, using triangles in your design as well as other geometric shapes.

Student identifications should reflect triangles in their choice of photos.

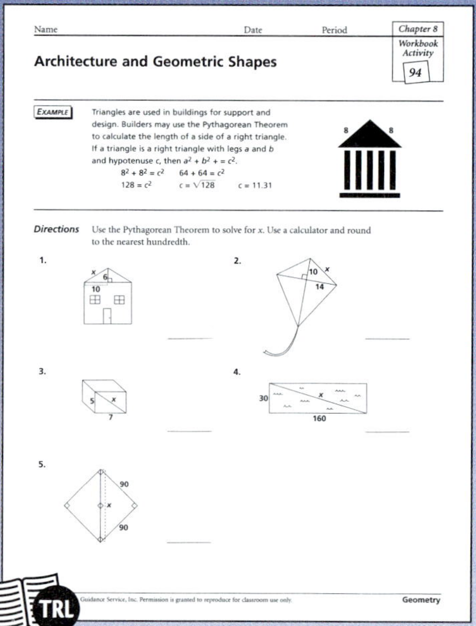

Name Date Period

Chapter 8 Workbook Activity 94

Architecture and Geometric Shapes

EXAMPLE Triangles are used in buildings for support and design. Builders may use the Pythagorean Theorem to calculate the length of a side of a right triangle. If a triangle is a right triangle with legs a and b and hypotenuse c, then $a^2 + b^2 + = c^2$.

$8^2 + 8^2 = c^2$ $64 + 64 = c^2$

$128 = c^2$ $c = \sqrt{128}$ $c = 11.31$

Directions Use the Pythagorean Theorem to solve for x. Use a calculator and round to the nearest hundredth.

1. _____
2. _____
3. _____
4. _____
5. _____

 Geometry

Workbook Activity 94

Find the length of the unknown side of each triangle. Use a calculator and round your answer to the nearest hundredth.

Example:

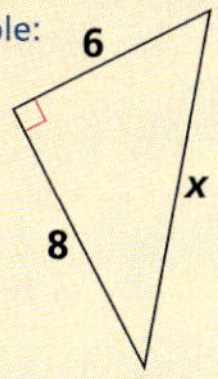

A. 14
B. 10
C. 7
D. 100

Solution: B; In a right triangle, $a^2 + b^2 = c^2$
Substitute the values for a and b.
$8^2 + 6^2 = x^2$
$64 + 36 = x^2$
$100 = x^2$
$\sqrt{100} = 10 = x$

1.

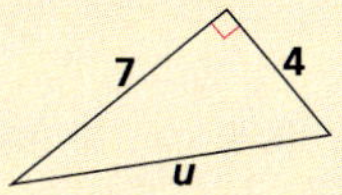

D

A. 65 **C.** 11
B. 80.6 **D.** 8.06

2.

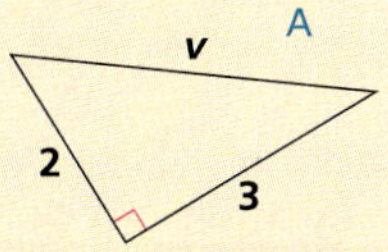

A

A. 3.61 **C.** 13
B. 5 **D.** 36.1

3.

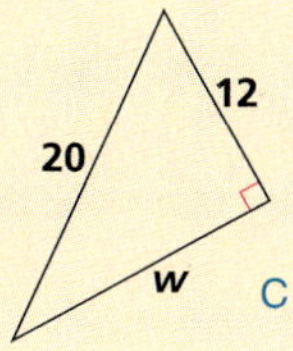

C

A. 28 **C.** 16
B. 32 **D.** 64.4

4.

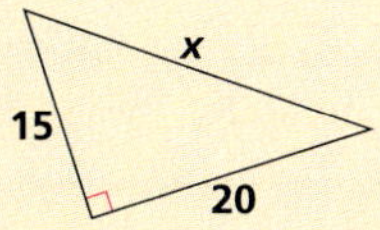

B

A. 6.25 **C.** 35
B. 25 **D.** 30

5.

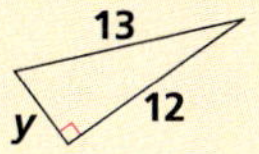

A

A. 5 **C.** 7
B. 12.5 **D.** 25

6.

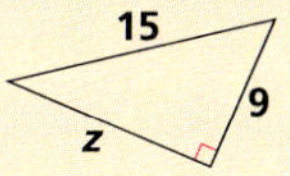

C

A. 17.49 **C.** 12
B. 24 **D.** 30.6

Name Date Period | *Chapter 8* *Mastery Test A* *page 1*

Chapter 8 Mastery Test A

Directions Use Plato's Formula to find Pythagorean Triples for the given integers. Use a calculator.

1. $m = 4$ __________
2. $m = 5$ __________
3. $m = 9$ __________
4. $m = 12$ __________
5. $m = 50$ __________

Directions Use the figure at the right to find the following quantities.

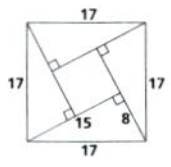

6. Area of the large square = __________
7. Area of each right triangle = __________
8. Side length and area of the small square = __________
9. Sum of the areas that make up the large square = __________
10. Does the area of the large square equal the sum of the areas of the four triangles plus the area of the small square? __________

Directions Use the figures to answer the questions.

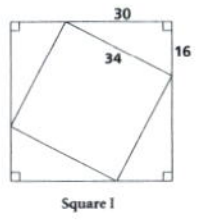

Square I

11. What is the area of each right triangle in Square I? __________
12. What is the area of the inner square in Square I? __________
13. What is the total area of Square I? __________
14. What is the sum of the areas of the four right triangles plus the area of the inner square? __________
15. Does the area of the large square equal the sum of the areas of the four triangles plus the area of the small square? __________

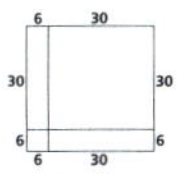

Square II

16. What is the area of the smaller square in Square II? __________
17. What is the area of the larger inside square? __________
18. What is the area of each rectangle in Square II? __________
19. What is the sum of the areas of the squares plus the areas of the rectangles? __________
20. How does the area of Square I compare to the area of Square II? __________

 Geometry

Name Date Period | *Chapter 8* *Mastery Test A* *page 2*

Chapter 8 Mastery Test A, continued

Directions Use the figure to answer the questions.

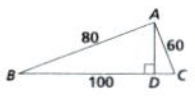

21. Name the three similar triangles. __________
22. What is the length of $\overline{AD}$? __________
23. What is the length of $\overline{BD}$? __________
24. What is the length of $\overline{CD}$? __________
25. What is the ratio between ΔABC, ΔDAC, and ΔDBA? __________

Directions Use the distance formula to find the distances between the given points. Circle the correct answer.

26. (–4, 2) and (2, 4) **A.** $\sqrt{32}$ **B.** $\sqrt{36}$ **C.** $\sqrt{40}$ **D.** $\sqrt{44}$
27. (1, –5) and (5, –2) **A.** $\sqrt{26}$ **B.** $\sqrt{5}$ **C.** $\sqrt{6}$ **D.** $\sqrt{24}$
28. (6, 1) and (1, 3) **A.** $\sqrt{29}$ **B.** $\sqrt{31}$ **C.** $\sqrt{27}$ **D.** $\sqrt{30}$
29. (–3, 3) and (–4, 4) **A.** $\sqrt{49}$ **B.** $\sqrt{50}$ **C.** $\sqrt{51}$ **D.** $\sqrt{52}$
30. (7, –1) and (–1, 7) **A.** $\sqrt{134}$ **B.** $\sqrt{132}$ **C.** $\sqrt{130}$ **D.** $\sqrt{128}$

Directions Use the converse of the Pythagorean Theorem to find the triples that could be the sides of a right triangle. Write the letter of the correct answer.

31. __________
A. 8, 16, 17 **B.** 4, 13, 15 **C.** 32, 255, 257 **D.** 10, 25, 27

32. __________
A. 9, 80, 82 **B.** 7, 48, 50 **C.** 11, 120, 122 **D.** 30, 225, 227

33. __________
A. 60, 80, 100 **B.** 6, 9, 11 **C.** 12, 6, 13 **D.** 90, 2022, 2024

34. __________
A. 18, 80, 84 **B.** 14, 49, 51 **C.** 9, 10, 11 **D.** 13, 5, 12

35. __________
A. 12, 14, 22 **B.** 5, 49, 51 **C.** 12, 16, 20 **D.** 40, 299, 301

 Geometry

TRL

Chapter 8 Mastery Test A

Chapter 8 Review

Each set of problems in the Chapter Review includes an example and solution to illustrate the concept. Use the given examples for reteaching the material in Chapter 8. For additional practice, refer to the Supplementary Problems for Chapter 8 (pages 410–411) and Additional Practice 47–53 in the Teacher's Resource Library.

Chapter 8 Mastery Test TRL

The Teacher's Resource Library includes parallel forms of the Chapter 8 Mastery Test. The difficulty level of the two forms is equivalent. You may wish to use one form as a pretest and the other form as a posttest. In addition, the Chapter 8 Cumulative Mastery Test covers content from Chapters 1–8.

Alternative Assessment

Alternative Assessment items correlate with student Goals for Learning at the beginning of this chapter.

- Have students use the Pythagorean Theorem to determine the length of an unknown side of a right triangle.

Construct four right triangles. For each triangle, assign lengths to two of the sides, leaving the length of the third side unknown. Exchange triangles with a partner and find the length of the unknown side.

- Have students determine the area of a trapezoid.

Draw five trapezoids assigning a length to each base and side. Exchange trapezoids with a partner and find the areas of your partner's trapezoids.

- Have students find the distance between points in the coordinate plane, using the distance formula.

Find the distance between the following points:

(2, 3) and (5, 6) ($\sqrt{18}$)
(4, 1) and (8, 4) ($\sqrt{25} = 5$)
(7, 3) and (14, 5) ($\sqrt{53}$)

Alternative Assessment

continued

▶ Have students identify Pythagorean Triples.

Let students work in groups or independently to determine which of the following are Pythagorean Triples.

21, 28, 35 (yes)

35, 84, 91 (yes)

15, 35, 39 (no)

More Review...

Calculate the area of each trapezoid.

Example:

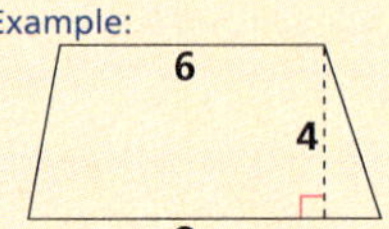

Solution: Area of a trapezoid = $h\frac{(a + b)}{2}$ where h is the height and a and b are the parallel sides. Substitute the given values in the formula, $h = 4$, $a = 6$, $b = 8$.

area = $4\frac{(6 + 8)}{2} = 4\frac{14}{2} = 4 \bullet 7$

area = 28

7.

40

8.

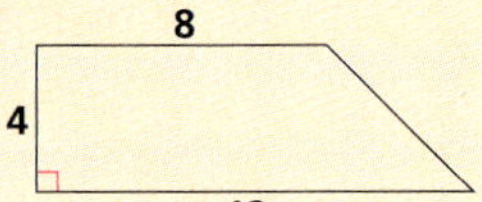

40

9.

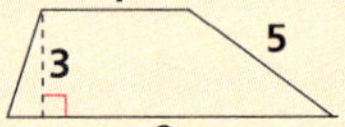

19.5

10.

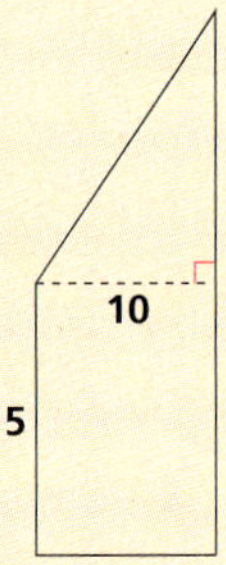

225

11.

77

12.

85

Name ______ Date ______ Period ______ *Chapter 8 Mastery Test B page 1*

Chapter 8 Mastery Test B

Directions Use Plato's Formula to find Pythagorean Triples for the given integers. Use a calculator.

1. $m = 6$ ______
2. $m = 7$ ______
3. $m = 8$ ______
4. $m = 13$ ______
5. $m = 100$ ______

Directions Use the figure at the right to find the following quantities.

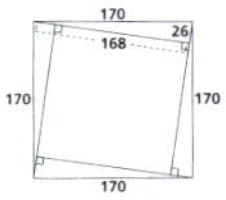

6. Area of the large square = ______
7. Area of each right triangle = ______
8. Side length and area of the small square = ______
9. Sum of the areas that make up the large square = ______
10. Does the area of the large square equal the sum of the areas of the four triangles plus the area of the small square? ______

Directions Use the figures to answer the questions.

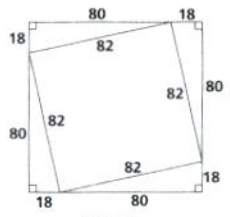

Square I

11. What is the area of each right triangle in Square I? ______
12. What is the area of the inner square in Square I? ______
13. What is the total area of Square I? ______
14. What is the sum of the areas of the four right triangles plus the area of the inner square? ______
15. Does the area of the large square equal the sum of the areas of the four triangles plus the area of the small square? ______
16. What is the area of the smaller square in Square II? ______
17. What is the area of the larger inside square? ______
18. What is the area of each rectangle in Square II? ______
19. What is the sum of the areas of the squares plus the areas of the rectangles? ______
20. How does the area of Square I compare to the area of Square II? ______

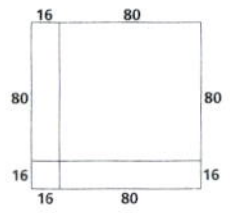

Square II

TRL

Geometry

Name ______ Date ______ Period ______ *Chapter 8 Mastery Test B page 2*

Chapter 8 Mastery Test B, continued

Directions Use the figure to answer the questions.

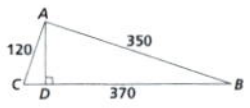

21. Name the three similar triangles. ______
22. What is the length of $\overline{AD}$? ______
23. What is the length of $\overline{BD}$? ______
24. What is the length of $\overline{CD}$? ______
25. What is the ratio between ΔABC, ΔDBA, and ΔDAC? ______

Directions Use the distance formula to find the distances between the given points. Circle the correct answer.

26. (5, 2) and (0, −4) **A.** $\sqrt{59}$ **B.** $\sqrt{60}$ **C.** $\sqrt{61}$ **D.** $\sqrt{62}$
27. (4, 6) and (5, 7) **A.** $\sqrt{1}$ **B.** $\sqrt{2}$ **C.** $\sqrt{3}$ **D.** $\sqrt{4}$
28. (−3, −2) and (0, 0) **A.** $\sqrt{10}$ **B.** $\sqrt{11}$ **C.** $\sqrt{12}$ **D.** $\sqrt{13}$
29. (0, 6) and (7, 8) **A.** $\sqrt{49}$ **B.** $\sqrt{51}$ **C.** $\sqrt{53}$ **D.** $\sqrt{55}$
30. (45, 3) and (49, 7) **A.** $\sqrt{32}$ **B.** $\sqrt{34}$ **C.** $\sqrt{38}$ **D.** $\sqrt{30}$

Directions Use the converse of the Pythagorean Theorem to find the triples that could be the sides of a right triangle. Write the letter of the correct answer.

31. ______
    A. 14, 48, 51 **B.** 13, 12, 5 **C.** 10, 8, 7 **D.** 100, 2499, 2500
32. ______
    A. 35, 37, 12 **B.** 34, 286, 288 **C.** 24, 144, 146 **D.** 11, 121, 123
33. ______
    A. 5, 10, 15 **B.** 14, 48, 50 **C.** 8, 14, 16 **D.** 80, 1599, 1600
34. ______
    A. 5, 6, 7 **B.** 1, 2, 3 **C.** 3, 4, 5 **D.** 6, 7, 8
35. ______
    A. 98, 2400, 2402 **B.** 77, 1481, 1483 **C.** 5, 4, 2 **D.** 20, 399, 400

TRL

Geometry

Chapter 8 Mastery Test B

Find the distance between the named points. Use graph paper or the formula for calculating distance. You may leave the distance in square root form.

Example: *A* and *B*

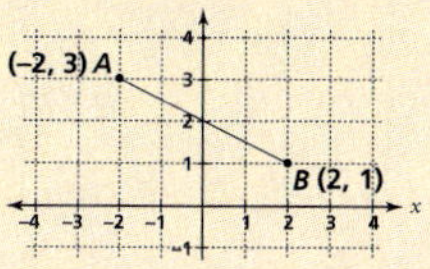

Solution: distance = $\sqrt{(x_2 - x_1)^2 + (y_2 - y_1)^2}$

Let (–2, 3) = (x_1, y_1) and let (2, 1) = (x_2, y_2)

then $d = \sqrt{(2 - (-2))^2 + (1 - 3)^2}$

$d = \sqrt{(4)^2 + (-2)^2}$

$d = \sqrt{16 + 4}$

$d = \sqrt{20} = 2\sqrt{5}$

13. *C* and *D* $\sqrt{41}$

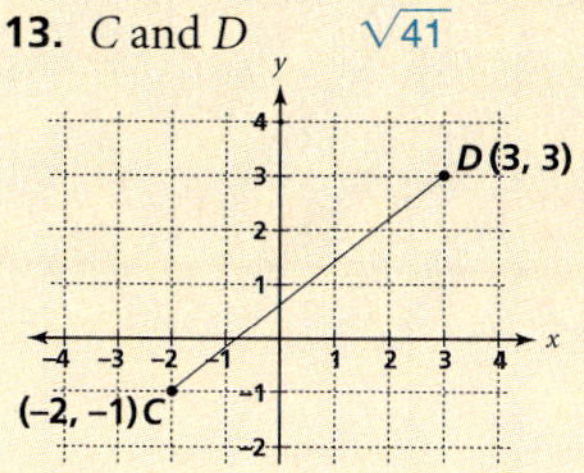

14. *E* and *F* $\sqrt{65}$

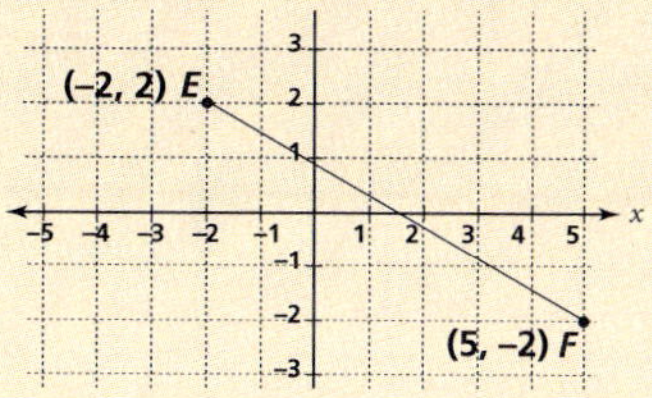

Tell whether or not these triples are sides of a right triangle. Answer *yes* or *no.*

Example: 25, 60, 65

Solution: Use the formula: $a^2 + b^2 = c^2$. Substitute the given values in the equation. $25^2 + 60^2 \stackrel{?}{=} 65^2$ $625 + 3{,}600 \stackrel{?}{=} 4{,}225$

4,225 = 4,225 Yes, the triple forms a right triangle.

15. 9, 12, 15 yes **17.** 12, 35, 37 yes **19.** 40, 400, 401 no

16. 7, 8, 11 no **18.** 35, 84, 91 yes **20.** 20, 99, 101 yes

Test-Taking Tip

When you read test directions, restate them in your own words. Tell yourself what you are expected to do.

Name Date Period

Chapter 8 — Cumulative Mastery Test — page 1

Chapter 8 Cumulative Mastery Test

Directions Graph each line using the slope and point given on a separate piece of paper. Also write the algebraic expression for each line.

1. Line *a*; $m = 3$, passes through (0, 4)
2. Line *b*; $m = \frac{1}{2}$, passes through (–3, 2)

Directions Complete the following constructions. Use a separate sheet of paper. You will need a straightedge and a compass.

3. Construct a scalene triangle using the lengths of segments *a*, *b*, and *c*. ___a___ ___b___ ___c___
4. Draw any equilateral triangle using sides the length of segment *z* and label the triangle *ABC*. Construct the altitudes, angle bisectors, and medians from *A*, *B*, and *C*. Label the points where each set of lines cross by their correct names. ___z___
5. Construct ΔABC with $\overline{AB} = \overline{AC}$ and $m\angle A = 90°$.
6. Draw any acute triangle and label it *ABC*. Construct a second triangle *A'B'C'* that is congruent to ΔABC using the AAS postulate.

Directions The pair of triangles is similar. Solve for the values of the unknowns.

7. a = ________
8. b = ________

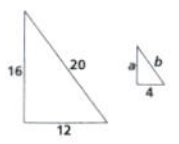

Directions Give the coordinates of the image quadrilateral under the following dilations. Use graph paper to graph the object and its images. (All dilations have (0, 0) as the center of the dilation.)

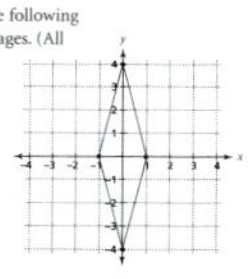

9. dilation of 2 ________ 13. dilation of $\frac{1}{3}$ ________
10. dilation of 3 ________ 14. dilation of $\frac{1}{4}$ ________
11. dilation of 4 ________ 15. dilation of $\frac{1}{5}$ ________
12. dilation of $\frac{1}{2}$ ________ 16. dilation of $\frac{1}{6}$ ________

TRL

 Geometry

Name Date Period

Chapter 8 — Cumulative Mastery Test — page 2

Chapter 8 Cumulative Mastery Test, continued

Directions Use Plato's Formula to find Pythagorean Triples for the given integers. Use a calculator.

17. $m = 17$ ________
18. $m = 9$ ________

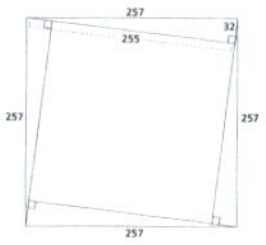

Directions Use the figure to find the following quantities.

19. Area of the large square = ________
20. Area of each right triangle = ________
21. Side length and area of the small square = ________
22. Sum of the areas that make up the large square = ________

Directions Use the figure to answer the questions.

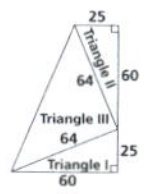

23. What are the areas of Triangle I and Triangle II? ________
24. What is the area of Triangle III? ________
25. What is the area of the entire trapezoid? ________

Directions Use the distance formula to find the distances between the given points. Circle the correct answer.

26. (0, 0) and (–4, –4) **A.** $\sqrt{30}$ **B.** $\sqrt{31}$ **C.** $\sqrt{32}$ **D.** $\sqrt{33}$
27. (8, 4) and (9, 2) **A.** $\sqrt{2}$ **B.** $\sqrt{3}$ **C.** $\sqrt{4}$ **D.** $\sqrt{5}$
28. (–4, –5) and (3, –1) **A.** $\sqrt{64}$ **B.** $\sqrt{65}$ **C.** $\sqrt{66}$ **D.** $\sqrt{6}$

Directions Use the converse of the Pythagorean Theorem to find the triples that could be the sides of a right triangle. Write the letter of the correct answer.

29. ________

A. 4, 5, 6 **B.** 7, 8, 9 **C.** 6, 8, 10 **D.** 8, 16, 18

29. ________

A. 24, 26, 10 **B.** 11, 120, 122 **C.** 14, 49, 51 **D.** 16, 65, 67

TRL

 Geometry

Chapter 8 Cumulative Mastery Test

Chapter 9

Planning Guide

Perimeter and Area

	Student Text Lesson						
	Student Pages	Vocabulary	Proof	Practice Exercises	Problem Solving	Calculator Practice	Solutions Key
Lesson 1 Perimeters of Polygons	284–286	✔		✔			✔
Lesson 2 Perimeter Formulas	287–290			✔	290		✔
Lesson 3 Perimeters and Diagonals	291–294			✔			✔
Lesson 4 Area: Squares, Rectangles, Parallelograms	295–299			✔	299		✔
Lesson 5 Areas of Trapezoids	300–303	✔		✔			✔
Lesson 6 Areas of Triangles	304–309			✔		307	✔
Application Tessellations	310			✔			✔

Chapter Activities

Teacher's Resource Library
Geometry on the Job 9: Manufacturing

Community Connection 9: Community Development

Teacher's Edition
Chapter 9 Project

Assessment Options

Student Text
Chapter 9 Review

Teacher's Resource Library
Chapter 9 Mastery Tests A and B

Chapter 9 Cumulative Mastery Test

Teacher's Edition
Chapter 9 Alternative Assessment

Student Text Features						Teaching Strategies								Learning Styles						Teacher's Resource Library		
Algebra Connection	Constructions	Experiencing Geometry	Technology Connection	Try This	Writing About Mathematics	Warm-Up Activity	Common Error	Algebra Connection	Mental Math	Manipulatives	Calculator	Applications	Group Problem Solving	Auditory/Verbal	Visual/Spatial	Body/Kinesthetic	Logical/Mathematical	Interpersonal/Group Learning	LEP/ESL	Workbook Activities	Additional Practice	Self-Study Guide
285						284		285											286	95	54	✔
287					290	287									290	289				96	55	✔
291						291								293			293	294		97	56	✔
297		296			298	295	298	297				297	299			298			298	98	57	✔
				302		300			301	301		303							302	99	58	✔
			308	309		304					308	308			309		307		306	100	59	✔
						310						310								101		✔

Manipulatives

Polygon Power Pack

Manipulatives Master 4
(Tessellations)

Chapter at a Glance

(Answer Keys for the Teacher's Resource Library begin on page 484 of this Teacher's Edition.)

Name Date Period — Chapter 9 *Geometry on the Job* 9

Manufacturing

Manufacturing is an industry that makes thousands of products from pencils to trucks. The word *manufacturing* comes from Latin and means "to make by hand." *Manufacturing* today means "to make by hand or by machine." Goods that are manufactured are divided into categories:

- heavy or light
- durable (lasts for a long time) or nondurable (a short- or immediate-use period)

The manufacturing process includes these steps—design and engineering of the product, purchasing the raw materials, making the product, and selling or distributing the product. Many different types of jobs are involved in each of these steps.

Investigate
Many tools designed to measure distance are manufactured. A simple tape measure, a cyclometer, or an odometer are examples of manufactured measurement devices

1. Choose a measurement device.
2. Tell about its history.
3. Tell who uses it and why.
4. Explain how to use the device to measure perimeter or area. Include any formulas you might need to complete the calculations.

 Geometry

Geometry on the Job 9

Name Date Period — Chapter 9 *Community Connection* 9

Community Development

There are probably thousands of separate plots of land in your community. Each one of these plots has a perimeter and an area. Planning for land use in communities started with the first community around 3500 B.C. Areas of the community were designated for housing, worship, or other activities. Today most communities have a master plan developed by a city planner or other city officials. The master plan identifies parts of the community for various functions such as residential, commercial, industrial, or public use such as parks. If you were to look at your town from above, you would probably be able to point out which areas are which.

The master plan for your community is laid out according to a set of laws called *zoning laws*. Generally, your community will have a board of elected or appointed officials who make decisions based on the community's zoning laws. These laws identify the type of buildings permitted in each part of the community. The laws also include regulations about size of lot, height of buildings, use of signs, and type of parking as well as general building and housing codes.

Directions
Find out about your community's master plan. Gather information and work together to make a report with drawings or maps to show which areas are residential, industrial, or commercial. Also include areas set aside for public use.

 Geometry

Community Connection 9

Chapter 9 Perimeter and Area

Maze gardens are amazing! Margam Country Park in West Glamorgan County, Wales, is a great example of one. There are walks of equal width. Thick parallel hedges separate one walk from another. The walks follow a series of paths. In maze gardens, the paths may lead to the exit or to the center of the garden. Walking through a complicated maze can be a fun way to spend an afternoon.

Imagine designing a maze garden. What would a designer need to know? Knowledge of geometry would be most helpful. The designer must determine the areas and perimeters of each walk as well as the whole garden. Also, the designer must consider the width and length of each hedge. With plans in hand, a construction crew can build the maze garden. Then walkers can enjoy making their way through it.

In Chapter 9, you will learn how to use formulas to calculate the perimeters and areas of polygons.

Goals for Learning

- To identify formulas for finding the perimeters of polygons
- To identify formulas for finding the areas of polygons
- To use the distance formula for finding perimeter
- To use formulas to find the perimeters and areas of polygons
- To use the Pythagorean Theorem in order to determine perimeter and area

283

SELF-STUDY GUIDE

Name ______

CHAPTER 9: Perimeter and Area

Goal 9.1 *To identify formulas for finding the perimeters of polygons*

Date	Assignment	Score
______	1: Read pages 284–285. Complete Exercises A–C on page 286.	______
______	2: Complete Workbook Activity 95.	______
______	3: Read pages 287–289. Complete Exercises A–D on pages 289–290.	______
______	4: Complete Workbook Activity 96.	______

Comments:

Goals 9.2 and 9.3 *To use the distance formula for finding perimeter; to use the Pythagorean Theorem in order to determine perimeter*

Date	Assignment	Score
______	5: Read pages 291–293. Complete Exercises A and B on pages 293–294.	______
______	6: Complete Workbook Activity 97.	______

Comments:

Goals 9.3 and 9.4 *To use the Pythagorean Theorem in order to determine perimeter; to identify formulas for finding the areas of polygons*

Date	Assignment	Score
______	7: Read pages 295–298. Complete Exercises A–E on pages 298–299.	______
______	8: Complete Workbook Activity 98.	______
______	9: Read pages 300–302. Complete Exercises A–C on pages 302–303.	______
______	10: Complete Workbook Activity 99.	______

Comments:

 Geometry

TRL

SELF-STUDY GUIDE

Name ______

CHAPTER 9: Perimeter and Area, continued

Goal 9.5 *To use formulas to find the perimeter and area of polygons*

Date	Assignment	Score
______	11: Read pages 304–307.	______
______	12: Read the Calculator Practice on page 307.	______
______	13: Complete Exercises A–C on pages 308–309.	______
______	14: Complete Workbook Activity 100.	______
______	15: Read and complete the Application Exercise on page 310.	______
______	16: Complete Workbook Activity 101.	______
______	17: Complete the Chapter 9 Review on pages 311–313.	______

Comments:

Student's Signature ______ Date ______

Instructor's Signature ______ Date ______

 Geometry

TRL

Chapter 9 Self-Study Guide

Introducing the Chapter

Use the photograph on these pages to help students distinguish between perimeter and area. Ask students to tell about any experiences they have had going through mazes. If students have no experiences to offer, invite them to speculate on what it would feel like to walk through a maze similar to the one in the photograph. Ask them what kinds of things would concern the maze designers. What would the designers have to keep in mind about the perimeter and area of the maze and the paths? Encourage students to be creative in their answers. Read the Goals for Learning and discuss each one with students.

Chapter Project

Challenge students to work in groups to create model mazes. Suggest that the overall dimensions of each maze be 4 ft by 4 ft. Indicate that the pathways and turns must be wide enough for a matchbox-sized car. Student mazes can be as simple or complex as they wish and may lead from one corner through the maze to a diagonal corner or from one edge to the center. For each maze, students can determine the lengths of the maze walls used, the outside perimeter of the maze, and the area of the maze.

Lesson at a Glance

Chapter 9 Lesson 1

Overview This lesson presents a formal definition of perimeter and formulas for finding the perimeters of polygons.

Objectives

- To find the perimeters of given polygons.
- To use the perimeter of a polygon to find the length of an unknown side.

Student Pages 284–286

Teacher's Resource Library TRL

Workbook Activity 95

Additional Practice 54

Mathematics Vocabulary

perimeter of a polygon

1. Warm-Up Activity

Invite several students to participate in a walk-around. Some can walk around a desk, counting paces. Others can walk around a room, the cafeteria, or schoolyard. Students should keep track of the number of paces they used in their walk-around. Record how many paces students used in walking around objects. Invite them to discuss how their walk-arounds are related to the perimeters of the objects. Point out that in this lesson students will learn some formal ways to describe and measure the distance around a polygon.

Lesson 1 Perimeters of Polygons

Perimeter of a polygon
Sum of the lengths of all the sides of the polygon

A common way to think about perimeter is to picture perimeter as the distance around a figure. This picture matches the more formal definition.

The **perimeter of a polygon** is the sum of the lengths of all the sides of the polygon.

EXAMPLES Perimeter = 5 + 6 + 7 = 18

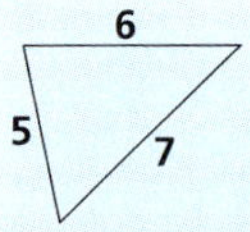

Perimeter = 10 + 15 + 13 + 20 = 58

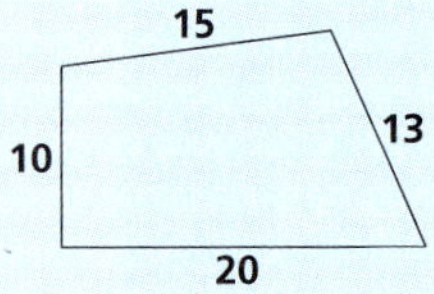

Perimeter =
10 + 30 + 20 + 8 + 15 = 83

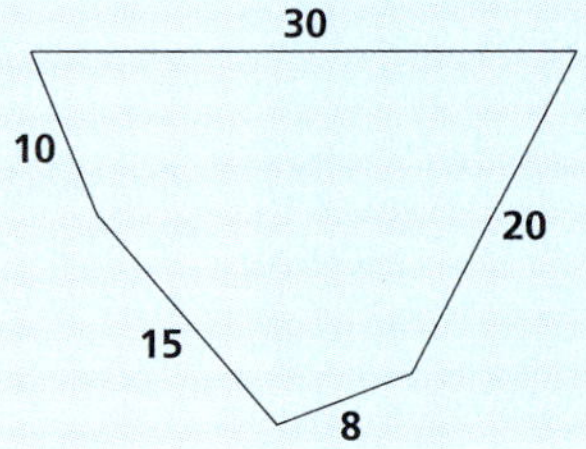

When a polygon has no equal sides, like those in the examples above, there is no formula or "shortcut" for finding the perimeter. The length of each side must be known.

$P = a + b + c$ no equal sides

$P = a + b + c + d$ no equal sides

$P = a + b + c + d + e$ no equal sides

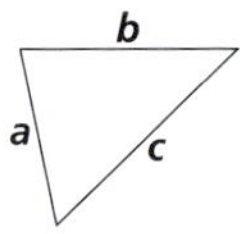

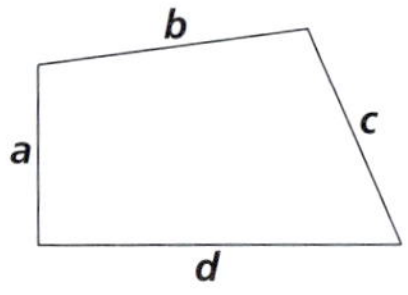

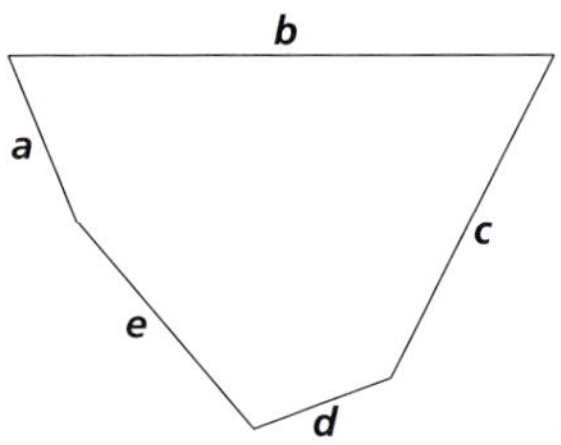

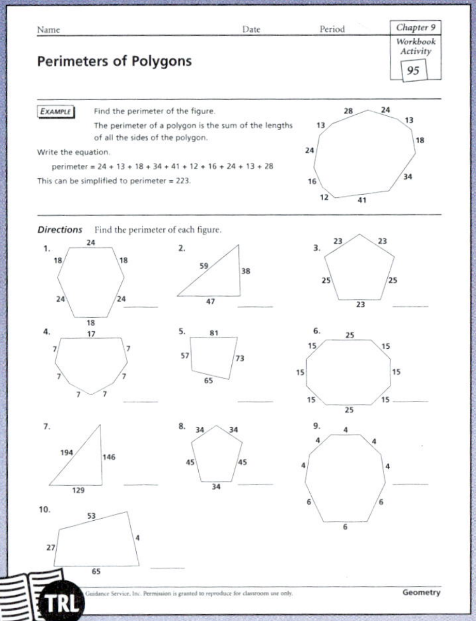
Name Date Period Chapter 9 Workbook Activity 95

Perimeters of Polygons

EXAMPLE Find the perimeter of the figure.
The perimeter of a polygon is the sum of the lengths of all the sides of the polygon.
Write the equation.
perimeter = 24 + 13 + 18 + 34 + 41 + 12 + 16 + 24 + 13 + 28
This can be simplified to perimeter = 223.

Directions Find the perimeter of each figure.

Geometry

Workbook Activity 95

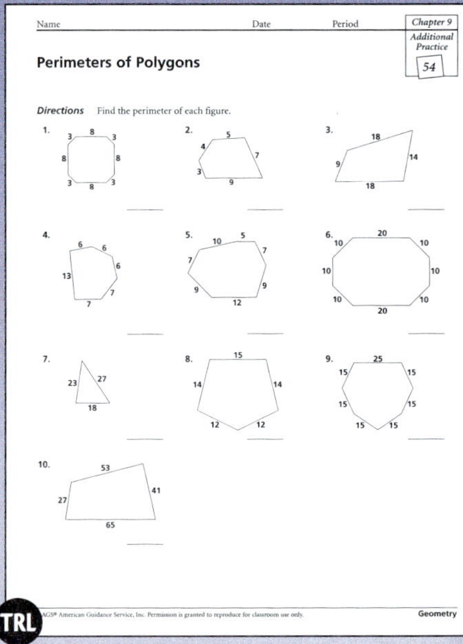
Name Date Period Chapter 9 Additional Practice 54

Perimeters of Polygons

Directions Find the perimeter of each figure.

Geometry

Additional Practice 54

If you know the perimeter of a polygon, then you can find the length of one missing side.

Algebra Connection

EXAMPLES

A quadrilateral has a perimeter of 60. If three of its sides are 10 ft, 12 ft, and 15 ft long, what is the length of the fourth side?

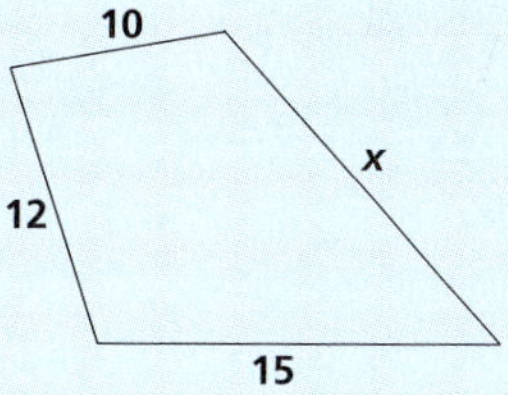

Draw a picture to help you solve the problem.

Let x = length of missing side

$10 + 12 + 15 + x = 60$

$37 + x = 60$

$x = 60 - 37$

$x = 23$ The missing side is 23 ft long.

The short side of a rectangular garden is 10 m, the long side is 15 m. What is the perimeter of the garden?

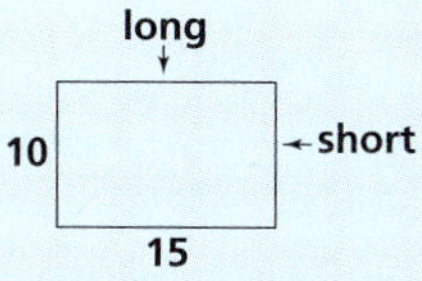

Draw a picture to help you solve the problem.

Perimeter = 2 short sides + 2 long sides

$P = 2(10) + 2(15)$

$P = 20 + 30 = 50$ The perimeter of the garden is 50 m.

Suppose you have 100 ft of wire fencing to make a square enclosure for your dog. How long will each side of the square be?

Draw a picture to help you solve the problem.

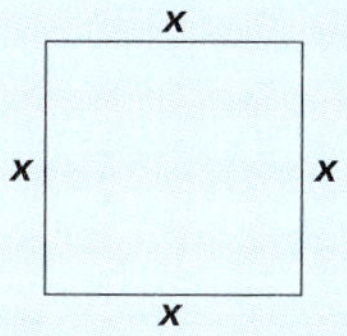

$P = 100$

$P = x + x + x + x = 4x$

$P = 4x = 100$

$x = 25$ Each side is 25 ft.

2. Teaching the Lesson

Work through the examples on page 284 with students. Point out that the perimeter is the distance around a polygon, whether that polygon is regular or irregular.

Ask a volunteer to copy the figure at the top of page 285 and to then explain how to find the length of the missing side. Ask two other volunteers to explain the process used in each of the remaining examples.

Algebra Connection

Direct students to draw five different polygons on a sheet of paper. For each polygon, students should supply the total perimeter and the lengths of all but one of the polygon's sides. Have students trade polygons with one another. Students can then write an algebraic equation for finding the missing side for each polygon. Have them exchange papers again. Direct them to determine if the algebraic equation—and its solution—accurately reflect the original polygon.

3. Reinforce and Extend

Learning Styles

LEP/ESL Direct students to work in small groups with students of varying language abilities to complete Exercise B. Ask students to explain to one another how they used perimeter to find the length of a missing side. As one student describes his or her thinking process, the others should record the steps used. Once the groups have completed the exercise, students can compare their processes and work together to write one version of the process that could work for finding the perimeters of polygons with any number of sides.

Selected Answers and Solutions begin on page 420.

Exercise A Find the perimeter of each polygon.

1.

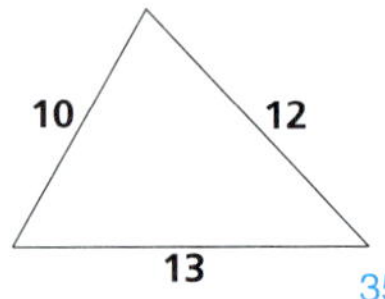

35

2.

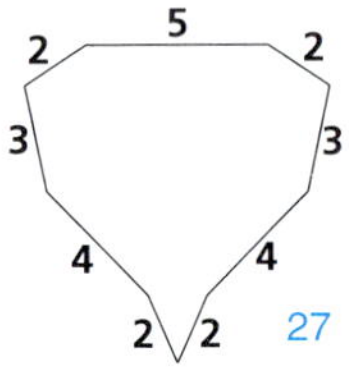

27

3.

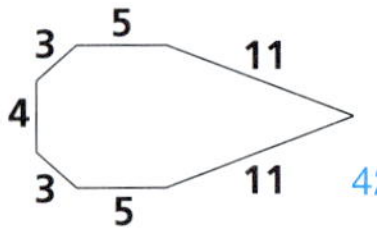

42

Exercise B Solve for the missing side. The perimeter is given.

4.

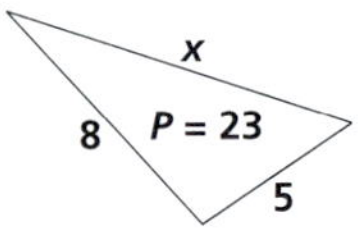

10

5.

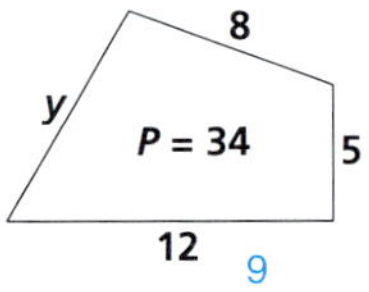

9

6.

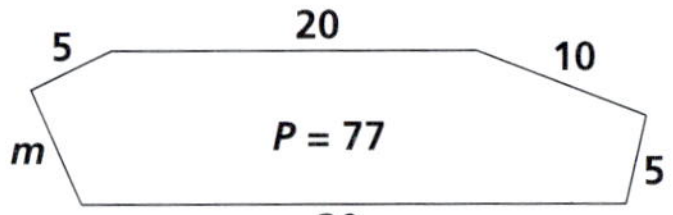

7

7.

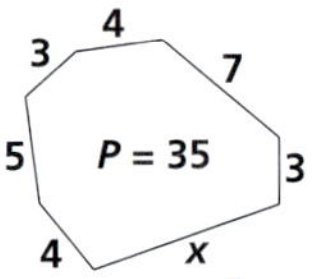

9

Exercise C Find the perimeter for each item.

8. A rectangle has a short side of 15 m and a long side of 25 m. What is the perimeter of the rectangle? $P = 80$ m

9. The short side of a rectangle is 23 ft, the long side of the same rectangle is 32 ft. What is the perimeter of the rectangle? $P = 110$ ft

10. The long side of a rectangle is 11 inches, the short side is 10 inches. What is the perimeter of the rectangle? $P = 42$ in.

Lesson 2 Perimeter Formulas

Algebra Connection

If a polygon has equal sides and equal angles, the polygon is a regular polygon. You can use a formula to find the perimeter. In the table below, s = the length of one side of a regular polygon.

Formulas for Perimeter		Examples	
Equilateral Triangle $P = s + s + s$ $P = 3s$	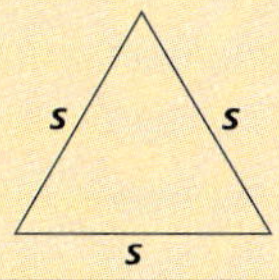	$P = 6 + 6 + 6$ $P = 3 \cdot 6 = 18$	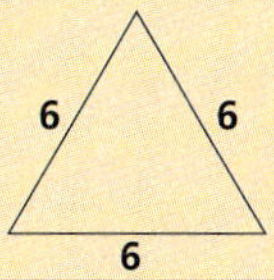
Rhombus $P = s + s + s + s$ $P = 4s$	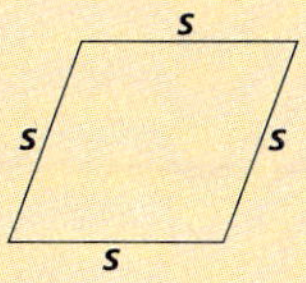	$P = 5 + 5 + 5 + 5$ $P = 4 \cdot 5 = 20$	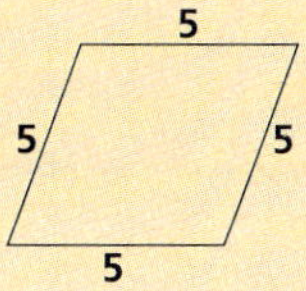
Square $P = s + s + s + s$ $P = 4s$	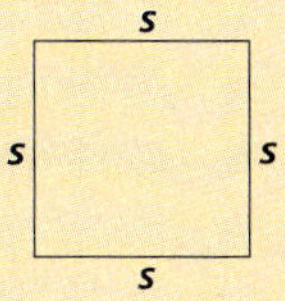	$P = 5 + 5 + 5 + 5$ $P = 4 \cdot 5 = 20$	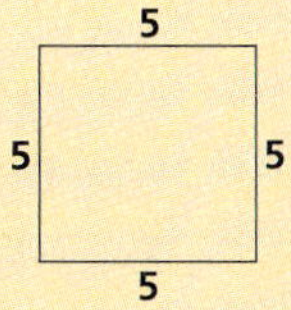
Regular Pentagon $P = s + s + s + s + s$ $P = 5s$	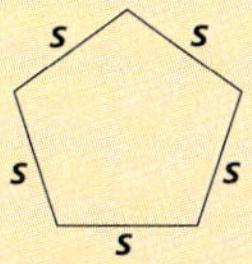	$P = 10 + 10 + 10 + 10 + 10$ $P = 5 \cdot 10 = 50$	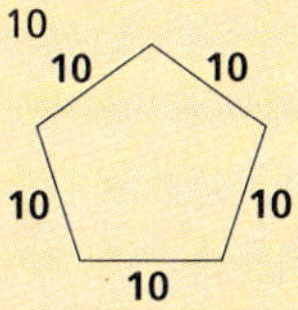
Regular Hexagon $P = s + s + s + s + s + s$ $P = 6s$	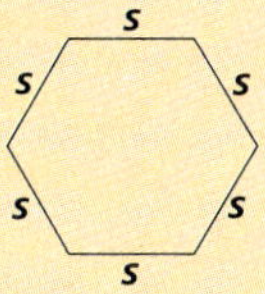	$P = 4 + 4 + 4 + 4 + 4 + 4$ $P = 6 \cdot 4 = 24$	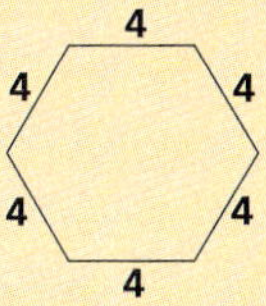
Regular n-sided Polygon $P = s + \ldots + s$ $P = ns$		P = (number of sides)(length of one side) $P = n \cdot s$	

Lesson at a Glance

Chapter 9 Lesson 2

Overview This lesson expands the concept of perimeter to regular polygons.

Objectives

- To find the perimeter of a regular polygon.
- To use the perimeter of a regular polygon and the number of sides to find the length of an unknown side.

Student Pages 287–290

Teacher's Resource Library TRL

Workbook Activity 96

Additional Practice 55

1. Warm-Up Activity

Ask students to recall and list the characteristics of a regular polygon. Ensure that students articulate that a regular polygon has sides of equal length and angles of equal measure. Ask students how they can use the term *similar* to describe regular polygons. Guide the discussion so that students characterize all regular polygons with a specified number of sides as similar.

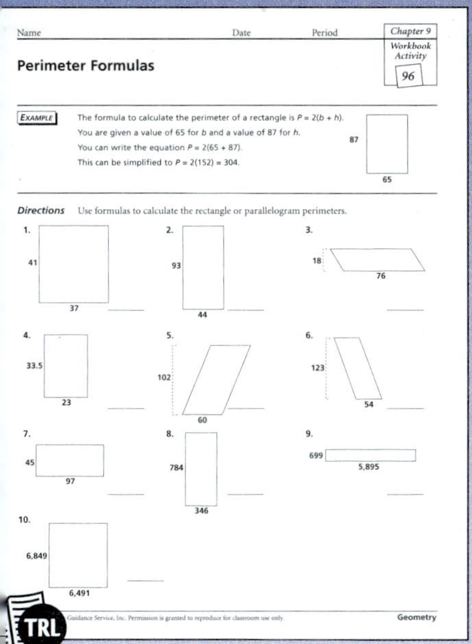

TRL Workbook Activity 96

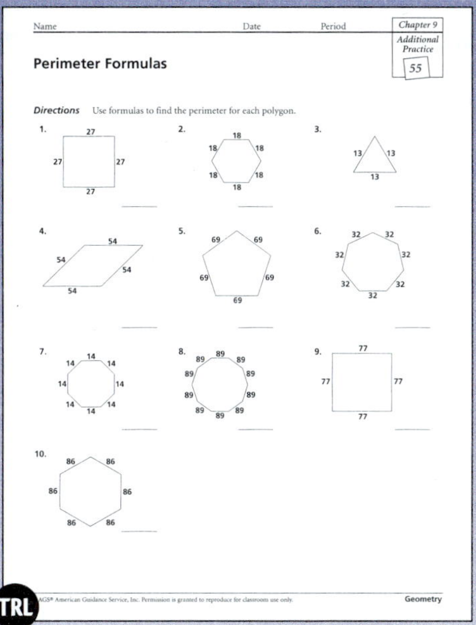

TRL Additional Practice 55

Perimeter and Area Chapter 9 287

2. Teaching the Lesson

Perform the following demonstration for students using a string, a ruler, and a rectangle suitable for use on an overhead projector. First, measure the length and width of the rectangle individually and complete the equations for perimeter shown in the left column on page 288.

To reinforce the concept of perimeter, wrap the string around the perimeter of the rectangle. Encourage students to predict the length of string needed to wrap around the edges of the rectangle. Confirm their predictions by cutting a string to the predicted length and wrapping it around the rectangle.

Invite a volunteer to perform a similar demonstration using a parallelogram.

The perimeter of a rectangle presents a special case. There are two pairs of equal sides, called the *length* and the *width*, or the *base* and the *height*.

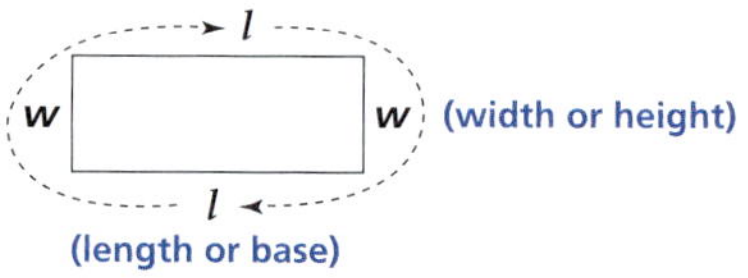

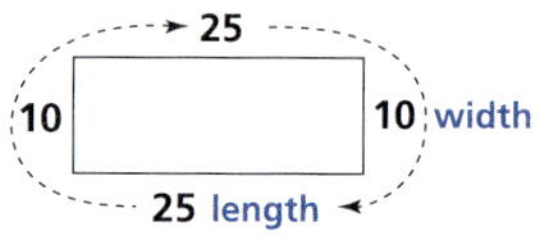

$P = l + w + l + w$

$P = l + l + w + w$

$P = 2l + 2w$

$P = 2(l + w)$

OR

$P = b + h + b + h$

$P = b + b + h + h$

$P = 2b + 2h$

$P = 2(b + h)$

$P = 25 + 10 + 25 + 10$

$P = 25 + 25 + 10 + 10$

$P = 2 \cdot 25 + 2 \cdot 10$

$P = 2(25 + 10) = 70$

FORMULAS: Perimeter of a rectangle
$P = 2l + 2w$ or $P = 2(l + w)$
$P = 2b + 2h$ or $P = 2(b + h)$

The formula for the perimeter of a parallelogram is the same as for the rectangle, except the terms used are *base* and *side*.

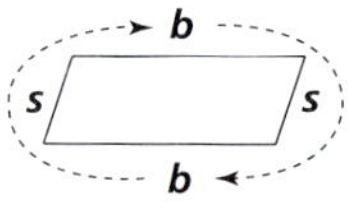

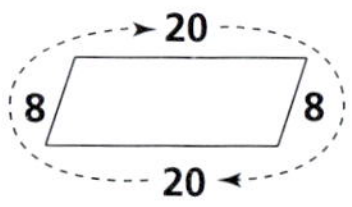

$P = b + s + b + s$

$P = b + b + s + s$

$P = 2b + 2s$

$P = 2(b + s)$

$P = 20 + 8 + 20 + 8$

$P = 20 + 20 + 8 + 8$

$P = 2 \cdot 20 + 2 \cdot 8$

$P = 2(20 + 8) = 56$

FORMULA: Perimeter of a parallelogram
$P = 2b + 2s$ or $P = 2(b + s)$

You can use these formulas to solve for a missing term.

EXAMPLES The perimeter of a rectangle is 200. Its length is 60. What is its width?

Substitute for P and l in the formula. Then solve for w, the unknown.

$$P = 2(l + w)$$
$$200 = 2(60 + w)$$
$$100 = 60 + w$$
$$40 = w$$

The perimeter of a rectangle is 600. The length of the rectangle is twice the width. How long is each side of the rectangle?

Let x = width, and $2x$ = length. Substitute in the formula. Then solve for x.

$$P = 2(l + w)$$
$$600 = 2(2x + x)$$
$$600 = 2(3x)$$
$$600 = 6x$$
$$100 = x$$ The width is 100, the length is 200.

Exercise A Use the formulas on page 287 to find the perimeter for each polygon.

1. equilateral triangle, $s = 10$

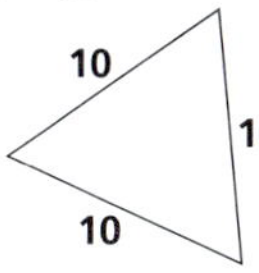

$P = 30$

2. rhombus, $s = 17$

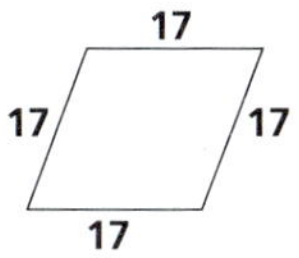

$P = 68$

3. square, $s = 14$

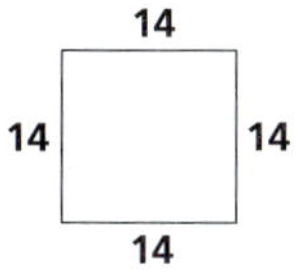

$P = 56$

4. regular pentagon, $s = 7$

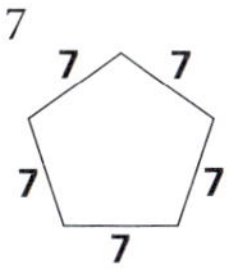

$P = 35$

5. regular hexagon, $s = 8$ $P = 48$
6. regular 10-sided polygon, $s = 12$ $P = 120$
7. regular 20-sided polygon, $s = 4$ $P = 80$
8. regular octagon, $s = 12$ $P = 96$

3. Reinforce and Extend

LEARNING STYLES

Body/Kinesthetic Present groups of students with an array of manipulative polygons such as those in the Polygon Power Pack or Pattern Blocks. Have students group like regular polygons together. Provide each group with a ruler or flexible tape measure. Ask students to measure and record the length of one side of the polygon. Then have them measure the perimeters of the polygons. Have them record their data in a table with headings such as square, rectangle, pentagon, octagon, and so on. Have students discuss their findings.

Learning Styles

Visual/Spatial Encourage pairs of students to work together to complete Exercise C. For each item, one student should draw a picture showing the information given in the problem. The second student should perform the needed calculations. Students should exchange roles after completing each item.

Writing About Mathematics

The square has the greater area. $P = 120$ for a rectangle implies $2(l + w) = 120$, or $l + w = 60$. Students can then test the rectangles that meet this condition. A rectangle 10×50 has an area of 500; a rectangle 20×40 has an area of 800, and a rectangle 30×30 ($l = w$, a square) has an area of 900.

Writing About Mathematics

Suppose a rectangle and a square each have a perimeter of 120 ft. Which figure has the greater area? (HINT: Choose a rectangle where $P =$ 120 and compare it to the square.)

Exercise B Use the formulas on page 288 to calculate the rectangle or parallelogram perimeters.

9. $l = 5, w = 2$	$P = 14$	**11.** $l = 6, w = 3$	$P = 18$
10. $l = 8, w = 4$	$P = 24$	**12.** $l = 11, w = 4$	$P = 30$

Exercise C Use what you know about perimeters to solve the following problems.

13. What is the length and width of a rectangle that has a perimeter of 160 and a length 3 times the width?

14. What are the side lengths of a rectangle whose length is 3 more than its width and whose perimeter is 84?

15. Two times the base of a parallelogram is equal to 3 times the side. The parallelogram has a perimeter of 800. What is its base? Side?

16. The side of a parallelogram is $\frac{1}{4}$ the length of its base. The parallelogram's perimeter is 125. What is the base and the side of the parallelogram?

17. The length of a rectangle is 6 times the width. The perimeter of the rectangle is 105. What is the length and width of the rectangle?

13. $w = 20$, $l = 60$
14. $w = 19.5$, $l = 22.5$
15. $b = 240$, $s = 160$
16. $b = 50$, $s = 12\frac{1}{2}$
17. $w = 7.5$, $l = 45$

Problem Solving

Exercise D Solve the following problems.

18. Delma is making a picture frame. She wants to glue gold braid around the edge of the frame. The frame is 9 in. × 14 in. What is the least amount of braid Delma will need? If the braid is sold by the yard and fractions of a yard, how much should Delma buy?

19. The El Sol community garden is in the shape of a regular octagon, 12 ft on a side. How much fencing is needed to enclose the garden?

20. Rhonda found 45 feet of fencing at a flea market. She wants to use the fencing to enclose her rectangular vegetable garden, which is 15 ft long. How wide can she make the garden and still be able to fence it in? Explain.

18. 46 in.; $\frac{46}{36}$ rounded to $1\frac{1}{3}$ yard
19. 96 ft
20. 7.5 ft ($P = 15 + 15 + x + x = 45$, $x = 7.5$)

Lesson 3 Perimeters and Diagonals

You'll find the Pythagorean Theorem is an important tool in finding the lengths of perimeters and diagonals.

Algebra Connection

EXAMPLES **Given:** a right-angled trapezoid with the measurements shown

Find: the perimeter

To find the perimeter, you'll need to add all the side lengths, including length x. To find x, you can use the Pythagorean Theorem.

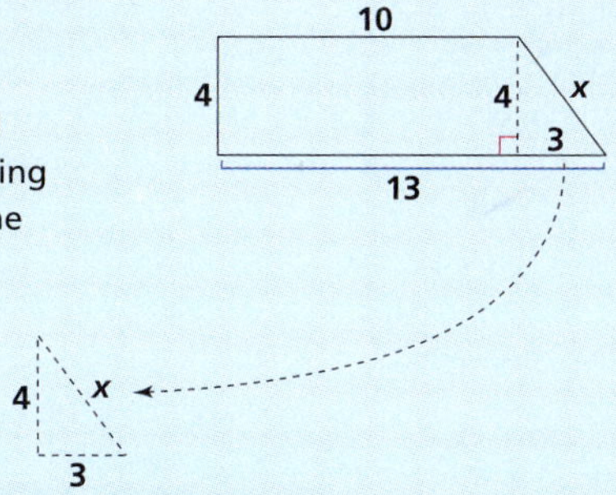

$P = 13 + x + 10 + 4$

$P = 27 + x$

$x^2 = 4^2 + 3^2$

$x^2 = 16 + 9 = 25$

$x = \sqrt{25} = 5$

$\therefore P = 27 + 5 = 32$

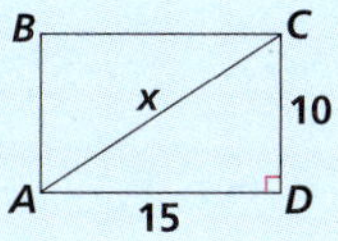

Given: rectangle $ABCD$ has side $\overline{AD} = 15$ and side $\overline{DC} = 10$

Find: the perimeter of $\triangle ADC$

To find the perimeter of the triangle, you will need to add the side lengths together, including the diagonal x, ($\overline{AC}$), the hypotenuse of the right $\triangle ADC$.

To find x, use the Pythagorean Theorem.

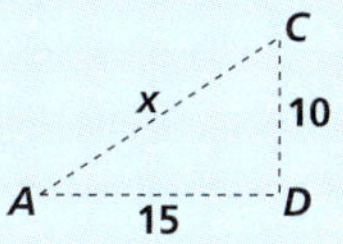

$P = 15 + 10 + x$

$x^2 = 10^2 + 15^2$

$x^2 = 100 + 225 = 325$

$x = \sqrt{325} = 18.0$

$\therefore P = 15 + 10 + 18 = 43$

Lesson at a Glance

Chapter 9 Lesson 3

Overview This lesson shows how the Pythagorean Theorem can be used to determine lengths of diagonals as well as perimeters.

Objectives

- To use the Pythagorean Theorem to find the length of a diagonal.
- To employ the distance formula and the Pythagorean Theorem to find the lengths of the sides of polygons on a coordinate grid.

Student Pages 291–294

Teacher's Resource Library TRL

Workbook Activity 97

Additional Practice 56

1. Warm-Up Activity

Reproduce the drawing of a right-angled trapezoid on the board or overhead projector. Display a variety of rectangles and triangles from a kit such as the Polygon Power Pack. Invite a volunteer to produce a right-angled trapezoid similar to your drawing. Repeat the activity, asking a volunteer to form a rectangle from two right triangles. Point out to students that in this lesson they may find it helpful to think of how polygons might fit together to form new polygons.

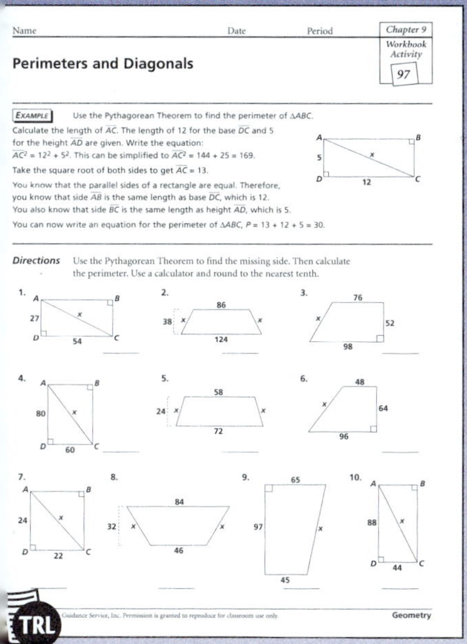

Name Date Period | *Chapter 9* *Workbook Activity* 97

Perimeters and Diagonals

EXAMPLE Use the Pythagorean Theorem to find the perimeter of $\triangle ABC$.

Calculate the length of $\overline{AC}$. The length of 12 for the base $\overline{DC}$ and 5 for the height $\overline{AD}$ are given. Write the equation: $\overline{AC}^2 = 12^2 + 5^2$. This can be simplified to $\overline{AC}^2 = 144 + 25 = 169$.

Take the square root of both sides to get $\overline{AC} = 13$.

You know that the parallel sides of a rectangle are equal. Therefore, you know that side $\overline{AB}$ is the same length as base $\overline{DC}$, which is 12. You also know that side $\overline{BC}$ is the same length as height $\overline{AD}$, which is 5.

You can now write an equation for the perimeter of $\triangle ABC$, $P = 13 + 12 + 5 = 30$.

Directions Use the Pythagorean Theorem to find the missing side. Then calculate the perimeter. Use a calculator and round to the nearest tenth.

1. 2. 3. 4. 5. 6. 7. 8. 9. 10.

 Geometry

Workbook Activity 97

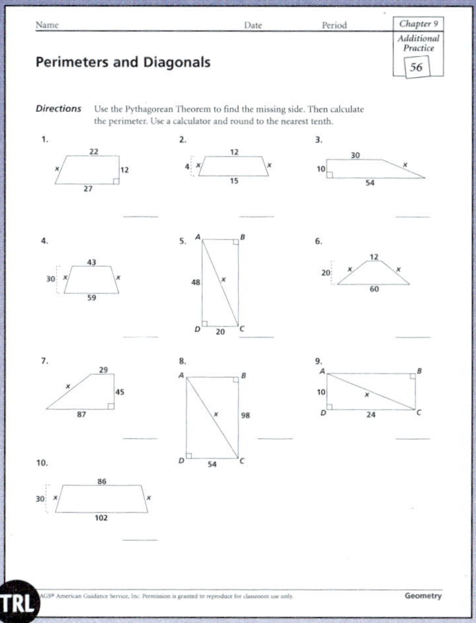

Name Date Period | *Chapter 9* *Additional Practice* 56

Perimeters and Diagonals

Directions Use the Pythagorean Theorem to find the missing side. Then calculate the perimeter. Use a calculator and round to the nearest tenth.

1. 2. 3. 4. 5. 6. 7. 8. 9. 10.

 Geometry

Additional Practice 56

2. Teaching the Lesson

Review with students the Pythagorean Theorem and its usefulness in determining the length of the hypotenuse of a right triangle. Draw students' attention to the example at the top of page 291. Point out that to find the perimeter of the trapezoid, students will need to know the lengths of all four sides, that is, they will have to find the length, x, in order to find the perimeter. Point out that this makes finding the perimeter a two-step process.

Ask students to explain each step for finding the perimeter of the right-angled trapezoid. Once students have found the value for x, suggest they write that value along the appropriate side of the trapezoid and then proceed to find the perimeter of the figure.

Invite a volunteer to explain each of the steps used to find the perimeter of the triangle on page 292. Once students have established the length of the hypotenuse, x, suggest they redraw the figure including the length of each side as part of the new diagram.

Remember that the absolute value of a number is equivalent to its distance from zero on the number line. $|n|$ is the symbol for the absolute value of n. For example, $|-3| = 3$.

EXAMPLE Calculate the perimeter of the triangle with these vertices in the coordinate plane.

$A = (1, 2)$

$B = (10, 2)$

$C = (1, 8)$

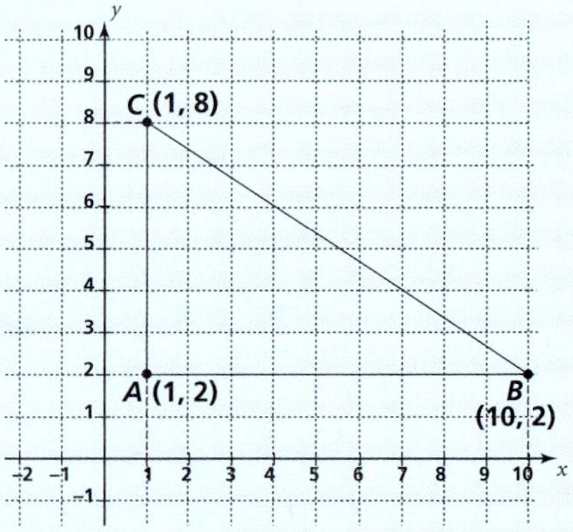

To find the perimeter, you will need to find the length of each side. You can use the Pythagorean Theorem to find the length of $\overline{BC}$. (Use a calculator and find lengths to the nearest tenth.)

$P = \overline{AC} + \overline{AB} + \overline{BC}$

$\overline{AC} = |\text{difference in the } y\text{-value}|$

$= |2 - 8| = 6 \text{ or } |8 - 2| = 6$

$\overline{AB} = |\text{difference in the } x\text{-value}|$

$= |1 - 10| = 9 \text{ or } |10 - 1| = 9$

$\overline{BC}^2 = 6^2 + 9^2$

$= 36 + 81$

$= 117$

$\overline{BC} = \sqrt{117} = 10.8$

$\therefore P = 6 + 9 + 10.8 = 25.8$

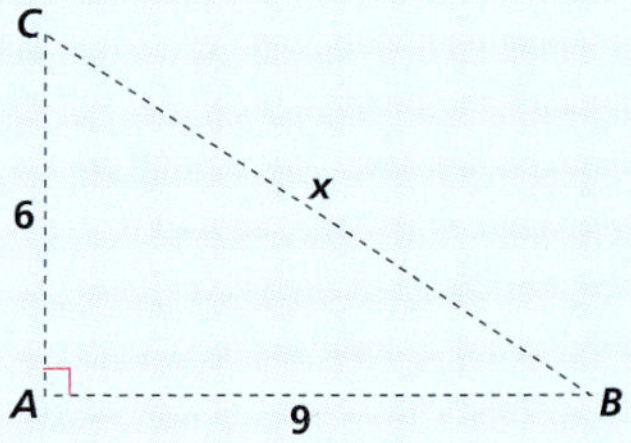

EXAMPLE Calculate the perimeter of $\triangle DEF$ whose vertices are

$D = (-2, 3)$

$E = (1, 6)$

$F = (3, 4)$

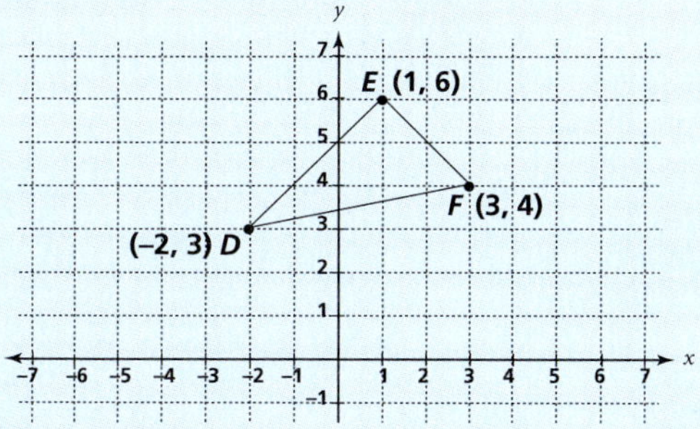

Use the distance formula you learned in Chapter 8 ($d = \sqrt{(x_2 - x_1)^2 + (y_2 - y_1)^2}$) to find the length of each of the three sides. Then add the lengths to find the perimeter. (Use a calculator and round to the nearest tenth.)

$P = \overline{DE} + \overline{EF} + \overline{FD}$

$\overline{DE}$	$\overline{EF}$	$\overline{FD}$
$D = (-2, 3), E = (1, 6)$	$E = (1, 6), F = (3, 4)$	$D = (-2, 3), F = (3, 4)$
$d = \sqrt{(-2 - 1)^2 + (3 - 6)^2}$	$d = \sqrt{(1 - 3)^2 + (6 - 4)^2}$	$d = \sqrt{(-2 - 3)^2 + (3 - 4)^2}$
$d = \sqrt{9 + 9} = \sqrt{18}$	$d = \sqrt{4 + 4} = \sqrt{8}$	$d = \sqrt{25 + 1} = \sqrt{26}$
$d = 4.2$	$d = 2.8$	$d = 5.1$

$\therefore P = 4.2 + 2.8 + 5.1 = 12.1$

Exercise A Use the Pythagorean Theorem to find the missing side. Then calculate the perimeter. (Use a calculator and round to the nearest tenth.)

1.

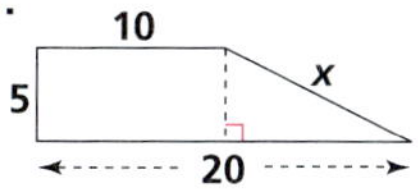

46.2

2.

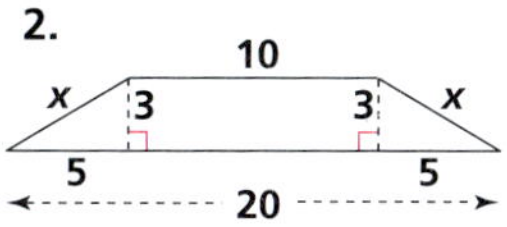

41.6

3.

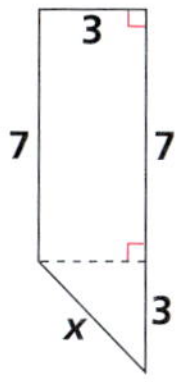

24.2

4. $P\triangle WYZ$

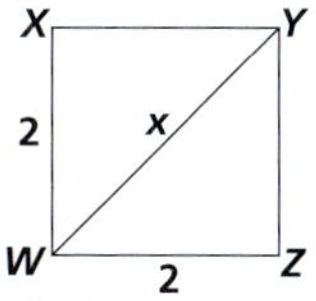

6.8

5. $P\triangle ABC$ 44.3

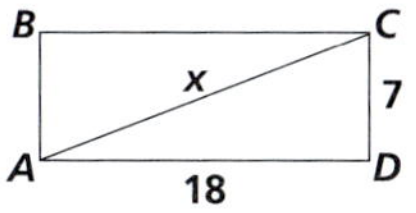

3. Reinforce and Extend

LEARNING STYLES

Logical/Mathematical Ask students to compare the example on page 292 with the one on page 293. Challenge them to explain why the distance formula, rather than the Pythagorean Theorem, is used in the example on page 293. (because the triangle is not a right triangle)

LEARNING STYLES

Auditory/Verbal Direct students to work in pairs to complete Exercise A. For each item, one student can explain to the other each of the two steps (finding the value of *x*, then using that value to find the perimeter) necessary to find the perimeter of the figure. Students can exchange speaking and listening roles after each item.

Learning Styles

Interpersonal/Group Learning Arrange students in groups of four. Provide each group with the following directions: Draw two triangles, rectangles, or trapezoids, making each side of the second figure three times as long as the corresponding side of the first.

Ask students to consider the following questions: What appears to be true about the two polygons? How do the perimeters of the two polygons compare? What generalization can you make about similar polygons and their perimeters? (The polygons are similar, the ratio of corresponding sides is 3:1, the ratio of the perimeters is also 3:1; for similar polygons, the ratio of the perimeters equals the ratio of the corresponding sides.)

Indicate to students that they should provide at least three instances for which their generalization applies.

Exercise B Find the perimeters of the following figures. Use the distance formula to find side lengths, then calculate the perimeter of each figure. (Use a calculator and round to the nearest tenth.)

6. 12

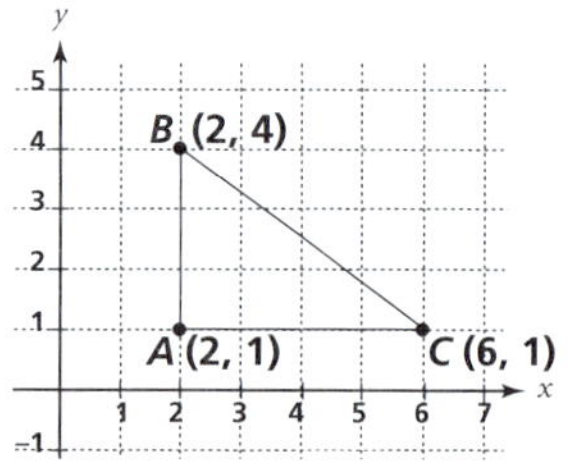

7. 22.4

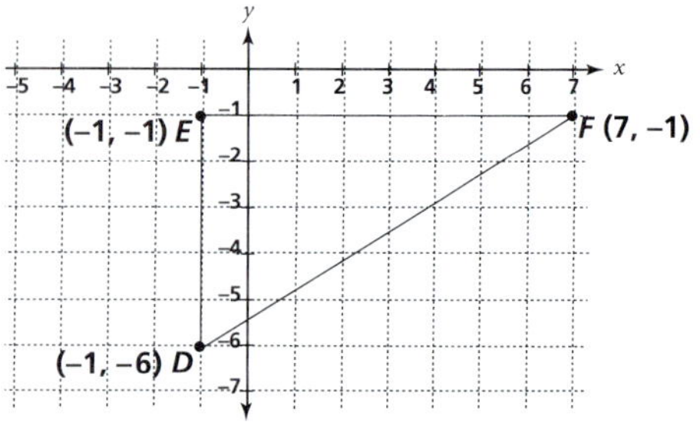

8. 22.5

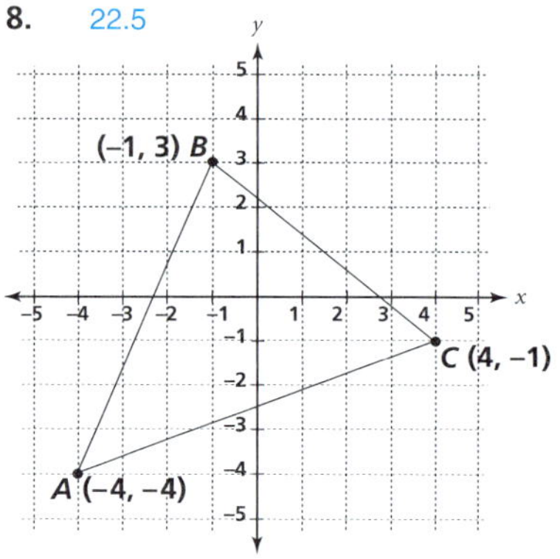

9. 21.7

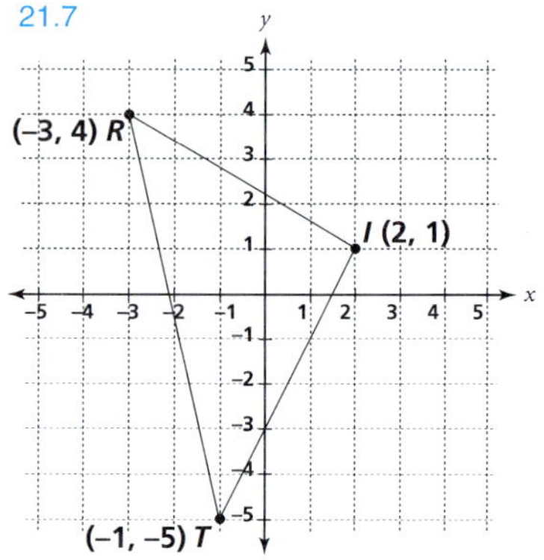

10. 24.4

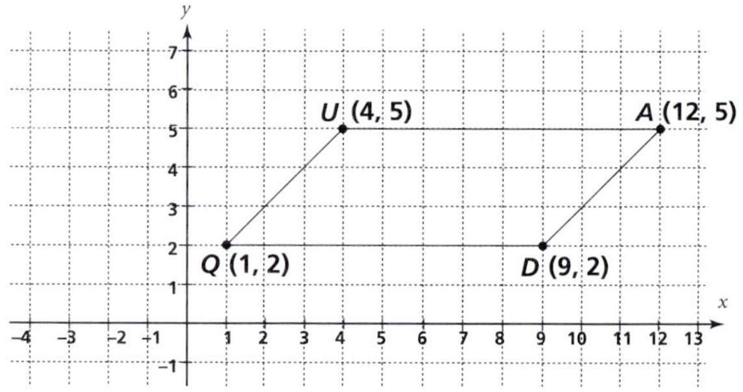

294 Chapter 9 Perimeter and Area

Lesson 4 Area: Squares, Rectangles, Parallelograms

Area is a measure of the region enclosed by a polygon. The basic unit of measurement is a square. Any square can be used. Here are some common examples.

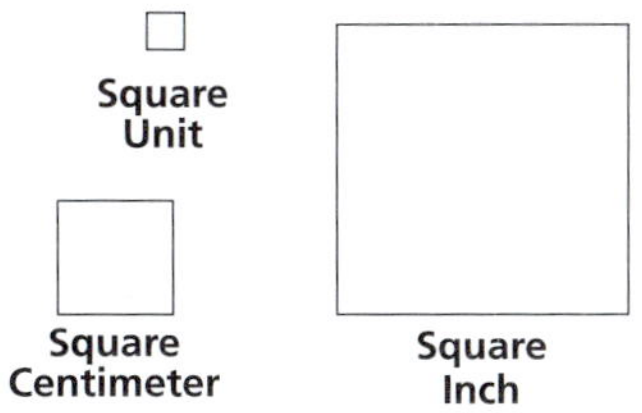

ABCD is a square with a side length equal to 10 units. The area of *ABCD* is measured in square units. Each unit is a square with a side length of 1. These units are called square units and can be written as units^2.

You can count the number of square units in square *ABCD* to find that its area is 100 square units. You can also calculate the area by multiplying the side length by the side width.

Area = (10 units)(10 units) = 100 units^2

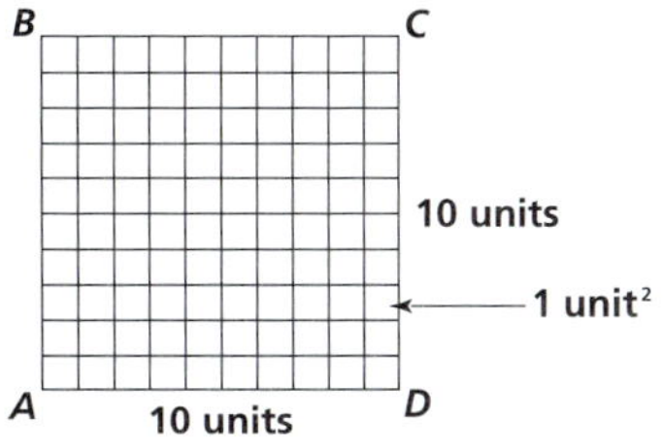

100 squares cover the entire area

EFGH is a rectangle with a width of 10 units and a length of 20 units.

You can count the number of square units in the rectangle—200 square units—or you can calculate the area by multiplying the length of the base times the length of the height.

Area = (20 units)(10 units) = 200 units^2

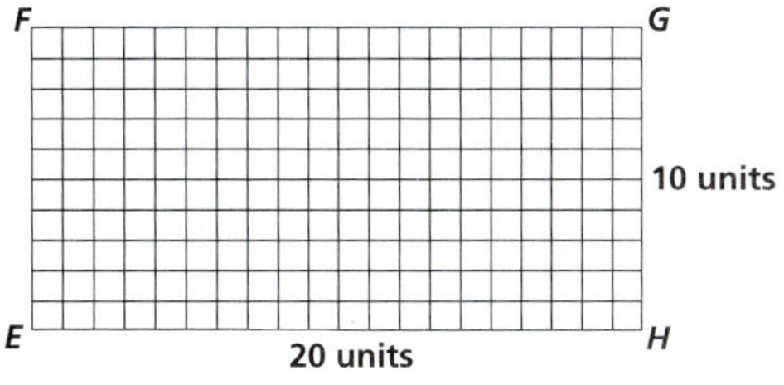

200 squares cover the entire area

FORMULAS: Area of square = s^2,
where s = length of a side

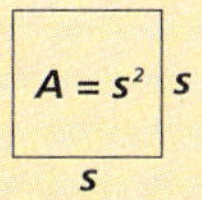

Area of rectangle = $b \bullet h$
where b = base and h = height

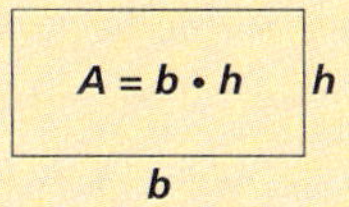

Lesson at a Glance

Chapter 9 Lesson 4

Overview This lesson introduces formulas for finding the areas of different-shaped polygons.

Objective

- To use appropriate formulas for finding areas of squares, rectangles, and parallelograms.

Student Pages 295–299

Teacher's Resource Library TRL

Workbook Activity 98

Additional Practice 57

1. Warm-Up Activity

Present students with a number of 1-in. square tiles. Ask students how they can use the tiles to model an array of 12 items. Encourage their thinking by modeling a 1 × 12 array, a 2 × 6 array, and a 3 × 4 array. Point out that in this lesson, students will learn that determining an area is much like forming an array—the number of objects in the array is equal to the number of rows times the number of items in each row. Ask students to speculate about how this might be related to the concept of *area.*

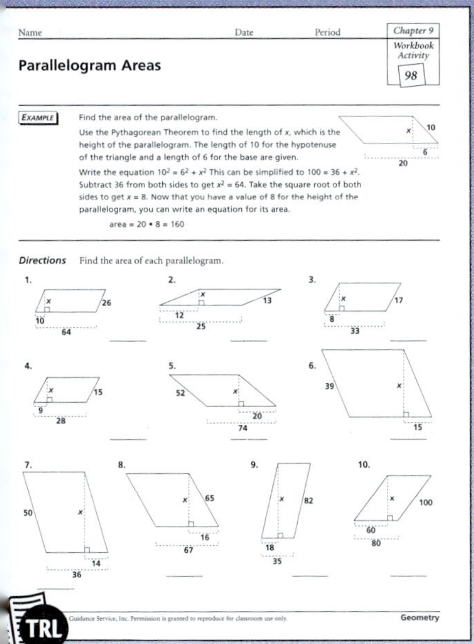

Name Date Period Chapter 9 Workbook Activity 98

Parallelogram Areas

EXAMPLE Find the area of the parallelogram.
Use the Pythagorean Theorem to find the length of x, which is the height of the parallelogram. The length of 10 for the hypotenuse of the triangle and a length of 6 for the base are given.
Write the equation $10^2 = 6^2 + x^2$. This can be simplified to $100 = 36 + x^2$. Subtract 36 from both sides to get $x^2 = 64$. Take the square root of both sides to get $x = 8$. Now that you have a value of 8 for the height of the parallelogram, you can write an equation for its area.
area = 20 • 8 = 160

Directions Find the area of each parallelogram.

Geometry

Workbook Activity 98

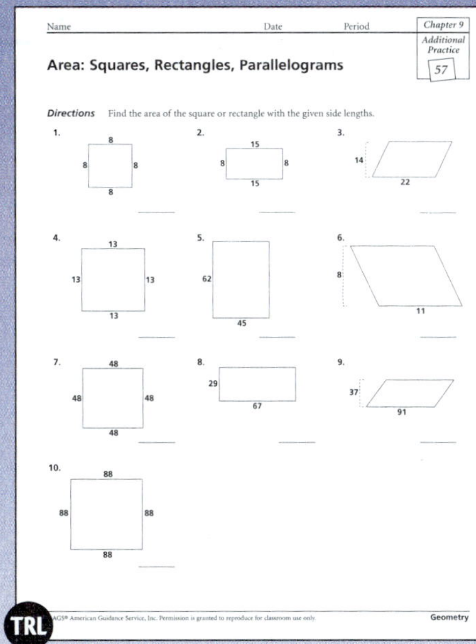

Name Date Period Chapter 9 Additional Practice 57

Area: Squares, Rectangles, Parallelograms

Directions Find the area of the square or rectangle with the given side lengths.

Geometry

Additional Practice 57

2. Teaching the Lesson

Point out that one way to think of area is to think of "tiling the plane." For example, an area 20 units by 10 units will take 200 1 × 1 units to cover. You might want to point to square ceiling tiles or square floor tiles as examples.

Use a rectangle and two similar right triangles to form a parallelogram. Then model the diagrams on page 296 to show that the area of the parallelogram is $b \cdot h$. Compare that area to the area of a rectangle, which is $l \cdot w$.

Experiencing Geometry

Model the activity for students. Then ask pairs of students to do the activity on their own. Each member of the pair should explain how to use the triangles from the rectangle to form a parallelogram.

As you work with the areas of polygons, you will meet these abbreviations:

sq = square
in. = inch
ft = foot
cm = centimeter
cm^2 = square centimeter

The area of a parallelogram is also $b \cdot h$. However, in this case, h is not the length of a side of the parallelogram but the length of the altitude of the parallelogram.

FORMULA: Area of parallelogram = $b \cdot h$

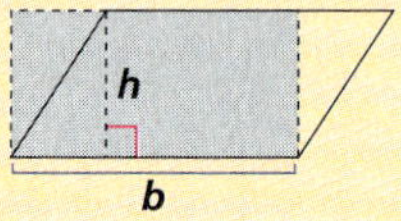

$b \cdot h$ calculates the area of the shaded rectangle. The triangles on the left and the right of the parallelogram are congruent. Therefore, the area of the shaded rectangle = the area of the parallelogram.

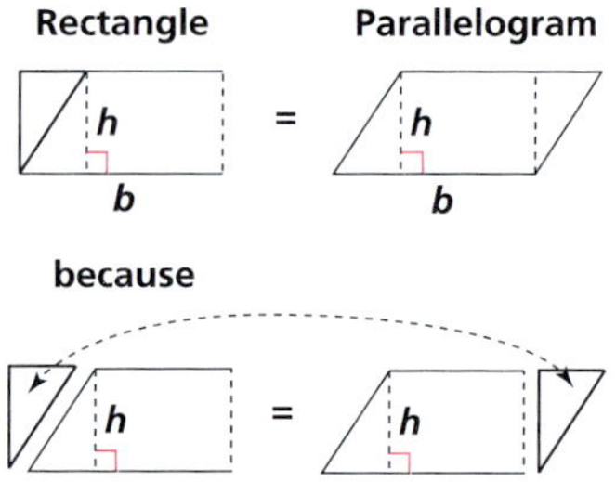

EXPERIENCING GEOMETRY

You can prove this for yourself using a construction-paper rectangle. Mark your rectangle with a dotted line and h as shown in the diagram. Next, cut the left triangle from the parallelogram as shown. Place it along the right side of your figure—the result is a parallelogram.

You can use these formulas to solve several different kinds of problems.

EXAMPLE The side of a square is 3 in. What is its area?

Area of square = s^2

= (3 in.)^2 = 9 sq in.

You can use the same formula, $A = s^2$, to find the side of a square if you know its area. You take the square root of each side of the equation.

$\sqrt{A} = s$

EXAMPLES The area of a square is 100 cm². What is the side length of the square?

$\sqrt{A} = s$

$\sqrt{100} = s$

$10 = s$ Each side is 10 cm.

1 foot = 12 inches. How many square inches in a square foot?

1 foot = 12 inches

(1 foot)(1 foot) = (12 in.)(12 in.) square both sides

1 sq ft = 144 sq in.

The height of a rectangle is 5 cm, its base is 10 cm. What is its area?

$A = b \bullet h$

$= (10 \text{ cm})(5 \text{ cm})$

$A = 50 \text{ cm}^2$

You can solve the formula $A = b \bullet h$ for either b or h.

$A = b \bullet h$ and $A = b \bullet h$

$\frac{A}{h} = b$ $\frac{A}{b} = h$

$A = b \bullet h$ $\frac{A}{h} = b$	h
b	

$A = b \bullet h$ $\frac{A}{b} = h$	h
b	

The area of a rectangle is 120 sq units, the base is 30 units. What is the height of the rectangle?

$A = b \bullet h$

$\frac{A}{b} = h$

$\frac{120}{30} = h$

$4 = h$ The height of the rectangle is 4 units.

Algebra Connection

3. Reinforce and Extend

ALGEBRA CONNECTION

Present students with the following statements. For each statement, direct students to give an equation for the unknown and then solve the equation.

Area = 30, $h = 6$, what is b?
(5)

Area = 90, $h = 30$, what is b?
(3)

Area = 81, $b = 9$, what is h?
(9)

Area = 256, $b = 16$, what is h?
(16)

Area = 256, $h = 8$, what is b?
(32)

APPLICATION

At Home

Point out to students that drawing on a grid can be helpful in deciding how to arrange furniture in a room. It is a technique used by architects and interior designers. Encourage students to take the measurements of a room in their home and to draw the measurements on grid paper, using a scale of 1 square = 6 inches. Then ask them to draw on the grid the locations of the furniture in the room. Remind them that the dimensions of the furniture should also use the scale 1 square = 6 in. Their drawing should include the area of the room as well as an estimate of the area covered by furniture. Post students' finished drawings.

Common Error

Often students dismiss units as unimportant or irrelevant. Stress to them that the units of area are square units such as cm^2, sq in., ft^2, and m^2.

Learning Styles

LEP/ESL Have students work in pairs to complete Exercise C. For each item, one student can give directions to the other showing how to solve for *h* or *b*; the second student can follow the directions. Students should exchange roles after each item.

Learning Styles

Body/Kinesthetic Reinforce the difference between area and perimeter with the following activity. Draw two large congruent polygons on the classroom floor. Ask students to illustrate the perimeter of one of the polygons by forming a line along the outline of the polygon. Ask students how they can use the second polygon to illustrate area. (Students can stand shoulder to shoulder inside the outlines of the polygon.)

Writing About Mathematics

Do these quadrilaterals have the same area? Explain.

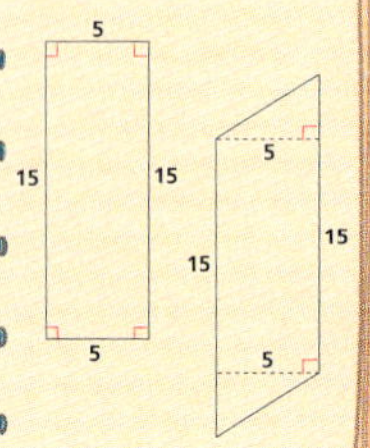

Example Find the area of this parallelogram.

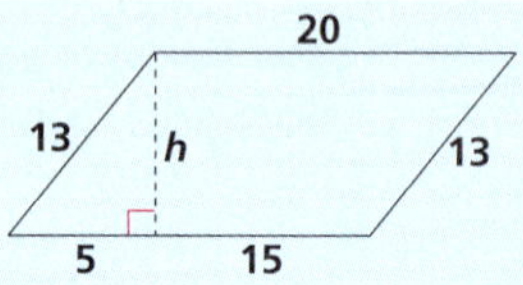

First you will need to use the Pythagorean Theorem to find the value of *h*.

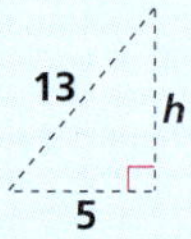

$A = b \bullet h$

$5^2 + h^2 = 13^2$

$25 + h^2 = 169$

$h^2 = 169 - 25$

$\sqrt{h^2} = \sqrt{144}$

$h = 12$

$A = 20 \bullet 12$

$A = 240$ sq units

Exercise A Find the area of the square or rectangle with the given side lengths.

1. $s = 15$ cm $225\ cm^2$
2. $s = 20$ ft 400 sq ft
3. $s = 9$ in. 81 sq in.
4. $s = 3$ km $9\ km^2$
5. $b = 8$ cm, $h = 6$ cm $48\ cm^2$
6. $b = 3$ in., $h = 4$ in. 12 sq in.
7. $b = 12$ km, $h = 5$ km $60\ km^2$
8. $b = 40$ m, $h = 15$ m $600\ m^2$

Exercise B Find the length of one side of a square with the given area.

9. $81\ cm^2$ 9 cm
10. $121\ m^2$ 11 m
11. 49 sq in. 7 in.
12. 60 sq ft 7.7 ft

Exercise C Find the missing dimension, either *h* or *b*, for rectangles with these areas.

13. $A = 110\ mi^2$, $h = 10$ mi

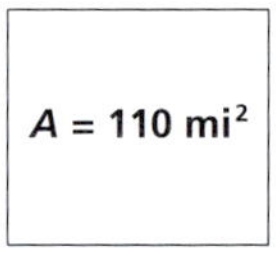

$b = 11$ mi

14. $A = 180\ km^2$, $b = 60$ km $h = 3$ km

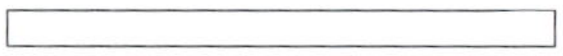

15. $A = 825\ m^2$, $b = 25$ m

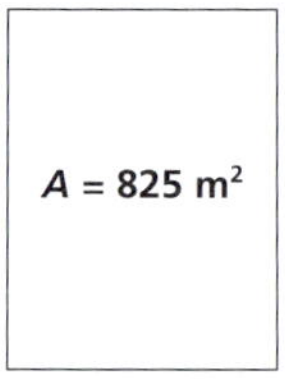

$h = 33$ m

Exercise D Find the area of each parallelogram.

16.

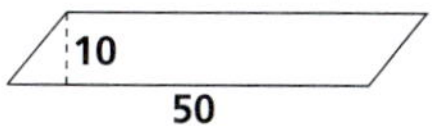

500

17.

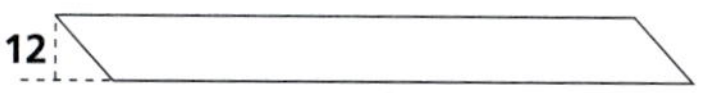

1,200

18.

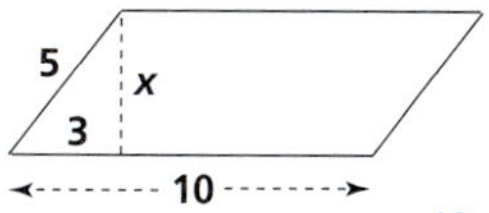

40

19.

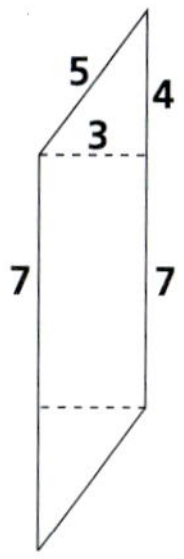

33

20.

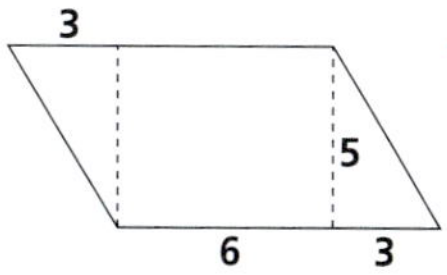

45

21.

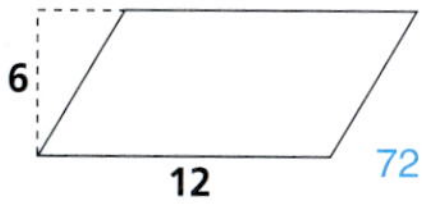

72

Problem Solving

Exercise E Solve the following problems.

22. The Lugo family is tiling their kitchen floor. How many square feet of tile will they need?

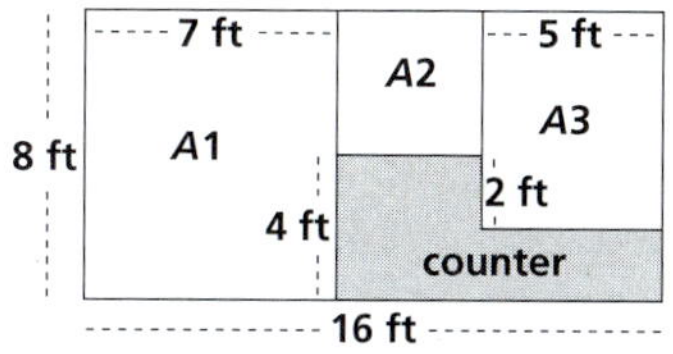

22. $A = A1 + A2 + A3 = (8 \bullet 7) + (4 \bullet 4) + (5 \bullet 6) = 56 + 16 + 30 = 102$ square feet
23. two; 1 application: $75 \bullet 30 = 2{,}250$; $5{,}000 \div 2{,}250 = 2.2$
24. Yes, the area he needs to paint is $120 \bullet 4 = 480$ sq ft.
25. 9 sq ft

23. Dharma purchased a bag of lawn food that covers 5,000 square feet. Her yard is 75 ft × 30 ft. How many complete applications does she have? Explain.

24. Raul plans to paint his room. There are four walls, each 10 ft wide by 12 ft high. The paint can says the paint will cover 500 square feet. Will he have enough paint to finish the job?

25. How many square feet are in one square yard? (Hint: Draw a picture.)

Group Problem Solving

Direct students to work in groups of four. Each group should be responsible for finding three polygons for which the following is true: The numeric value of the area is the same as the numeric value of the perimeter. For example, if a square is 4 cm on a side, its perimeter is 16 cm and its area is 16 square centimeters. Groups can present their findings as posters for a "Match the Number" contest. Students should explain how they determined their three matching numbers—trial and error or algebraically. Provide extra reinforcement for teams who solved the problem algebraically. ($2(l + w) = l \bullet w$ or $\frac{2w}{(w-2)} = l$.

Sample solutions: $l = 4$, $w = 4$; $l = 8$, $w = \frac{8}{3}$; $l = 10$, $w = \frac{5}{2}$)

Lesson at a Glance

Chapter 9 Lesson 5

Overview This lesson focuses on finding the area of a trapezoid.

Objectives

- To use a formula to find the area of a trapezoid.
- To find the lengths of the bases in a trapezoid.

Student Pages 300–303

Teacher's Resource Library TRL

Workbook Activity 99

Additional Practice 58

Mathematics Vocabulary

average base

1. Warm-Up Activity

Present students with a number of trapezoids with measurements marked on three sides. For example:

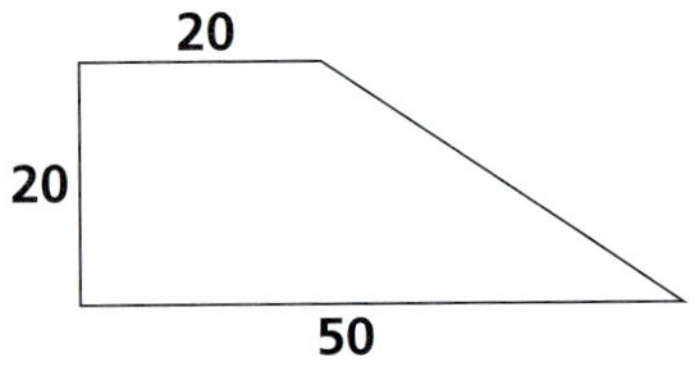

Before introducing the text on page 300, ask students to estimate the area of the above trapezoid. Lead them to the conclusion that 20 • 20 sq units is smaller than the actual area and that 20 • 50 square units is greater than the actual area. Ask students for suggestions on how to figure out the actual area. Encourage suggestions that propose finding an average base.

Lesson 5 Areas of Trapezoids

> ***Average base***
> *The line that connects the midpoints of the nonparallel sides of a trapezoid*

In Chapter 8, you used the formula for the area of a trapezoid to prove the Pythagorean Theorem. In this lesson, you will see how that formula came from that theorem.

Suppose you have a right-angled trapezoid and you want to know its area. You can think through a way to find the area using diagrams like these.

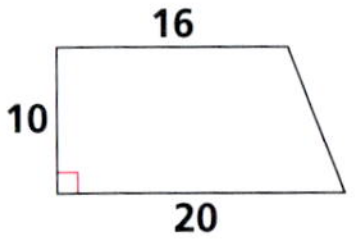

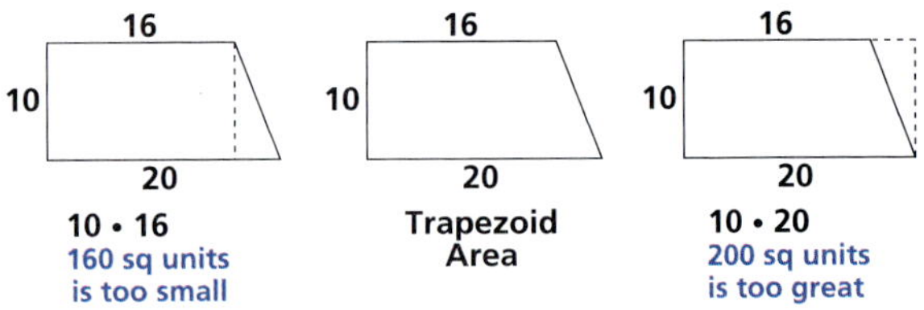

Based on the information in the diagram, you might decide to use the **average base** length.

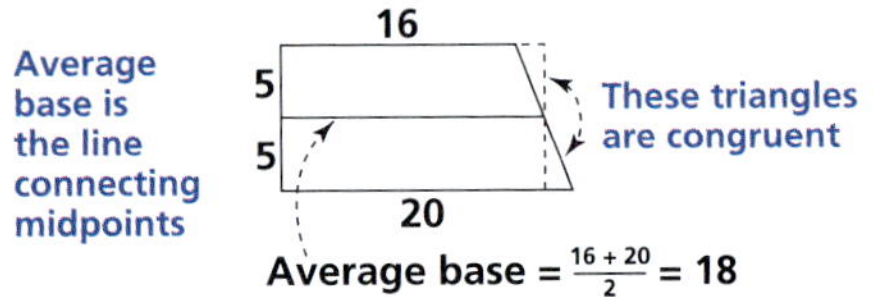

In this case you would find

$A = 10 \cdot 18$

$A = 180$ square units

Let's see if we can apply this to finding the area of any trapezoid. Remember, the average base connects the midpoints of the nonparallel sides.

> Remember, b = base. b_1 and b_2 are used to identify the two bases.

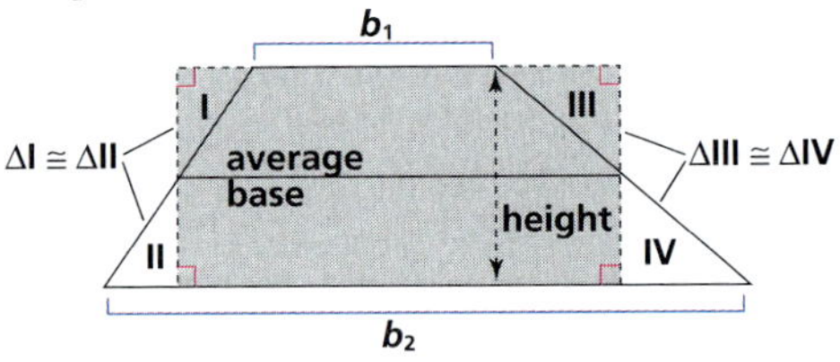

Shaded area = area of trapezoid

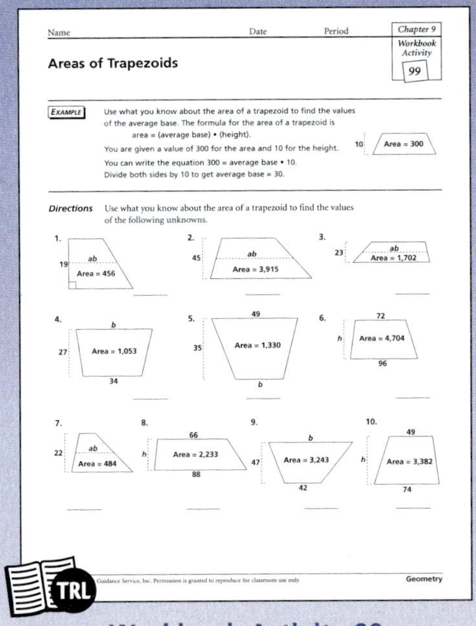

Name Date Period Chapter 9 Workbook Activity 99

Areas of Trapezoids

EXAMPLE Use what you know about the area of a trapezoid to find the values of the average base. The formula for the area of a trapezoid is area = (average base) • (height).
You are given a value of 300 for the area and 10 for the height.
You can write the equation 300 = average base • 10.
Divide both sides by 10 to get average base = 30.

Directions Use what you know about the area of a trapezoid to find the values of the following unknowns.

1. Area = 456 2. Area = 3,915 3. Area = 1,702 4. Area = 1,053 5. Area = 1,330 6. Area = 4,704 7. Area = 484 8. Area = 2,233 9. Area = 3,243 10. Area = 3,382

Geometry

Workbook Activity 99

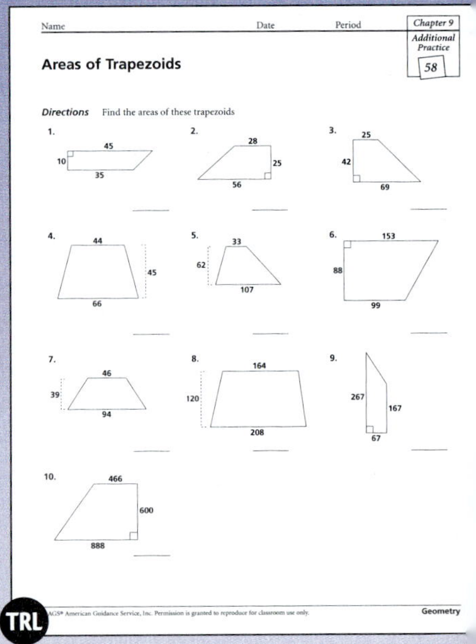

Name Date Period Chapter 9 Additional Practice 58

Areas of Trapezoids

Directions Find the areas of these trapezoids.

Geometry

Additional Practice 58

FORMULA: Area of trapezoid = (average base) • (height)

$= \frac{b_1 + b_2}{2}h$

You can use this formula to solve problems involving the areas of trapezoids.

EXAMPLES Calculate the area of the trapezoid shown.

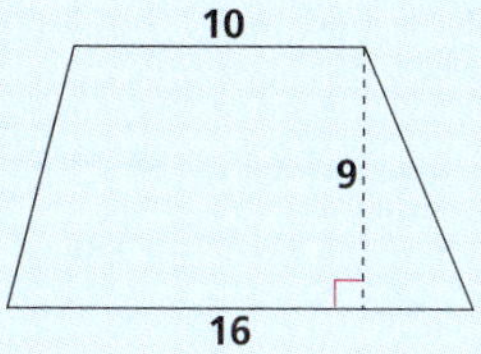

Use the formula $A = \frac{b_1 + b_2}{2}h$

Substitute the values from the diagram.

$b_1 = 10$, $b_2 = 16$, $h = 9$

$A = \frac{10 + 16}{2}9$

$= 13 \bullet 9 = 117$ square units

The area of a trapezoid is 120 square units and the average base is 10. What is the height of the trapezoid?

Solve the area formula for h. $A = (\text{average base}) \bullet (\text{height})$

$\frac{A}{\text{average base}} = h$

$\frac{120}{10} = h$

$12 = h$

The area of trapezoid $ABCD$ is 300 sq units and its height is 10. The lower base is twice as long as the upper base. How long is each base?

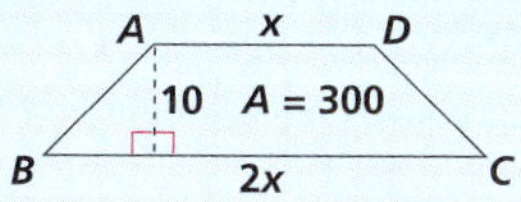

Draw a picture to help you.

Let $b_1 = x$ and $b_2 = 2x$, then substitute in the formula.

$A = \frac{b_1 + b_2}{2}h$

$300 = \frac{x + 2x}{2}10$

$300 = \frac{3x}{2}10$ simplify

$600 = (3x)(10)$ multiply by 2

$60 = 3x$ divide by 10

$20 = x$ divide by 3

∴ Upper base = 20, lower base = 40

2. Teaching the Lesson

Follow the text as you indicate to students that one way to find the area of a trapezoid is to find the average base length and multiply that number by h, the height of the trapezoid.

MENTAL MATH

Finding an average is an important step in finding the area of a trapezoid. Have students use mental math to find the averages of the following pairs of numbers.

3 and 5	(4)
6 and 10	(8)
12 and 24	(18)
13 and 25	(19)
13 and 24	$(18\frac{1}{2})$

MANIPULATIVES

M **Trapezoids**

Materials: Polygon Power Pack

Group Practice: Direct students to find a rectangle and right triangles they can use to model the trapezoid shown in the middle of page 300. Invite students to show how the triangles can be used with the rectangle to produce a trapezoid. Have them use the configuration to justify using the average base measurement to find the area of a trapezoid.

Student Practice: Encourage individuals to use the materials independently to create other trapezoids.

3. Reinforce and Extend

Learning Styles

LEP/ESL Invite students to work with an English proficient partner to complete Exercise A. Have one student explain how to find the area of a trapezoid. Students should indicate whether they need to use the Pythagorean Theorem to find an unknown measurement, or if they can use the formula

$A = \frac{b_1+b_2}{2}h$, or both.

Students can exchange roles after completing each item.

Try This...

Students should recognize that if the formula for the area of a trapezoid, $A = \frac{b_1+b_2}{2}h$, is applied to a rectangle, the factor in parentheses becomes either

$\frac{b_1 + b_1}{2} = b$ or $\frac{b_2 + b_2}{2} = b$,

so that the formula becomes $A = bh$, the formula for the area of a rectangle.

Example Calculate the area of the isosceles trapezoid *ISOL*.

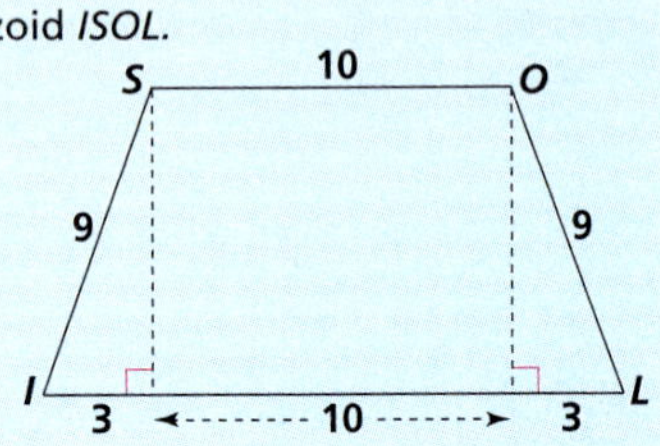

To be able to use the area formula, you must first find the height of *ISOL*.

Draw two perpendiculars from the upper base.

This shows you that the two right triangles formed must have short legs equal to (16 – 10) ÷ 2 = 3.

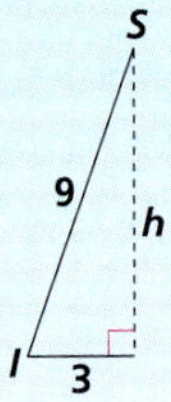

So, $9^2 = 3^2 + h^2$ by the Pythagorean Theorem

$81 = 9 + h^2$

$72 = h^2$

$\sqrt{72} = h = 8.5$ rounded to nearest tenth

Now you can substitute values in the area formula:

$A = \frac{b_1 + b_2}{2}h$

$A = \frac{10 + 16}{2}8.5$

$= (13)(8.5) = 110.5$ sq units

Exercise A Find the areas of these trapezoids.

1.
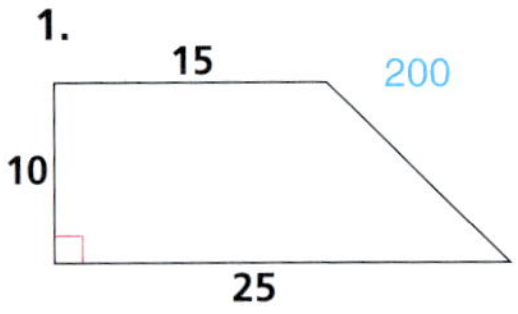

2.
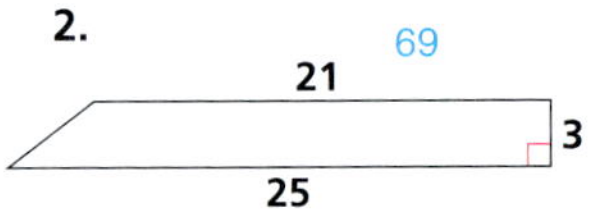

3.
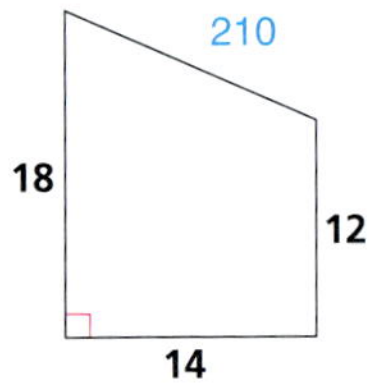

4.
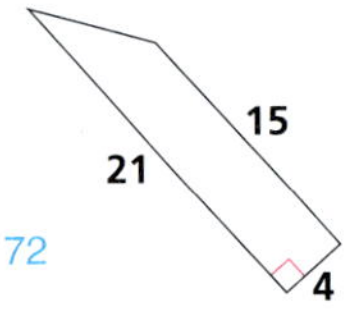

5.
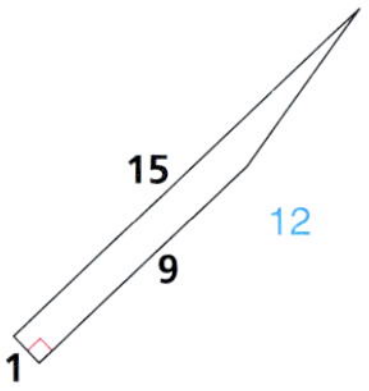

6.
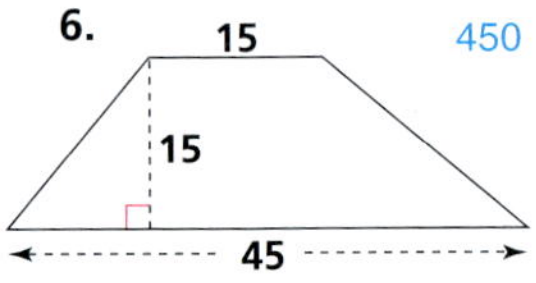

Try This...

Can you use the trapezoid formula to calculate the area of a rectangle? Try it and describe the results.

302 *Chapter 9 Perimeter and Area*

7.

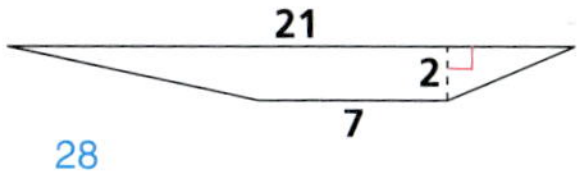

28

8.

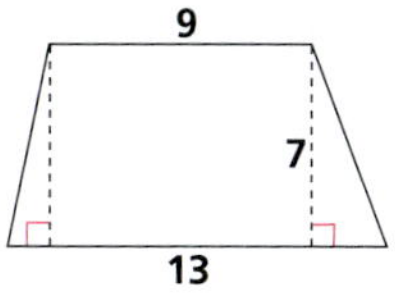

77

9.

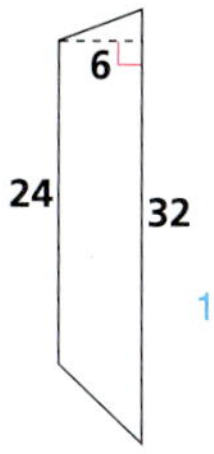

168

10.

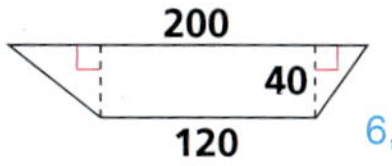

6,400

Exercise B Use what you know about the area of a trapezoid to find the values of the following unknowns.

11. average base

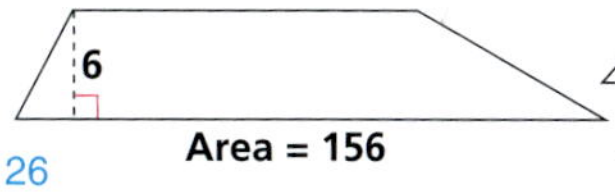

26

12.

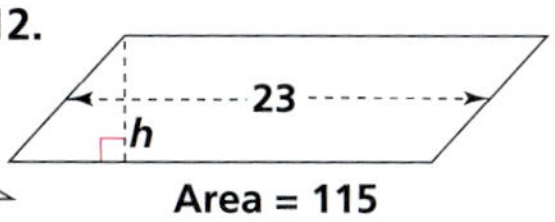

5

13.

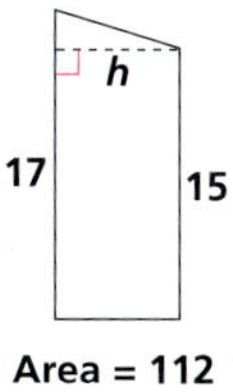

7

14. average base

8

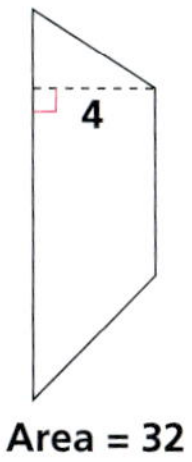

15. average base

5.5

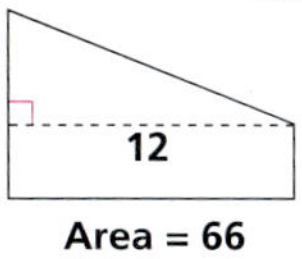

Exercise C Find the areas of these isosceles trapezoids. You may use a calculator. Round to the nearest tenth.

16.

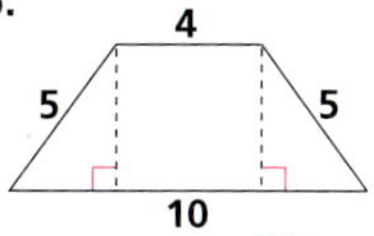

28

17.

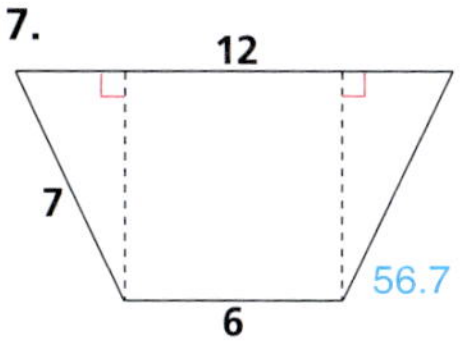

56.7

18.

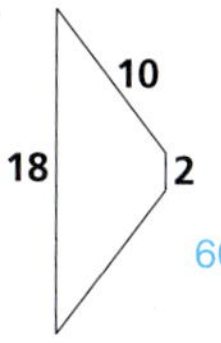

60

19.

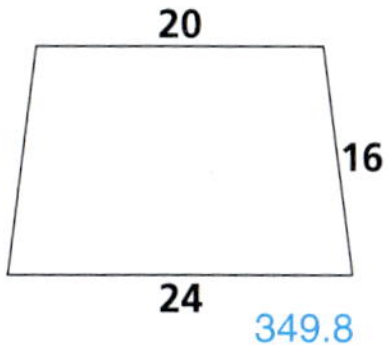

349.8

20.

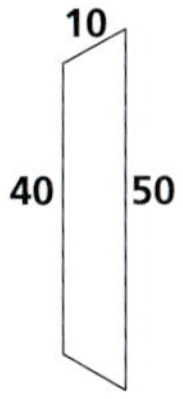

391.5

APPLICATION

Career Connection

Point out that much construction and design work revolves around the use of area and perimeter. Home builders need to know the area of a roof to order enough tiles to cover the roof. Interior decorators need to know the area of a room to order carpeting; they use the perimeter of the room when they order baseboards or decorative borders. Encourage interested students to conduct an interview with a carpenter, roofer, landscaper, interior designer, architect, or other professional or tradesperson involved in home building. Suggest that students make an appointment for the interview and prepare their questions in advance. Encourage them to ask about how geometry is useful in the profession. Students might want to ask what, if any, special training is needed, if employment is steady, and what people who are considering entering the profession should know. Students can share their interview results as part of a Career Corner segment on the evening news.

Lesson at a Glance

Chapter 9 Lesson 6

Overview This lesson demonstrates a variety of ways to find the area of a triangle.

Objectives

- To use a formula to find the area of a triangle.
- To use Heron's Formula to find the area of a triangle.

Student Pages 304–309

Teacher's Resource Library TRL

- **Workbook Activity** 100
- **Additional Practice** 59

1. Warm-Up Activity

Display a variety of plastic triangles on an overhead projector. Ask students to position pairs of right triangles so that they form either squares or rectangles. Ask students to speculate on how the area of the square or rectangle is related to the areas of the triangles. Elicit the observation that each triangle has half the area of a rectangle or square.

Lesson 6 Areas of Triangles

If you look closely at the areas of squares and rectangles, you may see a pattern for finding the areas of triangles.

Given a square

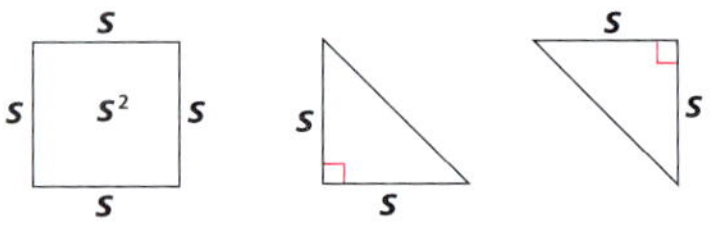

$A = s^2 \quad = \quad \frac{1}{2}s^2 \quad + \quad \frac{1}{2}s^2$

∴ Area of an isosceles right triangle with side $s = \frac{1}{2}s^2$

EXAMPLE Square Area of right triangles $= \frac{1}{2}(5)^2 = \frac{1}{2}25$

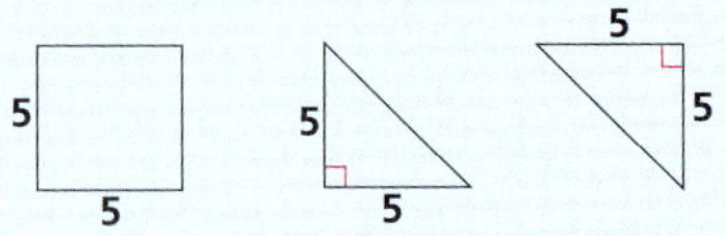

$A = 25 \quad = \quad \frac{1}{2}(25) \quad + \quad \frac{1}{2}(25)$

Given a rectangle

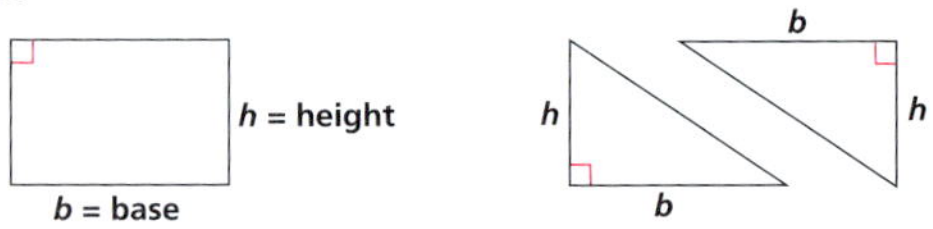

$A = bh \quad = \quad \frac{1}{2}bh \quad + \quad \frac{1}{2}bh$

∴ Area of any right triangle $= \frac{1}{2}bh$

EXAMPLE Rectangle

Area of right triangles $= \frac{1}{2}(10 \bullet 30) = 150$

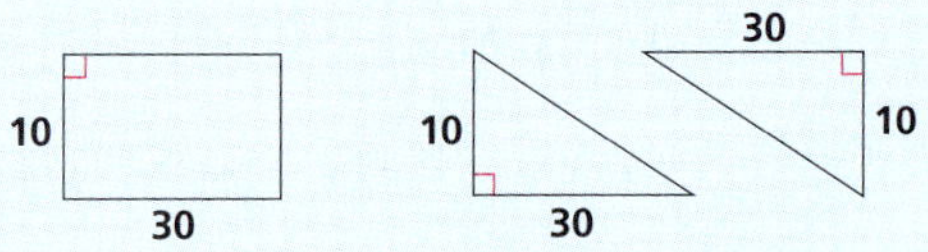

$A = 300 \quad = \quad \frac{1}{2}(300) \quad + \quad \frac{1}{2}(300)$

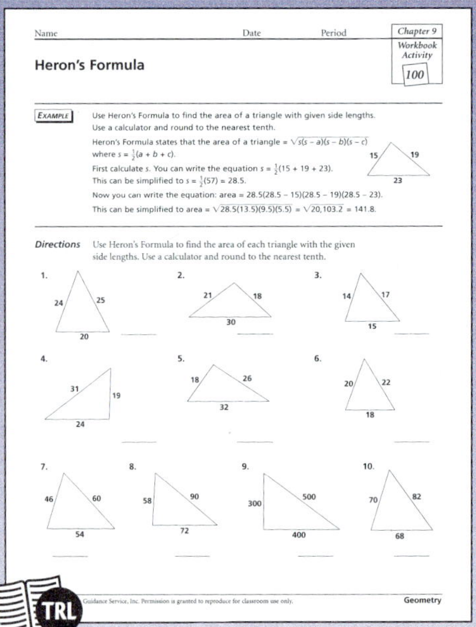

Name Date Period *Chapter 9 Workbook Activity 100*

Heron's Formula

EXAMPLE Use Heron's Formula to find the area of a triangle with given side lengths. Use a calculator and round to the nearest tenth.

Heron's Formula states that the area of a triangle $= \sqrt{s(s-a)(s-b)(s-c)}$ where $s = \frac{1}{2}(a + b + c)$.

First calculate s. You can write the equation $s = \frac{1}{2}(15 + 19 + 23)$. This can be simplified to $s = \frac{1}{2}(57) = 28.5$.

Now you can write the equation: area = 28.5(28.5 − 15)(28.5 − 19)(28.5 − 23).

This can be simplified to area $= \sqrt{28.5(13.5)(9.5)(5.5)} = \sqrt{20{,}103.2} = 141.8$.

Directions Use Heron's Formula to find the area of each triangle with the given side lengths. Use a calculator and round to the nearest tenth.

Workbook Activity 100

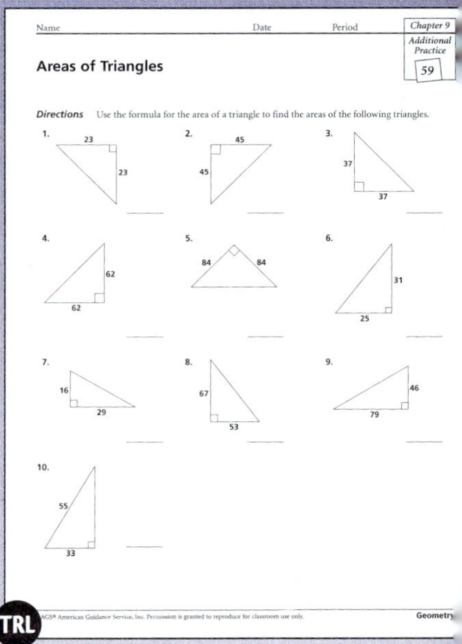

Name Date Period *Chapter 9 Additional Practice 59*

Areas of Triangles

Directions Use the formula for the area of a triangle to find the areas of the following triangles.

Additional Practice 59

Now we need to find the formula for the area of any triangle. We can start with the construction of a scalene triangle ABC.

EXAMPLE Let ΔABC be any scalene triangle with altitude $\overline{DB}$. The height of ΔABC equals $\overline{BD}$.

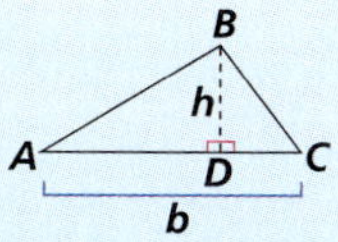

Construct a rectangle $AFEC$ with $b = \overline{AC}$ and $h = \overline{BD}$ on $\overline{AC}$. Area of rectangle $AFEC = b \bullet h$.

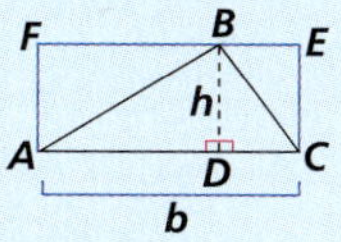

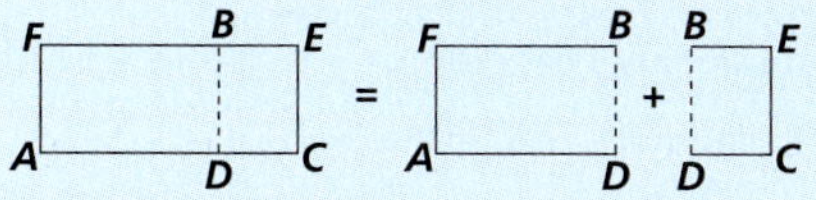

The whole is equal to the sum of its parts.

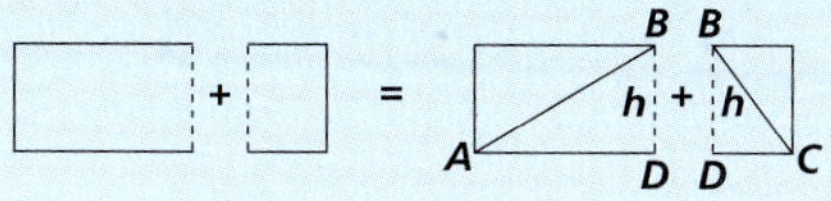

Draw diagonals to create right triangles.

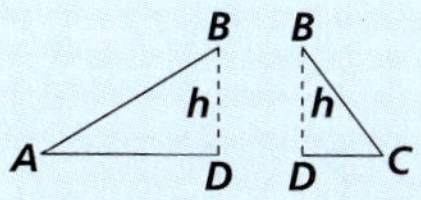

Areas of right triangles

$\frac{1}{2}(\overline{AD})h + \frac{1}{2}(\overline{DC})h$

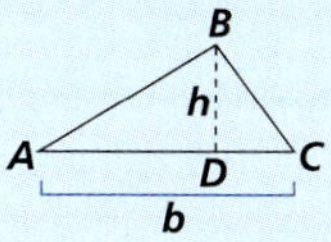

The whole equals the sum of its parts.

Area of scalene ΔABC

$= \frac{1}{2}(\overline{AD})h + \frac{1}{2}(\overline{DC})h$	Whole equals sum of its parts
$= \frac{1}{2}h(\overline{AD} + \overline{DC})$	Distributive Property
$= \frac{1}{2}(hb)$	Substitution $b = \overline{AD} + \overline{DC}$

Remember that for multiplication of real numbers $x(a + b) = xa + xb$.

FORMULA: Area of triangle = $\frac{1}{2}$(base)(height)

2. Teaching the Lesson

Make certain that students have assimilated the concept that the area of a triangle can be thought of as one-half the area of a square or rectangle. Then work through the example on page 305 with students. Be sure they can identify a justification or reason for each step. For example, in the second step "Construct a rectangle..." is possible because of Euclid's Postulate 1.

As an aside, review the Distributive Property of multiplication:

$x(a + b) = xa + xb$

Be sure students understand that the distributive property is used in simplifying $\frac{1}{2}(\overline{AD})h + \frac{1}{2}(\overline{DC})h$.

Common Error

Students often consider the lowest horizontal line of a triangle its "base." Point out that the base of a triangle can be any of the three sides, so that triangles in different orientations can have their bases in different orientations as well. Explain that this means an altitude can be to any one of the three sides of a triangle, usually whichever is most convenient for computation. However, emphasize that it is generally understood that in an isosceles triangle, the base is *not* one of the equal legs of the triangle.

3. Reinforce and Extend

Learning Styles

LEP/ESL Direct students to work in pairs, following the examples and computing the area of $\Delta A'B'C'$ with base 30 and altitude 8 **(120 sq units)**; $\Delta D'E'F'$ with base 50 and altitude 10 **(250 sq units)** and $\Delta G'H'I'$ with altitude 40 and base 12 **(240 sq units)**. In each case, students should explain each step they used to find the area.

You can use this formula to calculate the area of any triangle—all you need to know is the base and the height. Remember, the height of a triangle equals the height of the altitude to the side used as the base.

For ΔABC, the area can be calculated in three different ways.

Using $\overline{AC}$ as the base, $A = \frac{1}{2}(\overline{AC})h_1$.

Using $\overline{BC}$ as the base, $A = \frac{1}{2}(\overline{BC})h_2$.

Using $\overline{AB}$ as the base, $A = \frac{1}{2}(\overline{AB})h_3$.

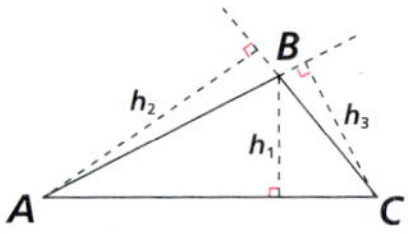

Examples

Calculate the area of ΔABC.

Area of triangle = $\frac{1}{2}bh$

$\frac{1}{2}(15)(8) = 60$ sq units

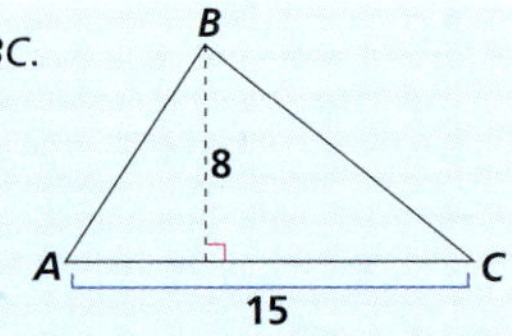

Calculate the area of ΔDEF.

Area of triangle = $\frac{1}{2}bh$

$\frac{1}{2}(25)(10) = 125$ sq units

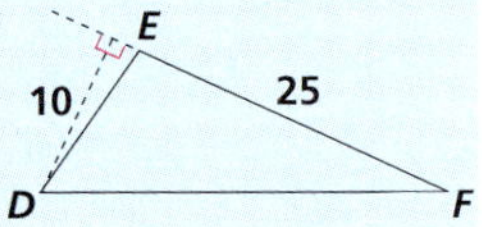

Calculate the area of ΔGHI.

Area of triangle = $\frac{1}{2}bh$

$\frac{1}{2}(12)(20) = 120$ sq units

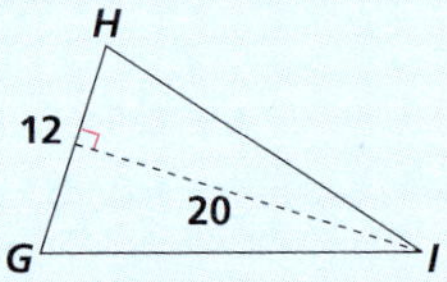

There is a way to find the area of a triangle even if the only information you have about the triangle is its sides.
This method for finding the area of a triangle was developed two thousand years ago by the Greek mathematician Heron of Alexandria.

Area of triangle = $\sqrt{s(s-a)(s-b)(s-c)}$

where a, b, and c are the sides of the triangle and $s = \frac{1}{2}(a + b + c)$

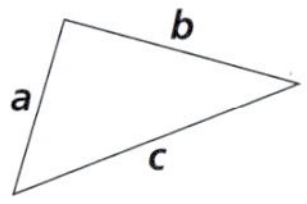

306 *Chapter 9 Perimeter and Area*

EXAMPLES Use Heron's Formula to find the area of a triangle whose side lengths are 3, 4, and 5.

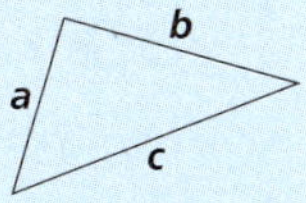

Area of triangle = $\sqrt{s(s-a)(s-b)(s-c)}$

Let $a = 3$, $b = 4$, $c = 5$

$s = \frac{(3+4+5)}{2} = 6$

$s - a = 6 - 3 = 3$; $s - b = 6 - 4 = 2$; $s - c = 6 - 5 = 1$

Area of triangle = $\sqrt{6(3)(2)(1)}$

$= \sqrt{36} = 6$

Use Heron's Formula to find the area of this triangle. (Use a calculator and round to the nearest tenth.)

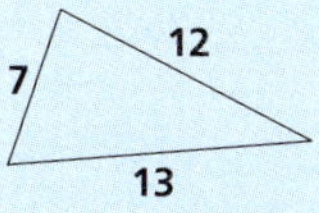

Area of triangle = $\sqrt{s(s-a)(s-b)(s-c)}$

Let $a = 7$, $b = 12$, $c = 13$

$s = \frac{(7+12+13)}{2} = \frac{32}{2} = 16$

$s - a = 16 - 7 = 9$; $s - b = 16 - 12 = 4$; $s - c = 16 - 13 = 3$

Area of triangle = $\sqrt{16(9)(4)(3)}$

$= \sqrt{1{,}728} = 41.6$ sq units

Calculator Practice

You can use the [√] key of your calculator to find the square root of a product. On some calculators, you need to press [2nd] [x^2] to find the square root of x.

EXAMPLE **Find:** square root of 6 • 7 • 8

Press 6 [×] 7 [×] 8 [×] [√] 18.3

or

Press 6 [×] 7 [×] 8 [=] [2nd] [x^2] 18.3

LEARNING STYLES

Logical/Mathematical Invite students to work in pairs. Suggest they use the example on page 307 to demonstrate the use of Heron's formula, which uses the perimeter of a triangle to find its area. Encourage students to explain each step of the process. Then have them draw three triangles, assigning a length to each side. Direct students to exchange triangles with one another. Students can then find the areas of the drawn triangles. Have students exchange triangles again. They can check for any algebraic or arithmetic errors. Once areas have been checked, triangles can be returned to the original artist.

Perimeter and Area *Chapter 9* **307**

CALCULATOR

Offer students additional opportunities to find the square root (to the nearest tenth) of products by presenting the following situations:

5 • 7 • 9	(17.7)
6 • 8 • 10	(21.9)
13 • 15 • 17	(57.6)
19 • 21 • 23	(95.8)

APPLICATION

In the Community

Draw students' attention to the *Technology Connection* at the bottom of page 308. Indicate to students that the four-color map problem seems simple at first glance, but its proof is rather complex. Invite students to find a map of the United States that uses only four colors. Have them indicate which colors represent which states. Alternately, students can find a globe in a library, classroom, or store. Have them report how many colors were used to color the countries and to color the oceans. What other conventions are used in coloring maps and globes? (Some maps and globes use different colors to represent different elevations or landforms. Note that some globes use six colors.)

Exercise A Use a calculator to find the square roots of these products. Round each answer to the nearest tenth.

1. 3 • 4 • 5 7.7
2. 12 • 13 • 7 33
3. 9 • 11 • 13 35.9
4. 6 • 8 • 9 20.8
5. 5 • 8 • 12 21.9

Exercise B Use the formula for the area of a triangle to find the areas of the following triangles.

6. 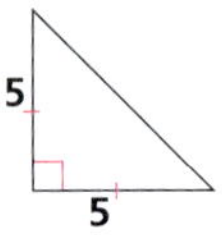

12.5

7. 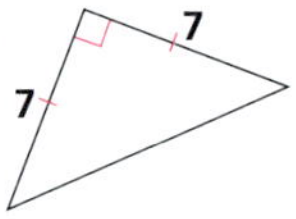

24.5

8. 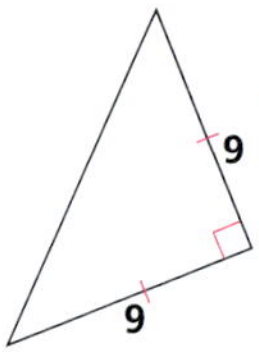

40.5

9. An isosceles right triangle with legs = 12 72

10. An isosceles right triangle with legs = 20 200

11. 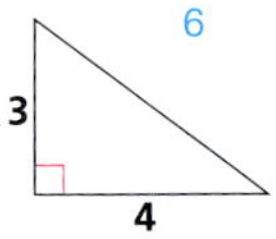

6

12. 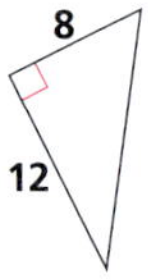

48

13. 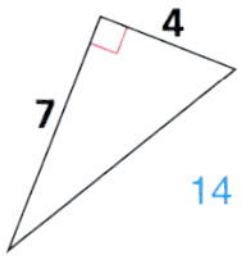

14

Technology Connection

In the mid-1800's, mathematicians formed the following statement concerning the number of colors needed on a map: Areas of any shape on a map can be distinguished from one another using no more than four colors. With the four colors, no states with the same color will touch each other. For more than one hundred years, mathematicians tried to prove this statement, known as the four-color map problem. Work on the proof led to the development of graph theory, the mathematics of complicated networks. Graph theory provides one of the basic tools used to design complicated telephone networks.

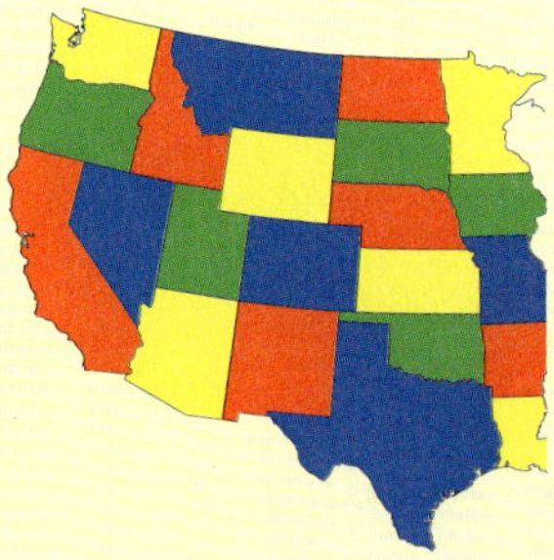

14. A right triangle with perpendicular sides of 10 and 20 100

15. A right triangle with perpendicular sides of 30 and 15 225

16.

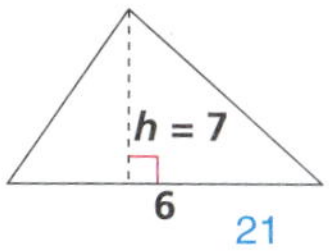

21

17.

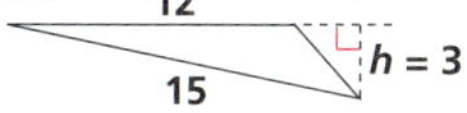

18

18.

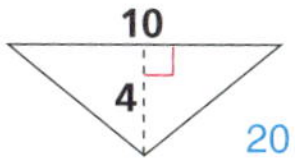

20

19. A scalene triangle with a base of 9 and a height of 5 22.5

20. A scalene triangle with a base of 15 and a height of 32 240

Exercise C Use Heron's Formula to find the areas of the triangles with the given side lengths. Use a calculator and round to the nearest tenth.

21. 6, 7, 8 20.3

22. 10, 15, 20 72.6

23. 11, 13, 16 71

24. 5, 6, 7 14.7

25. 10, 14, 22 51.9

TRY THIS...

Use Heron's Formula to find the area of an equilateral triangle whose sides measure x units.

Geometry Around Us

A bee's honeycomb is an example of hexagons in nature. Each cell is composed as a double layer of beeswax hexagons—polygons with six equal sides. Note that the hexagons fit closely together so that the honeycomb has no spaces between the cells of honey.

LEARNING STYLES

Visual/Spatial Encourage students to draw sketches for problems 14, 15, 19, and 20 of Exercise B. Have them exchange sketches with a partner to see if the sketches agree with one another and with the information in the text.

TRY THIS...

According to Heron's Formula,

Area $= \sqrt{s(s-x)(s-x)(s-x)}$ and

$s = \frac{1}{2}(x + x + x) = \frac{3x}{2}$. Substituting:

$$A = \sqrt{(\tfrac{3x}{2}) \cdot (\tfrac{3x}{2} - x) \cdot (\tfrac{3x}{2} - x) \cdot (\tfrac{3x}{2} - x)} =$$

$$A = \sqrt{(\tfrac{3x}{2})(\tfrac{x}{2})(\tfrac{x}{2})(\tfrac{x}{2})}$$

$$A = \sqrt{(\tfrac{3x}{2})(\tfrac{x^3}{8})}$$

$$A = \sqrt{(\tfrac{3x^4}{16})} = \frac{(x^2\sqrt{3})}{4}$$

Perimeter and Area *Chapter 9* **309**

Lesson at a Glance

Chapter 9 Application

Overview This lesson explores tessellations.

Objective

- To create a tessellation using a variety of regular polygons alone or in concert.

Student Page 310

Teacher's Resource Library TRL

Workbook Activity 101

1. Warm-Up Activity

Introduce tessellations by pointing out the pattern of ceiling tiles or floor tiles in the classroom. Then direct students' attention to the photograph on page 310.

2. Teaching the Lesson

Point out that for a figure, or a number of figures, to tessellate, they must form a pattern without gaps or overlaps. Discuss with students any tessellations they can see in their classroom or around the school. Help them recognize that tessellations may be as simple as the pattern of bricks in a wall or walkway or as complicated as a colorful mosaic.

3. Reinforce and Extend

APPLICATION

In the Community
Challenge students to find tessellations in their neighborhood. Encourage them to look for unusual patterns that use many-sided polygons. Students can sketch any tessellations they find, noting their location. Post sketched tessellations for students to share.

Application

Tessellations Look at your classroom floor or ceiling. Is either covered by a series of rectangles or squares placed one next to the other without gaps?

Each of these is an example of a *tessellation.* A tessellation is an arrangement of shapes that cover a plane surface without overlapping and without leaving holes. Simple tessellations can be made with regular polygons. More complex tessellations use other shapes. In the example shown, colorful tiles are used in a complex tessellation.

EXAMPLE Make a tessellation using the shape at the right.

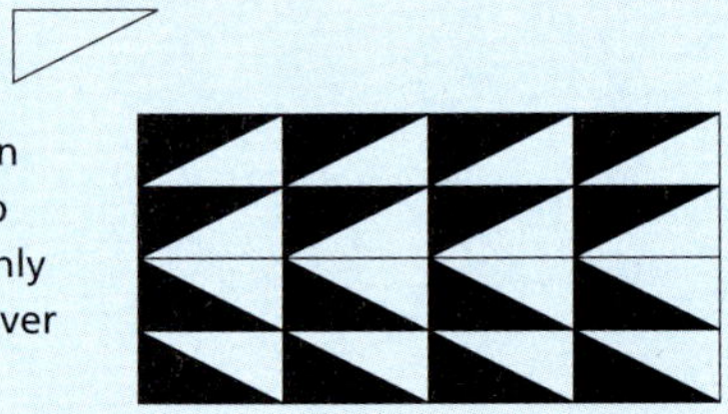

Note that to make the tessellation you may turn, rotate, slide, or flip the basic shape as needed. The only rule is to leave no gaps and to never overlap shapes.

Exercise

1. Test which of these shapes can tessellate. Draw each tessellation on a sheet of paper. When you are finished, compare your tessellations with those your classmates found.

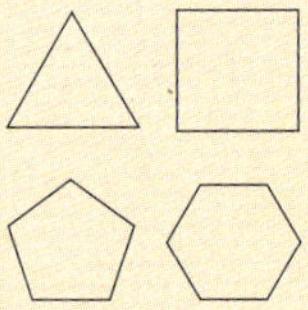

2. Draw a tessellation using regular octagons and squares.
3. Determine if you can make a tessellation using the octagon shown.

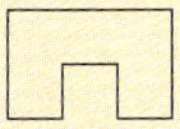

4. Draw a tessellation using triangles and rectangles.
5. Draw a tessellation using triangles and squares.

Student tessellations should reflect an understanding of the principle that tessellations cover a plane surface without overlapping and without leaving holes.

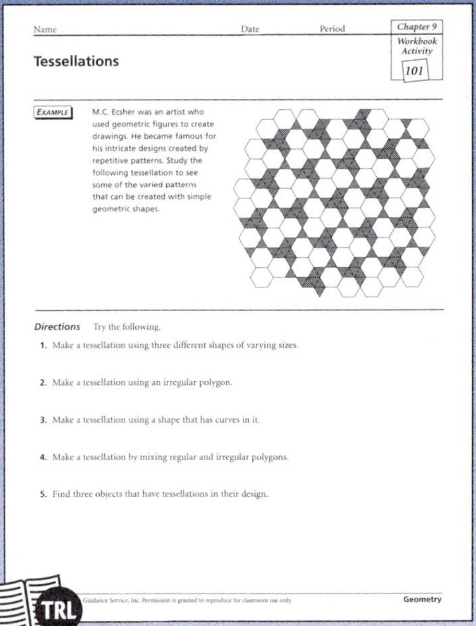

Name Date Period

Chapter 9 Workbook Activity 101

Tessellations

EXAMPLE M.C. Ecsher was an artist who used geometric figures to create drawings. He became famous for his intricate designs created by repetitive patterns. Study the following tessellation to see some of the varied patterns that can be created with simple geometric shapes.

Directions Try the following.

1. Make a tessellation using three different shapes of varying sizes.
2. Make a tessellation using an irregular polygon.
3. Make a tessellation using a shape that has curves in it.
4. Make a tessellation by mixing regular and irregular polygons.
5. Find three objects that have tessellations in their design.

Geometry

Workbook Activity 101

Chapter 9 Review

Match the polygon on the right with the formula for its area or perimeter on the left. Write the letter of the polygon. You may use some letters more than once.

Example: $A = l \cdot w$ — Solution: $l \cdot w$ is the formula for the area of a rectangle, choice B

1. $A = s^2$ C
2. $A = \frac{b_1 + b_2}{2}h$ D
3. $A = b \cdot h$ F
4. $A = \frac{1}{2}(b \cdot h)$ A
5. $A = \frac{1}{2}s^2$ E
6. $P = 4s$ C
7. $P = 2(l + w)$ B

A.

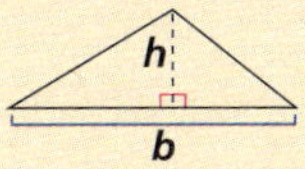

B.

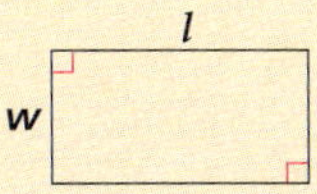

C.

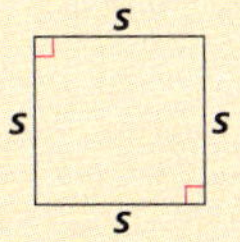

D.

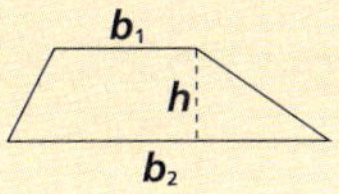

E.

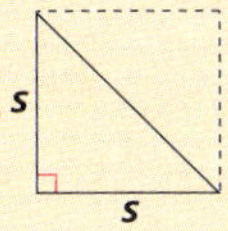

F.

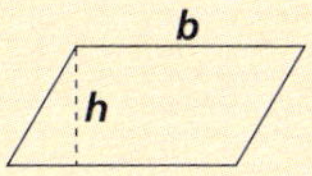

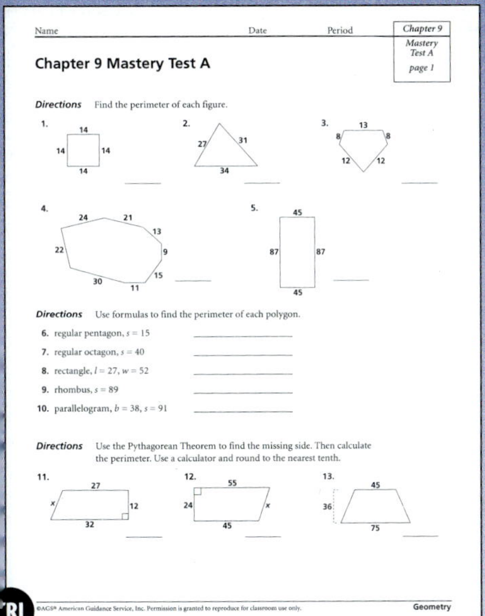

Name Date Period

Chapter 9 Mastery Test A page 1

Chapter 9 Mastery Test A

Directions Find the perimeter of each figure.

1. (square: 14, 14, 14, 14) ____
2. (triangle: 27, 31, 34) ____
3. (pentagon: 13, 8, 8, 12, 12) ____
4. (polygon: 24, 21, 13, 9, 15, 11, 30, 22) ____
5. (rectangle: 45, 87, 87, 45) ____

Directions Use formulas to find the perimeter of each polygon.

6. regular pentagon, $s = 15$ ____
7. regular octagon, $s = 40$ ____
8. rectangle, $l = 27$, $w = 52$ ____
9. rhombus, $s = 89$ ____
10. parallelogram, $b = 38$, $s = 91$ ____

Directions Use the Pythagorean Theorem to find the missing side. Then calculate the perimeter. Use a calculator and round to the nearest tenth.

11. (27, x, 12, 32) ____
12. (55, 24, x, 45) ____
13. (45, 36, 75) ____

 Geometry

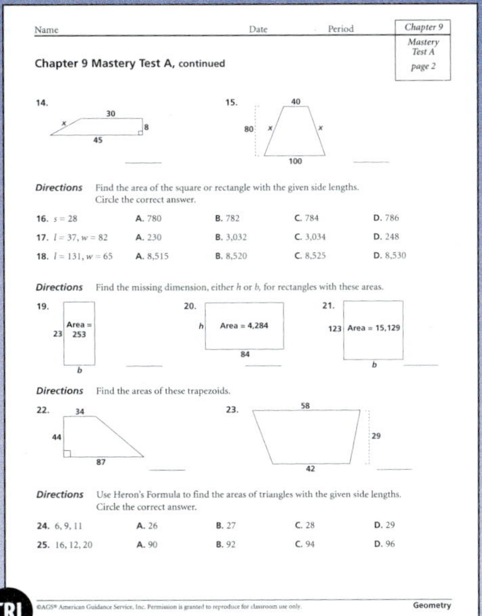

Name Date Period

Chapter 9 Mastery Test A page 2

Chapter 9 Mastery Test A, continued

14. (30, x, 8, 45) ____
15. (40, 80, x, x, 100) ____

Directions Find the area of the square or rectangle with the given side lengths. Circle the correct answer.

16. $s = 28$	A. 780	B. 782	C. 784	D. 786
17. $l = 37$, $w = 82$	A. 230	B. 3,032	C. 3,034	D. 248
18. $l = 131$, $w = 65$	A. 8,515	B. 8,520	C. 8,525	D. 8,530

Directions Find the missing dimension, either h or b, for rectangles with these areas.

19. (23; Area = 253; b) ____
20. (h; Area = 4,284; 84) ____
21. (123; Area = 15,129; b) ____

Directions Find the areas of these trapezoids.

22. (34, 44, 87) ____
23. (58, 29, 42) ____

Directions Use Heron's Formula to find the areas of triangles with the given side lengths. Circle the correct answer.

24. 6, 9, 11	A. 26	B. 27	C. 28	D. 29
25. 16, 12, 20	A. 90	B. 92	C. 94	D. 96

 Geometry

TRL

Chapter 9 Mastery Test A

Chapter 9 Review

Each set of problems in the Chapter Review includes an example and solution to illustrate the concept. Use the given examples for reteaching the material in Chapter 9. For additional practice, refer to the Supplementary Problems for Chapter 9 (pages 412–413) and Additional Practice 54–59 in the Teacher's Resource Library.

Chapter 9 Mastery Test TRL

The Teacher's Resource Library includes parallel forms of the Chapter 9 Mastery Test. The difficulty level of the two forms is equivalent. You may wish to use one form as a pretest and the other form as a posttest. In addition, the Chapter 9 Cumulative Mastery Test covers content from Chapters 1–9.

Alternative Assessment

Alternative Assessment items correlate with student Goals for Learning at the beginning of this chapter.

- Have students identify formulas for finding the perimeters of polygons.

Draw several polygons, both regular and irregular. For each, give a formula for finding the perimeter.

- Have students identify formulas for finding the areas of polygons.

Write a formula to find the area of each of the following polygons:

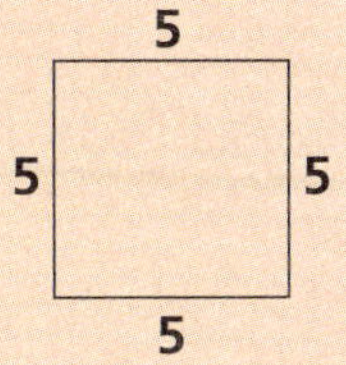

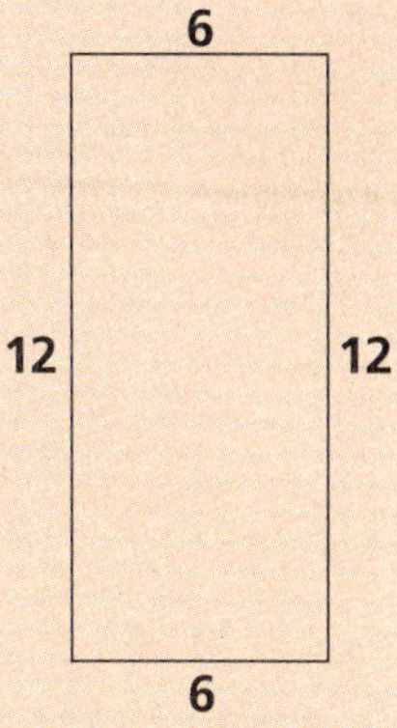

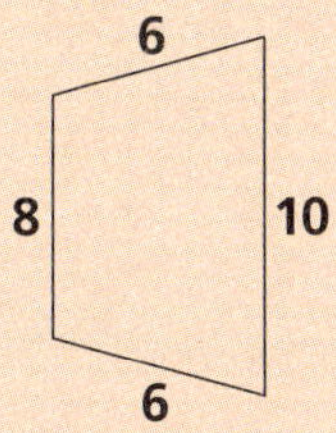

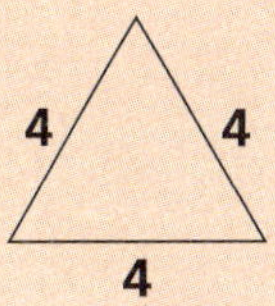

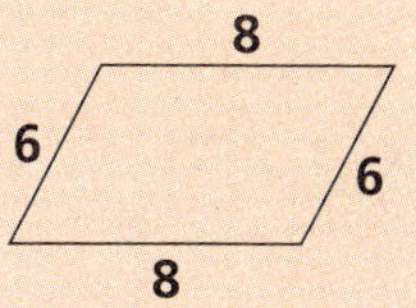

Alternative Assessment

continued

▶ Have students use the distance formula for finding perimeter.

Write the steps necessary to find the perimeter for the following figure. Give the perimeter.

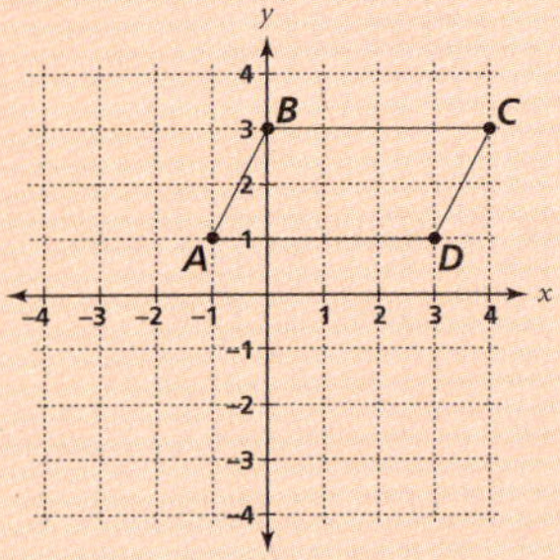

▶ Have students use formulas to find the perimeters and areas of polygons.

Find the perimeter and area of the shaded polygon.

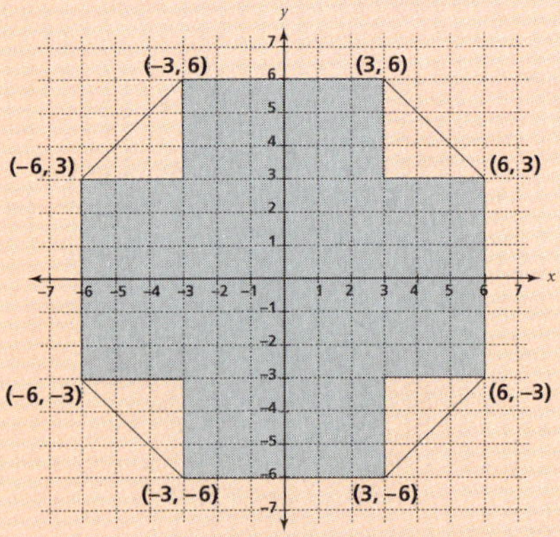

▶ Have students use the Pythagorean Theorem in order to determine perimeter and area.

Find the area and perimeter of the entire figure in item 4.

More Review...

Use the distance formula to find the perimeter of each polygon.

Example: $\triangle ABC$

Solution: $P = \overline{AB} + \overline{BC} + \overline{CA}$

$A = (-5, 4),\ B = (-1, 5)$

$d = \sqrt{(-5 - (-1))^2 + (4 - 5)^2}$

$d = \sqrt{16 + 1} = \sqrt{17}$

$d = 4.1$

$B = (-1, 5),\ C = (-1, 1)$

$d = \sqrt{(-1 - (-1))^2 + (5 - 1)^2}$

$d = \sqrt{0 + 16} = \sqrt{16}$

$d = 4$

$C = (-1, 1),\ A = (-5, 4)$

$d = \sqrt{(-1 - (-5))^2 + (1 - 4)^2}$

$d = \sqrt{16 + 9} = \sqrt{25}$

$d = 5$

$P = 4.1 + 4 + 5 = 13.1$

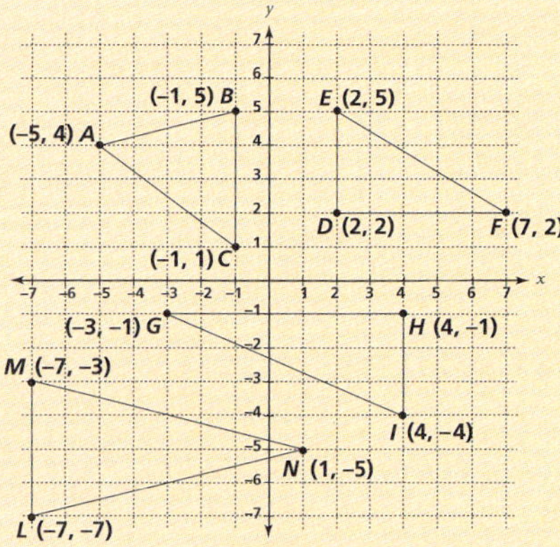

8. $\triangle DEF$ 13.8

9. $\triangle GHI$ 17.6

10. $\triangle LMN$ 20.4

Find the perimeter and area of each polygon.

Example:

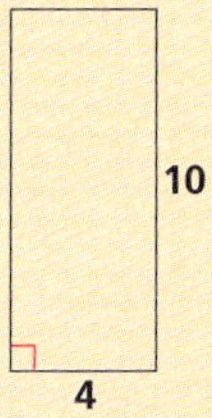

Solution: $P = 2(l + w) = 2(4 + 10) = 2(14) = 28$

$A = lw = (4)(10) = 40$ sq units

11.

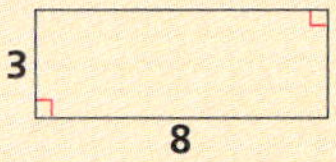

$P = 22$, $A = 24$ sq units

12.

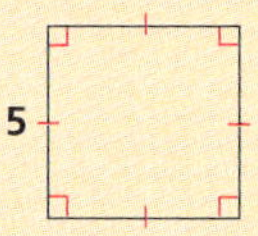

$P = 20$, $A = 25$ sq units

13.

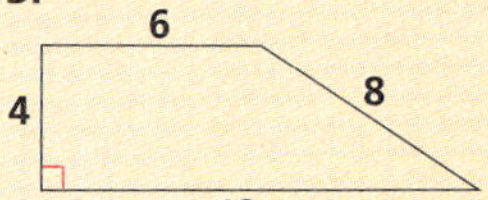

$P = 30$, $A = 36$ sq units

14.

$P = 38$, $A = 42$ sq units

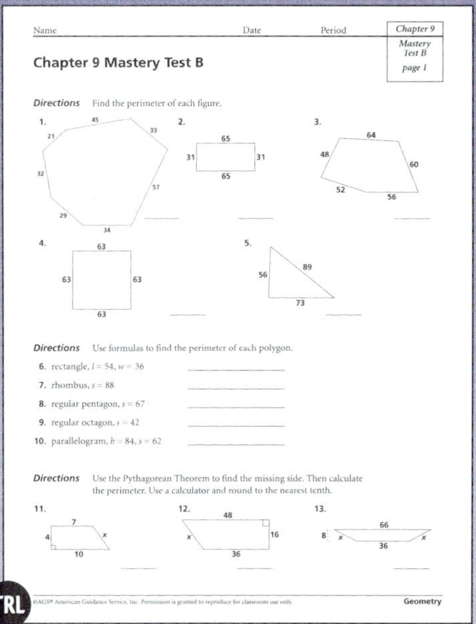

Name Date Period

Chapter 9 Mastery Test B page 1

Chapter 9 Mastery Test B

Directions Find the perimeter of each figure.

1. 2. 3. 4. 5.

Directions Use formulas to find the perimeter of each polygon.

6. rectangle, $l = 54$, $w = 36$ ______
7. rhombus, $s = 88$ ______
8. regular pentagon, $s = 67$ ______
9. regular octagon, $s = 42$ ______
10. parallelogram, $b = 84$, $s = 62$ ______

Directions Use the Pythagorean Theorem to find the missing side. Then calculate the perimeter. Use a calculator and round to the nearest tenth.

11. 12. 13.

TRL ©AGS® American Guidance Service, Inc. Permission is granted to reproduce for classroom use only. Geometry

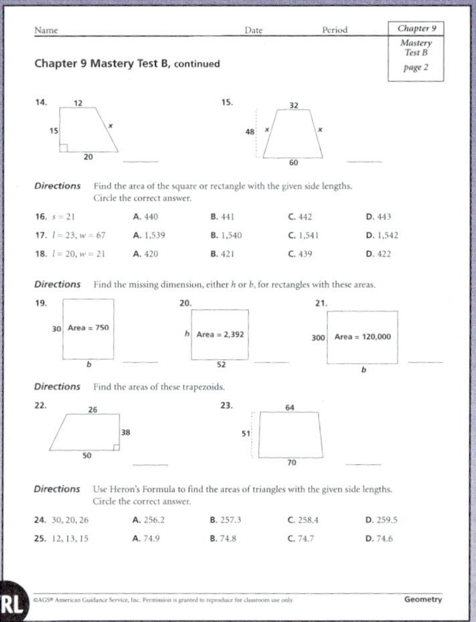

Name Date Period

Chapter 9 Mastery Test B page 2

Chapter 9 Mastery Test B, continued

14. 15.

Directions Find the area of the square or rectangle with the given side lengths. Circle the correct answer.

16. $s = 21$	A. 440	B. 441	C. 442	D. 443
17. $l = 23$, $w = 67$	A. 1,539	B. 1,540	C. 1,541	D. 1,542
18. $l = 20$, $w = 21$	A. 420	B. 421	C. 439	D. 422

Directions Find the missing dimension, either h or b, for rectangles with these areas.

19. Area = 750 20. Area = 2,392 21. Area = 120,000

Directions Find the areas of these trapezoids.

22. 23.

Directions Use Heron's Formula to find the areas of triangles with the given side lengths. Circle the correct answer.

24. 30, 20, 26	A. 256.2	B. 257.3	C. 258.4	D. 259.5
25. 12, 13, 15	A. 74.9	B. 74.8	C. 74.7	D. 74.6

TRL ©AGS® American Guidance Service, Inc. Permission is granted to reproduce for classroom use only. Geometry

Chapter 9 Mastery Test B

15.

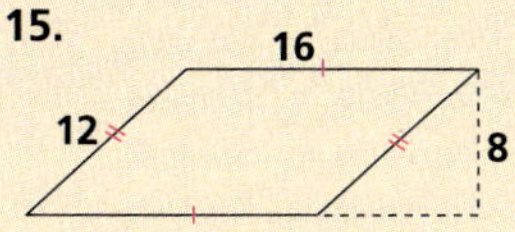

$P = 56$, $A = 128$ sq units

16.

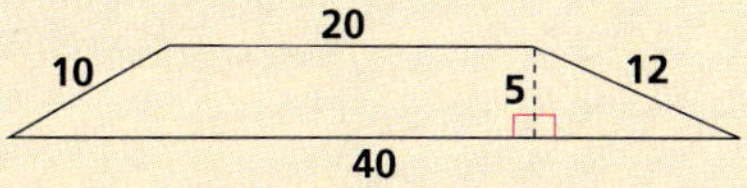

$P = 82$, $A = 150$ sq units

Use the Pythagorean Theorem to find the missing side. Then calculate the perimeter and the area of the *shaded* portion of each figure. (Use a calculator and round to the nearest tenth.)

Example:

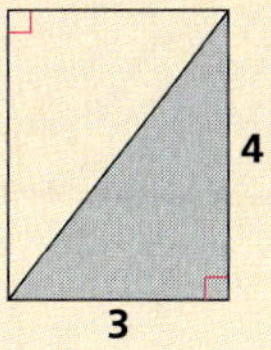

Solution: Let x = the missing side,

then $P = 3 + 4 + x$

$x^2 = 3^2 + 4^2 = 25$; $x = \sqrt{25} = 5$

substituting, $P = 3 + 4 + 5 = 12$

$A = \frac{1}{2}(bh) = \frac{1}{2}(3 \bullet 4) = \frac{1}{2}(12) = 6$ sq units

17.

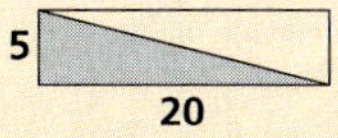

$P = 45.6$, $A = 50$ sq units

18.

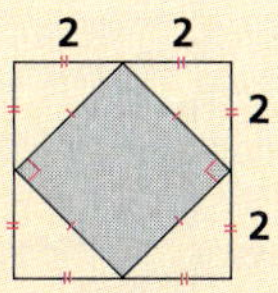

$P = 11.2$, $A = 7.8$ sq units

19.

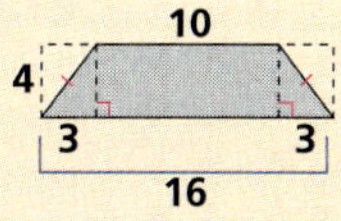

$P = 36$, $A = 52$ sq units

20.

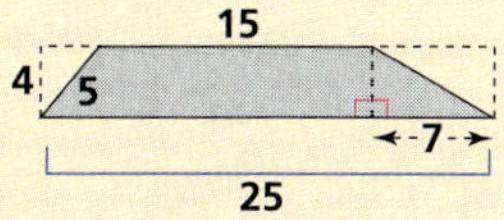

$P = 53.1$, $A = 80$ sq units

TEST-TAKING TIP

When taking a matching test, match all the items that you know go together for sure. Cross these items out. Then try to match the items that are left.

L

Name Date Period

Chapter 9
Cumulative Mastery Test
page 1

Chapter 9 Cumulative Mastery Test

Directions Graph each line on a separate sheet of paper using the slope and point given. Also write the algebraic expression for each line.

1. Line a; $m = \frac{1}{4}$, passes through $(5, -3)$
2. Line b, $m = 4$, passes through $(2, -3)$

Directions Complete the following constructions. Use a separate sheet of paper and a straightedge and a compass.

3. Construct ΔXYZ with $YZ = YX$ and $m\angle Y < 90°$.
4. Draw any right triangle and label it ABC. Construct a second triangle $A'B'C'$ that is congruent to ABC using the H–L theorem.

Directions Give the coordinates of the image quadrilateral under the following dilations. Graph the object and its images on a separate sheet of paper. All dilations have $(0, 0)$ as the center of the dilation.

5. dilation of 2
6. dilation of 4
7. dilation of $\frac{1}{2}$
8. dilation of $\frac{1}{4}$

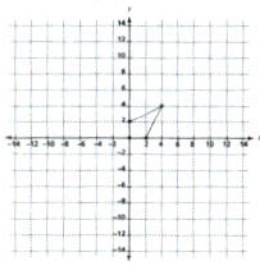

Directions Use Plato's Formula to find Pythagorean Triples for the given integers. Use a calculator.

9. m = 15 __________
10. m = 30 __________

Directions Use the figure to find the following quantities.

11. Area of each right triangle = __________
12. Side length and area of the small square = __________
13. Sum of the areas that make up the large square = __________

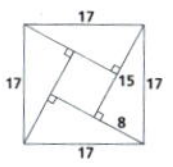

Geometry

TRL

Name Date Period

Chapter 9
Cumulative Mastery Test
page 2

Chapter 9 Cumulative Mastery Test, continued

Directions Use the distance formula to find the distances between the given points. Circle the correct answer.

14. $(-3, 1)$ and $(5, -2)$	A. 70	B. 71	C. 72	D. 73
15. $(3, 0)$ and $(-2, -2)$	A. 29	B. 30	C. 31	D. 2

Directions Use formulas to find the perimeter of each polygon.

16. rectangle, $l = 33$, $w = 54$ __________
17. pentagon, s = 232 __________

Directions Use the Pythagorean Theorem to find the missing side. Then calculate the perimeter. Use a calculator and round to the nearest tenth.

18.

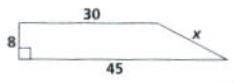

____

19.

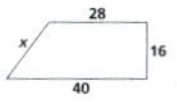

____

Directions Find the area of the square or rectangle with the given side lengths. Circle the correct answer.

20. $s = 45$	A. 2,025	B. 2,026	C. 2,027	D. 2,028
21. $s = 32$	A. 1,025	B. 1,024	C. 1,023	D. 1,022

Directions Find the missing dimension, either h or b, for rectangles with these areas.

22.

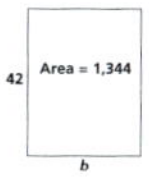

____

23.

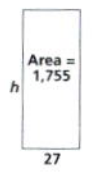

____

Directions Use Heron's Formula to find the areas of triangles with the given side lengths. Circle the correct answer.

24. 25, 35, 40	A. 433	B. 434	C. 435	D. 436
25. 17, 21, 32	A. 162.4	B. 162.5	C. 162.6	D. 162.7

Geometry

Chapter 9 Cumulative Mastery Test

Perimeter and Area *Chapter 9* **313**

Chapter 10

Planning Guide

Circles and Spheres

	Student Text Lesson						
	Student Pages	Vocabulary	Proof	Practice Exercises	Problem Solving	Calculator Practice	Solutions Key
Lesson 1 Definition of a Circle	316–318	✔		✔			✔
Lesson 2 The Ratio π	319–322	✔		✔	322		✔
Lesson 3 Approximating the Area of a Circle	323–325			✔			✔
Lesson 4 Area and Probability	326–327	✔		✔			✔
Lesson 5 Formula for the Area of a Circle	328–331			✔		330	✔
Lesson 6 Circles and Their Angles and Sectors	332–335	✔		✔			✔
Lesson 7 Tangents, Circumcircles, and Incircles	336–339	✔	339	✔			✔
Lesson 8 Sine, Cosine, and Tangent	340–341	✔		✔		341	✔
Lesson 9 Spheres	342–345	✔		✔			✔
Application Statistics	346			✔			✔

Chapter Activities

Teacher's Resource Library

Geometry on the Job 10: Astronomy

Community Connection 10: Around the Word

Teacher's Edition

Chapter 10 Project

Assessment Options

Student Text

Chapter 10 Review

Teacher's Resource Library

Chapter 10 Mastery Tests A and B

Chapter 10 Cumulative Mastery Test

Teacher's Edition

Chapter 10 Alternative Assessment

Student Text Features						Teaching Strategies								Learning Styles						Teacher's Resource Library		
Algebra Connection	Constructions	Experiencing Geometry	Technology Connection	Try This	Writing About Mathematics	Warm-Up Activity	Common Error	Algebra Connection	Mental Math	Manipulatives	Calculator	Applications	Group Problem Solving	Auditory/Verbal	Visual/Spatial	Body/Kinesthetic	Logical/Mathematical	Interpersonal/Group Learning	LEP/ESL	Workbook Activities	Additional Practice	Self-Study Guide
						316								318		317	318		318	102	60	✔
319		321			322	319		320	322			321					322			103	61	✔
					325	323	324								324			325		104	62	✔
						326	327								327					105	63	✔
						328				331	330			330						106, 107	64	✔
				335	335	332	334		333					334					335	108	65	✔
	339					336										339	338	338	338	109	66	✔
						340					341		341				341			110	67	✔
		342	345		345	342						344					345		343	111	68	✔
						346						346								112		✔

Manipulatives

Polygon Power Pack

Chapter at a Glance

Chapter 10: Circles and Spheres pages 314–349

Lessons

Teacher's Resource Library TRL

(Answer Keys for the Teacher's Resource Library begin on page 484 of this Teacher's Edition.)

Name Date Period | *Chapter 10* *Geometry on the Job* *10*

Astronomy

Astronomy is the study of the objects that make up the universe—stars, planets, and other objects in space. People who study astronomy are called *astronomers.* Many astronomers specialize and have special titles. An *astrophysicist* studies the physical and chemical properties of the universe. A *planetary astronomer* studies the planets. *Cosmologists* study the history and structure of the whole universe. *Observational astronomers* use telescopes to study astronomical bodies in the universe.

The study of astronomy began in ancient times. Early astronomers thought that Earth was the center of the solar system. In reality the sun is the center of the solar system and Earth is one of the nine planets that revolve around the sun. The solar system also includes asteroids, meteoroids, and comets, as well as dust and gas.

Try It Yourself
Use information from an encyclopedia or from the Internet to complete the chart. Calculate the radius by using the diameter.

Planets	Distance from Sun	Diameter	Radius
Mercury			
Venus			
Earth			
Mars			
Jupiter			
Saturn			
Uranus			
Neptune			
Pluto			

TRL Geometry

Geometry on the Job 10

Name Date Period | *Chapter 10* *Community Connection* *10*

Around the World

The earth is not actually a perfect sphere. Because of the force created by Earth's rotation around its own axis, it is squished out at the equator as if someone had pressed together the two poles slightly. However, from our frame of reference, it might as well be a sphere. Therefore, the globes that we use as maps to represent Earth three dimensionally are generally spherical.

Directions
Locate your hometown on the globe. Unless you live in a major city, it probably won't actually be on the globe itself, but you should be able to estimate its location fairly well using an atlas and measuring its relative distance and direction from the major cities around it. Choose two cities near it that are also on the globe and, on a transparency, form a triangle using your hometown and the two cities as endpoints. By measuring the angles created and the length of each of the sides, you should be able to copy the triangle (you will have to change the scale) onto the globe using the existing marks for the two cities. The point for your hometown won't be completely accurate because the lines drawn onto the globe were measured on a flat map and are being transferred to a curved surface. Label your town on the globe by using a piece of clay or clear tape that can be removed. After you have labeled it, try the following:

1. Locate the point directly opposite your town on the globe. Where is it? Name the continent or body of water. Tell what country the place is in if it is on land.
2. Locate points in the Northern Hemisphere and in the Southern Hemisphere that are equidistant from your town.
3. Measure a radius that is scaled to 500 miles on the globe. Draw a circle using this radius and your town as the center. Make a note of the places that lie on the circumference of the circle.

TRL Geometry

Community Connection 10

Chapter 10

Circles and Spheres

New Mexico is known as the Land of Enchantment. Its beautiful land and climate help it live up to its nickname. Visitors may also find its crop circles enchanting. These circles result from irrigation. Because of New Mexico's warm, dry climate, farmers must water their fields. Many farmers use a system of anchored sprinklers. These sprinklers have long arms that spread the life-giving water. The arms turn in circles so the water misses the corners of square fields. This leaves a pattern of circles dotting the land.

Each arm of the watering system is the radius of a circle. As a sprinkler turns, water covers a circular field completely. The resulting green fields turn gold as the grain and hay ripen. Geometry even plays a role in farming.

In Chapter 10, you will learn formulas for use with circles and spheres.

Goals for Learning

- To identify formulas for the radius, diameter, circumference, and area of a circle
- To find the circumference, diameter, and radius of a circle
- To solve problems involving area and probability
- To determine the areas of circles
- To construct circles
- To define trigonometric ratios
- To determine the volume and surface area of a sphere

Introducing the Chapter

Use the information on these pages to spark a discussion with students about circles, area of circles, and spheres. Discuss the photo and ask if any students have experience with smaller-scale circular sprinklers. Prompt students to describe a circle as the outer border, edge, or limit of the water's reach while describing the rest of the region as the area of the circle. Point out that in this chapter students will learn about the properties and characteristics of two-dimensional circles and their three-dimensional counterparts—spheres. Read the Goals for Learning with students and point out that this chapter will help them learn formulas for use with circles and spheres.

Chapter Project

Challenge students to design a set of pulleys or gears for which the following is true: 1 revolution of pulley A equals 3 revolutions of pulley B equals 9 revolutions of pulley C. Once students complete their design, encourage them to construct working models of their pulleys or gears.

As an alternative, students can identify, collect, and display a variety of gears and pulleys. For each item displayed, students should describe the relationship between the gears or pulleys.

SELF-STUDY GUIDE

Name ____________

CHAPTER 10: Circles and Spheres

Goal 10.1 To identify formulas for the radius, diameter, circumference, and area of a circle

Date	Assignment	Score
______	1: Read pages 316–317. Complete Exercises A and B on page 318.	______
______	2: Complete Workbook Activity 102.	______

Comments:

Goal 10.2 To find the circumference, diameter, and radius of a circle

Date	Assignment	Score
______	3: Read pages 319–321. Complete Exercises A–C on page 322.	______
______	4: Complete Workbook Activity 103.	______

Comments:

Goal 10.3 To solve problems involving area and probability

Date	Assignment	Score
______	5: Read pages 323–324. Complete Exercises A–C on page 325.	______
______	6: Complete Workbook Activity 104.	______
______	7: Read page 326. Complete Exercise A on page 327.	______
______	8: Complete Workbook Activity 105.	______

Comments:

Goal 10.4 To determine the areas of circles

Date	Assignment	Score
______	9: Read pages 328–329.	______
______	10: Read the Calculator Practice and complete Exercise A on page 330.	______
______	11: Complete Exercises B–D on page 331.	______
______	12: Complete Workbook Activities 106–107.	______

 Geometry

SELF-STUDY GUIDE

Name ____________

CHAPTER 10: Circles and Spheres, continued

Goal 10.4 To determine the areas of circles, continued

Date	Assignment	Score
______	13: Read pages 332–335. Complete Exercises A–C on page 335.	______
______	14: Complete Workbook Activity 108.	______

Comments:

Goal 10.5 To construct circles

Date	Assignment	Score
______	15: Read pages 336–339. Complete Exercises A and B on page 339.	______
______	16: Complete Workbook Activity 109.	______

Comments:

Goal 10.6 To use trigonometric ratios to solve problems

Date	Assignment	Score
______	17: Read pages 340–341.	______
______	18: Read the Calculator Practice and complete Exercise A on page 341.	______
______	19: Complete Workbook Activity 110.	______

Comments:

Goal 10.7 To determine the volume and surface area of a sphere

Date	Assignment	Score
______	20: Read pages 342–344. Complete Exercises A–C on pages 344–345.	______
______	21: Complete Workbook Activity 111.	______
______	22: Read and complete the Application Exercise on page 346.	______
______	23: Complete Workbook Activity 112.	______
______	24: Complete the Chapter 10 Review on pages 347–349.	______

Comments:

Student's Signature ____________ Date ____________

Instructor's Signature ____________ Date ____________

 Geometry

L TRL

Chapter 10 Self-Study Guide

Lesson at a Glance

Chapter 10 Lesson 1

Overview This lesson presents the idea of locus and the definition of a circle.

Objectives

- To identify a circle as the locus of points at the same distance from a given point.
- To identify the characteristics of a circle including its circumference, radius, diameter, quadrants, and chords.

Student Pages 316–318

Teacher's Resource Library TRL

Workbook Activity 102

Additional Practice 60

Mathematics Vocabulary

chord
circle
circumference
diameter
quadrant
radius

1. Warm-Up Activity

Ask students what they know about circles. Have them describe, or offer operational definitions of, terms such as *circle, radius, diameter,* and *circumference.* Accept all descriptions without prejudice or editing, recording them on an easel pad and save for use at the end of the lesson. Point out that in this lesson students will learn the terms they need to know to successfully complete the chapter.

Lesson 1 Definition of a Circle

Chord
Line segment joining two points on a circle

Circle
Locus of points at the same distance from a given point

Circumference
The complete length of a circle

Diameter
A chord that passes through the center of a circle

Quadrant
One-fourth of a circle

Radius
Distance between the center of a circle and any point on the circle

The plural of *radius* is *radii.*

If asked to point out a circle, you would not have a problem pointing to a number of items such as dinner plates, compact disks, or some clock faces—each of whose outline is a circle. But suppose someone asked you for a formal definition of a circle. You would need to provide a definition that uses the locations of points that satisfy certain conditions.

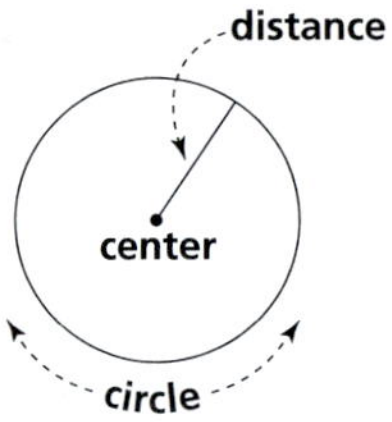

A **circle** is the locus of points equidistant (at the same distance) from a given point, called the center.

Remember that Euclid's Postulate 3 states that a circle may be drawn using any point as a center and any distance as a **radius.**

The radius, *r,* of a circle is the distance between the center and any point on the circle.

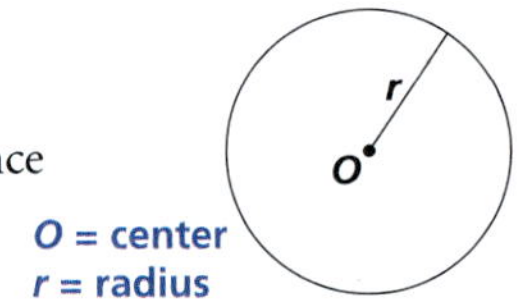

A **chord** of a circle is any line segment joining two points on the circle.

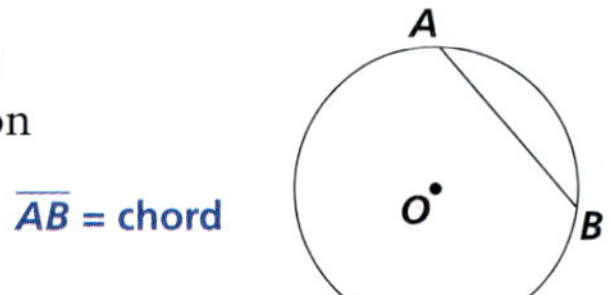

The **diameter** of a circle is a chord that passes through the center of a circle. The diameter of a circle is twice as long as a radius of the same circle. A diameter divides a circle into two equal parts called *semicircles.*

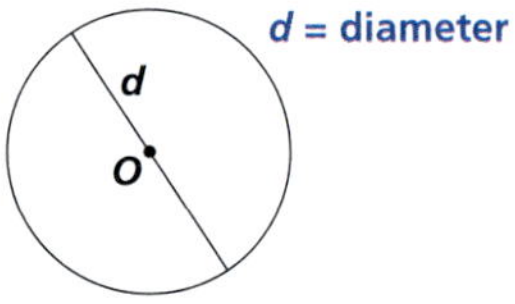

The **circumference** of a circle is the complete length of the circle. The circumference is $2\pi r$ where r is the radius of the circle. (The symbol π is the Greek letter *pi* and is pronounced "pie.") π is a real number whose value is approximately 3.14. You will learn more about π in the next lesson.

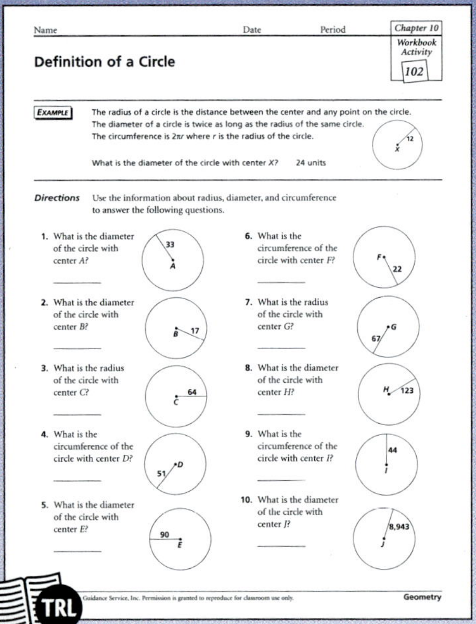

Name Date Period

Chapter 10 Workbook Activity 102

Definition of a Circle

EXAMPLE The radius of a circle is the distance between the center and any point on the circle. The diameter of a circle is twice as long as the radius of the same circle. The circumference is $2\pi r$ where r is the radius of the circle.

What is the diameter of the circle with center X? 24 units

Directions Use the information about radius, diameter, and circumference to answer the following questions.

1. What is the diameter of the circle with center A? ______
2. What is the diameter of the circle with center B? ______
3. What is the radius of the circle with center C? ______
4. What is the circumference of the circle with center D? ______
5. What is the diameter of the circle with center E? ______
6. What is the circumference of the circle with center F? ______
7. What is the radius of the circle with center G? ______
8. What is the diameter of the circle with center H? ______
9. What is the circumference of the circle with center I? ______
10. What is the diameter of the circle with center J? ______

Geometry

TRL

Workbook Activity 102

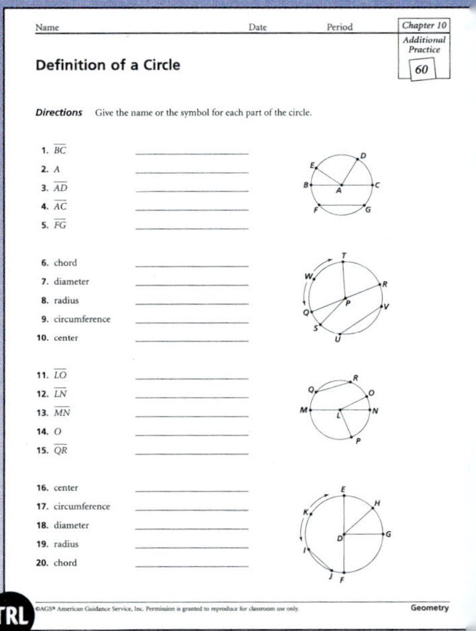

Name Date Period

Chapter 10 Additional Practice 60

Definition of a Circle

Directions Give the name or the symbol for each part of the circle.

1. $\overline{BC}$ ______
2. A ______
3. $\overline{AD}$ ______
4. $\overline{AC}$ ______
5. $\overline{FG}$ ______
6. chord ______
7. diameter ______
8. radius ______
9. circumference ______
10. center ______
11. $\overline{LO}$ ______
12. $\overline{LN}$ ______
13. $\overline{MN}$ ______
14. O ______
15. $\overline{QR}$ ______
16. center ______
17. circumference ______
18. diameter ______
19. radius ______
20. chord ______

Geometry

TRL

Additional Practice 60

The circumference of a circle is divided into 360°. Each circle can be divided into fourths of 90° each.

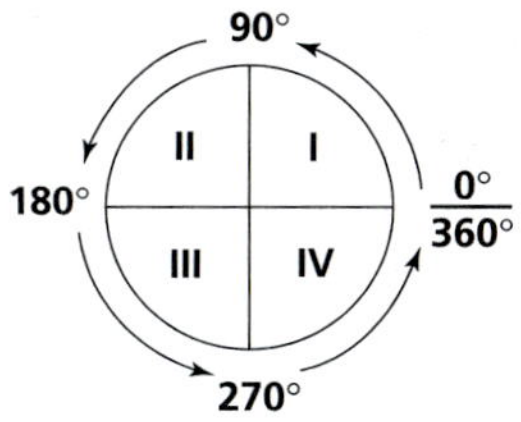

A **quadrant** of a circle is one-fourth of the circle. Quadrants of the circle are numbered from I to IV in a counterclockwise direction.

You can use these definitions to solve problems involving circles.

EXAMPLES Name the parts of the circle represented by O, $\overline{AB}$, $\overline{CD}$, $\overline{OE}$.

O = center
$\overline{AB}$ = chord
$\overline{CD}$ = diameter
$\overline{OE}$ = radius

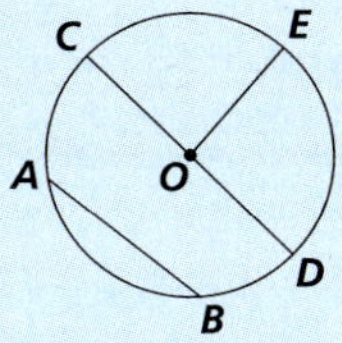

The radius of a circle is 10 units long. How long is the diameter of the circle? How long is the circumference?

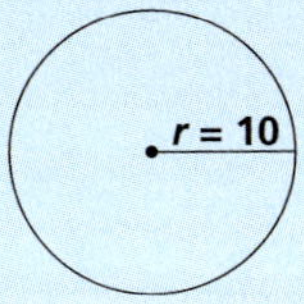

diameter = 2(radius) d = diameter
or $d = 2r$ r = radius
$d = 2(10) = 20$ units
circumference = $2\pi r$ (π approximately equal to 3.14)
$= 2\pi r$
$= 20\pi \approx 20(3.14) = 62.8$ units

The symbol ≈ means "approximately equal to."

2. Teaching the Lesson

Using a circle drawn on an overhead projector, illustrate each of the following: circle, diameter, radius. Then illustrate the terms *circumference* and *quadrants.* Point out to students that *circle* is the first term they have met whose definition involves the location (locus) of several points relative to a single point.

LEARNING STYLES

Body/Kinesthetic Have students work in groups of four. Give each group chalk and a string at least 4 feet long. Direct the students to use these materials to model each of the following terms: *circle, radius, chord,* and *diameter.* Indicate that students can use the chalk to draw directly on the floor. Circulate as students identify the terms they draw.

3. Reinforce and Extend

LEARNING STYLES

Logical/Mathematical Present the following inequalities to students. Ask whether they are always true, never true, or sometimes true. Elicit from students the criteria they need in order to say that something is never true. (a counterexample)

diameter > radius	(always true)
diameter > chord	(sometimes true)
radius > chord	(sometimes true)
diameter ≥ chord	(always true)

LEARNING STYLES

LEP/ESL Have students draw an example of each of this lesson's vocabulary terms. Suggest they label each drawing and use it as part of their picture dictionary. Encourage students to record relationships among the terms such as diameter = 2 radius or circumference = $2\pi r$. Direct students to compare their examples with the descriptions recorded at the beginning of the lesson. Have them determine the accuracy of their drawings as well as the earlier descriptions.

LEARNING STYLES

Auditory/Verbal Assign Exercise A to student pairs. For each item, one student can explain his or her answer, while the second student listens. Direct students to use complete sentences, offering the following model: "$\overline{XY}$ is a chord of the circle because it is a line segment joining two points on the circle." Students can exchange roles after each item.

Selected Answers and Solutions begin on page 420.

1. center
2. chord
3. diameter
4. radius
5. radius
6. $\overline{AO}$, $\overline{BO}$, $\overline{CO}$
7. $\overline{BC}$
8. $\overline{DE}$
9. C
10. O

Exercise A Give the name or the symbol for each part of the circle.

1. O

2. $\overline{XY}$

3. $\overline{LM}$

4. $\overline{OZ}$

5. $\overline{OX}$

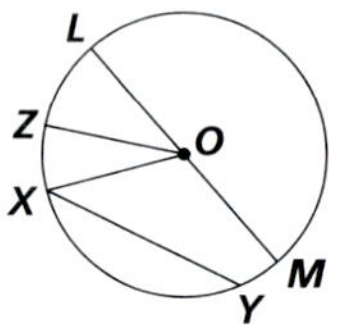

6. radius

7. diameter

8. chord

9. circumference

10. center

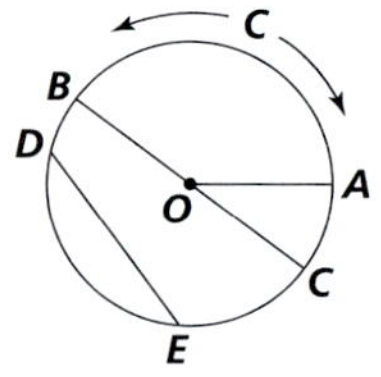

11. 12 units
12. 18π units
13. 7
14. 6
15. 12π units, 37.7 units

Exercise B Use the information about radius, diameter, and circumference to answer the following questions.

11. What is the diameter of the circle with center P?

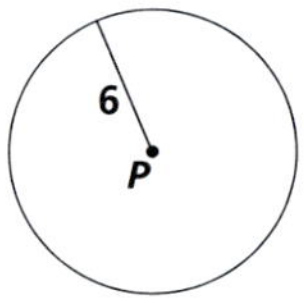

12. What is the circumference of the circle with center Q?

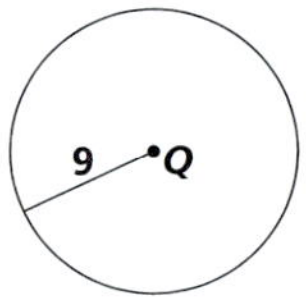

13. The circumference of a circle is 14π. What is the radius of the circle?

14. The diameter of a circle is 12. What is the radius of the circle?

15. The diameter of a circle equals 12 units. What is the exact circumference of the circle? What is the approximate circumference?

Lesson 2 The Ratio π

You have learned that triangles that have the same shape but are different sizes are similar.

Because the right triangles shown are similar, the corresponding sides are in proportion:

$$\frac{\text{altitude}}{\text{base}} = \frac{a}{b} = \frac{c}{d} = \frac{e}{f} \text{ and so on.}$$

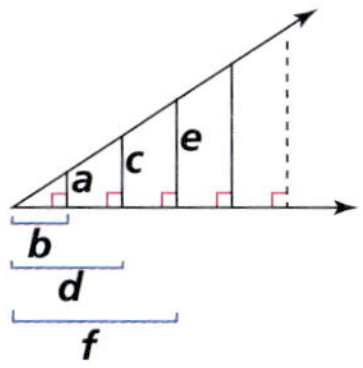

> ***Pi*** π
> *Ratio of the circumference of a circle to its diameter*

The same reasoning applies to circles. Although they may be different sizes, all circles are the same shape. This means that all circles are similar to one another. Because they are similar—

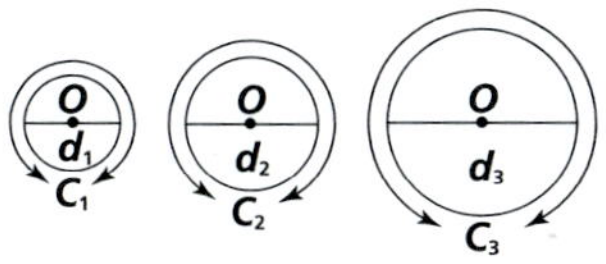

$$\frac{\text{circumference}}{\text{diameter}} = \frac{C_1}{d_1} = \frac{C_2}{d_2} = \frac{C_3}{d_3} \text{ and so on.}$$

These ratios always equal a single real number, **pi** or π.

$$\frac{\text{circumference}}{\text{diameter}} = \frac{C_1}{d_1} = \frac{C_2}{d_2} = \frac{C_3}{d_3} = \pi \text{ is true for any circle.}$$

π is a real number and can be placed on the real number line.

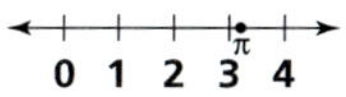

Because it is important in many formulas, π often has its own key on calculators. However, you can usually use algebra to work with formulas that include π without evaluating it or using a calculator.

Algebra Connection

Lesson at a Glance

Chapter 10 Lesson 2

Overview This lesson explores the constant π, the ratio of any circle's circumference to its radius.

Objectives

- To identify pi (π) as the ratio of a circle's circumference to its radius.
- To use the constant pi in appropriate formulas.
- To estimate a circle's circumference with an accuracy appropriate to the circumstance.

Student Pages 319–322

Teacher's Resource Library TRL

Workbook Activity 103

Additional Practice 61

Mathematics Vocabulary

Pi (π)

1. Warm-Up Activity

Write the symbol π, pi, on the board. Ask students if they are familiar with the symbol and what it stands for. Point out that in this lesson they will learn that pi represents a unique characteristic of circles.

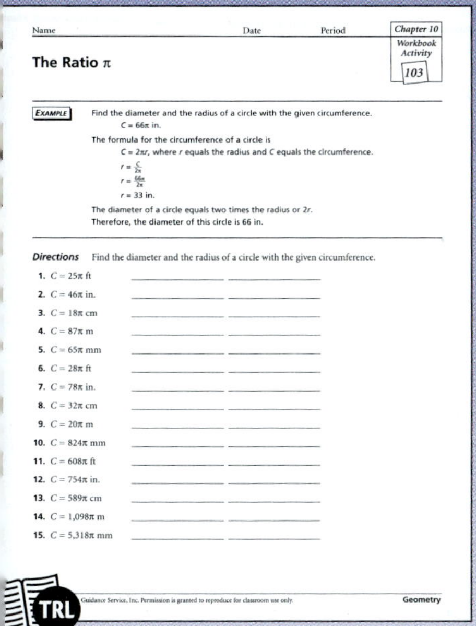

Name Date Period

Chapter 10 *Workbook Activity* *103*

The Ratio π

EXAMPLE Find the diameter and the radius of a circle with the given circumference.

$C = 66\pi$ in.

The formula for the circumference of a circle is

$C = 2\pi r$, where r equals the radius and C equals the circumference.

$r = \frac{C}{2\pi}$

$r = \frac{66\pi}{2\pi}$

$r = 33$ in.

The diameter of a circle equals two times the radius or $2r$.

Therefore, the diameter of this circle is 66 in.

Directions Find the diameter and the radius of a circle with the given circumference.

1. $C = 25\pi$ ft ______ ______
2. $C = 46\pi$ in. ______ ______
3. $C = 18\pi$ cm ______ ______
4. $C = 87\pi$ m ______ ______
5. $C = 65\pi$ mm ______ ______
6. $C = 28\pi$ ft ______ ______
7. $C = 78\pi$ in. ______ ______
8. $C = 32\pi$ cm ______ ______
9. $C = 20\pi$ m ______ ______
10. $C = 824\pi$ mm ______ ______
11. $C = 608\pi$ ft ______ ______
12. $C = 754\pi$ in. ______ ______
13. $C = 589\pi$ cm ______ ______
14. $C = 1{,}098\pi$ m ______ ______
15. $C = 5{,}318\pi$ mm ______ ______

 Geometry

TRL **Workbook Activity 103**

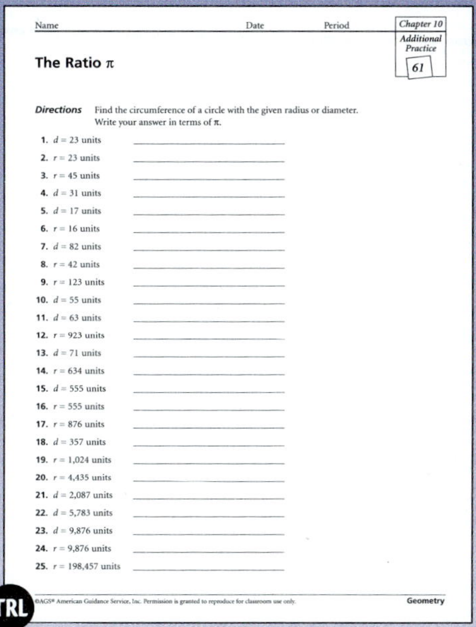

Name Date Period

Chapter 10 *Additional Practice* *61*

The Ratio π

Directions Find the circumference of a circle with the given radius or diameter. Write your answer in terms of π.

1. $d = 23$ units ______
2. $r = 23$ units ______
3. $r = 45$ units ______
4. $d = 31$ units ______
5. $d = 17$ units ______
6. $r = 16$ units ______
7. $d = 82$ units ______
8. $r = 42$ units ______
9. $r = 123$ units ______
10. $d = 55$ units ______
11. $d = 63$ units ______
12. $r = 923$ units ______
13. $d = 71$ units ______
14. $r = 634$ units ______
15. $d = 555$ units ______
16. $r = 555$ units ______
17. $r = 876$ units ______
18. $d = 357$ units ______
19. $r = 1{,}024$ units ______
20. $r = 4{,}435$ units ______
21. $d = 2{,}087$ units ______
22. $d = 5{,}783$ units ______
23. $d = 9{,}876$ units ______
24. $r = 9{,}876$ units ______
25. $r = 198{,}457$ units ______

 Geometry

TRL **Additional Practice 61**

2. Teaching the Lesson

Stress the uniqueness and constancy of pi—whether a circle is large or small, the ratio between its circumference and its diameter is always equal to π. Point out that even though the value of pi cannot be specified exactly, π can be placed on a real number line.

ALGEBRA CONNECTION

Use a calculator to give a value of π to seven places: 3.1415927 and point out that even this number does not give the exact value of π. Use the examples on page 320 to show that students can complete computations requiring π without actually evaluating it.

Work through the examples with students, indicating that for exact results they may leave π as a factor in an answer.

The ratio of the circumference of any circle to its diameter is called pi. In formula form:

$$\pi = \frac{\text{circumference}}{\text{diameter}} = \frac{C}{d}$$

Because $\pi = \frac{C}{d}$

$\pi d = C$ multiply both sides by d

and $d = \frac{C}{\pi}$ divide both sides by π

Also, because $d = 2r$, where r = radius:

$$\pi d = C$$

$$2\pi r = C$$

EXAMPLES

Calculate the circumference of a circle whose diameter is 100 cm.

Use the formula $C = \pi d$

$C = \pi(100 \text{ cm}) = 100\pi \text{ cm}$

Calculate the circumference of a circle whose radius is 25 in.

$C = \pi d = 2\pi r$ substitute $d = 2r$

$C = 2\pi(25 \text{ in.})$ substitute $r = 25$

$= 50\pi$ in.

Find the radius of a circle whose circumference is 50π in.

$C = \pi d$

$C = 2\pi r$ substitute $d = 2r$

$r = \frac{C}{2\pi}$ divide both sides by 2π

$r = \frac{50\pi}{2\pi}$ substitute $C = 50\pi$

$= 25$ in.

A circle with center *O* has a circumference of 42π. Its diameter is 42. What is the ratio of the diameter to the circumference?

$$\frac{42\pi}{42} = \pi$$

EXAMPLE The circumference of circle 1 is four times the circumference of circle 2. The diameter of circle 1 is 28. What is the diameter of circle 2?

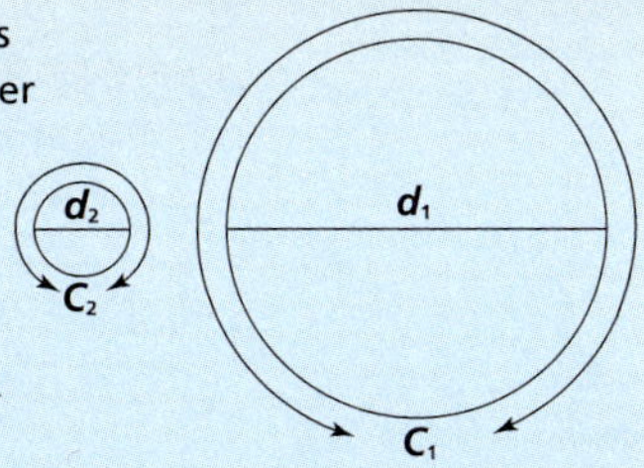

All circles are similar so

$$\frac{\text{circumference circle 1}}{\text{diameter circle 1}} = \frac{\text{circumference circle 2}}{\text{diameter circle 2}}$$

or

$\frac{C_1}{d_1} = \frac{C_2}{d_2}$	let x = diameter of circle 2
$\frac{C_1}{d_1} = \frac{C_2}{x}$	
$\frac{4C_2}{28} = \frac{C_2}{x}$	substitute $d_1 = 28$, $C_1 = 4C_2$
$(28)(C_2) = (4C_2)x$	product of means = product of extremes
$\frac{(28)(C_2)}{(4C_2)} = x$	divide by $4C_2$
$\frac{28}{4} = x$	simplify
$7 = x$	The diameter of circle 2 = 7 units.

EXPERIENCING GEOMETRY

Because pi is an irrational real number, it cannot be expressed as a terminating or repeating decimal expansion. Like other irrational numbers, such as $\sqrt{2}$ and $\sqrt{3}$, you can find an *approximate* value for pi.

Follow these steps to find an approximate value of π from different geometric models.

1. Collect a number of circular objects, such as a water glass, bucket, soda can, and dinner plate.
2. Use a string or flexible tape measure to measure each object's circumference.
3. Use a ruler to measure each circle's diameter (longest distance across). Be sure that measurements for a single object are in the same units.
4. Divide C by d.
5. Record your measurements and results in a table like the one below.

Object	Circumference, C	diameter, d	$\frac{C}{d}$

You should find that the values of $\frac{C}{d}$ are somewhat greater than 3.

Experiencing Geometry

If students have measured accurately, they should be able to derive a value of π accurate to hundredths, 3.14.

APPLICATION

At Home

Encourage students to recreate the Experiencing Geometry activity with objects they find at home. Indicate that they can use a variety of objects such as buttons, plates, pots, disks, and lamp shades. Suggest they record their measurements in a table similar to the one on page 321. Allow students to pool their findings and determine an average value of π based on their measurements.

3. Reinforce and Extend

MENTAL MATH

Work with the class as students complete Exercise A without pen, paper, or calculators. For further practice, ask students to use the formula $C = 2\pi r$ to give the exact circumference of circles with the following radii:

2, **(4π)**; 9, **(18π)**, 12.35, **(24.7π)**; 101.443 **(202.886π)**

Ask students to give the exact circumference of circles with the following diameters:

3.14, **(3.14π)**; 4.112, **(4.112π)**; 12.03, **(12.03π)**

LEARNING STYLES

Logical/Mathematical Have students find the value of pi used by the ancient Chinese. Suggest they compute its accuracy to today's value of pi.

($\frac{350}{113}$, 0.00000027)

Writing About Mathematics

Students should recognize that although pi is a real number, it is irrational. As a consequence, the value of pi cannot be stated as an exact decimal. Therefore, $n\pi$ will always be more accurate than the decimal equivalent.

Writing About Mathematics

Why is the circumference of a circle represented as a number times π ($n\pi$) more accurate than a decimal?

Exercise A Find the circumference of a circle with the given radius or diameter. Express your answer in terms of π.

1. $d = 12$ units $C = 12\pi$ units

2. $r = 3$ units $C = 6\pi$ units

3. $r = 15.5$ m $C = 31\pi$ m

4. $d = 8$ cm $C = 8\pi$ cm

5. $r = 10$ in. $C = 20\pi$ in.

6. $d = 9$ ft $C = 9\pi$ ft

7. $r = 9$ ft $C = 18\pi$ ft

8. $r = 33$ m $C = 66\pi$ m

Exercise B Find the diameter and the radius of a circle with the given circumference.

9. $C = 44\pi$ ft $d = 44$ ft, $r = 22$ ft

10. $C = 96\pi$ m $d = 96$ m, $r = 48$ m

11. $C = 12\pi$ in. $d = 12$ in., $r = 6$ in.

12. $C = 24\pi$ cm $d = 24$ cm, $r = 12$ cm

13. $C = 15\pi$ ft $d = 15$ ft, $r = 7\frac{1}{2}$ ft

14. $C = 43\pi$ mm $d = 43$ mm, $r = 21.5$ mm

15. $C = 22\pi$ cm $d = 22$ cm, $r = 11$ cm

16. $C = \pi$ in. $d = 1$ in., $r = \frac{1}{2}$ in.

Problem Solving

Exercise C Use what you know about similar circles, radius, diameter, and circumference to solve the following problems.

17. Kris used 23π feet of fencing to enclose his circular garden. How wide is the garden at its widest?
23 ft

18. Katja's trampoline has a diameter of 10 ft. Its circumference is twice as great as the circumference of Nelida's trampoline. What is the diameter of Nelida's trampoline?
5 ft

19. One sprinkler makes a circle with a diameter of 12 m. Another makes a circle with a diameter of 24 m. What is the ratio of the circumferences of the first circle to the second?
1 to 2 or $\frac{1}{2}$

20. The circumference of a circle is 13π. What is the ratio of its circumference to its diameter? Explain.
π, the ratio of a circle's circumference to its diameter is always π.

Lesson 3 Approximating the Area of a Circle

You already know that area can be measured by the number of square units it contains. We can use our knowledge of the area of a square to estimate, or approximate, the area of a circle.

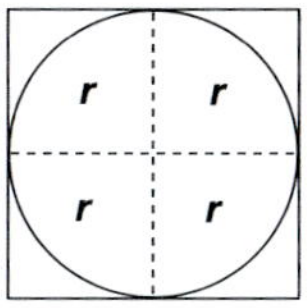

Draw any circle, radius r. Draw a square, side length $2r$, around the circle. Be sure the square's sides touch the circle. You can see that

the area of the circle < the area of the square, or

the area of the circle $< 4r^2$.

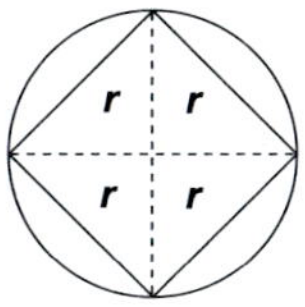

Now, draw a square inside the circle.

area of the small square < area of the circle, or

$\frac{1}{2}r^2 + \frac{1}{2}r^2 + \frac{1}{2}r^2 + \frac{1}{2}r^2 <$ area of the circle, or

$2r^2 <$ area of the circle

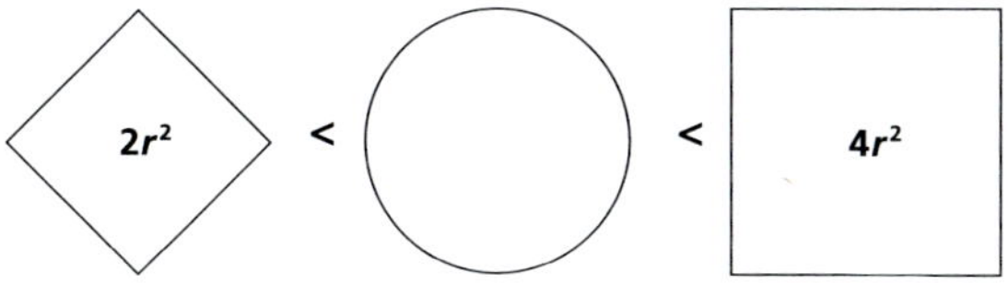

If we take the average of $2r^2$ and $4r^2$, we get $\frac{(2r^2 + 4r^2)}{2} = 3r^2$

$2r^2 \quad < \quad 3r^2 \quad < \quad 4r^2$

AREA ESTIMATION FORMULA: Area of a circle $\approx 3r^2$, where r = radius and $\approx$ means "approximately equal to."

The following example uses the Area Estimation Formula.

Lesson at a Glance

Chapter 10 Lesson 3

Overview This lesson coaches students on how to estimate or approximate the area of a circle.

Objectives

- To recognize that more exact approximations of pi (π) give more exact values of area.
- To estimate the area of a circle.

Student Pages 323–325

Teacher's Resource Library TRL

Workbook Activity 104

Additional Practice 62

1. Warm-Up Activity

Ask students to estimate some common lengths and widths in the classroom, for example, the height and width of a door, the width of a desk, the length of a bookcase. Have students describe situations in which an estimate is sufficient and situations in which exactness is required.

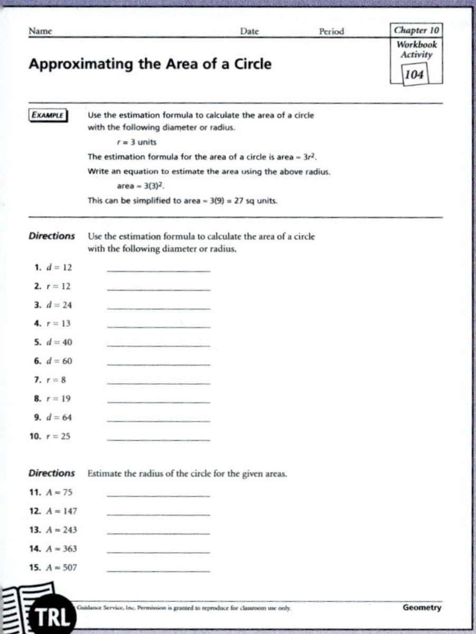

Name Date Period | Chapter 10 Workbook Activity 104

Approximating the Area of a Circle

EXAMPLE Use the estimation formula to calculate the area of a circle with the following diameter or radius.

$r = 3$ units

The estimation formula for the area of a circle is area $\approx 3r^2$.

Write an equation to estimate the area using the above radius.

area $\approx 3(3)^2$.

This can be simplified to area $\approx 3(9) = 27$ sq units.

Directions Use the estimation formula to calculate the area of a circle with the following diameter or radius.

1. $d = 12$ ______
2. $r = 12$ ______
3. $d = 24$ ______
4. $r = 13$ ______
5. $d = 40$ ______
6. $d = 60$ ______
7. $r = 8$ ______
8. $r = 19$ ______
9. $d = 64$ ______
10. $r = 25$ ______

Directions Estimate the radius of the circle for the given areas.

11. $A \approx 75$ ______
12. $A \approx 147$ ______
13. $A \approx 243$ ______
14. $A \approx 363$ ______
15. $A \approx 507$ ______

TRL Guidance Service, Inc. Permission is granted to reproduce for classroom use only. Geometry

Workbook Activity 104

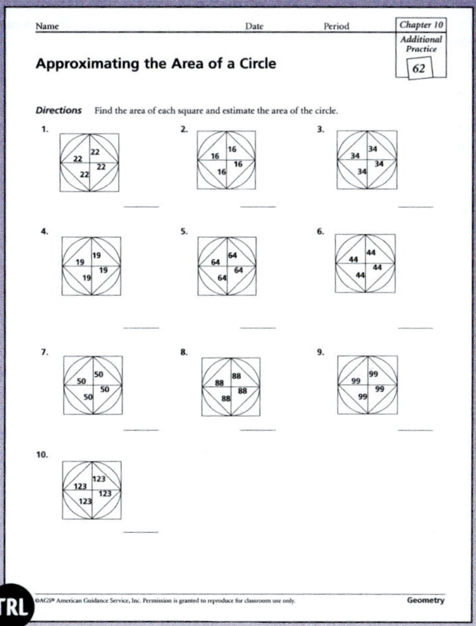

Name Date Period | Chapter 10 Additional Practice 62

Approximating the Area of a Circle

Directions Find the area of each square and estimate the area of the circle.

1. 22 22 22 22 ______
2. 16 16 16 16 ______
3. 34 34 34 34 ______
4. 19 19 19 19 ______
5. 64 64 64 64 ______
6. 44 44 44 44 ______
7. 50 50 50 50 ______
8. 88 88 88 88 ______
9. 99 99 99 99 ______
10. 123 123 123 123 ______

TRL ©AGS® American Guidance Service, Inc. Permission is granted to reproduce for classroom use only. Geometry

Additional Practice 62

2. Teaching the Lesson

Invite a volunteer to solve the formula area = $3r^2$ for r. Point out that the formula can be used to estimate the length of a radius or the circumference of a circle.

Common Error

Students may incorrectly conclude that the 3 in the formula area = $3r^2$, is the result of rounding pi (3.14) to the nearest whole number. Help students with the reasoning on page 323 so that they recognize that the factor 3 is the result of the application of a geometric principle.

Learning Styles

Visual/Spatial Invite teams of students to duplicate the diagram at the top of page 324 on grid paper three times. First have them use one grid square to represent one measurement unit, producing a center square 5 units on a side. Second, have them reproduce the diagram using two grid squares to represent one measurement unit. Finally, suggest students use three grid squares to represent one measurement unit. In the last case, the smaller square will be 15 units on a side.

For each diagram, have students count and record the number of complete grid squares in the larger square, the circle, and the smaller square, recording their results in a table similar to the one shown.

	Number of Grid Squares		
	larger square	**circle**	**smaller square**
trial 1			
trial 2			
trial 3			

Ask students to determine whether increasing the number of grid squares per measurement unit leads to a more accurate estimation of the area of the circle. Students should use data to support their conclusion.

Examples Estimate the area of a circle with a radius of 5 cm.

Use the Estimation Formula, area $\approx 3r^2$

$3(5)^2 = 3(25) = 75 \text{ cm}^2$

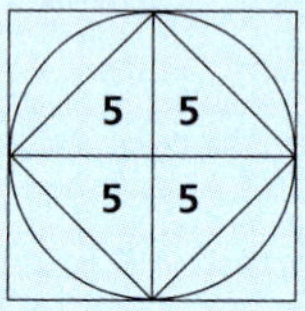

Here is a way to check the estimate:

Area of large square = $4r^2 = 4(5)^2 = 4(25) = 100 \text{ cm}^2$

Area of small square = $2r^2 = 2(5)^2 = 2(25) = 50 \text{ cm}^2$

Average of areas = $\frac{(100 + 50)}{2} = \frac{150}{2} = 75 \text{ cm}^2$

Estimate the area of a circle whose radius is 7 units.

Use the Estimation Formula, area $\approx 3r^2$

$3(7)^2 = 3(49) = 147$ sq units

The diameter of a circle is 20 in. Estimate the area of the circle.

First, find the radius: $d = 2r = 20$ in., $r = 10$ in.

Use the Estimation Formula, area $\approx 3r^2$

$3(10)^2 = 3(100) = 300$ sq in.

If you are given the area of a circle, you can use the Estimation Formula to estimate the length of the circle's radius.

Example The area of a circle is 75 sq units. Estimate its radius.

Use the Estimation Formula, area $\approx 3r^2$

$75 = 3r^2$

$25 = r^2$

$\sqrt{25} = r$

$5 = r$

Exercise A For each item, find the area of each square and estimate the area of the circle.

1. 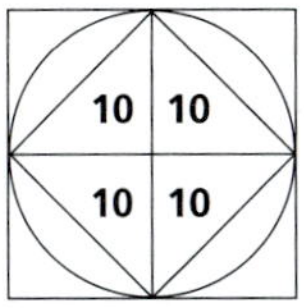

2. 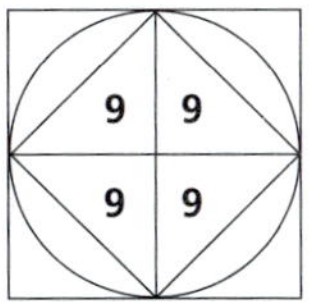

3. 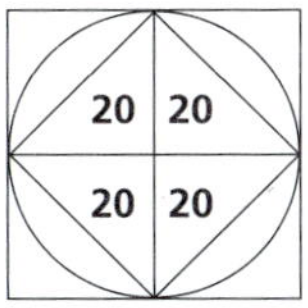

4. 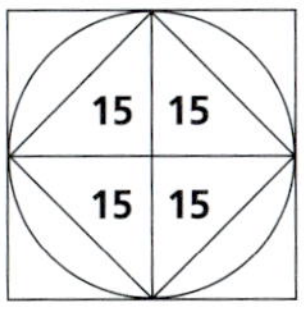

5. 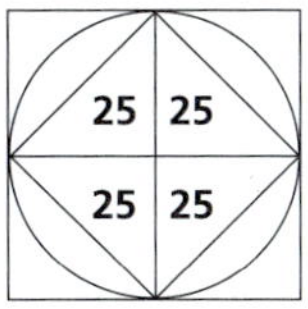

1. 400, 200, 300
2. 324, 162, 243
3. 1,600; 800; 1,200
4. 900, 450, 675
5. 2,500; 1250; 1,875

Exercise B Use the Estimation Formula to calculate the area of a circle with each diameter or radius.

6. $r = 4$ 48
7. $r = 6$ 108
8. $r = 9$ 243
9. $r = 17$ 867
10. $d = 14$ 147
11. $d = 24$ 432
12. $d = 36$ 972
13. $d = 18$ 243
14. $d = 30$ 675
15. $d = 42$ 1,323

Exercise C Estimate the radius of the circle with each given area, using the Estimation Formula.

16. $A = 48$ $r = 4$
17. $A = 108$ $r = 6$
18. $A = 192$ $r = 8$
19. $A = 300$ $r = 10$
20. $A = 147$ $r = 7$

Writing About Mathematics

In your own words, explain why the area of the smaller square is $\frac{1}{2}$ the area of the larger square.

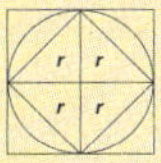

3. Reinforce and Extend

Learning Styles

Interpersonal/Group Learning Invite students to form groups of four. Allow groups time to discuss the following statement: "An estimate is a way to check on the accuracy of an answer." After a short discussion time, have a representative from each group explain to the class when and how estimation can be useful and when it can be used to check the accuracy of an answer.

Writing About Mathematics

Students should recognize that the area of the outside square can be thought of as the sum of the area of four smaller squares, each with an area of r^2 or $4r^2$. The area of the smaller square can be thought of as the sum of the area of four smaller squares, each with an area of $\frac{1}{2}r^2$ for a total area of $2r^2$. $\frac{2r^2}{4r^2} = \frac{1}{2}$.

Lesson at a Glance

Chapter 10 Lesson 4

Overview This lesson relies on the concept of area to introduce probability.

Objectives

- To identify probability as the ratio of the area of desired outcome to the total area.
- To compute probability from graphic representations.

Student Pages 326–327

Teacher's Resource Library TRL

Workbook Activity 105

Additional Practice 63

Mathematics Vocabulary

probability

1. Warm-Up Activity

Open a discussion of probability with a quote from a weather report similar to "There's an 80 percent chance of rain tomorrow." Ask students what the weather report means. Prompt a discussion about chance and probability. Ask students to explain in their own words what they think *probability* means. Let students know that in this lesson they will learn how to compute the probability of an event occurring.

2. Teaching the Lesson

Work through the examples with students. Tell students that finding the ratio of the area of the desired outcome to that of the total area is the same as the ratio of the desired outcome to the total number of outcomes possible.

Lesson 4 Area and Probability

> ***Probability***
> *The chance or likelihood of an event occurring*

Suppose you pick any point, at random, inside the circle on the right. What is the **probability** of picking a point in the top semicircle?

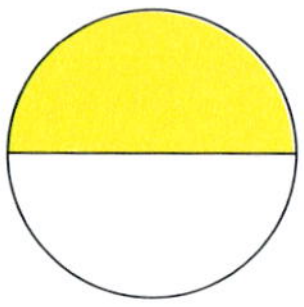

Since $\frac{\text{Area of Semicircle}}{\text{Area of Circle}} = \frac{1}{2}$,

you can conclude that the probability is $\frac{1}{2}$ or 50% by comparing the area of the desired outcome (top semicircle) to the whole circle.

$$\text{Probability} = \frac{\text{Area of Desired Outcome}}{\text{Total Area}}$$

EXAMPLE Points are chosen at random from the circle. The four quadrants are equal.

What is the probability of picking a point in QI?

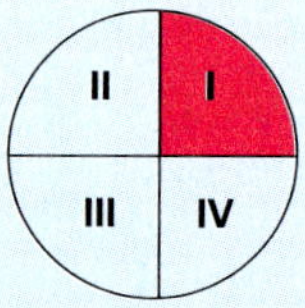

$P(\text{QI}) = ?$

$\frac{\text{Area I}}{\text{Area of Circle}} = \frac{1}{4}$ $\quad \frac{1}{4}$ of 100% = 25%

Suppose the given area is a square.

EXAMPLE The square is divided into three equal regions.

What is the probability of picking a point in the shaded region?

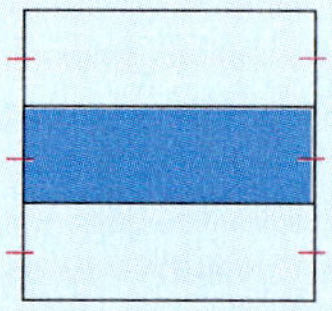

$P(\text{Shaded Region}) = ?$

$\frac{\text{Shaded Region}}{\text{Area of Square}} = \frac{1}{3}$ $\quad \frac{1}{3}$ of 100% = $33\frac{1}{3}$%

What is the probability of picking a point in the unshaded region?

$P(\text{Unshaded Region}) = ?$

$\frac{\text{Unshaded Region}}{\text{Area of Square}} = \frac{2}{3}$ $\quad \frac{2}{3}$ of 100% = $66\frac{2}{3}$%

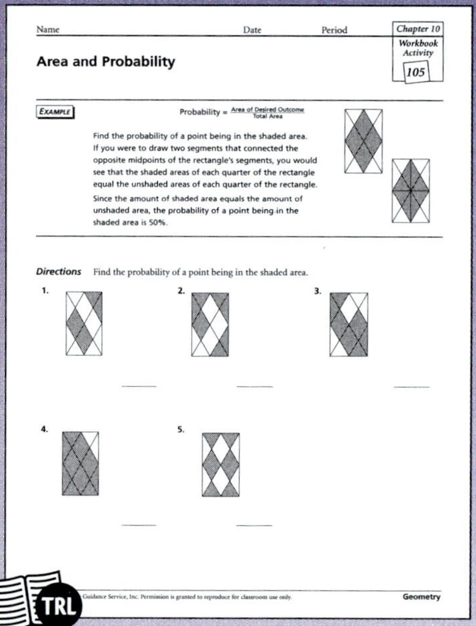

Name Date Period

Chapter 10 *Workbook Activity* 105

Area and Probability

EXAMPLE Probability = $\frac{\text{Area of Desired Outcome}}{\text{Total Area}}$

Find the probability of a point being in the shaded area. If you were to draw two segments that connected the opposite midpoints of the rectangle's segments, you would see that the shaded areas of each quarter of the rectangle equal the unshaded areas of each quarter of the rectangle.

Since the amount of shaded area equals the amount of unshaded area, the probability of a point being in the shaded area is 50%.

Directions Find the probability of a point being in the shaded area.

1. 2. 3.

4. 5.

 Geometry

Workbook Activity 105

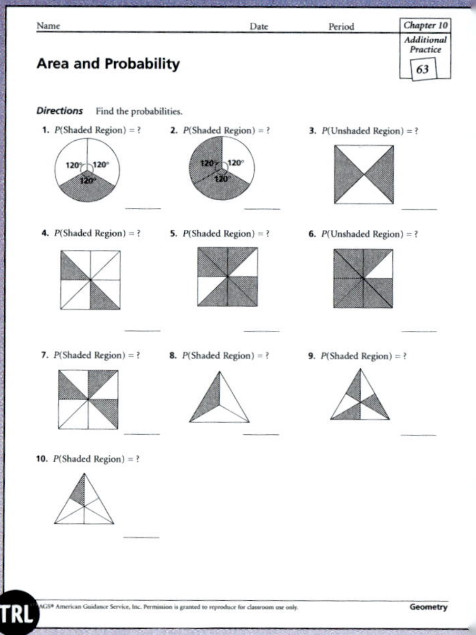

Name Date Period

Chapter 10 *Additional Practice* 63

Area and Probability

Directions Find the probabilities.

1. P(Shaded Region) = ?
2. P(Shaded Region) = ?
3. P(Unshaded Region) = ?
4. P(Shaded Region) = ?
5. P(Shaded Region) = ?
6. P(Unshaded Region) = ?
7. P(Shaded Region) = ?
8. P(Shaded Region) = ?
9. P(Shaded Region) = ?
10. P(Shaded Region) = ?

 Geometry

Additional Practice 63

Exercise A Find the probabilities.

1. P(QI or QII) = ?

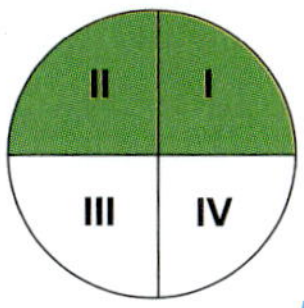

50%

2. P(Shaded Region) = ?

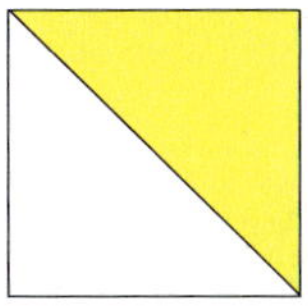

50%

3. P(QI or QII or QIII) = ?

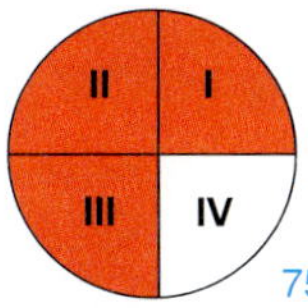

75%

4. P(Shaded Region) = ?

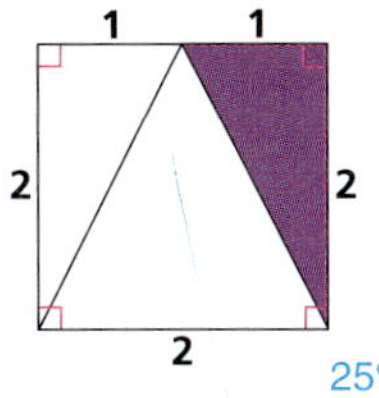

25%

5. P(Isosceles Triangle) = ?

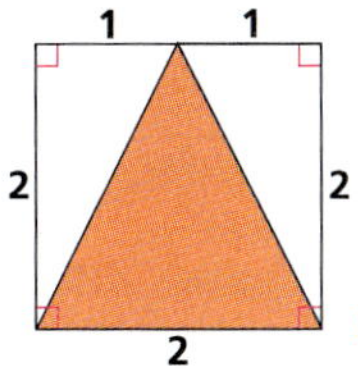

50%

Geometry Around Us

There are many reasons to know about the radius or circumference of a circle. For example, if you know the radius of your bicycle wheel, you know its circumference. If you know its circumference, you can tell how far your bike moved when the wheel completed one revolution. If you count the number of wheel revolutions, you can tell the distance you traveled. A car odometer works on this same principle.

COMMON ERROR

Some students may believe that a probability of 25 percent or one in four means that one in every four trials results in the desired outcome. Emphasize to students that probability is determined over a great number of trials.

3. Reinforce and Extend

LEARNING STYLES

Visual/Spatial Have students work in groups of four to design three dart boards with the following probabilities: 25 percent, 10 percent, and 5 percent. Allow time for groups to compare boards with one another.

Geometry Around Us

For example, a bicycle with a radius of 2 feet travels approximately 6.28 feet in one revolution and 628 feet in 100 revolutions. Students can determine how many revolutions will be required for the wheel to travel a mile. **(5,280 ft ÷ 6.28 = 840.8)**

Lesson at a Glance

Chapter 10 Lesson 5

Overview This lesson refines the estimate for finding the area of a circle.

Objectives

- To find the area of a circle using the formula πr^2.
- To use the formula for the area of a circle to find the value of an unknown radius.

Student Pages 328–331

Teacher's Resource Library TRL

Workbook Activities 106–107

Additional Practice 64

1. Warm-Up Activity

Remind students that in lesson three they used the average area of two squares to estimate the area of a circle. Invite a volunteer to reproduce the diagrams used to produce that estimation. Point out that in this lesson students will learn the formula for the exact area of a circle.

Lesson 5 Formula for the Area of a Circle

In Lesson 3, you used a formula for estimating the area of a circle. The exact area of a circle is given by the following formula.

> **FORMULA:** Area of circle = πr^2, where r is the radius of the circle.

The formula $A = \pi r^2$ gives a value slightly greater than that given by the Area Estimation Formula,

Area of circle $\approx 3r^2$.

Here is the reason for the formula. Suppose a circle were cut into many equal parts, as shown. As the number of parts increases, each part looks like an isosceles triangle.

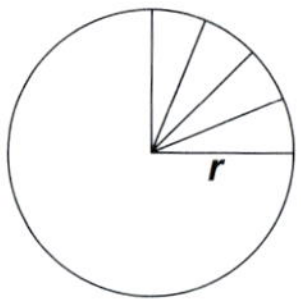

The height of each triangle begins to approximate the radius of the circle. At the same time, the sum of the bases of the triangles begins to approximate the circumference of the triangle.

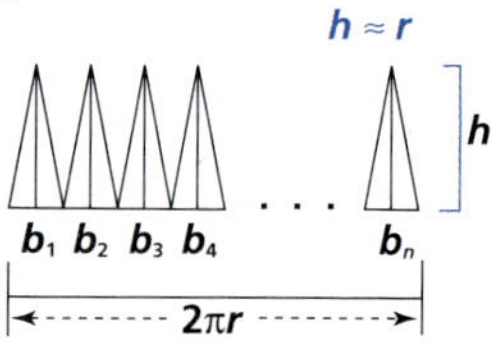

$$\begin{aligned}\text{Area of circle} = \text{total area} &= \tfrac{1}{2}b_1r + \tfrac{1}{2}b_2r + \tfrac{1}{2}b_3r + \tfrac{1}{2}b_4r \ldots + \tfrac{1}{2}b_nr\\ &= \tfrac{1}{2}r(b_1 + b_2 + b_3 + b_4 \ldots + b_n)\\ &= (\tfrac{1}{2}r)(\text{circumference})\\ &= (\tfrac{1}{2}r)(2\pi r)\\ &= \pi r^2\end{aligned}$$

$\therefore$ The formula for the area of any circle $= \pi r^2$, where r is the radius of the circle.

Because $2r = d$ d = diameter

$r = \frac{d}{2}$ can be substituted in the formula, giving

$$\begin{aligned}\text{Area of circle} &= \pi(\tfrac{d}{2})^2\\ &= \pi\tfrac{d^2}{4} = \tfrac{1}{4}\pi d^2\end{aligned}$$

> **FORMULA:** Area of circle = $\frac{1}{4}\pi d^2$, where d is the diameter of the circle.

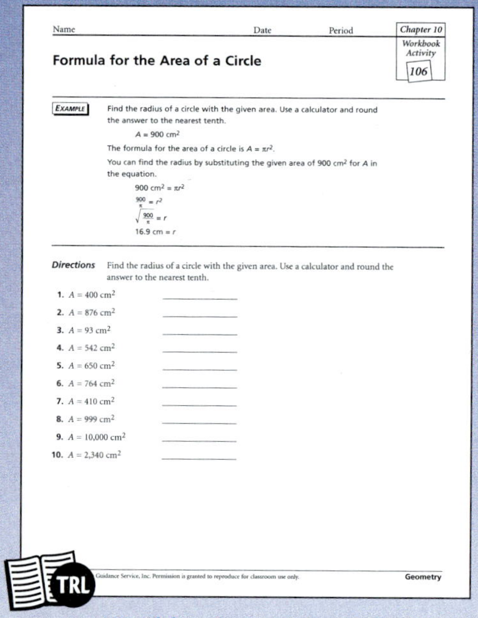

Name Date Period | Chapter 10 Workbook Activity 106

Formula for the Area of a Circle

EXAMPLE Find the radius of a circle with the given area. Use a calculator and round the answer to the nearest tenth.

$A = 900 \text{ cm}^2$

The formula for the area of a circle is $A = \pi r^2$.

You can find the radius by substituting the given area of 900 cm² for A in the equation.

$900 \text{ cm}^2 = \pi r^2$

$\frac{900}{\pi} = r^2$

$\sqrt{\frac{900}{\pi}} = r$

$16.9 \text{ cm} = r$

Directions Find the radius of a circle with the given area. Use a calculator and round the answer to the nearest tenth.

1. $A = 400 \text{ cm}^2$ ________
2. $A = 876 \text{ cm}^2$ ________
3. $A = 93 \text{ cm}^2$ ________
4. $A = 542 \text{ cm}^2$ ________
5. $A = 650 \text{ cm}^2$ ________
6. $A = 764 \text{ cm}^2$ ________
7. $A = 410 \text{ cm}^2$ ________
8. $A = 999 \text{ cm}^2$ ________
9. $A = 10{,}000 \text{ cm}^2$ ________
10. $A = 2{,}340 \text{ cm}^2$ ________

 Geometry

Workbook Activity 106

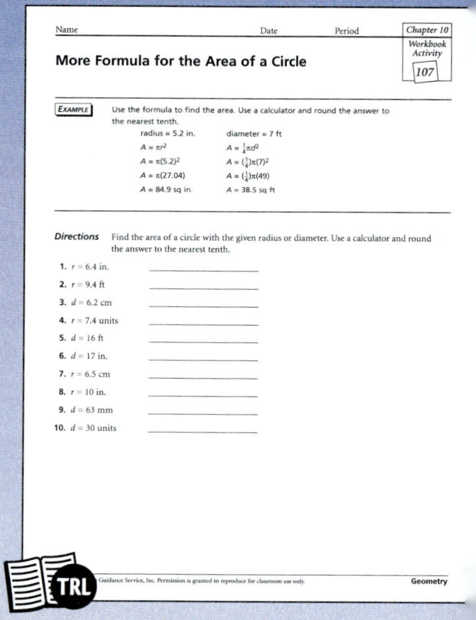

Name Date Period | Chapter 10 Workbook Activity 107

More Formula for the Area of a Circle

EXAMPLE Use the formula to find the area. Use a calculator and round the answer to the nearest tenth.

radius = 5.2 in.	diameter = 7 ft
$A = \pi r^2$	$A = \frac{1}{4}\pi d^2$
$A = \pi(5.2)^2$	$A = (\frac{1}{4})\pi(7)^2$
$A = \pi(27.04)$	$A = (\frac{1}{4})\pi(49)$
$A = 84.9$ sq in.	$A = 38.5$ sq ft

Directions Find the area of a circle with the given radius or diameter. Use a calculator and round the answer to the nearest tenth.

1. $r = 6.4$ in. ________
2. $r = 9.4$ ft ________
3. $d = 6.2$ cm ________
4. $r = 7.4$ units ________
5. $d = 16$ ft ________
6. $d = 17$ in. ________
7. $r = 6.5$ cm ________
8. $r = 10$ in. ________
9. $d = 63$ mm ________
10. $d = 30$ units ________

 Geometry

Workbook Activity 107

You can use these formulas to solve for the areas, radii, or diameters of circles.

EXAMPLES Find the area of a circle whose radius is 12 units. Give the answer in terms of π.

$$\text{Area of circle} = \pi r^2 = \pi(12)^2 = 144\pi \text{ sq units}$$

Find the area of a circle whose radius is 1.2 cm. Give the answer to the nearest tenth.

$$\text{Area of circle} = \pi r^2 = \pi(1.2)^2 = \pi(1.44)\text{ cm}^2 = 4.5\text{ cm}^2$$

Find the area of a circle whose diameter is 15 in. Calculate the answer to the nearest hundredth.

$$\text{Area of circle} = \tfrac{1}{4}\pi d^2 = (\tfrac{1}{4})\pi(15)^2 = (\tfrac{1}{4})\pi(225)\text{ sq in.} = 176.71\text{ sq in.}$$

The area of a circle is 200 sq units. Calculate the circle's radius to the nearest hundredth.

$$\text{Area of circle} = \pi r^2$$

$$200 = \pi r^2$$

$$\frac{200}{\pi} = r^2$$

$$\sqrt{\frac{200}{\pi}} = r$$

$$7.98\text{ units} = r$$

2. Teaching the Lesson

Lead students through the derivation of the formula given on page 328. Ask a volunteer to use algebra to show that $\pi r^2 = \frac{1}{4}\pi d^2$. Ask a second volunteer to show how to find the radius of a circle using the circle's area.

Invite volunteers to demonstrate, on the board or an overhead projector, each of the examples on page 329. Be sure to discuss with students the rules for rounding off decimals to the tenths or hundredths.

Ask students how they can use the formula for estimating the area of a circle to check the accuracy of their results using the formula $A = \pi r^2$.

Name Date Period — *Chapter 10* *Additional Practice* 64

Formula for the Area of a Circle

Directions Find the area of a circle with the given radius. Give your answer in terms of π.

1. $r = 4$ in. ____
2. $r = 6$ cm ____
3. $r = 8$ mm ____
4. $r = 11$ m ____
5. $r = 25$ in. ____
6. $r = 5$ in. ____
7. $r = 17$ cm ____
8. $r = 22$ mm ____
9. $r = 33$ m ____
10. $r = 50$ in. ____

Directions Find the area of a circle with the given radius. Use a calculator and round the answer to the nearest tenth.

11. $r = 7$ in. ____
12. $r = 9$ cm ____
13. $r = 13$ mm ____
14. $r = 41$ m ____
15. $r = 23$ in. ____
16. $r = 3$ in. ____
17. $r = 26$ cm ____
18. $r = 52$ mm ____
19. $r = 14$ m ____
20. $r = 99$ in. ____

 Geometry

Additional Practice 64

3. Reinforce and Extend

Calculator

Ask students to repeat Exercise A, this time giving the results to the nearest tenth. Have them determine which answers remained equally accurate. (1. 154; 2. 706.9; 3. 50.3; 4. 1,963.5; 5. 254.5; 9. 1.1. Answers for 6,7,8, and 10 remained equally accurate.)

Learning Styles

Auditory/Verbal Have students work in pairs. One student can offer step-by-step directions for finding the product or square root. His or her partner can follow the directions. Students can exchange roles after each item.

Calculator Practice

Some problems will require you to calculate an answer using a value for π. On some calculators, you will need to press a [2nd] or shift key in combination with the [π] key. Remember that you can find the square of a number using the [x^2] key. Using these special keys will help you calculate the areas of circles.

Examples Find the product of $5^2\pi$.

Press 5 [x^2] [×] [π] [=] 78.539816.

or Press 5 [x^2] [×] [2nd] [π] [=] 78.539816.

Often you will be instructed to give the answer to the nearest hundredth, 78.54, or to the nearest tenth, 78.5.

Remember that you can find the square root of a number using the [2nd] or the shift key in combination with the [x^2] key.

Find the square root of 2,209.

Press 2209 [√] [=] 47.

or Press 2209 [2nd] [x^2] [=] 47.

Exercise A Find the product or the square root. Use your calculator. Give the answer to the nearest hundredth.

1. $7^2\pi$	153.94	**6.** $\sqrt{2{,}704}$	52
2. $15^2\pi$	706.86	**7.** $\sqrt{625}$	25
3. $4^2\pi$	50.27	**8.** $\sqrt{12{,}321}$	111
4. $25^2\pi$	1,963.50	**9.** $\sqrt{1{,}234{,}321}$	1,111
5. $9^2\pi$	254.47	**10.** $\sqrt{10{,}201}$	101

Exercise B Find the area of a circle with the given radius. Give your answer in terms of π.

11. $r = 3$ in. 9π sq in.

12. $r = 23$ cm 529π cm^2

13. $r = 9$ m 81π m^2

14. $r = 40$ mi $1{,}600\pi$ mi^2

15. $r = 15$ km 225π km^2

Exercise C Find the area of a circle with the given radius or diameter. Use a calculator and give your answer to the nearest tenth.

16. $r = 12$ in. 452.4 sq in.

17. $r = 6$ ft 113.1 ft^2

18. $r = 13$ cm 530.9 cm^2

19. $r = 17$ mm 907.9 mm^2

20. $r = 20$ m 1,256.6 m^2

21. $d = 24$ km 452.4 km^2

22. $d = 100$ mi 7,854 mi^2

23. $d = 14$ ft 153.9 ft^2

24. $d = 40$ m 1,256.6 m^2

25. $d = 90$ mi 6,361.7 mi^2

Exercise D Find the radius of a circle with the given area. Use a calculator and give your answer to the nearest tenth.

26. $A = 210$ cm^2 $r = 8.2$ cm

27. $A = 400$ sq in. $r = 11.3$ in.

28. $A = 1{,}000$ m^2 $r = 17.8$ m

29. $A = 314$ km^2 $r = 10$ km

30. $A = 65$ mi^2 $r = 4.5$ mi

Manipulatives

M Polygons

Materials: Polygon Power Pack

Student Practice: Provide a variety of plastic circles and a ruler for each group of four students. Have students measure the radius of each circle and then compute the circumference and area of each. Students can record their data in a chart similar to the one shown below:

	radius	circumference	area
circle 1			
circle 2			
circle 3			

Have students find a variety of ratios using their data, for example, the ratio of the radius of circle 2 to that of circle 3; the ratio of the circumferences of two of the circles and the ratio of the areas of the circles. Encourage students to record their ratios in an orderly manner. Then ask what conclusions or generalizations they can make regarding the ratios they found. (The ratio of the radius, circumference, and area of any circle is the same as that of any other circle.)

Lesson at a Glance

Chapter 10 Lesson 6

Overview This lesson introduces arcs, sectors, and central angles.

Objectives

- To identify an arc, a central angle, and a sector of a circle.
- To use arcs, central angles, and/or sectors to solve problems.

Student Pages 332–335

Teacher's Resource Library TRL

Workbook Activity 108

Additional Practice 65

1. Warm-Up Activity

Draw a diagram of a piece of pie or pizza wedge on the board or an overhead projector. Point out to students that the diagram represents a portion of a circle. Ask them if they know any terms or measurements they can use to describe the wedge. Lead students to recognize the radius of the circle as the edge of the wedge. Point out to students that in this lesson they will learn about the curved edge of the wedge as well as the angle made by the radii.

Lesson 6 Circles and Their Angles and Sectors

Arc $\overset{\frown}{AB}$

A segment of a circle bounded by two endpoints

Central angle

An angle with its vertex at the center of a circle and the circle's radii as its sides

Sector

The area enclosed within a central angle of a circle

You are familiar with the shape of a slice of pizza or apple pie.

This shape is a part of a circle called a **sector.** A sector is contained in a **central angle** and its **arc.**

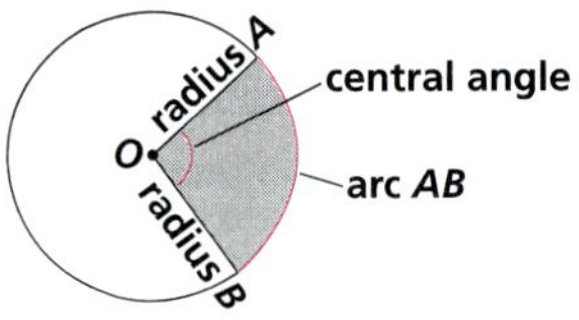

A segment of a circle bounded by two endpoints is called an arc.

The symbol for the arc between points A and B is $\overset{\frown}{AB}$. The shorter length is called the *minor arc,* the greater distance is the *major arc.* In this book, the symbol $\overset{\frown}{AB}$ refers to the minor arc.

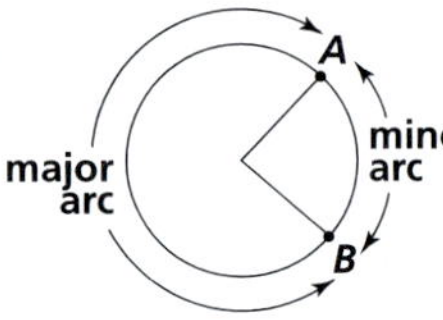

Arcs of a circle are measured in degrees. For example, for the circle above, $\overset{\frown}{AB} = 90°$. You can write "arc AB is 90°" in symbols as $\overset{\circ}{\overset{\frown}{AB}} = 90$.

A central angle of a circle has its vertex at the center of the circle and has sides formed by radii of the circle. Note that the central angle is always taken to be less than 180°. A sector of a circle is the area enclosed within a central angle of the circle.

You can form a central angle by drawing radii from the endpoints of an arc to the center of the circle. If you know the number of degrees in a circular arc, you automatically know the size of the central angle.

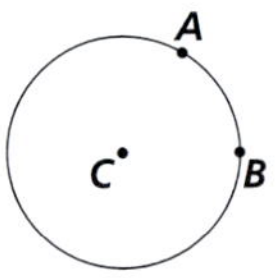

Given: $\overset{\frown}{AB} = 60°$

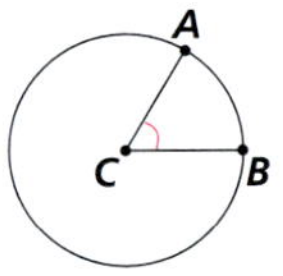

Draw radii AC and BC, forming central angle ACB. $m\angle ACB = 60°$

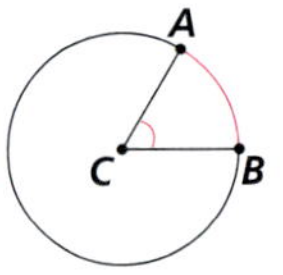

$\overset{\circ}{\overset{\frown}{AB}} = m\angle ACB$
arc degrees = central angle measure

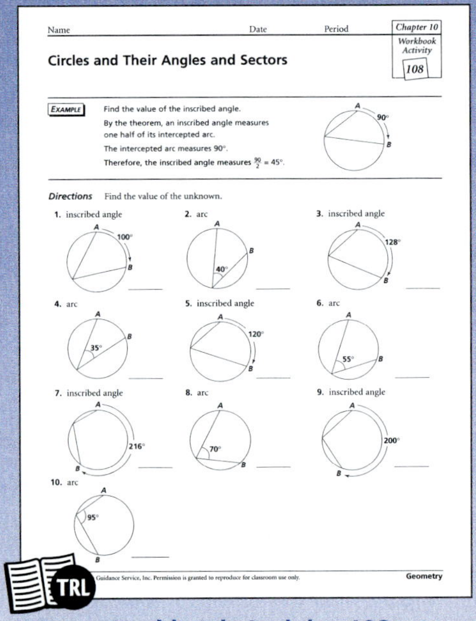

Name Date Period Chapter 10 Workbook Activity 108

Circles and Their Angles and Sectors

EXAMPLE Find the value of the inscribed angle.
By the theorem, an inscribed angle measures one half of its intercepted arc.
The intercepted arc measures 90°.
Therefore, the inscribed angle measures $\frac{90}{2} = 45°$.

Directions Find the value of the unknown.

1. inscribed angle
2. arc
3. inscribed angle
4. arc
5. inscribed angle
6. arc
7. inscribed angle
8. arc
9. inscribed angle
10. arc

Geometry

TRL

Workbook Activity 108

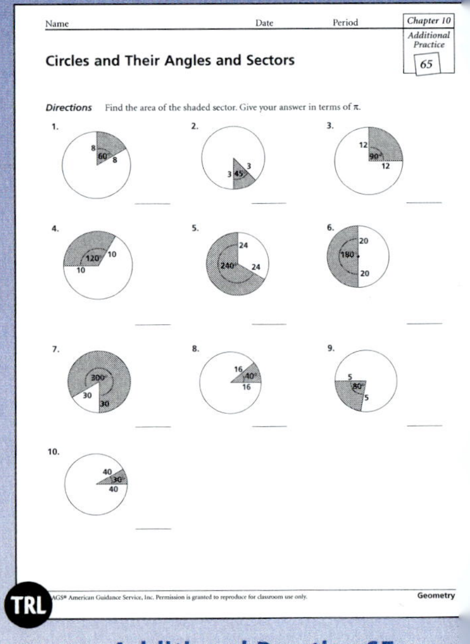

Name Date Period Chapter 10 Additional Practice 65

Circles and Their Angles and Sectors

Directions Find the area of the shaded sector. Give your answer in terms of π.

1. 2. 3. 4. 5. 6. 7. 8. 9. 10.

Geometry

TRL

Additional Practice 65

EXAMPLE **Given:** $\overset{\frown}{AB} = 30$, $\overset{\frown}{AC} = 130$

Find: $\overset{\frown}{BC}$, $m\angle AOB$, $m\angle AOC$, $m\angle BOC$

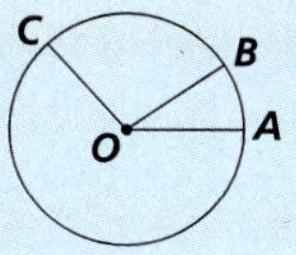

$\overset{\frown}{AC} - \overset{\frown}{AB} = \overset{\frown}{BC}$

$130 - 30 = 100 \;\therefore\; \overset{\frown}{BC} = 100$

m(central angle) = degrees in arc with radii as angle sides

$\overset{\frown}{AB} = 30$, so $m\angle AOB = 30°$

$\overset{\frown}{AC} = 130$, so $m\angle AOC = 130°$

$\overset{\frown}{BC} = 100$, so $m\angle BOC = 100°$

If you know the measure of the central angle, you can use the following ratio to determine what part of the circle's area is covered by the sector:

$$\frac{\text{central angle}}{360°} = \text{fraction of the circle covered by the sector}$$

With this information, you can then calculate the area of a sector.

EXAMPLE A circle has a radius of 5 units. What is the area of a sector whose central angle is 90°?

First, find the fraction of the circle's area covered by the sector:

$\frac{90°}{360°} = \frac{1}{4}$ of the circle

Area of circle $= \pi r^2$

$= 25\pi$ sq units

Area of sector $= \frac{1}{4}$ Area of circle

$= \frac{1}{4}(25\pi)$

$= \frac{25}{4}\pi$ sq units

2. Teaching the Lesson

Make certain students have a grasp of the terms. Ask them to answer the following: What is the sum of the major and minor arcs of a circle? **(360°)** How are the measures of a central angle and the arc it forms related? **(They're the same.)**

Point out to students that determining the area of a sector involves three steps. First, compare the central angle of the sector to the number of degrees in a circle. This ratio will give the ratio of the circle represented by the sector. Second, find the area of the entire circle using $A = \pi r^2$. Third, multiply the entire area of the circle by the fraction included in the sector.

MENTAL MATH

Ask students to give the measure of the major arc for the following minor arcs.

30°, **(330°)**; 60°, **(300°)**; 90°, **(270°)**; 180°, **(180°)**; 270°, **(90°)**

Circles and Spheres Chapter 10 **333**

Common Error

Students with poor reading habits may miss the difference between a central angle and an inscribed angle and the differences among a major arc, a minor arc, and an inscribed arc. Ask a volunteer to draw a diagram for one of these characteristics. Have a second volunteer identify the diagram. Repeat until all possible diagrams have been drawn and identified.

3. Reinforce and Extend

Learning Styles

Auditory/Verbal Direct pairs of students to make up a five-question multiple-choice quiz on central angles, major and minor arcs, inscribed angles and inscribed arcs. Encourage pairs to exchange and complete one another's quizzes. Invite students to post their quiz questions and allow class members to choose five questions they feel are a good study aid.

Prompt students with examples such as:

Central angles are formed by (a) diameters, (b) radii, (c) chords, (d) arcs. **(b)**

The diameter of a circle is also a(n) (a) radius, (b) circumference, (c) chord, (d) arc of a circle. **(c)**

The hands of a clock always form a(n) (a) diameter, (b) inscribed angle, (c) intercepted arc, (d) central angle. **(d)**

Examples A circle has a radius of 6 units, m(central angle) = 30°. Calculate the area of the sector.

First, find the fraction of the circle's area covered by the sector.

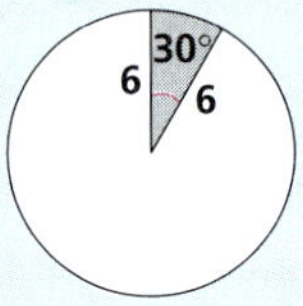

$\frac{30°}{360°} = \frac{1}{12}$ of the circle

Area of circle $= \pi r^2$

$= \pi(6)^2 = 36\pi$ sq units

Area of sector $= \frac{1}{12}$ Area of circle

$= \frac{1}{12}(36\pi)$

$= 3\pi$ sq units

A circle has a radius of 10 units, m(central angle) = 60°. What is the area of the circle not covered by the sector?

First, find the fraction of the circle's area covered by the sector.

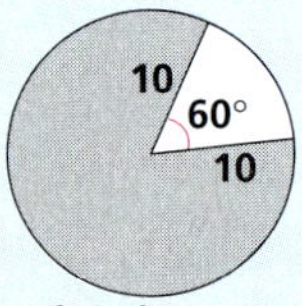

$\frac{60°}{360°} = \frac{1}{6}$ of the circle is covered by the sector, $\frac{5}{6}$ not covered

Area of circle $= \pi r^2$

$= \pi(10)^2 = 100\pi$

Area of circle not covered by sector $= \frac{5}{6}$ Area of circle

$= \frac{5}{6}(100\pi)$

$= \frac{250}{3}\pi$ sq units

In addition to a central angle, there can be another type of angle in a circle, an *inscribed angle.*

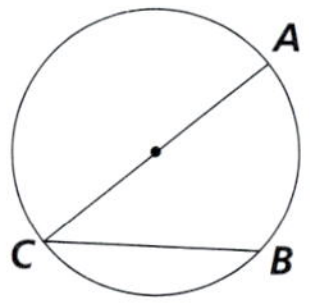

$\angle ACB$ is an example of an inscribed angle. Its sides, $\overline{AC}$ and $\overline{CB}$, are chords of the circle.

An inscribed angle has its vertex on the circle; its sides are chords of the circle. $\overset{\frown}{AB}$ is called the *intercepted arc.* We will assume the following theorem without proof.

Theorem: An inscribed angle is measured by $\frac{1}{2}$ its intercepted arc.

EXAMPLE An inscribed angle has an intercepted arc of 60°. What is the measure of the angle?

$m\angle C = (\frac{1}{2})60° = 30°$

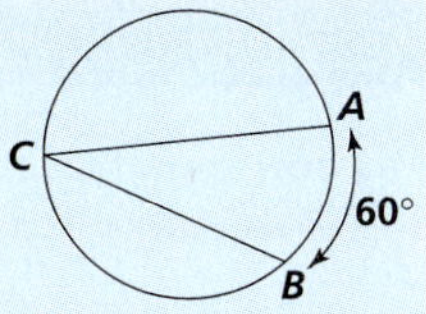

Exercise A Find the value for each of the circle components.

1. $m\angle ACB$ 85°
2. $\overset{\circ}{ED}$ 50°
3. $\overset{\circ}{BD}$ 60°
4. $\angle FCA$ 40°
5. minor arc FE 125°

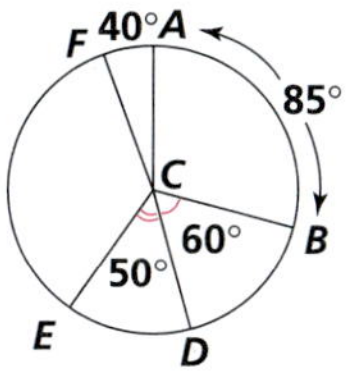

Exercise B Find the area of the shaded sector. Give your answer in terms of π.

6. 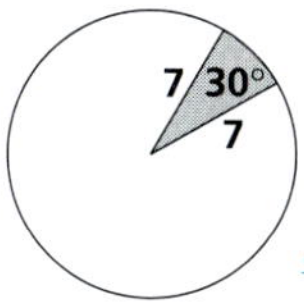

$\frac{49}{12}\pi$

7. 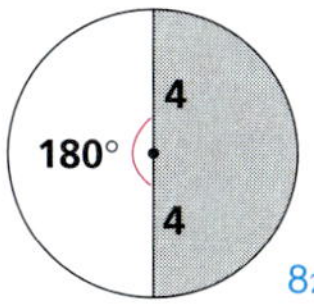

8π

8. 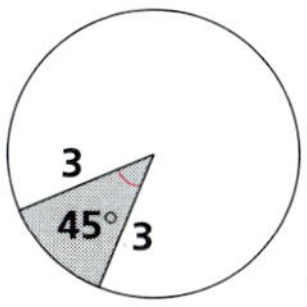

$\frac{9}{8}\pi$

Exercise C Find the value of the unknown.

9. inscribed angle 30°

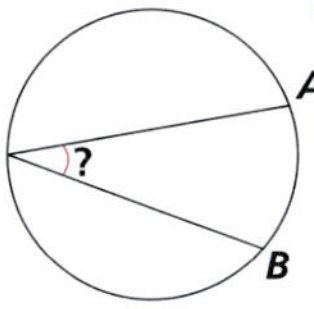

$\overset{\circ}{AB} = 60°$

10. arc 40°

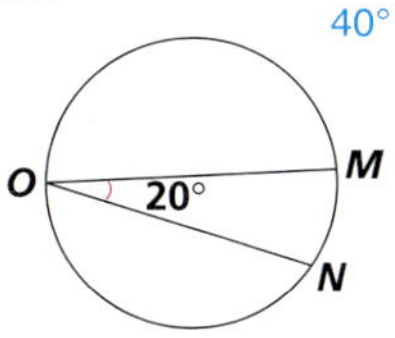

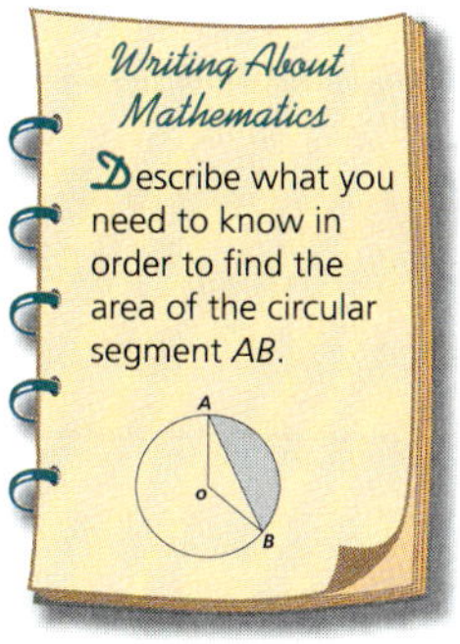

TRY THIS...

What is the size of any inscribed angle whose intercepted arc is a semicircle?

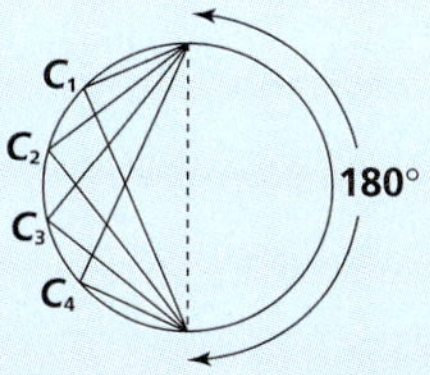

90°

LEARNING STYLES

LEP/ESL Have pairs of students complete Exercise A together. Once they agree on an answer, the students should work together to write step-by-step directions telling how to find the answer. Allow student pairs to exchange directions and offer suggestions on how to make the directions clearer or easier.

Writing About Mathematics

The area of circular segment AB is the area of the minor arc AB minus the area of ΔAOB. The following quantities are needed: the radius OA and the angle AOB are needed to find the area of the minor arc. If $\angle AOB$ is a right angle, $\overline{AB}$ can be computed and used to find the area of ΔAOB. If $\angle AOB$ is not a right angle, then the length of $\overline{AB}$ must also be known. The altitude, h, of ΔAOB must be computed and then used to compute the area of ΔAOB.

TRY THIS...

By the inscribed angle theorem, an inscribed angle is measured by $\frac{1}{2}$ its intercepted arc; $\frac{1}{2}(180°) = 90°$.

Lesson at a Glance

Chapter 10 Lesson 7

Overview This lesson presents additional characteristics of circles, including tangents, circumcircles, and incircles.

Objectives

- To identify angle bisectors and perpendicular bisectors as loci of points meeting specific conditions.
- To distinguish between incircles and circumcircles.
- To recognize a circumcenter and tangent as single points meeting specific requirements.

Student Pages 336–339

Teacher's Resource Library TRL

Workbook Activity 109

Additional Practice 66

Mathematics Vocabulary

angle bisector
circumcenter
circumcircle
incircle
perpendicular bisector
tangent

1. Warm-Up Activity

Remind students of their work with angle bisectors and perpendicular bisectors. Draw an angle or line on the board or an overhead projector. Invite a volunteer to construct the angle bisector. Have a second volunteer construct the perpendicular bisector.

Lesson 7 Tangents, Circumcircles, and Incircles

Angle bisector
Locus of points equidistant from the sides of an angle

Circumcenter
Center of a circumcircle and located at the intersection of the perpendicular bisectors of any two sides of a triangle

Circumcircle
Circle that passes through the three vertices of a triangle

Incircle
Circle in the interior of a triangle that is tangent to each side of the triangle

Perpendicular bisector
Locus of points equidistant from two given points

Tangent
Line that touches a circle at only one point

You now know that the measure of an inscribed angle is equal to half its intercepted arc.

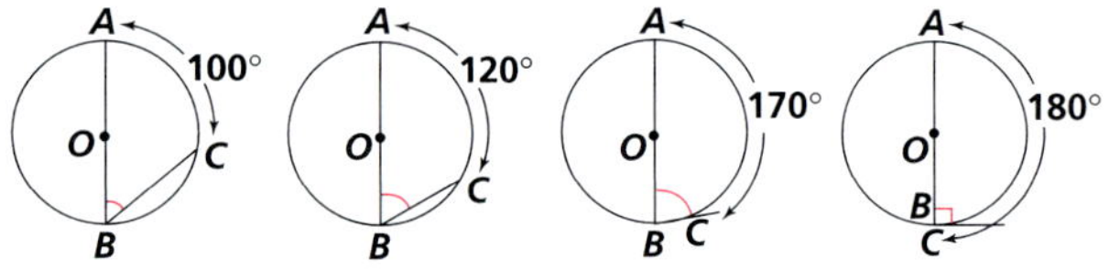

Arc → 180° Inscribed Angle → 90°

By studying the diagrams above, you can see that as point C moves closer to point B, the value of the intercepted arc moves toward 180°. The value of the inscribed angle moves toward 90°. When the arc is equal to 180°, the inscribed angle is 90°. Side BC is no longer a chord—it is now outside the circle and touches the circle only at point B.

In this case, $\overleftrightarrow{BC}$ is called a **tangent** to the circle. A tangent to a circle is a line that touches the circle at only one point. The point at which the tangent touches the circle is called the *point of tangency.*

In the diagram below, $\overline{OB}$ is a radius, B is the point of tangency, and $\overleftrightarrow{BD}$ is tangent to the circle. Note that *a radius drawn to the point of tangency of a line is perpendicular to the line.*

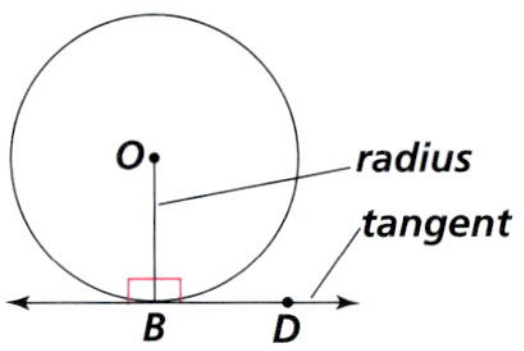

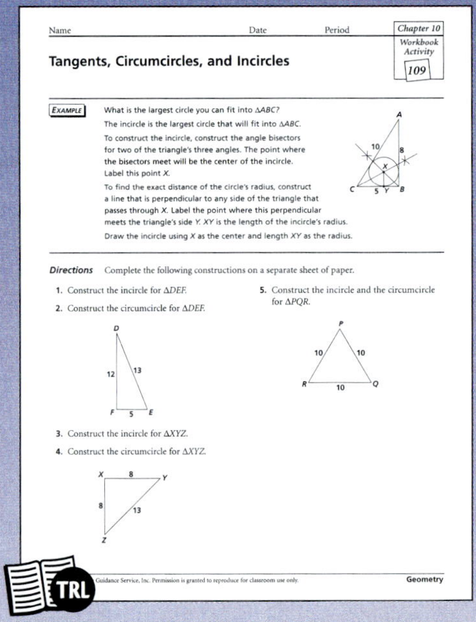

Name Date Period

Chapter 10 Workbook Activity 109

Tangents, Circumcircles, and Incircles

EXAMPLE What is the largest circle you can fit into ΔABC?

The incircle is the largest circle that will fit into ΔABC.

To construct the incircle, construct the angle bisectors for two of the triangle's three angles. The point where the bisectors meet will be the center of the incircle. Label this point X.

To find the exact distance of the circle's radius, construct a line that is perpendicular to any side of the triangle that passes through X. Label the point where this perpendicular meets the triangle's side Y. XY is the length of the incircle's radius. Draw the incircle using X as the center and length XY as the radius.

Directions Complete the following constructions on a separate sheet of paper.

1. Construct the incircle for ΔDEF.
2. Construct the circumcircle for ΔDEF.
3. Construct the incircle for ΔXYZ.
4. Construct the circumcircle for ΔXYZ.
5. Construct the incircle and the circumcircle for ΔPQR.

 Geometry

Workbook Activity 109

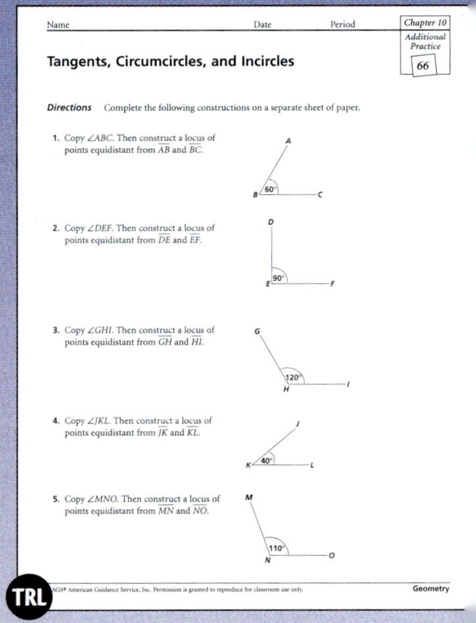

Name Date Period

Chapter 10 Additional Practice 66

Tangents, Circumcircles, and Incircles

Directions Complete the following constructions on a separate sheet of paper.

1. Copy $\angle ABC$. Then construct a locus of points equidistant from $\overline{AB}$ and $\overline{BC}$.
2. Copy $\angle DEF$. Then construct a locus of points equidistant from $\overline{DE}$ and $\overline{EF}$.
3. Copy $\angle GHI$. Then construct a locus of points equidistant from $\overline{GH}$ and $\overline{HI}$.
4. Copy $\angle JKL$. Then construct a locus of points equidistant from $\overline{JK}$ and $\overline{KL}$.
5. Copy $\angle MNO$. Then construct a locus of points equidistant from $\overline{MN}$ and $\overline{NO}$.

 Geometry

Additional Practice 66

Think back to the steps you use to construct a **perpendicular bisector** of any segment.

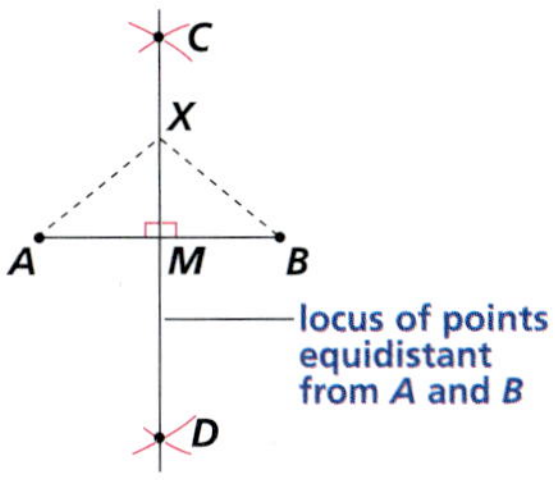

$\overline{CD} \perp$ bisector $\overline{AB}$

$M =$ midpoint of $\overline{AB}$

Also, any point, X, on $\overline{CD}$ is an equal distance from both point A and point B. We can indicate this by writing

$\Delta AXM \cong \Delta BXM$ Why? SAS Postulate

$\therefore\ \overline{AX} = \overline{XB}$ Why? Corresponding sides

> The word *equidistant* means "at an equal distance."

Because this will be true for any point on $\overline{CD}$, we can form the following definition.

The locus of points equidistant from two given points joined by a line segment is the perpendicular bisector of the segment that joins the two points.

This definition is useful for finding a point equidistant from three noncollinear (not on the same line) points.

> Remember, a *locus* definition is a definition that depends on, and specifies, the locations of points.

Given: points *A, B,* and *C* not on the same line

The locus of points equidistant from points A and B intersects the locus of points equidistant from points A and C at point D. Therefore, point D is equidistant from A, B, and C. Because of this, D can serve as the center of a circle that passes through points A, B, and C.

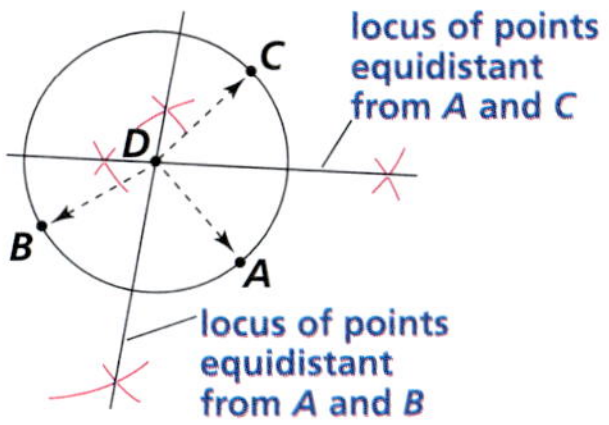

2. Teaching the Lesson

Help students realize that most of the definitions they have learned specify a single point. For example, only one point is the midpoint of a line, and a tangent to a circle touches the circle at only one point.

Indicate that in contrast, the definitions for *angle bisector* and *perpendicular bisector* specify many points. Draw the following diagram on the board or an overhead projector:

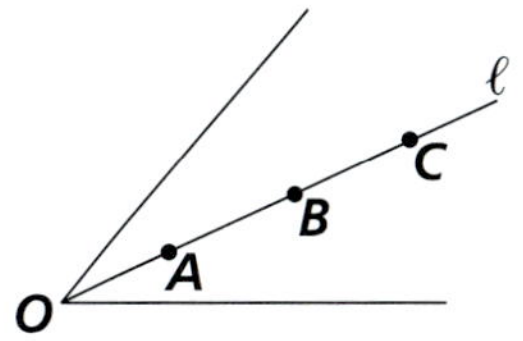

Point out that line ℓ is the bisector of angle O. Also point out that points A, B, C—and an infinite number of unnamed points—are each equidistant from the sides of the angle and lie on line ℓ. Indicate to students that one way to specify line ℓ is to call it the locus of points equidistant from the sides of the angle.

Ask students to follow your model of an angle bisector to explain the definition of a perpendicular bisector.

3. Reinforce and Extend

Learning Styles

LEP/ESL Give students an opportunity to reinforce the vocabulary for this lesson. Direct each student to write one definition on the face of an index card and pass the card to another student. Students receiving the cards can draw illustrations for the definitions on the backs of the cards. Encourage students to produce several cards showing incircles and circumcircles. Display cards, inviting students to choose those that best illustrate the mathematics vocabulary.

Learning Styles

Logical/Mathematical Invite a student to explain in his or her own words why constructing angle bisectors can produce a point that is equidistant from each side of a triangle.

Learning Styles

Interpersonal/Group Learning Direct students to work in groups of four. Instruct each group to draw several pairs of triangles with each pair on a separate sheet of paper. Students should draw two right, two acute, two isosceles, two equilateral, and two scalene triangles. Have the group find the angle bisectors for all three angles for one of each pair of triangles. Then have them find the perpendicular bisectors for all three sides of the second of each pair of triangles. Once constructions are complete, groups should make as many observations and generalizations as possible about incircles and circumcircles.

You already know that any three noncollinear points determine a triangle. And so, the construction results in a circle that passes through each of the three vertices of a triangle. This circle is called the *circumscribed circle* or **circumcircle.**

The circumcircle of a triangle is the circle that passes through the three vertices of the triangle. The center of this circle, called the **circumcenter,** is at the intersection of the perpendicular bisectors of any two sides of the triangle.

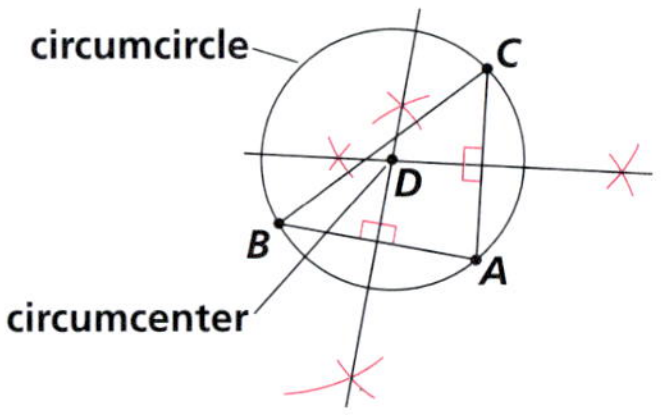

We can develop a similar definition based on an **angle bisector.** The points equidistant from the sides of an angle are on the bisector of the angle.

Example **Given:** any $\angle ABC$ with bisector $\overrightarrow{BD}$

Let X be any point on $\overrightarrow{BD}$. Draw a perpendicular from X to each side of angle ABC.

right $\Delta FXB \cong$ right ΔEXB
Why? AAS Theorem
$\therefore \overline{FX} = \overline{EX}$

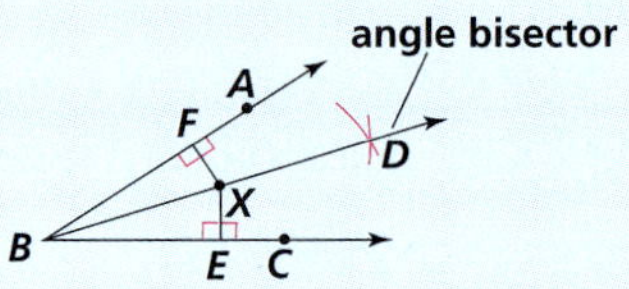

The locus of points equidistant from the sides of an angle is the angle bisector. This definition is useful for finding a point that is the center of a circle inside any triangle and tangent to each side of the triangle.

Look at ΔABC. Construct angle bisectors of $\angle A$ and $\angle B$. The point where the bisectors intersect is the center of the circle that touches each side of the triangle at one point. This circle is the *inscribed circle* or **incircle.**

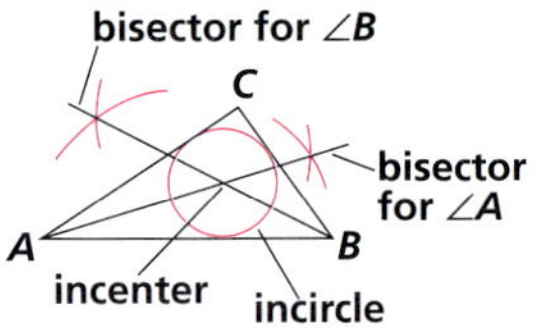

The incircle is a circle in the interior of a triangle that is tangent to each side of the triangle. The center of the circle, called the *in-center,* is the intersection of any two angle bisectors. The in-center is also the point that is the shortest distance from all three sides of the triangle.

CONSTRUCTION

Problem: What is the largest circle you can fit inside $\triangle ABC$?

Construct the incircle for the triangle. The circle will touch every side of the triangle. A larger circle would extend outside the triangle.

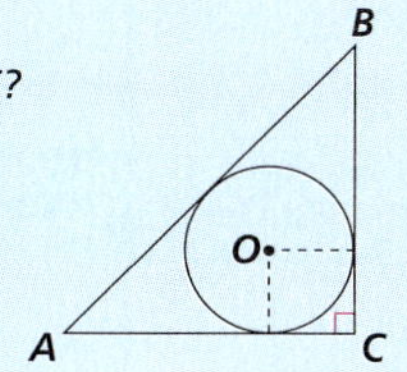

Exercise A Copy and complete the following proof.

Given: X, a point on $\overleftrightarrow{XD}$, a perpendicular bisector of $\overline{AB}$

To Prove: $\overline{AX} = \overline{BX}$

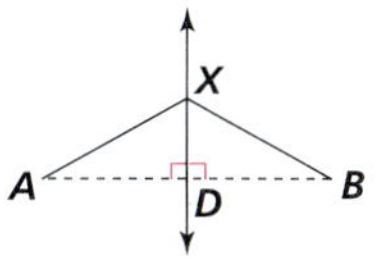

PROOF

Statement	Reason
1. X, a point on $\overleftrightarrow{XD}$	1. __1.__ Given.
2. D is the midpoint of __2.__ $\overline{AB}$	2. Given and definition of bisector.
3. $\overline{AD} = \overline{DB}$	3. __3.__ Definition of midpoint.
4. __4.__ $\overline{XD} = \overline{XD}$	4. A quantity is equal to itself.
5. __5.__ $m\angle ADX = m\angle BDX$	5. Given $\overleftrightarrow{XD}$ is $\perp$ to $\overline{AB}$, and all right angles are equal to each other.
6. $\triangle ADX \cong \triangle BDX$	6. __6.__ SAS Postulate.
7. $\therefore \overline{AX} = \overline{BX}$	7. __7.__ Corresponding sides of congruent triangles are equal.

Construction

Exercise B Complete the following constructions.

See Teacher's Edition page for answers.

8. Construct $\angle DEF$. Then construct a locus of points equidistant from $\overline{ED}$ and $\overline{EF}$.
9. Construct the circumcircle for any obtuse triangle.
10. Construct the incircle for any scalene triangle.

Construction

Students should recognize that they can find the largest incircle by constructing the angle bisectors for any two of the triangle's sides and using the point of intersection as the center of the circle.

LEARNING STYLES

Body/Kinesthetic Invite students to draw a variety of triangles on the board or classroom floor. For each triangle, have students construct a perpendicular bisector from one vertex to one opposite side of the triangle. Tell them to label the triangle similar to the triangle in the Exercise. Have students measure, and then compare the actual distances $\overline{AX}$ and $\overline{XB}$.

Answers to Exercise B

8.

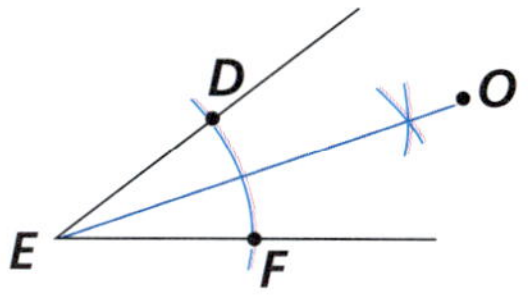

9.

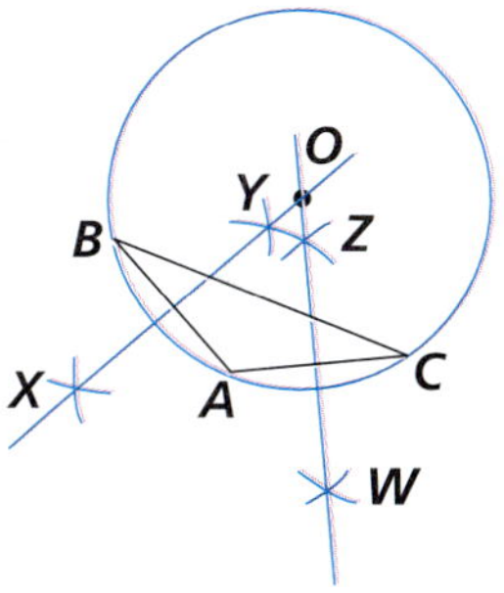

10.

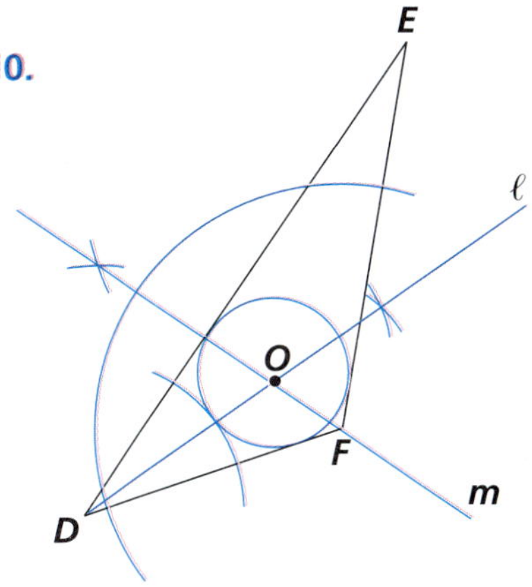

Lesson at a Glance

Chapter 10 Lesson 8

Overview This lesson introduces the ratios known as sine, cosine, and tangent.

Objectives

- To identify the ratios of sine, cosine, and tangent.
- To relate sine, cosine, and tangent to the ratios of sides of a right triangle.

Student Pages 340–341

Teacher's Resource Library TRL

Workbook Activity 110

Additional Practice 67

Mathematics Vocabulary

cosine (cos)
sine (sin)
tangent (tan)
unit circle

1. Warm-Up Activity

Remind students that they already know some of the characteristics of a right triangle. For example $a^2 + b^2 = c^2$ where a and b are the legs of a right triangle and c is the hypotenuse, or longest side. Point out to students that in this lesson they will learn about some other characteristics of right triangles.

2. Teaching the Lesson

Introduce each term, *sine, cosine,* and *tangent* by working through the examples in the text.

Lesson 8 Sine, Cosine, and Tangent

Cosine (cos)
For an angle of a right triangle (not the right angle), the ratio of the length of its adjacent leg to the length of the hypotenuse

Sine (sin)
For an angle of a right triangle (not the right angle), the ratio of the length of the opposite leg to the length of the hypotenuse

Tangent (tan)
For an angle of a right triangle (not the right angle), the ratio of the length of the opposite leg to the length of the adjacent leg

Unit circle
Circle whose radius is one

You can use a **unit circle,** a circle with radius = 1, to develop three important ratios for right triangles.

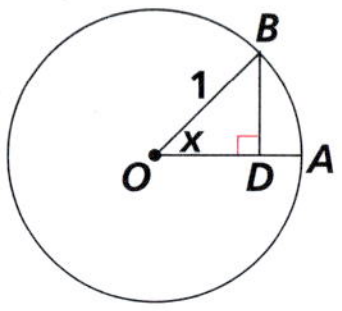

We can draw a central angle, $\angle x$, in the circle. We can then construct $\overline{BD}$ perpendicular to $\overline{OA}$. The result is right triangle ODB. Note that $\overline{OA}$ and $\overline{OB}$ equal 1, the radius of the circle.

EXAMPLE Form the ratio $\frac{BD}{OB}$, the side opposite $\angle x \div$ the hypotenuse of the right triangle.

$\frac{BD}{OB} = \frac{BD}{\text{radius}} = \frac{BD}{1} = BD$

This ratio is called the **sine** ratio of $\angle x$. It is abbreviated as sin x. For the unit circle,

$\sin x = \frac{BD}{1} = BD$

In general, $\sin x = \frac{\text{side opposite } \angle x}{\text{the hypotenuse}}$.

Note that any time you change the size of the central angle you also change the lengths of the sides and the ratios. This means that for every different angle x, there is a unique set of ratios.

EXAMPLE Form the ratio $\frac{OD}{OB}$, the side adjacent $\angle x \div$ the hypotenuse of the right triangle.

$\frac{OD}{OB} = \frac{OD}{\text{radius}} = \frac{OD}{1} = OD$

This ratio is called the **cosine** ratio of $\angle x$. It is abbreviated as cos x. For the unit circle,

$\cos x = \frac{OD}{1} = OD$

In general, $\cos x = \frac{\text{side adjacent } \angle x}{\text{the hypotenuse}}$.

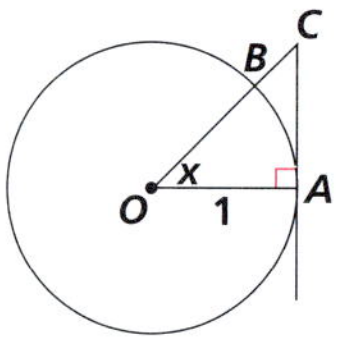

Using the unit circle, we can construct a tangent to the circle at point A. $\overline{OB}$ has been extended to meet the tangent at C. The result is right triangle OAC. Note that $\overline{OA}$ and $\overline{OB}$ equal 1, the radius of the circle. Right triangle OAC will be used for the **tangent** ratio.

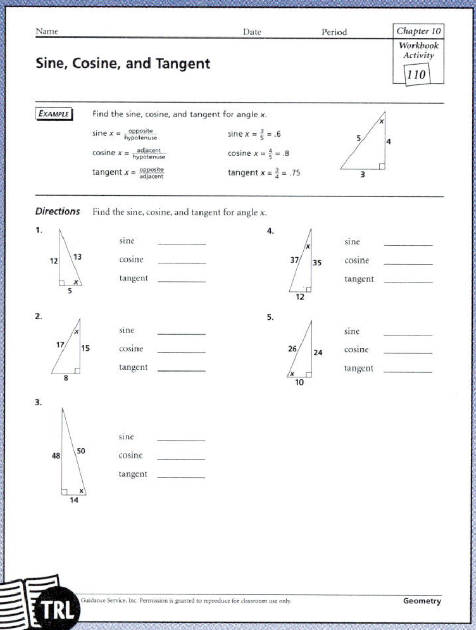

Name Date Period

Chapter 10 Workbook Activity 110

Sine, Cosine, and Tangent

EXAMPLE Find the sine, cosine, and tangent for angle x.

sine $x = \frac{\text{opposite}}{\text{hypotenuse}}$ sine $x = \frac{3}{5} = .6$

cosine $x = \frac{\text{adjacent}}{\text{hypotenuse}}$ cosine $x = \frac{4}{5} = .8$

tangent $x = \frac{\text{opposite}}{\text{adjacent}}$ tangent $x = \frac{3}{4} = .75$

Directions Find the sine, cosine, and tangent for angle x.

1. (12, 13, 5) sine ____ cosine ____ tangent ____
2. (17, 15, 8) sine ____ cosine ____ tangent ____
3. (48, 50, 14) sine ____ cosine ____ tangent ____
4. (37, 35, 12) sine ____ cosine ____ tangent ____
5. (26, 24, 10) sine ____ cosine ____ tangent ____

Guidance Service, Inc. Permission is granted to reproduce for classroom use only. Geometry

Workbook Activity 110

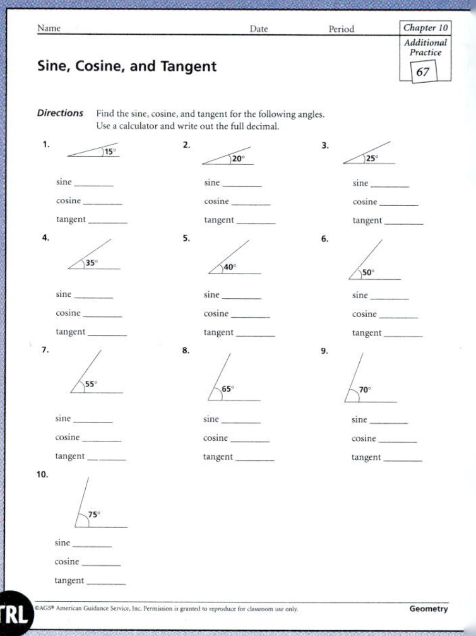

Name Date Period

Chapter 10 Additional Practice 67

Sine, Cosine, and Tangent

Directions Find the sine, cosine, and tangent for the following angles. Use a calculator and write out the full decimal.

1. 15° sine ____ cosine ____ tangent ____
2. 20° sine ____ cosine ____ tangent ____
3. 25° sine ____ cosine ____ tangent ____
4. 35° sine ____ cosine ____ tangent ____
5. 40° sine ____ cosine ____ tangent ____
6. 50° sine ____ cosine ____ tangent ____
7. 55° sine ____ cosine ____ tangent ____
8. 65° sine ____ cosine ____ tangent ____
9. 70° sine ____ cosine ____ tangent ____
10. 75° sine ____ cosine ____ tangent ____

 Geometry

Additional Practice 67

EXAMPLE Form the ratio $\frac{AC}{OA}$, the side opposite $\angle x$ ÷ the side adjacent angle x.

$\frac{AC}{OA} = \frac{AC}{\text{radius}} = \frac{AC}{1} = AC$

This ratio is called the tangent ratio of $\angle x$. It is abbreviated as tan x.

For the unit circle, $\tan x = \frac{AC}{1} = AC$

In general, $\tan x = \frac{\text{side opposite } \angle x}{\text{side adjacent } \angle x}$.

The sine, cosine, and tangent ratios are the basis of the branch of mathematics called *trigonometry.* These ratios provide values necessary to solve many different types of problems.

EXAMPLE Given an isosceles right triangle, find the sine, cosine, and tangent ratios of 45°.

Use the unit circle and the fact that the legs of the triangle are equal.

$\sin 45° = \frac{\text{opposite}}{\text{hypotenuse}} = \frac{BD}{OB} = \frac{x}{1} = x = 0.71$ (To find x, see below.)

$\cos 45° = \frac{\text{adjacent}}{\text{hypotenuse}} = \frac{OD}{OB} = \frac{x}{1} = x = 0.71$

$\tan 45° = \frac{\text{opposite}}{\text{adjacent}} = \frac{AC}{OA} = \frac{1}{1} = 1$

To find x: $x^2 + x^2 = 1$ Pythagorean Theorem

$2x^2 = 1 \quad x^2 = \frac{1}{2}$

$x = \sqrt{\frac{1}{2}} = 0.71$ Use a calculator and round to the nearest hundredth.

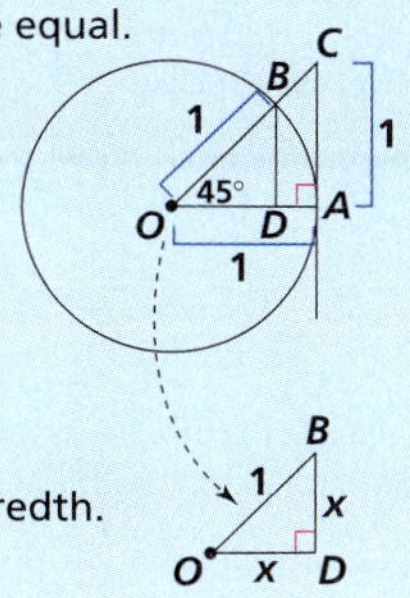

Calculator Practice

To find the sine, cosine, or tangent of a angle, use the SIN, COS, or TAN key on your calculator. Give the value to the nearest hundredth.

EXAMPLE Find the sine, cosine, and tangent for 22°.

Press 22 SIN display 0.37; press 22 COS display 0.93, press 22 TAN display 0.40.

Exercise A Use your calculator to find the sine, cosine, and tangent for each angle.

1. 0° 0, 1, 0
2. 30° 0.5, 0.87, 0.58
3. 45° 0.71, 0.71, 1
4. 60° 0.87, 0.5, 1.73
5. 90° 1, 0, undefined

3. Reinforce and Extend

LEARNING STYLES

Logical/Mathematical Present the following relationships between terms.

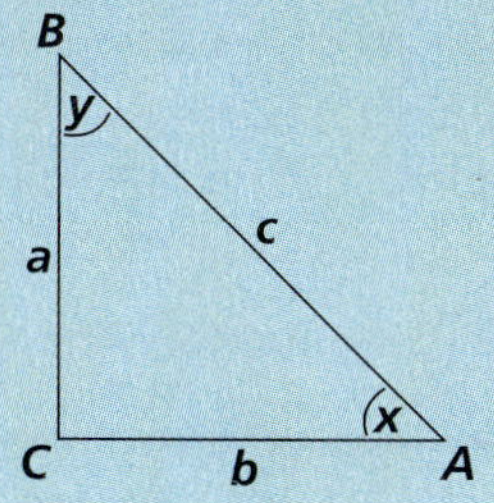

$\sin x = \frac{a}{c}$ $\sin y = \frac{b}{c}$

$\cos x = \frac{b}{c}$ $\cos y = \frac{a}{c}$

$\tan x = \frac{a}{b}$ $\tan y = \frac{b}{a}$

Stress the relationship between sin x and cos y as well as that of cos x and sin y. Ask students to use this information and algebra to show that $\frac{\sin x}{\cos x} = \tan x$.

GROUP PROBLEM SOLVING

Challenge students to justify each of the following equalities. Groups can prepare posters showing the algebraic steps or proofs using the Pythagorean Theorem. One student from each group should be prepared to explain the group's reasoning to the class.

$\frac{1}{\tan x} = \tan y$, $\frac{1}{\tan y} = \tan x$ and

$(\sin x)^2 + (\cos x)^2 = 1$

CALCULATOR

Offer further calculator practice by asking students to find the sine, cosine, and tangent of the following angles.

120° (0.87, –0.5, –1.73)
135° (0.71, –0.71, –1)
150° (0.5, –0.87, –0.58)
180° (0, –1, 0)

Lesson at a Glance

Chapter 10 Lesson 9

Overview This lesson introduces the sphere as a three-dimensional figure.

Objectives

- To describe a sphere as a locus of points at a given distance from a fixed point.
- To identify the center, radius, poles, and great circle of a given sphere.
- To compute the volume and surface area of given spheres.

Student Pages 342–345

Teacher's Resource Library TRL

Workbook Activity 111

Additional Practice 68

Mathematics Vocabulary

great circle
poles
sphere

1. Warm-Up Activity

Replicate the activity in *Experiencing Geometry* for students. Ask them to visualize the rotating circle as a sphere. If feasible, use an animating computer program to help students see the relationship between a circle and a sphere. Point out to students that the characteristics of a (three-dimensional) sphere are closely related to that of a (two-dimensional) circle.

Lesson 9 Spheres

Great circle
Circle on a sphere whose center is the center of the sphere and whose radius equals the radius of the sphere

Poles
Endpoints of the diameters of a sphere

Sphere
Locus of points in space equidistant from a fixed point

Circles and **spheres** are closely related.

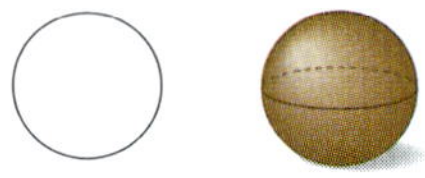

You can think of a sphere as the result of a circle spinning around one of its diameters.

A sphere is the locus of points in space that are at a given distance from a fixed point called the *center of the sphere.* The given distance is the *radius* of the sphere.

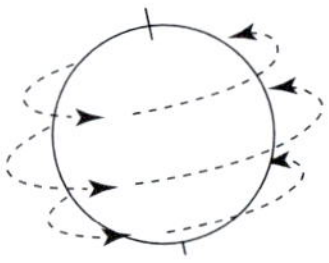

The diameter of a sphere is twice the radius, $d = 2r$, similar to the diameter of a circle.

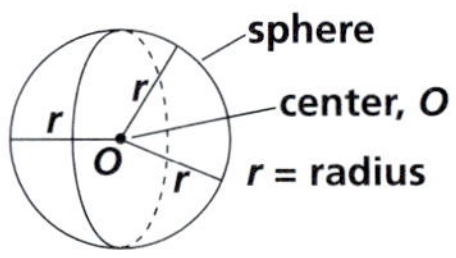

A **great circle** of a sphere is a circle on the sphere whose center is the center of the sphere and whose radius is equal to the radius of the sphere.

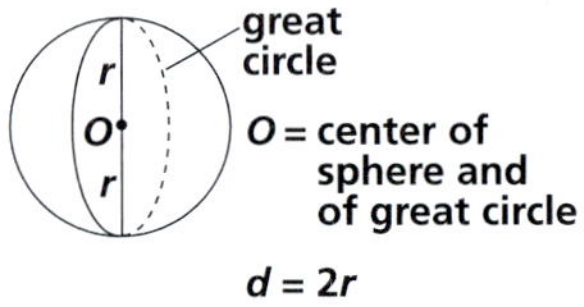

Experiencing Geometry

Cut two one-inch diameter circles from heavy construction paper or lightweight cardboard. Draw a diameter on each. Place a toothpick along the diameter of one circle so that its ends extend past each end of the diameter. Glue the toothpick in place. After the glue is dry, place a few drops of glue on the toothpick. Position the second circle on top of the toothpick so that its diameter lies along the toothpick. Allow the glue to dry on your paper-and-toothpick sandwich. Hold one end of the toothpick assembly between your thumb and forefinger. Roll the toothpick back and forth between your fingers. This will cause the paper circle to spin around the toothpick diameter. See if you can visualize the sphere that is formed as the circle spins.

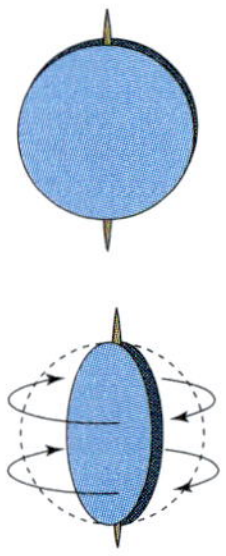

342 *Chapter 10 Circles and Spheres*

Name Date Period

Chapter 10 Workbook Activity 111

Spheres

EXAMPLE Find the surface area and the volume for a sphere with the given diameter. Use a calculator and round your answer to the nearest hundredth.

$d = 4$

The formula for the surface area of a sphere is $4\pi r^2$. You know that r = the radius and is $\frac{1}{2}$ of the diameter, d. The radius for this sphere is 2.

You can write the equation surface area = $4\pi(2^2)$. This can be reduced to surface area = $4\pi 4 = 16\pi$.

Using your calculator, you can multiply 16 times π to get 50.26.

The formula for the volume of a sphere is volume = $\frac{4}{3}\pi r^3$.

You can write the equation volume = $\frac{4}{3}\pi(2^3)$. This can be reduced to volume = $\frac{4}{3}\pi 8 = 6\pi = 18.85$.

Directions Find the surface area and the volume of spheres with the given diameters. Use a calculator and round your answer to the nearest hundredth.

1. $d = 10$ units $S =$ ______ $V =$ ______
2. $d = 6$ units $S =$ ______ $V =$ ______
3. $d = 8$ units $S =$ ______ $V =$ ______
4. $d = 2$ units $S =$ ______ $V =$ ______
5. $d = 14$ units $S =$ ______ $V =$ ______
6. $d = 16$ units $S =$ ______ $V =$ ______
7. $d = 18$ units $S =$ ______ $V =$ ______
8. $d = 20$ units $S =$ ______ $V =$ ______
9. $d = 50$ units $S =$ ______ $V =$ ______
10. $d = 100$ units $S =$ ______ $V =$ ______

 Geometry

TRL

Workbook Activity 111

Name Date Period

Chapter 10 Additional Practice 68

Spheres

Directions Find the surface area and the volume of spheres with the given radius. Give your answer in terms of π.

1. $r = 4$ units ______
2. $r = 5$ units ______
3. $r = 6$ units ______
4. $r = 8$ units ______
5. $r = 9$ units ______
6. $r = 2$ units ______
7. $r = 1$ unit ______
8. $r = 15$ units ______
9. $r = 13$ units ______
10. $r = 14$ units ______
11. $r = 16$ units ______
12. $r = 17$ units ______
13. $r = 19$ units ______
14. $r = 18$ units ______
15. $r = 20$ units ______
16. $r = 22$ units ______
17. $r = 25$ units ______
18. $r = 30$ units ______
19. $r = 50$ units ______
20. $r = 100$ units ______

 Geometry

TRL

Additional Practice 68

One way to picture a great circle is to picture the circle that divides the sphere into two equal halves called *hemispheres.*

The endpoints of the diameters of spheres are called **poles.**

The surface area of a sphere is equal to the area of four great circles, $4\pi r^2$, and the volume of a sphere is $\frac{4}{3}\pi r^3$.

Hemi- is a prefix that means "half," so *hemisphere* means "half a sphere."

FORMULAS: S = surface area of sphere = $4\pi r^2$
V = volume of sphere = $\frac{4}{3}\pi r^3$

EXAMPLES

What is the surface area of a sphere whose radius is 10 cm? Give your answer in terms of π.

S = surface area of sphere = $4\pi r^2$
Substitute the given radius.

S = surface area of sphere = $4\pi(10 \text{ cm})^2 = 400\pi \text{ cm}^2$

A sphere has a radius of 9 in. What is its volume? Use 3.14 for π.

V = volume of sphere = $\frac{4}{3}\pi r^3$
Substitute the given values.

V = volume of sphere = $\frac{4}{3}(3.14)(9 \text{ in.})^3$

$= \frac{4}{3}(3.14)(729 \text{ in.}^3)$ divide by 3

$= 4(3.14)(243) \text{ in.}^3$

$= 3{,}052.08 \text{ in.}^3$

Geometry Around Us

The planet Earth is often represented by a sphere. Which of the following correspond to great circles—lines of longitude, lines of latitude, the equator, or the prime meridian? Which "poles" are used on the geographic sphere?

2. Teaching the Lesson

Use a clear plastic sphere in conjunction with the example in the text to help students visualize the radius, surface area, and volume of a sphere.

LEARNING STYLES

LEP/ESL Ask students to work in small groups. Have the group members listen to one another as they describe the difference between the volume of a sphere and the surface area of a sphere in their own words. Have the group members come to an agreement upon definitions. Invite the groups to present their definitions of *volume* and *surface area* to the class.

Point out to students that a question such as *What happens to the surface area and the volume of the sphere if its radius is doubled?* is a question about changing variables. Impress upon students that one way to describe how the magnitudes change is to form a ratio comparing "before" and "after." Work through the example with students to show that doubling the radius gives a surface area four times greater and a volume eight times greater than the original.

3. Reinforce and Extend

APPLICATION

Career Connection

Ask students to speculate on why the relationship between surface area and volume is particularly important to package designers. Encourage students to inquire at local colleges and junior colleges about what courses might be most useful to a package designer. If feasible, students might contact a local food processor or packer and ask for an interview. Remind students to prepare several questions before the interview as well as to ask additional questions during the interview. Students can present their findings as a short Career Corner segment on the evening news. Indicate to students that their findings should include information about educational requirements and employment opportunities, as well as a specific example of a challenging packaging problem.

EXAMPLES A sphere has a radius of 10. What happens to the surface area and the volume of the sphere if its radius is doubled to 20?

Find the surface area and volume with radius 10, then with radius 20. Compare your results.

S = surface area of sphere = $4\pi r^2$

Substitute the given radius.

$r = 10 = 4\pi(10)^2$ $\quad$ $r = 20 = 4\pi(20)^2$

$= 4\pi(100)$ $\quad$ $= 4\pi(400)$

$= 400\pi$ $\quad$ $= 1{,}600\pi$

$\frac{1{,}600}{400} = \frac{4}{1}$

∴ Doubling the radius gives four times the surface area.

V = volume of sphere = $\frac{4}{3}\pi r^3$

Substitute the given radius.

$r = 10 = \frac{4}{3}\pi(10)^3$ $\quad$ $r = 20 = \frac{4}{3}\pi(20)^3$

$= \frac{4}{3}\pi(1{,}000)$ $\quad$ $= \frac{4}{3}\pi(8{,}000)$

$= \frac{(4{,}000)}{3}\pi$ $\quad$ $= \frac{(32{,}000)}{3}\pi$

$\frac{32{,}000}{4{,}000} = \frac{8}{1}$

∴ Doubling the radius gives eight times the volume.

Exercise A Find the surface area and the volume of a sphere with the given radius or diameter. Give your answer in terms of π.

1. $r = 3$ units $\quad S = 36\pi$ sq units; $V = 36\pi$ units^3

2. $r = 7$ units $\quad S = 196\pi$ sq units; $V = 457.3\pi$ units^3

3. $r = 4$ units $\quad S = 64\pi$ sq units; $V = 85.3\pi$ units^3

4. $d = 10$ units $\quad S = 100\pi$ sq units; $V = 166.67\pi$ units^3

5. $d = 12$ units $\quad S = 144\pi$ sq units; $V = 288\pi$ units^3

6. $d = 11$ units $\quad S = 121\pi$ sq units; $V = 221.83\pi$ units^3

344 *Chapter 10 Circles and Spheres*

Exercise B Find the surface area and the volume of a sphere with the given radius or diameter. Use a calculator and give your answer to the nearest hundredth.

7. $r = 2$ units — $S = 50.27$ sq units; $V = 33.51$ units^3
8. $r = 8$ units — $S = 804.25$ sq units; $V = 2{,}144.66$ units^3
9. $r = 11$ units — $S = 1{,}520.53$ sq units; $V = 5{,}575.28$ units^3
10. $d = 7$ units — $S = 153.94$ sq units; $V = 179.59$ units^3
11. $d = 20$ units — $S = 1{,}256.64$ sq units; $V = 4{,}188.79$ units^3
12. $d = 50$ units — $S = 7{,}853.98$ sq units; $V = 65{,}449.85$ units^3

Exercise C Copy the chart. Find the surface area and the volume of a sphere with the given radius. Fill in the blanks.

Radius	$S = 4\pi r^2$	$V = \frac{4}{3}\pi r^3$
13. 3	36π	36π
14. 6	144π	288π
15. 12	576π	$2{,}304\pi$

Writing About Mathematics

Review problems 14 and 15. Write a rule that describes what happens to the surface area and volume of a sphere when its radius is doubled.

TECHNOLOGY CONNECTION

Look closely at a soccer ball, and you will see that the surface of the sphere is composed of regular hexagons and pentagons. But the study of hexagons, pentagons, and spheres can lead to more than just soccer balls! Researchers have found that in addition to graphite and diamond, there is a third form of carbon—a molecule or cluster consisting of 60 atoms. The atoms in this carbon molecule occupy the positions represented by the vertices of the polygons on a soccer ball.

LEARNING STYLES

Logical/Mathematical

Challenge students to write three volume and surface area problems. The given in each problem can be the surface area or volume, the unknown in each problem can be the radius or diameter. Students can exchange and solve one another's problems.

Writing About Mathematics

Students should be able to interpret the data and conclude that doubling the radius gives a surface area four times greater and a volume eight times greater than the original.

Technology Connection

The carbon molecule formed of 60 carbon atoms is called *buckminsterfullerene* after R. Buckminster Fuller, the engineer and architect who developed the geodesic dome.

Lesson at a Glance

Chapter 10 Application

Overview This lesson introduces statistics and ways to visually present statistical information.

Objectives

- To identify circle graphs as those that show the relationship of the part to the whole.
- To identify line graphs as those that show changes over time.
- To identify bar graphs as a means to compare data.

Student Page 346

Teacher's Resource Library TRL

Workbook Activity 112

1. Warm-Up Activity

Ask students to identify information presented to them as statistics. For example, "three out of four dentists choose...," "40 percent of teenagers have tried...."

2. Teaching the Lesson

Indicate to students that there are many ways to present the same or similar data. Invite students to describe the information shown in each graph or chart. Engage them in a discussion of the usefulness of graphs in presenting numerical information. Encourage students to describe the different kinds of information each type of graph presents.

3. Reinforce and Extend

APPLICATION

In the Community

Have students scan local and national newspapers and other periodicals for articles containing tables, charts, and graphs. Students can cull graphs and tables from these articles and present them to the class. For each item presented students can discuss why the data is in chart, table, or graph form, and how the visual does or does not help them understand the data.

Application

Statistics Every day you are exposed to many different kinds of information. Increasingly, that information is in the form of statistics—numerical facts about people, places, or things. One way to understand data is to present them in the form of graphs, including circle graphs or pie charts.

EXAMPLES The data in the table give you information about the numbers of students in grades 9–12 in a school district during the years 1997 to 2000.

Students in Grades 9–12

	1997	1998	1999	2000
Grade 9	3056	3169	3352	3604
Grade 10	2945	2868	3027	3131
Grade 11	2749	2629	2656	2748
Grade 12	2650	2473	2431	2843
Total	11,400	11,139	11,466	12,326

The circle graph shows the same information for 2000 in a different format. A good way to show how a total population is divided is to use the various sectors of a circle graph.

Students in Grades 9–12, 2000

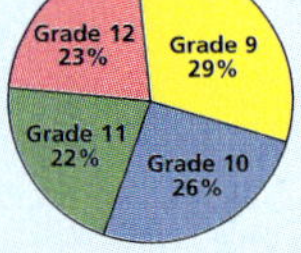

Exercise Each graph below represents some characteristic of the data about students in grades 9–12 from 1997 to 2000. Match each graph with its title. Explain your choice. Then answer the questions.

A.

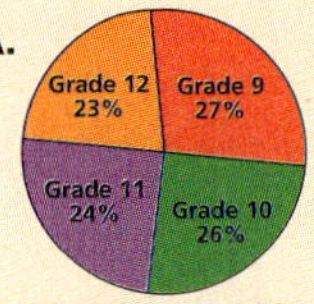

School Year, 1997

B.

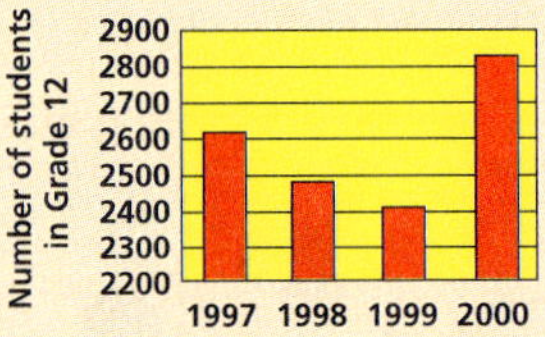

C.

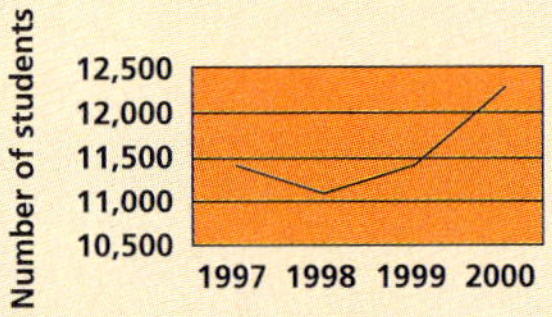

1. Changes in Enrollment, Grades 9–12, 1997–2000 C
2. Students in Grade 12, 1997–2000 B
3. Students in Grades 9–12, 1997 A
4. Which graph shows the relationship between the parts and the whole? A
5. Which graph shows how the student population in grades 9–12 changed over time? C

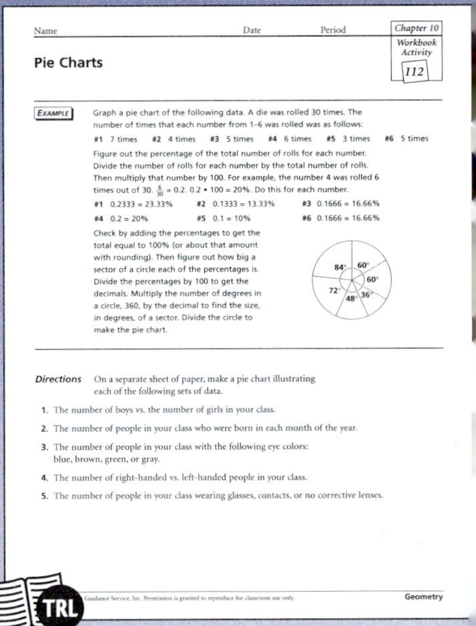

Name Date Period

Chapter 10 Workbook Activity 112

Pie Charts

EXAMPLE Graph a pie chart of the following data. A die was rolled 30 times. The number of times that each number from 1-6 was rolled was as follows:

#1 7 times #2 4 times #3 5 times #4 6 times #5 3 times #6 5 times

Figure out the percentage of the total number of rolls for each number. Divide the number of rolls for each number by the total number of rolls. Then multiply that number by 100. For example, the number 4 was rolled 6 times out of 30. $\frac{6}{30} = 0.2$. $0.2 \cdot 100 = 20\%$. Do this for each number.

#1 0.2333 = 23.33% #2 0.1333 = 13.33% #3 0.1666 = 16.66%

#4 0.2 = 20% #5 0.1 = 10% #6 0.1666 = 16.66%

Check by adding the percentages to get the total equal to 100% (or about that amount with rounding). Then figure out how big a sector of a circle each of the percentages is. Divide the percentages by 100 to get the decimals. Multiply the number of degrees in a circle, 360, by the decimal to find the size, in degrees, of a sector. Divide the circle to make the pie chart.

Directions On a separate sheet of paper, make a pie chart illustrating each of the following sets of data.

1. The number of boys vs. the number of girls in your class.
2. The number of people in your class who were born in each month of the year.
3. The number of people in your class with the following eye colors: blue, brown, green, or gray.
4. The number of right-handed vs. left-handed people in your class.
5. The number of people in your class wearing glasses, contacts, or no corrective lenses.

 Geometry

TRL

Workbook Activity 112

Chapter 10 Review

Tell which quantity each formula represents. Write the letter of the correct answer.

Example: $\frac{C}{2\pi}$ Solution: A; $C = 2\pi r$, solve for r, $r = \frac{C}{2\pi}$

A. radius of a circle C. circle circumference

B. volume of a sphere D. area of a circle

1. πr^2 D

A. area of a sphere **C.** circle circumference

B. volume of a sphere **D.** area of a circle

2. $\frac{4}{3}\pi r^3$ A

A. volume of a sphere **C.** circle circumference

B. surface area of a sphere **D.** diameter

3. $\sqrt{\frac{A}{\pi}}$ A

A. radius of a circle **C.** circle circumference

B. radius of a sphere **D.** area of a circle

4. $4\pi r^2$ C

A. radius of a circle **C.** surface area of a sphere

B. volume of a sphere **D.** diameter

5. $2r$ D

A. radius of a circle **C.** volume of a sphere

B. area of a circle **D.** diameter

Find the circumference and the area of each circle. Use a calculator and give your answer to the nearest tenth.

Example:

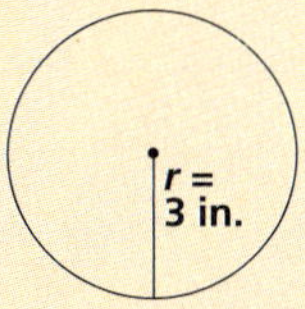

Solution: $C = 2\pi r$, substitute for r.
$C = 2\pi(r) = 2\pi(3) = 18.8$ in.
Use the formula for area,
$A = \pi r^2$. Substitute for r.
$A = \pi(3)^2 = 9\pi = 28.3$ sq in.

Name Date Period

Chapter 10 Mastery Test A page 1

Chapter 10 Mastery Test A

Directions Give the name for each part of the circle.

1. $\overline{PR}$ ______
2. $\overline{QS}$ ______
3. Q ______
4. $\overline{UV}$ ______
5. $\overline{QP}$ ______

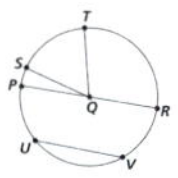

Directions Find the diameter and the radius of a circle with the given circumference.

6. $C = 30\pi$ ft $d =$ ______ $r =$ ______
7. $C = 46\pi$ ft $d =$ ______ $r =$ ______
8. $C = 22\pi$ ft $d =$ ______ $r =$ ______
9. $C = 50\pi$ ft $d =$ ______ $r =$ ______
10. $C = 82\pi$ ft $d =$ ______ $r =$ ______

Directions Use the estimation formula to calculate the area of a circle with the following radius.

11. $r = 10$ ______
12. $r = 12$ ______
13. $r = 15$ ______
14. $r = 20$ ______
15. $r = 50$ ______

Directions Find the probabilities of a point being in the shaded area.

16. 17. 18.

 Geometry

Name Date Period

Chapter 10 Mastery Test A page 2

Chapter 10 Mastery Test A, continued

Directions Find the area of a circle with the given radius. Use a calculator and round the answer to the nearest tenth.

19. $r = 8$ ______
20. $r = 10$ ______
21. $r = 5$ ______
22. $r = 22$ ______
23. $r = 15$ ______

Directions Find the area of the shaded sector. Circle the correct answer.

24.

A. 108π B. 12π C. 14π D. 16π

25.

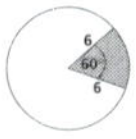

A. 4π B. 5π C. 6π D. 7π

Directions Complete the following constructions on a separate sheet of paper.

26. Construct $\angle ABC$. Then construct a locus of points equidistant from AB and BC.
27. Construct the circumcircle for any acute triangle.
28. Construct the incircle for any obtuse triangle.
29. Construct the circumcircle for any scalene triangle.
30. Construct the incircle for any right triangle.

Directions Find the surface area and the volume of spheres with the given diameter. Circle the correct answer.

31. $d = 6$ A. $S = 36\pi; V = 27\pi$ B. $S = 36\pi; V = 20.25\pi$ C. $S = 27\pi; V = 36\pi$ D. $S = 20.25\pi; V = 36\pi$
32. $d = 8$ A. $S = 64\pi; V = 48\pi$ B. $S = 48\pi; V = 64\pi$ C. $S = 68\pi; V = 44\pi$ D. $S = 46\pi; V = 84\pi$
33. $d = 10$ A. $S = 93.75\pi; V = 100\pi$ B. $S = 125\pi; V = 100\pi$ C. $S = 100\pi; V = 93.75\pi$ D. $S = 100\pi; V = 125\pi$
34. $d = 16$ A. $S = 384\pi; V = 256\pi$ B. $S = 284\pi; V = 356\pi$ C. $S = 356\pi; V = 284\pi$ D. $S = 256\pi; V = 384\pi$
35. $d = 20$ A. $S = 400\pi; V = 750\pi$ B. $S = 750\pi; V = 400\pi$ C. $S = 450\pi; V = 700\pi$ D. $S = 700\pi; V = 450\pi$

TRL Geometry

Chapter 10 Mastery Test A

Chapter 10 Review

Each set of problems in the Chapter Review includes an example and solution to illustrate the concept. Use the given examples for reteaching the materials in Chapter 10. For additional practice, refer to the Supplementary Problems for Chapter 10 (pages 414–415) and Additional Practice 60–68 in the Teacher's Resource Library.

Chapter 10 Mastery Test TRL

The Teacher's Resource Library includes parallel forms of the Chapter 10 Mastery Test. The difficulty level of the two forms is equivalent. You may wish to use one form as a pretest and the other form as a posttest. In addition, the Chapter 10 Cumulative Mastery Test covers content from Chapters 1–10.

Alternative Assessment

Alternative Assessment items correlate with student Goals for Learning at the beginning of this chapter.

- Have students identify formulas for the radius, diameter, circumference, and area of a circle.

Write the formulas for finding the circumference and area of a circle. Indicate the relationship between the diameter and the radius of a circle. Use algebra to solve those equations for the radius of a circle.

- Have students find the circumference, diameter, and radius of a circle.

Find a circular object (can, jar) and measure and compute its circumference, diameter, and radius.

- Have students solve problems involving area and probability.

Draw diagrams representing probabilities of 50, 75, 85, and 95 percent.

- Have students determine the areas of circles.

Construct three circles of different radius. Find the area and circumference of each.

Alternative Assessment

continued

- Have students construct circles.

Construct ΔABC and its circumcircle and its incircle.

- Have students use trigonometric ratios to solve problems.

Use a calculator and sin, cos, or tan to find the value of the unknown side for each of the following:

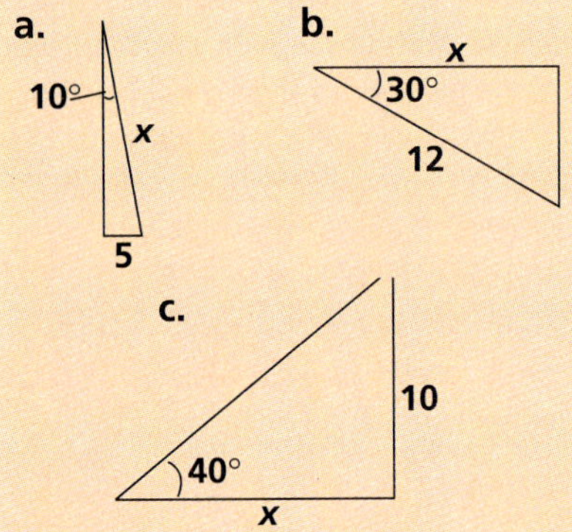

(a. $\sin 10 = \frac{5}{x}$, $x = 29.4$;
b. $\cos 30 = \frac{x}{12}$, $x = 10.44$;
c. $\tan 40 = \frac{10}{x} = 11.9$)

- Have students determine the volume and surface area of a sphere.

Find the volume and surface area of a sphere with radius = r and radius = 5.

($SA = 4\pi r^2$, $V = \frac{4}{3}\pi r^3$; 100π; 166.67π)

More Review...

6. $C = 31.4$ m; $A = 78.5$ m^2
7. $C = 44$ units; $A = 153.9$ sq units
8. $C = 25.1$ units; $A = 50.3$ sq units
9. $C = 37.7$ cm; $A = 113.1$ cm^2
10. $C = 62.8$ km; $A = 314.2$ km^2

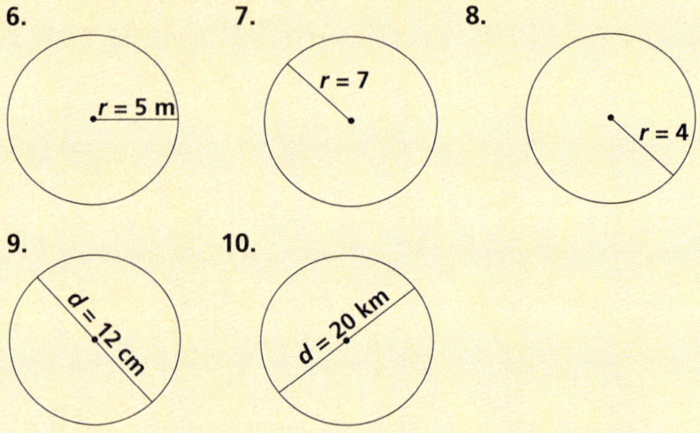

Find the area of the shaded part of each circle. Give your answer in terms of π.

Example:

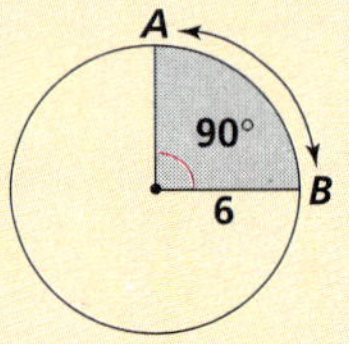

Solution: Find what part of the circle is shaded: $\frac{90}{360} = \frac{1}{4}$
Area of whole circle = $\pi r^2 = \pi(6)^2 = \pi 36$
Area of sector = $\frac{1}{4}$ area of whole circle = $\frac{\pi 36}{4} = 9\pi$

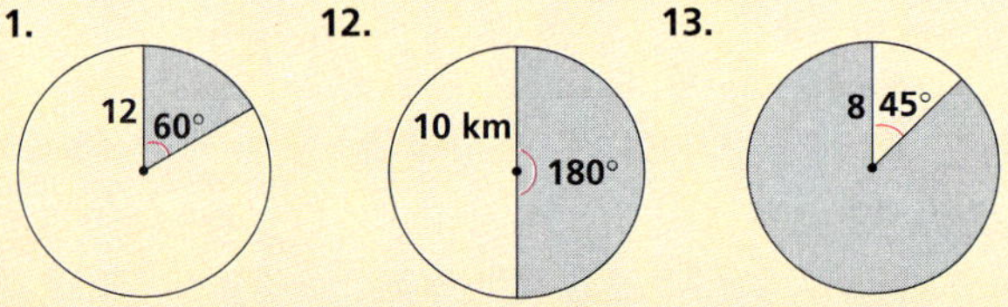

11. $A = 24\pi$
12. $A = 12.5\pi$ km^2
13. $A = 56\pi$
14. 0.54, 0.84, 0.65
15. 0.72, 0.69, 1.04

Use your calculator to find the sine, cosine, and tangent for each angle. Round to the nearest hundredth.

Example: 60° Solution: Press 60 SIN, 0.87; Press 60 COS, 0.5; Press 60 TAN, 1.73

14. 33° **15.** 46°

Name Date Period

Chapter 10 Mastery Test B page 1

Chapter 10 Mastery Test B

Directions Give the name for each part of the circle.

1. $\overline{HI}$ ____
2. C ____
3. $\overline{CD}$ ____
4. $\overline{DE}$ ____
5. $\overline{DG}$ ____

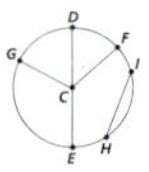

Directions Find the diameter and the radius of a circle with the given circumference.

6. $C = 25\pi$ ft $d =$ ____ $r =$ ____
7. $C = 80\pi$ ft $d =$ ____ $r =$ ____
8. $C = 56\pi$ ft $d =$ ____ $r =$ ____
9. $C = 34\pi$ ft $d =$ ____ $r =$ ____
10. $C = 98\pi$ ft $d =$ ____ $r =$ ____

Directions Use the estimation formula to calculate the area of a circle with the following radius.

11. $r = 14$ ____
12. $r = 9$ ____
13. $r = 17$ ____
14. $r = 25$ ____
15. $r = 100$ ____

Directions Find the probabilities of a point being in the shaded area.

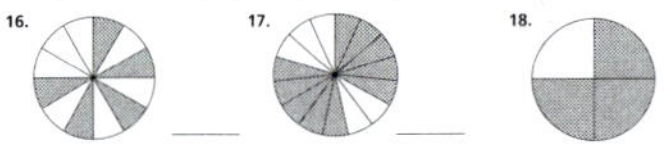

TRL

 Geometry

Name Date Period

Chapter 10 Mastery Test B page 2

Chapter 10 Mastery Test B, continued

Directions Find the area of a circle with the given radius. Use a calculator and round the answer to the nearest tenth.

19. $r = 9$ ____
20. $r = 17$ ____
21. $r = 7$ ____
22. $r = 21$ ____
23. $r = 28$ ____

Directions Find the area of the shaded sector. Circle the correct answer.

24.

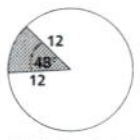

A. 19π **B.** 19.1π **C.** 19.2π **D.** 19.4π

25.

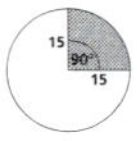

A. 56.25π **B.** 56.75π **C.** 56.5π **D.** 56π

Directions Complete the following constructions on a separate sheet of paper.

26. Construct the incircle for any right triangle.
27. Construct the circumcircle for any acute triangle.
28. Construct $\angle XYZ$. Then construct a locus of points equidistant from XY and YZ.
29. Construct the circumcircle for any obtuse triangle.
30. Construct the incircle for any scalene triangle.

Directions Find the surface area and the volume of spheres with the given diameter. Circle the correct answer.

31. $d = 4$ **A.** $S = 6\pi$; $V = 16\pi$ **B.** $S = 60\pi$; $V = 16\pi$ **C.** $S = 16\pi$; $V = 6\pi$ **D.** $S = 16\pi$; $V = 60\pi$
32. $d = 12$ **A.** $S = 162\pi$; $V = 144\pi$ **B.** $S = 144\pi$; $V = 162\pi$ **C.** $S = 44\pi$; $V = 62\pi$ **D.** $S = 62\pi$; $V = 44\pi$
33. $d = 24$ **A.** $S = 576\pi$; $V = 1{,}296\pi$ **B.** $S = 1{,}296\pi$; $V = 576\pi$ **C.** $S = 1{,}276\pi$; $V = 596\pi$ **D.** $S = 596\pi$; $V = 1{,}276\pi$
34. $d = 48$ **A.** $S = 10{,}386\pi$; $V = 2{,}304\pi$ **B.** $S = 10{,}386\pi$; $V = 2{,}034\pi$ **C.** $S = 2{,}034\pi$; $V = 10{,}386\pi$ **D.** $S = 2{,}304\pi$; $V = 10{,}368\pi$
35. $d = 60$ **A.** $S = 20{,}250\pi$; $V = 3{,}600\pi$ **B.** $S = 20{,}600\pi$; $V = 3{,}250\pi$ **C.** $S = 3{,}600\pi$; $V = 20{,}250\pi$ **D.** $S = 3{,}250\pi$; $V = 20{,}600\pi$

TRL

 Geometry

Chapter 10 Mastery Test B

Find the surface area and the volume of each sphere. Give your answer in terms of π.

Example: $r = 3$ in. Solution: Use the formula for surface area,
$S = 4\pi r^2$. Substitute for r.
$S = 4\pi(3)^2 = 36$ sq in.
Use the formula for volume,
$V = \frac{4}{3}\pi r^3$. Substitute for r.
$V = \frac{4}{3}\pi(3)^3 = (\frac{4}{3})\pi(27) = 36\pi$ cu in.

16. $r = 6$

17. $r = 10$

18. $r = 5$ m

19. $r = 7$

20. $r = 4$ km

21. $d = 6$

16. $S = 144\pi$ sq in.; $V = 288\pi$ cu in.
17. $S = 400\pi$; $V = \frac{4{,}000\pi}{3} = 1{,}333.3\pi$
18. $S = 100\pi$ m^2; $V = \frac{500\pi}{3} = 166.7$ m^3
19. $S = 196\pi$; $V = \frac{1{,}372\pi}{3} = 457.3\pi$
20. $S = 64\pi$ km^2; $V = \frac{256\pi}{3} = 85.3\pi$ km^3
21. $S = 36\pi$; $V = 36\pi$

Complete the following constructions.

Example: Construct the angle bisector for an acute angle. Solution:

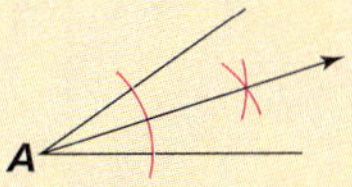

22. Construct the circumcircle for a scalene triangle.

23. Construct the incircle for an isosceles triangle.

See Teacher's Edition page for answers.

Find the following probabilities for a point randomly chosen from the inside of each circle.

Example: P(Unshaded Region)

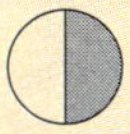

Solution: $P = \frac{\text{Unshaded Region}}{\text{Area of Square}} = \frac{1}{2}$ of 100% = 50%

24. P(Shaded Region) 75%

25. P(Unshaded Region) 25%

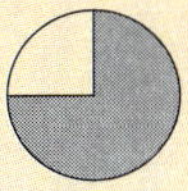

TEST-TAKING TIP

Effective studying takes place over a period of time. Spend time studying new material for several days or weeks before a test. Don't try to learn new material the night before a test.

Answers to Problems

22.

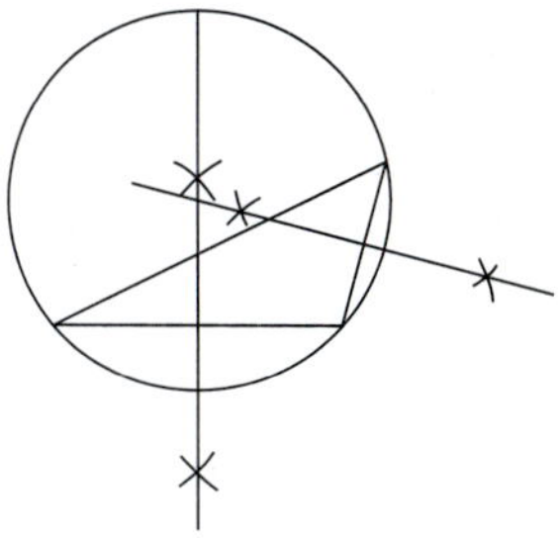

23.

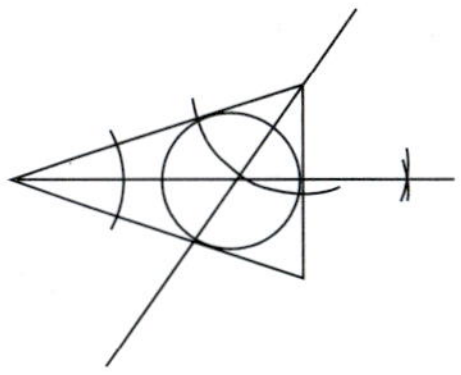

Name Date Period

Chapter 10 Cumulative Mastery Test page 1

Chapter 10 Cumulative Mastery Test

Directions Complete the following constructions. Use a separate sheet of paper and a straightedge and a compass.

1. Construct ΔABC with $\overline{AB} = \overline{AC}$ and $m\angle A > 90°$.
2. Draw any right triangle and label it ABC. Construct a second triangle $A'B'C'$ that is congruent to ABC using the SSS Theorem.

Directions Give the coordinates of the image quadrilateral under the following dilations. Use graph paper to graph the object and its images. (All dilations have (0, 0) as the center of the dilation.)

3. dilation of 3 ________
4. dilation of $\frac{1}{2}$ ________

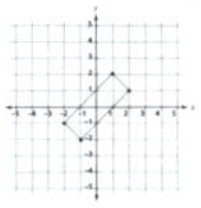

Directions Use the figure to find the following quantities.

5. Area of each right triangle = ________
6. Side length and area of the small square = ________
7. Sum of the areas that make up the large square = ________

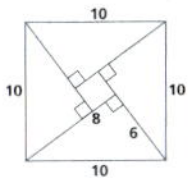

Directions Use formulas to find the perimeter of each polygon.

8. rectangle, $l = 56$, $w = 34$ ________
9. hexagon, $s = 356$ ________

Directions Find the area of the square or rectangle with the given side lengths. Write the letter of the correct answer.

10. $s = 36$ ________
    A. 1,292 B. 1,296 C. 1,300 D. 1,304
11. $l = 15$, $w = 45$ ________
    A. 660 B. 665 C. 670 D. 675

RL

 Geometry

Name Date Period

Chapter 10 Cumulative Mastery Test page 2

Chapter 10 Cumulative Mastery Test, continued

Directions Find the missing dimension, b, for the rectangle shown.

12. ________

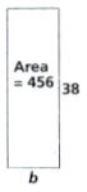

Directions Use Heron's Formula to find the areas of triangles with the given side lengths. Write the letter of the correct answer.

13. 37, 42, 45 ________
    A. 6275.62 B. 6274.62 C. 6276.62 D. 6273.62
14. 10, 16, 14 ________
    A. 69.3 B. 70.3 C. 71.3 D. 72.3

Directions Find the diameter and the radius of a circle with the given circumference.

15. $C = 43\pi$ ft ________
16. $C - 66\pi$ ft ________

Directions Find the area of a circle with the given radius. Use a calculator and round your answer to the nearest tenth.

17. $r = 11$ ________
18. $r = 19$ ________

Directions Find the area of the shaded sector. Circle the correct answer.

19.

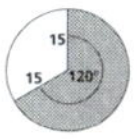

A. 75π B. 65π C. 80π D. 70π

Directions Find the surface area and the volume of the sphere with the given diameter. Circle the correct answer.

20. $d = 13$
    A. $S = 1{,}647.75\pi$; $V = 676\pi$
    B. $S = 676\pi$; $V = 1{,}675.47\pi$
    C. $S = 1{,}675.47\pi$; $V = 676\pi$
    D. $S = 676\pi$; $V = 1{,}647.75\pi$

TRL

 Geometry

Chapter 10 Cumulative Mastery Test

Chapter 11

Planning Guide

Solid Geometric Figures and Their Measures

	Student Text Lesson						
	Student Pages	Vocabulary	Proof	Practice Exercises	Problem Solving	Calculator Practice	Solutions Key
Lesson 1 Basic Volume Formulas	352–355	✔		✔	355		✔
Lesson 2 Volumes of Pyramids and Cones	356–359	✔		✔			✔
Lesson 3 Surface Areas of Prisms and Cylinders	360–363	✔		✔			✔
Lesson 4 Surface Areas of Pyramids and Cones	364–367			✔	367	367	✔
Lesson 5 Measurements	368–369	✔		✔			✔
Application Origami	370			✔			✔

Chapter Activities

Teacher's Resource Library
Geometry on the Job 11: Plumbing

Community Connection 11: Water Supply

Teacher's Edition
Chapter 11 Project

Assessment Options

Student Text
Chapter 11 Review

Teacher's Resource Library
Chapter 11 Mastery Tests A and B

Chapter 11 Cumulative Mastery Test

Teacher's Edition
Chapter 11 Alternative Assessment

Student Text Features						Teaching Strategies								Learning Styles						Teacher's Resource Library		
Algebra Connection	Constructions	Experiencing Geometry	Technology Connection	Try This	Writing About Mathematics	Warm-Up Activity	Common Error	Algebra Connection	Mental Math	Manipulatives	Calculator	Applications	Group Problem Solving	Auditory/Verbal	Visual/Spatial	Body/Kinesthetic	Logical/Mathematical	Interpersonal/Group Learning	LEP/ESL	Workbook Activities	Additional Practice	Self-Study Guide
354						352	354	355						353		353	354	355		113	69	✔
		356	359			356			358	358		359					358		357	114	70	✔
						360	361					362			361				363	115	71	✔
365				367	367	364					367		366						365	116	72	✔
						368						369							369	117	73	✔
						370									370					118		✔

Manipulatives

Geometric Volume Set

GeoSolids

Chapter at a Glance

Chapter 11: Solid Geometric Figures and Their Measures pages 350–373

(Answer Keys for the Teacher's Resource Library begin on page 484 of this Teacher's Edition.)

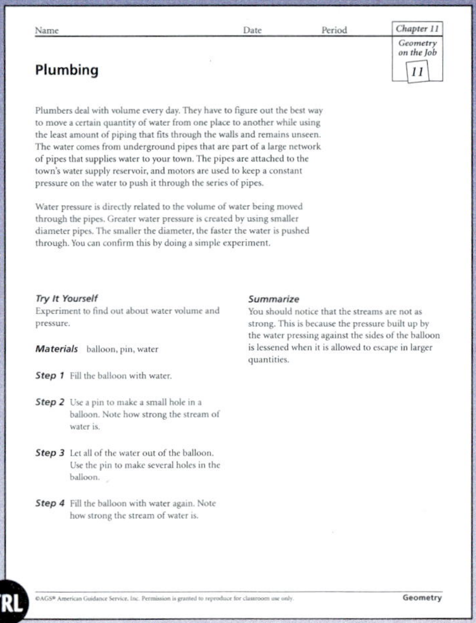

Name Date Period

Chapter 11 Geometry on the Job 11

Plumbing

Plumbers deal with volume every day. They have to figure out the best way to move a certain quantity of water from one place to another while using the least amount of piping that fits through the walls and remains unseen. The water comes from underground pipes that are part of a large network of pipes that supplies water to your town. The pipes are attached to the town's water supply reservoir, and motors are used to keep a constant pressure on the water to push it through the series of pipes.

Water pressure is directly related to the volume of water being moved through the pipes. Greater water pressure is created by using smaller diameter pipes. The smaller the diameter, the faster the water is pushed through. You can confirm this by doing a simple experiment.

Try It Yourself
Experiment to find out about water volume and pressure.

Materials balloon, pin, water

Step 1 Fill the balloon with water.

Step 2 Use a pin to make a small hole in a balloon. Note how strong the stream of water is.

Step 3 Let all of the water out of the balloon. Use the pin to make several holes in the balloon.

Step 4 Fill the balloon with water again. Note how strong the stream of water is.

Summarize
You should notice that the streams are not as strong. This is because the pressure built up by the water pressing against the sides of the balloon is lessened when it is allowed to escape in larger quantities.

 Geometry

TRL Geometry on the Job 11

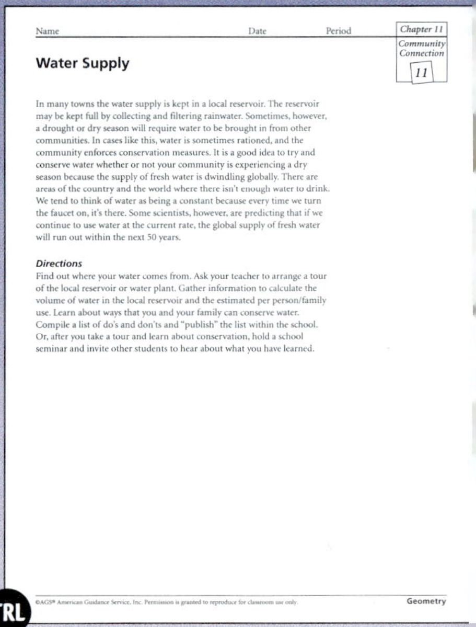

Name Date Period

Chapter 11 Community Connection 11

Water Supply

In many towns the water supply is kept in a local reservoir. The reservoir may be kept full by collecting and filtering rainwater. Sometimes, however, a drought or dry season will require water to be brought in from other communities. In cases like this, water is sometimes rationed, and the community enforces conservation measures. It is a good idea to try and conserve water whether or not your community is experiencing a dry season because the supply of fresh water is dwindling globally. There are areas of the country and the world where there isn't enough water to drink. We tend to think of water as being a constant because every time we turn the faucet on, it's there. Some scientists, however, are predicting that if we continue to use water at the current rate, the global supply of fresh water will run out within the next 50 years.

Directions
Find out where your water comes from. Ask your teacher to arrange a tour of the local reservoir or water plant. Gather information to calculate the volume of water in the local reservoir and the estimated per person/family use. Learn about ways that you and your family can conserve water. Compile a list of do's and don'ts and "publish" the list within the school. Or, after you take a tour and learn about conservation, hold a school seminar and invite other students to hear about what you have learned.

 Geometry

TRL Community Connection 11

Chapter 11

Solid Geometric Figures and Their Measures

Cleveland, Ohio, is home to the Rock and Roll Hall of Fame. Here the hall is seen reflected in the harbor waters of Lake Erie. Since its construction in the mid-1990s, this building has become a landmark. The hall's architect I. M. Pei is well-known for his use of geometric shapes. He designed the hall with a central pyramid of glass and steel. Cylinders and rectangular solids reach out from the pyramid. Pei's modern design seems perfect for the hall. It's as much fun as the music it honors.

In Chapter 11, you will learn formulas for use with geometric solids.

Goals for Learning

- To identify formulas for determining the volume or area of solids
- To determine the volumes of solids
- To find the surface areas of solids
- To use formulas to determine a missing dimension of a solid given its surface area or volume
- To use customary and metric units of measurement

351

Introducing the Chapter

Use the photograph on these pages to spark students' interest in a variety of shapes and volumes. Discuss with students the sensations associated with entering a room with ceilings a foot or so above their heads. Have them contrast those sensations with sensations they experience when entering a building with an atrium—a roof or ceiling several stories high. Encourage students to realize that the shape and enclosed volume of buildings can vary as much as the shape and volume of ordinary containers. Ask volunteers to read the Goals for Learning and discuss each one with students.

Chapter Project

Groups of students can research and build models of existing pyramids. Encourage students to use a variety of research tools such as reference books, architectural magazines, and the Internet. Exhibits should include Khufu's pyramid in Egypt, Quetzalcóatl in Mexico, the Louvre addition in France, and the Transamerica building in San Francisco, as well as other pyramids of students' choosing. Information gathered should include the shape and dimensions of each pyramid and computations of the lateral surface area and the volume enclosed by the pyramid. Groups should choose an appropriate scale and build models of each pyramid they research.

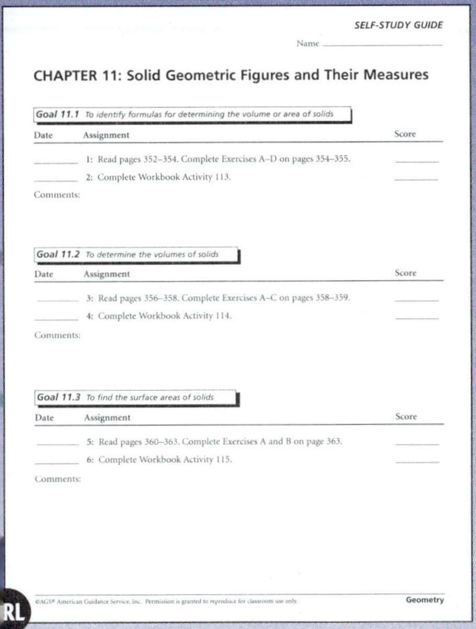

SELF-STUDY GUIDE

Name ____________

CHAPTER 11: Solid Geometric Figures and Their Measures

Goal 11.1 *To identify formulas for determining the volume or area of solids*

Date	Assignment	Score
______	1: Read pages 352–354. Complete Exercises A–D on pages 354–355.	______
______	2: Complete Workbook Activity 113.	______

Comments:

Goal 11.2 *To determine the volumes of solids*

Date	Assignment	Score
______	3: Read pages 356–358. Complete Exercises A–C on pages 358–359.	______
______	4: Complete Workbook Activity 114.	______

Comments:

Goal 11.3 *To find the surface areas of solids*

Date	Assignment	Score
______	5: Read pages 360–363. Complete Exercises A and B on page 363.	______
______	6: Complete Workbook Activity 115.	______

Comments:

©AGS® American Guidance Service, Inc. Permission is granted to reproduce for classroom use only. Geometry

TRL

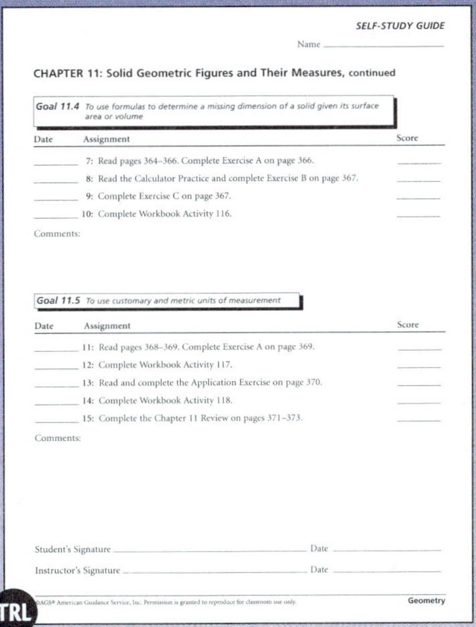

SELF-STUDY GUIDE

Name ____________

CHAPTER 11: Solid Geometric Figures and Their Measures, continued

Goal 11.4 *To use formulas to determine a missing dimension of a solid given its surface area or volume*

Date	Assignment	Score
______	7: Read pages 364–366. Complete Exercise A on page 366.	______
______	8: Read the Calculator Practice and complete Exercise B on page 367.	______
______	9: Complete Exercise C on page 367.	______
______	10: Complete Workbook Activity 116.	______

Comments:

Goal 11.5 *To use customary and metric units of measurement*

Date	Assignment	Score
______	11: Read pages 368–369. Complete Exercise A on page 369.	______
______	12: Complete Workbook Activity 117.	______
______	13: Read and complete the Application Exercise on page 370.	______
______	14: Complete Workbook Activity 118.	______
______	15: Complete the Chapter 11 Review on pages 371–373.	______

Comments:

Student's Signature ____________ Date ______

Instructor's Signature ____________ Date ______

©AGS® American Guidance Service, Inc. Permission is granted to reproduce for classroom use only. Geometry

TRL

Chapter 11 Self-Study Guide

Lesson at a Glance

Chapter 11 Lesson 1

Overview This lesson presents the volume formulas for rectangular and triangular prisms as well as cylinders.

Objectives

- To identify the volume of a solid as (area of its base)(height).
- To compute the volumes of a variety of solids including rectangular and triangular prisms and cylinders.

Student Pages 352–355

Teacher's Resource Library TRL

Workbook Activity 113

Additional Practice 69

Mathematics Vocabulary

prism

1. Warm-Up Activity

Ask students what they know about volume. If feasible, present students with two or three differently shaped containers that have the same capacity, for instance, a tall, thin cylinder and a short, wide cylinder. Ask students how they could compare how much each container could hold. Encourage them to think beyond filling one container and emptying it into the others. Point out to students that in this lesson they will learn formulas for volume that will help them compare how much different containers can hold.

Lesson 1 Basic Volume Formulas

Prism

A three-dimensional figure that has congruent polygons as bases in parallel planes and whose sides are parallelograms

Volume

The measure of the number of unit cubes in a solid figure

In the last chapter, you created a sphere by spinning a circular area around its diameter. In fact, you can create any **volume** by moving an area through space.

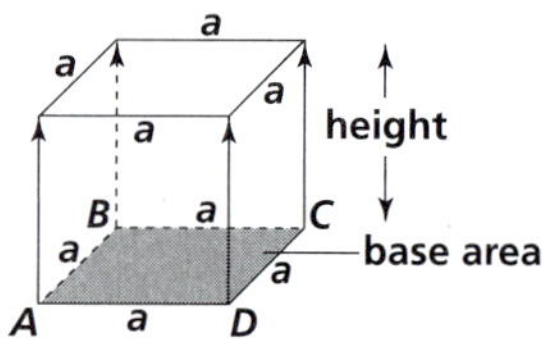

Think of a square, *ABCD*. Picture what would happen if you moved the square straight up. You could call the distance you moved the square the *height*. You will have created a volume that has a base congruent to the original square *ABCD*. In addition, the figure now has some height, *h*, the distance you moved the square. You have now created a **prism.** The size of the volume is calculated by multiplying the area of the base by the height the base moved.

Start with any area as a base, and the same formula will apply.

Rectangular Box

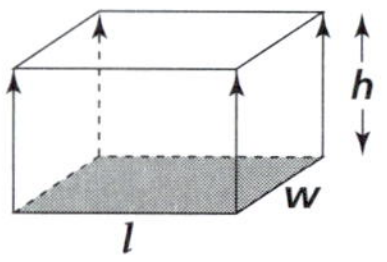

V = (area of base)(height)
$V = l \bullet w \bullet h$

Triangular Prism

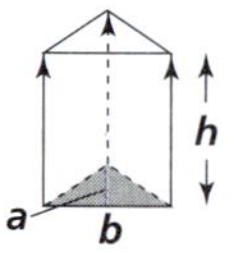

V = (area of base Δ)(height)
$V = (\frac{1}{2}ba)(h)$ where a = altitude of the base Δ

Cylinder

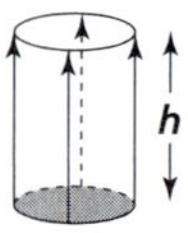

V = (area of base circle)(height)
$V = (\pi r^2)(h)$

FORMULA: Volume of a solid = area of the base times the height

EXAMPLE What is the volume of a cube with a side of 10 units?

Volume = area of base • height
= area of a square • height
Substitute values:
= (10 • 10)10
= 1,000 cubic units

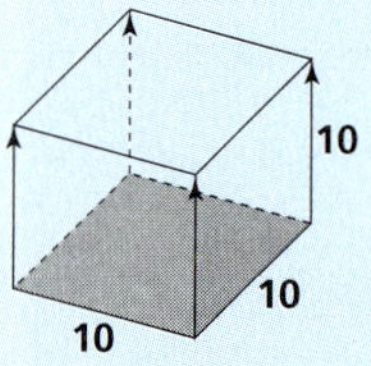

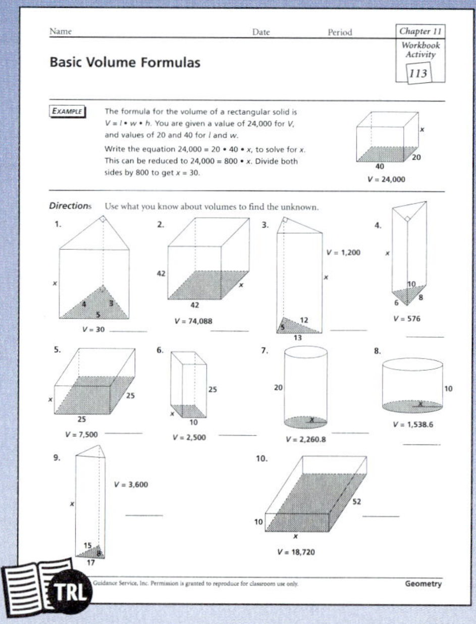

Name Date Period *Chapter 11 Workbook Activity 113*

Basic Volume Formulas

EXAMPLE The formula for the volume of a rectangular solid is $V = l \bullet w \bullet h$. You are given a value of 24,000 for V, and values of 20 and 40 for l and w.
Write the equation 24,000 = 20 • 40 • x, to solve for x. This can be reduced to 24,000 = 800 • x. Divide both sides by 800 to get x = 30.

Directions Use what you know about volumes to find the unknown.

 Geometry

TRL

Workbook Activity 113

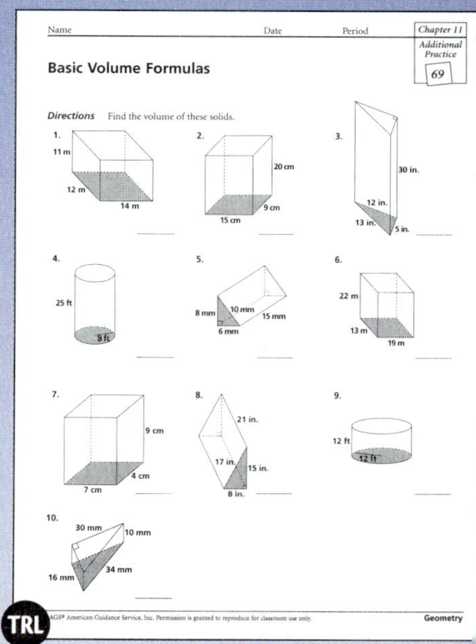

Name Date Period *Chapter 11 Additional Practice 69*

Basic Volume Formulas

Directions Find the volume of these solids.

 Geometry

TRL

Additional Practice 69

For three-dimensional measure, the word *cubic* means the basic measure is a unit cube. The 10^3 cube contains 1,000 of these cubes, so its volume is 1,000 cubic units.

Volume is the measure of the number of unit cubes contained in a solid figure.

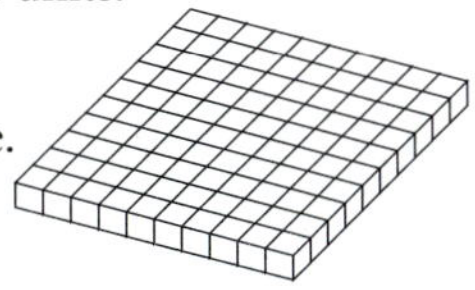

*C*ubic units may be indicated using the abbreviation *cu,* as in *cu units,* or using the exponent 3, as in *unit*^{3}.

EXAMPLES What is the volume of a rectangular box whose base measures 3 units by 10 units and whose height is 20 units?

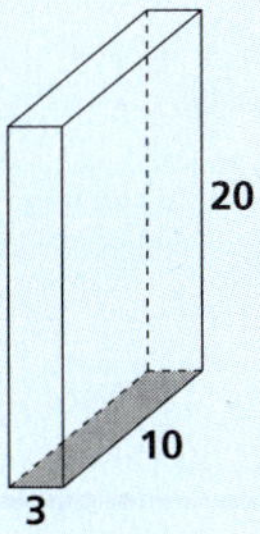

V = area of base • height
Substitute values:

= (3 • 10)20

= 600 cubic units

A triangular prism is a solid whose base is a triangle. For the prism at right, the area of the base is 15 cm^2 and the height is 30 cm. What is the volume of the prism?

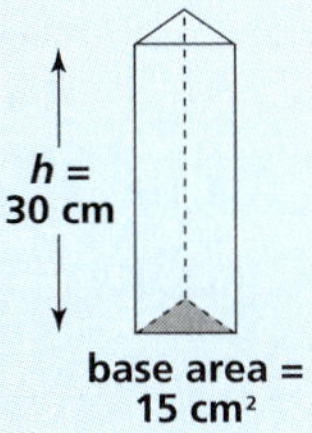

V = area of base • height
Substitute values:

= (15 cm^2)30 cm

= 450 cm^3

A cylindrical tank has a radius of 10 ft and a height of 25 ft. What is its volume? Give your answer in terms of π.

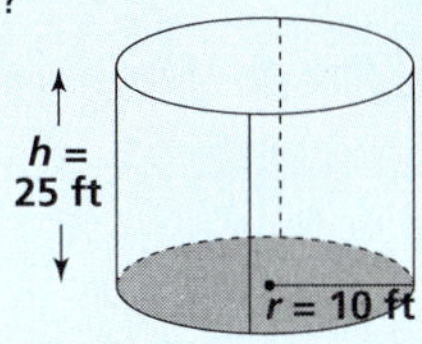

V = area of base • height
Substitute values:

= $(\pi r^2)h$

= $\pi(10^2)(25)$

= $2{,}500\pi$ cu ft

2. Teaching the Lesson

Model each of the examples on pages 352 and 353 for students. For each, ask a volunteer to identify—and if appropriate, compute—the area of each solid.

LEARNING STYLES

Body/Kinesthetic If feasible, present students with a selection of unit cubes, tens rods, and hundreds squares. Encourage students to use the materials to construct a cube which has a length, width, and height of 10 units. Ask students to give the volume of their completed solid. Have them explain their answer.

LEARNING STYLES

Auditory/Verbal Ask students to explain what is similar about the four examples. Prompt their answer by asking: *How are the formulas for the volume of a cube, a rectangular box, a cylindrical tank, and a triangular prism alike?* If necessary, write the formula for each (area of base)(height) on the board or overhead projector.

3. Reinforce and Extend

Learning Styles

Logical/Mathematical Ask students to explain how using different estimates for π can affect the computation of the volume of a given solid. Encourage students to draw illustrations to support their explanations.

Common Error

Students may be most comfortable identifying the bottommost, horizontal portion of a solid as its base. As they study the diagrams in Exercises A–D, point out to students that this identification depends on the orientation of the object. Turning the object upside down or merely rotating it 90° allows for another side to be considered the base of the object.

You can now use any approximation you wish for π. If your answer needs to be accurate to the nearest hundred, you might choose $\pi \approx 3$. If you need an answer to the nearest tenth, you might choose $\pi \approx 3.14$. If you need a more accurate answer, you might choose $\pi \approx 3.1416$.

Let $\pi \approx 3$	Let $\pi \approx 3.14$	Let $\pi \approx 3.1416$
$V = 2{,}500 \cdot 3$	$V = 2{,}500 \cdot 3.14$	$V = 2{,}500 \cdot 3.1416$
$V = 7{,}500$	$V = 7{,}850$	$V = 7{,}854$

Algebra Connection

Example The volume of a rectangular box is 2,400 cu units. The measurements of the base are 60 x 20. What is the height of the box?

First, draw a sketch. Include the information that is given. Let h be the unknown. Substitute the given values and solve the formula for h.

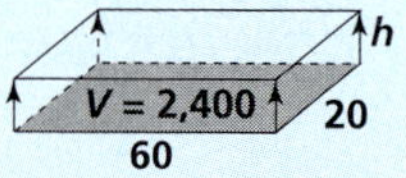

V = area of base • height

$2{,}400 = (60)(20)h$

$2{,}400 = 1{,}200h$

$2 = h$ The height of the box is 2.

Selected Answers and Solutions begin on page 420.

Exercise A Find the volume of each rectangular solid.

1.

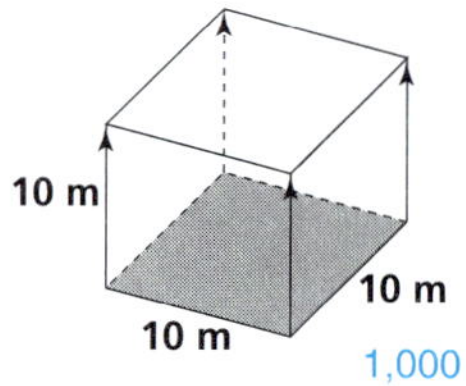

1,000 m^3

2.

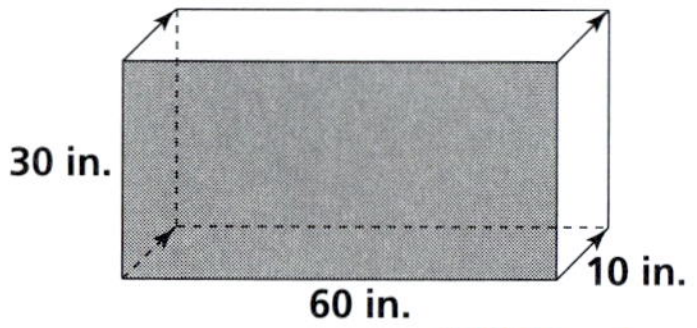

18,000 cu in.

3.

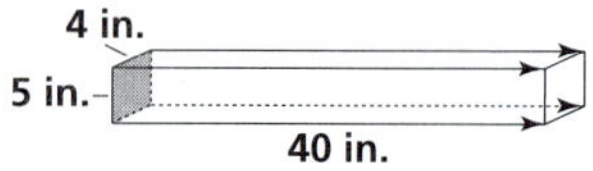

800 cu in.

4.

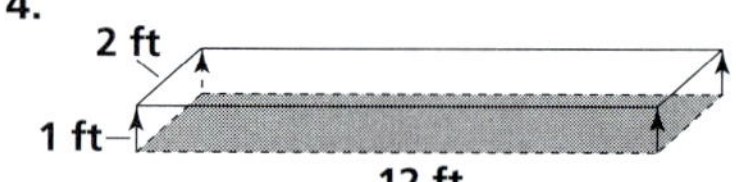

24 cu ft

Exercise B Find the volume of each triangular prism.

5.

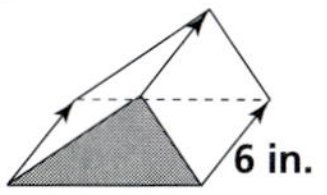

Base area = 24 sq in.

144 cu in.

6.

Base area = 12 m²

180 m³

7.

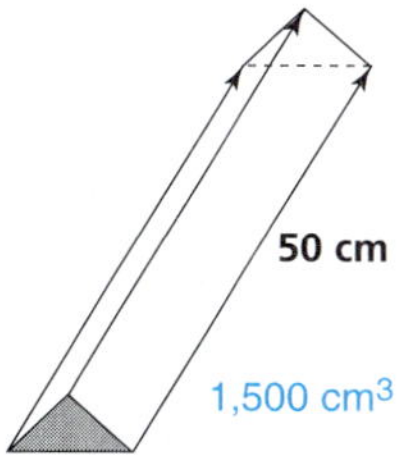

1,500 cm³

Base area = 30 cm²

Exercise C Find the volume of each cylinder. Give your answer in terms of π. Then use 3, 3.1, and 3.14 as approximations for π.

	V in terms of π	π = 3	π = 3.1	π = 3.14
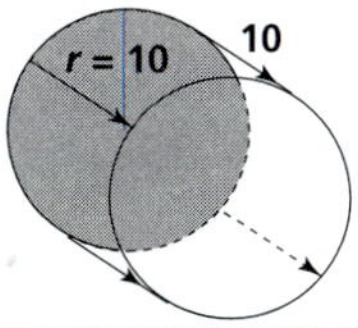	**8.** ___ 1,000π	**9.** ___ 3,000	**10.** ___ 3,100	**11.** ___ 3,140
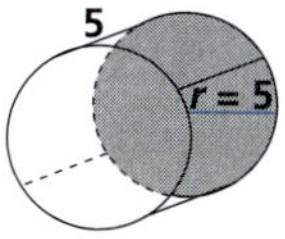	**12.** ___ 125π	**13.** ___ 375	**14.** ___ 387.5	**15.** ___ 392.5

Problem Solving

Exercise D Use what you know about volumes to find the unknown in each item.

16. $h = ?$

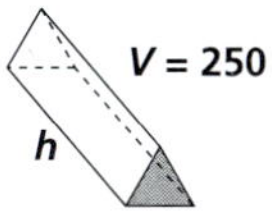

17. $r = ?$

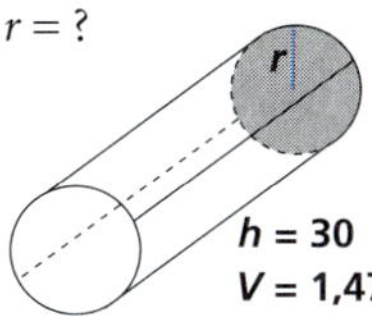

18. $l = ?$

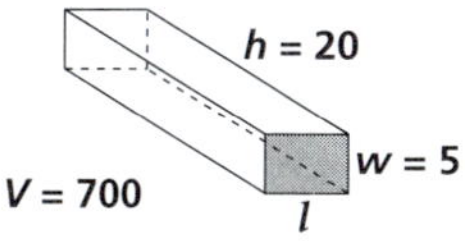

19. $s = ?$

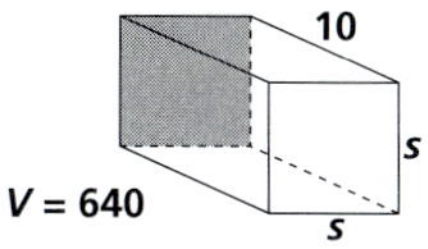

20. $s = ?$

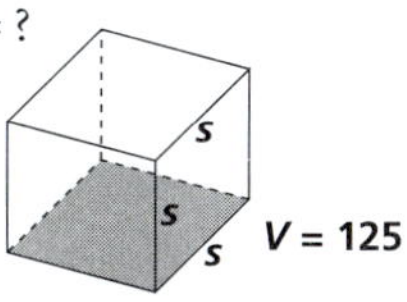

16. $h = 10$
17. $r = 7$
18. $l = 7$
19. $s = 8$
20. $s = 5$

Learning Styles

Interpersonal/Group Learning Have student pairs complete two more columns of the chart for Exercise C, one using π = 3.141 and the second using π = 3.1416. Ask students to give a rounded value for the quantities in items 10 to 14. Have students compare their new column entries with the rounded quantities. Invite them to use their data to support rounding to the nearest tenth for π and for computations using π.

Algebra Connection

Have students work in pairs. Ask each pair to write an algebraic equation for each unknown in Exercise D. Pairs of students can exchange results with one another. Allow time for pairs to revise their equations if necessary.

16. $h = V \div \text{(base area)}$
17. $r = \sqrt{\frac{V}{\pi h}}$
18. $l = \frac{V}{hw}$
19. $s = \sqrt{\frac{V}{l}}$
20. $s = \sqrt[3]{V}$

Lesson at a Glance

Chapter 11 Lesson 2

Overview This lesson presents the formulas for the volume of pyramids and cones.

Objectives

- To identify the volume of a pyramid or cone as the area of its base times its height.
- To compute the volume of solids such as a triangular, square, or rectangular pyramid.
- To compute the volume of a cone.

Student Pages 356–359

Teacher's Resource Library TRL

Workbook Activity 114

Additional Practice 70

Mathematics Vocabulary

cone
pyramid

1. Warm-Up Activity

Ask students to speculate on how the volume of a cone compares with the volume of a cylinder with the same height and radius. Accept and record all responses. If feasible, direct students to complete the *Experiencing Geometry* activity. Discuss their results. Point out to students that in this lesson they will learn formulas for finding the volume of pyramids and cones.

Lesson 2 Volumes of Pyramids and Cones

Cone
A three-dimensional figure whose base is a circle and all segments that join the circle to a vertex, V, not in the plane of the circle

Pyramid
A three-dimensional figure with a single base and sides that are triangles

A **pyramid** is a three-dimensional figure with a single base and sides that are triangles. The sides of a pyramid meet at a vertex opposite the base. A pyramid can be named by the shape of its base.

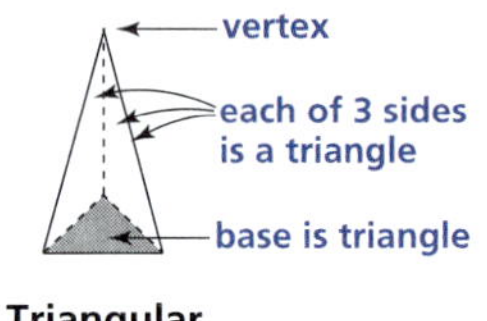

Triangular Pyramid

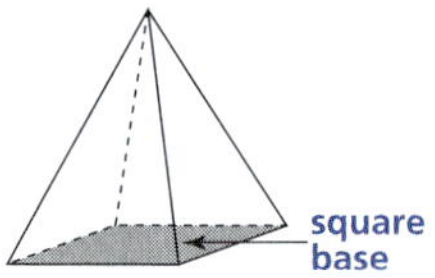

Square Pyramid

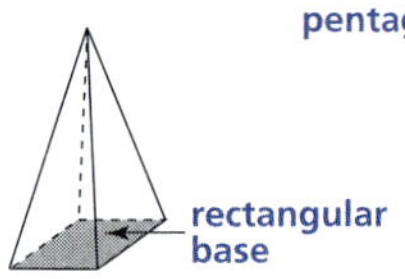

Rectangular Pyramid

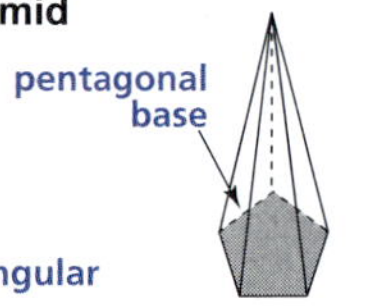

Pentagonal Pyramid

If the base of a pyramid is a circle, the pyramid is called a **cone.**

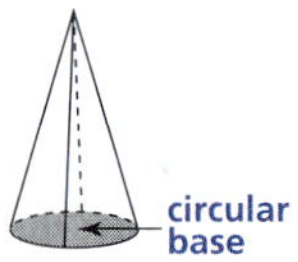

Cone

Remember, you found the volume of a rectangular prism and a cylinder. How do you think the volume of a pyramid might compare to the volume of a cylinder of similar height?

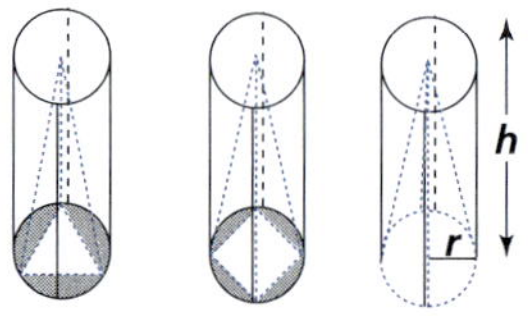

Look at the illustrations. The volumes of the pyramids are completely contained inside the cylinders. The same is true for the volume of the cone—it can be completely contained in a cylinder. The volume of a cone with height h and radius r is less than the volume of a cylinder with the same height and radius. Use this experiment to figure out how the volumes of a cone and a cylinder are related.

Experiencing Geometry

For this experiment you will need a cone and a cylinder that have the same radius and height. You may find it easier to choose a cylindrical drinking glass, can, or plastic container and then roll a sheet of paper into a cone shape having the same radius and height as your container. Fill the cone with rice or small beans such as lentils. Empty the cone into the cylinder. Find how many cones full of rice or beans it takes to fill the cylinder.

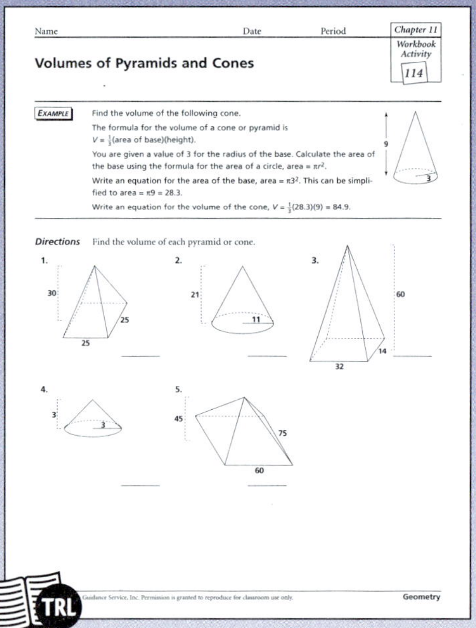
Name Date Period | Chapter 11 Workbook Activity 114

Volumes of Pyramids and Cones

EXAMPLE Find the volume of the following cone.
The formula for the volume of a cone or pyramid is
$V = \frac{1}{3}$(area of base)(height).
You are given a value of 3 for the radius of the base. Calculate the area of the base using the formula for the area of a circle, area = πr^2.
Write an equation for the area of the base, area = $\pi 3^2$. This can be simplified to area = $\pi 9 = 28.3$.
Write an equation for the volume of the cone, $V = \frac{1}{3}(28.3)(9) = 84.9$.

Directions Find the volume of each pyramid or cone.

1. 2. 3. 4. 5.

Geometry

Workbook Activity 114

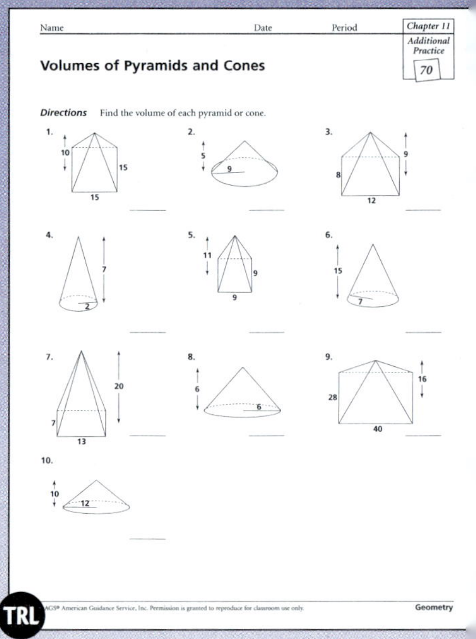
Name Date Period | Chapter 11 Additional Practice 70

Volumes of Pyramids and Cones

Directions Find the volume of each pyramid or cone.

1. 2. 3. 4. 5. 6. 7. 8. 9. 10.

Geometry

Additional Practice 70

From your experiment you can conclude that the volume of the cone is about one-third of the volume of the cylinder. The correct formula is given here.

FORMULA: Volume of cone = $\frac{1}{3}$(area of base)(height) = $\frac{1}{3}\pi r^2 h$

EXAMPLE What is the volume of a cone whose height is 10 units and whose base has radius 5? Express your answer in terms of π.

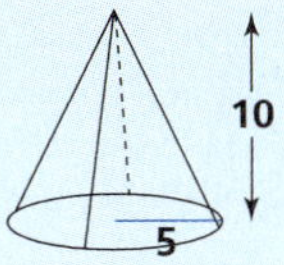

First, make a sketch that shows the given information. Then use the formula.

Volume of cone = $\frac{1}{3}\pi r^2 h$ Substitute values for *r* and *h*:

$= (\frac{1}{3})(\pi 5^2)(10)$

$= \frac{1}{3}(25\pi)(10)$

$= (\frac{250}{3})\pi$ cubic units

Once you have this value, you can use any approximation of π to obtain a decimal value. You can let $\pi = 3$, $\pi = 3.14$, or $\pi = 3.1416$.

Because you already know that a cone is a pyramid with a circular base, you can reason that the formula for the volume of a pyramid will be much like the formula for the volume of a cone.

FORMULA: Volume of pyramid = $\frac{1}{3}$(area of base)(height)

EXAMPLE What is the volume of a square pyramid whose height is 15 units and the length of each side of the square is 5 units?

Make a sketch that shows the given information.

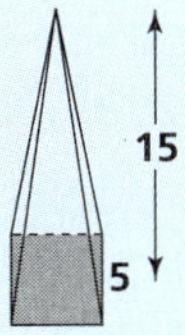

Volume of pyramid = $\frac{1}{3}$(area of base)(height)

area of base = $5^2 = 25$

Volume of pyramid = $\frac{1}{3}$(area of base)(height)

Substitute values for area and *h*:

$= \frac{1}{3}(25)(15)$

$= (25)(5)$

$= 125$ cubic units

2. Teaching the Lesson

Work through the examples with students, making certain they understand that the formula $V = \frac{1}{3}\pi r^2 h$ is equivalent to the area of the base times the height. Ask a volunteer to follow the model and find the area of a cone with radius = 4 units, whose height is 10 units. **($\frac{160\pi}{3}$ cubic units)** Invite a second volunteer to find the volume of a square pyramid whose height is 15 units and whose base is 10 units on a side. **(500 cubic units)**

LEARNING STYLES

LEP/ESL Provide poster board, scissors, and tape or glue to groups of four students. Invite students to create several isosceles triangles, some with bases of equal length, others of unequal length. Invite students to use the triangles to form triangular, square, rectangular, pentagonal, and hexagonal pyramids. Have students trace the base of each pyramid onto paper to confirm the base of the pyramid. Using their constructed pyramids as models, have students draw a diagram of each, label the diagram, and include it in their picture dictionary.

3. Reinforce and Extend

Mental Math

Invite the class to use mental math to find the volume of the following cones.

area 15, height 7; $V = 35$ cu units

area 8, height 21; $V = 56$ cu units

area 5, height 33; $V = 55$ cu units

radius = 2, h = 6; $V = 8\pi$ cu units

radius = 5, h = 30; $V = 250\pi$ cu units

Manipulatives

M | Working with Formulas

Materials: Geometric Volume Set

Student Practice: Invite students to confirm the formulas in this lesson using the Geometric Volume Set and sand, dried beans, or rice to compare the volumes of pyramids, prisms, cones, and cylinders.

Learning Styles

Logical/Mathematical Have students work together to complete Exercise A. Ask them to complete a fourth column of the chart, using the π key on their calculator. Direct students to compare this value with the other entries in the table. Encourage students to discuss which entry they think is most accurate. Which do they think offers a suitable estimate for most work?

EXAMPLES What is the volume of a triangular pyramid whose base is a 3, 4, 5 right triangle and whose height is 20 units?

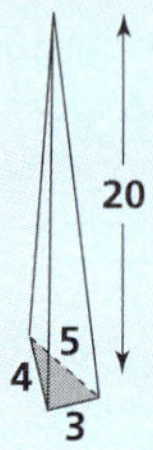

Make a sketch.

Volume of pyramid = $\frac{1}{3}$(area of base)(height)

area of base = $\frac{1}{2}bh = \frac{1}{2}(3)(4)$

= 6 sq units

Volume of pyramid = $\frac{1}{3}$(area of base)(height)

Substitute values for area and h:

= $(\frac{1}{3})(6)(20)$

= (2)(20)

= 40 cubic units

What is the height of a cone whose volume is 120 cubic units and whose radius is 5? Use $\pi = 3.14$.

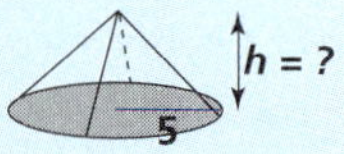

Make a sketch.

Volume of cone = $\frac{1}{3}$(area of base)(height)

$120 = \frac{1}{3}\pi(25)h$

Substitute values in the formula:

$120 = (\frac{25}{3})\pi h$ multiply

$120 = \frac{25}{3}(3.14)h$ use $\pi = 3.14$

$120 = 26.17h$ solve for h

$h = 4.59$

Exercise A Find the volume of each cone. Give your answer in terms of π. Then use 3, 3.1, and 3.14 as approximations for π.

	V in terms of π	$\pi = 3$	$\pi = 3.1$	$\pi = 3.14$
$r = 10, h = 10$	1. 333.3π	2. 1,000	3. 1,033.3	4. 1,046.67
$r = 6, h = 12$	5. 144π	6. 432	7. 446.4	8. 452.16
$r = 3, h = 4$	9. 12π	10. 36	11. 37.2	12. 37.7

Exercise B Find the volume of each pyramid. Be sure to make and label a sketch for each.

13. $s = 10$ in., $h = 10$ in.

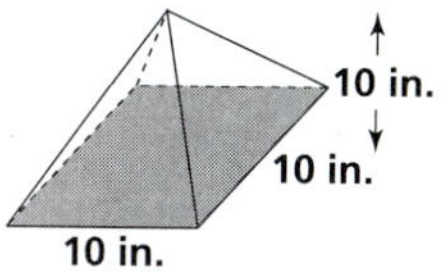

333.3 cu in.

14. $l = 8$ ft, $w = 5$ ft, $h = 15$ ft

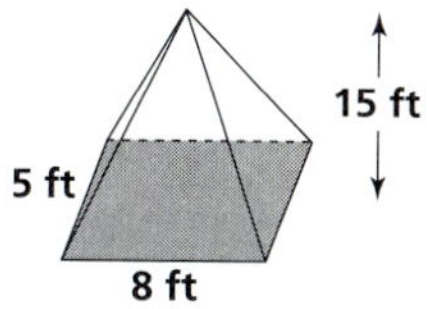

200 cu ft

15. base area = 20 cm^2, $h = 12$ cm

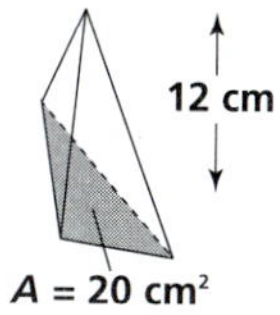

80 cm^3

16. Square pyramid: $s = 5$ in., $h = 50$ in.

416.67 cu in.

17. Triangular pyramid: base area = 15 m^2, $h = 20$ m

100 m^3

Exercise C Use the formula for the volume of a pyramid or cone to find the value of the unknown. Use $\pi = 3.1$.

18. $V = 80$ cm^3, $w = ?$

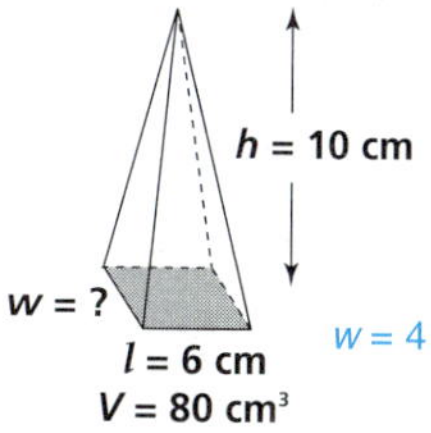

$w = 4$ cm

19. $V = 8{,}906.3$ cu in., $r = ?$

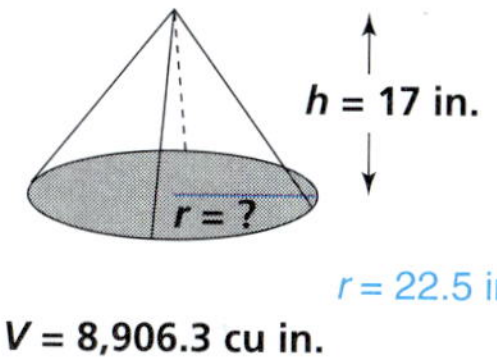

$r = 22.5$ in

20. $V = 700$ cu ft, $w = ?$

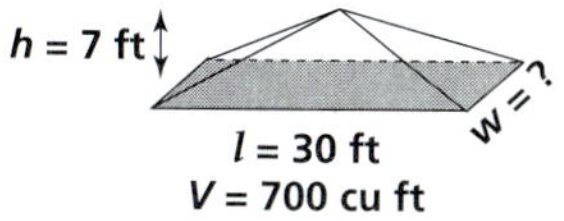

$w = 10$ ft

Technology Connection

Engineers and others who test aircraft for safety and resistance to vibrations often test the aircraft in an anechoic chamber. An anechoic chamber is designed to deaden or eliminate vibrations and reverberations. The inside surface of some anechoic chambers is covered by foam pyramids embedded with a vibration-absorbing material such as carbon.

Technology Connection

Aerospace engineers are not the only people interested in deadening vibrations. Audio engineers for recording studios must also be certain that extra vibrations do not make their way onto recordings.

Application

Career Connection

Suggest students gather information about acoustic engineering or audio (sound) engineering. Students can learn about Pythagoras' studies as well as the acoustic design of ancient stadiums and coliseums and comparing them to today's designs for auditoriums and theaters. Students can investigate what knowledge and experience is necessary to be an acoustic engineer and where such engineers might find employment. Alternatively, they can learn about audio engineering and the requirements needed to work in the recording industry. Findings can be presented as a two-minute recruitment speech for Can You Hear?, an audio engineering firm.

Lesson at a Glance

Chapter 11 Lesson 3

Overview This lesson introduces computations needed to find the surface areas of prisms and cylinders.

Objectives

- To recognize that the surface area of a prism or cylinder is the sum of the surface areas of the faces plus the surface area of the base.
- To compute the surface area of selected prisms and cylinders.

Student Pages 360–363

Teacher's Resource Library TRL

Workbook Activity 115

Additional Practice 71

Mathematics Vocabulary

right prism

1. Warm-Up Activity

Display a variety of similar-sized prisms and cylinders. Ask students how they could determine how much paint—or paper—they would need to cover each. Point out that in this lesson they will use their intuition to determine formulas for finding the surface areas of some solids.

Geometry Around Us

Point out to students that if a second right prism is placed in the band of colors, the light will bend again and will appear as a single beam of white light as it leaves the second prism. Indicate that under certain circumstances, raindrops in the air can act like the prism to form rainbows.

Lesson 3 Surface Areas of Prisms and Cylinders

Right prism

Prism in which the base and the sides are at right angles

In addition to volume, three-dimensional figures also have another measurable characteristic—surface area. The surface areas of three-dimensional geometric figures can be computed by adding the areas of the sides, the base, and the top.

Cube

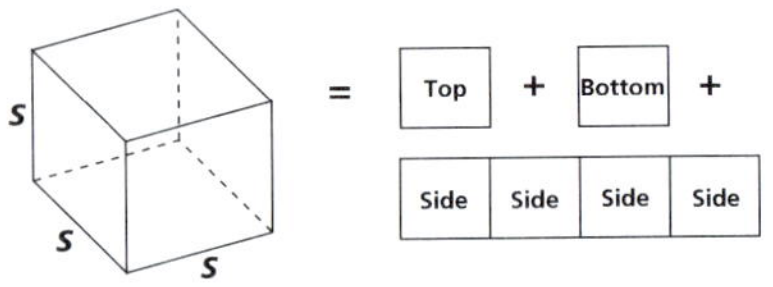

Total surface area of cube = area of 6 squares
Total surface area = 2(area of base) + areas of four sides

FORMULA: Surface area of cube = $6s^2$

Rectangular Prism

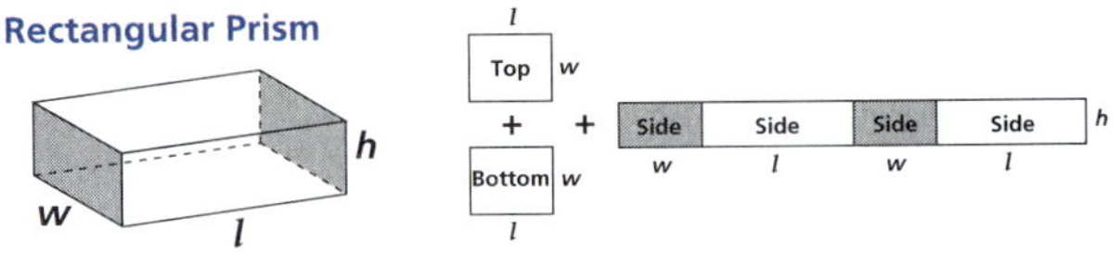

$= 2(lw) \quad + \quad h(l + w + l + w)$

$= 2(lw) \quad + \quad h(2l + 2w)$

$= 2lw \quad + \quad 2hl \quad + \quad 2hw$

Area = base and top + two larger sides + two smaller sides

FORMULA: Surface area of rectangular prism = $2(lw + hl + hw)$

Geometry Around Us

Prisms are used to change the direction of a beam of light in optical instruments such as binoculars and periscopes. This triangular prism bends the light as it passes through and splits the light into a band of colors.

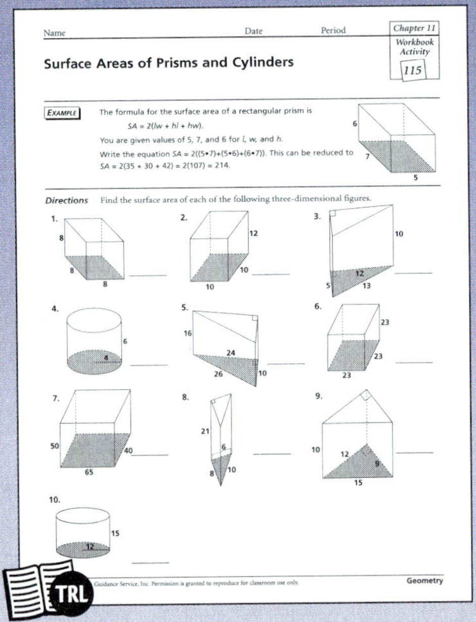
Name Date Period | *Chapter 11 Workbook Activity 115*

Surface Areas of Prisms and Cylinders

EXAMPLE The formula for the surface area of a rectangular prism is
$SA = 2(lw + hl + hw)$.
You are given values of 5, 7, and 6 for l, w, and h.
Write the equation $SA = 2((5 \cdot 7)+(5 \cdot 6)+(6 \cdot 7))$. This can be reduced to $SA = 2(35 + 30 + 42) = 2(107) = 214$.

Directions Find the surface area of each of the following three-dimensional figures.

1. 2. 3. 4. 5. 6. 7. 8. 9. 10.

Geometry

Workbook Activity 115

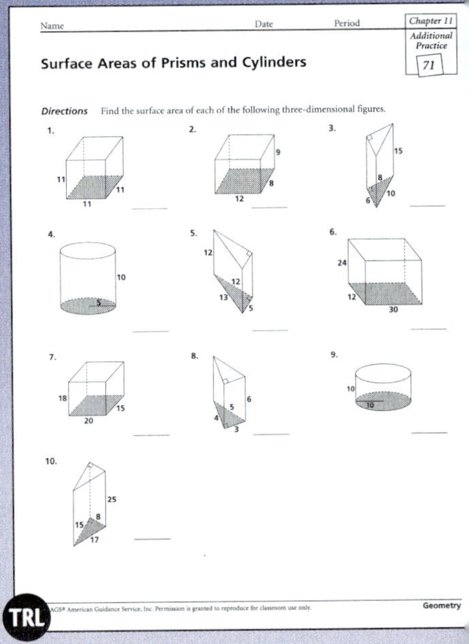
Name Date Period | *Chapter 11 Additional Practice 71*

Surface Areas of Prisms and Cylinders

Directions Find the surface area of each of the following three-dimensional figures.

1. 2. 3. 4. 5. 6. 7. 8. 9. 10.

Geometry

Additional Practice 71

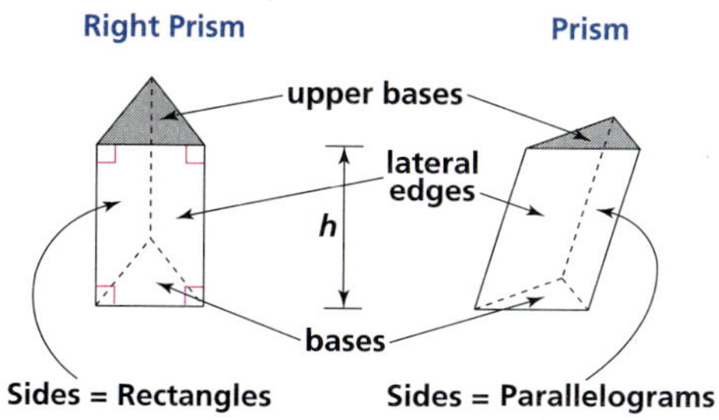

Cubes and rectangular prisms can be grouped as **right prisms.** In a right prism, the side surfaces are perpendicular to the base and the upper surface. The total surface area is the sum of the areas of the base, upper surface (top), and the sides. The side surfaces are rectangles. In prisms that are not right prisms, the sides are parallelograms.

Right Prism | **Surfaces are two triangles and three rectangles**

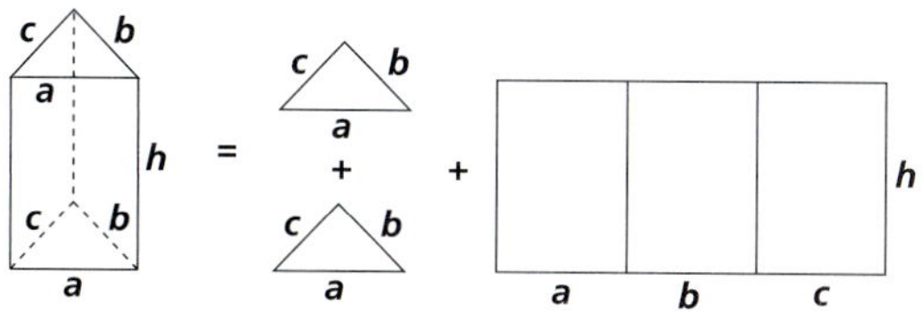

FORMULA: Surface area of right prism = 2(area of triangle) + $h(a + b + c)$

If the side surfaces of a prism are not perpendicular to the base and top, the sides are parallelograms.

Triangular Prism | **Surfaces are two triangles and three parallelograms**

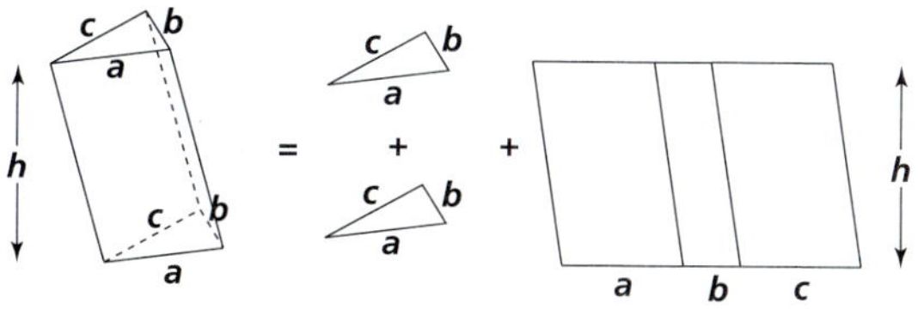

2. Teaching the Lesson

Use plastic models or the GeoSolids of each of the solids mentioned in the examples. With a marker, number or *X* each side as you include it in the total surface area. After completing each example, allow students to inspect the model. Be sure they understand that each side was included once in the computation of the surface area.

Learning Styles

Visual/Spatial Invite teams of students to use poster board or construction paper to duplicate each of the solids described on pages 360 and 361. Suggest they also make a model for a cylinder. Have students follow the diagrams to construct each of the solids. Then have them draw, on separate sheets of paper, each of the surfaces used to make up the solid. Encourage them to produce diagrams similar to those shown in the text.

Ask students to label their solids to match those in the text. Challenge them to use their solids to derive the formulas for finding surface area.

Common Error

When working with prisms, students often forget to include the base as part of the total surface area.

3. Reinforce and Extend

Application

At Home

Encourage students to collect several empty boxes, such as cereal or shoe boxes, from home. Encourage them to carefully slit the edges of the box, unfolding it so that it matches the rectangular—or square—diagrams shown in the text. Indicate that a test of such a diagram is that it can be folded again so that it encloses the original space. Mention to students that they may need to make more than one attempt to know which edges to slit. Invite students to share their successful attempts with the class. Have the class compute the surface area of each diagram presented by students.

A cylinder whose parallel bases are circles requires a special formula.

Cylinder | **Surfaces are two circles and one rectangle**

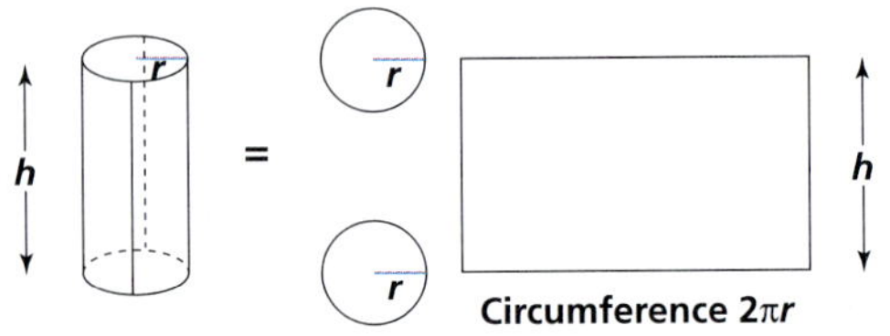

Formula: Surface area of cylinder = $2\pi r^2 + (2\pi r)h$

You can use these formulas to calculate the surface areas of some of the most common three-dimensional objects.

Examples

What is the total surface area of a cube whose side is 5 units long?

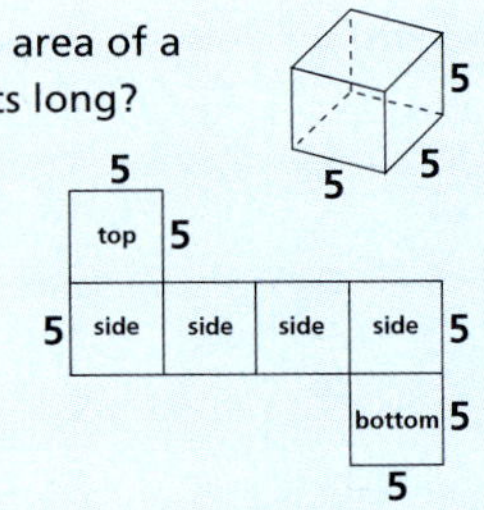

Sketch all the surfaces of the cube.

Surface area of cube = $6s^2$

$= 6(5^2) = 6(25)$

$= 150$ sq units

What is the total surface area of a box that is 10 in. on a side and has no lid?

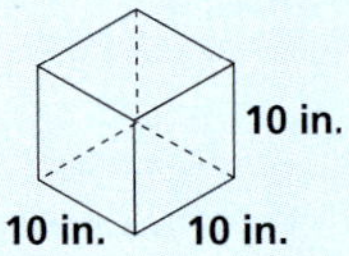

Sketch all the surfaces of the box.

Surface area of cube = $5s^2$

$= 5(10^2) = 5(100)$

$= 500$ sq in.

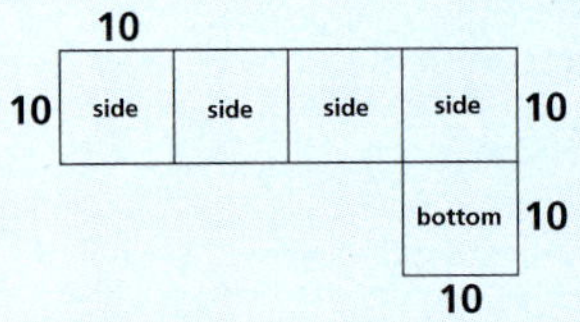

EXAMPLES Calculate the total surface area of a rectangular box whose dimensions are 10 • 8 • 5.

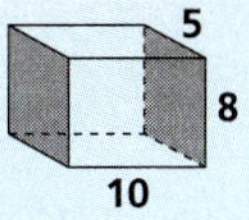

Sketch all the surfaces of the box.

Surface area of rectangular prism = $2(lw + hl + hw)$

let $l = 10$, $w = 5$, $h = 8$

Surface area = $2[(8 \cdot 10) + (5 \cdot 10) + (8 \cdot 5)]$

= 160 + 100 + 80

= 340 sq units

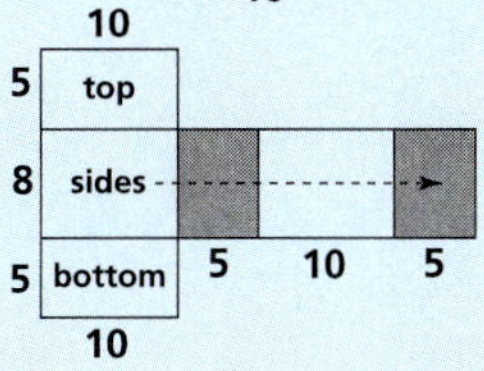

What size label is needed to cover the sides of a can whose radius is 2 in. and whose height is 7 in.?

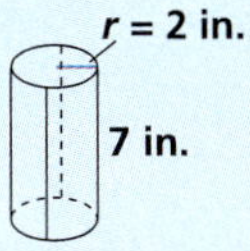

Sketch each surface of the cylinder.

surface area = $l \cdot w$

$w = 7$

l = circumference = $2\pi r$

$= 2\pi 2 = 4\pi$

surface area = $7(4\pi)$

$= 28\pi$ sq in.

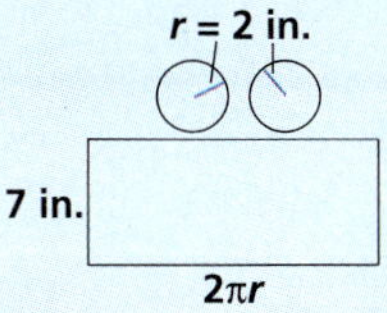

Exercise A Find the surface area of each of the following three-dimensional figures. Some answers may include radicals or π.

1.

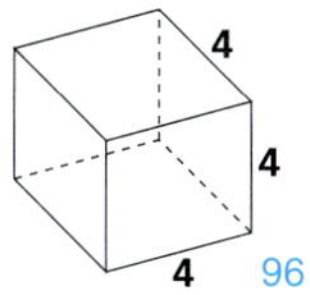

96 sq units

2.

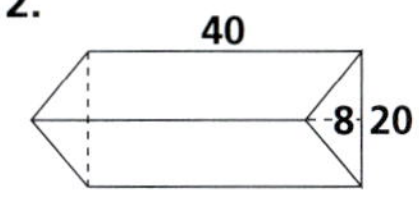

$960 + 160\sqrt{41}$ sq units

3.

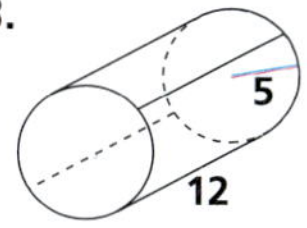

170π sq units

Exercise B Solve the following problems. You may leave some answers in terms of radicals or π.

4. What is the surface area of a 12-in.-tall cylindrical vase whose radius is 4 in.?

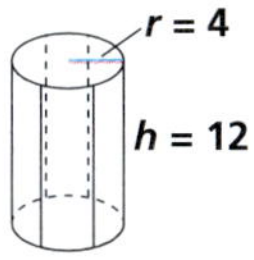

112π sq in.

5. Juanita plans to paint the walls and ceiling of a room bright green. The room is 10 ft wide and 12 ft long. The ceilings are 9 ft above the floor. She has one can of paint that covers 500 sq ft. Will it be enough? Why?

No, she's planning to paint 516 sq ft.

LEARNING STYLES

LEP/ESL Trios of students can work together to complete Exercise A. For each item, one member of the group can produce a sketch, showing the surfaces to be included in the total surface area. A second student can write out step-by-step directions for computing the total surface area. The third student can follow the directions. Teams can compare final products with one another, discussing any differences in sketches or directions.

Lesson at a Glance

Chapter 11 Lesson 4

Overview This lesson describes methods for finding the surface area of pyramids and cones.

Objectives

- To recognize that the surface area of pyramids and cones is the sum of the surface areas of the sides plus the surface area of the base.
- To compute the surface area of selected pyramids and cones.

Student Pages 364–367

Teacher's Resource Library TRL

Workbook Activity 116

Additional Practice 72

1. Warm-Up Activity

Display a variety of poster board or cardboard pyramids and cones. Make certain that the insides of each solid are a color different from the outside. Take each pyramid and cone apart for students. Ask students to speculate on how they might use the parts of the solids to determine their surface areas.

Lesson 4 Surface Areas of Pyramids and Cones

In the previous lesson, you found the area of each side of a three-dimensional figure and then added the areas. You can use the same method to find the surface areas of pyramids and cones. First, separate the base of the pyramid (a polygon) from its sides (triangles). Here is an example using a square pyramid.

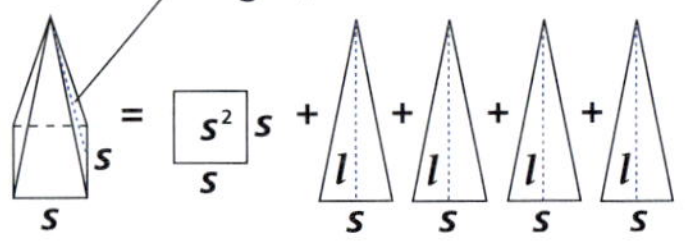

The area of each of the four side triangles is found using the base of the triangle and a measurement called the *slant height* of the pyramid. Think of the slant height as the straight-line distance you would have to climb from the midpoint of a base to the top of the pyramid. The symbol for slant height is l.

$$= s \cdot s \quad + \tfrac{1}{2}sl + \tfrac{1}{2}sl + \tfrac{1}{2}sl + \tfrac{1}{2}sl$$
$$= s^2 \quad + \quad 2sl$$

FORMULA: Surface area of pyramid = (area of base) + areas of four triangles = $s^2 + 2sl$

The Transamerica Building in San Francisco is shaped like a pyramid.

The surface area of a cone is found in much the same way. Think of the surface of the cone cut from the base to the vertex.

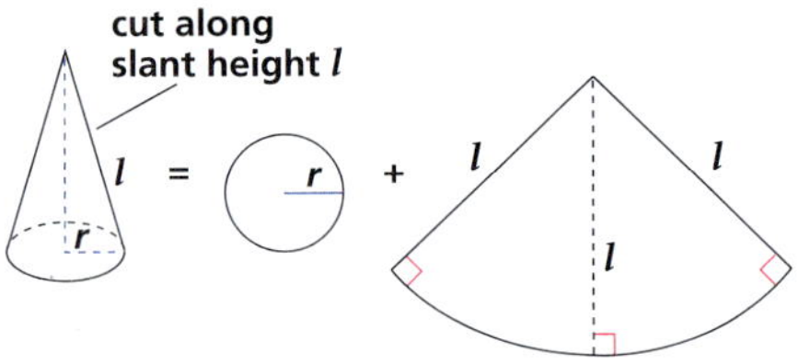

Then flatten the surface. You will see that the surface of the cone is made up of the surfaces of two areas. One is a circle with radius r. The other is like a triangle with a base that is the circumference of the circle and a height that is the slant height of the cone.

FORMULA: Surface area of cone = (area of circular base) + area of triangle $= \pi r^2 + \frac{1}{2}bl$

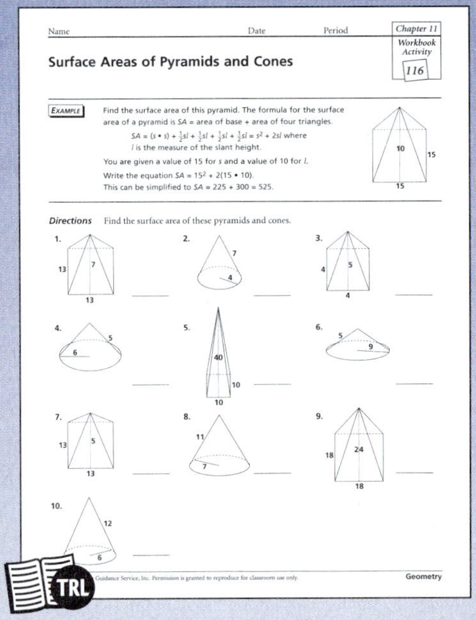

Name Date Period | Chapter 11 Workbook Activity 116

Surface Areas of Pyramids and Cones

EXAMPLE Find the surface area of this pyramid. The formula for the surface area of a pyramid is SA = area of base + area of four triangles.

$SA = (s \cdot s) + \frac{1}{2}sl + \frac{1}{2}sl + \frac{1}{2}sl + \frac{1}{2}sl = s^2 + 2sl$ where l is the measure of the slant height.

You are given a value of 15 for s and a value of 10 for l.

Write the equation $SA = 15^2 + 2(15 \cdot 10)$.

This can be simplified to $SA = 225 + 300 = 525$.

Directions Find the surface area of these pyramids and cones.

1. 2. 3. 4. 5. 6. 7. 8. 9. 10.

Guidance Service, Inc. Permission is granted to reproduce for classroom use only. Geometry

Workbook Activity 116

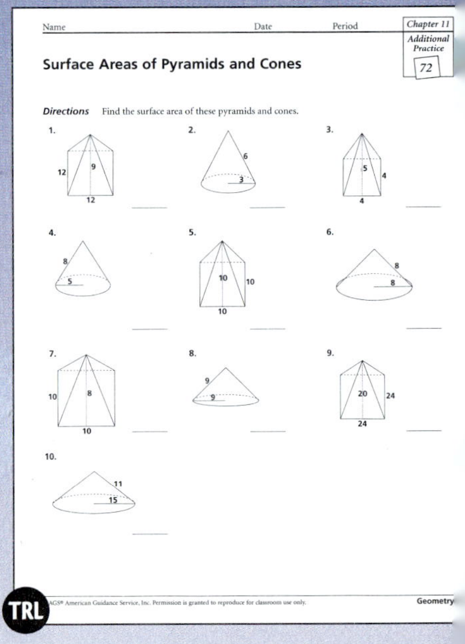

Name Date Period | Chapter 11 Additional Practice 72

Surface Areas of Pyramids and Cones

Directions Find the surface area of these pyramids and cones.

1. 2. 3. 4. 5. 6. 7. 8. 9. 10.

AGS® American Guidance Service, Inc. Permission is granted to reproduce for classroom use only. Geometry

Additional Practice 72

In this case, the base of the triangle, b, equals the circumference of the circle. The height of the triangle is the slant height, l, of the side's edge.

$$= \pi r^2 \quad + \quad \tfrac{1}{2}(2\pi r \cdot l)$$
$$= \pi r^2 \quad + \quad \pi r l$$
$$= \pi r(r + l)$$

Algebra Connection

Each of these formulas can be used to find the total surface areas of pyramids and cones.

EXAMPLES Find the areas of the sides and then the total surface area of the square pyramid with the given dimensions.

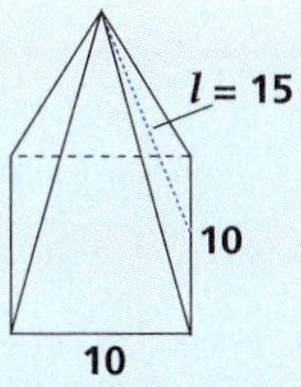

Sketch all the surfaces of the pyramid.

Surface area of pyramid = (area of base) + areas of four triangles

Surface area of pyramid sides = areas of four triangles

$= 4(\tfrac{1}{2})(bl)$ $\quad b = 10, l = 15$

$= 4(\tfrac{1}{2})(10)(15) = 2(150)$

$= 300$ sq units

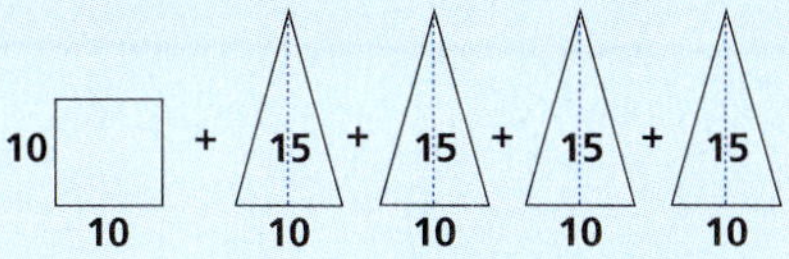

Surface area of base = 10 • 10 = 100 sq units

Surface area of pyramid = 100 sq units + 300 sq units

= 400 sq units

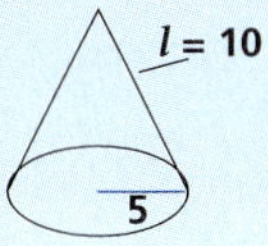

What is the surface area of a cone whose radius is 5 and whose slant height is 10?

Sketch the surface.

Surface area of cone = (area of circular base) + area of triangle

$= \pi r^2 + \tfrac{1}{2}bl$

b = the circumference of the circle = $2\pi r = 2\pi 5$, $l = 10$

$= \pi 5^2 \quad + \quad \tfrac{1}{2}(2\pi 5)\ (10)$

$= \pi 25 \quad + \quad \pi 50$

$= 75\pi$

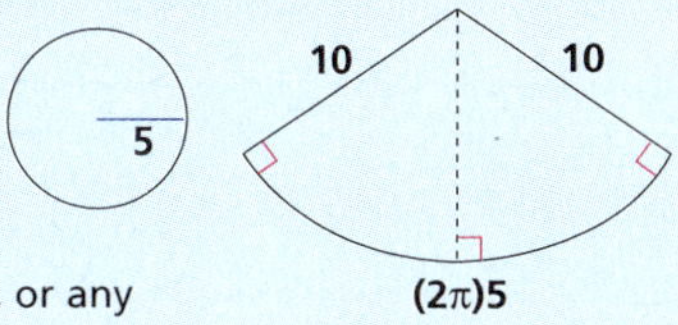

As with cylinders, you can substitute 3, 3.1, 3.14, or any other approximation for π.

2. Teaching the Lesson

Demonstrate how to find the surface area of a pyramid. Following the example on page 364, take apart a paper pyramid and complete the diagram for the solid. Do the same for the cone example at the bottom of the page. Allow a volunteer to reassemble the two-part model, closing the cut with tape.

Help students recognize that finding the surface area of a pyramid or cone is the same as finding the surface area of a prism or cylinder—one must find the surface area of each part or component.

LEARNING STYLES

LEP/ESL Give groups of four students a selection of model pyramids and cones. Group members should identify the following: faces, edges, vertices, the base, lateral faces, and how they would find slant height. Group members can be responsible for correcting any errors.

3. Reinforce and Extend

Group Problem Solving

Engage students in a discussion of soft-drink sizes available at fast-food restaurants. Have students work in groups of four. If feasible, encourage students to bring in a selection of paper cups used for soft drinks, noting the price, without tax, for each size. Challenge students to determine how to find the volume of the cups (most are frustums of cones). Groups should investigate how many cubic centimeters of liquid each cup holds, what is the price per cubic centimeters of a soft drink, and how much paper is needed to make the cup. Students can present their findings as a Consumer Corner segment of the evening news.

Example Find the total surface area of the triangular pyramid whose base measures 10 units on a side and whose slant height is 25 units.

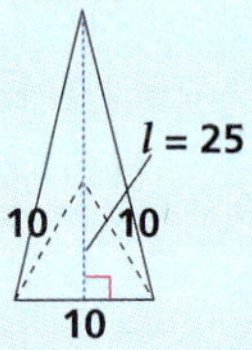

Sketch all the surfaces of the pyramid.

Surface area of pyramid = (area of base) + areas of three triangles

Surface area of pyramid sides = areas of three triangles

$= 3(\frac{1}{2})(bl) \qquad b = 10, l = 25$

$= 3(\frac{1}{2})(10)(25) = 3(5 \bullet 25)$

$= 375$ sq units

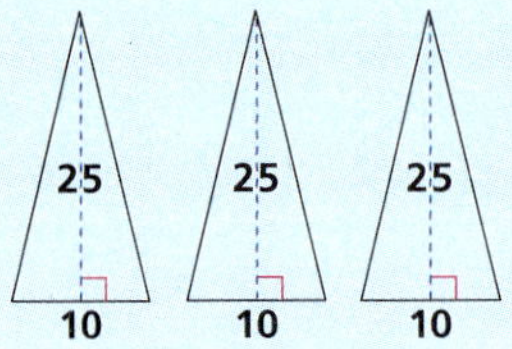

Surface area of base $= \frac{1}{2}(bh)$

$b = 10$ but you must find the value of h:

Using the Pythagorean Theorem:

$10^2 = 5^2 + h^2$

$100 = 25 + h^2$

$75 = h^2$

$\sqrt{3 \bullet 25} = h$

$5\sqrt{3} = h$

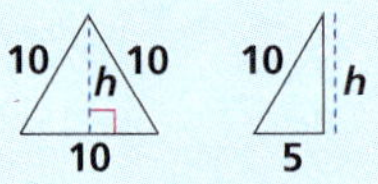

Surface area of base $= \frac{1}{2}(bh) = \frac{1}{2}(10)(5\sqrt{3})$

$= 25\sqrt{3}$

Surface area of pyramid = (area of base) + areas of three triangles

$= 25\sqrt{3} + 375$ sq units

Exercise A Find the surface areas of these pyramids and cones.

1.

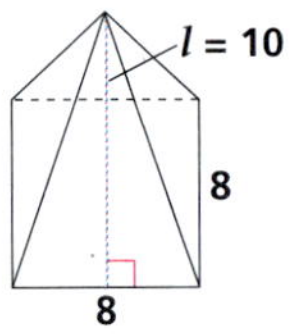

224 sq units

2.

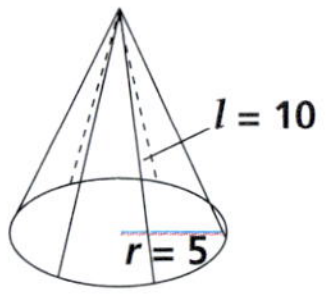

75π sq units

3.

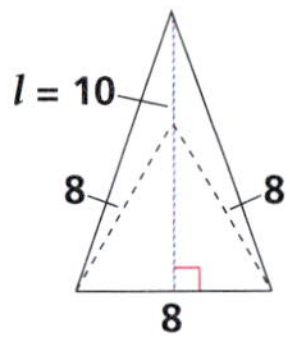

$120 + 16\sqrt{3}$ sq units

Calculator Practice

When calculating the area of the base of a cone, you may find it helpful to use the x^2 key on your calculator.

EXAMPLE Find the value of 17^2.

Press 17 x^2 289.

Exercise B Find the squares of the following numbers.

4. 16 256 **5.** 18 324 **6.** 24 576

7. 23 529 **8.** 19 361

Problem Solving

Exercise C Solve the following problems.

9. Roger has to make a model tepee for social studies class. He has a triangle with the dimensions shown. Can he make a wigwam whose base radius is 8 ft? Explain.

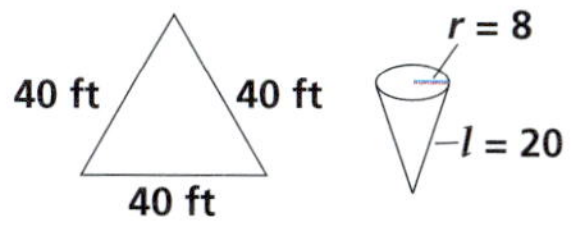

No, he needs a triangle whose base ≥ 50 ft, the circumference of the base.

10. The largest of the Egyptian pyramids, built for Pharaoh Khufu, is a square pyramid with the dimensions shown. What is the surface area of the sides of Khufu's pyramid?

86,394 m²

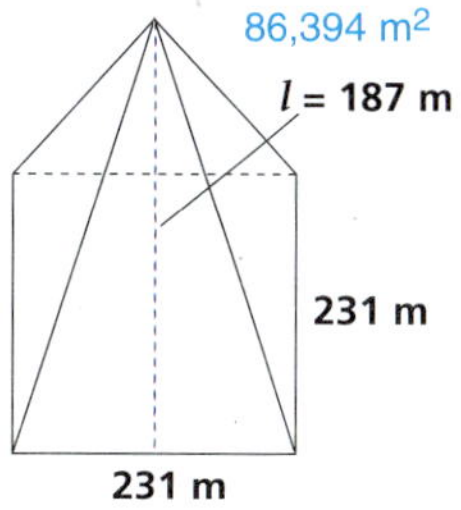

TRY THIS...

Find the surface area of a pyramid whose sides have a slant height of 20 units and whose base is a rectangle 10 units long and 5 units wide.

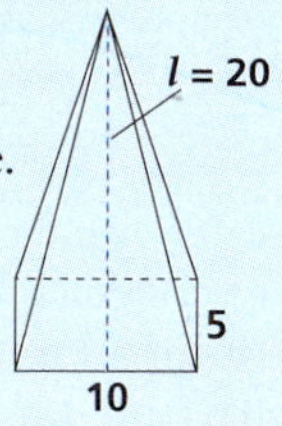

350 sq units

Writing About Mathematics

Are the areas of the sides of these two pyramids equal? Explain. (Hint: You may use numbers to test your idea.)

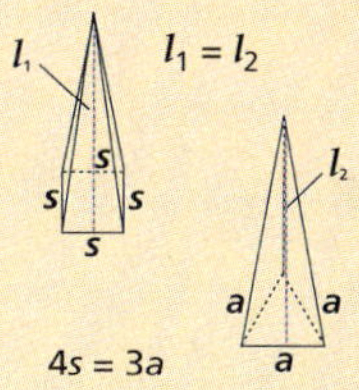

CALCULATOR

Point out that another useful key is the $\sqrt{\ }$ key which takes the square root of a number. Have students use their calculators to find the following square roots.

$\sqrt{625}$	25
$\sqrt{169}$	13
$\sqrt{225}$	15
$\sqrt{100}$	10

TRY THIS...

Students should recognize that they need to find the area of each side of the pyramid:

Surface area of pyramid = (area of base) + area of four triangles

$(l \cdot w) + \frac{1}{2}(l)(h) + \frac{1}{2}(l)(h) + \frac{1}{2}(w)(h) + \frac{1}{2}(w)(h)$

$5 \cdot 10 + (10 \cdot 20) + (5 \cdot 20)$

$50 + 200 + 100 = 350$ units

Lesson at a Glance

Chapter 11 Lesson 5

Overview This lesson presents length, weight, and capacity units for the customary system and length, mass, and capacity units for the metric system.

Objectives

- To recognize that there is more than one accepted standard of measurement.
- To use the relationships between units to convert one measure to another within the same system of measurement.

Student Pages 368–369

Teacher's Resource Library TRL

Workbook Activity 117

Additional Practice 73

Mathematics Vocabulary

capacity
customary
metric

1. Warm-Up Activity

Remind students that they have dealt with lengths such as inches, feet, centimeters, and meters throughout their study of geometry. Invite students to come to the board or overhead projector and draw lines that they estimate to be one inch, one centimeter, and one foot. Point out that in this lesson students will learn about the relationships among units of measure in the same system.

Lesson 5 Measurements

Capacity
The amount a container will hold when full

Customary
Usual or common, ordinary; system of measurement used mostly in the United States; inches and cups are basic units

Metric
System of measurement based on powers of ten; gram and meter are basic units

In your study of surface areas and volumes, you used measurements from two different systems. In the United States, most measurements are given in **customary** units—the units of length are inch (in.), foot (ft), yard (yd), and mile (mi).

To convert one measurement to another, you need to know the relationship between the measurements.

1 ft = 12 in.   1 yd = 3 ft   1 mi = 1,760 yd
1 yd = 36 in.   1 mi = 5,280 ft

The customary system has a special unit for area, the acre.

1 acre = 43,560 sq ft = 4,840 sq yd

There are customary units for **capacity** and weight as well.

Capacity	1 cup = 8 fluid ounces	2 pints = 1 quart
	2 cups = 1 pint	4 quarts = 1 gallon
Weight	1 pound = 16 ounces	2,000 pounds = 1 ton

You can use these relationships to change one unit of measurement to another.

EXAMPLE A car has a total length of 2 yd 1 ft. How many feet long is the car? How many inches?

2 yd = 2 • 3 ft = 6 ft   so 2 yd 1 ft = 7 ft
7 ft = 7 • 12 in. = 84 in.

Most of the world uses the **metric** measurement system. In the metric system, the basic unit is the meter (m). Other units in the metric system are formed by multiplying the meter by positive or negative powers of ten. Some of the other units of length are centimeter (cm), decimeter (dm), and kilometer (km).

A centimeter is about the width of your little finger.

To convert one measurement to another, you need to know the relationship between the measurements.

10 cm = 1 dm   10 dm = 1 m   1,000 m = 1 km
100 cm = 1 m

Name ____ Date ____ Period ____ Chapter 11 Workbook Activity 117

Measurements

EXAMPLE Complete the following statement.
14 pints = ____ gallons
You know that a gallon equals 4 quarts. You also know that a quart equals 2 pints. Therefore, a gallon = 4 quarts = 8 pints.
If you substitute 8 pints for gallons, you can write the following equation.
14 pints = x(8 pints). Divide both sides by 8 pints to get x = 1.75 gallons.

Directions Complete each statement.

1. 5 ft = ____ in.
2. 13 yd = ____ ft
3. 56 in. = ____ ft
4. 7 mi = ____ yd
5. 58,080 ft = ____ mi
6. 78 gallons = ____ qt
7. 28 cups = ____ pints
8. 304 ounces = ____ lb
9. 15 cups = ____ fl oz
10. 90,000 lb = ____ tons
11. 66 kg = ____ g
12. 57 l = ____ ml
13. 92 cm = ____ mm
14. 843 cm = ____ m
15. 88 km = ____ cm
16. 85 ft = ____ in.
17. 36 yd = ____ ft
18. 1,008 in. = ____ ft
19. 12 mi = ____ yd
20. 290,400 ft = ____ mi

 Geometry

Workbook Activity 117

Name ____ Date ____ Period ____ Chapter 11 Additional Practice 73

Measurements

Directions Complete each statement.

1. 3 ft = ____ in.
2. 5 yd = ____ ft
3. 48 in. = ____ ft
4. 3 mi = ____ yd
5. 31,680 ft = ____ mi
6. 3 gallons = ____ qt
7. 12 cups = ____ pints
8. 128 ounces = ____ lb
9. 7 cups = ____ fl oz
10. 60,000 lb = ____ tons
11. 23 kg = ____ g
12. 34 l = ____ ml
13. 17 cm = ____ mm
14. 99 cm = ____ m
15. 45 km = ____ cm
16. 14 ft = ____ in.
17. 64 yd = ____ ft
18. 96 in. = ____ ft
19. 8 mi = ____ yd
20. 174,240 ft = ____ mi
21. 15 gallons = ____ qt
22. 242 cups = ____ pints
23. 384 ounces = ____ lbs
24. 41 cups = ____ fl oz
25. 444,000 lb = ____ tons
26. 49 kg = ____ g
27. 77 l = ____ ml
28. 152 cm = ____ mm
29. 375 cm = ____ m
30. 564 km = ____ cm

 Geometry

Additional Practice 73

The metric system also has units for capacity—the liter—and for mass—the gram. (Mass is the amount of matter in an object.)

With customary units you need to remember many different ratios for measurements of length, capacity, and weight. In the metric system, you need to remember the prefix that names the power of ten used in the measurement.

Prefix	milli-	centi-	deci-	Basic Unit	kilo-
power of ten	10^{-3} 0.001	10^{-2} 0.01	10^{-1} 0.1	10^{0} 1	10^{3} 1,000
example	millimeter (mm)	centimeter (cm)	decimeter (dm)	meter (m)	kilometer (km)
example	milliliter (ml)	centiliter (cl)	deciliter (dl)	liter (l)	kiloliter (kl)
example	milligram (mg)	centigram (cg)	decigram (dg)	gram (g)	kilogram (kg)

EXAMPLE A fund-raising race is advertised as a 5K race. The K stands for kilometers (usually abbreviated km). How many meters long is the race? How many centimeters?

1 km = 1,000 m, so 5 km = 5 • 1,000 m = 5,000 m

1 m = 100 cm, so 5,000 m = 5,000 • 100 cm = 500,000 cm

Exercise A Complete each statement.

	Answer		Answer
1. 12 yd = ___ in.	432	**6.** 13 cm = ___ mm	130
2. 14 pounds = ___ ounces	224	**7.** 15 kg = ___ g	15,000
3. 1 mi = ___ in.	63,360	**8.** 22 l = ___ ml	22,000
4. 8 pints = ___ gal	1	**9.** 10 km = ___ cm	1,000,000
5. 1 sq yd = ___ sq ft	9	**10.** 12.5 cm = ___ mm	125

2. Teaching the Lesson

LEARNING STYLES

LEP/ESL Allow students to work in groups of four to create a weights and measure chart that shows the relationship between the units of measure in one system. The chart should be designed so that the entries complete a sentence. For example,

to change	to	
feet	inches	divide by 12

3. Reinforce and Extend

APPLICATION

In the Community

Direct students to make a survey of items which they see or purchase on a regular basis. How many give measurements in the customary system only? How many in metric only? How many give both metric and customary measures? Students can present the results of their survey in a circle graph.

Lesson at a Glance

Chapter 11 Application

Overview This lesson introduces origami, the Japanese art of paper folding.

Objective

- To fold a 2-dimensional sheet of paper into a three-dimensional object.

Student Page 370

Teacher's Resource Library TRL

Workbook Activity 118

1. Warm-Up Activity

Ask students if they are familiar with origami. Invite them to present to the class any origami pieces they have made.

2. Teaching the Lesson

Tell students that the gray in the diagrams indicates the underside of the sheet of paper. Demonstrate the example for students. Paper that is dark on one side, light on the other will amplify the demonstration.

3. Reinforce and Extend

Learning Styles

Visual/Spatial Allow interested students to find examples of frogs, cranes, and other animals made by folding paper.

Application

Origami You already know that you can bend a sheet of paper to form a three-dimensional surface such as a cylinder or cone. You probably also realize that you can fold paper to make many geometric as well as fanciful three-dimensional figures. The Japanese art of paper folding, often called *origami*, offers a way to make hundreds of different figures each using only a single sheet of paper.

EXAMPLE Make a rectangular box following these eight steps.

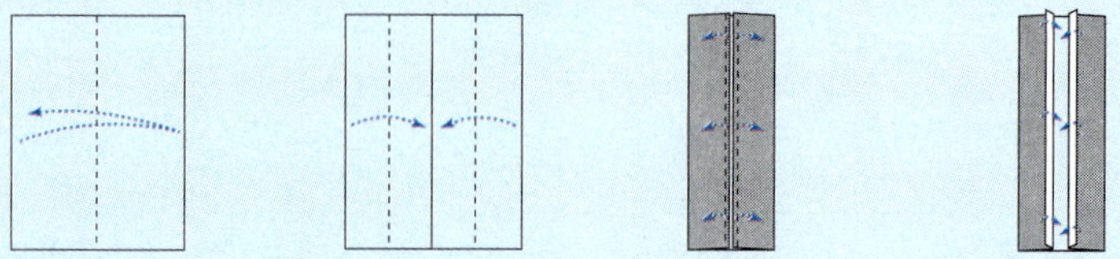

1. Fold paper in half, crease, and open.
2. Fold each edge to meet center crease.
3. Make a small outward fold along long edges. Crease.
4. Undo fold from Step 3.

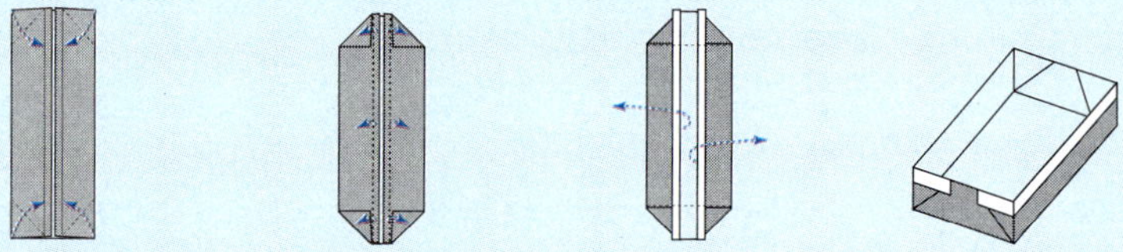

5. Make 45° fold at each corner.
6. Open fold from Step 3.
7. Lift flaps and form box.
8. You may need to square the corners.

Exercise Construct another rectangular box.

1. Find the area of the sheet of paper you plan to use for your box.
2. Fold the paper to make the box.
3. Find the outside surface area of your completed box.
4. Find the ratio of the surface area of the completed box to the surface area of the original sheet of paper.
5. Compare results with your classmates. Results will vary based on constructions of students.

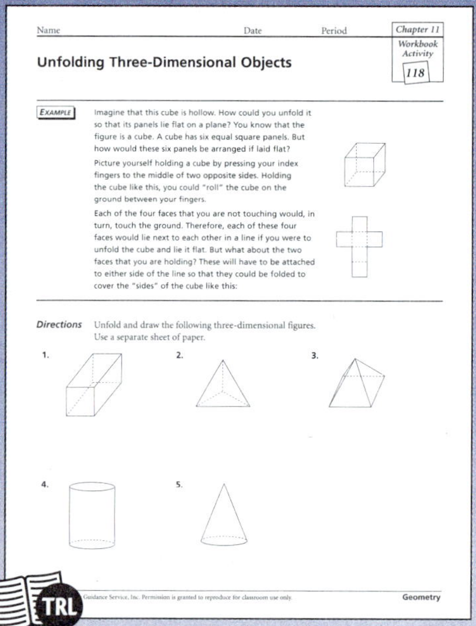

Name Date Period

Chapter 11 *Workbook Activity* *118*

Unfolding Three-Dimensional Objects

EXAMPLE Imagine that this cube is hollow. How could you unfold it so that its panels lie flat on a plane? You know that the figure is a cube. A cube has six equal square panels. But how would these six panels be arranged if laid flat?

Picture yourself holding a cube by pressing your index fingers to the middle of two opposite sides. Holding the cube like this, you could "roll" the cube on the ground between your fingers.

Each of the four faces that you are not touching would, in turn, touch the ground. Therefore, each of these four faces would lie next to each other in a line if you were to unfold the cube and lie it flat. But what about the two faces that you are holding? These will have to be attached to either side of the line so that they could be folded to cover the "sides" of the cube like this:

Directions Unfold and draw the following three-dimensional figures. Use a separate sheet of paper.

1.

2.

3.

4.

5.

 Geometry

Workbook Activity 118

Chapter 11 Review

Write the letter of the correct description.

Example: The volume of a rectangular solid equals

A. (area)^2 C. base area + areas of four sides

B. lwh D. $\frac{1}{2}$(area + height)

Solution: B; lwh

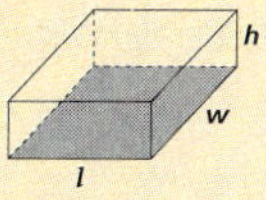

1. The volume of a cube equals B

A. (side)(side) **C.** base area + areas of four sides

B. $s \cdot s \cdot s$ **D.** $\frac{1}{2}$(area + height)

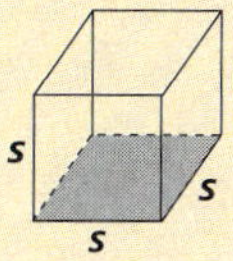

2. The surface area of a cube equals C

A. $(s^2)(h)$ **C.** $6s^2$

B. $4(s^2)(h)$ **D.** $6(s^2)(h)$

3. The volume of a triangular prism equals C

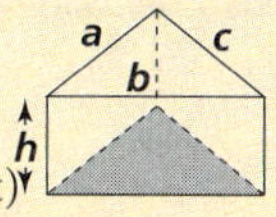

A. (base area)(areas of sides) **C.** (base area)(height)

B. $\frac{1}{3}$(base)(height) **D.** $\frac{1}{2}$(base)(height)

4. The surface of a triangular prism equals D

A. 2(base area)(areas of sides) **C.** $\frac{1}{2}$(base)(height)

B. $\frac{1}{3}$(base)(height) **D.** 2(base area)+ $h(a + b + c)$

5. The volume of a cylinder equals D

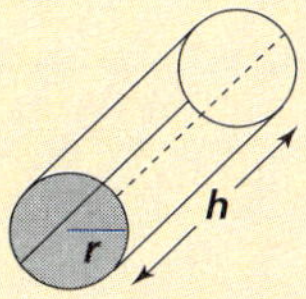

A. $(2\pi r^2)(h)$ **C.** $\pi r^2 + h$

B. $\frac{1}{3}$(base)(height) **D.** $(\pi r^2)(h)$

6. The surface area of a cylinder equals C

A. $(2\pi r^2)(h)$ **C.** $2\pi r^2 + 2\pi rh$

B. $\frac{1}{3}$(base)(height) **D.** $(\pi r^2)(h)$

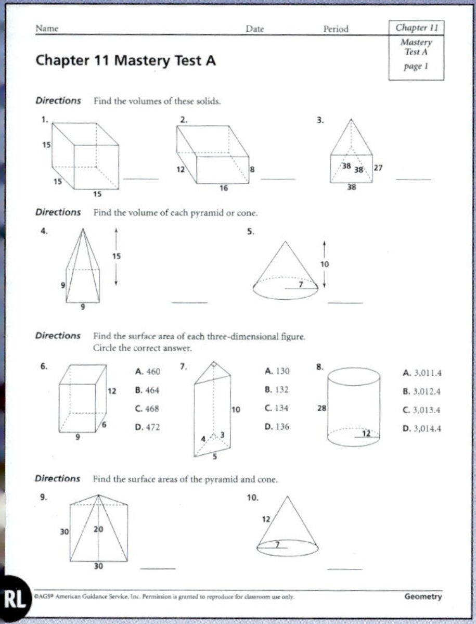

Name Date Period | *Chapter 11* *Mastery Test A* *page 1*

Chapter 11 Mastery Test A

Directions Find the volumes of these solids.

1. 2. 3.

Directions Find the volume of each pyramid or cone.

4. 5.

Directions Find the surface area of each three-dimensional figure. Circle the correct answer.

6. A. 460 B. 464 C. 468 D. 472

7. A. 130 B. 132 C. 134 D. 136

8. A. 3,011.4 B. 3,012.4 C. 3,013.4 D. 3,014.4

Directions Find the surface areas of the pyramid and cone.

9. 10.

©AGS® American Guidance Service, Inc. Permission is granted to reproduce for classroom use only. Geometry

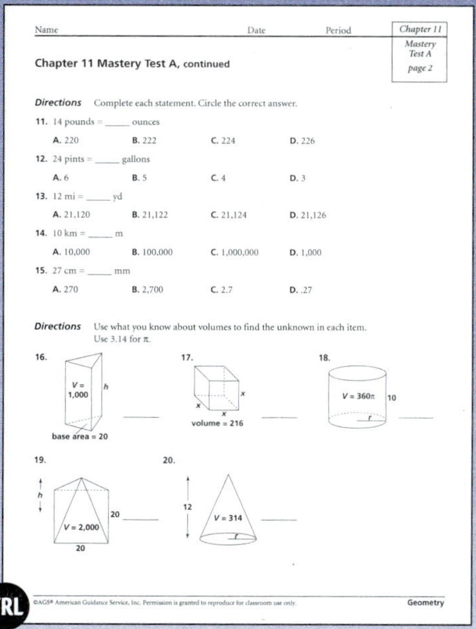

Name Date Period | *Chapter 11* *Mastery Test A* *page 2*

Chapter 11 Mastery Test A, continued

Directions Complete each statement. Circle the correct answer.

11. 14 pounds = _____ ounces
A. 220 B. 222 C. 224 D. 226

12. 24 pints = _____ gallons
A. 6 B. 5 C. 4 D. 3

13. 12 mi = _____ yd
A. 21,120 B. 21,122 C. 21,124 D. 21,126

14. 10 km = _____ m
A. 10,000 B. 100,000 C. 1,000,000 D. 1,000

15. 27 cm = _____ mm
A. 270 B. 2,700 C. 2.7 D. .27

Directions Use what you know about volumes to find the unknown in each item. Use 3.14 for π.

16. 17. 18.

19. 20.

©AGS® American Guidance Service, Inc. Permission is granted to reproduce for classroom use only. Geometry

Chapter 11 Mastery Test A

Chapter 11 Review

Each set of problems in the Chapter Review includes an example and solution to illustrate the concept. Use the given examples for reteaching the material in Chapter 11. For additional practice, refer to the Supplementary Problems for Chapter 11 (pages 416–417) and Additional Practice 69–73 in the Teacher's Resource Library.

Chapter 11 Mastery Test TRL

The Teacher's Resource Library includes parallel forms of the Chapter 11 Mastery Test. The difficulty level of the two forms is equivalent. You may wish to use one form as a pretest and the other form as a posttest. In addition, the Chapter 11 Cumulative Mastery Test covers content from Chapters 1–11.

Alternative Assessment

Alternative Assessment items correlate with student Goals for Learning at the beginning of this chapter.

- Have students identify the formulas for determining the volume or area of solids.

Write the formulas for finding the volume of a rectangular box, a cube, and a cylinder. Draw a diagram appropriate for each formula.

- Have students determine the volumes of solids.

Choose three regular, but differently shaped, solids in the classroom. Find the volume of each.

- Have students find the surface areas of solids.

Find the surface area of each of three solid forms found in the classroom.

Alternative Assessment

continued

- Have students use formulas to determine a missing dimension of a solid given its surface area or volume.

Use algebra to solve for the unknown in each of the following volume formulas. Draw the solid for each of the formulas.

$V = \frac{1}{2}bh$ for b $\quad \frac{2V}{h}$

$V = \pi r^2 h$ for r $\quad \sqrt{\frac{V}{\pi h}}$

$V = \frac{1}{3}s^2 h$ for s $\quad \sqrt{\frac{3v}{h}}$

$V = \frac{1}{3}\pi r^2 h$ for h $\quad \frac{3V}{\pi r^2}$

- Have students use customary and metric units of measurement.

Measure the length of four items in the classroom and measure the capacity of two items. Give the measurements in both customary and metric units.

More Review...

7. The volume of a cone equals A

A. $\frac{1}{3}(\pi r^2)(h)$ **B.** $(2\pi r^2)(h)$

C. (base area)(circumference)

D. (circumference)(slant height)

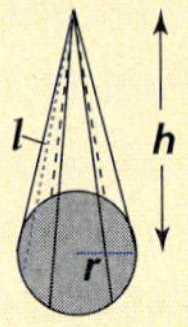

8. The surface area of a cone equals D

A. $(2\pi r^2)(h)$ **C.** $\frac{1}{3}(\pi r^2)(h)$

B. (base area)(circumference) **D.** $\pi r^2 + \frac{1}{2}(2\pi rl)$

9. The volume of a square pyramid equals D

A. $(s^2)(h)$ **C.** (base area)(areas of sides)

B. $4(s^2)(h)$ **D.** $\frac{1}{3}(s^2)(h)$

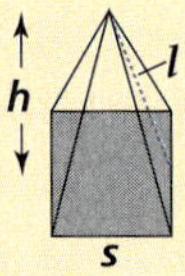

10. The surface of a square pyramid equals B

A. $(s^2)(h)$ **C.** (base area)(areas of sides)

B. $(s^2) + 4(\frac{1}{2}bl)$ **D.** $\frac{1}{3}(s^2)(h)$

Find the volume and total surface area for each figure. You may give your answers in terms of π.

Example:

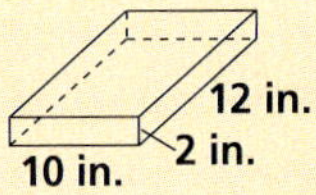

Solution: Volume

$l \bullet w \bullet h = 12 \bullet 10 \bullet 2 = 240$ cu in.

Total Surface Area

$2(lw \bullet hl \bullet hw) = 2(10 \bullet 12 + 2 \bullet 12 + 2 \bullet 10)$

$2(120 + 24 + 20) = 2(164) = 328$ sq in.

11. Volume 72,000 cm³

12. Total Surface Area 12,000 cm²

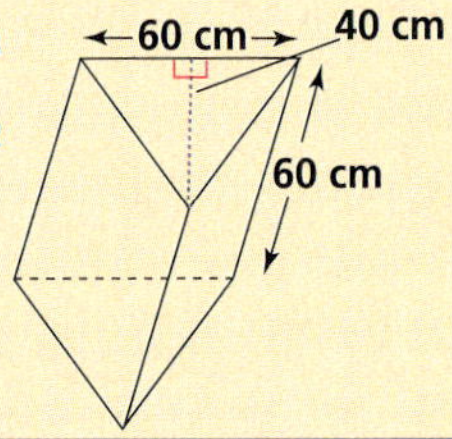

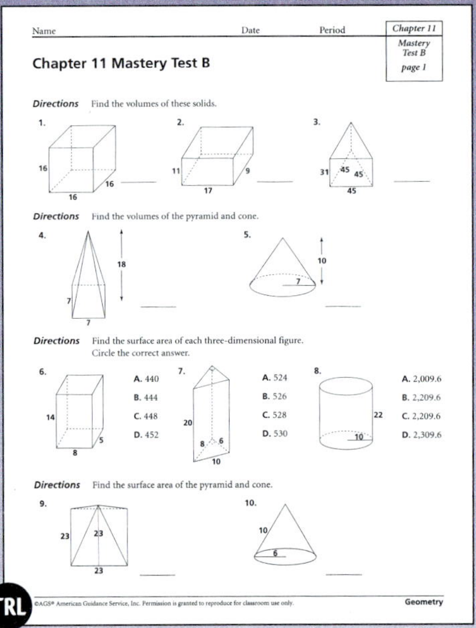

Name Date Period | *Chapter 11 Mastery Test B page 1*

Chapter 11 Mastery Test B

Directions Find the volumes of these solids.

1. (16, 16, 16) ______ 2. (11, 9, 17) ______ 3. (31, 45, 45, 45) ______

Directions Find the volumes of the pyramid and cone.

4. (18, 7, 7) ______ 5. (10, 7) ______

Directions Find the surface area of each three-dimensional figure. Circle the correct answer.

6. (14, 5, 8) **A.** 440 **B.** 444 **C.** 448 **D.** 452

7. (20, 8, 6, 10) **A.** 524 **B.** 526 **C.** 528 **D.** 530

8. (22, 10) **A.** 2,009.6 **B.** 2,209.6 **C.** 2,209.6 **D.** 2,309.6

Directions Find the surface area of the pyramid and cone.

9. (23, 23, 23) ______ 10. (10, 6) ______

TRL ©AGS® American Guidance Service, Inc. Permission is granted to reproduce for classroom use only. **Geometry**

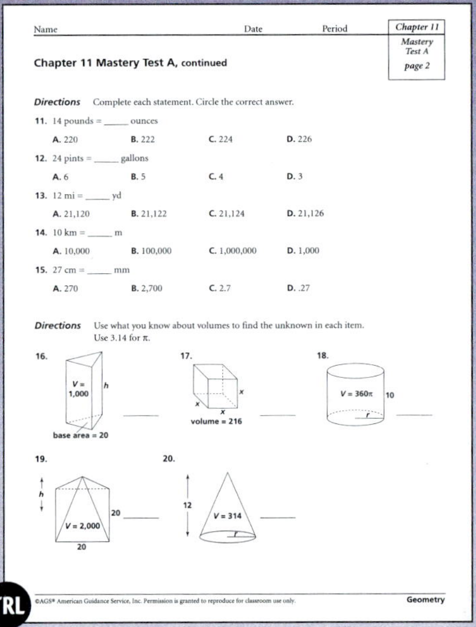

Name Date Period | *Chapter 11 Mastery Test A page 2*

Chapter 11 Mastery Test A, continued

Directions Complete each statement. Circle the correct answer.

11. 14 pounds = _____ ounces

A. 220 **B.** 222 **C.** 224 **D.** 226

12. 24 pints = _____ gallons

A. 6 **B.** 5 **C.** 4 **D.** 3

13. 12 mi = _____ yd

A. 21,120 **B.** 21,122 **C.** 21,124 **D.** 21,126

14. 10 km = _____ m

A. 10,000 **B.** 100,000 **C.** 1,000,000 **D.** 1,000

15. 27 cm = _____ mm

A. 270 **B.** 2,700 **C.** 2.7 **D.** .27

Directions Use what you know about volumes to find the unknown in each item. Use 3.14 for π.

16. (V = 1,000, h, base area = 20) ______ 17. (x, x, x, volume = 216) ______ 18. ($V = 360\pi$, 10, r) ______

19. (h, V = 2,000, 20, 20) ______ 20. (12, V = 314, r) ______

TRL ©AGS® American Guidance Service, Inc. Permission is granted to reproduce for classroom use only. **Geometry**

Chapter 11 Mastery Test B

13. Volume 1536π cu ft

14. Total Surface Area 512π sq ft

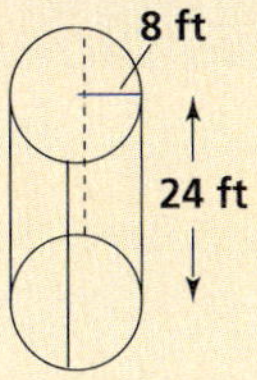

15. Volume 270π cu in.

16. Total Surface Area 202.5π sq in.

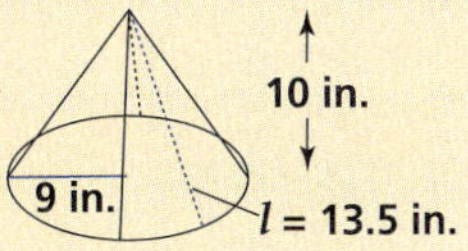

Use what you know about surface areas and volumes to find the unknowns.

17. 10 ft, 120 in.
18. 5.8 cm
19. 10 cm, 1 dm
20. 20 units

Example: The topless box has a surface area of 125 sq in. What is the length of a side, *s*, of the box?
Solution: The topless box has 5 sides. Divide the area by 5. 125 ÷ 5 = 25. Take the square root of 25 to find the length of side *s*. $\sqrt{25} = 5$ in.

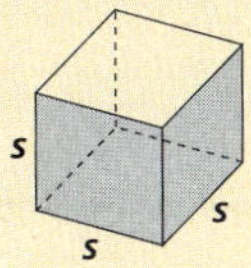

17. A rectangular box with length 6 ft and width 6 ft has a volume of 360 cubic ft. What is the height of the box in feet? What is the height in inches?

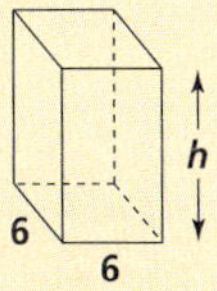

18. The volume of a cylinder is 600π cm^3, its height is 18 cm. What is the approximate radius of the cylinder?

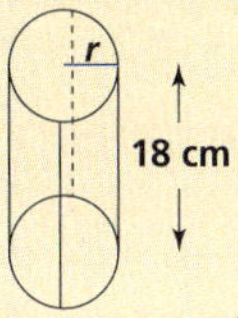

19. The volume of a cone is 600π cm^3, its height is 18 cm. What is the radius of the cone in cm? What is the radius in dm?

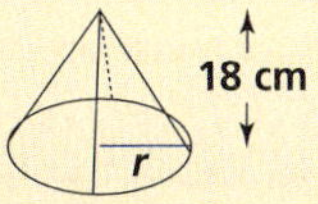

20. The surface area of a triangular pyramid is 692 sq units. The height of each triangle is 17.3 units. How many units is the base of each face of the pyramid?

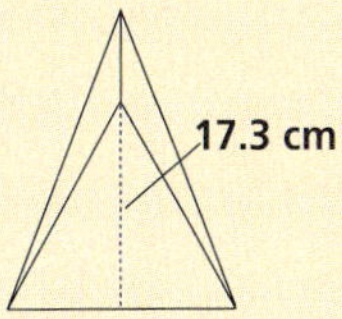

Test-Taking Tip

Before you begin a test, look it over quickly. Try to set aside enough time to complete each section.

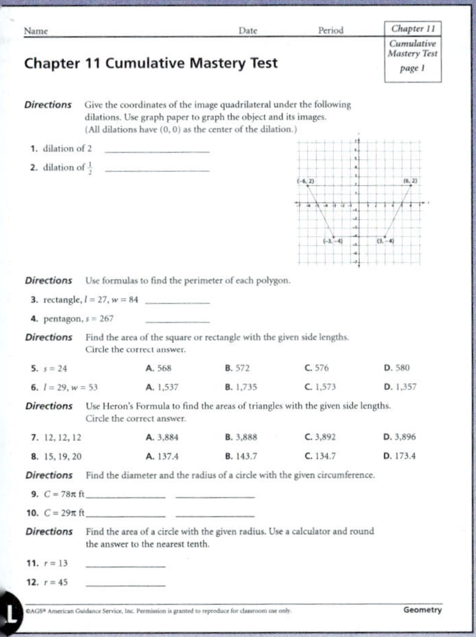

Name Date Period

Chapter 11 Cumulative Mastery Test page 1

Chapter 11 Cumulative Mastery Test

Directions Give the coordinates of the image quadrilateral under the following dilations. Use graph paper to graph the object and its images. (All dilations have (0, 0) as the center of the dilation.)

1. dilation of 2 ________
2. dilation of $\frac{1}{2}$ ________

Directions Use formulas to find the perimeter of each polygon.

3. rectangle, $l = 27$, $w = 84$ ________
4. pentagon, $s = 267$ ________

Directions Find the area of the square or rectangle with the given side lengths. Circle the correct answer.

5. $s = 24$	A. 568	B. 572	C. 576	D. 580
6. $l = 29$, $w = 53$	A. 1,537	B. 1,735	C. 1,573	D. 1,357

Directions Use Heron's Formula to find the areas of triangles with the given side lengths. Circle the correct answer.

7. 12, 12, 12	A. 3,884	B. 3,888	C. 3,892	D. 3,896
8. 15, 19, 20	A. 137.4	B. 143.7	C. 134.7	D. 173.4

Directions Find the diameter and the radius of a circle with the given circumference.

9. $C = 78\pi$ ft ________ ________
10. $C = 29\pi$ ft ________ ________

Directions Find the area of a circle with the given radius. Use a calculator and round the answer to the nearest tenth.

11. $r = 13$ ________
12. $r = 45$ ________

 Geometry

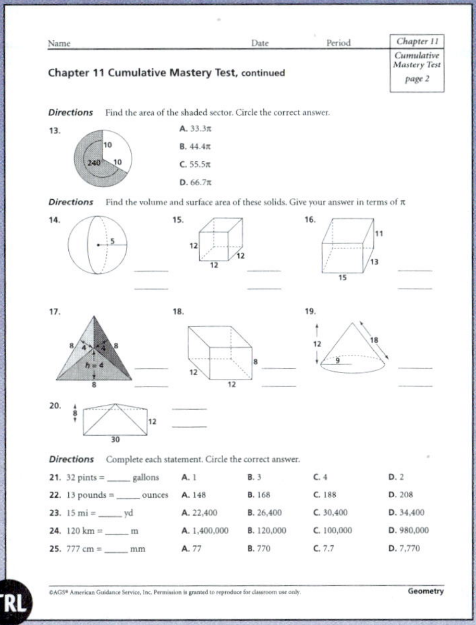

Name Date Period

Chapter 11 Cumulative Mastery Test page 2

Chapter 11 Cumulative Mastery Test, continued

Directions Find the area of the shaded sector. Circle the correct answer.

13. A. 33.3π B. 44.4π C. 55.5π D. 66.7π

Directions Find the volume and surface area of these solids. Give your answer in terms of π

14. ________ ________
15. ________ ________
16. ________ ________
17. ________ ________
18. ________ ________
19. ________ ________
20. ________ ________

Directions Complete each statement. Circle the correct answer.

21. 32 pints = ____ gallons	A. 1	B. 3	C. 4	D. 2
22. 13 pounds = ____ ounces	A. 148	B. 168	C. 188	D. 208
23. 15 mi = ____ yd	A. 22,400	B. 26,400	C. 30,400	D. 34,400
24. 120 km = ____ m	A. 1,400,000	B. 120,000	C. 100,000	D. 980,000
25. 777 cm = ____ mm	A. 77	B. 770	C. 7.7	D. 7,770

 Geometry

L TRL

Chapter 11 Cumulative Mastery Test

Chapter 12

Planning Guide

Geometry and Imagination

	Student Text Lesson						
	Student Pages	Vocabulary	Proof	Practice Exercises	Problem Solving	Calculator Practice	Solutions Key
Lesson 1 Lines and Planes in Space	376–378			✔	378		✔
Lesson 2 Loci in the Coordinate Plane	379–382			✔		382	✔
Lesson 3 Compound Loci	383–385			✔			✔
Application Spirals	386			✔			✔

Chapter Activities

Teacher's Resource Library

Geometry on the Job 12: Mathematicians

Community Connection 12: Meeting Places

Teacher's Edition

Chapter 12 Project

Assessment Options

Student Text

Chapter 12 Review

Teacher's Resource Library

Chapter 12 Mastery Tests A and B

Chapter 12 Cumulative Mastery Test

Chapters 1–12 Final Mastery Test

Teacher's Edition

Chapter 12 Alternative Assessment

Student Text Features						Teaching Strategies								Learning Styles						Teacher's Resource Library		
Algebra Connection	Constructions	Experiencing Geometry	Technology Connection	Try This	Writing About Mathematics	Warm-Up Activity	Common Error	Algebra Connection	Mental Math	Manipulatives	Calculator	Applications	Group Problem Solving	Auditory/Verbal	Visual/Spatial	Body/Kinesthetic	Logical/Mathematical	Interpersonal/Group Learning	LEP/ESL	Workbook Activities	Additional Practice	Self-Study Guide
		377			377	376						378			377			376	378	119	74	✔
379			382			379		379			382		381	381						120	75	✔
	384			385		383										385				121	76	✔
						386						386			386		386			122		✔

Chapter at a Glance

Chapter 12: Geometry and Imagination pages 374–389

(Answer Keys for the Teacher's Resource Library begin on page 484 of this Teacher's Edition.)

Name Date Period | Chapter 12 *Geometry on the Job* 12

Mathematicians

The study of mathematics can lead to many different careers. Algebra, geometry, trigonometry, and calculus are used in many jobs. Some mathematicians get their undergraduate degree and become teachers at the high school level and with advanced degrees can become teachers at the college and university level.

Many mathematicians work in business, government, or industry. Some become accountants, statisticians, or computer operators. Mathematicians can work as computer programmers or system analysts. Insurance companies hire mathematicians to calculate tables for various costs for policies and life expectancy charts. Other mathematicians use their training to conduct research for transportation, energy, manufacturing, or communications industries.

Try It Yourself
Look around you. Where do you see mathematics at work. Who uses the basic principles of mathematics in your school, at home, in the community, and around the world? How does this affect your life? What changes does it create?

Make a chart about mathematics and mathematicians in your world. Tell what mathematics is used, who uses it, and why.

TRL Geometry

Geometry on the Job 12

Name Date Period | Chapter 12 *Community Connection* 12

Meeting Places

The concept of a locus of points may seem rather theoretical to you. However, you use similar ideas all the time without realizing it. For instance, let's say that you and your friends are planning a trip to a game and will be driving in one car. Everyone lives in different parts of town, and it would take a lot of time for the person who will be driving to drive to everyone's house and pick everyone up. So you decide to meet at the person's house who is most central and easy for everyone else to get to. If you were to draw a large circle around each house on a map, this meeting spot would be in the middle of where most of the circles converge. Maybe no one in the group actually lives in this exact spot, but there will be a convenient place to meet.

Directions
Choose several friends and pretend that the group is to meet at a central location for a field trip. Draw the circles on a map of your community and decide where the central location will be. Repeat the process with other groups to find the central location.

TRL Geometry

Community Connection 12

Chapter 12

Geometry and Imagination

The fair comes to town. You can feel the excitement. Workers raise the Ferris wheel and other rides. You're looking forward to a weekend of fun. Take a close look. You'll see geometry in action. The Ferris wheel is a good example. The wheel is made up of two large parallel circles, each in its own plane. Radii span from a central point and result in the equidistant loci of the circles. The wheel's cars dangle between the two outer circles. They connect to each circle at only one point. As a result, they can be considered tangent to each one. Climb aboard a Ferris wheel car. As the wheel turns, you can enjoy a bird's-eye view of your community and the fair. Everywhere you look, you will see geometry as part of a festival of fun and excitement.

In Chapter 12, you will learn to imagine the ways geometric figures intersect in a plane and in space.

Goals for Learning

- To identify how geometric figures intersect in space
- To write equations for loci
- To construct simple and compound loci
- To describe compound loci

375

Introducing the Chapter

Use the photograph to introduce students to the concept of loci, the location of points that fulfill specific requirements. Ask students who have experienced a Ferris wheel ride to describe the ride. Ask them to describe their motion. Indicate that they could use geometric terms to describe their motion. Ease students to the discovery that they traveled in a circle around the center (hub) of the Ferris wheel. At every point in their ride they were a constant distance from the center of the Ferris wheel. Lead them to indicate that this distance was the radius of the wheel. Point out that in this chapter students will study more about geometric figures that are described by the location of points. Read the Goals for Learning and discuss each one with students.

Chapter Project

Groups of students can work together to design a board game based on the descriptions of the interaction of geometric figures. The game must include descriptions of geometric figures, several kinds of interactions, and the results of the interactions. Encourage students to use their creativity to determine how the descriptions are used and how players advance along a board. Of course, game rules must also specify the conditions for winning.

SELF-STUDY GUIDE

Name ______

CHAPTER 12: Geometry and Imagination

Goal 12.1 *To identify how geometric figures intersect in space*

Date	Assignment	Score
______	1: Read pages 376–377. Complete Exercises A–C on pages 377–378.	______
______	2: Complete Workbook Activity 119.	______

Comments:

Goal 12.2 *To write equations for loci*

Date	Assignment	Score
______	3: Read pages 379–381. Complete Exercises A and B on pages 381–382.	______
______	4: Read the Calculator Practice and complete Exercise C on page 382.	______
______	5: Complete Workbook Activity 120.	______

Comments:

 Geometry

SELF-STUDY GUIDE

Name ______

CHAPTER 12: Geometry and Imagination, continued

Goals 12.3 and 12.4 *To construct simple and compound loci; to describe compound loci*

Date	Assignment	Score
______	6: Read pages 383–384. Complete Exercises A and B on pages 384–385.	______
______	7: Complete Workbook Activity 121.	______
______	8: Read and complete the Application Exercise on page 386.	______
______	9: Complete Workbook Activity 122.	______
______	10: Complete the Chapter 12 Review on pages 387–389.	______

Comments:

Student's Signature ______ Date ______

Instructor's Signature ______ Date ______

 Geometry

TRL

Chapter 12 Self-Study Guide

Lesson at a Glance

Chapter 12 Lesson 1

Overview This lesson introduces students to the figures formed when a sphere and a line interact.

Objectives

- To visualize the possible interactions of a line and a sphere and a plane and a sphere.
- To draw or sketch selected three-dimensional intersections.

Student Pages 376–378

Teacher's Resource Library TRL

Workbook Activity 119

Additional Practice 74

1. Warm-Up Activity

Begin the lesson by recreating the interactions of circles and lines as shown on the right of page 376. If feasible, model the interactions using a long dowel or rod and a flat paper ring. Model the interactions a second time replacing the flat paper ring with a flat paper circle.

Learning Styles

Interpersonal/Group Learning Have students work in groups of three or four to plan, devise, and build models. The models should show the loci of points equidistant from a specific point in space and intersections of lines and spheres or lines and cylinders in space. Encourage students to be creative in the use of the materials they use. Each group should provide a written description of each model and the plan for devising it. Ask students to present their models and explain how they serve as examples of specific intersections.

Lesson 1 Lines and Planes in Space

Remember from Chapter 10 that the locus of points *describes the location of points. The plural of* locus *is* loci.

You may find it easy to draw geometric figures in a plane. However, it is more difficult to show three-dimensional figures with a drawing on a flat sheet of paper. To understand some geometric figures you need to be able to imagine how those figures look—and how they act in space.

In the Plane

Recall that the *locus of points* at equal distances from a given point is a *circle*. The center is the given point; *r*, the radius, is the given distance.

In Space

The locus of points at equal distances from a given point is a *sphere*. The center is the given point; *r*, the radius, is the given distance.

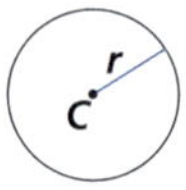

Circle

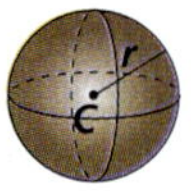

Sphere

Now, think about any line. In how many ways can the circle—or the sphere—and the line relate to one another?

No intersection

Tangent at a single point

Intersects in two points

If the chord formed equals the diameter of the circle, you know the line must pass through the center of the circle.

If the chord formed equals the diameter of the sphere, you know the line must pass through the center of the sphere.

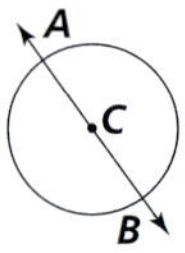

$\overline{AB}$ = diameter

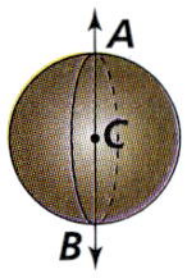

$\overline{AB}$ = diameter
A and B are poles.

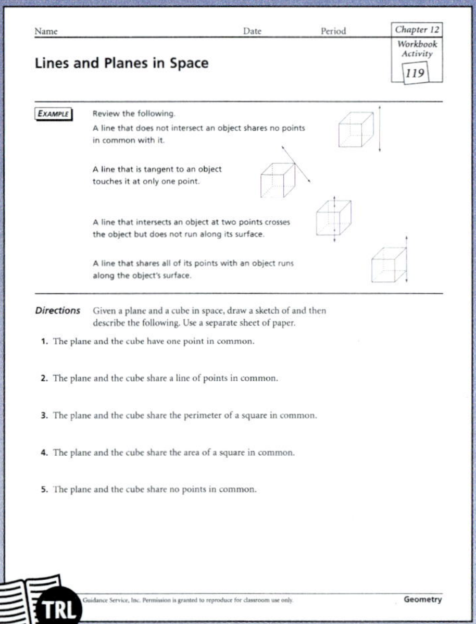

Name Date Period

Chapter 12 Workbook Activity 119

Lines and Planes in Space

EXAMPLE Review the following.

A line that does not intersect an object shares no points in common with it.

A line that is tangent to an object touches it at only one point.

A line that intersects an object at two points crosses the object but does not run along its surface.

A line that shares all of its points with an object runs along the object's surface.

Directions Given a plane and a cube in space, draw a sketch of and then describe the following. Use a separate sheet of paper.

1. The plane and the cube have one point in common.
2. The plane and the cube share a line of points in common.
3. The plane and the cube share the perimeter of a square in common.
4. The plane and the cube share the area of a square in common.
5. The plane and the cube share no points in common.

Geometry

Workbook Activity 119

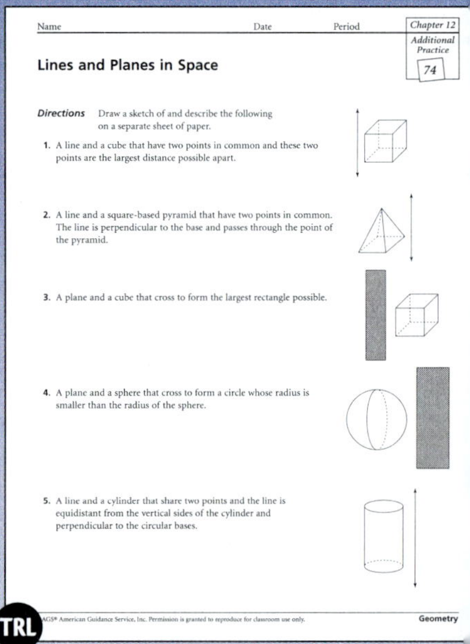

Name Date Period

Chapter 12 Additional Practice 74

Lines and Planes in Space

Directions Draw a sketch of and describe the following on a separate sheet of paper.

1. A line and a cube that have two points in common and these two points are the largest distance possible apart.
2. A line and a square-based pyramid that have two points in common. The line is perpendicular to the base and passes through the point of the pyramid.
3. A plane and a cube that cross to form the largest rectangle possible.
4. A plane and a sphere that cross to form a circle whose radius is smaller than the radius of the sphere.
5. A line and a cylinder that share two points and the line is equidistant from the vertical sides of the cylinder and perpendicular to the circular bases.

Geometry

Additional Practice 74

You have visualized how a line and other geometric figures intersect. Now, try to imagine, or visualize, how a plane and a sphere might intersect.

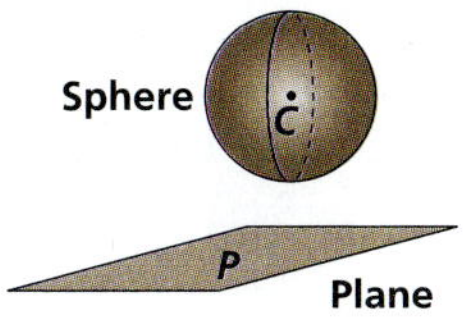

No intersection

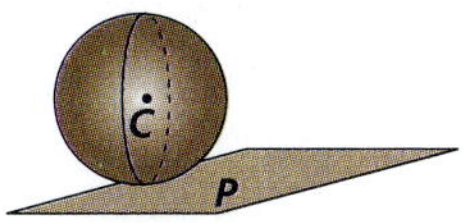

Tangent at a single point

Intersection is a circle. The center of the circle of intersection is not the center of the sphere, so the circle is not a great circle.

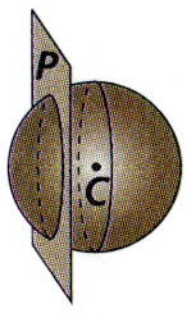

A plane passing through the center of a sphere intersects in a great circle. The center of the circle of intersection is the center of the sphere, and both lie in plane *P*, so the circle is a great circle. A great circle divides the sphere into two hemispheres.

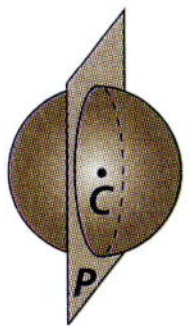

Exercise A Given a cylinder and a line in space, draw a sketch and then describe the following. (You may wish to use an empty can and a pencil to model each item.)

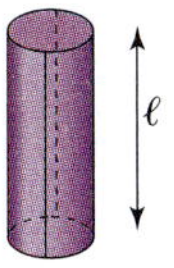

1. The line and the cylinder have one point in common.
   The line is tangent to the cylinder.
2. The line and the cylinder have two points in common.
   The line passes through the cylinder.
3. The line is part of the cylinder, completely.
   The line lies on the surface of the cylinder.
4. The line is at a fixed distance from the cylinder. The distance is larger than the radius of the cylinder.
   The locus is a new cylinder at a fixed distance from the given cylinder.

Remember that a sphere is the locus of points, in space, at equal distances from a given point.

Experiencing Geometry

Try the following intersections. You will need a round ball of soft clay for the sphere. You can use a piece of cardboard as the plane. Use your sphere and plane to model each of these possibilities: not touching, tangent, intersecting to form a circle, and intersecting to form a great circle.

☞ *Selected Answers and Solutions begin on page 420.*

Writing About Mathematics

An arrow hits a circular target at a single point. Is this an example of a line tangent to a circle? Explain why or why not.

2. Teaching the Lesson

Model the interactions shown on pages 376 and 377 using a log rod or dowel to represent a line, and a sheet of cardboard or poster board to represent the plane. Extruded-foam balls (Styrofoam) can be used to represent the sphere. Indicate to students that they can use these materials to model some of the interactions of geometric figures, but they will have to imagine or visualize other interactions.

Learning Styles

Visual/Spatial Allow groups of four to work together to complete the *Experiencing Geometry* activity. Groups can label and display their results. The class can consider each group's efforts during a walk-around critique. Groups should be allowed time to revise their models as necessary.

Writing About Mathematics

Because the target circle is a plane, an arrow hitting the target represents a point in the plane rather than a tangent to a circle.

3. Reinforce and Extend

Learning Styles

LEP/ESL Direct students to work with a partner. Have the partners use an empty juice can and a sheet of paper to model items 5–10. One student can create the model, and the second student can form the description of the model. Students should exchange roles after each item.

Application

Career Connection

Chefs and their assistants use a number of cutting styles for vegetables, meats, and desserts they serve. For example, green beans cut in the French style have a resulting oval-shaped cut whereas straight-cut beans have a circular cut. Arrange for a visit to a local restaurant or catering service where a chef can demonstrate how different foods are cut at various angles. Encourage students to ask about the resulting shapes and the reasons why foods are cut differently.

5. The plane makes a vertical slice through the cylinder.
6. The plane makes a horizontal slice through the cylinder.
7. The plane must "rest against" the side of the cylinder.
8. No, the plane needs to intersect with an edge of the cylinder.

Exercise B Given a cylinder and a plane in space, draw a sketch and then describe the following. (You may wish to use an empty can and a sheet of paper to model each item.)

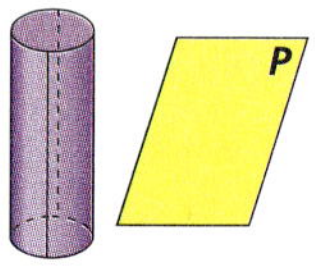

5. Describe how the intersection can be a pair of parallel lines.

6. How can the plane and the cylinder intersect to form a circle?

7. How can the plane and the cylinder meet to form a single straight line?

8. If the cylinder is infinitely long, can the cylinder and the plane ever meet in just one point? Explain.

Problem Solving

Exercise C Solve the following problems.

9. A violent tornado sends a two-by-four board through a tree trunk. If the board were a line, describe the situation using geometric terms.

A line passed through a circle forming a chord.

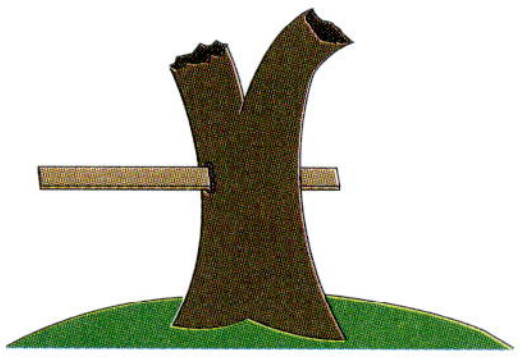

10. The illustration shows several beads strung on a wire. What geometric figures can be used to describe the illustration? Where are the holes for the wire located?

A line, spheres, and cylinders; The wire passes through the middle of the cylinders and forms the diameter of the spheres.

Geometry Around Us

Look around you for examples of the interaction of geometric figures. For instance, a knife (plane) is used to cut a cucumber (cylinder) to form slices (circles).

Lesson 2 Loci in the Coordinate Plane

You can use the idea of loci, the location of points, to solve many problems in algebra. Here are some examples.

Algebra Connection

EXAMPLES

Given: line $y = 3$

Find: the locus of points one unit away from $y = 3$

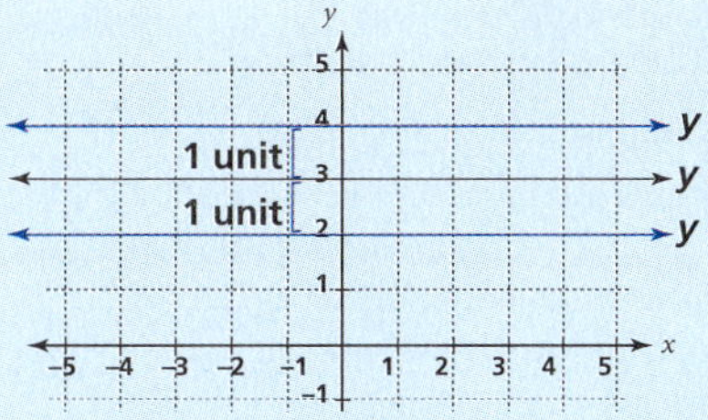

The locus will be two parallel lines. Each line will be a distance of 1 unit from $y = 3$. The equations of the lines are $y = 4$ and $y = 2$.

Given: line $y = -x + 4$

Find: the locus of points equidistant from the given line and passing through $y = 3$ and $y = 5$

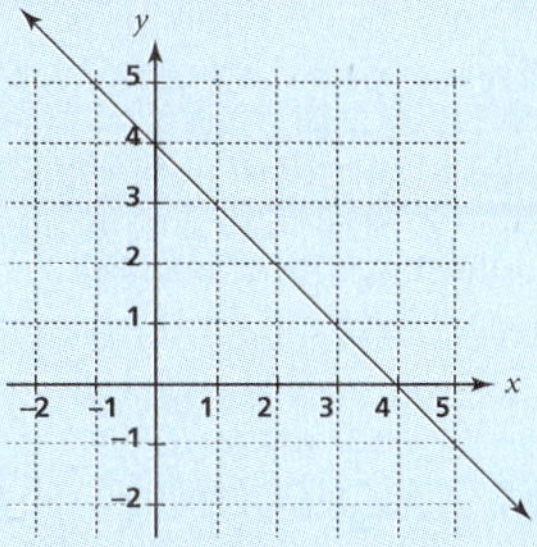

The locus will be two parallel lines—one passing through a point one unit above the y-intercept of 4 and one passing one unit below the y-intercept of 4. Because the lines are parallel, they have equal slope. Therefore, the lines are

$y = -x + 5$ one unit above the given line

$y = -x + 4$ the given line

$y = -x + 3$ one unit below the given line

Make a sketch to help you check whether your answer is reasonable.

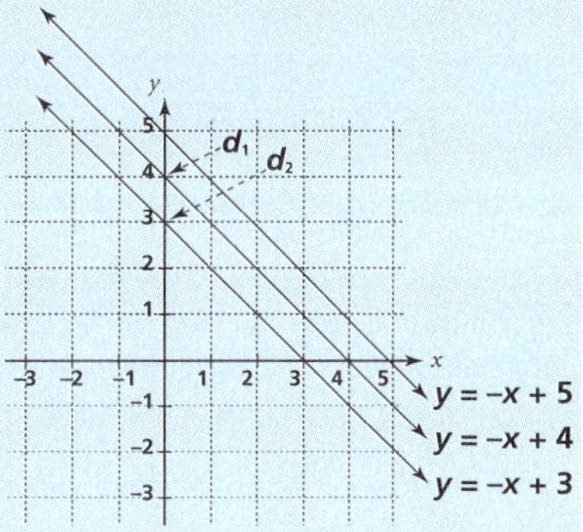

Note: The perpendicular distances, d_1 and d_2, are equal because the two right triangles formed are congruent. (AAS: right angles are equal; vertical angles 1 and 2 are equal; distances along y-axis are equal.)

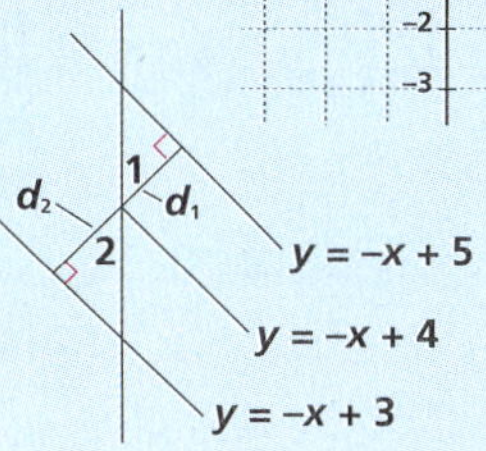

Lesson at a Glance

Chapter 12 Lesson 2

Overview This lesson models for students how to solve problems using the loci and the coordinate plane.

Objectives

- To represent loci on the coordinate plane.

Student Pages 379–382

Teacher's Resource Library TRL

Workbook Activity 120

Additional Practice 75

1. Warm-Up Activity

Remind students that the equation of a line is usually given in the form $y = mx + b$, where m gives the slope of the line and b is the y-intercept.

ALGEBRA CONNECTION

Review with students some of the characteristics of the equations of lines: parallel lines have the same slope, if x = any number, the line is parallel to the x-axis, if y = any number, the line is parallel to the y-axis.

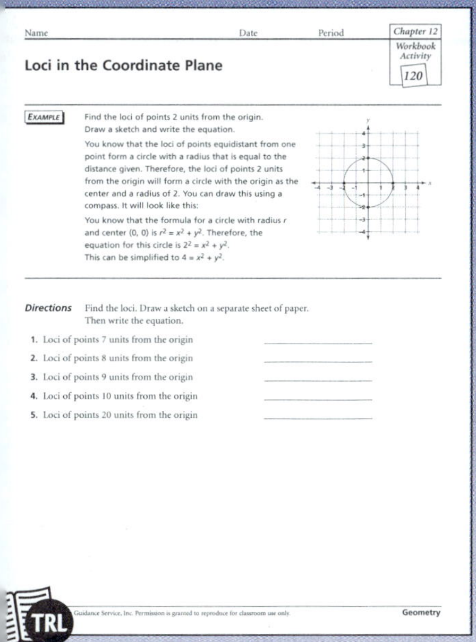

Name Date Period | Chapter 12 Workbook Activity 120

Loci in the Coordinate Plane

EXAMPLE Find the loci of points 2 units from the origin. Draw a sketch and write the equation.

You know that the loci of points equidistant from one point form a circle with a radius that is equal to the distance given. Therefore, the loci of points 2 units from the origin will form a circle with the origin as the center and a radius of 2. You can draw this using a compass. It will look like this:

You know that the formula for a circle with radius r and center (0, 0) is $r^2 = x^2 + y^2$. Therefore, the equation for this circle is $2^2 = x^2 + y^2$. This can be simplified to $4 = x^2 + y^2$.

Directions Find the loci. Draw a sketch on a separate sheet of paper. Then write the equation.

1. Loci of points 7 units from the origin ____
2. Loci of points 8 units from the origin ____
3. Loci of points 9 units from the origin ____
4. Loci of points 10 units from the origin ____
5. Loci of points 20 units from the origin ____

Geometry

TRL

Workbook Activity 120

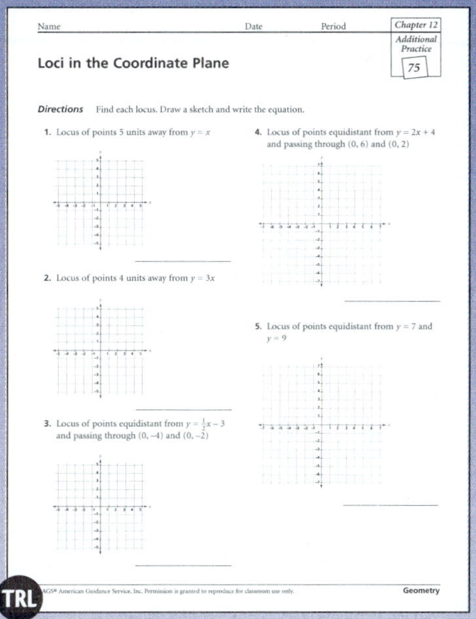

Name Date Period | Chapter 12 Additional Practice 75

Loci in the Coordinate Plane

Directions Find each locus. Draw a sketch and write the equation.

1. Locus of points 5 units away from $y = x$
2. Locus of points 4 units away from $y = 3x$
3. Locus of points equidistant from $y = \frac{1}{2}x - 3$ and passing through (0, −4) and (0, −2)
4. Locus of points equidistant from $y = 2x + 4$ and passing through (0, 6) and (0, 2)
5. Locus of points equidistant from $y = 7$ and $y = 9$

Geometry

TRL

Additional Practice 75

2. Teaching the Lesson

Work through the examples on pages 379 and 380. Invite student volunteers to demonstrate the examples on page 380. Be sure students understand how to find the midpoint of the perpendicular line connecting the two y-intercepts.

In the second example, ask students how they could determine that $(\frac{1}{2}, 4)$ is equidistant from $(3, 4)$ and $(-2, 4)$. How could they determine that $(\frac{1}{2}, 7)$ is equidistant from the two points? **(Count over and up and see if the counts are the same or use the distance formula.)**

You can use the information in the note on page 379 to solve a related problem.

EXAMPLES

Given: two parallel lines: $y = 2x + 1$ and $y = 2x + 5$

Find: the locus of points equidistant from the two lines

First, find the midpoint of the perpendicular line connecting $y = 1$ and $y = 5$. That midpoint is $y = 3$. Then write an equation for the line parallel to the given lines and passing through $y = 3$.

$y = 2x + 3$ is the locus.

Make a sketch to check if your answer is reasonable.

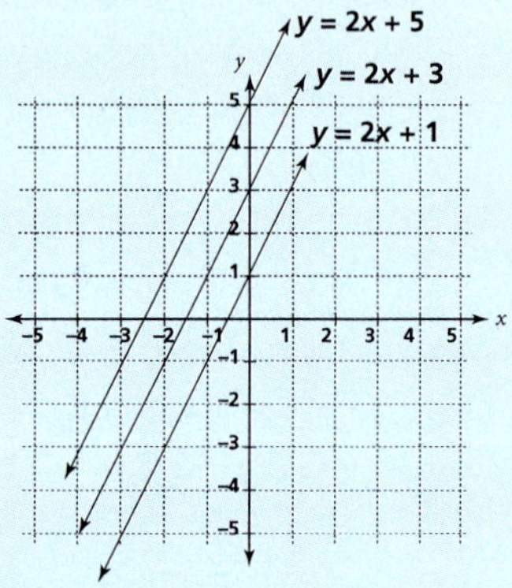

Given: the pair of points (3, 4) and (–2, 4) and the line segment connecting the two points

Write the equation for the perpendicular bisector to the line segment.

$x = \frac{1}{2}$ is the locus.

Make a sketch to check if your answer is reasonable.

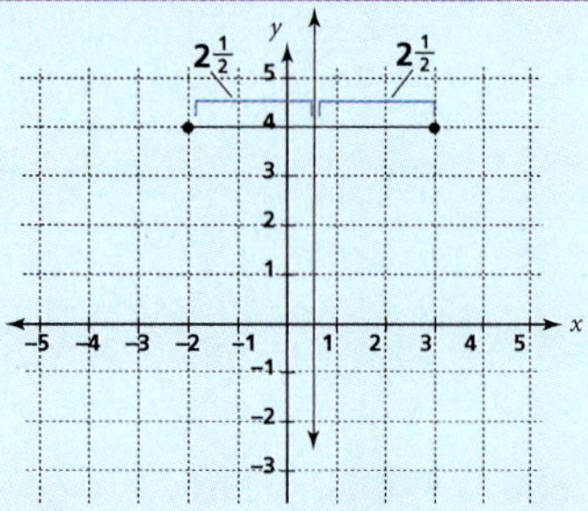

What is the locus of points 5 units from the origin in the coordinate plane?

From your knowledge of geometric definitions, you know that the locus is a circle with center (0, 0) and a radius of 5. You can find the equation for the circle using algebra.

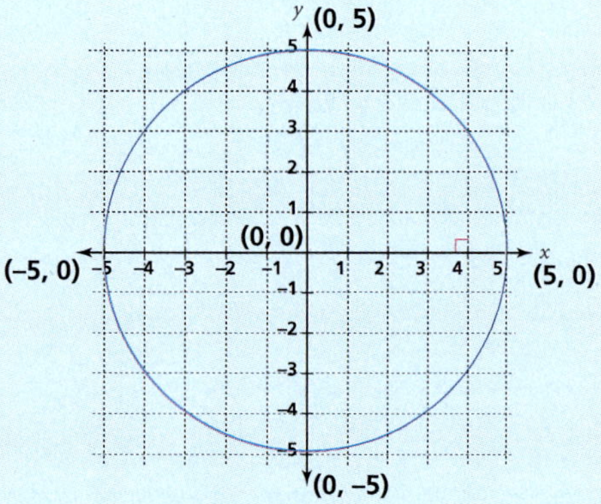

EXAMPLES *(continued)*

Let (x, y) be any point on the circle. Using the Pythagorean Theorem you know that $5^2 = x^2 + y^2$. This is the equation of the locus, a circle with radius 5 and a center at (0, 0). In general, the formula for a circle with radius r and center (0, 0) is $r^2 = x^2 + y^2$.

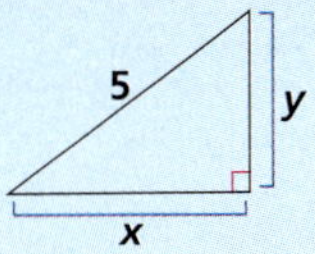

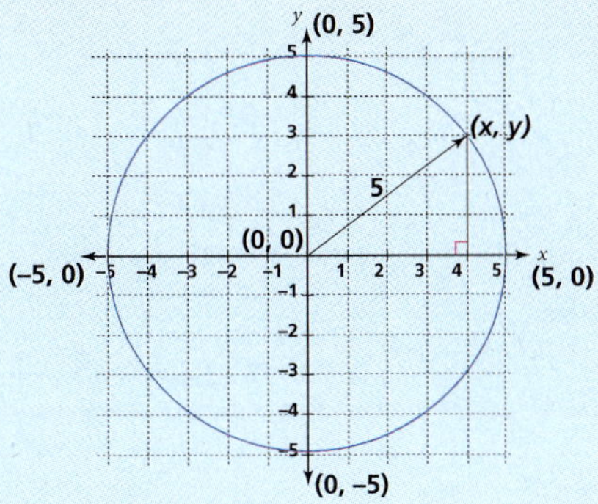

Exercise A Find each locus. Draw a sketch and write the equation.

1. Locus of points 3 units away from $y = 5$

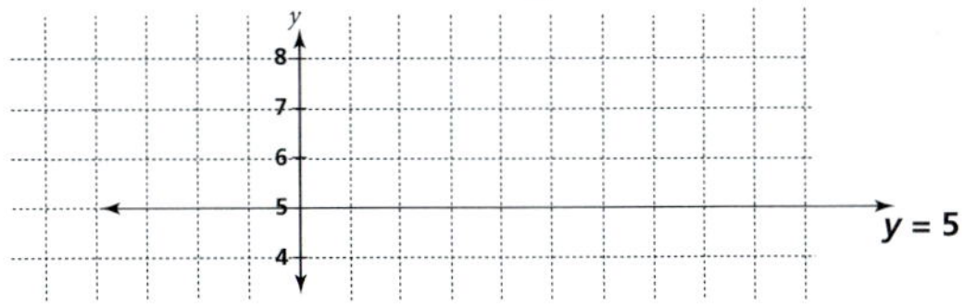

2. Locus of points equidistant from $y = x + 3$ and passing through $y = 1$ and $y = 5$

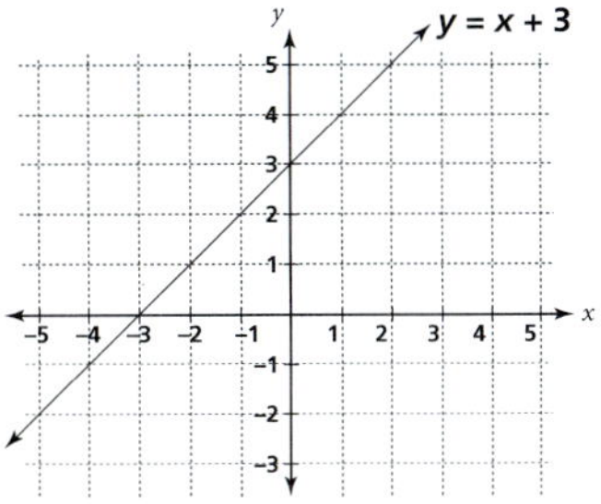

1. $y = 8$ and $y = 2$
2. $y = x + 1$ and $y = x + 5$
3. $y = -\frac{1}{2}x + 1$ and $y = -\frac{1}{2}x + 3$

3. Locus of points equidistant from $y = -\frac{1}{2}x + 2$ and passing through $y = 1$ and $y = 3$

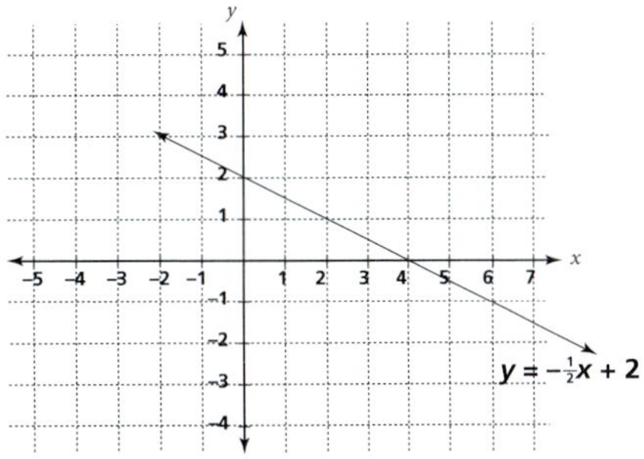

3. Reinforce and Extend

Learning Styles

Auditory/Verbal Students can work in pairs to complete Exercise A. One student can solve the problem, thinking out loud. The second student can execute the solution steps described by the first student. Pairs should examine the results of their work before they move to the next item. Students should exchange roles after each item.

Group Problem Solving

Have small groups of students work together to write two problems similar to those in Exercise A. The groups must write the problems and prepare a solution key. The key should include a sketch and an equation for each problem. The key might also provide step-by-step instructions for solving the problems. Groups then exchange and solve one another's problems. They check their solutions, using the solution keys.

CALCULATOR

Have students consider the results of items 11–15 and the following items. Have them describe any patterns they observe. Ask students to find the squares of the following numbers: 707, 505, 303. (499,849; 255,025; 91,809; the first and last numbers are the square of the first (or last) digit, the center numbers are double the square of the first (or last) digit.)

Technology Connection

Discuss with students the advantage of being able to visualize a slice of a person's brain or heart.

4. Locus of points equidistant from $x = 2$ and $x = -3$ — $x = -\frac{1}{2}$
5. Locus of points equidistant from $y = 3$ and $y = -5$ — $y = -1$
6. Locus of points equidistant from (3, 2) and (–4, 2) — $x = -\frac{1}{2}$

Exercise B Find each locus. Draw a sketch and write the equation.

7. Locus of points 6 units from origin — $36 = x^2 + y^2$
8. Locus of points 4 units from origin — $16 = x^2 + y^2$
9. Locus of points $\frac{1}{2}$ unit from origin — $0.25 = x^2 + y^2$
10. Locus of points 3 units from origin — $9 = x^2 + y^2$

Calculator Practice

You can use the x^2 key on your calculator to compute r^2 as you write the equation of a circle.

EXAMPLE Find 15^2.

Press 15 x^2. The display reads 225.

Exercise C Find the squares of following numbers.

11. 111 12,321 **13.** 101 10,201 **15.** 606 367,236

12. 1,111 1,234,321 **14.** 202 40,804

TECHNOLOGY CONNECTION

The CAT scan is one of many tools doctors use to learn about a person's health. A CAT scan produces a series of x-ray pictures. Each picture shows a "slice" of a person's body as if a plane were passed through it. Doctors can study the pictures to learn the location of a tumor or other abnormality.

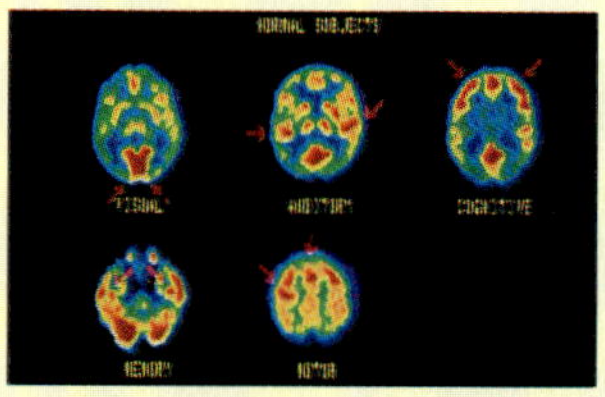

382 *Chapter 12 Geometry and Imagination*

Lesson 3 Compound Loci

You've learned how to visualize loci and how to use the idea of loci to solve many problems in algebra. In this lesson, you will be looking for points that satisfy more than one condition at a time. Here are some examples.

EXAMPLES **Given:** lines ℓ and m, points A and B

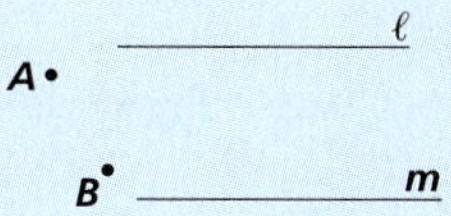

What is the locus of points equidistant from A and B and the parallels ℓ and m?

You will need to find each locus separately and then determine the loci's intersection.

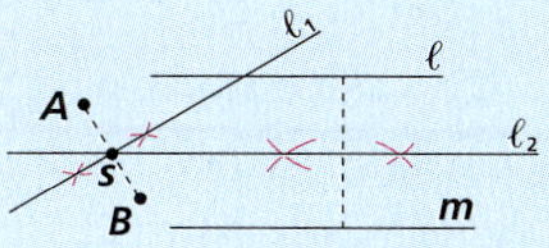

First, find the locus of points equidistant from A and B by finding the perpendicular bisector of $\overline{AB}$. Label this line ℓ_1.

Next, find the locus of points equidistant from ℓ and m. Label this line ℓ_2.

The two loci, ℓ_1 and ℓ_2, intersect at s. Therefore, s is both equidistant from A and B and equidistant to the parallels ℓ and m.

Given: line ℓ, $\overline{AB}$, and point C

Find: the locus of points at a distance of $\overline{AB}$ from C and a distance of $\overline{AB}$ from ℓ.

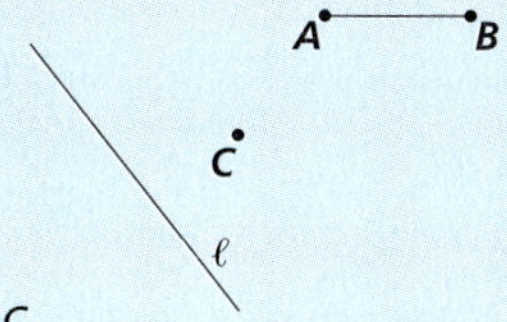

First, find the locus of points a distance of $\overline{AB}$ from C. This is a circle with radius $\overline{AB}$ and center C.

Next, draw two lines, each parallel to ℓ and each a distance of $\overline{AB}$ from ℓ.

Lesson at a Glance

Chapter 12 Lesson 3

Overview This lesson introduces compound loci—points that satisfy more than one condition at a time.

Objectives

- To construct compound loci.
- To describe compound loci.

Student Pages 383–385

Teacher's Resource Library TRL

Workbook Activity 121

Additional Practice 76

1. Warm-Up Activity

Ask all students with the letter *a* in their first name to stand at the front of the room. Ask all those who have the letter *e* in their first name to stand at the back of the room. Ask any students with both and *a* and an *e* in their first name to stand in the center of the room. Explain that the students, if any, in the center of the room have names that fulfill two conditions: having an *a* and an *e*. Point out that in this lesson students will learn how to describe points that fulfill more than one condition.

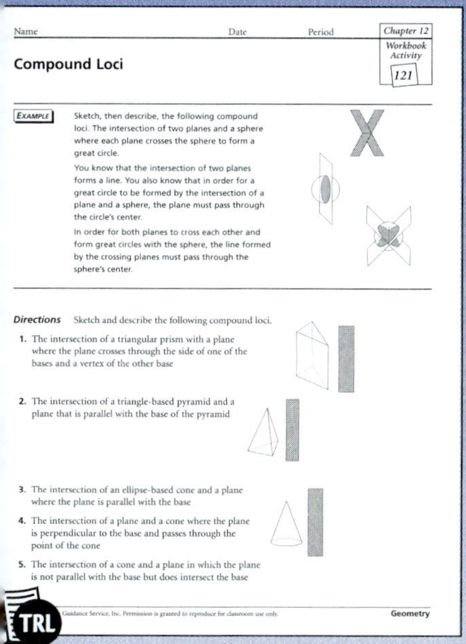
Name Date Period Chapter 12 Workbook Activity 121

Compound Loci

EXAMPLE Sketch, then describe, the following compound loci. The intersection of two planes and a sphere where each plane crosses the sphere to form a great circle.

You know that the intersection of two planes forms a line. You also know that in order for a great circle to be formed by the intersection of a plane and a sphere, the plane must pass through the circle's center.

In order for both planes to cross each other and form great circles with the sphere, the line formed by the crossing planes must pass through the sphere's center.

Directions Sketch and describe the following compound loci.

1. The intersection of a triangular prism with a plane where the plane crosses through the side of one of the bases and a vertex of the other base
2. The intersection of a triangle-based pyramid and a plane that is parallel with the base of the pyramid
3. The intersection of an ellipse-based cone and a plane where the plane is parallel with the base
4. The intersection of a plane and a cone where the plane is perpendicular to the base and passes through the point of the cone
5. The intersection of a cone and a plane in which the plane is not parallel with the base but does intersect the base

Geometry

Workbook Activity 121

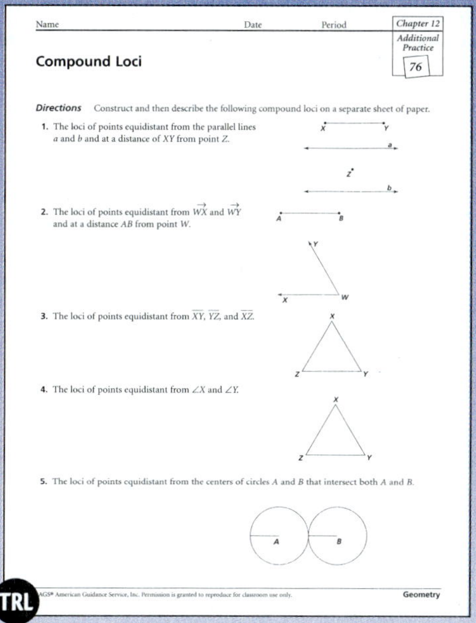
Name Date Period Chapter 12 Additional Practice 76

Compound Loci

Directions Construct and then describe the following compound loci on a separate sheet of paper.

1. The loci of points equidistant from the parallel lines a and b and at a distance of XY from point Z.
2. The loci of points equidistant from $\overrightarrow{WX}$ and $\overrightarrow{WY}$ and at a distance AB from point W.
3. The loci of points equidistant from $\overline{XY}$, $\overline{YZ}$, and $\overline{XZ}$.
4. The loci of points equidistant from $\angle X$ and $\angle Y$.
5. The loci of points equidistant from the centers of circles A and B that intersect both A and B.

Geometry

Additional Practice 76

2. Teaching the Lesson

Work through the first example on page 383 with students. Point out that they must first find the points that satisfy the first condition—those are on ℓ_1. Then they must find the points that satisfy the second condition—those are on ℓ_2. Lead them to the conclusion that points that satisfy both conditions must be on both lines. Stress that in this instance there is only one such point.

Ask a volunteer to work through the second example using the text as a model.

Answers to Exercise A

1.

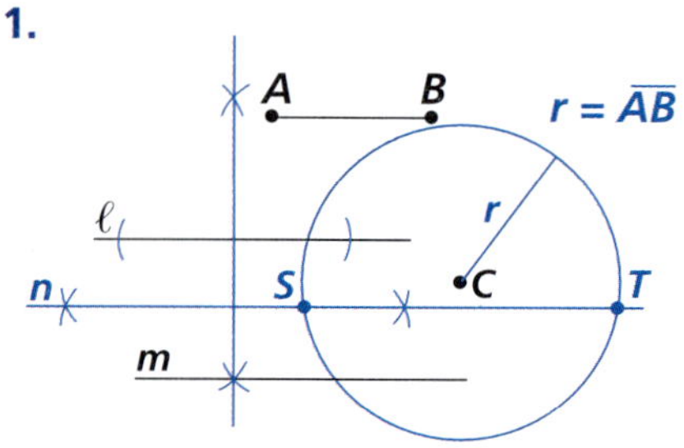

2.

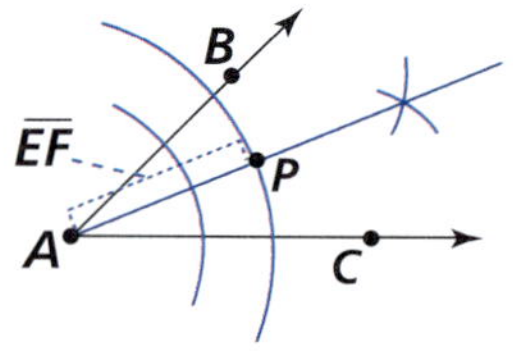

3.

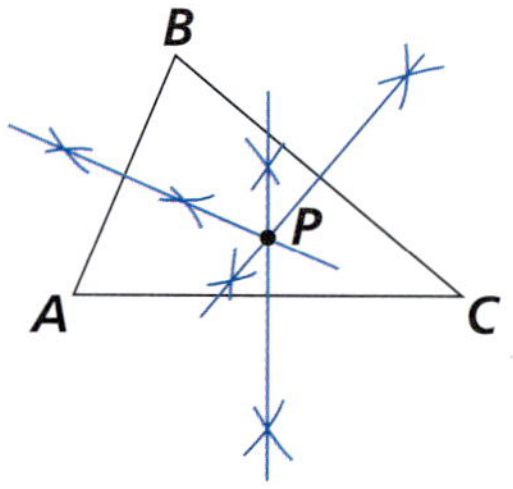

4.

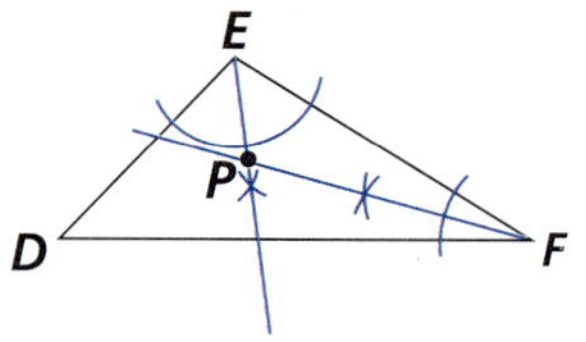

EXAMPLES *(continued)*

Find where the two loci intersect and label these points *S* and *T*.

S and *T* are the only two points that are a distance of $\overline{AB}$ both from point *C* and from line ℓ.

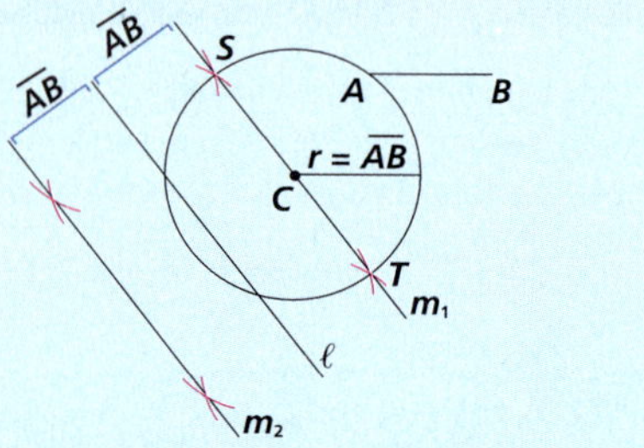

Describe and then sketch the following compound loci in space.

(a) points equidistant from a given point *C*

(b) points equidistant from a given line

(c) points equidistant from a point *C* and a line, with the point being on the given line

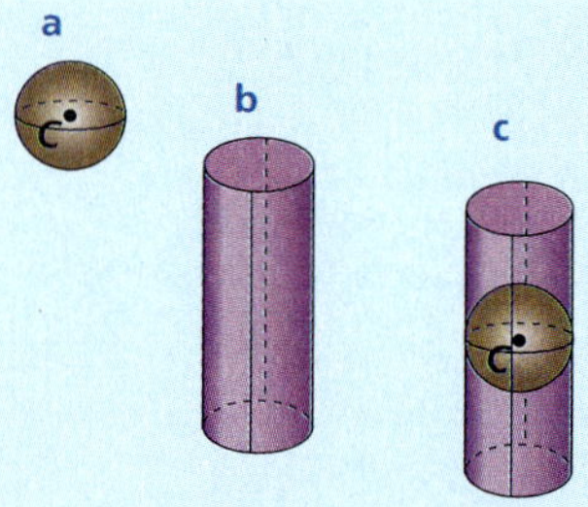

The compound locus is a great circle where the sphere touches the cylinder.

Exercise A Construct and then describe the following compound loci.

1. The locus of points equidistant from the parallel lines ℓ and *m* and at a distance of $\overline{AB}$ from *C*.

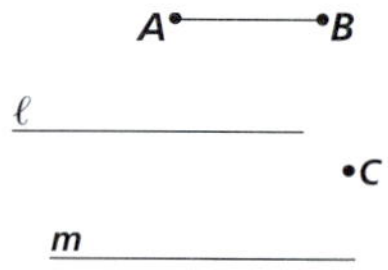

two points *S* and *T* on the circle and equidistant from ℓ and *m*

2. The locus of points equidistant from $\overrightarrow{AB}$ and $\overrightarrow{AC}$ and at a distance of $\overline{EF}$ from *A*.

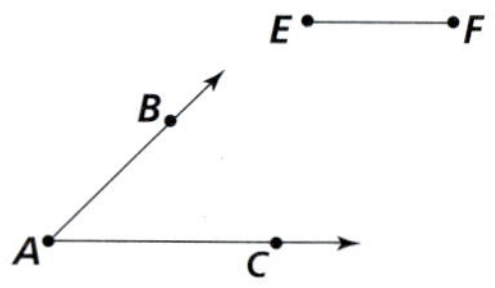

one point that is both along the angle bisector and on the circle with radius *EF*

Answers to Exercise B

5.

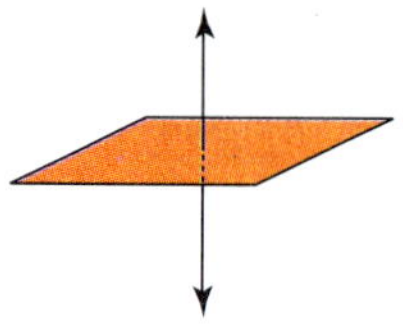

6.

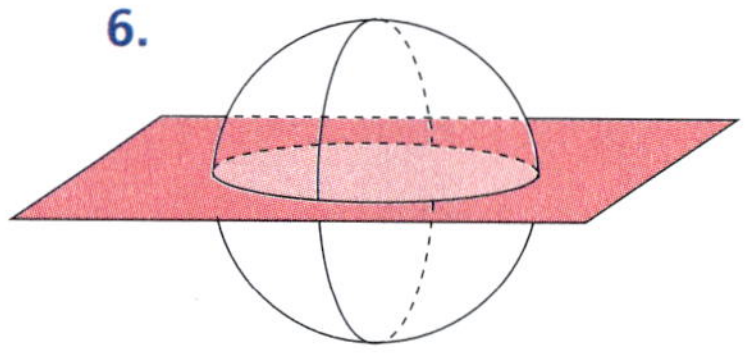

7.

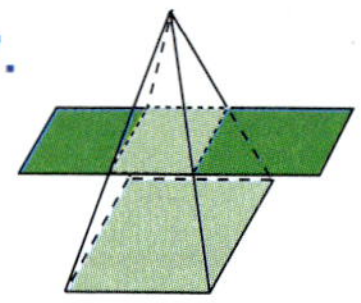

8.

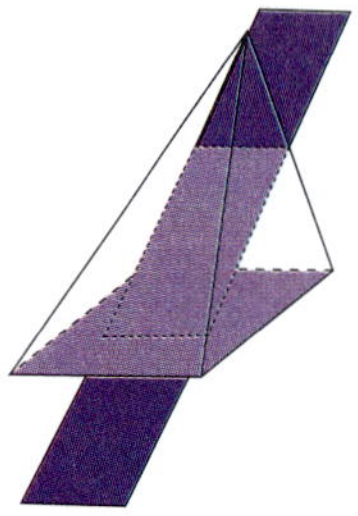

9.

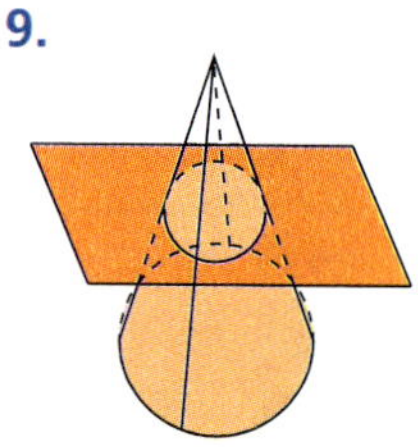

10.

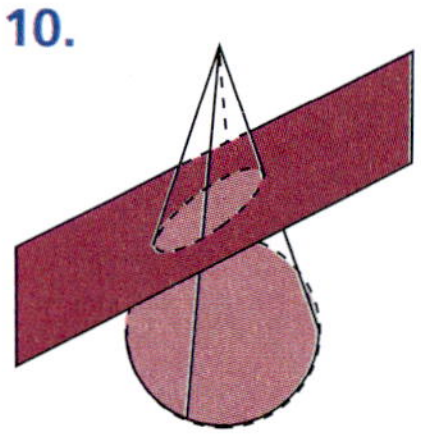

3. The locus of points equidistant from vertices A, B, and C.

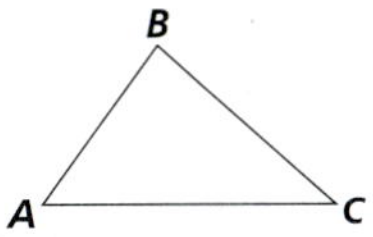

the intersection of the perpendicular bisectors of each side of the triangle

4. The locus of points equidistant from $\angle E$ and $\angle F$.

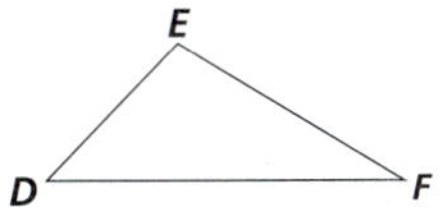

the intersection of the angle bisectors of each angle

Exercise B Sketch and then describe the following compound loci.

5. The intersection of a line with a plane. a point
6. The intersection of a plane with a sphere. a circle
7. The intersection of a plane and a square pyramid in which the plane is parallel to the base of the pyramid. a square smaller than the base
8. The intersection of a plane and a square pyramid in which the plane is not parallel to the base but does intersect the base. a parallelogram
9. The intersection of a plane and a cone in which the plane is parallel to the base of the cone. a circle with a radius smaller than the radius of the base

10. The intersection of a plane and a cone in which the plane is not parallel to the base. an oval (or ellipse)

TRY THIS...

Make a solid box 2 in. wide by 2 in. high by 8 in. long from clay. Use a piece of cardboard to act as a plane. Form the following figures, describing the intersection of the plane and the prism: a square, a rectangle, and a triangle.

3. Reinforce and Extend

LEARNING STYLES

Body/Kinesthetic Have pairs of students use materials such as clay balls, empty juice cans, and sheets of cardboard and paper to model each item in Exercise B. Invite pairs to share their models with the class. Allow pairs to revise their models as necessary.

TRY THIS...

Students can repeat the *Try This* using a clay triangle 8 inches long with each side 1 inch × 8 inch.

Lesson at a Glance

Chapter 12 Application

Overview This lesson introduces students to an equiangular spiral.

Objective

- To draw an equiangular spiral using golden ratio squares as a guide.

Student Page 386

Teacher's Resource Library TRL

Workbook Activity 122

1. Warm-Up Activity

If feasible, provide an actual nautilus shell, pine cone, and sunflower seed head for students to examine.

2 Teaching the Lesson

Learning Styles

Visual/Spatial Some students may benefit from producing the exercise on large sheets of craft paper. In this case, they may use a string and pen to draw the arcs of the spiral.

Learning Styles

Logical/Mathematical Challenge students to write a mathematical description of their spiral. For example, the spiral is the locus of points in which the radius changes every quarter circle according to the following pattern:

$r = 1, r = 1, r = 2, r = 3, r = 5, r = 8, \ldots$
$r_n = r_n - 2 + r_n - 1$.

Application

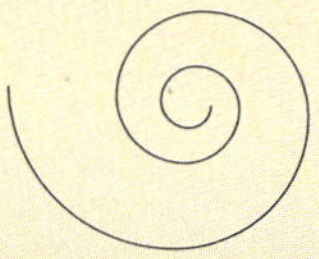

Spirals Suppose you start a construction with a center and a radius. However, this time, instead of keeping the radius constant, you increase the radius as you draw. What figure will you draw?

Many plants and animals—from the seed head of a sunflower to a parrot's beak—exhibit a spiral shape. The chambered nautilus, for example, produces a shell that has a spiral structure. As the animal grows, the shell grows as well. The chamber the animal lives in becomes larger, yet it remains the same shape.

EXAMPLE One of the characteristics of the nautilus shell is that as it intersects the outer shell, a radius drawn from the center of the shell to the outer shell always forms the same angle. This type of spiral is called an *equiangular spiral*.

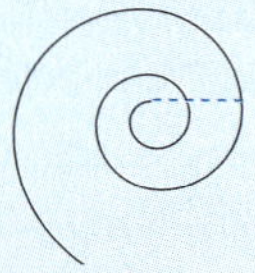

Exercise Follow these instructions to produce an equiangular spiral.

First, prepare a guide you can use to draw the spiral.

1. Draw a 1-unit square.
2. Draw a second 1-unit square next to the first.
3. Attach a 2-unit square along a matching side length.
4. Attach a 3-unit square along a matching side length.

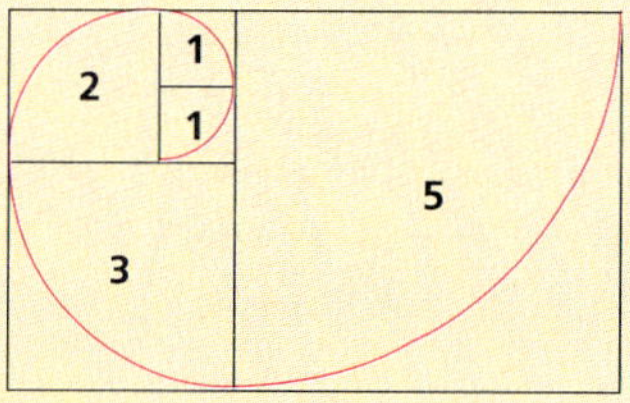

5. Continue adding squares to the guide. Each square will have a side length equal to the sum of the two previous side lengths: $2 + 3 = 5$, $3 + 5 = 8$, $5 + 8 = 13$, $8 + 13 = 21$ and so on to $r_n = r_{n-2} + r_{n-1}$.

Using the side length of each square as the radius, draw a quarter-circle arc in each square. Draw the arc so that it connects the opposite corners of the square and connects with the previously drawn arc. Write a locus definition for the spiral you have drawn.

Example: The spiral is the locus of points such that the radius changes every quarter-circle according to this pattern: $r = 1$, $r = 1$, $r = 2$, $r = 3$, $r = 5$, $r = 8$, $\ldots r_n = r_{n-2} + r_{n-1}$.

3. Reinforce and Extend

Application

In the Community Direct interested students to use references including the library and Internet to investigate the geometry of the Golden Rectangle or Fibonacci sequence. Students can report their findings in a Math Minute segment of the evening news.

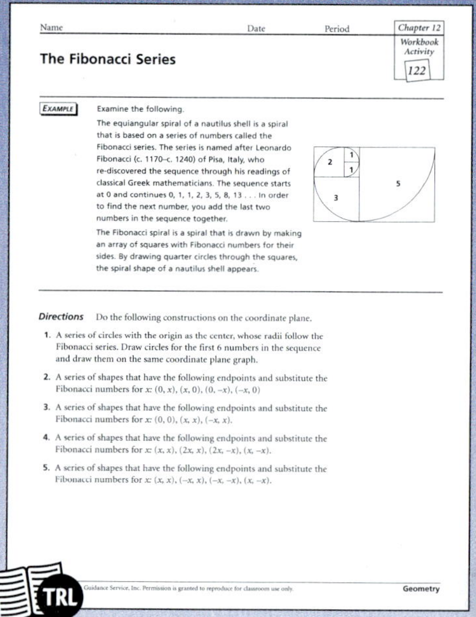

Name Date Period | Chapter 12 Workbook Activity 122

The Fibonacci Series

EXAMPLE Examine the following.

The equiangular spiral of a nautilus shell is a spiral that is based on a series of numbers called the Fibonacci series. The series is named after Leonardo Fibonacci (c. 1170–c. 1240) of Pisa, Italy, who re-discovered the sequence through his readings of classical Greek mathematicians. The sequence starts at 0 and continues 0, 1, 1, 2, 3, 5, 8, 13 . . . In order to find the next number, you add the last two numbers in the sequence together.

The Fibonacci spiral is a spiral that is drawn by making an array of squares with Fibonacci numbers for their sides. By drawing quarter circles through the squares, the spiral shape of a nautilus shell appears.

Directions Do the following constructions on the coordinate plane.

1. A series of circles with the origin as the center, whose radii follow the Fibonacci series. Draw circles for the first 6 numbers in the sequence and draw them on the same coordinate plane graph.
2. A series of shapes that have the following endpoints and substitute the Fibonacci numbers for x: $(0, x), (x, 0), (0, -x), (-x, 0)$
3. A series of shapes that have the following endpoints and substitute the Fibonacci numbers for x: $(0, 0), (x, x), (-x, x)$.
4. A series of shapes that have the following endpoints and substitute the Fibonacci numbers for x: $(x, x), (2x, x), (2x, -x), (x, -x)$.
5. A series of shapes that have the following endpoints and substitute the Fibonacci numbers for x: $(x, x), (-x, x), (-x, -x), (x, -x)$.

 Geometry

Workbook Activity 122

Chapter 12 Review

Sketch and describe the following.

Example: A line and a circle have one point in common.

Solution: The line is tangent to the circle.

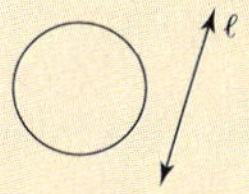

1. A line and circle have two points in common.

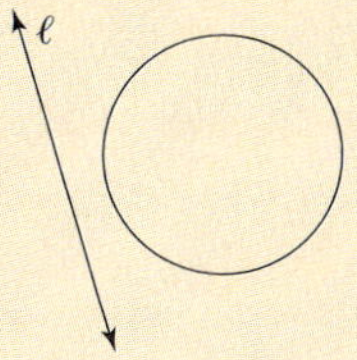

2. A line and a sphere have one point in common.
3. A line and a sphere have two points in common.

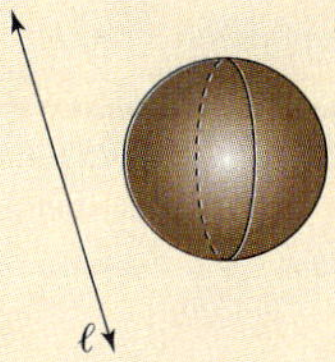

4. A line and a cylinder have one point in common.
5. A line and a cylinder have infinitely many points in common.

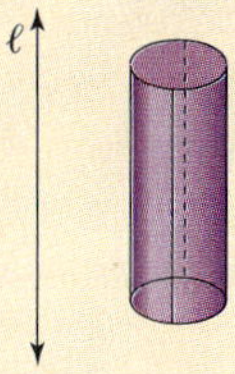

6. A plane passes through the center of a sphere.
7. A plane is tangent to a sphere.

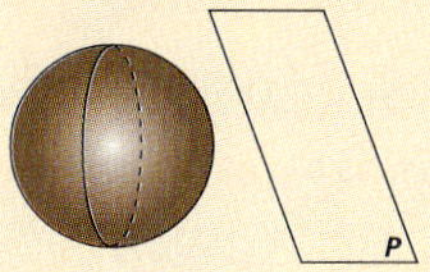

1. The line passes through the circle, forming a chord.
2. The line is tangent to the sphere.
3. The line passes through the sphere.
4. The line is tangent to the cylinder.
5. The line is on the surface of the cylinder.
6. A great circle is formed; the center of the great circle is the center of the sphere.
7. The plane and sphere meet at one point only.

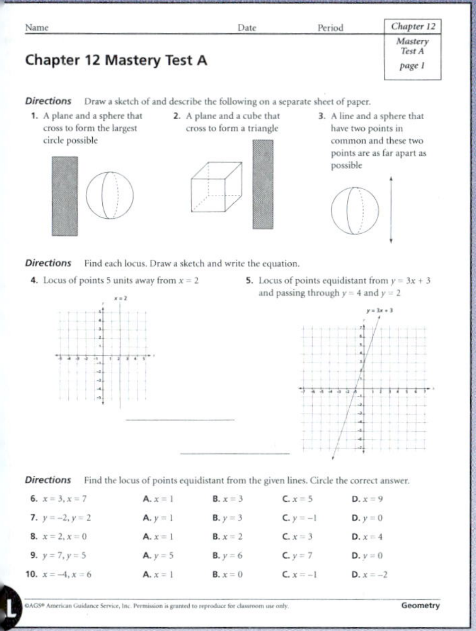

Name Date Period

Chapter 12 Mastery Test A page 1

Chapter 12 Mastery Test A

Directions Draw a sketch of and describe the following on a separate sheet of paper.

1. A plane and a sphere that cross to form the largest circle possible
2. A plane and a cube that cross to form a triangle
3. A line and a sphere that have two points in common and these two points are as far apart as possible

Directions Find each locus. Draw a sketch and write the equation.

4. Locus of points 5 units away from $x = 2$
5. Locus of points equidistant from $y = 3x + 3$ and passing through $y = 4$ and $y = 2$

Directions Find the locus of points equidistant from the given lines. Circle the correct answer.

6. $x = 3, x = 7$	A. $x = 1$	B. $x = 3$	C. $x = 5$	D. $x = 9$
7. $y = -2, y = 2$	A. $y = 1$	B. $y = 3$	C. $y = -1$	D. $y = 0$
8. $x = 2, x = 0$	A. $x = 1$	B. $x = 2$	C. $x = 3$	D. $x = 4$
9. $y = 7, y = 5$	A. $y = 5$	B. $y = 6$	C. $y = 7$	D. $y = 0$
10. $x = -4, x = 6$	A. $x = 1$	B. $x = 0$	C. $x = -1$	D. $x = -2$

©AGS® American Guidance Service, Inc. Permission is granted to reproduce for classroom use only. Geometry

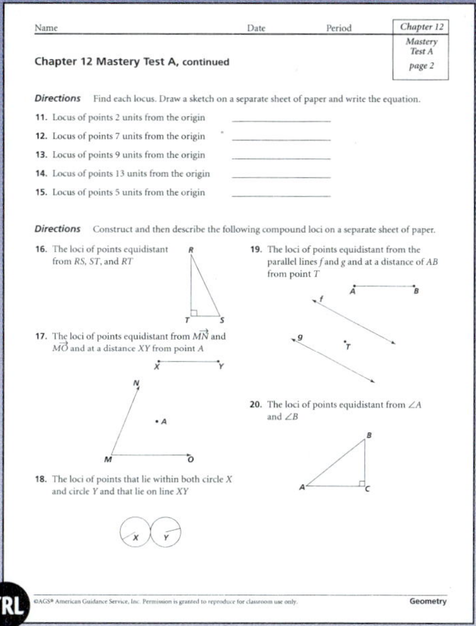

Name Date Period

Chapter 12 Mastery Test A page 2

Chapter 12 Mastery Test A, continued

Directions Find each locus. Draw a sketch on a separate sheet of paper and write the equation.

11. Locus of points 2 units from the origin ______
12. Locus of points 7 units from the origin ______
13. Locus of points 9 units from the origin ______
14. Locus of points 13 units from the origin ______
15. Locus of points 5 units from the origin ______

Directions Construct and then describe the following compound loci on a separate sheet of paper.

16. The loci of points equidistant from RS, ST, and RT
17. The loci of points equidistant from $\overrightarrow{MN}$ and $\overrightarrow{MO}$ and at a distance XY from point A
18. The loci of points that lie within both circle X and circle Y and that lie on line XY
19. The loci of points equidistant from the parallel lines f and g and at a distance of AB from point T
20. The loci of points equidistant from $\angle A$ and $\angle B$

©AGS® American Guidance Service, Inc. Permission is granted to reproduce for classroom use only. Geometry

TRL

Chapter 12 Mastery Test A

Chapter 12 Review

Each set of problems in the Chapter Review includes an example and solution to illustrate the concept. Use the given examples for reteaching the material in Chapter 12. For additional practice, refer to the Supplementary Problems for Chapter 12 (pages 418–419) and Additional Practice 74–76 in the Teacher's Resource Library.

Chapter 12 Mastery Test TRL

The Teacher's Resource Library includes parallel forms of the Chapter 12 Mastery Test. The difficulty level of the two forms is equivalent. You may wish to use one form as a pretest and the other form as a posttest. In addition, the Chapter 12 Cumulative Mastery Test covers content from Chapters 1–12.

Alternative Assessment

Alternative Assessment items correlate with student Goals for Learning at the beginning of this chapter.

- Have students identify how geometric figures intersect in space.

Use a juice or soup can and a sheet of paper to model how a cylinder and plane might interact. Model and describe at least three interactions.

- Have students write equations for loci.

Write equations for and graph the loci of the following:

points 3 units from $y = -4$

points 4 units from $x = -3$

points equidistant from $x = 3$ and $x = 5$

Alternative Assessment

continued

- Have students construct simple and compound loci.

Use ordinary materials to model the points equidistant from a point C in space, points equidistant from a given line, and points that are equidistant from C on line ℓ that are also equidistant from ℓ.

- Have students describe compound loci.

Describe the loci modeled in the previous item.

Chapters 1–12 Final Mastery Test

TRL

The Teacher's Resource Library includes the Final Mastery Test. This test is pictured on pages 482–483 of the Teacher's Edition. The Final Mastery Test assesses the major learning objectives of this text, with emphasis on Chapters 7–12.

Answers to Problems 10–18

10.

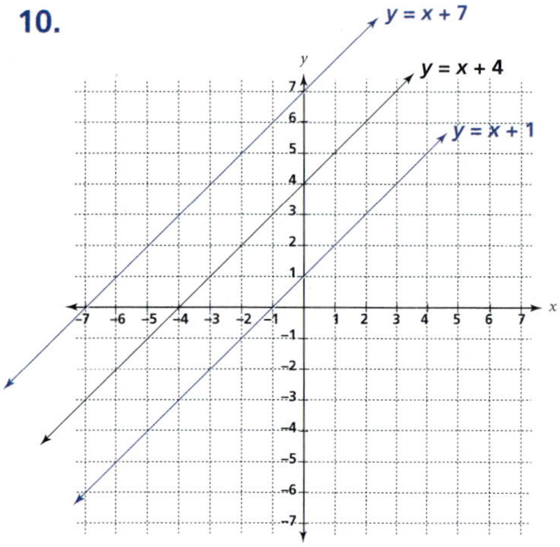

11.

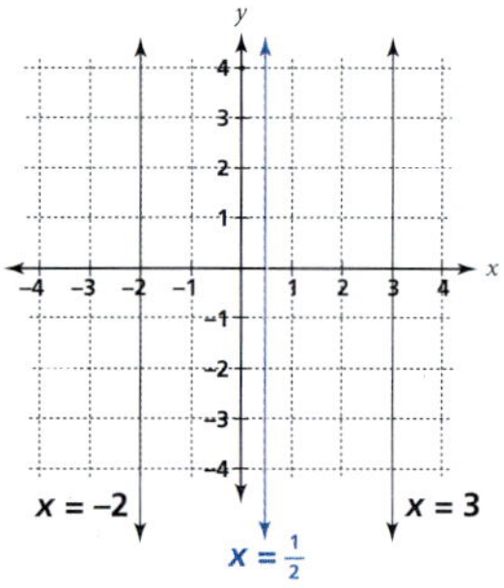

More Review...

8. Two parallel lines are formed.
9. The plane and the cylinder intersect to form a circle.

8. A plane perpendicular to the base of a cylinder passes through the cylinder.

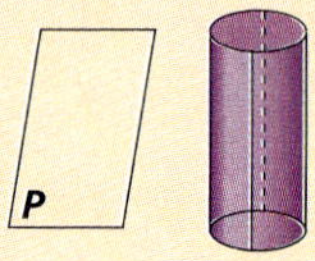

9. A plane parallel to the base of a cylinder passes through the cylinder.

Draw a sketch and write the equation for each locus.

Example: Locus of points 2 units away from $y = 4$.

Solution: The locus is two lines parallel to $y = 4$ and two units away from $y = 4$. The equations are $y = 2$ and $y = 6$.

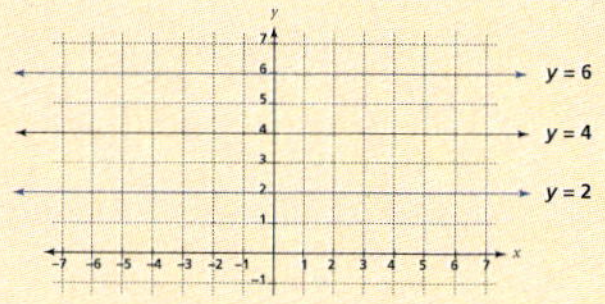

10. $y = x + 1$ and $y = x + 7$
11. $x = \frac{1}{2}$
12. $y = -1$
13. $y = 1$
14. $25 = x^2 + y^2$
15. $4 = x^2 + y^2$
16. $9 = x^2 + y^2$

10. Locus of points equidistant from $y = x + 4$ and passing through $y = 1$ and $y = 7$

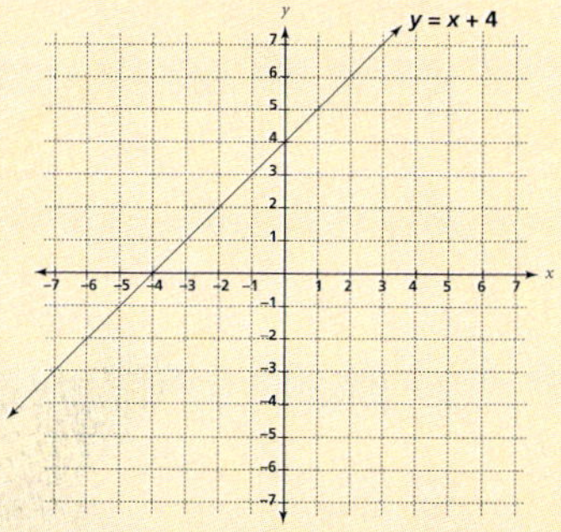

11. Locus of points equidistant from $x = -2$ and $x = 3$

12. Locus of points equidistant from $y = 2$ and $y = -4$

13. Locus of points equidistant from (3, 4) and (3, −2)

14. Locus of points 5 units from origin

15. Locus of points 2 units from origin

16. Locus of points 3 units from origin

Name Date Period | *Chapter 12* *Mastery Test B* *page 1*

Chapter 12 Mastery Test B

Directions Draw a sketch of and describe the following on a separate sheet of paper.

1. A plane and a pentagonal-based pyramid that cross to form a smaller similar pentagon
2. A line and an equilateral triangle-based pyramid that have two points in common
3. A plane and a square-based pyramid that cross to form a trapezoid

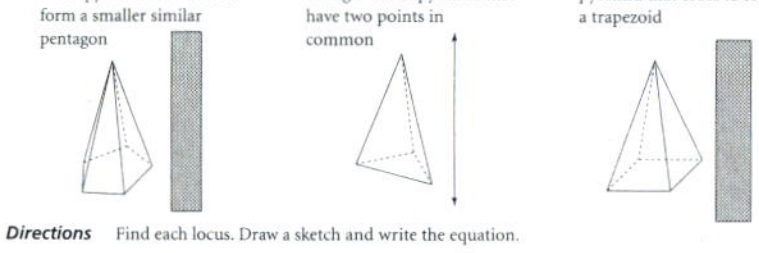

Directions Find each locus. Draw a sketch and write the equation.

4. Locus of points 6 units away from $x = -1$
5. Locus of points equidistant from $y = \frac{1}{3}x + 1$ and passing through $y = 0$ and $y = 2$

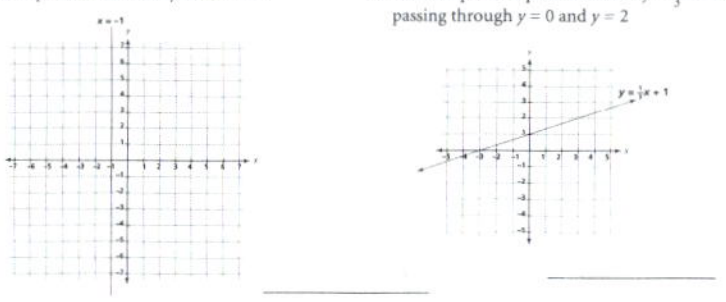

Directions Find the locus of points equidistant from the given lines. Circle the correct answer.

6. $y = 4, y = 2$	**A.** $y = 1$	**B.** $y = 3$	**C.** $y = -1$	**D.** $y = 0$
7. $x = 2, x = 6$	**A.** $x = 1$	**B.** $x = 2$	**C.** $x = 3$	**D.** $x = 4$
8. $x = 4, x = 6$	**A.** $x = 1$	**B.** $x = 3$	**C.** $x = 5$	**D.** $x = 9$
9. $y = 7, y = 3$	**A.** $y = 5$	**B.** $y = 6$	**C.** $y = 7$	**D.** $y = 0$
10. $x = -4, x = 2$	**A.** $x = 1$	**B.** $x = 0$	**C.** $x = -1$	**D.** $x = -2$

TRL

 Geometry

Name Date Period | *Chapter 12* *Mastery Test B* *page 2*

Chapter 12 Mastery Test B, continued

Directions Find each locus. Draw a sketch on a separate sheet of paper and write the equation.

11. Locus of points 3 units from the origin ______
12. Locus of points 12 units from the origin ______
13. Locus of points 14 units from the origin ______
14. Locus of points 33 units from the origin ______
15. Locus of points 4 units from the origin ______

Directions Construct and then describe the following compound loci on a separate sheet of paper.

16. The loci of points equidistant from the parallel lines l and m and at a distance of FG from point W

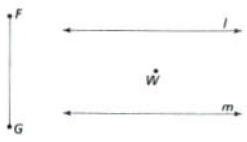

17. The loci of points equidistant from $\overrightarrow{DE}$ and $\overrightarrow{DF}$ and at a distance BC from point S

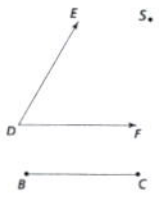

18. The loci of points equidistant from AB, BC, and AC

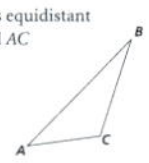

19. The loci of points equidistant from $\angle E$ and $\angle F$

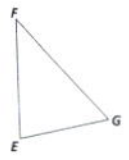

20. The loci of points that lie within both circle C and circle D and that lie on line CD

TRL

 Geometry

Chapter 12 Mastery Test B

Construct and describe the following compound loci.

Example: The locus of points equidistant from parallel lines ℓ and m and equidistant from A and B

Solution: Find each locus separately. First, find the locus of points. Line p_1 is equidistant from both ℓ and m. Every point on line p_2 is equidistant from A and B. p_1 and p_2 intersect at T. Therefore, T is equidistant from ℓ and m and A and B.

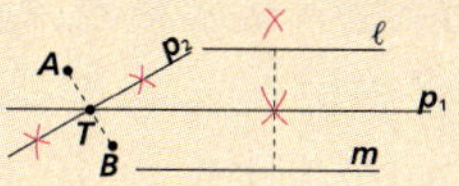

17. The locus of points at a distance of $\overline{CD}$ from P and at a distance of $\overline{CD}$ from line ℓ.

two points S and T, each on the circle with center P and radius $\overline{CD}$ and at a distance of $\overline{CD}$ from ℓ

18. The locus of points equidistant from the parallel lines m and n and a distance of $\overline{EF}$ from a point, P.

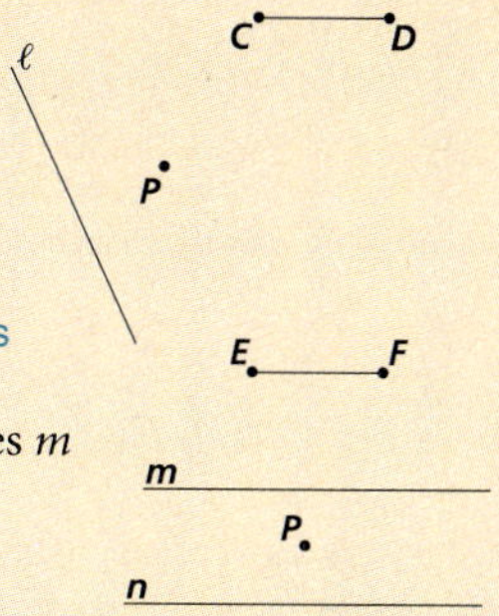

Identify the compound loci formed.

Example: The intersection of a plane and a hexagonal pyramid in which the plane is parallel to the base of the pyramid.

A. point **B.** triangle **C.** hexagon **D.** none

Solution: C; a hexagon smaller than the base

18. two points, each on the circle with center P and radius $\overline{EF}$ and midway between lines m and n

19. The intersection of a plane and cube in which the plane is parallel to the base.

A. point **C.** square

B. rectangle **D.** none

C

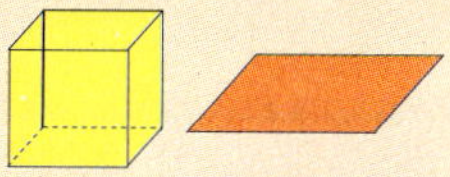

20. The intersection of a plane and cube in which the plane is perpendicular to the diagonal of the base.

A. point **C.** square

B. rectangle **D.** none

B

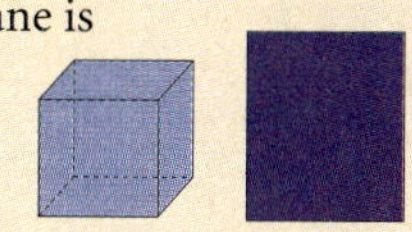

TEST-TAKING TIP

Always read directions more than once. Solve the problems and remember to give the units if needed.

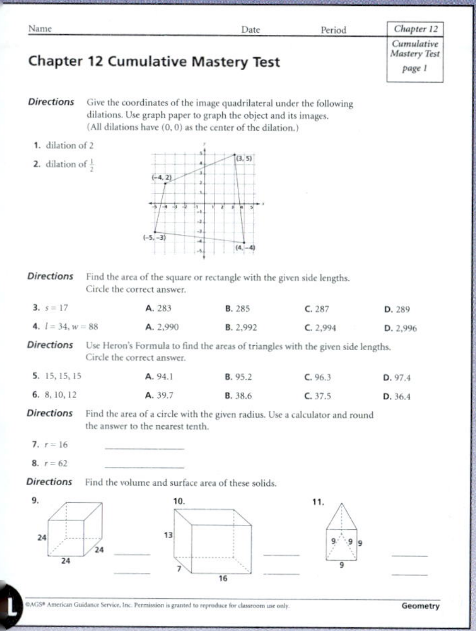

Name Date Period *Chapter 12 Cumulative Mastery Test page 1*

Chapter 12 Cumulative Mastery Test

Directions Give the coordinates of the image quadrilateral under the following dilations. Use graph paper to graph the object and its images. (All dilations have (0, 0) as the center of the dilation.)

1. dilation of 2
2. dilation of $\frac{1}{2}$

Directions Find the area of the square or rectangle with the given side lengths. Circle the correct answer.

	A.	B.	C.	D.
3. $s = 17$	A. 283	B. 285	C. 287	D. 289
4. $l = 34$, $w = 88$	A. 2,990	B. 2,992	C. 2,994	D. 2,996

Directions Use Heron's Formula to find the areas of triangles with the given side lengths. Circle the correct answer.

	A.	B.	C.	D.
5. 15, 15, 15	A. 94.1	B. 95.2	C. 96.3	D. 97.4
6. 8, 10, 12	A. 39.7	B. 38.6	C. 37.5	D. 36.4

Directions Find the area of a circle with the given radius. Use a calculator and round the answer to the nearest tenth.

7. $r = 16$
8. $r = 62$

Directions Find the volume and surface area of these solids.

9. 10. 11.

©AGS® American Guidance Service, Inc. Permission is granted to reproduce for classroom use only. Geometry

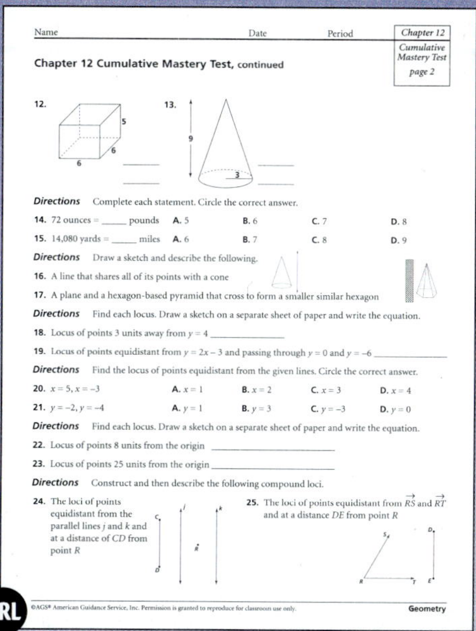

Name Date Period *Chapter 12 Cumulative Mastery Test page 2*

Chapter 12 Cumulative Mastery Test, continued

12. 13.

Directions Complete each statement. Circle the correct answer.

	A.	B.	C.	D.
14. 72 ounces = ____ pounds	A. 5	B. 6	C. 7	D. 8
15. 14,080 yards = ____ miles	A. 6	B. 7	C. 8	D. 9

Directions Draw a sketch and describe the following.

16. A line that shares all of its points with a cone
17. A plane and a hexagon-based pyramid that cross to form a smaller similar hexagon

Directions Find each locus. Draw a sketch on a separate sheet of paper and write the equation.

18. Locus of points 3 units away from $y = 4$
19. Locus of points equidistant from $y = 2x - 3$ and passing through $y = 0$ and $y = -6$

Directions Find the locus of points equidistant from the given lines. Circle the correct answer.

	A.	B.	C.	D.
20. $x = 5$, $x = -3$	A. $x = 1$	B. $x = 2$	C. $x = 3$	D. $x = 4$
21. $y = -2$, $y = -4$	A. $y = 1$	B. $y = 3$	C. $y = -3$	D. $y = 0$

Directions Find each locus. Draw a sketch on a separate sheet of paper and write the equation.

22. Locus of points 8 units from the origin
23. Locus of points 25 units from the origin

Directions Construct and then describe the following compound loci.

24. The loci of points equidistant from the parallel lines j and k and at a distance of CD from point R
25. The loci of points equidistant from $\overrightarrow{RS}$ and $\overrightarrow{RT}$ and at a distance DE from point R

©AGS® American Guidance Service, Inc. Permission is granted to reproduce for classroom use only. Geometry

Chapter 12 Cumulative Mastery Test

Answers to Problems continued

12.

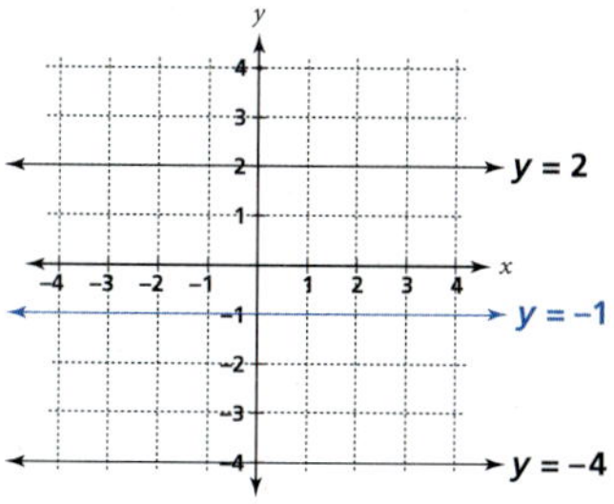

13.

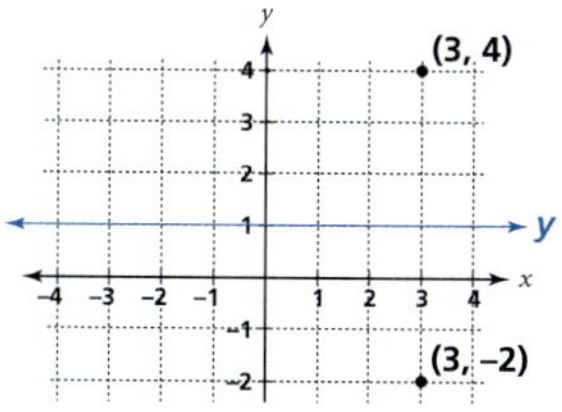

14.

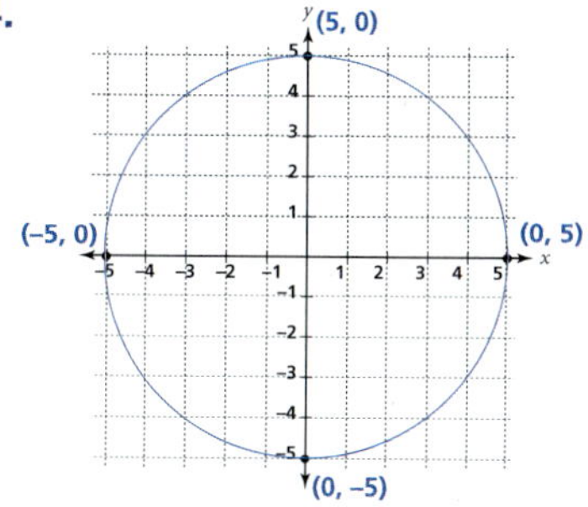

15.

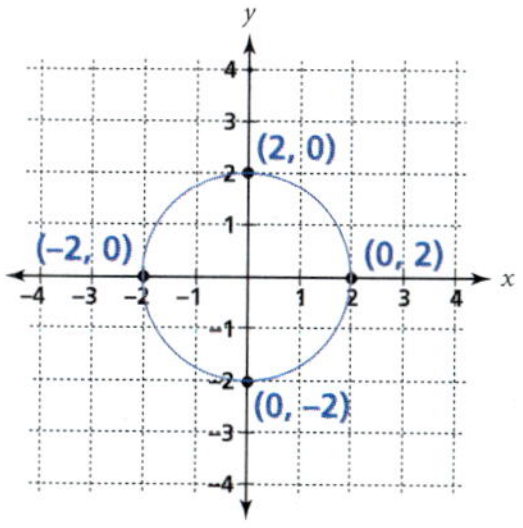

16.

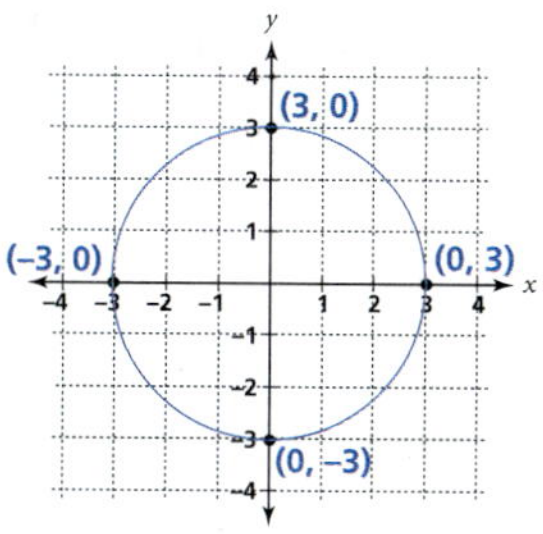

17.

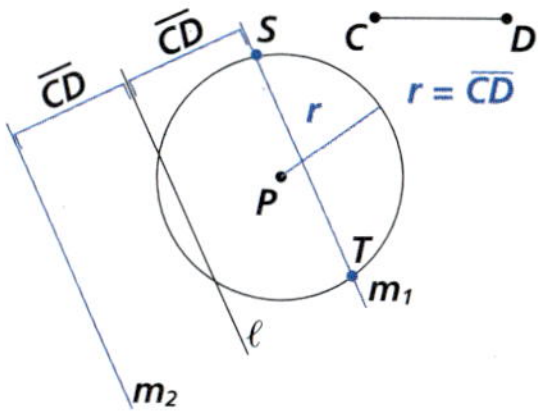

18.

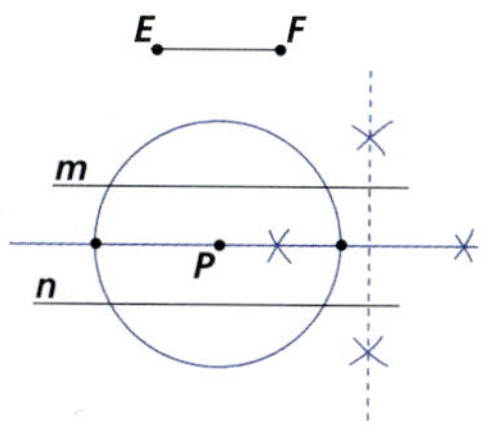

Chapter 1 Supplementary Problems

Tell whether each figure is a point, a line, a line segment, or a ray. Then use symbols to name each figure.

1.

line, $\overleftrightarrow{XY}$

2.

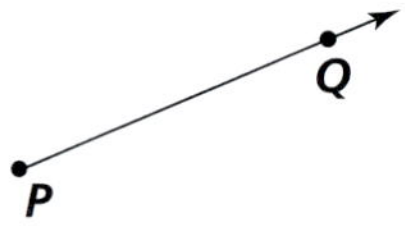

ray, $\overrightarrow{PQ}$

3.

line, *m*

4.

point, *B*

5.

line segment, $\overline{ST}$

6.

line, *y*

Use the Ruler Postulate to name the real number or letter corresponding to the point named.

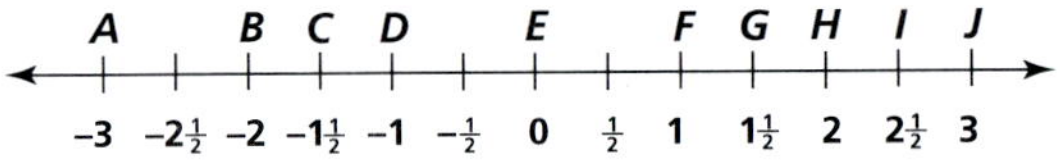

7. *A* −3 **8.** −2 *B* **9.** *H* 2 **10.** $2\frac{1}{2}$ I **11.** $-1\frac{1}{2}$ *C*

12. *J* 3 **13.** 1 *F* **14.** *E* 0 **15.** *G* $1\frac{1}{2}$ **16.** −1 *D*

Find the distance between the points.

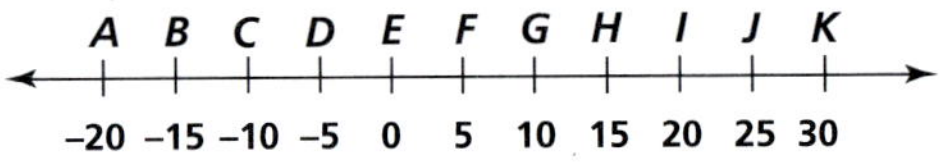

17. *B* and E 15 **18.** *G* and *I* 10 **19.** *E* and *F* 5 **20.** *F* and *H* 10 **21.** *C* and *D* 5

22. *D* and *F* 10 **23.** *C* and *E* 10 **24.** *B* and *G* 25 **25.** *A* and *D* 15 **26.** *B* and *I* 35

390 *Supplementary Problems* *Geometry*

Chapter 1 Supplementary Problems

Write the letter of the answer to each question.

27. How many points are used to name a line? C

A. none **B.** one **C.** two

28. How many dimensions does a plane have? C

A. none **B.** one **C.** two

29. What geometric term can you use to describe beads on a string? B

A. points on a ray **B.** points on a line segment **C.** points on a line

30. What are points on the same line called? A

A. collinear points **B.** parallel points **C.** endpoints

Name each figure in another way.

31. $\overleftrightarrow{AC}$ $\overleftrightarrow{CA}$

32. $\overrightarrow{XZ}$ $\overrightarrow{XY}$

33. $\overline{QR}$ $\overline{RQ}$

34. $\angle E$ $\angle DEF$

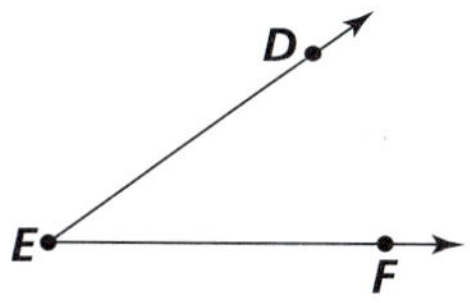

Copy each angle. Then bisect each angle using a compass and straightedge.

35.

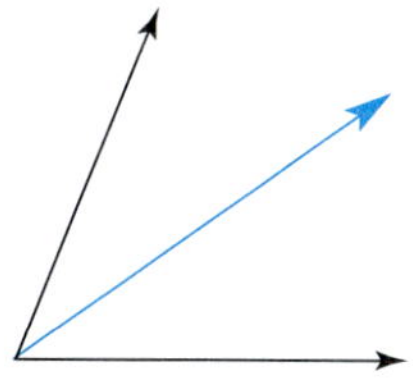

36.

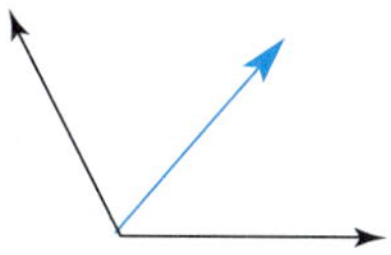

Student constructions should match angles with bisectors shown here.

Geometry Supplementary Problems 391

Chapter 1 Supplementary Problems

Classify each angle. Write *acute, right, obtuse,* or *straight.*

37. $m\angle DEF = 90°$

38. $m\angle Q = 120°$

39. $m\angle XYZ < 90°$

40. $m\angle 2 = 180°$

41.

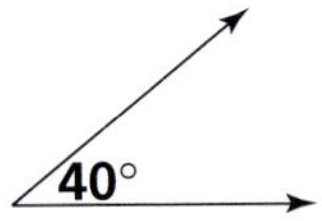

42.

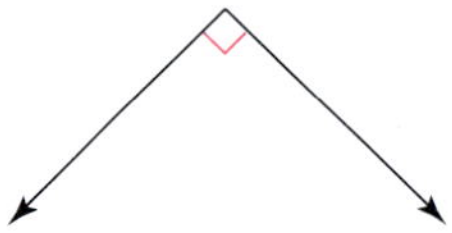

37. right
38. obtuse
39. acute
40. straight
41. acute
42. right

Find the measure of each angle.

43. $m\angle 2$ 43. 90°

44. $m\angle 6$ 44. 55°

45. $m\angle 4$ 45. 35°

46. $m\angle 3$ 46. 55°

47. $m\angle VXS$ 47. 180°

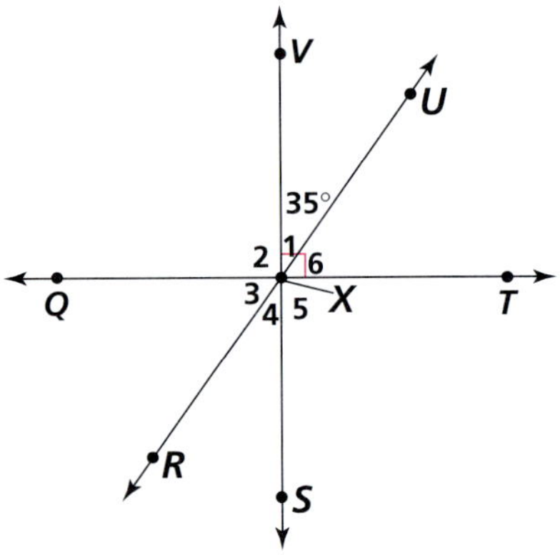

Solve for the missing angle(s).

48.

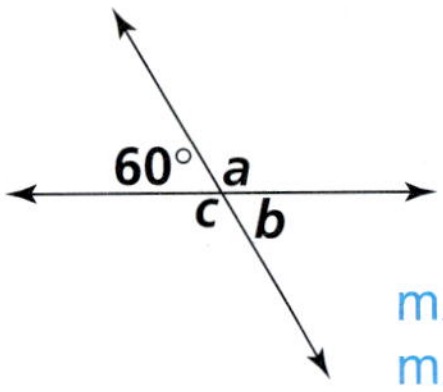

$m\angle a = 120°$
$m\angle b = 60°$
$m\angle c = 120°$

49.

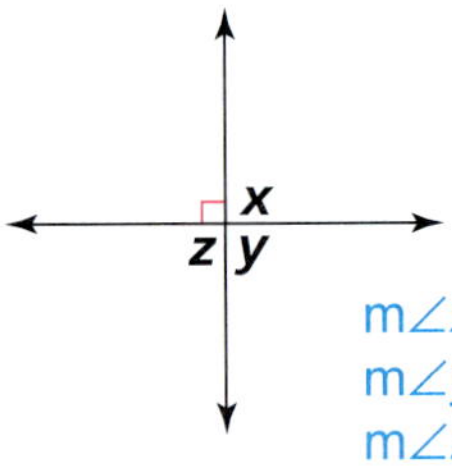

$m\angle x = 90°$
$m\angle y = 90°$
$m\angle z = 90°$

50.

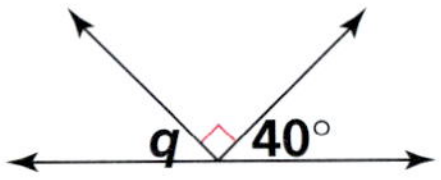

$m\angle q = 50°$

392 *Supplementary Problems* *Geometry*

Chapter 2 Supplementary Problems

Copy each statement. Draw one line under the hypothesis. Circle the conclusion.

1. If an angle is a right angle, then its measure is 90°.
2. If an angle is a straight angle, then its measure is 180°.
3. If a figure is a quadrilateral, then it has 4 sides.
4. If two angles have equal measures, then they are congruent.
5. If an animal is an insect, then it has 6 legs.

Write the converse of each conditional and tell whether it is true or false.

6. If a figure has three sides, then it is a triangle.
7. If two angles are supplementary, then the sum of their measures is 180°.
8. If it is raining, then the sun is not shining.
9. If you are a citizen of the United States, then you must pay taxes.

6. If a figure is a triangle, then it has three sides; true
7. If the sum of the measures of two angles is 180°, then the angles are supplementary; true
8. If the sun is not shining, then it is raining; false
9. If you must pay taxes, then you are a citizen of the United States; false

Decide which postulate allows the construction.

10. Connect the points with line segments to form a rectangle.

11. Draw circle *P* with radius 4.

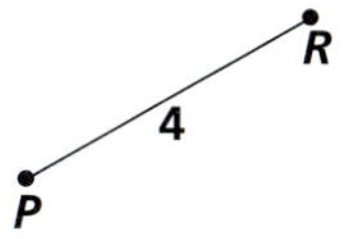

12. Draw 2 right angles equal to one another.

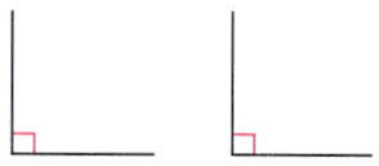

10. Postulate 1; A straight line can be drawn from any point to any point.
11. Postulate 3; A circle may be described with any center and distance.
12. Postulate 4; All right angles are equal to one another.

Geometry *Supplementary Problems* **393**

Decide which postulate allows the conclusion to be made.

13.

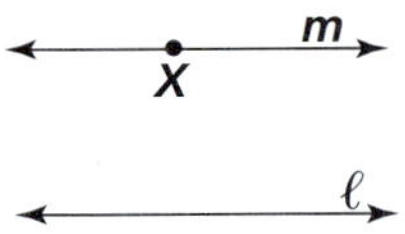

Conclusion: Line m is the only line that passes through point X and is parallel to line ℓ.

Parallel Postulate; If there is line ℓ and a point P not on ℓ, then there is only one line that passes through P and is parallel to ℓ.

14.

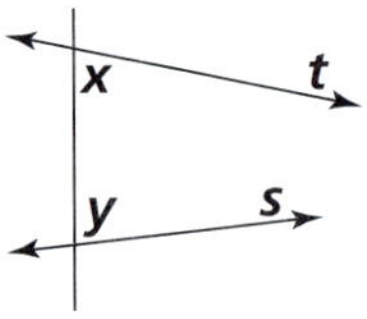

$m\angle x + m\angle y < 180°$

Conclusion: Lines t and s are intersecting lines.

Postulate 5; If one straight line crosses two other straight lines, four angles are formed. If the two inside angles together measure less than two right angles, then the two lines will, if extended indefinitely, meet on the same side as the two angles.

15.

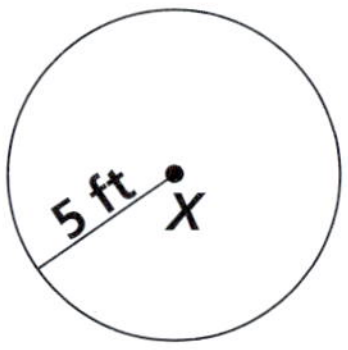

Conclusion: Circle X is shown with a radius of 5 feet.

Postulate 3; A circle may be described with any center and distance.

Answer each question. Tell which postulate or axiom you used.

16. Is $m\angle A = m\angle E$?

Yes; Postulate 4

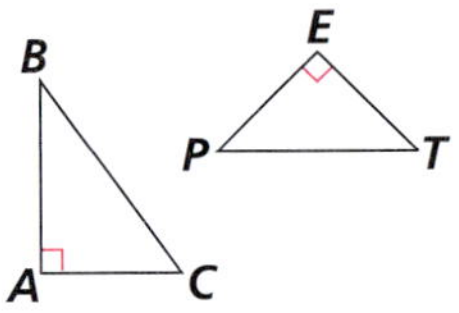

17. What is the measure of $\angle A$?

$m\angle A = 30°$; Axiom 1

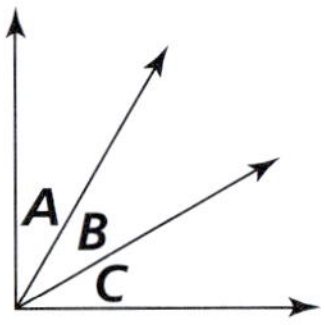

$m\angle A = m\angle C$
$m\angle B = m\angle C$
$m\angle B = 30°$

18. Which angles, or sum of angles, have a measure less than 90°?

angles a, b, and c; $m\angle a + m\angle b$; $m\angle a + m\angle c$; $m\angle b + m\angle c$; Axiom 5

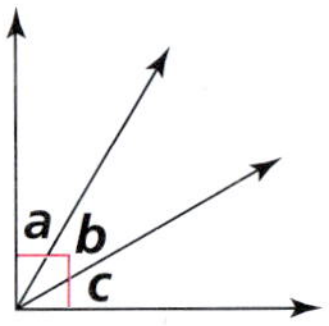

394 Supplementary Problems Geometry

Chapter 2 Supplementary Problems

19. If $m\angle a + m\angle b = m\angle b + m\angle c$, does $m\angle a = m\angle c$?

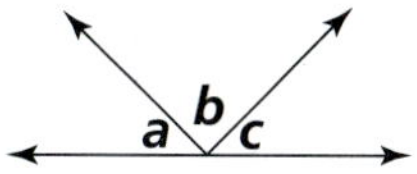

Yes; Axiom 2

20. If $4x + 16 = 56$, does $4x = 40$?

Yes; Axiom 3

Find the measures of angles Q, R, and S.

21.

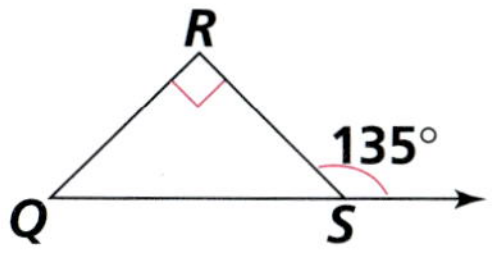

$m\angle Q = m\angle S$

$m\angle R = 90°$, $m\angle S = 45°$, $m\angle Q = 45°$

22.

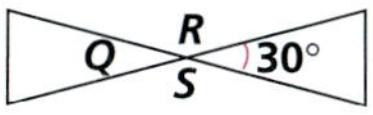

$m\angle Q = 30°$, $m\angle R = 150°$, $m\angle S = 150°$

23.

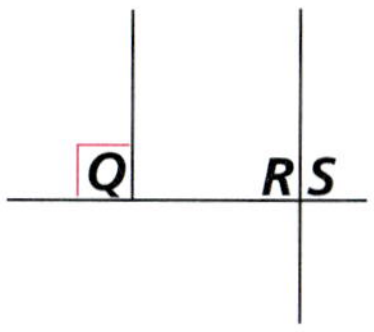

$m\angle Q = 90°$, $m\angle R = 90°$, $m\angle S = 90°$

$m\angle Q = m\angle R$

Give a reason for each of the following statements. Use the diagram at the right.

24. $m\angle e = m\angle c$

25. $m\angle a = m\angle c$

26. $m\angle e = 60°$

27. $m\angle b - m\angle c = m\angle b - m\angle a$

28. $m\angle f < 180°$

29. $m\angle c + m\angle d = 180°$

30. $m\angle e + m\angle f < 180°$

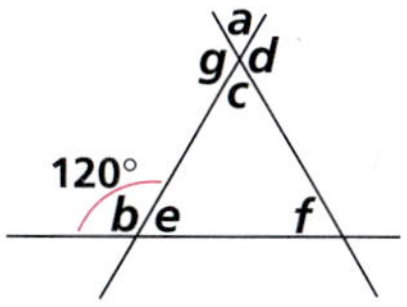

$\angle e = \angle c$

24. Given
25. Vertical angles are equal.
26. Definition of supplementary angles.
27. Axiom 3
28. Axiom 5
29. Definition of supplementary angles.
30. Postulate 5

Geometry *Supplementary Problems* **395**

Chapter 3 Supplementary Problems

Use the figure to identify lines or angles as described below.

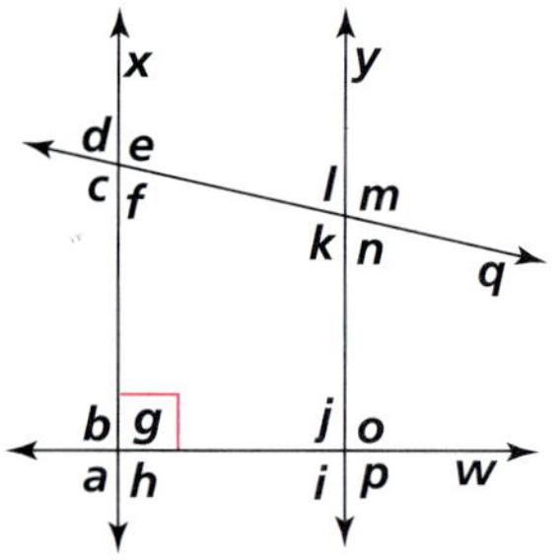

1. parallel lines lines x and y

2. transversal line q, w, x, or y

3. a pair of acute, vertical angles
$\angle d$ and $\angle f$, or $\angle l$ and $\angle n$

4. a pair of obtuse, vertical angles
$\angle k$ and $\angle m$, or $\angle c$ and $\angle e$

5. intersecting lines
lines x and q, or x and w, or y and q, q and w, or y and w

6. a pair of acute, alternate interior angles
$\angle f$ and $\angle l$

Classify each pair of angles. Write *alternate interior, alternate exterior, corresponding,* or *supplementary.*

$r \parallel s$

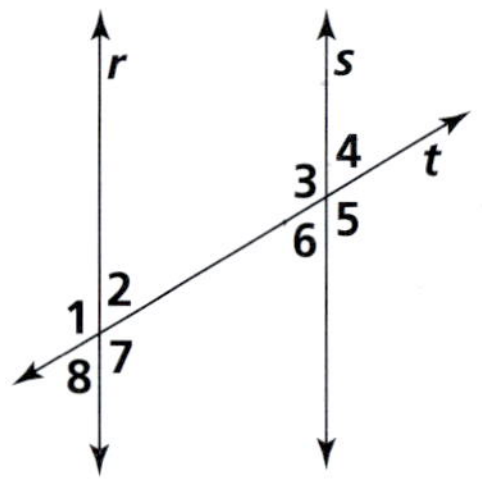

7. supplementary
8. alternate interior
9. alternate exterior
10. alternate interior
11. supplementary
12. alternate exterior
13. corresponding
14. supplementary
15. corresponding
16. supplementary

7. $\angle 1$ and $\angle 2$

8. $\angle 3$ and $\angle 7$

9. $\angle 4$ and $\angle 8$

10. $\angle 2$ and $\angle 6$

11. $\angle 6$ and $\angle 7$

12. $\angle 1$ and $\angle 5$

13. $\angle 6$ and $\angle 8$

14. $\angle 2$ and $\angle 3$

15. $\angle 1$ and $\angle 3$

16. $\angle 1$ and $\angle 4$

396 *Supplementary Problems* *Geometry*

Find the measure of each angle.

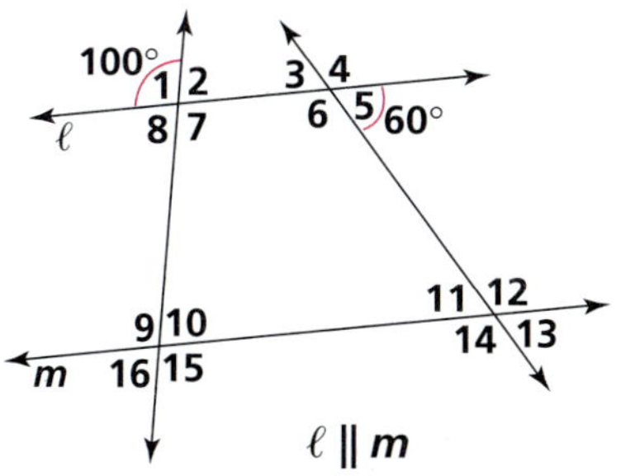

17. ∠2 80°	**21.** ∠11 60°	**25.** ∠8 80°
18. ∠9 100°	**22.** ∠12 120°	**26.** ∠10 80°
19. ∠4 120°	**23.** ∠16 80°	**27.** ∠3 60°
20. ∠6 120°	**24.** ∠13 60°	**28.** ∠14 120°

Use the theorems about parallel lines to find the measure of each angle.

29. The measure of $\angle b$ is twice that of $\angle c$.

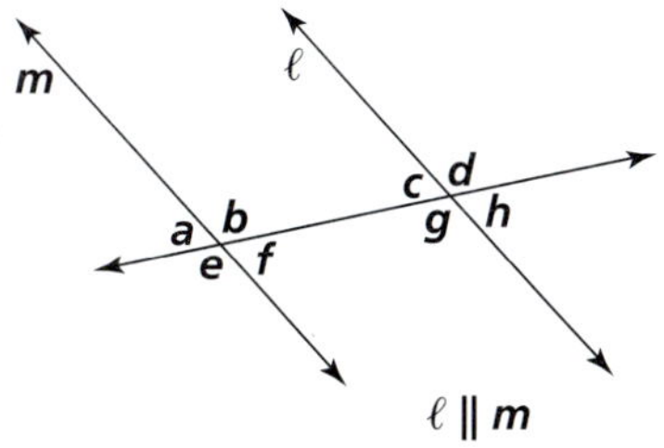

$m\angle b =$ ___

$m\angle c =$ ___

$m\angle b = 120°$; $m\angle c = 60°$

30. The measure of $\angle w$ is $\frac{1}{3}$ that of $\angle y$.

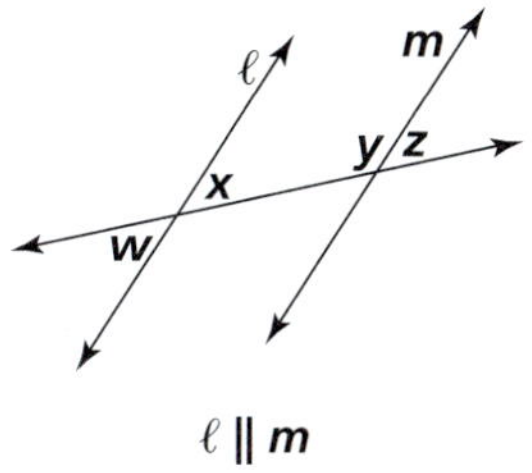

$m\angle w =$ ___

$m\angle y =$ ___

$m\angle w = 45°$; $m\angle y = 135°$

Geometry *Supplementary Problems* **397**

Chapter 3 Supplementary Problems

Use definitions and theorems to complete the statements.

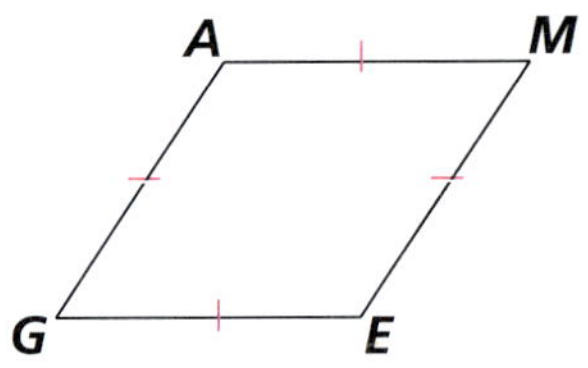

31. $\overline{GA} \parallel$ ______ $\overline{EM}$

32. $\overline{EM} \cong$ ______ $\overline{EG}, \overline{GA}, \overline{AM}$

33. $\overline{AM} \parallel$ ______ $\overline{GE}$

34. $m\angle A + m\angle$ ______ $= 180°$ *G* or *M*

35. $m\angle M + m\angle$ ______ $= 180°$ *E* or *A*

36. $m\angle E \cong m\angle$ ______ *A*

Write *parallel* or *not parallel* for each pair of lines crossed by a third line.

37.

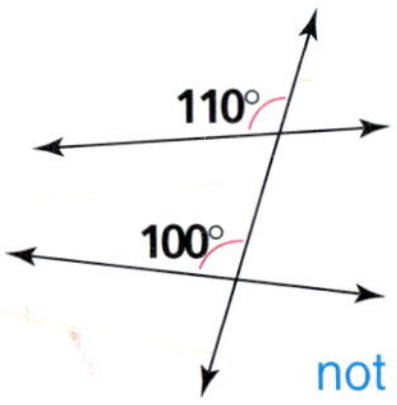

not parallel

38.

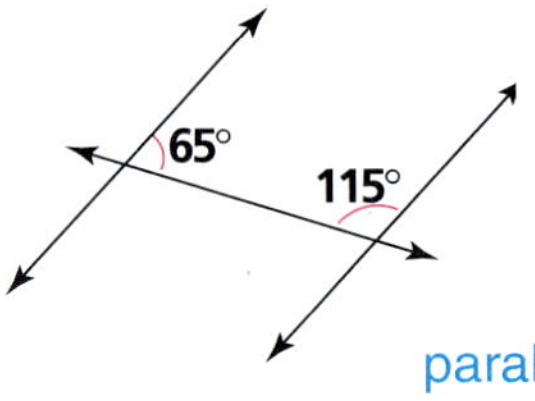

parallel

39.

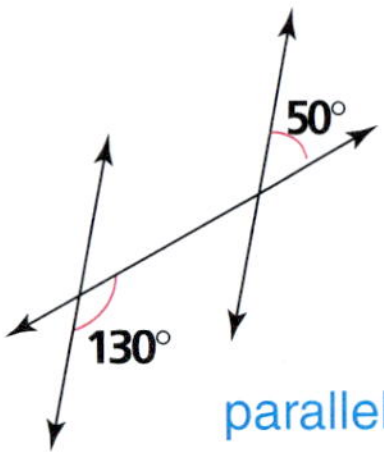

parallel

40.

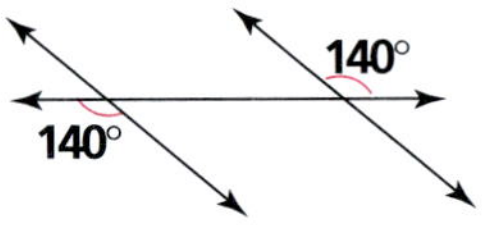

parallel

398 *Supplementary Problems* *Geometry*

Name the ordered pair that corresponds to each point.

1. *G* (1, 2)

2. *A* (–4, 1)

3. *C* (–2, –3)

4. *J* (4, 0)

5. *F* (0, –4)

6. *I* (3, 4)

7. *D* (–1, 0)

8. *H* (2, –3)

9. *B* (–3, 0)

10. E (0, 3)

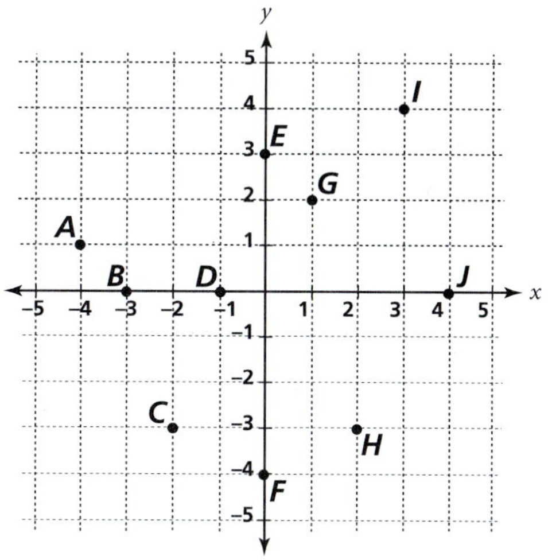

Name the point located at each ordered pair.

11. (1, 0) S

12. (4, –3) W

13. (3, 0) U

14. (–3, 1) N

15. (–2, –2) O

16. (1, 3) R

17. (4, 4) V

18. (0, 2) P

19. (1, –3) T

20. (0, –2) Q

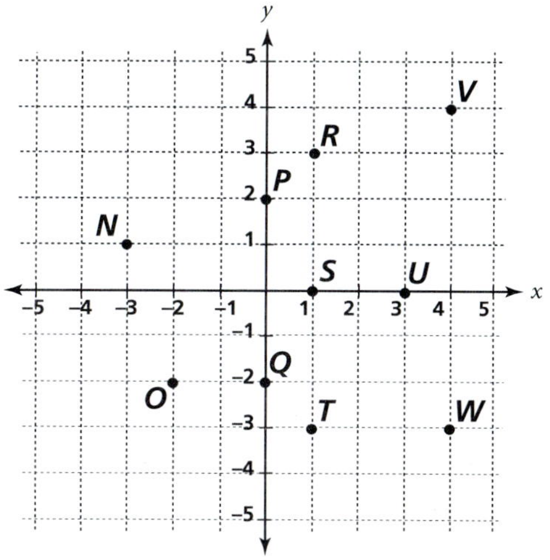

Geometry *Supplementary Problems* **399**

Find the slope, m, of each line.

21. line ℓ $m = 3$

22. line s $m = \frac{2}{3}$

23. line n $m = -2$

24. line p $m = -1$

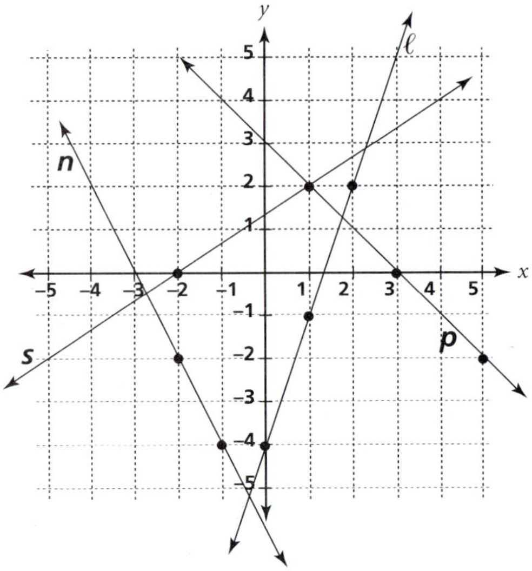

Find the slope, m, of the line that passes through the given points.

25. (1, 2) and (0, 0) $m = 2$

26. (3, 0) and (2, 0) $m = 0$

27. (8, 3) and (4, 1) $m = \frac{1}{2}$

28. (4, 0) and (2, 2) $m = -1$

29. (–3, –3) and (5, 2) $m = \frac{5}{8}$

30. (10, –6) and (1, –6) $m = 0$

Write the equation of the line that passes through each pair of points in problems 25–30 above. Use the form $y = mx + b$.

31. problem 25

32. problem 26

33. problem 27

34. problem 28

35. problem 29

36. problem 30

31. $y = 2x$

32. $y = 0$

33. $y = \frac{1}{2}x - 1$

34. $y = -x + 4$

35. $y = \frac{5}{8}x - \frac{9}{8}$

36. $y = -6$

400 *Supplementary Problems* *Geometry*

Chapter 4 Supplementary Problems

Find another point on each line using the slope and point given.

37. l; $m = 3$; passes through (–2, 4)

38. t; $m = \frac{2}{5}$; passes through (1, –4)

39. r; $m = -2$; passes through (–1, 3)

40. q; $m = -\frac{3}{4}$; passes through (0, 0)

37. (–1, 7)
38. (6, –2)
39. (0, 1)
40. (4, –3)

Make a grid and graph each of the lines described in problems 37–40 above. Connect the points with a line.

41. line l (–2, 4), (–1, 7)

42. line t (1, –4), (6, –2)

43. line r (–1, 3), (0, 1)

44. line q (0, 0), (4, –3)

See page 450 for graphs

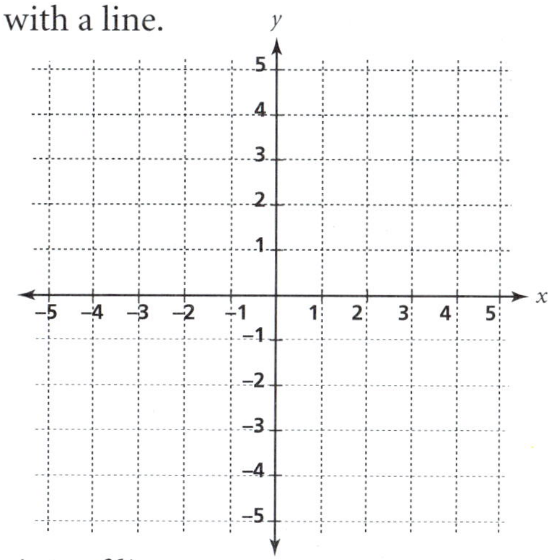

Use the midpoint formula to find the midpoints of line segments having the following endpoints.

45. (1, 1) and (5, 7) (3, 4)

46. (3, 4) and (–1, 2) (1, 3)

47. (8, –3) and (–2, 5) (3, 1)

48. (12, 8) and (6, –4) (9, 2)

49. (0, 6) and (4, 0) (2, 3)

50. (1, 5) and (–6, 7) $(-\frac{5}{2}, 6)$

Geometry *Supplementary Problems* **401**

Chapter 5 Supplementary Problems

Name the polygon as precisely as you can.

1.

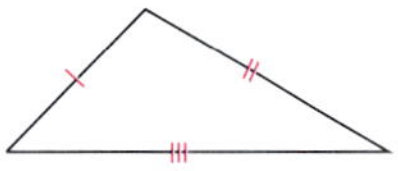

scalene triangle

2.

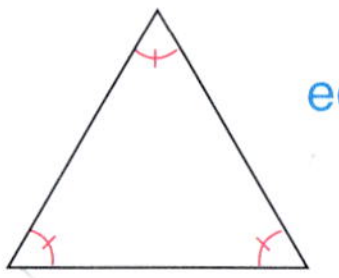

equilateral triangle

3.

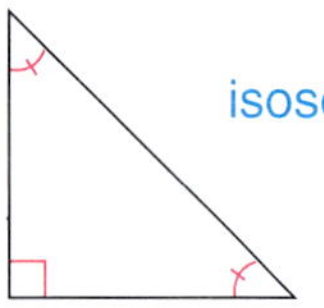

isosceles right triangle

4.

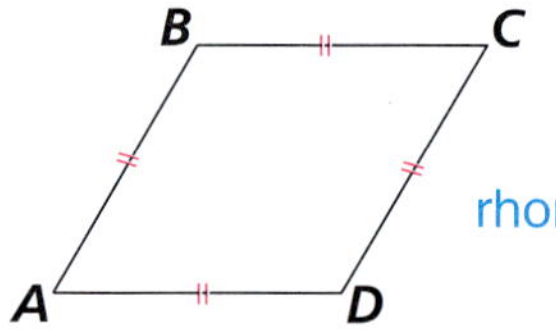

rhombus

5.

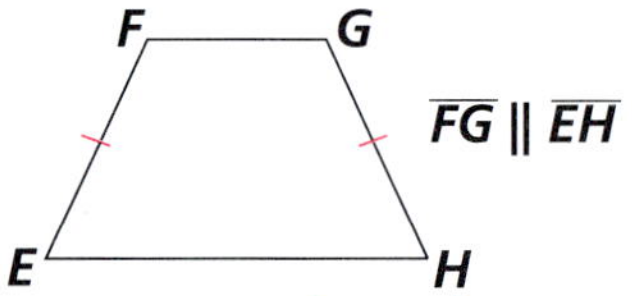

isosceles trapezoid

6.

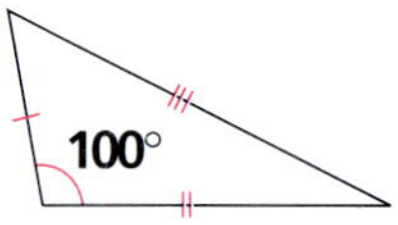

obtuse scalene triangle

Complete the following constructions on a separate sheet of paper. Use only a straightedge and a compass.

See pages 450-451 for answers.

7. Construct an equilateral triangle.

8. Draw any large scalene triangle and label the angles X, Y, and Z. Construct the altitude from X to $\overline{YZ}$.

9. Construct a scalene triangle using the following lengths for the sides.

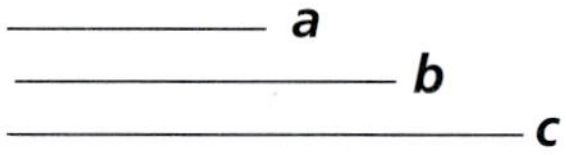

10. Construct an isosceles triangle. Label the angles XYZ and identify which sides are equal.

11. Construct ΔLMN with $\overline{LM} = \overline{LN}$ and $m\angle L > 90°$.

402 *Supplementary Problems Geometry*

Find the measure of $\angle x$.

12.

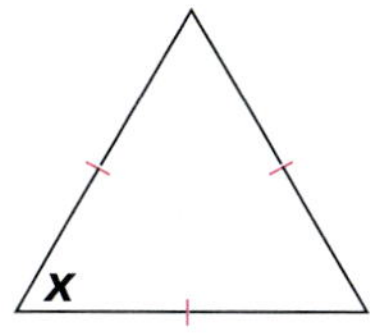

60°

13.

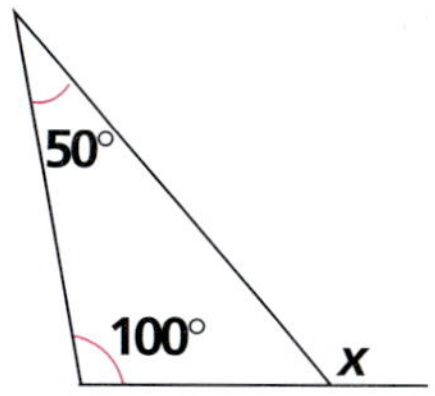

150°

14.

80°

15.

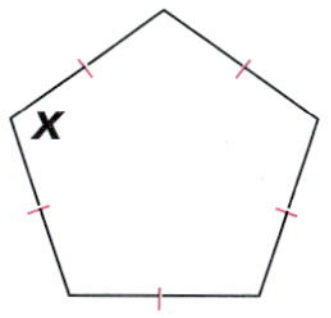

108°

16.

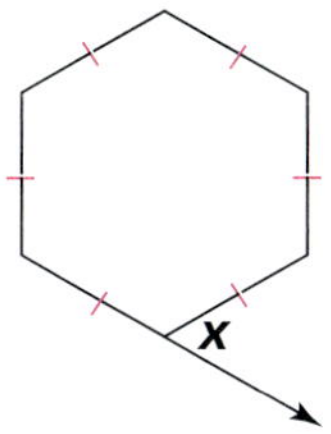

60°

17.

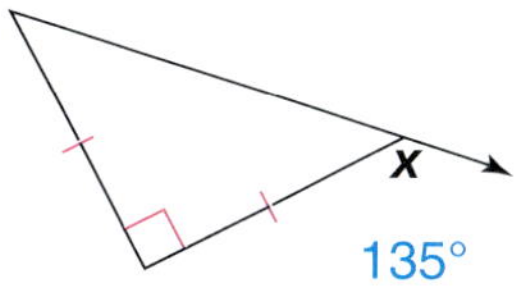

135°

Use the information given to find the angles' measures.

18. m$\angle$2 40°

19. m$\angle$3 110°

20. m$\angle$8 70°

21. m$\angle$5 70°

22. m$\angle$4 40°

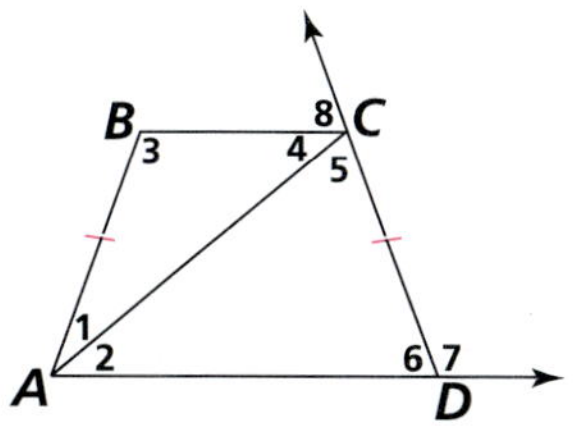

ABCD **is an isosceles trapezoid**
m$\angle$7 = 110°
m$\angle$1 = 30°

Geometry *Supplementary Problems* **403**

Chapter 5 Supplementary Problems

Answer the questions.

23. What is an altitude?

24. What is the sum of the measures of the interior angles of a regular octagon?

25. What is the measure of each angle in an equilateral triangle?

26. What is the measure of each interior angle of a 9-sided regular polygon?

27. How can you find the measure of an exterior angle of a triangle?

28. Is a trapezoid a parallelogram? Why or why not?

29. How can you find the number of triangles formed in a polygon by the diagonals from one vertex?

30. What is a median?

23. A line from the vertex of a triangle and perpendicular to the opposite side
24. 1,080
25. 60°
26. 140°
27. Find the sum of the two nonadjacent interior angles.
28. No, because it has only one pair of parallel sides
29. Subtract 2 from the number of sides the polygon has.
30. A line from the vertex of a triangle to the midpoint of the opposite side

Use the information given to answer the questions.

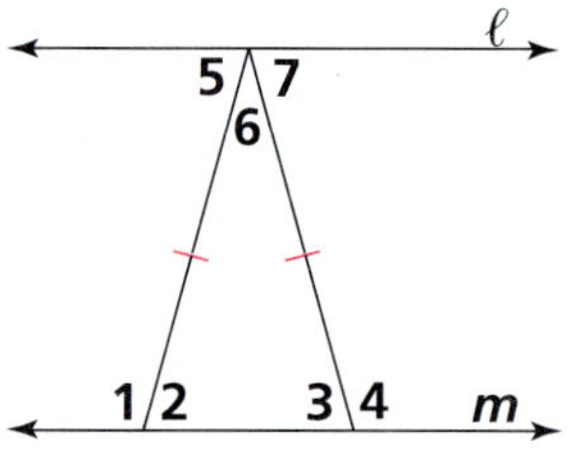

$\ell \parallel m$

31. Which angles are congruent to $\angle 2$?

32. If $m\angle 6 = 40°$, what is the sum of $m\angle 5$ and $m\angle 7$?

33. If $m\angle 2 = 70°$ and $m\angle 4 = 110°$, what is $m\angle 6$?

34. Which angle measure equals the sum of $m\angle 6$ and $m\angle 2$?

35. Are $\angle 1$ and $\angle 4$ exterior angles? Why or why not?

31. $\angle 3$, $\angle 5$, $\angle 7$
32. 140°
33. 40°
34. $m\angle 1$ or $m\angle 4$
35. yes; An exterior angle is formed by extending one side of the triangle along any vertex.

404 *Supplementary Problems* *Geometry*

Chapter 6 Supplementary Problems

Tell whether the triangles in each pair are congruent. Write *yes* or *no.* If your answer is *yes,* name the postulate that supports your answer. If your answer is *no,* explain why.

1.

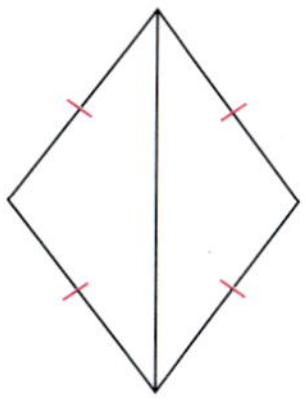

yes; SSS

2.

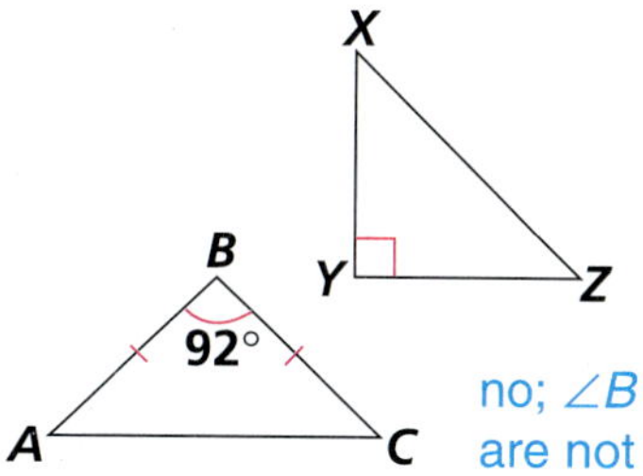

no; $\angle B$ and $\angle Y$ are not congruent.

3.

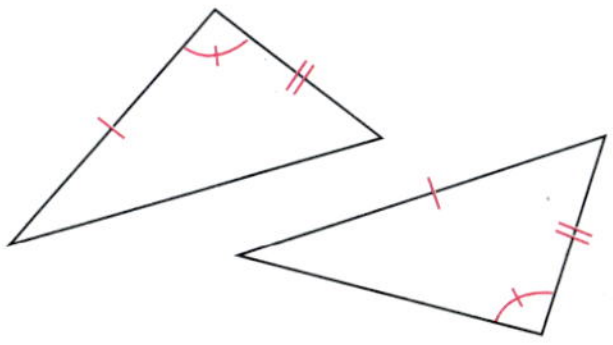

no; for SAS, the angle must be the included angle.

4.

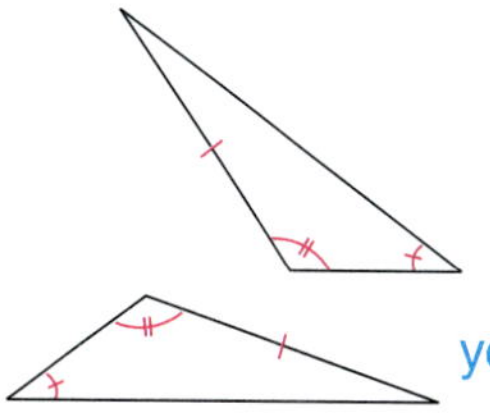

yes; AAS

Tell whether you can construct a triangle with the given sides. Write *yes* or *no*.

5.

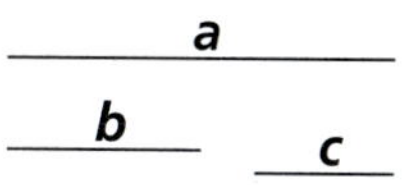

no

6.

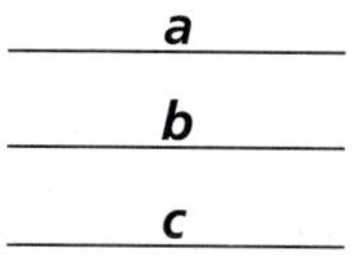

yes

7. $a = 6, b = 3, c = 7$ yes

8. $a = 8, b = 3, c = 4$ no

9. $a = 14, b = 12, c = 5$ yes

10. $a = 5, b = 6, c = 18$ no

11. $a = 5, b = 5, c = 10$ no

12. $a = 9, b = 9\frac{1}{2}, c = 18$ yes

Geometry *Supplementary Problems* **405**

Chapter 6 Supplementary Problems

Solve the following the problems.

13. $x + 2x + 3x = 180°$ $x = 30°$

14. $3x + 5x = 120°$ $x = 15°$

15. If m∠1 is half m∠2, what is m∠1? ∠2?

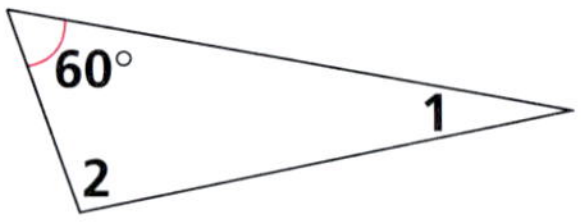

m∠2 = 80°; m∠1 = 40°

16. If m∠2 = m∠3, what is m∠2? ∠3?

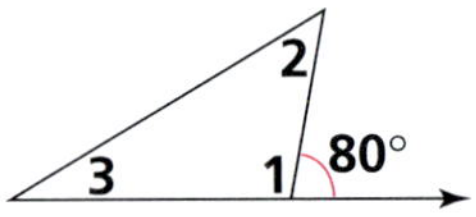

m∠2 = 40°, m∠3 = 40°

Complete the proof.

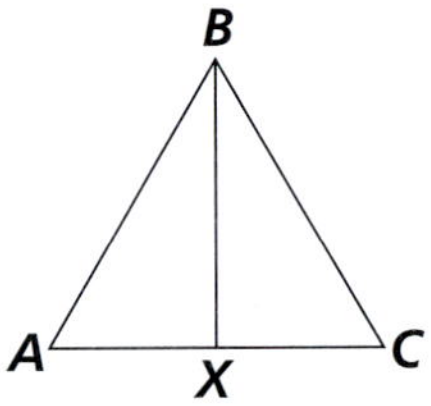

Given: ΔABC is an equilateral triangle.

$\overline{BX}$ is a perpendicular bisector.

Prove: $\Delta ABX \cong \Delta CBX$

PROOF

Statement	Reason
1. $\overline{BX}$ is a perpendicular bisector.	1. **17.** ______
2. $\overline{AB} = \overline{CB}$	2. **18.** ______
3. $\overline{AX} = \overline{XC}$	3. **19.** ______
4. $\overline{BX} = \overline{BX}$	4. **20.** ______
5. $\Delta ABX = \Delta CBX$	5. **21.** ______

17. Given.
18. Definition of equilateral triangle.
19. Definition of perpendicular bisector.
20. Any item is equal to itself.
21. SSS

406 *Supplementary Problems* *Geometry*

Chapter 6 Supplementary Problems

Tell how many lines of symmetry each figure has.

22. 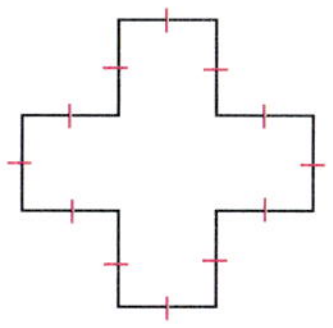

4

23. 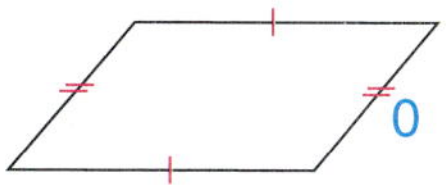

0

24. 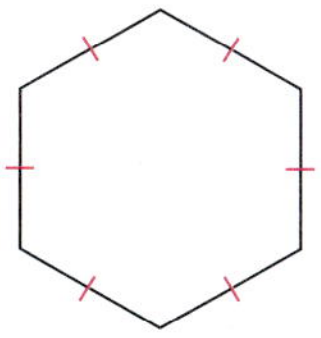

6

25. 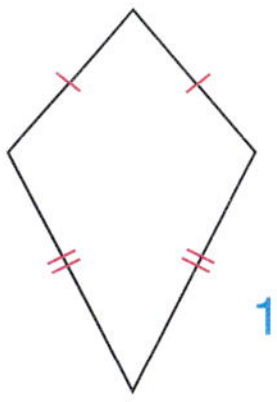

1

Identify the coordinate of the image of each point under the translation (3, –2).

26. $F = (8, 0)$ (11, –2)

27. $H = (12, -10)$ (15, –12)

28. $T = (9, -7)$ (12, –9)

29. $S = (-12, 3)$ (–9, 1)

30. $R = (3, 2)$ (6, 0)

31. $B = (-5, -2)$ (–2, –4)

32. $P = (-28, -45)$ (–25, –47)

33. $D = (-3, 4)$ (0, 2)

Reflect each image over the indicated axis. Give the coordinates of the image vertices.

34. *y*-axis

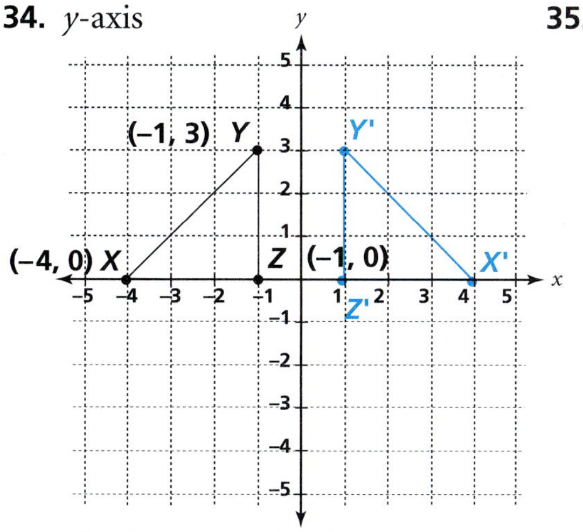

$X' = (4, 0)$, $Y' = (1, 3)$, $Z' = (1, 0)$

35. *x*-axis

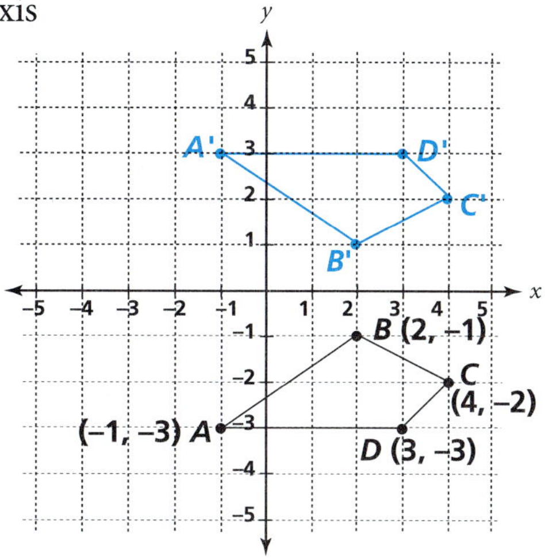

$A' = (-1, 3)$, $B' = (2, 1)$, $C' = (4, 2)$, $D' = (3, 3)$

Geometry Supplementary Problems 407

Chapter 7 Supplementary Problems

Are the following ratios equal? Write *yes* or *no*.

1. $\frac{6}{7}$ and $\frac{12}{14}$ yes

2. $\frac{1}{3}$ and $\frac{4}{11}$ no

3. $\frac{2}{5}$ and $\frac{7}{17}$ no

4. $\frac{6}{7}$ and $\frac{12}{14}$ yes

5. $\frac{9}{12}$ and $\frac{2}{3}$ no

6. $\frac{5}{8}$ and $\frac{10}{16}$ yes

7. $\frac{11}{10}$ and $\frac{110}{100}$ yes

8. $\frac{8}{3}$ and $\frac{24}{9}$ yes

9. $\frac{7}{16}$ and $\frac{21}{32}$ no

10. $\frac{3}{25}$ and $\frac{1}{5}$ no

Tell whether the triangles in each pair are similar. Answer *yes* or *no*. Give reasons for your answer.

11.

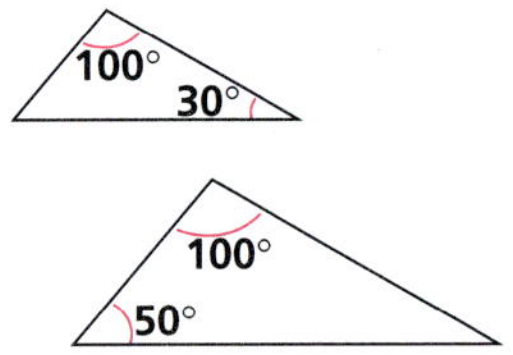

yes; 3 equal angles

12.

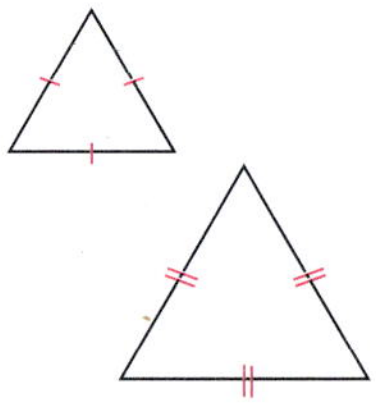

Yes; all angles are 60°

Name the corresponding angles and sides of the similar triangles.

13. $m\angle BDE = ?$ $m\angle A$

14. $m\angle C = ?$ $m\angle BED$

15. $\frac{DB}{AB} = \frac{BE}{?}$ $\overline{BC}$

16. $\frac{AC}{?} = \frac{BC}{?}$ $\overline{DE}$, $\overline{BE}$

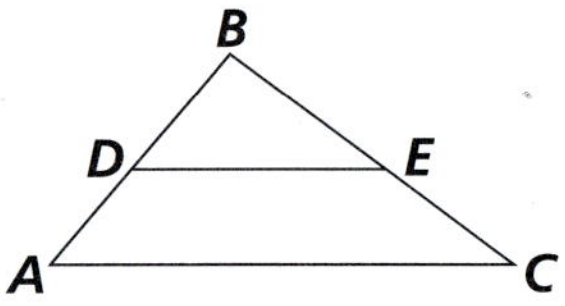

$\overline{DE} \parallel \overline{AC}$

408 *Supplementary Problems* *Geometry*

Chapter 7 Supplementary Problems

Solve for the unknown value in each pair of similar triangles.

17. x 20

18. y 10

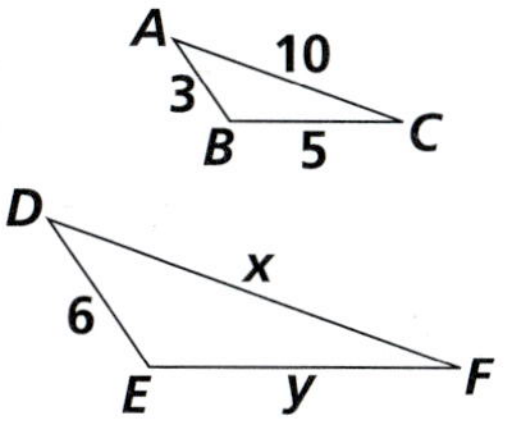

19. x 100°

20. y 50°

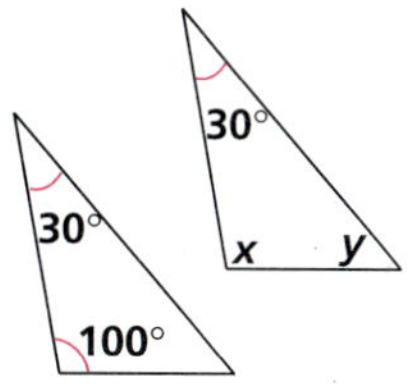

Find the perimeter of each polygon.

21. The ratio of similarity is 3:2

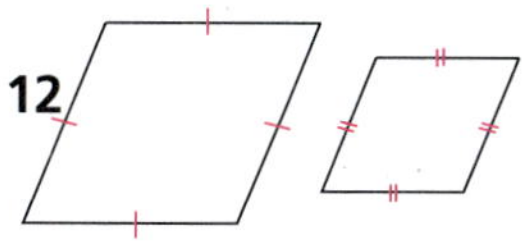

48, 32

22. The ratio of similarity is 3:1

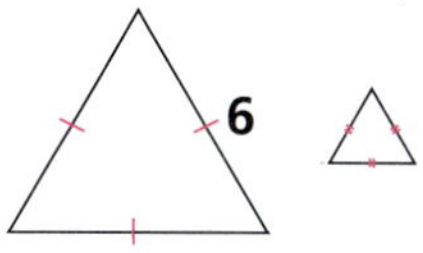

18, 6

23. The ratio of similarity is 4:5

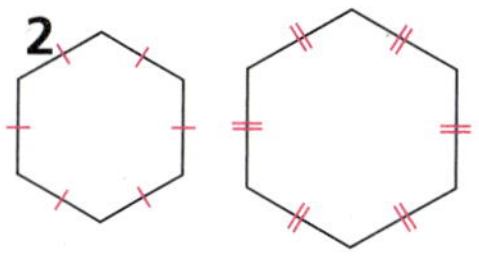

12, 15

Solve the following problems.

24. To find the height of a building, an engineer used similar triangles and the height of a smaller building to draw the diagram shown. What is the height of the taller building? 480 feet

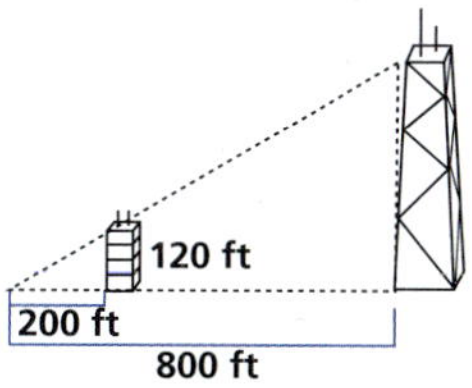

25. Jonas is making a poster by making a 3 × 5-inch photo 4 times larger. He drew the size of the photo on a grid as shown. What will the coordinates of the poster be if Jonas draws its image on the grid?

(–6, 10)(6, 10)(–6, –10)(6, –10)

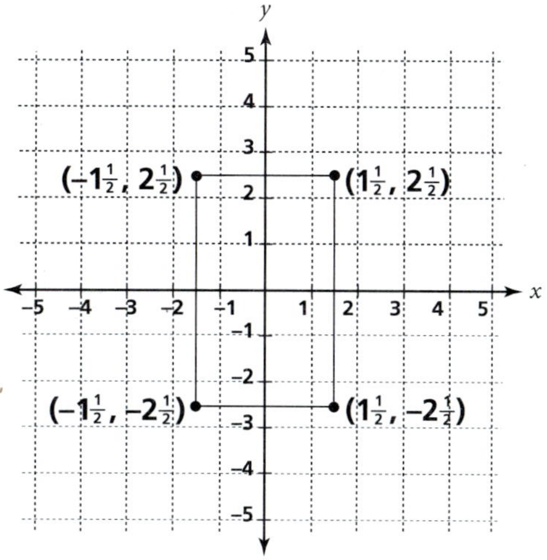

Geometry *Supplementary Problems* **409**

Chapter 8 Supplementary Problems

Find additional triples using the given triples and multiplier.

1. triple (3, 4, 5) by 2 6, 8, 10

2. triple (3, 4, 5) by 8 24, 32, 40

3. triple (3, 4, 5) by 10 30, 40, 50

4. triple (5, 12, 13) by 2 10, 24, 26

5. triple (5, 12, 13) by 3 15, 36, 39

6. triple (5, 12, 13) by 10 50, 120, 130

Find the length of the unknown side of each right triangle.

7.

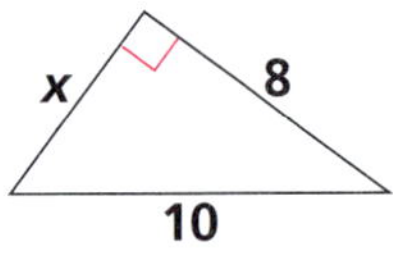

$x = 6$

8.

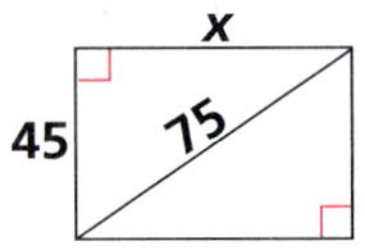

$x = 60$

9.

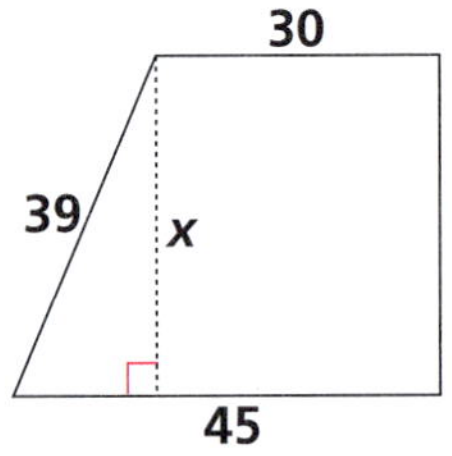

$x = 36$

Find the area of each trapezoid.

10.

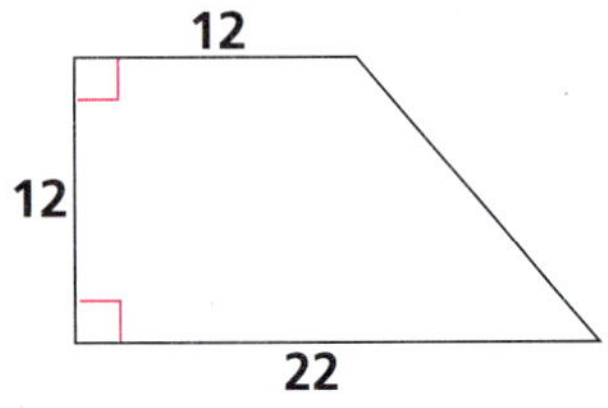

$A = 204$

11.

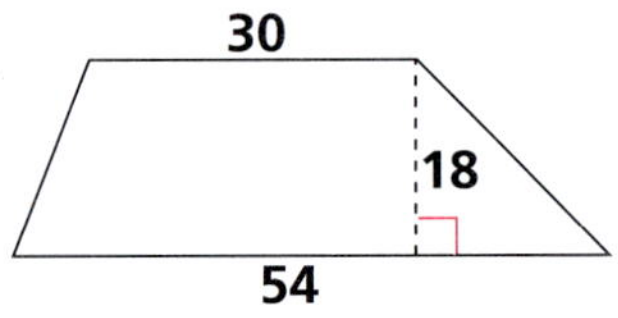

$A = 756$

Solve the following problems.

12. A window washer uses a ladder to reach a window 15 feet high. The base of the ladder must be placed at least 4 feet from the wall for safety reasons. Will a 16-foot ladder reach the window?

yes

410 *Supplementary Problems* *Geometry*

Chapter 8 Supplementary Problems

13. Charlotte cut a triangular piece of glass to fit in a 90° corner of the frame of a large stained-glass window. After cutting the piece, she measures the lengths of the sides and finds they are $3\frac{1}{2}$ ft, 6 ft, and $7\frac{1}{4}$ ft. Will the piece of glass fit in the corner? Explain your answer.

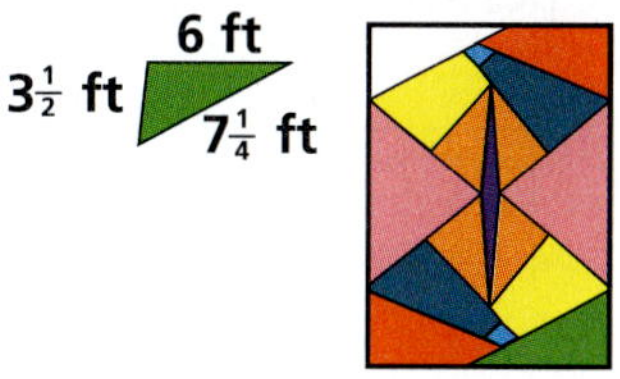

no, the cut piece is not a right triangle

Tell whether or not each triangle is a right triangle. Answer *yes* or *no.* If it is, name the right angle.

14.

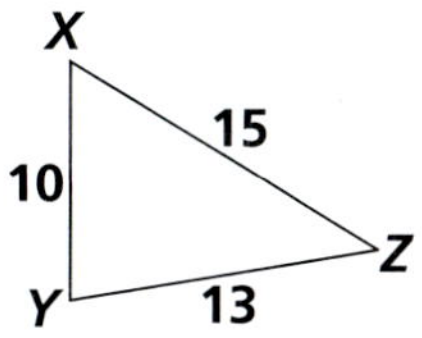

no

15.

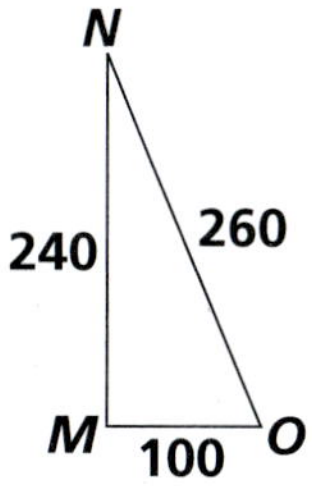

yes; $\angle M$

16.

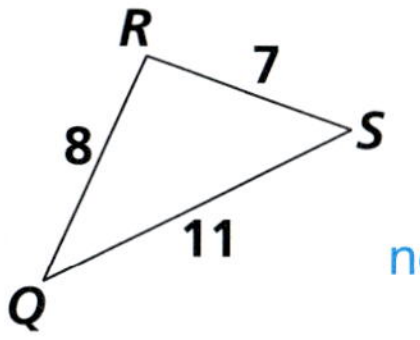

no

17.

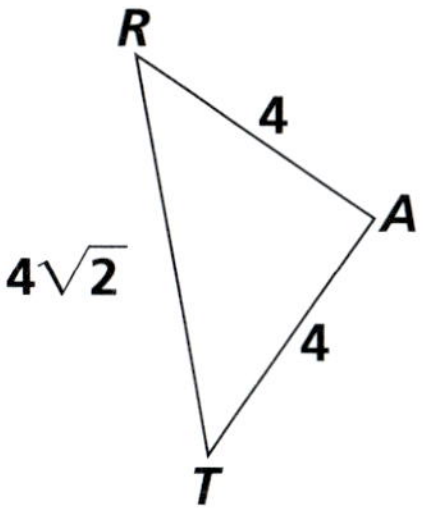

yes; $\angle A$

Use the distance formula to find the distances between the given points. Round to the nearest tenth.

18. (–4, –6) and (8, 10) $d = 20.0$

19. (8, 10) and (20, 32) $d = 25.1$

20. (9, –5) and (20, –10) $d = 12.1$

Geometry *Supplementary Problems* **411**

Find the perimeter of each polygon.

1.

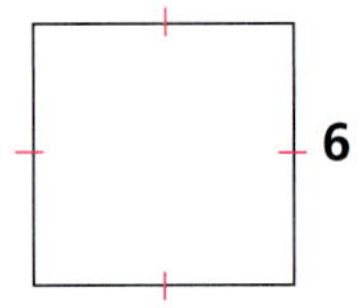

$P = 24$

2.

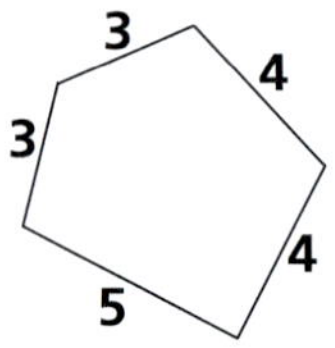

$P = 19$

3.

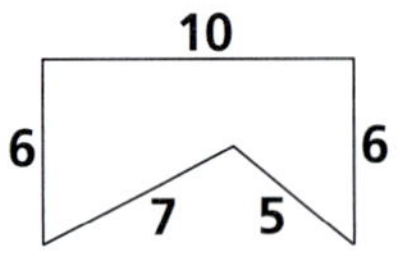

$P = 34$

4.

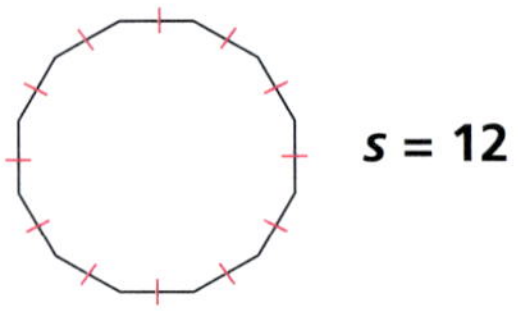

$P = 144$

5. rhombus, $s = 125$ $P = 500$

6. rectangle, $l = 14$, $w = 10$ $P = 48$

Solve the following problems.

7. What are the length and width of a rectangle that has a perimeter of 60 and a length 2 times the width? $l = 20$, $w = 10$

8. Three times the base of a parallelogram is equal to 4 times the height. The parallelogram has a perimeter of 42. What is its base? Height? $b = 12$, $h = 9$

9. What is the length of the diagonal of a square with an area of 64? $8\sqrt{2}$

10. The length of a rectangle is 4 times the width. The perimeter of the rectangle is 120. What are the length and width of the rectangle? $w = 12$, $l = 48$

Find the length of the unknown side.

11.

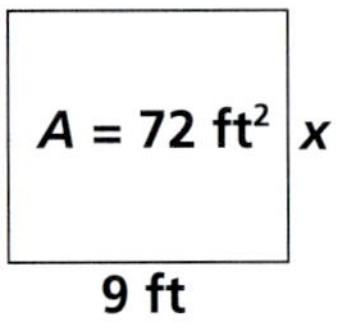

$x = 8$

12.

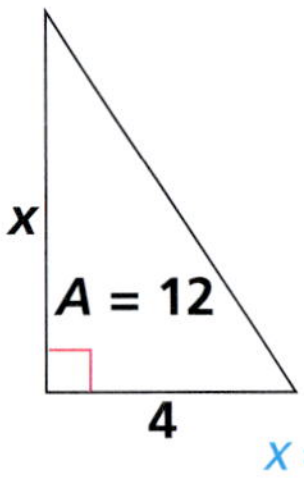

$x = 6$

13.

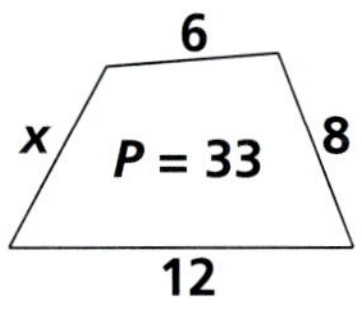

$x = 7$

412 *Supplementary Problems* *Geometry*

Chapter 9 Supplementary Problems

Find the area of each quadrilateral.

14.

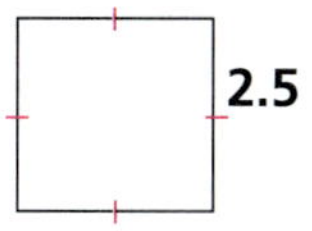

$A = 6.25$

15.

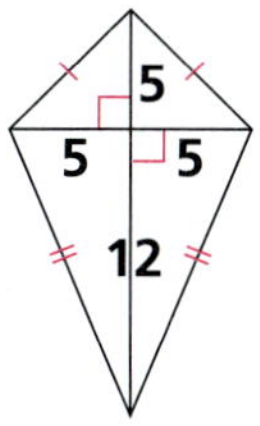

$A = 85$

16.

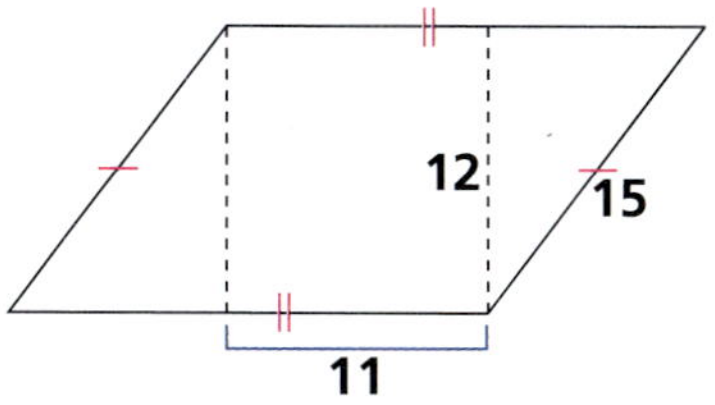

$A = 240$

17.

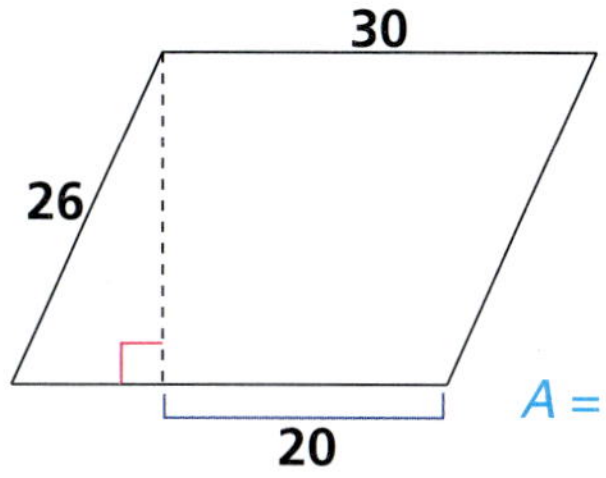

$A = 720$

Use Heron's Formula to find the areas of triangles with the given side lengths. Use a calculator and round to the nearest tenth.

18. 9, 10, 13 $A = 44.9$

19. 4, 5, 7 $A = 9.8$

20. 10, 10, 16 $A = 48.0$

21. 12, 12, 12 $A = 62.4$

Find the area of each triangle.

22.

$A = 270$

23.

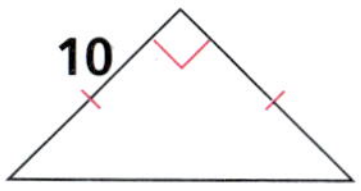

$A = 50$

24.

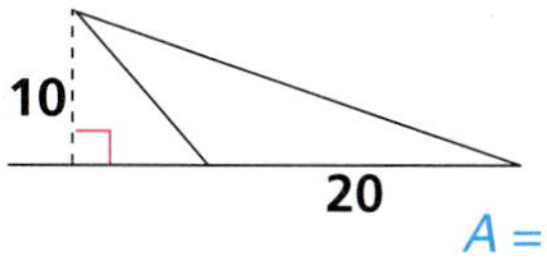

$A = 100$

25.

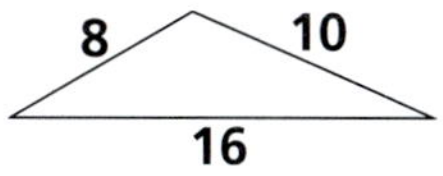

$A = 32.7$

Geometry *Supplementary Problems* **413**

Chapter 10 Supplementary Problems

Estimate the circumference and the area of a circle with the given radius or diameter. Use 3.1 as an approximate value for π.

		Circumference		Area
$r = 1$ in.	**1.** ______	6.2 in.	**2.** ______	3.1 sq in.
$d = 1$ cm	**3.** ______	3.1 cm	**4.** ______	0.78 cm^2
$r = 20$ ft	**5.** ______	124 ft	**6.** ______	1,240 sq ft
$d = 30$ m	**7.** ______	93 m	**8.** ______	697.5 m^2
$r = \frac{1}{2}$ in.	**9.** ______	3.1 in.	**10.** ______	0.78 sq in.

Find the area of the shaded sector. Give your answer in terms of π.

11.

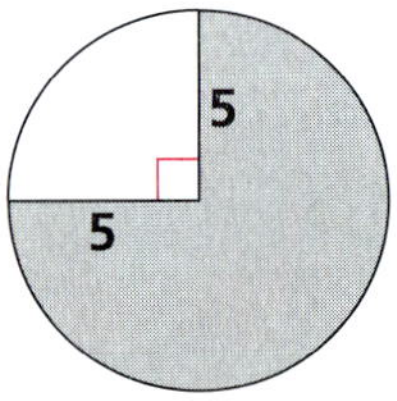

$\frac{75}{4}\pi$

12.

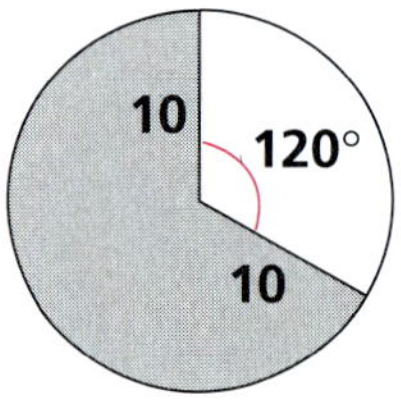

$\frac{200}{3}\pi$

13.

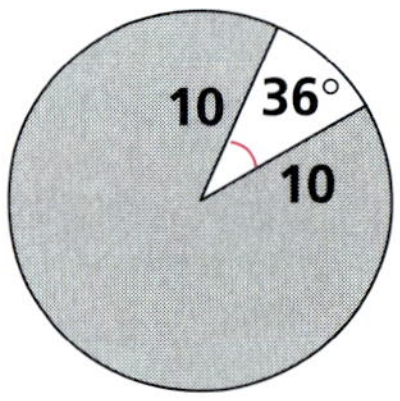

90π

Find the value of the unknown.

14. inscribed angle 60°

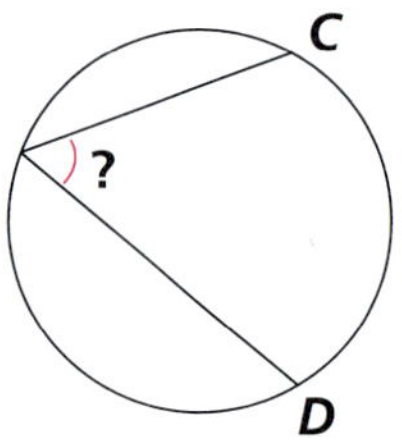

15. $\overset{\frown}{PR} = ?$ 160°

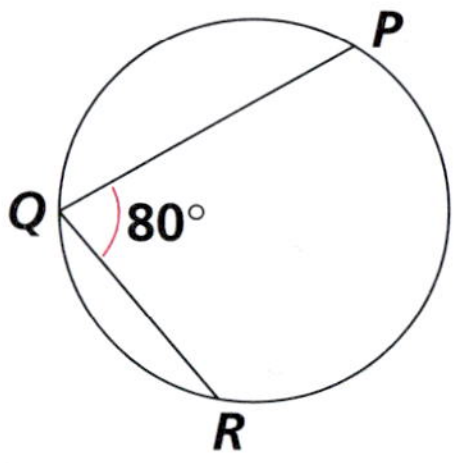

414 *Supplementary Problems* *Geometry*

Find the probabilities.

16. P(shaded area)

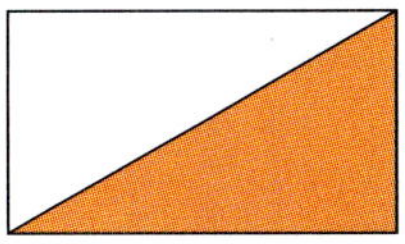

50%

17. P(shaded area)

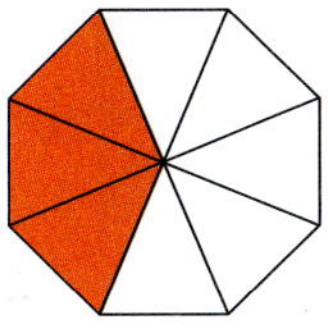

37.5%

18. P(shaded area)

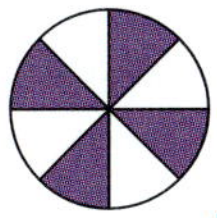

50%

Find the surface area and the volume for each sphere. Give your answer in terms of π.

	Surface Area	Volume
$r = 1$ mi	**19.** ______ 4π sq mi	**20.** ______ $\frac{4}{3}\pi$ cu mi
$d = 4$ ft	**21.** ______ 16π sq ft	**22.** ______ $\frac{32}{3}\pi$ cu ft
$r = 20$ ft	**23.** ______ $1{,}600\pi$ sq ft	**24.** ______ $\frac{32{,}000}{3}\pi$ cu ft
$d = 30$ m	**25.** ______ 900π m^2	**26.** ______ $4{,}500\pi$
$r = \frac{1}{2}$ in.	**27.** ______ 1π sq in.	**28.** ______ $\frac{1}{6}\pi$ cu in.

Choose the correct term to complete each sentence.

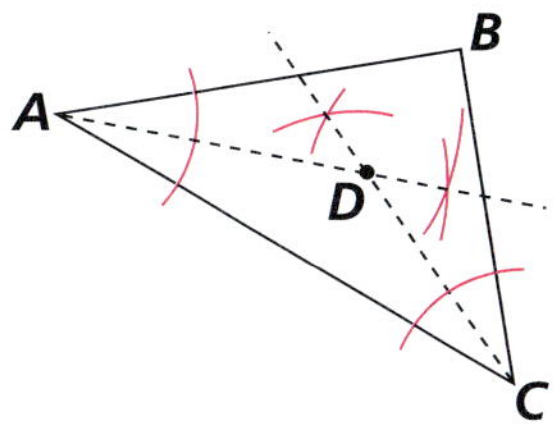

29. In ΔABC, the point D where the angle bisectors meet is the center of the triangle's ____. B

A. circumcenter **B.** incircle **C.** circumcircle

30. An incircle of ΔABC touches _____. B

A. each vertex of the triangle **B.** each side of the triangle

C. each perpendicular bisector

Geometry *Supplementary Problems* **415**

Chapter 11 Supplementary Problems

Find the volumes of these solids.

1.

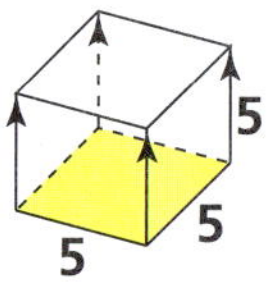

125 cu units

2.

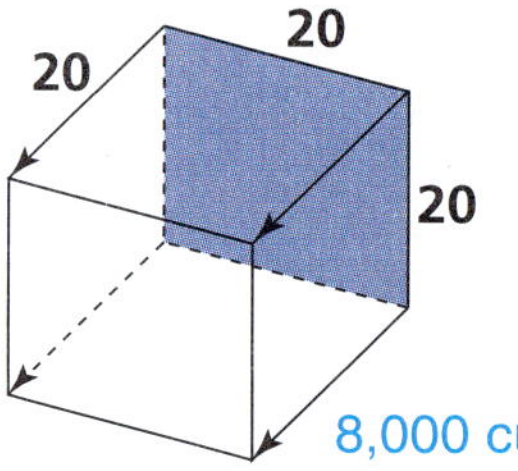

8,000 cu units

3.

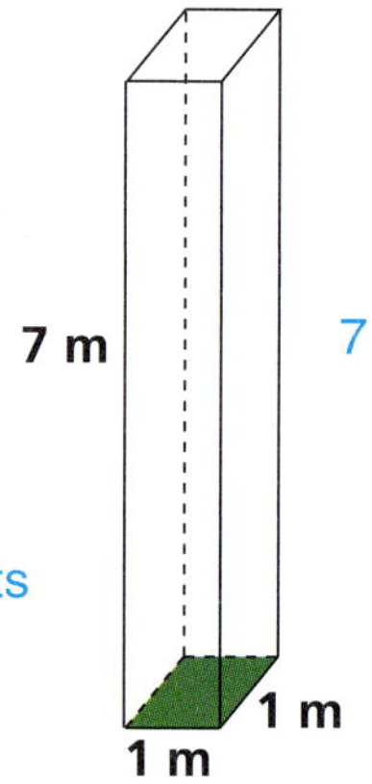

7 m³

4.

432 cm³

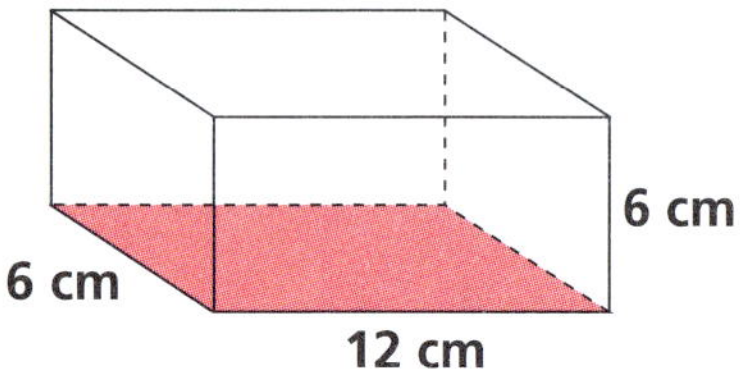

5.

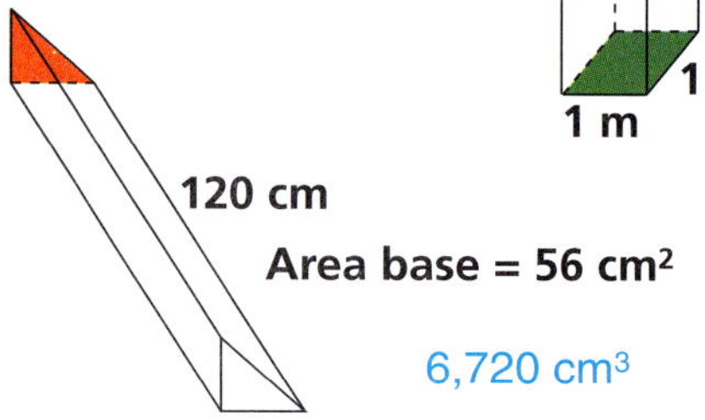

6,720 cm³

Find the surface area of each figure. Some answers may be in terms of π.

6.

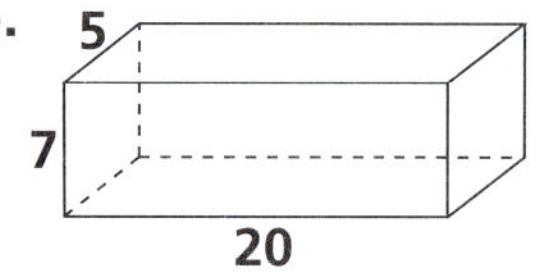

550 square units

7.

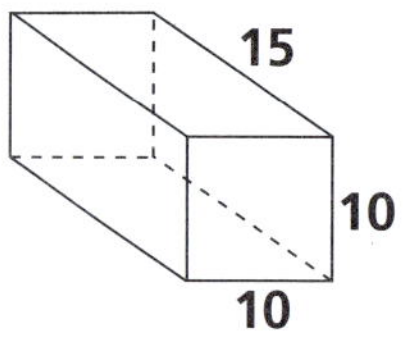

800 square units

8.

248 square units

9.

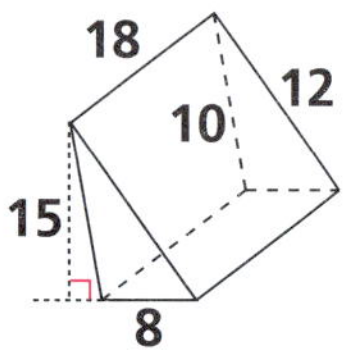

660 square units

10.

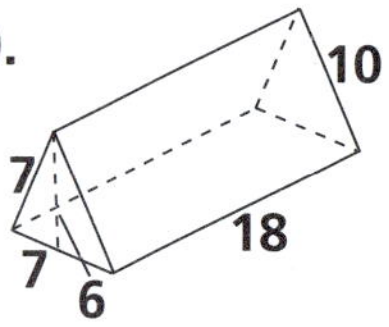

474 square units

11.

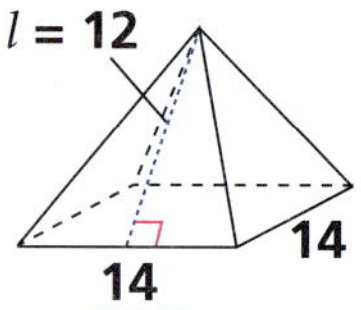

532 square units

12.

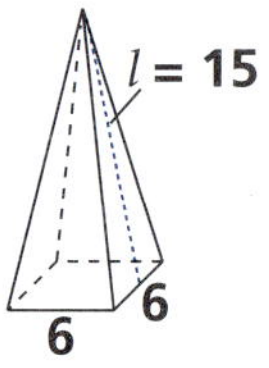

216 square units

13.

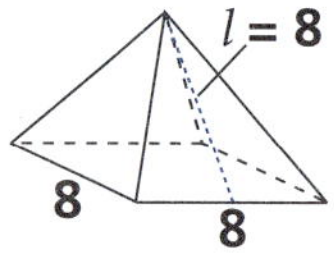

192 square units

14.

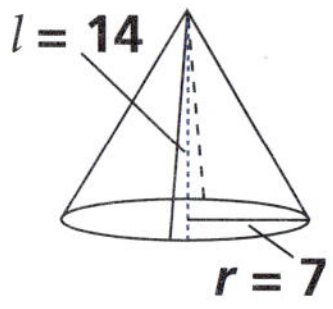

147π square units

15.

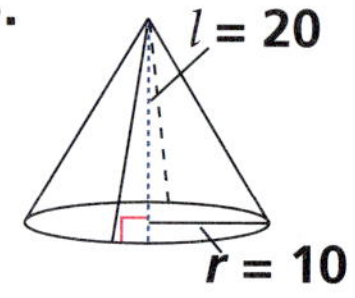

300π square units

416 *Supplementary Problems* *Geometry*

Chapter 11 Supplementary Problems

Find the volume and total surface area for each figure. You may give your answers in terms of π.

16.

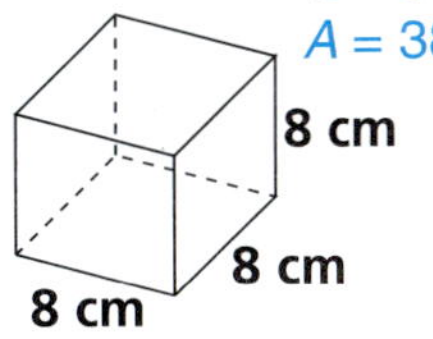

$V = 512 \text{ cm}^3$
$A = 384 \text{ cm}^2$

17.

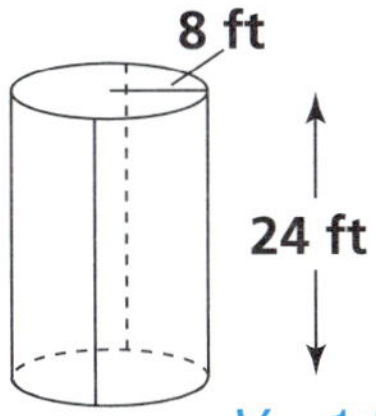

$V = 1{,}536\pi$ cu ft
$A = 512$ sq ft

18.

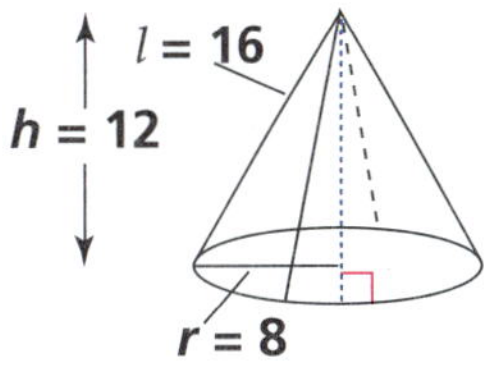

$V = 256\pi$ cubic units
$A = 192\pi$ square units

19.

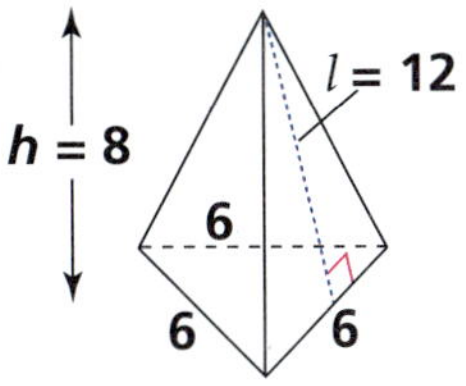

20.

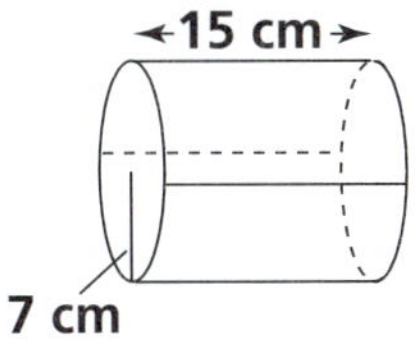

19. $V = 24\sqrt{3}$ cubic units
$A = 108 + 9\sqrt{3}$ square units
20. $V = 735\pi \text{ cm}^3$
$A = 308\pi \text{ cm}^2$

Solve the following problems. You may leave some answers in terms of π.

21. What is the surface area of a triangular door stop whose base measures 6 in. × 6 in. × 6 in. and whose height is 12 in.?

$216 + 18\sqrt{3}$

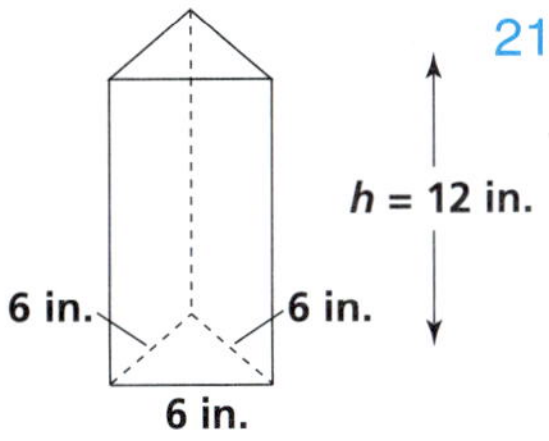

22. Deonne plans to cover the sides and roof of the birdhouse below with copper sheeting. How much copper will he need? 642 cm^2

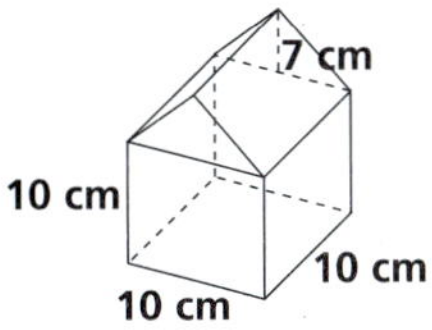

23. What is the surface area of an open aquarium that is 24 in. wide, 14 in. long, and 18 in. deep? 1,704 sq in.

24. Mr. Joungston needs to paint the sides but not the dome of the old corn silo. How many square feet of metal will he be painting?

$1{,}296\pi$ sq ft

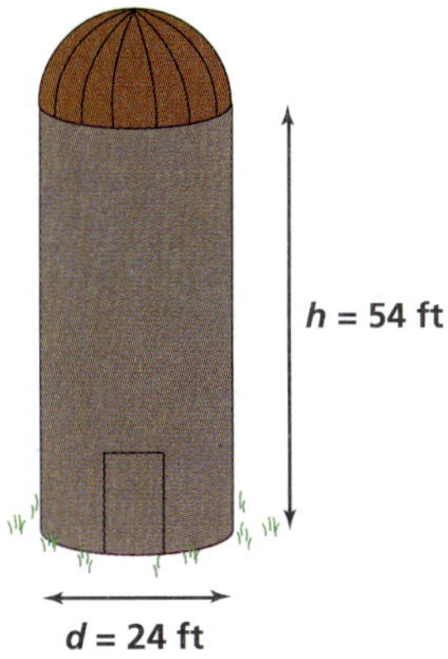

25. How much plastic shrink-wrap is needed to wrap the case of 24 cans of water?

912 sq in.

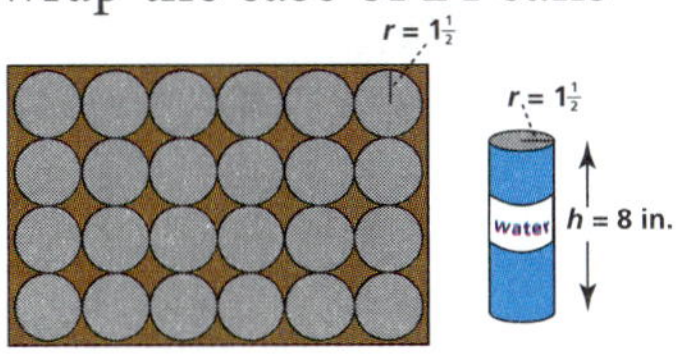

Geometry Supplementary Problems 417

Chapter 12 Supplementary Problems

Sketch and describe the locus of the intersection of the following pairs of figures.

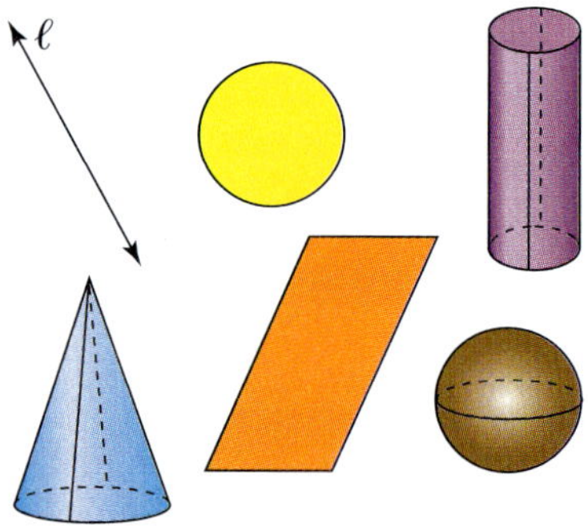

1. A line and a circle have one point in common.
2. A line and a circle have two points in common.
3. A plane and a cone form a circle.
4. A plane and a cone form an oval.
5. Two parallel lines and a circle form 4 points.
6. Two parallel planes and a sphere produce a circle and a great circle.
7. Each of two planes forms a distinct great circle with a sphere.
8. Two parallel planes and a cylinder produce 2 circles.
9. Two parallel planes and a cylinder produce two ovals.
10. Two parallel planes and a cylinder produce 2 lines.

See pages 452-453 for answers.

Draw a sketch and write an equation for each locus.

11. Locus of points 2 units away from $y = \frac{1}{2}$. $y = 2.5$ and $y = -1.5$
12. Locus of points 4 units away from $x = -2$.

$x = -6$ and $x = 2$

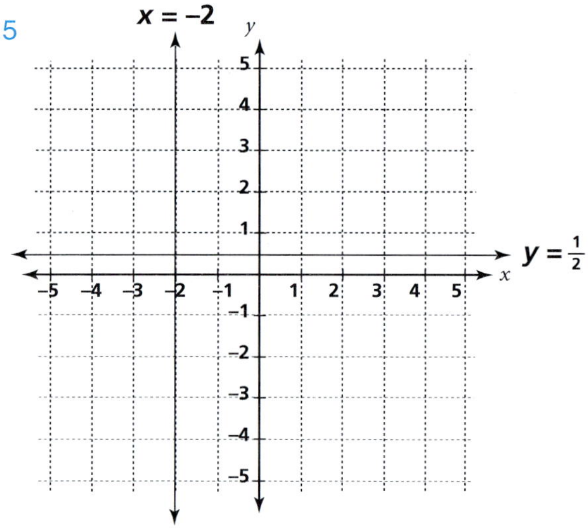

418 *Supplementary Problems* *Geometry*

13. Locus of points equidistant from $x = 2$ and $x = 5$.

$x = 3.5$

See page 453 for answers.

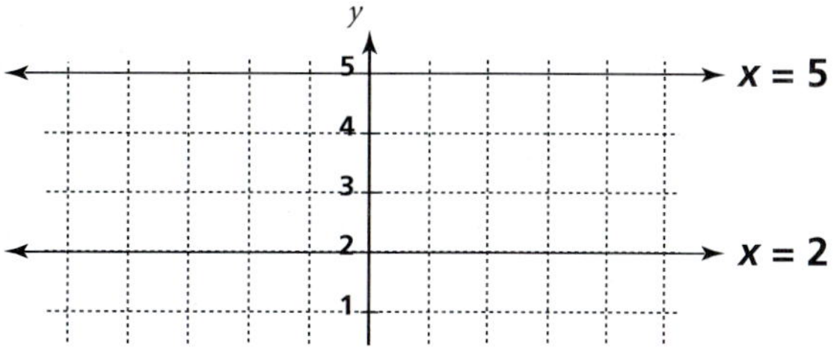

14. Locus of points equidistant from $y = -x + 2$ and passing through $y = 4$ and $y = 0$.

$y = -x + 4$ and $y = -x$

15. Locus of points equidistant from $y = -\frac{1}{3}x + 3$ and passing through $y = 1$ and $y = 5$.

$y = -\frac{1}{3}x + 1$ and $y = -\frac{1}{3}x + 5$

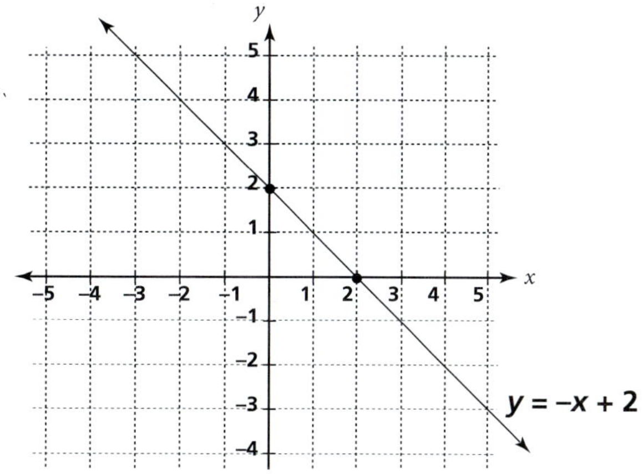

Sketch and describe these compound loci.

16. The locus of points equidistant from $\overrightarrow{XW}$ and $\overrightarrow{XZ}$ and at a distance of $\overline{GH}$ from X.

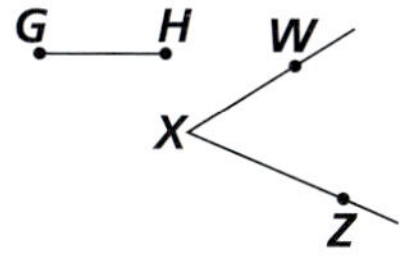

16. one point along the angle bisector and on the circle with radius *GH* 17. the intersection of the angle bisectors of each angle of the triangle

17. The locus of points equidistant from $\overline{XY}$, $\overline{YZ}$, and $\overline{ZX}$.

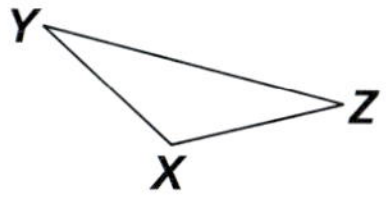

18. The locus of points equidistant from $\angle H$, $\angle I$, and $\angle J$.

the intersection of the perpendicular bisectors of each side

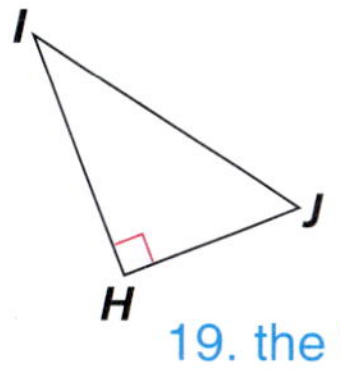

19. The locus of points equidistant from points A and B and equidistant from points C and D.

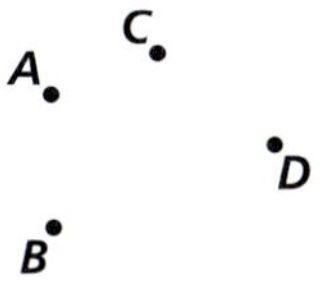

19. the intersection of the perpendicular bisectors of $\overline{AB}$ and $\overline{CD}$

20. The locus of points equidistant from the parallel lines ℓ and m and at a distance of $\overline{GH}$ from P.

two points on the circle with radius *GH* and center *P* and on the perpendicular bisector of a line perpendicular to both ℓ and m

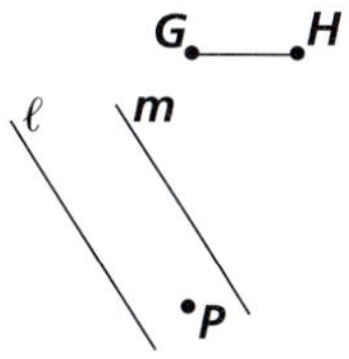

Geometry *Supplementary Problems* **419**

CHAPTER 1

Lesson 1, pages 4–6

1. *A, F, C;* Use a straightedge to draw a line through any three points that appear to be on the same line. If the line goes through all three, then they are collinear. **3.** *A, B, C; A, D, E; E, F, G;* and *G, H, I* **4.** ray; A ray has one endpoint and one end that extends with no end. **5.** line **7.** ray **9.** line segment **10.** $\overleftrightarrow{GH}$, $\overleftrightarrow{HG}$, and *m*; A line can be named by any two of its labeled points in either order and by a lowercase script letter. **11.** two; A line segment has two endpoints by definition. **13.** one **15.** points **17.** ray **18.** one; Only one line segment can be drawn between two points because a line segment needs two endpoints. **19.** three

Try This

n = number of sides or points; $n \cdot \frac{n-1}{2}$ or count down from $n - 1$ to 1, adding each possibility; square = 4 $\rightarrow$ 3 + 2 + 1 = 6 segments; 5 points $\rightarrow$ 4 + 3 + 2 + 1 = 10 segments

Lesson 2, pages 9–10

1. *A* –3, *B* –1, *C* 2, *D* 3.5, *E* 6; The Ruler Postulate states that for every point on the line, there is exactly one real number. **2.** –3 *A*, $-\frac{1}{2}$ *B*, 0 *C*, $\frac{3}{4}$ *D*, $\sqrt{2}$ *E*, π *F*, 4 *G* **3.** 3 units; The Ruler Placement Postulate states that given two points, *A* and *B*, on a line, the number line can be chosen so that *A* is at zero and *B* is at a positive number. **5.** 2 inches; Place the ruler so that the left edge of the line segment is at zero and the line segment lines up with the edge of the ruler. Then take a reading of where the right endpoint of the line segment falls on the ruler. **7.** 1 cm; See directions for problem 5. **9.** 3 cm **10.** $\overline{AB} = 2$, $\overline{BC} = 4$, $\overline{AC} = 6$; $\overline{AB} + \overline{BC} = 2 + 4 = 6$; *B* is between *A* and *C*. **11.** $\overline{AB} = 4$, $\overline{BC} = 8$, $\overline{AC} = 12$; $\overline{AB} + \overline{BC} = 4 + 8 = 12$; *B* is between *A* and *C*. **12.** Ruler Postulate; The Ruler Postulate states that the points on a line can be placed in a one-to-one correspondence with the real numbers so that for every point on the line, there is exactly one real number. **13.** Ruler Placement Postulate **15.** no; If the points are not in a straight line, $AB + BC \neq AC$.

Lesson 3, page 14

1. ∠*Q*, ∠*RQP;* An angle can be named by its vertex point or by its vertex point and a labeled point on each ray with the vertex point in the middle. **3.** ∠*QRS*, ∠*SRQ* **4.** Use the steps on page 12.

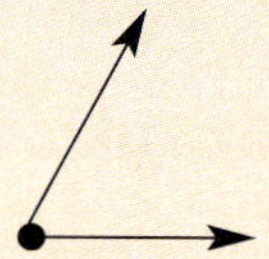

5.

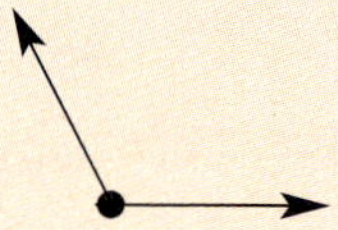

7. Use the steps on page 13.

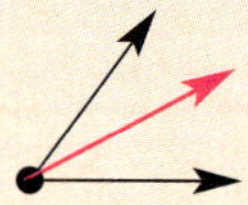

9.

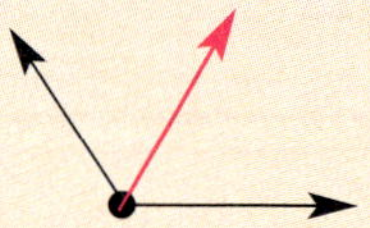

10. Yes, common point *B* could be the vertex of an angle.

420 *Selected Answers and Solutions* *Geometry*

Lesson 4, pages 18–19

1. acute; Any angle that measures less than 90° is acute. **3.** straight **5.** acute **7.** obtuse **9.** Angle measurements will vary but should be less than 90°. **11.** Angle measurements will vary but should be greater than 90°. **12.** $\ell \perp m$; Lines that form right angles are said to be perpendicular. **13.** $\overline{AB} \perp \overline{BC}$ **15.** When two perpendicular lines intersect, they form four right angles.

Lesson 5, pages 22–23

1. supplementary; If the sum of two angle measures equals 180°, they are supplementary. **3.** supplementary **5.** 80°; Press 180 [–] 100 [=]. The answer is 80. **7.** 118° **9.** 127° **11.** 99° **13.** $\angle m$ and $\angle p$, $\angle o$ and $\angle n$; Vertical angles have a common vertex and their sides are formed by the same lines. **15.** $\angle t$ and $\angle y$, $\angle u$ and $\angle x$, $\angle v$ and $\angle w$, $\angle s$ and $\angle z$ **17.** 90°; f and c are vertical angles; $c = 90°$, so $f = 90°$ **19.** 45°

Lesson 6, pages 26–27

1. $m\angle a = 100°$, $m\angle b = m\angle c = 80°$; $\angle a$ is a vertical angle to the angle labeled 100°. Therefore, $\angle a = 100°$. $\angle b$ is supplementary to $\angle a$. $180° - 100° = 80°$. Therefore, $\angle b = 80°$. $\angle c$ is a vertical angle to $\angle b$. Therefore, $\angle c = 80°$. **3.** $m\angle w = 150°$, $m\angle z = m\angle q = 30°$ **4.** 48°; $\angle 1$ is complementary to $\angle 2$, which equals 42°. Since the sum of two complementary angles must be equal to 90°, $90° - 42° = 48°$. **5.** 90° **7.** 42° **9.** 180°; $\angle 3$, or $\angle EBF$, is supplementary to $\angle CBE$, which is a right angle. Right angles are equal to 90°. $\angle 6$ and $\angle 3$ are vertical angles, so $\angle 6 = 90°$; $90° + 90° = 180°$. **11.** $\angle 1$, $\angle 4$; $\angle 2$, $\angle 5$; and $\angle 3$, $\angle 6$ **13.** For any pair of vertical angles, the angles have equal measures. **15.** 36° and 144° **17.** 72° and 108° **19.** 30° and 150°

Application, page 28

1. yes; *D* to *C* to *B* to *D* to *A* to *B*, or start at *B* and reverse. The network is traversable because you can trace a route along its points without repeating a section. **3.** yes; *D* to *A* to *C* to *E* to *A* to *B* to *C* to *D* to *E* **5.** yes; *B* to *D* to *C* to *B* to *A* to *D* to *F* to *B* to *G* to *F* to *E* to *D*

Chapter Review, pages 29–31

1. line segment, $\overline{AB}$ **3.** line, $\overleftrightarrow{RS}$ **5.** ray, $\overrightarrow{ML}$ **7.** line, m **9.** line, $\overleftrightarrow{RQ}$ **11.** B **13.** C **15.** 1 inch **17.** 5 cm **19.** $\angle S$, $\angle TSR$ **21.** $\angle YXW$, $\angle X$, or $\angle a$ **23.** right **25.** acute **27.** obtuse **29.** 60° **31.** 65° **33.** 140° **35.** Angle measure should be 180°.

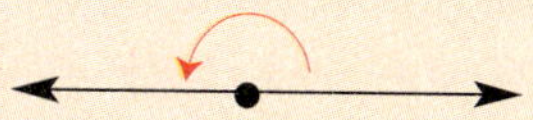

CHAPTER 2

Lesson 1, pages 37–38

1. *a figure is a triangle*, hypothesis; *it has three sides*, conclusion; The statement that follows the "If" is the hypothesis. The statement that follows the "then" is the conclusion. **3.** *the sum of the measures of two angles is 90°*, hypothesis; *the angles are complementary*, conclusion **5.** *you use SuperBright detergent*, hypothesis; *your clothes will be clean*, conclusion. **6.** True; Right angles always

Geometry *Selected Answers and Solutions* **421**

equal 90°. Since the conclusion meets the hypothesis, then the conditional is considered true. **7.** True **9.** False **C11.** If an angle measure is $90° < m < 180°$, the angle is obtuse. The converse of a conditional is the reverse of the hypothesis and conclusion. **C13.** If two angles are complementary, then the sum of their measures is 90°. **C15.** If an animal has four legs, then it is a dog. **D11.** The conditional and its converse are true. An obtuse angle must measure more than 90° and less than 180°. Conversely, any angle that measures more than 90° and less than 180° must be obtuse. **D13.** The conditional and the converse are true. **D15.** The conditional is true and the converse is false.

Try This

Answers will vary.

Lesson 2, pages 42–43

1. Postulate 3, a circle may be described with any center and distance.
3. Postulate 5, Parallel Postulate
5. Postulate 4, all right angles are equal to one another. **7.** Postulate 1
9. Postulate 3 **10.** Lines n and m are parallel; $m\angle d = 90°$ because $\angle a$ and $\angle d$ are supplementary and $m\angle b = 90°$ because $\angle b$ and $\angle c$ are supplementary. Then, since $m\angle c + m\angle d = 180°$, the lines do not meet the requirement for intersection and so they must be parallel.

Lesson 3, pages 45–48

1. Postulate 1, a straight line can be drawn from any point to any point.
3. Postulate 2 **5.** Postulate 2
7. Postulate 3 **9.** Postulates 1 and 3
11. Postulate 1; Construction is completed by connecting points K and L with a straightedge. Postulate 1 states that a straight line can be drawn from any point to any point.

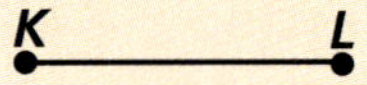

13. Postulate 4, all four angles are equal.

15. Postulate 5 (Parallel Postulate)

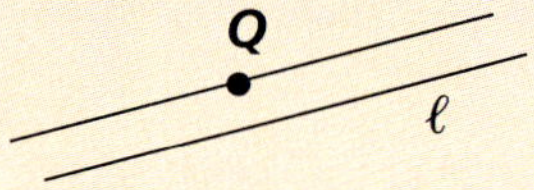

17. Postulate 1

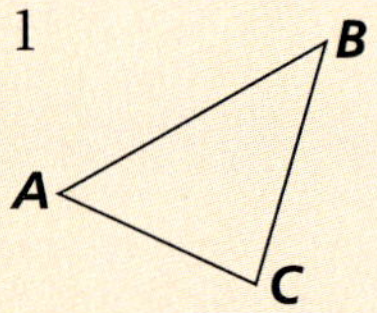

19. Postulate 5 (Parallel Postulate)

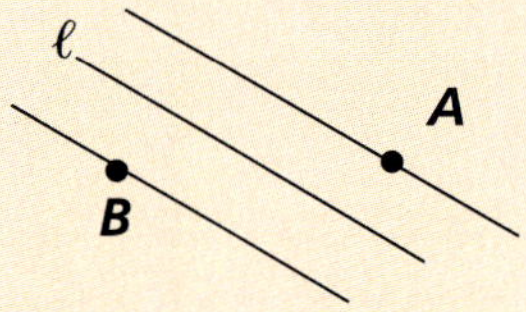

Lesson 4, pages 51–52

1. Axiom 2, which states that if equals are added to equals, the sums are equal.
3. Axiom 3 **5.** Axiom 2 **6.** $\angle BAC < 90°$; $\angle CAD < 90°$. Axiom 5, the whole is greater than the sum of the parts. Since $\angle CAD$ and $\angle BAC$ are parts of $\angle BAD$, which measures 90°, then they must each

422 *Selected Answers and Solutions* *Geometry*

measure less than 90°. **7.** Yes; Axiom 1, substitution for equals. **9.** Angles 1, 2, 3, and 1 + 2, 1 + 3, and 2 + 3 are all less than 180°; Axiom 5, whole is greater than parts **11.** 8 pt; Axiom 1, substitution; If 1 quart equals 2 pints, and 1 gallon equals 4 quarts, then 1 gallon = 4 × 2 pints = 8 pints.
13. 16 oz; Axiom 1, substitution
15. 3,650 days; Axiom 1, substitution

Lesson 5, pages 54–55

1. Given; The first statement in a proof is given because it is part of the conditional being proven. **3.** $\angle a$ and $\angle b$ are adjacent on ℓ and are supplementary. **5.** Axiom 3, if equals are subtracted from equals, the differences are equal. **6.** $x = 60°$, $y = 120°$, $z = 60°$; The vertical angle theorem states that vertical angles are equal. Therefore, $m\angle y$ is equal to its vertical angle, which is equal to 120°. Angles x and z are both supplementary to the angle labeled 120°. Since 180° – 120° = 60°, both x and z are equal to 60°. **7.** $x = 160°$, $y = 20°$, $z = 160°$ **9.** $x = 89°$, $y = 91°$, $z = 89°$ **11.** Angles a and b are vertical angles. The theorem states that vertical angles are equal. **13.** $b + d = 180°$, so $b < 180°$; Axiom 5, whole is greater than parts **15a.** Angles x and w are vertical angles. The theorem states that vertical angles are equal. Angles x and w are right angles. Postulate 4, all right angles are equal to one another.
15b. Vertical angles are equal. Postulate 4, all right angles are equal to one another. **15c.** Postulate 4, right angles are equal; Axiom 1, substitution.

Application, page 56

1. Artwork will vary. **3.** Answers will vary. **5.** Answers will vary.

Chapter Review, pages 57–59

1. An angle is acute. **3.** A figure has three sides. **5.** It measures less than 90°. **7.** It is a triangle. **9.** If two angles are complementary, then the sum of their measures is 90°. **11.** Postulate 1
13. Postulate 3 **15.** Axiom 5 **17.** $m\angle a + m\angle c = 180°$ **19.** $m\angle a + m\angle c = m\angle a + m\angle b$ **21.** A **23.** A **25.** A

CHAPTER 3

Lesson 1, pages 63–64

1. True; Lines AE and BD are both perpendicular to line FC and therefore never meet. Since they never meet and they are both on the same plane, they are parallel lines. **3.** True **5.** True
7. True **9.** True **11.** $\ell \nparallel m$ **13.** $p \parallel q$
15. $r \parallel s \parallel t$

Try This

The lines are skew because they are not in the same plane.

Lesson 2, page 67

1. w, x, s, r; Exterior angles are outside the parallel lines. Angles w, x, s, and r are outside the parallel lines ℓ and m.
3. w and p, x and q, z and s, y and r
5. w and x, z and y, w and z, x and y, p and q, s and r, p and s, q and r
6. alternate exterior; f and k are exterior angles on opposite sides of the transversal. **7.** exterior
9. corresponding **11.** supplementary
13. corresponding **15.** supplementary

Geometry *Selected Answers and Solutions* **423**

16. yes; The corresponding angles should appear to be the same exact measure. **17.** no **19.** yes

Lesson 3, pages 70–72

1. 180°; The theorem states that interior angles on the same side of the transversal are supplementary. Since $\angle a$ and $\angle c$ are interior angles on the same side of the transversal, they are supplementary. Therefore, their sum must equal 180°. **3.** $m\angle d$ **5.** $m\angle e$ **7.** $m\angle g$ **9.** $m\angle a$ **11.** $m\angle c$ **13.** 120°; $\angle q$ is supplementary with the angle whose measure is 60°; 180° – 60° = 120°; $\angle q$ = 120°. **15.** 120°; $\angle q$ and $\angle s$ are vertical angles and therefore equal. **17.** 60°; $\angle x$ and the 60° angle are corresponding angles and therefore equal. **19.** 120°; $\angle z$ and $\angle x$ are supplementary. $m\angle x$ = 60°; 180° – 60° = 120°. **20.** $\angle x$ and $\angle y$ are not supplementary; By indirect proof assumptions, the opposite of the "To prove" statement, which is that $\angle x$ and $\angle y$ are supplementary. **21.** Definition of not supplementary angles **23.** Euclid's Postulate 5 **24.** 90°; corresponding angles are equal **25.** $m\angle x = m\angle y = m\angle i = m\angle j = 45°$ and $m\angle w = m\angle z = m\angle h = m\angle k = 135°$ **26.** 128°; The sum of two supplementary angles is 180°. Press 180° [–] 52° [=] 128°. **27.** 152° **29.** 117°

Lesson 4, pages 75–76

1. $m\angle z$ = 135°, $m\angle y$ = 45°, interior angles are supplementary; $\angle z$ and $\angle y$ are interior angles on the same side of the transversal and therefore are supplementary; $\angle z + \angle y$ = 180°. $\angle z = 3\angle y$; $3\angle y + \angle y$ = 180°. $3\angle y + \angle y = 4\angle y$ = 180°. Divide both sides by 4; $\angle y$ = 45°. $\angle z = 3\angle y$, or 3 × 45° = 135°. Therefore, $\angle z$ = 135° and $\angle y$ = 45°. **3.** $m\angle a = m\angle z$, corresponding angles are equal; $m\angle z = m\angle a$ = 162°; $m\angle w$ = 18° **5.** $m\angle b = m\angle x$, corresponding angles are equal; $m\angle x = m\angle b$ = 45° ($m\angle a$ = 135°) **7.** 105°; $\angle o$ is supplementary with the 75° angle. 180° – 75° = 105°. **9.** 105° **11.** 75° **13.** 105° **15.** 75° **17.** 105°

18.

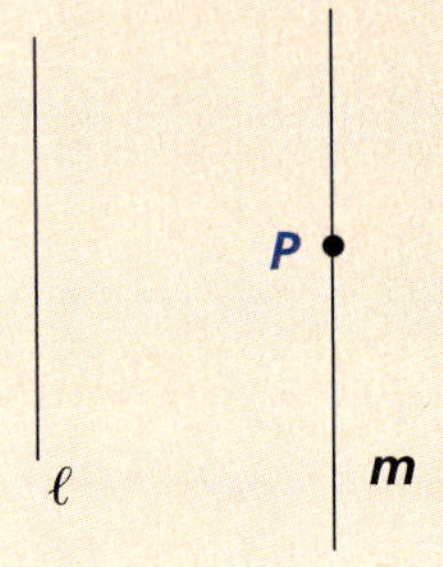

19.

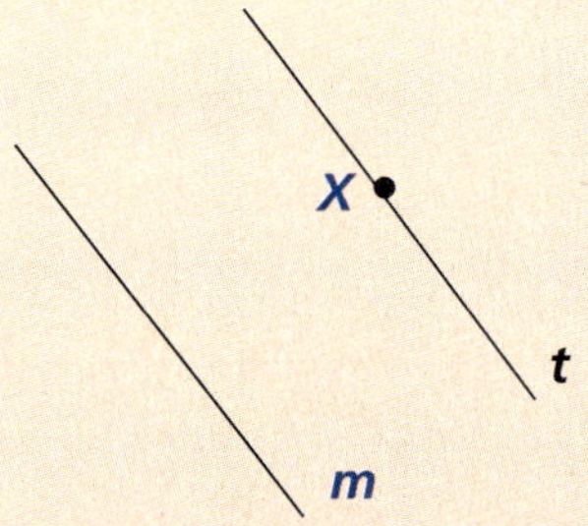

Lesson 5, pages 80–82

1. $\overline{CD}$; A parallelogram is a quadrilateral whose opposite sides are parallel. The opposite side to $\overline{AB}$ is $\overline{CD}$. **3.** $\overline{AD}$, $\overline{BC}$ **5.** 180° **7.** 180° **9.** $\overline{HG}$ **11.** $\overline{FG}$ **13.** 180° **15.** definition of a parallelogram; A parallelogram is a quadrilateral whose opposite sides are parallel. $\overline{AB}$ and $\overline{CD}$ are opposite each other; $\overline{AD}$ and $\overline{BC}$ are also opposite sides. **17.** Theorem, interior angles on the same side of the transversal are supplementary.

424 *Selected Answers and Solutions* *Geometry*

19. definition of a trapezoid; A trapezoid is a quadrilateral with exactly one pair of parallel sides. Since $\overline{WZ}$ and $\overline{XY}$ are not parallel, then $\overline{WX}$ and $\overline{YZ}$ must be parallel. **21.** Theorem, interior angles on the same side of the transversal are supplementary. **23.** True. A rhombus is a parallelogram with four equal sides. By theorem, interior angles on the same side of a transversal are supplementary; the sum of two supplementary angles is 180°. If one angle is a right angle and measures 90°, then 180° – 90° = 90°. The other sides are also parallel, so the above process would apply to those sides as well. **25.** True. Rectangle has four right angles; square also has four right angles.

Lesson 6, pages 87–89

1. parallel, equal alternate interior angles; The angles shown as equal to 130° are alternate interior angles. By the Alternate Interior Angles Postulate, if a transversal intersects two lines so that the alternate interior angles are equal, then the lines are parallel. **3.** parallel; equal corresponding angles **5.** not parallel; unequal corresponding angles and not alternate interior angles
7. parallel; equal corresponding angles
9. parallel; sum of interior angles on same side of transversal equals 180°
11. parallel; alternate interior angles equal **13.** equal alternate interior angles; $\angle d$ and $\angle e$ are interior angles on opposite sides of the transversal and are therefore alternate interior angles. The Alternate Interior Angles Postulate states that if a transversal intersects two lines so that the alternate interior angles are equal, then the lines are parallel.
15. alternate interior angles equal
17. corresponding angles equal
19. interior angles on the same side of the transversal are supplementary
21. corresponding angles equal
23. alternate interior angles equal
25. corresponding angles equal
26.

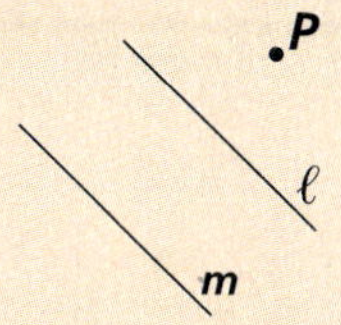

27.

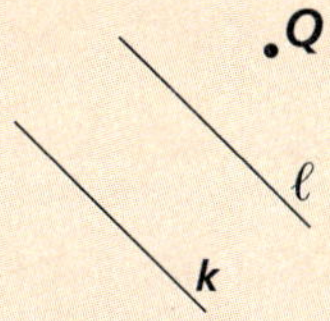

29. The corresponding angles always equal 45°, so the lines will always be parallel. The converse states that if corresponding angles are equal, then the lines are parallel.

Application, page 90

1. not a polygon; A polygon has line segments that intersect at their endpoints. This figure's segments do not meet at their endpoints. **3.** not a polygon; not all segments **5.** polygon
7. polygon **9.** polygon

Geometry *Selected Answers and Solutions* **425**

Chapter Review, pages 91–93

1. A **3.** B **5.** acute, interior **7.** adjacent, supplementary **9.** acute, corresponding **11.** 135° **13.** 45° **15.** 135° **17.** $\angle a + \angle b \neq 180°$ **19.** Euclid's Postulate 5 **21.** C **23.** parallelogram **25.** Line m should be parallel to ℓ. Point X is not on line ℓ.

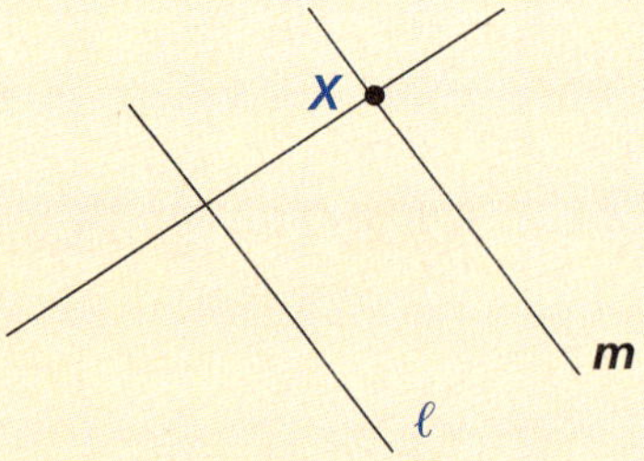

CHAPTER 4

Lesson 1, pages 99–100

1. (1, 1); A point is named by using the x- and y-values together. The x-value is 1 and the y-value is 1. **3.** (2, 3) **5.** (−2, 1) **7.** (−1, −3) **9.** (2, −2) **11.** Q; A point is named by using the x- and y-values together. **13.** T **15.** U **17.** R **19.** W **21.** (−2, 0); The x-value is −2 and the y-value is 0. **23.** (0, 0) **25.** (2, 0) **26.** (0, 2); The x-value is 0 and the y-value is 2. **27.** (0, 1) **29.** (0, −1) **31.** The x-values are on the same vertical line, $x = 3$; $-3 \geq y \leq 2$. **33.** $x \leq 0$ **35.** Each square can be specified by using a letter-number combination. The coordinate system names intersections or points; the chess system names specific areas.

Lesson 2, pages 102–103

1. $y = 3$, x is any real number; Any point on line ℓ has a y-value of 3; since line ℓ extends in both directions, x is any real number. **3.** $y = 1$, x is any real number **5.** $y = -2$, x is any real number **6.** $(x, 3)$; x = any real number and $y = 3$. **7.** $(x, 2)$ **9.** $(x, -1)$

11.

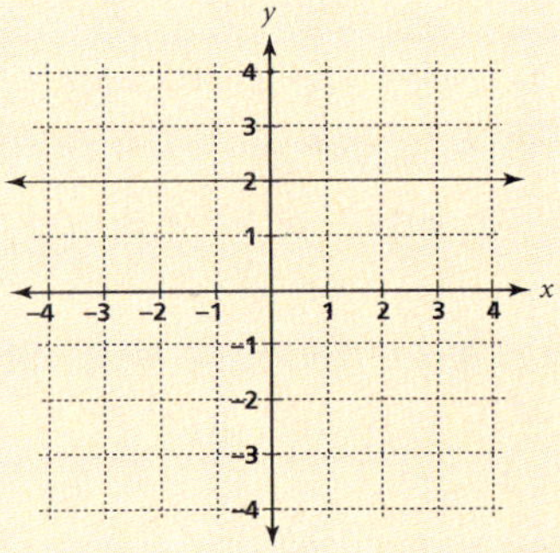

13.

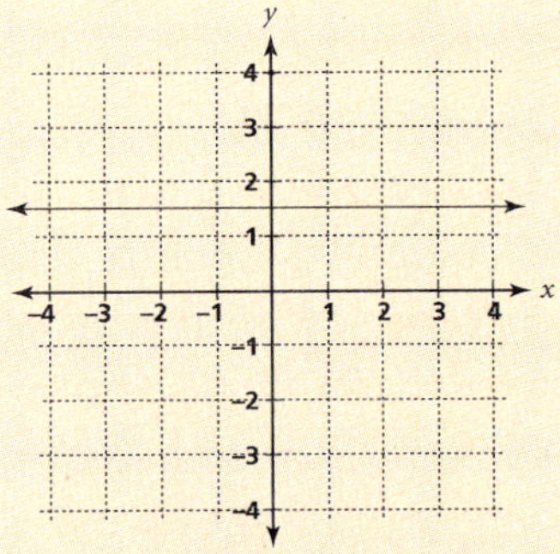

15.

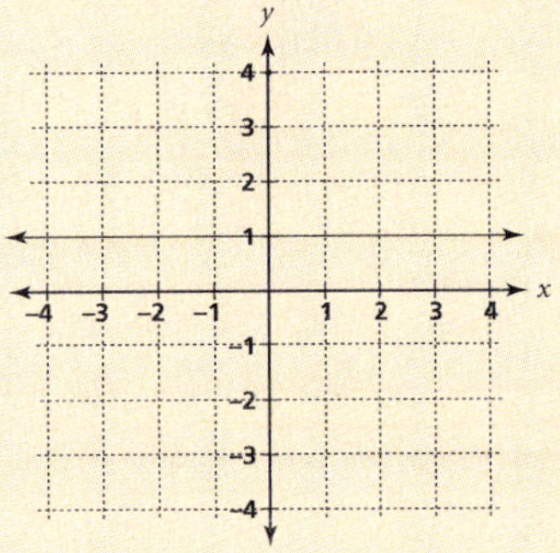

426 *Selected Answers and Solutions* *Geometry*

17.

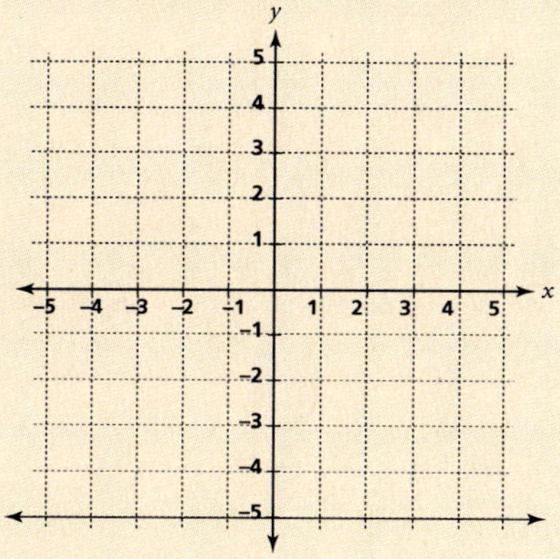

19. $y = 0$, x is any real number; The x-axis is a line perpendicular to the y-axis. It passes through point 0 on the y-axis; its y-value is 0.

Lesson 3, page 106

1. $x = -3$, y is any real number; Any point on line ℓ has an x-value of -3; since line ℓ extends in both directions, y is any real number. **3.** $x = -1$, y is any real number **5.** $x = 2$, y is any real number **6.** $(-3, y)$; $x = -3$ and $y =$ any real number **7.** $(-2, y)$ **9.** $(3, y)$

11.

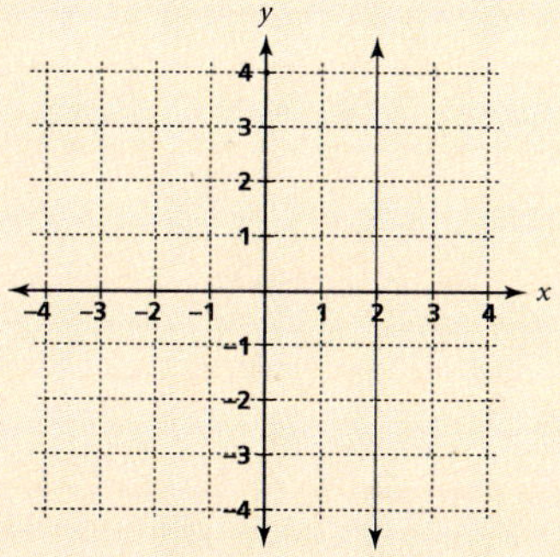

13.

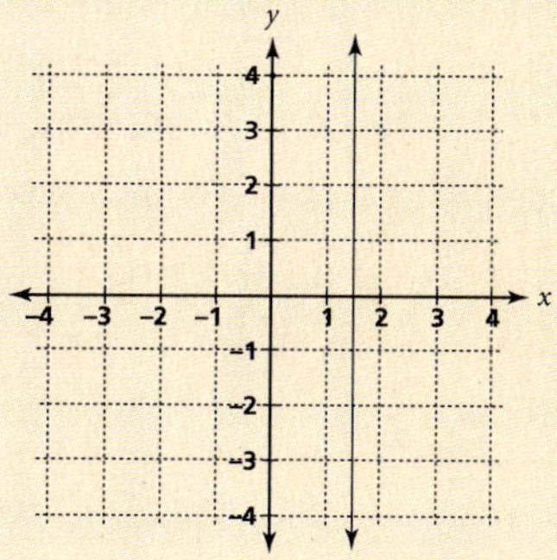

15.

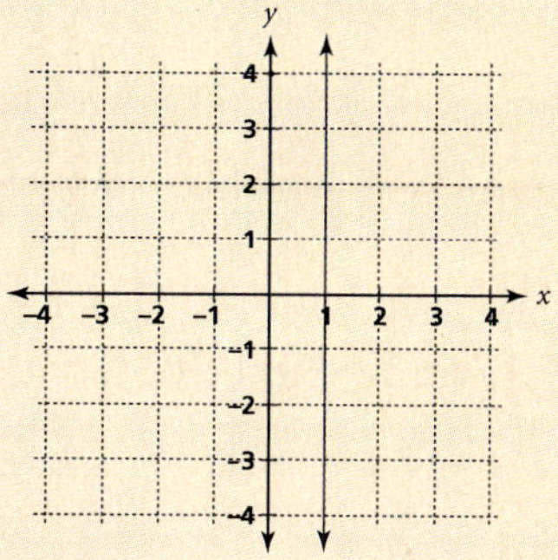

17.

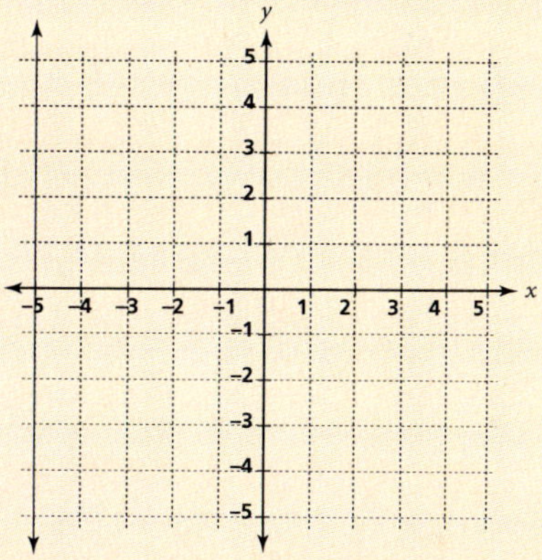

19. $x = 0$, y is any real number; The y-axis is a line perpendicular to the x-axis. It passes through point 0 on the x-axis; its x-value is 0.

Geometry *Selected Answers and Solutions* **427**

Lesson 4, pages 111–112

1. 0.5; Press (6 − 3), then press + (4 − 2) =. The answer is 0.5. **3.** 0.5 **5.** –0.5 **6.** $m = \frac{1}{3}$; Use the two points listed on the line to formulate the equation to find the slope; $\frac{y_1 - y_2}{x_1 - x_2} = \frac{4-3}{5-2} = \frac{1}{3}$. **7.** $m = \frac{3}{7}$ **9.** $m = 1$ **10.** $m = 3$; Use the two points to find the slope; $\frac{y_1 - y_2}{x_1 - x_2} = \frac{16-4}{7-3} = \frac{12}{4} = 3$. **11.** $m = \frac{6}{5}$ **13.** $m = -1$ **14.** Slope of ℓ = slope of m = 1; Use the two points on line ℓ to find the slope; $\frac{y_1 - y_2}{x_1 - x_2} = \frac{6-(-2)}{2-(-6)} = \frac{8}{8} = 1$. Use the two points on line m to find the slope; $\frac{y_1 - y_2}{x_1 - x_2} = \frac{0-(-3)}{3-0} = \frac{3}{3} = 1$ **15.** slope of ℓ = slope of $m = -2$

Lesson 5, pages 117–118

1. $x = 2, y = 1$ and $x = 1, y = 2$

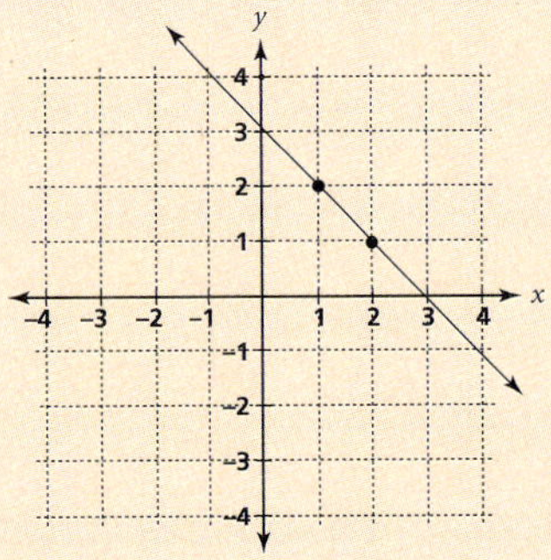

3.

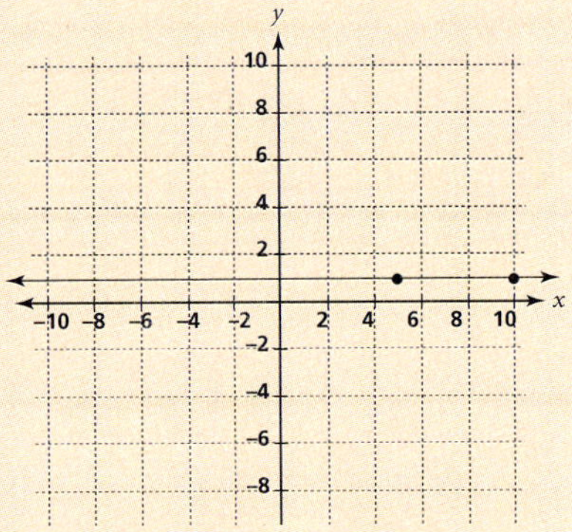

5.

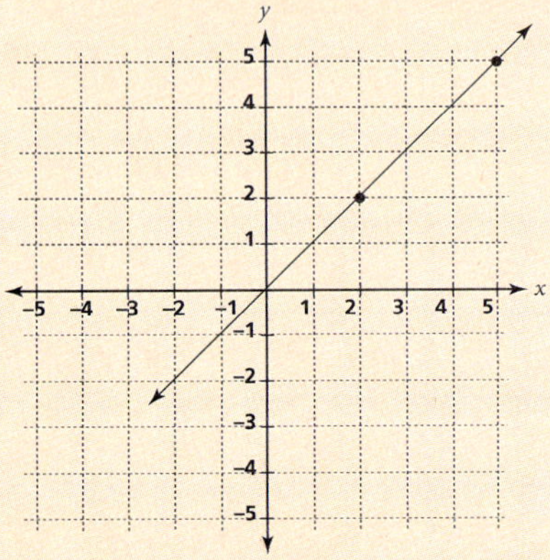

7.

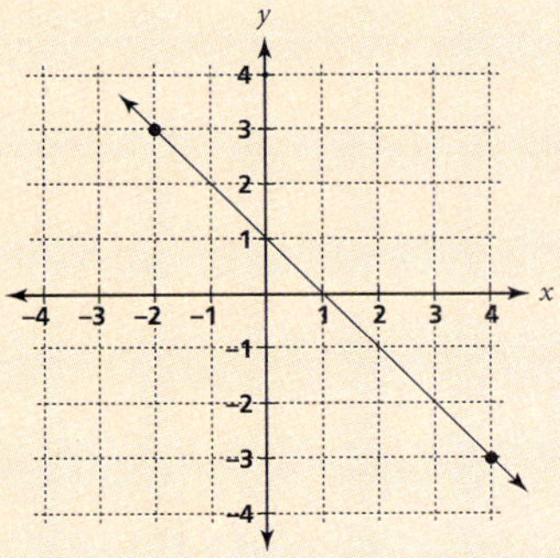

8. Domain: $1 \le x \le 2$; Range: $1 \le y \le 2$

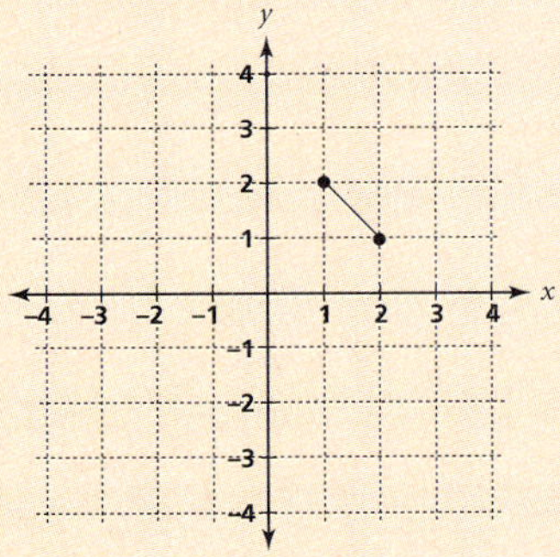

428 *Selected Answers and Solutions* *Geometry*

9. Domain: $-2 \leq x \leq 1$; Range: $1 \leq y \leq 6$

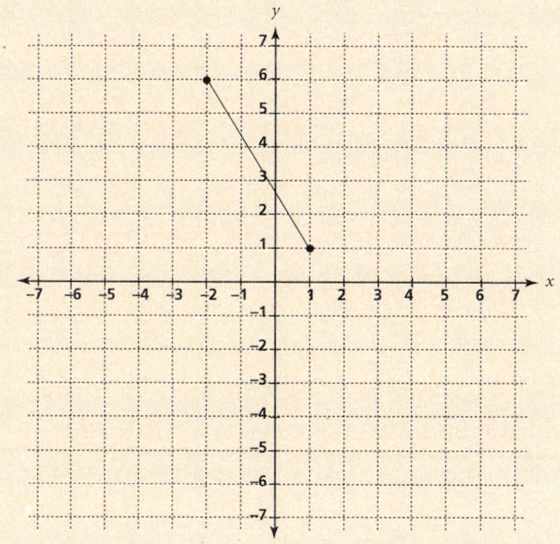

11. Domain: $-1 \leq x \leq 4$; Range: $y = 2$

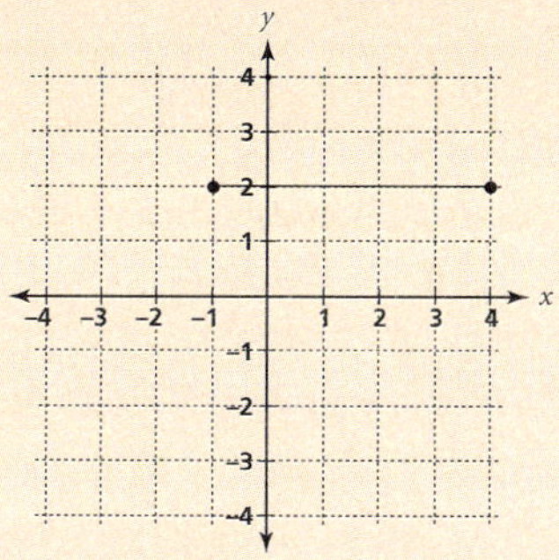

13. $y = -x + 3$; First calculate the slope; slope $= -1$. Then plug in the x- and y-values for one of the points listed along with the slope to get the value of b. $1 = -1(2) + b$; $1 = -2 + b$; $b = 3$. Then represent the equation using letter values for x and y and real values for b and m. **15.** $y = 1$ **17.** $y = x$ **18.** slope $= \frac{7}{10}$ or $-\frac{7}{10}$; Staircase is the y-axis and ground is the x-axis; the measurements result in ordered pairs of (0, 7) and (10, 0); $\frac{y_1 - y_2}{x_1 - x_2} = \frac{7 - 0}{0 - 10} = -\frac{7}{10}$ **19.** slope $= \frac{1}{2}$

Try This

Answers will vary.

Lesson 6, page 122

1. Slope equals 1 or $\frac{1}{1}$; add 1 to both the x- and y-values to graph point B. Connect points A and B to graph the line.

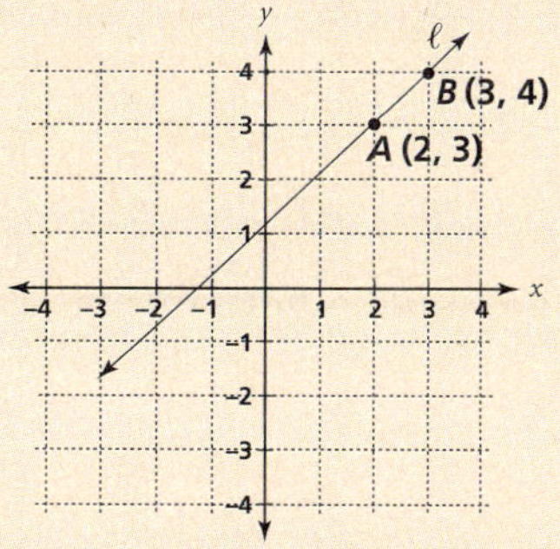

3.

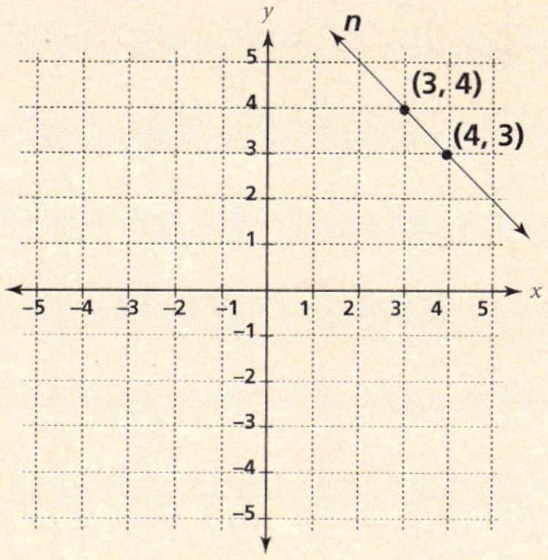

5.

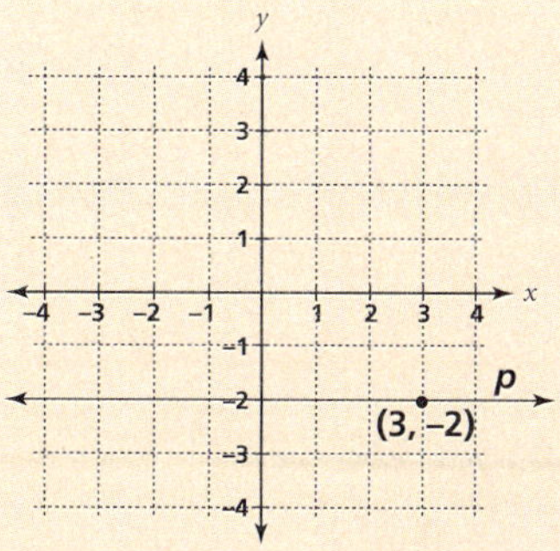

Geometry Selected Answers and Solutions 429

7.

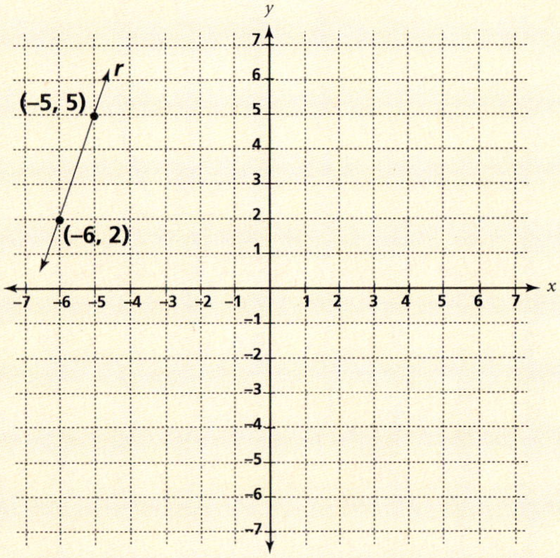

9.

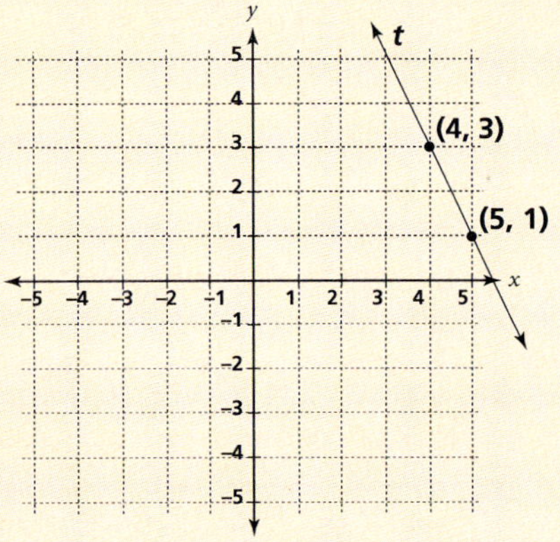

11. $y = x + 1$; Take the slope and a value for x and y; plug these into the slope equation to yield a value of 1 for b. $3 = 1(2) + b$; $3 = 2 + b$; $b = 1$. Then represent the equation with letters for x and y and the m and b values as numbers. **13.** $y = -x + 7$ **15.** $y = -2$, $x =$ any real number **17.** $y = 3x + 20$ **19.** $y = -2x + 11$

Lesson 7, pages 126–127

1. $(9, 13\frac{1}{2})$; $(9, 13.5)$; $\frac{x_1 + x_2}{2} = \frac{12 + 6}{2} = 9$ and $\frac{y_1 + y_2}{2} = \frac{18 + 9}{2} = 13.5$ **3.** $(-2, \frac{5}{2})$; $(-2, 2.5)$ **5.** $(-5, 7\frac{1}{2})$; $(-5, 7.5)$

7.

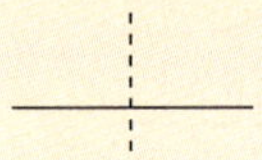

9.

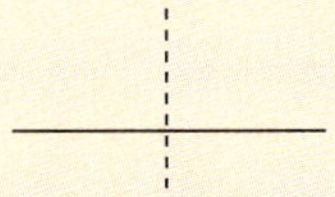

11. Constructions will vary but should include the arcs necessary to find the midpoints. 12. (4, 3); Use the midpoint formula and plug in the two points given in the problem; $\frac{10 + -2}{2} = 4$ and $\frac{3 + 3}{2} = 3$. This gives the midpoint x- and y-values of 4 and 3. **13.** (1, –3) **15.** (4, –2)

Application, page 128

1. slope = 1.34 **3.** slope = 1.73 **5.** slope = 1.33 **7.** slope = 1 **9.** Beast

Chapter Review, pages 129–131

1. a **3.** c **5.** d **7.** d **9.** c **11.** $y = -x + 5$ **13.** $y = -\frac{1}{3}x + 6$ **15.** $y = x + 3$ **17.** (–4, 8) **19.** (–3, 7)

21.

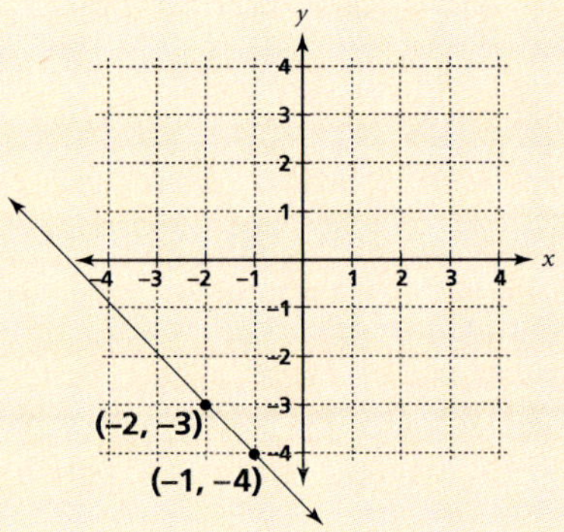

430 *Selected Answers and Solutions* *Geometry*

23.

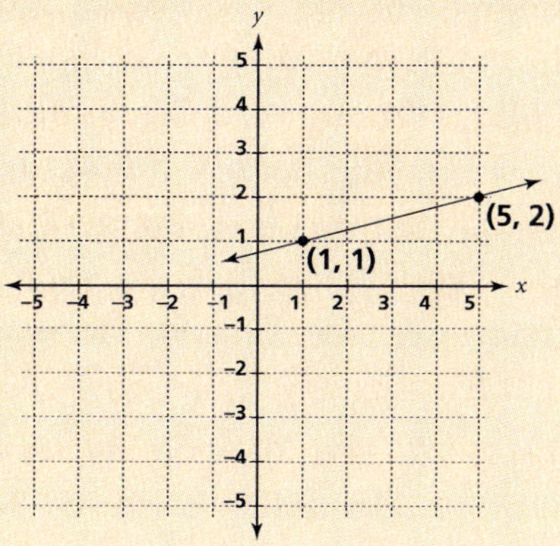

25.

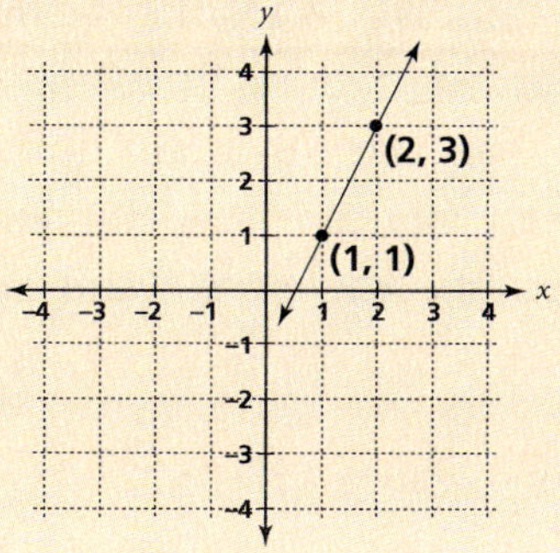

CHAPTER 5

Lesson 1, page 138

1. isosceles; according to the definition of an isosceles triangle. **3.** scalene **5.** isosceles **7.** isosceles **9.** isosceles

11.

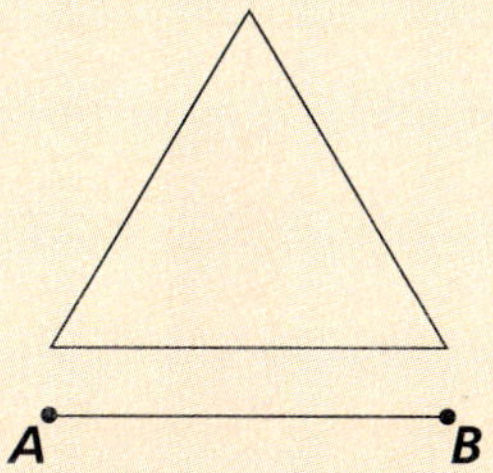

13.

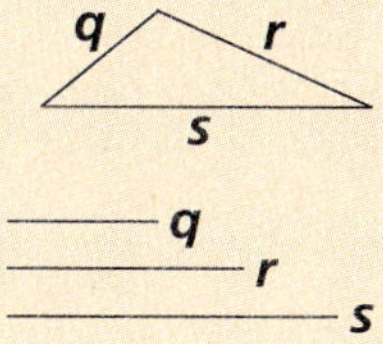

15.

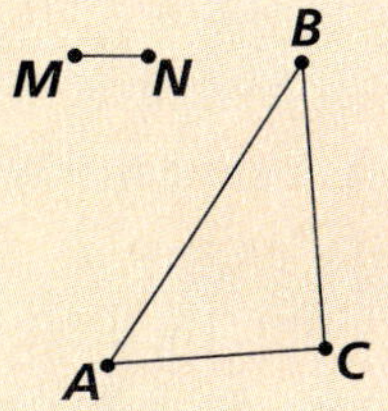

Lesson 2, page 142

1. right triangle; By definition, a triangle with one right angle **3.** acute triangle **5.** equilateral triangle **7.** acute equilateral triangle **9.** right scalene triangle

11. right isosceles triangle

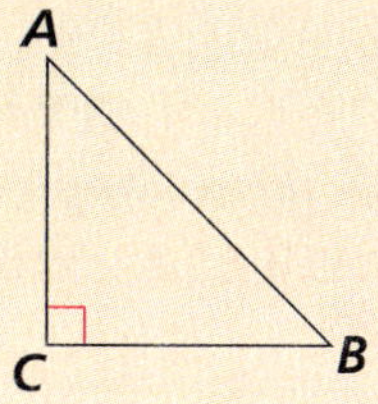

13. obtuse isosceles triangle

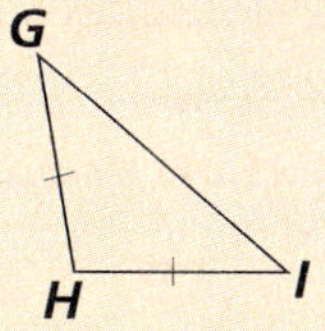

15. right scalene triangle

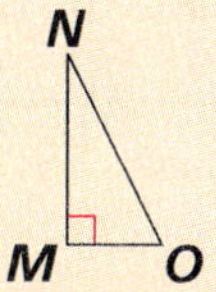

Try This

Constructions will vary. A right triangle cannot be equilateral because it does not have three 60° angles.

Geometry *Selected Answers and Solutions* **431**

Lesson 3, page 145

1. False; A rectangle is also a parallelogram but not a square. **3.** False; isosceles trapezoid **5.** False; not all rhombuses have equal angles **7.** True **8.** Figure with four sides; By definition, a rhombus has four sides **9.** parallel and equal to each other and the other pair of sides as well **11.** rhombus; Four equal sides with angles that are not equal to each other **13.** quadrilateral **15.** rectangle

Lesson 4, page 148

1. definition of parallelogram; It is given that *ABCD* is a parallelogram. By definition, opposite sides are equal and parallel. $\overline{BC}$ and $\overline{AD}$ are opposite sides and therefore parallel. **3.** definition of parallelogram **5.** angle 4; Angle 4 and angle 1 are alternate interior angles on the transversal *WY* crossing parallel lines *XY* and *WZ*. By theorem, alternate interior angles are equal. **7.** 25°; Angle 4 and angle 1 are alternate interior angles on the transversal $\overline{EG}$. By theorem, alternate interior angles are equal. **9.** 65° **11.** 65° **12.** 55°; Angle 4 and angle 1 are alternate interior angles on the transversal *JL*. By theorem, alternate interior angles are equal. **13.** 55° **15.** 35°

Lesson 5, page 153

1. 180°; The figure shown with one pair of parallel lines at the top and bottom. The left side is a transversal. Angles 1 and 2 are interior angles on the same side of this transversal. By theorem, interior angles on the same side of the transversal are supplementary and equal 180°. **3.** 360° **5.** 180° **7.** 40°; The angle x is equal to 180° minus the measures of the other two angles. One of these angles is a right angle that measures 90° and the other is 50°. Therefore, $x = 180° - (90° + 50°)$ or $x = 180° - 140° = 40°$. **9.** 80° **11.** 50° **13.** 30° **15.** 30° **17.** 20° **18.** no; This would give an angle sum greater than 180°; An obtuse angle, by definition, is an angle that measures more than 90°. A right angle measures 90°. Added together, the sum will be more than 180°. By definition, a triangle has three angles whose sum equals 180°. **19.** yes; one angle > 90° and two equal sides

Lesson 6, pages 158–159

1. 5; By definition, a diagonal is a line that is not part of the original figure, so $n = 5$. **3.** 1,080° **5.** 7 **7.** 7 **9.** 8 **11.** 44 **13.** 1,620° **15.** 54 **17.** regular hexagon; All the angles are equal; only three diagonals are shown. **19.** 720°

Lesson 7, page 162

1.

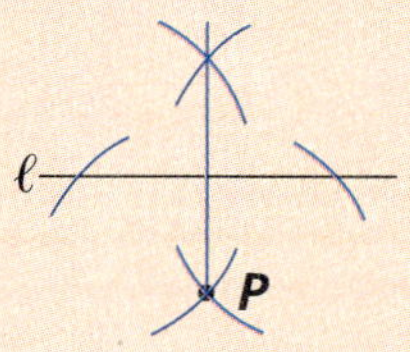

3.

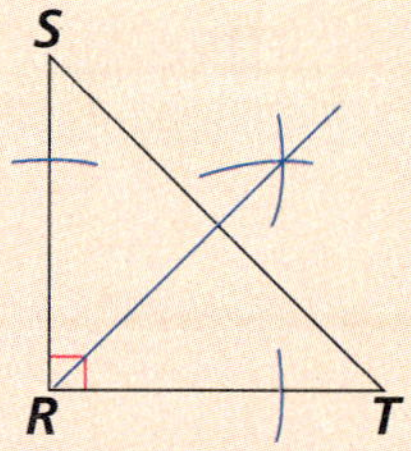

432 *Selected Answers and Solutions Geometry*

5. yes; $\overline{TP} \parallel \overline{RA}$, alternate angles are equal. Therefore, $\angle RXT = \angle XTP = 90°$, and $\overline{XT}$ is perpendicular to $\overline{RA}$ and $\overline{TP}$.

Try This

Triangles will vary. All perpendicular bisectors divide lines in half and form right angles with them.

Lesson 8, pages 165–166

1.

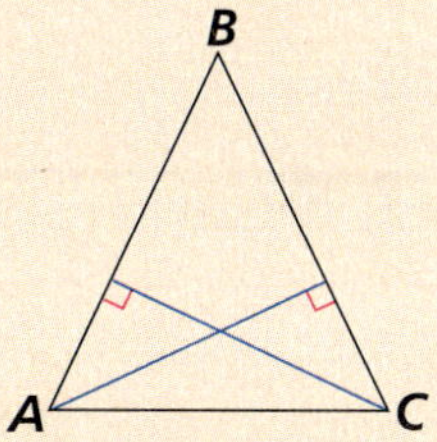

3.

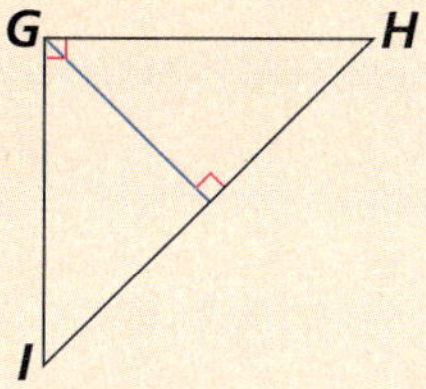

5.

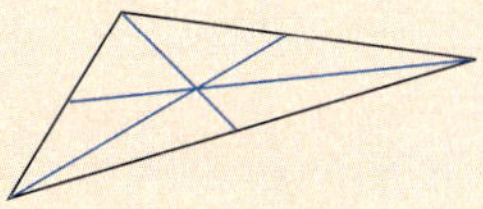

7.

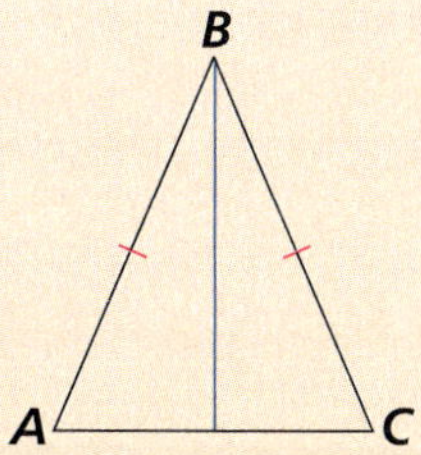

9.

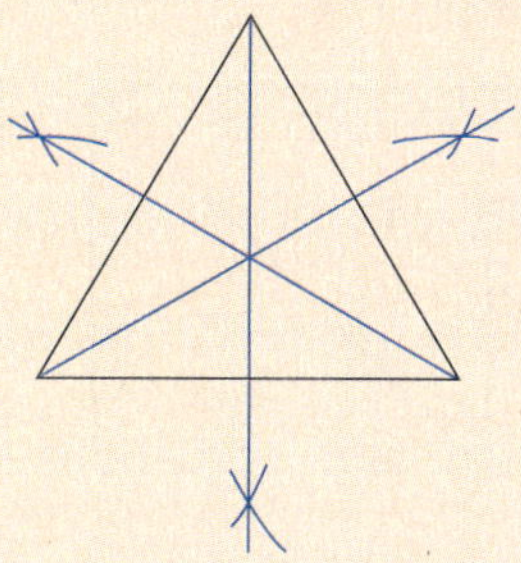

11. They are the triangle's sides, or legs. The sides that make the right angle are perpendicular to each other. Since the definition of an altitude is that it is perpendicular to one of the triangle's sides, then the sides themselves are the altitudes. **13.** They seem to be the same. **15.** The intersection of the angle bisectors appears to be equidistant from each side of the triangle.

Lesson 9, pages 170–171

1. 30°; Angle 1 is an adjacent angle with the angles labeled 70° and 80° whose exterior sides form a straight line that is equivalent to a straight angle that measures 180°. Therefore, $180° - (70° + 80°) = 180° - 150° = 30°$. **3.** 80° **5.** 160° **7.** 180° **9.** 160° **11.** 160° **13.** True; $\angle d$ is an exterior angle and supplementary to $\angle c$; $m\angle d = 180° - m\angle c$. Since the sum of a triangle's angles equal 180°, then the sum of $m\angle a + m\angle b = 180° - m\angle c$. Therefore, $m\angle a + m\angle b = 180° - m\angle c = m\angle d$. **15.** True **17.** False **19.** 60°; If the angles are all equal then they must equal 180° ÷ 3 angles; $180° \div 3 = 60°$. **21.** 70°; Press 43 [+] 27 [=]. The answer is 70. **23.** 97° **25.** 102°

Geometry Selected Answers and Solutions **433**

Application, page 172

1.

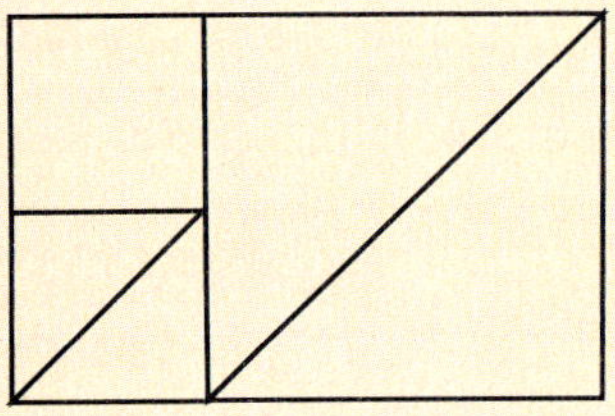

3.

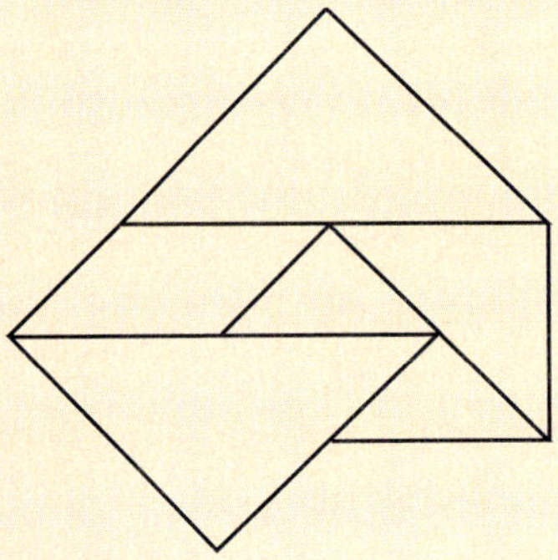

5. Answers will vary.

Chapter Review, pages 173–175

1. True **3.** False; A right triangle can also be scalene. **5.** False; There cannot be an obtuse right triangle. **7.** 35° **9.** 55° **11.** 55° **13.** 180° **15.** 80° **17.** 120° **19.** D **21.** B

23.

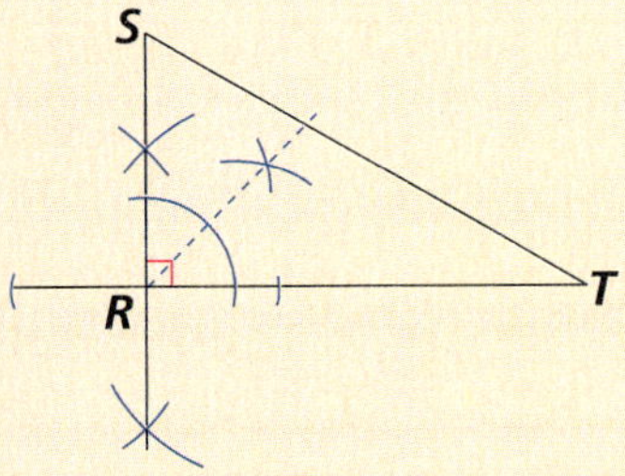

25.

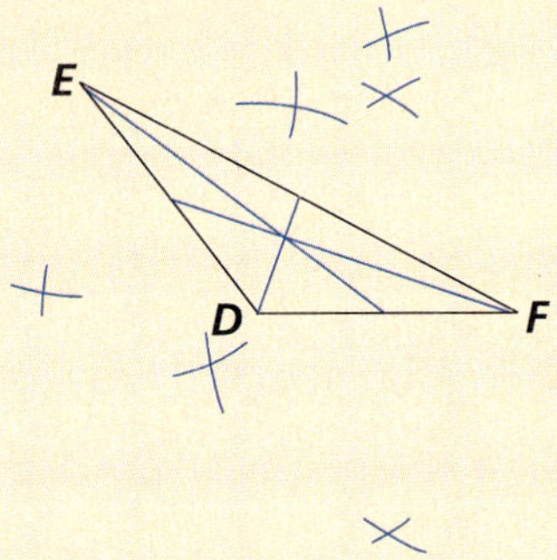

CHAPTER 6

Lesson 1, pages 182–183

1. $\overline{AB} \cong \overline{DE}$, $\angle A \cong \angle D$, $\overline{AC} \cong \overline{DF}$; SAS postulate; A single tick mark on one pair of corresponding sides, double tick marks on the other pair of corresponding sides, and the included angles are marked with arcs **3.** $\overline{HI}$ and $\overline{HK}$, $\overline{HJ}$ and $\overline{HG}$; The included angles between these two pairs of corresponding sides are marked congruent and are equal because they are vertical. For the triangles to be congruent by SAS, the above sides must be congruent. **5.** $\overline{MN}$ and $\overline{ON}$, $\overline{NL}$ and $\overline{NL}$ **7.** Given; This is given as a feature of the figure. **9.** $\angle ACB$ **11.** Given **13.** $\angle EFG \cong \angle HIK$ **15.** 40 paces, SAS congruency; The two triangles are congruent according to the SAS Postulate. If the triangles are congruent, then corresponding sides are congruent. So $\overline{AB}$ and $\overline{A'B'}$ are congruent.

434 *Selected Answers and Solutions Geometry*

Lesson 2, page 188

1. no, $b + c = a$; Because $b + c$ is not greater than a but only equal to it.
3. yes, $a + b > c$, $b + c > a$, $a + c > b$
5. yes; $14 > 7$, $11 > 10$, $17 > 4$ **7.** yes; $20 > 12$, $23 > 9$, $21 > 11$ **9.** yes; $6 > 5$, $8 > 3$
11. >; Because $\overline{PQ}$ is not a perpendicular, it cannot be the shortest distance. **13.** = **15.** = **17.** >
18.

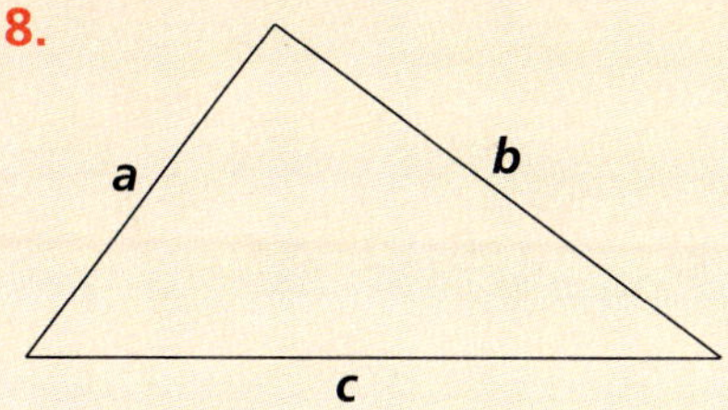

19.

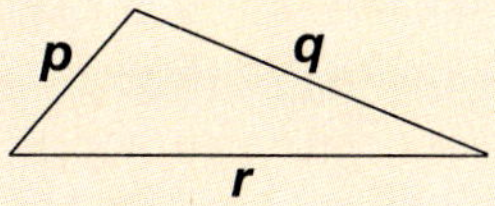

Lesson 3, pages 192–193

1. yes, SSS; The two outer pairs of sides are marked congruent by the single and double tick marks. Both triangles share the third side, which is equal to itself. Therefore, all three sides are equal.
3. Yes, ASA or SAS **5.** Yes, SSS **6.** Given
7. Given **9.** $\angle ABE = \angle DBC$ **11.** Given; Given as a feature of the figure.
13. Given **15.** $\overline{GI} = \overline{GI}$
17.

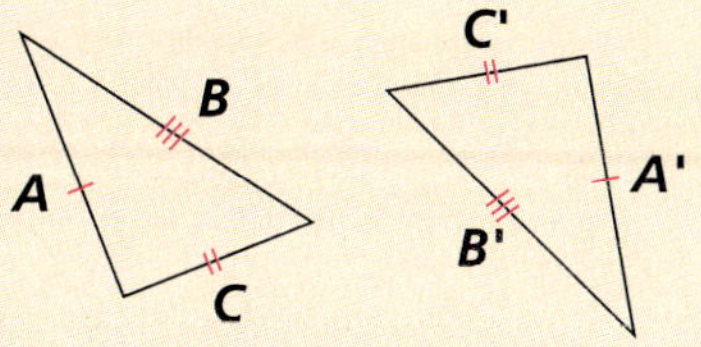

19. no; The problem describes two equilateral triangles. All equilateral triangles have equal angles because all of the angles equal 60°. However, two equilateral triangles can have different length sides.

Lesson 4, pages 198–199

1. SAS; Two pairs of sides and the included angle are shown to be congruent. **3.** AAS **5.** Given; Definition of perpendicular lines. **7.** H-L Theorem
9. Given; Parallel lines have equal corresponding angles. **11.** $m\angle 1 = m\angle 3$
13. $m\angle 1 = 30°$, $m\angle 2 = 60°$; The acute angles in a right triangle are complementary and therefore equal to 90°. $\angle 2 = 2 \cdot \angle 1$; $\angle 2 + \angle 1 = 90°$; $(2 \cdot \angle 1) + \angle 1 = 90°$; $3 \cdot \angle 1 = 90°$. Divide both sides by 3. $\angle 1 = 30°$; $90° - 30° = 60°$. $\angle 2$ is equal to 60°. **15.** $m\angle 3 = 50°$, $m\angle 4 = 40°$ **16.** $x = 20°$; if $5x + x = 120°$, then $6x = 120°$. Divide both sides by 6 to get $x = 20°$. **17.** $x = 20°$ **19.** $x = 24°$

Lesson 5, page 204

1. $A' = (3, -5)$; By definition, a reflection over the x-axis has the same x-value and a negative of the y-value. **3.** $C' = (-3, -5)$ **5.** $E' = (2, 0)$ **7.** $G' = (-4, 2)$; By definition, a reflection over the y-axis has the same y-value and a negative of the x-value. **9.** $I' = (4, 2)$ **11.** $K' = (-3, 0)$ **13.** $A' = (1, 2)$, $B' = (-1, 6)$, $C' = (-2, 2)$, $D' = (-1, -2)$; Since the line of reflection is vertical, the reflected figures' points have the same y-value as the original figure. To find the x-value of each reflected point, subtract 2 from the

Geometry *Selected Answers and Solutions* **435**

original x-value. This gives the distance from the line of reflection for the original point. This distance has to be exactly reflected. To find the x-value for the reflected point, subtract this distance from the x-value of the line of reflection, or 2. **15.** $L' = (0, -6)$, $M' = (-2, -2)$, $N' = (-6, -1)$, $O' = (-1, -10)$

Lesson 6, page 206

1. no; The given line does not create coinciding images when the figure is folded along it and is not a line of symmetry. **3.** no **5.** yes **6.** 3; The figure is a triangle. **7.** 4 **9.** It must bisect an interior angle.

Try This

Any diameter seems to be a line of symmetry. There are an infinite number of diameters, so there must be an infinite number of lines of symmetry.

Lesson 7, page 209

1. $A' = (0, 8)$; A slide to the left by any number is made by subtracting the number from the original x-value; $3 - 3 = 0$ = new x-value. A slide up by any number is made by adding the number to the original y-value; $6 + 2 = 8$ = new y-value. **3.** $C' = (0, -4)$ **5.** $E' = (-3, 7)$ **7.** $G' = (-3, 2)$ **8.** $A' = (-3, 5)$, $B' = (2, 3)$, $C' = (8, 8)$; To slide the triangle 4 right, add 4 to each x-value for each endpoint. To slide the triangle 2 up, add 2 to each y-value. **9.** $E' = (-5, 1)$, $F' = (0, 5)$, $G' = (0, 0)$ **11.** $(4, 4)$; The point given has an x-value of 3 and a y-value of -1. To find the new values, plug these old values into $x + 1$ and $y + 5$. **13.** $(1, -5)$ **15.** $(7, -1)$ **16.** $(x - 4, y + 4)$; To get new value of x, subtract distance of 4 from old value. To get new value of y, add distance of 4 to old value. **17.** $(x + 3, y + 3)$ **19.** $(x + 2, y - 4)$

Lesson 8, pages 212–213

1.

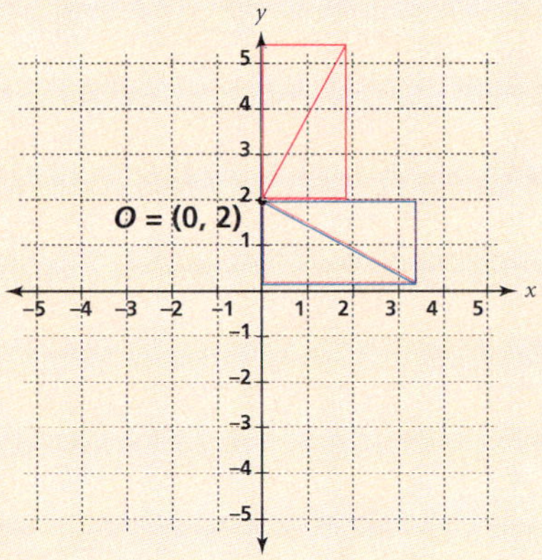

3.

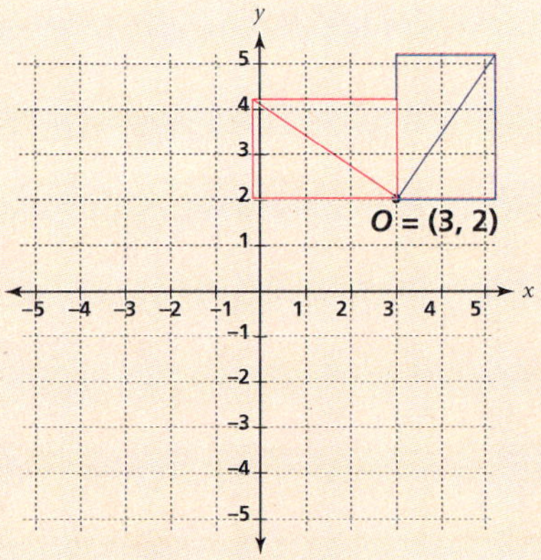

436 *Selected Answers and Solutions* *Geometry*

5.

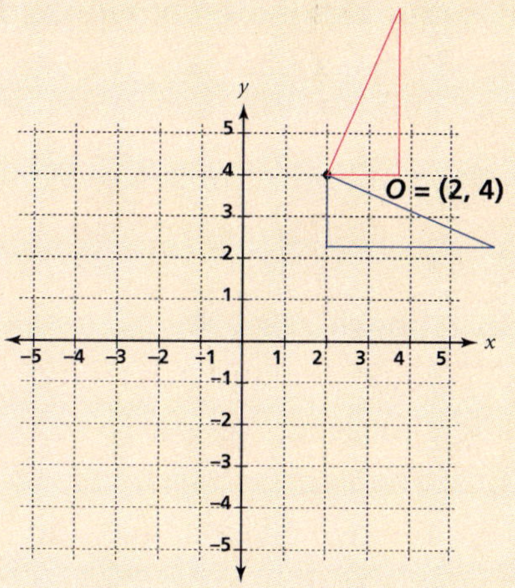

7.

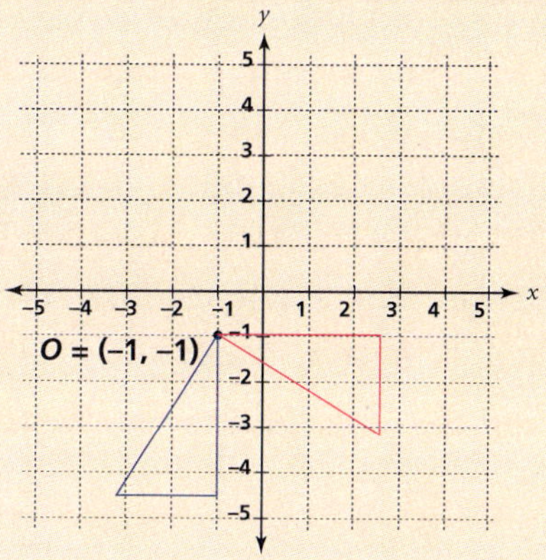

8.

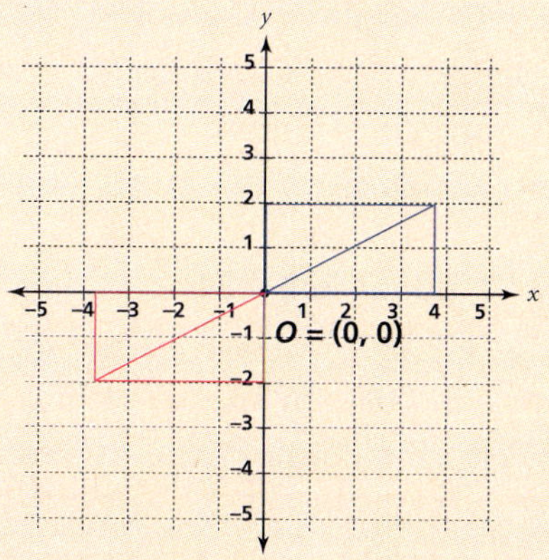

9.

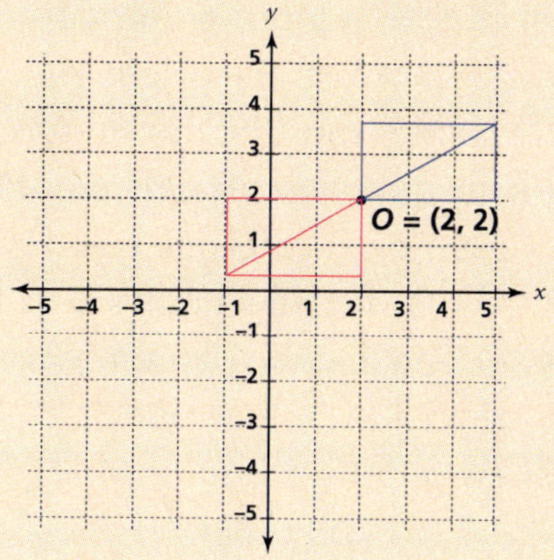

11.

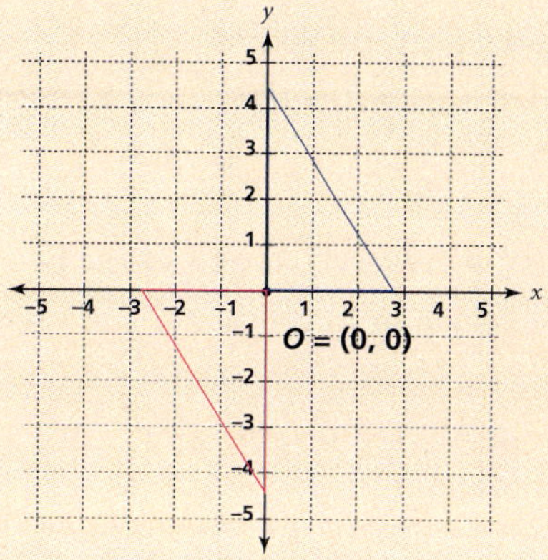

13.

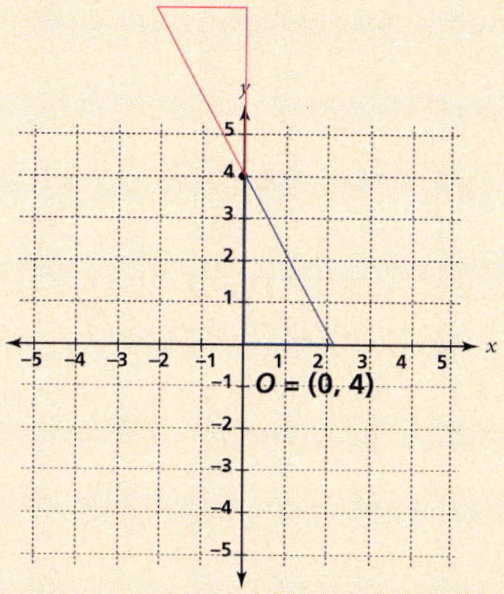

Geometry *Selected Answers and Solutions* **437**

15. (–4, 5); To rotate a point 180°, the distance between the origin and the x- and y-values must be moved in the opposite direction. The distance from the origin to the original point's x-value is 4 – 0 = 4; the new point's x-value is 0 – 4 = –4. The distance from the origin to the original point's y-value is –5 – 0 = –5; the new point's y-value is 0 – (–5) = 5. **17.** (–7, –1) **19.** (–8, 3)

Application, page 214

1. Choices will vary. **3.** Symmetry will vary. **5.** Answers will vary.

Chapter Review, pages 215–217

1. C **3.** A **5.** C **7.** A **9.** 35° **11.** 45° **13.** yes **15.** no **17.** reflection **19.** rotation through 180° counterclockwise

CHAPTER 7

Lesson 1, pages 224–225

1. 6; To solve for x, multiply the extremes together and the means together; 2 • 9 = 3x; 18 = 3x; 6 = x. **3.** 3.55 **5.** 40 **6.** 10; Sides AC and DF are similar sides of the similar triangles that act as one side of the equation as $\frac{20}{10}$. $\overline{AB}$ and $\overline{DE}$ are also similar. The other side of the equation is $\frac{DE}{5}$. The whole equation is $\frac{20}{10} = \frac{DE}{5}$. Multiply extremes and means; 20 • 5 = 10$\overline{DE}$; 100 = 10$\overline{DE}$; 10 = $\overline{DE}$ **7.** 24 **9.** 4.2 **11.** 8 **13.** 15.4 **15.** 13.33 **16.** no; For two ratios to be equal, the extremes multiplied together and the means multiplied together must be equal; 3 • 16 = 48; 5 • 10 = 50. The products are not equal, so the ratios are not equal. **17.** no **19.** yes **21.** no **23.** no **25.** yes

Try This

Give one counterexample (ΔABC with sides 2, 2, 3 and ΔDEF with sides 4, 4, 5).

Lesson 2, pages 229–230

1. yes, 3 equal angles; Each triangle has two angle measures given. To solve for the third angle in each triangle, add together the two given angles and subtract this sum from 180°. The first triangle's given angles' sum is 93°; 180° – 93° = 87°. The second triangle's given angles' sum is 148°; 180° – 148° = 32°. Both triangles have angle measures of 61°, 32°, and 87°. **3.** yes, sides are in proportion **5.** yes, angle sum and equal base angles **6.** m$\angle D$; $\angle A$ and $\angle D$ are alternate interior angles and are equal. **7.** m$\angle E$ **9.** $\overline{ED}$, $\overline{DC}$ **11.** m$\angle RTS$ **13.** m$\angle TSR$ **15.** $\overline{RS}$, $\overline{TS}$ **16.** 16; Use the ratio formula; $\frac{11}{22} = \frac{8}{x}$; 11$x$ = 176; x = 16. **17.** 12 **19.** 12 **21.** 7.5 **23.** 16 **24.** 343 inches; $\frac{84}{450} = \frac{64}{x}$; 84$x$ = 28,800; x = 342.85, which can be rounded up to 343 inches. **25.** 24 feet

Lesson 3, pages 233–234

1. 60°; 180° – $\frac{360°}{3}$ = x; 180° – 120° = x; 60° = x. **3.** 108° **5.** 135° **7.** 150° **9.** 168° **11.** 10; Both triangles are equilateral triangles. One side of the ratio formula is $\frac{3}{5}$. The other side is the perimeter of the first triangle, 6, over the unknown perimeter of the second triangle, x. The whole equation is $\frac{3}{5} = \frac{6}{x}$; 3$x$ = 30; x = 10. **13.** 75 **15.** 56

438 *Selected Answers and Solutions* *Geometry*

Lesson 4, pages 238–239

1. $A' = (2, 4)$, $B' = (4, 8)$, $C' = (10, 10)$

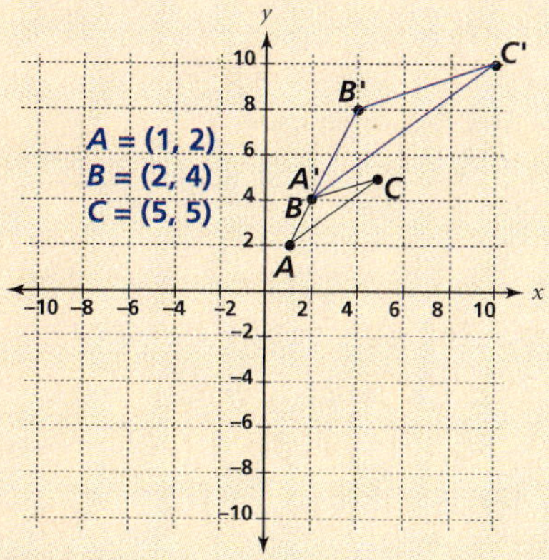

3. $A' = (4, 8)$, $B' = (8, 16)$, $C' = (20, 20)$

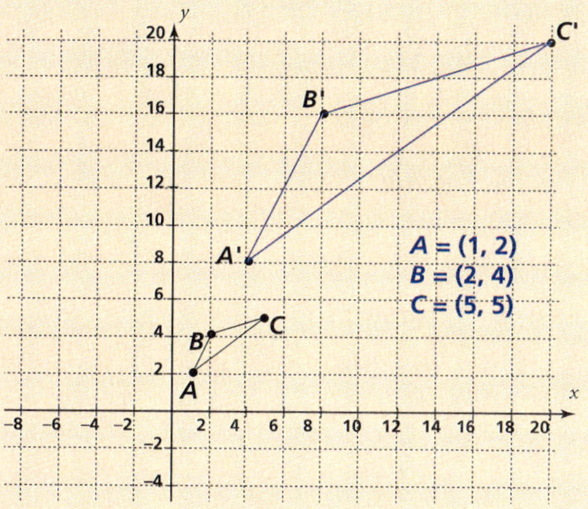

5. $D' = (-6, 6)$, $E' = (9, 3)$, $F' = (6, -6)$

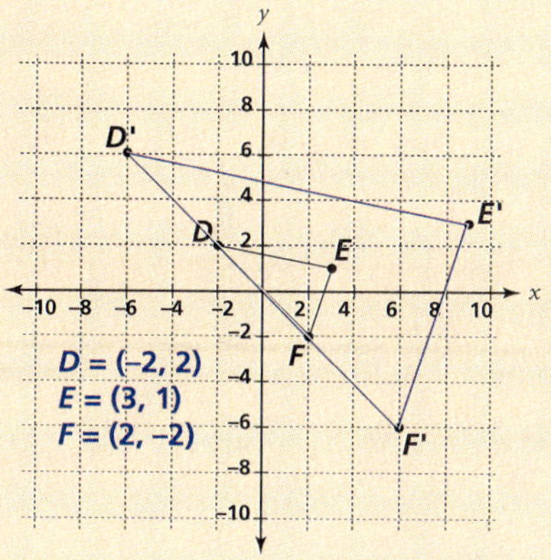

7. $J' = (-9, 3)$, $K' = (-6, 12)$, $L' = (0, 0)$

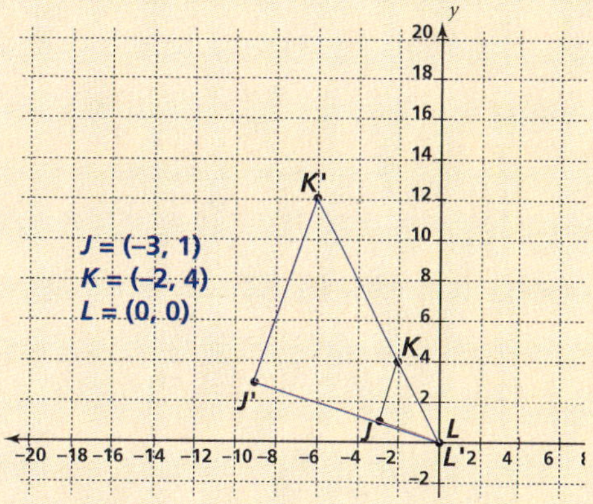

8. $A' = (0, 0)$, $B' = (0, 4)$, $C' = (4, 4)$, $D' = (4, 0)$

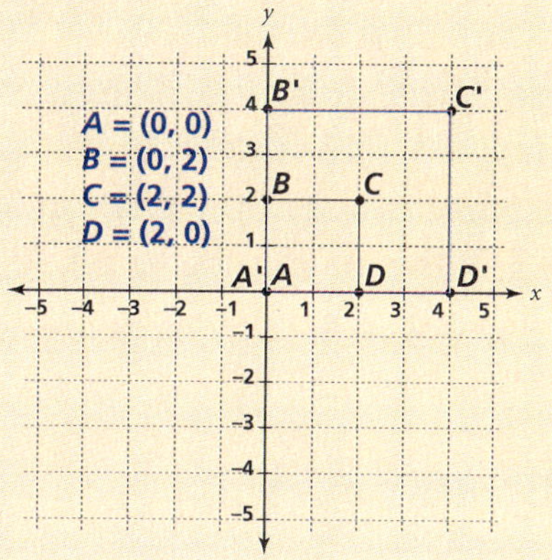

9. $A' = (0, 0)$, $B' = (0, 6)$, $C' = (6, 6)$, $D' = (6, 0)$

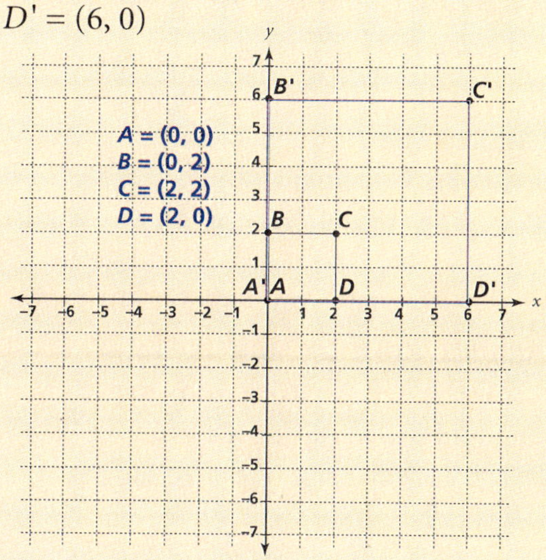

Geometry *Selected Answers and Solutions* **439**

11. $A' = (-4, -4)$, $B' = (-8, 2)$, $C' = (2, 2)$, $D' = (6, -4)$

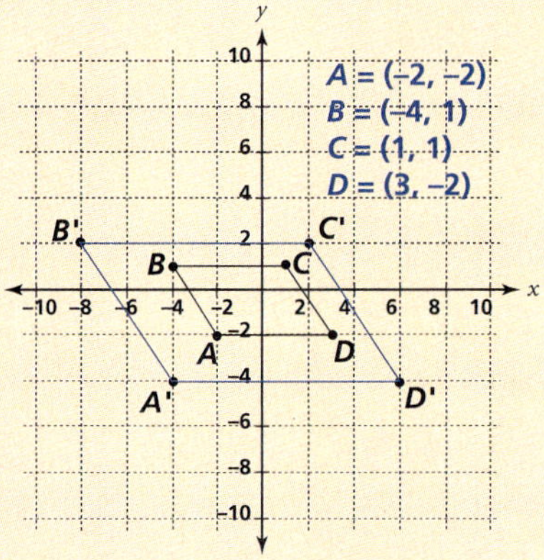

13. 18 inches × 12 inches; For the posters to be 3 times the size of the flyers, multiply each side by the dilation amount; 6 • 3 = 18 and 4 • 3 = 12.
15. The magnitude of the dilation to get 4 × 5's is 2; the magnitude to get 10 × 8's is 4. The vertices have been multiplied by 2 and 4.

Lesson 5, pages 242–243

1. $A' = (-2, -2)$, $B' = (-2, 2)$, $C' = (2, 0)$; Multiply each of the original values by $\frac{1}{2}$ to get the coordinates shown. **3.** $D' = (1, -2)$, $E' = (0, 2)$, $F' = (2, 1)$ **5.** $S' = (0, 0)$, $T' = (0, 1)$, $U' = (1, 1)$, $V' = (1, 0)$; Multiply each of the original values by $\frac{1}{6}$ to get the coordinates shown. **7.** $S' = (0, 0)$, $T' = (0, 4)$, $U' = (4, 4)$, $V' = (4, 0)$ **9.** $S' = (0, 0)$, $T' = (0, 3)$, $U' = (3, 3)$, $V' = (3, 0)$ **11.** 1,120 cm or about 11 m; Put the ratio of 1:80 into the ratio formula as $\frac{1}{80}$. 1 corresponds to the length of the scaled-down car, or 14 cm. To solve for the size of the actual car, or x, write the equation as $\frac{1}{80} = \frac{14}{x}$; $x = 14 \cdot 80$ or $x = 1{,}120$. **13.** 45 inches **15.** 2,750 gallons in 10 minutes, 16,500 in an hour

Application, page 244

1. a, it has the largest scale **3.** 2,310 miles **5.** $\frac{1}{2}$ inch

Chapter Review, pages 245–247

1. C **3.** D **5.** A **7.** A **9.** 8 **11.** 24 **13.** $7\frac{1}{3}$ **15.** 11 **17.** 56 **19.** $G' = (-6, -3)$, $H' = (-9, 6)$, $I' = (-3, 3)$ **21.** $A' = (-1, 0)$, $B' = (-1, 1)$, $C' = (0, 1)$, $D' = (0, 0)$ **23.** $A' = (-2, -4)$, $B' = (-2, 8)$, $C' = (8, 8)$, $D' = (8, -4)$ **25.** $A' = (-3, -6)$, $B' = (-3, 12)$, $C' = (12, 12)$, $D' = (12, -6)$

CHAPTER 8

Try This, page 252

$3m$, $4m$, $5m$, where $m > 1$

Lesson 1, page 253

1. 169; The square of 5 is 25. The square of 12 is 144; 25 + 144 = 169. **3.** 400 **5.** 100 **6.** 12, 16, 20; The triple given is 3, 4, 5 and the multiplier is 4; 3 • 4 = 12, 4 • 4 = 16, 5 • 4 = 20 **7.** 15, 20, 25 **9.** 21, 28, 35 **11.** 20, 48, 52 **13.** 30, 72, 78 **15.** 100, 240, 260 **16.** no; Use the formula: $a^2 + b^2 = c^2$. $3^2 + 6^2 = 9^2$; 9 + 36 = 81 is a false statement because 9 + 36 = 45, not 81. **17.** yes **19.** no **21.** 6, 8, 10; Plato's formula is $(2m)^2 + (m^2 - 1)^2 = (m^2 + 1)^2$: Use 3 for m; $(2 \cdot 3)^2 + (3^2 - 1)^2 = (3^2 + 1)^2$; $6^2 + (9 - 1)^2 = (9 + 1)^2$; $6^2 + 8^2 = 10^2$, which gives the triple 6, 8, 10. **23.** 20, 99, 101 **25.** 40, 399, 401

440 Selected Answers and Solutions Geometry

Lesson 2, page 256

1. 169 sq units; The square measures 13 inches per side. $A = lw$; $13 \cdot 13 = 169$ **3.** 7, 49 **5.** yes **6.** 9, 12, 15; Longest side is 15; shortest side is 9; square's side is 3, $3 + 9 = 12$ **7.** 54 **9.** Subtract area of small square from area of large square; divide by 4; $225 - 9 = 216$; $\frac{216}{4} = 54$; area of each triangle = 54 sq units

Lesson 3, pages 260–261

1. 25; Press 625 [√] 25. **3.** 63 **5.** 30; Area of any right triangle is $\frac{1}{2}bh$. The base of each triangle is 5 and the height is 12; $\frac{1}{2}(5 \cdot 12) = \frac{1}{2}(60) = 30$ **7.** 289 **9.** 25 **11.** 60 **13.** the same **14.** 13; For a right triangle, $b^2 + h^2 = \text{hypotenuse}^2$; $5^2 + 12^2 = \text{hypotenuse}^2$; $25 + 144 = 169 = \text{hypotenuse}^2$; $\sqrt{169} = 13$ **15.** 41 **17.** 3.6 **19.** 2

Lesson 4, pages 264–265

1. ΔFGH, ΔFIG, ΔGIH; A perpendicular with the hypotenuse creates two smaller triangles that are similar to the first triangle. **3.** $\angle I = \angle G$, $\angle H = \angle H$, $\angle IGH = \angle F$ **5.** yes, AA Similarity Postulate **6.** 16.64 cm; Clock hands form the base and height of a right triangle; use the formula $b^2 + h^2 = \text{hypotenuse}^2$; $9^2 + 14^2 = \text{hypotenuse}^2$; $81 + 196 = 277 = \sqrt{277} \approx 16.64$ **7.** 45 feet **9.** 95.39 paces **11.** 14.42 inches **13.** Yes, the opening has a diagonal of 5 feet, which is wide enough. **15.** 721.11 m

Lesson 5, pages 268–269

1. 84; Formula for area of a trapezoid is $(\frac{a+b}{2})h$; $(\frac{10+18}{2})6 = (\frac{28}{2})6 = 14 \cdot 6 = 84$ **3.** 67.5 **5.** 105 **7.** 21 **9.** 19 **10.** 6; Area of any right triangle is $\frac{1}{2}bh$; $\frac{3 \cdot 4}{2} = \frac{12}{2} = 6$ **11.** 6 **13.** 24.5 **15.** 24 **17.** 132.5 **19.** 24 **21.** 98 **23.** 480 **25.** 2,312

Lesson 6, pages 273–274

1. $\sqrt{26}$; $d = \sqrt{(x_2 - x_1)^2 + (y_2 - y_1)^2}$; $d = \sqrt{(-4 - (-5))^2 + (-3 - 2)^2}$; $d = \sqrt{(1)^2 + (-5)^2}$; $d = \sqrt{1 + 25} = \sqrt{26}$ **3.** $\sqrt{52}$ **5.** 11 **7.** $\sqrt{388} = 2\sqrt{97}$ **9.** $\sqrt{325} = 5\sqrt{13}$ **11.** 10; $d = \sqrt{(x_2 - x_1)^2 + (y_2 - y_1)^2}$; $d = \sqrt{(4 - (-2))^2 + (5 - (-3))^2}$; $d = \sqrt{(6)^2 + (8)^2}$; $d = \sqrt{36 + 64} = \sqrt{100} = 10$ **13.** $\sqrt{29}$ **15.** $\sqrt{50} = 5\sqrt{2}$ **16.** $\sqrt{5{,}165}$; $d = \sqrt{(x_2 - x_1)^2 + (y_2 - y_1)^2}$; $d = \sqrt{(30 - (-8))^2 + (-11 - 50)^2}$; $d = \sqrt{(38)^2 + (-61)^2} = \sqrt{1{,}444 + 3{,}721} = \sqrt{5{,}165}$ **17.** $\sqrt{125} = 5\sqrt{5}$ **19.** 5

Lesson 7, page 277

1. yes; For ΔABC, if $\overline{BC}^2 = \overline{AB}^2 + \overline{AC}^2$, then ΔABC is a right triangle. Given sides 3, 4, and 5; $5^2 = 3^2 + 4^2$; $25 = 9 + 16$. Statement is true, so triangle is a right triangle. **3.** no **5.** no **7.** no **9.** no **11.** 12; $a^2 + b^2 = c^2$; $5^2 + \overline{BD}^2 = 13^2$; $25 + \overline{BD}^2 = 169$; subtract 25 from both sides to get $\overline{BD}^2 = 144$; take the square root of each side to get $\overline{BD} = 12$. **13.** For isosceles triangle, a line from the vertex perpendicular to the base bisects the base. **15.** Both angles are right angles.

Geometry *Selected Answers and Solutions* **441**

Application, page 278

1. Answers will vary but should reflect triangles in photos. **3.** Answers will vary but should reflect triangles in photos. **5.** Answers will vary but should include triangles in designs.

Chapter Review, pages 279–281

1. D **3.** C **5.** A **7.** 40 **9.** 19.5 **11.** 77 **13.** $\sqrt{41}$ **15.** yes **17.** yes **19.** no

CHAPTER 9

Lesson 1, page 286

1. 35; $10 + 12 + 13 = 35$ **3.** 42 **4.** 10; Let x = the missing side; $8 + 5 + x = 23$; $13 + x = 23$; $x = 10$ **5.** 9 **7.** 9 **8.** $P = 80$ m; $2(15) + 2(25) = P$; $30 + 50 = 80$ **9.** $P = 110$ ft

Lesson 2, pages 289–290

1. $P = 30$; $P = 3s$: $P = 3(10) = 30$ **3.** 56 **5.** 48 **7.** 80 **9.** 14; $P = 2(l + w)$: $P = 2(5 + 2)$; $P = 2(7) = 14$ **11.** 18 **13.** $w = 20$, $l = 60$; $P = 2(l + w)$: l is 3 times w; $P = 2(3w + w)$; $160 = 2(3w + w)$ or $160 = 2(4w)$; $160 = 8w$; $w = 20$. Since l is 3 times w: $3(20) = 60$. **15.** $b = 240$, $s = 160$ **17.** $w = 7.5$, $l = 45$ **18.** 46 in.; $\frac{46}{36}$ rounded to $1\frac{1}{3}$ yards; $P = 2(l + w)$: $P = 2(9 + 14)$; $P = 2(23)$; $P = 46$; (1 yard = 36 inches) **19.** 96 ft

Lesson 3, pages 293–294

1. 46.2; Let x = the missing side: $5^2 + 10^2 = x^2$; $25 + 100 = 125$; $x = \sqrt{125} = 11.2$; $P = 5 + 10 + 20 + 11.2 = 46.2$ **3.** 24.2 **5.** 44.3 **6.** 12; $P = \overline{AB} + \overline{BC} + \overline{CA}$;
$d = \sqrt{(x_2 - x_1)^2 + (y_2 - y_1)^2}$;
$\overline{AB} = \sqrt{(2 - 2)^2 + (1 - 4)^2}$;
$\overline{AB} = \sqrt{(0)^2 + (3)^2} = \sqrt{9} = 3$
$\overline{BC} = \sqrt{(6 - 2)^2 + (1 - 4)^2}$;
$\overline{BC} = \sqrt{(4)^2 + (3)^2}$;
$\overline{BC} = \sqrt{16 + 9} = \sqrt{25} = 5$
$\overline{CA} = \sqrt{(6 - 2)^2 + (1 - 1)^2}$;
$\overline{CA} = \sqrt{(4)^2 + 0} = \sqrt{16} = 4$;
$3 + 4 + 5 = 12$ **7.** 22.4 **9.** 21.7

Lesson 4, pages 298–299

1. 225 cm^2; $A = s^2$; $15^2 = 225$ **3.** 81 sq in. **5.** 48 cm^2 **7.** 60 km^2 **9.** 9 cm; $\sqrt{A} = s$; $\sqrt{81} = 9$ **11.** 7 in. **13.** $b = 11$ mi; $A = bh$; $110 = 10 \cdot b$; $\frac{110}{10} = b = 11$ **15.** 33 m **16.** 500; $A = bh$; $A = 10 \cdot 50 = 500$ **17.** 1,200 **19.** 33 **21.** 72 **22.** 102 sq ft; Area of $A1 = 8 \cdot 7 = 56$; Area of $A2 = 4 \cdot 4 = 16$; Area of $A3 = 5 \cdot 6 = 30$; $56 + 16 + 30 = 102$ **23.** two; one application area is $75 \cdot 30 = 2{,}250$; $\frac{5{,}000}{2{,}250} = 2.2$ **25.** 9 sq ft

Lesson 5, pages 302–303

1. 200; $A = \frac{b_1 + b_2}{2} h$; $\frac{15 + 25}{2} 10 = \frac{40}{2} 10$; $(20)(10) = 200$ **3.** 210 **5.** 12 **7.** 28 **9.** 168 **11.** 26; $A = (\text{average base}) \cdot h$; $156 = \text{average base} \cdot 6$; $\frac{156}{6} = \text{average base} = 26$ **13.** 7 **15.** 5.5 **16.** 28; $(10 - 4) \div 2 = 3$; $a^2 + b^2 = c^2$; $3^2 + b^2 = c^2$; $9 + b^2 = 25$; $b^2 = 16$; $b = 4$; $A = \frac{b_1 + b_2}{2} h$; $\frac{10 + 4}{2} 4 = \frac{14}{2} 4 = (7)(4) = 28$ **17.** 56.7 **19.** 349.8

Try This

Yes. Because $b_1 = b_2$ in a rectangle, the average base will be either $\frac{(b_1 + b_1)}{2}$ or $\frac{(b_2 + b_2)}{2} = b_1 = b_2$, giving the results $b \cdot h$ for area, which is equal to $l \cdot w$.

Lesson 6, pages 308–309

1. 7.7; $3 \cdot 4 = 12$; $12 \cdot 5 = 60$; $\sqrt{60} \approx 7.7$ **3.** 35.9 **5.** 21.9 **6.** 12.5; $A = \frac{1}{2}bh$; $(5 \cdot 5) = 25$; $\frac{1}{2}25 = 12.5$. **7.** 24.5 **9.** 72 **11.** 6 **13.** 14 **15.** 225 **17.** 18 **19.** 22.5 **21.** 20.3; $A = \sqrt{s(s-a)(s-b)(s-c)}$ where $s = \frac{1}{2}(a + b + c)$; $s = \frac{1}{2}(6 + 7 + 8) = \frac{1}{2}(21) = 10.5$; $A = \sqrt{10.5(10.5-6)(10.5-7)(10.5-8)} = \sqrt{10.5(4.5)(3.5)(2.5)} = \sqrt{413.4} = 20.3$ **23.** 71 **25.** 51.9

Try This

$$A = \sqrt{\frac{3x}{2} \cdot \left(\frac{3x}{2} - x\right) \cdot \left(\frac{3x}{2} - x\right) \cdot \left(\frac{3x}{2} - x\right)}$$

$$= \sqrt{\frac{3x^4}{16}} = \frac{x^2\sqrt{3}}{4} = (0.4)x^2$$

Application, page 310

1. Drawings should reflect the principle that tessellations cover a plane surface without overlapping and without leaving holes. **3.** Yes **5.** Drawings should reflect the principle that tessellations cover a plane surface without overlapping and without leaving holes.

Chapter Review, pages 311–313

1. C **3.** F **5.** E **7.** B **9.** 17.6 **11.** $P = 22$, $A = 24$ sq units **13.** $P = 30$, $A = 36$ sq units **15.** $P = 56$, $A = 128$ sq units **17.** $P = 45.6$, $A = 50$ sq units **19.** $P = 36$, $A = 52$ sq units

CHAPTER 10

Lesson 1, page 318

1. center; O is the center of the circle. **3.** diameter **5.** radius **7.** $\overline{BC}$ **9.** C **11.** 12 units; $d = 2r$; $d = 2 \cdot 6 = 12$ **13.** 7 **15.** 12π units, 37.7 units

Lesson 2, page 322

1. 12π units; $C = \pi d$; $C = 12\pi$ **3.** 31π m **5.** 20π in. **7.** 18π ft **9.** $d = 44$ ft, $r = 22$ ft; $C = d\pi = 2r\pi$; $44\pi = d\pi$ and $44\pi = 2r\pi$; $d = 44$ and $2r = 44$; $r = 22$ **11.** $d = 12$ in., $r = 6$ in. **13.** $d = 15$ ft, $r = 7.5$ ft **15.** $d = 22$ cm, $r = 11$ cm **17.** 23 ft; $C = d\pi$; fence $= C$; widest part of garden $= d$; $23\pi = d\pi$; $d = 23$ **19.** 1 to 2 or $\frac{1}{2}$

Lesson 3, page 325

1. large square = 400 sq units, interior square = 200 sq units, estimated area of circle = 300 sq units; A of large square $= 4r^2 = 4(10)^2 = 4(100) = 400$; A of interior square $= 2r^2 = 2(10)^2 = 2(100) = 200$; A of circle $= \frac{400 + 200}{2} = 300$ **3.** large square = 1,600 sq units; interior square = 800 sq units; estimated area of circle = 1,200 sq units **5.** large square = 2,500 sq units; interior square = 1,250 sq units; estimated area of circle = 1,875 sq units **6.** 48; $A \approx 3r^2$; $A = 3(4^2) = 3(16) = 48$ **7.** 108 **9.** 867 **11.** 432 **13.** 243 **15.** 1,323 **16.** $r = 4$; $A \approx 3r^2$; $48 = 3r^2$; $16 = r^2$; $\sqrt{16} = r = 4$ **17.** $r = 6$ **19.** $r = 10$

Lesson 4, page 327

1. 50%; $\frac{\text{area of desired outcome}}{\text{total area}} = \frac{2}{4} = \frac{1}{2} = 50\%$ **3.** 75% **5.** 50%

Lesson 5, pages 330–331

1. 153.94; Press [π] [×] 7 [x^2] [=] 153.94. **3.** 50.27 **5.** 254.47 **7.** 25 **9.** 1,111 **11.** 9π sq in.; $A = \pi r^2$; $A = \pi(3)^2 = 9\pi$ **13.** 81π m^2 **15.** 225π km^2 **16.** 452.4 sq in.; $A = \pi r^2$; $A = \pi 12^2 = 144\pi$; $144 \cdot \pi \approx 452.4$ **17.** 113.1 ft^2 **19.** 907.9 mm^2 **21.** 452.4 km^2

Geometry *Selected Answers and Solutions* **443**

23. 153.9 ft^2 **25.** 6,361.7 mi^2
26. $r = 8.2$ cm; $A = \pi r^2$; $210 = \pi r^2 = 66.7 = r^2$; $\sqrt{66.7} = 8.2$ **27.** $r = 11.3$ in.
29. $r = 10.0$ km

Lesson 6, page 335

1. 85°; The measure of any central angle is equal to the arc that its endpoints make on the circle; $\angle ACB$ = arc AB.
3. 60° **5.** 125° **6.** $\frac{49}{12}\pi$; The shaded area has a central angle of 30°. A circle has 360°; $\frac{360}{30} = 12$; the shaded area is $\frac{1}{12}$ of the circle. $A = \pi r^2$; $A = \pi 7^2 = 49\pi$; $\frac{1}{12}(\frac{49}{12}\pi) = \frac{49}{12}\pi$ **7.** 8π **9.** 30°; An inscribed angle is measured by $\frac{1}{2}$ its intercepted arc; $\frac{1}{2}(60) = 30$

Try This

90°

Lesson 7, page 339

1. Given; This statement is part of the given information about the proof.
3. Definition of midpoint. **5.** $m\angle ADX = m\angle BDX$ **7.** Corresponding sides of congruent triangles are equal.

8.

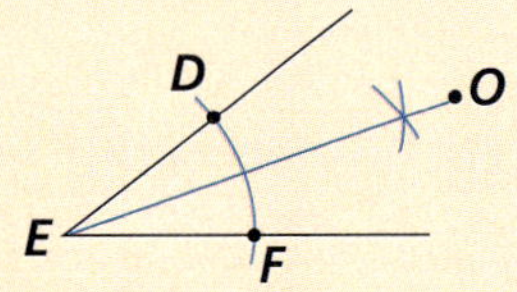

9.

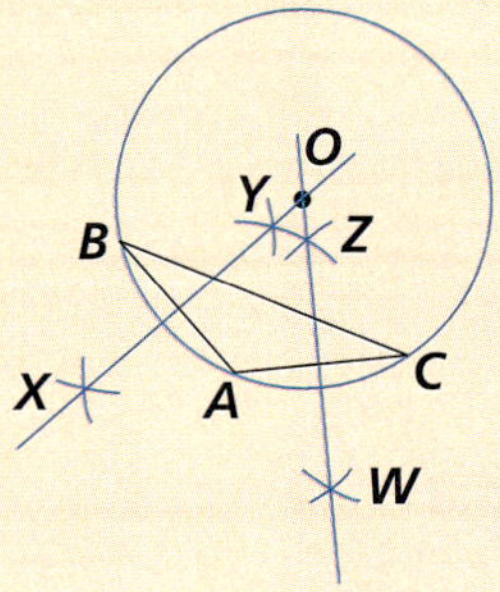

Lesson 8, page 341

1. 0, 1, 0; Press 0 [SIN] display 0; Press 0 [COS] display 1; Press 0 [TAN] display 0.
3. 0.71, 0.71, 1 **5.** 1, 0, undefined

Lesson 9, pages 344–345

1. $S = 36\pi$ sq units, $V = 36\pi$ $units^3$; Formula for surface area is $S = 4\pi r^2$; $S = 4\pi(3)^2 = 4\pi(9) = 36\pi$. Formula for volume is $V = \frac{4}{3}\pi r^3$; $V = \frac{4}{3}\pi(3)^3 = \frac{4}{3}\pi 27 = 36\pi$ **3.** $S = 64\pi$ sq units, $V = 85.3\pi$ $units^3$
5. $S = 144\pi$ sq units, $V = 288\pi$ $units^3$
7. $S = 50.27$ sq units, $V = 33.51$ $units^3$; $S = 4\pi r^2$; $S = 4\pi(2)^2 = 4\pi(4) = 16\pi$; using a calculator: $16 \cdot 3.14 \approx 50.27$. $V = \frac{4}{3}\pi r^3$; $V = \frac{4}{3}\pi(2)^3 = \frac{4}{3}\pi(8) = 10.7\pi$; using a calculator: $10.7 \cdot 3.14 \approx 33.51$ $units^3$.
9. $S = 1{,}520.53$ sq units, $V = 5{,}575.28$ $units^3$ **11.** $S = 1{,}256.64$ sq units, $V = 4{,}188.79$ $units^3$ **13.** 36π, 36π; $S = 4\pi r^2$; $S = 4\pi(3)^2 = 36\pi$; $V = \frac{4}{3}\pi r^3$; $V = \frac{4}{3}\pi(3)^3 = \frac{4}{3}\pi 27 = 36\pi$ **15.** 576π, $2{,}304\pi$

Application, page 346

1. C; Graph shows total number of students in grades 9–12 by year. **3.** A; Graph shows data for 1997 only. **5.** C

Chapter Review, pages 347–349

1. D **3.** A **5.** D **7.** $C = 44.0$ units, $A = 153.9$ sq units **9.** $C = 37.7$ cm, $A = 113.1$ cm^2 **11.** $A = 24\pi$ **13.** $A = 56\pi$ **15.** 0.72, 0.69, 1.04 **17.** $S = 400\pi$, $V = 1{,}333.3\pi$ **19.** $S = 196\pi$, $V = 457.3\pi$ **21.** $S = 36\pi$, $V = 36\pi$

444 *Selected Answers and Solutions* *Geometry*

23.

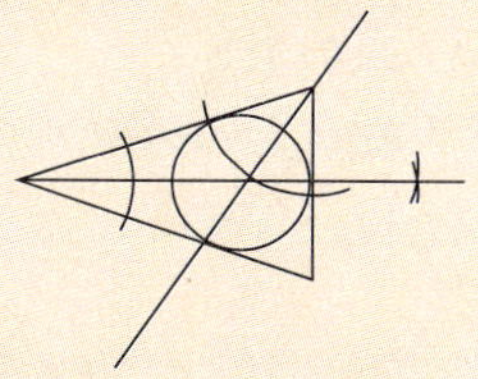

25. 25%

CHAPTER 11

Lesson 1, pages 354–355

1. 1,000 m^3; V = area of base • height; area of base = lw; area of base = 10 • 10 = 100. V = 100 • height; V = 100 • 10 = 1,000 **3.** 800 cu in. **5.** 144 cu in.; V = area of base • height; $V = 24 \cdot 6 = 144$ **7.** 1,500 cm^3 **8.** $1{,}000\pi$; V = area of base • height; (base is circle with radius of 10) $A = \pi r^2 = \pi(10)^2 = 100\pi$. $V = 100\pi(10) = 1{,}000\pi$ **9.** 3,000 **11.** 3,140 **13.** 375 **15.** 392.5 **16.** $h = 10$; V = area of base • height; $250 = 25 \cdot h$; $h = 10$ **17.** $r = 7$ **19.** $s = 8$

Lesson 2, pages 358–359

1. 333.3π; Formula for volume of cone is $V = \frac{1}{3}\pi r^2 h$; $V = (\frac{1}{3})(\pi 10^2)(10) = \frac{1}{3}(100\pi)(10) = \frac{1}{3}(1{,}000\pi) = 333.3\pi$ **3.** 1,033.3 **5.** 144π **7.** 446.4 **9.** 12π **11.** 37.2 **13.** 333.3 cu in.; Formula for volume of pyramid is $V = \frac{1}{3}$(area of base)(height); area = $10^2 = 100$; $V = \frac{1}{3}(100)(10) = \frac{1}{3}(1{,}000) = 333.3$ **15.** 80 cm^3 **17.** 100m^3 **18.** $w = 4$ cm; Formula for volume of pyramid is $V = \frac{1}{3}$(area of base)(height); area of base = lw; $80 = \frac{1}{3}(6w)(10)$; $80 = 20w$; $w = 4$ **19.** $r = 22.5$ in.

Lesson 3, page 363

1. 96 sq units; Formula for surface area of cube is $S = 6s^2$; $S = 6(4^2) = 6(16) = 96$ **3.** 170π sq units **4.** 112π; Formula for surface area of cylinder is $S = 2\pi r^2 + (2\pi r)h$; (subtract the surface area of one circular base from the equation) $S = \pi r^2 + (2\pi r)h$; $S = \pi 4^2 + 2\pi(4)(12) = 16\pi + 96\pi = 112\pi$ **5.** No, she is planning to paint 516 sq ft.

Lesson 4, pages 366–367

1. 224 sq units; Formula for surface area of pyramid is S = (area of base) + area of each of the triangles. Area of the square base is $A = s^2$; $A = 8^2 = 64$. Formula for area of triangle is $A = \frac{1}{2}(bh)$; $A = \frac{1}{2}(8)(10) = \frac{1}{2}(80) = 40$. $S = 64 + 4(40) = 64 + 160 = 224$ **3.** $120 + 16\sqrt{3}$ sq units **4.** 256; Press 16 [x^2]. The display reads 256. **5.** 324 **7.** 529 **9.** No, he needs a triangle with a base ≥ 50.24 feet; $C = 2\pi r = 2\pi 8 = 16\pi \approx 50.24$

Try This

350 sq units

Lesson 5, page 369

1. 432 in.; A yard is 36 inches long; $12 \times 36 = 432$ **3.** 63,360 **5.** 9 **7.** 15,000 **9.** 1,000,000

Application, page 370

1. Results will vary based on size of paper. **3.** Results will vary based on size of box. **5.** Results will vary based on constructions.

Geometry *Selected Answers and Solutions* **445**

Chapter Review, pages 371–373

1. B **3.** C **5.** D **7.** A **9.** D **11.** 72,000 cm^3 **13.** $1{,}536\pi$ cu ft **15.** 270π cu in. **17.** 10 ft, 120 in. **19.** 10 cm, 1 dm

CHAPTER 12

Lesson 1, pages 377–378

1. The line is tangent to the cylinder; By definition, a tangent line can share only one point with an object. **3.** The line lies on the surface of the cylinder. **5.** The plane makes a vertical slice through the cylinder. **7.** The plane must "rest against" the side of the cylinder.
9. A line passed through a circle forming a chord; The circumference of the trunk is comparable to a circle and the board inside the trunk to a line segment.

Lesson 2, pages 381–382

1. $y = 8$ and $y = 2$; For points in a line to be equidistant from a given line, they must form a parallel line. Two lines are parallel to $y = 5$ and they are points 3 units away. They are $y = 8$ and $y = 2$.
3. $y = \frac{1}{2}x + 1$ and $y = -\frac{1}{2}x + 3$ **5.** $y = -1$
7. $36 = x^2 + y^2$; Points that are equidistant from one point such as the origin create a circle. Each point on the circle has an x- and y-value. A right triangle can be drawn using the radius as the hypotenuse and the axis as the base and height; hypotenuse$^2 = b^2 + h^2$; $6^2 = x^2 + y^2$ with x equal to the x-value of the point and y equal to the y-value. **9.** $0.25 = x^2 + y^2$ **11.** 12,321; Press 111 [x^2]. The display reads 12,321. **13.** 10,201 **15.** 367,236

Lesson 3, pages 384–385

1. Two points S and T on the circle and equidistant from ℓ and m

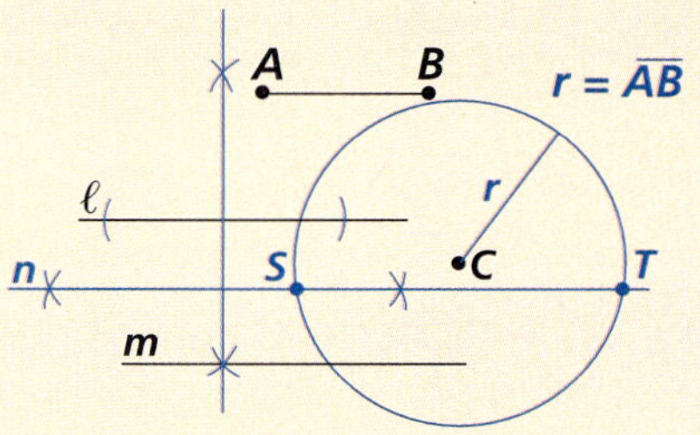

3. The intersection of the perpendicular bisectors of each side of the triangle

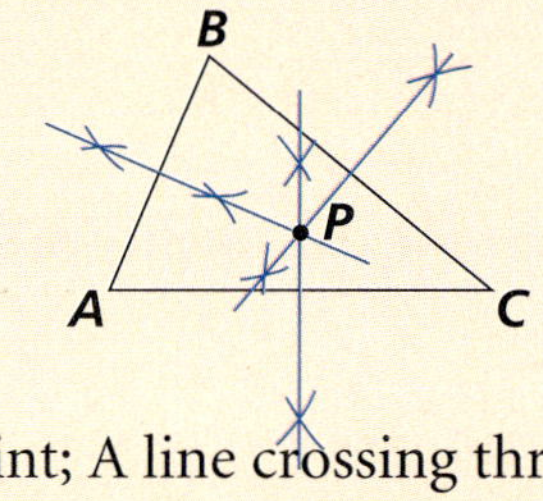

5. a point; A line crossing through a plane will cross at only one point.
7. a square smaller than the base
9. a circle with radius smaller than the radius of the base

Try This

square: intersect perpendicular to length; rectangle: intersect parallel to length; triangle: a line intersecting the length diagonally forms a triangle on the surface of the rectangle

446 Selected Answers and Solutions Geometry

Application, page 386

1. Results will vary based on unit size of constructions, but each square should have 4 equal sides and 90° angles.
3. Results will vary based on unit size of constructions. **5.** Results will vary based on unit size of constructions. Example: the spiral is the locus of points such that the radius changes every quarter-circle according to the pattern $r_n = r_{n-1} + r_{n-1}$.

Chapter Review, pages 387–389

1. The line passes through the circle, forming a chord.

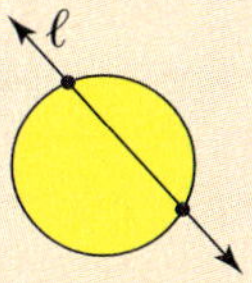

3. The line passes through the sphere.

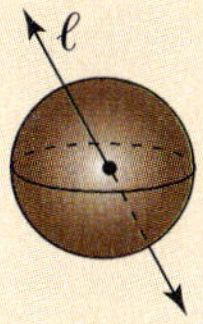

5. The line is on the surface of the cylinder.

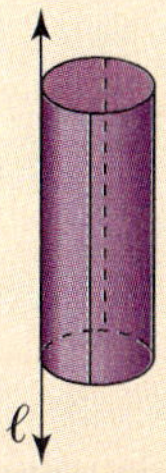

7. The plane and the sphere meet at one point only.

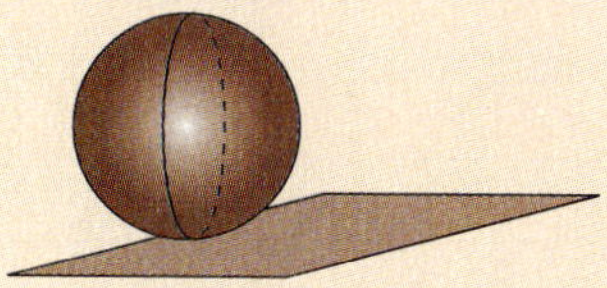

9. The plane and the cylinder intersect to form a circle.

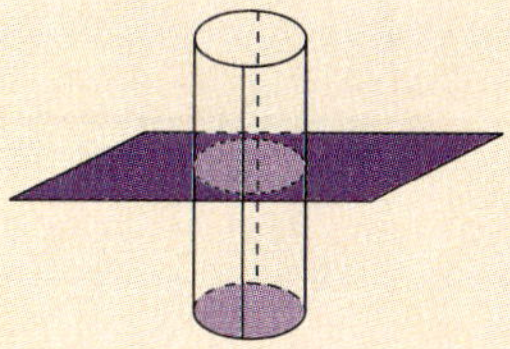

11. $x = \frac{1}{2}$

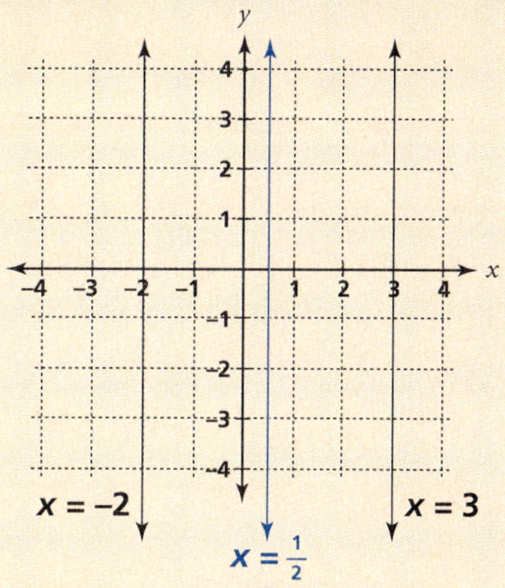

Geometry *Selected Answers and Solutions* **447**

13. $y = 1$

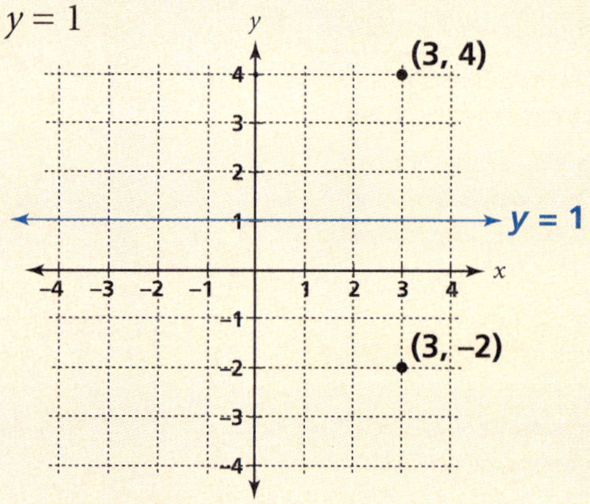

15. $4 = x^2 + y^2$

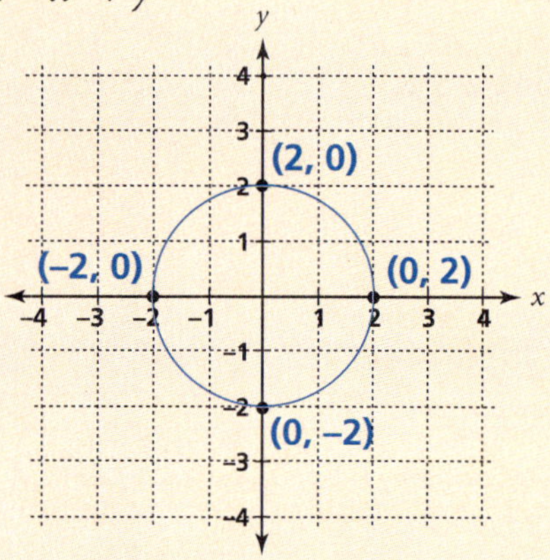

17. two points S and T, each on the circle with center P and radius CD and at a distance of $\overline{CD}$ from ℓ

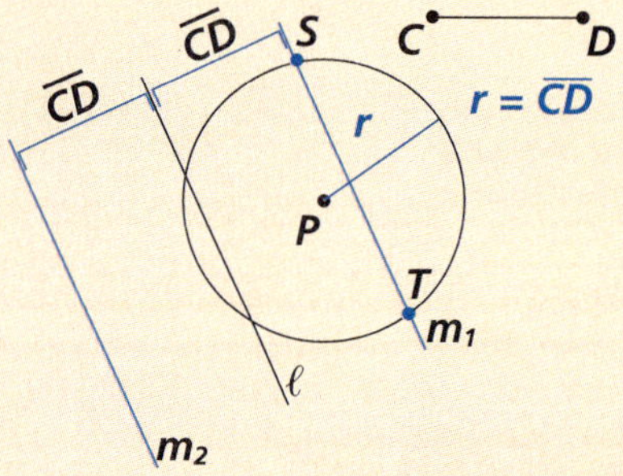

19. C

448 *Selected Answers and Solutions* *Geometry*

Supplementary Problems

CHAPTER 1

Pages 390–392

1. line, $\overleftrightarrow{XY}$ **2.** ray, $\overrightarrow{PQ}$ **3.** line, m **4.** point, B **5.** line segment, $\overline{ST}$ **6.** line, y **7.** –3 **8.** B **9.** 2 **10.** I **11.** C **12.** 3 **13.** F **14.** 0 **15.** $1\frac{1}{2}$ **16.** D **17.** 15 **18.** 10 **19.** 5 **20.** 10 **21.** 5 **22.** 10 **23.** 10 **24.** 25 **25.** 15 **26.** 35 **27.** C **28.** C **29.** B **30.** A **31.** $\overleftrightarrow{CA}$ **32.** $\overrightarrow{XY}$ **33.** $\overline{RQ}$ **34.** $\angle DEF$

35.

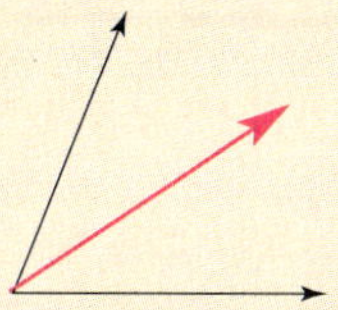

36.

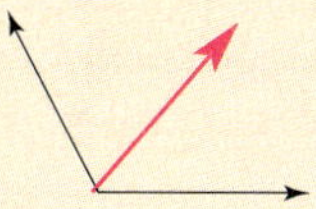

37. right **38.** obtuse **39.** acute **40.** straight **41.** acute **42.** right **43.** 90° **44.** 55° **45.** 35° **46.** 55° **47.** 180° **48.** $m\angle a = 120°$, $m\angle b = 60°$, $m\angle c = 120°$ **49.** $m\angle x = 90°$, $m\angle y = 90°$, $m\angle z = 90°$ **50.** $m\angle q = 50°$

CHAPTER 2

Pages 393–395

1. *an angle is a right angle*, hypothesis; *its measure is 90°*, conclusion **2.** *an angle is a straight angle*, hypothesis; *its measure is 180°*, conclusion **3.** *a figure is a quadrilateral*, hypothesis; *it has 4 sides*, conclusion **4.** *two angles have equal measures*, hypothesis; *they are congruent*, conclusion **5.** *an animal is an insect*, hypothesis; *it has 6 legs*, conclusion **6.** If a figure is a triangle, then it has three sides; true **7.** If the sum of the measures of two angles is 180°, then the angles are supplementary; true **8.** If the sun is not shining, then it is raining; false **9.** If you must pay taxes, then you are a citizen of the United States; false **10.** Postulate 1 **11.** Postulate 3 **12.** Postulate 4 **13.** Parallel Postulate **14.** Postulate 5 **15.** Postulate 3 **16.** yes; Postulate 4 **17.** $m\angle A = 30°$; Axiom 1 **18.** angles a, b, and c; $m\angle a + m\angle b$; $m\angle a + m\angle c$; $m\angle b + m\angle c$; Axiom 5 **19.** yes; Axiom 2 **20.** yes; Axiom 3 **21.** $m\angle R = 90°$, $m\angle S = 45°$, $m\angle Q = 45°$ **22.** $m\angle Q = 30°$, $m\angle R = 150°$, $m\angle S = 150°$ **23.** $m\angle Q = 90°$, $m\angle R = 90°$, $m\angle S = 90°$ **24.** Given **25.** Vertical angles are equal. **26.** Definition of supplementary angles. **27.** Axiom 3 **28.** Axiom 5 **29.** Definition of supplementary angles. **30.** Postulate 5

CHAPTER 3

Pages 396–398

1. lines x and y **2.** line q, w, x, or y **3.** $\angle d$ and $\angle f$, or $\angle l$ and $\angle n$ **4.** $\angle k$ and $\angle m$, or $\angle c$ and $\angle e$ **5.** lines x and q, or x and w, or y and q, q and w, or y and w **6.** $\angle f$ and $\angle l$ **7.** supplementary **8.** alternate interior **9.** alternate exterior **10.** alternate interior **11.** supplementary **12.** alternate exterior **13.** corresponding **14.** supplementary **15.** corresponding **16.** supplementary **17.** 80° **18.** 100° **19.** 120° **20.** 120° **21.** 60° **22.** 120° **23.** 80° **24.** 60° **25.** 80° **26.** 80° **27.** 60° **28.** 120° **29.** $m\angle b = 120°$; $m\angle c = 60°$

Geometry Selected Answers and Solutions 449

30. $m\angle w = 45°$; $m\angle y = 135°$ **31.** $\overline{EM}$
32. $\overline{EG}, \overline{GA}, \overline{AM}$ **33.** $\overline{GE}$ **34.** G or M
35. E or A **36.** A **37.** not parallel
38. parallel **39.** parallel **40.** parallel

CHAPTER 4

Pages 399–401

1. (1, 2) **2.** (–4, 1) **3.** (–2, –3) **4.** (4, 0) **5.** (0, –4) **6.** (3, 4) **7.** (–1, 0) **8.** (2, –3) **9.** (–3, 0) **10.** (0, 3) **11.** S **12.** W **13.** U **14.** N **15.** O **16.** R **17.** V **18.** P **19.** T **20.** Q **21.** $m = 3$ **22.** $m = \frac{2}{3}$ **23.** $m = -2$ **24.** $m = -1$ **25.** $m = 2$ **26.** $m = 0$ **27.** $m = \frac{1}{2}$ **28.** $m = -1$ **29.** $m = \frac{5}{8}$ **30.** $m = 0$ **31.** $y = 2x$ **32.** $y = 0$ **33.** $y = \frac{1}{2}x - 1$ **34.** $y = -x + 4$ **35.** $y = \frac{5}{8}x - \frac{9}{8}$ **36.** $y = -6$ **37.** (–1, 7) **38.** (6, –2) **39.** (0, 1) **40.** (4, –3)
41. (–2, 4), (–1, 7)

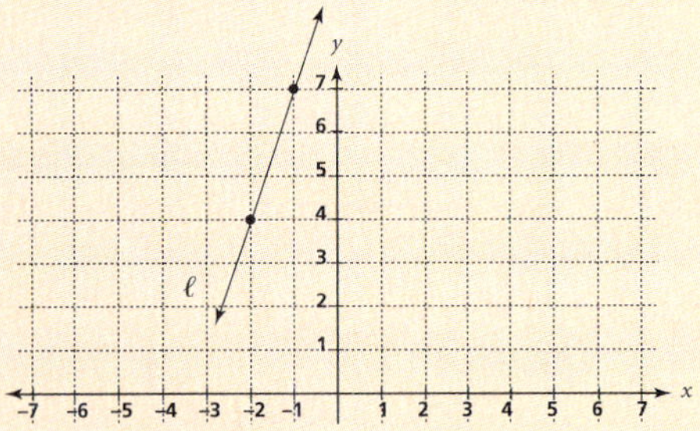

42. (1, –4), (6, –2)

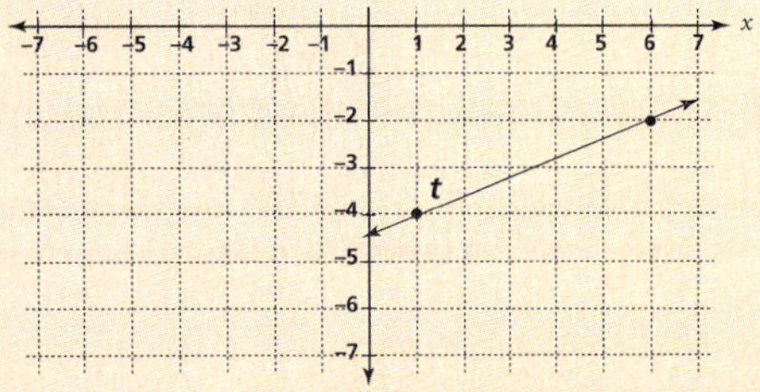

43. (–1, 3), (0, 1)

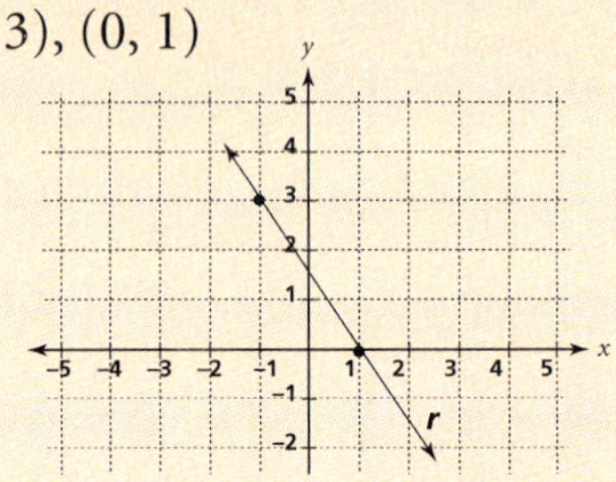

44. (0, 0), (4, –3)

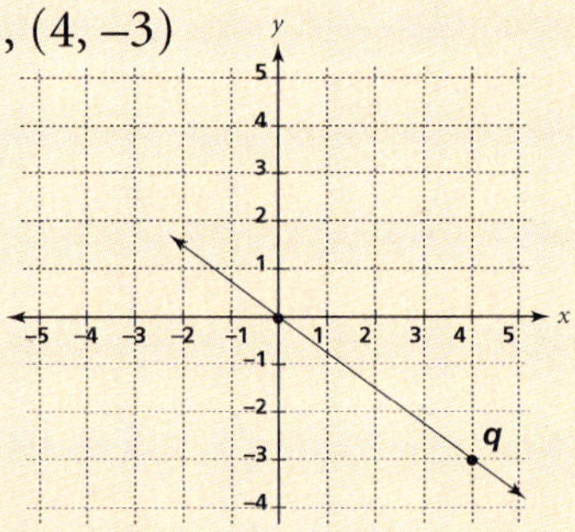

45. (3, 4) **46.** (1, 3) **47.** (3, 1) **48.** (9, 2) **49.** (2, 3) **50.** $(-\frac{5}{2}, 6)$

CHAPTER 5

Pages 402–404

1. scalene triangle **2.** equilateral triangle **3.** isosceles right triangle **4.** rhombus **5.** isosceles trapezoid **6.** obtuse scalene triangle

7.

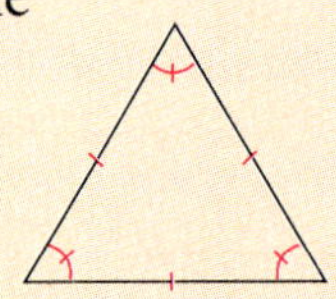

8.

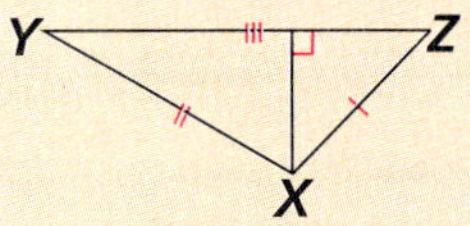

9.

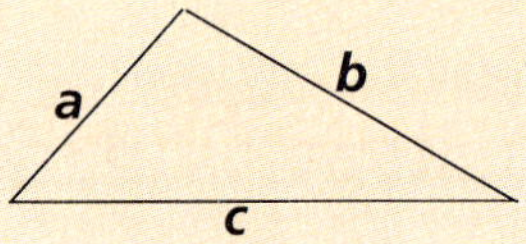

10.

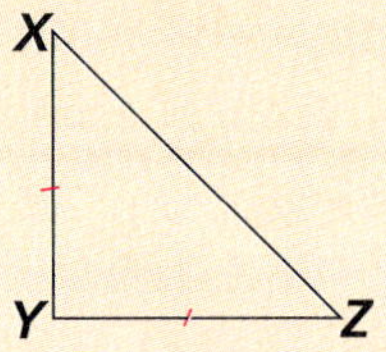

450 Selected Answers and Solutions Geometry

11.

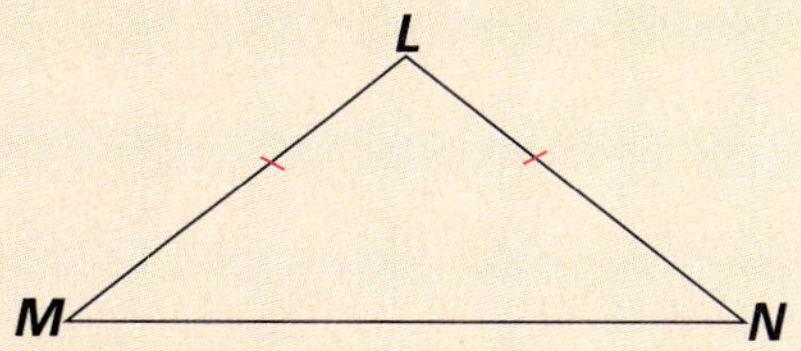

12. 60° **13.** 150° **14.** 80° **15.** 108° **16.** 60° **17.** 135° **18.** 40° **19.** 110° **20.** 70° **21.** 70° **22.** 40° **23.** A line from the vertex of a triangle and perpendicular to the opposite side **24.** 1,080 **25.** 60° **26.** 140° **27.** Find the sum of the two nonadjacent interior angles. **28.** No, because it has only one pair of parallel sides **29.** Subtract 2 from the number of sides the polygon has. **30.** A line from the vertex of a triangle to the midpoint of the opposite side **31.** ∠3, ∠5, ∠7 **32.** 140° **33.** 40° **34.** m∠1 or m∠4 **35.** yes; An exterior angle is formed by extending one side of the triangle along any vertex.

CHAPTER 6

Pages 405–407

1. yes; SSS **2.** no; ∠*B* and ∠*Y* are not congruent. **3.** no; for SAS, the angle must be the included angle. **4.** yes; AAS **5.** no **6.** yes **7.** yes **8.** no **9.** yes **10.** no **11.** no **12.** yes **13.** $x = 30°$ **14.** $x = 15°$ **15.** m∠2 = 80°; m∠1 = 40° **16.** m∠2 = 40°, m∠3 = 40° **17.** Given. **18.** Definition of equilateral triangle. **19.** Definition of perpendicular bisector. **20.** Any item is equal to itself. **21.** SSS **22.** 4 **23.** 0 **24.** 6 **25.** 1 **26.** (11, –2) **27.** (15, –12) **28.** (12, –9) **29.** (–9, 1) **30.** (6, 0) **31.** (–2, –4) **32.** (–25, –47) **33.** (0, 2)

34. *X*' = (4, 0), *Y*' = (1, 3), *Z*' = (1, 0)

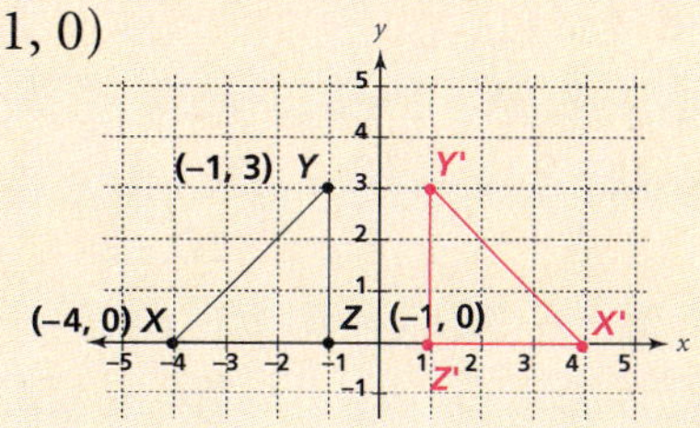

35. *A*' = (–1, 3), *B*' = (2, 1), *C*' = (4, 2), *D*' = (3, 3)

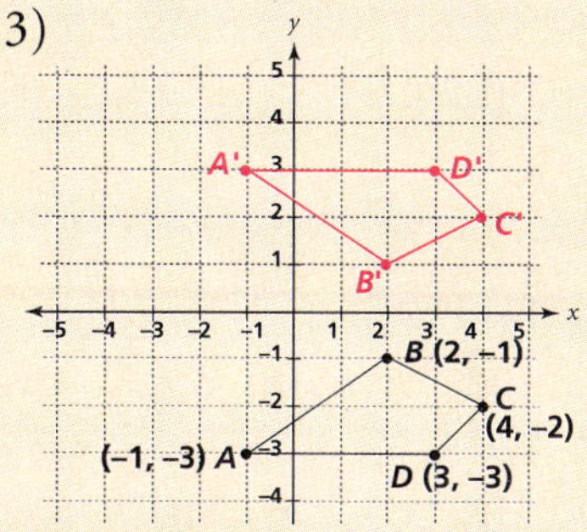

CHAPTER 7

Pages 408–409

1. yes **2.** no **3.** no **4.** yes **5.** no **6.** yes **7.** yes **8.** yes **9.** no **10.** no **11.** yes; 3 equal angles **12.** yes; all angles are 60° **13.** m∠*A* **14.** m∠*BED* **15.** $\overline{BC}$ **16.** $\overline{DE}$, $\overline{BE}$ **17.** 20 **18.** 10 **19.** 100° **20.** 50° **21.** 48, 32 **22.** 18, 6 **23.** 12, 15 **24.** 480 feet **25.** (–6, 10) (6, 10) (–6, –10) (6, –10)

CHAPTER 8

Pages 410–411

1. 6, 8, 10 **2.** 24, 32, 40 **3.** 30, 40, 50 **4.** 10, 24, 26 **5.** 15, 36, 39 **6.** 50, 120, 130 **7.** $x = 6$ **8.** $x = 60$ **9.** $x = 36$ **10.** $A = 204$ **11.** $A = 756$ **12.** yes **13.** no, the cut piece is not a right triangle **14.** no **15.** yes; ∠*M* **16.** no **17.** yes; ∠*A* **18.** $d = 20.0$ **19.** $d = 25.1$ **20.** $d = 12.1$

Geometry Selected Answers and Solutions **451**

CHAPTER 9

Pages 412–413

1. $P = 24$ **2.** $P = 19$ **3.** $P = 34$ **4.** $P = 144$ **5.** $P = 500$ **6.** $P = 48$ **7.** $l = 20, w = 10$ **8.** $b = 12, h = 9$ **9.** $8\sqrt{2}$ **10.** $w = 12, l = 48$ **11.** $x = 8$ **12.** $x = 6$ **13.** $x = 7$ **14.** $A = 6.25$ **15.** $A = 85$ **16.** $A = 240$ **17.** $A = 720$ **18.** $A = 44.9$ **19.** $A = 9.8$ **20.** $A = 48.0$ **21.** $A = 62.4$ **22.** $A = 270$ **23.** $A = 50$ **24.** $A = 100$ **25.** $A = 32.7$

CHAPTER 10

Pages 414–415

1. 6.2 in. **2.** 3.1 sq in. **3.** 3.1 cm **4.** 0.78 cm^2 **5.** 124 ft **6.** 1,240 sq ft **7.** 93 m **8.** 697.5 m^2 **9.** 3.1 in. **10.** 0.78 sq in. **11.** $\frac{75}{4}\pi$ **12.** $\frac{200}{3}\pi$ **13.** 90π **14.** 60° **15.** 160° **16.** 50% **17.** 37.5% **18.** 50% **19.** 4π sq mi **20.** $\frac{4}{3}\pi$ cu mi **21.** 16π sq ft **22.** $\frac{32}{3}\pi$ cu ft **23.** $1{,}600\pi$ sq ft **24.** $\frac{32{,}000}{3}\pi$ cu ft **25.** 900π m^2 **26.** $4{,}500\pi$ **27.** 1π sq in. **28.** $\frac{1}{6}\pi$ cu in. **29.** B **30.** B

CHAPTER 11

Pages 416–417

1. 125 cu units **2.** 8,000 cu units **3.** 7 m^3 **4.** 432 cm^3 **5.** 6,720 cm^3 **6.** 550 square units **7.** 800 square units **8.** 248 square units **9.** 660 square units **10.** 474 square units **11.** 532 square units **12.** 216 square units **13.** 192 square units **14.** 147π square units **15.** 300π square units **16.** $V = 512$ cm^3, $A = 384$ cm^2 **17.** $V = 1{,}536\pi$ cu ft, $A = 512$ sq ft **18.** $V = 256\pi$ cubic units, $A = 192\pi$ square units **19.** $V = 24\sqrt{3}$ cubic units, $A = 108 + 9\sqrt{3}$ square units **20.** $V = 735\pi$ cm^3, $A = 308\pi$ cm^2 **21.** $216 + 18\sqrt{3}$ sq in. **22.** 642 cm^2 **23.** 1,704 sq in. **24.** $1{,}296\pi$ sq ft **25.** 348 sq in.

CHAPTER 12

Pages 418–419

1.

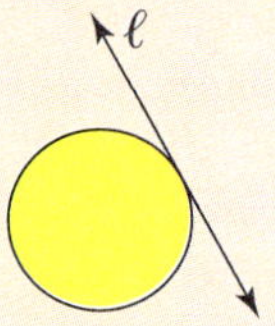

2.

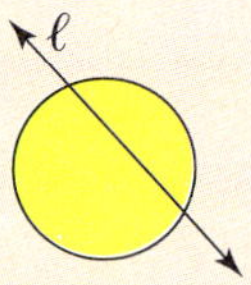

3.

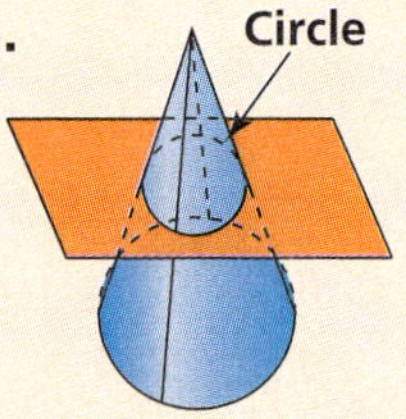

4.

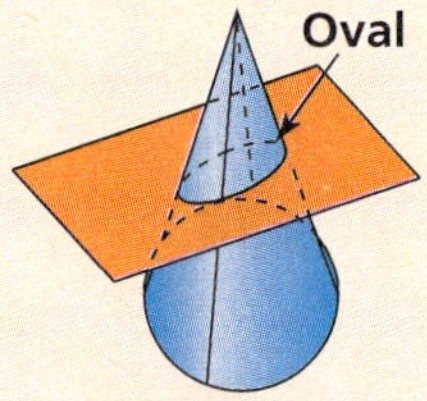

5.

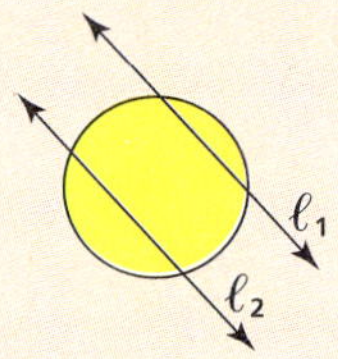

6.

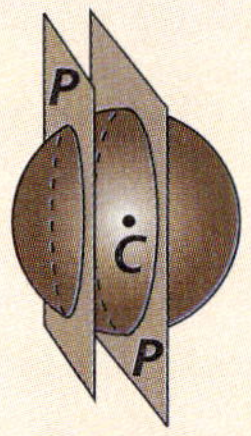

7.

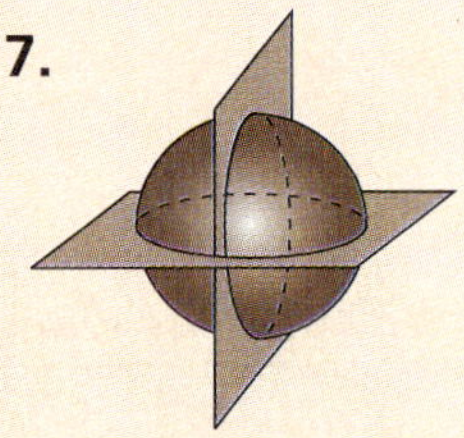

8.

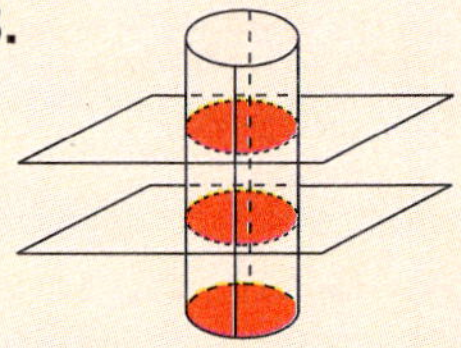

9.

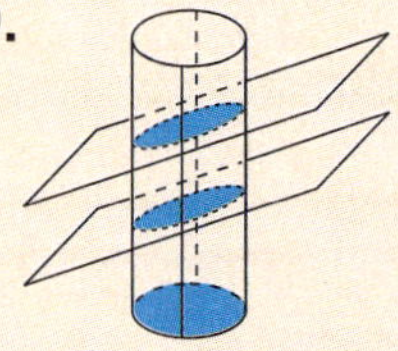

10.

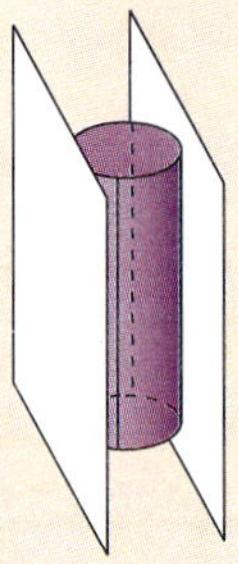

452 Selected Answers and Solutions Geometry

11. $y = 2.5$ and $y = -1.5$

12. $x = -6$ and $x = 2$

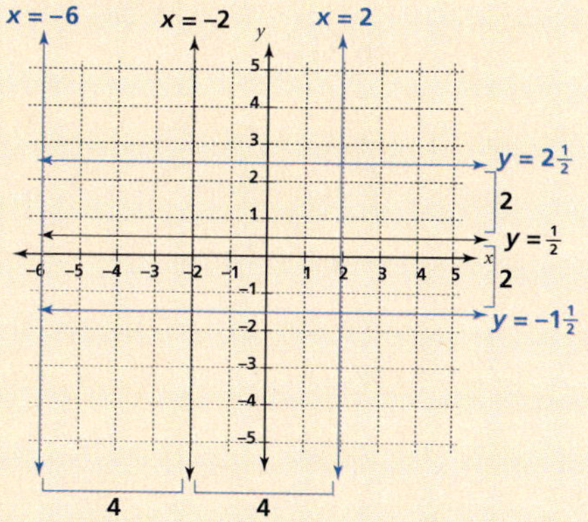

13. $x = 3.5$

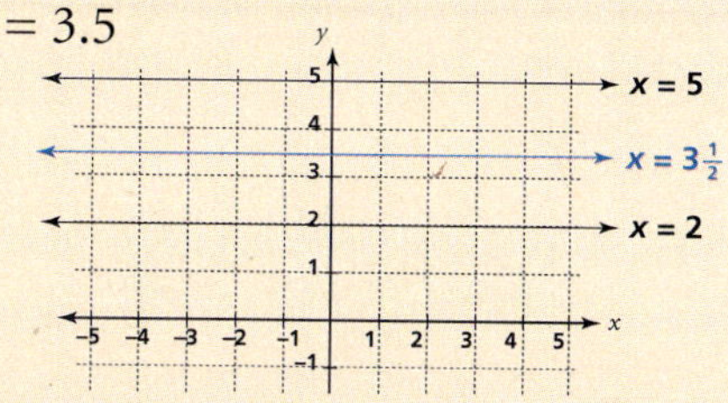

14. $y = -x + 4$ and $y = -x$

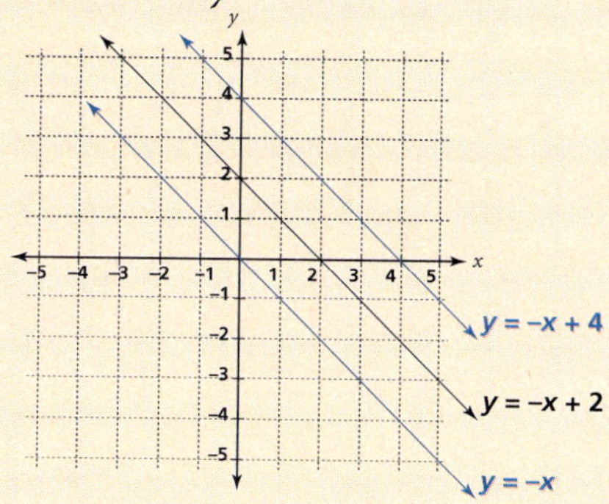

15. $y = -\frac{1}{3}x + 1$ and $y = -\frac{1}{3}x + 5$

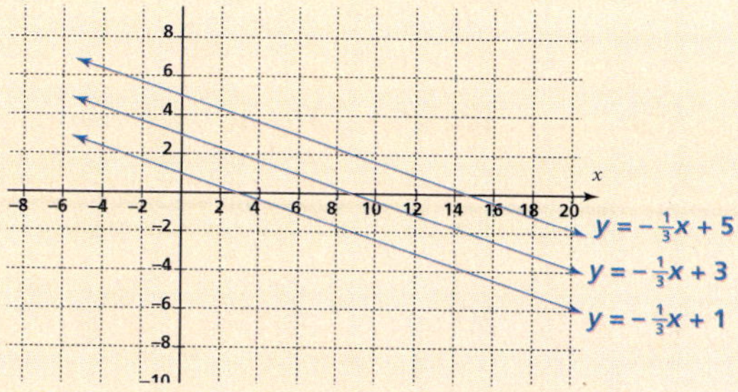

16. one point along the angle bisector and on the circle with radius GH **17.** the intersection of the angle bisectors of each angle of the triangle **18.** the intersection of the perpendicular bisectors of each side **19.** the intersection of the perpendicular bisectors of $\overline{AB}$ and $\overline{CD}$ **20.** two points on the circle with radius GH and center P and on the perpendicular bisector of a line perpendicular to both ℓ and m

Algebra Review

Page 454

1. 10 **2.** 23 **3.** 20 **4.** 35 **5.** 19 **6.** 4 **7.** 3 **8.** 12 **9.** 5 **10.** 1 **11.** 7 **12.** $1\frac{1}{2}$ **13.** 17 **14.** $6\frac{1}{2}$ **15.** −3 **16.** −4 **17.** $\frac{1}{2}$ **18.** −9 **19.** $3\frac{1}{2}$ **20.** −17

Page 455

1. 2 **2.** 4 **3.** 5 **4.** 15 **5.** 40 **6.** 2 **7.** 4 **8.** 6 **9.** 91 **10.** 20 **11.** −2 **12.** −2 **13.** −6 **14.** −10 **15.** −5 **16.** −4 **17.** −4 **18.** −5 **19.** −20 **20.** −30

Page 457

1. 7 **2.** 3 **3.** 10 **4.** 3 **5.** −2 **6.** $-6\frac{2}{3}$ **7.** $16\frac{1}{2}$ **8.** $1\frac{1}{5}$ **9.** $33\frac{1}{3}$ **10.** $-9\frac{1}{3}$ **11.** 3 **12.** 7 **13.** $6\frac{1}{4}$ **14.** 5 **15.** −16 **16.** 6 **17.** 2 **18.** 7 **19.** 0 **20.** 9 **21.** −7 **22.** 3 **23.** 10 **24.** 6 **25.** 5 **26.** −3 **27.** −8 **28.** 4 **29.** 20 **30.** 8

Geometry *Selected Answers and Solutions* **453**

Solution of Equations Using Addition/Subtraction

EXAMPLES Solve for x when $x - 5 = 7$

Solution:

$x - 5 = 7$	given
$+5 = +5$	add 5 to both sides
$x = 12$	answer

Check: Substitute 12 for x in $x - 5 = 7$

$12 - 5 = 7$ is True, $x = 12$ is correct

Solve for x when $x + 5 = 7$

Solution:

$x + 5 = 7$	given
$-5 = -5$	subtract 5 from both sides (or add –5)
$x = 2$	answer

Check: Substitute 2 for x in $x + 5 = 7$

$2 + 5 = 7$ is True, $x = 2$ is correct

Exercise Solve for x and check.

1. $x - 3 = 7$ 10
2. $x - 10 = 13$ 23
3. $x - 4 = 16$ 20
4. $x - 15 = 20$ 35
5. $x - 9 = 10$ 19
6. $x + 3 = 7$ 4
7. $x + 10 = 13$ 3
8. $x + 4 = 16$ 12
9. $x + 15 = 20$ 5
10. $x + 9 = 10$ 1
11. $x - 2 = 5$ 7
12. $x - \frac{1}{2} = 1$ $1\frac{1}{2}$
13. $x - 13 = 4$ 17
14. $x - 1\frac{1}{2} = 5$ $6\frac{1}{2}$
15. $x - 7 = -10$ –3
16. $x + 7 = 3$ –4
17. $x + \frac{1}{2} = 1$ $\frac{1}{2}$
18. $x + 13 = 4$ –9
19. $x + 1\frac{1}{2} = 5$ $3\frac{1}{2}$
20. $x + 7 = -10$ –17

454 *Algebra Review* *Geometry*

Solution of Equations Using Multiplication/Division

EXAMPLES Solve for x when $3x = 12$

Solution:

$3x = 12$ given

$3x \div 3 = 12 \div 3$ divide both sides by 3 (or multiply by $\frac{1}{3}$)

$x = 4$ answer

Check: Substitute 4 for x in $3x = 12$

$3 \bullet 4 = 12$ is True, $x = 4$ is correct

Solve for x when $5x = -30$

Solution:

$5x = -30$ given

$\frac{1}{5}(5x) = \frac{1}{5}(-30)$ multiply both sides by $\frac{1}{5}$ (or divide by 5)

$x = -\frac{30}{5} = -6$ answer

Check: Substitute -6 for x in $5x = -30$

$5(-6) = -30$ is True, $x = -6$ is correct

Exercise Solve for x and check.

1. $3x = 6$ 2

2. $4x = 16$ 4

3. $7x = 35$ 5

4. $10x = 150$ 15

5. $15x = 600$ 40

6. $5x = 10$ 2

7. $15x = 60$ 4

8. $8x = 48$ 6

9. $10x = 910$ 91

10. $40x = 800$ 20

11. $3x = -6$ -2

12. $5x = -10$ -2

13. $6x = -36$ -6

14. $10x = -100$ -10

15. $15x = -75$ -5

16. $7x = -28$ -4

17. $6x = -24$ -4

18. $9x = -45$ -5

19. $16x = -320$ -20

20. $40x = -1,200$ -30

Geometry *Algebra Review* **455**

Solving Equations

EXAMPLES Solve for x when $3x + 5 = 17$

Solution:

Remember, the strategy is to get x = some number.

$3x + 5 = 17$	given
$3x = 12$	subtract 5 from both sides
$x = 4$	divide both sides by 3
$x = 4$	answer

Check: Substitute 4 for x in $3x + 5 = 17$

$(3 \bullet 4) + 5 = 17$ is True, $x = 4$ is correct

Solve for x when $\frac{3}{2}x + 2 = -9$

Solution:

$\frac{3}{2}x + 2 = -9$	given
$\frac{3}{2}x = -11$	subtract 2 from both sides
$\frac{2}{3}(\frac{3}{2}x) = \frac{2}{3}(-11)$	multiply both sides by $\frac{2}{3}$ to make $1x$
$x = -\frac{22}{3}$	answer

Check: Substitute $-\frac{22}{3}$ for x in $\frac{3}{2}x + 2 = -9$

$\frac{3}{2}(-\frac{22}{3}) + 2 = -9$

$-11 + 2 = -9$ is True, $x = -\frac{22}{3}$ is correct

456 *Algebra Review* *Geometry*

Algebra Review

Solving Equations *(continued)*

Exercise Solve for x and check.

1. $2x + 5 = 19$ 7
2. $6x + 3 = 21$ 3
3. $4x + 9 = 49$ 10
4. $3x - 2 = 7$ 3
5. $7x + 2 = -12$ –2
6. $\frac{3}{2}x + 1 = -9$ $-6\frac{2}{3}$
7. $\frac{2}{3}x - 1 = 10$ $16\frac{1}{2}$
8. $\frac{5}{6}x + 1 = 2$ $1\frac{1}{5}$
9. $\frac{3}{10}x - 7 = 3$ $33\frac{1}{3}$
10. $\frac{3}{4}x - 3 = -10$ $-9\frac{1}{3}$
11. $3x + 4 = 13$ 3
12. $5x + 6 = 41$ 7
13. $\frac{4}{5}x + (-5) = 0$ $6\frac{1}{4}$
14. $2x - 3 = 7$ 5
15. $\frac{3}{4}x + 12 = 0$ –16
16. $2x - 6 = 6$ 6
17. $4x - 1 = 7$ 2
18. $2x + 3 = 17$ 7
19. $-9x - 4 = -4$ 0
20. $3x - 4 = 23$ 9
21. $-\frac{2}{7}x - 6 = -4$ –7
22. $9x - 25 = 2$ 3
23. $3x - 25 = 5$ 10
24. $-6x + 20 = -16$ 6
25. $4x + 5 = 25$ 5
26. $-4x + 2 = 14$ –3
27. $-\frac{1}{2}x + 2 = 6$ –8
28. $2x - (-10) = 18$ 4
29. $3x + 40 = 100$ 20
30. $2x - 8 = 8$ 8

Geometry *Algebra Review* **457**

Axioms

Axiom 1 Things that are equal to the same thing are equal to each other. (p. 49)

Axiom 2 If equals are added to equals, the sums are equal. (p. 49)

Axiom 3 If equals are subtracted from equals, the differences are equal. (p. 49)

Axiom 4 Things that are alike or coincide with one another are equal to one another. (p. 49)

Axiom 5 The whole, or sum, is greater than the parts. (p. 49)

Postulates

Ruler Postulate The points on a line can be placed in a one-to-one correspondence with the real numbers so that

1. for every point on the line, there is exactly one real number.
2. for every real number, there is exactly one point on the line.
3. the distance between any two points is the absolute value of the difference of the corresponding real numbers. (p. 7)

Ruler Placement Postulate Given two points, A and B on a line, the number line can be chosen so that A is at zero and B is at a positive number. (p. 8)

Segment Addition Postulate If B is between A and C, then $\overline{AB} + \overline{BC} = \overline{AC}$. (p. 8)

Euclid's Postulate 1 A straight line can be drawn from any point to any point. (p. 39)

Euclid's Postulate 2 A finite straight line can be extended continuously in a straight line. (p. 40)

Euclid's Postulate 3 A circle may be described with any center and distance. (p. 40)

458 *Axioms, Postulates, and Theorems* *Geometry*

Euclid's Postulate 4 All right angles are equal to one another. (p. 40)

Euclid's Postulate 5 If two lines ℓ and m are cut by a third line t, and the two inside angles, a and b, together measure less than two right angles, then the two lines ℓ and m, if extended, will meet on the same side as the two angles a and b. (p. 41)

Parallel Postulate 5 If there is a line ℓ and a point P not on ℓ, then there is only one line that passes through P and is parallel to ℓ. (p. 41)

Alternate Interior Angles Postulate If a transversal intersects two lines so that the alternate interior angles are equal, then the lines are parallel. (pp. 73, 84)

SAS (Side-Angle-Side) Postulate If two sides and the included angle of one triangle are congruent to the corresponding two sides and included angle of a second triangle, the triangles are congruent. (p. 179)

SSS (Side-Side-Side) Postulate If three sides of a triangle are congruent to three sides of a second triangle, then the triangles are congruent. (p. 189)

ASA (Angle-Side-Angle) Postulate If two angles and the included side of a triangle are congruent to the corresponding angles and included side of a second triangle, then the triangles are congruent. (p. 190)

AA Similarity Postulate If two angles of a triangle have the same measures as two angles of a second triangle, then the two triangles are similar. (p. 228)

Theorems

Theorem If angles are vertical angles, then their measures are equal. (p. 53)

Theorem If two lines are parallel, then the interior angles on the same side of the transversal are supplementary. (p. 68)

Geometry *Axioms, Postulates, and Theorems* **459**

Theorem If two lines are parallel, then the corresponding angles are equal. (p. 69)

Theorem If two lines are parallel, then the alternate interior angles are equal. (p. 70)

Theorem If a figure is a parallelogram, then its angle sum is 360°. (p. 81)

Theorem If a figure is a trapezoid, then its angle sum is 360°. (p. 82)

Theorem If corresponding angles are equal, then the lines are parallel. (p. 84)

Theorem If two lines are cut by a transversal so that the angles on the same side of the transversal are supplementary, then the lines are parallel. (p. 85)

Angle Sum Theorem The sum of the angle measures in any triangle is 180°. (p. 168)

Corollary to the Angle Sum Theorem If ΔABC is any triangle, then an exterior angle of ABC has the same measure as the sum of the measures of the two nonadjacent interior angles. (p. 168)

Shortest Distance Theorem The perpendicular segment from a point to a line is the shortest segment from that point to the line. The length of the perpendicular segment is called the distance from the point to the line. (p. 186)

Triangle Inequality Theorem The sum of the lengths of any two sides of a triangle must be greater than the length of the third side. (p. 186)

AAS (Angle-Angle-Side) Theorem If two angles and one side not included between those angles are congruent to two angles and the corresponding side of a second triangle, then the two triangles are congruent. (p. 194)

Theorem The acute angles in any right triangle are complementary. (p. 195)

460 *Axioms, Postulates, and Theorems* *Geometry*

Axioms, Postulates, and Theorems

Hypotenuse-Leg (H-L) Theorem Two right triangles are congruent if the hypotenuse and leg of one triangle are congruent to the hypotenuse and corresponding leg of a second triangle. (p. 196)

Theorem In any proportion, the product of the extremes equals the product of the means. (p. 222)

Theorem If one acute angle of a right triangle has the same measure as one acute angle of a second right triangle, then the two right triangles are similar. (p. 228)

Pythagorean Theorem If a triangle is a right triangle with legs a and b and hypotenuse c, then $a^2 + b^2 = c^2$. (p. 255)

In any right triangle ABC, with $\overline{AC}$ the hypotenuse, $\overline{AC}^2 = \overline{AB}^2 + \overline{BC}^2$. (p. 262)

Converse of the Pythagorean Theorem If for ΔABC, $\overline{BC}^2 = \overline{AB}^2 + \overline{AC}^2$, then ΔABC is a right triangle. (p. 275)

Theorem An inscribed angle is measured by $\frac{1}{2}$ its intercepted arc. (p. 334)

Note: The theorems in this list are those that are featured as the main subjects of the lessons. Other theorems appear throughout the text.

Geometry *Axioms, Postulates, and Theorems* **461**

Geometric Symbols and Formulas

Symbols

$\angle A$	angle *A*
$m\angle A$	measure of angle *A* in degrees
$\overleftrightarrow{AB}$	line through points *A* and *B*
$\overline{AB}$	segment *AB* with endpoints *A* and *B*
ΔABC	triangle *ABC*
$\overrightarrow{AB}$	ray *AB*, with endpoint *A* and passing through *B*
$\cong$	is congruent to
$\ncong$	is not congruent to
$\perp$	is perpendicular to
$\not\perp$	is not perpendicular to
$\parallel$	is parallel to
$\nparallel$	is not parallel to
n°	*n* degrees
$\therefore$	therefore
$a \Rightarrow b$	If *a* then *b*
π	$\frac{\text{circumference}}{\text{diameter}}$ of a circle
$\odot$	circle

Formulas

$C = \pi d$	C = circumference d = diameter
$d = 2r$	r = radius
$\pi = \frac{c}{d}$	π as a ratio

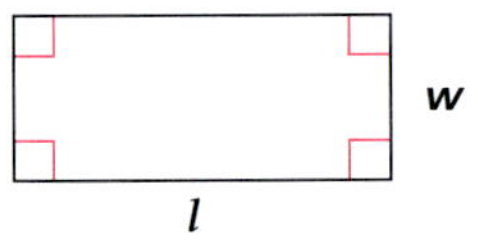

Perimeter of a rectangle = $2l + 2w$

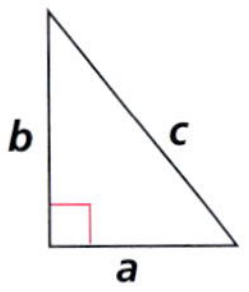

$a^2 + b^2 = c^2$ Pythagorean Formula

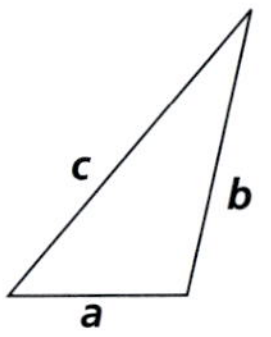

Perimeter of a scalene $\Delta = a + b + c$

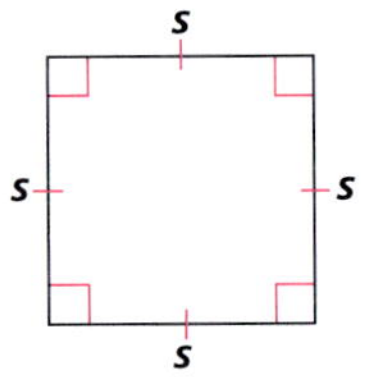

Perimeter of a square = $4s$

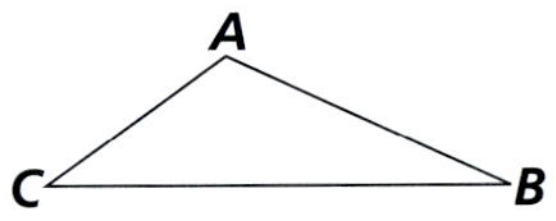

$m\angle A + m\angle B + m\angle C = 180^\circ$

462 *Symbols and Formulas* *Geometry*

Geometric Symbols and Formulas

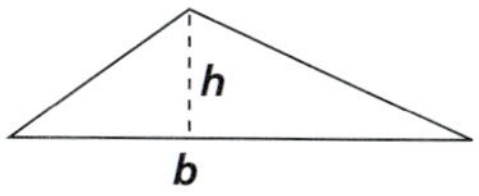

Area of a $\Delta = \frac{1}{2}bh$

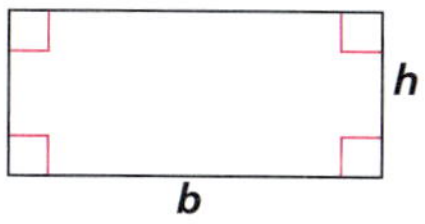

Area of a rectangle = bh

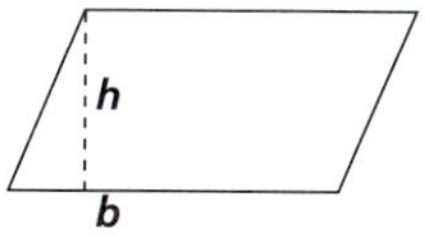

Area of a parallelogram = bh

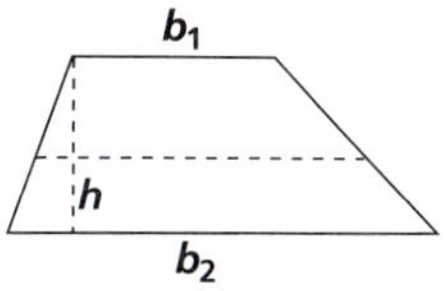

Area of a trapezoid = $h(\frac{b_1 + b_2}{2})$

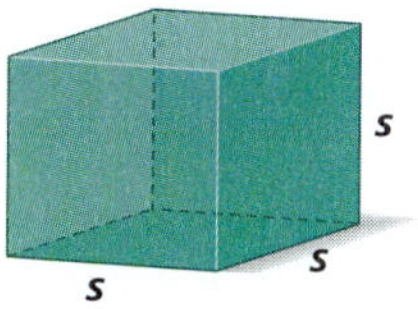

Volume of a cube = s^3

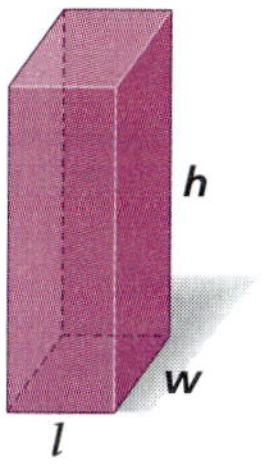

Volume of a rectangular box = lwh

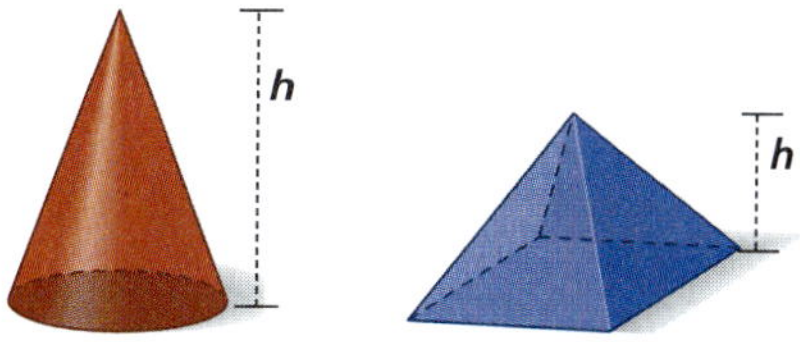

Volume of a cone or a pyramid
$= \frac{1}{3}h$(area of base)

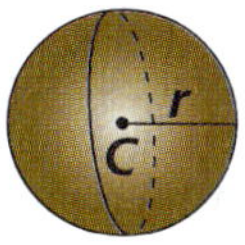

Volume of a sphere = $\frac{4}{3}\pi r^3$

Geometry *Symbols and Formulas* **463**

Measurement Conversion Factors

Metric Measures

Length

1,000 meters (m) = 1 kilometer (km)
100 centimeters (cm) = 1 m
10 decimeters (dm) = 1 m
1,000 millimeters (mm) = 1 m
10 cm = 1 decimeter (dm)
10 mm = 1 cm

Area

100 square millimeters (mm^2) = 1 square centimeter (cm^2)
10,000 cm^2 = 1 square meter (m^2)
10,000 m^2 = 1 hectare (ha)

Volume

1,000 cubic meters (m^3) = 1 cubic centimeter (cm^3)
100 cm^3 = 1 cubic decimeter (dm^3)
1,000,000 cm^3 = 1 cubic meter (m^3)

Capacity

1,000 milliliters (mL) = 1 liter (L)
1,000 L = 1 kiloliter (kL)

Mass

1,000 kilograms (kg) = 1 metric ton (t)
1,000 grams (g) = 1 kg
1,000 milligrams (mg) = 1 g

Temperature Degrees Celsius (°C)

0°C = freezing point of water
37°C = normal body temperature
100°C = boiling point of water

Customary Measures

Length

12 inches (in.) = 1 foot (ft)
3 ft = 1 yard (yd)
36 in. = 1 yd
5,280 ft = 1 mile (mi)
1,760 yd = 1 mi
6,076 feet = 1 nautical mile

Area

144 square inches (sq in.) = 1 square foot (sq ft)
9 sq ft = 1 square yard (sq yd)
43,560 sq ft = 1 acre (A)

Volume

1,728 cubic inches (cu in.) = 1 cubic foot (cu ft)
27 cu ft = 1 cubic yard (cu yard)

Capacity

8 fluid ounces (fl oz) = 1 cup (c)
2 c = 1 pint (pt)
2 pt = 1 quart (qt)
4 qt = 1 gallon (gal)

Weight

16 ounces (oz) = 1 pound (lb)
2,000 lb = 1 ton (T)

Temperature Degrees Fahrenheit (°F)

32°F = freezing point of water
98.6°F = normal body temperature
212°F = boiling point of water

Time

60 seconds (sec) = 1 minute (min)
60 min = 1 hour (hr)
24 hr = 1 day

464 *Symbols and Formulas* *Geometry*

Measurement Conversion Factors

To change	To	Multiply by	To change	To	Multiply by
centimeters	inches	.3937	meters	feet	3.2808
centimeters	feet	.03281	meters	miles	.0006214
cubic feet	cubic meters	.0283	meters	yards	1.0936
cubic meters	cubic feet	35.3145	metric tons	tons (long)	.9842
cubic meters	cubic yards	1.3079	metric tons	tons (short)	1.1023
cubic yards	cubic meters	.7646	miles	kilometers	1.6093
feet	meters	.3048	miles	feet	5,280
feet	miles (nautical)	.0001645	miles (statute)	miles (nautical)	.8684
feet	miles (statute)	.0001894	miles/hour	feet/minute	88
feet/second	miles/hour	.6818	millimeters	inches	.0394
gallons (U.S.)	liters	3.7853	ounces avdp	grams	28.3495
grams	ounces avdp	.0353	ounces	pounds	.0625
grams	pounds	.002205	pecks	liters	8.8096
hours	days	.04167	pints (dry)	liters	.5506
inches	millimeters	25.4000	pints (liquid)	liters	.4732
inches	centimeters	2.5400	pounds advp	kilograms	.4536
kilograms	pounds avdp	2.2046	pounds	ounces	16
kilometers	miles	.6214	quarts (dry)	liters	1.1012
liters	gallons (U.S.)	.2642	quarts (liquid)	liters	.9463
liters	pecks	.1135	square feet	square meters	.0929
liters	pints (dry)	1.8162	square meters	square feet	10.7639
liters	pints (liquid)	2.1134	square meters	square yards	1.1960
liters	quarts (dry)	.9081	square yards	square meters	.8361
liters	quarts (liquid)	1.0567	yards	meters	.9144

Geometry Symbols and Formulas **465**

Glossary

A

Acute angle—An angle whose measure is less than 90° (p. 15)

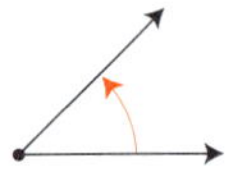

Acute triangle—A triangle whose angles are all less than 90° (p. 139)

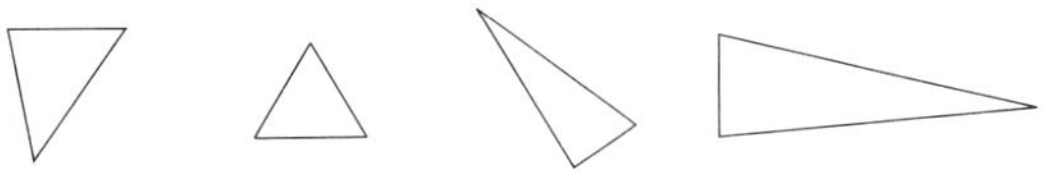

Adjacent—Sharing a common side (p. 146)

Adjacent angles—Angles with a common vertex and one common side (p. 20)

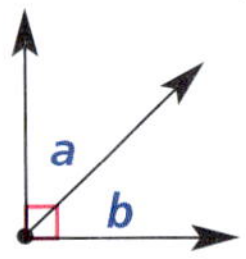

Adjacent vertices—Vertices that share a common side (p. 146)

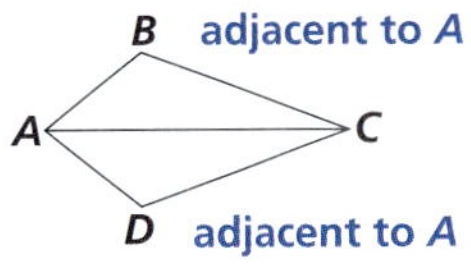

Alternate exterior angles—Pairs of exterior angles on opposite sides of a transversal (p. 65)

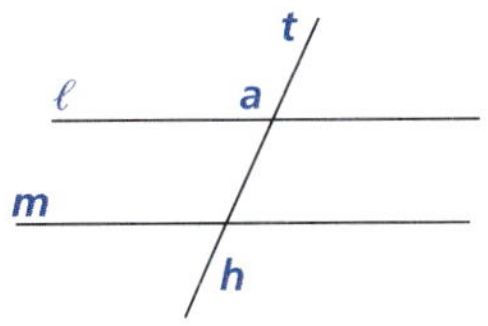

Alternate interior angles—Pairs of interior angles on opposite sides of a transversal (p. 65)

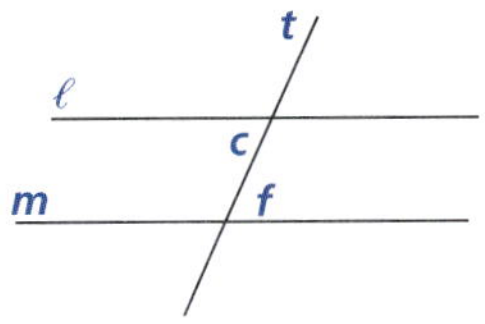

Altitude—Line from the vertex of a triangle and perpendicular to the opposite side (p. 163)

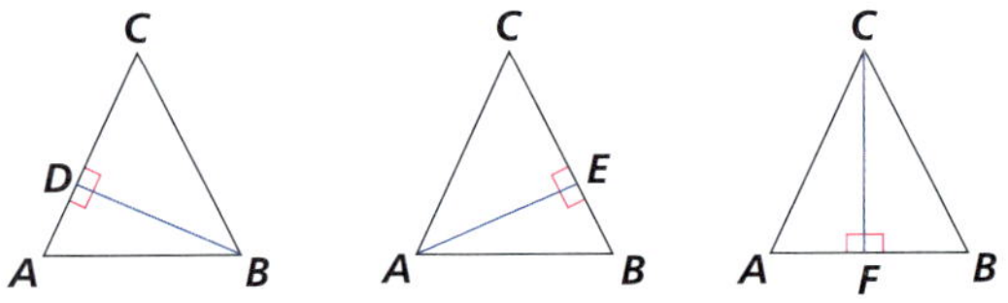

Angle—A figure made up of two sides, or rays, with a common endpoint (p. 11)

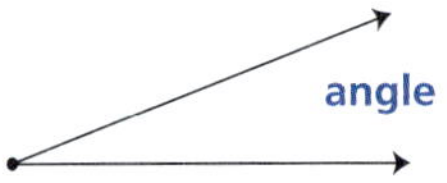

Angle bisector—Ray that divides an angle into two equal parts (p. 11); locus of points equidistant from the sides of an angle (p. 336)

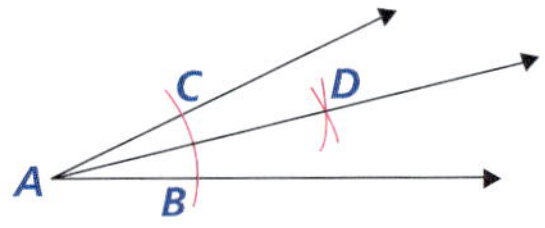

Arc—Part of a circle (p. 11); a segment of a circle bounded by two endpoints (p. 332)

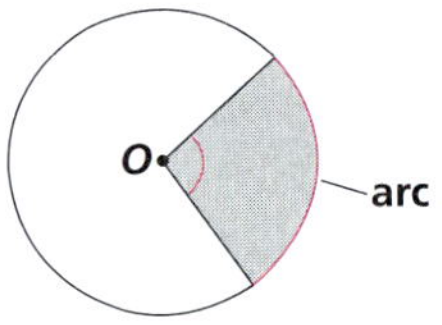

Area—The number of square units inside a closed region (p. 254)

5 • 5 = 25 sq units

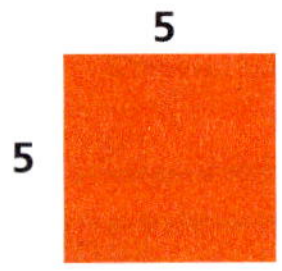

Argument—Course of reasoning (p. 68)

Assumption—Something taken to be true without proof (p. 68)

Average—A number found by dividing the sum of all the quantities by the number of quantities (p. 266)

$\frac{(a + b)}{2}$ = average of a and b

466 *Glossary* *Geometry*

Average base—The line that connects the midpoints of the nonparallel sides of a trapezoid (p. 300)

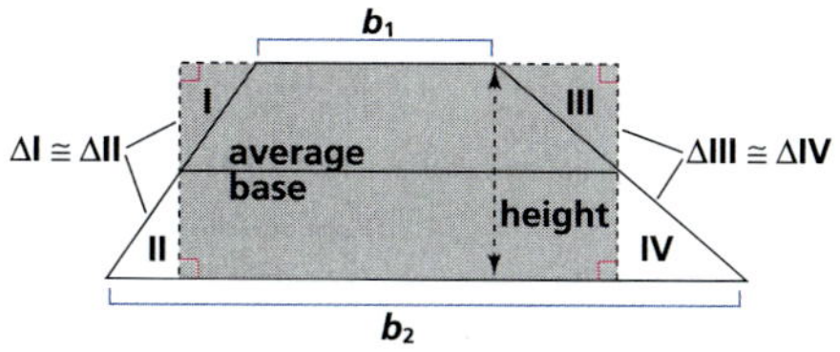

Axiom—A common notion taken to be true without proof (p. 49)

B

Bisect—To divide into two equal parts (pp. 11, 123)

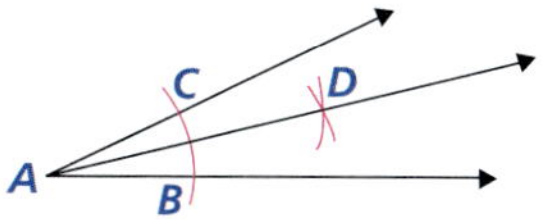

C

Capacity—The amount a container will hold when full (p. 368)

Central angle—An angle with its vertex at the center of a circle and the circle's radii as its sides (p. 332)

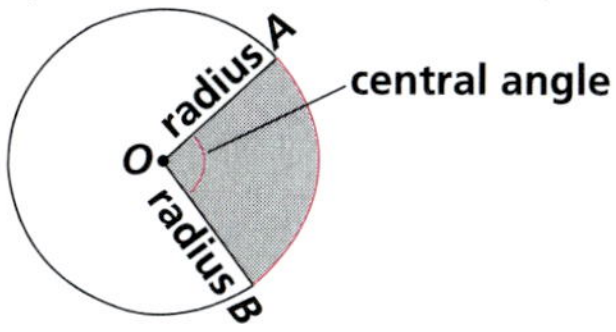

Centroid—Point where the three medians of a triangle intersect (p. 163)

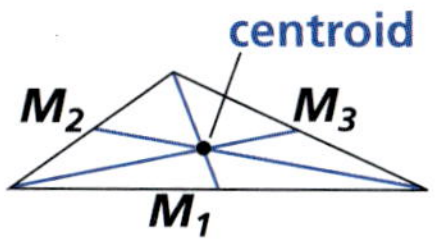

Characteristics—Special features or qualities (p. 266)

Chord—Line segment joining two points on a circle (p. 316)

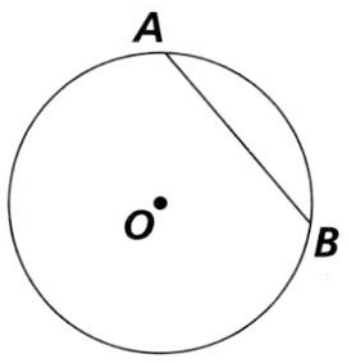

Circle—Locus of points at the same distance from a given point (p. 316)

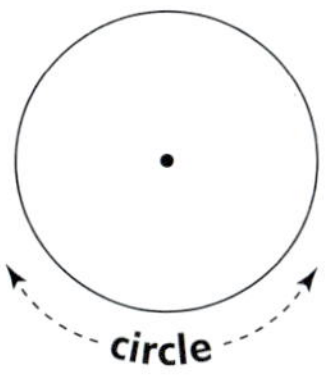

Circumcenter—Center of a circumcircle and located at the intersection of the perpendicular bisectors of any two sides of a triangle (p. 336)

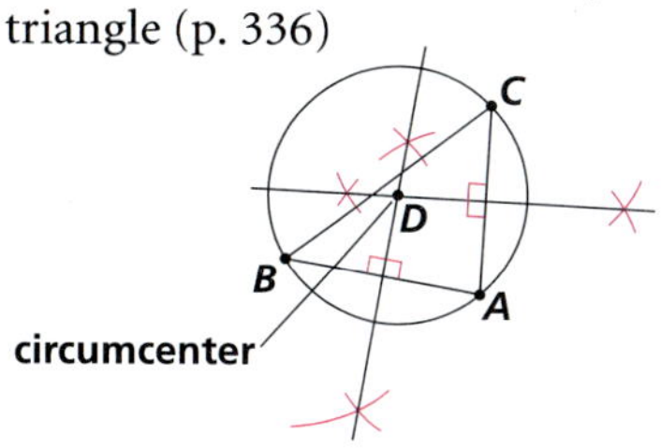

Circumcircle—Circle that passes through the three vertices of a triangle (p. 336)

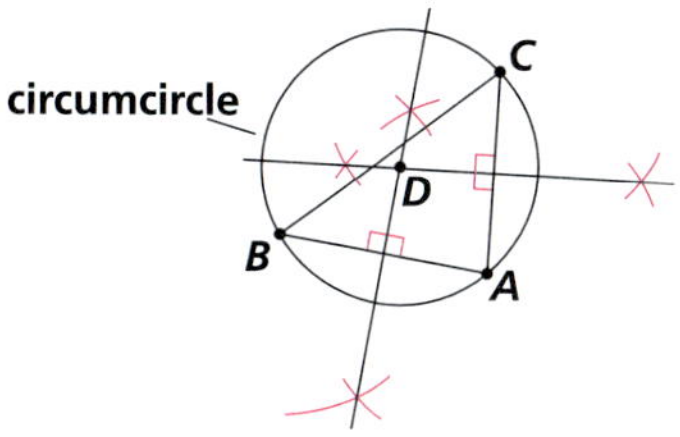

Circumference—The complete length of a circle (p. 316)

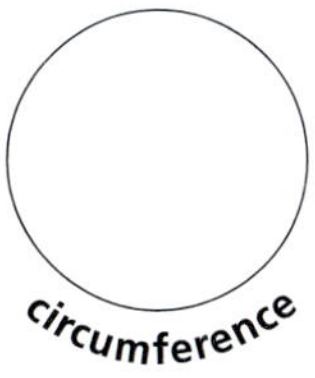

Coincide—To be alike or occupy the same position at the same time (pp. 49, 149)

Collinear—Points on the same line (p. 2)

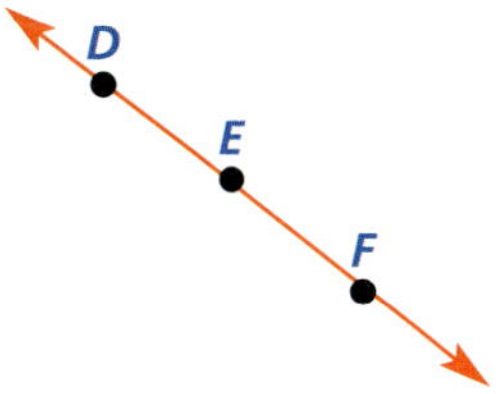

Geometry Glossary **467**

Compass—A tool used to draw circles and parts of circles called arcs (p. 11)

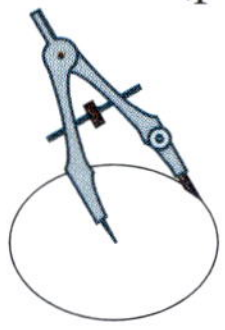

Complementary angles—Two angles whose measures add to 90° (p. 20)

Concave polygon—Polygon for which a segment joining any two points in the interior does not lie entirely within the figure (p. 154)

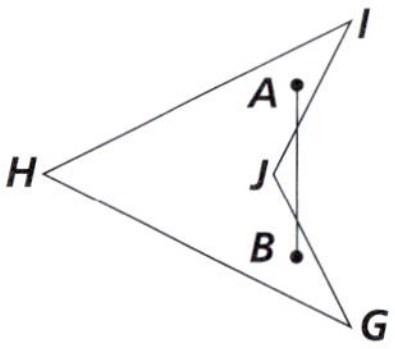

Conclusion—The then . . . part of a conditional (p. 34)

Conditional—A statement in the If . . . then . . . form (p. 34)

Cone—A three-dimensional figure whose base is a circle and all segments that join the circle to a vertex, *V*, not in the plane of the circle (p. 356)

Congruent angles—Angles that have equal sides and equal measures (p. 178)

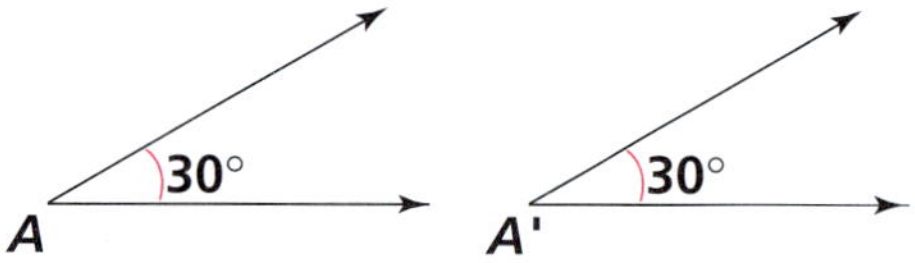

Congruent segments—Segments that have the same length (p. 178)

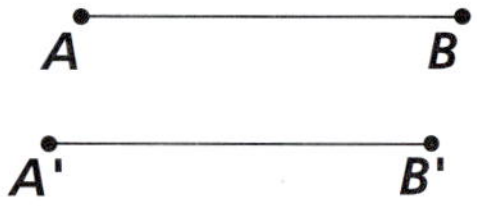

Congruent triangles—Triangles whose corresponding angles and sides are congruent (p. 178)

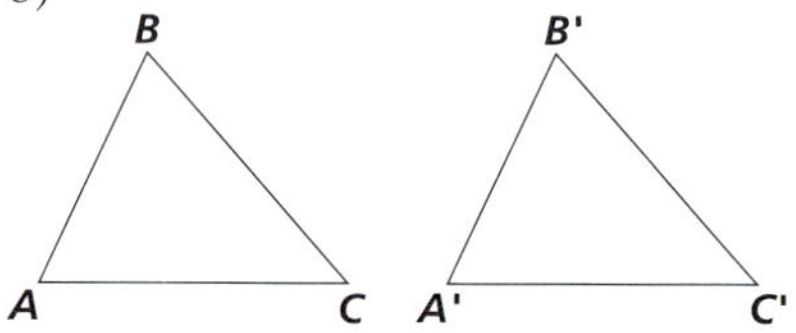

Construction—Process of making a line, angle, or figure according to specific requirements (p. 11)

Contradiction—A statement that can be shown to be true and false (p. 68)

Converse—A new conditional formed from a given conditional in which the hypothesis and conclusion are switched (p. 34)

Convex polygon—Polygon for which any line segment joining any two points in the interior lies entirely within the figure (p. 154)

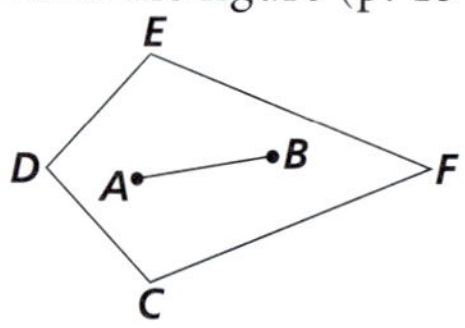

Coordinate plane—Plane formed by placing two real number lines at right angles (p. 96)

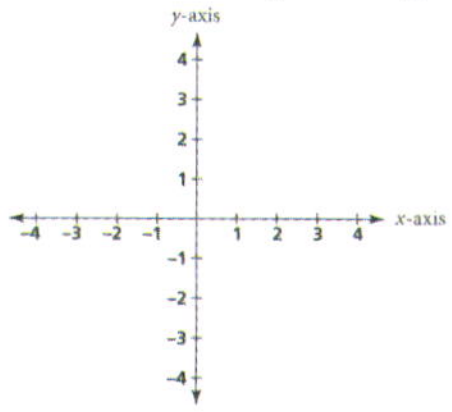

Coplanar—Lines or points in the same plane (p. 62)

Corollary—A conclusion that follows easily from a given theorem (p. 167)

Corresponding angles—Pairs of angles in similar positions (pp. 65, 178)

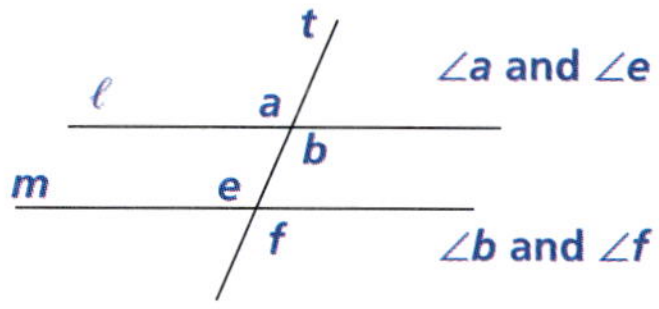

468 *Glossary* *Geometry*

Cosine (cos)—For an angle of a right triangle (not the right angle), the ratio of the length of its adjacent leg to the length of the hypotenuse (p. 340)

Counterexample—A single example that proves a statement false (p. 143)

Customary—Usual or common, ordinary; system of measurement used mostly in the United States; inches and cups are basic units (p. 368)

D

Deductive (Deductive Thinking)—Logical reasoning in which a conclusion necessarily follows from the propositions stated (see pp. 53–54 for first example of proof)

Degree—A unit of angle measure (p. 15)

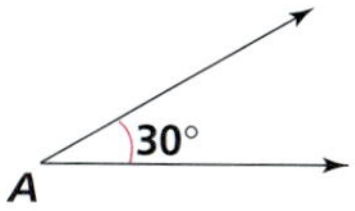

Diagonal—A line segment connecting any two nonadjacent vertices of a polygon (p. 146)

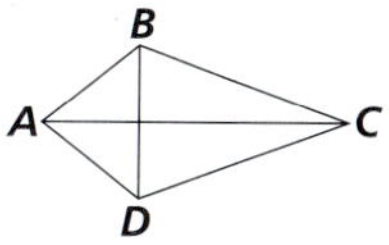

Diameter—Distance across a circle through the center (p. 220); a chord that passes through the center of a circle (p. 316)

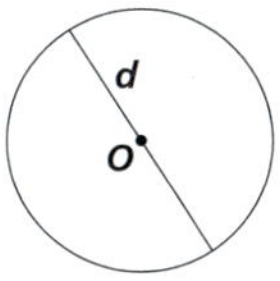

Dilation—Process by which a geometric figure is enlarged or shrunk (p. 235)

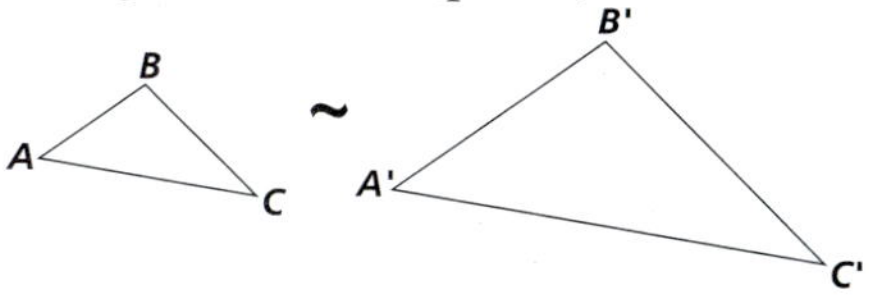

Domain—The *x*-values of a function (p. 113)

E

Endpoints—The points *A* and *B* at the ends of a line segment (p. 2)

Equilateral triangle—A triangle with three equal sides (p. 134)

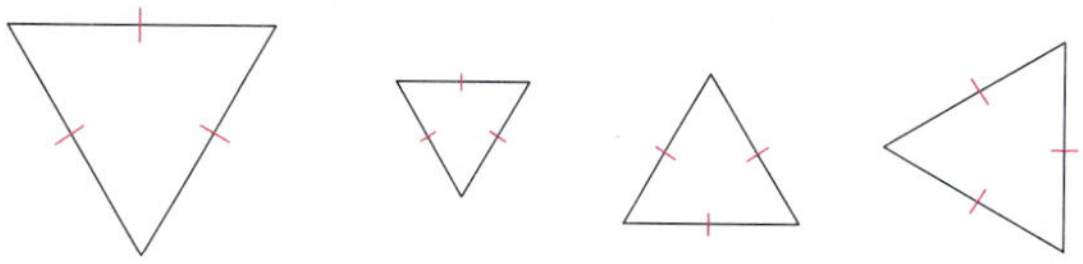

Exterior angle—Angle outside parallel lines (p. 65); angle formed by extending one side of a triangle through any vertex (p. 167)

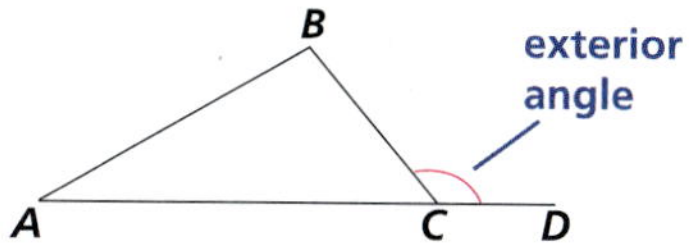

F

Function—A rule that pairs every *x*-value with one and only one *y*-value (p. 113)

G

Geometry—The study of points, lines, angles, surfaces, and solids (p. 2)

Graph—A diagram in the coordinate plane (p. 96)

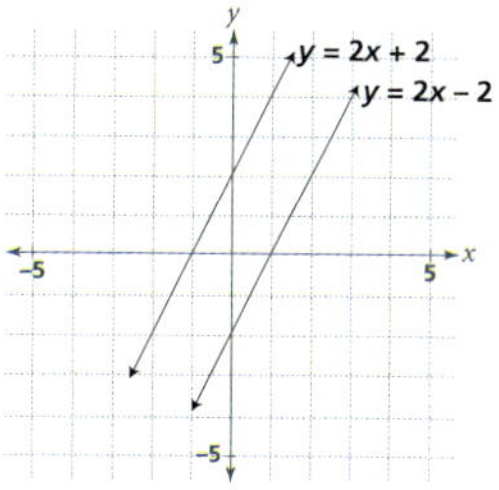

Great circle—Circle on a sphere whose center is the center of the sphere and whose radius equals the radius of the sphere (p. 342)

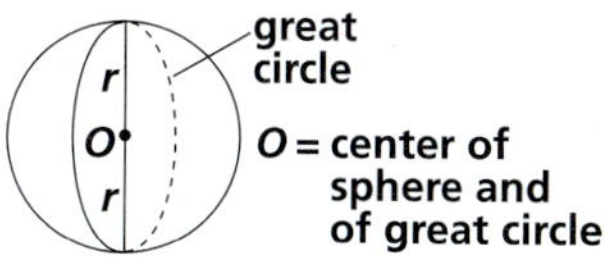

Geometry Glossary **469**

H

Hexagon—a six-sided polygon (p. 232)

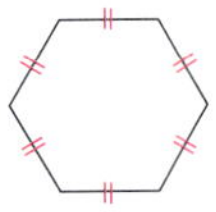

Horizontal—Parallel to the horizon (p. 96)

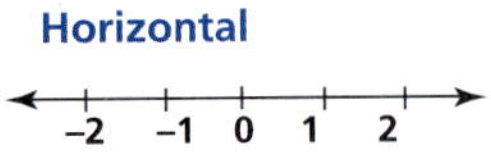

Hypotenuse—The side opposite the right angle in a right triangle (pp. 139, 184)

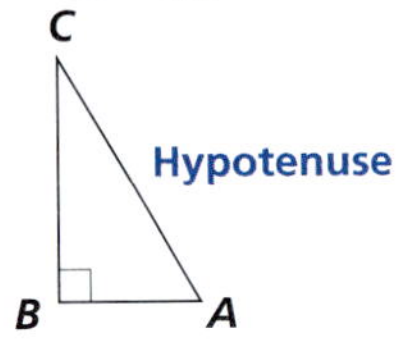

Hypothesis—The given or If . . . part of a conditional (p. 34)

I

Image—Reflection of an object (p. 200)

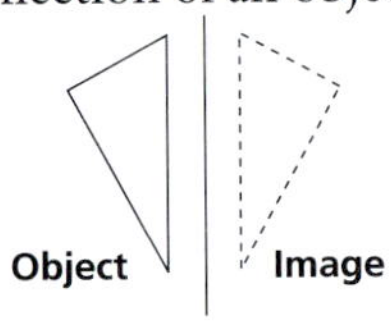

In-center—The point where the angle bisectors of the angles of a triangle intersect (p. 163)

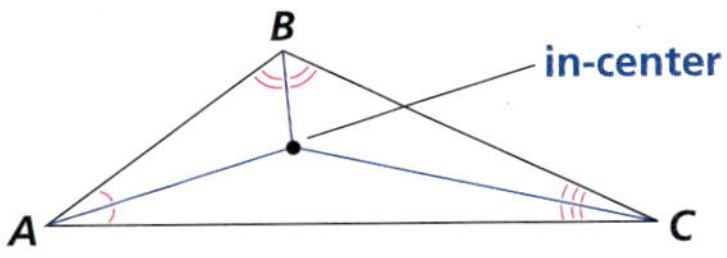

Incircle—Circle in the interior of a triangle that is tangent to each side of the triangle (p. 336)

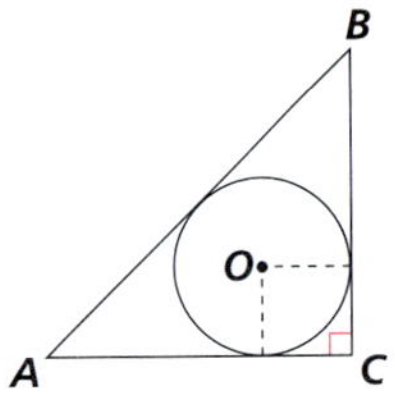

Indirect proof—A proof that shows that the hypothesis cannot be false because accepted facts would be contradicted (p. 68)

Inductive—Using the similarities of particular cases to make a general statement (pp. 80, 149)

Integer—One of the set of positive and negative whole numbers including zero (p. 250)

. . . –3, –2, –1, 0, 1, 2, 3, . . .

Interior—Inside (p. 154); Segment *AB* is in the interior of the polygon.

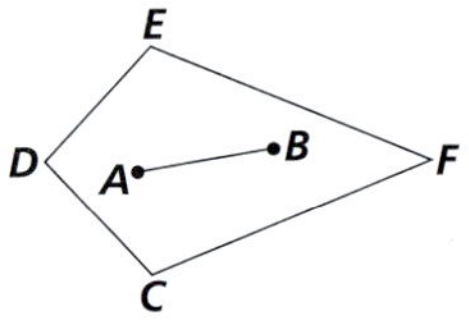

Interior angle—Angle inside parallel lines (p. 65)

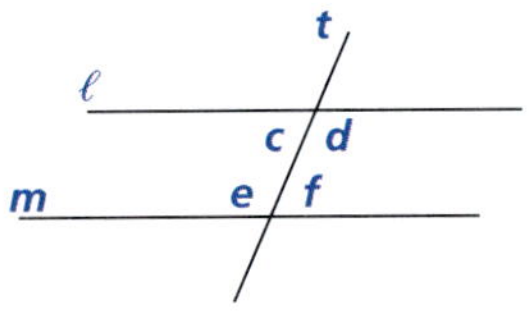

Intersect—To cross each other; point at which two or more lines cross each other (p. 20)

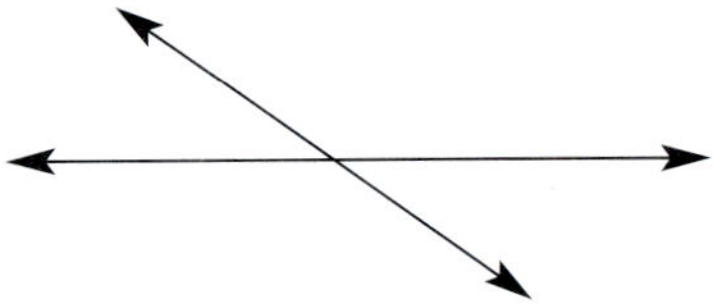

Isosceles—A geometric figure with two sides of equal length (p. 77)

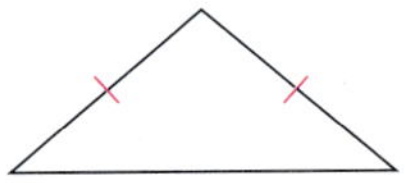

Isosceles triangle—A triangle with two equal sides (p. 134)

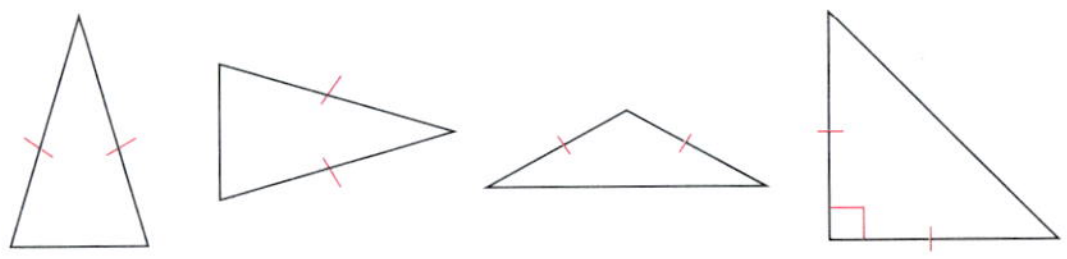

470 *Glossary* *Geometry*

L

Line—A set of many points that extend in opposite directions without ending (p. 2)

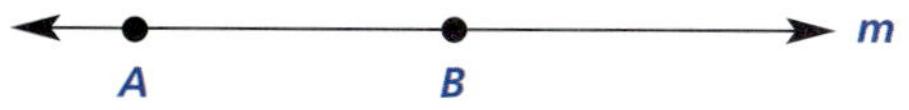

Line of symmetry—Reflection line of an object for which the object coincides with the image (p. 205)

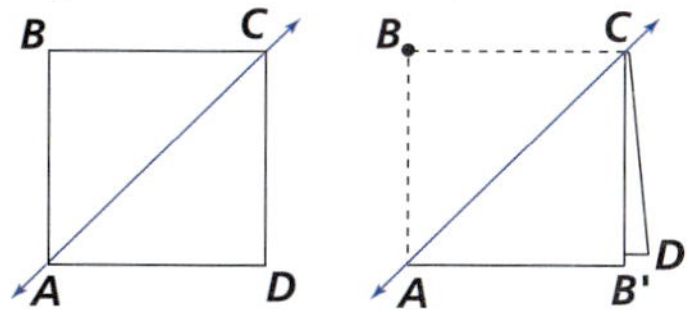

Line segment—The set of points *A*, *B*, and all the points between *A* and *B* (p. 2)

Locus of points—Location of points (plural *loci*) (pp. 335, 376)

M

Median—Line from a vertex of a triangle to the midpoint of the opposite side (p. 163)

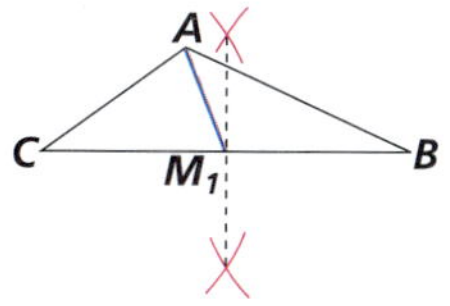

Metric—System of measurement based on powers of ten; gram and meter are basic units (p. 368)

Midpoint—Point that divides a line segment into two equal parts (p. 123)

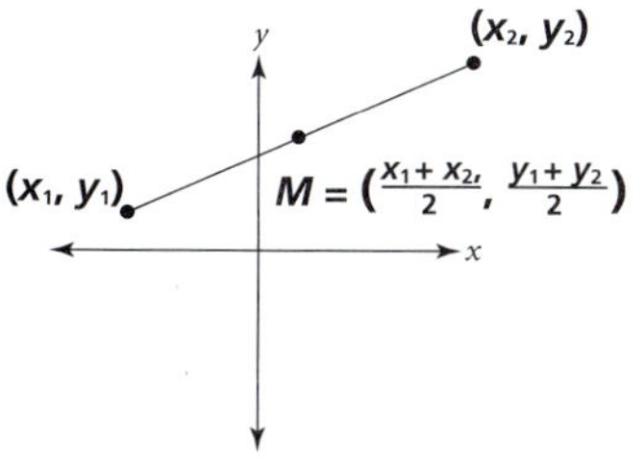

N

Nonadjacent—Not having a common side (p. 146)

Nonadjacent vertices—Vertices that do not share a common side (p. 146)

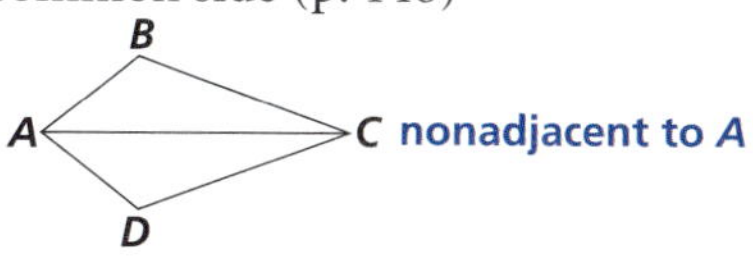

Notion—Idea (p. 49)

O

Obtuse angle—An angle whose measure is greater than 90° but less than 180° (p. 15)

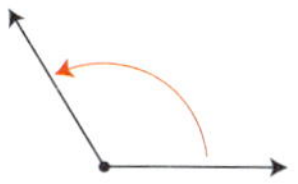

Obtuse triangle—A triangle having one angle greater than 90° (p. 139)

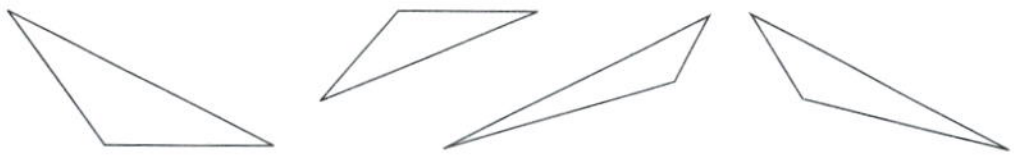

Octagon—an eight-sided polygon (p. 233)

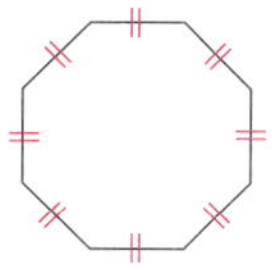

Ordered pair—A set of two real numbers that locate a point on a plane (p. 96)

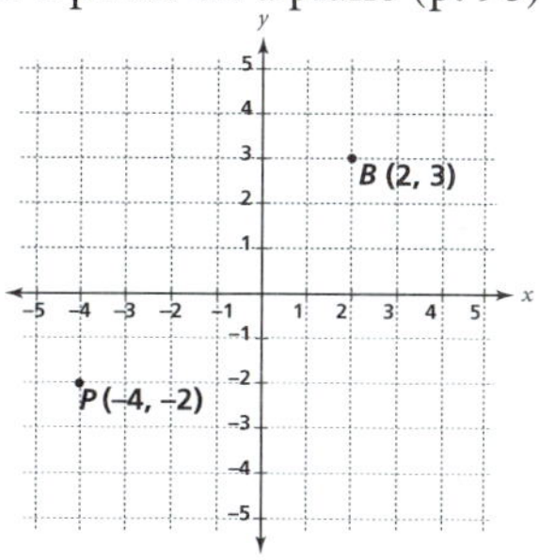

Origin—The point at which the *x*-axis and *y*-axis in the coordinate system intersect (p. 96)

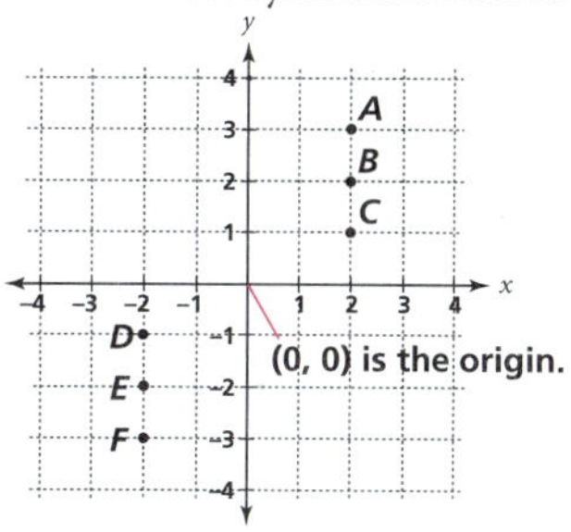

Geometry Glossary **471**

Orthocenter—The point where the altitudes of a triangle meet (p. 163)

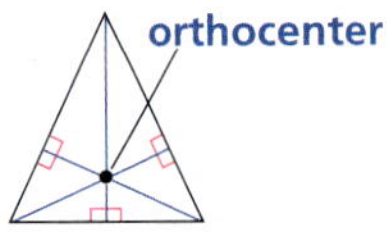

P

Parallel lines—Lines in the same plane that never meet (p. 62)

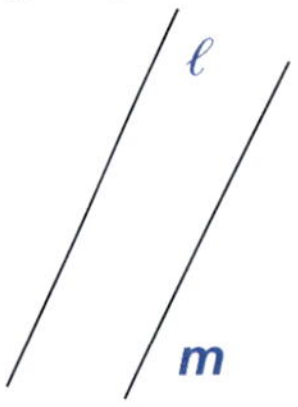

Parallelogram—A quadrilateral whose opposite sides are parallel (pp. 77, 143)

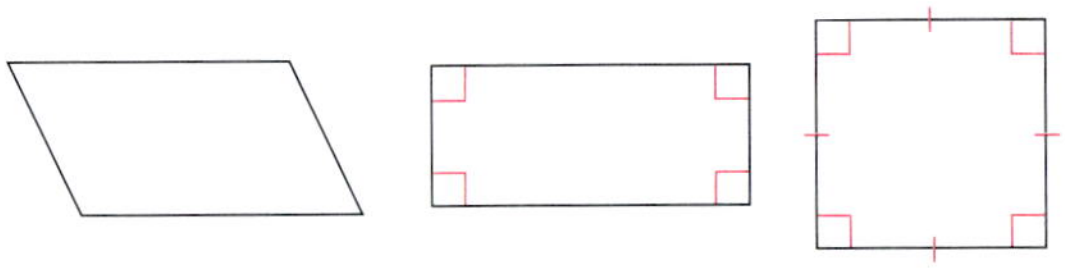

Pentagon—a five-sided polygon (p. 231)

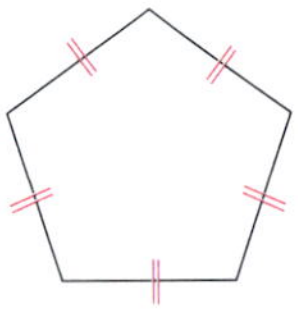

Perimeter—The distance around the outside of a shape (p. 231)

$P = a + b + c + d + e$

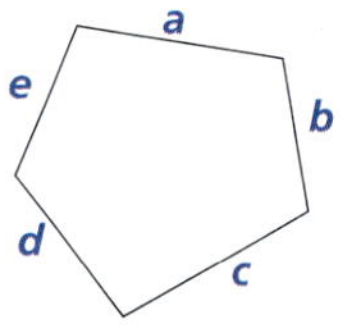

Perimeter of a polygon—Sum of the lengths of all the sides of the polygon (p. 284)

Perimeter = 10 + 15 + 13 + 20 = 58

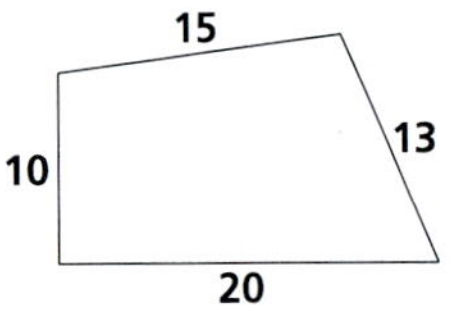

Perpendicular bisector—Line that bisects a given line segment and forms a right angle with it (p. 160); locus of points equidistant from two given points (p. 336)

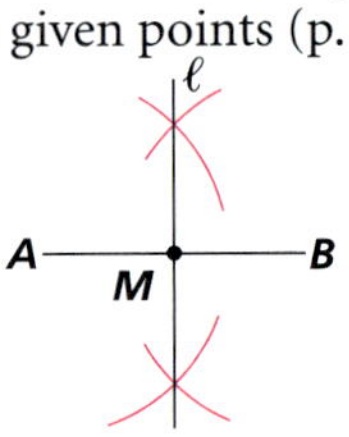

Perpendicular lines—Lines that form right angles (p. 15)

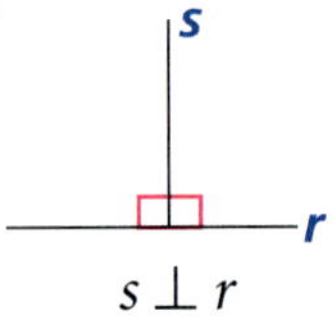

Pi (π)—Ratio of the circumference of a circle to its diameter (p. 319)

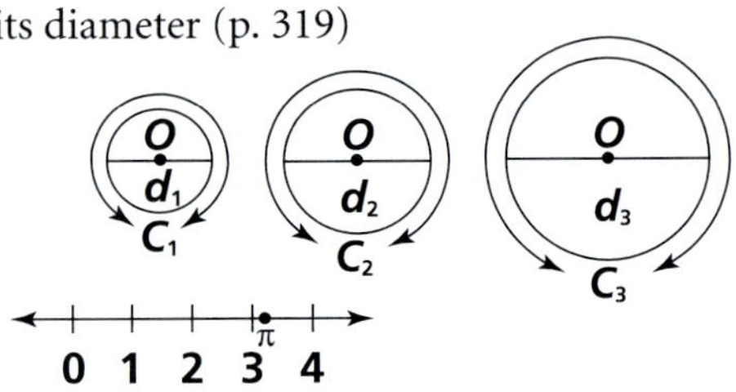

Plane—A two-dimensional flat surface (p. 2)

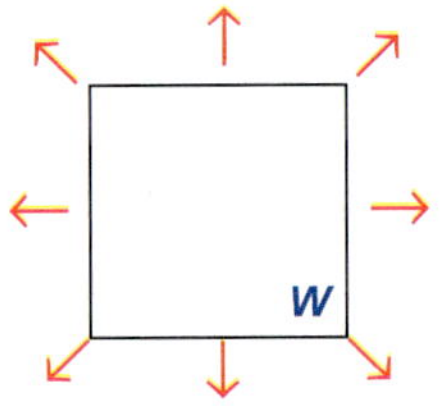

Point—A location in space represented by a dot (p. 3)

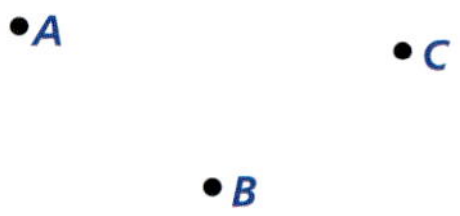

Poles—Endpoints of the diameters of a sphere (p. 342)

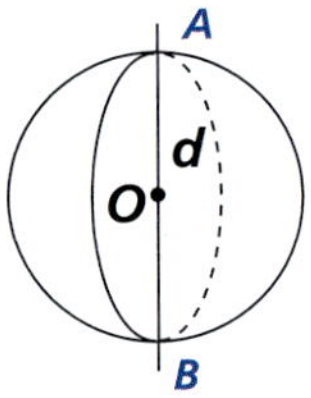

472 Glossary Geometry

Polygon—A closed, many-sided figure (pp. 90, 154)

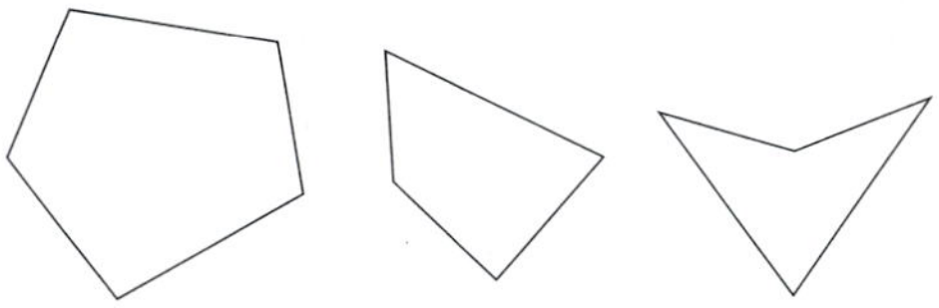

Postulate—A statement about geometric figures accepted as true without proof (p. 7)

Prism—A three-dimensional figure that has congruent polygons as bases in parallel planes and whose sides are parallelograms (p. 352)

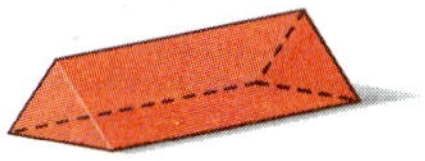

Probability—The chance or likelihood of an event occurring (p. 326)

Proof—A series of true statements leading to a desired conclusion (p. 53)

Proportion—The equality of two ratios (p. 220)

$$\frac{a}{b} = \frac{c}{d}$$

Protractor—A tool used to draw or measure angles (p. 15)

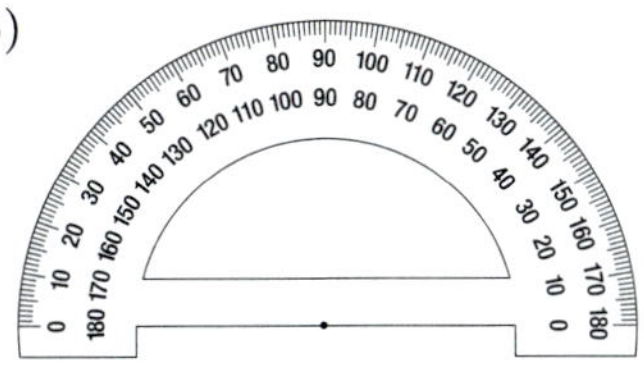

Pyramid—A three-dimensional figure with a single base and sides that are triangles (p. 356)

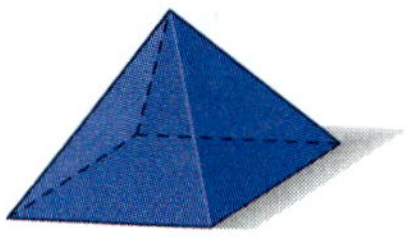

Pythagorean Triple—Integers a, b, c such that $a^2 + b^2 = c^2$ (p. 250)

Q

Quadrant—One-fourth of the coordinate plane (p. 96); one-fourth of a circle (p. 316)

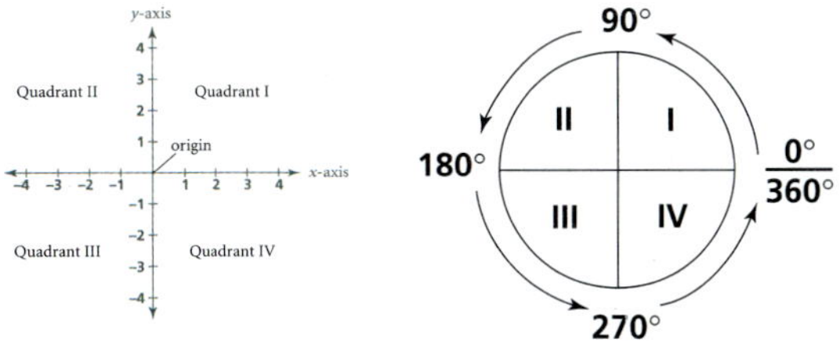

Quadrilateral—A geometric figure with four sides (p. 77)

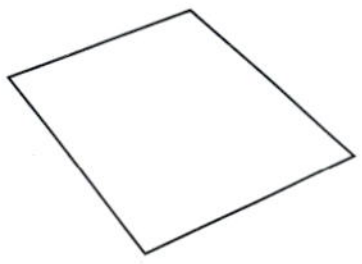

R

Radius—The distance from the center of a circle to the circle (pp. 231, 316)

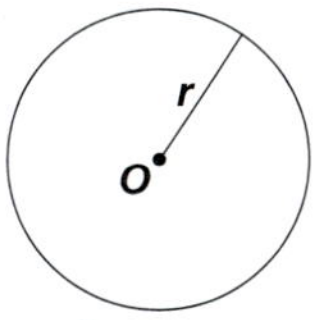

Range—The y-values of a function (p. 113)

Ratio—Comparison of two like quantities (p. 220)

$$\frac{a}{b}$$

Ray—A set of points that is part of a line; a ray has one endpoint and extends in one direction with no end (p. 3)

Rectangle—A parallelogram with four right angles (pp. 77, 143)

Geometry Glossary **473**

Regular polygon—Polygon with sides of equal length and angles of equal measure (p. 154)

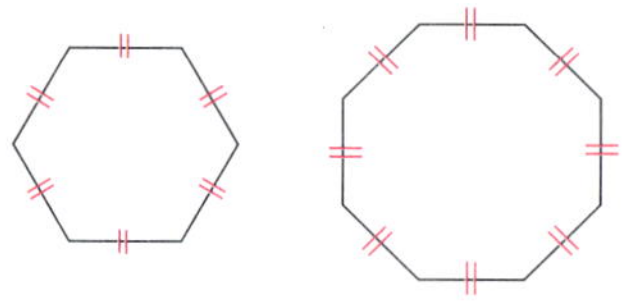

Rhombus—A parallelogram with four equal sides (pp. 77, 143)

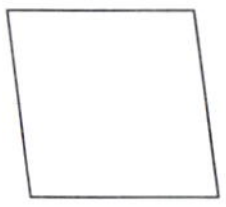

Right angle—An angle whose measure is 90° (p. 15)

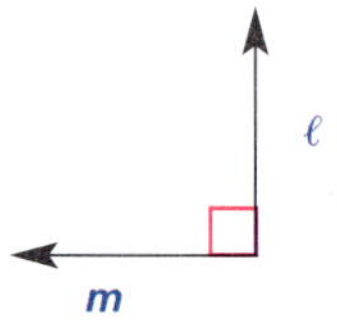

Right prism—Prism in which the base and the sides are at right angles (p. 360)

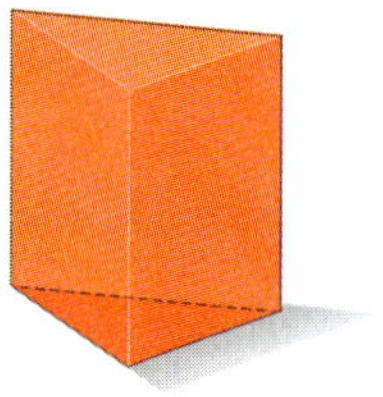

Right triangle—A triangle having one right angle (p. 139)

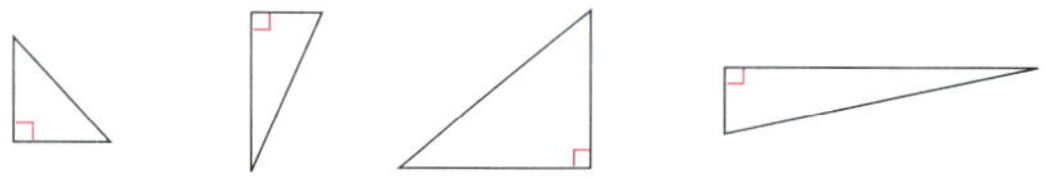

Rotation—Transformation in which a geometric figure turns around a center without affecting its size or shape (p. 210)

S

Scalene—A geometric figure with sides of unequal lengths (p. 77)

Scalene triangle—A triangle with no equal sides (p. 134)

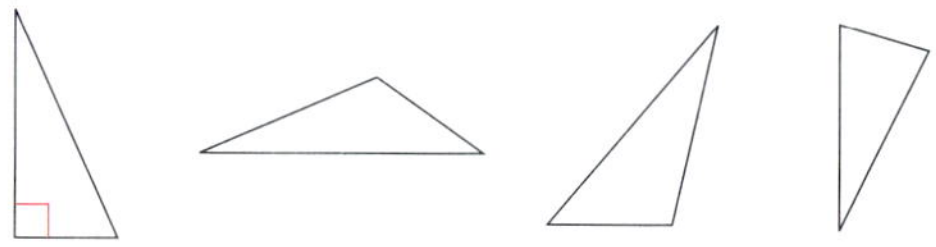

Sector—The area enclosed within a central angle of a circle (p. 332)

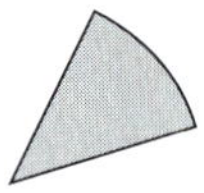

Set—A collection of particular things, like the set of points on a line segment (p. 3)

Similar—Having the same shape but a different size (p. 220)

Similar triangles—Triangles whose corresponding angles are equal and whose corresponding sides are proportional (p. 226)

Sine (sin)—For an angle of a right triangle (not the right angle), the ratio of the length of the opposite leg to the length of the hypotenuse (p. 340)

Skew lines—Lines in space that are not coplanar and never meet (p. 62)

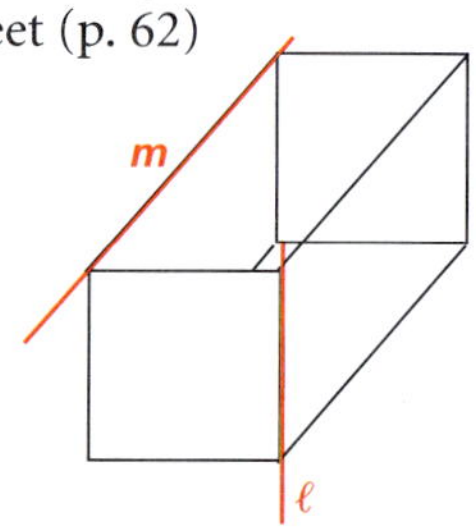

Slope of a line—Ratio of the difference in y-values to the difference in x-values between two points (p. 107)

$$\text{slope} = \frac{\text{rise}}{\text{run}}$$

Sphere—Locus of points in space equidistant from a fixed point (p. 342)

474 *Glossary* *Geometry*

Square—A rectangle with sides of equal length (p. 77)

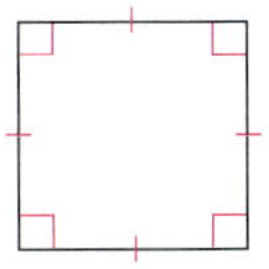

Straight angle—An angle whose measure is 180° (p. 15)

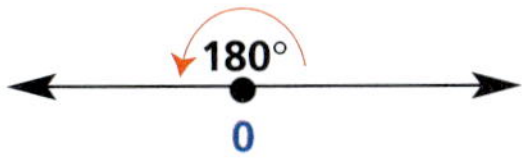

Strict—Exact and precise (p. 34)

Supplementary angles—Two angles whose measures add to 180° (p. 20)

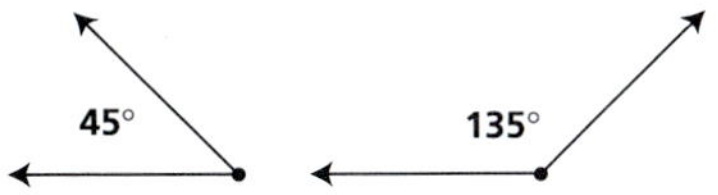

T

Tangent—Line that touches a circle at only one point (p. 336)

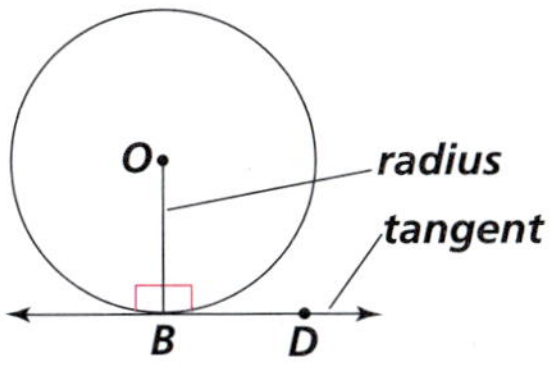

Tangent (tan)—For an angle of a right triangle (not the right angle), the ratio of the length of the opposite leg to the length of the adjacent leg (p. 340)

Tessellation—an arrangement of shapes that cover a plane surface without overlapping and without leaving holes (p. 310)

Theorem—A statement that can be proven true (p. 53)

Transformation—Movement of a geometric figure from one location to another (p. 207)

Translation—Transformation in which a geometric figure slides from one location to another without affecting its size or shape (p. 207)

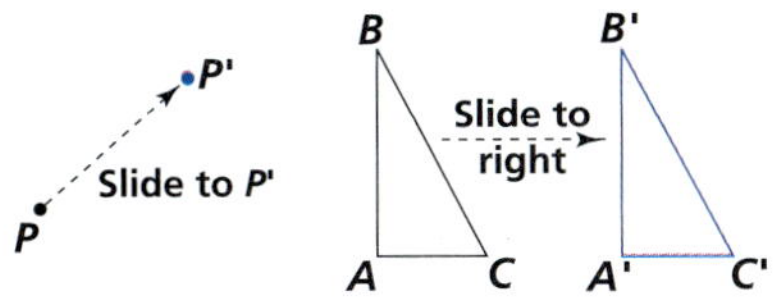

Transversal—A line that crosses parallel lines (p. 65)

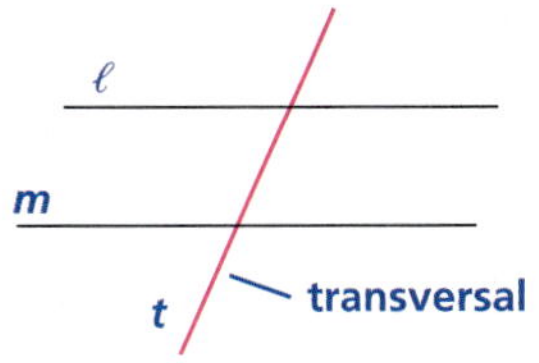

Trapezoid—A quadrilateral with exactly one pair of parallel sides (pp. 77, 143)

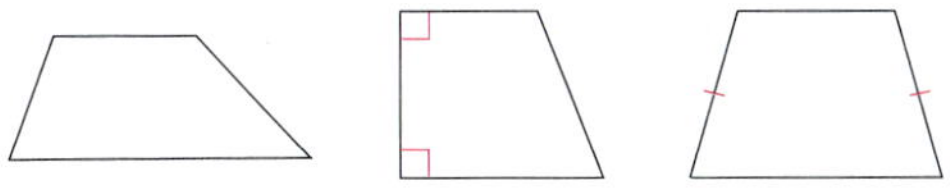

U

Unit circle—Circle whose radius is one (p. 340)

V

Valid—Based on evidence or fact (p. 68)

Vertex—The point common to both sides of an angle (p. 11)

Vertical—Straight up and down (p. 96)

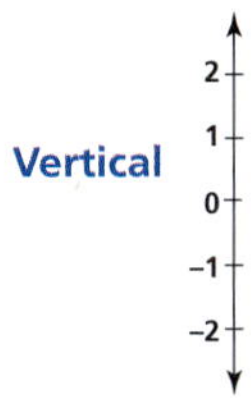

Vertical angles—Angles that have a common vertex and whose sides are formed by the same lines (p. 20)

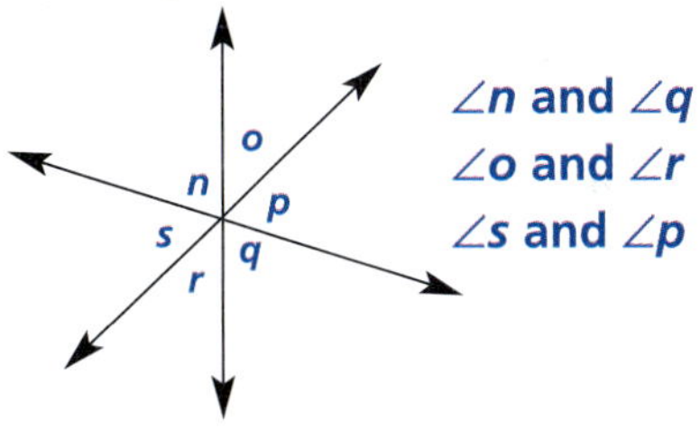

Vertices—Plural of *vertex* (p. 146)

Volume—The measure of the number of unit cubes in a solid figure (p. 352)

(3 • 10) 20 = 600 cubic units

Y

***Y*-intercept**—The *y*-value of the point where a line crosses the *y*-axis (when $x = 0$) (p. 113)

Geometry Glossary **475**

Index

476 *Index* *Geometry*

478 *Index* *Geometry*

Q

R

S

T

U

V

W

X

Y

480 *Index* *Geometry*

Midterm Mastery Test

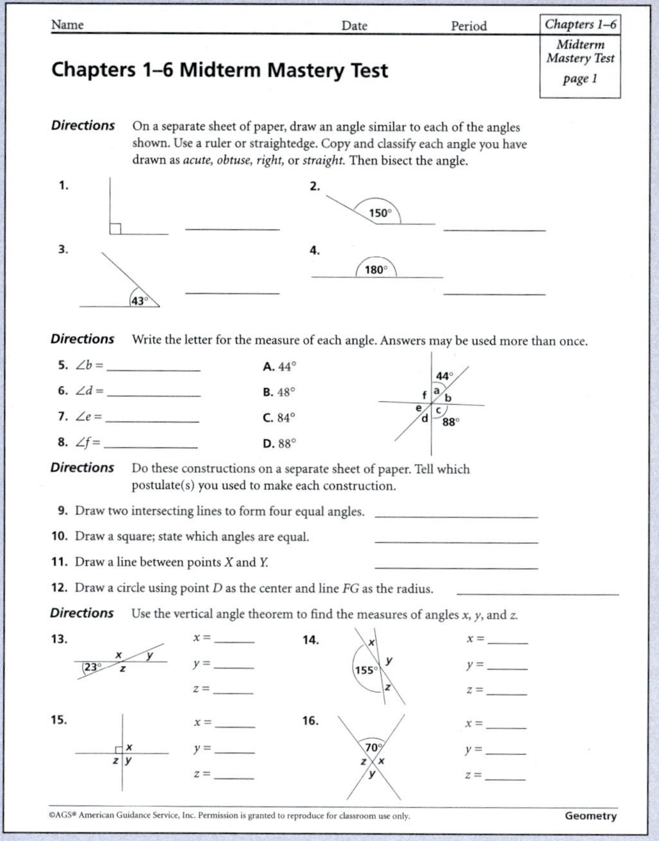

Name Date Period

Chapters 1–6 *Midterm Mastery Test* *page 1*

Chapters 1–6 Midterm Mastery Test

Directions On a separate sheet of paper, draw an angle similar to each of the angles shown. Use a ruler or straightedge. Copy and classify each angle you have drawn as *acute, obtuse, right,* or *straight.* Then bisect the angle.

1. ____________ **2.** 150° ____________

3. 43° ____________ **4.** 180° ____________

Directions Write the letter for the measure of each angle. Answers may be used more than once.

5. $\angle b =$ ____________ **A.** 44°

6. $\angle d =$ ____________ **B.** 48°

7. $\angle e =$ ____________ **C.** 84°

8. $\angle f =$ ____________ **D.** 88°

44° a b c d e f 88°

Directions Do these constructions on a separate sheet of paper. Tell which postulate(s) you used to make each construction.

9. Draw two intersecting lines to form four equal angles. ____________

10. Draw a square; state which angles are equal. ____________

11. Draw a line between points *X* and *Y*. ____________

12. Draw a circle using point *D* as the center and line *FG* as the radius. ____________

Directions Use the vertical angle theorem to find the measures of angles *x, y,* and *z.*

13. 23° x y z $x =$ ______ $y =$ ______ $z =$ ______

14. 155° x y z $x =$ ______ $y =$ ______ $z =$ ______

15. x y z $x =$ ______ $y =$ ______ $z =$ ______

16. 70° x y z $x =$ ______ $y =$ ______ $z =$ ______

 Geometry

Midterm Mastery Test Page 1

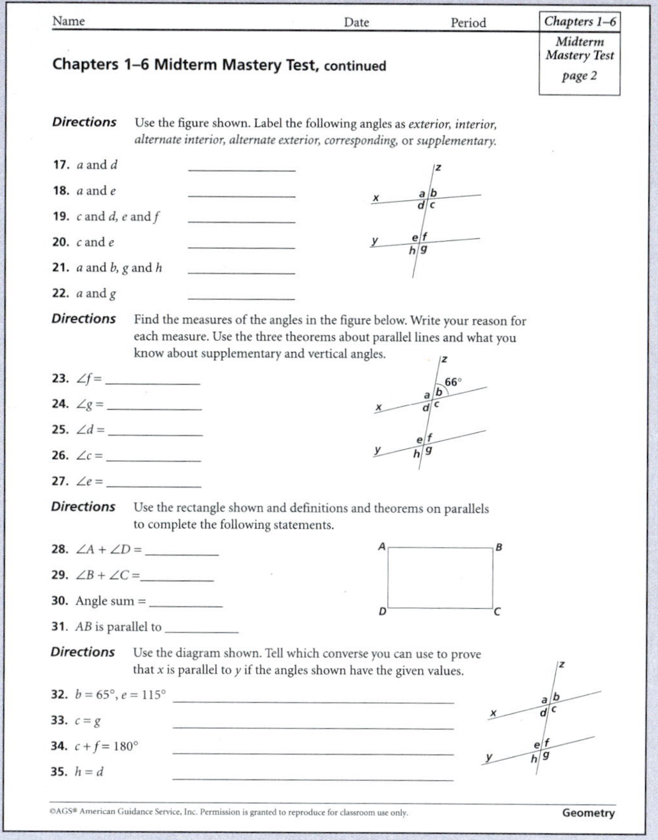

Name Date Period

Chapters 1–6 *Midterm Mastery Test* *page 2*

Chapters 1–6 Midterm Mastery Test, continued

Directions Use the figure shown. Label the following angles as *exterior, interior, alternate interior, alternate exterior, corresponding,* or *supplementary.*

17. *a* and *d* ____________

18. *a* and *e* ____________

19. *c* and *d, e* and *f* ____________

20. *c* and *e* ____________

21. *a* and *b, g* and *h* ____________

22. *a* and *g* ____________

z x y a b d c e f h g

Directions Find the measures of the angles in the figure below. Write your reason for each measure. Use the three theorems about parallel lines and what you know about supplementary and vertical angles.

23. $\angle f =$ ____________

24. $\angle g =$ ____________

25. $\angle d =$ ____________

26. $\angle c =$ ____________

27. $\angle e =$ ____________

z 66° x y a b d c e f h g

Directions Use the rectangle shown and definitions and theorems on parallels to complete the following statements.

28. $\angle A + \angle D =$ ____________

29. $\angle B + \angle C =$ ____________

30. Angle sum = ____________

31. *AB* is parallel to ____________

A B D C

Directions Use the diagram shown. Tell which converse you can use to prove that *x* is parallel to *y* if the angles shown have the given values.

32. $b = 65°, e = 115°$ ____________

33. $c = g$ ____________

34. $c + f = 180°$ ____________

35. $h = d$ ____________

z x y a b d c e f h g

 Geometry

Midterm Mastery Test Page 2

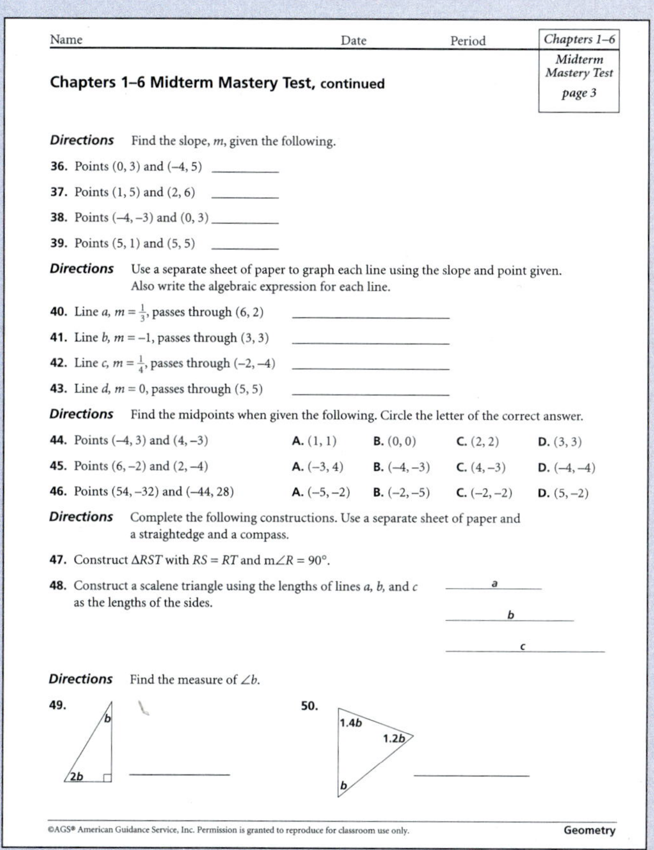

Name Date Period

Chapters 1–6 *Midterm Mastery Test* *page 3*

Chapters 1–6 Midterm Mastery Test, continued

Directions Find the slope, *m*, given the following.

36. Points (0, 3) and (–4, 5) ____________

37. Points (1, 5) and (2, 6) ____________

38. Points (–4, –3) and (0, 3) ____________

39. Points (5, 1) and (5, 5) ____________

Directions Use a separate sheet of paper to graph each line using the slope and point given. Also write the algebraic expression for each line.

40. Line *a*, $m = \frac{1}{3}$, passes through (6, 2) ____________

41. Line *b*, $m = -1$, passes through (3, 3) ____________

42. Line *c*, $m = \frac{1}{4}$, passes through (–2, –4) ____________

43. Line *d*, $m = 0$, passes through (5, 5) ____________

Directions Find the midpoints when given the following. Circle the letter of the correct answer.

44. Points (–4, 3) and (4, –3) **A.** (1, 1) **B.** (0, 0) **C.** (2, 2) **D.** (3, 3)

45. Points (6, –2) and (2, –4) **A.** (–3, 4) **B.** (–4, –3) **C.** (4, –3) **D.** (–4, –4)

46. Points (54, –32) and (–44, 28) **A.** (–5, –2) **B.** (–2, –5) **C.** (–2, –2) **D.** (5, –2)

Directions Complete the following constructions. Use a separate sheet of paper and a straightedge and a compass.

47. Construct ΔRST with $RS = RT$ and $m\angle R = 90°$.

48. Construct a scalene triangle using the lengths of lines *a, b,* and *c* as the lengths of the sides.

a b c

Directions Find the measure of $\angle b$.

49. b 2b ____________

50. 1.4b 1.2b b ____________

 Geometry

Midterm Mastery Test Page 3

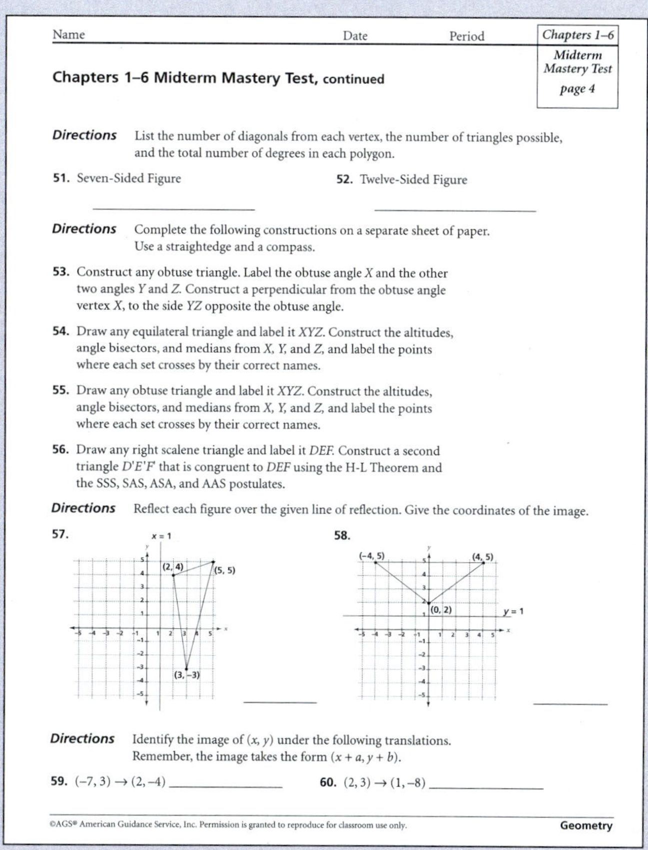

Name Date Period

Chapters 1–6 *Midterm Mastery Test* *page 4*

Chapters 1–6 Midterm Mastery Test, continued

Directions List the number of diagonals from each vertex, the number of triangles possible, and the total number of degrees in each polygon.

51. Seven-Sided Figure ____________ **52.** Twelve-Sided Figure ____________

Directions Complete the following constructions on a separate sheet of paper. Use a straightedge and a compass.

53. Construct any obtuse triangle. Label the obtuse angle *X* and the other two angles *Y* and *Z*. Construct a perpendicular from the obtuse angle vertex *X*, to the side *YZ* opposite the obtuse angle.

54. Draw any equilateral triangle and label it *XYZ*. Construct the altitudes, angle bisectors, and medians from *X, Y,* and *Z,* and label the points where each set crosses by their correct names.

55. Draw any obtuse triangle and label it *XYZ*. Construct the altitudes, angle bisectors, and medians from *X, Y,* and *Z,* and label the points where each set crosses by their correct names.

56. Draw any right scalene triangle and label it *DEF*. Construct a second triangle *D'E'F* that is congruent to *DEF* using the H-L Theorem and the SSS, SAS, ASA, and AAS postulates.

Directions Reflect each figure over the given line of reflection. Give the coordinates of the image.

57. x = 1 (2, 4) (5, 5) (3, –3) ____________

58. (–4, 5) (4, 5) (0, 2) y = 1 ____________

Directions Identify the image of (*x, y*) under the following translations. Remember, the image takes the form $(x + a, y + b)$.

59. $(-7, 3) \rightarrow (2, -4)$ ____________ **60.** $(2, 3) \rightarrow (1, -8)$ ____________

 Geometry

Midterm Mastery Test Page 4

Final Mastery Test

Name Date Period

Chapters 1–12 *Final Mastery Test page 1*

Chapters 1–12 Final Mastery Test

Directions On a separate sheet of paper, draw an angle similar to each of the angles shown. Use a ruler or straightedge. Copy and classify each angle you have drawn as *acute, obtuse, right,* or *straight.* Then bisect the angle.

1. 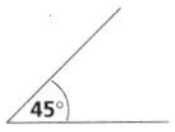

2. 

Directions Do these constructions on a separate sheet of paper. Tell which postulate(s) you used to make the construction.

3. Draw a circle using point *X* as the center and line *RS* as the radius. ______

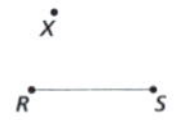

4. Draw two intersecting lines to form four equal angles. ______

Directions Find the measures of the angles in the figure below. Write your reason for each measure. Use the three theorems about parallel lines and what you know about supplementary and vertical angles.

5. $\angle g$ ______

6. $\angle e$ ______

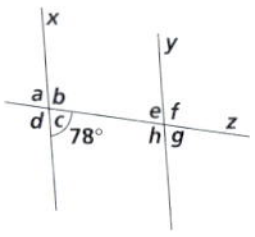

Directions Graph each line using the slope and point given. Use a separate sheet of paper. Also write the algebraic expression for each line.

7. Line *a*, $m = 3$, passes through (0, 4) ______

8. Line *b*, $m = \frac{1}{2}$, passes through (5, 2) ______

Directions Complete the following constructions. Use a separate sheet of paper and a straightedge and a compass.

9. Construct ΔABC with $AB = AC$ and $m\angle A < 90°$.

10. Construct a scalene triangle using the lengths of segments *a*, *b*, and *c* as the lengths of the sides.

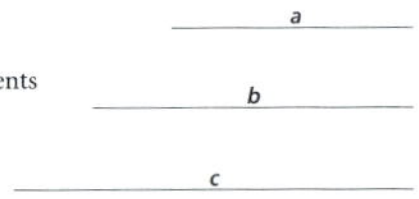

 Geometry

Final Mastery Test Page 1

Name Date Period

Chapters 1–12 *Final Mastery Test page 2*

Chapters 1–12 Final Mastery Test, continued

Directions Find the measure of $\angle x$.

11. 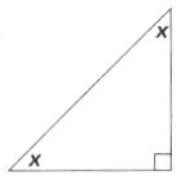

______

12. 

______

Directions Complete the following constructions on a separate sheet of paper. Use a straightedge and a compass.

13. Draw any right scalene triangle and label it *ABC*. Construct a second triangle *A′B′C′* that is congruent to *ABC* using the H-L Theorem and the SSS, SAS, ASA, and AAS postulates.

14. Draw any right triangle and label it *ABC*. Construct the altitudes, angle bisectors, and medians from *A*, *B*, and *C*, and label the points where each set crosses by their correct names.

15. Draw any equilateral triangle and label it *ABC*. Construct the altitudes, angle bisectors, and medians from *A*, *B*, and *C*, and label the points where each set crosses by their correct names.

Directions Reflect each figure over the given line of reflection. Give the coordinates of the image.

16. 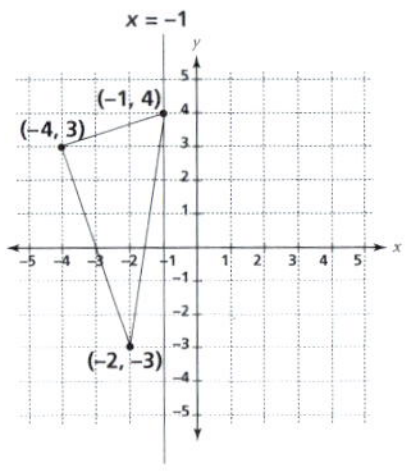

______

17. 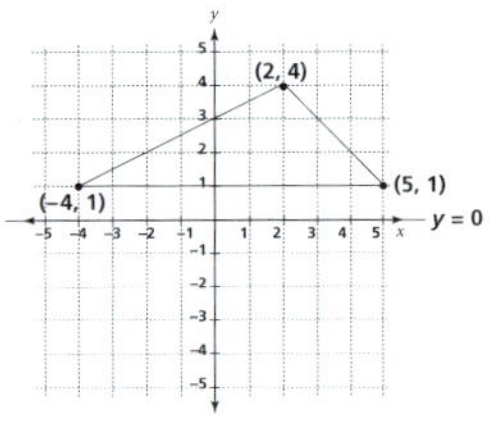

______

 Geometry

Final Mastery Test Page 2

Name Date Period

Chapters 1–12 *Final Mastery Test page 3*

Chapters 1–12 Final Mastery Test, continued

Directions Identify the image of (*x*, *y*) under the following translations. Remember, the image takes the form $(x + a, y + b)$.

18. $(5, -2) \rightarrow (3, -4)$ ______

19. $(7, 4) \rightarrow (2, 9)$ ______

Directions The triangles are similar. Solve for the values of the unknowns.

20. $s =$ ______

21. $r =$ ______

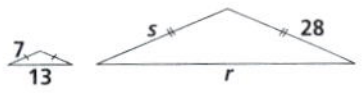

Directions Give the coordinates of the image triangle under the following dilations. Use graph paper to graph the object and its images. (All dilations have (0, 0) as the center of the dilation.)

22. dilation of 2

23. dilation of $\frac{1}{2}$

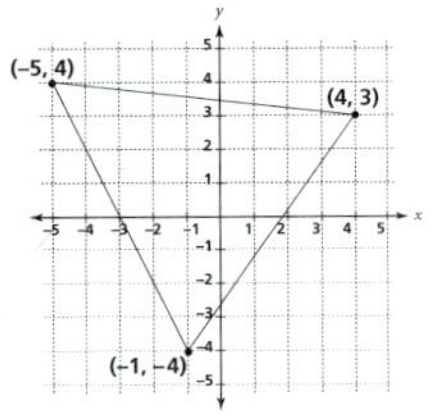

Directions Use Plato's Formula to find Pythagorean Triples for the given integers. Use a calculator.

24. $m = 6$ ______

25. $m = 9$ ______

Directions Use the distance formula to find the distances between the given points. Circle the correct answer.

26. (8, 3) and (6, 7) **A.** 12 **B.** 16 **C.** 20 **D.** 24

27. (–4, 0) and (1, 3) **A.** 32 **B.** 34 **C.** 36 **D.** 38

 Geometry

Final Mastery Test Page 3

Name Date Period

Chapters 1–12 *Final Mastery Test page 4*

Chapters 1–12 Final Mastery Test, continued

Directions Use the converse of the Pythagorean Theorem to find the triples that could be the sides of a right triangle from the given choices. Write the letter of the correct answer on the line. There will be only one correct answer for each problem.

28. ______ **A.** 12, 16, 18 **B.** 15, 17, 4 **C.** 40, 399, 400 **D.** 12, 16, 20

29. ______ **A.** 40, 75, 85 **B.** 60, 80, 90 **C.** 13, 12, 12 **D.** 51, 49, 7

Directions Find the perimeter of each figure.

30. 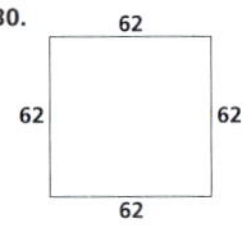

______

31. 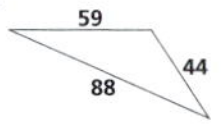

______

Directions Use formulas to find the perimeter of each polygon.

32. regular septagon, $s = 57$ ______

33. regular decagon, $s = 17$ ______

Directions Use the Pythagorean Theorem to find the missing side. Then calculate the perimeter. Use a calculator and round to the nearest tenth.

34. 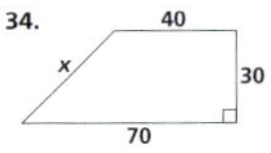

______

35. 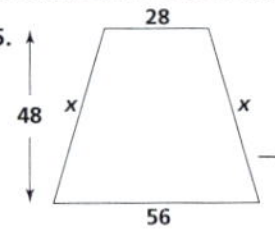

______

Directions Find the area of the square or rectangle with the given side lengths. Circle the correct answer.

36. $s = 37$ **A.** 9,163 **B.** 1,936 **C.** 3,916 **D.** 1,369

37. $l = 44$, $w = 66$ **A.** 9,204 **B.** 2,904 **C.** 2,409 **D.** 2,094

Directions Find the missing dimension, either *h* or *b*, for rectangles with these areas.

38. 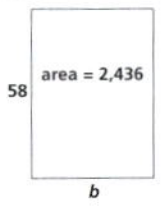

______

39. 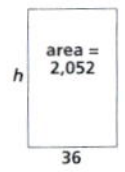

______

 Geometry

Final Mastery Test Page 4

Final Mastery Test

Name Date Period

Chapters 1–12
Final Mastery Test
page 5

Chapters 1–12 Final Mastery Test, continued

Directions Find the areas of these trapezoids.

34. 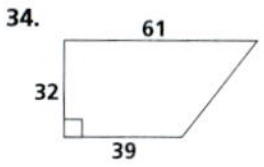

____

35. 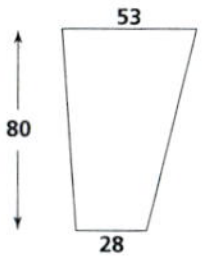

____

Directions Use Heron's Formula to find the areas of triangles with the given side lengths. Circle the correct answer.

42. 22, 28, 30 **A.** 218 **B.** 318 **C.** 281 **D.** 2.18

43. 17, 18, 19 **A.** 149.3 **B.** 143.9 **C.** 139.4 **D.** 134.9

Directions Find the diameter and the radius of a circle with the given circumference. Then calculate the area of the circle. You may state your answer in terms of π.

44. $C = 94\pi$ ft $d =$ __________ $r =$ __________

45. $C = 58\pi$ ft $d =$ __________ $r =$ __________

Directions Find the area of the shaded sector. Circle the correct answer.

46. 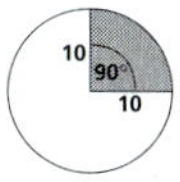

A. 50π
B. 25π
C. 100π
D. 10π

47. 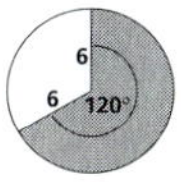

A. 12π
B. 48π
C. 36π
D. 24π

Directions Complete the following constructions on a separate sheet of paper. Use a straightedge and a compass.

48. Construct the circumcircle for any obtuse triangle.

49. Construct the incircle for any right triangle.

Directions Find the surface area and the volume of spheres with the given diameter.

50. $d = 12$ $S=$ __________ $V =$ __________

51. $d = 14$ $S=$ __________ $V =$ __________

 Geometry

Final Mastery Test Page 5

Name Date Period

Chapters 1–12
Final Mastery Test
page 6

Chapters 1–12 Final Mastery Test, continued

Directions Find the volume of these quadrilateral solids.

52. 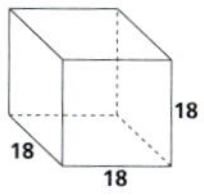

____

53. 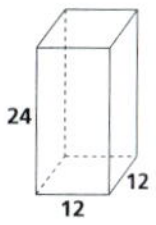

____

Directions Find the volume of each pyramid or cone.

54. 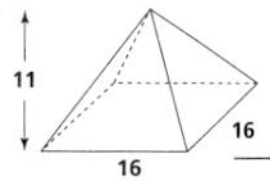

____

55. 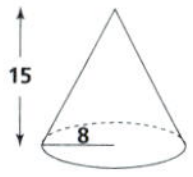

____

Directions Find the surface area of each of the following figures.

56. 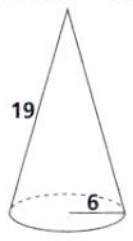

____

57. 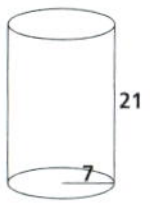

____

Directions On a separate sheet of paper, draw a sketch and describe the following.

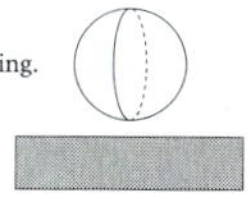

58. A plane and a sphere which cross to form a circle that has a smaller radius than the radius of the sphere

Directions On a separate sheet of paper, construct and then describe the following compound loci.

59. The locus of points equidistant from $\overline{XY}$, $\overline{YZ}$, and $\overline{XZ}$.

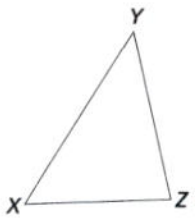

60. The loci of points equidistant from the parallel lines *c* and *d* and at a distance of *EF* from point *G*.

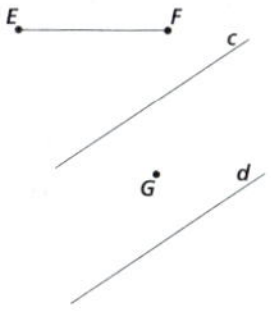

 Geometry

Final Mastery Test Page 6

Teacher's Resource Library Answer Key

Additional Practice

Additional Practice 1—Points and Lines in the Plane

1. $\overrightarrow{AB}$ **2.** $\overline{NT}$ **3.** n **4.** $\overleftrightarrow{BC}$ **5.** $\overrightarrow{LM}$ **6.** $\overline{YZ}$ **7.** $\overleftrightarrow{DE}$, ED, t **8.** $\overleftrightarrow{ML}$, $\overleftrightarrow{OL}$, $\overleftrightarrow{OM}$, $\overleftrightarrow{LM}$, $\overleftrightarrow{LO}$, $\overleftrightarrow{MO}$ **9.** $\overleftrightarrow{SR}$, $\overleftrightarrow{TR}$, $\overleftrightarrow{TS}$, $\overleftrightarrow{RS}$, $\overleftrightarrow{RT}$, $\overleftrightarrow{ST}$, c **10.** z **11.** one **12.** none **13.** two **14.** four **15.** none **16.** $\overline{AB}$ **17.** $\overline{AB}$, $\overline{BC}$, and $\overline{AC}$ **18.** $\overline{AB}$, $\overline{BC}$, $\overline{CD}$, $\overline{AC}$, $\overline{AD}$, $\overline{BD}$ **19.** $\overline{AC}$, $\overline{AD}$, $\overline{AE}$, $\overline{AB}$, $\overline{BC}$, $\overline{CD}$, $\overline{DE}$, $\overline{BD}$, $\overline{BE}$, $\overline{CE}$ **20.** $\overline{AB}$, $\overline{BC}$, $\overline{CD}$, $\overline{DE}$, $\overline{EF}$, $\overline{AC}$, $\overline{AD}$, $\overline{AE}$, $\overline{AF}$, $\overline{BD}$, $\overline{BE}$, $\overline{BF}$, $\overline{DF}$, $\overline{CE}$, $\overline{CF}$

Additional Practice 2—Ruler Postulates

1. $A = 0, B = 2, C = 3, D = 4$ **2.** $L = -4, M = 1, N = 2.3, O = 3.14, P = 5$ **3.** 2.5 in. **4.** 4 in. **5.** 6 in. **6.** 5 in. **7.** 40 mm **8.** 100 mm **9.** 140 mm **10.** 50 mm **11.** 8 **12.** 4 **13.** $3 + 3 = 6$ **14.** $6 + 3 = 9$ **15.** $5 + 2 = 7$

Additional Practice 3—Copying and Bisecting Angles

1. $\angle ABC$ or $\angle CBA$ **2.** $\angle STU$ or $\angle UTS$ **3.** $\angle MNO$ or $\angle r$ **4.** $\angle EFG$ or $\angle z$ **5.–8.** Students should draw and copy angles with the measures of the angles shown. **9.** Students should draw two adjacent angles of 30° each. **10.** Students should draw two adjacent angles of 60° each. **11.** Students should draw two adjacent angles of 75° each. **12.** Students should draw two adjacent angles of 50° each. **13.** yes **14.** no **15.** yes

Additional Practice 4—Angle Measurement

1. acute **2.** obtuse **3.** acute **4.** straight **5.** right **6.** acute **7.** right **8.** obtuse **9.** straight **10.** acute **11.** Students should draw a right angle. **12.** Students should draw an obtuse angle. **13.** Students should draw a straight angle. **14.** Students should draw an acute angle. **15.** x and y **16.** $\overline{AB}$ and $\overline{BC}$ **17.** $\overline{TU}$ and $\overline{TS}$ **18.** Students should draw two perpendicular lines. **19.** right angles **20.** four 90° angles

Additional Practice 5—Complementary and Supplementary Angles

1. 50° **2.** 15° **3.** 70° **4.** 25° **5.** 125° **6.** 60° **7.** 150° **8.** $180° - z°$ **9.** 85° **10.** supplementary **11.** complementary **12.** complementary **13.** supplementary and adjacent **14.** $\angle s$ and $\angle u$, $\angle t$ and $\angle v$ **15.** $\angle d$ and $\angle f$, $\angle e$ and $\angle g$ **16.** $\angle q$ and $\angle n$, $\angle l$ and $\angle o$, $\angle p$ and $\angle m$ **17.** 90° **18.** 45° **19.** 45° **20.** 45°

Additional Practice 6—Algebra and Angles

1. $b = 95°, c = 85°, d = 95°$ **2.** $k = 66°, m = 114°, l = 66°$ **3.** 36° **4.** 90° **5.** 36° **6.** 54° **7.** 90° **8.** Students should draw two supplementary angles, one 135° and the other 45°. **9.** 80°, 10° **10.** 160°, 20°

Additional Practice 7—If . . . Then Statements

1. If a figure is a triangle, then it has three angles. **2.** If a figure is a square, then it contains four right angles. **3.** If the sum of angles A and B is 180°, then angles A and B are supplementary. **4.** If angle Z measures 91°, then angle Z is obtuse. **5.** If angle M measures 60°, then angle M is acute. **6.** True **7.** True **8.** True **9.** True **10.** False **11.** If a given angle is a right angle, then it measures 90°; True, True **12.** If a given angle is an obtuse angle, then it measures more than 90°; True, True **13.** If a figure is a square, then it has four sides and two of its angles are right angles; False, False **14.** If three angles are complementary, then the sum of their measures is 90°; False, False **15.** If a figure is a triangle, then it has three sides; True, True

Additional Practice 8—Euclid's Five Postulates

1.

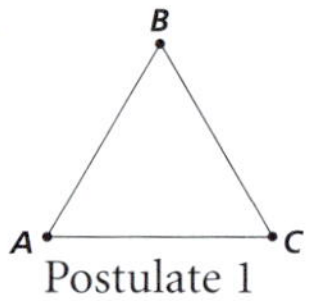

Postulate 1

2.

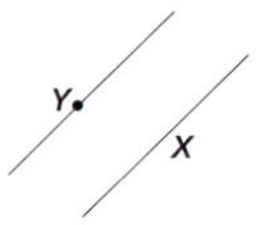

Postulate 2

3.

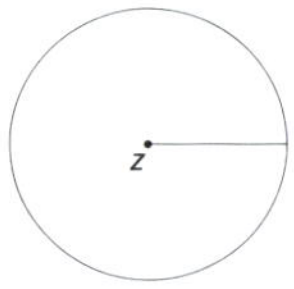

Postulate 3

4.

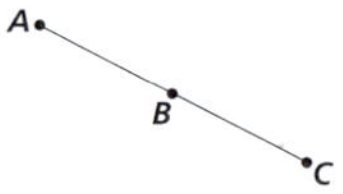

Postulate 2

5. Postulate 3 **6.** Postulate 1 **7.** Postulate 2 **8.** Postulate 4 **9.** True; a circle has only one distance for radius. **10.** True; Postulate 2

Additional Practice 9—Using Euclid's Postulates

1.

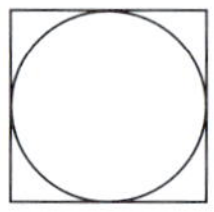

Postulates 3 and 1

2.

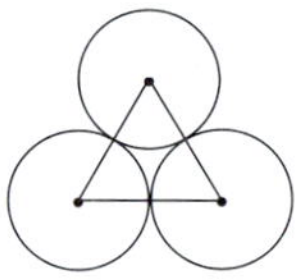

Postulate 3

3.

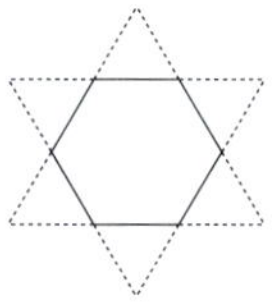

Postulate 2

4.

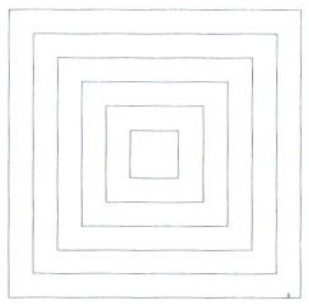

Postulate 2

5. Postulate 4 **6.** Postulate 2 **7.** Postulate 2 **8.** Postulate 4 **9.** Postulate 1 **10.** Postulate 5

Additional Practice 10—Axioms or Common Notions

1. Axiom 3 **2.** Axiom 2 **3.** Axiom 3 **4.** Axiom 2 **5.** equal; Axiom 4 **6.** equal; Axiom 4 **7.** yes; Axiom 1 **8.** They're smaller than the whole; Axiom 5 **9.** 12; Axiom 1 **10.** 5,280; Axiom 1

Additional Practice 11—A Simple Proof

1. $x = 91°, y = 89°, z = 91°$ **2.** $x = 150°, y = 30°, z = 150°$ **3.** $x = 80°, y = 100°, z = 80°$ **4.** $x = 90°, y = 90°, z = 90°$ **5.** $x = 6°, y = 174°, z = 6°$ **6.** vertical angles are equal **7.** $m\angle a + m\angle d = 180°$; Axiom 5, sum is greater than the parts **8.** $m\angle c + m\angle b = 180°$; Axiom 5, sum is greater than the parts **9.** Axiom 3, equals subtracted from equals give equal differences **10.** Euclid's Postulate 5

Additional Practice 12—Parallel Lines

1. True **2.** True **3.** True **4.** True **5.** True **6.** False **7.** True **8.** False **9.** True **10.** False **11.** $a \nparallel b$ **12.** $x \parallel y$ **13.** $s \perp t$ **14.** $\nparallel$ **15.** $\perp$

Additional Practice 13—Transversals

1. $\angle a$ and $\angle e$, $\angle c$ and $\angle g$, $\angle d$ and $\angle h$, $\angle b$ and $\angle f$ **2.** $\angle a$, $\angle b$, $\angle g$, $\angle h$ **3.** $\angle d$ and $\angle f$, $\angle c$ and $\angle e$ **4.** $\angle c$, $\angle d$, $\angle e$, $\angle f$ **5.** $\angle a$ and $\angle b$, $\angle b$ and $\angle c$, $\angle c$ and $\angle d$, $\angle a$ and $\angle d$, $\angle e$ and $\angle f$, $\angle f$ and $\angle g$, $\angle g$ and $\angle h$, $\angle h$ and $\angle e$ **6.** supplementary **7.** corresponding **8.** alternate exterior **9.** alternate interior **10.** interior **11.** exterior **12.** corresponding **13.** alternate interior **14.** supplementary **15.** alternate exterior

Additional Practice 14—Theorems About Parallel Lines

1. $\angle f$ **2.** $\angle e$ **3.** $\angle a$ **4.** $\angle b$ **5.** $\angle h$ **6.** $\angle g$ **7.** $\angle f$ **8.** $\angle e$ **9.** $\angle c$, vertical angles are equal; $\angle e$, corresponding angles are equal; $\angle g$, alternate exterior angles are equal **10.** $\angle d$, vertical angles are equal; $\angle f$, corresponding angles are equal; $\angle h$, alternate interior angles are equal **11.** $\angle a$, vertical angles are equal; $\angle g$, corresponding angles are equal; $\angle e$, alternate interior angles are equal **12.** $\angle b$, vertical angles are equal; $\angle h$, corresponding angles are equal; $\angle f$, alternate exterior angles are equal **13.** $\angle g$, vertical angles are equal; $\angle a$, corresponding angles are equal; $\angle c$, alternate interior angles are equal **14.** $\angle h$, vertical angles are equal; $\angle b$, corresponding angles are equal; $\angle d$, alternate exterior angles are equal **15.** $\angle e$, vertical angles are equal; $\angle c$, corresponding angles are equal; $\angle a$, alternate exterior angles are equal

Additional Practice 15—Constructions and Problem Solving

1. $m\angle a = 120°$, $m\angle b = 60°$; supplementary straight angle **2.** $m\angle a = 100°$, $m\angle b = 80°$; supplementary straight angle **3.** $m\angle a = 120°$, $m\angle b = 60°$, $m\angle c = 60°$; supplementary straight angle and vertical angles are equal **4.** $m\angle a = 162°$, $m\angle b = 18°$, $m\angle c = 162°$; supplementary straight angle and vertical angles are equal **5.** 135° **6.** 45° **7.** 135° **8.** 45° **9.** 135° **10.** 45° **11.** 135° **12.** 45° **13.** 135° **14.** 45° **15.** 135°

Additional Practice 16—Quadrilaterals and Parallels

1. $\overline{ZY}$ **2.** $\overline{XY}$ **3.** $\overline{WZ}, \overline{XY}$ **4.** 180° **5.** 360° **6.** 90° **7.** 90° **8.** 180° **9.** 360° **10.** 180° **11.** $\overline{AD}, \overline{BC}$ **12.** Given: definition of a rhombus. **13.** Theorem 1, interior angles and supplementary **14.** Theorem 1, interior angles and supplementary **15.** Postulate 1, equals added to equals = 360° are equal

Additional Practice 17—Proving Lines Parallel

1. parallel; Converse 1 **2.** parallel; Converse 2 **3.** parallel; Converse 1 **4.** parallel; Converse 2 **5.** parallel; Converse 3 **6.** not parallel **7.** corresponding angles are equal **8.** corresponding angles are equal **9.** alternate interior angles are equal **10.** interior angles on the same side of the transversal are supplementary **11.** interior angles on the same side of the transversal are supplementary **12.** corresponding angles are equal **13.** corresponding angles are equal **14.** alternate interior angles are equal **15.** interior angles on the same side of the transversal are supplementary

Additional Practice 18—Naming Points

1. (2, 2) **2.** (−3, 4) **3.** (3, 1) **4.** (−3, −2) **5.** (2, −1) **6.** (5, −1) **7.** (−4, 1) **8.** (−5, −5) **9.** (−1, 4) **10.** (1, −1) **11.** *X* **12.** *T* **13.** *W* **14.** *S* **15.** *Y* **16.** *Q* **17.** *Z* **18.** *R* **19.** *U* **20.** *V* **21.** (0, −2) **22.** (2, 0) **23.** (−1, 0) **24.** (0, 0) **25.** (0, 1)

Additional Practice 19—Horizontal Lines

1. $y = -2$, x is any real number **2.** $y = 2$, x is any real number **3.** $y = 1$, x is any real number **4.** $y = -3$, x is any real number **5.** $y = 3$, x is any real number **6.** $y = -4$, x is any real number **7.** $y = -1$, x is any real number **8.** $y = 4$, x is any real number **9.** $y = -5$, x is any real number **10.** $y = 5$, x is any real number **11.** $(x, 4)$ **12.** $(x, -3)$ **13.** $(x, 1)$ **14.** $(x, -4)$ **15.** $(x, 3)$ **16.** $(x, -1)$ **17.** $(x, 5)$ **18.** $(x, -5)$ **19.** $(x, 2)$ **20.** $(x, -2)$

Additional Practice 20—Vertical Lines

1. $x = -5$, y is any real number **2.** $x = 4$, y is any real number **3.** $x = 1$, y is any real number **4.** $x = 3$, y is any real number **5.** $x = -3$, y is any real number **6.** $x = -4$, y is any real number **7.** $x = 5$, y is any real number **8.** $x = 2$, y is any real number **9.** $x = -2$, y is any real number **10.** $x = -1$, y is any real number **11.** $(1, y)$ **12.** $(2, y)$ **13.** $(-1, y)$ **14.** $(-2, y)$ **15.** $(3, y)$ **16.** $(-3, y)$ **17.** $(5, y)$ **18.** $(-5, y)$ **19.** $(-4, y)$ **20.** $(4, y)$

Additional Practice 21—Slope of a Line

1. $m = -\frac{1}{2}$ **2.** $m = -\frac{5}{7}$ **3.** $m = \frac{1}{2}$ **4.** $m = 1$ **5.** $m =$ undefined

Additional Practice 22—Using Algebra: $y = mx + b$

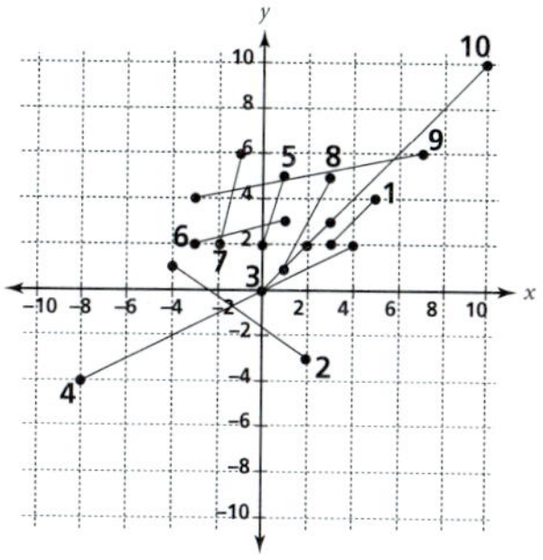

1. range = $2 \le y \le 4$; domain = $3 \le x \le 5$; $y = x - 1$ **2.** range = $-3 \le y \le 1$; domain = $-4 \le x \le 2$; $y = -\frac{2}{3}x - \frac{5}{3}$ **3.** range = $0 \le y \le 3$; domain = $0 \le x \le 3$; $y = x$ **4.** range = $-4 \le y \le 2$; domain = $-8 \le x \le 4$; $y = \frac{1}{2}x$ **5.** range = $2 \le y \le 5$; domain = $0 \le x \le 1$; $y = 3x + 2$ **6.** range = $2 \le y \le 3$; domain = $-3 \le x \le 1$; $y = \frac{1}{4}x + 2\frac{3}{4}$ **7.** range = $2 \le y \le 6$; domain = $-2 \le x \le -1$; $y = 4x + 10$ **8.** range = $1 \le y \le 5$; domain = $1 \le x \le 3$; $y = 2x - 1$ **9.** range = $4 \le y \le 6$; domain = $-3 \le x \le 7$; $y = \frac{1}{5}x + 4\frac{3}{5}$ **10.** range = $2 \le y \le 10$; domain = $2 \le x \le 10$; $y = x$

Additional Practice 23—The Point-Slope Formula

1. graph with line through (1, 4) and (2, 6) **2.** graph with line through (−2, 3) and (−1, 7) **3.** graph with line through (−1, −3) and (0, −4) **4.** graph with line through (3, −2) and (6, −1) **5.** graph with line through (4, 5) and (5, 5) **6.** graph with line through (−2, 0) and (−1, −3) **7.** graph with line through (−3, 4) and (−1, 5) **8.** graph with line through (2, 2) and (3, 2) **9.** graph with line through (0, 0) and (4, −1) **10.** graph with line through (−4, −3) and (−3, 0) **11.** $y = 2x + 2$ **12.** $y = 4x + 11$ **13.** $y = x - 4$ **14.** $y = \frac{1}{3}x - 3$ **15.** $y = 5$, x is any real number **16.** $y = -3x - 6$ **17.** $y = \frac{1}{2}x + 5\frac{1}{2}$ **18.** $y = 2$, x is any real number **19.** $y = -\frac{1}{4}x$ **20.** $y = 3x + 9$

Additional Practice 24—Finding the Midpoint of a Segment

1. (0, 0) **2.** $(1, 4\frac{1}{2})$ **3.** (1, 6) **4.** $(1, -3\frac{1}{2})$ **5.** (4, −2) **6.** (−3, 0) **7.** (3, 4) **8.** (0, 3) **9.** (−3, 1) **10.** (3, 3) **11.** midpoint at 2 inches, should include arcs necessary to find midpoint **12.** midpoint at 1.5 inches, should include arcs necessary to find midpoint **13.** midpoint at 3.5 inches, should include arcs necessary to find midpoint **14.** midpoint at 3 centimeters, should include arcs necessary to find midpoint **15.** midpoint at 4 centimeters, should include arcs necessary to find midpoint

Additional Practice 25—Naming Triangles Using Their Sides

1. equilateral **2.** scalene **3.** isosceles **4.** scalene **5.** isosceles **6.** equilateral **7.** scalene **8.** isosceles **9.** equilateral **10.** scalene

11.

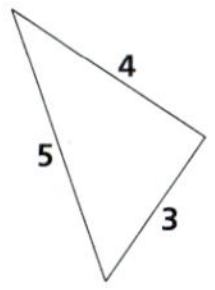

12.

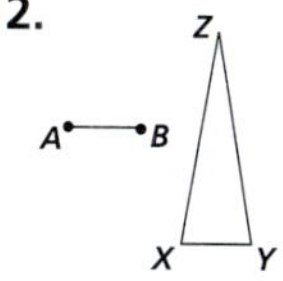

13.

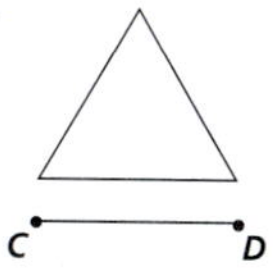

14.

15.

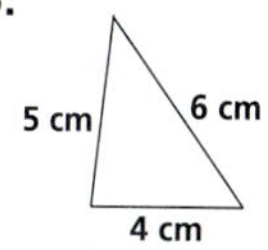

Additional Practice 26—Naming Triangles Using Their Angles

1. acute equilateral triangle **2.** right isosceles triangle **3.** right scalene triangle **4.** acute scalene triangle **5.** obtuse triangle **6.** obtuse isosceles triangle **7.** right triangle **8.** acute isosceles triangle **9.** acute equilateral triangle **10.** right scalene triangle

11.
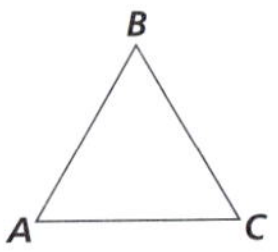

12.
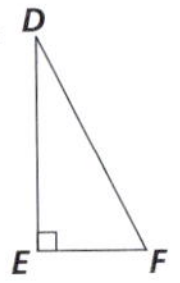

13.
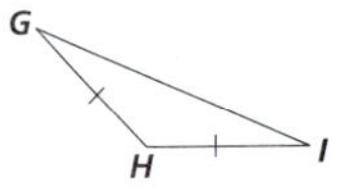

14.
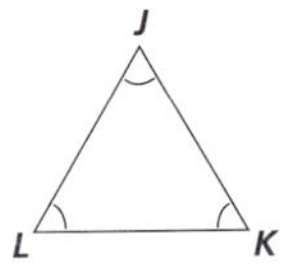

15.
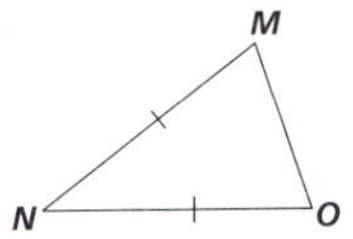

Additional Practice 27—Special Quadrilaterals

1. True **2.** False, no trapezoids are rhombuses **3.** False, rhombus **4.** True **5.** False, only squares are rectangles and rhombuses **6.** False, rectangles **7.** isosceles trapezoid **8.** quadrilateral **9.** square, rhombus **10.** irregular quadrilateral, scalene trapezoid

Additional Practice 28—Quadrilaterals and Their Diagonals

1. 45° **2.** 45° **3.** 45° **4.** 90° **5.** 90° **6.** 80° **7.** 10° **8.** 80° **9.** 90° **10.** 90° **11.** 40° **12.** 50° **13.** 40° **14.** 90° **15.** 90°

Additional Practice 29—Angle Sum of Any Triangle

1. 180° **2.** 180° **3.** 360° **4.** 100° **5.** 80° **6.** 180° **7.** 180° **8.** 360° **9.** 120° **10.** 90°

Additional Practice 30—Concave and Convex Polygons

1. 9 **2.** 54 **3.** 10 **4.** 1,800° **5.** 1 **6.** 2 **7.** 2 **8.** 360° **9.** 5 **10.** 20 **11.** 6 **12.** 1,080° **13.** 7 **14.** 35 **15.** 8 **16.** 1,440° **17.** No **18.** No **19.** Yes, it would create a regular hexagon **20.** Yes, it is a regular octagon

Additional Practice 31—Constructing Perpendiculars

1. Student constructions should reflect correct use of construction steps for constructing a line perpendicular to a given line either from a point not on the line or through a point on the line. **2.** Lines m and o are parallel because the interior angles formed by the transversal n are all right angles. **3.–5.** Student constructions should reflect correct use of construction steps for constructing a line perpendicular to a given line either from a point not on the line or through a point on the line.

Additional Practice 32—Altitudes, Angle Bisectors, and Medians

1.–3. Student constructions should reflect correct use of construction steps. **4.** The altitudes, medians, and angle bisectors are the same in an equilateral triangle. **5.** Student constructions should reflect correct use of construction steps; no, they are not the same.

Additional Practice 33—Proof of the Angle Sum Theorem

1. 35° **2.** 145° **3.** 180° **4.** 80° **5.** 35° **6.** 65° **7.** True **8.** False **9.** True **10.** False

Additional Practice 34—Corresponding Parts: SAS Postulate

1. $\overline{AB} \cong \overline{DE}, \overline{BC} \cong \overline{EF}, \angle B \cong \angle E$ **2.** $\overline{BC} \cong \overline{EF}, \overline{CA} \cong \overline{FD}, \angle C \cong \angle F$ **3.** $\overline{CA} \cong \overline{FD}, \overline{AB} \cong \overline{DE}, \angle A \cong \angle D$ **4.** $\overline{AC} \cong \overline{DF}, \overline{BC} \cong \overline{EF}, \angle C \cong \angle F$ **5.** $\overline{BC} \cong \overline{EF}, \overline{AC} \cong \overline{DF}, \angle C \cong \angle F$ **6.** $\overline{QR}$ and $\overline{SR}$ or $\overline{TR}$; $\overline{PR}$ and $\overline{TR}$ or $\overline{SR}$ **7.** $\overline{AC}$ and $\overline{DC}$ or $\overline{EC}$; $\overline{BC}$ and $\overline{EC}$ or $\overline{DC}$ **8.** $\overline{AB}$ and $\overline{AD}$ **9.** $\overline{YZ}$ and $\overline{XZ}$ **10.** $\overline{AB}$ and $\overline{XY}$; $\overline{BC}$ and $\overline{YZ}$

Additional Practice 35—SSS Construction and Triangle Inequality

1.–6. Constructions should reflect given line lengths. **7.** Yes, $a + b > c, a + c > b, b + c > a$ **8.** No, $a + b < c$ **9.** No, $a + b = c$ **10.** Yes, $a + b > c, a + c > b, b + c > a$

Additional Practice 36—SSS and ASA Congruencies

1. Yes, SSS **2.** Yes, ASA **3.** No **4.** No **5.** Yes, SAS **6.** Yes, SSS **7.** Yes, SAS **8.** No **9.** Yes, SSS **10.** Yes, SSS

Additional Practice 37—Special Congruencies: Right Triangles

1. SAS **2.** H-L **3.** AAS **4.** AAS **5.** AAS **6.** Yes, SSS **7.** Yes, SAS **8.** SSS, SAS, AAS, H-L **9.** SSS, SAS **10.** Yes, SSS

Additional Practice 38—Reflections in the Coordinate Plane

1. $A' = (4, -1)$ **2.** $B' = (1, 6)$ **3.** $C' = (-5, 0)$ **4.** $D' = (7, 3)$ **5.** $E' = (-2, -1)$ **6.** $A' = (-4, 5)$ **7.** $A' = (-3, -2)$ **8.** $H' = (6, 0)$ **9.** $I' = (-9, 4)$ **10.** $J' = (3, -3)$ **11.** $K' = (-4, 1)$ **12.** $L' = (-5, 3)$ **13.** $M' = (2, 0)$ **14.** $N' = (7, 4)$ **15.** $O' = (1, 1)$ **16.** $P' = (3, 3)$ **17.** $Q' = (-4, -5)$ **18.** $R' = (0, -5)$ **19.** $S' = (-3, -2)$ **20.** $T' = (7, 0)$

Additional Practice 39—Special Reflections: Symmetries

1. Yes **2.** Yes **3.** Yes **4.** No **5.** Yes **6.** Yes **7.** No **8.** No **9.** No **10.** Yes **11.** No **12.** Yes **13.** Yes **14.** No **15.** Yes

Additional Practice 40—Slides and Translations

1. $A' = (5, -3)$ **2.** $B' = (3, 2)$ **3.** $C' = (1, -1)$ **4.** $D' = (11, -9)$ **5.** $E' = (8, 0)$ **6.** $J' = (-2, 1)$ **7.** $K' = (1, 4)$ **8.** $L' = (5, 2)$ **9.** $M' = (-5, 3)$ **10.** $N' = (4, -3)$ **11.** $A' = (7, 7), B' = (0, 9)$ **12.** $C' = (-4, -5), D' = (0, -9)$ **13.** $E' = (7, -7), F' = (12, -2)$ **14.** $G' = (-5, 8), H' = (1, 4)$ **15.** $I' = (1, -3), J' = (8, 4)$

Additional Practice 41—Rotations

1. new coordinate $P' = (-3, 3)$ **2.** new coordinate $P' = (-2, 0)$ **3.** new coordinate $P' = (-1, 5)$ **4.** new coordinate $P' = (5, 8)$ **5.** new coordinate $P' = (-4, 5)$

Additional Practice 42—Similar Figures and Proportions

1. 12 **2.** 10 **3.** 5 **4.** 4 **5.** 13 **6.** 13 **7.** 7.5 **8.** 12 **9.** 10 **10.** 8 **11.** 4 **12.** $1\frac{1}{3}$ **13.** $15\frac{5}{8}$ **14.** $40\frac{5}{8}$ **15.** $3\frac{2}{3}$

Additional Practice 43—Similar Triangles

1. No, noncongruent angles **2.** Yes, all angles congruent **3.** Yes, all angles congruent **4.** Yes, all angles congruent **5.** No, sides not in proportion **6.** No, all angles not congruent **7.** Yes, sides all in proportion **8.** Yes, all angles congruent **9.** Yes, all angles congruent **10.** No, sides not in proportion

Additional Practice 44—Angle Measure in Regular Polygons

1. 135° **2.** 147.27° **3.** 108° **4.** 128.6° **5.** 90° **6.** 140° **7.** 152.3° **8.** 120° **9.** 60° **10.** 144°

Additional Practice 45—Enlarging Geometric Figures

1. $A' = (6, 2)$ $B' = (8, 10)$ $C' = (-6, 2)$ **2.** $A' = (9, 3)$ $B' = (12, 15)$ $C' = (-9, 3)$ **3.** $A' = (12, 4)$ $B' = (16, 20)$ $C' = (-12, 4)$ **4.** $A' = (15, 5)$ $B' = (20, 25)$ $C' = (-15, 5)$ **5.** $A' = (18, 6)$ $B' = (24, 30)$ $C' = (-18, 6)$ **6.** $A' = (-2, 2)$ $B' = (2, 2)$ $C' = (-2, -2)$ **7.** $A' = (-3, 3)$ $B' = (3, 3)$ $C' = (-3, -3)$ **8.** $A' = (-4, 4)$ $B' = (4, 4)$ $C' = (-4, -4)$ **9.** $A' = (-5, 5)$ $B' = (5, 5)$ $C' = (-5, -5)$ **10.** $A' = (-6, 6)$ $B' = (6, 6)$ $C' =$

(–6, –6) **11.** $A' = (-6, -12)$ $B' = (8, -6)$ $C' = (0, 0)$ **12.** $A' = (-9, -18)$ $B' = (12, -9)$ $C' = (0, 0)$ **13.** $A' = (-12, -24)$ $B' = (16, -12)$ $C' = (0, 0)$ **14.** $A' = (-15, -30)$ $B' = (20, -15)$ $C' = (0, 0)$ **15.** $A' = (-18, -36)$ $B' = (24, -18)$ $C' = (0, 0)$ **16.** $A' = (-4, 6)$ $B' = (6, 8)$ $C' = (2, 0)$ **17.** $A' = (-6, 9)$ $B' = (9, 12)$ $C' = (3, 0)$ **18.** $A' = (-8, 12)$ $B' = (12, 16)$ $C' = (4, 0)$ **19.** $A' = (-10, 15)$ $B' = (15, 20)$ $C' = (5, 0)$ **20.** $A' = (-12, 18)$ $B' = (18, 24)$ $C' = (6, 0)$

Additional Practice 46—Shrinking Geometric Figures

1. $A' = (4\frac{1}{2}, 4)$ $B' = (-3\frac{1}{2}, 2\frac{1}{2})$ $C' = (0, -3)$ **2.** $A' = (3, 2\frac{2}{3})$ $B' = (-2\frac{1}{3}, 1\frac{2}{3})$ $C' = (0, -2)$ **3.** $A' = (2\frac{1}{4}, 2)$ $B' = (-1\frac{3}{4}, 1\frac{1}{4})$ $C' = (0, -1\frac{1}{2})$ **4.** $A' = (1\frac{4}{5}, 1\frac{3}{5})$ $B' = (-1\frac{2}{5}, 1)$ $C' = (0, -1\frac{1}{5})$ **5.** $A' = (1\frac{1}{2}, 1\frac{1}{3})$ $B' = (-1\frac{1}{6}, \frac{5}{6})$ $C' = (0, -1)$ **6.** $A' = (-5, 5)$ $B' = (5, 5)$ $C' = (-5, -5)$ **7.** $A' = (-4, 4)$ $B' = (4, 4)$ $C' = (-4, -4)$ **8.** $A' = (-3, 3)$ $B' = (3, 3)$ $C' = (-3, -3)$ **9.** $A' = (-2, 2)$ $B' = (2, 2)$ $C' = (-2, -2)$ **10.** $A' = (-1, 1)$ $B' = (1, 1)$ $C' = (-1, -1)$ **11.** $A' = (4\frac{1}{2}, -2\frac{1}{2})$ $B' = (4, 2\frac{1}{2})$ $C' = (0, 0)$ **12.** $A' = (6, -3\frac{1}{3})$ $B' = (5\frac{1}{3}, 3\frac{1}{3})$ $C' = (0, 0)$ **13.** $A' = (6\frac{3}{4}, -3\frac{3}{4})$ $B' = (6, 3\frac{3}{4})$ $C' = (0, 0)$ **14.** $A' = (7\frac{1}{5}, -4)$ $B' = (6\frac{2}{5}, 4)$ $C' = (0, 0)$ **15.** $A' = (7\frac{1}{2}, -4\frac{1}{6})$ $B' = (6\frac{2}{3}, 4\frac{1}{6})$ $C' = (0, 0)$

Additional Practice 47—Pythagorean Triples

1. 24, 32, 40 **2.** 27, 36, 45 **3.** 30, 40, 50 **4.** 33, 44, 55 **5.** 36, 48, 60 **6.** 39, 52, 65 **7.** 42, 56, 70 **8.** 45, 60, 75 **9.** 48, 64, 80 **10.** 75, 100, 125 **11.** 40, 96, 104 **12.** 45, 108, 117 **13.** 50, 120, 130 **14.** 55, 132, 143 **15.** 60, 144, 156 **16.** 65, 156, 169 **17.** 70, 168, 182 **18.** 75, 180, 195 **19.** 80, 192, 208 **20.** 125, 300, 325

Additional Practice 48—Pythagorean Triples and a Proof

1. 100 **2.** 24 **3.** 4 **4.** 100 **5.** Yes **6.** 400 **7.** 96 **8.** 16 **9.** 400 **10.** Yes **11.** 1,369 **12.** 210 **13.** 529 **14.** 1,369 **15.** Yes

Additional Practice 49—Pythagorean Demonstration

1. 60 **2.** 17 **3.** 289 **4.** 529 **5.** 529 **6.** 64 **7.** 225 **8.** 120 **9.** 529 **10.** they are equal **11.** 24 **12.** 10 **13.** 100 **14.** 196 **15.** 196

Additional Practice 50—Pythagorean Theorem and Similar Triangles

1. 46.2 **2.** 110.8 **3.** 19.2 **4.** ΔABC, ΔABD, ΔADC **5.** $\angle A = \angle D$; $\angle B = \angle B$; $\angle C = \angle A$ **6.** $\angle D = \angle D$, $\angle A = \angle B$; $\angle C = \angle A$ **7.** $\angle A = \angle D$; $\angle B = \angle A$; $\angle C = \angle C$ **8.** $\overline{CB}$ **9.** $\overline{AB}$ **10.** $\overline{CA}$ **11.** Yes, AAS **12.** 13:12 **13.** 13:5 **14.** 12:5 **15.** 300 **16.** ~116 **17.** ~277 **18.** 116:277:300 **19.** 5:12:13 **20.** Yes

Additional Practice 51—Pythagorean Proof and Trapezoids

1. 56 **2.** 65 **3.** 84 **4.** 35 **5.** 54 **6.** 1,161 **7.** 100 **8.** 6,321 **9.** 128 **10.** 2,112 **11.** 1,000 **12.** 968 **13.** 10 **14.** 4,984 **15.** 5,056

Additional Practice 52—Distance Formula: Pythagorean Theorem

1. $\sqrt{29}$ **2.** $\sqrt{53}$ **3.** $\sqrt{61}$ **4.** 5 **5.** $\sqrt{90}$ **6.** $\sqrt{170}$ **7.** $\sqrt{104}$ **8.** $\sqrt{68}$ **9.** $\sqrt{98}$ **10.** $\sqrt{26}$ **11.** $\sqrt{10}$ **12.** $\sqrt{29}$ **13.** $\sqrt{116}$ **14.** $\sqrt{45}$ **15.** $\sqrt{65}$ **16.** $\sqrt{130}$ **17.** 5 **18.** $\sqrt{65}$ **19.** $\sqrt{80}$ **20.** $\sqrt{65}$

Additional Practice 53—Converse of the Pythagorean Theorem

1. Yes **2.** No **3.** No **4.** Yes **5.** No **6.** No **7.** Yes **8.** Yes **9.** No **10.** Yes **11.** Yes **12.** Yes **13.** No **14.** Yes **15.** No **16.** Yes **17.** Yes **18.** No **19.** No **20.** Yes

Additional Practice 54—Perimeters of Polygons

1. 44 **2.** 28 **3.** 59 **4.** 45 **5.** 59 **6.** 100 **7.** 68 **8.** 67 **9.** 115 **10.** 186

Additional Practice 55—Perimeter Formulas

1. 108 **2.** 108 **3.** 39 **4.** 216 **5.** 345 **6.** 224 **7.** 112 **8.** 890 **9.** 308 **10.** 516

Additional Practice 56—Perimeters and Diagonals

1. $x = 13$, $P = 74$ **2.** $x = 4.3$, $P = 35.5$ **3.** $x = 26$, $P = 120$ **4.** $x = 31$, $P = 164$ **5.** $x = 52$, $P = 136$ **6.** $x = 31.2$, $P = 134.5$ **7.** $x = 73.4$, $P = 234.4$ **8.** $x = 111.9$, $P = 304$ **9.** $x = 26$, $P = 68$ **10.** $x = 31$, $P = 250$

Additional Practice 57—Area: Squares, Rectangles, Parallelograms

1. 64 **2.** 120 **3.** 308 **4.** 169 **5.** 2,790 **6.** 88 **7.** 2,304 **8.** 1,943 **9.** 3,367 **10.** 7,744

Additional Practice 58—Areas of Trapezoids

1. 400 **2.** 1,050 **3.** 1,974 **4.** 2,475 **5.** 4,340 **6.** 11,088 **7.** 2,730 **8.** 22,320 **9.** 14,539 **10.** 406,200

Additional Practice 59—Areas of Triangles

1. 264.5 **2.** 1,012.5 **3.** 684.5 **4.** 1,922 **5.** 3,528 **6.** 387.5 **7.** 232 **8.** 1,775.5 **9.** 1,817 **10.** 726

Additional Practice 60—Definition of a Circle

1. diameter **2.** center **3.** radius **4.** radius **5.** chord **6.** $\overline{UV}$ or $\overline{QR}$ **7.** $\overline{QR}$ **8.** $\overline{PQ}$, $\overline{PR}$, $\overline{PS}$, $\overline{PT}$ **9.** W **10.** P **11.** radius **12.** radius **13.** diameter or chord **14.** center **15.** chord **16.** D **17.** K **18.** $\overline{EF}$ **19.** $\overline{DE}$, $\overline{DF}$, $\overline{DG}$, $\overline{DH}$ **20.** $\overline{IJ}$ or $\overline{EF}$

Additional Practice 61—The Ratio π

1. 23π units **2.** 46π units **3.** 90π units **4.** 31π units **5.** 17π units **6.** 32π units **7.** 82π units **8.** 84π units **9.** 246π units **10.** 55π units **11.** 63π units **12.** $1{,}846\pi$ units **13.** 71π units **14.** $1{,}268\pi$ units **15.** 555π units **16.** $1{,}110\pi$ units **17.** $1{,}752\pi$ units **18.** 357π units **19.** $2{,}048\pi$ units **20.** $8{,}870\pi$ units **21.** $2{,}087\pi$ units **22.** $5{,}783\pi$ units **23.** $9{,}876\pi$ units **24.** $19{,}752\pi$ units **25.** $396{,}914\pi$ units

Additional Practice 62—Approximating the Area of a Circle

1. large square = 1,936; small square = 968; circle estimate = 1,452 **2.** large square = 1,024; small square = 512; circle estimate = 768 **3.** large square = 4,624; small square = 2,312; circle estimate = 3,468 **4.** large square = 1,444; small square = 722; circle estimate = 1,083 **5.** large square = 16,384; small square = 8,192; circle estimate = 12,288 **6.** large square = 7,744; small square = 3,872; circle estimate = 5,808 **7.** large square = 10,000; small square = 5,000; circle estimate = 7,500 **8.** large square = 30,976; small square = 15,488; circle estimate = 23,232 **9.** large square = 39,204; small square = 19,602; circle estimate = 29,403 **10.** large square = 60,516; small square = 30,258; circle estimate = 45,387

Additional Practice 63—Area and Probability

1. $33\frac{1}{3}\%$ **2.** $66\frac{2}{3}\%$ **3.** 50% **4.** 25% **5.** 75% **6.** $87\frac{1}{2}\%$ **7.** $37\frac{1}{2}\%$ **8.** $33\frac{1}{3}\%$ **9.** 50% **10.** $16\frac{2}{3}\%$

Additional Practice 64—Formula for the Area of a Circle

1. 16π sq in. **2.** 36π sq cm **3.** 64π sq mm **4.** 121π sq m **5.** 625π sq in. **6.** 25π sq in. **7.** 289π sq cm **8.** 484π sq mm **9.** $1{,}089\pi$ sq m **10.** $2{,}500\pi$ sq in. **11.** 153.9 sq in. **12.** 254.5 cm^2 **13.** 530.9 mm^2 **14.** 5,280.9 m^2 **15.** 1,661.9 sq in. **16.** 28.3 sq in. **17.** 2,123.7 cm^2 **18.** 8,494.6 mm^2 **19.** 615.7 m^2 **20.** 30,789.8 sq in.

Additional Practice 65—Circles and Their Angles and Sectors

1. $\frac{64}{6}\pi$ **2.** $\frac{9}{8}\pi$ **3.** 36π **4.** $\frac{100}{3}\pi$ **5.** 384π **6.** 200π **7.** 750π **8.** $\frac{256}{9}\pi$ **9.** $\frac{25}{4.5}\pi$ **10.** $\frac{1600}{12}\pi$

Additional Practice 66—Tangents, Circumcircles, and Incircles

1. 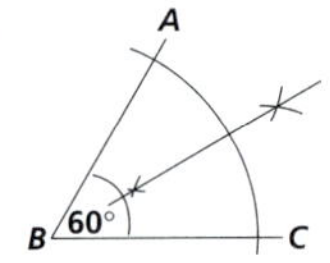

2. 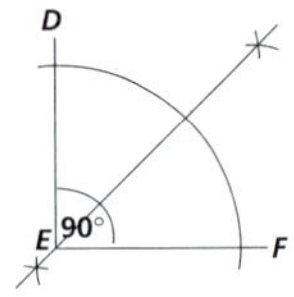

3. 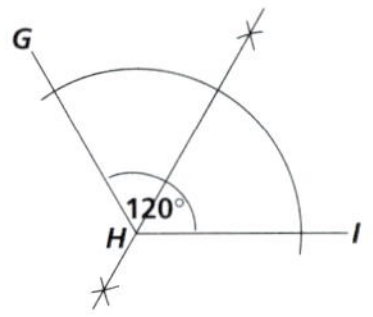

4. 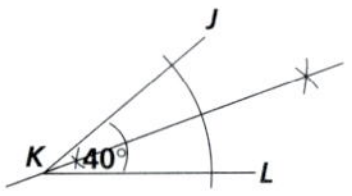

5. 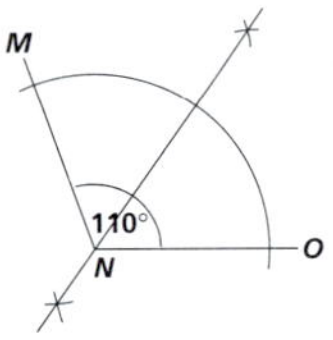

Additional Practice 67—Sine, Cosine, and Tangent

1. 0.26, 0.97, 0.27 **2.** 0.34, 0.94, 0.36 **3.** 0.42, 0.91, 0.47 **4.** 0.57, 0.82, 0.70 **5.** 0.64, 0.77, 0.84 **6.** 0.77, 0.64, 1.19 **7.** 0.82, 0.57, 1.43 **8.** 0.91, 0.42, 2.14 **9.** 0.94, 0.34, 2.75 **10.** 0.97, 0.26, 3.73

Additional Practice 68—Spheres

1. $S = 64\pi$ sq units; $V = 85.33\pi$ units^3 **2.** $S = 100\pi$ sq units; $V = 166.67\pi$ units^3 **3.** $S = 144\pi$ sq units; $V = 288\pi$ units^3 **4.** $S = 256\pi$ sq units; $V = 682.67\pi$ units^3 **5.** $S = 324\pi$ sq units; $V = 972\pi$ units^3 **6.** $S = 16\pi$ sq units; $V = 10.67\pi$ units^3 **7.** $S = 4\pi$ sq units; $V = 1.33\pi$ units^3 **8.** $S = 900\pi$ sq units; $V = 4{,}500\pi$ units^3 **9.** $S = 676\pi$ sq units; $V = 2{,}929.33\pi$ units^3 **10.** $S = 784\pi$ sq units; $V = 3{,}658.67\pi$ units^3 **11.** $S = 1{,}024\pi$ sq units; $V = 5{,}461.33\pi$ units^3 **12.** $S = 1{,}156\pi$ sq units; $V = 6{,}550.67\pi$ units^3 **13.** $S = 1{,}444\pi$ sq units; $V = 9{,}145.33\pi$ units^3 **14.** $S = 1{,}296\pi$ sq units; $V = 7{,}776\pi$ units^3 **15.** $S = 1{,}600\pi$ sq units; $V = 10{,}666.67\pi$ units^3 **16.** $S = 1{,}936\pi$ sq units; $V = 14{,}197.33\pi$ units^3 **17.** $S = 2{,}500\pi$ sq units; $V = 20{,}833.33\pi$ units^3 **18.** $S = 3{,}600\pi$ sq units; $V = 36{,}000\pi$ units^3 **19.** $S = 10{,}000\pi$ sq units; $V = 166{,}666.67\pi$ units^3 **20.** $S = 40{,}000\pi$ sq units; $V = 1{,}333{,}333.33\pi$ units^3

Additional Practice 69—Basic Volume Formulas

1. 1,848 m^3 **2.** 2,700 cm^3 **3.** 900 in.^{3} **4.** 5,026.4 ft^3 **5.** 360 mm^3 **6.** 5,434 m^3 **7.** 252 cm^3 **8.** 1,260 in.^{3} **9.** 5,425.92 ft^3 **10.** 2,400 mm^3

Additional Practice 70—Volumes of Pyramids and Cones

1. 750 m^3 **2.** 424.10 cm^3 **3.** 288 ft^3 **4.** 29.32 in.^{3} **5.** 297 cm^3 **6.** 769.67 ft^3 **7.** 606.67 in.^{3} **8.** 226.19 mm^3 **9.** 5,973.33 ft^3 **10.** 1,507.92 m^3

Additional Practice 71—Surface Areas of Prisms and Cylinders

1. 726 cm^2 **2.** 552 ft^2 **3.** 408 **4.** 471 m^2 **5.** 420 **6.** 2,736 cm^2 **7.** 1,860 sq in. **8.** 84 mm^2 **9.** 1,257 m^2 **10.** 1,120

Additional Practice 72—Surface Areas of Pyramids and Cones

1. 360 in.^{2} **2.** 84.82 cm^2 **3.** 56 ft^2 **4.** 204.20 m^2 **5.** 300 yd^2 **6.** 402.34 mm^2 **7.** 260 ft^2 **8.** 508.92 cm^2 **9.** 1,728 mm^2 **10.** 1,225.19 in.^{2}

Additional Practice 73—Measurements

1. 36 **2.** 15 **3.** 4 **4.** 5,280 **5.** 6 **6.** 12 **7.** 6 **8.** 8 **9.** 56 **10.** 30 **11.** 23,000 **12.** 34,000 **13.** 170 **14.** 0.99 **15.** 4,500,000 **16.** 168 **17.** 192 **18.** 8 **19.** 14,080 **20.** 33 **21.** 60 **22.** 121 **23.** 24 **24.** 328 **25.** 222 **26.** 49,000 **27.** 77,000 **28.** 1,520 **29.** 3.75 **30.** 56,400,000

Additional Practice 74—Lines and Planes in Space

1. The line passes through diagonal corners of the cube. **2.** The line passes through the point of the pyramid and the center of the square base. **3.** The plane intersects the cube at the upper edge of one face and the lower edge of the opposite face. **4.** The plane crosses the sphere to form any circle that is not a great circle. **5.** The line passes through the center of each base of the cylinder. The line is also perpendicular to each base.

Additional Practice 75—Loci in the Coordinate Plane

1. 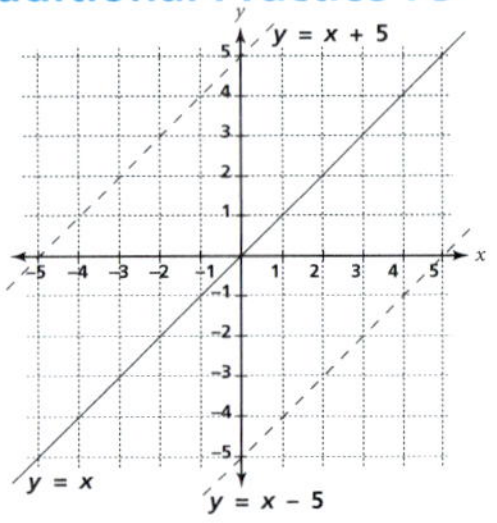

2. 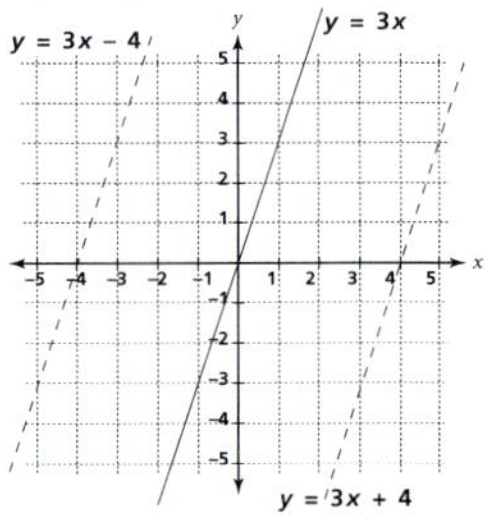

3. 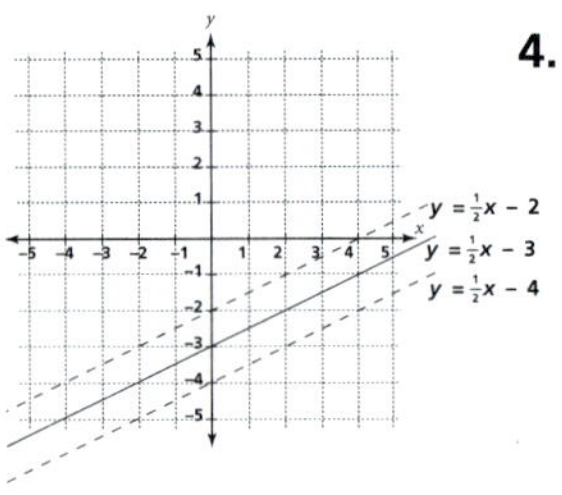

4. 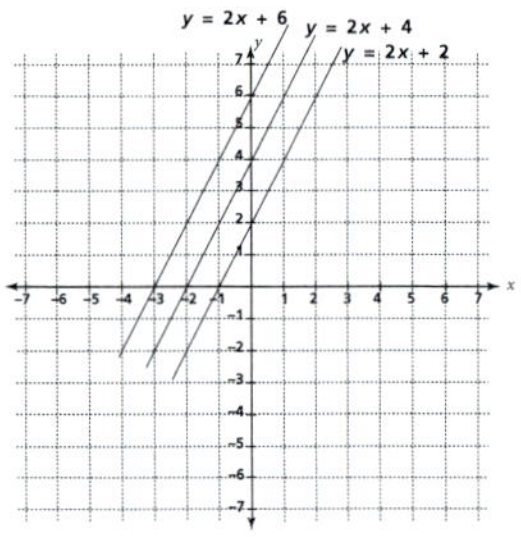

5. 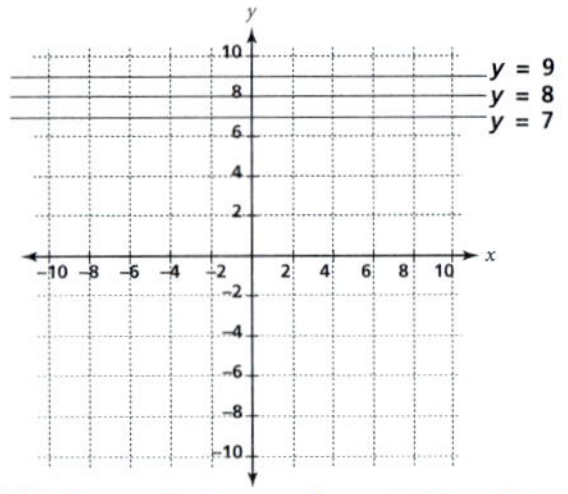

Additional Practice 76—Compound Loci

1. 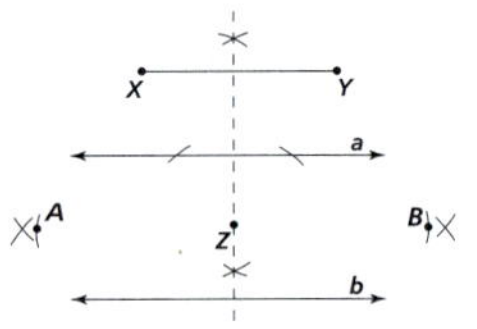

2. 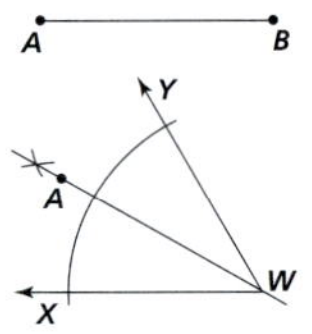

3. 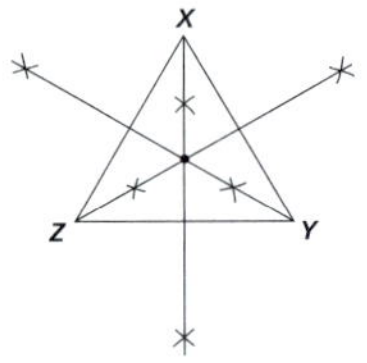

4. 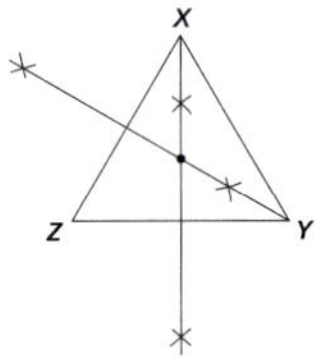

5. 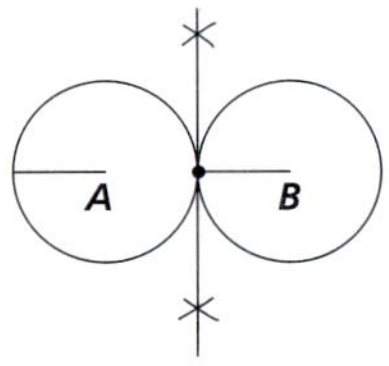

Workbook Activities

Workbook Activity 1—Points and Lines in the Plane

1. two **2.** three **3.** one **4.** four **5.** four **6.** three **7.** four **8.** five **9.** ray **10.** line **11.** line segment **12.** ray **13.** line **14.** point **15.** line segment

Workbook Activity 2—Measuring Line Segments

1. 5 in. **2.** 3 in. **3.** 6 in. **4.** 2 in. **5.** 1 in. **6.** 13 cm **7.** 16 cm **8.** 6 cm **9.** 10 cm **10.** 11 cm

Workbook Activity 3—Ruler Postulates

1. 3 units **2.** 7 units **3.** 9 units **4.** 7 units **5.** $\overline{AB} = 3$, $\overline{BC} = 5$, $\overline{AC} = 8$; $\overline{AB} + \overline{BC} = 3 + 5 = 8$; B is between A and C.

Workbook Activity 4—Copying and Bisecting Angles

1.–5. Students should copy the angles shown.

6.

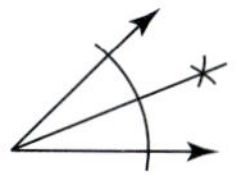

7.

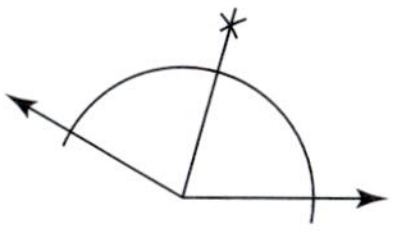

8.

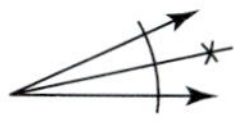

9.

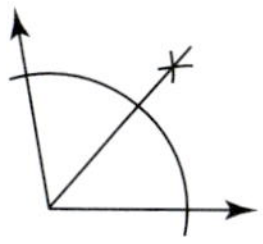

10.

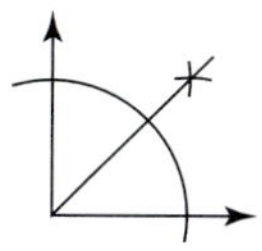

Workbook Activity 5—Angle Measurement

1. 15° **2.** 30° **3.** 110° **4.** 90° **5.** 140° **6.** 170° **7.** obtuse **8.** straight **9.** right **10.** acute

Workbook Activity 6—Complementary and Supplementary Angles

1. Angle should be 50°. **2.** Angle should be 20°. **3.** Angle should be 60°. **4.** Angle should be 40°. **5.** Angle should be 15°. **6.** Angle should be 105°. **7.** Angle should be 140°. **8.** Angle should be 55°. **9.** Angle should be 40°. **10.** Angle should be 135°.

Workbook Activity 7—Algebra and Angles

1. $m\angle r = 76°$, $m\angle s = 104°$, $m\angle t = 76°$ **2.** $m\angle l = 42°$, $m\angle m = 138°$, $m\angle n = 42°$ **3.** $m\angle k = 70°$, $m\angle m = 110°$, $m\angle l = 70°$ **4.** $m\angle a = 80°$, $m\angle b = 100°$, $m\angle c = 80°$ **5.** $m\angle f = 145°$, $m\angle g = 35°$, $m\angle h = 145°$ **6.** 46° **7.** 90° **8.** 44° **9.** 46° **10.** 90°

Workbook Activity 8—Drawing Networks

1. traversable **2.** not traversable; crosses a section twice **3.** traversable **4.** traversable **5.** transportation engineer, electric or telephone company

Workbook Activity 9—Conditionals

1. True; If an angle is less than 90°, then the angle is an acute angle. **2.** True; If two circles have the same radius, then they are equal to one another. **3.** False; If a figure has five sides, then it must contain at least one right angle. **4.** True; If a closed figure has four angles, then the sum of their measures must equal 360°. **5.** True; If the measure of one angle created by the intersection of two lines is 90°, then all four of the angles created by the intersection of the two lines will measure 90°. **6.** False; If your computer doesn't work when you turn it on, then there must be something wrong with the processor. **7.** True; If it is snowing outside, then the temperature must be less than 32° Fahrenheit. **8.** False; If it is raining outside, then you will get wet. **9.** If two angles add to 180°, then they are supplementary. **10.** If two circles have equal radii, then they are equal.

Workbook Activity 10—Converses

1. False; If an angle is right, then it is less than 90°. False **2.** False; If a circle and square are equal, then the circle has the same diameter as the length of the side of a square. False **3.** True; If the sum of two pairs of angles equals 360°, then they are supplementary. True. **4.** False; If two circles are equal, then they have the same center but different radii. False **5.** True; If the figure is a rectangle, then all four angles are 90°. True **6.** True; If the measures of two right angles are added together, then the sum is equal to the measures of two pairs of complementary angles. True **7.–10.** Answers will vary. Check students' responses. Possible answers: **7.** If an angle equals 90°, then it is right. **8.** If it is below 32°F, then it is snowing. **9.** If you go out in the rain without an umbrella, you get wet. **10.** If it is red, then it is blue.

Workbook Activity 11—Lines and Euclid's Postulates

1. Postulate 1 **2.** Postulate 2 **3.** Postulate 2 **4.** True, Postulate 1 **5.** False, Postulate 2

Workbook Activity 12—Circles and Right Angles

1.

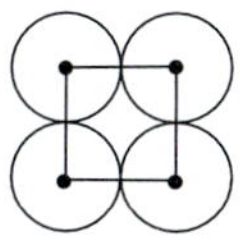

2. Student drawings should show a square with four right angles. **3.** circle; Postulate 1 **4.** circle; Postulate 3 **5.** square, right angle; Postulate 4

Workbook Activity 13—Using Euclid's Postulates

1. Postulates 2 and 3 **2.** Postulate 1 **3.** Postulates 2 and 4 **4.** Postulates 1 and 5 **5.** Postulates 2 and 3

Workbook Activity 14—Making Constructions

1. Postulate 1 **2.** Postulates 1, 3, and 4 **3.** Postulates 1 and 3 **4.** Postulate 2 **5.** Postulate 3 **6.** Postulate 4

7.

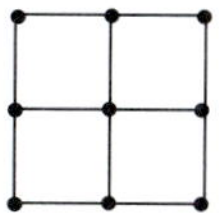

8.

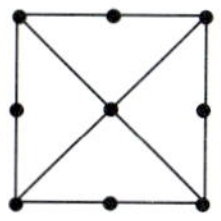

9.

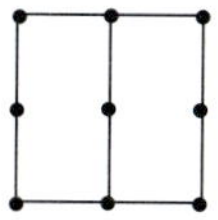

10.

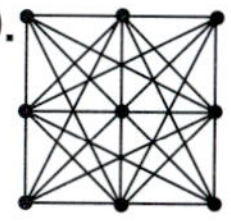

Workbook Activity 15—Axioms and Equals

1. Axiom 2 **2.** Axiom 3 **3.** Axiom 3 **4.** Axiom 2 **5.** Axiom 3 **6.** Axiom 3 **7.** $m\angle x = 55°$; Axiom 3 **8.** $m\angle x = 125°$; Axiom 2 **9.** $x = 180°$; Axiom 2 **10.** $m\angle x = 90°$; Axiom 3

Workbook Activity 16—Axioms and Figures

1. circles are equal; Axiom 4 **2.** each half is smaller than the whole; Axiom 5 **3.** triangles are equal; Axiom 4 **4.** all angles are less than 180°; Axiom 5 **5.** circles are equal; Axiom 4

Workbook Activity 17—Theorems

1. $m\angle x = 105°$, adjacent and supplementary angles; $m\angle y = 75°$, vertical angles equal; $m\angle z = 105°$, vertical angles equal **2.** $m\angle x = 75°$, adjacent and supplementary angles; $m\angle y = 105°$, vertical angles equal; $m\angle z = 75°$, vertical angles equal **3.** $m\angle x = 155°$, adjacent and supplementary angles; $m\angle y = 25°$, vertical angles equal; $m\angle z = 155°$, vertical angles equal **4.** $m\angle x = 140°$, adjacent and supplementary angles; $m\angle y = 40°$, vertical angles equal; $m\angle z = 140°$, vertical angles equal **5.** $m\angle x = 50°$, adjacent and supplementary angles; $m\angle y = 130°$, vertical angles equal; $m\angle z = 50°$, vertical angles equal

Workbook Activity 18—Reasoning

1. vertical angles are equal **2.** Axiom 5, whole is greater than the parts **3.** Axiom 5, whole is greater than the parts **4.** Axiom 1, substitution of equals **5.** vertical angles and right angles are equal **6.** vertical angles and right angles are equal **7.** adjacent and supplementary angles and things that are equal to the same thing are equal to each other **8.** vertical angles are equal and equals subtracted from equals

Workbook Activity 19—Art and Geometry

1.–5. Answers will vary. Check students' responses.

Workbook Activity 20—Figures with Parallel Lines

1. a and b: red; c and d: red **2.** a and b: green **3.** m, n, and o: red; d: green **4.** ℓ and m: green **5.** r and s: red; t: green

Workbook Activity 21—More Figures with Parallel Lines

1. 4 **2.** 12 **3.** 3 **4.** 4 **5.** Answers will vary. Check students' responses.

Workbook Activity 22—Transversals

1. c **2.** b **3.** a **4.** c **5.** Constructions will vary. Corresponding, alternate interior, and alternate exterior angles are equal. Interior angles or any two adjacent angles or exterior angles on the same side of the transversal are supplementary.

Workbook Activity 23—More Transversals

1. $\angle k$, $\angle l$, $\angle m$, $\angle n$ **2.** $\angle i$, $\angle j$, $\angle o$, $\angle p$ **3.** $\angle i$ and $\angle j$, $\angle j$ and $\angle k$, $\angle k$ and $\angle l$, $\angle l$ and $\angle i$, $\angle m$ and $\angle n$, $\angle n$ and $\angle o$, $\angle o$ and $\angle p$, $\angle p$ and $\angle m$ **4.** $\angle i$ and $\angle m$, $\angle j$ and $\angle n$, $\angle k$ and $\angle o$, $\angle l$ and $\angle p$ **5.** $\angle k$ and $\angle m$, $\angle l$ and $\angle n$ **6.** corresponding **7.** alternate interior **8.** alternate exterior **9.** exterior **10.** interior

Workbook Activity 24—Theorems Using Parallel Lines

1. 180° **2.** 180° **3.** $m\angle y$ **4.** $m\angle z$ **5.** $m\angle u$ **6.** $m\angle s$ **7.** $m\angle x$ **8.** $m\angle v$ **9.** $m\angle w$ **10.** $m\angle y$

Workbook Activity 25—Solving Problems with Theorems and Parallel Lines

1. 30°; $m\angle a + m\angle b = 180°$; $m\angle a = 5b$; $5b + m\angle b = 180°$; $6b = 180°$; $b = 30°$ **2.** 150°; $\angle a$ supplementary to $\angle b$ **3.** 150°; $\angle c$ vertical angle to $\angle a$ **4.** 30°; $\angle d$ vertical angle to $\angle b$ **5.** 150°; $\angle e$ corresponding angle to $\angle a$ **6.** 30°; $\angle f$ corresponding angle to $\angle b$; or $\angle f$ supplementary to $\angle e$; or $\angle f$ alternate interior angle to $\angle d$ **7.** 150°; $\angle g$ corresponding angle to $\angle c$; or $\angle g$ supplementary to $\angle f$; or $\angle g$ vertical angle to $\angle e$ **8.** 30°; $\angle g$ corresponding angle to $\angle d$; or $\angle h$ supplementary to $\angle g$ or $\angle e$; or $\angle h$ vertical angle to $\angle f$ **9.** 20° **10.** 60°

Workbook Activity 26—Constructions

1.–10. Student constructions will vary but should show alternate interior angles with given measures.

Workbook Activity 27—Quadrilaterals and Parallels

1. $\overline{DC}$ **2.** $\overline{BC}$ **3.** $\overline{BC}$ and $\overline{AD}$ **4.** $\overline{AB}$ and $\overline{DC}$ **5.** 180° **6.** 180° **7.** 180° **8.** 360° **9.** $\overline{AB}$ and $\overline{DC}$ **10.** 115°; 180°; 360°

Workbook Activity 28—Parallel Lines

1. Converse 2 **2.** Converse 1 **3.** Converse 3 **4.** Converse 2 **5.** Converse 2 **6.** Converse 3 **7.** Converse 2 **8.** Converse 1 **9.** Converse 3 **10.** Converse 2

Workbook Activity 29—Polygons

1. polygon **2.** polygon **3.** not a polygon; its segments intersect more than two other segments **4.** not a polygon; not line segments **5.** not a polygon; not a closed figure **6.** polygon **7.** not a polygon; not line segments **8.** not a polygon; not closed, contains other segments inside **9.** not a polygon; not line segments **10.** not a polygon; its segments intersect more than two other segments

Workbook Activity 30—Graphing Ordered Pairs

1.–10.

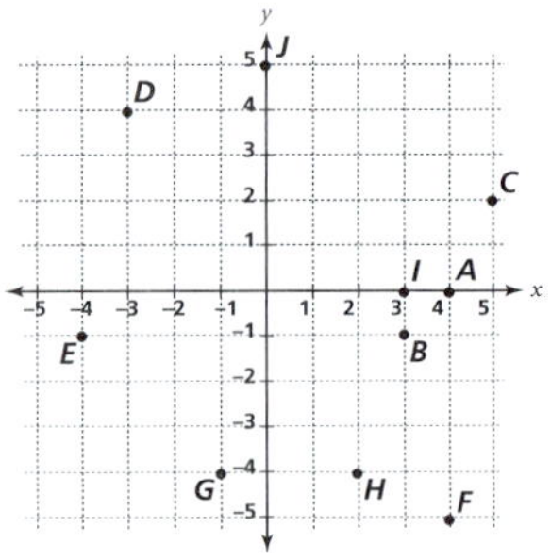

11.–20.

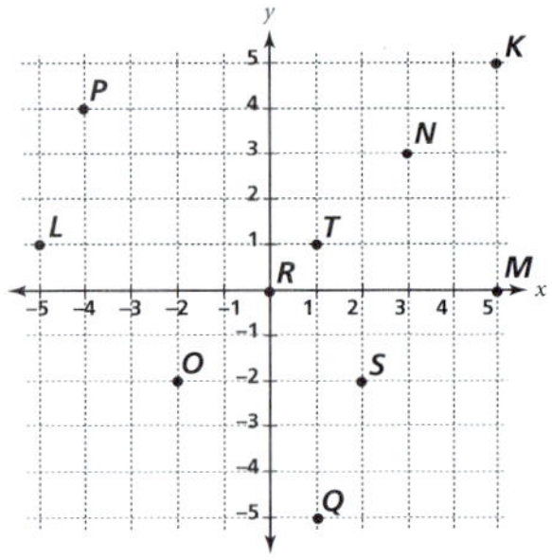

Workbook Activity 31—Using Formulas to Graph Ordered Pairs

1. $y = 1$ **2.** $y = 3$ **3.** $x = 2$ **4.** $y = -5$ **5.** $y = -4$ **6.** $y = 6$ **7.** $x = 1$ **8.** $y = -9$ **9.** $x = -2$ **10.** $y = 0$

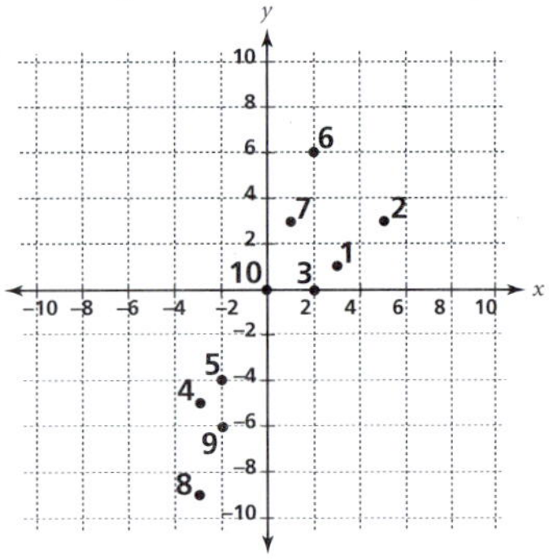

Workbook Activity 32—Graphing Horizontal Lines

1. horizontal line crossing the y-axis at 3 **2.** horizontal line crossing the y-axis at –1 **3.** horizontal line crossing the y-axis at 4 **4.** horizontal line crossing the y-axis at –2 **5.** horizontal line crossing the y-axis at –5 **6.** horizontal line crossing the y-axis at 0 **7.** horizontal line crossing the y-axis at 2 **8.** horizontal line crossing the y-axis at 3 **9.** horizontal line crossing the y-axis at 5 **10.** horizontal line crossing the y-axis at –4

Workbook Activity 33—Graphing Ordered Pairs

1. horizontal line crossing the y-axis at –3 **2.** horizontal line crossing the y-axis at 1 **3.** horizontal line crossing the y-axis at –2 **4.** horizontal line crossing the y-axis at –5 **5.** horizontal line crossing the y-axis at –4 **6.** horizontal line crossing the y-axis at 0 **7.** horizontal line crossing the y-axis at –5 **8.** horizontal line crossing the y-axis at 2 **9.** horizontal line crossing the y-axis at 4 **10.** horizontal line crossing the y-axis at –1

Workbook Activity 34—Graphing Vertical Lines

1. vertical line crossing the x-axis at –2 **2.** vertical line crossing the x-axis at –1 **3.** vertical line crossing the x-axis at 3 **4.** vertical line crossing the x-axis at –4 **5.** vertical line crossing the x-axis at 5 **6.** vertical line crossing the x-axis at 0 **7.** vertical line crossing the x-axis at 2 **8.** vertical line crossing the x-axis at –5 **9.** vertical line crossing the x-axis at –3 **10.** vertical line crossing the x-axis at 4

Workbook Activity 35—Graphing Ordered Pairs

1. vertical line crossing the x-axis at 4 **2.** vertical line crossing the x-axis at 1 **3.** vertical line crossing the x-axis at 2 **4.** vertical line crossing the x-axis at –5 **5.** vertical line crossing the x-axis at 3 **6.** vertical line crossing the x-axis at 0 **7.** vertical line crossing the x-axis at –2 **8.** vertical line crossing the x-axis at –1 **9.** vertical line crossing the x-axis at 5 **10.** vertical line crossing the x-axis at –3

Workbook Activity 36—Finding the Slope of a Line

1. $m = -\frac{5}{6}$ **2.** $m = -\frac{1}{2}$ **3.** $m = -1$ **4.** $m = 2$ **5.** $m = 2$ **6.** $m = -2$ **7.** $m = \frac{1}{2}$ **8.** m = undefined **9.** $m = 5$ **10.** $m = \frac{1}{2}$ **11.** $m = 1$ **12.** $m = 0$ **13.** $m = -5$ **14.** $m = 1$ **15.** $m = -3$

Workbook Activity 37—The Slope of Parallel Lines

1. $m = 4$ **2.** $m = 3$ **3.** $m = -5$ **4.** $m = -\frac{1}{3}$

Workbook Activity 38—$y = mx + b$

1. $y = x + 2$ **2.** $y = 5x$ **3.** $y = -3x - 7$ **4.** $y = -x + 1$ **5.** $x = 1$, y is any real number

Workbook Activity 39—Using Slope in Real-Life Examples

1. domain = 300 ft, range = 100 ft, $m = \frac{1}{3}$ **2.** 8 in. **3.** domain = 4, range = 8 **4.** domain = 10 ft, range = 4 ft, $m = \frac{2}{5}$ **5.** $y = \frac{1}{3}x$; $y = 2x$; $y = \frac{2}{5}x$

Workbook Activity 40—The Point-Slope Formula

1. line a passes through (2, –1) and (4, 0) **2.** line b passes through (0, –3) and (1, –5) **3.** line c passes through (3, 5) and (7, 2) **4.** line d passes through (–5, 1) and (–4, 4) **5.** line e passes through (2, 0) and (4, 5) **6.** $y = \frac{1}{2}x - 2$ **7.** $y = -2x - 3$ **8.** $y = -\frac{3}{4}x + 7\frac{1}{4}$ **9.** $y = 3x + 16$ **10.** $y = \frac{5}{2}x - 5$

Workbook Activity 41—More Point Slope

1. line a passes through (–3, –2) and (1, –7) **2.** line b passes through (0, 0) and (1, 3) **3.** line c passes through (–5, 5) and (–4, 1) **4.** line d passes through (–2, –4) and (5, –1) **5.** line e passes through (1, 1) and (3, –2) **6.** $y = -\frac{5}{4}x - 5\frac{3}{4}$ **7.** $y = 3x$ **8.** $y = -4x - 15$ **9.** $y = \frac{3}{7}x - 3\frac{1}{7}$ **10.** $y = -\frac{3}{2}x + 2\frac{1}{2}$

Workbook Activity 42—The Midpoint of a Segment

1. (0, 0) **2.** $(1, -\frac{1}{2})$ **3.** (4, –3) **4.** (1, 5) **5.** (–2, 1)

Workbook Activity 43—Graphing Line Segments

1. (0, –6); graph line segment between (4, 6) and (0, –6) with midpoint (2, 0) **2.** (1, 3); graph line segment between (–3, –5) and (1, 3) with midpoint (–1, –1) **3.** (1, –2); graph line segment between (5, 4) and (1, –2) with midpoint (3, 1) **4.** (–5, 1); graph line segment between (3, –5) and (–5, 1) with midpoint (–1, –2) **5.** (4, –4); graph line segment between (–2, 6) and (4, –4) with midpoint (1, 1) **6.** (–3, –1); graph line segment between (–5, –3) and (–3, –1) with midpoint (–4, –2) **7.** (6, –8); graph line segment between (4, 4) and (6, –8) with midpoint (5, –2) **8.** (5, –3); graph line segment between (1, 7) and (5, –3) with midpoint (3, 2) **9.** (5, 2); graph line segment between (–3, –4) and (5, 2) with midpoint (1, –1) **10.** (3, –1); graph line segment between (1, 1) and (3, –1) with midpoint (2, 0)

Workbook Activity 44—Water Slides

Slide One: the first drop slope is $\frac{1}{2}$, the second drop slope is $\frac{1}{3}$, the third drop slope is 1, and the end slope is $\frac{1}{4}$
Slide Two: the first drop slope is 1, the second drop slope is $\frac{1}{2}$, the third drop slope is 2, and the end slope is $\frac{1}{3}$

Workbook Activity 45—Triangle Sides

1. equilateral **2.** scalene **3.** isosceles **4.** isosceles **5.** scalene **6.** equilateral **7.** scalene **8.** isosceles **9.** equilateral **10.** scalene **11.** isosceles **12.** equilateral **13.** isosceles or scalene **14.** isosceles **15.** equilateral

Workbook Activity 46—Naming Triangles by Their Angles

1. obtuse isosceles triangle **2.** acute scalene triangle **3.** acute equilateral triangle **4.** obtuse scalene triangle **5.** right isosceles triangle **6.** right scalene triangle **7.** acute isosceles triangle **8.** obtuse scalene triangle **9.** equilateral triangle **10.** right scalene triangle

Workbook Activity 47—Constructing Triangles Using Their Angles

1. Students' triangle should have angles of 90°, 45°, and 45°.
2. Students' triangle should have angles of 120°, 30°, and 30°.
3. Students' triangle should have angles of 90°, 80°, and 10°.
4. Students' triangle should have angles of 60°, 60°, and 60°.
5. Students' triangle should have angles of 60°, 60°, and 60°.

Workbook Activity 48—Special Quadrilaterals

1. True, rhombus **2.** True, isosceles trapezoid **3.** False **4.** True, irregular quadrilateral **5.** True, isosceles trapezoid **6.** True, parallelogram **7.** True, right trapezoid **8.** True, quadrilateral (shaped like an arrowhead) **9.** False **10.** False

Workbook Activity 49—Diagonals

1. Definition of a rectangle. **2.** Alternate interior angles. **3.** Definition of a rectangle. **4.** Alternate interior angles. **5.** Definition of a rectangle. **6.** Substitution of equals. **7.** Division of equals. **8.** Substitution of equals. **9.** Definition of a parallelogram. **10.** Alternate interior angles.

Workbook Activity 50—More Diagonals

1. 80° **2.** 10° **3.** 80° **4.** 10° **5.** 60° **6.** 30° **7.** 60° **8.** 30° **9.** 85° **10.** 5°

Workbook Activity 51—Triangle Angles

1. $x = 55°$ **2.** $x = 60°$ **3.** $x = 45°$ **4.** $x = 120°$ **5.** $x = 70°$ **6.** $x = 97°$ **7.** $x = 33°$ **8.** $x = 48°$ **9.** $x = 13°$ **10.** $x = 48°$

Workbook Activity 52—Triangle Angles and Algebra

1. $x = 15°$ **2.** $x = 30°$ **3.** $x = 10°$ **4.** $x = 20°$ **5.** $x = 60°$ **6.** $x = 65°$ **7.** $x = 30°$ **8.** $x = 20°$ **9.** $x = 60°$ **10.** $x = 10°$

Workbook Activity 53—Constructing Regular Polygons

1. Students' construction should show proper arcs and interior angles of 108°. **2.** Students' construction should show proper arcs and interior angles of 135°. **3.** Students' construction should show proper arcs and interior angles of 144°. **4.** Students' construction should show proper arcs and interior angles of 128.6°. **5.** Students' construction should show proper arcs and interior angles of 120°

Workbook Activity 54—Geometric Patterns

1. 1 diagonal

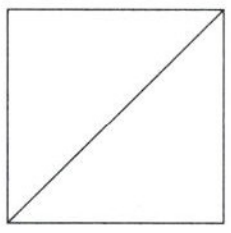

2. 2 diagonals

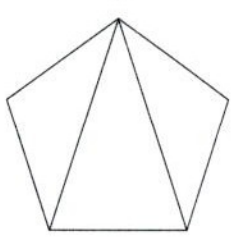

3. 7 diagonals

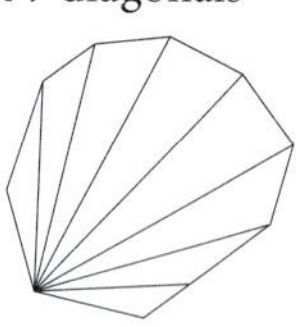

4. 5 diagonals

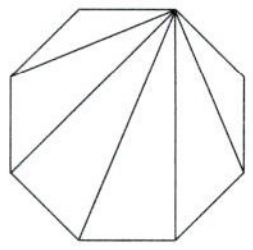

5. 6 diagonals

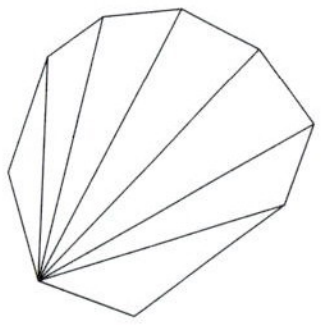

Workbook Activity 55—Constructing Perpendiculars

1.–5. Students' constructions should reflect correct use of construction steps.

Workbook Activity 56—More Perpendiculars

1.–10. Students' constructions should include the necessary arcs and perpendiculars.

Workbook Activity 57—Angle Bisectors and Medians

1.

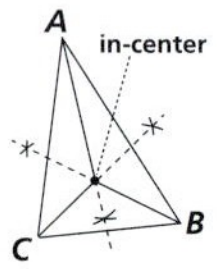

2.

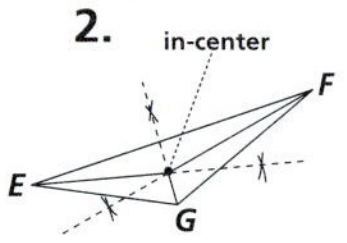

3.

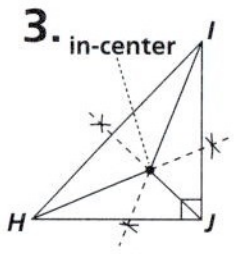

4.

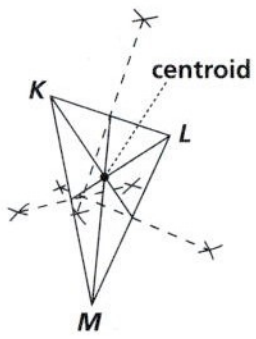

5.

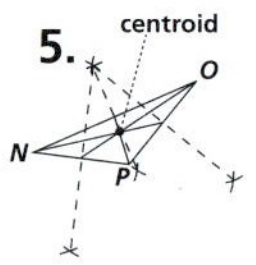

Workbook Activity 58—Altitudes

1. Students' construction should show orthocenter inside an acute triangle. **2.** Students' construction should show orthocenter outside an obtuse triangle. **3.** Students' construction should show orthocenter at vertex of a right angle. **4.** orthocenter will be inside the triangle **5.** orthocenter moves closer and closer to triangle's edge; orthocenter moves onto vertex of right angle; orthocenter moves outside of triangle

Workbook Activity 59—Angle Sum Theorem

1. 90° **2.** 60° **3.** 30° **4.** 40° **5.** 80° **6.** 60° **7.** 30° **8.** 30° **9.** 120° **10.** 180°

Workbook Activity 60—More of the Angle Sum Theorem

1. 80° **2.** 40° **3.** 60° **4.** 80° **5.** 20° **6.** 80° **7.** 60° **8.** 30° **9.** 90° **10.** 180°

Workbook Activity 61—Geometric Puzzles

1.–5. Shapes will vary.

Workbook Activity 62—Proving Triangles Congruent by SAS

Reason

1. Definition of a square. **2.** Definition of a square. **3.** Definition of a square. **4.** SAS postulate.

Reason

1. Definition of a parallelogram. **2.** Definition of a parallelogram. **3.** Definition of a parallelogram. **4.** Interior angles on the same side of a transversal are supplementary. **5.** Definition of a parallelogram. **6.** Interior angles on the same side of a transversal are supplementary.

Workbook Activity 63—Proving Triangles Congruent by SSS

Reason

1. Definition of a parallelogram. **2.** Definition of a parallelogram. **3.** Any quantity is equal to itself. **4.** SSS postulate.

Reason

1. Definition of an isosceles triangle. **2.** Any quantity is equal to itself. **3.** Definition of a perpendicular. **4.** Definition of a perpendicular. **5.** Definition of a right triangle. **6.** Pythagorean Theorem.

Workbook Activity 64—Proving Triangles Congruent by ASA

Reason

1. Definition of a square. **2.** Definition of a transversal. **3.** Alternate interior angles are equal. **4.** Definition of a square. **5.** Definition of a transversal. **6.** Alternate interior angles are equal. **7.** Any quantity is equal to itself. **8.** ASA postulate.

Workbook Activity 65—Hypotenuse-Leg Theorem

Reason

1. Given: definition of an isosceles triangle. **2.** Any quantity is equal to itself. **3.** Given: definition of an altitude. **4.** Given: definition of an altitude. **5.** All right angles are equal. **6.** Hypotenuse-Leg Theorem.

Reason

1. Given: definition of a square. **2.** Any quantity is equal to itself. **3.** Given: definition of a square. **4.** Given: definition of a square.

Workbook Activity 66—Reflections

1. $A' = (1, -1)$, $B' = (1, -5)$, $C' = (5, -5)$, $D' = (5, -1)$ **2.** $E' = (-2, 3)$, $F' = (-5, 3)$, $G' = (-5, -4)$, $H' = (-2, -4)$ **3.** $I' = (-2, -1)$, $J' = (0, -4)$, $K' = (4, -4)$, $L' = (4, -1)$ **4.** $M' = (-3, -2)$, $N' = (-1, -6)$, $O' = (3, -6)$, $P' = (1, -2)$ **5.** $Q' = (-1, 1)$, $R' = (-1, 5)$, $S' = (-3, 7)$, $T' = (-5, 5)$, $U' = (-5, 1)$

Workbook Activity 67—Reflections Over Different Lines

1. $A' = (2, 2)$, $B' = (0, 2)$, $C' = (0, -3)$, $D' = (2, -3)$ **2.** $E' = (-2, 2)$, $F' = (0, 2)$, $G' = (2, 7)$, $H' = (0, 7)$ **3.** $I' = (-3, -3)$, $J' = (-3, 3)$, $K' = (-5, 2)$, $L' = (-5, -2)$ **4.** $X' = (4, 4)$, $R' = (4, 6)$ **5.** $Q' = (-3, -3)$, $R' = (-3, -6)$, $S' = (-1, -6)$, $T' = (-1, -3)$

Workbook Activity 68—Alphabet Symmetry

1. one vertical line **2.** one vertical line **3.** one vertical line, one horizontal line **4.** one horizontal line **5.** one horizontal line **6.** one vertical line **7.** one vertical line, one horizontal line **8.** one horizontal line **9.** no lines **10.** one vertical line, one horizontal line All lines are centered on the letters.

Workbook Activity 69—Identifying Image Translations

1. (0, 6) **2.** (5, 3) **3.** (–2, 1) **4.** (7, 5) **5.** (–4, 2) **6.** (–7, 0) **7.** (2, 4) **8.** (–4, 4) **9.** (17, –1) **10.** (–8, 2) **11.** (–2, 0) **12.** (– 10, 4) **13.** $(x - 2, y - 3)$ **14.** $(x + 3, y + 6)$ **15.** $(x + 5, y + 7)$

Workbook Activity 70—Rotating Images

1. rotated image's endpoints: (2, 1), (6, 1), (2, –2) **2.** rotated image's endpoints: (–3, 2), (0, –1), (–2, –3) **3.** rotated image's endpoints: (–4, –5), (3, –7), (0, –11) **4.** rotated image's endpoints: (–3, 1), (1, 3), (0, –7) **5.** rotated image's endpoints are (–4, –2), (–1, –4), (–2, –8)

Workbook Activity 71—3-D Symmetry

Answers will vary. Representative answers are shown.

1.

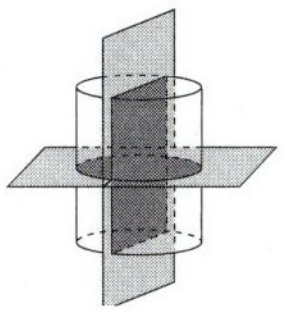

2.

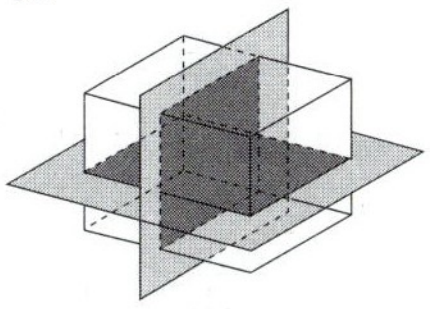

3.

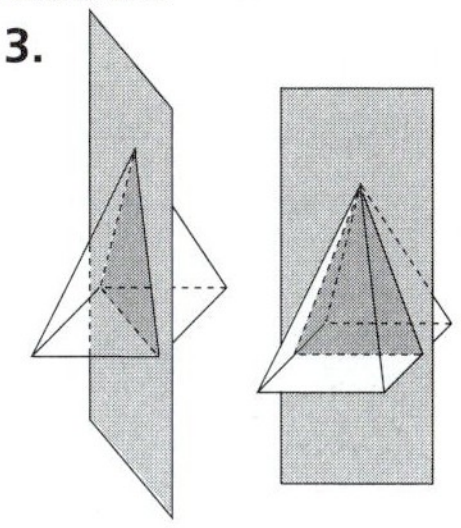

4.

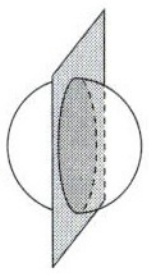

5.

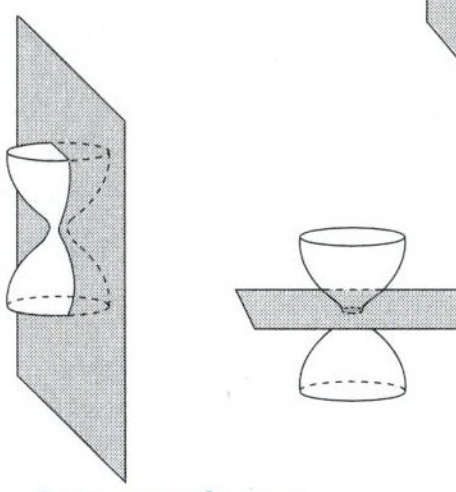

Workbook Activity 72—Proportions

1. $x = 9$ **2.** $x = 16$ **3.** $x = 92$ **4.** $x = 1$ **5.** $x = 14$ **6.** $x = 6$ **7.** $x = 20$ **8.** $x = 39$ **9.** $x = 11\frac{1}{3}$ **10.** $x = 15$ **11.** $x = 23\frac{3}{4}$ **12.** $x = 24$ **13.** $x = 16$ **14.** $x = 25$ **15.** $x = 8\frac{2}{3}$

Workbook Activity 73—Determining Equal Ratios

1. No **2.** Yes **3.** Yes **4.** Yes **5.** No **6.** Yes **7.** No **8.** No **9.** Yes **10.** Yes **11.** No **12.** Yes **13.** No **14.** Yes **15.** Yes

Workbook Activity 74—Corresponding Angles and Sides

1. $m\angle S$ **2.** $m\angle T$ **3.** $m\angle SRT$ **4.** $\overline{TS}, \overline{TR}$ **5.** $\overline{SR}, \overline{TS}$ **6.** $m\angle D$ **7.** $m\angle E$ **8.** $m\angle F$ **9.** $\overline{DE}, \overline{EF}$ **10.** $\overline{DF}, \overline{DE}$ **11.** $m\angle D$ **12.** $m\angle E$ **13.** $m\angle F$ **14.** $\overline{DE}, \overline{EF}$ **15.** $\overline{DF}, \overline{DE}$

Workbook Activity 75—Using the AA Postulate

1. 9 **2.** 9 **3.** 35° **4.** 55° **5.** 16 **6.** 9.6 **7.** 12.8 **8.** 7.2 **9.** 3 **10.** 3 **11.** 5 **12.** 5 **13.** 12 **14.** 5 **15.** 8

Workbook Activity 76—The Ratio of Similarity

1. 96 **2.** 20 **3.** 51 **4.** 45 **5.** 64

Workbook Activity 77—Angle Measure in Regular Polygons

1. 108° **2.** 144° **3.** 140° **4.** 147.27° **5.** 150°

Workbook Activity 78—Finding the Ratio of Similarity

1. 1:3 **2.** 3:4 **3.** 17:50 **4.** 1:2 **5.** 2:5

Workbook Activity 79—Dilating Images

1. $A' = (2, 2), B' = (0, 2), C' = (0, 0), D' = (2, 0)$ **2.** $A' = (3, 3), B' = (0, 3), C' = (0, 0), D' = (3, 0)$ **3.** $A' = (4, 4), B' = (0, 4), C' = (0, 0), D' = (4, 0)$ **4.** $A' = (5, 5), B' = (0, 5), C' = (0, 0), D' = (5, 0)$ **5.** $A' = (6, 6), B' = (0, 6), C' = (0, 0), D' = (6, 0)$ **6.** $A' = (-6, 2), B' = (-4, -4), C' = (4, -4), D' = (6, 2)$ **7.** $A' = (-9, 3), B' = (-6, -6), C' = (6, -6), D' = (9, 3)$ **8.** $A' = (-12, 4), B' = (-8, -8), C' = (8, -8), D' = (12, 4)$ **9.** $A' = (-15, 5), B' = (-10, -10), C' = (10, -10), D' = (15, 5)$ **10.** $A' = (-18, 6), B' = (-12, -12), C' = (12, -12), D' = (18, 6)$

Workbook Activity 80—More Dilations

1. $A' = (2, 2), B' = (6, 2), C' = (8, 6), D' = (4, 6)$ **2.** $A' = (3, 3), B' = (9, 3), C' = (12, 9), D' = (6, 9)$ **3.** $A' = (4, 4), B' = (12, 4), C' = (16, 12), D' = (8, 12)$ **4.** $A' = (5, 5), B' = (15, 5), C' = (20, 15), D' = (10, 15)$ **5.** $A' = (6, 6), B' = (18, 6), C' = (24, 18), D' = (12, 18)$ **6.** $A' = (-6, 2), B' = (-2, 2), C' = (-2, 6), D' = (-6, 6)$ **7.** $A' = (-9, 3), B' = (-3, 3), C' = (-3, 9), D' = (-9, 9)$ **8.** $A' = (-12, 4), B' = (-4, 4), C' = (-4, 12), D' = (-12, 12)$ **9.** $A' = (-15, 5), B' = (-5, 5), C' = (-5, 15), D' = (-15, 15)$ **10.** $A' = (-18, 6), B' = (-6, 6), C' = (-6, 18), D' = (-18, 18)$

Workbook Activity 81—Shrinking Images

1. $A' = (0, 0), B' = (3, 0), C' = (3, 3), D' = (0, 3)$ **2.** $A' = (0, 0), B' = (2, 0), C' = (2, 2), D' = (0, 2)$ **3.** $A' = (0, 0), B' = (1\frac{1}{2}, 0), C' = (1\frac{1}{2}, 1\frac{1}{2}), D' = (0, 1\frac{1}{2})$ **4.** $A' = (0, 0), B' = (1\frac{1}{5}, 0), C' = (1\frac{1}{5}, 1\frac{1}{5}), D' = (0, 1\frac{1}{5})$ **5.** $A' = (0, 0), B' = (1, 0), C' = (1, 1), D' = (0, 1)$ **6.** $A' = (-5, -5), B' = (-5, 5), C' = (5, 5), D' = (5, -5)$ **7.** $A' = (-4, -4), B' = (-4, 4), C' = (4, 4), D' = (4, -4)$ **8.** $A' = (-3, -3), B' = (-3, 3), C' = (3, 3), D' = (3, -3)$ **9.** $A' = (-2, -2), B' = (-2, 2), C' = (2, 2), D' = (2, -2)$ **10.** $A' = (-1, -1), B' = (-1, 1), C' = (1, 1), D' = (1, -1)$

Workbook Activity 82—More Shrinking

1. $(-3, \frac{1}{2}), (0, 3), (3, 3), (0, \frac{1}{2})$ **2.** $(-4, \frac{2}{3}), (0, 4), (4, 4), (0, \frac{2}{3})$ **3.** $(-4\frac{1}{2}, \frac{3}{4}), (0, 4\frac{1}{2}), (4\frac{1}{2}, 4\frac{1}{2}), (0, \frac{3}{4})$ **4.** $(-4\frac{4}{5}, \frac{4}{5}), (0, 4\frac{4}{5}), (4\frac{4}{5}, 4\frac{4}{5}), (0, \frac{4}{5})$ **5.** $(-5, \frac{5}{6}), (0, 5), (5, 5), (0, \frac{5}{6})$ **6.** $(0, 0), (0, 2), (2, 2), (2, 0)$ **7.** $(0, 0), (0, 4), (4, 4), (4, 0)$ **8.** $(0, 0), (0, 6), (6, 6), (6, 0)$ **9.** $(0, 0), (0, 8), (8, 8), (8, 0)$ **10.** $(0, 0), (0, 1), (1, 1), (1, 0)$

Workbook Activity 83—Zooming In

Answers will vary.

Workbook Activity 84—Checking Triples

1. No **2.** No **3.** Yes **4.** No **5.** No **6.** Yes **7.** No **8.** Yes **9.** No **10.** Yes **11.** Yes **12.** No **13.** No **14.** Yes **15.** No

Workbook Activity 85—Plato's Formula

1. 8, 15, 17 **2.** 10, 24, 26 **3.** 16, 63, 65 **4.** 18, 80, 82 **5.** 22, 120, 122 **6.** 24, 143, 145 **7.** 26, 168, 170 **8.** 28, 195, 197 **9.** 32, 255, 257 **10.** 34, 288, 290 **11.** 36, 323, 325 **12.** 38, 360, 362 **13.** 42, 440, 442 **14.** 44, 483, 485 **15.** 60, 899, 901

Workbook Activity 86—Pythagorean Triples and Proofs

1. 289 **2.** 60 **3.** 49 **4.** 289 **5.** Yes **6.** 1,156 **7.** 240 **8.** 196 **9.** 1,156 **10.** Yes

Workbook Activity 87—Pythagorean Demonstration

1. 120 **2.** 26 **3.** 676 **4.** 1,156 **5.** 1,156 **6.** 100 **7.** 576 **8.** 240 **9.** 1,156 **10.** They are equal.

Workbook Activity 88—Pythagorean Theorem and Similar Triangles

1. 48 **2.** 64 **3.** 36 **4.** $\overline{XY}$ **5.** $\overline{WX}$ **6.** $\overline{WY}$ **7.** Yes, $<Y + <YWZ = 90° = <X + <XWZ$ **8.** 5:4 **9.** 5:3 **10.** 4:3 **11.** 240 **12.** 144 **13.** 192 **14.** 5:4:3 **15.** Yes, they are the same.

Workbook Activity 89—More with Similar Triangles

1. 14.14 **2.** 14.14 **3.** 14.14 **4.** $\overline{BC}$ **5.** $\overline{AC}$ **6.** $\overline{AB}$ **7.** Yes, AAS **8.** 28.8:20 **9.** 28.8:20 **10.** 1:1 **11.** 68.28 **12.** 48.28 **13.** 48.28 **14.** 68.28:48.28:48.28 **15.** Yes, they are the same.

Workbook Activity 90—Pythagorean Proof and Trapezoids

1. 312 **2.** 132 **3.** 48 **4.** 5,760 **5.** 20 **6.** 2,738 **7.** 344 **8.** 966 **9.** 2,375 **10.** 2,144

Workbook Activity 91—Distance Formula

1. $\sqrt{34}$ **2.** $\sqrt{53}$ **3.** $\sqrt{130}$ **4.** $\sqrt{37}$ **5.** $\sqrt{29}$ **6.** $\sqrt{116}$ **7.** $\sqrt{97}$ **8.** $\sqrt{29}$ **9.** $\sqrt{52}$ **10.** $\sqrt{65}$

Workbook Activity 92—More Distance Formula

1. $\sqrt{29}$ **2.** $\sqrt{85}$ **3.** $\sqrt{5}$ **4.** $\sqrt{146}$ **5.** $\sqrt{29}$ **6.** $\sqrt{97}$ **7.** $\sqrt{98}$ **8.** $\sqrt{8}$ **9.** $\sqrt{85}$ **10.** $\sqrt{65}$ **11.** $\sqrt{394}$ **12.** $\sqrt{85}$ **13.** $\sqrt{65}$ **14.** $\sqrt{72}$ **15.** $\sqrt{52}$

Workbook Activity 93—Converse of the Pythagorean Theorem

1. Yes **2.** Yes **3.** No **4.** No **5.** No **6.** No **7.** Yes **8.** No **9.** No **10.** Yes **11.** Yes **12.** No **13.** No **14.** No **15.** Yes

Workbook Activity 94—Architecture and Geometric Shapes

1. $x = 11.66$ **2.** $x = 17.20$ **3.** $x = 8.60$ **4.** $x = 162.79$ **5.** $x = 127.28$

Workbook Activity 95—Perimeters of Polygons

1. 126 **2.** 144 **3.** 119 **4.** 59 **5.** 276 **6.** 140 **7.** 469 **8.** 192 **9.** 38 **10.** 186

Workbook Activity 96—Perimeter Formulas

1. 156 **2.** 274 **3.** 188 **4.** 113 **5.** 324 **6.** 354 **7.** 284 **8.** 2,260 **9.** 13,188 **10.** 26,680

Workbook Activity 97—Perimeters and Diagonals

1. 141.4 **2.** 295 **3.** 279.2 **4.** 240 **5.** 180 **6.** 288 **7.** 78.6 **8.** 204.4 **9.** 306 **10.** 230.4

Workbook Activity 98—Parallelogram Areas

1. 1,536 **2.** 125 **3.** 495 **4.** 336 **5.** 3,552 **6.** 1,584 **7.** 1,728 **8.** 4,221 **9.** 2,800 **10.** 6,400

Workbook Activity 99—Areas of Trapezoids

1. $ab = 24$ **2.** $ab = 87$ **3.** $ab = 74$ **4.** $b = 44$ **5.** $b = 27$ **6.** $h = 56$ **7.** $ab = 22$ **8.** $h = 29$ **9.** $b = 96$ **10.** $h = 55$

Workbook Activity 100—Heron's Formula

1. 223.4 **2.** 186 **3.** 99.7 **4.** 227.9 **5.** 233.9 **6.** 169.7 **7.** 1,189.3 **8.** 2,085 **9.** 60,000 **10.** 2,274.7

Workbook Activity 101—Tessellations

Answers will vary.

Workbook Activity 102—Definition of a Circle

1. 66 **2.** 34 **3.** 64 **4.** 320.44 **5.** 180 **6.** 138.23 **7.** 67 **8.** 246 **9.** 276.46 **10.** 17,886

Workbook Activity 103—The Ratio π

1. $d = 25$ ft, $r = 12.5$ ft **2.** $d = 46$ in., $r = 23$ in. **3.** $d = 18$ cm, $r = 9$ cm **4.** $d = 87$ m, $r = 43.5$ m **5.** $d = 65$ mm, $r = 32.5$ mm **6.** $d = 28$ ft, $r = 14$ ft **7.** $d = 78$ in., $r = 39$ in. **8.** $d = 32$ cm, $r = 16$ cm **9.** $d = 20$ m, $r = 10$ m **10.** $d = 824$ mm, $r = 412$ mm **11.** $d = 608$ ft, $r = 304$ ft **12.** $d = 754$ in., $r = 377$ in. **13.** $d = 589$ cm, $r = 294.5$ cm **14.** $d = 1{,}098$ m, $r = 549$ m **15.** $d = 5{,}318$ mm, $r = 2{,}659$ mm

Workbook Activity 104—Approximating the Area of a Circle

1. 108 **2.** 432 **3.** 432 **4.** 507 **5.** 1,200 **6.** 2,700 **7.** 192 **8.** 1,083 **9.** 3,072 **10.** 1,875 **11.** 5 **12.** 7 **13.** 9 **14.** 11 **15.** 13

Workbook Activity 105—Area and Probability

1. 37.5% **2.** 50% **3.** 62.5% **4.** 87.5% **5.** 50%

Workbook Activity 106—Formula for the Area of a Circle

1. 11.3 cm **2.** 16.7 cm **3.** 5.4 cm **4.** 13.1 cm **5.** 14.4 cm **6.** 15.6 cm **7.** 11.4 cm **8.** 17.8 cm **9.** 56.4 cm **10.** 27.3 cm

Workbook Activity 107—More Formula for the Area of a Circle

1. 128.7 sq in. **2.** 277.6 ft^2 **3.** 30.2 cm^2 **4.** 172 units^2 **5.** 201.1 ft^2 **6.** 227 sq in. **7.** 132.7 cm^2 **8.** 314.2 sq in. **9.** 3,117.2 mm^2 **10.** 706.9 units^2

Workbook Activity 108—Circles and Their Angles and Sectors

1. 50° **2.** 80° **3.** 64° **4.** 70° **5.** 60° **6.** 110° **7.** 108° **8.** 140° **9.** 100° **10.** 190°

Workbook Activity 109—Tangents, Circumcircles, and Incircles

1.–2.

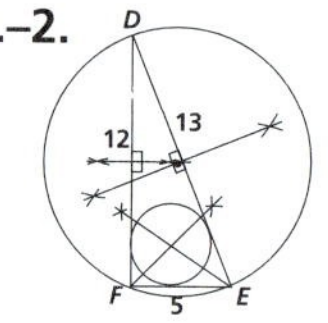

3.–4.

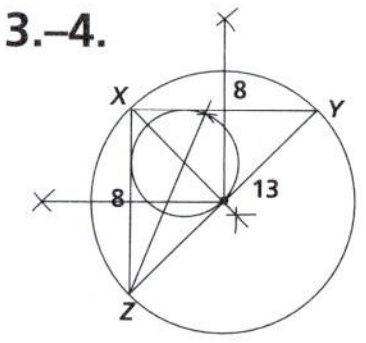

5.

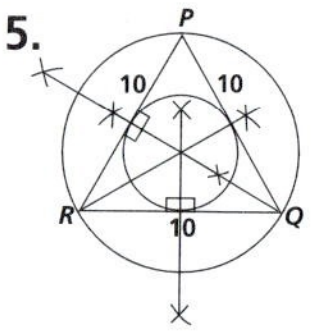

Workbook Activity 110—Sine, Cosine, and Tangent

1. 0.92, 0.38, 2.4 **2.** 4.7, 0.88, 0.53 **3.** 0.96, 0.28, 3.43 **4.** 0.32, 0.94, 0.34 **5.** 0.92, 0.38, 2.4

Workbook Activity 111—Spheres

1. 314.16 sq units; 523.60 units^3 **2.** 113.10 sq units; 113.10 units^3 **3.** 201.06 sq units; 268.08 units^3 **4.** 12.57 sq units; 4.19 units^3 **5.** 615.75 sq units; 1,436.76 units^3 **6.** 804.25 sq units; 2,144.66 units^3 **7.** 1,017.88 sq units; 3,053.63 units^3 **8.** 1,256.64 sq units; 4,188.79 units^3 **9.** 7,853.98 sq units; 65,449.85 units^3 **10.** 31,415.93 sq units; 523,598.78 units^3

Workbook Activity 112—Pie Charts

Answers will vary.

Workbook Activity 113—Basic Volume Formulas

1. $x = 5$ **2.** $x = 42$ **3.** $x = 40$ **4.** $x = 24$ **5.** $x = 12$ **6.** $x = 10$ **7.** $x = 6$ **8.** $x = 7$ **9.** $x = 60$ **10.** $x = 36$

Workbook Activity 114—Volumes of Pyramids and Cones

1. 6,250 **2.** 2,660.93 **3.** 8,960 **4.** 28.27 **5.** 67,500

Workbook Activity 115—Surface Area of Prisms and Cylinders

1. 384 **2.** 680 **3.** 360 **4.** 251.32 **5.** 1,200 **6.** 3,174 **7.** 15,700 **8.** 552 **9.** 468 **10.** 2,035.69

Workbook Activity 116—Surface Areas of Pyramids and Cones

1. 351 **2.** 138.23 **3.** 56 **4.** 207.34 **5.** 900 **6.** 395.83 **7.** 299 **8.** 395.83 **9.** 1,188 **10.** 339.28

Workbook Activity 117—Measurements

1. 60 **2.** 39 **3.** $4\frac{2}{3}$ **4.** 12,320 **5.** 11 **6.** 312 **7.** 14 **8.** 19 **9.** 120 **10.** 45 **11.** 66,000 **12.** 57,000 **13.** 920 **14.** 8.43 **15.** 8,800,000 **16.** 1,020 **17.** 108 **18.** 84 **19.** 21,120 **20.** 55

Workbook Activity 118—Unfolding Three-Dimensional Objects

1. will look like a square unfolded except middle panels are rectangular **2.** will look like a large triangle divided into four smaller equal triangles, one in the middle with the other three pointing off of its sides **3.** will look like a square with four equal triangles, one off of each side **4.** will look like a rectangle with two equal circles attached to each longer side, one at the left end and one at the right end **5.** will look like an isosceles triangle with a rounded bottom edge and a circle attached to that edge

Workbook Activity 119—Lines and Planes in Space

1. The plane touches the cube at one corner. **2.** The plane rests against one edge of the cube. **3.** The plane rests against one face of the cube. **4.** The plane passes through the cube and is parallel to two faces of the cube. **5.** The plane does not touch the cube, even if the plane were extended in all directions.

Workbook Activity 120—Loci in the Coordinate Plane

1. $49 = x^2 + y^2$ **2.** $64 = x^2 + y^2$ **3.** $81 = x^2 + y^2$ **4.** $100 = x^2 + y^2$ **5.** $400 = x^2 + y^2$

Workbook Activity 121—Compound Loci

1. The plane passes through one vertex of the top triangle and through one side of the base triangle. A triangle is formed. **2.** The plane, parallel to the pyramid base, intersects the pyramid to form a triangle similar to the pyramid base. **3.** The plane, parallel to the base of the cone, intersects to form an ellipse similar to the ellipse at the base of the cone. **4.** The plane passes through the vertex of the cone and is perpendicular to the base of the cone. A triangle is formed. **5.** The plane intersects the base but is not parallel to the base. An ellipse is formed.

Workbook Activity 122—The Fibonacci Series

1. Six concentric circles with the origin at the center, whose radii are the first six numbers (1, 2, 3, 5, 8, 13) of the Fibonacci series. **2.** Concentric diamonds with these vertices: (0, 1), (1, 0), (0, –1), (–1, 0); (0, 2), (2, 0), (0, –2), (–2, 0); (0, 3), (3, 0), (0, –3), (–3, 0); (0, 5), (5, 0), (0, –5), (–5, 0); (0, 8), (8, 0), (0, –8), (–8, 0) **3.** Concentric triangles with the origin as one vertex and these additional vertices: (1, 1), (–1, 1); (2, 2), (–2, 2); (3, 3), (–3, 3); (5, 5), (–5, 5); (8, 8), (–8, 8) **4.** Overlapping rectangles with these vertices: (1, 1), (2, 1), (2, –1), (1, –1); (2, 2), (4, 2), (4, –2), (2, –2); (3, 3), (6, 3), (6, –3), (3, –3); (5, 5), (10, 5), (10, –5), (5, –5); (8, 8), (16, 8), (16, –8), (8, –8) **5.** Concentric squares with these vertices: (1, 1), (–1, 1), (–1, –1), (1, –1); (2, 2), (–2, 2), (–2, –2), (2, –2); (3, 3), (–3, 3), (–3, –3), (3, –3); (5, 5), (–5, 5), (–5, –5), (5, –5); (8, 8), (–8, 8), (–8, –8), (8, –8)

Community Connection

Community Connection 1—12

Answers will vary. Check students' responses.

Geometry on the Job

Geometry on the Job 1—12

Answers will vary. Check students' responses.

Mastery Test A

Chapter 1 Mastery Test A

1. Ray $\overrightarrow{AB}$, AB **2.** Line CD, Line $\overleftrightarrow{DC}$, $\overleftrightarrow{CD}$, DC **3.** Line segment EF, Line segment FE, $\overline{EF}$, $\overline{FE}$ **4.** Line z, Line GH, Line HG, $\overleftrightarrow{GH}$, $\overleftrightarrow{HG}$ **5.** A **6.** D **7.** C **8.** B **9.** 6 units **10.** 8 units **11.** 3 units **12.** 5 units **13.** 5 **14.** 5 **15.** 10 **16.** 3 **17.–20.** Students should copy each angle shown. **21.–24.** Students should copy and bisect each angle shown. **25.** straight **26.** right **27.** acute **28.** obtuse **29.** Students should draw adjacent angles whose measures are 45° and 45°. **30.** Students should draw adjacent angles whose measures are 45° and 135°. **31.** Students should draw adjacent angles whose measures are 85° and 5°. **32.** Students should draw adjacent angles whose measures are 5° and 175°. **33.** B **34.** D **35.** C

Chapter 2 Mastery Test A

1. If a figure is a square, then it has four sides. **2.** If an angle measures less than 90°, then it is acute. **3.** If a figure is a line segment, then it has two endpoints. **4.** If an angle measures more than 90°, then it is obtuse. **5.** If two lines intersect, then two pairs of vertical angles are formed. **6.** True; If a figure is a rectangle, then it has four sides and four right angles; True. **7.** False; If an animal is a horse, then it has four legs; True. **8.** True; If a circle's diameter is larger than a square's sides, then the circle encloses the square; True. **9.** True; If each of two angles measures less than 90°, then the two angles are complementary; False. **10.** True; If the sum of the interior angles formed between two lines equals 360°, then the two lines are parallel; True. **11.** Postulate 3 **12.** Postulate 2 **13.** Postulate 4 **14.** Postulate 1 **15.** Postulate 5 **16.** Axiom 2 **17.** Axiom 3 **18.** Axiom 3 **19.** Axiom 3 **20.** They are equal; Axiom 4 **21.** $\angle ABC$, $\angle CBD$, $\angle DBE$, $\angle ABD$, $\angle CBE$; Axiom 5 **22.** They are supplementary; Axiom 1 **23.** Yes; Axiom 1 **24.** 144; Axiom 1 **25.** 36,500; Axiom 1 **26.** 100; Axiom 1 **27.** $\angle x =$ C, $\angle y =$ B, $\angle z =$ C **28.** $\angle x =$ A, $\angle y =$ D, $\angle z =$ A **29.** $\angle x =$ A, $\angle y =$ A, $\angle z =$ A **30.** $\angle x =$ B, $\angle y =$ D, $\angle z =$ B

Chapter 3 Mastery Test A

1. True **2.** True **3.** True **4.** False **5.** True **6.** False **7.** True **8.** False **9.** B **10.** C **11.** B **12.** C **13.** C **14.** C **15.** B **16.** A **17.** 135°; supplementary to $\angle g$ **18.** 135°; supplementary to $\angle g$, vertical to $\angle h$ **19.** 45°; vertical to $\angle g$, supplementary to $\angle h$, supplementary to $\angle f$ **20.** 45°; corresponding to $\angle g$ **21.** 135°; alternate exterior to $\angle f$, corresponding to $\angle h$, supplementary to $\angle c$ **22.** 135°; corresponding to $\angle f$, alternate interior to $\angle h$, vertical to $\angle d$, supplementary to $\angle c$ **23.** 45°; vertical to $\angle c$, supplementary to $\angle b$ or $\angle d$, corresponding to $\angle e$, alternate exterior to $\angle g$ **24.–25.** Students should construct parallel lines and points as instructed. **26.** $\angle a = 135°$ and $\angle b = 45°$; interior angles on the same side of the transversal are supplementary **27.** $\angle w = 72°$ and $\angle x = 108°$; interior angles on the same side of the transversal are supplementary **28.** $\angle r = 160°$ and $\angle s = 20°$; two lines are cut by a transversal so that the exterior angles on the same side of the transversal are supplementary **29.** $\overline{DC}$ **30.** $\overline{BC}$; **31.** 180° **32.** interior angles on the same side of the transversal are supplementary **33.** alternate interior angles are equal **34.** corresponding angles are equal **35.** $\angle h$ is vertical with $\angle f$; $\angle f$ and $\angle c$ are interior angles on the same side of the transversal

Chapter 4 Mastery Test A

1. (3, 6) **2.** (–4, 1) **3.** (2, –6) **4.** (0, 4) **5.** (–5, –3) **6.** (1, 0) **7.** (4, 2) **8.** (–2, –5) **9.** graph shows a horizontal line at $y = 3$; ordered pair $(x, 3)$ **10.** graph shows a horizontal line at $y = -2$; ordered pair $(x, -2)$ **11.** graph shows a vertical line at $x = -2$; ordered pair $(-2, y)$ **12.** graph shows a horizontal line at $y = 4$; ordered pair $(x, 4)$ **13.** graph shows a vertical line at $x = 5$; ordered pair $(5, y)$ **14.** graph shows a vertical line at $x = 0$, which is the y-axis; ordered pair $(0, y)$ **15.** graph shows a vertical line at $x = -4$; ordered pair $(-4, y)$ **16.** graph shows a horizontal line at $y = 0$, which is the x-axis; ordered pair $(x, 0)$ **17.** slope $= -\frac{2}{3}$ **18.** slope = 1 **19.** slope $= \frac{1}{4}$ **20.** slope = 3 **21.** slope = –4 **22.** slope $= \frac{5}{4}$ **23.** slope $= -\frac{1}{3}$

24. slope = 0
25. $y = -4x + 20$
26. $y = 2x - 7$

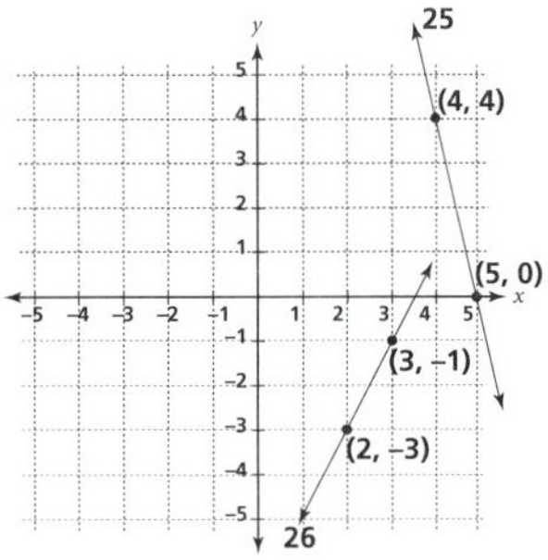

27. $y = \frac{4}{3}x$
28. $x = -4$, y is any real number

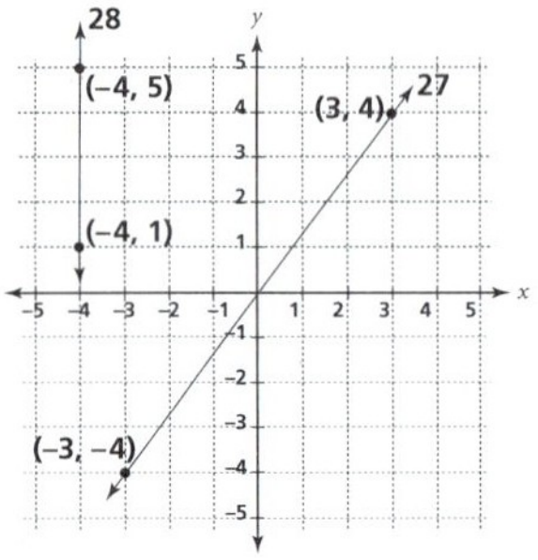

29. $y = 3x - 11$
30. $y = -x + 1$
31. $y = \frac{2}{3}x + 5\frac{1}{3}$

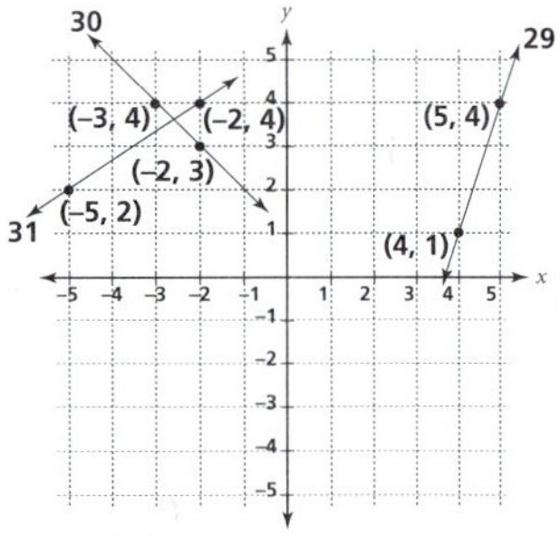

32. C **33.** A **34.** B **35.** D

Chapter 5 Mastery Test A

1.–5. Student constructions should reflect triangles with the given side lengths and/or angle measures. **6.** acute equilateral triangle **7.** obtuse isosceles triangle **8.** right scalene triangle **9.** acute scalene triangle **10.** right isosceles triangle **11.** False, isosceles trapezoid **12.** True **13.** False, a rectangle where $w > l$ or $l > w$ **14.** True **15.** False, rectangles, rhombuses **16.** 55° **17.** 35° **18.** 55° **19.** 90° **20.** 90° **21.** D **22.** B **23.** A **24.** C **25.** D **26.** 3, 4, 720° **27.** 2, 3, 540° **28.–29.** Students should construct perpendicular lines as instructed. **30.** $\angle 1 = 60°$, $\angle 2 = 30°$ and $\angle 3 = 90°$

Chapter 6 Mastery Test A

1. Given: definition of an isosceles triangle. **2.** Given. **3.** Definition of angle bisector. **4.** Any quantity is equal to itself. **5.** SAS postulate. **6.** No, $9 < 10$ **7.** Yes, $9 > 7$, $10 > 6$, $13 > 3$ **8.** Yes, $30 > 13$, $28 > 15$ **9.** No, $8 < 9$ **10.** Yes, $12 > 1$, $7 > 6$ **11.** Student construction should reflect triangle with the given line lengths. **12.–15.** Student answers will vary. Check responses. **16.** B **17.** C **18.** D **19.** (4, –2) **20.** (3, 5) **21.** (–2, 0) **22.** (–4, 1) **23.** Yes **24.** No **25.** Yes **26.** $(x - 7, y - 3)$ **27.** $(x + 4, y - 6)$ **28.** $(x + 1, y)$ **29.** $(x + 3, y + 8)$ **30.** $(x - 7, y - 9)$ **31.** (–3, –5) **32.** (2, –6) **33.** (5, 4) **34.** (0, 4) **35.** (–7, 0)

Chapter 7 Mastery Test A

1. $x = 16$ **2.** $x = 30$ **3.** $x = 1$ **4.** $x = 6$ **5.** 12 **6.** 12 **7.** 6 **8.** 7 **9.** Yes, 3 equal angles **10.** No, sides not in proportion **11.** Yes, isosceles triangles have equal base angles **12.** Yes, 3 equal angles **13.** 7 **14.** $11\frac{2}{3}$ **15.** 6 **16.** 9 **17.** 156° **18.** 128.6° **19.** 90° **20.** 162.9° **21.** image endpoints: (6, 4), (2, 0), (–4, 4), (–2, –6), (4, –4) **22.** image endpoints: (9, 6), (3, 0), (–6, 6), (–3, –9), (6, –6) **23.** image endpoints: (12, 8), (4, 0), (–8, 8), (–4, –12), (8, –8) **24.** image endpoints: (15, 10), (5, 0), (–10, 10), (–5, –15), (10, –10) **25.** image endpoints: (18, 12), (6, 0), (–12, 12), (–6, –18), (12, –12) **26.** (–6, 1), (0, 5), (6, 2), (5, –4), (–6, –4) **27.** $(-4, \frac{2}{3})$, $(0, 3\frac{1}{3})$, $(4, 1\frac{1}{3})$, $(3\frac{1}{3}, -2\frac{2}{3})$, $(-4, 2\frac{2}{3})$ **28.** $(-3, \frac{1}{2})$, $(0, 2\frac{1}{2})$, $(3, 1)$, $(2\frac{1}{2}, -2)$, $(-3, -2)$ **29.** $(-2\frac{2}{5}, \frac{2}{5})$, $(0, 2)$, $(2\frac{2}{5}, \frac{4}{5})$, $(2, -1\frac{3}{5})$, $(-2\frac{2}{5}, -1\frac{3}{5})$ **30.** $(-2, \frac{1}{3})$, $(0, 1\frac{2}{3})$, $(2, \frac{2}{3})$, $(1\frac{2}{3}, -1\frac{1}{3})$, $(-2, -1\frac{1}{3})$

Chapter 8 Mastery Test A

1. 8, 15, 17 **2.** 10, 24, 26 **3.** 18, 80, 82 **4.** 24, 143, 145 **5.** 100, 2499, 2501 **6.** 289 **7.** 60 **8.** 7, 49 **9.** 289 **10.** Yes **11.** 240 **12.** 1,156 **13.** 2,116 **14.** 2,116 **15.** Yes **16.** 256 **17.** 900 **18.** 480 **19.** 2,116 **20.** They are equal. **21.** ΔABC, ΔDBA, ΔDAC **22.** 48 **23.** 64 **24.** 36 **25.** 5:3:4 **26.** C **27.** C **28.** A **29.** B **30.** D **31.** C **32.** B **33.** A **34.** D **35.** C

Chapter 9 Mastery Test A

1. 56 **2.** 92 **3.** 53 **4.** 145 **5.** 264 **6.** 75 **7.** 320 **8.** 158 **9.** 356 **10.** 258 **11.** 84 **12.** 150 **13.** 198 **14.** 100 **15.** 310.8 **16.** C **17.** C **18.** A **19.** $b = 11$ **20.** $h = 51$ **21.** $b = 123$ **22.** 2,662 **23.** 1,450 **24.** B **25.** D

Chapter 10 Mastery Test A

1. diameter or chord **2.** radius **3.** center **4.** chord **5.** radius **6.** $d = 30$ ft; $r = 15$ ft **7.** $d = 46$ ft; $r = 23$ ft **8.** $d = 22$ ft; $r = 11$ ft **9.** $d = 50$ ft; $r = 25$ ft **10.** $d = 82$ ft; $r = 41$ ft **11.** 300 **12.** 432 **13.** 675 **14.** 1,200 **15.** 7,500 **16.** 25% **17.** 40% **18.** $33\frac{1}{3}$% **19.** 201.1 **20.** 314.2 **21.** 78.5 **22.** 1,520.5 **23.** 706.8 **24.** B **25.** C **26.–30.** Student constructions will vary. **31.** B **32.** A **33.** D **34.** D **35.** A

Chapter 11 Mastery Test A

1. 3,375 **2.** 1,536 **3.** 16,877.7 **4.** 405 **5.** 512.86 **6.** C **7.** B **8.** D **9.** 2,100 **10.** 417.82 **11.** C **12.** D **13.** A **14.** A **15.** A **16.** $h = 50$ m **17.** $x = 6$ **18.** $r = 6$ **19.** $h = 15$ **20.** $r = 5$

Chapter 12 Mastery Test A

1. The plane passes through the center of the sphere forming a great circle. **2.** The plane moves from one vertex of the cube to the vertices along the opposite edge. **3.** The line passes through the center of the sphere. **4.** $x = 7$, $x = -3$ **5.** $y = 3x + 2$, $y = 3x + 4$ **6.** C **7.** D **8.** A **9.** B **10.** A **11.** $4 = x^2 + y^2$ **12.** $49 = x^2 + y^2$ **13.** $81 = x^2 + y^2$ **14.** $169 = x^2 + y^2$ **15.** $25 = x^2 + y^2$ **16.** intersection of the perpendicular bisectors of each side of the triangle **17.** point that is distance $\overline{XY}$ from A and is part of the angle bisector for $\angle NMO$ **18.** point at which the circles touch **19.** two points on the circle with T as the center that lie on a line parallel to f and g and halfway between them **20.** the intersection of the angle bisectors of $\angle A$ and $\angle B$

Mastery Test B

Chapter 1 Mastery Test B

1. $\overleftrightarrow{XY}$ or $\overleftrightarrow{YX}$ **2.** Line z **3.** ray **4.** line segment **5.** line **6.** –4 **7.** –2 **8.** 1 **9.** 2.3 **10.** 3.14 **11.–15.** Students should copy and bisect each angle shown. **16.** A **17.** B **18.** B **19.** C **20.** B **21.** 43° **22.** 158° **23.** 84° **24.** 9° **25.** 176° **26.** 72° **27.** 37° **28.** 72° **29.** 71° **30.** D **31.** A **32.** C **33.** B **34.** C **35.** A

Chapter 2 Mastery Test B

1. If a figure is a ray, then it has one endpoint **2.** If a figure is a triangle, then it has three sides **3.** If a figure is an angle, then it has two rays **4.** If an angle is straight, then it measures 180° **5.** If an angle measures 90°, then it is a right angle **6.** False; If a figure is a circle, then it has no line segments; True. **7.** True; If a figure is a rectangle, then it has four sides and four right angles; True. **8.** True; If an angle measures 48°, then the angle is complementary to an angle measuring 42°; True. **9.** True; If each of two angles measures less than 180°, then the two angles are supplementary; False. **10.** True; If all angles were right angles, then all angles would equal 90°; True. **11.** Postulate 1 **12.** Postulate 3 **13.** Postulate 2 **14.** Postulate 4 **15.** Postulate 5 **16.** Postulate 2 **17.** Axiom 3 **18.** Axiom 2 **19.** Axiom 2 **20.** Axiom 3 **21.** $\angle t + \angle u = \angle v + \angle w$; Axiom 2 **22.** Yes; Axiom 1 **23.** $\angle xwy$, $\angle ywz$; Axiom 5 **24.** They are equal; Axiom 4 **25.** $\angle m$ = D, $\angle n$ = A, $\angle o$ = D **26.** $\angle m$ = C, $\angle n$ = A, $\angle o$ = C **27.** $\angle m$ = D, $\angle n$ = B, $\angle o$ = D **28.** $\angle m$ = B, $\angle n$ = B, $\angle o$ = B **29.** $\angle m$ = B, $\angle n$ = C, $\angle o$ = B **30.** $\angle m$ = D, $\angle n$ = A, $\angle o$ = D

Chapter 3 Mastery Test B

1. False **2.** True **3.** False **4.** True **5.** False **6.** True **7.** True **8.** True **9.** C **10.** B **11.** C **12.** B **13.** C **14.** C **15.** B **16.** A **17.** 120°; supplementary to $\angle y$ **18.** 120°; supplementary to $\angle y$, vertical angle to $\angle z$ **19.** 60°; vertical angle to $\angle y$, supplementary to $\angle x$ or $\angle z$ **20.** 60°; corresponding angle to $\angle y$, alternate interior angle to $\angle w$ **21.** 120°; alternate interior angle to $\angle x$, supplementary to $\angle w$, corresponding angle to $\angle z$ **22.** 120°; corresponding angle to $\angle x$, supplementary to $\angle u$, vertical angle to $\angle v$, alternate exterior angle to $\angle z$ **23.** 60°; corresponding angle to $\angle w$, supplementary to $\angle t$ or $\angle v$, vertical angle to $\angle u$, alternate exterior angle to $\angle y$ **24.–25.** Students should construct parallel lines and points as instructed. **26.** $\angle y = 120°$ and $\angle x = 60°$; interior angles on the same side of the transversal are supplementary **27.** $\angle a = 150°$ and $\angle b = 30°$; interior angles on the same side of the transversal are supplementary **28.** $\angle c = 60°$ and $\angle d = 120°$; interior angles on the same side of the transversal are supplementary **29.** $\overline{ZY}$ **30.** $\overline{WX}$ and $\overline{ZY}$ **31.** 180° **32.** corresponding angles are equal **33.** corresponding angles are equal **34.** alternate interior angles are equal **35.** corresponding angles are equal

Chapter 4 Mastery Test B

1. (1, 3) **2.** (–6, 0) **3.** (5, –2) **4.** (–4, –4) **5.** (5, –1) **6.** (–5, 0) **7.** (0, 4) **8.** (0, 0) **9.** graph shows a vertical line at $x = 5$; ordered pair $(5, y)$ **10.** graph shows a horizontal line at $y = 4$; ordered pair $(x, 4)$ **11.** graph shows a horizontal line at $y = -1$; ordered pair $(x, -1)$ **12.** graph shows a vertical line at $x = 2$; ordered pair $(2, y)$ **13.** graph shows a horizontal line at $y = -5$; ordered pair $(x, -5)$ **14.** graph shows a vertical line at $x = -3$; ordered pair $(-3, y)$ **15.** graph shows a horizontal line at $y = 1$; ordered pair $(x, 1)$ **16.** graph shows a horizontal line at $y = -3$; ordered pair $(x, -3)$ **17.** slope $= -\frac{5}{3}$ **18.** slope $= 1\frac{1}{2}$ **19.** slope $= \frac{1}{2}$ **20.** slope = 1 **21.** slope = –1 **22.** slope = 2 **23.** slope = undefined **24.** slope = 0

25. $y = x + 1$
26. $y = 2x + 2$

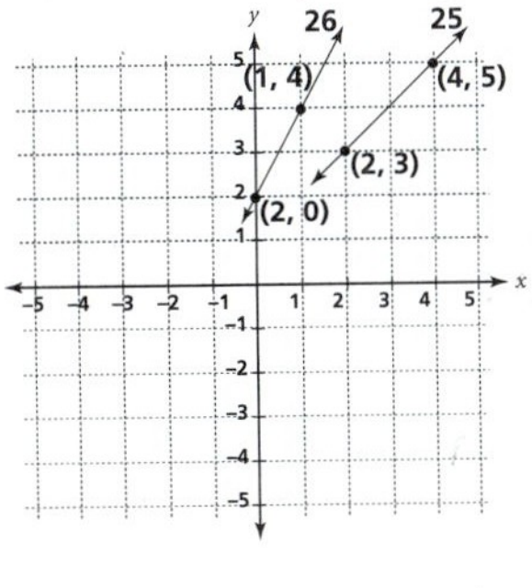

27. $y = \frac{1}{5}x + 4\frac{4}{5}$
28. $y = 5$, x is any real number

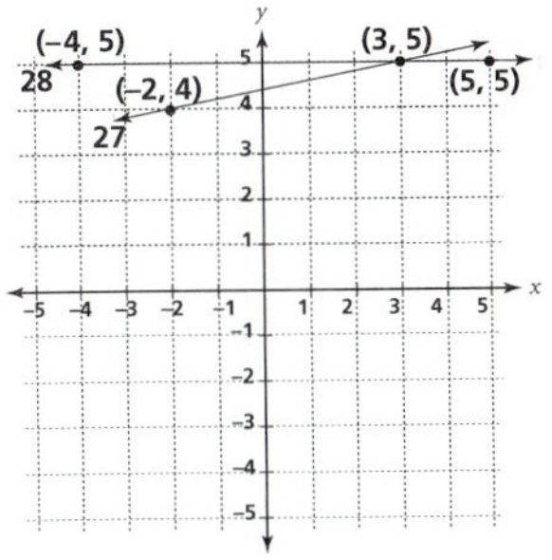

29. $y = x$
30. $y = 2x$
31. $y = \frac{1}{3}x + 1$

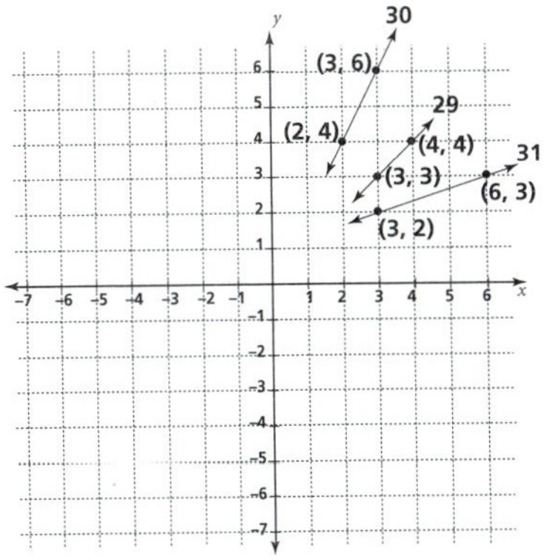

32. B **33.** D **34.** C **35.** D

Chapter 5 Mastery Test B

1.–5. Student constructions should reflect triangles with the given lengths and/or angle measures. **6.** obtuse isosceles triangle **7.** right isosceles triangle **8.** acute scalene triangle **9.** acute equilateral triangle **10.** right scalene triangle **11.** True **12.** False, no parallelograms are trapezoids. **13.** False, a rectangle where $w > l$ or $l > w$ **14.** False, scalene or right trapezoid **15.** False, rhombuses with one pair of obtuse and one pair of acute angles **16.** 40° **17.** 40° **18.** 40° **19.** 180° **20.** 180° **21.** B **22.** C **23.** D **24.** A **25.** C **26.** 1, 2, 360° **27.** 9, 10, 1,800° **28.–30.** Students should construct figures and perpendiculars as instructed.

Chapter 6 Mastery Test B

1. Given. **2.** Given. **3.** Definition of angle bisector. **4.** Any quantity is equal to itself. **5.** SAS postulate. **6.** Yes; $6 > 5, 8 > 3$ **7.** Yes; $12 > 6$ **8.** No; $8 < 13$ **9.** Yes; $8 > 4, 9 > 3, 7 > 5$ **10.** No; $2 < 3$ **11.–15.** Student constructions should reflect triangles with the given side lengths and/or congruent sides and angles. **16.** C **17.** A **18.** D **19.** $A' = (3, 6)$ **20.** $B' = (-1, 7)$ **21.** $C' = (4, -5)$ **22.** $D' = (2, 3)$ **23.** one vertical line and one horizontal line through the center **24.** 8 that go through the midpoints of opposite segments, 8 that go through opposite vertices **25.** an infinite number of diameters **26.** $(x - 2, y + 11)$ **27.** $(x - 5, y + 8)$ **28.** $(x - 3, y - 5)$ **29.** $(x, y - 6)$ **30.** $(x - 11, y)$ **31.** (9, –3) **32.** (–3, 4) **33.** (2, –6) **34.** (–1, 0) **35.** (0, –5)

Chapter 7 Mastery Test B

1. $x = 9\frac{1}{3}$ **2.** $x = 22.3$ **3.** $x = 3$ **4.** $x = 40$ **5.** 18 **6.** 15 **7.** 6 **8.** 8 **9.** Yes, isosceles triangles have equal base angles. **10.** Yes, the sides are in proportion. **11.** Yes, 3 equal angles **12.** No, the sides are not in proportion. **13.** 9 **14.** 10 **15.** 12 **16.** 21.8 **17.** 150° **18.** 154.3° **19.** 108° **20.** 165.6° **21.** image endpoints: (2, 2), (0, 4), (–2, 2), (–2, –2), (2, –2) **22.** image endpoints: (3, 3), (0, 6), (–3, 3), (–3, –3), (3, –3) **23.** image endpoints: (4, 4), (0, 8), (–4, 4), (–4, –4), (4, –4) **24.** image endpoints: (5, 5), (0, 10), (–5, 5), (–5, –5), (5, –5) **25.** image endpoints: (6, 6), (0, 12), (–6, 6), (–6, –6), (6, –6) **26.** image endpoints: (6, 4), (0, 6), (–6, 4), (–5, –3), (5, –3)

27. image endpoints: $(4, 2\frac{2}{3})$, $(0, 4)$, $(-4, 2\frac{2}{3})$, $(-3\frac{1}{3}, -2)$, $(3\frac{1}{3}, -2)$

28. image endpoints: $(3, 2)$, $(0, 3)$, $(-3, 2)$, $(-2\frac{1}{2}, -1\frac{1}{2})$, $(2\frac{1}{2}, -1\frac{1}{2})$

29. image endpoints: $(2\frac{2}{5}, 1\frac{3}{5})$, $(0, 2\frac{2}{5})$, $(-2\frac{2}{5}, 1\frac{3}{5})$, $(-2, -1\frac{1}{5})$, $(2, -1\frac{1}{5})$

30. image endpoints: $(2, 1\frac{1}{3})$, $(0, 2)$, $(-2, 1\frac{1}{3})$, $(-1\frac{2}{3}, -1)$, $(1\frac{2}{3}, -1)$

Chapter 8 Mastery Test B

1. 12, 35, 37 **2.** 14, 48, 50 **3.** 16, 63, 65 **4.** 26, 168, 170 **5.** 200; 9,999; 10,001 **6.** 28,900 **7.** 2,184 **8.** 142; 20,164 **9.** 28,900 **10.** Yes **11.** 720 **12.** 6,724 **13.** 9,604 **14.** 9,604 **15.** Yes **16.** 324 **17.** 6,400 **18.** 1,440 **19.** 9,604 **20.** They are equal. **21.** ΔABC, ΔABD, ΔACD **22.** 113.5 **23.** 331.1 **24.** 38.9 **25.** 37:35:12 **26.** C **27.** B **28.** D **29.** C **30.** A **31.** A **32.** A **33.** B **34.** C **35.** A

Chapter 9 Mastery Test B

1. 251 **2.** 192 **3.** 280 **4.** 252 **5.** 218 **6.** 180 **7.** 352 **8.** 335 **9.** 336 **10.** 292 **11.** 26 **12.** 120 **13.** 136 **14.** 64 **15.** 192 **16.** B **17.** C **18.** A **19.** $b = 25$ **20.** $h = 46$ **21.** $b = 400$ **22.** 1,444 **23.** 3,417 **24.** A **25.** B

Chapter 10 Mastery Test B

1. chord **2.** center **3.** radius **4.** diameter or chord **5.** radius **6.** $d =$ 25 ft; $r = 12.5$ ft **7.** $d = 80$ ft; $r = 40$ ft **8.** $d = 56$ ft; $r = 28$ ft **9.** $d =$ 34 ft; $r = 17$ ft **10.** $d = 98$ ft; $r = 49$ ft **11.** 588 **12.** 243 **13.** 867 **14.** 1,875 **15.** 30,000 **16.** 41.66% **17.** 66.67% **18.** 75% **19.** 254.5 **20.** 907.9 **21.** 153.9 **22.** 1,385.4 **23.** 2,462.9 **24.** C **25.** A **26.–30.** Constructions will vary. **31.** D **32.** B **33.** A **34.** D **35.** C

Chapter 11 Mastery Test B

1. 4,096 **2.** 1,683 **3.** 9,060.53 **4.** 294 **5.** 513.11 **6.** B **7.** C **8.** A **9.** 1,587 **10.** 301.58 **11.** D **12.** A **13.** C **14.** B **15.** B **16.** $h = 105$ **17.** $x = 9$ **18.** $r = 8$ **19.** $h = 90$ **20.** $r = 10$

Chapter 12 Mastery Test B

1. The plane crosses through the pyramid parallel to the base. **2.** The line passes through the point of the pyramid and the center of the triangle base. **3.** The plane crosses through the pyramid so that it is not parallel to any side or to the base. **4.** $x = -7$, $x = 5$ **5.** $y = \frac{1}{3}x$, $y = \frac{1}{3}x + 2$ **6.** B **7.** D **8.** C **9.** A **10.** C **11.** $9 = x^2 + y^2$ **12.** $144 = x^2 + y^2$ **13.** $196 = x^2 + y^2$ **14.** $1{,}089 = x^2 + y^2$ **15.** $16 = x^2 + y^2$ **16.** Two points along a line midway between l and m and also distance $\overline{FG}$ from W **17.** Two points along the bisector of $\angle EDF$ and also distance $\overline{BC}$ from S **18.** Point at the intersection of perpendicular bisectors **19.** Point at the intersection of the two angle bisectors **20.** Line segment within overlapping area with left endpoint on circle D and right endpoint on circle C

Cumulative Mastery Test

Chapter 1 Cumulative Mastery Test

1. $\overline{XY}$ **2.** $\overrightarrow{AB}$ **3.** $\overleftrightarrow{EF}$ **4.** $\overline{RS}$ **5.** $\overrightarrow{CD}$ **6.** 3 units **7.** 6 units **8.** 8 units **9.** 4 units **10.** 9 units **11.–15.** Students should copy and bisect each angle shown. **16.** B **17.** C **18.** B **19.** C **20.** 37°, 127° **21.** 13°, 103° **22.** 48°, 138° **23.** 57°, 147° **24.** 84°, 174° **25.** C **26.** A **27.** C **28.** B **29.** B **30.** C

Chapter 2 Cumulative Mastery Test

1. 9 units **2.** 3 units **3.** 6 units **4.** 9 units **5.** 6 units **6.–10.** Students should copy and bisect the angle shown. **6.** acute **7.** acute **8.** obtuse **9.** obtuse **10.** right **11.** C **12.** C **13.** C **14.** B **15.** 70° **16.** 20° **17.** 90° **18.** 70° **19.** 20° **20.** 90° **21.** True; If an angle is supplementary to an angle measuring 80°, then the angle measures 100°; True. **22.** True; If each angle measures less than 90°, then the two angles are complementary; False. **23.** False; If a figure is a triangle, then it has line segments; True. **24.** Axiom 2 **25.** Axiom 3 **26.** Axiom 3 **27.** $\angle m = 136°$, $\angle n = 44°$, $\angle o = 136°$ **28.** $\angle m = 92°$, $\angle n = 88°$, $\angle o = 92°$ **29.** $\angle m = 103°$, $\angle n = 77°$, $\angle o =$ 103° **30.** $\angle m = 36°$, $\angle n = 144°$, $\angle o = 36°$

Chapter 3 Cumulative Mastery Test

1. Students should copy and bisect obtuse angle shown. **2.** Students should copy and bisect acute angle shown. **3.** Students should copy and bisect straight angle shown. **4.** Students should copy and bisect acute angle shown. **5.** C **6.** D **7.** B **8.** D **9.** Euclid's Postulate 3; Constructions will vary. **10.** Euclid's Postulate 4; Constructions will vary. **11.** Euclid's Postulate 1; Constructions will vary. **12.** Euclid's Postulates 2 and 4; Constructions will vary. **13.** $\angle p = 90°$, $\angle q = 90°$, $\angle r = 90°$ **14.** $\angle p = 145°$, $\angle q = 35°$, $\angle r = 145°$ **15.** $\angle p = 32°$, $\angle q =$ 148°, $\angle r = 32°$ **16.** $\angle p = 120°$, $\angle q = 60°$, $\angle r = 120°$ **17.** alternate interior **18.** exterior **19.** interior **20.** supplementary **21.** alternate exterior **22.** corresponding **23.** 70°; supplementary to $\angle h$ **24.** 110°; vertical angle to $\angle h$, supplementary to $\angle g$ **25.** 110°; alternate interior angle to $\angle f$, corresponding angle to $\angle h$ **26.** 110°; corresponding angle to $\angle f$, vertical angle to $\angle d$, alternate exterior angle to $\angle h$ **27.** 70°; corresponding angle to $\angle g$, supplementary to $\angle d$ or $\angle b$ **28.** 180° **29.** 180° **30.** 360° **31.** $\overline{ON}$ **32.** corresponding angles are equal **33.** interior angles on the same side of the transversal are supplementary **34.** $\angle s$ is vertical to $\angle q$; $\angle q$ and $\angle n$ are interior angles on the same side of the transversal **35.** corresponding angles are equal

Chapter 4 Cumulative Mastery Test

1. A; Students should copy and bisect the angle shown. **2.** C; Students should copy and bisect the angle shown. **3.** B; Students should copy and bisect the angle shown. **4.** A; Students should copy and bisect the angle shown. **5.** 45° **6.** 45° **7.** 90° **8.** 45° **9.** $\angle a =$ 118°, $\angle b = 62°$, $\angle c = 118°$ **10.** $\angle a = 22°$, $\angle b = 158°$, $\angle c = 22°$ **11.** $\angle a = 90°$, $\angle b = 90°$, $\angle c = 90°$ **12.** $\angle a = 130°$, $\angle b = 50°$, $\angle c =$ 130° **13.** 97°; supplementary to $\angle w$ **14.** 83°; vertical angle to $\angle w$ or supplementary to $\angle x$ **15.** 83°; alternate interior angle to $\angle w$ or corresponding to $\angle y$ **16.** 83°; corresponding angle to $\angle w$ or vertical to $\angle u$ or alternate exterior to $\angle y$ **17.** $(3, -5)$ **18.** $(-4, 1)$ **19.** $(2, -2)$ **20.** $(0, 0)$ **21.** graph shows a horizontal line at $y = 2$; ordered pair $(x, 2)$ **22.** graph shows a vertical line at $x = -3$; ordered pair $(-3, y)$ **23.** graph shows a horizontal line at $y = 1$; ordered pair $(x, 1)$ **24.** graph shows a vertical line at $x = 5$; ordered pair $(5, y)$ **25.** slope $= -1$ **26.** slope $= \frac{1}{2}$ **27.** slope $= 2$ **28.** slope $= 4$

29. $y = 3x - 12$

30. $y = -2x + 11$

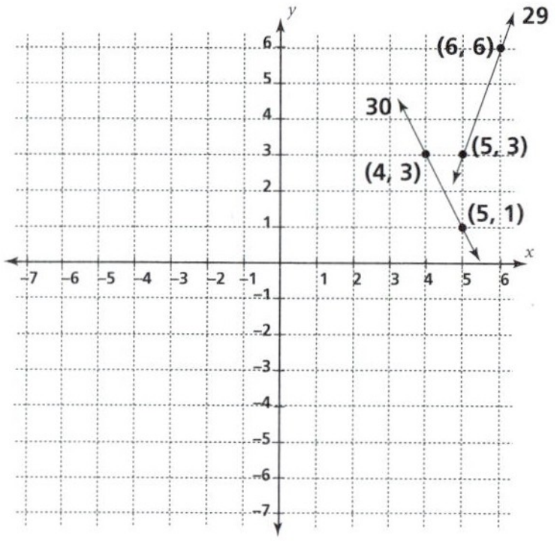

31. $y = \frac{1}{2}x + 4\frac{1}{2}$
32. $y = x$

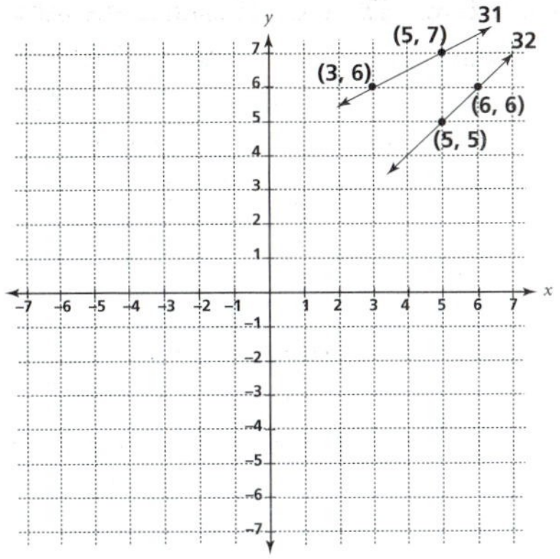

33. (0, 4) **34.** (4, 3) **35.** (–8, 14)

Chapter 5 Cumulative Mastery Test

1. C **2.** A **3.** C **4.** $\angle x = 125°$, $\angle y = 55°$, $\angle z = 125°$ **5.** $\angle x = 57°$, $\angle y = 123°$, $\angle z = 57°$ **6.** $\angle x = 89°$, $\angle y = 91°$, $\angle z = 89°$ **7.** D; interior angles on the same side of the transversal are supplementary **8.** D; corresponding angle to $\angle d$ or supplementary to $\angle e$ **9.** C; supplementary to $\angle h$ or vertical to $\angle e$ **10.** graph shows horizontal line at $y = 5$; ordered pair $(x, 5)$ **11.** graph shows vertical line at $x = -6$; ordered pair $(-6, y)$ **12.** slope $= -\frac{2}{3}$, midpoint = (–1, 2) **13.** slope = 2, midpoint = (4, 2) **14.** slope = undefined, midpoint = $(0, 6\frac{1}{2})$ **15.** $y = -4x - 10$; graph shows line passing through (–3, 2) and (–2, –2) **16.** $y = \frac{3}{4}x + 7$; graph shows line passing through (–8, 1) and (–4, 4) **17.–18.** Constructions should reflect given line lengths and angle measures. **19.** $\angle a = 30°$ **20.** $\angle a = 45°$ **21.** 5; 6; 1,080° **22.** 17; 18; 3,240° **23.–25.** Constructions should include all terms labeled correctly.

Chapter 6 Cumulative Mastery Test

1. $\angle x$ = D, $\angle y$ = B, $\angle z$ = D **2.** $\angle x$ = A, $\angle y$ = D, $\angle z$ = A **3.** 108°; corresponding angle to $\angle d$ **4.** 72°; supplementary to $\angle d$ **5.** 72°; supplementary to $\angle h$ or alternate exterior to $\angle a$ **6.** slope $= -\frac{1}{3}$, midpoint $= (-1\frac{1}{2}, 5\frac{1}{2})$ **7.** slope = 1, midpoint = (–1, 5) **8.** slope = 6, midpoint $= (4\frac{1}{2}, -2)$ **9.** $y = 2x - 9$; graph shows line passing through (5, 1) and (6, 3) **10.** $y = \frac{1}{4}x - 1$; graph shows line passing through (4, 0) and (8, 1) **11.** Construction shows equilateral triangle with sides of length x. **12.** Constructions will vary but should show an acute isosceles triangle. **13.** $x = 10°$ **14.** $x = 36°$ **15.–20.** Constructions will vary. **21.** image coordinates: (–3, 1), (–7, 1), (–7, 4) **22.** image coordinates: (–4, 1), (–6, –1), (–1, 0) **23.** $(x + 5, y + 9)$ **24.** $(x - 3, y + 1)$ **25.** image coordinates: (–1, –4), (–4, 2), (3, –5), (6, 2)

Chapter 7 Cumulative Mastery Test

1. 123°; corresponding to $\angle z$ **2.** 57°; supplementary to $\angle v$ **3.** $y = \frac{2}{3}x + \frac{5}{3}$; graph shows line passing through (2, 3) and (5, 5) **4.** $y = 3x + 14$; graph shows line passing through (–3, 5) and (–2, 8) **5.–10.** Constructions will vary. **11.** A' = (5, –3), B' = (–3, –3), C' = (1, –6) **12.** $13\frac{1}{3}$ **13.** $16\frac{2}{3}$ **14.** $4\frac{2}{3}$ **15.** 6 **16.** A' = (4, 0), B' = (0, 4), C' = (–4, 0), D' = (0, –4) **17.** A' = (6, 0), B' = (0, 6), C' = (–6, 0), D' = (0, –6) **18.** A' = (8, 0), B' = (0, 8), C' = (–8, 0), D' = (0, –8) **19.** A' = (10, 0), B' = (0, 10), C' = (–10, 0), D' = (0, –10) **20.** A' = (12, 0), B' = (0, 12), C' = (–12, 0), D' = (0, –12) **21.** A' = (0, 0), B' = (3, 0), C' = (3, 3), D' = (0, 3) **22.** A' = (0, 0), B' = (2, 0), C' = (2, 2), D' = (0, 2) **23.** A' = (0, 0), $B' = (1\frac{1}{2}, 0)$, $C' = (1\frac{1}{2}, 1\frac{1}{2})$, $D' = (0, 1\frac{1}{2})$ **24.** A' = (0, 0), $B' = (1\frac{1}{5}, 0)$, $C' = (1\frac{1}{5}, 1\frac{1}{5})$, $D' = (0, 1\frac{1}{5})$ **25.** A' = (0, 0), B' = (1, 0), C' = (1, 1), D' = (0, 1)

Chapter 8 Cumulative Mastery Test

1. $y = 3x + 4$; graph shows line passing through (0, 4) and (1, 7) **2.** $y = \frac{1}{2}x + 3\frac{1}{2}$; graph shows line passing through (–3, 2) and (–1, 3) **3.–6.** Constructions should reflect triangles with given side lengths and/or angle measures. **7.** $5\frac{1}{3}$ **8.** $6\frac{2}{3}$ **9.** image endpoints: (2, 0), (0, 8), (–2, 0), (0, –8) **10.** image endpoints: (3, 0), (0, 12), (–3, 0), (0, –12) **11.** image endpoints: (4, 0), (0, 16), (–4, 0), (0, –16) **12.** image endpoints: $(\frac{1}{2}, 0)$, (0, 2), $(-\frac{1}{2}, 0)$, (0, –2) **13.** image endpoints: $(\frac{1}{3}, 0)$, $(0, 1\frac{1}{3})$, $(-\frac{1}{3}, 0)$, $(0, -1\frac{1}{3})$ **14.** image endpoints: $(\frac{1}{4}, 0)$, (0, 1), $(-\frac{1}{4}, 0)$, (0, –1) **15.** image endpoints: $(\frac{1}{5}, 0)$, $(0, \frac{4}{5})$, $(-\frac{1}{5}, 0)$, $(0, -\frac{4}{5})$ **16.** $(\frac{1}{6}, 0)$, $(0, \frac{2}{3})$, $(-\frac{1}{6}, 0)$, $(0, -\frac{2}{3})$ **17.** 34, 288, 290 **18.** 18, 80, 82 **19.** 66,049 **20.** 4,080 **21.** 223; 49,729 **22.** 66,049 **23.** 750 **24.** 2,112.5 **25.** 3,612.5 **26.** C **27.** D **28.** B **29.** C **30.** A

Chapter 9 Cumulative Mastery Test

1. $y = \frac{1}{4}x - \frac{17}{4}$; graph shows line passing through (5, –3) and (9, –2) **2.** $y = 4x - 11$; graph shows line passing through (2, –3) and (3, 1) **3.–4.** Constructions should reflect triangles with given line lengths and/or angle measures. **5.** image endpoints: (8, 8), (0, 4), (0, 0), (4, 0) **6.** image endpoints: (16, 16), (0, 8), (0, 0), (8, 0) **7.** image endpoints: (2, 2), (0, 1), (0, 0), (1, 0) **8.** image endpoints: (1, 1), $(0, \frac{1}{2})$, (0, 0), $(\frac{1}{2}, 0)$ **9.** 30, 224, 226 **10.** 60, 899, 901 **11.** 60 **12.** 7, 49 **13.** 289 **14.** D **15.** A **16.** 174 **17.** 1,160 **18.** $x = 17$, $P = 100$ **19.** $x = 20$, $P = 104$ **20.** A **21.** B **22.** $b = 32$ **23.** $h = 65$ **24.** A **25.** D

Chapter 10 Cumulative Mastery Test

1.–2. Constructions should create triangles with given line lengths and/or angle measures. **3.** image endpoints: (3, 6), (6, 3), (–3, –6), (–6, –3) **4.** image endpoints: $(\frac{1}{2}, 1)$, $(1, \frac{1}{2})$, $(-\frac{1}{2}, -1)$, $(-1, -\frac{1}{2})$ **5.** 24 **6.** 2, 4 **7.** 100 **8.** 180 **9.** 2,136 **10.** B **11.** D **12.** $b = 12$ **13.** C **14.** A **15.** $d = 43$ ft; $r = 21.5$ ft **16.** $d = 66$ ft; $r = 33$ ft **17.** 380.1 **18.** 1,134.1 **19.** A **20.** B

Chapter 11 Cumulative Mastery Test

1. image endpoints: (12, 4), (–12, 4), (–6, –8), (6, –8) **2.** image endpoints: (3, 1), (–3, 1), $(-1\frac{1}{2}, -2)$, $(1\frac{1}{2}, -2)$ **3.** 222 **4.** 1,335 **5.** C **6.** A **7.** B **8.** C **9.** $d = 78$ ft; $r = 39$ ft **10.** $d = 29$ ft; $r = 14.5$ ft **11.** 530.9 **12.** 6,361.5 **13.** D **14.** $S = 100\pi$; $V = 166.7\pi$ **15.** $S = 864$; $V = 1{,}728$ **16.** $S = 1{,}006$; $V = 2{,}145$ **17.** $S = 75.71$; $V = 36.95$ **18.** $S = 672$; $V = 1{,}152$ **19.** $S = 763.38$; $V = 1{,}017.85$ **20.** $S = 696$; $V = 960$ **21.** C **22.** D **23.** B **24.** B **25.** D

Chapter 12 Cumulative Mastery Test

1. image endpoints: (6, 10), (–8, 4), (–10, –6), (8, –8) **2.** image endpoints: $(1\frac{1}{2}, 2\frac{1}{2})$, (–2, 1), $(-2\frac{1}{2}, -1\frac{1}{2})$, (2, –2) **3.** D **4.** B **5.** D **6.** A **7.** 803.8 **8.** 12,070.2 **9.** $S = 3{,}456$; $V = 13{,}824$ **10.** $S = 822$; $V = 1{,}456$ **11.** $S = 278.07$; $V = 105.2$ **12.** $S = 192$; $V = 180$ **13.** $S = 117.81$; $V = 84.82$ **14.** A **15.** C **16.** The line lies on the surface of the cone. **17.** The plane crosses through the pyramid parallel to the base. **18.** $y = 7$, $y = 1$ **19.** $y = 2x$, $y = 2x - 6$ **20.** A **21.** C **22.** $64 = x^2 + y^2$; circle $r = 8$ **23.** $625 = x^2 + y^2$; circle $r = 25$ **24.** Two points on a line midway between and parallel to j and k that are distance $\overline{CD}$ above and below R **25.** One point on the angle bisector of $\angle SRT$ that is also distance $\overline{DE}$ from R

Midterm Mastery Test—Chapters 1–6

1. right **2.** obtuse **3.** acute **4.** straight **5.** B **6.** A **7.** B **8.** D

9. Euclid's Postulate 4

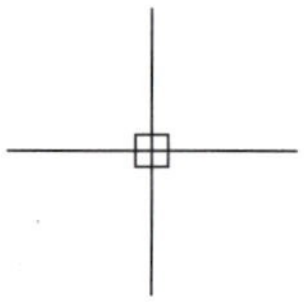

10. Euclid's Postulate 4

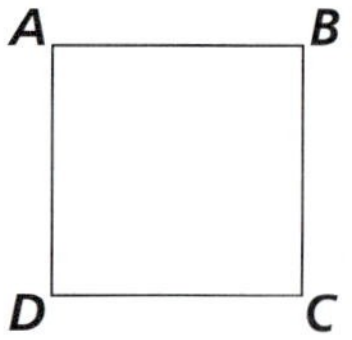

11. Euclid's Postulate 1

12. Euclid's Postulate 3

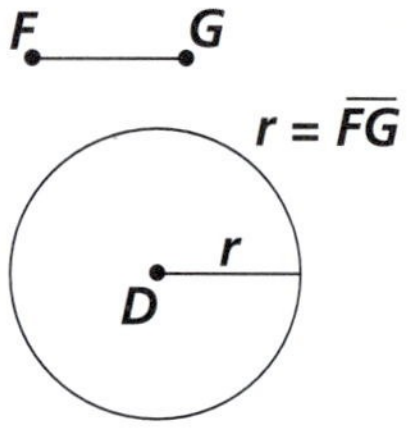

13. 157°, 23°, 157° **14.** 25°, 155°, 25° **15.** 90°, 90°, 90° **16.** 110°, 70°, 110° **17.** supplementary **18.** corresponding **19.** interior **20.** alternate interior **21.** exterior **22.** alternate exterior **23.** 66°; corresponding to $\angle b$ **24.** 114°; supplementary to $\angle f$ **25.** 66°; vertical angle to $\angle b$ and alternate interior to $\angle f$ **26.** 114°; supplementary to $\angle b$, supplementary to $\angle d$, and corresponding to $\angle g$ **27.** 114°; alternate interior to $\angle c$, vertical to $\angle g$, supplementary to $\angle f$ **28.** 180° **29.** 180° **30.** 360° **31.** $\overline{DC}$ **32.** If sum of interior angles on same side of transversal = 180°, then $x \parallel y$. **33.** If corresponding angles are equal, then $x \parallel y$. **34.** If sum of interior angles on same side of transversal = 180°, then $x \parallel y$. **35.** If corresponding angles are equal, then $x \parallel y$. **36.** slope $= -\frac{1}{2}$ **37.** slope = 1 **38.** slope $= 1\frac{1}{2}$ **39.** slope = undefined **40.** line passes through (9, 3) and (6, 2); $y = \frac{1}{3}x$ **41.** line passes through (0, 6), (3, 3) and (6, 0); $y = -x + 6$ **42.** line passes through (–2, –4) and $(0, -\frac{7}{2})$; $y = \frac{1}{4}x - \frac{7}{2}$ **43.** horizontal line crosses x-axis at 5; $y = 5$ **44.** B **45.** C **46.** D **47.–48.** Student constructions should reflect triangles with the given side lengths and/or angle measures. **49.** $b = 30°$ **50.** $b = 50°$ **51.** 4, 5, 900° **52.** 9, 10, 1,800° **53.** Constructions will vary. **54.–56.** Student constructions will vary but should reflect understanding of how to construct requested elements. **57.** image coordinates: (0, 4), (–1, –3), (–3, 5) **58.** image coordinates: (–4, –3), (4, –3), (0, 0) **59.** $(x + 9, y - 7)$ **60.** $(x - 1, y - 11)$

Final Mastery Test—Chapters 1–12

1. acute; Students should copy and bisect the angle shown. **2.** straight; Students should copy and bisect the angle shown. **3.** Student constructions should show circle with center and radius as indicated; Euclid's Postulate 3 **4.** Student constructions should show two perpendicular intersecting lines forming four 90° angles; Euclid's Postulates 2 and 4 **5.** 78°; corresponding angles **6.** 78°; alternate interior angles

7. $y = 3x + 4$

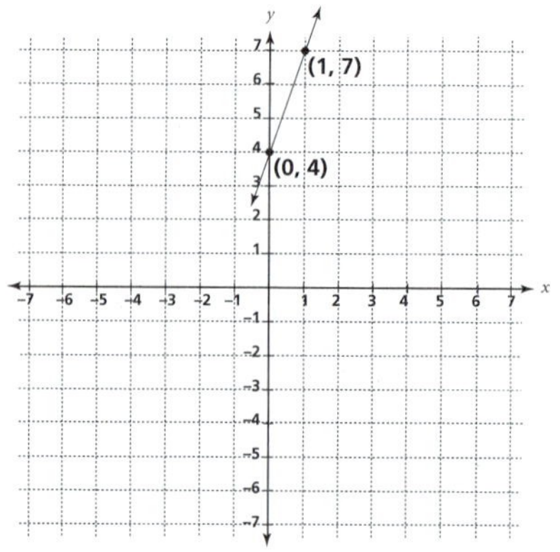

8. $y = \frac{1}{2}x - \frac{1}{2}$

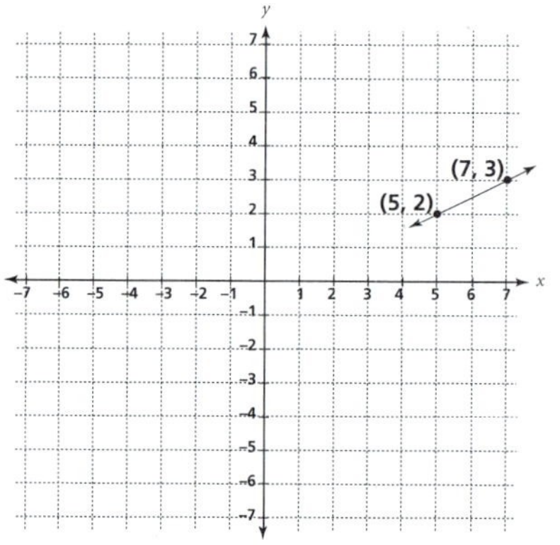

9.–10. Student constructions should reflect triangles with the given side lengths and/or angle measures. **11.** $x = 45°$ **12.** $x = 30°$ **13.–15.** Student constructions will vary but should reflect understanding of how to construct requested elements. **16.** image coordinates: (–1, 4), (2, 3), (0, –3) **17.** image coordinates: (2, –4), (5, –1), (–4, –1) **18.** $(x - 2, y - 2)$ **19.** $(x - 5, y + 5)$ **20.** 28 **21.** 52 **22.** image endpoints: (–10, 8), (8, 6), (–2, –8) **23.** image endpoints: $(-2\frac{1}{2}, 2)$, $(2, 1\frac{1}{2})$, $(-\frac{1}{2}, -2)$ **24.** 12, 35, 37 **25.** 18, 80, 82 **26.** C **27.** B **28.** D **29.** A **30.** 248 **31.** 191 **32.** 399 **33.** 170 **34.** 182.4 **35.** 184 **36.** D **37.** B **38.** $b = 42$ **39.** $h = 57$ **40.** 1,600 **41.** 3,240 **42.** B **43.** C **44.** 94 ft; 47 ft; 2,209π sq ft **45.** 58 ft; 29 ft; 841π sq ft **46.** B **47.** A **48.–49.** Answers will vary. **50.** 452.16, 904.32 **51.** 615.44, 1,436.03 **52.** 5,832 **53.** 3,456 **54.** 471 **55.** 1,230.88 **56.** 471 **57.** 923.16 **58.** The plane crosses through the sphere to form a circle but does not pass through the sphere's center. **59.** Point at the intersection of the angle bisectors **60.** Two points along the perpendicular bisector of the line connecting lines c and d and at a distance of $\overline{EF}$ from G